PERIODIC TABLE OF THE ELEMENTS

(Modified from a table published and copyrighted by Sargent-Welch Scientific Company, Skokie, Illinois, and used with their permission)

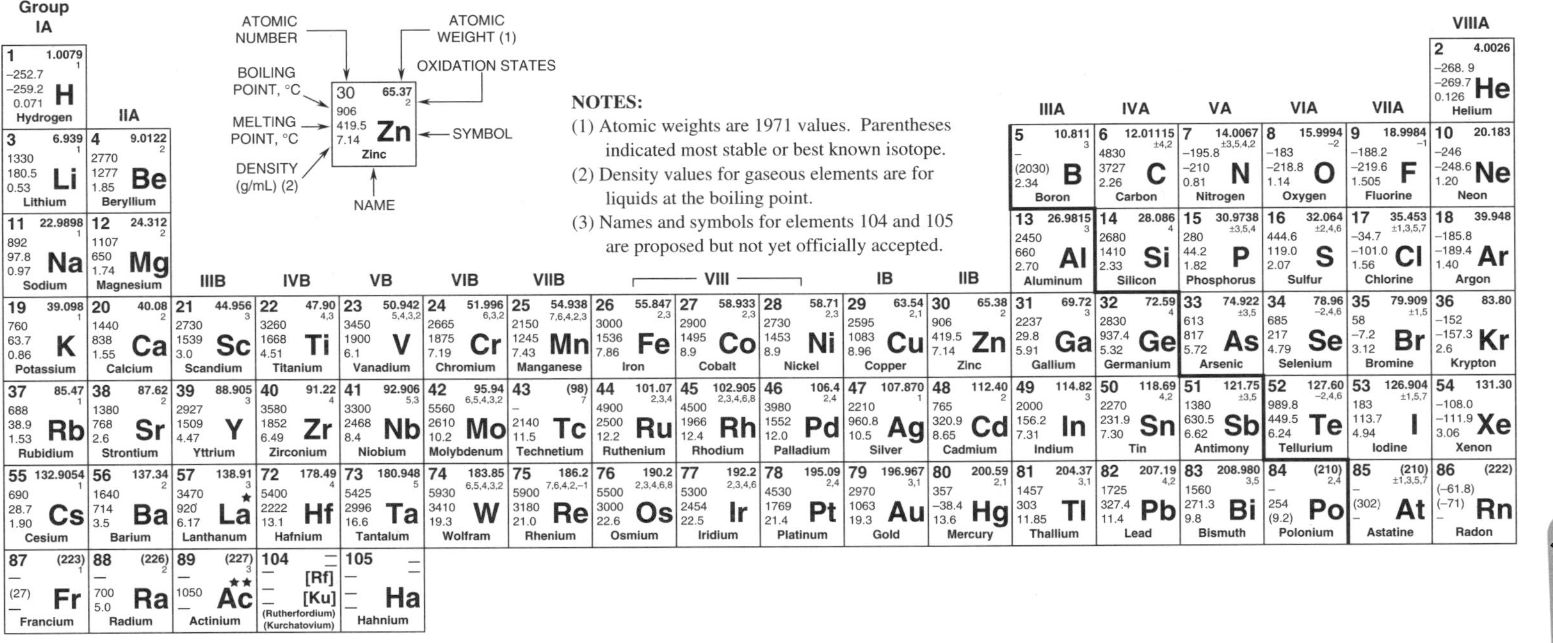

Key: ATOMIC NUMBER (30), ATOMIC WEIGHT (1) (65.37), OXIDATION STATES (2), BOILING POINT, °C (906), MELTING POINT, °C (419.5), DENSITY (g/mL) (2) (7.14), SYMBOL (Zn), NAME (Zinc)

NOTES:

(1) Atomic weights are 1971 values. Parentheses indicated most stable or best known isotope.

(2) Density values for gaseous elements are for liquids at the boiling point.

(3) Names and symbols for elements 104 and 105 are proposed but not yet officially accepted.

Group	Atomic number	Symbol	Name	Atomic weight	Oxidation states	Boiling point, °C	Melting point, °C	Density (g/mL)
IA	1	H	Hydrogen	1.0079	1	−252.7	−259.2	0.071
VIIIA	2	He	Helium	4.0026		−268. 9	−269.7	0.126
IA	3	Li	Lithium	6.939	1	1330	180.5	0.53
IIA	4	Be	Beryllium	9.0122	2	2770	1277	1.85
IIIA	5	B	Boron	10.811	3	–	(2030)	2.34
IVA	6	C	Carbon	12.01115	±4,2	4830	3727	2.26
VA	7	N	Nitrogen	14.0067	±3,5,4,2	−195.8	−210	0.81
VIA	8	O	Oxygen	15.9994	−2	−183	−218.8	1.14
VIIA	9	F	Fluorine	18.9984	−1	−188.2	−219.6	1.505
VIIIA	10	Ne	Neon	20.183		−246	−248.6	1.20
IA	11	Na	Sodium	22.9898	1	892	97.8	0.97
IIA	12	Mg	Magnesium	24.312	2	1107	650	1.74
IIIA	13	Al	Aluminum	26.9815	3	2450	660	2.70
IVA	14	Si	Silicon	28.086	4	2680	1410	2.33
VA	15	P	Phosphorus	30.9738	±3,5,4	280	44.2	1.82
VIA	16	S	Sulfur	32.064	±2,4,6	444.6	119.0	2.07
VIIA	17	Cl	Chlorine	35.453	±1,3,5,7	−34.7	−101.0	1.56
VIIIA	18	Ar	Argon	39.948		−185.8	−189.4	1.40
IA	19	K	Potassium	39.098	1	760	63.7	0.86
IIA	20	Ca	Calcium	40.08	2	1440	838	1.55
IIIB	21	Sc	Scandium	44.956	3	2730	1539	3.0
IVB	22	Ti	Titanium	47.90	4,3	3260	1668	4.51
VB	23	V	Vanadium	50.942	5,4,3,2	3450	1900	6.1
VIB	24	Cr	Chromium	51.996	6,3,2	2665	1875	7.19
VIIB	25	Mn	Manganese	54.938	7,6,4,2,3	2150	1245	7.43
VIII	26	Fe	Iron	55.847	2,3	3000	1536	7.86
VIII	27	Co	Cobalt	58.933	2,3	2900	1495	8.9
VIII	28	Ni	Nickel	58.71	2,3	2730	1453	8.9
IB	29	Cu	Copper	63.54	2,1	2595	1083	8.96
IIB	30	Zn	Zinc	65.38	2	906	419.5	7.14
IIIA	31	Ga	Gallium	69.72	3	2237	29.8	5.91
IVA	32	Ge	Germanium	72.59	4	2830	937.4	5.32
VA	33	As	Arsenic	74.922	±3,5	613	817	5.72
VIA	34	Se	Selenium	78.96	−2,4,6	685	217	4.79
VIIA	35	Br	Bromine	79.909	±1,5	58	−7.2	3.12
VIIIA	36	Kr	Krypton	83.80		−152	−157.3	2.6
IA	37	Rb	Rubidium	85.47	1	688	38.9	1.53
IIA	38	Sr	Strontium	87.62	2	1380	768	2.6
IIIB	39	Y	Yttrium	88.905	3	2927	1509	4.47
IVB	40	Zr	Zirconium	91.22	4	3580	1852	6.49
VB	41	Nb	Niobium	92.906	5,3	3300	2468	8.4
VIB	42	Mo	Molybdenum	95.94	6,5,4,3,2	5560	2610	10.2
VIIB	43	Tc	Technetium	(98)	7	–	2140	11.5
VIII	44	Ru	Ruthenium	101.07	2,3,4	4900	2500	12.2
VIII	45	Rh	Rhodium	102.905	2,3,4,6,8	4500	1966	12.4
VIII	46	Pd	Palladium	106.4	2,4	3980	1552	12.0
IB	47	Ag	Silver	107.870	1	2210	960.8	10.5
IIB	48	Cd	Cadmium	112.40	2	765	320.9	8.65
IIIA	49	In	Indium	114.82	3	2000	156.2	7.31
IVA	50	Sn	Tin	118.69	4,2	2270	231.9	7.30
VA	51	Sb	Antimony	121.75	±3,5	1380	630.5	6.62
VIA	52	Te	Tellurium	127.60	−2,4,6	989.8	449.5	6.24
VIIA	53	I	Iodine	126.904	±1,5,7	183	113.7	4.94
VIIIA	54	Xe	Xenon	131.30		−108.0	−111.9	3.06
IA	55	Cs	Cesium	132.9054	1	690	28.7	1.90
IIA	56	Ba	Barium	137.34	2	1640	714	3.5
IIIB	57	La ★	Lanthanum	138.91	3	3470	920	6.17
IVB	72	Hf	Hafnium	178.49	4	5400	2222	13.1
VB	73	Ta	Tantalum	180.948	5	5425	2996	16.6
VIB	74	W	Wolfram	183.85	6,5,4,3,2	5930	3410	19.3
VIIB	75	Re	Rhenium	186.2	7,6,4,2,−1	5900	3180	21.0
VIII	76	Os	Osmium	190.2	2,3,4,6,8	5500	3000	22.6
VIII	77	Ir	Iridium	192.2	2,3,4,6	5300	2454	22.5
VIII	78	Pt	Platinum	195.09	2,4	4530	1769	21.4
IB	79	Au	Gold	196.967	3,1	2970	1063	19.3
IIB	80	Hg	Mercury	200.59	2,1	357	−38.4	13.6
IIIA	81	Tl	Thallium	204.37	3,1	1457	303	11.85
IVA	82	Pb	Lead	207.19	4,2	1725	327.4	11.4
VA	83	Bi	Bismuth	208.980	3,5	1560	271.3	9.8
VIA	84	Po	Polonium	(210)	2,4	–	254	(9.2)
VIIA	85	At	Astatine	(210)	±1,3,5,7	–	(302)	–
VIIIA	86	Rn	Radon	(222)		(−61.8)	(−71)	–
IA	87	Fr	Francium	(223)	1	–	(27)	–
IIA	88	Ra	Radium	(226)	2	–	700	5.0
IIIB	89	Ac ★★	Actinium	(227)	3	–	1050	–
IVB	104	[Rf] [Ku]	(Rutherfordium) (Kurchatovium)	–	–	–	–	–
VB	105	Ha	Hahnium	–	–	–	–	–

★ Lanthanides

Atomic number	Symbol	Name	Atomic weight	Oxidation states	Boiling point, °C	Melting point, °C	Density (g/mL)
58	Ce	Cerium	140.12	3,4	3468	795	6.67
59	Pr	Praseodymium	140.907	3,4	3127	935	6.77
60	Nd	Neodymium	144.24	3	3027	1024	7.00
61	Pm	Promethium	(147)	3	–	(1027)	–
62	Sm	Samarium	150.35	3,2	1900	1072	7.54
63	Eu	Europium	151.96	3,2	1439	826	5.26
64	Gd	Gadolinium	157.25	3	3000	1312	7.89
65	Tb	Terbium	158.924	3,4	2800	1356	8.27
66	Dy	Dysprosium	162.50	3	2600	1407	8.54
67	Ho	Holmium	164.930	3	2600	1461	8.80
68	Er	Erbium	167.26	3	2900	1497	9.05
69	Tm	Thulium	168.934	3	1727	1545	9.33
70	Yb	Ytterbium	173.04	3,2	1427	824	6.98
71	Lu	Lutetium	174.97	3	3327	1652	9.84

★★ Actinides

Atomic number	Symbol	Name	Atomic weight	Oxidation states	Boiling point, °C	Melting point, °C	Density (g/mL)
90	Th	Thorium	232.04	4	3850	1750	11.7
91	Pa	Protactinium	(231)	5,4	–	(1230)	15.4
92	U	Uranium	238.036	6,5,4,3	3818	1132	19.07
93	Np	Neptunium	(237)	6,5,4,3	–	637	19.5
94	Pu	Plutonium	(242)	6,5,4,3	3235	640	–
95	Am	Americium	(243)	6,5,4,3	–	–	11.7
96	Cm	Curium	(247)	3	–	–	–
97	Bk	Berkelium	(247)	4,3	–	–	–
98	Cf	Californium	(249)	3	–	–	–
99	Es	Einsteinium	(254)		–	–	–
100	Fm	Fermium	(253)		–	–	–
101	Md	Mendelevium	(256)		–	–	–
102	No	Nobelium	(254)		–	–	–
103	Lr	Lawrencium	(257)		–	–	–

INTRODUCTION TO ENVIRONMENTAL ENGINEERING

McGraw-Hill Series in Water Resources and Environmental Engineering

Consulting Editor

George Tchobanoglous, *University of California, Davis*

Bailey and Ollis: *Biochemical Engineering Fundamentals*
Bouwer: *Groundwater Hydrology*
Canter: *Environmental Impact Assessment*
Chanlett: *Environmental Protection*
Chow, Maidment, and Mays: *Applied Hydrology*
Crites and Tchobanoglous: *Small and Decentralized Wastewater Management*
Davis and Cornwell: *Introduction to Environmental Engineering*
deNevers: *Air Pollution Control Engineering*
Eckenfelder: *Industrial Water Pollution Control*
LaGrega, Buckingham, and Evans: *Hazardous Waste Management*
Linsley, Franzini, Freyberg, and Tchobanoglous: *Water Resources and Engineering*
McGhee: *Water Supply and Sewerage*
Mays and Tung: *Hydrosystems Engineering and Management*
Metcalf & Eddy, Inc.: *Wastewater Engineering: Collection and Pumping of Wastewater*
Metcalf & Eddy, Inc.: *Wastewater Engineering: Treatment, Disposal, Reuse*
Peavy, Rowe, and Tchobanoglous: *Environmental Engineering*
Sawyer and McCarty: *Chemistry for Environmental Engineering*
Tchobanoglous, Theisen, and Vigil: *Integrated Solid Waste Management: Engineering Principles and Management Issues*
Wentz: *Hazardous Waste Management*
Wentz: *Safety, Health, and Environmental Protection*

INTRODUCTION TO ENVIRONMENTAL ENGINEERING

Third Edition

Mackenzie L. Davis
Michigan State University

David A. Cornwell
Environmental Engineering & Technology, Inc.

Boston Burr Ridge, IL Dubuque, IA Madison, WI New York
San Francisco St. Louis Bangkok Bogotá Caracas Kuala Lumpur
Lisbon London Madrid Mexico City Milan Montreal New Delhi
Santiago Seoul Singapore Sydney Taipei Toronto

McGraw-Hill Higher Education

A Division of The ***McGraw-Hill*** *Companies*

INTRODUCTION TO ENVIRONMENTAL ENGINEERING

This book is printed on acid-free paper.

7 8 9 0 DOC/DOC 0 9 8 7 6 5 4 3 2

ISBN 0-07-015918-1

Editorial director: Tom Casson
Sponsoring editor: Erica M. Munson
Project manager: Karen Nelson
Production supervisor: Melonie Salvati
Compositor: Publication Services
Typeface: 10.5/12 Times Roman
Printer: R. R. Donnelley & Sons

Library of Congress Cataloging-in-Publication Data

Davis, Mackenzie Leo (date)
Introduction to Environmental Engineering/Mackenzie L. Davis, David A. Cornwell./3rd ed.
p. cm.
Includes bibliographical references and index.
ISBN 0-07-015918-1
1. Environmental engineering. I. Cornwell, David A., 1948– . II. Title.
III. McGraw-Hill series in water resources and environmental engineering.
TD145 .D26 1998
628–dc21 97-21847

http://www.mhhe.com

To our students,
who make it worthwhile

ABOUT THE AUTHORS

Mackenzie L. Davis is an Associate Professor of Environmental Engineering at Michigan State University. He received all his degrees from the University of Illinois. From 1968 to 1971 he served as a Captain in the U.S. Army Medical Service Corps. During his military service he conducted air pollution surveys at Army ammunition plants. From 1971 to 1973 he was Branch Chief of the Environmental Engineering Branch at the U.S. Army Construction Engineering Research Laboratory. His responsibilities included supervision of research on air, noise, and water pollution control and solid waste management for Army facilities. In 1973 he joined the faculty at Michigan State University. He teaches and conducts research in the areas of air pollution control and hazardous waste management.

In 1987 and 1989–92, under an Intergovernmental Personnel Assignment with the Office of Solid Waste of the U.S. Environmental Protection Agency, Dr. Davis performed technology assessments of treatment methods used to demonstrate the regulatory requirements for the land disposal restrictions ("land ban") promulgated under the Hazardous and Solid Waste Amendments.

Dr. Davis is a member of the following professional organizations: American Chemical Society; American Institute of Chemical Engineers; American Society for Engineering Education; American Meteorological Society; American Society of Civil Engineers; Air & Waste Management Association; Association of Environmental Engineering Professors; and the Water Environment Federation. Currently, he serves as chairperson of the editorial review boards of the *Journal of the Air & Waste Management Association* and *EM.*

His honors and awards include the State-of-the-Art award from A.S.C.E., Chapter Honor Member of Chi Epsilon, Sigma Xi, and election as a Diplomate in the American Academy of Environmental Engineers with certification in hazardous waste management. He has received teaching awards from the American Society of Civil Engineers Students Chapter, Michigan State University College of Engineering, North Central Section of the American Society for Engineering Education, Great Lakes Region of Chi Epsilon, and the AMOCO Corporation. He is a registered professional engineer in Illinois and Michigan.

David A. Cornwell is president of the consulting firm Environmental Engineering & Technology, Inc. He received his Ph.D. in Civil/Environmental Engineering from the University of Florida. He was an Associate Professor in the Civil and Environmental Engineering Department at Michigan State University prior to entering the consulting field. He has chaired many American Water Works Association committees and activities, including serving as a trustee to the Research Division, and he received the 1986 AWWA Engineering Division Best Paper Award. He is also an author in the latest edition of AWWA's *Water Quality and Treatment* book.

PREFACE

This text is for use in an introductory sophomore-level engineering course. It is written in a language and style that will appeal to sophomores. (Some pundits claim it is puerile; we claim that it is at least sophomoric.) We assume that the book will be used in one of the first engineering courses encountered by the student. In the 50 offerings of the course material in this format, we have found that mature college seniors in allied fields—such as biology, chemistry, resource development, fisheries and wildlife, microbiology, and soils science—have no difficulty with the material. Likewise, we suspect that any bright science/mathematics-oriented high school senior would find the material easy to master. Senior level engineering students will find the material a snap, and perhaps beneath their dignity.

We have emphasized concepts, definitions, descriptions, abundant illustrations (to the horror of our beloved publisher), and a tad of humor. In the mathematical presentations we have provided only a few derivations. Engineering professors will complain (it must be confessed with justice) of insufficient rigor. But to this question there are two sides; however important it may be to maintain a uniformly high standard in pure mathematics, the engineering professional may occasionally do well to rest content with the result of the argument. To our minds, the more stringent procedures of the pure mathematician may yield a result that is not more but less demonstrative—and even confusing—to the beginning engineering student. And to insist on the highest standard would mean exclusion of many important subjects altogether in view of the space that would be required.

Two themes are carried through the text. The first is an introduction to the concept of mass balance as a tool for problem solving. The concept is introduced in the first chapter and then applied for conservative systems in hydrology (hydrologic cycle, development of the rational formula, and reservoir design). This theme is expanded to include chemical reaction kinetics, reactor design, and sludge mass balance in Chapter 3. The DO sag curve is developed using a mass balance approach. The design equations for a completely mixed activated sludge system and a more elaborate sludge mass balance are developed in Chapter 5. Mass balance is used to account for the production of sulfur dioxide from the combustion of coal and in the development of absorber design equations in Chapter 6. In Chapter 9, a mass balance approach is used for waste audit.

The second theme is the concept that pollution control begins with the minimization of the generation of waste. This is introduced in Chapter 2 under the heading of water conservation. It is addressed again in the sludge management section of the water treatment chapter; in air pollution control, resource conservation, and recovery in solid waste management; and in the reduction of hazardous waste generation rates.

A solution manual is available for qualified instructors. Sample course outlines and solved exams are included in the solution manual. Please inquire with your McGraw-Hill representative. We appreciate any comments, suggestions, corrections, and contributions of problems for future revisions.

As it stands in the curriculum at Michigan State University, the course bearing the title of this book provides the foundation for four follow-on senior level environmental engineering courses. We believe in and support the philosophy of the Association of Environmental Engineering Professors that the undergraduate environmental engineering curriculum must be expanded if we are to maintain modern, rigorous graduate programs. The fact that our better high schools provide opportunities for first-year calculus and college level chemistry, physics, and computer courses allows us to do this.

ACKNOWLEDGMENTS

As with any other text, the number of individuals who have made it possible far exceeds those whose names grace the cover. At the hazard of leaving someone out, we would like to explicitly thank the following individuals for their contribution.

The following students helped to solve problems, proofread text, prepare illustrations, raise embarrassing questions, and generally make sure that other students could understand it: Stephanie Albert, Deb Allen, Mark Bishop, Kristen Brandt, Jeff Brown, Nicole Chernoby, Linda Clowater, Shauna Cohen, John Cooley, Ted Coyer, Marcia Curran, Kimberly Doherty, Bobbie Dougherty, Lisa Egleston, Craig Fricke, Elizabeth Fry, Edith Hooten, Kathy Hulley, Angela Ilieff, Gary Lefko, Lynelle Marolf, Lisa McClanahan, Tim McNamara, Becky Mursch, Cheryl Oliver, Lynnette Payne, Jim Peters, Kristie Piner, Christine Pomeroy, Erica Rayner, Bob Reynolds, Laurene Rhyne, Sandra Risley, Lee Sawatzki, Mary Stewart, and Rick Wirsing. To them a hearty thank you!

We would also like to thank the following reviewers for their many helpful comments and suggestions: Wayne Chudyk, Tufts University; John Cleasby, Iowa State University; Michael J. Humenick, University of Wyoming; Tim C. Keener, University of Cincinnati; Paul King, Northeastern University; Susan Masten, Michigan State University; R. J. Murphy, University of South Florida; Thomas G. Sanders, Colorado State University; and Ron Wukasch, Purdue University.

To John Eastman, our esteemed friend and former colleague, we offer our sincere appreciation. His contribution of Chapter 4, as well as constructive criticism and "independent" testing of the material, have been exceptionally helpful.

And last, but certainly not least, we wish to thank our families, who have put up with the nonsense of book writing.

Mackenzie L. Davis
David F. Cornwell

CONTENTS

Preface ix

1 Introduction 1

1-1 What Is Environmental Engineering? 2
1-2 Introduction to Environmental Engineering 4
1-3 Environmental Systems Overview 7
1-4 Environmental Legislation and Regulation 16
1-5 Environmental Ethics 33
1-6 A Materials Balance Approach to Problem Solving 35
1-7 Chapter Review 41
1-8 Problems 42
1-9 Discussion Questions 43
1-10 Additional Reading 45

2 Hydrology 46

2-1 Fundamentals 47
2-2 Rainfall Analysis 57
2-3 Runoff Analysis 62
2-4 Storage of Reservoirs 84
2-5 Groundwater and Wells 87
2-6 Waste Minimization and Water Conservation 110
2-7 Chapter Review 110
2-8 Problems 112
2-9 Discussion Questions 128
2-10 Additional Reading 130

3 Water Treatment 131

3-1 Introduction 132
3-2 Coagulation 172
3-3 Softening 178
3-4 Reactors 199
3-5 Mixing and Flocculation 201

3-6 Sedimentation 211
3-7 Filtration 228
3-8 Disinfection 240
3-9 Adsorption 249
3-10 Water Plant Waste Management 250
3-11 Chapter Review 265
3-12 Problems 267
3-13 Discussion Questions 281
3-14 Additional Reading 282

4 Water Quality Management 283

4-1 Introduction 284
4-2 Water Pollutants and Their Sources 285
4-3 Water Quality Management in Rivers 288
4-4 Water Quality Management in Lakes 320
4-5 Chapter Review 330
4-6 Problems 332
4-7 Discussion Questions 336
4-8 Additional Reading 337

5 Wastewater Treatment 338

5-1 Wastewater Microbiology 339
5-2 Characteristics of Wastewater 351
5-3 On-Site Disposal Systems 354
5-4 Municipal Wastewater Treatment Systems 361
5-5 Unit Operations of Pretreatment 364
5-6 Primary Treatment 372
5-7 Unit Processes of Secondary Treatment 374
5-8 Disinfection 411
5-9 Advanced Wastewater Treatment 411
5-10 Land Treatment 415
5-11 Sludge Treatment 418
5-12 Sludge Disposal 441
5-13 Chapter Review 445
5-14 Problems 447
5-15 Discussion Questions 457
5-16 Additional Reading 457

6 Air Pollution 459

6-1 Physical and Chemical Fundamentals 460
6-2 Air Pollution Perspective 463
6-3 Air Pollution Standards 463
6-4 Effects of Air Pollutants 465
6-5 Origin and Fate of Air Pollutants 475
6-6 Micro and Macro Air Pollution 481
6-7 Air Pollution Meteorology 491
6-8 Atmospheric Dispersion 500
6-9 Indoor Air Quality Model 509
6-10 Air Pollution Control of Stationary Sources 511
6-11 Air Pollution Control of Mobile Sources 535

6-12 Waste Minimization 542
6-13 Chapter Review 542
6-14 Problems 544
6-15 Discussion Questions 549
6-16 Additional Reading 549

7 Noise Pollution 550

7-1 Introduction 551
7-2 Effects of Noise on People 564
7-3 Rating Systems 580
7-4 Community Noise Sources and Criteria 584
7-5 Transmission of Sound Outdoors 590
7-6 Traffic Noise Prediction 599
7-7 Noise Control 610
7-8 Chapter Review 624
7-9 Problems 625
7-10 Discussion Questions 629
7-11 Additional Reading 629

8 Solid Waste Management 630

8-1 Perspective 631
8-2 Collection 639
8-3 Interroute Transfer 654
8-4 Disposal by Sanitary Landfill 658
8-5 Waste to Energy 677
8-6 Resource Conservation and Recovery 682
8-7 Chapter Review 691
8-8 Problems 693
8-9 Discussion Questions 701
8-10 Additional Reading 701

9 Hazardous Waste Management 702

9-1 The Hazard 703
9-2 Risk 707
9-3 Definition and Classification of Hazardous Waste 726
9-4 RCRA and HSWA 732
9-5 CERCLA and SARA 745
9-6 Hazardous Waste Management 750
9-7 Treatment Technologies 755
9-8 Land Disposal 783
9-9 Groundwater Contamination and Remediation 789
9-10 Chapter Review 798
9-11 Problems 800
9-12 Discussion Questions 808
9-13 Additional Reading 809

10 Ionizing Radiation 810

10-1 Fundamentals 811
10-2 Biological Effects of Ionizing Radiation 825
10-3 Radiation Standards 834

10-4 Radiation Exposure 837
10-5 Radiation Protection 841
10-6 Radioactive Waste 850
10-7 Chapter Review 863
10-8 Problems 864
10-9 Discussion Questions 865

Appendix A Properties of Air, Water, and Selected Chemicals 867

Appendix B Noise Computation Tables and Nomographs 879

Appendix C EPA Hazardous Waste Codes 883

Index 893

CHAPTER 1

INTRODUCTION

1-1 WHAT IS ENVIRONMENTAL ENGINEERING?
Professions, Learned and Otherwise
And What Is Engineering?
On to Environmental Engineering
1-2 INTRODUCTION TO ENVIRONMENTAL ENGINEERING
Where Do We Start?
A Short Outline of This Book
Le Système International d'Unités
1-3 ENVIRONMENTAL SYSTEMS OVERVIEW
Systems as Such
Water Resource Management System
Air Resource Management System
Solid Waste Management
Multimedia Systems
1-4 ENVIRONMENTAL LEGISLATION AND REGULATION
Water Quality Management
Air Quality Management
Noise Pollution Control
Solid Waste
Hazardous Wastes
Atomic Energy and Radiation
Environmental Legislation and Regulation in the 1980s
1-5 ENVIRONMENTAL ETHICS
Case 1: To Add or Not to Add
Case 2: Too Close for Comfort
1-6 A MATERIALS BALANCE APPROACH TO PROBLEM SOLVING

1-7 CHAPTER REVIEW
1-8 PROBLEMS
1-9 DISCUSSION QUESTIONS
1-10 ADDITIONAL READING

1-1 WHAT IS ENVIRONMENTAL ENGINEERING?

Professions, Learned and Otherwise

Webster's dictionary defines the learned professions as law, medicine, and theology. It has been suggested that engineers may not be learned enough to rank among these because the study of law, medicine, or theology requires considerably more than four years of undergraduate work. There was a time, some hundred years ago, when the four-year engineering program was two years longer than those of the learned professions! At any rate, *Webster's* is willing to concede that engineering, along with teaching and writing, is a profession even if it is not "learned." At a minimum, a profession is an occupation that requires advanced training in the liberal arts or sciences and mental rather than manual work. This definition excludes professional athletes, police, firefighters, politicians, actors, and soldiers.

But being a professional is more than being in or of a profession. True professionals are those who pursue their learned art in a spirit of public service.[1] True professionalism is defined by the following seven characteristics:[2]

1. Professional decisions are made by means of general principles, theories, or propositions that are independent of the particular case under consideration.
2. Professional decisions imply knowledge in a specific area in which the person is expert. The professional is an expert only in his or her profession and not an expert at everything.
3. The professional's relations with his or her clients are objective and independent of particular sentiments about them.
4. A professional achieves status and financial reward by accomplishment, not by inherent qualities such as birth order, race, religion, sex, or age or by membership in a union.
5. A professional's decisions are assumed to be on behalf of the client and to be independent of self-interest.
6. The professional relates to a voluntary association of professionals and accepts only the authority of those colleagues as a sanction on his or her own behavior.

---

[1]American Society of Civil Engineers (ASCE), *Official Record,* 1973.

[2]Edgar H. Schein, *Organizational Socialization and the Profession of Management,* 3rd Douglas Murray McGregor Memorial Lecture to the Alfred P. Sloan School of Management, Massachusetts Institute of Technology, 1968.

AMERICAN SOCIETY OF CIVIL ENGINEERS
CODE OF ETHICS

Fundamental Principles

Engineers uphold and advance the integrity, honor and dignity of the engineering profession by:

1. using their knowledge and skill for the enhancement of human welfare;
2. being honest and impartial and serving with fidelity the public, their employers and clients;
3. striving to increase the competence and prestige of the engineering profession; and
4. supporting the professional and technical societies of their disciplines.

Fundamental Canons

1. Engineers shall hold paramount the safety, health and welfare of the public in the performance of their professional duties.
2. Engineers shall perform services only in areas of their competence.
3. Engineers shall issue public statements only in an objective and truthful manner.
4. Engineers shall act in professional matters for each employer or client as faithful agents or trustees, and shall avoid conflicts of interest.
5. Engineers shall build their professional reputation on the merit of their services and shall not compete unfairly with others.
6. Engineers shall act in such a manner as to uphold and enhance the honor, integrity, and dignity of the engineering profession.
7. Engineers shall continue their professional development throughout their careers, and shall provide opportunities for the professional development of those engineers under their supervision.

FIGURE 1-1
American Society of Civil Engineers code of ethics.

7. A professional is someone who knows better what is good for clients than do the clients. The professional's expertise puts the client into a very vulnerable position. This vulnerability has necessitated the development of strong professional codes and ethics, which serve to protect the client. Such codes are enforced through the colleague peer group.

The branch of engineering called civil engineering, from which environmental engineering is primarily, but not exclusively, derived, has an established code of ethics that embodies these principles. The 1977 revision of the code is summarized in Figure 1-1.

And What Is Engineering?

The Engineer's Council for Professional Development (ECPD) has published a definition of engineering to which we subscribe:

> Engineering is the profession in which a knowledge of the mathematical and natural sciences gained by study, experience, and practice is applied with judgment to

> develop ways to utilize economically the materials and forces of nature for the benefit of mankind.[3]

This definition implies that there are fundamental differences between scientists and engineers. The key is not so much in the individual parts of the definition, but rather in the integration of the parts. It is inherent in the professional development of the engineer that he or she must attain experience, practice, and judgment under the tutelage of an experienced engineer.[4] Engineering has at least this much in common with the learned professions!

Engineers are frequently pressed to explain why they are different from scientists. Consider the following distinction: "Scientists discover things. Engineers make them work."[5]

On to Environmental Engineering

The Environmental Engineering Division of the American Society of Civil Engineers (ASCE) has published the following statement of purpose:

> Environmental engineering is manifest by sound engineering thought and practice in the solution of problems of environmental sanitation, notably in the provision of safe, palatable, and ample public water supplies; the proper disposal of or recycle of wastewater and solid wastes; the adequate drainage of urban and rural areas for proper sanitation; and the control of water, soil, and atmospheric pollution, and the social and environmental impact of these solutions. Furthermore it is concerned with engineering problems in the field of public health, such as control of arthropod-borne diseases, the elimination of industrial health hazards, and the provision of adequate sanitation in urban, rural, and recreational areas, and the effect of technological advances on the environment.

Thus, we may consider what environmental engineering is not. It is not concerned primarily with heating, ventilating, or air conditioning (HVAC), nor is it concerned primarily with landscape architecture. Neither should it be confused with the architectural and structural engineering functions associated with built environments, such as homes, offices, and other workplaces.

1-2 INTRODUCTION TO ENVIRONMENTAL ENGINEERING

Where Do We Start?

We have used the ASCE definition of an environmental engineer as a basis for this book. Given the constraints of time and space, we have limited ourselves to the

---

[3] We might add "and womankind."

[4] Although the Engineer-in-Training (E.I.T.) exam may be taken upon completion of the B.S., the Professional Engineers (P.E.) exam may not be taken until after four years of engineering experience.

[5] Robert MacVicar, Vice-President, Oklahoma State University, Commencement Address, June 1983.

following topics from the definition:

1. Provision of safe, palatable, and ample public water supplies
2. Proper disposal of or recycling of wastewater and solid wastes
3. Control of water, soil, and atmospheric pollution (including noise as an atmospheric pollutant)

A Short Outline of This Book

Hydrology (Chapter 2) provides a beginning point for our discussion as we look toward providing ample water from either surface water or groundwater. Since hydrology is concerned with flooding as well as with droughts, we also touch on the adequate drainage portion of the definition. The discussion of the physics of groundwater movement will give you the tools you need to understand problems of groundwater pollution.

In Chapter 3 we turn from water quantity to water quality. First, we review some basic chemistry concepts and calculations; then we examine some characteristics of water that affect its quality. Finally, we explain how to treat water for public consumption.

In Chapter 4 we consider the effects of various materials on water quality. In particular, we spend a good deal of time examining the effects of organic pollution on the levels of dissolved oxygen in the water. Dissolved oxygen is required for higher forms of aquatic life, such as fish, to survive.

Wastewater treatment is the subject of Chapter 5. Here, we look at how we can remove pollutants that reduce the quality of the lake or stream. Our emphasis is on municipal wastewater treatment.

In Chapters 6 and 7, we turn to the control of atmospheric pollution and noise control. After a brief introduction to the health effects and other environmental impacts of air pollutants and noise, we examine transport processes that carry pollutants from their source to people, as well as some methods of control.

Solid waste is the topic of Chapter 8. Collection, disposal, and recycling of solid waste are fundamental needs of our complex urban society. This chapter will present some of the tools for understanding and solving problems in solid waste management.

Hazardous waste is the topic of Chapter 9. Methods of dealing with abandoned hazardous waste sites and managing the wastes we are continually generating are discussed. We examine some alternatives for treatment of these wastes as an application of the technologies addressed in earlier chapters.

The final chapter is a brief examination of ionizing radiation. A brief introduction to health effects of radiation is followed by a discussion of management techniques for both radioactive waste and x-rays.

Le Système International d'Unités

Since this text is written using the International System of Units (SI), we thought it would be appropriate to include a brief discussion of the evolution of SI at this point. The following was extracted from the American Society for Testing Materials (ASTM) publication numbered E380:

> The decimal system of units was first conceived of in the 16th century, when there was a great confusion and jumble of units of weights and measures. It was not until 1790, however, that the French National Assembly requested the French Academy of Sciences to work out a system of units suitable for adoption by the entire world. This system, based on the meter as a unit of length and the gram as a unit of mass, was adapted as a practical measure to benefit industry and commerce. Physicists soon realized its advantages and it was adopted also in scientific and technical circles. The importance of the regulation of weights and measures was recognized in Article 1, Section 8 of the United States Constitution. Although the Constitution was written in 1787, the metric system was not legalized in this country until 1866. In 1893, the international meter and kilogram became the fundamental standards of length and mass in the United States, both for metric and customary weights and measures.
>
> Meanwhile, international standardization began with a meeting in 1870 of 15 nations in Paris that led to the International Metric Convention of May 20, 1875. It was at this convention that a permanent International Bureau of Weights and Measures was established near Paris. A General Conference on Weights and Measures (CGPM) was also constituted to handle all international matters concerning the metric system. The CGPM nominally meets every sixth year in Paris. Its function is to control the International Bureau of Weights and Measures. The Bureau's duties include preserving the metric standards, comparing national standards with them, and conducting research to establish new standards. The National Bureau of Standards represents the United States in these activities.
>
> The original metric system provided a coherent set of units for the measurement of length, area, volume, capacity, and mass based on two fundamental units: the meter and the kilogram. Measurement of additional quantities required for science and commerce has necessitated development of additional fundamental and derived units. Numerous other systems based on these two metric units have been used. A unit of time was added to produce the centimeter-gram-second (CGS) system adopted in 1881 by the International Congress of Electricity. By about 1900 practical measurements in metric units began to be based on the meter-kilogram-seconds (MKS) system. In 1935 Professor Giovanni Giorgi recommended that the MKS system of mechanics be linked with the electromagnetic system of units by adoption of one of the units—ampere, coulomb, ohm, or volt—for the fourth base unit. This recommendation was accepted and in 1950 the ampere, the unit of electric current, was established as a base unit to form the MKSA system.
>
> In 1954 the 10th CGPM adopted a rationalized and coherent system of units based on the four MKSA units, plus the degree Kelvin as the unit of temperature and the candela as the unit of luminous intensity. In 1960 the 11th CGPM formally gave the system the full title of International System of Units, for which the abbreviation is "SI" in all languages. Thirty-six countries, including the United States, participated in the 1960 conference. In 1964 the 12th CGPM made some refinements, and in 1967 the 13th CGPM redefined the second, renamed the unit of temperature as the kelvin (K), and revised the definition of the candela. In 1971 the 14th CGPM added a seventh base unit, the mole, and approved the pascal (Pa) as a special name for the SI unit of pressure or stress, the newton per square metre, the siemens (S) as a special name for the unit of electrical conductance, and the reciprocal ohm (mho) or the ampere per volt.

SI is a rationalized selection of units from the metric system which individually are not new. It includes a unit of force (the newton) which was introduced in place of the kilogram-force to indicate by its name that it is a unit of force and not of mass. SI is a coherent system with seven base units for which names, symbols, and precise definitions have been established. Many derived units are defined in terms of the base units, symbols assigned to each, and, in some cases, given names, as for example, the newton (N).

Originally (1795) the litre was intended to be identical with the cubic decimetre. The third General Conference on Weights and Measures, meeting in 1901, decided to define the litre as the volume occupied by the mass of one kilogram of pure water at its maximum density under normal atmospheric pressure. Careful determinations subsequently established the litre so defined as being equivalent to 1.000028 dm^3. This conference declared that the word "litre" was a special name for the cubic decimetre. Thus its use is permitted in SI, but is discouraged, since it creates two units for the same quantity, and its use in precision measurements might conflict with measurements recorded under the old definition.

The SI naming convention for factors of ten is included on the inside cover of this book.

1-3 ENVIRONMENTAL SYSTEMS OVERVIEW

Systems as Such

Before we begin in earnest, we thought it worth taking a look at the problems to be discussed in this text in a larger perspective. Engineers like to call this the "systems approach," that is, looking at all the interrelated parts and their effects on one another. In environmental systems it is doubtful that mere mortals can ever hope to identify all the interrelated parts, to say nothing of trying to establish their effects on one another. The first thing the systems engineer does, then, is to simplify the system to a tractable size that behaves in a fashion similar to the real system. The simplified model does not behave in detail as the system does, but it gives a fair approximation of what is going on.

We have followed this pattern of simplification in our description of three environmental systems: the water resource management system, the air resource management system, and the solid waste management system. Pollution problems that are confined to one of these systems are called single-medium problems if the medium is either air, water, or soil. Many important environmental problems are not confined to one of these simple systems but cross the boundaries from one to the other. These problems are referred to as *multimedia* pollution problems.

Water Resource Management System

Water supply subsystem. The nature of the water source commonly determines the planning, design, and operation of the collection, purification, transmission, and distribution works. The two major sources used to supply community and industrial needs are referred to as *surface water* and *groundwater.* Streams, lakes, and rivers are the surface water sources. Groundwater sources are those pumped from wells.

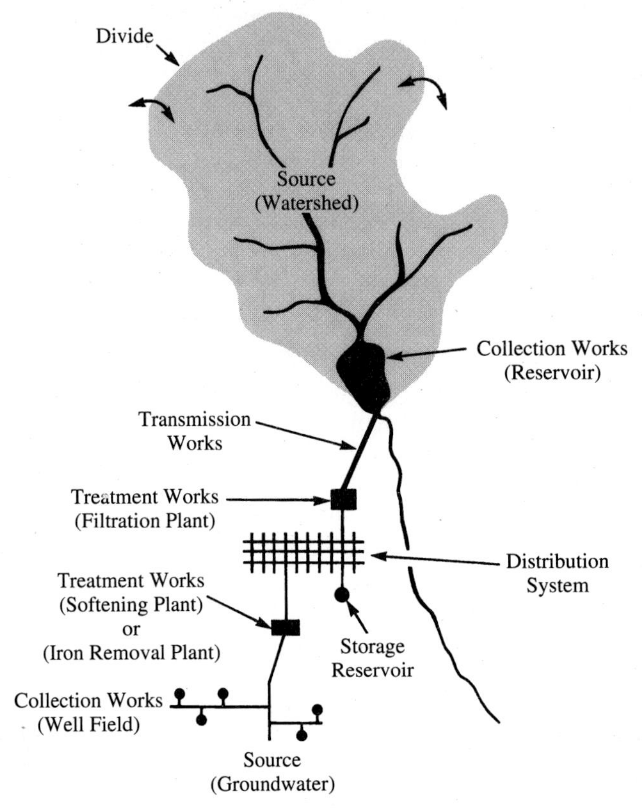

FIGURE 1-2
An extension of the water supply resource system.

Figure 1-2 depicts an extension of the water resource system to serve a small community. The source in each case determines the type of collection works and the type of treatment works.[6] The pipe network in the city is called the distribution system. The pipes themselves are often referred to as water mains. Water in the mains generally is kept at a pressure between 200 and 860 kilopascals (kPa). Excess water produced by the treatment plant during periods of low demand[7] (usually the nighttime hours) is held in a storage reservoir. The storage reservoir may be elevated (the ubiquitous water tower), or it may be at ground level. The stored water is used to meet high demand during the day. Storage compensates for changes in demand and allows a smaller treatment plant to be built. It also provides emergency backup in case of a fire.

---

[6] *Works* is a noun used in the plural to mean "engineering structures." It is used in the same sense as *art works.*

[7] *Demand* is the use of water by consumers. This use of the word derives from the economic term meaning "the desire for a commodity." The consumers express their desire by opening the faucet or flushing the water closet (W.C.).

Population and water-consumption patterns are the prime factors that govern the quantity of water required and hence the source and the whole composition of the water resource system. One of the first steps in the selection of a suitable water-supply source is determining the demand that will be placed on it. The essential elements of water demand include average daily water consumption and peak rate of demand. Average daily water consumption must be estimated for two reasons: (1) to determine the ability of the water source to meet continuing demands over critical periods when surface flows are low or groundwater tables are at minimum elevations, and (2) for purposes of estimating quantities of stored water that would satisfy demands during these critical periods. The peak demand rates must be estimated in order to determine plumbing and pipe sizing, pressure losses, and storage requirements necessary to supply sufficient water during periods of peak water demand.

Many factors influence water use for a given system. For example, the mere fact that water under pressure is available stimulates its use, often excessively, for watering lawns and gardens, for washing automobiles, for operating air-conditioning equipment, and for performing many other activities at home and in industry. The following factors have been found to influence water consumption in a major way:

1. Industrial activity
2. Meterage
3. System management
4. Standard of living
5. Climate

The following factors also influence water consumption to a lesser degree: extent of sewerage, system pressure, water price, and availability of private wells.

The influence of industry is to increase per capita[8] water demand. Small rural and suburban communities will use less water per person than industrialized communities. Industry is probably the largest single factor influencing per capita water use.

The second most important factor in water use is whether individual consumers have water meters. Meterage imposes a sense of responsibility not found in unmetered residences and businesses. This sense of responsibility reduces per capita water consumption because customers repair leaks and make more conservative water-use decisions almost regardless of price. Because water is so cheap, price is not much of a factor.

Following meterage closely is the aspect called system management. If the water distribution system is well managed, per capita water consumption is less than if it is not well managed. Well-managed systems are those in which the managers know when and where leaks in the water mains occur and have them repaired promptly.

Industrial activity, meterage, and system management are more significant factors controlling water consumption than are either the standard of living or the

[8] *Per capita* is a Latin term that means "by heads." Here it means "per person." This assumes that each person has one head (on the average).

TABLE 1-1
Examples of variation in per capita water consumption

		Percent of per capita consumption		
Location	**Lpcd**	**Industry**	**Commercial**	**Residential**
Lansing, MI	479	17.3	40.2	42.5
East Lansing, MI	310	0	5	95
Michigan State University	307	0	1	99

Data from local treatment plants, 1994.

climate. The rationale for the latter two factors is straightforward. Per capita water use increases with an increased standard of living. Highly developed countries use much more water than the less developed nations. Likewise, higher socioeconomic status implies greater per capita water use than lower socioeconomic status. Higher average annual temperature implies higher per capita water use, whereas areas of high rainfall experience lower water use.

The average national value for fresh water consumptive use in the United States in 1990 was estimated to be 1,411 liters per capita per day (Lpcd).[9] A similar study conducted in 1970 yielded a value of 628 Lpcd. The average single family residence uses about 400 Lpcd. The variation in demand is normally reported as a factor of the average day. For metered dwellings the factors are as follows: maximum day $= 2.2 \times$ average day; peak hour $= 5.3 \times$ average day.[10] Some average daily figures and the contribution of various sectors to demand are shown in Table 1-1.

Wastewater disposal subsystem. Safe disposal of all human wastes is necessary to protect the health of the individual, the family, and the community, and also to prevent the occurrence of certain nuisances. To accomplish satisfactory results, human wastes must be disposed of so that:

1. They will not contaminate any drinking water supply.
2. They will not give rise to a public health hazard by being accessible to insects, rodents, or other possible carriers that may come into contact with food or drinking water.
3. They will not give rise to a public health hazard by being accessible to children.
4. They will not cause violation of laws or regulations governing water pollution or sewage disposal.

[9]P. H. Gleick, ed., *Water in Crisis: A Guide to the World's Fresh Water Resources,* New York: Oxford University Press, p. 388, 1993.

[10]F. P. Linaweaver, Jr., J. C. Geyer, and J. B. Wolff, "Summary Report on the Residential Water Use Research Project," *Journal of the American Water Works Association,* vol. 59, p. 267, 1967.

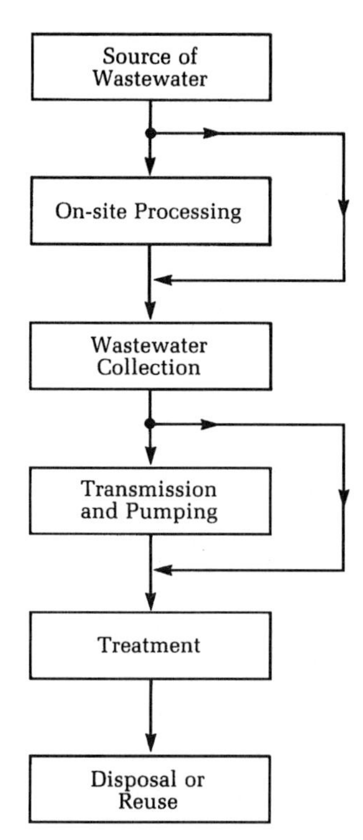

FIGURE 1-3
Wastewater management subsystem. (*Source:* R.K. Linsley and J.B. Fanzini, *Water Resources Engineering,* New York: McGraw-Hill, 1979. Reprinted by permission.)

5. They will not pollute or contaminate the waters of any bathing beach, shellfish-breeding ground, or stream used for public or domestic water-supply purposes, or for recreational purposes.
6. They will not give rise to a nuisance due to odor or unsightly appearance.

These criteria can best be met by the discharge of domestic sewage to an adequate public or community sewerage system.[11] Where no community sewer system exists, on-site disposal by an approved method is mandatory.

In its simplest form the wastewater management subsystem is composed of six parts (Figure 1-3). The source of wastewater may be either industrial wastewater or domestic sewage or both.[12] Industrial wastewater may be subject to some pretreatment on site if it has the potential to upset the municipal wastewater treatment plant (WWTP). Federal regulations refer to municipal wastewater treatment systems as publicly owned treatment works, or POTWs.

---

[11]U.S. Department of Health Education and Welfare, *Manual of Septic Tank Practice* (Public Health Service Publication No. 526), Washington, DC: U.S. Government Printing Office, 1970.

[12]Domestic sewage is sometimes called sanitary sewage, although it is far from being sanitary!

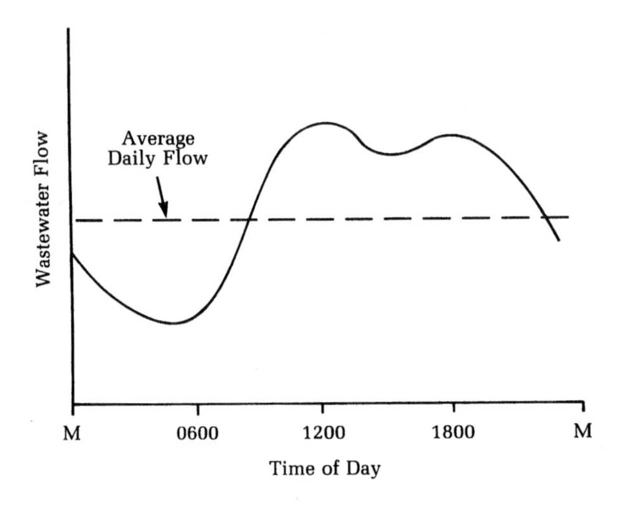

FIGURE 1-4
Typical variation in daily wastewater flow.

The quantity of sewage flowing to the WWTP varies widely throughout the day in response to water usage. A typical daily variation is shown in Figure 1-4. Most of the water used in a community will end up in the sewer. Between 5 and 10 percent of the water is lost in lawn watering, car washing, and other consumptive uses. Consumptive use may be thought of as the difference between the average rate that water flows into the distribution system and the average rate that wastewater flows into the WWTP (excepting the effects of leaks in the pipes).

The quantity of wastewater, with one exception, depends on the same factors that determine the quantity of water required for supply. The major exception is that underground water (groundwater) conditions may strongly affect the quantity of water in the system because of leaks. Whereas the drinking water distribution system is under pressure and is relatively tight, the sewer system is gravity operated and is relatively open. Thus, groundwater may *infiltrate,* or leak into, the system. When manholes lie in low spots, there is the additional possibility of *inflow* through leaks in the manhole cover. Other sources of inflow include direct connections from roof gutters and downspouts, as well as sump pumps used to remove water from basement footing tiles. Infiltration and inflow (I & I) are particularly important during rainstorms. The additional water from I & I may hydraulically overload the sewer causing sewage to back up into houses as well as to reduce the efficiency of the WWTP. New construction techniques and materials have made it possible to reduce I & I to insignificant amounts.

Sewers are classified into three categories: sanitary, storm, and combined. *Sanitary sewers* are designed to carry municipal wastewater from homes and commercial establishments. With proper pretreatment, industrial wastes may also be discharged into these sewers. *Storm sewers* are designed to handle excess rainwater to prevent flooding of low areas. While sanitary sewers convey wastewater to treatment facilities, storm sewers generally discharge into rivers and streams. *Combined sewers* are expected to accommodate both municipal wastewater and stormwater. These systems are designed so that during dry periods the wastewater is carried to a treatment facility. During rain storms, the excess water is discharged directly into a river, stream, or lake without treatment. Unfortunately the stormwater is mixed with untreated sewage. Modern design practice discourages the building of combined sewers, and the continued improvement of our natural water bodies will probably require extensive replacement of combined sewers with separate systems for sanitary and storm flow.

When gravity flow is not possible or when sewer trenches become uneconomically deep, the wastewater may be pumped. When the sewage is pumped vertically to discharge into a higher-elevation gravity sewer, the location of the sewage pump is called a *lift station.*

Sewage treatment is performed at the WWTP to stabilize the waste material, that is, to make it less putrescible. The effluent from the WWTP may be discharged into an ocean, lake, or river (called the receiving body). Alternatively, it may be discharged onto (or into) the ground, or be processed for reuse. The by-product sludge from the WWTP also must be disposed of in an environmentally acceptable manner.

Whether the waste is discharged onto the ground or into a receiving body, care must be exercised not to overtax the assimilative capacity of the ground or receiving body. The fact that the wastewater effluent is cleaner than the river into which it flows does not justify the discharge if it turns out to be the proverbial "straw that breaks the camel's back."

In summary, water resource management is the process of managing both the quantity and the quality of the water used for human benefit without destroying its availability and purity.

Air Resource Management System

Our air resource differs from our water resource in two important aspects. The first is in regard to quantity. Whereas engineering structures are required to provide an adequate water supply, air is delivered free of charge in whatever quantity we desire. The second aspect is in regard to quality. Unlike water, which can be treated before we use it, it is impractical to go about with a gas mask on to treat impure air and with ear plugs in to keep out the noise.

The balance of cost and benefit to obtain a desired quality of air is termed *air resource management*. Cost-benefit analyses can be problematic for at least two reasons. First is the question of what is desired air quality. The basic objective is, of course, to protect the health and welfare of people. But how much air pollution can we stand? We know the tolerable limit is something greater than zero, but tolerance varies from person to person. Second is the question of cost versus benefit. We know that we don't want to spend the entire Gross National Product to ensure that no individual's health or welfare is impaired, but we do know that we want to spend some amount. Although the cost of control can be reasonably determined by standard engineering and economic means, the cost of pollution is still far from being quantitatively assessed.

Air resource management programs are instituted for a variety of reasons. The most defensible reasons are that (1) air quality has deteriorated and there is a need for correction, and (2) the potential for a future problem is strong.

In order to carry out an air resource management program effectively, all of the elements shown in Figure 1-5 must be employed. (Note that with the appropriate

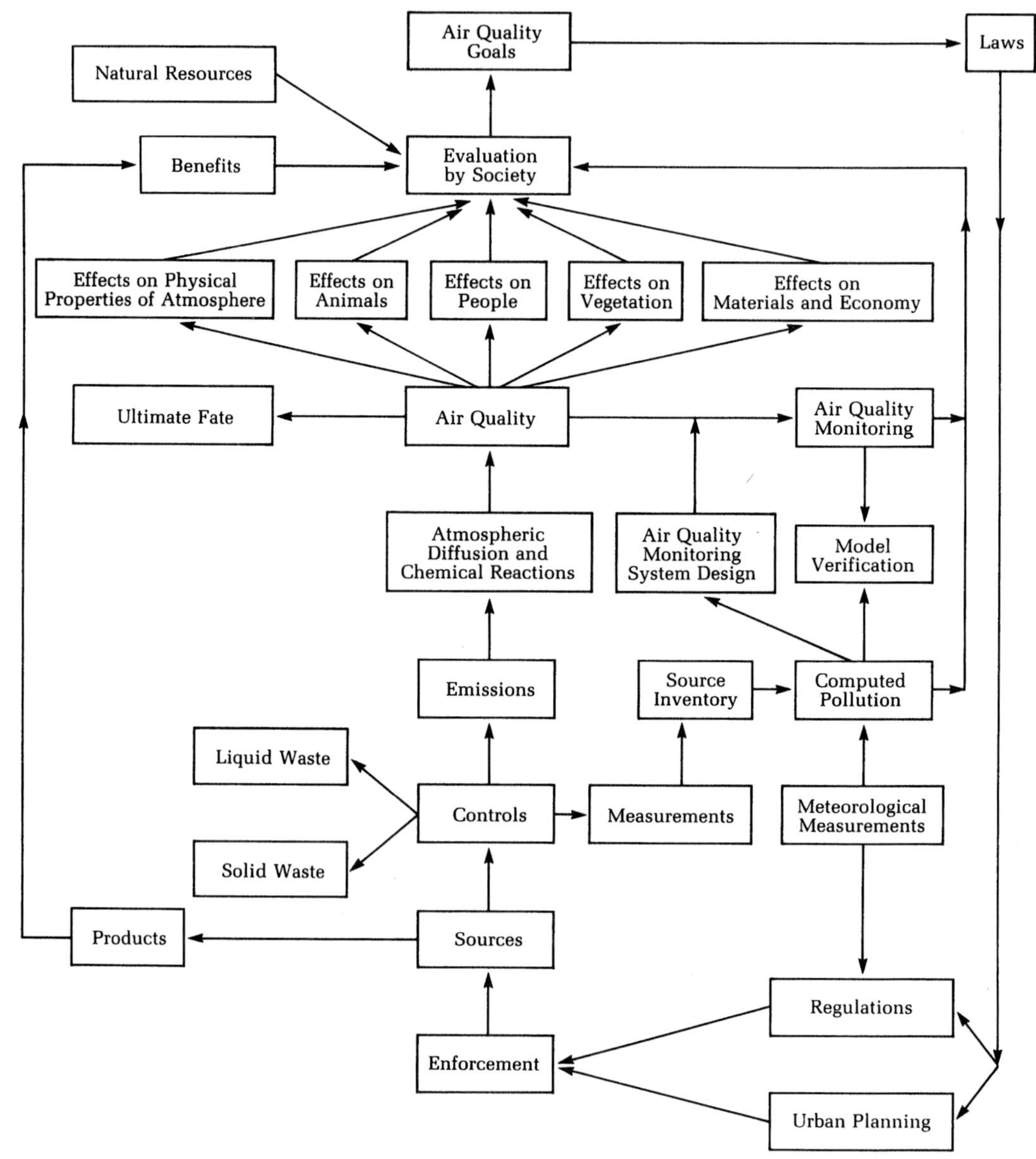

FIGURE 1-5
A simplified block diagram of an air resource management system.

substitution of the word *water* for *air,* these elements apply to management of water resources as well.)

Solid Waste Management

In the past, solid waste was considered a resource, and we will examine its current potential as a resource. Generally, however, solid waste is considered a problem to

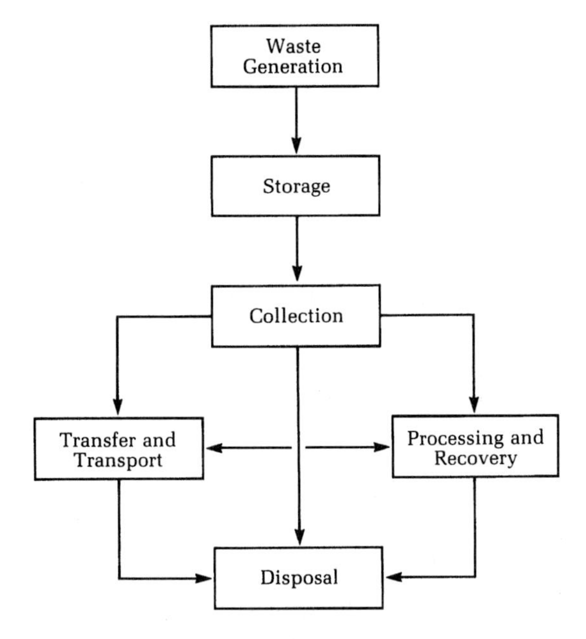

FIGURE 1-6
A simplified block diagram of a solid waste management system. (*Source:* G. Tchobanoglous, H. Theisen, and R. Eliassen, *Solid Wastes,* New York: McGraw-Hill, 1977. Reprinted by permission.)

be solved as cheaply as possible rather than a resource to be recovered. A simplified block diagram of a solid waste management system is shown in Figure 1-6.

While typhoid and cholera epidemics of the mid-1800s spurred water resource management efforts, and while air pollution episodes have prompted better air resource management, we have yet to feel the impact of material or energy shortages severe enough to encourage modern solid waste management. The landfill "crisis" of the 1980s appears to have abated in the early 1990s due to new or expanded landfill capacity and to many initiatives to reduce the amount of solid waste generated. In 1992, 39 states and the District of Columbia had some form of recycling law. Some states have waste-reduction goals, some have recycling goals, and some have both. Some of these goals are mandatory while others are voluntary.

Multimedia Systems

Many environmental problems cross the air-water-soil boundary. An example is acid rain that results from the emission of sulfur oxides and nitrogen oxides into the atmosphere. These pollutants are washed out of the atmosphere, thus cleansing it, but in turn polluting water and changing the soil chemistry, which ultimately results in the death of fish and trees. Thus, our historic reliance on the natural cleansing processes of the atmosphere in designing air-pollution-control equipment has failed to deal with the multimedia nature of the problem. Likewise, disposal of solid waste by incineration results in air pollution, which in turn is controlled by scrubbing with water, resulting in a water pollution problem.

Three lessons have come to us from our experience with multimedia problems. First, it is dangerous to develop models that are too simplistic. Second, environmental engineers must use a multimedia approach and, in particular, work with a multidisciplinary team to solve environmental problems. Third, the best solution to environmental pollution is waste minimization—if waste is not produced, it does not need to be treated or disposed of.

1-4 ENVIRONMENTAL LEGISLATION AND REGULATION

Water Quality Management

Drinking water. The Safe Drinking Water Act of 1974 was the first legislation aimed at providing safe water for consumption rather than for discharge into waterways. In this legislation, Congress directed the EPA to establish drinking water standards for public water systems serving 25 or more people or having 15 or more service connections. In 1975, the EPA published the Interim Drinking Water Regulations.

In 1980, Congress added a provision to prohibit underground injection of wastes without a permit. Major modifications were added in 1986 with the passage of the Safe Drinking Water Act Amendments. The Amendments converted the interim regulations to permanent status; established enforcement responsibility; prohibited the use of lead pipes, solder, and fluxes; and moved to protect sole source aquifers.

The Safe Drinking Water Act Amendments of 1996 strengthened and expanded the protection of drinking water by providing grants for compliance and enforcement, enhanced water-system capacity, operator training, and development of solutions to source pollution. In addition, it provided for public notification of violations within 24 hours (rather than 2 weeks under the old act), and annual reporting of levels of regulated contaminants to consumers. Relief from analysis of contaminants that have never been found and are unlikely to occur was given to reduce analytical costs. EPA was funded to conduct research on health effects and treatment for arsenic, radon, and Crytosporidium. In addition, EPA was required to develop a screening program to identify the risks posed by substances that have an effect similar to that produced by naturally occurring estrogen and to screen pesticides and other chemicals for estrogenic effects.

Water pollution control. (The discussion on water pollution control follows that of L. Ortolano in *Environment Planning and Design.*[13]) Prior to 1948, the principal responsibility for controlling water pollution was assumed by the states and by various local and regional agencies. The first institutions to deal with water pollution problems were created soon after the "sanitary awakening" of the 1850s, when waterborne diseases reached epidemic proportions.

The federal role in water pollution control began with the Public Health Service Act of 1912. This act established the Streams Investigation Station at Cincinnati to carry out water pollution research. The Oil Pollution Act was passed in 1924 to

[13] L. Ortolano, *Environmental Planning and Design,* New York: Wiley, 1984.

TABLE 1-2
Federal laws controlling water pollution

Year	Title	Selected elements of legislation[a]
1948	Water Pollution Control Act	Funds for state water pollution control agencies Technical assistance to states Limited provisions for legal action against polluters
1958	Federal Water Pollution Control Act (FWPCA)	Funds for water pollution research and training Construction grants to municipalities Three-stage enforcement process
1965	Water Quality Act	States set water quality standards States prepare implementation plans
1972	FWPCA Amendments	Zero discharge of pollutants goal BPT and BAT effluent limitations NPDES permits Enforcement based on permit violations
1977	Clean Water Act	BAT requirements for toxic substances BCT requirements for conventional pollutants
1981	Municipal Waste Treatment Construction Grants Amendments	Reduced federal share in construction grants program

[a]The table entries include only the new policies and programs established by each of the laws. Often these provisions were carried forward in modified form as elements of subsequent legislation.

Legend:
BPT = Best Practical Treatment
BAT = Best Available Treatment
NPDES = National Pollution Discharge Elimination System
BCT = Best Conventional Treatment

prevent oily discharges on coastal waters. During the 1930s and 1940s, there was a continuing debate over whether the federal government should take a greater role in controlling water pollution. This debate led to the limited expansion of federal powers expressed in the Water Pollution Control Act of 1948 (Table 1-2). The Federal Water Pollution Control Act (FWPCA) of 1956 was the cornerstone of early federal efforts to reduce pollution. Key elements of the act included a new program of subsidies for municipal treatment plant construction and an expanded basis for federal legal action against polluters. Increased funding for state water pollution control efforts and new support for research and training activities were also provided. Each of these programs was continued in the many amendments to the Federal Water Pollution Control Acts in the 1960s and 1970s.

The 1956 legislation mandated a program of construction grants (subsidies to help pay for municipal treatment facilities) as a response to pressures from municipalities. The cities felt that federal funds should be used to help pay for the treatment plants required by federal law. Grants made under the 1956 act could subsidize as much as one-third of the construction cost for a municipal plant, but the maximum grant for a single project was limited to $600,000. A total of $50 million per year was authorized for the program.

The Water Quality Act of 1965 carried forward many provisions of the earlier federal legislation, generally with an increase in levels of funding. The 1965 act also introduced important new requirements for states to establish ambient water quality standards and detailed plans indicating how the standards would be met. The act also shifted responsibility for administering the federal water quality program from the U.S. Public Health Service to a separate agency, the Federal Water Pollution Control Administration, within the Department of Health, Education, and Welfare (HEW). This was not a permanent change. In 1970, a presidential reorganization order placed the water pollution control activities and several other federal environmental programs in the newly created Environmental Protection Agency (EPA).

In amending the Federal Water Pollution Control Act in 1972, Congress introduced (1) national water quality goals, (2) technology-based effluent limitations, (3) a national discharge permit system, and (4) federal court actions against sources violating permit conditions.

The 1972 amendments aimed to restore and maintain "the chemical, physical and biological integrity of the nation's waters." The amendments specified, as a national goal, that the "discharge of pollutants into navigable waters be eliminated by 1985." This also included an interim goal:

> [W]herever attainable, an interim goal of water quality which provides for the protection and propagation of fish, shellfish and wildlife and provides for recreation in and on the water [should] be achieved by July 1, 1983.

The EPA administrator was required to set effluent restrictions that met the following general requirements of the 1972 amendments: By 1977, all dischargers were to achieve "best practicable control technology currently available" (BPT); and by 1983, all dischargers were to have the "best available technology economically achievable" (BAT). After delays caused by numerous legal challenges to the EPA administrator's effluent limitations guidelines, the BPT provisions were implemented. However, the BAT requirements were so heavily disputed that Congress modified them in the Clean Water Act of 1977.

The principal criticism of the original BAT effluent limitations was that the costs of the very high required percentage reductions in residuals would be much greater than the benefits. In defining BAT, costs were considered, but only in the general context of affordability by industry. Computations of the social benefits of stringent effluent controls were not a central factor. Congress presumed the benefits of eliminating water pollutants would be substantial. Congressional insistence on very strict effluent limitations can also be interpreted as an effort to guarantee the rights of Americans to high-quality waters.

In 1977, Congress responded to critics of BAT by requiring it only for toxic substances. A different requirement was introduced for "conventional pollutants," such as biochemical oxygen demand and suspended solids. The effluent limitations guidelines for these pollutants were to be based on the "best conventional pollutant control technology" (BCT).

The Clean Water Act of 1977 strongly endorsed the view that waterborne toxic substances must be controlled. The text of the act included a list of 65 substances, or classes of substances, to be used as the basis for defining toxics. This list resulted from a 1976 settlement of a legal action in which several environmental organizations sued the EPA administrator for failing to issue toxic pollutant standards. This

TABLE 1-3
EPA's priority pollutant list

1. Antimony
2. Arsenic
3. Beryllium
4. Cadmium
5a. Chromium (III)
5b. Chromium (VI)
6. Copper
7. Lead
8. Mercury
9. Nickel
10. Selenium
11. Silver
12. Thallium
13. Zinc
14. Cyanide
15. Asbestos
16. 2,3,7,8-TCDD (Dioxin)
17. Acrolein
18. Acrylonitrile
19. Benzene
20. Bromoform
21. Carbon tetrachloride
22. Chlorobenzene
23. Chlorodibromomethane
24. Chloroethane
25. 2-Chloroethylvinyl ether
26. Chloroform
27. Dichlorobromomethane
28. 1,1-Dichloroethane
29. 1,2-Dichloroethane
30. 1,1-Dichloroethylene
31. 1,2-Dichloropropane
32. 1,3-Dichloropropylene
33. Ethylbenzene
34. Methyl bromide
35. Methyl chloride
36. Methylene chloride
37. 1,2,2,2-Tetrachloroethane
38. Tetrachloroethylene
39. Toluene
40. 1,2-trans-dichloroethylene
41. 1,1,1-Trichloroethane
42. 2,4 Dichlorophenol
43. Trichloroethylene
44. Vinyl chloride
45. 2-Chlorophenol
46. 2,4-Dichlorophenol
47. 2,4-Dimethylphenol
48. 2-Methyl-4-chlorophenol
49. 2,4-Dinitrophenol
50. 2-Nitrophenol
51. 4-Nitrophenol
52. 3-Methyl-4-chlorophenol
53. Pentachlorophenol
54. Phenol
55. 2,4,6-Trichlorophenol
56. Acenaphthene
57. Acenaphthylene
58. Anthracene
59. Benzidine
60. Benzo(a)anthracene
61. Benzo(a)pyrene
62. Benzo(a)fluoranthene
63. Benzo(ghi)perylene
64. Benzo(k)fluoranthene
65. bis(2-Chloroethoxy)methane
66. bis(2-Chloroethyl)ether
67. bis(2-Chloroisopropyl)ether
68. bis(2-Ethylhexyl)phthalate
69. 4-Bromophenyl phenyl ether
70. Butylbenzyl phthalate
71. 2-Chloronaphthalene
72. 4-Chlorophenyl phenyl ether
73. Chrysene
74. Dibenzo(a,h)anthracene
75. 1,2-Dichlorobenzene
76. 1,3-Dichlorobenzene
77. 1,4-Dichlorobenzene
78. 3,3-Dichlorobenzidine
79. Diethyl phthalate
80. Dimethyl phthalate
81. Di-n-butyl phthalate
82. 2,4-Dinitrotoluene
83. 2,6-Dinitrotoluene
84. Di-n-octyl phthalate
85. 1,2-Diphenylhydrazine
86. Fluoranthene
87. Fluorene
88. Hexachlorobenzene
89. Hexachlorobutadiene
90. Hexachlorocyclopentadiene
91. Hexachloroethane
92. Indeno(1,2,3-cd)pyrene
93. Isophorone
94. Naphthalene
95. Nitrobenzene
96. N-Nitrosodimethylamine
97. N-Nitrosodi-n-propylamine
98. N-Nitrosodiphenylamine
99. Phenanthrene
100. Pyrene
101. 1,2,4-Trichlorobenzene
102. Aldrin
103. alpha-BHC
104. beta-BHC
105. gamma-BHC
106. delta-BHC
107. Chlordane
108. 4,4′-DDT
109. 4,4′-DDE
110. 4,4′-DDD
111. Dieldrin
112. alpha-Endosulfan
113. beta-Endosulfan
114. Endosulfan sulfate
115. Endrin
116. Endrin aldehyde
117. Heptachlor
118. Heptachlor epoxide
119. PCB-1242
120. PCB-1254
121. PCB-1221
122. PCB-1232
123. PCB-1248
124. PCB-1260
125. PCB-1016
126. Toxaphene

Source: 40 CFR 131.36, July 1, 1993.

list was subsequently expanded by EPA to include 127 "priority pollutants" (Table 1-3).

Effluent limitations required by the FWPCA amendments of 1972 (and later the Clean Water Act of 1977) formed the basis for issuing "National Pollutant Discharge Elimination System" (NPDES) permits. The permit system idea stemmed from actions taken by the Department of Justice in the late 1960s. With the support of a favorable interpretation by the Supreme Court, attorneys for the United States relied on the 1899 River and Harbor Act to prosecute industrial sources of water pollution. The 1899 act, which was drafted originally to prohibit deposits of refuse in navigable waters to keep them clear for boat traffic, was interpreted in the 1960s as applying to liquid waste as well. In December 1970, the EPA administrator issued an executive order calling for a water quality management program using permits and penalties based on the River and Harbor Act of 1899. Although this program was delayed by court challenges in 1971, Congress made it a central part of the federal strategy embodied in the FWPCA amendments of 1972.

Air Quality Management

(The discussion on air quality management follows that of L. Ortolano in *Environmental Planning and Design.*[14]) The earliest programs to manage air quality regulated emissions from smokestacks using nuisance law municipal ordinances. The first antismoke ordinance in the United States was issued by Chicago in 1881. Little progress was made in air pollution control in the first half of the twentieth century.

During the 1950s, there was a shift away from nuisance law and municipal ordinances as bases for air quality management. Two factors stimulated the development of additional air pollution control strategies. One was a tragic air pollution episode at Donora, Pennsylvania, which caused death to 20 people and illness to several thousand. The second factor was the growing recognition of the linkage between automobile exhausts and photochemical smog.

The pattern of reliance on local agencies began to change with the passage of the Federal Air Pollution Control Act of 1955 (Table 1-4). This act established a program of federally funded research grants administered by the U.S. Public Health Service. Although the act represented an expansion of the federal role, it was a very limited one. The legislative history of the act indicates congressional intent to restrain federal involvement and to respect the rights and responsibilities of states, counties, and cities in controlling air pollution.

The federal role was further extended by the Clean Air Act of 1963, which allowed direct federal intervention to reduce interstate pollution. The form of intervention followed the enforcement process in the Federal Water Pollution Control Act of 1956.

The first federal restrictions on auto emissions came with the Motor Vehicle Air Pollution Control Act of 1965. Based on earlier auto emission control efforts in California, the 1965 act gave the Secretary of the Department of Health, Education, and Welfare authority to establish permissible emission levels for new automobiles beginning with the 1968 model year. The control of emissions from older vehicles was left to individual states.

---

[14]L. Ortolano, *Environmental Planning and Design,* New York: Wiley, 1984.

TABLE 1-4
Federal laws controlling air pollution

Year	Title	Selected elements of legislation[a]
1955	Air Pollution Control Act	Funds for air pollution research
1960	Motor Vehicle Exhaust Act	Funds for research on vehicle emissions
1963	Clean Air Act	Three-stage enforcement process Funds for state and local air pollution control agencies
1965	Motor Vehicle Air Pollution Control Act	Emission regulations for cars beginning with 1968 models
1967	Air Quality Act	Federally issued criteria documents Federally issued control technique documents Air quality and control regions (AQCRs) defined Requirements for states to set ambient standards for AQCRs Requirements for state implementation plans
1970	Clean Air Act Amendments	National ambient air quality standards New source performance standards Technology forcing auto emission standards Transportation control plans
1977	Clean Air Act Amendments	Relaxation of previous auto emission requirements Vehicle inspection and maintenance programs Prevention of significant deterioration areas Emission offsets for nonattainment areas Study ozone depletion National emission standards for hazardous air pollutants (NESHAP)
1980	Acid Precipitation Act	Development of a long-term research plan
1986	Radon Gas and Indoor Air Quality Research Act	Research program to gather data and to coordinate and assess federal action
1990	Clean Air Act Amendments	Sets attainment dates for criteria air pollutants Imposes new requirements for auto emissions and establishes clean fuels program Identifies 189 hazardous air pollutants to be regulated Establishes SO_2 allowances for acid rain control Establishes a national permit system Sets schedule for phase-out of ozone-depleting compounds

[a]The table entries include only the new policies and programs established by each of the laws. Often these provisions were carried forward in modified form as elements of subsequent legislation.

The Air Quality Act of 1967 borrowed concepts from the Water Quality Control Act of 1965 by requiring states to develop ambient air quality standards and state implementation plans (SIPs) to achieve the standards. Implementation plans were to include emission requirements for controlling air pollution and a timetable for meeting the requirements. Deadlines were set for submitting ambient standards, which were to be established on a region-wide basis.

Although the Clean Air Act Amendments of 1970 continued many of the research and state aid programs established by prior legislation, several aspects of the amendments represented dramatic changes in strategy. These involved (1) the requirement that the administrator of EPA set national ambient air quality standards (NAAQS) and emission standards for selected categories of new industrial facilities, and (2) the explicit delineation (by Congress) of auto emission standards. Another manifestation of the expanded role of the federal government was the requirement of the 1970 amendments that the EPA administrator issue new source performance standards (NSPS). These standards were to control new stationary sources categorized by the administrator as contributing significantly to air pollution.

The Clean Air Act Amendments of 1977 relaxed the emission requirements somewhat and extended the compliance deadlines into the early 1980s. They also defined a concept of Prevention of Significant Deterioration (PSD) areas and required that an area that meets the national ambient standards for a given air pollutant be declared a PSD area for that pollutant. The amendments also defined three classes of PSD areas. For each class, numerical limits indicated the maximum permissible increment of air quality degradation from all new (or modified) stationary sources of pollution in an area.

The 1977 amendments also indicated that significant new sources of pollution could locate in areas that did not meet the NAAQS, but only if certain conditions were satisfied. The amendments required that a significant new source locating in a nonattainment area (one which has not achieved the NAAQS) had to meet strict emission-reduction requirements developed by the EPA administrator. In addition, discharges from the new source had to be more than offset by reductions in emissions from other sources in the region.

In 1979, the EPA extended the concept of emission offsets, as used in nonattainment areas, to a different context: multiple sources of air pollution generated at a single site. This extension, known as the bubble policy, is illustrated in Figure 1-7. The figure depicts a firm that must control releases from smokestacks at two adjacent plants. Before the bubble policy, the firm had to comply with emission standards that allowed only 100 Mg/d from each plant.[15] The total discharge was 200 Mg/d. The unit cost of emission controls for Plant A was much higher than that for Plant B, but the emission requirements were insensitive to these cost differences. Using the bubble policy, the firm is free to decide how to reduce residuals at each plant. The only restriction is that its total discharge must be no greater than 200 Mg/d. Imagine that a bubble surrounds the two plants. The policy allows the firm to make choices within the bubble, but the total discharge from the bubble is restricted. In the early 1980s, the original bubble policy was extended to include plants that were not at the same location (multiplant bubbles).

The Clean Air Act Amendments of 1990 (CAAA) mandate that the EPA promulgate more than 175 new regulations, 30 guidance documents, 35 studies, and 50 new research initiatives. The Congressional mandates are categorized under eleven

---

[15] Mg/d = megagram per day. 1 Mg = 1,000 kg.

Without Bubble
Total Allowed Emissions = 200 Mg/d
Control Cost = $20 Million

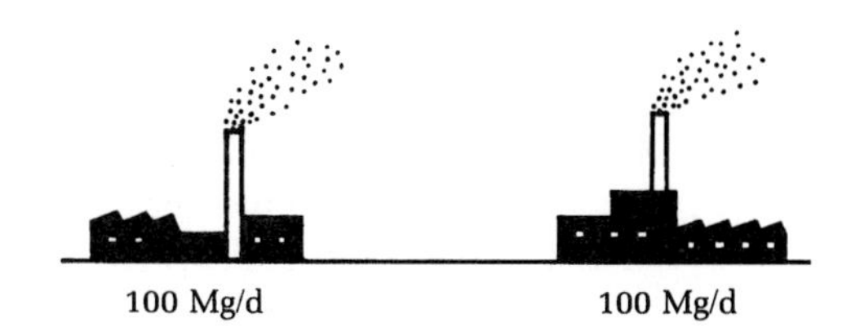

With Bubble
Total Allowed Emissions = 200 Mg/d
Control Cost = $15 Million

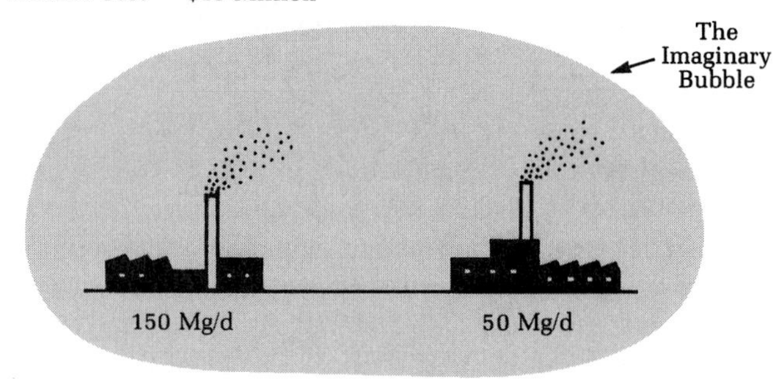

FIGURE 1-7
Illustration of bubble concept.

"Titles" in the Act. It has become common to refer to the requirements of the CAAA by title number.

In light of the fact that three previous deadlines for attainment had come and gone, Title I establishes sixteen new deadlines. Although these are primarily aimed at ozone, there are also classifications for carbon monoxide and fine particulates.

Provisions relating to mobile sources are spelled out in Title II. Cars are required to have dashboard warning lights that signal whether or not pollution control equipment is working. These devices frequently have impregnated chemicals that react with the pollutants. The life expectancy of these devices, in terms of miles driven, is specified as 100,000 miles, rather than the previous requirement of 50,000 miles. Auto makers are required to produce some cars that use clean fuels such as alcohol and some that are powered by electricity. In addition, inspection and maintenance (I/M) programs for metropolitan areas have been expanded.

Because the previous legislation establishing national emission standards for hazardous pollutants (NESHAPs) based on health risk proved too cumbersome, Title III established an initial list of 189 hazardous air pollutants (HAPs) shown in Table 1-5 and directed EPA to establish emission standards based on technology. These standards are to be the maximum achievable control technology (MACT) for a given source category.

TABLE 1-5
Hazardous air pollutants (HAPs)

Acetaldehyde	1,3-Dichloropropene
Acetamide	Dichlorvos
Acetonitrile	Diethanolamine
Acetophenone	N,N-Diethyl aniline (N,N-Dimethylaniline)
2-Acetylaminofluorene	Diethyl sulfate
Acrolein	3,3-Dimethoxybenzidine
Acrylamide	Dimethyl aminoazobenzene
Acrylic acid	3,3′-Dimethyl benzidine
Acrylonitrile	Dimethyl carbamoyl chloride
Allyl chloride	Dimethyl formamide
4-Aminobiphenyl	1,1-Dimethyl hydrazine
Aniline	Dimethyl phthalate
o-Anisidine	Dimethyl sulfate
Asbestos	4,6-Dinitro-o-cresol, and salts
Benzene (including benzene from gasoline)	2,4-Dinitrophenol
Benzidine	2,4-Dinitrotoluene
Benzotrichloride	1,4-Dioxane (1,4-Diethyleneoxide)
Benzyl chloride	1,2-Diphenylhydrazine
Biphenyl	Epichlorohydrin (1-chloro-2,3-epoxypropane)
Bis(2-ethylhexyl)phthalate (DEHP)	1,2-Epoxybutane
Bis(chloromethyl)ether	Ethyl acrylate
Bromoform	Ethyl benzene
1,3-Butadiene	Ethyl carbamate (Urethane)
Calcium cyanamide	Ethyl chloride (Chloroethane)
Caprolactam	Ethylene dibromide (Dibromoethane)
Captan	Ethylene dichloride (1,2-Dichloroethane)
Carbaryl	Ethylene glycol
Carbon disulfide	Ethylene imine (Aziridine)
Carbon tetrachloride	Ethylene oxide
Carbonyl sulfide	Ethylene thiourea
Catechol	Ethylidene dichloride (1,1-Dichloroethane)
Chloramben	Formaldehyde
Chlordane	Heptachlor
Chlorine	Hexachlorobenzene
Chloroacetic acid	Hexachlorobutadiene
2-Chloroacetophenone	Hexachlorocyclopentadiene
Chlorobenzene	Hexachloroethane
Chlorobenzilate	Hexamethylene-1,6-diisocyanate
Chloroform	Hexamethylphosphoramide
Chloromethyl methyl ether	Hexane
Chloroprene	Hydrazine
Cresols/Cresylic acid (isomers and mixture)	Hydrochloric acid
o-Cresol	Hydrogen fluoride (Hydrofluoric acid)
m-Cresol	Hydrogen sulfide
p-Cresol	Hydroquinone
Cumene	Isophorone
2,4-D, salts and esters	Lindane (all isomers)
DDE	Maleic anhydride
Diazomethane	Methanol
Dibenzofurans	Methoxychlor
1,2-Dibromo-3-chloropropane	Methyl bromide (Bromomethane)
Dibutylphthalate	Methyl chloride (Chloromethane)
1,4-Dichlorobenzene(p)	Methyl chloroform (1,1,1-Trichloroethane)
3,3-Dichlorobenzidene	Methyl ethyl ketone (2-Butanone)
Dichloroethyl ether [Bis(2-chloroethyl)ether]	Methyl hydrazine

TABLE 1-5
Hazardous air pollutants (HAPs) (*continued*)

Methyl iodide (Iodomethane)
Methyl isobutyl ketone (Hexone)
Methyl isocyanate
Methyl methacrylate
Methyl tert butyl ether
4,4-Methylene bis(2-chloroaniline)
Methylene chloride (Dichloromethane)
Methylene diphenyl diisocyanate (MDI)
4,4′-Methylenedianiline
Naphthalene
Nitrobenzene
4-Nitrobiphenyl
4-Nitrophenol
2-Nitropropane
N-Nitroso-N-methylurea
N-Nitrosodimethylamine
N-Nitrosomorpholine
Parathion
Pentachloronitrobenzene (Quintobenzene)
Pentachlorophenol
Phenol
p-Phenylenediamine
Phosgene
Phosphine
Phosphorus
Phthalic anhydride
Polychlorinated biphenyls (Aroclors)
1,3-Propane sultone
beta-Propiolactone
Propionaldehyde
Propoxur (Baygon)
Propylene dichloride (1,2-Dichloropropane)
Propylene oxide
1,2-Propylenimine (2-Methyl aziridine)
Quinoline
Quinone
Styrene
Styrene oxide
2,3,7,8-Tetrachlorodibenzo-p-dioxin
1,1,2,2-Tetrachloroethane
Tetrachloroethylene (Perchloroethylene)
Titanium tetrachloride
Toluene
2,4-Toluene diamine
2,4-Toluene diisocyanate
o-Toluidine
Toxaphene (chlorinated camphene)
1,2,4-Trichlorobenzene
1,1,2-Trichloroethane
Trichloroethylene
2,4,5-Trichlorophenol
2,4,6-Trichlorophenol
Triethylamine
Trifluralin
2,2,4-Trimethylpentane
Vinyl acetate
Vinyl bromide
Vinyl chloride
Vinylidene chloride (1,1-Dichloroethylene)
Xylenes (isomers and mixture)
o-Xylenes
m-Xylenes
p-Xylenes
Antimony compounds
Arsenic compounds (inorganic, including arsine)
Beryllium compounds
Cadmium compounds
Chromium compounds
Cobalt compounds
Coke oven emissions
Cyanide compounds[1]
Glycol ethers[2]
Lead compounds
Manganese compounds
Mercury compounds
Fine mineral fibers[3]
Nickel compounds
Polycylic organic matter[4]
Radionuclides (including radon)[5]
Selenium compounds

NOTE: For all listings above which contain the word "compounds" and for glycol ethers, the following applies: Unless otherwise specified, these listings are defined as including any unique chemical substance that contains the named chemical (i.e., antimony, arsenic, etc.) as part of that chemical's infrastructure.

[1]X′CN where X = H′ or any other group where a formal dissociation may occur. For example KCN or $Ca(CN)_2$

[2]Includes mono- and di- ethers of ethylene glycol, diethylene glycol, and triethylene glycol R-$(OCH2CH2)_n$-OR′ where
n = 1, 2, or 3
R = alkyl or aryl groups
R′ = R, H, or groups which, when removed, yield glycol ethers with the structure: R-$(OCH2CH)_n$-OH. Polymers are excluded from the glycol category.

[3]Includes mineral fiber emissions from facilities manufacturing or processing glass, rock, or slag fibers (or other mineral derived fibers) of average diameter 1 micrometer or less.

[4]Includes organic compounds with more than one benzene ring, and which have a boiling point greater than or equal to 100°C.

[5]A type of atom which spontaneously undergoes radioactive decay.

Source: Public Law 101-549, Nov. 15, 1990.

TABLE 1-6
Schedule for phasing out production of ozone-destroying chemicals*

Date	Carbon tetrachloride	Methyl chloroform	Other Class I substances
1991	100	100	85
1992	90	100	80
1993	80	90	75
1994	70	85	65
1995	15	70	50
1996	15	50	40
1997	15	50	15
1998	15	50	15
1999	15	50	15
2000	—	20	—
2001	—	20	—

*It is unlawful for any person to produce an annual quantity greater than the percentage specified in the table. The percentages refer to the quantity in the baseline year.

Source: Public Law 101-549, Nov. 15, 1990.

Under Title IV, the Act outlines a new nationwide approach to the problem of acid rain. The law sets up a market-based system to lower sulfur dioxide emissions. EPA will issue emission allowances to power plants listed in the act. The allowances are set below current emission levels. Plants may meet the allowances by installing control technology or by purchasing allowances from plants that have emissions below their allowance. For example, in November of 1994, Niagara Mohawk, which serves upstate New York, and the Arizona Public Service Co. traded emission allowances for carbon monoxide and sulfur dioxide.

Unlike the Clean Water Act, no provision for permits was included in the original Clean Air Act (1963). Title V remedies this deficiency by making it unlawful to operate one of the sources listed in the Act except by compliance with a permit.

Depletion of the ozone layer is addressed in Title VI of the Act. A schedule for phasing out the production of ozone-destroying chemicals was promulgated in the Act with provision that EPA could accelerate the schedule. In 1993, EPA established the accelerated schedule shown in Table 1-6.

Noise Pollution Control

The discussion on noise pollution control follows C. R. Bragdon, *Inter-noise,* and Council on Environmental Quality Reports.[16] The city of Boston was the first to regulate noise, beginning in 1850. This ordinance was primarily concerned with noise occurring in public places that would disturb the peace. Consequently, restrictions

[16]C. R. Bragdon, *Inter-noise 74,* Oct. 1974, and Council on Environmental Quality Reports subsequent to 1974.

were placed on the use of public spaces for conducting noisy activities. It was not until the beginning of the 20th century that other cities became interested in regulating noise.

In the late 1920s the Health Commissioner of New York City established a Noise Abatement Commission. This commission, the first of its kind, sought to identify city noise sources, assess their impact, and recommend solutions. The commission's report, *City Noise,* published in 1930, was unique in many ways. An acoustical inventory of noise sources was undertaken along with a survey of public opinion and recommendations for laws to control noise. The legislation that this widely circulated report urged was subsequently adopted by many other cities. Primary noise provisions included muffler requirements for vehicles, building construction restrictions in residential areas, the prohibition of horns and whistles, the regulation of peddlers and vendors, and the prohibition of noise from mechanical or electrical sound-making or reproducing equipment.

In 1948 the National Institute of Municipal Law Officers (NIMLO) prepared a research report to guide municipalities in controlling noise. This report, subsequently referred to as the NIMLO model ordinance, was to be widely disseminated and adopted. Until recently this ordinance has been the basis for most laws.

It was not until 1952 that specific noise emission levels appeared. These first laws contained vehicle noise level requirements expressed in decibels. Seattle, Washington, and Cincinnati, Ohio, adopted these first quantitative ordinances. Following the control of mobile noise sources were laws establishing specific fixed source limits that used the zoning ordinance as the legal basis. Chicago, Illinois, in 1955 adopted the first law that restricted land use activity that produced noise. It represented a new approach to zoning. Restrictions were not placed on the classification of industry by type (that is, light manufacturing or heavy manufacturing) and by permitted uses, but rather on their performance in terms of noise emission. Many cities adopted provisions similar or identical to those used in Chicago.

In 1970 NIMLO modified their earlier nuisance-type ordinance, adding decibel provisions as an alternative. The League of California Cities prepared a model document that contained sound levels for zoning districts in rural, suburban, and urban areas.

In 1973 there were seventeen states that had types of noise laws concerned with zoning and land use, motor vehicles, and aircraft and building codes.

Prior to 1970 the federal government's activities in noise abatement had no central focus. The emphasis was on specific activities regulated separately by individual agencies. Each noise problem was considered in isolation.

The landmark legislation in the area of occupational noise abatement was enacted in 1942 and is known as the Walsh-Healey Public Contracts Act. This act established minimum working conditions for employees of contractors who supply the federal government with materials, supplies, and equipment in excess of $10,000. However, it was not until 1969 that the Secretary of Labor interpreted this as applicable to noise! (Note: These applied only to supply contracts and not to construction contracts.)

In the 1962 amendments to the Federal Aid Highways Act, economic, social, and environmental impacts were included as requirements for consideration in the development of plans for construction. The following year, the Federal Housing Administration implemented minimum standards for airborne noise and building structure noise permitted for mortgage underwriting.

The Department of Transportation Act (1966) included provisions to promote research on noise abatement with particular attention to aircraft. This was followed by the 1968 amendments to the Federal Aviation Administration Act that directed the Secretary of Transportation to prescribe rules for control and abatement of aircraft noise.

1970 was a banner year for environmental legislation and rule making. The Federal Aid Highways Act was amended to direct the Secretary of Transportation to develop and promulgate standards for highway noise levels compatible with different land uses. The Occupational Safety and Health Act of 1970 (OSHA) enabled the Secretary of Labor to apply the Walsh-Healey standards with new meaning. Walsh-Healey merely excluded from bidding on federal contracts those suppliers who failed to meet minimum work condition standards. OSHA provided penalties for those suppliers, including civil and criminal law sanctions. Construction noise was brought under federal consideration in the Construction Safety Act of 1970. This act carried the Walsh-Healey provisions to the supply of construction contracts.

Also in 1970 Congress added Title IV to the Clean Air Act amendments. This act was entitled "Noise Pollution and Abatement Act of 1970," and it set up the Office of Noise Abatement and Control in the EPA. The major provision of the act was to give the EPA the discretion to identify and classify causes and sources of noise and to assess effects. Authority for actual abatement of federal sources of noise was also included.

Solid Waste

(The discussion on solid waste follows G. Tchobanoglous, *et al., Solid Wastes.*[17]) Modern solid waste legislation dates from 1965 when the Solid Waste Disposal Act, Title II of Public Law 89-272, was enacted by Congress. The intent of this act was to

1. Promote the demonstration, construction, and application of solid waste management and resource recovery systems which preserve and enhance the quality of air, water, and land resources.
2. Provide technical and financial assistance to states and local governments and interstate agencies in the planning and development of resource recovery and solid waste disposal programs.
3. Promote a national research and development program for improved management techniques, more effective organizational arrangements, and new and improved methods of collection, separation, recovery, and recycling of solid wastes, and the environmentally safe disposal of nonrecoverable residues.

[17] G. Tchobanoglous, H. Theisen, and R. Eliassen, *Solid Wastes,* New York: McGraw-Hill, p. 39, 1977.

4. Provide for the promulgation of guidelines for solid waste collection, transport, separation, recovery, and disposal systems.
5. Provide training grants in occupations involving the design, operation, and maintenance of solid waste disposal systems.

Enforcement of this act became the responsibility of the U.S. Public Health Service (USPHS) and the Bureau of Mines. The USPHS had responsibility for most of the municipal wastes. The Bureau of Mines was charged with supervision of solid wastes from mining activities and the fossil-fuel solid wastes from power plants and industrial steam plants.

The Solid Waste Disposal Act of 1965 was amended by Public Law 95-512, the Resources Recovery Act of 1970. The act directed that the emphasis of the national solid waste management program be shifted from disposal as its primary objective to that of recycling and reuse of recoverable materials in solid wastes or to the conversion of wastes to energy.

Another feature of the 1970 act was the mandate of Congress to the Secretary of Health, Education, and Welfare to prepare a report on the treatment and disposal of hazardous wastes, including radioactive, toxic chemical, biological, and other wastes of significance to the public health and welfare. Previously, the Atomic Energy Act of 1954 had authorized the U.S. Atomic Energy Commission to manage all radioactive wastes generated by the commission and the nuclear power industry.

Hazardous Wastes

(Part of this discussion on hazardous wastes is excerpted from J. Quarles, *Federal Regulation of Hazardous Wastes.*[18]) For many years little attention was paid to where waste materials, including the contaminants removed from the air and the water, went. These contaminants often returned to the environment through the disposal of waste. Waste disposal practices were discovered to have caused environmental damage in many cases, especially in contamination of groundwater used as a drinking-water supply. This led Congress to enact the Resource Conservation and Recovery Act of 1976, commonly known as RCRA (and pronounced "rick-rah").

RCRA addresses the handling of hazardous waste at facilities currently operating and at those yet to be constructed. The act was designed in large part to meet disposal needs resulting from the Clean Air Act and Clean Water Act, which require industries to remove hazardous substances from their air emissions and their water discharges. Neither statute, however, ensures that the ultimate disposition of waste materials will be environmentally sound. RCRA was intended to provide that ensurance. RCRA does not, however, deal directly with abandoned sites or closed facilities where hazardous wastes have been handled or disposed of in the past. These

---

[18] J. Quarles, *Federal Regulation of Hazardous Wastes,* Washington, DC: The Environmental Law Institute, 1982.

locations are covered by the Comprehensive Environmental Response, Compensation, and Liability Act (CERCLA, pronounced "sir-klah"), commonly referred to as "Superfund," enacted by Congress in 1980. Finally, RCRA also does not control the disposition of hazardous substances within the productive stream of commerce. Such substances include chemicals covered by the Toxic Substances Control Act; pesticides regulated under the Federal Insecticide, Fungicide, and Rodenticide Act; or other hazardous products subject to the Hazardous Materials Transportation Act and to other types of federal regulation.

The five major elements in the federal approach to hazardous waste management are:

1. Federal classification of hazardous waste
2. Cradle-to-grave manifest (record-keeping) system
3. Federal standards for safeguards to be followed by generators, transporters, and facilities that treat, store, or dispose of hazardous waste
4. Enforcement of federal standards for facilities through a permit program
5. Authorization of state programs to operate in lieu of the federal program

The act directs the U.S. Environmental Protection Agency to promulgate regulations necessary to put the federal program into full effect.

Unhappy with the progress in implementing RCRA, Congress in 1984 passed the Hazardous and Solid Waste Amendments (HSWA, pronounced "hiss-wah"). The scope of RCRA was significantly increased. Under the legislation:

1. Waste minimization was established as the preferred method for managing hazardous waste.
2. Untreated hazardous waste was banned from land disposal and EPA was directed to establish treatment standards for land disposal.
3. New technology standards, such as double liners, leachate collection systems, and extensive groundwater monitoring, were established for land disposal facilities.
4. New requirements were established for small quantity generators.
5. The EPA was directed to establish standards for underground storage tanks.
6. The EPA was directed to evaluate criteria for municipal solid waste landfills and upgrade monitoring requirements.

The Comprehensive Environmental Response, Compensation, and Liability Act of 1980 (CERCLA) provided authority for removal of hazardous substances from improperly constructed or operated active sites not in compliance with RCRA and from inactive disposal sites.

The most fundamental feature of CERCLA is that it provides basic operating authority to the federal government to take direct action to remove hazardous substances from dangerous inactive disposal sites and to assist with cleaning up emergency spills. This includes authority to carry out investigations, testing, and monitoring of disposal sites. It also includes authority to implement remedial measures to remove contaminants in the groundwater.

CERCLA earned its nickname, "Superfund," from the provision of a $1.6 billion Hazardous Substance Response Trust Fund. Seven-eighths of the money is to be provided by industry through taxes on crude oil, certain petroleum products, and 42 chemical feedstocks; one-eighth is to be provided by government through appropriations from general revenues.

In cases where responsibility for the wastes that cause contamination can be traced to companies with financial resources, CERCLA places financial responsibility for the cleanup on those companies. The statute establishes a new and far-reaching set of federal laws under which liability can be imposed on such companies even when they are only indirectly involved in the ownership or operation of the facilities where the wastes were disposed. After the government has identified a site as a threat to the environment, it may call upon those liable companies to undertake the cleanup at their own cost. Alternatively, if such companies refuse to assume responsibility for the cleanup, the government can carry out the remedial program using money from the fund and then bring suit against the companies for reimbursement.

A National Contingency Plan (NCP) establishes the rules for how EPA will use its authority and spend its money. To qualify for expenditure of CERCLA funds, a site must appear on the National Priorities List (NPL). The EPA developed the Hazardous Ranking System (HRS) as a method of assigning a site to the NPL. As of 1989, 981 sites had been placed on the NPL. In addition CERCLA contains notice requirements for all releases (spills) of reportable quantities of hazardous substances and creates a Post-Closure Liability Fund for qualified disposal facilities.

The Superfund Amendments and Reauthorization Act (SARA) of 1986 extended the provisions of CERCLA. In addition to establishing an $8.5 billion fund for cleanup, SARA establishes that EPA:

1. Revise the NPL and the HRS on which it is based.
2. Revise the NCP.
3. Is authorized to subpoena documents and witnesses.
4. Can spend money to investigate sites and design remedies, and can permit private parties to conduct cleanup.
5. Has broad enforcement authority to require private parties to undertake cleanup.
6. Must impose the more stringent of federal standards or state standards.
7. May use mixed funding, that is, both federal and private money.
8. Develop an administrative record of decisions.

The Toxic Substances Control Act (TSCA) of 1976 is unique in hazardous waste legislation in that it requires disclosure of information about the toxicity of new materials before they enter into commercial manufacture. It deals with hazardous waste in only one instance: polychlorinated biphenyls (PCBs). At the federal level, rules for the disposal of PCBs are set under TSCA (pronounced "tos-ka") rather than RCRA or CERCLA.

Atomic Energy and Radiation

Laws and regulations to manage radioactive materials and radiation exposure began with the Atomic Energy Act of 1946. The act established the Atomic Energy Commission (AEC) and directed it to conduct research and development on peaceful applications of fissionable and radioactive materials. The Atomic Energy Act of 1954 provided for control of uranium and thorium ("source material" for nuclear reactors), plutonium and enriched uranium (classified as special nuclear material because of their potential use in atomic weapons), and other by-products of the nuclear industry. The Energy Reorganization Act of 1974 divided the developmental and regulatory functions of the AEC between two agencies: the Energy and Research and Development Administration (ERDA) and the Nuclear Regulatory Commission (NRC). In restructuring the administration of energy-related matters after the Arab oil boycott, the Energy Organization Act of 1977 replaced ERDA with the Department of Energy (DOE). The NRC was given jurisdiction over reactor construction and operation. It regulates the possession, use, transportation, handling, and disposal of radioactive materials and wastes. The DOE is responsible for research and development and will operate defense and high-level waste repositories.

The diminishing space at low-level disposal sites led to the enactment of the Low-Level Waste Policy Act (LLWPA) in 1980. Each state is responsible for providing for the availability of capacity either within or outside the state for disposal of low-level radioactive waste generated within its borders. States were encouraged to enter into *compacts* with their neighbors to more efficiently manage the waste. The law allowed the compacts to exclude wastes from other regions and allowed existing disposal sites to impose surcharges for disposal of wastes from regions without sites. The surcharge was to be used for site development. Difficulties in negotiating the compacts prompted the enactment of the Low-Level Radioactive Waste Policy Act Amendments of 1985 (LLRWPAA). It stipulated that the three existing commercial sites remain open for use by all states through 1992. Annual and total limits on the volume of waste that can be sent from reactors were established. DOE is responsible for overseeing the compact arrangements with authority to allocate additional emergency capacity to reactors. The NRC can authorize emergency access to existing sites.

The Nuclear Waste Policy Act of 1982 directed DOE to develop a plan for storage of high-level radioactive waste. Following the requirements of the law, DOE began investigation of nine sites in the west and two in the east. Under the act, the EPA established standards that specified release limits for 1,000 and 10,000 years after disposal.

Because of loudly voiced concern over the direction of the DOE's mission plan and the decision to abandon the search for a repository site in the east, Congress passed the Nuclear Waste Policy Act Amendments of 1987. The amendments restructured DOE's high-level waste program. The only site that would be considered would be Yucca Flats, Nevada. Furthermore, spent fuel would be required to be shipped in NRC-approved packages after notification of state and local governments. During the years 2007-2010, DOE is to study the need for a second repository.

For mixed wastes, that is, both hazardous and radioactive, RCRA and HSWA apply to the hazardous characteristic. As of 1987, disposal rules must comply with both NRC rules for radioactivity and EPA rules for hazardous constituents. Before then, only the NRC rules applied. Likewise, for leaking disposal sites, CERCLA and SARA rules apply as well as the NRC rules.

Radiation exposure from x-rays and medical diagnosis and treatment are regulated under the Radiation Control for Health and Safety Act of 1968.

Environmental Legislation and Regulation in the 1980s

Under the administration of President Ronald Reagan, major environmental legislation stagnated. Furthermore, the Reagan administration was noteworthy for unprecedented contraction of the environmental movement of the 1960s and 1970s. Mismanagement and contempt for Congress marked the attitude of Mr. Reagan's first EPA administrator and her appointed (not civil service) staff. Unethical practices forced the wholesale removal of the top echelon of the EPA administration in early 1983. This leads us to our next topic—environmental ethics.

1-5 ENVIRONMENTAL ETHICS

The birth of environmental ethics as a force is partly a result of concern for our own long-term survival, as well as our realization that humans are but one form of life, and that we share our earth with other forms of life.[19]

The difficulty of using self-modification to arrive at environmentally ethical decisions is best illustrated by the following two examples. We have not attempted to provide pat solutions, but rather have left them for you and your instructor to resolve.

Case 1: To Add or Not to Add

A friend of yours has discovered that his firm is adding nitrites and nitrates to bacon to help preserve it. He also has read that these compounds are precursors to cancer-forming chemicals that are produced in the body. On the other hand, he realizes that certain disease organisms such as those that manufacture botulism toxin have been known to grow in bacon that had not been treated. He asks you whether he should (a) protest to his superiors knowing he might get fired; (b) leak the news to the press; (c) remain silent because the risk of dying from cancer is less than the absolute certainty of dying from botulism.

Note: The addition of nitrite to bacon is approved by the Food and Drug Administration.

---

[19]P. A. Vesilind, *Environmental Pollution and Control,* Ann Arbor, MI: Ann Arbor Science, p. 214, 1975.

Case 2: Too Close for Comfort

As the only engineer in the shop who has had any training in noise pollution, you have been asked, on your third day on the job, to review a manufacturer's bid for noise control devices on aeration blowers for a wastewater treatment plant. After reading the bid proposal and making a few calculations, you conclude that the noise silencers will protect the workers. However, your reading of *Introduction to Environmental Engineering* leads you to surmise that the noise level for nearby neighbors will be excessive at night. (The city has no noise ordinance.) You know that if you ask to go back out on bid for better noise control equipment, construction of the plant will be delayed 90 days. During these 90 days untreated sewage will enter the river. What do you recommend to your new boss?

We think it is important to point out that many environmentally related decisions such as those described above are much more difficult than the problems presented in the remaining chapters of this book. Frequently these problems are related more to ethics than to engineering. The problems arise when there are several courses of action with no *a priori* certainty as to which is best. Decisions related to safety, health, and welfare are easily resolved. Decisions as to which course of action is in the best interest of the public are much more difficult to resolve. Furthermore, decisions as to which course of action is in the best interest of the environment are at times in conflict with those which are in the best interest of the public. Whereas decisions made in the public interest are based on professional ethics, decisions made in the best interest of the environment are based on environmental ethics.

Ethos, the Greek word from which "ethic" is derived, means the character of a person as described by his or her actions. This character was developed during the evolutionary process and was influenced by the need for adapting to the natural environment. Our ethic is our way of doing things. Our ethic is a direct result of our natural environment. During the latter stages of the evolutionary process, *Homo sapiens* began to modify the environment rather than submit to what, millennia later, became known as Darwinian natural selection. As an example, consider the cave dweller who, in the chilly dawn of prehistory, realized the value of the saber-toothed tiger's coat and appropriated it for personal use. Inevitably a pattern of appropriation developed, and our ethic became more self-modified than environmentally adapted. Thus, we are no longer adapted to our natural environment but rather to our self-made environment. In the ecological context, such maladaptation results in one of two consequences: (1) the organism (*Homo sapiens*) dies out; or (2) the organism evolves to a form and character that is once again compatible with the natural environment.[20] Assuming that we choose the latter course, how can this change in character (ethic) be brought about? Each individual must change his or her character or ethic, and the social system must change to become compatible with the global ecology.

---

[20]Ibid., p. 216.

The acceptable system is one in which we learn to share our exhaustible resources—to regain a balance. This requires that we reduce our needs and that the materials we use must be replenishable. We must treat all of the earth as a sacred trust to be used so that its content is neither diminished nor permanently changed; we must release no substances that cannot be reincorporated without damage to the natural system. The recognition of the need for such adaptation (as a means of survival) has developed into what we now call the *environmental ethic.*[21]

1-6 A MATERIALS BALANCE APPROACH TO PROBLEM SOLVING

You are familiar with the concept that matter can neither be created nor destroyed but that it can be changed in form. This concept serves as a basis for describing and analyzing environmental engineering problems. The concept is called a *materials balance,* or a *mass balance.* In its simplest form it may be viewed as an accounting procedure. You perform a form of mass balance each time you balance your check book:

$$\text{Balance} = \text{Deposit} - \text{Withdrawal} \tag{1-1}$$

In an environmental system or subsystem, the equation would be written:

$$\text{Accumulation} = \text{Input} - \text{Output} \tag{1-2}$$

where Accumulation, Input, and Output refer to the mass quantities accumulating in the system or flowing into or out of the system.

In the mass-balance approach, we begin solving the problem by drawing a flowchart of the process or a conceptual diagram of the environmental subsystem. All of the known inputs, outputs, and accumulation are converted to the same mass units and placed on the diagram. Unknown inputs, outputs, and accumulation are also marked on the diagram. This helps us define the problem. System boundaries (imaginary blocks around the process or part of the process) are drawn in such a way that calculations are made as simple as possible. We then write materials-balance equations to solve for unknown inputs, outputs, or accumulations or to demonstrate that we have accounted for all of the components by demonstrating that the materials balance "closes," that is, the accounting balances.

Example 1-1. Mr. and Mrs. Konzzumer have no children. In an average week they purchase and bring into their house approximately 50 kg of consumer goods (food, magazines, newspapers, appliances, furniture, and associated packaging). Of this amount, 50 percent is consumed as food. Half of the food is used for biological maintenance and ultimately released as CO_2; the remainder is discharged to the sewer system.

[21] Ibid., p. 215.

Approximately 1 kg accumulates in the house. The Konzzumers recycle approximately 25 percent of the solid waste that is generated. Estimate the amount of solid waste they place at the curb each week.

Solution

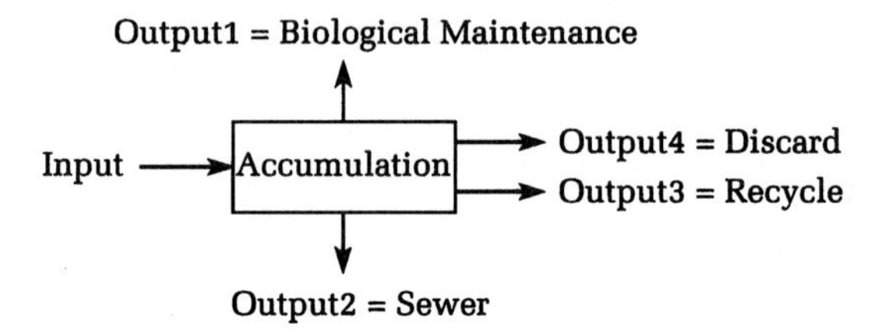

Write the mass balance equation.

$$\text{Input} = \text{Output1} + \text{Output2} + \text{Output3} + \text{Output4} + \text{Accumulation}$$

Now we need to calculate known outputs.

One half of input is food $= (0.5)(50 \text{ kg}) = 25 \text{ kg}$

One half of food is used for biological maintenance $= \text{Output1} = (0.5)(25 \text{ kg}) = 12.5 \text{ kg}$

One half of the food is lost to the sewer system $= \text{Output2} = (0.5)(25 \text{ kg}) = 12.5 \text{ kg}$

The recycled amount is 25 percent of what remains of input after food and accumulation is removed $= \text{Output3} = .25(\text{Input} - \text{Output1} - \text{Output2} - \text{Accumulation}) = .25(50 - 12.5 - 12.5 - 1) = 6 \text{ kg}$

Now we can solve for Output4:

$$\begin{aligned}\text{Output4} &= \text{Input} - \text{Output1} - \text{Output2} - \text{Output3} - \text{Accumulation} \\ &= 50 - 12.5 - 12.5 - 6 - 1 \\ &= 18 \text{ kg}\end{aligned}$$

For many environmental problems time is an important factor in establishing the degree of severity of the problem or in designing a solution. In these instances Equation 1-2 is modified to the following form:

$$\text{Rate of accumulation} = \text{Rate of input} - \text{Rate of output} \tag{1-3}$$

where *rate* is used to mean per unit of time. In the calculus this may be written as:

$$\frac{dM}{dt} = \frac{d(\text{In})}{dt} - \frac{d(\text{Out})}{dt} \tag{1-4}$$

where M refers to the mass accumulated and (In) and (Out) refer to the mass flowing in or out. As part of the description of the problem, a convenient time interval that is meaningful for the system must be chosen.

Example 1-2. Truely Clearwater is filling her bathtub but she forgot to put the plug in. If the volume of water for a bath is 0.350 cubic meters and the tap is flowing at 1.32 L/min and the drain is running at 0.32 L/min, how long will it take to fill the tub to bath level? How much water will be wasted? Assume the density of water is 1,000 kg/m^3.

Solution

Q_{in} = 1.32 L/min → $\forall_{accumulation}$ → Q_{out} = 0.32 L/min

Solving without integration

$$\text{mass} = (\text{volume})(\text{density}) = (\forall)(\rho)$$

$$\text{volume} = (\text{flow rate})(\text{time}) = (Q)(t)$$

So for the mass balance equation, noting that $0.350 \text{ m}^3 = 350 \text{ L}$,

$$(\forall_{ACC})(\rho) = (Q_{in})(\rho)(t) - (Q_{out})(\rho)(t)$$

$$\forall_{ACC} = Q_{in}(t) - Q_{out}(t)$$

$$\forall_{ACC} = 1.32t - 0.32t$$

$$350 \text{ L} = (1.00 \text{ L/min})(t)$$

$$t = 350 \text{ min}$$

Solving using integration

Using Equation 1-4,

$$\frac{dM}{dt} = \frac{d(\text{In})}{dt} - \frac{d(\text{Out})}{dt}$$

dM/dt = the mass accumulated over time. In this case, since the density is a constant, the integral would look like

$$\int_0^t \forall \, dt$$

We want to see how long it takes to fill, so at time t, the bath is full; hence

$$\int_0^t \forall \, dt = 350 \text{ L}$$

Integrating the other side of the equation,

$$350 \text{ L} = \int_0^t (Q_{in})\, dt - \int_0^t (Q_{out})\, dt$$

$$= (Q_{in})(t)\Big|_0^t - (Q_{out})(t)\Big|_0^t$$

$$= [(Q_{in})(t) - (Q_{in})(0)] - [(Q_{out})(t) - (Q_{out})(0)]$$

$$= (Q_{in})(t) - (Q_{out})(t)$$

So

$$350 \text{ L} = (1.32 \text{ L/min})(t) - (0.32 \text{ L/min})(t)$$

$$350 \text{ L} = (1.0 \text{ L/min})(t)$$

$$t = 350 \text{ min}$$

Equation 1-4 is applicable when there is no chemical or biological reaction, or radioactive decay of the substances in the mass balance. In these instances the substance is said to be *conserved.* Examples of conservative substances include salt in water and carbon dioxide in air. Examples of nonconservative substances include decomposing organic matter and particulate matter that is settling from the air.

The state of mixing in the system is an important consideration in the application of Equation 1-4. Consider a coffee cup containing approximately 200 mL of black coffee (or another beverage of your choice). If we add a dollop (about 20 mL) of cream and immediately take a sample (or a sip), we would not be surprised to find that the cream was not evenly distributed throughout the coffee. If, on the other hand, we mixed the coffee and cream vigorously and then took a sample, it would not matter if we sipped from the left or right of the cup, or, for that matter, put a valve in the bottom and sampled from there, we would expect the cream to be distributed evenly. In terms of a mass balance on the coffee cup system, the cup itself would define the system boundary. If the coffee and cream were not mixed well, then the place we take the sample would strongly affect the value of $d(\text{Out})/dt$ in Equation 1-4. On the other hand, if the coffee and cream were instantaneously well mixed, then any place we take the sample would yield the same result. That is, any output would look exactly like the contents of the cup. This system is called a completely mixed system. A more formal definition is that *completely mixed systems* are those in which every drop of fluid is homogeneous with every other drop, that is, every drop of fluid contains the same concentration of material or physical property (for example, temperature). If a system is completely mixed, then we may assume that the output from the system (concentration, temperature, etc.) is the same as the contents within the system boundary. Although we frequently make use of this assumption to solve mass-balance problems, it is often very difficult to achieve in real systems. This means that solutions to mass-balance problems that utilize this assumption must be taken as approximations to reality.

When a system has operated in such a way that the rate of input and the rate of output are constant and equal, then, of course, the rate of accumulation is zero (i.e. $dM/dt = 0$). This condition is called *steady state.* In solving mass-balance problems, it is often convenient to make an assumption that steady-state conditions have been achieved.

Example 1-3. A storm sewer is carrying snow melt containing 1.200 g/L of sodium chloride into a small stream. The stream has a naturally occurring sodium chloride concentration of 20 mg/L. If the storm sewer flow rate is 2,000 L/min and the stream flow rate is 2.0 m^3/s, what is the concentration of salt in the stream? Assume that the sewer flow and the stream flow are completely mixed, that the salt is a conservative substance, and that the system is at steady state.

Solution. The first step is to draw a mass balance diagram as shown below.

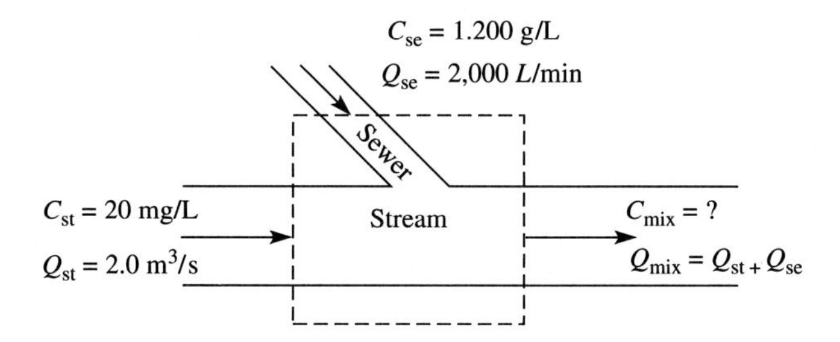

Note that the mass of salt may be calculated as:

$$\text{mass/time} = (\text{concentration})(\text{flow rate})$$

or

$$\text{mass/time} = (\text{mg/L})(\text{L/min}) = \text{mg/min}$$

Using the notation in the diagram where the subscript "st" refers to the stream and the subscript "se" refers to the sewer, the mass balance may be written as:

$$\text{Rate of Accumulation} = [C_{st}Q_{st} + C_{se}Q_{se}] - C_{mix}Q_{mix}$$

$$\text{where } Q_{mix} = Q_{st} + Q_{se}$$

Since we may assume steady state, the rate of accumulation equals zero and

$$C_{mix}Q_{mix} = [C_{st}Q_{st} + C_{se}Q_{se}]$$

Solving for C_{mix}

$$C_{mix} = \frac{[C_{st}Q_{st} + C_{se}Q_{se}]}{Q_{st} + Q_{se}}$$

Before substituting in the values, the units are converted as follows:

$$C_{se} = (1.200\text{ g/L})(1{,}000\text{ mg/g}) = 1{,}200\text{ mg/L}$$

$$Q_{st} = (2.0\text{ m}^3\text{/s})(1{,}000\text{ L/m}^3)(60\text{ s/min}) = 120{,}000\text{ L/min}$$

$$C_{mix} = \frac{[(20\text{ mg/L})(120{,}000\text{ L/min})] + [(1{,}200\text{ mg/L})(2{,}000\text{ L/min})]}{120{,}000\text{ L/min} + 2{,}000\text{ L/min}}$$

$$C_{mix} = 39.34 \text{ or } 39\text{ mg/L}$$

In most systems of environmental interest, transformations occur within the system: byproducts are formed (e.g., sludge) or compounds are destroyed (e.g., ozone). Because many environmental reactions do not occur instantaneously, the time dependence of the reaction must be taken into account. Equation 1-3 may be written to account for time-dependent transformation as follows:

$$\begin{matrix}\text{Accumulation} \\ \text{Rate}\end{matrix} = \begin{matrix}\text{Input} \\ \text{Rate}\end{matrix} - \begin{matrix}\text{Output} \\ \text{Rate}\end{matrix} \pm \begin{matrix}\text{Transformation} \\ \text{Rate}\end{matrix} \qquad (1\text{-}5)$$

Time-dependent reactions are called *kinetic reactions*. The rate of transformation, or *reaction rate (r)*, is used to describe the rate of formation or disappearance of a substance or chemical species. In the calculus:

$$\frac{dM}{dt} = \frac{d(\text{In})}{dt} - \frac{d(\text{Out})}{dt} \pm r \tag{1-6}$$

The reaction rate is often some complex function of temperature, pressure, the reacting components, and/or products of reaction.

A convenient model for the decay of nonconservative substances is the first-order reaction. In this model it is assumed that the rate of loss of the substance is proportional to the amount of substance present at any given time, t. In the calculus:

$$\frac{dC}{dt} = -kC \tag{1-7}$$

where k = reaction rate constant, s^{-1} or d^{-1}
C = concentration of substance

The differential equation may be integrated to yield either

$$\ln \frac{C}{C_0} = -kt \tag{1-8}$$

or

$$C = C_0 e^{-kt} \tag{1-9}$$

where C_0 = initial concentration.

For simple systems, we assume the substance is distributed uniformly throughout the volume. Thus, the total mass of substance (M) is equal to the product of the concentration and volume ($C\forall$) and, when $\forall$ is a constant, the mass rate of decay of the substance is

$$\frac{dM}{dt} = \frac{d(C\forall)}{dt} = \forall \frac{dC}{dt} \tag{1-10}$$

Thus, for decay of a substance, we can rewrite Equation 1-6 as

$$\frac{dM}{dt} = \frac{d(\text{In})}{dt} - \frac{d(\text{Out})}{dt} - kC\forall \tag{1-11}$$

Example 1-4. A well-mixed sewage lagoon is receiving 430 m^3/d of sewage. The lagoon has a surface area of 10 hectares (ha) and a depth of 1.0 m. The pollutant concentration in the raw sewage is 180 mg/L. The organic matter in the sewage degrades biologically (decays) in the lagoon according to first-order kinetics. The reaction rate constant (decay coefficient) is 0.70 d^{-1}. Assuming no other water losses or gains (evaporation, seepage, or rainfall) and that the lagoon is completely mixed, find the steady-state concentration of the pollutant in the effluent.

Solution. We begin by drawing the mass-balance diagram.

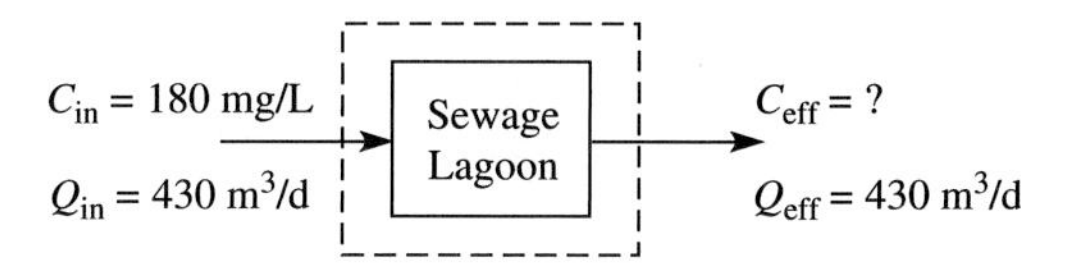

Assuming steady-state conditions, the mass-balance equation may be written as:

$$\text{Input rate} = \text{Output rate} + \text{Decay rate}$$

Input rate is

$$(430 \text{ m}^3/\text{d})(1{,}000 \text{ L/m}^3)(180 \text{ mg/L}) = 77{,}400{,}000 \text{ mg/d}$$

With a volume of

$$(10 \text{ ha})(10^4 \text{ m}^2/\text{ha})(1 \text{ m}) = 100{,}000 \text{ m}^3$$

and the decay coefficient of 0.70 d^{-1}, the decay rate is

$$kC\forall = (0.70 \text{ d}^{-1})(100{,}000 \text{ m}^3)(1{,}000 \text{ L/m}^3)(C_{\text{lagoon}}) = 70{,}000{,}000\ (C_{\text{lagoon}})$$

Now we take advantage of the assumption that the lagoon is completely mixed and assume that $C_{\text{eff}} = C_{\text{lagoon}}$. Thus,

$$kC\forall = 70{,}000{,}000\ (C_{\text{eff}})$$

Substituting into the mass-balance equation

$$\text{Output rate} = 77{,}400{,}000 \text{ mg/d} - 70{,}000{,}000 \text{ L/d}(C_{\text{eff}})$$

or

$$C_{\text{eff}}(430 \text{ m}^3/\text{d})(1{,}000 \text{ L/m}^3) = 77{,}400{,}000 \text{ mg/d} - 70{,}000{,}000 \text{ L/d}(C_{\text{eff}})$$

Solving for C_{eff}

$$C_{\text{eff}} = \frac{77{,}400{,}000 \text{ mg/d}}{70{,}430{,}000 \text{ L/d}} = 1.10 \text{ mg/L}$$

1-7 CHAPTER REVIEW

When you have completed studying this chapter, you should be able to do the following without the aid of your textbooks or notes:

1. Sketch and label a water resource system including (*a*) source; (*b*) collection works; (*c*) transmission works; (*d*) treatment works; and (*e*) distribution works.
2. State the proper general approach to treatment of a surface water and a groundwater (see Figure 1-2).
3. Define the word "demand" as it applies to water.
4. List the five most important factors contributing to water consumption and explain why each has an effect.

5. State the rule-of-thumb water requirement for an average city on a per-person basis and calculate the average daily water requirement for a city of a stated population.
6. Define the acronyms WWTP and POTW.
7. Explain why separate storm sewers and sanitary sewers are preferred over combined sewers.
8. Explain the purpose of a lift station.
9. Identify and explain the following acronyms and concepts found in environmental legislation: BPT, BAT, BCT, NPDES, HAP, MACT, "bubble policy," NIMLO, Walsh-Healey, OSHA, RCRA, and Superfund.
10. Define the law of conservation of mass.
11. Write the general mass balance equations for systems with and without transformation.
12. Define the following terms: conservative pollutants, reactive chemicals, steady-state conditions, completely mixed systems.
13. Write the mathematical expression for the decay of a substance by first-order kinetics with respect to the substance.

1-8 PROBLEMS

1-1. Estimate the average daily water consumption for the United States in 1970. The population was 203,302,031.

1-2. Estimate the population of the United States in 1990 and then estimate the average daily water consumption. Use the following population data.

Year	Population
1940	132,164,569
1950	151,325,798
1960	179,323,175
1970	203,302,031
1980	226,542,203

1-3. A small residential development of 28 houses is being planned. Assume that the average residential consumption applies, and that each house has three residents. Estimate the additional average daily water production in L/d that will have to be supplied by the city.

1-4. Repeat Problem 1-3 for 83 houses, but assume that low-flush valves reduce water consumption by 18 percent.

1-5. Using the data in Problem 1-3 and assuming that the houses are metered, determine what additional demand will be made at the peak hour.

1-6. If a faucet is dripping at a rate of one drop per second and each drop contains 0.150 milliliters, calculate how much water (in liters) will be lost in one year.

1-7. Savabuck University has installed standard pressure-operated flush valves on their water closets. When flushing, these valves deliver 130.0 L/min. If the delivered water costs $0.26 per cubic meter, what is the monthly cost of not repairing a broken valve which flushes continuously?

1-8. A sanitary landfill has available space of 16.2 ha at an average depth of 10 m. Seven hundred sixty-five (765) cubic meters of solid waste are dumped at the site five days per week. This waste is compacted to twice its delivered density. Draw a mass-balance diagram and estimate the expected life of the landfill in years.

1-9. Each month the Speedy Dry Cleaning Company buys one barrel (0.160 m^3) of carbon tetrachloride dry cleaning fluid. Ninety percent of the fluid is lost to the atmosphere and 10 percent remains as residue to be disposed of. The density of carbon tetrachloride is 1.5940 g/mL. Draw a mass-balance diagram and estimate the monthly mass emission rate to the atmosphere (kg/mo.).

1-10. The Rappahannock River near Warrenton, VA, has a flow rate of 3.00 m^3/s. Tin Pot Run (a pristine stream) discharges into the Rappahannock at a flow rate of 0.05 m^3/s. To study mixing of the stream and river, a conservative tracer is to be added to the stream. If the instruments that can measure the tracer can detect a concentration of 1.0 mg/L, what minimum concentration must be achieved in the stream so that 1.0 mg/L of tracer can be measured after the river and stream mix? Assume that the 1.0 mg/L of tracer is to be measured after complete mixing of the stream and Rappahannock has been achieved and that no tracer is in Tin Pot Run or the Rappahannock above the point where the two streams mix. What mass rate (kg/d) of tracer must be added to the stream?

1-11. A sewage lagoon that has a surface area of 10 ha and a depth of 1 m is receiving 8,640 m^3/d of sewage containing 100 mg/L of biodegradable contaminant. At steady state, the effluent from the lagoon must not exceed 20 mg/L of biodegradable contaminant. Assuming the lagoon is well mixed and that there are no losses or gains of water in the lagoon other than the sewage input, what biodegradation reaction rate coefficient (d^{-1}) must be achieved?

1-9 DISCUSSION QUESTIONS

1-1. Would you expect the demand for water to drop in half if the price ($/L) doubled? Explain your reasoning.

1-2. The water supply for the City of Peoria, AZ, is from wells. Other than disinfection, no water treatment is provided. A filtration plant would be appropriate to improve the quality of the water. True or False? If the answer is false, revise the statement so that it is true.

1-3. The water treatment plant for the town of Gettysburg, PA, was built 20 years ago. Over the last few years, there has been difficulty in maintaining water pressure in the system over the 4th of July weekend. In some parts of town only a trickle of water flows from the tap during early morning and late evening hours. There are no problems during the remainder of the year. Explain why the town may be having water pressure problems.

1-4. The town of West Lafayette, IN, is considering two proposals for a new water-treatment plant. West Lafayette's average daily demand is 8,640 m^3/d. Proposal A is to build a plant that will produce 360 m^3/h and a storage reservoir to hold 2,000 m^3 of water. Proposal B is to build a plant that will produce 1,080 m^3/h. but no water storage reservoir will be provided. Which proposal do you recommend? Explain why.

1-5. Homeowners in the town of Rolla, MO, have connected their downspouts and the sump pumps from their footing drains to the sanitary sewer system. The rainwater and sump

water entering the sewer is called (choose one):
(a) Infiltration
(b) Inflow

These connections to the sanitary sewer, in effect, make it a (choose one):
(c) Storm sewer
(d) Combined sewer

Explain why you have made your choices.

1-6. The Shiny Plating Company is using about 2,000.0 kg/wk of organic solvent for vapor degreasing of metal parts before they are plated. The Air Pollution Engineering and Testing Company (APET) has measured the air in the workroom and in the stack which vents the degreaser. APET has determined that 1,985.0 kg/wk is being vented up the stack and that the workroom environment is within occupational standards. The 1,985.0 kg/wk is well above the allowable emission rate of 11.28 kg/wk.

Elizabeth Fry, the plant superintendent, has asked J.R. Injuneer, the plant engineer, to review two alternative control approaches offered by APET and to recommend one of them.

The first method is to purchase a pollution control device to put on the stack. This control system will reduce the solvent emission to 1.0 kg/wk. Approximately 1,950.0 kg of the solvent which is captured each week can be recycled back to the degreaser. Approximately 34.0 kg of the solvent must be discharged to the wastewater treatment plant (WWTP). J. R. has determined that this small amount of solvent will not adversely affect the performance of the WWTP. In addition, the capital cost of the pollution control equipment will be recovered in about two years as a result of savings from recovering lost solvent.

The second method is to substitute a solvent that is not on the list of regulated emissions. The price of the substitute is about 10 percent higher than the solvent currently in use. J. R. has estimated that the substitute solvent loss will be about 100.0 kg/wk. The substitute collects moisture and loses its effectiveness in about a month's time. The substitute solvent cannot be discharged to the WWTP because it will adversely affect the WWTP performance. Consequently, about 2,000 kg must be hauled to a hazardous waste disposal site for storage each month. Because of the lack of capital funds and the high interest rate for borrowing, J. R. recommends that the substitute solvent be used. Do you agree with this recommendation? Explain your reasoning.

1-7. Ted Terrific is the manager of a leather tanning company. In part of the tanning operation a solution of chromic acid is used. It is company policy that the spent chrome solution is put in 0.20-m^3 drums and shipped to a hazardous waste disposal facility.

On Thursday the 12th, the day shift miscalculates the amount of chrome to add to a new batch and makes it too strong. Since there is not enough room in the tank to adjust the concentration, Abe Lincoln, the shift supervisor, has the tank emptied and a new one prepared and makes a note to the manager that the bad batch needs to be reworked.

On Monday the 16th, Abe Lincoln looks for the bad batch and cannot find it. He notifies Ted Terrific that it is missing. Upon investigation, Ted finds that Rip Van-Winkle, the night-shift supervisor, dumped the batch into the sanitary sewer at 3:00 a.m. on Friday the 13th. Ted makes discreet inquiries at the wastewater plant and finds that they have had no process upsets. After Ted severely disciplines Rip, he should: (Choose the correct answer and explain your reasoning.)

A. Inform the city and state authorities of the illegal discharge as required by law even though no apparent harm resulted.
B. Keep the incident quiet because it will cause trouble for the company without doing the public any good. No harm was done and the shift supervisor has been punished.
C. Advise the president and board of directors and let them decide whether to follow A or B.

1-10 ADDITIONAL READING

A. S. Goodman, *Principles of Water Resources Planning,* Englewood Cliffs, NJ: Prentice Hall, 1984.
R. J. Bibbero and I. G. Young, *Systems Approach to Air Pollution Control,* New York: Wiley, 1974.
W. H. Rodgers, Jr., *Environmental Law, 2nd ed.,* St. Paul, MN: West Publishing Co., 1994.
Selected Environmental Law Statutes, 1992–93 Educational Edition, St. Paul, MN: West Publishing Co., 1992.

CHAPTER 2

HYDROLOGY

2-1 FUNDAMENTALS
The Hydrologic Cycle
Surface Water Hydrology
Groundwater Hydrology
Common Units of Measurement
The Hydrologic Equation
2-2 RAINFALL ANALYSIS
Point Precipitation Analysis
2-3 RUNOFF ANALYSIS
Estimation of Amount of Runoff
Estimation of Time of Arrival
Estimation of Probability of Occurrence
2-4 STORAGE OF RESERVOIRS
Classification of Reservoirs
Volume of Reservoirs
2-5 GROUNDWATER AND WELLS
Construction of Wells
Cone of Depression
Definition of Terms
Well Hydraulics
Groundwater Contamination
2-6 WASTE MINIMIZATION AND WATER CONSERVATION
2-7 CHAPTER REVIEW
2-8 PROBLEMS
2-9 DISCUSSION QUESTIONS
2-10 ADDITIONAL READING

2-1 FUNDAMENTALS

The Hydrologic Cycle

The global system that supplies and removes water from the earth's surface is known as the hydrologic cycle (Figure 2-1). Water is transferred to the earth's atmosphere through two processes: (1) *evaporation* and (2) *transpiration.*[1] As moist air rises, it cools. Eventually enough moisture accumulates and the mass cools sufficiently to nucleate (form small crystals) on microscopic particles. Sufficient growth causes the droplets or snowflakes to become heavy enough to fall as precipitation. As they fall on the earth's surface, the droplets either run over the ground into streams and rivers (*surface runoff,* or just *runoff*) or percolate into the ground to form groundwater.

Surface Water Hydrology

Precipitation. Surface water hydrology begins before the precipitate hits the ground. The form the precipitate takes (rain, sleet, hail, or snow) is important. For example, it takes about 10 mm of snow to make the equivalent of 1 mm of rain. Other factors of importance are the size of the area over which the precipitation falls, the intensity of the precipitation, and its duration.

Once the precipitation hits the ground, a number of things can happen. It can evaporate promptly. This is especially true if the surface is hot and impervious. If the

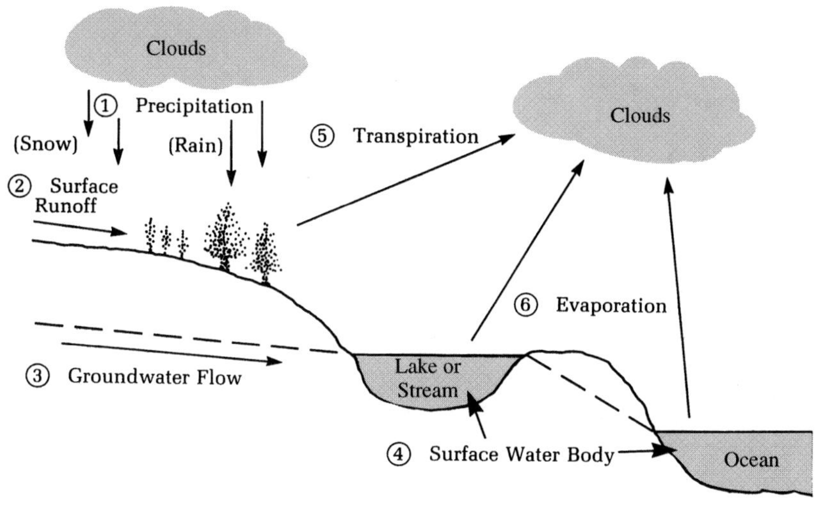

FIGURE 2-1
The hydrologic cycle.

[1]Transpiration is the process whereby plants give off water vapor through the pores of their leaves. The moisture comes from the roots through capillary action.

soil is dry and/or porous, the precipitate may *infiltrate* into the ground or it may only wet the surface. This process and the process of wetting leaves and blades of grass is called *interception.* The precipitate may be trapped in small depressions or puddles. It may remain there until it evaporates or until the depressions fill and overflow. And last but not least, it may run off directly to the nearest stream or lake to become surface water. The four factors (evaporation, infiltration, interception, and trapping) that reduce the amount of direct runoff are called *abstractions.*

Streamflow. The water that makes up our streams and rivers is derived from two sources: direct runoff and groundwater exfiltration, or base flow, as it is more commonly called. Direct runoff is a consequence of precipitation. Base flow is the dry weather flow that results from the seepage of groundwater out of stream banks.

The amount of water that reaches a stream is a function of the abstractions mentioned above and the catchment area or watershed that feeds the stream. The watershed, or *basin,* is defined by the surrounding topography (Figure 2-2). The perimeter of the watershed is called a *divide.* It is the highest elevation surrounding the watershed. All of the water that falls on the inside of the divide has the potential to be shed into the streams of the basin encompassed by the divide. Water falling outside of the divide is shed to another basin.

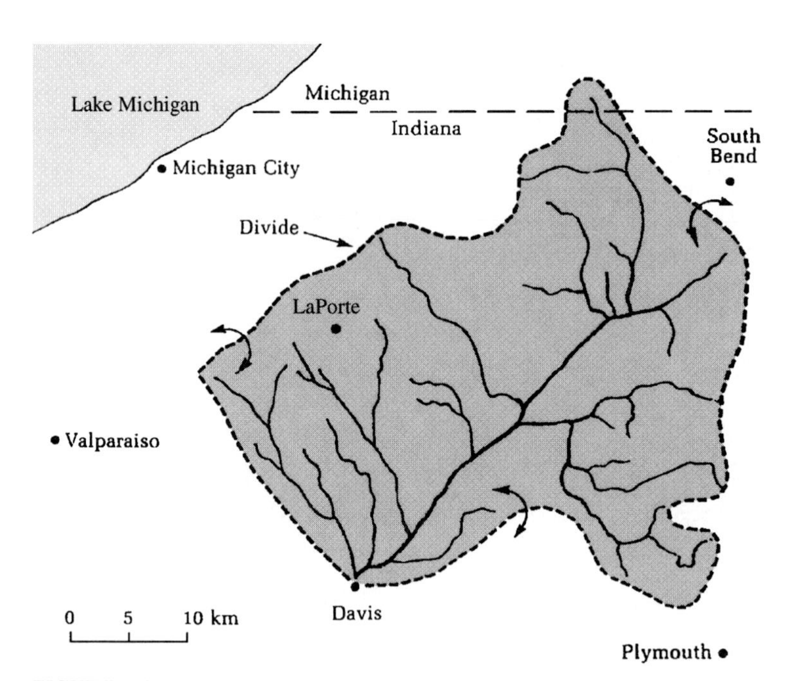

FIGURE 2-2
The Kankakee River Basin above Davis, IN. *Note:* Arrows indicate that precipitation falling inside the dashed line is in the Davis watershed, while that falling outside is in another watershed. The dashed line then "divides" the watersheds.

Groundwater Hydrology

Water table (unconfined) aquifer. As we mentioned earlier, part of the precipitation that falls on the soil may infiltrate. This water replenishes the soil moisture or is used by growing plants and returned to the atmosphere by transpiration. Water that drains downward below the root zone finally reaches a level at which all of the openings or voids in the earth's materials are filled with water. This zone is known as the *zone of saturation.* Water in the zone of saturation is referred to as groundwater. The geologic formation that bears the water is called an *aquifer.* The upper surface of the zone of saturation, if not confined by impermeable material, is called the *water table* (Figure 2-3). The aquifer is called a *water table aquifer* or an *unconfined aquifer.* Water will rise to the level of the water table in an unpumped water table well.

The smaller void spaces in the porous material just above the water table may contain water as a result of capillarity. This zone is referred to as the *capillary fringe* (Figure 2-3*b*). It is not a source of supply since the water held will not drain freely by gravity. The region from the saturated zone to the surface is also called the *vadose zone.*

Springs. Because of the irregularities in underground deposits and in surface topography, the water table occasionally intersects the surface of the ground or the bed of a stream, lake, or ocean. At these points of intersection, groundwater moves out

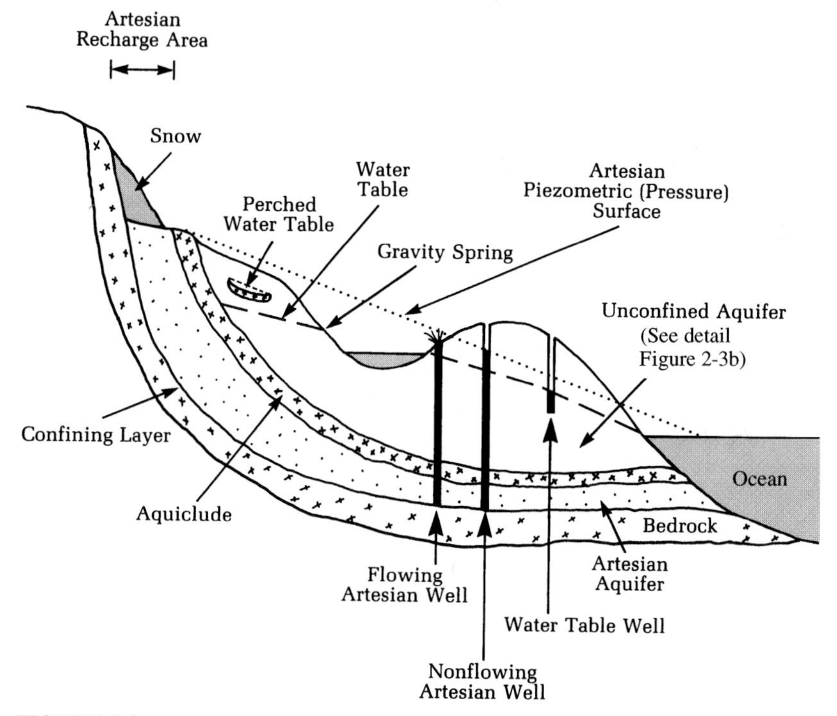

FIGURE 2-3a
Schematic of groundwater aquifers.

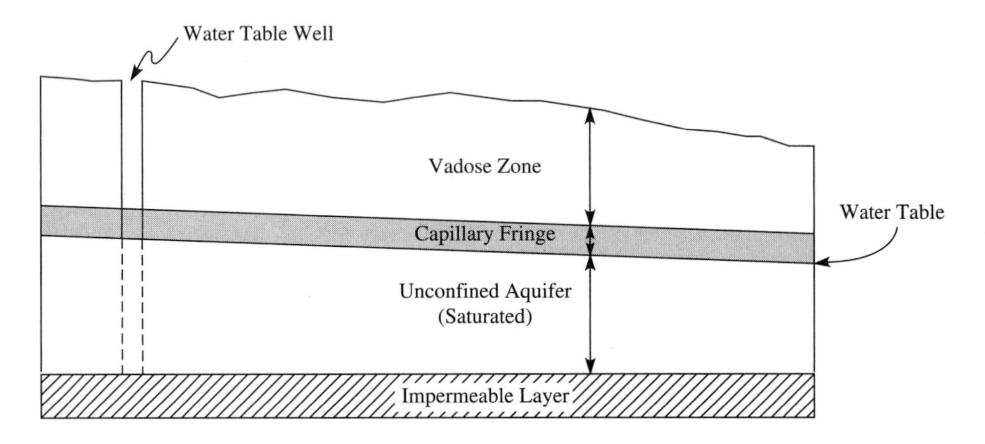

FIGURE 2-3b
Detail of the unconfined aquifer.

of the aquifer. The place where the water table breaks the ground surface is called a *gravity* or *seepage spring* (Figure 2-3).

Perched water table. A perched water table is a lens of water held above the surrounding water table by an impervious layer. It may cover an area from a few hundred square meters to several square kilometers.

Artesian aquifer. As water percolates into an aquifer and flows downhill, the lower layers come under pressure. This pressure is the result of the mass of water in the upper layers pressing on the water in the lower layers, much as deep sea divers are under greater and greater pressure as they go deeper and deeper into the sea. The system is analogous to a manometer (Figure 2-4). When there is no constriction in the manometer, the water level in each leg rises to the same height. If the left leg is raised, the increased water pressure in that leg pushes the water up in the right leg until the levels are equal again. If the right leg is clamped shut then, of course, the water will not rise to the same level. However, at the point where the clamp is placed, the water pressure will increase. This pressure is the result of the height of water in the left leg.

A special type of groundwater system occurs when an overlying impermeable formation and an underlying impermeable formation restrict the water, much as the walls of a manometer. The impermeable layers are called *confining layers.* Other names given to these layers are *aquicludes* if they are essentially impermeable, or *aquitards* if they are less permeable than the aquifer but not truly impermeable. An aquifer between impermeable layers is called a *confined aquifer.* If the water in the aquifer is under pressure, it is called an *artesian aquifer* (Figure 2-3). The name "artesian" comes from the French province of Artois (*Artesium* in Latin) where, in the days of the Romans, water flowed to the surface of the ground from a well.

Water enters an artesian aquifer at some location where the confining layers intersect the ground surface. This is usually in an area of geological uplift. The ex-

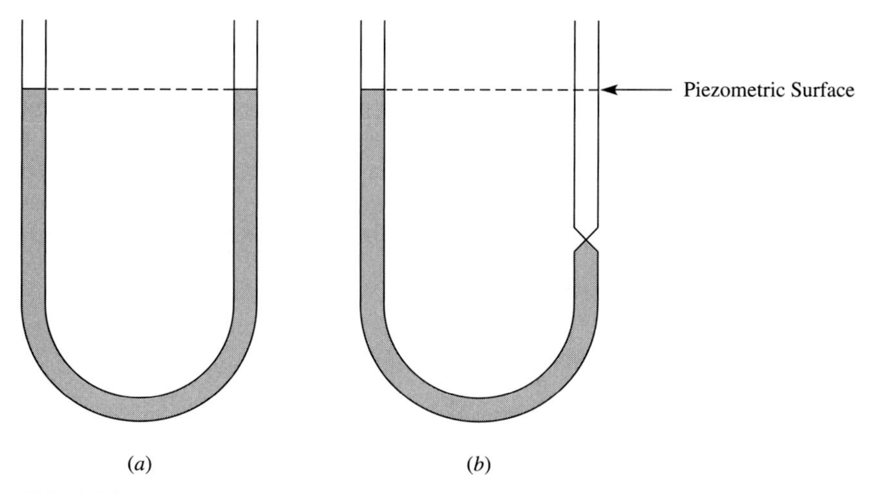

FIGURE 2-4
Manometer analogy to water in an aquifer. Manometer "a" is analogous to an unconfined aquifer. Manometer "b" is analogous to a confined aquifer.

posed surface of the aquifer is called the *recharge area.* The artesian aquifer is under pressure for the same reason that the pinched manometer is under pressure, that is, because the recharge area is higher than the bottom of the top aquiclude and, thus, the height of the water above the aquiclude causes pressure in the aquifer. The greater the vertical distance between the recharge area and the bottom of the top aquiclude, the higher the height of the water, and the higher the pressure.

Piezometric surfaces. If we place small tubes (*piezometers*) into an artesian aquifer along its length, the water pressure will cause water to rise in the tubes much as the water in the legs of a manometer rises to a point of equilibrium. The height of the water above the bottom of the aquifer is a measure of the pressure in the aquifer. An imaginary plane drawn through the points of equilibrium is called a *piezometric surface.* In an unconfined aquifer, the piezometric surface is the water table.

If the piezometric surface of a confined aquifer lies above the ground surface, a well penetrating into the aquifer will flow naturally without pumping. If the piezometric surface is below the ground surface, the well will not flow without pumping.

Common Units of Measurement

Precipitation, evaporation, and transpiration. These quantities are reported as either depths or rates. In SI units[2] depths are reported in millimeters (mm) and rates are reported in millimeters per hour (mm/h).

Speed and flow. The speed[3] of water movement in rivers is reported in meters per second (m/s). Beneath the ground, water movement is more likely to be on the order

[2]"Le Système International d'Unités" or "The International System of Units."

[3]Note that this is speed. Velocity is speed in a specified direction.

of meters per day (m/d). Flow is the volume of water moving past a point in a unit of time (m^3/s). Common symbols for speed and flow are u and Q respectively.

Volume. The common unit of volume is the cubic meter (m^3). Other equivalent units may also be used. For example, the depth of rainfall over a specified area is a volume. A hectare-meter (ha-m) is equivalent to a volume of water one meter deep covering one hectare (10,000 m^2) of land. Another example is the volume produced by a flow of 1 m^3/s in a 24-hour period. This volume is called a second-meter-day (smd). A common symbol for volume is $\forall$.

The Hydrologic Equation

Because the total quantity of water available to the earth is finite, the global hydrologic system is considered to be a closed system; that is, it is self-contained or in mass balance. As such it may be described by a simple mass-balance equation:

$$\forall_P(\rho) - \forall_S(\rho) - \forall_R(\rho) - \forall_G(\rho) - \forall_E(\rho) - \forall_T(\rho) = 0 \qquad (2\text{-}1)$$

where $\forall$ refers to the volume and the subscripts are defined as follows:

P = precipitation
S = storage
R = runoff
G = groundwater infiltration
E = evaporation
T = transpiration

The ρ refers to the density of water so that the product of volume times density yields mass. Averaged over the globe, the density is often assumed to be a constant. Thus, the equation is often written in terms of the volumes alone as the density terms cancel out.

In contrast to the global system, most of the problems we are interested in are open systems. We can modify the basic hydrologic equation by considering any hydrologic subsystem as a mass-balancing problem. Let us consider the system shown in Figure 2-5*a* as an example. A simplified mass-balance diagram of this system is shown in Figure 2-5*b*. Using the generalized form of the mass-balance equation from Chapter 1, we would write

$$\text{Rate of accumulation} = \text{Rate of input} - \text{Rate of output}$$

or in more sophisticated terms:

$$\frac{dS}{dt} = \frac{d(\text{In})}{dt} - \frac{d(\text{Out})}{dt} \qquad (2\text{-}2)$$

This says that the difference between all of the inputs and all of the outputs is equal to the change in storage in the subsystem. As shown in Figure 2-5*a*, precipitation (P_{in}), river inflow (R_{in}), and groundwater inflow (G_{in}) are inputs. Evaporation (E_{out}), transpiration (T_{out}), river outflow (R_{out}), and groundwater outflow (G_{out}) are outputs. As in the global mass balance, we often assume that the density of the water is constant throughout the system and thus ignore it in writing hydrologic mass-balance equations.

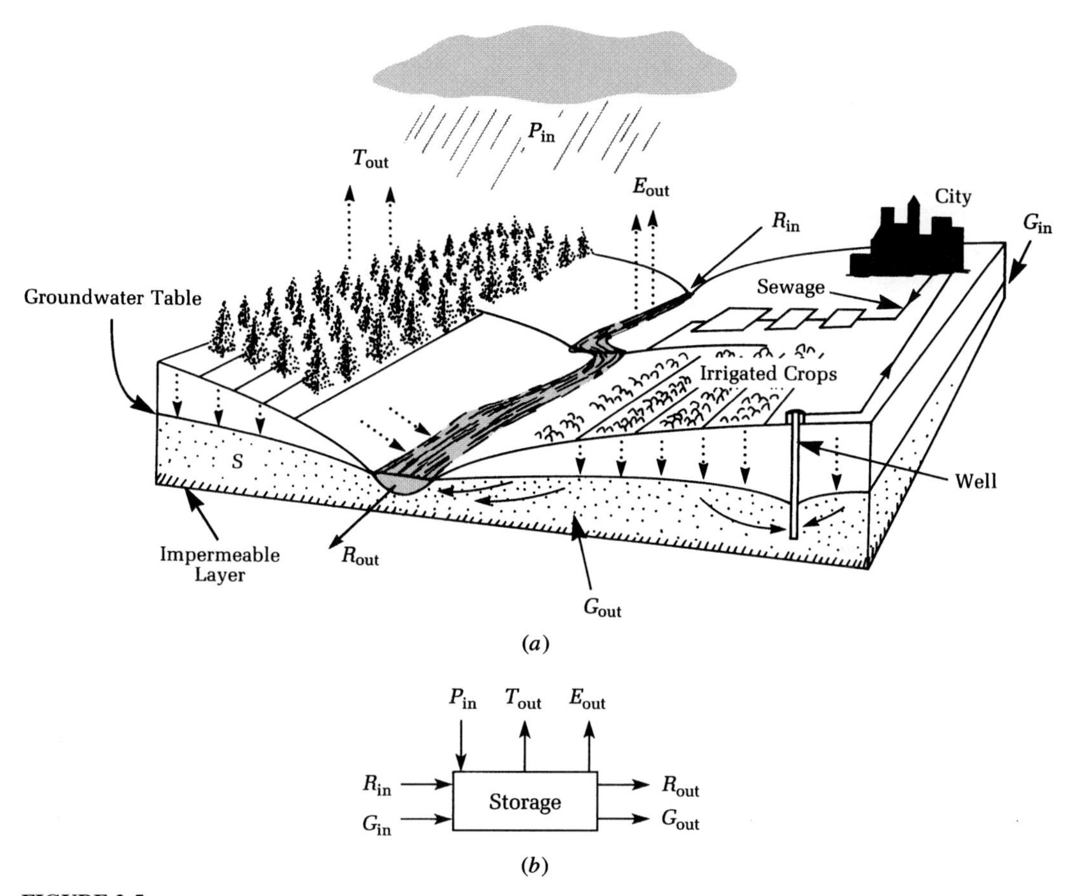

FIGURE 2-5
(*a*) Schematic diagram of a hydrologic subsystem; (*b*) mass balance diagram of hydrologic subsystem.

The terms of the mass-balance equation for the hydrologic equation may be expanded to show their functional relationship to other physical phenomena. For example, the amount of runoff is a function of the characteristics of the surface (paved, cultivated, flat, steep-sloped). The amount of storage, for example, is a function of the type of soil or geological formation. These two aspects of the hydrologic equation are discussed in Sections 2-3 and 2-5. In the following paragraphs, we wish to elaborate on the behavior of the other terms.

Infiltration. Of the numerous equations developed to describe infiltration, Horton's equation is useful to examine because it characterizes three phenomena of interest. Horton[4] expressed the infiltration rate as:

$$f = f_c + (f_o - f_c)e^{-kt} \quad (2\text{-}3)$$

[4]R. E. Horton, *Surface Runoff Phenomena: Part I, Analysis of the Hydrograph,* Horton Hydrologic Lab Pub. 101, Ann Arbor, MI, Edward Bros., 1935.

where f = infiltration rate, mm/h
f_c = equilibrium or final infiltration rate, mm/h
f_o = initial infiltration rate, mm/h
k = empirical constant, h^{-1}
t = time, h

This expression assumes that the rate of precipitation is greater than the rate of inflation.

Infiltration rate is a function of the properties of the soil; thus, the values for f_o, f_c, and k are, as you might expect, a function of the soil type. Some examples are (in mm/h and h^{-1}):

	f_o	f_c	k
Dothan loamy sand	88	67	1.4
Fuquay pebbly loamy sand	159	61	4.7

Soil moisture content, vegetative cover, organic matter, and season affect these values.

The second property of interest is that the infiltration rate is an inverse exponential function of time. If the rate of precipitation exceeds the rate of infiltration, a plot of infiltration rate versus time will reveal that as rainfall continues, the rate at which the ground soaks it up decreases because the pore spaces in the soil fill up with water. Since typical values for f_o and f_c are greater than prevailing rainfall intensity this may lead to calculated decreases in infiltration even though there is capacity to accept precipitation at higher rates.

The third property, which is directly related to hydrologic balances, is that the area under the infiltration curve represents the volume of water that infiltrates. Integration of Horton's equation yields the volume:

$$\forall = f_c t + \frac{f_o - f_c}{k}(1 - e^{-kt}) \tag{2-4}$$

Evaporation. The loss of water from the surface of a lake or other water body is a function of solar radiation, air and water temperature, wind speed, and the difference in vapor pressures at the water surface and in the overlying air. As with estimates of infiltration rate, there are numerous methods for estimating evaporation. Dalton[5] first expressed the fundamental relationship in the form:

$$E = (e_s - e_a)(a + bu) \tag{2-5}$$

[5]J. Dalton, "Experimental Essays on the Constitution of Mixed Gases; on the Force of Steam or Vapor from Waters and Other Liquids, Both in a Torricellian Vacuum and in Air; on Evaporation; and on the Expansion of Gases by Heat," *Mem. Proc. Manchester Lit. Phil. Soc.*, vol. 5, pp. 535–602, 1802.

where E = evaporation rate, mm/d
e_s = saturation vapor pressure, kPa
e_a = vapor pressure in overlying air, kPa
a, b = empirical constants
u = wind speed, m/s

Empirical studies at Lake Hefner, Oklahoma, yielded a similar relationship:

$$E = 1.22(e_s - e_a)u \tag{2-6}$$

From these expressions, it is apparent that high wind speeds and low humidities (vapor pressure in the overlying air) result in large evaporation rates. You may note that the units for these expressions do not make much sense. This is because these are empirical expressions developed from field data. The constants have implied conversion factors in them. In applying empirical expressions, care must be taken to use the same units as those used by the author of the expression.

Evapotranspiration. Water loss from plants (transpiration) is difficult to separate from losses from the soil surface or root zone. For mass-balance calculations, these are often lumped together under the term evapotranspiration. The rate of evapotranspiration is a function of soil moisture, soil type, plant type, wind speed, and temperature. Plant types may affect evapotranspiration rates dramatically. For example, an oak tree may transpire as much as 160 L/d while a corn plant may transpire only about 1.9 L/d.

Example 2-1. Silk's Lake has a surface area of 70.8 ha. For the month of April the inflow was 1.5 m^3/s. The dam regulated the outflow (discharge) from Silk's Lake to be 1.25 m^3/s. If the precipitation recorded for the month was 7.62 cm and the storage volume increased by an estimated 650,000 m^3, what is the estimated evaporation in m^3 and cm? Assume that no water infiltrates out of the bottom of Silk's Lake.

Solution. Begin by drawing the mass-balance diagram:

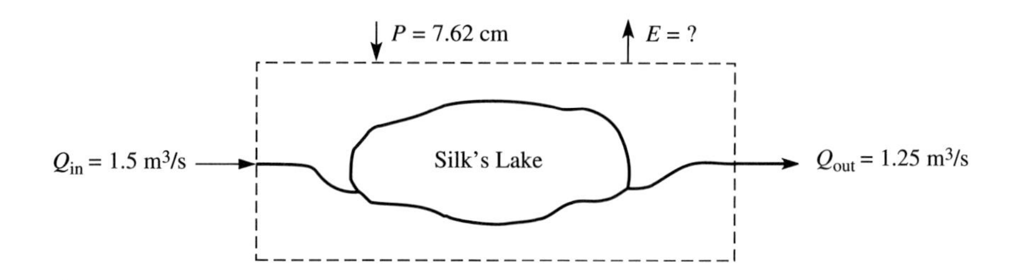

The mass-balance equation is:

$$\text{Accumulation} = \text{Input} - \text{Output}$$

The accumulation is given as 650,000 m^3. The input consists of the inflow and the precipitation. The product of the precipitation depth and the area on which it fell (70.8 ha) will yield a volume. The output consists of outflow plus evaporation.

$$\Delta S = [(Q_{in})(t) + (P)(\text{area})]_{\text{input}} - [(Q_{out})(t) + E]_{\text{output}}$$

Noting that April has 30 days and making the appropriate units conversions:

$$\begin{aligned}650{,}000\ \text{m}^3 &= (1.5\ \text{m}^3/\text{s})(30\ \text{d})(86{,}400\ \text{s/d})\\ &\quad + (7.62\ \text{cm})(70.8\ \text{ha})(10^4\ \text{m}^2/\text{ha})(1\text{m}/100\ \text{cm})\\ &\quad - (1.25\ \text{m}^3/\text{s})(30\ \text{d})(86{,}400\ \text{s/d}) - E\end{aligned}$$

Solving for E:

$$E = 3.89 \times 10^6\ \text{m}^3 + 5.39 \times 10^4\ \text{m}^3 - 3.24 \times 10^6\ \text{m}^3 - 6.50 \times 10^5\ \text{m}^3$$

$$E = 5.39 \times 10^4\ \text{m}^3$$

For an area of 70.8 ha, the evaporation depth is:

$$E = \frac{5.39 \times 10^4\ \text{m}^3}{(70.8\ \text{ha})(10^4\ \text{m}^2/\text{ha})} = 0.076\ \text{m or } 7.6\ \text{cm}$$

Example 2-2. During April, the wind speed over Silk's Lake was estimated to be 4.0 m/s. The air temperature averaged 20 °C and the relative humidity was 30%. The water temperature averaged 10 °C. Estimate the evaporation rate using the empirical relationship in Equation 2-6.

Solution. From the water temperature and Table 2-1, the saturation vapor pressure is estimated as $e_s = 1.227$ kPa. The vapor pressure in the air may be estimated as the product of the relative humidity and the saturation vapor pressure at the air temperature:

$$e_a = (2.337\ \text{kPa})(0.30) = 0.70\ \text{kPa}$$

The daily evaporation rate is then estimated to be:

$$E = 1.22(1.227 - 0.70)(4.0\ \text{m/s}) = 2.57\ \text{mm/d}$$

The monthly evaporation would then be estimated to be:

$$E = (2.56\ \text{mm/d})(30\ \text{d}) = 76.8\ \text{mm or } 7.7\ \text{cm}$$

TABLE 2-1
Water vapor pressures at various temperatures

Temperature, °C	Vapor pressure, kPa
0	0.611
5	0.872
10	1.227
15	1.704
20	2.337
25	3.167
30	4.243
35	5.624
40	7.378
50	12.34

2-2 RAINFALL ANALYSIS

Of the many variables of rainfall that might be of interest, we are concerned primarily with four:

1. Space: the average rainfall over the area
2. Intensity: how hard it rains
3. Duration: how long it rains at any given intensity
4. Frequency: how often it rains at any given intensity and duration

Point Precipitation Analysis

Data from a single nearby rain gage are often sufficiently representative to allow their use in the design of small projects. The analysis of data from a single gage is called point precipitation analysis. Spatial analysis is much more complex and is left for more advanced courses.

Rain gages and rain gage records. There are three types of rain gage in use: the U.S. Weather Bureau standard, the weighing bucket, and the tipping bucket. The standard gage is used for manually recording 24-hour accumulations of precipitation. The weighing bucket provides a continuous strip chart record of accumulated precipitation (Figure 2-6). The tipping bucket records precipitation by logging the number of times the cup tips. The cup is designed to tip when 0.25 mm of precipitation has accumulated.

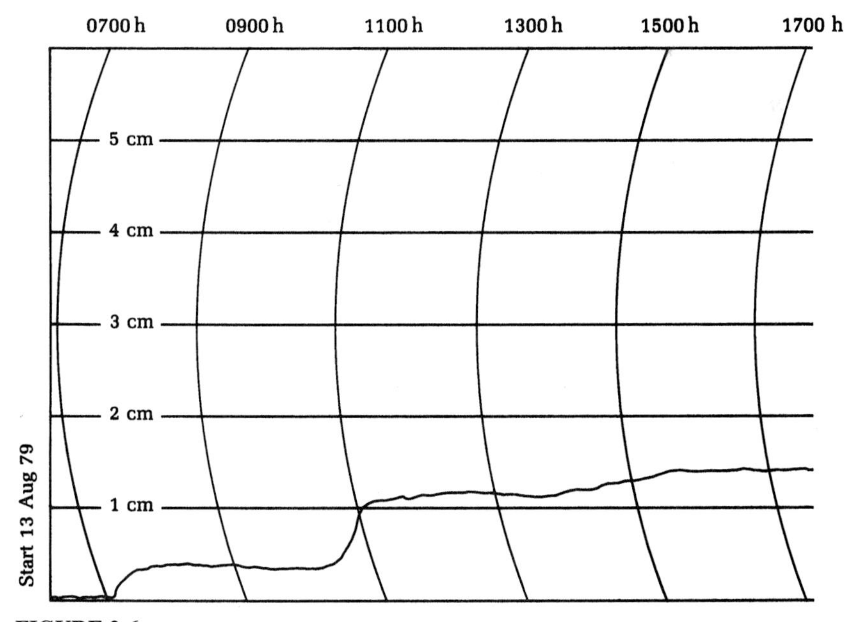

FIGURE 2-6
Strip chart record from weighing-bucket rain gage.

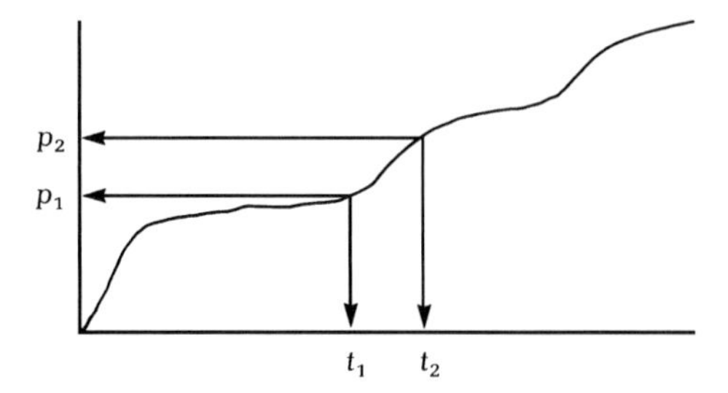

FIGURE 2-7
Computation of rainfall intensity from weighing-bucket strip chart record.

Interpreting rain gage records. The analysis of standard and tipping-bucket measurements is fairly straightforward. Weighing-bucket charts require a slight degree of interpretation.

The weighing-bucket chart is wound on a drum for data collection. The drum rotates at a constant speed while the pen inscribes a line of accumulated precipitation (actually mass converted to depth). Thus, when there is no rain the line on the chart is horizontal. When it rains, the line is sloped upward; the steeper the slope, the more intense the rain.

Intensity is computed from weighing-bucket records by determining the slope of the line (Figure 2-7):

$$\frac{\Delta p}{\Delta t} = \frac{(p_2 - p_1)}{(t_2 - t_1)}$$

where p_1 and p_2 are the accumulated precipitation at times t_1 and t_2.

The duration of precipitation for a given intensity is t. The normal procedure is to select fixed time intervals and calculate Δp at each interval. Thus, it is possible for one rainfall event to produce data for several durations.[6] For example, a 15-minute rain could yield the following pieces of data: one 15-minute duration event, one 10-minute duration event, and three 5-minute duration events.

Intensity-duration-frequency curves (IDF). The family of curves that depicts the relationship between the intensity, duration, and frequency of precipitation at a point is a fundamental part of the rational method of design for storm water disposal. "In practice, the collection and analysis of basic rainfall data by the designer are limited to extensive projects. Rainfall data compiled and processed by the U.S. Weather Bureau, the Department of Agriculture, and similar government agencies are used widely where the size of the project or lack of local records does not justify complete local statistical analysis."[7]

---

[6] An *event* is any continuous period of precipitation.

[7] Joint Committee of the American Society of Civil Engineers and the Water Pollution Control Federation, *Design and Construction of Sanitary and Storm Sewers* (ASCE Manuals and Reports on Engineering Practice No. 37, or WPCF Manual of Practice No. 9), New York: American Society of Civil Engineers, p. 46, 1969.

Table 2-2 is the compilation of a partial series of rainfall events. Rather than a record of all rainfalls, it is a record of rainfall intensities above some practical minimum. It gives the frequency or number of times that a rainfall of given intensity and duration will be equaled or exceeded for the period of record. For example, looking at the first row in Table 2-2, one would expect seven rainfall events with an intensity of 160.0 mm/h or more and a duration of five minutes to occur in any 45-year period ($1968 - 1923 = 45$ years) in the Dismal Swamp.

You should note two other facts about the table. First, the numbers in the table are also ranks. If the rainfall events for five-minute duration storms are arranged in descending order of intensity, then the 7th storm in the sequence or "the 7th-ranked" storm has an intensity of 160.0 mm/h or more. We assume that the ranks are spaced evenly between the recorded ranks, that is, that intensity and rank are linear. Thus, for the 5th-ranked storm of 5-minute duration, by interpolation, we can estimate that it would have an intensity of 170.0 mm/h or more.

The second fact that you should note is the ranks may be used to infer the probability that a given intensity storm will be equaled or exceeded. Again, using the seventh-ranked storm, we may infer that rainfall intensities of 160.0 mm/h or greater will occur with a frequency of seven times in 45 years. An annual average probability of occurrence would be $\frac{7}{45} = 0.16$ or 16 percent. Hydrologists and engineers often use the reciprocal of annual average probability because it has some temporal significance. The reciprocal is called the average return period or average recurrence interval (T):

$$T = \frac{1}{\text{Annual average probability}} \tag{2-7}$$

For the case of our seventh-ranked storm, the average return period of a 160.0 mm/h, five-minute storm is 6.25 years. This means that we would expect a storm of 160.0 mm/h or greater once every 6.25 years on the average.

Because the amount of reliable data available is limited to the last 100 years or less,[8] it is customary to use Weibull's formula for calculating return period:[9]

$$T = \frac{n + 1}{m} \tag{2-8}$$

where T = average return period in years
n = number of years of record
m = rank of storm, with most intense storm given a rank of 1

Weibull's formula allows for a small correction when the number of years of record is small. At larger values of n it closely approximates $T = \frac{n}{m}$.

[8] Systematic measurement of precipitation was begun by the Surgeon General of the Army in 1819, while streamflow data collection did not begin until 1888.

[9] W. Weibull, "A Statistical Theory of the Strength of Materials," *Ing. Vetenskapsakad. Akad. Handl,* (Stockholm), vol. 151, p. 15, 1939.

TABLE 2-2
Rainfall record for the Dismal Swamp (1 Oct. 1923–30 Sep. 1968)

	Number of storms of stated intensity or more										
Duration	Intensity (mm/h)										
(min)	20.0	30.0	40.0	60.0	80.0	100.0	120.0	140.0	160.0	180.0	200.0
5						245	49	16	7	3	2
10					256	64	15	7	4	1	
15				241	94	18	6	3	2		
20		240	80	36	10	4	2	1			
30	202	44	17	9	2	2	1				
40	76	31	8	1							
50	30	12	3								
60	9	2									

Example 2-3. Prepare a table of plotting points for an IDF curve for a 5-year storm at the Dismal Swamp. Compute points for each duration given in Table 2-2.

Solution. Since Table 2-2 is a table of ranks, we need to determine the rank of the 5-year storm. First, we rearrange Weibull's formula:

$$m = \frac{n+1}{T}$$

where

$$n = 1968 - 1923 = 45 \text{ y}$$

$$T = 5 \text{ y}$$

thus

$$m = \frac{46}{5} = 9.2$$

Starting with the 5-minute duration, we note that the 9.2-ranked storm lies between the 16th- and 7th-ranked storm; that is,

Intensity (mm/h)		
140.0		160.0
16	9.2	7

We also note that the ranks increase from right to left while the intensities increase from left to right. Keeping this in mind, and recalling that we assume a linear relationship between intensity and rank, we may interpolate by simple proportions:

$$\frac{9.2 - 7}{16 - 7}(160.0 - 140.0) = 4.89$$

Thus, the 9.2-ranked storm is 4.89 mm/h less than 160.0 mm/h:

$$160.0 - 4.89 = 155.11 \text{ or } 155.1 \text{ mm/h}$$

The completed table would appear as follows:

Intensity and duration values for a five-year storm at Dismal Swamp

Duration (min)	Intensity (mm/h)
5	155.1
10	134.5
15	114.7
20	82.7
30	59.5
40	39.5
50	33.1
60	—

Note that a similar table could be constructed for each intensity given in Table 2-2. This would give us twice as many points to use for fitting the curve.

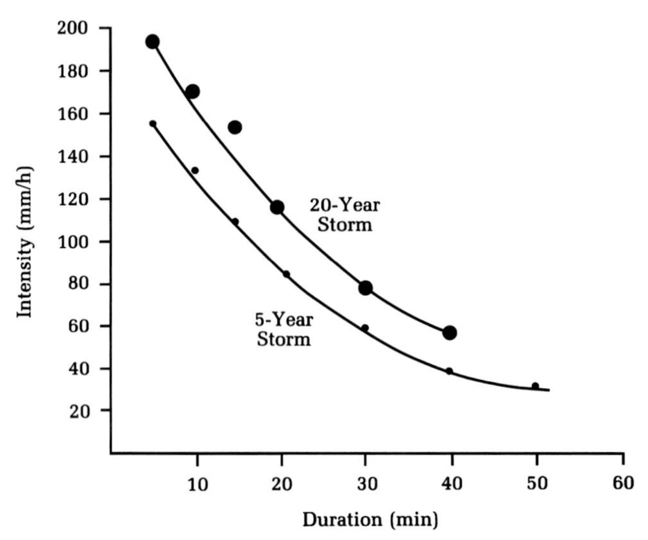

FIGURE 2-8
Intensity-duration-frequency curves for the Dismal Swamp.

The IDF curve for Example 2-3 and the curve for a return period of 20 years are plotted in Figure 2-8. You should note that the frequency curves join occurrences that are not necessarily from the same storm. They represent the average intensity expected for a given duration. They do not represent a sequence of intensities during a single storm.

2-3 RUNOFF ANALYSIS

Three runoff questions are of interest:

1. How much of the rain that falls on a watershed reaches the stream or storm sewer draining it?
2. How long does it take for the runoff to reach the stream or storm sewer?
3. How often does the runoff cause a flood?

Estimation of Amount of Runoff

Stream gages. Streamflow measurements are made by recording the height of the surface of the water above a reference datum. The elevation (*stage*) readings are calibrated in terms of streamflow (*discharge*). At manual recording stations, readings are made from a marked rod (*staff gage*) placed in the stream (Figure 2-9). At automatic recording stations, a float and cable system is used to drive a pen on a strip chart recorder (Figure 2-10). A *stilling well* (Figure 2-11) is used to minimize the effects of wave action and to protect the float from floating logs and other materials. For small streams, a dam with a weir plate (Figure 2-12) may be installed. This system increases the change in elevation for small changes in streamflow and makes readings more precise and accurate.

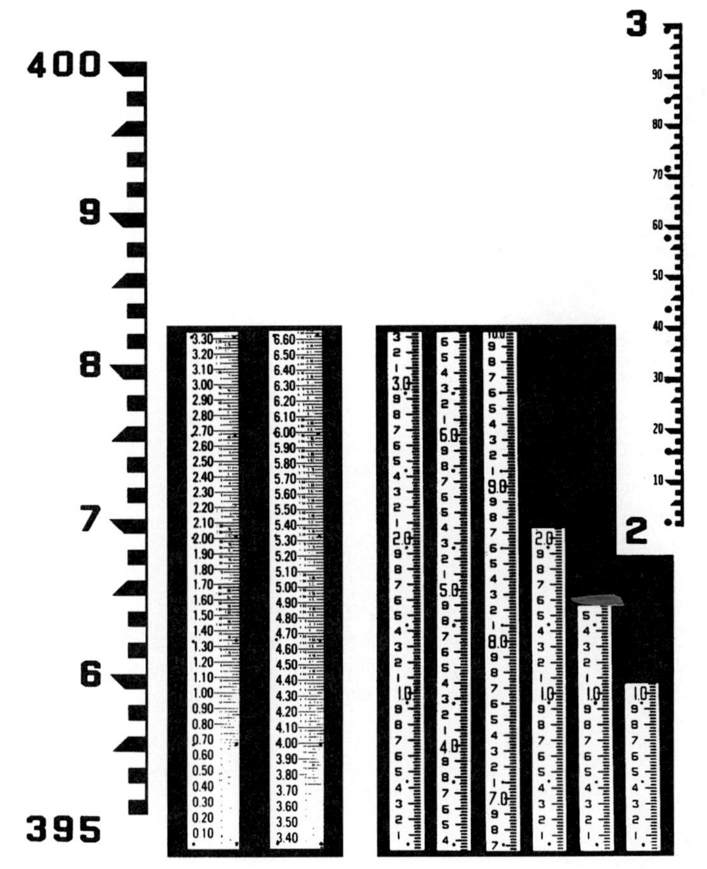

FIGURE 2-9
Staff gages for measurement of stream stage. (Courtesy of Leupold & Stevens, Inc.)

Hydrographs. A graphical representation of the discharge of a stream at a single gaging station is called a *hydrograph* (Figure 2-13). As we mentioned earlier, during the period between storms the base flow is a result of exfiltration of groundwater from the banks of the stream. Discharge from precipitation excess, that is, that which remains after abstractions, causes a hump in the hydrograph. This hump is called the *direct runoff hydrograph* (DRH).

Obviously, any precipitation excess that occurs at the extremities of a watershed will not be recorded at the basin outlet until some time lapse has occurred. As precipitation continues, enough time elapses for the more distant areas to add to the outlet discharge. The lag time of the peak and the shape of the DRH depend on the precipitation pattern and the characteristics of the basin (size, slope, shape, and channel storage capacity).

Unit hydrograph method. A unit hydrograph (UH) is a DRH that results from a unit of precipitation excess over a watershed for a unit period of time. Although any unit depth may be selected (fathoms, furlongs, feet, hands, or cubits all would do), we

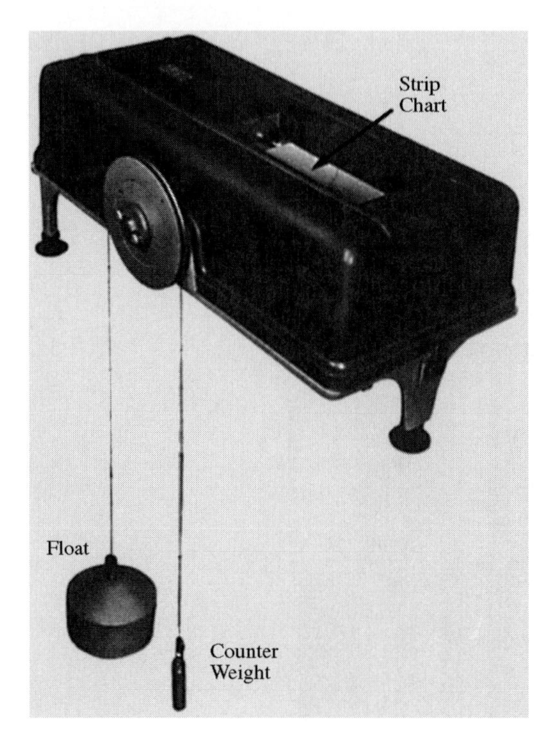

FIGURE 2-10
Float system and strip chart recorder for continuous stage measurement. (Courtesy of Leupold & Stevens, Inc.)

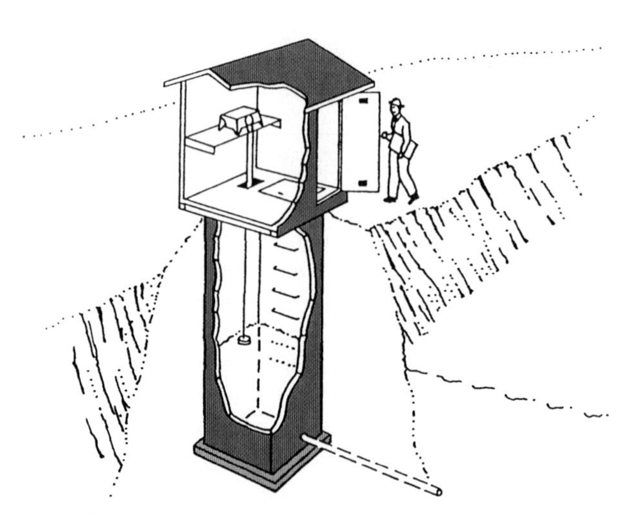

FIGURE 2-11
Stilling well. (Courtesy of Leupold & Stevens, Inc.)

FIGURE 2-12
Dam and weir with float system for stage measurement. (Courtesy of Leupold & Stevens, Inc.)

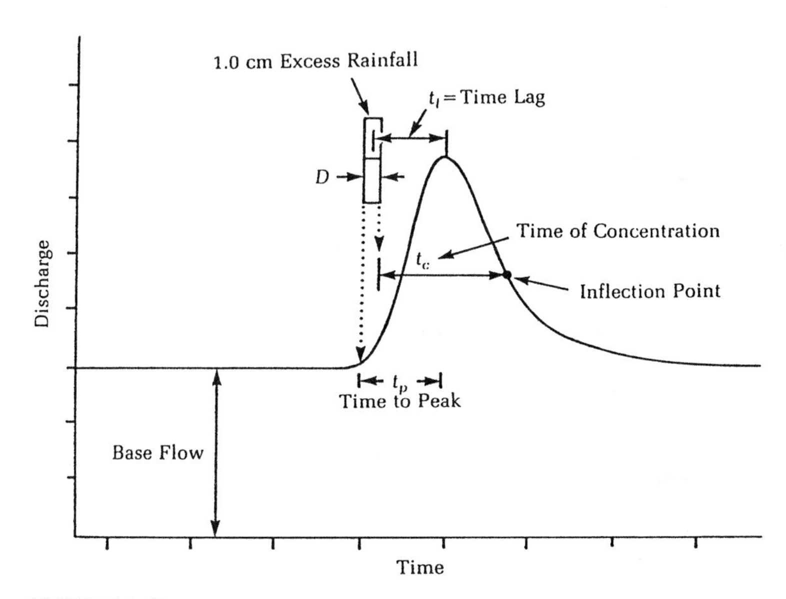

FIGURE 2-13
An idealized hydrograph showing a uniform base flow and a superimposed direct runoff hydrograph resulting from 1.0 cm of rainfall excess.

have selected an excess of 1.0 cm after abstractions as a workable unit depth.[10] The presumption is that if you can determine an average UH, then you can approximate the DRH for any other rainfall excess over the same unit time by multiplying the UH ordinates by the amount of the rainfall excess. For example, a 2.0-cm rainfall excess would yield a DRH with ordinates twice as large as a 1.0-cm UH. The method is limited to watersheds between 3,000- and 4,000-square kilometers in area.[11] Since the intent of the UH is to portray discharge caused by direct runoff so we can use it to predict the DRH for other storms, the first step in constructing a UH is to remove the groundwater contribution. This step is called hydrograph separation. There are a number of graphical procedures for hydrograph separation.

The second step in the construction of the UH is to estimate the total volume of water that occurs as direct runoff. Since the hydrograph is a plot of discharge versus time, the area under the DRH is equal to the volume of direct runoff. The volume is computed by numerical integration of the area under the curve. This is simply a summation of the products of an arbitrary unit of time (dt) and the height of the DRH ordinate at the center of the selected time interval.

The third step is to convert the volume of direct runoff to a storm depth of runoff. This is done by dividing the volume of direct runoff by the area of the watershed in square meters and then multiplying by a conversion factor of 100 cm/m.

The fourth step is to divide the ordinates of the DRH by the storm depth computed in step three. The quotients are the ordinates of the UH. They have units of $m^3/s \cdot cm$.

The unit duration of the UH is determined from the *hyetograph* (time-rainfall graph) of the storm that was used to develop the UH. Since all of the precipitation does not result in direct runoff, an effective duration of excess precipitation must be estimated. The effective duration becomes the UH unit duration.

Example 2-4. Determine the unit hydrograph ordinates for the Triangle River hydrograph shown in Figure 2-14. The area of the watershed is 16.2 square kilometers.

Solution. The first step is to determine the depth of the storm precipitation spread over the watershed. The depth is equivalent to the volume of water divided by the area. The volume is equal to the area under the hydrograph. Because of the rather symmetrical shape of this particular hydrograph, it would be easy to find the area from the principles of geometry. However, in the interest of developing a technique that will also be applicable to more customary hydrographs, we will numerically integrate the area under the curve. We do this by taking a convenient slice or *dt* and multiplying it by

---

[10]In the original development of the UH by Sherman, a unit depth was defined as 1.0 inch of rainfall excess. L. K. Sherman, "Stream-Flow from Rainfall by the Unit-Graph Method," *Engineering News Record*, vol. 108, pp. 501–505, 1932.

[11]W. Viessman, J. W. Knapp, G. L. Lewis, and T. Harbaugh, *Introduction to Hydrology*, 2nd ed, New York: Harper & Row, p. 117, 1977.

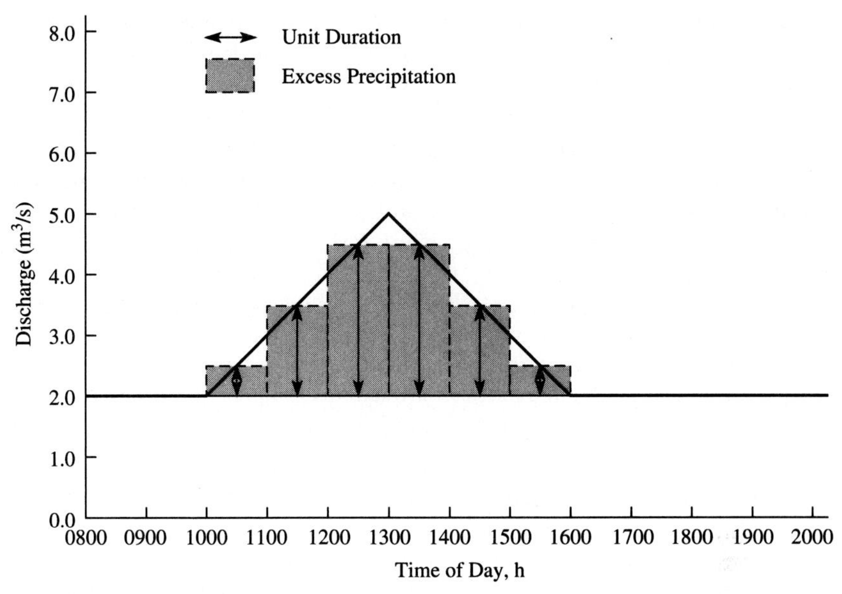

FIGURE 2-14
Triangle River hydrograph.

the height of the direct runoff (DRH) ordinate. The direct runoff ordinate is simply the difference between the total ordinate and the base ordinate. In this particular instance the base ordinate is, by observation, 2.0 m^3/s for all time periods. Using a convenient time interval of 1 hour, the following tabular computations are used to numerically integrate the area under the curve:

Time interval (h)	Total ordinate (m^3/s)	Base ordinate (m^3/s)	DRH ordinate (m^3/s)	Volume increment (m^3)
10–11	2.5	2.0	0.5	1,800
11–12	3.5	2.0	1.5	5,400
12–13	4.5	2.0	2.5	9,000
13–14	4.5	2.0	2.5	9,000
14–15	3.5	2.0	1.5	5,400
15–16	2.5	2.0	0.5	1,800
				$\sum$ = 32,400

The volume increment is calculated as follows: First, the difference between the total ordinate and the base ordinate is found for the time increment selected. In the first row, for the time period from 10 AM to 11 AM the total ordinate is read from the hydrograph (Figure 2-14) as 2.5 m^3/s:

$$\text{Total ordinate} - \text{Base ordinate} = \text{DRH ordinate}$$

$$2.5\ \text{m}^3/\text{s} \quad - \quad 2.0\ \text{m}^3/\text{s} \quad = \quad 0.5\ \text{m}^3/\text{s}$$

To find the area (volume) represented by this slice, the flow rate is multiplied by the time interval selected (1 h) with appropriate units conversions:

$$(0.5 \text{ m}^3/\text{s})(1 \text{ h})(3{,}600 \text{ s/h}) = 1{,}800 \text{ m}^3$$

This process is continued for all the slices shown in Figure 2-14. The total volume (area under the curve) is estimated as 32,400 m^3. We can verify this by using the geometry of the triangle:

$$1/2(\text{base})(\text{height}) = (0.5)(6 \text{ h})(5.0 \text{ m}^3/\text{s} - 2.0 \text{ m}^3/\text{s})(3{,}600 \text{ s/h}) = 32{,}400 \text{ m}^3$$

Since we wish to construct a *unit hydrograph,* we need to determine whether or not this storm produced 1.0 cm of rainfall excess over the watershed. If it did, then we may use the ordinates directly. If not, then we must adjust the ordinates so that they would be equivalent to that produced by a 1.0 cm rainfall excess. We can determine whether or not this storm produced 1.0 cm by dividing the volume of rainfall by the area of the watershed (given as 16.2 km^2):

$$\frac{32{,}400 \text{ m}^3}{(16.2 \text{ km}^2)(1 \times 10^6 \text{ m}^2/\text{km}^2)} \times 100 \text{ cm/m} = 0.20 \text{ cm}$$

It is obvious that the storm is too small and, hence, the ordinates are too small. By dividing the ordinates by the storm depth, we can synthesize ordinates for a unit hydrograph. For example, for the first DRH ordinate:

$$\frac{\text{DRH ordinate}}{\text{Storm depth}} = \frac{0.5 \text{ m}^3/\text{s}}{0.2 \text{ cm}} = 2.5 \text{ m}^3/\text{s} \cdot \text{cm}$$

This ordinate would be located at the center of the slice that was used to establish it, i.e., halfway between 1000 and 1100 hours (see the arrows in Figure 2-14), i.e., 1030. For a generic hydrograph starting at a time equal to zero, the plotting point would be 0.5 h. The remaining unit hydrograph ordinates are tabulated below.

Triangle River plotting time (h)	Generic plotting time (h)	UH ordinate ($m^3/s \cdot cm$)
1030	0.5	2.5
1130	1.5	7.5
1230	2.5	12.5
1330	3.5	12.5
1430	4.5	7.5
1530	5.5	2.5

The unit "$m^3/s \cdot cm$" is read as

$$\frac{\text{m}^3}{(\text{s})(\text{cm})}$$

This means if we multiply a UH ordinate by the cm of excess rainfall, we will get units of m^3/s for the ordinate.

We can check our logic by calculating the area under a similar triangle using these new ordinates.

Time interval (h)	DRH ordinate (m^3/s)	Volume increment (m^3)
10–11	2.5	9,000
11–12	7.5	27,000
12–13	12.5	45,000
13–14	12.5	45,000
14–15	7.5	27,000
15–16	2.5	9,000
		$\sum$ = 162,000

Recalculating our storm depth:

$$\frac{162{,}000 \text{ m}^3}{(16.2 \text{ km}^2)(1 \times 10^6 \text{ m}^2/\text{km}^2)} \times 100 \text{ cm/m} = 1.00 \text{ cm}$$

The unit hydrograph may be applied to a sequence of storms that have the same unit duration. There are two fundamental assumptions in the technique. The first is that storms of the same unit duration have ordinates that are in proportion to the unit hydrograph ordinates. Thus, simple ratios can account for differences in runoff excess. The second assumption is that a sequence of storms may be approximated by superimposing one hydrograph over another (with appropriate time lag) and adding the ordinates together. This is illustrated in the next example.

Example 2-5. Using the hyetograph in Figure 2-15, and the unit hydrograph ordinates from Example 2-4, determine the DRH ordinates and compound runoff.

Solution. The tabular computations are shown below. The explanation follows the table.

Time interval	Time (h)	Rainfall excess (cm)	DRH ordinates 1	DRH ordinates 2	DRH ordinates 3	Compound runoff (m^3/s)
1	0–1	0.5	1.25	N/A	N/A	1.25
2	1–2	2.0	3.75	5.0	N/A	8.75
3	2–3	1.0	6.25	15.0	2.5	23.75
4	3–4	0.0	6.25	25.0	7.5	38.75
5	4–5	0.0	3.75	25.0	12.5	41.25
6	5–6	0.0	1.25	15.0	12.5	28.75
7	6–7	0.0	0.0	5.0	7.5	12.5
8	7–8	0.0	0.0	0.0	2.5	2.5

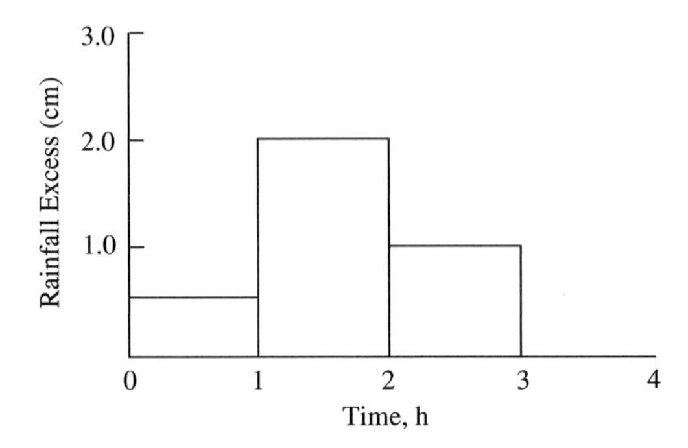

FIGURE 2-15
Hyetograph for Triangle River basin.

The time interval is simply an enumeration of the segments. For the first hour, from the hyetograph in Figure 2-15, the rainfall excess is 0.5 cm. For the second and third hours, the rainfall excesses are 2.0 and 1.0 cm, respectively. No rain falls after the end of the third hour. The column labeled DRH 1 refers to the ordinates that are generated from the rainfall excess (0.5 cm) occurring in the first hour. Likewise, the DRH 2 refers to the ordinates resulting from the 2.0-cm rainfall excess in the second hour.

The first set of ordinates is obtained by multiplying the rainfall excess by each of the UH ordinates, that is:

$$\text{(Rainfall excess)(UH Ordinate)} = \text{DRH ordinate}$$

Using the UH ordinates from Example 2-4:

$$(0.5 \text{ cm})(2.5 \text{ m}^3/\text{s} \cdot \text{cm}) = 1.25 \text{ m}^3/\text{s}$$
$$(0.5 \text{ cm})(7.5 \text{ m}^3/\text{s} \cdot \text{cm}) = 3.75 \text{ m}^3/\text{s}$$
$$(0.5 \text{ cm})(12.5 \text{ m}^3/\text{s} \cdot \text{cm}) = 6.25 \text{ m}^3/\text{s}$$
$$(0.5 \text{ cm})(12.5 \text{ m}^3/\text{s} \cdot \text{cm}) = 6.25 \text{ m}^3/\text{s}$$
$$(0.5 \text{ cm})(7.5 \text{ m}^3/\text{s} \cdot \text{cm}) = 3.75 \text{ m}^3/\text{s}$$
$$(0.5 \text{ cm})(2.5 \text{ m}^3/\text{s} \cdot \text{cm}) = 1.25 \text{ m}^3/\text{s}$$

The values for the second DRH start an hour later. Thus, under the column DRH 2, the first row is not applicable (N/A) since the rain that falls in the second hour (time interval 2) cannot reach the stream in the first hour. Likewise, under the column DRH 3, the first and second rows are N/A because rain that falls in the third hour cannot reach the stream in the first or second hour.

The DRH ordinates for the second hour of rainfall excess are obtained in the same fashion as those for the first, that is by multiplying the rainfall excess by each of the UH ordinates:

$$(2.0 \text{ cm})(2.5 \text{ m}^3/\text{s} \cdot \text{cm}) = 5.0 \text{ m}^3/\text{s}$$
$$(2.0 \text{ cm})(7.5 \text{ m}^3/\text{s} \cdot \text{cm}) = 15.0 \text{ m}^3/\text{s}$$
$$(2.0 \text{ cm})(12.5 \text{ m}^3/\text{s} \cdot \text{cm}) = 25.0 \text{ m}^3/\text{s}$$
$$(2.0 \text{ cm})(12.5 \text{ m}^3/\text{s} \cdot \text{cm}) = 25.0 \text{ m}^3/\text{s}$$
$$(2.0 \text{ cm})(7.5 \text{ m}^3/\text{s} \cdot \text{cm}) = 15.0 \text{ m}^3/\text{s}$$
$$(2.0 \text{ cm})(2.5 \text{ m}^3/\text{s} \cdot \text{cm}) = 5.0 \text{ m}^3/\text{s}$$

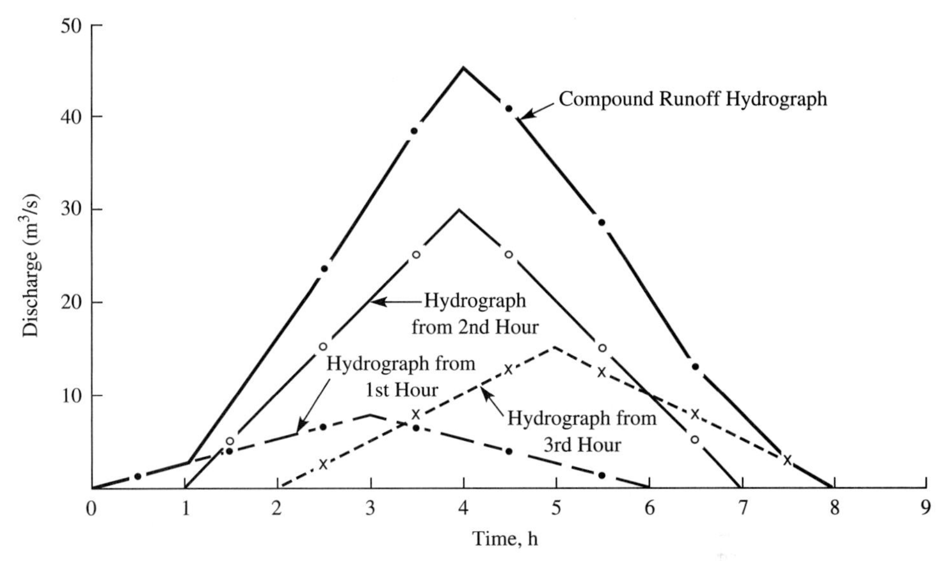

FIGURE 2-16
Compound runoff hydrograph for Triangle River. *Note:* Base flow is not shown.

You should note that the table is carried beyond the last rainfall period in the hyetograph until all of the ordinates are used since it takes some finite length of time for the last drop of rainfall excess to reach the stream.

The compound runoff is the sum of the DRH ordinates for each of the time intervals. For example:

$$1.25 + \text{N/A} + \text{N/A} = 1.25$$
$$3.75 + 5.0 + \text{N/A} = 8.75$$
$$6.25 + 15.0 + 2.5 = 23.75$$

To plot the compound runoff hydrograph, the compound runoff ordinates are plotted at 1.0-h intervals, starting 0.5 h from time zero in accordance with the plotting position of the UH ordinates specified earlier. A plot of the individual hydrographs for each of the storms, their superposition, and the resulting compound hydrograph are shown in Figure 2-16.

Rational method. This method of determining runoff is one of the simplest applications of the hydrologic equation. A good example for us to look at is a paved parking lot (Figure 2-17). A mass-balance equation of the form of Equation 2-2 applies in this case. The only input is precipitation. The only output is direct runoff. Assuming that the density of water is constant, the mass-balance equation is:

$$\frac{\text{Storage}}{\text{Unit of time}} = \frac{\text{Volume of precipitation}}{\text{Unit of time}} - \frac{\text{Volume of runoff}}{\text{Unit of time}}$$

or

$$\frac{dS}{dt} = \frac{\forall_P}{dt} - \frac{\forall_R}{dt} \tag{2-9}$$

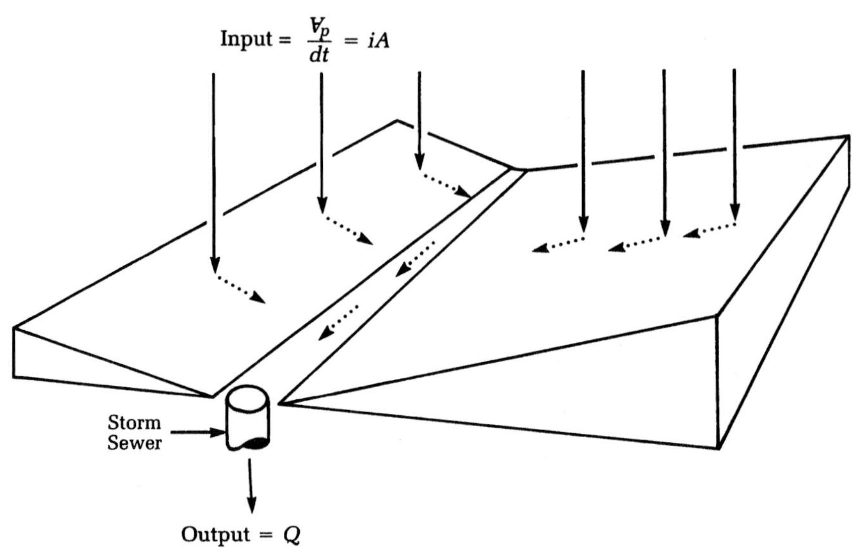

FIGURE 2-17
The application of the hydrologic equation to a parking lot having area = A.

If the rainfall on the lot continues for a long enough period at a constant intensity, at some time the system will reach *steady state.* At steady state each drop of water that falls on the watershed conceptually displaces a drop through the outlet. Thus, further rainfall at the same intensity does not increase the discharge at the outlet. The hyetograph and corresponding DRH for this situation are shown in Figure 2-18. The time that it takes for steady state to be achieved is called the *time of concentration* (t_c). The time of concentration is primarily a function of the basin geometry, surface conditions, and slope.

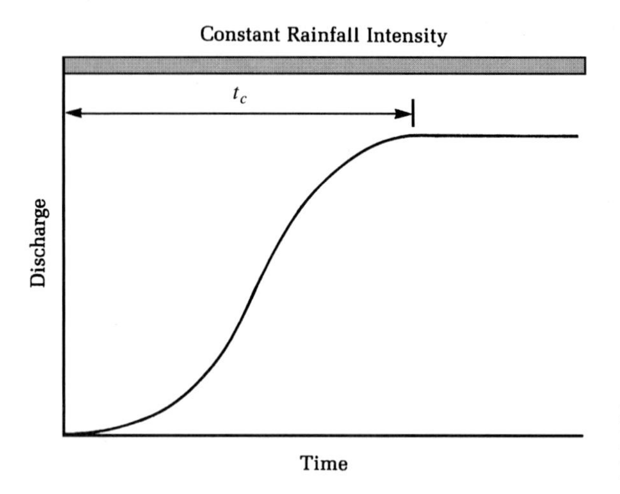

FIGURE 2-18
Hyetograph and hydrograph for a parking lot.

At steady state the storage term (dS/dt) in Equation 2-9 is equal to zero. The equation then reduces to the form

$$\frac{\forall_P}{dt} = \frac{\forall_R}{dt} \tag{2-10}$$

The input ($\forall_P/dt$) is a volume-per-unit time that may easily be shown to be equal to the product of the rainfall intensity (i) and the area of the watershed (A):

$$\frac{\forall_P}{dt} = iA \tag{2-11}$$

The output volume per unit time ($\forall_R/dt$) is direct runoff. It is equal to the discharge (Q). Since few natural or artificial surfaces are completely impervious (remember the chuck holes!), not all of the precipitation reaches the outlet. We can account for this loss by assuming that only some fraction (C) of the rainfall makes it to the outlet. By substituting all of these into Equation 2-10 and rearranging so that output is on the left, we obtain the rational formula

$$Q = 0.0028\, CiA \tag{2-12}$$

where Q = peak runoff rate, m^3/s
C = runoff coefficient
i = average rainfall intensity, mm/h
A = area of watershed, ha
0.0028 = conversion factor, $m^3 \cdot h/mm \cdot ha \cdot s$

The original derivation of the rational method was in English units. In the English system the use of intensity in inches/h and area in acres yields a runoff in ft^3/s without any conversion factor. Hence the name "rational" because the units work out rationally! Although the basic principles of the rational method are applicable to large watersheds, an upper limit of 13 square kilometers is recommended.[12] A selected list of runoff coefficients is given in Table 2-3.

The coefficients in Table 2-3 are applicable for storms of 5- to 10-year return period. Less frequent, higher intensity storms will require the use of higher coefficients because infiltration and other losses have a proportionally smaller effect on runoff. The coefficients are based on the assumption that the design storm does not occur when the ground is frozen.

Example 2-6. What is the peak discharge from the grounds of the Beauregard Long Ashby High School during a 5-year storm? The school grounds encompass a 16.2 ha plot that is 1.3 km east of the Dismal Swamp rain gage. Assume that the average time

[12] Joint Committee of the American Society of Civil Engineers and the Water Pollution Control Federation, *Design and Construction of Sanitary and Storm Sewers,* p. 43.

TABLE 2-3
Selected runoff coefficients

Description of area or character of surface	Runoff coefficient	Description of area or character of surface	Runoff coefficient
Business		Railroad yard	0.20 to 0.35
Downtown	0.70 to 0.95	Unimproved	0.10 to 0.30
Neighborhood	0.50 to 0.70	Pavement	
Residential		Asphaltic and concrete	0.70 to 0.95
Single-family	0.30 to 0.50	Brick	0.70 to 0.85
Multi-units, detached	0.40 to 0.60	Roofs	0.75 to 0.95
Multi-units, attached	0.60 to 0.75	Lawns, sandy soil	
Residential (suburban)	0.25 to 0.40	Flat, 2 percent	0.05 to 0.10
Apartment	0.50 to 0.70	Average, 2 to 7 percent	0.10 to 0.15
Industrial		Steep, 7 percent	0.15 to 0.20
Light	0.50 to 0.80	Lawns, heavy soil	
Heavy	0.60 to 0.90	Flat, 2 percent	0.13 to 0.17
Parks, cemeteries	0.10 to 0.25	Average, 2 to 7 percent	0.18 to 0.22
Playgrounds	0.20 to 0.35	Steep, 7 percent	0.25 to 0.35

Source: Joint Committee of the American Society of Civil Engineers and the Water Pollution Control Federation, *Design and Construction of Sanitary and Storm Sewers,* p. 51. See Note 6, supra.

of concentration of the grounds is 53 minutes. The composition of the grounds is as follows:

Character of surface	Area (m^2)	Runoff coefficient
Parking lot, asphaltic	11,150	0.85
Building	10,800	0.75
Lawns, heavy soil		
2.0% slope	35,000	0.17
6.0% slope	105,050	0.20
	$\sum$ = 162,000	

Solution. We begin by computing the weighted runoff coefficient, that is, the product of the fraction of the area and its runoff coefficient.

$$\begin{aligned} AC &= (11{,}150)(0.85) + (10{,}800)(0.75) + (35{,}000)(0.17) + (105{,}050)(0.20) \\ &= 44{,}537.5 \text{ m}^2 \text{ or } 4.45 \text{ ha} \end{aligned}$$

Since the Dismal Swamp rain gage is only 1.3 km away, we shall use the IDF curve obtained in Example 2-3 to determine the intensity. By definition, the peak discharge for a watershed occurs when the duration of the storm equals the time of concentration. Thus, we select a duration of 53 minutes and read a value of 30 mm/h at the 5-year storm curve in Figure 2-8.

The peak runoff is then

$$Q = (0.0028)(4.45)(30) = 0.37 \text{ m}^3/\text{s}$$

Thus, a storm sewer large enough to handle 0.37 m^3/s of flow is required to carry storm water away from the BLAHS grounds.

Estimation of Time of Arrival

In addition to the quantity of discharge, it is often desirable to know when the peak flow will arrive at the watershed outlet or at some point along the discharge channel. This is particularly important when analyzing a series of watersheds that contribute to a river or sewer at various distances downstream from the headwater. The coincident arrival of two peaks would influence the design dramatically.

Lag time. The time of arrival of the peak discharge is determined inherently in the UH method of estimating runoff. The lag time is the time from the midpoint of excess rainfall to the peak discharge.

Time of concentration. The time of concentration (t_c) is the time required for direct runoff to flow from the hydraulically most remote part of the drainage area to the watershed outlet. One of the major assumptions of the rational method is that the average rainfall intensity used in Equation 2-12 has continued for a period long enough to establish direct runoff and that rainfall has continued long enough to equal or exceed t_c. Thus, it is impossible to use the rational formula without being able to estimate t_c.

Although there are several methods for estimating t_c, the Federal Aviation Agency formula appears to be the easiest to use:[13]

$$t_c = \frac{1.8(1.1 - C)\sqrt{3.28D}}{\sqrt[3]{S}} \tag{2-13}$$

where t_c = time of concentration, min
C = runoff coefficient
D = overland flow distance, m
S = slope, %

Example 2-7. Estimate t_c for the BLAHS 6-percent-slope lawn in Example 2-6. Assume that the overland flow distance was 300.0 m.

Solution. From Example 2-6 we use the same value of C, namely 0.20. Thus,

$$t_c = \frac{1.8(1.1 - 0.20)\sqrt{(3.28)(300.0)}}{\sqrt[3]{6.0}}$$

$$t_c = \frac{50.82}{1.82} = 27.97 \text{ or } 28.0 \text{ min}$$

---

[13] Federal Aviation Agency, Department of Transportation, *Airport Drainage* (Advisory Circular A/C 150-5320-5B), Washington, DC: U.S. Government Printing Office, 1970.

Estimation of Probability of Occurrence

The frequency of occurrence ($1/T$) used in the design of water supply or storm water control projects should be a function of the cost of the project and the benefits to be obtained from it. That is, the benefit-cost ratio should be greater than 1.0 to justify the project on economic grounds. While the cost of construction can be estimated in a straightforward manner, the environmental cost may be impossible to estimate. Likewise, the benefit beyond a cheaper supply of water or a reduced amount of flood damage is difficult to quantify.

The following paragraphs provide some guidance on the selection of a design frequency. They were obtained from the ASCE sewer design manual with their permission. [14]

In practice, benefit-cost studies usually are not conducted for the ordinary urban storm drainage project. Judgment supported by records of performance in other similar areas is usually the basis of selecting the design frequency.

The range of frequencies used in engineering offices is as follows:

1. For storm sewers in residential areas, 2 to 15 years, with 5 years most commonly reported.
2. For storm sewers in commercial and high-value districts, 10 to 50 years, depending on economic justification.
3. For flood protection works and reservoir design, economics usually dictate a 50-year return period.

Other factors that may affect choice of design frequency include:

1. Use of greater return periods for design of those parts of the system not economically susceptible to future relief.
2. Use of longer recurrence intervals for design of combined sewers than for separate storm sewers because of basement flooding and consequent greater damage which may occur with overloaded combined sewers.
3. Use of greater recurrence intervals for design of special structures, such as expressway drainage pumping systems, where runoff exceeding capacity would seriously disrupt an important facility. Design frequencies of 50 years or more may be justified in such cases, particularly in small drainage areas, even though the project may be located in a district justifying only 5-year frequency for normal drainage.
4. Adoption of shorter return periods than normal but commensurate with available funds so that some degree of protection can be provided.

---

[14] Joint Committee of the American Society of Civil Engineers and the Water Pollution Control Federation, *Design and Construction of Sanitary and Storm Sewers,* p. 44.

The cost of storm sewers is not directly proportional to design frequency. Rousculp cites studies of effects of various factors on sewer cost and shows that sewer systems designed for 10-year storms may cost only about 6 to 11 percent more than systems designed for 5-year storms, depending on the sewer slope.[15] The lesser increase applies to steeper sewers.

If the peak flow is estimated by the rational method, it is assumed that the return period (inverse probability) of the peak flow is the same as that of the rainfall used to obtain it. If you use the UH method to determine the discharge, you must resort to some other method of estimating the probability of occurrence.

Annual series. Extreme-value analysis is a probability analysis of the largest or smallest values in a data set. Each of the extreme values is selected from an equal time interval. For example, if the largest value in each year of record is used, the extreme-value analysis is called an *annual maxima* series. If the smallest value is used, it is called an *annual minima* series.

Because of the climatic effects on most hydrologic phenomena, a water year or hydrologic year is adopted instead of a calendar year. The U.S. Geological Survey (U.S.G.S.) has adopted the 12-month period from October 1 to September 30 as the hydrologic year for the United States. This period was chosen for two reasons: "(1) to break the record during the low-water period near the end of the summer season, and (2) to avoid breaking the record during the winter, so as to eliminate computation difficulties during the ice period."[16]

The procedure for an annual maxima or minima analysis is as follows:

1. Select the minimum or maximum value in each 12-month interval (October to September) over the period of record.
2. Rank each value starting with the highest (for annual maxima) or lowest (for annual minima) as rank number one.
3. Compute a return period using Equation 2-8.
4. Plot the annual maxima series on a special probability paper known as *Gumbel paper* or flood data paper. Although the same paper may be used for annual minima series, Gumbel recommends a log extremal probability paper (axis of ordinates is log scale) for droughts.[17]

From the Gumbel plot, the return period for a flood or drought of any magnitude may be determined. Conversely, for any magnitude of flood or drought, you may determine how frequently it will occur.

---

[15]J. A. Rousculp, "Relation of Rainfall and Runoff to Cost of Sewers," *Transactions of the American Society of Civil Engineers,* vol. 104, p. 1473, 1939.

[16]M. C. Boyer, "Streamflow Measurement," in *Handbook of Applied Hydrology,* V. T. Chow, Ed., New York: McGraw-Hill, pp. 15–41, 1964.

[17]E. J. Gumbel, "Statistical Theory of Droughts," *Proceedings of the American Society of Civil Engineers,* (Sep. No. 439), vol. 80, pp. 1–19, May 1954.

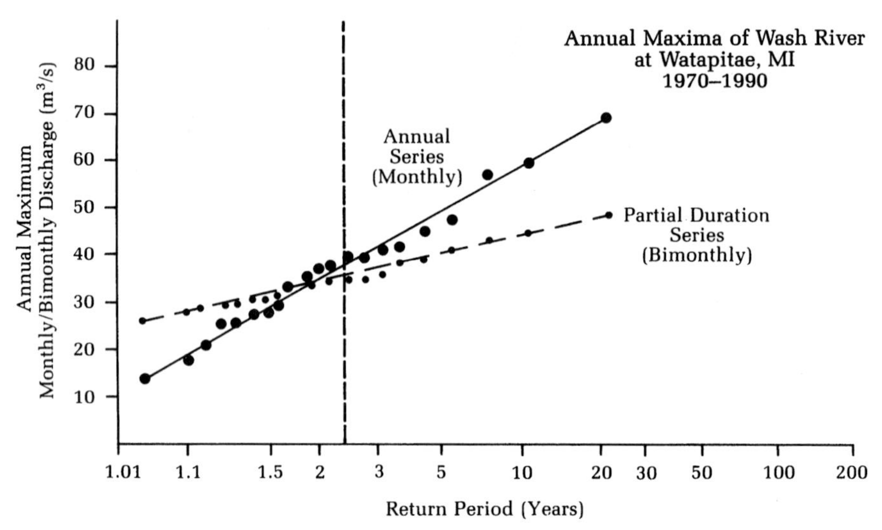

FIGURE 2-19
Gumbel plot of annual maxima of Wash River at Watapitae, MI.

In statistical parlance a Gumbel plot is a linearization of a Type I probability distribution. The logarithmically transformed version of the Type I distribution is called a log-Pearson Type III distribution. The return period of the mean (X) of the Type I distribution occurs at $T = 2.33$ years. Thus, the U.S.G.S. takes the return period of the mean annual flood to be 2.33 years. This is marked by a vertical dashed line on Gumbel paper (Figure 2-19).

The data in Table 2-4 were used to plot the annual maxima line in Figure 2-19. The computations are explained in Example 2-8.

Example 2-8. Perform an annual maxima extreme-value analysis on the data in Table 2-4. Determine the recurrence interval of monthly flows equal to or greater than 58.0 m^3/s. Also determine the discharge of the mean monthly annual flood.

Solution. To begin we select the maximum discharge in each hydrologic year. The first nine months of 1969 and the last three months of 1990 cannot be used because they are not complete hydrologic years. After selecting the maximum value in each year, we rank the data and compute the return period. The 1970 water year begins in October 1969.

The computations are summarized in Table 2-5. The return period and flows are plotted as the solid line in Figure 2-19. From Figure 2-19 we find that the return period for a flood of 58.0 m^3/s is about 9.2 years. The mean annual flood is about 37 m^3/s.

Partial-duration series. It often happens that the second largest or second smallest flow in a water year is larger or smaller than the maxima or minima from a different water year. To take these events into consideration, a partial series of the data is examined. The theoretical relationship between an annual series and partial series

TABLE 2-4
Average monthly discharge of the Wash River at Watapitae, MI (discharge in m^3/s)

Year	J	F	M	A	M	J	J	A	S	O	N	D
1969	2.92	5.10	1.95	4.42	3.31	2.24	1.05	0.74	1.02	1.08	3.09	7.62
1970	24.3	16.7	11.5	17.2	12.6	7.28	7.53	3.03	10.2	10.9	17.6	16.7
1971	15.3	13.3	14.2	36.3	13.5	3.62	1.93	1.83	1.93	3.29	5.98	12.7
1972	11.5	4.81	8.61	27.0	4.19	2.07	1.15	2.04	2.04	2.10	3.12	2.97
1973	11.1	7.90	41.1	6.77	8.27	4.76	2.78	1.70	1.46	1.44	4.02	4.45
1974	2.92	5.10	28.7	12.2	7.22	1.98	0.91	0.67	1.33	2.38	2.69	3.03
1975	7.14	10.7	9.63	21.1	10.2	5.13	3.03	10.9	3.12	2.61	3.00	3.82
1976	7.36	47.4	29.4	14.0	14.2	4.96	2.29	1.70	1.56	1.56	2.04	2.35
1977	2.89	9.57	17.7	16.4	6.83	3.74	1.60	1.13	1.13	1.42	1.98	2.12
1978	1.78	1.95	7.25	24.7	6.26	8.92	3.57	1.98	1.95	3.09	3.94	12.7
1979	13.8	6.91	12.9	11.3	3.74	1.98	1.33	1.16	0.85	2.63	6.49	5.52
1980	4.56	8.47	59.8	9.80	6.06	5.32	2.14	1.98	2.17	3.40	8.44	11.5
1981	13.8	29.6	38.8	13.5	37.2	22.8	6.94	3.94	2.92	2.89	6.74	3.09
1982	2.51	13.1	27.9	22.9	16.1	9.77	2.44	1.42	1.56	1.83	2.58	2.27
1983	1.61	4.08	14.0	12.8	33.2	22.8	5.49	4.25	5.98	19.6	8.5	6.09
1984	21.8	8.21	45.1	6.43	6.15	10.5	3.91	1.64	1.64	1.90	3.14	3.65
1985	8.92	5.24	19.1	69.1	26.8	31.9	7.05	3.82	8.86	5.89	5.55	12.6
1986	6.20	19.1	56.6	19.5	20.8	7.73	5.75	2.95	1.49	1.69	4.45	4.22
1987	15.7	38.4	14.2	19.4	6.26	3.43	3.99	2.79	1.79	2.35	2.86	10.9
1988	21.7	19.9	40.0	40.8	11.7	13.2	4.28	3.31	9.46	7.28	14.9	26.5
1989	31.4	37.5	29.6	30.8	11.9	5.98	2.71	2.15	2.38	6.03	14.2	11.5
1990	29.2	20.5	34.9	35.3	13.5	5.47	3.29	3.14	3.20	2.11	5.98	7.62

TABLE 2-5
Tabulated computations of annual maxima for the Wash River at Watapitae, MI

Year	Discharge (m^3/s)	Rank	$T = \frac{n+1}{m}$
1970	24.3	18	1.22
1971	36.3	11	2.00
1972	27.0	16	1.38
1973	41.1	6	3.67
1974	28.7	14	1.57
1975	21.1	19	1.16
1976	47.4	4	5.50
1977	17.7	20	1.10
1978	24.7	17	1.29
1979	13.8	21	1.05
1980	59.8	2	11.00
1981	38.8	8	2.75
1982	27.9	15	1.47
1983	33.2	13	1.69
1984	45.1	5	4.40
1985	69.1	1	22.00
1986	56.6	3	7.33
1987	38.4	9	2.44
1988	40.8	7	3.14
1989	37.5	10	2.20
1990	35.3	12	1.83

TABLE 2-6
Theoretical relationship between partial series and annual series return periods

Partial series	Annual series
0.5	1.18
1.0	1.58
1.45	2.08
2.0	2.54
5.0	5.52
10.0	10.5
50.0	50.5
100.0	100.5

Source: W. B. Langbein, "Annual Floods and Partial Duration Series," *Transactions of the American Geophysical Union,* vol. 30, 1949.

is shown in Table 2-6. The partial series is approximately equal to the annual series for return periods greater than ten years.[18]

If the time period over which the event occurs is also taken into account, the analysis is termed a partial-duration series. While it is fairly easy to define a flood as "any" flow that exceeds the capacity of the drainage system, in order to properly define a drought we must specify the low flow and its duration. For example, if a roadway is covered with water for ten minutes, we can say that it is flooded. In contrast, if the flow in a river is below our demand for ten minutes, we certainly would not declare it a drought! Thus, a partial-duration series is particularly relevant for low-flow conditions.

From an environmental engineering point of view, three low-flow durations are of particular interest. The 10-year return period of seven days of low flow has been selected by many states as the critical flow for water pollution control. Wastewater treatment plants must be designed to provide sufficient treatment to allow effluent discharge without driving the quality of the receiving stream below the standard when the dilution capacity of the stream is at a ten-year low.

A longer duration low flow and longer return period are selected for water supply. In the Midwest, durations of 1 to 5 years and return periods of 25 to 50 years are used in the design of water-supply reservoirs. Where water supply is by direct draft (withdrawal) from a river, the duration selected may be on the order of 30 to 90 days with a ten-year return period.

The procedure for performing a partial-duration series analysis is very similar to that used for an annual series.

Complete series. All of the observed data are used in a complete series analysis. This analysis is usually presented in one of two forms: as a duration curve (Figure 2-20), or as a cumulative probability distribution function (CDF) (Figure 2-21). In either form the analysis shows the percent of time that a given flow will be equaled or exceeded. The percent of time is interpreted as the probability that a watershed will yield a given flow over a long period of time. Thus it is sometimes called a yield analysis.

The procedure for preparing the data for plotting is as follows:

1. Establish a series of class intervals that cover the range of discharge observations in increasing order of magnitude.
2. Tabulate the number of observations in each class interval.
3. Cumulatively sum the number of observations in each class interval starting with the highest flow.
4. Compute the percentage of all the observations that appear in each class.

---

[18]W. B. Langbein, "Annual Flood and Partial Duration Series," *Transactions of the American Geophysical Union*, vol. 30, pp. 879–881, 1949.

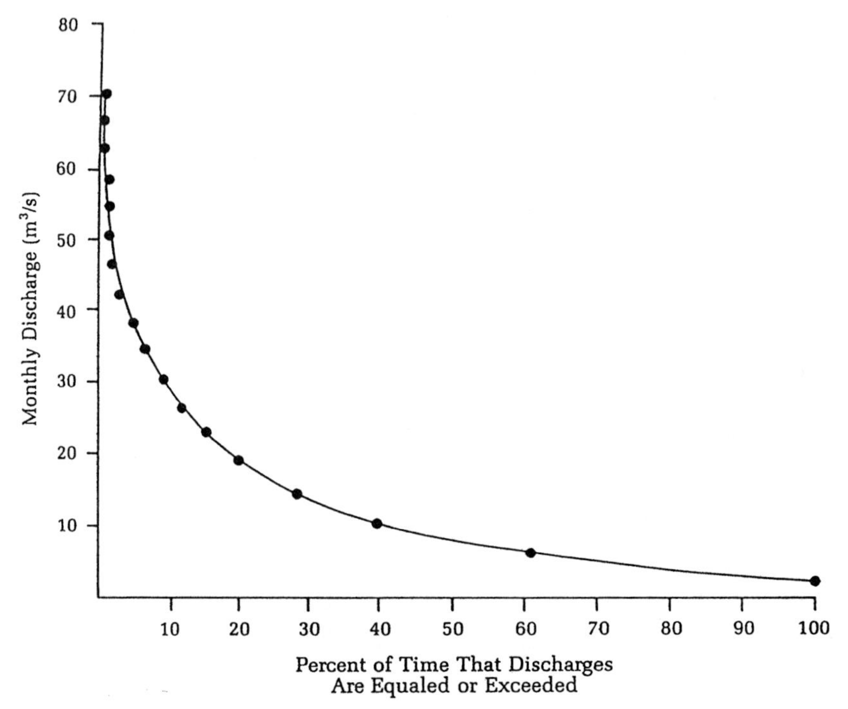

FIGURE 2-20
Duration curve for Wash River at Watapitae, MI.

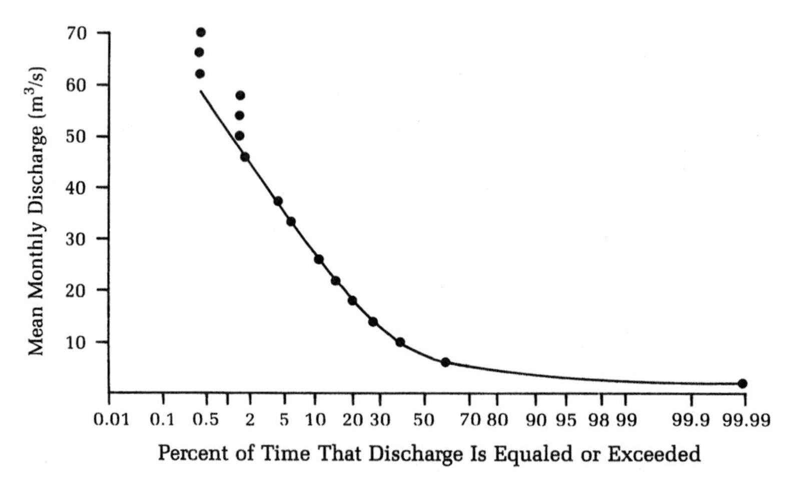

FIGURE 2-21
Cumulative probability distribution for Wash River at Watapitae, MI.

TABLE 2-7
Tabulation of computations for yield analysis

Class mean (m^3/s)	Class interval[a] (m^3/s)	Total	Accumulated	Percent
2	0–4	106	264	100.0
6	4–8	54	158	59.9
10	8–12	30	104	39.4
14	12–16	23	74	28.0
18	16–20	13	51	19.3
22	20–24	8	38	14.4
26	24–28	6	30	11.4
30	28–32	8	24	9.1
34	32–36	4	16	6.1
38	36–40	5	12	4.6
42	40–44	2	7	2.6
46	44–48	1	5	1.9
50	48–52	0	4	1.5
54	52–56	0	4	1.5
58	56–60	3	4	1.5
62	60–64	0	1	0.4
66	64–68	0	1	0.4
70	68–72	1	1	0.4

[a]Class intervals were actually 0.00–4.00, 4.01–8.00, 8.01–12.00, and so on. Zeros were eliminated to save space.

Example 2-9. Perform a complete series analysis on the data in Table 2-4. Determine the percent of time that a monthly flow of 28.0 m^3/s is equaled or exceeded. Also determine the mean monthly discharge.

Solution. Eighteen class intervals were selected for the analysis. These are shown in Table 2-7. The numbers in the column labeled "Total" are the number of observations in the class interval. The cumulative sum starting at the bottom of the column is in the next column. Thus, "Accumulated" means sum of observations equal to or greater than the class interval. "Percent" is the percent of readings equal to or greater than the class interval, computed by dividing the cumulative sum for each class interval by the total number of observations and multiplying by 100 percent.

The mean of each class interval is plotted against "Percent" in Figures 2-20 and 2-21. From these figures we find that a monthly flow of 28.0 m^3/s is equaled or exceeded about 10 percent of the time. The mean monthly discharge is the flow that is equaled or exceeded 50 percent of the time. From either Figure 2-20 or 2-21 we find that the mean monthly discharge is about 8 m^3/s.

When to use which series. You should not expect that the probability of occurrence ($1/T$) computed from an annual series will be the same as that found from a complete series. There are many reasons for this difference: Among the most obvious is the fact that in an annual series we treat $1/12$ of the data as if it were all of the data when, in fact, it is not even a representative sample. It is only the extreme end of the possible range of values.

The following guidelines can be used to decide when to use which analysis:

1. Use an annual series to predict the size of flood that a storm sewer or drainage channel must handle.
2. Use a partial series to predict low-flow conditions for wastewater dilution and water supply.
3. Use a complete series to determine the long-time reliability (yield) for water supply or power generation.

In practice the complete-series analysis can be performed to decide whether or not it is worth doing a partial series for water supply. If the complete series indicates that the mean monthly flow will not supply the demand, then computation of a partial series to determine the storage requirement is not worth the trouble, since it would be impossible to store enough water.

2-4 STORAGE OF RESERVOIRS

Classification of Reservoirs

For our purpose we can classify reservoirs either by size or by use. The size of the reservoir is used to establish the degree of safety to be incorporated into the design of the dam and spillway. The use or uses of the reservoir are a basis for evaluating the benefit-cost ratio.

Major dams (reservoir capacity greater than 6×10^7 m^3) are designed to withstand the maximum probable flood. Intermediate-sized dams (1×10^6 to 6×10^7 m^3) are designed to handle the discharge from the most severe storm considered to be reasonably characteristic of the watershed. For minor reservoirs (less than 1×10^6 m^3) the dams are designed to handle floods with return periods of 50 to 100 years.

Some of the benefits derived from reservoirs include the following: (1) flood control; (2) hydroelectric power; (3) irrigation; (4) water supply; (5) navigation; (6) preservation of aquatic life; and (7) recreation. The multipurpose or multiuse reservoir is the rule rather than the exception. Very seldom is it possible to justify the cost of a major reservoir on the basis of a single use.

Volume of Reservoirs

Mass diagram. The techniques for determining the storage volume required for a reservoir are dependent both on the size and use of the reservoir. We shall discuss the simplest procedure, which is quite satisfactory for small water-supply impoundments, storm-water retention ponds, and wastewater equalization basins. It is called the mass diagram or Rippl method.[19] The main disadvantage of the Rippl method is that it assumes that the sequence of events leading to a drought or flood will be

---

[19]W. Rippl, "The Capacity of Storage Reservoirs for Water Supply," *Proceedings of the Institution of Civil Engineers* (London), vol. 71, p. 270, 1883.

the same in the future as it was in the past. More sophisticated techniques have been developed to overcome this disadvantage, but these techniques are left for more advanced classes.

The Rippl procedure for determining the storage volume is an application of the mass-balance approach (Equation 2-2). In this case it is assumed that the only input is the flow into the reservoir (Q_{in}) and that the only output is the flow out of the reservoir (Q_{out}). Therefore,

$$\frac{dS}{dt} = \frac{d(\text{In})}{dt} - \frac{d(\text{Out})}{dt}$$

becomes

$$\frac{dS}{dt} = Q_{in} - Q_{out} \tag{2-14}$$

with the assumption that the density term cancels out because the change in density across the reservoir is negligible.

If we multiply both sides of the equation by dt, the inflow and outflow become volumes (flow rate $\times$ time $=$ volume), that is,

$$dS = (Q_{in})(dt) - (Q_{out})(dt) \tag{2-15}$$

By substituting finite time increments (Δt), the change in storage is then

$$(Q_{in})(\Delta t) - (Q_{out})(\Delta t) = \Delta S \tag{2-16}$$

By cumulatively summing the storage terms, we can estimate the size of the reservoir. If the reservoir design is for water supply, then Q_{out} is the demand, and zero or positive values of storage (ΔS) indicate there is enough water to meet the demand. If the storage is negative, then the reservoir must have a capacity equal to the absolute value of cumulative storage to meet the demand. If the reservoir design is for flood protection, then Q_{out} is the capacity of the downstream river to hold water, and zero or negative values indicate the river is below flood stage. If the storage is positive, then the reservoir must have capacity equal to the cumulative storage to prevent flooding.

Example 2-10. Using the data in Table 2-4, determine the storage required to meet a demand of 2.0 m^3/s for the period from August 1976 through December 1978.

Solution. The computations are summarized in the table below.

Month	Q_{in} (m^3/s)	$Q_{in}(\Delta t)$ (10^6 m^3)	Q_{out} (m^3/s)	$Q_{out}(\Delta t)$ (10^6 m^3)	ΔS (10^6 m^3)	$\sum(\Delta S)$ (10^6 m^3)
1976						
Aug	1.70	4.553	2.0	5.357	−0.8035	−0.8035
Sep	1.56	4.043	2.0	5.184	−1.140	−1.944
Oct	1.56	4.178	2.0	5.357	−1.178	−3.122
Nov	2.04	5.287	2.0	5.184	0.1036	−3.019
Dec	2.35	6.294	2.0	5.357	0.9374	−2.081

(continued)

Month	Q_{in} (m^3/s)	$Q_{in}(\Delta t)$ (10^6 m^3)	Q_{out} (m^3/s)	$Q_{out}(\Delta t)$ (10^6 m^3)	ΔS (10^6 m^3)	$\sum(\Delta S)$ (10^6 m^3)
1977						
Jan	2.89	7.741	2.0	5.357	2.384	
Feb	9.57					
Mar	17.7					
Apr	16.4					
May	6.83					
Jun	3.74					
Jul	1.60	4.285	2.0	5.357	−1.071	−1.071
Aug	1.13	3.027	2.0	5.357	−2.330	−3.401
Sep	1.13	2.929	2.0	5.184	−2.255	−5.657
Oct	1.42	3.803	2.0	5.357	−1.553	−7.210
Nov	1.98	5.132	2.0	5.184	−0.052	−7.262
Dec	2.12	5.678	2.0	5.357	0.3214	−6.940
1978						
Jan	1.78	4.768	2.0	5.357	−0.5892	−7.530
Feb	1.95	4.717	2.0	4.838	−0.121	−7.651
Mar	7.25	19.418	2.0	5.357	14.061	
Apr	24.7					
May	6.26					
Jun	8.92					
Jul	3.57					
Aug	1.98	5.303	2.0	5.357	−0.0536	−0.0536
Sep	1.95	5.054	2.0	5.184	−0.1296	−0.1832
Oct	3.09	8.276	2.0	5.357	2.919	
Nov	3.94					
Dec	12.7					

The data in the first and second columns of the table were extracted from Table 2-4.

The third column is the product of the second column and the time interval for the month. For example, for August (31 d) and September (30 d), 1976:

$$(1.70\ \text{m}^3/\text{s})(31\ \text{d})(86{,}400\ \text{s/d}) = 4{,}553{,}280\ \text{m}^3$$

$$(1.56\ \text{m}^3/\text{s})(30\ \text{d})(86{,}400\ \text{s/d}) = 4{,}043{,}520\ \text{m}^3$$

The fourth column is the demand given in the problem statement.

The fifth column is the product of the demand and the time interval for the month. For example, for August and September 1976:

$$(2.0\ \text{m}^3/\text{s})(31\ \text{d})(86{,}400\ \text{s/d}) = 5{,}356{,}800\ \text{m}^3$$

$$(2.0\ \text{m}^3/\text{s})(30\ \text{d})(86{,}400\ \text{s/d}) = 5{,}184{,}000\ \text{m}^3$$

The sixth column (ΔS) is the difference between the third and fifth columns. For example, for August and September 1976:

$$4{,}553{,}280\ \text{m}^3 - 5{,}356{,}800\ \text{m}^3 = -803{,}520\ \text{m}^3$$

$$4{,}043{,}520\ \text{m}^3 - 5{,}184{,}000\ \text{m}^3 = -1{,}140{,}480\ \text{m}^3$$

The last column ($\sum(\Delta S)$) is the sum of the last value in that column and the value in the sixth column. For August 1976, it is $-803{,}520$ m^3 since this is the first value.

For September 1976, it is

$$(-803{,}520 \text{ m}^3) + (-1{,}140{,}480 \text{ m}^3) = -1{,}944{,}000 \text{ m}^3$$

The following logic is used in interpreting the table. From August through December 1976, the demand exceeds the flow, and storage must be provided. The maximum storage required for this interval is 3.122×10^6 m^3. In January 1977, the storage (ΔS) exceeds the deficit ($\sum(\Delta S)$) from December 1976. If we view the deficit as the volume of water in a virtual reservoir with a total capacity of 3.122×10^6 m^3, then in December 1976, the volume of water in the reservoir is 1.041×10^6 m^3 ($3.122 \times 10^6 - 2.081 \times 10^6$). The January 1977 inflow exceeds the demand and fills the reservoir deficit of 2.081×10^6 m^3.

Since the inflow (Q_{in}) exceeds the demand (2.0 m^3/s) for the months of February through June 1977, no storage is required during this period. Hence, no computations were performed.

From July 1977 through February 1978, the demand exceeds the inflow, and storage is required. The maximum storage required is 7.651×10^6 m^3. Note that the computations for storage did not stop in December 1977, even though the inflow exceeded the demand. This is because the storage was not sufficient to fill the reservoir deficit. The storage was sufficient to fill the reservoir deficit in March 1978.

You should note that these tabulations are particularly well suited to spreadsheet-type programs.

The storage volume determined by the Rippl method must be increased to account for water lost through evaporation and volume lost through the accumulation of sediment.

Application of the mass diagram. When existing storm sewers cannot handle runoff, consideration is given to providing a retention basin as an alternative to replacing the existing storm sewer. The retention basin serves two functions. First, it delays the coincidence of peak flows, and second, it can be designed to release storm water at a rate the storm sewer can handle. The procedure for applying the mass diagram to retention basins is basically the same as it is for water supply. The DRH for the design storm is substituted for the monthly discharge readings. The allowable discharge to the existing sewer is substituted for Q_{out}.

Wastewater treatment plants function best when the flow into the plant is constant. The general daily fluctuation in water use in most municipalities and industries results in an alternating pattern of high and low wastewater discharge. The peaks and valleys of flow can be evened out with an equalization basin. The volume of the equalization basin is chosen to shave off the peaks and fill in the valleys of the inflow mass diagram.

2-5 GROUNDWATER AND WELLS

Although the portion of the population of the United States supplied by surface water is twice that supplied by groundwater, the number of communities supplied by groundwater is four times that supplied by surface water (Table 2-8). The reason

TABLE 2-8
People served by groundwater and surface-water systems

Size (number of people served)	Groundwater systems		Surface-water systems	
	Number of systems	Number of people	Number of systems	Number of people
25–100	16,140	934,000	1,160	69,000
101–500	15,950	3,906,000	2,261	657,000
501–1,000	4,980	3,651,000	1,227	925,000
1,001–3,300	5,814	10,774,000	2,504	4,924,000
3,301–10,000	2,374	13,769,000	1,711	10,262,000
10,001–25,000	914	14,482,000	939	15,117,000
25,001–50,000	361	12,882,000	446	15,945,000
50,001–75,000	99	5,954,000	161	9,900,000
75,001–100,000	45	3,871,000	76	6,552,000
100,001–500,000	81	15,382,000	186	38,437,000
500,001–1,000,000	7	5,079,000	27	18,395,000
More than 1,000,000	1	1,705,000	13	27,344,000

Active community water systems as of August 31 1993 (*Source:* Federal Data Reporting System)

for this pattern is that larger cities are supplied by surface water while many small communities use groundwater.

Groundwater has several characteristics that make it desirable as a water supply source. First, the groundwater system provides natural storage, which eliminates the cost of impoundment works. Second, since the supply frequently is available at the point of demand, the cost of transmission is reduced significantly. Third, because groundwater is filtered by the natural geologic strata, groundwater is clearer to the eye than surface water.

Construction of Wells

Modern wells consist of more than a simple hole in the ground (Figure 2-22). A steel pipe called a *casing* is placed in the well hole to maintain the integrity of the hole. The casing is sealed to the surrounding soil with a cement grout, and a screen is placed at the bottom of the casing to allow water in and to keep soil material out. Two types of pump may be used. In the diagram shown, the pump motor is at the ground surface and the pump itself is placed down in the well above the well screen. The alternative is a submersible pump. In this case both the pump and the motor are lowered into the casing; water is pumped out of the well through a discharge pipe or drop pipe.

Sanitary considerations. The penetration of a water-bearing formation by a well provides a direct route for possible contamination of the groundwater. Although there are different types of wells and well construction, there are basic sanitary aspects

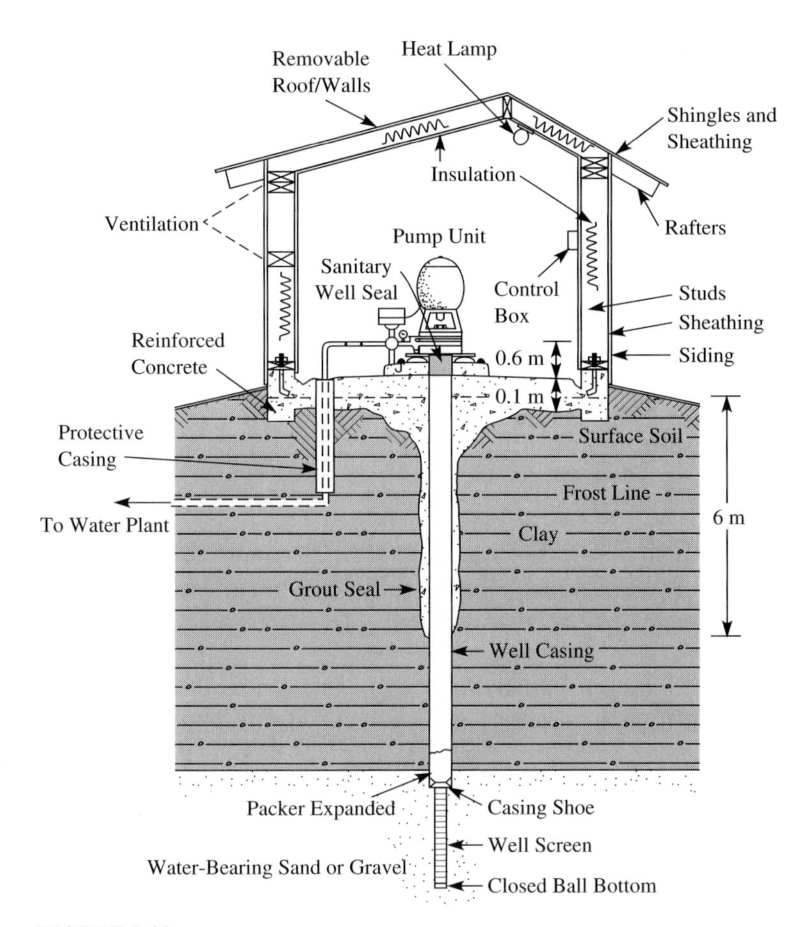

FIGURE 2-22
Pumphouse. [*Source:* U.S. Environmental Protection Agency, *Manual of Individual Water Supply Systems,* (Publication No. EPA-430-9-73-003), Washington, DC: U.S. Government Printing Office, 1973.]

that must be considered and followed (refer to Figure 2-23):[20]

1. The annular space outside the casing should be filled with a watertight cement grout or puddled clay from the surface to the depth necessary to prevent entry of contaminated water. A minimum of 6 m is recommended.
2. For artesian aquifers, the casing should be sealed into the overlying impermeable formations so as to retain the artesian pressure.
3. When a water-bearing formation containing water of poor quality is penetrated, the formation should be sealed off to prevent the infiltration of water into the well and aquifer.

---

[20]U.S. Environmental Protection Agency, *Manual of Individual Water Supply Systems* (Publication No. EPA-430-9-73-003), Washington, DC: U.S. Government Printing Office, pp. 45–50, 107–109, 1973.

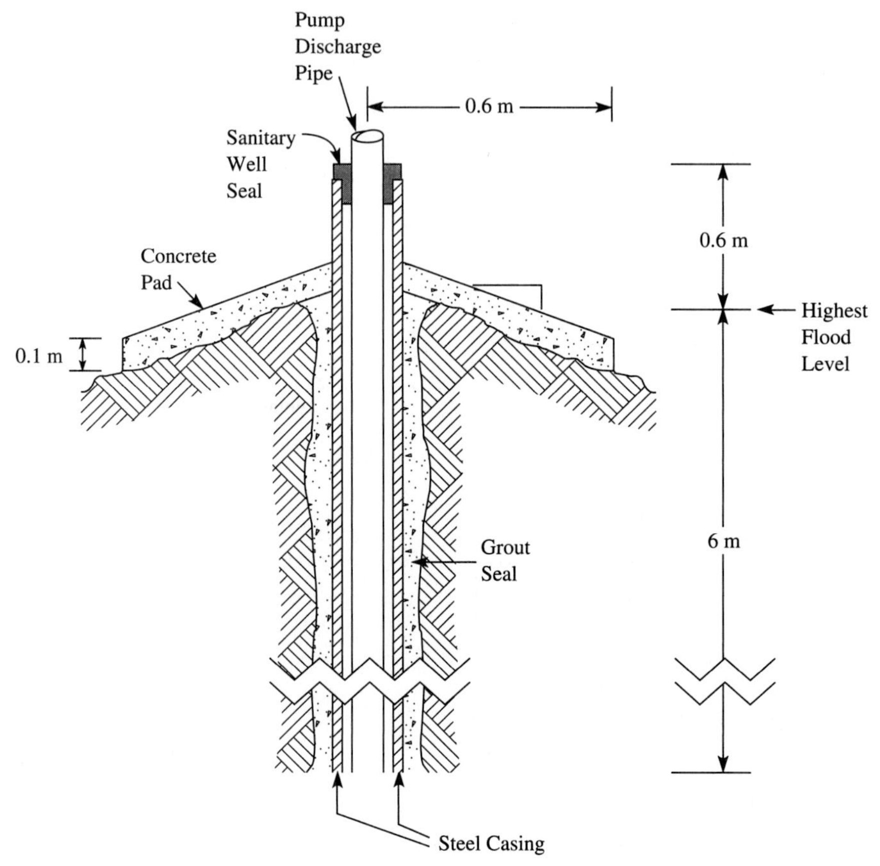

FIGURE 2-23
Sanitary considerations in well construction.

4. A sanitary well-seal with an approved vent should be installed at the top of the well casing to prevent the entrance of contaminated water or other objectionable material.

Well covers and seals. Every well should be provided with an overlapping, tight-fitting cover at the top of the casing or pipe sleeve to prevent contaminated water or other material from entering the well.

The seal in a well that is exposed to possible flooding should be elevated at least 0.6 m above the highest known flood level. When this is not possible, the seal should be watertight and equipped with a vent line whose opening to the atmosphere is at least 0.6 m above the highest known flood level.

Well covers and pump platforms should be elevated above the adjacent finished ground level. Pumproom floors should be constructed of reinforced, watertight concrete sloped away from the well so that surface and wastewater cannot stand near the well. The minimum thickness of such a slab or floor should be 0.1 m. Concrete slabs or floors should be poured separately from the grout formation seal and, where

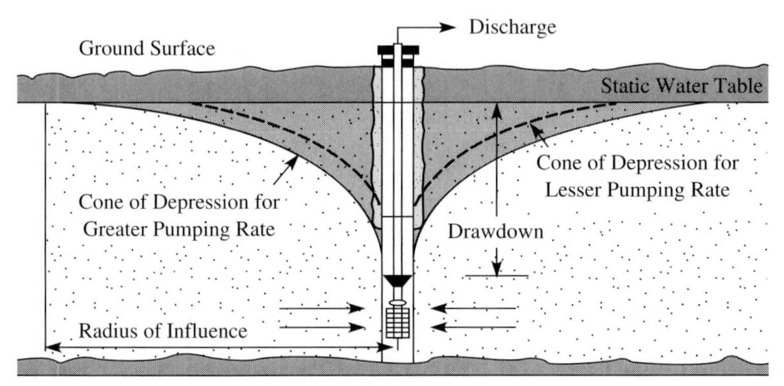

FIGURE 2-24
Effect of pumping rate on cone of depression. (*Source:* U.S. Environmental Protection Agency, *Manual of Individual Water Supply Systems.* See Note 20, supra.)

the threat of freezing exists, insulated from it and the well casing by a plastic or mastic coating or sleeve to prevent bonding of the concrete to either.

All water wells should be readily accessible at the top for inspection, servicing, and testing. This requires that any structure over the well be easily removable to provide full, unobstructed access for well-servicing equipment.

Disinfection of wells. All newly constructed wells should be disinfected to neutralize contamination from equipment, material, or surface drainage introduced during construction. Every well should be disinfected promptly after construction or repair.

Pumphousing. A pumphouse installed above the surface of the ground should be used. It should be unnecessary to use an underground discharge connection if an insulated, heated pumphouse is provided. For individual installations in rural areas, two 60-watt light bulbs, a thermostatically controlled electric heater, or a heating cable will generally provide adequate protection when the pumphouse is properly insulated. Since power failures may occur, an emergency gasoline-driven power supply or pump should be considered.

Cone of Depression[21]

When a well is pumped, the level of the piezometric surface in the vicinity of the well will be lowered (Figure 2-24). This lowering, or drawdown, causes the piezometric surface to take the shape of an inverted cone called a cone of depression. Since the water level in a pumped well is lower than that in the aquifer surrounding it, the water flows from the aquifer into the well. At increasing distances from the well, the drawdown decreases until the slope of the cone merges with the static water table. The distance from the well at which this occurs is called the radius of

[21]U.S. Environmental Protection Agency, *Manual of Individual Water Supply Systems*, pp. 28–30, 48–50, 107–109, 1973.

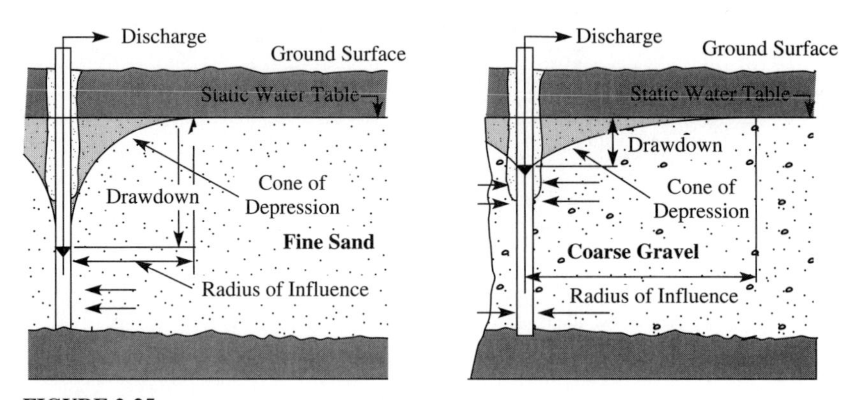

FIGURE 2-25
Effect of aquifer material on cone of depression. (*Source:* U.S. Environmental Protection Agency, *Manual of Individual Water Supply Systems.* See Note 20, supra.)

influence. The radius of influence is not constant but tends to expand with continued pumping. At a given pumping rate, the shape of the cone of depression depends on the characteristics of the water-bearing formation. Shallow and wide cones will form in aquifers composed of coarse sands or gravel. Deeper and narrower cones will form in fine sand or sandy clay (Figure 2-25). As the pumping rate increases, the drawdown increases. Consequently the slope of the cone steepens. When other conditions are equal for two wells, it may be expected that pumping costs will be higher for the well surrounded by the finer material because of greater drawdown.

When the cones of depression overlap, the local water table will be lowered (Figure 2-26). This requires additional pumping lifts to obtain water from the interior portion of the group of wells. A wider distribution of the wells over the groundwater basin will reduce the cost of pumping and will allow the development of a larger quantity of water. One rule of thumb is that two wells should be placed no closer together than two times the thickness of the water-bearing strata. For more than two wells they should be spaced at least 75 meters apart.

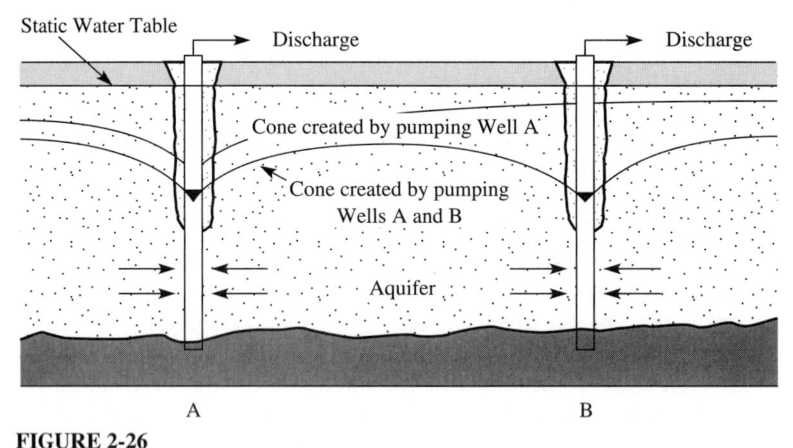

FIGURE 2-26
Effect of overlapping cones of depression. (*Source:* U.S. Environmental Protection Agency, *Manual of Individual Water Supply Systems.* See Note 20, supra.)

TABLE 2-9
Typical values of aquifer parameters

Aquifer material	Porosity (%)	Specific yield (%)	Hydraulic conductivity (m/s)
Unconsolidated			
Clay	55	3	1.2×10^{-6}
Loam	35	5	6.4×10^{-6}
Fine sand	45	10	3.5×10^{-5}
Medium sand	37	25	1.5×10^{-4}
Coarse sand	30	25	6.9×10^{-4}
Sand and gravel	20	16	6.1×10^{-4}
Gravel	25	22	6.4×10^{-3}
Consolidated			
Shale	<5	3	1.2×10^{-12}
Granite	<1	0	1.2×10^{-10}
Sandstone	15	5	5.8×10^{-7}
Limestone	15	2	5.8×10^{-6}
Fractured rock	5	2	5.8×10^{-5}

Definition of Terms

The aquifer parameters identified and defined in this section are those relevant to determining the available volume of water and the ease of its withdrawal.

Porosity. The ratio of the volume of voids (open spaces) in the soil to the total volume is called *porosity.* It is a measure of the amount of water that can be stored in the spaces between soil particles. It does not indicate how much of this water is available for development. Some typical values are shown in Table 2-9.

Specific yield. The percentage of water that is free to drain from the aquifer under the influence of gravity is defined as *specific yield* (Figure 2-27). Specific yield is not equal to porosity because the molecular and surface tension forces in the pore spaces keep some of the water in the voids. Specific yield reflects the amount of water available for development. Some average values are shown in Table 2-9.

Storage coefficient (S). This parameter is akin to specific yield. The *storage coefficient* is the volume of available water resulting from a unit decline in the piezometric surface over a unit horizontal cross-sectional area. It has units of m^3 of water/m^3 of aquifer or, in essence, no units at all! Storage coefficient and specific yield may be used interchangeably for unconfined aquifers. Values of S for unconfined aquifers range from 0.01 to 0.35. For confined aquifers the values of S vary from 1×10^{-3} to 1×10^{-5}.

Hydraulic gradient and head. The slope of the piezometric surface is called the *hydraulic gradient.* It is measured in the direction of the steepest slope of the piezometric surface. Groundwater flows in the direction of the hydraulic gradient and at a rate proportional to the slope.

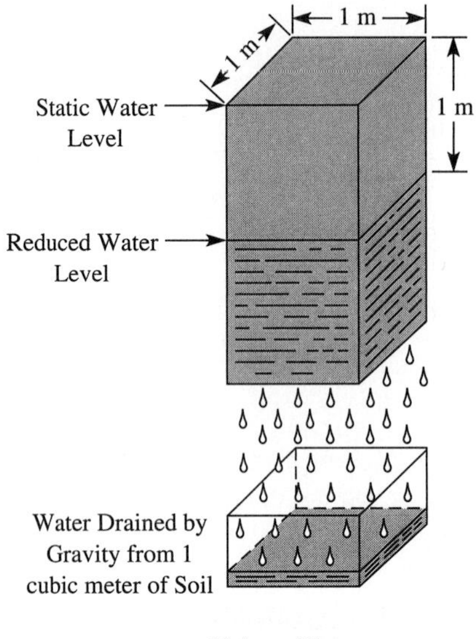

FIGURE 2-27
Specific yield. (*Source:* Johnson Division, UOP, *Ground Water and Wells,* St. Paul, MN: Johnson Division, UOP, 1975. Reprinted by permission.)

The vertical distance from a reference plane (zero datum) to the bottom of a well or piezometer (Figure 2-28) is called the *elevation head.* The height to which water will rise in the piezometer is called the *pressure head.* The sum of the elevation head and the pressure head is called the *total head.*

Using Figure 2-28 and the assumption that the groundwater is flowing in the plane of the page, we can define the gradient to be:

$$\text{Hydraulic gradient} = \frac{\text{Change in head}}{\text{Horizontal distance}} \tag{2-17}$$

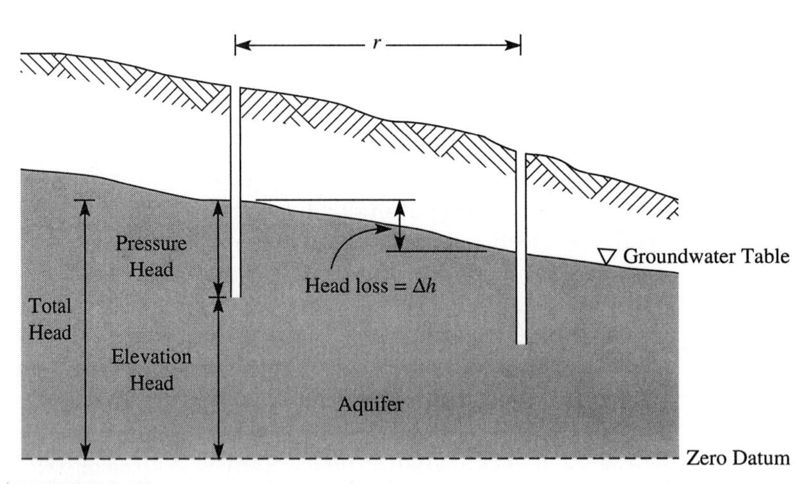

FIGURE 2-28
Geometry for definition of head and hydraulic gradient.

or in the differential sense:

$$\text{Hydraulic gradient} = \frac{\Delta h}{r} = \frac{dh}{dr} \tag{2-18}$$

We generally require three wells, one of which is out of plane with the other two, to define the hydraulic gradient. Heath[22] suggests the following graphical procedure for finding the hydraulic gradient between three closely spaced wells:

1. Draw a line between the two wells with the highest and lowest head and divide it into equal intervals. By interpolation, find the place on the line between these two wells where the head is equal to the head of the third well.
2. Draw a line from the third well to the point on the line between the first two wells where the head is the same as that in the third well. This line is called an *equipotential line.* This means that the head anywhere along the line should be constant. Groundwater will flow in a direction perpendicular to this line.
3. Draw a line perpendicular to the equipotential line through the well with the lowest or highest head. The groundwater flow is in a direction parallel to this line. It is called a *flow line.*
4. Calculate the gradient as the difference in head between the head on the equipotential line and the head at the lowest or highest well divided by the distance from the equipotential line to that well.

Example 2-11. For the wells shown in plan view below, determine the direction of flow and the hydraulic gradient. The total head is given for each well as follows:

Well A = 10.4 m
Well B = 10.0 m
Well C = 9.9 m

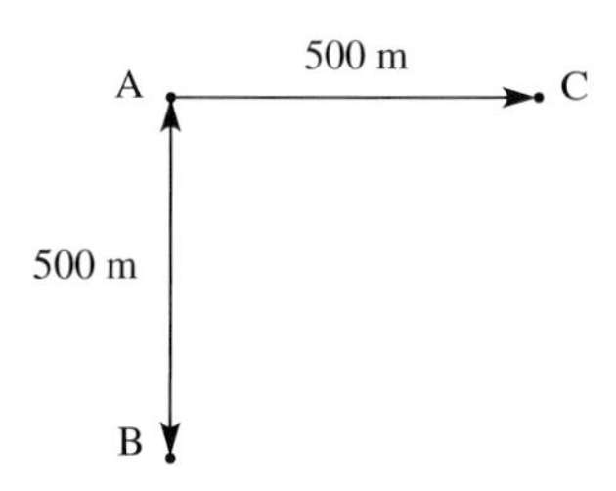

[22]R. C. Heath, *Basic Ground-Water Hydrology,* U.S. Geological Survey Water-Supply Paper 2220, U.S. Government Printing Office, 1983.

Solution. The stepwise graphical solution procedure is shown below.

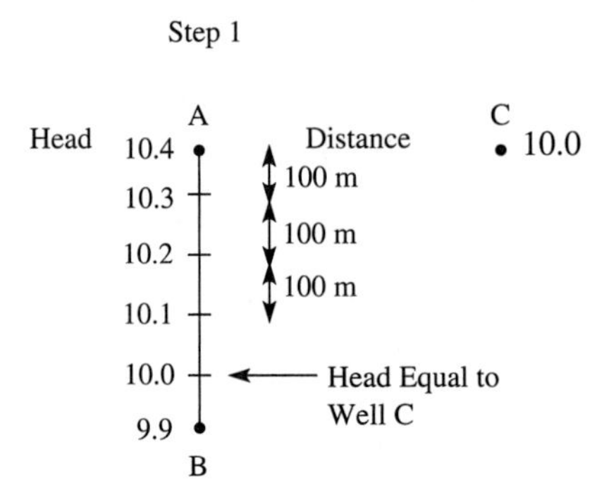

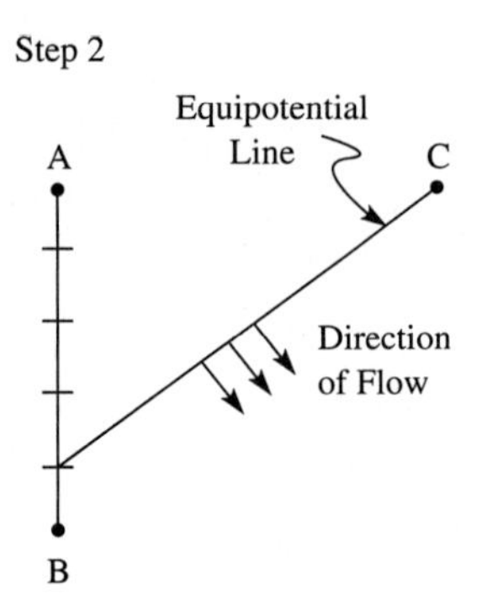

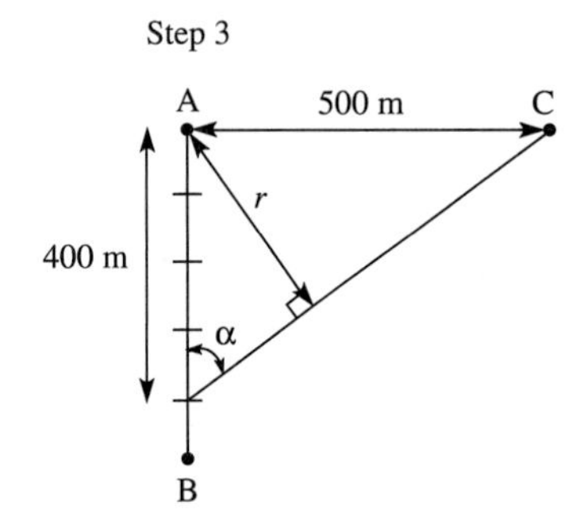

The distance r must be determined in order to calculate the hydraulic gradient. From the plan view, we may note that the wells form a right triangle with legs of 400 m and 500 m. The angle α may be computed as:

$$\tan^{-1}(\alpha) = \frac{500}{400}$$

and $\alpha = 51.34°$
The distance r is

$$r = (400)\sin\alpha = 400\sin 51.34 = 312.35 \text{ m}$$

The hydraulic gradient is then:

$$\text{Hydraulic gradient} = \frac{10.4 \text{ m} - 10.0 \text{ m}}{312.35 \text{ m}} = 0.00128$$

Note that the hydraulic gradient has no units.

Hydraulic conductivity (K). The property of an aquifer that is a measure of its ability to transmit water under a sloping piezometric surface is called *hydraulic conductivity.* It is defined as the discharge that occurs through a unit cross section of aquifer (Figure 2-29) under a hydraulic gradient of 1.00. It has units of speed (m/s). Typical values are given in Table 2-9.

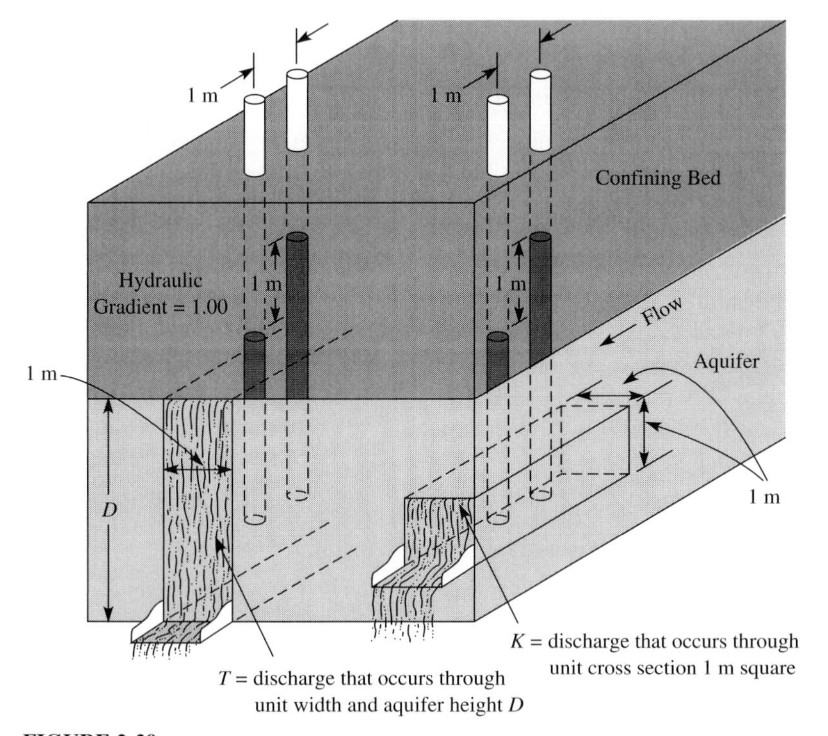

FIGURE 2-29
Illustration of definition of hydraulic conductivity (K) and transmissibility (T). (*Source:* Johnson Division, UOP, *Ground Water and Wells.* Reprinted by permission. See source note for Figure 2-26, supra.)

Transmissibility (T). The coefficient of transmissibility (T) is a measure of the rate at which water will flow through a unit width vertical strip of aquifer extending through its full saturated thickness (Figure 2-29) under a unit hydraulic gradient. It has units of m^2/s. Values of the transmissibility coefficient range from 1.0×10^{-4} to 1.5×10^{-1} m^2/s.

Well Hydraulics

Equations for calculating the discharge that results when the piezometric surface is lowered are based on the work of the French hydrologist Henri Darcy. In 1856 he discovered that the velocity of water flow in a porous aquifer was proportional to the hydraulic gradient. He proposed the following equation, which is now called Darcy's law:[23]

$$v = K\frac{dh}{dr} \tag{2-19}$$

[23]Henri Darcy, *Les Fontaines Publiques de la Ville de Dijon*, Paris: Victor Dalmont, pp. 570, 590–594, 1856.

where v = velocity, m/s
K = permeability, m/s
dh/dr = slope of the hydraulic gradient, m/m

The *Darcy velocity* is not a real velocity because it assumes that the full cross-sectional area of the aquifer is available for water to flow through. However, much of the cross-sectional area is soil material; thus, the actual area through which the water flows is much less. As a result, the actual linear velocity of the groundwater is considerably faster than the Darcy velocity.

We may determine the actual average linear velocity by taking into account the fraction of the cross-sectional area filled with soil. Let us define the flow (Q) to be the product of the total cross-sectional area and the Darcy velocity, that is $Q = Av$. The actual average linear velocity (v') through the voids times the area of the voids (A') is also equal to the flow (Q). Thus, the actual velocity is related to the Darcy velocity:

$$A'v' = Av = Q \tag{2-20}$$

If we solve this expression for the average linear velocity, v', we find:

$$v' = \frac{Av}{A'} \tag{2-21}$$

If we multiply the top and bottom by a unit length (L), then we have

$$v' = \frac{ALv}{A'L} \tag{2-22}$$

The product AL is the total volume of the soil. The product $A'L$ is the void volume. The ratio of the void volume to the total volume is the definition of porosity. The actual linear velocity may then be defined in terms of the porosity as

$$v' = \frac{\text{Darcy velocity}}{\text{Porosity}} = \frac{v}{\eta} \tag{2-23}$$

or in terms of the hydraulic gradient

$$v' = \frac{K(dh/dr)}{\eta} \tag{2-24}$$

The gross discharge is the product of the velocity of flow and the area (A) through which it flows.

$$Q = vA = KA\frac{dh}{dr} \tag{2-25}$$

This equation has been solved for steady state and nonsteady or *transient flow*. Steady state is a condition under which no changes occur with time. It will seldom, if ever, occur in practice, but may be approached after very long periods of pumping. Transient-flow equations include a factor of time. The derivation of these equations is based on the following assumptions:

1. The well is pumped at a constant rate.
2. Flow towards the well is radial and uniform.

3. Initially the piezometric surface is horizontal.
4. The well fully penetrates the aquifer and is open for the entire height of the aquifer.
5. The aquifer is homogeneous in all directions and is of infinite horizontal extent.
6. Water is released from the aquifer in immediate response to a drop in the piezometric surface.

Steady flow in a confined aquifer. The equation describing steady, confined aquifer flow was first presented by Dupuit in 1863 and subsequently extended by Theim in 1906.[24] It may be written as follows (refer to Figure 2-30 for an

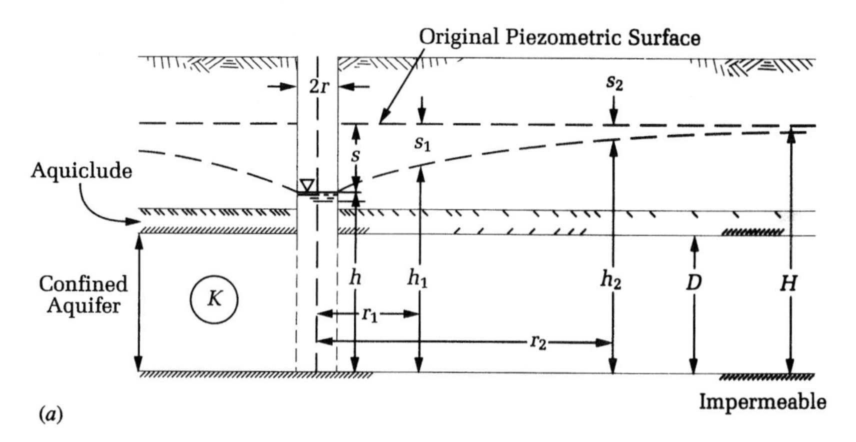

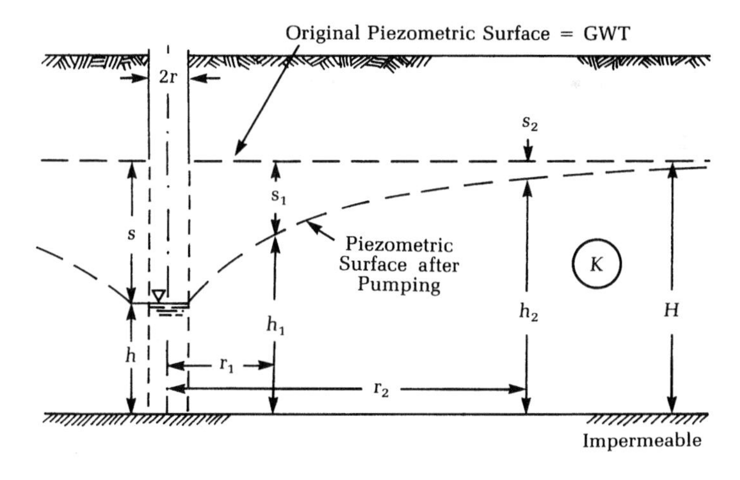

FIGURE 2-30
Geometry and symbols for a pumped well in (*a*) confined aquifer and (*b*) unconfined aquifer. (*Source:* H. Bouwer, *Groundwater Hydrology,* New York: McGraw-Hill, 1978. Reprinted by permission.)

[24] J. Dupuit, *Etudes Théoriques et Pratiques sur le Mouvement des Eaux dans les Canaux Découverts et à Travers les Terrains Perméables*, Paris: Dunod, 1863; Gunther Theim, *Hydrologische Methoden*, Leipzig, Germany: J. M. Gebhart, 1906.

explanation of the notation):

$$Q = \frac{2\pi T(h_2 - h_1)}{\ln(r_2/r_1)} \tag{2-26}$$

where
$T = KD =$ transmissibility, m^2/s
$D =$ thickness of artesian aquifer, m
$h_1, h_2 =$ height of piezometric surface above confining layer, m
$r_1, r_2 =$ radius from pumping well, m
$\ln =$ logarithm to base e

Example 2-12. An artesian aquifer 10.0 m thick with a piezometric surface 40.0 m above the bottom confining layer is being pumped by a fully penetrating well. The aquifer is a medium sand with a hydraulic conductivity of 1.50×10^{-4} m/s. Steady state drawdowns of 5.00 m and 1.00 m are observed at two nonpumping wells located 20.0 m and 200.0 m, respectively, from the pumped well. Determine the discharge at the pumped well.

Solution. First we determine h_1 and h_2:

$$h_1 = 40.0 - 5.00 = 35.0 \text{ m}$$

$$h_2 = 40.0 - 1.00 = 39.0 \text{ m}$$

so

$$Q = \frac{(2\pi)(1.50 \times 10^{-4})(10.0)(39.0 - 35.0)}{\ln(200/20)}$$

$$Q = 0.0164 \text{ or } 0.016 \text{ m}^3/\text{s}$$

Steady flow in an unconfined aquifer. For unconfined aquifers the factor D in Equation 2-26 is replaced by the height of the water table above the lower boundary of the aquifer. The equation then becomes

$$Q = \frac{\pi K(h_2^2 - h_1^2)}{\ln(r_2/r_1)} \tag{2-27}$$

Example 2-13. A 0.50 m diameter well fully penetrates an unconfined aquifer which is 30.0 m thick. The drawdown at the pumped well is 10.0 m and the hydraulic conductivity of the gravel aquifer is 6.4×10^{-3} m/s. If the flow is steady and the discharge is 0.014 m^3/s, determine the drawdown at a site 100.0 m from the well.

Solution. First we calculate h_1

$$h_1 = 30.0 - 10.0 = 20.0 \text{ m}$$

Then we apply Equation 2-27 and solve for h_2. Note that $r_1 = 0.50 \text{ m}/2 = 0.25$ m.

$$0.014 = \frac{\pi(6.4 \times 10^{-3})(h_2^2 - (20.0)^2)}{\ln(100/0.25)}$$

$$h_2^2 - 400.0 = \frac{(0.014)(5.99)}{\pi(6.4 \times 10^{-3})}$$

$$h_2 = (4.17 + 400.0)^{1/2}$$

$$h_2 = 20.10 \text{ m}$$

The drawdown is then

$$s_2 = H - h_2 = 30.0 - 20.10 = 9.90 \text{ m}$$

Unsteady flow in a confined aquifer. A solution for the transient-flow problem was developed by Theis in 1935.[25] Using heat-flow theory as an analogy, he found the following for an infinitesimally small diameter well with radial flow:

$$s = \frac{Q}{4\pi T} \int_u^\infty \left(\frac{e^{-u}}{u}\right) du \tag{2-28}$$

where s = drawdown ($H - h$), m

$u = \dfrac{r^2 S}{4Tt}$

r = distance between pumping well and observation well, or radius of pumping well, m

S = storage coefficient

T = transmissibility, m^2/s

t = time since pumping began, s

Some explanations of the terms may be of use here. The lower case s refers to the drawdown at some time, t, after the start of pumping. The time does not appear explicitly in Equation 2-28 but is used to compute the value of u to be used in the integration. The transmissibility and storage coefficient also are used to calculate u. The transmissibility may be determined from the permeability and the thickness of the aquifer as it was for steady-state flow. Field pumping tests may also be used to define T. You should note that the r term used to calculate the value of u may take on values ranging upward from the radius of the well. Thus, you could, if you wished, calculate every point on the cone of depression (i.e., value of s) by iterating the calculation with values of r from the well radius to infinity. If you with to calculate the drawdown at a specific distance from the pumping well, then, of course, you must use that distance for r. The integral in Equation (2-28) is called the well function of u and is evaluated by the following series expansion:

$$W(u) = -0.577216 - \ln u + u - \frac{u^2}{2 \cdot 2!} + \frac{u^3}{3 \cdot 3!} - \cdots \tag{2-29}$$

A table of values of $W(u)$ was prepared by Ferris, *et al.*[26] It is reproduced in Table 2-10.

---

[25] C. V. Theis, "The Relation Between the Lowering of the Piezometric Surface and the Rate and Duration of Discharge of a Well Using Ground Water Storage," *Transactions of the American Geophysical Union*, vol. 16, pp. 519–524, 1935.

[26] J. G. Ferris, *et al., Theory of Aquifer Tests* (U.S. Geological Survey Water-Supply Paper No. 1536-E), Washington, DC: U.S. Government Printing Office, 1962, pp. 69–174.

TABLE 2-10
Values of $W(u)$

N \ u	$N \times 10^{-15}$	$N \times 10^{-14}$	$N \times 10^{-13}$	$N \times 10^{-12}$	$N \times 10^{-11}$	$N \times 10^{-10}$	$N \times 10^{-9}$	$N \times 10^{-8}$	$N \times 10^{-7}$	$N \times 10^{-6}$	$N \times 10^{-5}$	$N \times 10^{-4}$	$N \times 10^{-3}$	$N \times 10^{-2}$	$N \times 10^{-1}$	N
1.0	33.9616	31.6590	29.3564	27.0538	24.7512	22.4486	20.1460	17.8435	15.5409	13.2383	10.9357	8.6332	6.3315	4.0379	1.8229	0.2194
1.1	33.8662	31.5637	29.2611	26.9585	24.6559	22.3533	20.0507	17.7482	15.4456	13.1430	10.8404	8.5379	6.2363	3.9436	1.7371	0.1860
1.2	33.7792	31.4767	29.1741	26.8715	24.5689	22.2663	19.9637	17.6611	15.3586	13.0560	10.7534	8.4509	6.1494	3.8576	1.6595	0.1584
1.3	33.6992	31.3966	29.0940	26.7914	24.4889	22.1863	19.8837	17.5811	15.2785	12.9759	10.6734	8.3709	6.0695	3.7785	1.5889	0.1355
1.4	33.6251	31.3225	29.0199	26.7173	24.4147	22.1122	19.8096	17.5070	15.2044	12.9018	10.5993	8.2968	5.9955	3.7054	1.5241	0.1162
1.5	33.5561	31.2535	28.9509	26.6483	24.3458	22.0432	19.7406	17.4380	15.1354	12.8328	10.5303	8.2278	5.9266	3.6374	1.4645	0.1000
1.6	33.4916	31.1890	28.8864	26.5838	24.2812	21.9786	19.6760	17.3735	15.0709	12.7683	10.4657	8.1634	5.8621	3.5739	1.4092	0.08631
1.7	33.4309	31.1283	28.8258	26.5232	24.2206	21.9180	19.6154	17.3128	15.0103	12.7077	10.4051	8.1027	5.8016	3.5143	1.3578	0.07465
1.8	33.3738	31.0712	28.7686	26.4660	24.1634	21.8608	19.5583	17.2557	14.9531	12.6505	10.3479	8.0455	5.7446	3.4581	1.3089	0.06471
1.9	33.3197	31.0171	28.7145	26.4119	24.1094	21.8068	19.5042	17.2016	14.8990	12.5964	10.2939	7.9915	5.6906	3.4050	1.2649	0.05620
2.0	33.2684	30.9658	28.6632	26.3607	24.0581	21.7555	19.4529	17.1503	14.8477	12.5451	10.2426	7.9402	5.6394	3.3547	1.2227	0.04890
2.1	33.2196	30.9170	28.6145	26.3119	24.0093	21.7067	19.4041	17.1015	14.7969	12.4964	10.1938	7.8914	5.5907	3.3069	1.1829	0.04261
2.2	33.1731	30.8705	28.5679	26.2653	23.9628	21.6602	19.3576	17.0550	14.7524	12.4498	10.1473	7.8449	5.5443	3.2614	1.1454	0.03719
2.3	33.1286	30.8261	28.5235	26.2209	23.9183	21.6157	19.3131	17.0106	14.7080	12.4054	10.1028	7.8004	5.4999	3.2179	1.1099	0.03250
2.4	33.0861	30.7835	28.4809	26.1783	23.8758	21.5732	19.2706	16.9680	14.6654	12.3628	10.0603	7.7579	5.4575	3.1763	1.0762	0.02844
2.5	33.0453	30.7427	28.4401	26.1375	23.8349	21.5323	19.2298	16.9272	14.6246	12.3220	10.0194	7.7172	5.4167	3.1365	1.0443	0.02491
2.6	33.0060	30.7035	28.4009	26.0983	23.7957	21.4931	19.1905	16.8880	14.5854	12.2828	9.9802	7.6779	5.3776	3.0983	1.0139	0.02185
2.7	32.9683	30.6657	28.3631	26.0606	23.7580	21.4554	19.1528	16.8502	14.5476	12.2450	9.9425	7.6401	5.3400	3.0615	0.9849	0.01918
2.8	32.9319	30.6294	28.3268	26.0242	23.7216	21.4190	19.1164	16.8138	14.5113	12.2087	9.9061	7.6038	5.3037	3.0261	0.9573	0.01686
2.9	32.8968	30.5943	28.2917	25.9891	23.6865	21.3839	19.0813	16.7788	14.4762	12.1736	9.8710	7.5687	5.2687	2.9920	0.9309	0.01482
3.0	32.8629	30.5604	28.2578	25.9552	23.6526	21.3500	19.0474	16.7449	14.4423	12.1397	9.8371	7.5348	5.2349	2.9591	0.9057	0.01305
3.1	32.8302	30.5276	28.2250	25.9224	23.6198	21.3172	19.0146	16.7121	14.4095	12.1069	9.8043	7.5020	5.2022	2.9273	0.8815	0.01149
3.2	32.7984	30.4958	28.1932	25.8907	23.5880	21.2855	18.9829	16.6803	14.3777	12.0751	9.7726	7.4703	5.1706	2.8965	0.8583	0.01013
3.3	32.7676	30.4651	28.1625	25.8599	23.5573	21.2547	18.9521	16.6495	14.3470	12.0444	9.7418	7.4395	5.1399	2.8668	0.8361	0.008939
3.4	32.7378	30.4352	28.1326	25.8300	23.5274	21.2249	18.9223	16.6197	14.3171	12.0145	9.7120	7.4097	5.1102	2.8379	0.8147	0.007891
3.5	32.7088	30.4062	28.1036	25.8010	23.4985	21.1959	18.8933	16.5907	14.2881	11.9855	9.6830	7.3807	5.0813	2.8099	0.7942	0.006970
3.6	32.6806	30.3780	28.0755	25.7729	23.4703	21.1677	18.8651	16.5625	14.2599	11.9574	9.6548	7.3526	5.0532	2.7827	0.7745	0.006160
3.7	32.6532	30.3506	28.0481	25.7455	23.4429	21.1403	18.8377	16.5351	14.2325	11.9300	9.6274	7.3252	5.0259	2.7563	0.7554	0.005448
3.8	32.6266	30.3240	28.0214	25.7188	23.4162	21.1136	18.8110	16.5085	14.2059	11.9033	9.6007	7.2985	4.9993	2.7306	0.7371	0.004820
3.9	32.6006	30.2980	27.9954	25.6928	23.3902	21.0877	18.7851	16.4825	14.1799	11.8773	9.5748	7.2725	4.9735	2.7056	0.7194	0.004267
4.0	32.5753	30.2727	27.9701	25.6675	23.3649	21.0623	18.7598	16.4572	14.1546	11.8520	9.5495	7.2472	4.9482	2.6813	0.7024	0.003779
4.1	32.5506	30.2480	27.9454	25.6428	23.3402	21.0376	18.7351	16.4325	14.1299	11.8273	9.5248	7.2225	4.9236	2.6576	0.6859	0.003349
4.2	32.5265	30.2239	27.9213	25.6187	23.3161	21.0136	18.7110	16.4084	14.1058	11.8032	9.5007	7.1985	4.8997	2.6344	0.6700	0.002969
4.3	32.5029	30.2004	27.8978	25.5952	23.2926	20.9900	18.6874	16.3884	14.0823	11.7797	9.4771	7.1749	4.8762	2.6119	0.6546	0.002633
4.4	32.4800	30.1774	27.8748	25.5722	23.2696	20.9670	18.6644	16.3619	14.0593	11.7567	9.4541	7.1520	4.8533	2.5899	0.6397	0.002336
4.5	32.4575	30.1519	27.8523	25.5497	23.2471	20.9446	18.6420	16.3394	14.0368	11.7342	9.4317	7.1295	4.8310	2.5684	0.6253	0.002073
4.6	32.4355	30.1329	27.8303	25.5277	23.2252	20.9226	18.6200	16.3174	14.0148	11.7122	9.4097	7.1075	4.8091	2.5474	0.6114	0.001841
4.7	32.4140	30.1114	27.8088	25.5062	23.2037	20.9011	18.5985	16.2959	13.9933	11.6907	9.3882	7.0860	4.7877	2.5268	0.5979	0.001635
4.8	32.3929	30.0904	27.7878	25.4852	23.1826	20.8800	18.5774	16.2748	13.9723	11.6697	9.3671	7.0650	4.7667	2.5068	0.5848	0.001453
4.9	32.3723	30.0697	27.7672	25.4646	23.1620	20.8594	18.5568	16.2542	13.9516	11.6491	9.3465	7.0444	4.7462	2.4871	0.5721	0.001291
5.0	32.3521	30.0495	27.7470	25.4444	23.1418	20.8392	18.5366	16.2340	13.9314	11.6289	9.3263	7.0242	4.7261	2.4679	0.5598	0.001148
5.1	32.3323	30.0297	27.7271	25.4246	23.1220	20.8194	18.5168	16.2142	13.9116	11.6091	9.3065	7.0044	4.7064	2.4491	0.5478	0.001021
5.2	32.3129	30.0103	27.7077	25.4051	23.1026	20.8000	18.4974	16.1948	13.8922	11.5896	9.2871	6.9850	4.6871	2.4306	0.5362	0.0009086
5.3	32.2939	29.9913	27.6887	25.3861	23.0835	20.7809	18.4783	16.1758	13.8732	11.5706	9.2681	6.9659	4.6681	2.4126	0.5250	0.0008086
5.4	32.2752	29.9726	27.6700	25.3674	23.0648	20.7622	18.4596	16.1571	13.8545	11.5519	9.2494	6.9473	4.6495	2.3948	0.5140	0.0007198

TABLE 2-10

Values of $W(u)$ (*continued*)

N \ u	N × 10^{-15}	N × 10^{-14}	N × 10^{-13}	N × 10^{-12}	N × 10^{-11}	N × 10^{-10}	N × 10^{-9}	N × 10^{-8}	N × 10^{-7}	N × 10^{-6}	N × 10^{-5}	N × 10^{-4}	N × 10^{-3}	N × 10^{-2}	N × 10^{-1}	N
5.5	32.2568	29.9542	27.6516	25.3491	23.0465	20.7439	18.4413	16.1387	13.8361	11.5336	9.2310	6.9289	4.6313	2.3775	0.5034	0.0006409
5.6	32.2388	29.9362	27.6336	25.3310	23.0285	20.7259	18.4233	16.1207	13.8181	11.5155	9.2130	6.9109	4.6134	2.3604	0.4930	0.0005708
5.7	32.2211	29.9185	27.6159	25.3133	23.0108	20.7082	18.4056	16.1030	13.8004	11.4978	9.1953	6.8932	4.5958	2.3437	0.4830	0.0005085
5.8	32.2037	29.9011	27.5985	25.2959	22.9934	20.6908	18.3882	16.0856	13.7830	11.4804	9.1779	6.8758	4.5785	2.3273	0.4732	0.0004532
5.9	32.1866	29.8840	27.5814	25.2789	22.9763	20.6737	18.3711	16.0685	13.7659	11.4633	9.1608	6.8588	4.5615	2.3111	0.4637	0.0004039
6.0	32.1608	29.8672	27.5646	25.2620	22.9595	20.6569	18.3543	16.0517	13.7491	11.4465	9.1440	6.8420	4.5448	2.2953	0.4544	0.0003601
6.1	32.1533	29.8507	27.5481	25.2455	22.9429	20.6403	18.3378	16.0352	13.7326	11.4300	9.1275	6.8254	4.5283	2.2797	0.4454	0.0003211
6.2	32.1370	29.8344	27.5318	25.2293	22.9267	20.6241	18.3215	16.0189	13.7163	11.4138	9.1112	6.8092	4.5122	2.2645	0.4366	0.0002864
6.3	32.1210	29.8184	27.5158	25.2133	22.9107	20.6081	18.3055	16.0029	13.7003	11.3978	9.0952	6.7932	4.4963	2.2494	0.4280	0.0002555
6.4	32.1053	29.8027	27.5001	25.1975	22.8949	20.5923	18.2898	15.9872	13.6846	11.3820	9.0795	6.7775	4.4806	2.2346	0.4197	0.0002279
6.5	32.0898	29.7872	27.4846	25.1820	22.8794	20.5768	18.2742	15.9717	13.6691	11.3665	9.0640	6.7620	4.4652	2.2201	0.4115	0.0002034
6.6	32.0745	29.7719	27.4693	25.1667	22.8641	20.5616	18.2590	15.9564	13.6538	11.3512	9.0487	6.7467	4.4501	2.2058	0.4036	0.0001816
6.7	32.0595	29.7569	27.4543	25.1517	22.8491	20.5465	18.2439	15.9414	13.6388	11.3362	9.0337	6.7317	4.4351	2.1917	0.3959	0.0001621
6.8	32.0446	29.7421	27.4395	25.1369	22.8343	20.5317	18.2291	15.9265	13.6240	11.3214	9.0189	6.7169	4.4204	2.1779	0.3883	0.0001448
6.9	32.0300	29.7275	27.4249	25.1223	22.8197	20.5171	18.2145	15.9119	13.6094	11.3608	9.0043	6.7023	4.4059	2.1643	0.3810	0.0001293
7.0	32.0156	29.7131	27.4105	25.1079	22.8053	20.5027	18.2001	15.8976	13.5950	11.2924	8.9899	6.6879	4.3916	2.1508	0.3738	0.0001155
7.1	32.0015	29.6989	27.3963	25.0937	22.7911	20.4885	18.1860	15.8834	13.5808	11.2782	8.9757	6.6737	4.3775	2.1376	0.3668	0.0001032
7.2	31.9875	29.6849	27.3823	25.0797	22.7771	20.4746	18.1720	15.8694	13.5668	11.2642	8.9617	6.6598	4.3636	2.1246	0.3599	0.00009219
7.3	31.9737	29.6711	27.3685	25.0659	22.7633	20.4608	18.1582	15.8556	13.5530	11.2504	8.9479	6.6460	4.3500	2.1118	0.3532	0.00008239
7.4	31.9601	29.6575	27.3549	25.0523	22.7497	20.4472	18.1446	15.8420	13.5394	11.2368	8.9343	6.6324	4.3364	2.0991	0.3467	0.00007364
7.5	31.9467	29.6441	27.3415	25.0389	22.7363	20.4337	18.1311	15.8286	13.5260	11.2234	8.9209	6.6190	4.3231	2.0867	0.3403	0.00006583
7.6	31.9334	29.6308	27.3282	25.0257	22.7231	20.4205	18.1179	15.8153	13.5127	11.2102	8.9076	6.6057	4.3100	2.0744	0.3341	0.00005886
7.7	31.9203	29.6178	27.3152	25.0126	22.7100	20.4074	18.1048	15.8022	13.4997	11.1971	8.8946	6.5927	4.2970	2.0623	0.3280	0.00005263
7.8	31.9074	29.6048	27.3023	24.9997	22.6971	20.3945	18.0919	15.7893	13.4868	11.1842	8.8817	6.5798	4.2842	2.0503	0.3221	0.00004707
7.9	31.8947	29.5921	27.2895	24.9869	22.6844	20.3818	18.0792	15.7766	13.4740	11.1714	8.8689	6.5671	4.2716	2.0386	0.3163	0.00004210
8.0	31.8821	29.5795	27.2769	24.9744	22.6718	20.3692	18.0666	15.7640	13.4614	11.1589	8.8563	6.5545	4.2591	2.0269	0.3106	0.00003767
8.1	31.8697	29.5671	27.2645	24.9619	22.6594	20.3568	18.0542	15.7516	13.4490	11.1464	8.8439	6.5421	4.2468	2.0155	0.3050	0.00003370
8.2	31.8574	29.5548	27.2523	24.9497	22.6471	20.3445	18.0419	15.7393	13.4367	11.1342	8.8317	6.5298	4.2346	2.0042	0.2996	0.00003015
8.3	31.8453	29.5427	27.2401	24.9375	22.6350	20.3324	18.0298	15.7272	13.4246	11.1220	8.8195	6.5177	4.2226	1.9930	0.2943	0.00002699
8.4	31.8333	29.5307	27.2282	24.9256	22.6230	20.3204	18.0178	15.7152	13.4126	11.1101	8.8076	6.5057	4.2107	1.9820	0.2891	0.00002415
8.5	31.8215	29.5189	27.2163	24.9137	22.6112	20.3086	18.0060	15.7034	13.4008	11.0982	8.7957	6.4939	4.1990	1.9711	0.2840	0.00002162
8.6	31.8098	29.5072	27.2046	24.9020	22.5995	20.2969	17.9943	15.6917	13.3891	11.0865	8.7840	6.4822	4.1874	1.9604	0.2790	0.00001936
8.7	31.7982	29.4957	27.1931	24.8905	22.5879	20.2853	17.9827	15.6801	13.3776	11.0750	8.7725	6.4707	4.1759	1.9498	0.2742	0.00001733
8.8	31.7868	29.4842	27.1816	24.8790	22.5765	20.2739	17.9713	15.6687	13.3661	11.0635	8.7610	6.4592	4.1646	1.9393	0.2694	0.00001552
8.9	31.7755	29.4729	27.1703	24.8678	22.5652	20.2626	17.9600	15.6574	13.3548	11.0523	8.7497	6.4480	4.1534	1.9290	0.2647	0.00001390
9.0	31.7643	29.4618	27.1592	24.8566	22.5540	20.2514	17.9488	15.6462	13.3437	11.0411	8.7386	6.4368	4.1423	1.9187	0.2602	0.00001245
9.1	31.7533	29.4507	27.1481	24.8455	22.5429	20.2404	17.9378	15.6352	13.3326	11.0300	8.7275	6.4258	4.1313	1.9087	0.2557	0.00001115
9.2	31.7424	29.4398	27.1372	24.8346	22.5320	20.2294	17.9268	15.6243	13.3217	11.0191	8.7166	6.4148	4.1205	1.8987	0.2513	0.000009988
9.3	31.7315	29.4290	27.1264	24.8238	22.5212	20.2186	17.9160	15.6135	13.3109	11.0083	8.7058	6.4040	4.1098	1.8888	0.2470	0.000008948
9.4	31.7208	29.4183	27.1157	24.8131	22.5105	20.2079	17.9053	15.6028	13.3002	10.9976	8.6951	6.3934	4.0992	1.8791	0.2429	0.000008018
9.5	31.7103	29.4077	27.1051	24.8025	22.4999	20.1973	17.8948	15.5922	13.2896	10.9870	8.6845	6.3828	4.0887	1.8695	0.2387	0.000007185
9.6	31.6998	29.3972	27.0946	24.7920	22.4895	20.1869	17.8843	15.5817	13.2791	10.9765	8.6740	6.3723	4.0784	1.8599	0.2347	0.000006439
9.7	31.6894	29.3868	27.0843	24.7817	22.4791	20.1765	17.8739	15.5713	13.2688	10.9662	8.6637	6.3620	4.0681	1.8505	0.2308	0.000005771
9.8	31.6792	29.3766	27.0740	24.7714	22.4688	20.1663	17.8637	15.5611	13.2585	10.9559	8.6534	6.3517	4.0579	1.8412	0.2269	0.000005173
9.9	31.6690	29.3664	27.0639	24.7613	22.4587	20.1561	17.8535	15.5509	13.2483	10.9458	8.6433	6.3416	4.0479	1.8320	0.2231	0.000004637

Source: J. G. Ferris, et al., *Theory of Aquifer Tests* (U.S. Geologic Survey Water-Supply Paper No. 1536–E) Washington, DC: U.S. Government Printing Office, 1962, p. 96.

Example 2-14. If the storage coefficient is 2.74×10^{-4} and the transmissibility is 2.63×10^{-3} m²/s, calculate the drawdown that will result at the end of 100 days of pumping a 0.61-m-diameter well at a rate of 2.21×10^{-2} m³/s.

Solution. Begin by computing u. The radius is

$$r = \frac{0.61 \text{ m}}{2} = 0.305 \text{ m}$$

and

$$u = \frac{(0.305 \text{ m})^2(2.74 \times 10^{-4})}{4(2.63 \times 10^{-3} \text{ m}^2/\text{s})(100 \text{ d})(86{,}400 \text{ s/d})} = 2.80 \times 10^{-10}$$

The factor of 86,400 is to convert days to seconds.

From table of $W(u)$ versus u find that at 2.8×10^{-10}, $W(u) = 21.4190$.

Compute s:

$$s = \frac{2.21 \times 10^{-2} \text{ m}^3/\text{s}}{4(3.14)2.63 \times 10^{-3} \text{ m}^3/\text{s}}(21.4190)$$

$$= 14.33 \text{ or } 14 \text{ m}$$

Unsteady flow in an unconfined aquifer. There is no exact solution to the transient-flow problem for unconfined aquifers because T changes with time and r as the water table is lowered. Furthermore, vertical-flow components near the well invalidate the assumption of radial flow that is required to obtain an analytical solution. If the unconfined aquifer is very deep in comparison to the drawdown, the transient-flow solution for a confined aquifer may be used for an approximate solution. In general, however, numerical methods yield more satisfactory solutions.

Determining the hydraulic properties of a confined aquifer. The estimation of the transmissibility and storage coefficient of an aquifer is based on the results of a pumping test. The preferred situation is one in which one or more observation wells located at a distance from the pumping well are used to gather the data.

The transmissibility may be determined in the steady-state condition by using Equation 2-26. If we define drawdown as $s = H - h$, then the rearrangement of Equation 2-26 yields

$$T = \frac{Q \ln(r_2/r_1)}{2\pi(s_1 - s_2)} \tag{2-30}$$

where s_1 = drawdown at radius r_1
s_2 = drawdown at radius r_2

In the transient state we cannot solve for T and S directly. We have selected the Cooper and Jacob method from the several indirect methods that are available.[27]

[27] H. H. Cooper, Jr. and C. E. Jacob, "A Generalized Graphical Method for Evaluating Formation Constants and Summarizing Well Field History," *Transactions American Geophysical Union*, vol. 27, pp. 520–534, 1946.

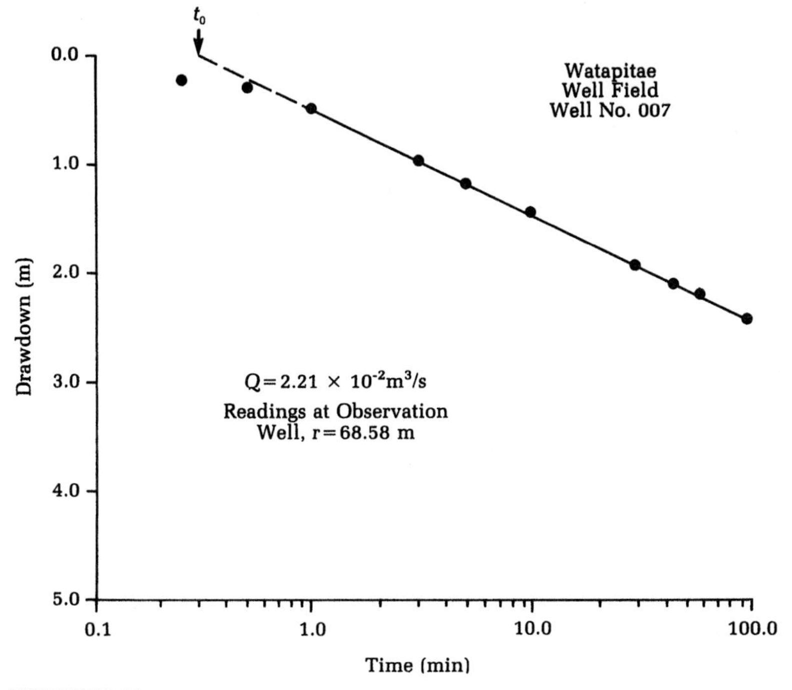

FIGURE 2-31
Pumping test results.

For values of u less than 0.01, they found that Equation 2-28 could be rewritten as follows:

$$s = \frac{Q}{4\pi T} \ln \frac{2.25Tt}{r^2 S} \tag{2-31}$$

A semilogarithmic plot of s versus t (log scale) from the results of a pumping test (Figure 2-31) enables a direct calculation of T from the slope of the line. From Equation 2-31, the difference in drawdown at two points in time may be shown to be

$$s_2 - s_1 = \frac{Q}{4\pi T} \ln \frac{t_2}{t_1} \tag{2-32}$$

Solving for T we find

$$T = \frac{Q}{4\pi(s_2 - s_1)} \ln \frac{t_2}{t_1} \tag{2-33}$$

Cooper and Jacob showed that an extrapolation of the straight-line portion of the plot to the point where $s = 0$ yields a "virtual" (imaginary) starting time (t_0). At this virtual time, Equation 2-31 may be solved for the storage coefficient, S, as follows:

$$S = \frac{2.25Tt_0}{r^2} \tag{2-34}$$

The calculus implies that the distance to the observation well (r) may be as little as the radius of the pumping well itself. This means that drawdown measured in the pumping well may be used as a source of data.

Example 2-15. Determine the transmissibility and storage coefficient for the Watapitae Wells based on the pumping test data plotted in Figure 2-31.

Solution. Using Figure 2-31 we find $s_1 = 0.49$ m at $t_1 = 1.0$ min and at $t_2 = 10.0$ min we find $s_2 = 1.43$ m. Thus,

$$T = \frac{2.21 \times 10^{-2}}{4(3.14)(1.43 - 0.49)} \ln \frac{10.0}{1.0}$$

$$= (1.87 \times 10^{-3})(2.30) = 4.31 \times 10^{-3} \text{ m}^2/\text{s}$$

From Figure 2-31 we find that the extrapolation of the straight portion of the graph yields $t_0 = 0.30$ min. Using the distance between the pumping well and the observation ($r = 68.58$ m) we find

$$S = \frac{(2.25)(4.31 \times 10^{-3})(0.30)(60)}{(68.58)^2}$$

$$= 3.7 \times 10^{-5}$$

The factor of 60 is to convert minutes into seconds. Now we should check to see if our implicit assumption that u is less than 0.01 was true. We use $t = 10.0$ min for the check.

$$u = \frac{(68.58)^2(3.7 \times 10^{-5})}{4(4.31 \times 10^{-3})(10.0)(60)} = 0.017$$

This is a bit high. However, it is obvious that at 100 minutes it would be acceptable. Since the slope does not change, we will take this as a reasonable solution.

Determining the hydraulic properties of an unconfined aquifer. Transmissibility may be determined under steady-state conditions in a fashion similar to that for confined aquifers using Equation 2-30.

Providing that the basic assumptions can be met, the transient-state equations for a confined aquifer may be applied to an unconfined aquifer. Because the unconfined layer often does not release water immediately after a drop in the piezometric surface, the transient equations often are not valid. Thus, great care should be taken in using them.

Pumping test results from nonhomogeneous aquifers. Of course, our assumption that an aquifer is homogeneous seldom applies in real life. Under a few circumstances the inhomogeneities even out to yield average aquifer parameters. Under many other circumstances more complex techniques are required.

Two special cases of inhomogeneity are of particular interest. The first case is where the cone of depression intercepts a barrier. This is shown in Figure 2-32. The net result is that a greater pumping lift is required to achieve the same well yield as would have been obtained if the barrier had not existed.

The second case is where the cone of depression intersects a recharge area (Figure 2-33) such as a lake, stream, or underground stream. The net result is to

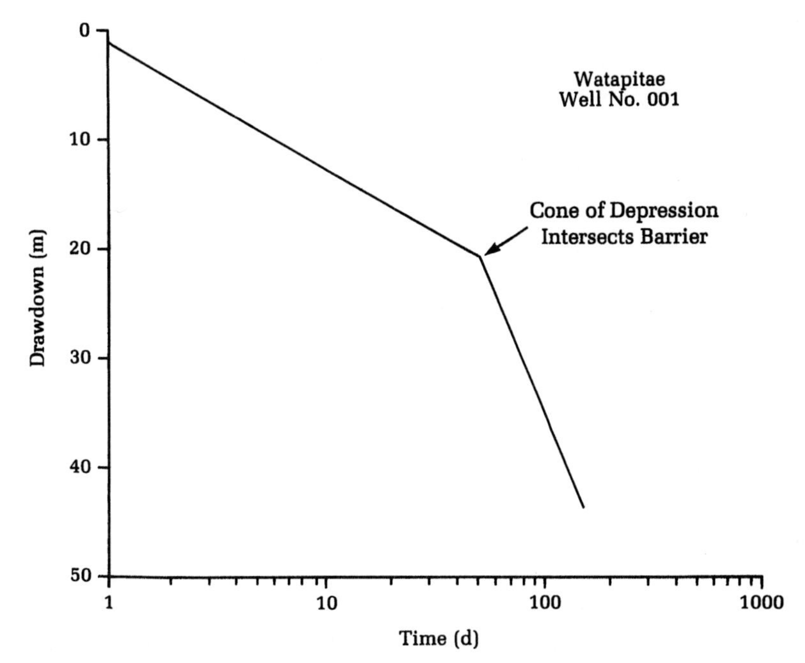

FIGURE 2-32
Pumping test curve showing effect of barrier at day 50.

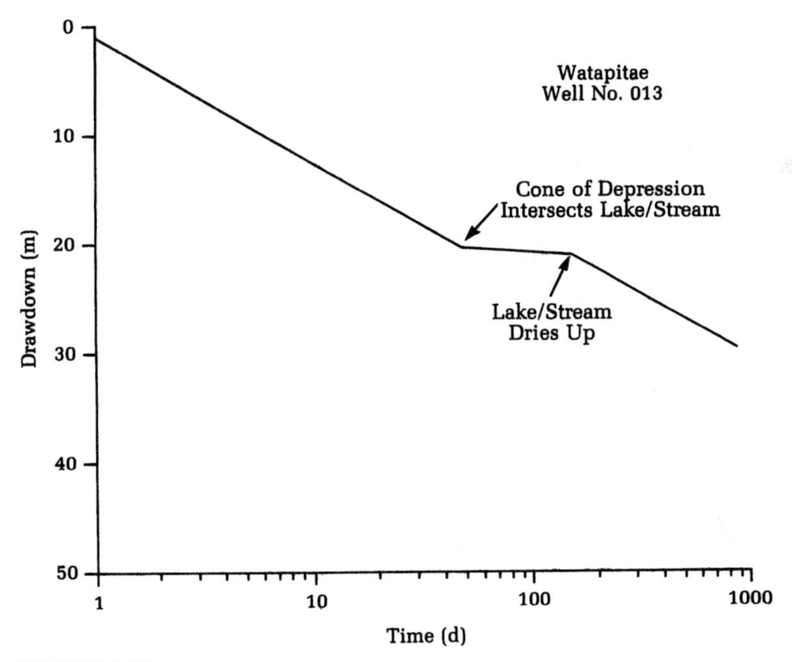

FIGURE 2-33
Pumping test curve showing effect of recharge from day 50 to day 150.

reduce the lift required to maintain the desired yield of the well. Of course, it is possible to dry up the lake or stream if the pumping rate is large enough. This is shown by the increase in slope after the plateau.

Calculating interference. As we mentioned earlier, the cones of depression of wells located close together will overlap; this interference will reduce the potential yield of both wells. In severe circumstances, well interference can cause drawdowns that leave shallow wells dry.

A solution to the well interference problem can be achieved by the method of superposition. This method assumes that the drawdown at a particular location is equal to the sum of the drawdowns from all of the influencing wells. Mathematically this can be represented as follows:

$$s_r = \sum_{i=1}^{n} s_i \tag{2-35}$$

where s_i = individual drawdown caused by well i at location r.

Example 2-16. Three wells are located at 75-m intervals along a straight line. Each well is 0.50 m in diameter. The coefficient of transmissibility is 2.63×10^{-3} m^2/s and the storage coefficient is 2.74×10^{-4}. Determine the drawdown at each well if each well is pumped at 4.42×10^{-2} m^3/s for 10 days.

Solution. The drawdown at each well will be the sum of the drawdown of each well pumping by itself plus the interference from each of the other two wells. Since each well is the same diameter and pumps at the same rate, we may compute one value of the term $Q/(4\pi T)$ and apply it to each well.

$$\frac{Q}{4\pi T} = \frac{4.42 \times 10^{-2}}{4(3.14)(2.63 \times 10^{-3})} = 1.34$$

In addition, since each well is identical, the individual drawdowns of the wells pumping by themselves will be equal. Thus, we may compute one value of u and apply it to each well.

$$u = \frac{(0.25)^2(2.74 \times 10^{-4})}{4(2.63 \times 10^{-3})(10)(86{,}400)} = 1.88 \times 10^{-9}$$

Using $u = 1.9 \times 10^{-9}$ and referring to Table 2-10 we find $W(u) = 19.5042$. The drawdown of each individual well is then

$$s = (1.34)(19.5042) = 26.14 \text{ m}$$

Before we begin calculating interference, we should label the wells so that we can keep track of them. Let us call the two outside wells A and C and the inside well B. Let us now calculate interference of well A on well B, that is, the increase in drawdown at well B as a result of pumping well A. Because we have pumped only for ten days, we must use the transient-flow equations and calculate u at 75 m.

$$u_{75} = \frac{(75)^2(2.74 \times 10^{-4})}{4(2.63 \times 10^{-3})(10)(86{,}400)} = 1.70 \times 10^{-4}$$

From Table 2-10, $W(u) = 8.1027$. The interference of well A on B is then

$$s_{\text{A on B}} = (1.34)(8.1027) = 10.86 \text{ m}$$

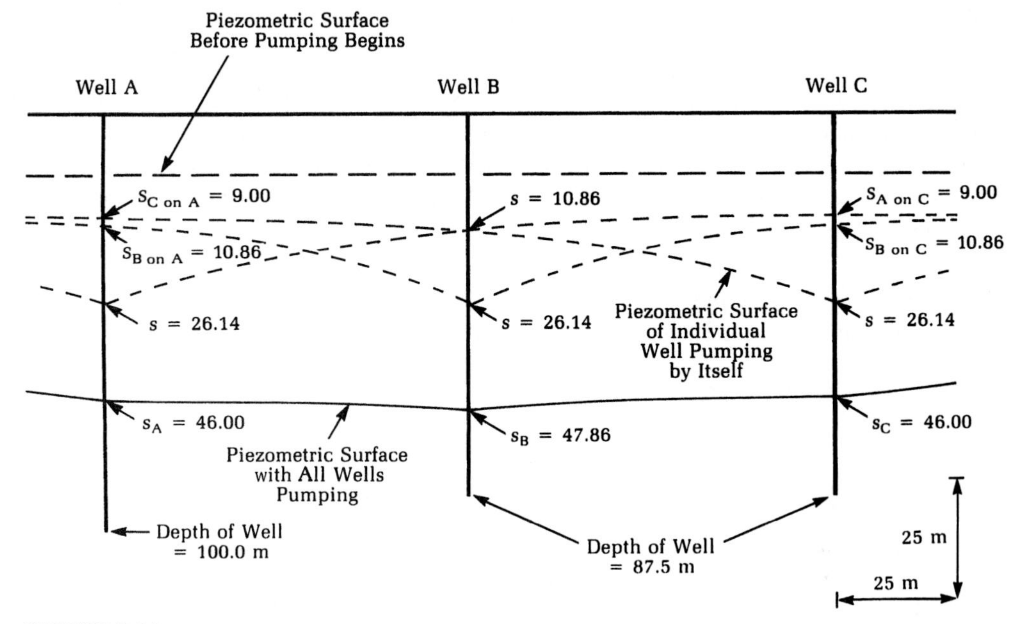

FIGURE 2-34
Interference drawdown of three wells.

In a similar fashion we calculate the interference of well A on well C.

$$u_{150} = (150)^2(3.0145 \times 10^{-8}) = 6.78 \times 10^{-4}$$

and $W(u) = 6.7169$

$$s_{\mathrm{A\,on\,C}} = (1.34)(6.7169) = 9.00 \text{ m}$$

Since our well arrangement is symmetrical, the following equalities may be used:

$$s_{\mathrm{A\,on\,B}} = s_{\mathrm{B\,on\,A}} = s_{\mathrm{B\,on\,C}} = s_{\mathrm{C\,on\,B}}$$

and

$$s_{\mathrm{A\,on\,C}} = s_{C\,\mathrm{on}\,A}$$

The total drawdown at each well is computed as follows:

$$s_{\mathrm{A}} = s + s_{\mathrm{B\,on\,A}} + s_{\mathrm{C\,on\,A}}$$
$$s_{\mathrm{A}} = 26.14 + 10.86 + 9.00 = 46.00 \text{ m}$$
$$s_{\mathrm{B}} = s + s_{\mathrm{A\,on\,B}} + s_{\mathrm{C\,on\,B}}$$
$$s_{\mathrm{B}} = 26.14 + 10.86 + 10.86 = 47.86 \text{ m}$$
$$s_{\mathrm{C}} = s_{\mathrm{A}} = 46.00 \text{ m}$$

Drawdowns are measured from the undisturbed piezometric surface. Note that if the wells are pumped at different rates, the symmetry would be destroyed and the value $Q/(4\pi T)$ would have to be calculated separately for each case. Likewise, if the distances were not symmetric, then separate u values would be required. The results of these calculations have been plotted in Figure 2-34.

Groundwater Contamination

The well hydraulics equations may be used to design purge wells to remove contaminated groundwater. The lowering of the piezometric surface will cause contaminants to flow to the well where they may be removed to the surface for treatment. In some cases, a well field may be required to intercept the contaminant plume. In these cases, the well interference effects may be put to good use in designing the distribution of the wells.

2-6 WASTE MINIMIZATION AND WATER CONSERVATION

"Water: too much, too little, too dirty."[28] These are the conditions that prompt water resource management. Often the incentive for increased control follows a major disaster, namely, flood, drought, fish kill, or epidemic. Historically, solutions to water resource problems in general, and water supply in particular, have been structural in nature. Building reservoirs, dredging channels, and driving new and deeper wells are traditional ways that civil engineers have solved these problems. Modern water resource management also considers nonstructural means to achieve water resource goals.

A key nonstructural means to achieve water resource goals is to minimize the use of water. Water that is not used has reduced potential for being contaminated or lost through evaporation/transpiration. In addition to conserving water, a reduced municipal water demand reduces the cost of water supply works, water treatment and distribution, and wastewater collection and disposal facilities. Many examples can be cited where simple changes in habits can reduce water consumption and, in turn, pollution of natural waters. For example, washing of streets with high-pressure hoses can be replaced with wet sweeping, and spray irrigation can be replaced with trickle irrigation. The installation of water meters in Boulder, Colorado, reduced the average daily demand from 802 Lpcd to 635 Lpcd.[29] The use of water-saving devices such as high-pressure, low-volume shower heads and low-volume flush toilets can reduce per capita household demand by as much as 16 percent.[30]

2-7 CHAPTER REVIEW

When you have completed studying this chapter you should be able to do the following without the aid of your textbooks or notes:

1. Sketch and explain the hydrologic cycle, labeling the parts as in Figure 2-1.
2. List and explain the four factors that reduce the amount of direct runoff.

---

[28] D. P. Loucks, J. R. Stedinger, and D. A. Haith, *Water Resource Systems Planning and Analysis*, Englewood Cliffs, NJ: Prentice-Hall, p. 3, 1981.

[29] S. H. Hanke, "Demand for Water Under Dynamic Conditions," *Water Resources Research*, vol. 6, No. 5, 1970.

[30] J. E. Flack, "Achieving Urban Water Conservation," *Water Resources Bulletin*, vol. 16, No. 1, 1980.

3. Explain the difference between streamflow that results from direct runoff and that which results from baseflow.
4. Define evaporation, transpiration, runoff, baseflow, watershed, basin, and divide.
5. Sketch the groundwater hydrologic system, labeling the parts as in Figure 2-3.
6. Explain why water in an artesian aquifer is under pressure and why water may rise above the surface in some instances and not others.
7. Explain why infiltration rates decrease with time.
8. Explain return period or recurrence interval in terms of probability that an event will take place.
9. Explain what a unit hydrograph is.
10. Explain the purpose of the unit hydrograph and explain how it might be applied in the analysis of a storm sewer design or a stream flood-control project.
11. Using a sketch, show how the groundwater contribution to a hydrograph is identified.
12. Define the rational formula and identify the units as used in this text.
13. Explain why the rational method may yield inadequate results when applied by inexperienced engineers, while experienced engineers may get adequate results.
14. Define time of concentration and explain how it is used in conjunction with the rational method.
15. Distinguish between an annual series, a partial series, and a complete series analysis and explain when to use each.
16. Sketch a graph of a hydrologic year for rainfall or stream-flow events and a calendar year for rainfall or stream-flow events and explain why a hydrologic year is chosen for data representation instead of a calendar year.
17. Sketch a subsurface cross-section from the results of a well boring log and identify pertinent hydrogeologic features.
18. Sketch a well and label the major sanitary protection features according to this text.
19. Sketch a piezometric profile for a single well pumping at a high rate, and sketch a profile for the same well pumping at a low rate.
20. Sketch a piezometric profile for two or more wells located close enough together to interfere with one another.
21. Sketch a well-pumping test curve which shows (*a*) the interception of a barrier, and (*b*) the interception of a recharge area.
22. Give two examples of methods to minimize water use.

With the use of this text you should be able to do the following:

1. Compute mass balances for open and closed hydrologic systems.
2. Compute infiltration rates by Horton's method and estimate the volume of infiltration.

3. Estimate the volume of water loss through transpiration given the air and water temperature, wind speed, and relative humidity.
4. Use Horton's equation and/or some form of Dalton's equation to solve complex mass-balance problems.
5. Construct an intensity-duration-frequency curve from a compilation of intense rainfall occurrences.
6. Construct a unit hydrograph for a given stream-gaging station if you are provided with a rainfall and total flow at the gaging station.
7. Apply a given unit hydrograph to construct a compound hydrograph if you are provided with an observed rainfall and an estimate of the rainfall losses due to infiltration and evaporation.
8. Determine the peak flow (Q in the rational formula) and time of arrival (t_c) resulting from a rainfall of specified intensity and duration in a well-defined watershed.
9. Perform an extreme-value analysis on average monthy discharge data and interpret the analysis in terms of return period.
10. Perform a complete-series analysis on average monthly discharge data and interpret the analysis in terms of reliability for water supply.
11. Using the mass-balance method, determine the volume of a reservoir or retention basin for a given demand or flood control given appropriate discharge data.
12. Calculate the drawdown at a pumped well or observation well if you are given the proper input data.
13. Calculate the transmissibility and storage coefficient for an aquifer if you are provided with the results of a pumping test.
14. Calculate the interference effects of two or more wells.

2-8 PROBLEMS

2-1. Lake Kickapoo, TX, is approximately 12 km in length by 2.5 km in width. The inflow for the month of March is 3.26 m^3/s and the outflow is 2.93 m^3/s. The total monthly precipitation is 15.2 cm and the evaporation is 10.2 cm. The seepage is estimated to be 2.5 cm. Estimate the change in storage during the month of March.

Answer: 1.63×10^6 m^3

2-2. A 4,000-km^2 watershed receives 102 cm of precipitation in one year. The average flow of the river draining the watershed is 34.2 m^3/s. Infiltration is extimated to be 5.5×10^{-7} cm/s and evapotranspiration is estimated to be 40 cm/y. Determine the change in storage in the watershed over one year. The ratio of runoff (in cm) to precipitation is termed the runoff coefficient. Compute the runoff coefficient for this watershed.

2-3. Using the values of f_o, f_c, and k for a Dothan loamy sand (see Section 2-1), find the infiltration rate at times of 12, 30, 60, and 120 minutes. Compute the total volume of infiltration over 120 minutes.

2-4. Infiltration data from an experiment yield an initial infiltration rate of 4.70 cm/h and a final equilibrium infiltration rate of 0.70 cm/h after 60 minutes of steady precipitation. The value of k was estimated to be 10.85 min^{-1}. Determine the total volume of infiltration for the following storm sequence: 30 mm/h for 30 minutes, 53 mm/h for 30 minutes, 23 mm/h for 30 minutes.

2-5. Using the empirical equation developed for Lake Hefner, estimate the evaporation from a lake on a day that the air temperature is 30 °C, the water temperature is 15 °C, the wind speed is 9 m/s, and the relative humidity is 30%.

2-6. The Dalton-type evaporation equation implies that there is a limiting relative humidity above which evaporation will be nil regardless of the wind speed. Using the Lake Hefner empirical equation, estimate the relative humidity at which evaporation will be nil if the water temperature is 10 °C and the air temperature is 25 °C.

2-7. Prepare an IDF curve for a 2-year storm at Dismal Swamp using the data in Table 2-2. *Hint:* Curve should intersect 98.7 mm/h at 15 minutes duration.

2-8. Prepare an IDF curve for a 10-year storm at Dismal Swamp using the data in Table 2-2.

2-9. Prepare an intensity-duration-frequency curve for a 5-year storm with the data shown in the following table.

Annual maximum intensity (mm/h)

Year	30-minute duration	60-minute duration	90-minute duration	120-minute duration
1960	122.3	100.3	81.1	55.3
1961	104.6	82.9	64.5	39.7
1962	81.0	60.8	41.5	16.5
1963	145.1	123.7	104.7	81.7
1964	83.0	61.5	40.7	16.0
1965	70.1	51.1	30.1	11.3
1966	94.7	71.0	51.7	26.7
1967	63.7	41.5	21.8	10.1
1968	57.9	35.7	17.1	9.7
1969	71.5	50.0	31.3	15.9

Hint: Curve should intersect 96.8 mm/h at 60 minutes duration.

2-10. Prepare an intensity-duration-frequency curve for a 2-year storm with the data from Problem 2-9.

2-11. Determine the unit hydrograph ordinates for the Tursiops River with the streamflow data shown in the table below that resulted from a 5-hour storm of uniform intensity. The basin area is 100.0 square kilometers.

Time (h)	Flow (m^3/s)	Time (h)	Flow (m^3/s)	Time (h)	Flow (m^3/s)
0	0.55	35	5.77	65	1.64
5	0.50	40	5.02	70	1.10
10	0.45	45	4.29	75	0.79
15	1.98	50	3.51	80	0.47
20	4.82	55	2.72	85	0.25
25	6.24	60	2.19	90	0.25
30	6.86				

2-12. Determine the unit hydrograph ordinates for the Orca River with the streamflow data shown in the table below. The basin area is 310.0 square kilometers. The unit duration of the storm was five hours.

Time (h)	Flow (m^3/s)	Time (h)	Flow (m^3/s)	Time (h)	Flow (m^3/s)
0	1.60	35	21.55	70	5.15
5	1.73	40	18.13	75	3.46
10	1.57	45	15.77	80	2.48
15	1.41	50	13.48	85	1.48
20	6.22	55	11.03	90	0.79
25	15.14	60	8.55	95	0.77
30	19.60	65	6.88	100	0.77

2-13. Determine the unit hydrograph ordinates for the Isoceles River with the streamflow data shown in Figure P-2-13 below. The basin area is 14.40 square kilometers, and the unit duration of the storm is one hour. For ease of computation, locate ordinates at hourly intervals, that is, 1500, 1600, and 1700 hours. Note that the time is military time, that is, 1500 h = 3 P.M., 2000 h = 8 P.M., etc.

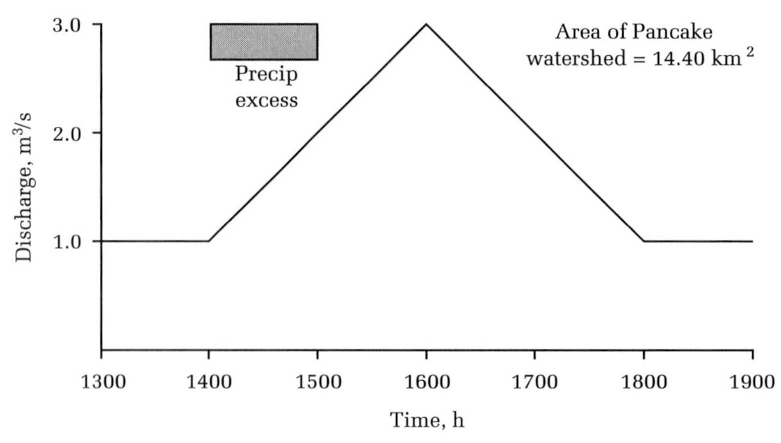

FIGURE P-2-13
Storm hydrograph for Isoceles River.

2-14. Using the unit hydrograph developed in Problem 2-11, determine the streamflow hydrograph resulting from the storm shown in Figure P-2-14 below. Note that the time is military time, that is, 1500 h = 3 P.M., 2000 h = 8 P.M., etc.

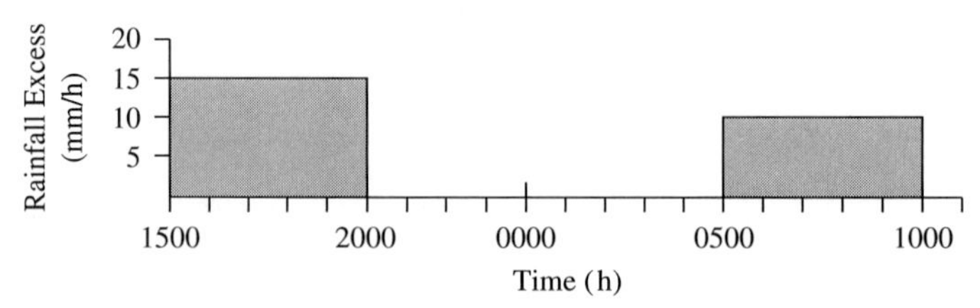

FIGURE P-2-14
Plot of rainfall during storm.

2-15. Using the unit hydrograph developed in Problem 2-12, determine the streamflow hydrograph resulting from the storm shown in Figure P-2-15 below.

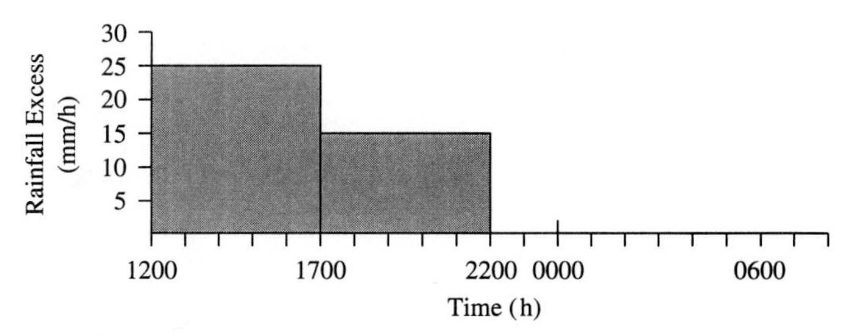

FIGURE P-2-15
Plot of rainfall during storm.

2-16. Using the unit hydrograph developed in Problem 2-13, determine the streamflow hydrograph resulting from the storm shown in Figure P-2-16 below.

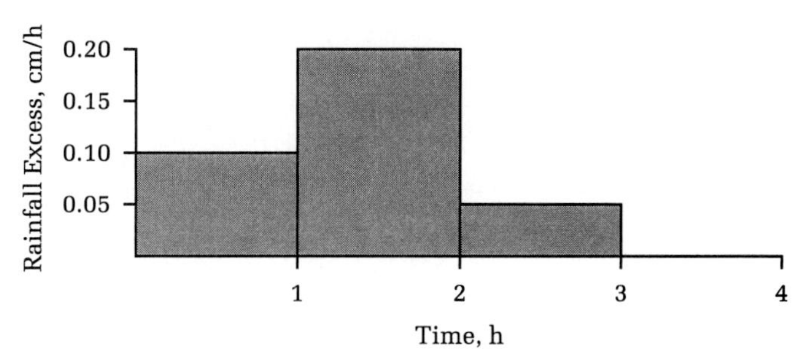

FIGURE P-2-16
Plot of rainfall during storm.

2-17. Mechanicsville has obtained a grant under the Federal Program for Urban Development of Greenspace and the Environment to build a condominium complex for the retired. The total land set aside is 74,010 m^2. Of this area 15,831 m^2 will be used for Swiss chalet condominiums with slate roofs. Public streets and drives will occupy 18,886 m^2. The remainder will be lawns. The area is flat and has a sandy soil. Using the most conservative estimates of C, find the peak discharge from the development for a 2-year storm. Assume an overland flow distance of 272 m and use the IDF curves given in Figure P-2-17.

2-18. Paul Revere is planning to build a shopping mall and parking lot on the north side of Hindry Road as shown in Figure P-2-18. The existing culvert near the southwest corner of the proposed parking lot was designed for a 5-year storm. Determine the frequency that the capacity of the culvert will be exceeded if it is not enlarged when the Revere mall is constructed. Also determine the peak discharge that must be handled if the design criteria are changed to specify a 10-year storm. Use the IDF curves in Figure P-2-17.

2-19. Dr. Florence Nightengale is planning to build a clinic on the east side of Okemos Road as shown in Figure P-2-19. The existing culvert (No. 481) was designed for a 5-year storm. Determine if the capacity of the culvert will be exceeded if it is not enlarged when the clinic is built. Also determine the peak discharge that must be handled if the design criteria are changed to specify a 10-year storm. Use the IDF curves in Figure P-2-17.

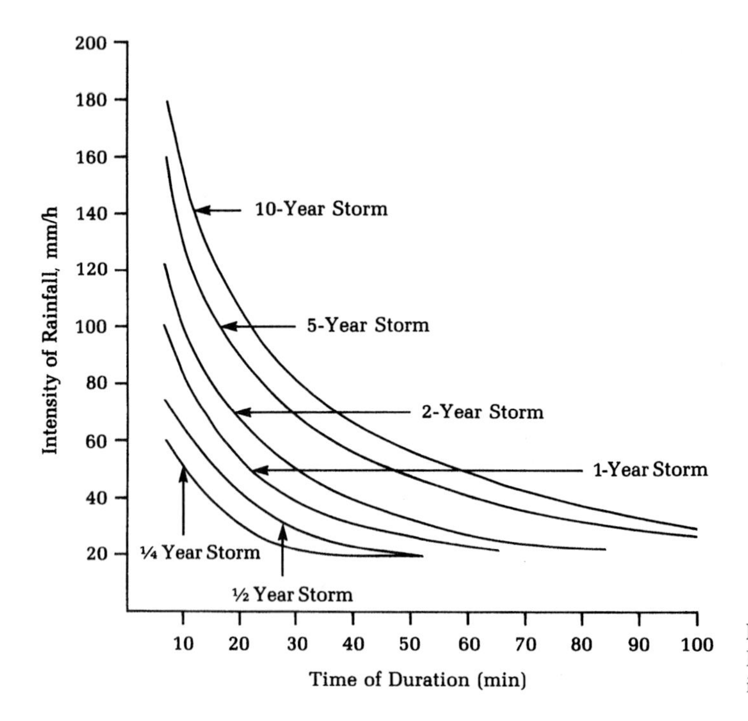

FIGURE P-2-17
IDF curve for Mechanicsville.

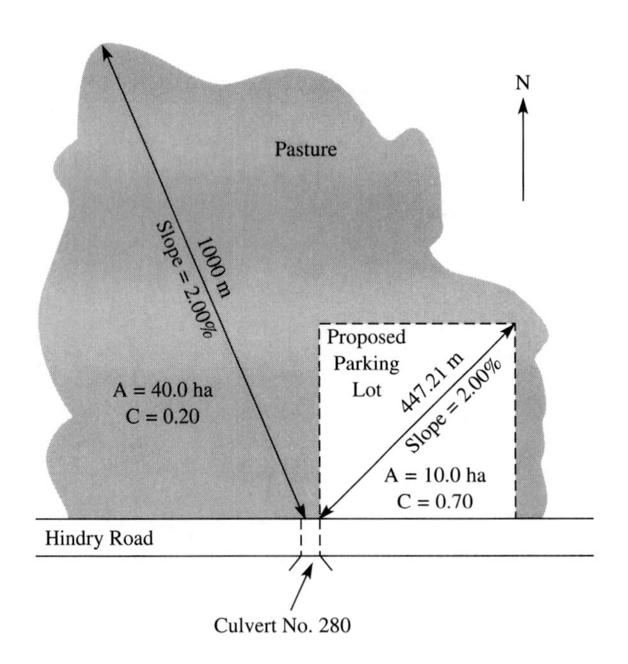

FIGURE P-2-18
Sketch of proposed shopping mall.

2-20. Perform an annual minima extreme-value analysis on the Menominee River data given in the table on page 118. Determine the return period of a flow less than or equal to 40.0 m^3/s. Provide your tabulated results as well as your plot. Explain how you determined the return period.

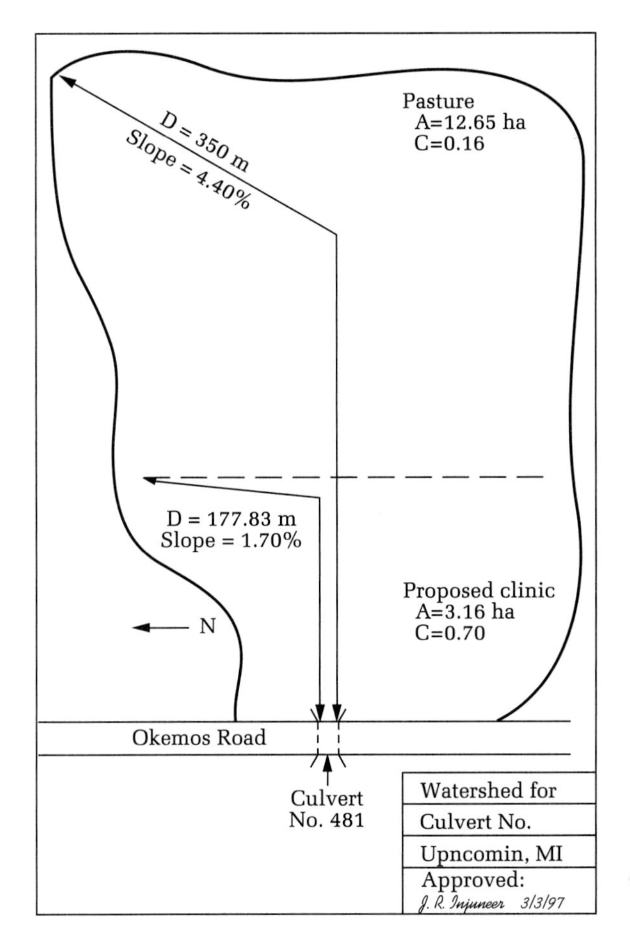

FIGURE P-2-19
Sketch of proposed clinic.

2-21. Perform an annual minima extreme-value analysis on the Hoko River data given in the table on page 119. Determine the flow for a return period of eight years. Provide your tabulated results as well as your plot.

2-22. Perform an annual minima value analysis on the Rappahannock River data given in the table on page 120. Determine the return period of a flow less than or equal to 1.0 m^3/s. Give your tabulated results as well as your plot. Explain how you determined the return period.

2-23. Perform a complete-series analysis of the Spokane River data given in the table on page 121. Can the river supply 50 m^3/s for 90 percent or more of the time?

2-24. Perform a complete-series analysis of the Squannacook River data given in the table on page 122. Can the river supply 2.0 m^3/s for 90 percent or more of the time?

2-25. Perform a complete-series analysis of the Clear Fork Trinity River data given in the table on page 123. Can the river supply 1.5 m^3/s for 90 percent or more of the time?

2-26. Construct a mass balance for the Spokane River (Problem 2-23) and determine the storage required to meet a demand of 175 m^3/s for the period from May 1953 to April 1954. Use a spreadsheet program you have written to solve this problem.

Menominee River below Koss, MI (Problem 2-20)

Mean monthly discharge (m^3/s)

Year	Jan	Feb	Mar	Apr	May	June	July	Aug	Sept	Oct	Nov	Dec
1946	76.3	69.8	149.0	106.0	85.3	132.0	89.5	62.2	58.5	56.6	76.7	56.8
1947	54.0	53.0	60.1	157.0	164.0	103.0	72.1	55.5	52.7	51.0	59.3	47.1
1948	49.2	35.4	72.5	103.0	82.3	48.5	42.2	45.8	39.6	34.8	56.1	49.2
1949	45.9	47.9	57.4	87.7	84.7	62.0	102.0	51.7	57.2	57.0	57.8	58.6
1950	59.1	57.6	61.1	199.0	243.0	101.0	69.2	65.8	49.0	43.4	48.9	48.4
1951	50.5	44.5	68.0	267.0	171.0	143.0	159.0	93.3	122.0	147.0	117.0	87.0
1952	76.9	75.6	66.5	219.0	94.0	86.7	153.0	97.0	58.6	45.6	50.2	51.9
1953	59.4	62.5	103.0	167.0	133.0	166.0	174.0	87.5	67.0	56.1	53.4	63.6
1954	57.0	66.7	68.5	182.0	184.0	138.0	73.2	61.0	91.1	123.0	84.6	68.5
1955	66.2	61.0	70.5	254.0	108.0	106.0	48.7	56.1	38.0	59.7	63.0	57.4
1956	59.0	54.8	49.7	270.0	108.0	88.4	116.0	83.4	61.5	48.8	53.1	53.6
1957	49.4	46.1	69.7	130.0	93.0	65.0	41.4	36.3	52.3	52.7	73.3	59.8
1958	54.0	51.3	61.8	123.0	65.9	62.0	132.0	47.0	58.5	47.7	68.9	48.4
1959	46.7	43.1	55.0	110.0	105.0	56.7	48.3	78.0	142.0	155.0	122.0	78.2
1960	82.3	71.0	62.4	242.0	373.0	135.0	83.4	72.1	80.8	60.5	102.0	68.2
1961	52.6	49.2	77.0	158.0	186.0	82.5	60.8	53.8	48.9	57.9	67.6	63.1
1962	53.9	52.4	69.2	168.0	168.0	107.0	59.3	52.1	73.8	65.3	54.7	51.4
1963	46.8	43.8	56.1	87.7	120.0	99.1	43.6	40.6	38.0	34.2	35.6	35.7
1964	37.2	33.8	41.8	70.2	131.0	63.2	42.1	56.9	65.1	55.6	69.8	54.2
1965	49.3	44.1	50.9	173.0	361.0	83.6	51.7	45.6	56.2	67.5	76.6	87.2
1966	78.5	66.6	131.0	157.0	117.0	113.0	44.6	63.0	43.5	62.6	63.7	59.2
1967	59.8	64.0	65.2	295.0	133.0	135.0	100.0	66.3	51.6	86.0	108.0	63.3
1968	50.5	58.0	72.6	134.0	108.0	168.0	141.0	78.1	155.0	93.8	90.2	82.7
1969	89.9	90.0	84.4	229.0	157.0	118.0	87.1	51.1	42.6	65.3	68.9	58.2
1970	62.7	51.0	58.8	114.0	109.0	157.0	56.9	46.6	49.1	64.1	122.0	97.5
1971	72.4	64.7	89.4	284.0	155.0	93.3	67.5	51.0	47.0	92.4	88.2	80.2
1972	65.9	56.7	68.5	194.0	254.0	88.6	68.2	108.0	91.0	134.0	138.0	77.2
1973	85.3	73.8	226.0	240.0	290.0	111.0	72.6	80.3	69.0	66.9	82.5	66.3
1974	62.8	65.4	73.4	142.0	107.0	111.0	61.6	86.4	77.8	58.8	106.0	75.5
1975	66.3	66.0	68.4	193.0	209.0	116.0	54.8	41.5	63.2	43.4	69.2	86.3
1976	67.5	69.2	96.9	298.0	147.0	79.1	41.2	36.9	30.3	—	—	—

Hoko River near Sekiu, WA (Problem 2-21)

Mean monthly discharge (m^3/s) 1963–1973

Year	Jan	Feb	Mar	Apr	May	June	July	Aug	Sept	Oct	Nov	Dec
1963	12.1	15.0	8.55	9.09	5.78	1.28	2.59	1.11	.810	13.3	26.1	20.3
1964	27.3	12.2	18.0	8.21	4.08	3.62	4.53	2.44	4.28	7.67	13.3	14.7
1965	27.4	27.3	5.01	5.61	6.68	1.38	.705	.830	.810	7.31	16.8	19.6
1966	29.8	11.3	17.6	5.18	2.67	2.10	1.85	.986	1.54	10.3	17.0	39.0
1967	35.6	26.8	18.5	6.51	3.43	1.46	.623	.413	.937	25.7	14.2	27.8
1968	34.2	22.4	15.7	9.20	3.68	2.65	1.72	1.55	9.12	16.8	16.5	25.2
1969	17.2	18.5	12.9	12.8	3.74	2.23	1.19	.810	6.15	7.84	9.15	15.9
1970	17.3	12.1	8.50	17.7	3.85	1.32	.932	.708	4.22	7.96	13.9	25.4
1971	32.7	21.0	21.1	8.13	3.43	2.83	1.83	.932	2.22	10.7	22.7	22.0
1972	27.4	26.9	25.4	14.6	3.00	1.00	5.32	.841	2.00	1.14	11.8	37.8
1973	28.0	9.23	11.3	4.13	5.30	4.93	1.63	.736	.810	13.1	29.8	31.5

Rappahannock River near Warrenton, VA (Problem 2-22)

Mean monthly discharge (m^3/s)

Year	Jan	Feb	Mar	Apr	May	June	July	Aug	Sept	Oct	Nov	Dec
1943	7.08	9.23	10.5	8.55	6.12	3.31	1.18	0.294	0.269	0.057	1.38	1.06
1944	4.67	3.14	8.44	5.66	3.71	1.88	0.334	0.450	4.13	4.62	1.87	5.27
1945	4.11	5.44	4.81	3.62	3.79	2.94	4.22	9.68	13.0	3.77	4.05	8.27
1946	8.44	7.90	7.36	5.38	9.94	7.84	2.60	4.70	2.02	3.14	2.35	2.38
1947	6.94	3.79	6.77	3.57	4.84	3.74	2.92	2.39	1.40	1.04	6.60	2.74
1948	3.85	6.09	7.28	9.77	11.5	4.05	4.16	13.1	2.86	7.73	10.4	14.1
1949	13.8	11.0	8.95	11.9	13.2	5.10	4.39	3.79	1.67	2.03	2.37	2.94
1950	2.76	7.45	7.73	4.42	7.56	5.44	4.25	1.60	5.75	4.19	5.63	20.0
1951	5.72	12.8	11.0	12.1	5.44	9.40	2.89	1.19	0.447	0.453	1.93	5.13
1952	7.62	9.74	12.5	18.0	9.32	3.43	2.21	1.98	2.50	0.951	9.80	7.28
1953	11.3	7.25	12.6	8.24	10.6	4.39	1.36	0.685	0.343	0.357	0.773	2.25
1954	2.47	2.54	5.69	5.72	4.39	1.95	0.875	0.572	0.131	1.90	1.80	3.57
1955	2.42	4.11	8.21	5.07	3.85	4.16	0.801	20.3	3.54	2.47	2.19	1.62
1956	2.21	6.54	7.31	6.23	2.46	1.06	10.3	3.60	2.17	4.42	6.34	3.91
1957	4.33	8.07	8.95	8.72	3.31	2.15	0.402	0.008	0.391	1.02	1.66	6.43
1958	10.1	6.82	12.0	11.8	9.03	2.97	5.07	3.57	1.33	1.56	1.81	1.89
1959	3.45	3.06	4.76	7.08	3.28	5.04	0.804	0.513	0.759	2.68	2.05	3.88
1960	4.11	9.71	7.70	13.3	11.3	9.97	2.97	1.85	2.77	1.10	1.23	1.31
1961	3.31	15.4	9.85	15.5	11.1	6.82	3.23	2.24	1.70	1.16	1.77	4.25
1962	5.44	5.61	16.8	10.7	5.27	6.88	3.57	1.51	0.855	0.932	4.73	3.60
1963	6.51	4.19	13.6	4.45	2.55	3.20	0.496	0.136	0.160	0.121	2.28	2.21
1964	9.97	8.18	9.63	11.8	7.25	1.37	1.44	0.660	0.697	1.72	1.83	3.82
1965	6.51	13.8	15.0	6.31	3.74	1.34	5.27	0.365	3.09	0.459	0.379	0.450
1966	0.694	5.83	4.45	4.45	6.63	1.70	0.225	0.135	4.47	3.51	3.31	4.42
1967	5.63	5.86	13.6	3.82	4.53	1.57	1.04	7.42	2.27	3.82	2.57	10.0
1968	12.9	6.68	9.40	4.56	4.02	4.59	2.64	1.64	1.03	0.971	4.93	3.00
1969	4.47	4.84	5.07	3.51	1.93	1.70	1.54	1.93	2.12	1.80	3.37	6.12
1970	5.92	11.4	6.14	9.83	4.30	2.34	3.62	1.62	0.413	—	—	—

Spokane River near Otis Orchards, WA (Problem 2-23)

Mean monthly discharge (m^3/s)

Year	Jan	Feb	Mar	Apr	May	June	July	Aug	Sept	Oct	Nov	Dec
1950	—	—	—	—	—	—	—	—	—	59.9	123.0	257.0
1951	217.0	492.0	200.0	422	460	156.0	36.3	4.25	8.44	75.6	98.7	145.0
1952	115.0	140.0	107.0	491	624	165.0	52.4	9.49	33.3	36.1	44.7	50.5
1953	169.0	311.0	163.0	246	525	358.0	32.3	12.8	29.8	52.2	53.0	97.8
1954	116.0	190.0	254.0	398	657	414.0	68.0	28.13	28.3	82.1	106.0	93.1
1955	60.0	76.3	62.4	257	544	468.0	82.3	19.5	26.0	86.0	166.0	354.0
1956	292.0	141.0	195.0	685	809	391.0	47.8	22.0	27.9	64.7	78.9	101.0
1957	99.0	61.0	278.0	461	792	329.0	33.6	12.5	15.7	55.5	66.9	73.0
1958	80.0	245.0	234.0	408	548	152.0	29.5	4.50	24.4	36.8	153.0	240.0
1959	356.0	233.0	192.0	465	351	410.0	35.5	11.6	45.2	84.4	224.0	172.0
1960	117.0	154.0	202.0	600	470	266.0	35.4	16.0	45.0	47.8	57.6	88.5
1961	93.0	454.0	406.0	389	559	352.0	8.86	4.96	19.1	24.5	55.9	50.7
1962	107.0	142.0	124.0	534	535	232.0	39.9	11.0	13.3	50.3	96.6	186.0
1963	152.0	268.0	208.0	353	303	92.1	26.9	5.32	10.1	21.2	68.0	71.2
1964	76.5	97.6	81.6	328	594	619.0	66.5	34.0	92.7	43.0	88.6	408.0
1965	282.0	314.0	268.0	475	602	158.0	57.3	28.9	23.2	32.5	82.7	76.7
1966	105.0	63.7	175.0	451	388	126.0	38.3	10.3	12.0	45.6	60.1	122.0
1967	205.0	282.0	191.0	272	504	413.0	40.8	13.3	30.8	49.3	68.5	93.6
1968	104.0	276.0	348.0	226	232	155.0	42.3	22.3	40.5	92.9	166.8	198.0
1969	254.0	138.0	162.0	638	651	220.0	50.7	25.2	40.5	45.8	49.0	60.8
1970	86.8	225.0	214.0	260	512	382.0	51.3	32.5	35.6	—	—	—

Squannacook River near West Groton, MA (Problem 2-24)

Mean monthly discharge (m^3/s)

Year	Jan	Feb	Mar	Apr	May	June	July	Aug	Sept	Oct	Nov	Dec
1951	3.48	8.18	6.63	6.63	3.20	2.38	2.40	1.49	1.06	1.82	6.60	4.47
1952	6.68	5.07	6.51	9.94	5.44	3.71	.87	1.01	.69	.45	1.05	3.37
1953	4.79	6.77	11.44	9.80	6.34	1.21	.52	.42	.29	.51	1.34	3.65
1954	2.06	3.20	4.67	4.53	9.71	2.75	1.21	1.05	6.94	2.27	6.26	7.16
1955	2.92	3.06	5.41	6.17	2.77	1.44	.46	1.63	.61	8.38	8.61	2.17
1956	9.15	3.29	3.82	14.56	5.21	2.50	.77	.40	.50	.54	1.14	2.33
1957	2.92	2.63	4.22	3.99	2.65	.87	.37	.22	.22	.29	.97	3.91
1958	5.89	3.48	6.60	12.40	5.35	1.29	.81	.49	.45	.62	1.13	1.49
1959	2.07	2.05	5.41	8.67	2.37	1.22	1.17	.59	.82	2.55	4.08	5.55
1960	3.51	3.96	3.03	14.73	5.52	2.41	1.09	1.21	2.71	2.18	3.34	2.49
1961	1.57	3.09	7.28	11.10	4.67	2.31	1.03	.80	1.23	.99	2.06	1.73
1962	3.14	1.80	5.47	10.93	3.71	1.25	.56	.69	.50	2.95	4.73	4.30
1963	2.19	1.76	6.83	7.53	2.66	.77	.38	.24	.25	.35	1.52	1.98
1964	3.77	2.57	7.33	6.57	1.85	.59	.38	.25	.21	.27	.36	.79
1965	.65	1.33	2.38	3.79	1.47	.59	.23	.20	.19	.27	.45	.64
1966	.61	1.96	5.55	2.92	2.46	.80	.26	.18	.27	.52	1.75	1.35
1967	1.68	1.53	2.64	10.62	6.29	3.17	2.22	.72	.47	.60	1.07	3.03
1968	2.02	2.14	9.60	3.79	3.82	4.79	1.92	.61	.48	.46	1.88	4.33
1969	2.21	2.17	5.81	10.70	2.80	1.01	.58	1.03	.93	.52	5.24	5.83

Clear Fork Trinity River at Fort Worth, TX (Problem 2-25)

Mean monthly discharge (m^3/s)

Year	Jan	Feb	Mar	Apr	May	June	July	Aug	Sept	Oct	Nov	Dec
1940	—	—	—	—	—	—	—	—	—	0.00	5.63	15.4
1941	4.59	23.8	7.50	6.91	10.2	17.0	2.07	2.29	0.20	1.71	0.631	0.926
1942	0.697	0.595	0.614	58.93	24.1	9.09	0.844	0.714	1.21	10.0	2.38	1.87
1943	1.33	1.00	3.99	3.71	8.38	3.77	0.140	0.00	1.33	0.014	0.00	0.139
1944	0.311	4.93	2.83	2.25	13.3	1.68	0.210	0.609	4.11	0.985	0.515	1.47
1945	3.06	30.38	35.23	28.85	5.69	21.7	2.14	0.230	0.162	0.971	0.617	0.541
1946	1.88	5.75	3.54	1.89	6.57	5.86	0.153	1.45	4.02	0.906	12.2	10.3
1947	4.64	2.62	4.87	5.13	2.27	4.25	0.292	0.054	0.535	0.371	0.331	3.51
1948	3.99	16.9	9.06	1.91	2.64	1.11	1.22	0.00	0.00	0.00	0.00	0.003
1949	0.309	4.19	9.94	4.16	55.21	11.1	1.38	0.450	0.447	4.53	0.711	0.614
1950	3.28	14.7	3.26	12.7	15.1	2.50	3.60	2.44	10.6	1.12	0.711	0.801
1951	0.708	0.994	0.719	0.527	1.37	6.20	0.980	0.00	0.00	0.00	0.006	0.090
1952	0.175	0.413	0.297	1.93	3.65	0.210	0.003	0.029	0.007	0.00	0.368	0.167
1953	0.099	0.080	0.134	0.671	0.934	0.008	0.286	0.249	0.041	0.546	0.182	0.066
1954	0.108	0.092	0.114	0.088	0.278	0.017	0.021	0.015	0.008	0.047	0.024	0.063
1955	0.091	0.153	0.317	0.145	0.464	0.640	0.049	0.050	0.119	0.104	0.055	0.058
1956	0.069	0.218	0.026	0.306	1.35	0.30	0.026	0.019	0.029	0.266	0.030	0.170
1957	0.065	0.300	0.385	12.8	23.6	59.9	6.97	1.36	0.501	0.476	0.855	1.55
1958	1.65	1.61	4.59	5.69	28.1	0.589	0.524	0.456	0.549	0.572	0.490	0.566
1959	0.759	0.776	0.120	0.261	0.097	0.685	0.379	0.668	0.473	9.03	1.64	3.65
1960	11.8	4.45	3.26	1.42	0.631	0.379	0.660	0.566	0.467	0.498	0.241	0.648
1961	2.05	1.92	3.40	1.02	0.306	2.34	0.821	0.816	1.08	0.824	0.297	0.504
1962	0.311	2.03	0.467	0.759	0.459	0.236	0.745	1.41	6.94	1.31	0.405	0.767
1963	0.345	0.268	0.379	1.74	3.79	1.48	0.527	0.586	0.331	0.277	0.249	0.266
1964	0.416	0.266	1.16	0.813	1.02	0.374	0.535	0.963	3.96	0.351	1.47	0.886
1965	2.13	14.6	4.16	2.28	20.5	2.45	1.22	0.821	0.776	0.394	0.476	0.213
1966	0.169	0.354	0.462	6.40	23.1	18.5	5.32	0.951	0.294	1.37	0.15	0.134
1967	0.244	0.244	0.198	0.688	1.04	3.65	0.354	0.068	0.697	0.473	0.394	0.558
1968	1.64	3.85	23.2	1.89	15.7	6.12	0.583	0.144	0.220	0.419	0.396	0.206
1969	0.259	0.555	2.66	12.1	21.2	0.745	0.674	0.30	1.56	0.917	0.459	1.94
1970	5.78	6.37	27.0	3.31	15.0	1.03	0.521	0.697	1.23	—	—	—

2-27. If the demand for a water supply from the Squannacook River (Problem 2-24) is 3.71 m^3/s, what size reservoir (in m^3) is required to provide this flow during the drought that occurs between June and December 1953? Work this problem by the mass balance method.

2-28. For the Clear Fork Trinity River (Problem 2-25) determine the storage required to prevent a flood of 10 m^3/s for the period December 1958 to November 1960. Assume the downstream discharge is constant at 1 m^3/s. Use a spreadsheet program you have written to solve this problem.

2-29. For the Clear Fork Trinity River (Problem 2-25) determine the storage required to prevent a flood of 15 m^3/s for the period from January 1953 to December 1953. Assume the downstream discharge is constant at 2 m^3/s. Use a spreadsheet program you have written to solve this problem.

2-30. Four monitoring wells have been placed around a leaking underground storage tank. The wells are located at the corners of a 100-ha square. The total piezometric head in each of the wells is as follows: NE corner, 30.0 m; SE corner, 30.0 m; SW corner, 30.6 m; NW corner, 30.6 m. Determine the magnitude and direction of the hydraulic gradient.

2-31. After a long wet spell, the water levels in the wells described in Problem 2-30 were measured and found to be the following distances from the ground surface: NE corner, 3.0 m; SE corner, 3.0 m; SW corner, 3.6 m; NW corner, 3.4 m. Assume that the ground surface is at the same elevation for each of the wells. Determine the magnitude and direction of the hydraulic gradient.

2-32. A gravelly sand has a hydraulic conductivity of 6.9×10^{-4} m/s, a hydraulic gradient of 0.00141, and a porosity of 20%. Determine the Darcy velocity and the average linear velocity.

2-33. Two piezometers have been placed along the direction of flow in a confined aquifer that is 30.0 m thick. The piezometers are 280 m apart. The difference in piezometric head between the two is 1.4 m. The aquifer hydraulic conductivity is 50 m/d and the porosity is 20%. Estimate the travel time for water to flow between the two piezometers.

2-34. A fully penetrating well in a 28.0 m thick artesian aquifer pumps at a rate of 0.00380 m^3/s for 1,941 days and causes a drawdown of 64.05 m at an observation well 48.00 m from the pumping well. How much drawdown will occur at an observation well 68.00 m away? The original piezometric surface was 94.05 m above the bottom confining layer. The aquifer material is sandstone. Report your answer to two decimal places.

Answer: $s_2 = 51.08$ m

2-35. It is undesirable to lower the piezometric surface of a confined aquifer below the aquiclude because this will destroy the structural integrity of the aquifer formation. Determine the maximum rate of pumping that would be permissible for the case described in Example 2-12 if the following conditions prevail:

1. The observation well at 200.0 m maintained the same drawdown.
2. Another observation well 2.0 m from the pumped well was used to observe lowering of the piezometric surface to the bottom of the aquiclude.

Report answer to two decimal places.

2-36. An artesian aquifer 5 m thick with a piezometric surface 65 m above the bottom confining layer is being pumped by a fully penetrating well. The aquifer is a mixture of sand and gravel. A steady-state drawdown of 7 m is observed at a nonpumping well located

10 m away. If the pumping rate is 0.020 m^3/s, how far away is a second nonpumping well that has an observed drawdown of 2 m?

2-37. For an unconfined aquifer, there is a possibility that drawdown will lower the piezometric surface to the bottom of the well and that the water will stop flowing. For the case described in Example 2-13, determine the maximum pumping rate that can be sustained indefinitely if the drawdown at the observation well 100.0 m from the pumped well is 9.90 m and the drawdown at the pumped well is limited by the depth of the aquifer, that is, 30.0 m. Report your answer to two decimal places.

Answer: $Q = 1.36\ m^3/s$

2-38. A contractor is trying to estimate the distance to be expected of a drawdown of 4.81 m from a pumping well under the following conditions:

Pumping rate = 0.0280 m^3/s
Pumping time = 1,066 d
Drawdown in observation well = 9.52 m
Observation well is located 10.00 m from the pumping well
Aquifer material = medium sand
Aquifer thickness = 14.05 m

Assume that the well is fully penetrating in an unconfined aquifer. Report your answer to two decimal places.

2-39. A well with a 0.25 m diameter fully penetrates an unconfined aquifer that is 20 m thick. The well has a discharge of 0.015 m^3/s and a drawdown of 8 m. If the flow is steady and the permeability is 1.5×10^{-4}, what is the height of the piezometric surface above the confining layer at a site 80 m from the well?

2-40. An aquifer yields the following results from pumping a 0.61 m diameter well at 0.0303 m^3/s: $s = 0.98$ m in 8 min; $s = 3.87$ m in 24 h. Determine its transmissibility. Report your answer to two decimal places.

Answer: $T = 4.33 \times 10^{-3}\ m^2/s$

2-41. An aquifer yields a drawdown of 1.04 m at an observation well 96.93 m from a well pumping at 0.0170 m^3/s after 80 min of pumping. The virtual time is 0.6 min and the transmissibility is $5.39 \times 10^{-3}\ m^2/s$. Determine the storage coefficient.

2-42. Using the data from Problem 2-41, find the drawdown at the observation well 80 days after pumping begins.

2-43. If the transmissibility is $2.51 \times 10^{-3}\ m^2/s$ and the storage coefficient is 2.86×10^{-4}, calculate the drawdown that will result at the end of two days of pumping a 0.5 m diameter well at a rate of 0.0194 m^3/s.

2-44. Determine the storage coefficient for an artesian aquifer from the pumping test results shown in the table below. The measurements were made at an observation well 300.00 m away from the pumping well. The pumping rate was 0.0350 m^3/s.

Time (min)	Drawdown (m)
100.0	3.10
500.0	4.70
1700.0	5.90

Answer: $S = 1.9 \times 10^{-5}$

2-45. Rework Problem 2-44, but assume that the data were obtained at an observation well 100.0 m away from the pumping well.

2-46. Determine the storage coefficient for an artesian aquifer from the pumping test results shown in the table below. The measurements were made at an observation well 100.00 m away from the pumping well. The pumping rate was 0.0221 m^3/s.

Time (min)	Drawdown (m)
10.0	1.35
100.0	3.65
1440.0	6.30

2-47. Rework Problem 2-46, but assume that the data were obtained at an observation well 60.0 m away from the pumping well.

2-48. Two wells located 106.68 m apart are both pumping at the same time. Well A pumps at 0.0379 m^3/s and well B pumps at 0.0252 m^3/s. The diameter of each well is 0.460 m. The transmissibility is 4.35×10^{-3} m^2/s and the storage coefficient is 4.1×10^{-5}. What is the interference of well A on well B after 365 days of pumping? Report your answer to two decimal places.

Answer: Interference of well A on B is 9.29 m.

2-49. Using the data from Problem 2-48, find the total drawdown in well B after 365 days of pumping. Report your answer to two decimal places.

2-50. If two wells, No. 12 and No. 13, located 100.0 m apart, are pumping at rates of 0.0250 m^3/s and 0.0300 m^3/s, respectively, what is the interference of well No. 12 on well No. 13 after 280 days of pumping? The diameter of each well is 0.500 m. The transmissibility is 1.766×10^{-3} m^2/s and the storage coefficient is 6.675×10^{-5}. Report your answer to two decimal places.

2-51. Using the data from Problem 2-50, find the total drawdown in well 13 after 280 days of pumping. Report your answer to two decimal places.

2-52. For the well field layout shown in Figure P-2-52, determine the effect of adding a sixth well. Is there any potential for adverse effects on the well or the aquifer? Assume all wells are pumped for 180 days and that each well is 0.914 m in diameter. Well data are given in the table below. Aquifer data are shown below Figure P-2-52.

Chug-a-lug Brewery Well Field No. 1

Well no.	Pumping rate (m^3/s)	Depth of well
1	0.0426	169.0
2	0.0473	170.0
3	0.0426	170.5
4	0.0404	168.84
5	0.0457	170.0
6 (proposed)	0.0473	170.0

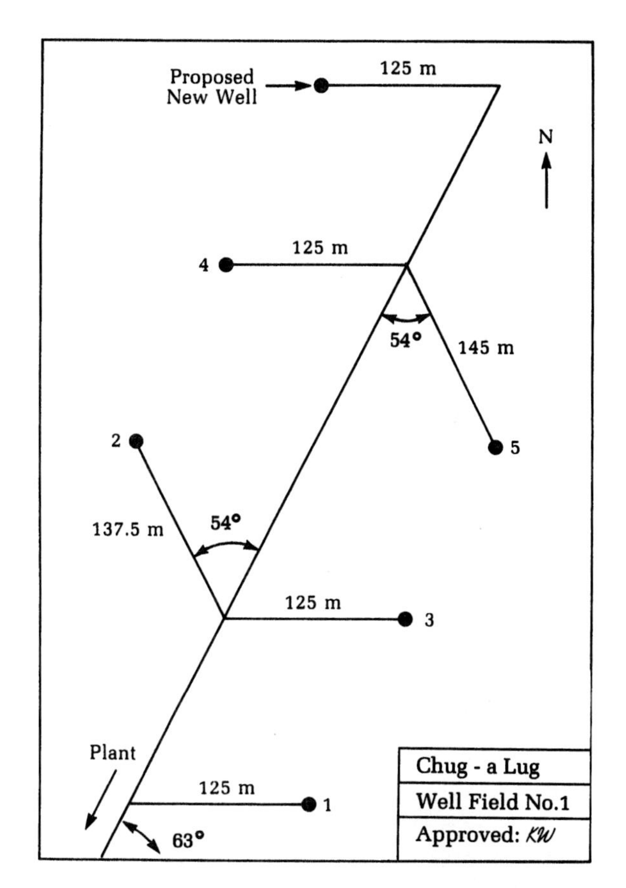

FIGURE P-2-52
Layout for well field.

The aquifer characteristics are as follows:

Storage coefficient $= 2.80 \times 10^{-5}$
Transmissibility $= 1.79 \times 10^{-3}$ m^2/s
Nonpumping water level = 7.60 m below grade
Depth to top of artesian aquifer = 156.50 m

2-53. What pumping rate, pumping time, or combination thereof, can be sustained by the new well in Problem 2-52 if all of the well diameters are enlarged to 1.80 m?

2-54. For the well field layout shown in Figure P-2-54, determine the effect of adding a sixth well. Is there any potential for adverse effects on the well or the aquifer? Assume all wells are pumped for 100 days and that each well is 0.300 m in diameter. Aquifer data are given below. Well data are shown below Figure P-2-54.

The aquifer characteristics are as follows:

Storage coefficient $= 2.11 \times 10^{-6}$
Transmissibility $= 4.02 \times 10^{-3}$ m^2/s
Nonpumping water level = 9.50 m below grade
Depth to top of artesian aquifer = 50.1 m

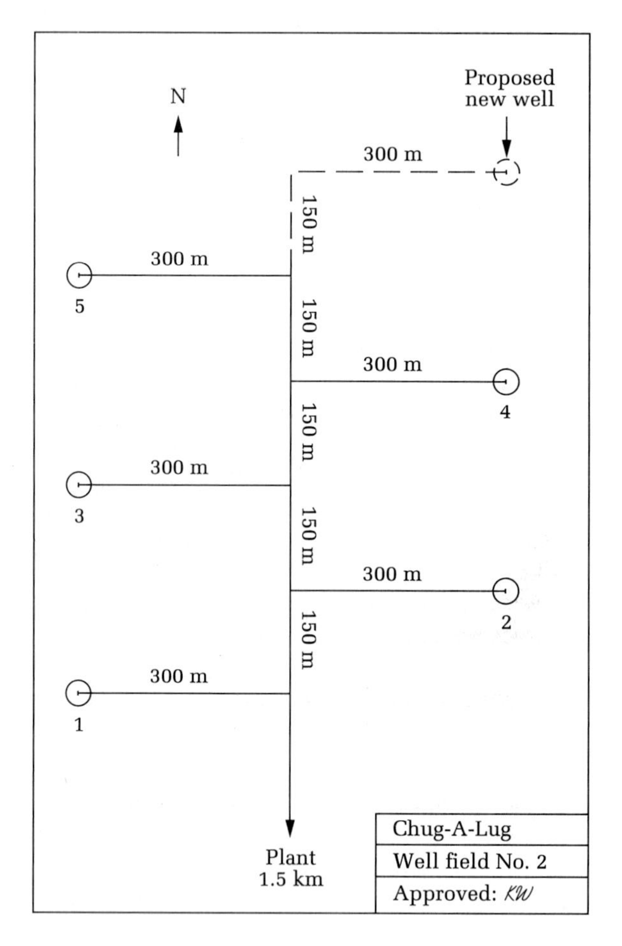

FIGURE P-2-54
Layout for well field.

Chug-a-lug Brewery Well Field No. 2

Well no.	Pumping rate (m^3/s)	Depth of well
1	0.020	105.7
2	0.035	112.8
3	0.020	111.2
4	0.015	108.6
5	0.030	113.3
6 (proposed)	0.025	109.7

2-55. What pumping rate, pumping time, or combination thereof, can be sustained by the new well in Problem 2-54 if all of the well diameters are enlarged to 1.50 m?

2-9 DISCUSSION QUESTIONS

2-1. An artesian aquifer is under pressure because of the weight of the overlying geologic strata. Is this sentence true or false? If it is false, rewrite the sentence to make it true.

2-2. Identify the base flow in the following hydrographs.

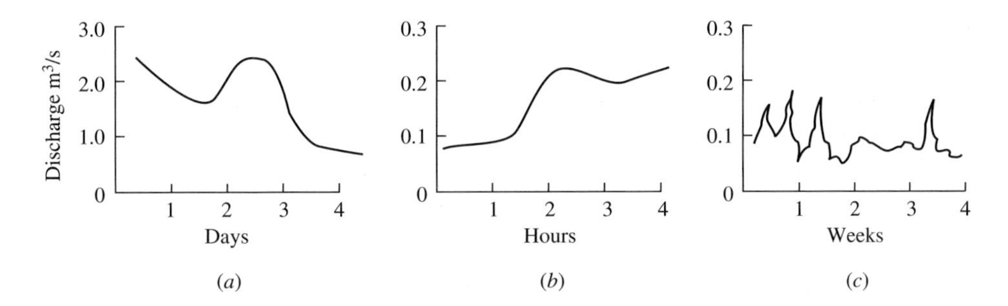

2-3. As a field engineer you have been asked to estimate how long you would have to measure the discharge from a mall parking lot before the maximum discharge would be achieved. What data would you have to gather to make the estimate?

2-4. Explain why it is impossible to use the rational formula without being able to estimate the time of concentration.

2-5. When a flood has a recurrence interval (return period) of 5 years, it means that the chance of another flood of the same or less severity occurring next year is 5 percent. Is this sentence true or false? If it is false, rewrite the sentence to make it true.

2-6. The hydrologic year used for data representation of rainfall or runoff flow events is from September 30 to October 1. Is this sentence true or false? If it is false, rewrite the sentence to make it true.

2-7. For the following well boring log, identify the pertinent hydrogeologic features. The well screen is set at 6.0–8.0 m and the static water level after drilling is 1.8 m from the ground surface.

Strata	Depth, m	Remarks
Top soil	0.0–0.5	
Sandy till	0.5–6.0	Water encountered at 1.8 m
Sand	6.0–8.0	
Clay	8.0–9.0	
Shale	9.0–10.0	Well terminated

2-8. For the following well boring log (Bracebridge, Ontario, Canada), identify the pertinent hydrogeologic features. The well screen is set at 48.0–51.8 m and the static water level after drilling is 10.2 m from the ground surface.

Strata	Depth, m	Remarks
Sand	0.0–6.1	
Gravelly clay	6.10–8.6	
Fine sand	8.6–13.7	
Clay	13.7–17.5	Casing sealed
Fine sand	17.5–51.8	
Bedrock	51.8	Well terminated

2-9. Sketch the piezometric profiles for two wells that interfere with one another. Well A pumps at 0.028 m^3/s and well B pumps at 0.052 m^3/s. Show the ground water table before pumping, the drawdown curve of each well pumping alone, and the resultant when both wells are operated together.

2-10 ADDITIONAL READING

Books

H. Bouwer, *Groundwater Hydrology,* New York: McGraw-Hill; 1978.

V. T. Chow, D. R. Maidment, and L. W. Mays, *Applied Hydrology,* New York: McGraw-Hill, 1988.

V. T. Chow, *Handbook of Applied Hydrology,* New York: McGraw-Hill, 1964.

Johnson Division, UOP, *Ground Water and Wells,* St. Paul, MN: Johnson Division, UOP, 1975.

R. K. Linsley, Jr., and J. B. Franzini, *Water Resources Engineering,* New York: McGraw-Hill, 1979.

R. K. Linsley, Jr., M. A. Kohler, and J. L. E. Paulhus, *Hydrology for Engineers,* New York: McGraw-Hill, 1975.

R. H. McCuen, *Hydrologic Analysis and Design,* Englewood Cliffs, NJ: Prentice-Hall, 1989.

V. M. Ponce, *Engineering Hydrology Principles and Practices,* Englewood Cliffs, NJ: Prentice-Hall, 1989.

W. Viessman, G. L. Lewis, and J. W. Knapp, *Introduction to Hydrology,* 3rd ed., New York: Harper & Row, 1989.

Journals

Water Resources Bulletin

Water Resources Research

CHAPTER 3

WATER TREATMENT

3-1 INTRODUCTION
Water Chemistry
Reaction Kinetics
Water Quality
Physical Characteristics
Chemical Characteristics
Microbiological Characteristics
Radiological Characteristics
Water Quality Standards
Water Classification and Treatment Systems
3-2 COAGULATION
Colloid Stability
Colloid Destabilization
Coagulation
3-3 SOFTENING
Lime-Soda Softening
More Advanced Concepts in Lime-Soda Softening
Ion-Exchange Softening
3-4 REACTORS

3-5 MIXING AND FLOCCULATION
Rapid Mix
Flocculation
3-6 SEDIMENTATION
Overview
Sedimentation Concepts
Determination of v_s
Determination of v_o
3-7 FILTRATION
Grain Size Characteristics
Filter Hydraulics
3-8 DISINFECTION
Disinfection Kinetics
Chlorine Reactions in Water
Chlorine-Disinfecting Action
Chlorine/Ammonia Reactions
Practices of Water Chlorination
Chlorine Dioxide
Ozonation
Ultraviolet Radiation
Advanced Oxidation Processes (AOPs)
3-9 ADSORPTION
3-10 WATER PLANT WASTE MANAGEMENT
Sludge Production and Characteristics
Minimization of Sludge Generation
Sludge Treatment
Ultimate Disposal
3-11 CHAPTER REVIEW
3-12 PROBLEMS
3-13 DISCUSSION QUESTIONS
3-14 ADDITIONAL READING

3-1 INTRODUCTION

Approximately 80 percent of the United States population turns their taps on every day to take a drink of publicly supplied water. They all assume that when they take a drink it is safe. They probably never even think of safety.

In the United States there are 215,000 water systems that are classified as public systems. Of these, 57,000 are community systems and 158,000 are noncommunity systems. Community water systems include, for example, cities, townships, subdivisions, and trailer parks. Noncommunity water systems include restaurants, motels, campgrounds, service stations, and other systems serving transient populations. Community systems serve 91 percent of the U.S. population.

In the period 1985 through 1994, fewer than 25 waterborne disease outbreaks occurred each year.[1] In developing nations, clean water is the exception rather than the rule.

> . . . Something like 40 percent of the human race does not have adequate access to safe water. Waterborne diseases are estimated to kill more than 25,000 people daily. Schistosomiasis and filariasis, the world's largest causes of blindness, affect—according to one estimate—some 450 million people in more than 70 nations. There are, economist Barbara Ward has said, cities in the developing world where 60 percent of the children born die of infantile gastritis before the age of five . . .[2]

The fact that the United States and developed countries have an outstanding water supply record is no accident. Since the early 1900s sanitary engineers have been working in the United States to reduce waterborne disease. Figure 3-1 shows the incidence of typhoid cases in Philadelphia from 1890 to 1935. This is typical of the decrease in waterborne disease as public water supply treatment has increased. Philadelphia received its water from rivers and distributed the water untreated until about 1906, when slow sand filters were put into use. An immediate reduction

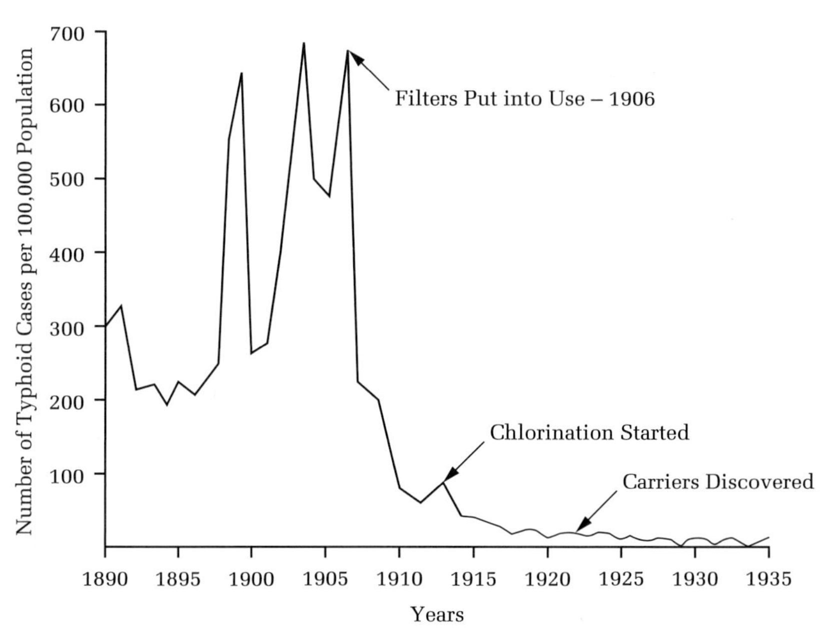

FIGURE 3-1
Typhoid fever cases per 100,000 population from 1890 to 1935, Philadelphia.

---

[1] M.H. Kramer, *et al.*, "Waterborne Disease: 1993 and 1994," *Journal of American Water Works Association,* vol. 88, no. 3, pp. 66–80, March, 1996.

[2] Excerpted from remarks by former EPA Administrator Russel E. Train that were delivered before the Los Angeles World Affairs Forum, December 16, 1976.

in typhoid fever was realized. Disinfection of the water by the addition of chlorine further decreased the number of typhoid cases. A still greater decrease was accomplished after 1920 by careful control over infected persons who had become carriers. Since 1952 the death rate from typhoid fever in the United States has been less than 1 per 1,000,000. Many countries and organizations have committed their technological and financial resources to developing nations in recognition of the principle that reasonable access to safe and adequate drinking water is a fundamental right of all people.

A water that can be consumed in any desired amount without concern for adverse health effects is termed a *potable* water. Potable does not necessarily mean that the water tastes good. This is in contrast to a *palatable* water, which is one that is pleasing to drink, but not necessarily safe. We have learned that we must provide a water that is both potable and palatable, for if it is not palatable people will turn to untreated water that may not be potable. The widespread availability of potable water in the United States does not mean that there are no operational or control deficiencies in water systems. As populations increase, so must water production. Often increased production requires the use of new water sources that contain higher contaminant levels. As production increases while plant capacity remains the same, the task of producing potable water becomes increasingly difficult. The scientific community is continually making advances in identifying contaminants and discovering potential long-term health effects of constituents that had not been previously identified. At the same time that some scientists are assessing existing pollutants, other scientists are producing new chemicals at a tremendous rate.

Water Chemistry

An understanding of the fundamentals of water chemistry is essential to the comprehension of the chapters on water quality and wastewater treatment. You should plan to spend ample time working the examples in this section.

Physical properties of water. The basic physical properties of water relevant to water treatment are density and viscosity. Density is a measure of the concentration of matter and is expressed in three ways:

1. Mass density, ρ. *Mass density* is mass per unit volume and is measured in units of kg/m^3. Appendix A, Table A-1, shows the variation of density with temperature for pure water free from air. Dissolved impurities change the density in direct proportion to their concentration and their own density. In environmental engineering applications, it is common to ignore the density increase due to impurities in the water. However, environmental engineers do not ignore the density of the matter when dealing with high concentrations, such as thickened sludge or commercial liquid chemicals.
2. Specific weight, γ. *Specific weight* is weight (force) per unit volume, measured in units of kN/m^3. The specific weight of a fluid is related to its density by the acceleration of gravity, g, which is 9.81 m/s^2.

$$\gamma = \rho g \tag{3-1}$$

3. Specific gravity, S. Specific gravity is given by

$$S = \rho/\rho_0 = \gamma/\gamma_0 \tag{3-2}$$

where the subscript zero denotes the density of water at 3.98°C, 1,000 kg/m^3, and the specific weight of water, 9.81 kN/m^3.

For quick approximations, the density of water at normal temperature is taken as 1,000 kg/m^3 (which is conveniently 1 kg/L) with a specific gravity = 1.

All substances, including liquids, exhibit a resistance to movement, an internal friction. The higher the friction, the harder it is to pump the liquid. A measure of the friction is viscosity. Viscosity is presented in one of two ways:

1. Dynamic viscosity, or absolute viscosity, μ, has dimensions of mass per unit length per time, with units of Pa · s.

2. Kinematic viscosity, ν, is found by

$$\nu = \mu/\rho \tag{3-3}$$

and has dimensions of length squared per time with the corresponding units m^2/s.

States of solution impurities. From an environmental engineering point of view, substances can exist in water in one of three classifications—*suspended, colloidal,* or *dissolved.*

A dissolved substance is one which is truly in solution. The substance is homogeneously dispersed in the liquid. Dissolved substances can be simple atoms or complex molecular compounds. Dissolved substances are in the liquid, that is, there is only one phase present. The substance cannot be removed from the liquid without accomplishing a phase change such as distillation, precipitation, adsorption, or extraction. In *distillation* either the liquid or the substance itself is changed from a liquid phase to a gas phase in order to achieve separation. In *precipitation* the substance in the liquid phase combines with another chemical to form a solid phase, thus achieving separation from the water. *Adsorption* also involves a phase change, wherein the dissolved substance reacts with a solid particle to form a solid particle-substance complex. *Liquid extraction* can separate a substance from water by extracting it into another liquid, hence a phase change from water to a different liquid. But under no circumstances can physical methods such as filtration, sedimentation, or centrifugation remove dissolved substances.

Suspended solids are large enough to settle out of solution or be removed by filtration. In this case there are two phases present, the liquid water phase and the suspended-particle solid phase. The lower size range of this class is 0.1 to 1.0μm, about the size of bacteria. In environmental engineering, suspended solids are defined as those solids that can be filtered by a glass fiber filter disc and are properly called filterable solids. Suspended solids can be removed from water by physical methods such as sedimentation, filtration, and centrifugation.

Colloidal particles are in the size range between dissolved substances and suspended particles. They are in a solid state and can be removed from the liquid by physical means such as very high-force centrifugation or filtration through membranes with very small pore spaces. However, the particles are too small to be removed by sedimentation or by normal filtration processes. Colloidal particles exhibit the Tyndall effect; that is, when light passes through a liquid containing colloidal particles, the light is reflected by the particles. The degree to which a colloidal suspension reflects light at a 90° angle to the entrance beam is measured by *turbidity.* Turbidity is a relative measure, and there are various standards against which a sample is compared. The different standards employed are similar, although not exactly the same. For our purposes we will simply refer to the measure of turbidity as a turbidity unit (TU). For a given particle size, the higher the turbidity, the higher the concentration of colloidal particles.

Another useful term in environmental engineering that is used to describe a solution state is *color*. Color is not separate from the above three categories, but rather is a combination of dissolved and colloidal materials. Color is widely used in environmental engineering because it, in itself, can be measured. However, it is very difficult to distinguish "dissolved color" from "colloidal color." Some color is caused by colloidal iron or manganese complexes, although the most common cause of color is from complex organic compounds that originate from the decomposition of organic matter. One common source of color is the degradation of soil humus, which produces humic acids. Humic acids impart reddish-brown color to the water. Humic acids have molecular weights between 800 and 50,000, the lower being dissolved and the greater, colloidal. Most color seems to be between 3.5 and 10μm, which is colloidal. Color is measured by the ability of the solution to absorb light. Color particles can be removed by the methods discussed for dissolved or colloidal particles, depending upon the state of the color.

Figure 3-2[3] presents an overview by size of the types of particles that are often dealt with in water treatment. A relatively new technique that is being used in water treatment to help evaluate water quality is *particle counting*. A particle counter counts the number of particles in a water sample and reports the results by particle size, generally from 1 to 30μm. Figure 3-3 shows a sample count comparing the distribution of particles in a raw water to that of the finished water. While particle counting does not indicate anything about the kind of particle, it can be useful in assessing overall treatment efficiency as well as characterizing water sources.

Chemical units. Since solutes in solution are often analyzed by weight, the terms *weight percent* and *milligram per liter* are used. In order to perform stoichiometric calculations,[4] it is necessary to convert to common units, and the terms *molarity* and *normality* are used.

Weight percent, P, is sometimes employed to express approximate concentrations of commercial chemicals or of solid concentrations of sludges. The term

[3] N. McTigue, and D. Cornwell, "The Use of Particle Counting for the Evaluation of Filter Performance," AWWA Seminar Proceedings, *Filtration: Meeting New Standards,* AWWA Conference, 1988.

[4] *Stoichiometry* is the part of chemistry concerned with measuring the proportions of elements involved in a reaction. Stoichiometric calculations are an application of the principle of conservation of mass to chemical reactions.

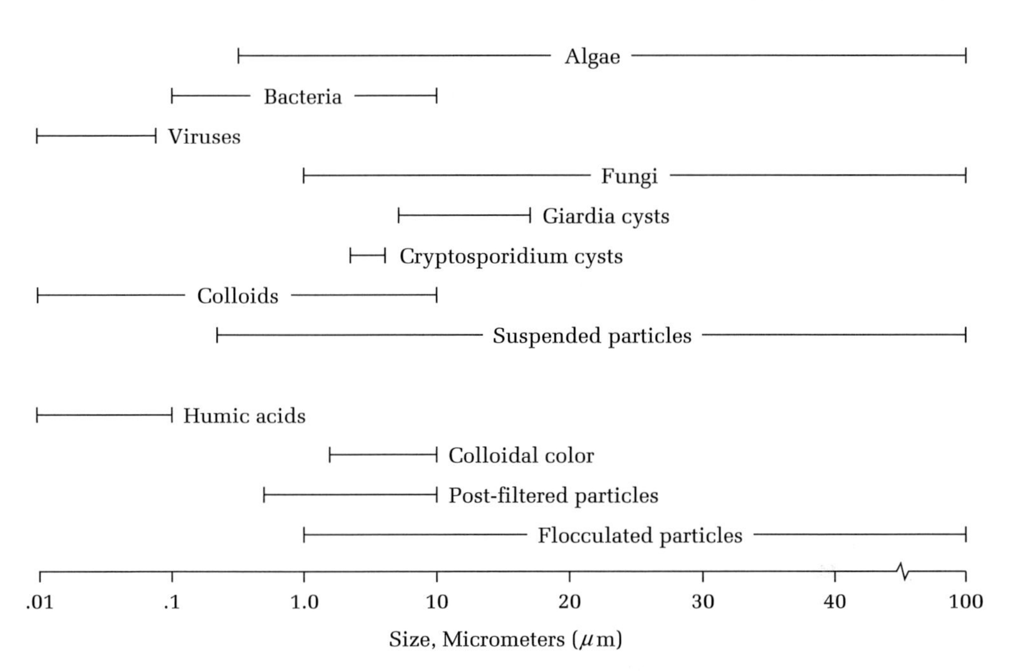

FIGURE 3-2
Particulates in water.

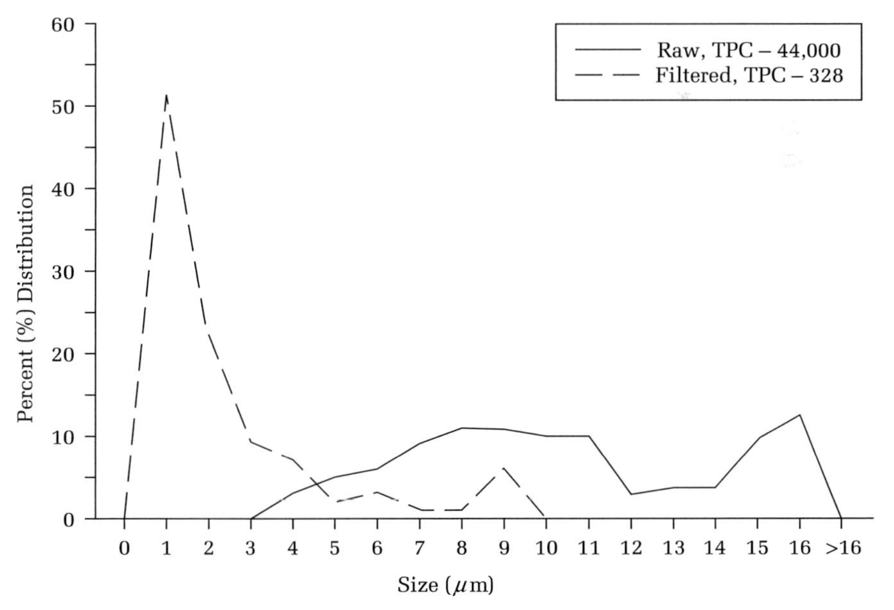

FIGURE 3-3
Particle distribution changes through treatment.

specifies the grams of substance per 100 grams of solution and is mathematically expressed as

$$P = \frac{W}{W + W_0} \times 100\% \tag{3-4}$$

where P = percent of substance by weight
W = grams of substance
W_0 = grams of solution

Analysts generally give results directly in mass per volume (concentration), and the units are mg/L. In environmental engineering it is often assumed that the substance does not change the density of the water. This is generally untrue, but it does make for some useful conversions, and the assumption is not too inaccurate for dilute concentrations. If such an assumption is made and we recall that 1 mL of water weighs 1 g (again an approximation), then

$$\frac{1 \text{ mg}}{\text{L}} = \frac{10^{-3} \text{ mL}}{\text{L}} = \frac{10^{-3} \text{ mL}}{10^3 \text{ mL}} = \frac{1 \text{ mL}}{10^6 \text{ mL}} = 1 \text{ ppm} \tag{3-5}$$

or 1 mg/L equals 1 part per million (ppm). If the same assumptions are made, then the weight percent of 1 mg/L can be determined:

$$P = \frac{W}{W + W_0} \times 100 = \frac{1 \text{ mg}(100)}{1 \text{ L}} = \frac{10^{-3} \text{ g}(100)}{10^3 \text{ g}} = 1 \times 10^{-4}\% \tag{3-6}$$

or 1 mg/L equals $1 \times 10^{-4}\%$, which can be translated into 1% = 10,000 mg/L.

In order to work with chemical reactions it is necessary to convert weight concentrations to molarity or normality. A *mole* is 6.02×10^{23} molecules of a substance. Chemical reactions are expressed in integral numbers of moles. A mole of a substance has a relative weight called its *molecular weight* (MW). Molecular weight is the sum of the atomic weights. A table of atomic weights is given inside the front cover of this book. *Molarity* is the number of moles in a liter of solution. A 1-molar (1 *M*) solution has 1 mole of substance per liter of solution. Molarity is related to mg/L by

$$\begin{aligned} \text{mg/L} &= \text{molarity} \times \text{molecular weight} \times 10^3 \\ &= (\text{moles/L})\ (\text{g/mole})\ (10^3 \text{ mg/g}) \end{aligned} \tag{3-7}$$

A second unit, *equivalent weight* (EW), is frequently used in softening and redox reactions. The equivalent weight is the molecular weight divided by the number (n) of electrons transferred in redox reactions or the number of protons transferred in acid/base reactions.

The value of n depends on how the molecule reacts. In this text we are concerned with molecules that react in acid/base reactions or precipitation reactions. **In an acid/base reaction, *n* is the number of hydrogen ions that the molecule transfers. That is, an acid gives up an EW of hydrogen ions, and a base accepts an EW of hydrogen ions. In a precipitation reaction, *n* is the valence of the element in question. For compounds, *n* is equal to the number of hydrogen ions that would be required to replace the cation; that is, for $CaCO_3$ it would take two hydrogen**

ions to replace the calcium, therefore, $n = 2$. In oxidation/reduction reactions, n is equal to the change in oxidation number that the compound undergoes in the reaction. Obviously, it is difficult to recognize reaction capacity without the context of the reaction. Common valence states of elements found in water are listed in Appendix A.

Normality (N) is the number of equivalent weights per liter of solution and is related to molarity (M) by

$$N = Mn \tag{3-8}$$

Example 3-1. Commercial sulfuric acid, H_2SO_4, is often purchased as a 93 weight percent solution. Find the mg/L of H_2SO_4 and the molarity and normality of the solution. Sulfuric acid has a specific gravity of 1.839.

Solution. Since 1 L of water weighs 1,000 g, 1 L of 100% H_2SO_4 weighs

$$1{,}000(1.839) = 1{,}839 \text{ g}$$

93% (1,839) g $= 1{,}710$ g of H_2SO_4, or 1.7×10^6 mg/L of H_2SO_4 in a 93% solution. The molecular weight of H_2SO_4 is found by looking up the atomic weights on the inside cover of this book:

$$\begin{aligned} 2\text{H} &= 2(1) &&= 2 \\ \text{S} &= && 32 \\ 4\text{O} &= 4(16) &&= \underline{64} \\ & && 98 \text{ g/mole} \end{aligned}$$

The molarity is found by using Equation 3-7:

$$\frac{1{,}710 \text{ g/L}}{98 \text{ g/mole}} = 17.45 \text{ mole/L or } 17.45\ M$$

The normality is found from Equation 3-8, realizing that H_2SO_4 can give up two hydrogen ions and therefore $n = 2$ equivalents/mole:

$$N = 17.45 \text{ mole/L } (2 \text{ equiv/mole}) = 34.9 \text{ equiv/L}$$

Example 3-2. Find the weight of sodium bicarbonate, $NaHCO_3$, necessary to make a 1 M solution. Find the normality of the solution.

Solution. The molecular weight of $NaHCO_3$ is 84; therefore by using Equation 3-7:

$$\text{mg/L} = (1 \text{ mole/L})(84 \text{ g/mole})(10^3 \text{ mg/g}) = 84{,}000$$

HCO_3^- is able to give or accept only one proton; therefore $n = 1$, and the normality is the same as the molarity.

Example 3-3. Find the equivalent weight of each of the following: Ca^{2+}, CO_3^{2-}, $CaCO_3$.

Solution. Equivalent weight was defined as

$$\text{EW} = \frac{\text{Atomic or molecular weight}}{n}$$

The units of EW are grams/equivalent (g/eq) or milligrams/milliequivalent (mg/meq).

For calcium, n is equal to the valence or oxidation state in water, so $n = 2$. From the table on the inside cover of the book, the atomic weight of Ca^{2+} is 40.08. The equivalent weight is then

$$EW = \frac{40.08}{2} = 20.04 \text{ g/eq or } 20.04 \text{ mg/meq}$$

For the carbonate ion (CO_3^{2-}) the oxidation state of 2^- is used for n since the base CO_3^{2-} can potentially accept 2 hydrogen ions (H^+). The molecular weight is

$$\begin{aligned} C &= && 12.01 \\ 3O &= 3(15.9994) &=& \underline{48.00} \\ &&& 60.01 \end{aligned}$$

and the equivalent weight is

$$EW = \frac{60.01}{2} = 30.00 \text{ g/eq or } 30.00 \text{ mg/meq}$$

In $CaCO_3$, $n = 2$ since it would take two hydrogen ions to replace the cation (Ca^{2+}) to form carbonic acid, H_2CO_3. Its molecular weight is the sum of the atomic weights of Ca^{2+} and CO_3^{2-} and is, therefore, equal to $40.08 + 60.01 = 100.09$. Its equivalent weight is

$$EW = \frac{100.09}{2} = 50.04 \text{ g/eq or mg/meq}$$

Chemical reactions. There are four principal types of reactions of importance in environmental engineering: precipitation, acid/base, ion-association, and oxidation/reduction.

Dissolved ions can react with each other and form a solid compound. This phase-change reaction of dissolved to solid state is called a precipitation reaction. Typical of a precipitation reaction is the formation of calcium carbonate when a solution of calcium is mixed with a solution of carbonate:

$$Ca^{2+} + CO_3^{2-} \rightleftharpoons CaCO_3(s) \tag{3-9}$$

The (s) in the above reaction denotes that the $CaCO_3$ is in the solid state. When no symbol is used to designate state, it is assumed to be dissolved. The arrows in the reaction imply that the reaction is reversible and so could proceed to the right (that is, the ions are combining to form a solid) or to the left (that is, the solid is dissociating into the ions).

Often, out of convenience, we talk about compounds when in reality a compound does not exist in water. Take, for example, a water containing sodium chloride and calcium sulfate. We would say that the water has NaCl and $CaSO_4$ in it, but no implication is made regarding the association of Na and Cl or Ca and SO_4. The following reactions occur:

$$CaSO_4(s) \rightleftharpoons Ca^{2+} + SO_4^{2-} \tag{3-10}$$

and

$$NaCl(s) \rightleftharpoons Na^+ + Cl^- \tag{3-11}$$

such that the water consists of four unassociated ions: Na^+, Ca^{2+}, Cl^-, and SO_4^{2-}. Don't make the mistake of thinking that the sodium and chloride are together.

Acid/base reactions are a special type of ionization when a hydrogen ion is added to or removed from solution. An acid could be added to water to produce a hydrogen ion, as by the addition of hydrochloric acid to water with the reaction

$$HCl \rightleftharpoons H^+ + Cl^- \tag{3-12}$$

The above reaction is simplified in that it is assumed that water is present. The reaction is properly written

$$HCl + H_2O \rightleftharpoons H_3O^+ + Cl^- \tag{3-13}$$

A hydrogen ion could also be removed from water, as by the addition of a base:

$$NaOH + H_3O^+ \rightleftharpoons 2H_2O + Na^+ \tag{3-14}$$

In some cases, ions may exist in water complexed with other ions. Formation of dissolved complexes are ion-association reactions. In this case, the ions are "tied" together in the solution. The complex could be a neutral compound, such as soluble mercuric chloride:

$$Hg^{2+} + 2Cl^- \rightleftharpoons HgCl_2 \tag{3-15}$$

More often, the soluble complex has a charge and is itself an ion. Metal ion complexes are common examples:

$$Al^{3+} + OH^- \rightleftharpoons AlOH^{2+} \tag{3-16}$$

The $AlOH^{2+}$ is still soluble, but acts differently than did the individual species before complexation.

Oxidation/reduction reactions involve valence changes and the transfer of electrons. When iron metal corrodes, it releases electrons:

$$Fe^0 \rightleftharpoons Fe^{2+} + 2e^- \tag{3-17}$$

If one element releases electrons, then another must be available to accept the electrons. In iron pipe corrosion, hydrogen gas is often produced:

$$2H^+ + 2e^- \rightleftharpoons H_2(g) \tag{3-18}$$

where the symbol (g) indicates the hydrogen is in the gas phase.

Precipitation reactions. All complexes are soluble in water to a certain extent. Likewise, all complexes are limited by how much can be dissolved in water. Some compounds, such as NaCl, are very soluble; other compounds, such as AgCl, are very insoluble—only a small amount will go into solution. Visualize a solid compound being placed in distilled water. Some of the compound will go into solution. At some time no more of the compound will dissolve, and equilibrium will be reached. The time to reach equilibrium may be seconds or centuries. The solubility reaction is written as follows:

$$A_aB_b(s) \rightleftharpoons aA^{b+} + bB^{a-} \tag{3-19}$$

For example,

$$Ca_3(PO_4)_2(s) \rightleftharpoons 3Ca^{2+} + 2PO_4^{3-} \tag{3-20}$$

TABLE 3-1
Selected solubility constants at 25° C

Substance	Equilibrium equation	pK_s	Application
Aluminum hydroxide	$Al(OH)_3(s) \rightleftharpoons Al^{3+} + 3OH^-$	32.9	Coagulation
Aluminum phosphate	$AlPO_4(s) \rightleftharpoons Al^{3+} + PO_4^{3-}$	22.0	Phosphate removal
Calcium carbonate	$CaCO_3(s) \rightleftharpoons Ca^{2+} + CO_3^{2-}$	8.305	Softening, corrosion control
Ferric hydroxide	$Fe(OH)_3(s) \rightleftharpoons Fe^{3+} + 3OH^-$	38.57	Coagulation, iron removal
Ferric phosphate	$FePO_4(s) \rightleftharpoons Fe^{3+} + PO_4^{3-}$	21.9	Phosphate removal
Magnesium hydroxide	$Mg(OH)_2(s) \rightleftharpoons Mg^{2+} + 2OH^-$	11.25	Softening

Interestingly, the product of the activity of the ions (approximated by the molar concentration) is always a constant for a given compound at a given temperature. This constant is called the solubility constant, K_s. In the general form it is written as

$$K_s = [A]^a[B]^b \tag{3-21}$$

where, in this text, [] denotes *molar* concentrations. **Do not use mg/L!** A table of constants is shown in Table 3-1 and in Appendix A. K_s values are often reported as pK_s, where

$$pK_s = -\log K_s \tag{3-22}$$

The constant works equally well whether we are dissolving a solid (reaction going to the right) or precipitating ions (reaction going to the left). If we place $A_aB_b(s)$ in water, for every a moles of A that dissolve, b moles of B will dissolve until equilibrium is reached. But kinetically[5] it might take years to happen.

When precipitating ions, it is possible to have a higher concentration of ions in solution than dictated by the solubility product. This is called a supersaturated solution.

Example 3-4. How many mg/L of PO_4^{3-} would be in solution at equilibrium with $AlPO_4(s)$?

Solution. The pertinent reaction is

$$AlPO_4(s) \rightleftharpoons Al^{3+} + PO_4^{3-}$$

The associated pK_s is found in Table 3-1 as 22.0 and calculated as

$$K_s = 10^{-22} = [Al][PO_4]$$

For every mole of $AlpO_4$ that dissolves, one mole of Al^{3+} and one mole of PO_4^{3-} are released into solution. At equilibrium, the molar concentration of Al^{3+} and PO_4^{3-} in

[5] *Kinetics* is the part of chemistry concerned with rates of reactions and factors that affect them.

solution will be equal, so we may say

$$[Al^{3+}] = [PO_4^{3-}] = X$$

Substituting X for each compound in the K_s expression,

$$10^{-22} = X^2$$

Solving for X (which is equal to PO_4^{3-}), we find $PO_4^{3-} = 10^{-11}$ moles per liter in solution. The molecular weight is 95 g/mole, so the concentration in mg/L is

$$(95 \text{ g/mole})(10^3 \text{ mg/g})(10^{-11} \text{ moles/L}) = 9.5 \times 10^{-7} \text{ mg/L}$$

Example 3-5. If 50.0 mg of CO_3^{2-} and 50.0 mg of Ca^{2+} are present in 1 L of water, what will be the final (equilibrium) concentration of Ca^{2+}?

Solution. The molecular weight of Ca^{2+} is 40.08 and that of CO_3^{2-} is 60.01, resulting in initial molar concentrations of 1.25×10^{-3} moles/L and 8.33×10^{-4} moles/L for Ca^{2+} and CO_3^{2-} respectively.

$$K_s = 10^{-pK_s} = 10^{-8.305} = [Ca^{2+}][CO_3^{2-}]$$

For every mole of Ca^{2+} that is removed from solution, one mole of CO_3^{2-} is removed from solution. If the amount removed is given by Z, then

$$10^{-8.305} = 4.95 \times 10^{-9} = [1.25 \times 10^{-3} - Z][8.33 \times 10^{-4} - Z]$$

$$1.04 \times 10^{-6} - (2.08 \times 10^{-3})\,Z + Z^2 = 0$$

$$Z = \frac{-b \pm \sqrt{b^2 - 4ac}}{2a}$$

$$= \frac{2.08 \times 10^{-3} \pm \sqrt{4.34 \times 10^{-6} - 4(1.04 \times 10^{-6})}}{2}$$

$$= 8.28 \times 10^{-4}$$

so that the final Ca^{2+} concentration is

$$[Ca^{2+}] = 1.25 \times 10^{-3} - 8.28 \times 10^{-4} = 4.22 \times 10^{-4}\, M$$

or

$$(4.22 \times 10^{-4} \text{ moles/L})(40 \text{ g/mole})(10^3 \text{ mg/g}) = 16.9 \text{ mg/L}$$

Acid/base reactions. For the purposes of this text, acids are defined as those compounds that release protons. Bases are those compounds that accept protons.

The simple reaction for the release of a proton is

$$HA \rightleftharpoons H^+ + A^- \qquad (3\text{-}23)$$

In order for HA to release the proton (H^+), something must accept the proton. Often that something is water, that is,

$$H^+ + H_2O \rightleftharpoons H_3O^+ \qquad (3\text{-}24)$$

resulting in the net reaction

$$HA + H_2O \rightleftharpoons H_3O^+ + A^- \tag{3-25}$$

It is understood that water is generally present. Hence Equation 3-23 is used in place of Equation 3-25. In the case of Equation 3-25, water is acting as the base; that is, it accepts the proton. If a base is added to water, the water can act as an acid.

$$B^- + H_2O \rightleftharpoons HB + OH^- \tag{3-26}$$

In the above reaction the base (B^-) accepts a proton from water. If a compound is a stronger acid than water, then water will act as a base. If a compound is a stronger base than water, then water will act as an acid.

You can quickly see that acid/base chemistry centers on water and that it is important to know how strong an acid water is. Water itself is ionized in water by the equation

$$H_2O \rightleftharpoons H^+ + OH^- \tag{3-27}$$

The degree of ionization of water is very small and can be measured by what is called the ion product of water, K_w. It is found by

$$K_w = [OH^-][H^+] \tag{3-28}$$

and has a value of 10^{-14} ($pK_w = 14$) at 25°C. A solution is said to be acidic if $[H^+]$ is greater than $[OH^-]$, neutral if equal, and basic if $[H^+]$ is less than $[OH^-]$. If the solution is neutral, then $[H^+] = [OH^-] = 10^{-7}$ M. If the solution is acidic, H^+ is greater than 10^{-7} M. A convenient expression for the hydrogen ion concentration is pH, given by

$$pH = -\log [H^+] \tag{3-29}$$

Therefore, a neutral solution at 25°C has a pH of 7 (written pH 7), an acidic solution has a pH < 7, and a basic solution has a pH > 7.

Acids are classified as strong acids or weak acids. *Strong acids* have a tendency to donate their protons to water. For example,

$$HCl \rightarrow H^+ + Cl^- \tag{3-30}$$

which we recall is the simplified form of

$$HCl + H_2O \rightarrow H_3O^+ + Cl^- \tag{3-31}$$

A list of important strong acids is in Table 3-2. Note the use of the single arrow to signify that, for practical purposes, we may assume that the reaction proceeds completely to the right.

Example 3-6. If 100 mg of H_2SO_4 (MW = 98) is added to 1 L of water, what is the final pH?

Solution. Using the molecular weight of sulfuric acid we find

$$\left(\frac{100 \text{ mg}}{1 \text{ L } H_2O}\right)\left(\frac{1}{98 \text{ g/mole}}\right)\left(\frac{1}{10^3 \text{ mg/g}}\right) = 1.02 \times 10^{-3} \text{ mole/L}$$

The reaction is

$$H_2SO_4 \rightarrow 2H^+ + SO_4^{2-}$$

and therefore $2(1.02 \times 10^{-3})M$ H^+ is produced. The pH is

$$pH = -\log(2.04 \times 10^{-3}) = 2.69$$

Weak acids are acids that do not completely dissociate in water. An equilibrium exists between the dissociated ions and undissociated compound. The reaction of a weak acid is

$$HW \rightleftharpoons H^+ + W^- \tag{3-32}$$

An equilibrium constant exists that relates the degree of dissociation:

$$K_a = \frac{[H^+][W^-]}{[HW]} \tag{3-33}$$

As with other K values,

$$pK_a = -\log K_a \tag{3-34}$$

A list of important weak acids in water and in wastewater treatment is in Table 3-3. By knowing the pH of a solution (which can be easily found with a pH meter) it is possible to get a rough idea of the degree of dissociation of the acid. For example, if

TABLE 3-2
Strong acids

Substance	Equilibrium equation	Significance
Hydrochloric acid	$HCl \rightarrow H^+ + Cl^-$	pH adjustment
Nitric acid	$HNO_3 \rightarrow H^+ + NO_3^-$	Analytical techniques
Sulfuric acid[a]	$H_2SO_4 \rightarrow 2H^+ + SO_4^{2-}$	pH adjustment, coagulation

[a] Dissociation of the second proton, $HSO_4^- \rightleftharpoons H^+ + SO_4^{2-}$, is actually a weak acid reaction with a pK_a of 1.92. As long as the pH of the solution is above 2.5, the release of both protons may be considered complete.

TABLE 3-3
Selected weak acid dissociation constants at 25°C

Substance	Equilibrium equation	pK_a	Significance
Acetic acid	$CH_3COOH \rightleftharpoons H^+ + CH_3COO^-$	4.75	Anaerobic digestion
Carbonic acid	$H_2CO_3\ (CO_2 + H_2O) \rightleftharpoons H^+ + HCO_3^-$	6.35	Corrosion, coagulation,
	$HCO_3^- \rightleftharpoons H^+ + CO_3^{2-}$	10.33	softening, pH control
Hydrogen sulfide	$H_2S \rightleftharpoons H^+ + HS^-$	7.2	Aeration, odor control,
	$HS^- \rightleftharpoons H^+ + S^{2-}$	11.89	corrosion
Hypochlorous acid	$HOCl \rightleftharpoons H^+ + OCl^-$	7.54	Disinfection
Phosphoric acid	$H_3PO_4 \rightleftharpoons H^+ + H_2PO_4^-$	2.12	Phosphate removal
	$H_2PO_4 \rightleftharpoons H^+ + HPO_4^{2-}$	7.20	plant nutrient,
	$HPO_4^{2-} \rightleftharpoons H^+ + PO_4^{3-}$	12.32	analytical

the pH is equal to the pK_a (that is, $[H^+] = K_a$), then from Equation 3-33, $[HW] = [W^-]$ and the acid is 50 percent dissociated. If the $[H^+]$ is two orders of magnitude (100 times) less than the K_a, then $100[H^+] = K_a$ (or pH $\gg$ pK).

$$100[H^+] = \frac{[H^+][W^-]}{[HW]}$$

or $100\,[HW] = [W^-]$. We would conclude that essentially all the acid is dissociated ($W^- \gg HW$). Correspondingly, if pH $\ll$ pK then $[HW] \gg [W^-]$, and none of the acid is dissociated.[6]

Example 3-7. If 15 mg/L of HOCl is added to a potable water for disinfection and the final measured pH is 7.0, what percent of the HOCl is not dissociated? Assume the temperature is 25°C.

Solution. The reaction is

$$HOCl \rightleftharpoons H^+ + OCl^-$$

From Table 3-3, we find the pK_a is 7.54 and

$$K_a = 10^{-7.54} = 2.88 \times 10^{-8}$$

Writing the equilibrium constant expression in the form of Equation 3-33

$$K_a = \frac{[H^+][OCl^-]}{[HOCl]}$$

and substituting the values for K_a and $[H^+]$

$$2.88 \times 10^{-8} = \frac{[10^{-7}][OCl^-]}{[HOCl]}$$

Solving for the HOCl concentration

$$[HOCl] = 3.47[OCl^-]$$

Since the fraction of HOCl that has not dissociated plus the OCl^- that was formed by the dissociation must, by the law of conservation of mass, equal 100% of the original HOCl added:

$$[HOCl] + [OCl^-] = 100\% \text{ (of the total HOCl added to the solution)}$$

then

$$3.47[OCl^-] + [OCl^-] = 100\%$$

$$4.47[OCl^-] = 100\%$$

$$[OCl^-] = \frac{100\%}{4.47} = 22.37\%$$

---

[6]If $[H^+] < K_a$, then pH $>$ pK. The symbol $\gg$ means greater by two orders of magnitude.

and

$$[HOCl] = 3.47(22.37\%) = 77.6\%$$

Buffer solutions. A solution that resists large changes in pH when an acid or base is added or when the solution is diluted is called a *buffer* solution. A solution containing a weak acid and its salt is an example of a buffer. Atmospheric carbon dioxide (CO_2) produces a natural buffer through the following reactions:

$$CO_2(g) \rightleftharpoons CO_2 + H_2O \rightleftharpoons H_2CO_3 \rightleftharpoons H^+ + HCO_3^- \rightleftharpoons 2H^+ + CO_3^{2-} \quad (3\text{-}35)$$

where H_2CO_3 = carbonic acid
HCO_3^- = bicarbonate ion
CO_3^{2-} = carbonate ion

This is perhaps the most important buffer system in water and wastewater treatment. We will be referring to it several times in this and subsequent chapters as the *carbonate buffer system.*

As depicted in Equation 3-35, the CO_2 in solution is in equilibrium with atmospheric $CO_2(g)$. Any change in the system components to the right of CO_2 causes the CO_2 either to be released from solution or to dissolve.

We can examine the character of the buffer system in resisting a change in pH by assuming the addition of an acid or a base and applying the law of mass action (Le Chatelier's principle). For example, if an acid is added to the system, it unbalances it by increasing the hydrogen ion concentration. Therefore, the carbonate combines with it to form bicarbonate. Bicarbonate reacts to form more carbonic acid, which in turn dissociates to CO_2 and water. The excess CO_2 can be released to the atmosphere in a thermodynamically open system. Alternatively, the addition of a base consumes hydrogen ions and the system moves to the right with the CO_2 being replenished from the atmosphere. When CO_2 is bubbled into the system or is removed by passing an inert gas such as nitrogen through the liquid (a process called *stripping*), the pH will change more dramatically because the atmosphere is no longer available as a source or sink for CO_2. Figure 3-4 summarizes the four general responses of the carbonate buffer system. The first two cases are common in natural settings when the reactions proceed over a relatively long period of time. In a water treatment plant, we can alter the reactions more quickly than the CO_2 can be replenished from the atmosphere. The second two cases are not common in natural settings. They are used in water treatment plants to adjust the pH.

In natural waters in equilibrium with atmospheric CO_2, the amount of CO_3^{2-} in solution is quite small in comparison to the HCO_3^- in solution. The presence of Ca^{2+} in the form of limestone rock or other naturally occurring sources of calcium results in the formation of calcium carbonate ($CaCO_3$), which is very insoluble. As a consequence, it precipitates from solution. The reaction of Ca^{2+} with CO_3^{2-} to form a precipitate is one of the fundamental reactions used to soften water.

Alkalinity. Alkalinity is defined as the sum of all titratable bases down to about pH 4.5. It is found by experimentally determining how much acid it takes to lower the pH of water to 4.5. In most waters the only significant contributions to alkalinity are

Case I
Acid is added to carbonate buffer system[a]

Reaction shifts to the left as $H_2CO_3^*$ is formed when H^+ and HCO_3^- combine[b]
CO_2 is released to the atmosphere
pH is lowered slightly because the availability of free H^+ (amount depends on buffering capacity)

Case II
Base is added to carbonate buffer system

Reaction shifts to the right
CO_2 from the atmosphere dissolves into solution
pH is raised slightly because H^+ combines with OH^- (amount depends on buffering capacity)

Case III
CO_2 is bubbled into carbonate buffer system

Reaction shifts to the right because $H_2CO_3^*$ is formed when CO_2 and H_2O combine
CO_2 dissolves into solution
pH is lowered

Case IV
Carbonate buffer system is stripped of CO_2

Reaction shifts to the left to form more $H_2CO_3^*$ to replace that removed by stripping
CO_2 is removed from solution
pH is raised

[a]Refer to Equation 3-35
[b]The asterisk * in the H_2CO_3 is used to signify the sum of CO_2 and H_2CO_3 in solution.

FIGURE 3-4
Behavior of the carbonate buffer system with the addition of acids and bases or the addition and removal of CO_2.

the carbonate species and any free H^+ or OH^-. The total H^+ that can be taken up by a water containing primarily carbonate species is

$$\text{Alkalinity} = [HCO_3^-] + 2[CO_3^{2-}] + [OH^-] - [H^+] \tag{3-36}$$

where [] refers to concentrations in moles/L. In most natural water situations (pH 6 to 8), the OH^- and H^+ are negligible, such that

$$\text{Alkalinity} = [HCO_3^-] + 2[CO_3^{2-}] \tag{3-37}$$

Note that $[CO_3^{2-}]$ is multiplied by two because it can accept two protons. The pertinent acid/base reactions are

$$H_2CO_3 \rightleftharpoons H^+ + HCO_3^- \qquad pK_{a1} = 6.35 \text{ at } 25°C \tag{3-38}$$

$$HCO_3^- \rightleftharpoons H^+ + CO_3^{2-} \qquad pK_{a2} = 10.33 \text{ at } 25°C \tag{3-39}$$

From the pK values, some useful relationships can be found. The more important ones are as follows:

1. Below pH of 4.5, essentially all of the carbonate species are present as H_2CO_3, and the alkalinity is negative (due to the H^+).
2. At a pH of 8.3 most of the carbonate species are present as HCO_3^-, and the alkalinity equals HCO_3^-.
3. Above a pH of 12.3, essentially all of the carbonate species are present as CO_3^-, and the alkalinity equals $2[CO_3^-] + [OH^-]$. The $[OH^-]$ may not be insignificant at this pH.

From Equation 3-37 and our discussion of buffer solutions, it can be seen that alkalinity serves as a measure of buffering capacity. The greater the alkalinity, the greater the buffering capacity. In environmental engineering, then, we differentiate between alkaline water and water having high alkalinity. Alkaline water has a pH greater than 7, while a water with high alkalinity has a high buffering capacity. An alkaline water may or may not have a high buffering capacity. Likewise, a water with a high alkalinity may or may not have a high pH.

By convention, alkalinity is not expressed in molarity units as shown in the above equations, but rather in mg/L as $CaCO_3$. In order to convert species to mg/L as $CaCO_3$, multiply mg/L as the species by the ratio of the equivalent weight of $CaCO_3$ to the species equivalent weight:

$$\text{mg/L as } CaCO_3 = (\text{mg/L as species})\left(\frac{EW_{CaCO_3}}{EW_{species}}\right) \tag{3-40}$$

The alkalinity is then found by adding all the carbonate species and the hydroxide, and then subtracting the hydrogen ions. When using the units "mg/L as $CaCO_3$," the terms are added directly. The multiple of two for CO_3^{2-} has already been accounted for in the conversion.

Example 3-8. A water contains 100.0 mg/L CO_3^{2-} and 75.0 mg/L HCO_3^- at a pH of 10. Calculate the alkalinity exactly at 25°C. Approximate the alkalinity by ignoring $[OH^-]$ and $[H^+]$.

Solution. First, convert CO_3^{2-}, HCO_3^-, OH^-, and H^+ to mg/L as $CaCO_3$.

The equivalent weights are

$$CO_3^{2-}: \text{MW} = 60, n = 2, \text{EW} = 30$$
$$HCO_3^-: \text{MW} = 61, n = 1, \text{EW} = 61$$
$$H^+: \text{MW} = 1, n = 1, \text{EW} = 1$$
$$OH^-: \text{MW} = 17, n = 1, \text{EW} = 17$$

and the concentration of H^+ and OH^- is calculated as follows: pH = 10; therefore $[H^+] = 10^{-10}$ M. Using Equation 3-7,

$$\text{mg/L} = (10^{-10}\ \text{moles/L})(1\ \text{g/mole})(10^3\ \text{mg/g}) = 10^{-7}$$

Using Equation 3-28,

$$[OH^-] = \frac{K_w}{[H^+]} = \frac{10^{-14}}{10^{-10}} = 10^{-4}\ \text{moles/L}$$

and

$$\text{mg/L} = (10^{-4}\ \text{moles/L})(17\ \text{g/mole})(10^3\ \text{mg/g}) = 1.7$$

Now, the mg/L as $CaCO_3$ is found by using Equation 3-40 and taking the equivalent weight of $CaCO_3$ to be 50:

$$CO_3^{2-} = 100.0\left(\frac{50}{30}\right) = 167$$

$$HCO_3^- = 75.0\left(\frac{50}{61}\right) = 61$$

$$H^+ = 10^{-7}\left(\frac{50}{1}\right) = 5 \times 10^{-6}$$

$$OH^- = 1.7\left(\frac{50}{17}\right) = 5.0$$

The exact alkalinity (in mg/L) is found by

$$\begin{aligned}\text{Alkalinity} &= 61 + 167 + 5.0 - (5 \times 10^{-6})\\ &= 233\ \text{mg/L as } CaCO_3\end{aligned}$$

It is approximated by 61 + 167 = 228 mg/L as $CaCO_3$. This is a 2.2 percent error.

Activity Coefficients. To this point our discussions have assumed that the solutions being analyzed were dilute. That is, the total ion concentrations were low (generally less than 10^{-2} M). For dilute solutions, the ions in solution can be considered to act independently from one another. As the concentration of ions in solution increases, the interaction of their electric charges affects their equilibrium relationships. This interaction is measured in terms of *ionic strength.* To account for high ionic strength, the equilibrium relationships are modified by incorporating *activity coefficients.* These are symbolized by γ(ion). Activity is then the product of the molar concentration of the species and its activity coefficient. For example, the solubility product of $CaCO_3$ would be

$$K_s = \{\gamma(Ca^{2+}) \times [Ca^{2+}]\}\{\gamma(CO_3) \times [CO_3]\} \tag{3-41}$$

Reaction Kinetics

Many reactions that occur in the environment do not reach equilibrium quickly. Some examples include disinfection of water, gas transfer into and out of water, removal of organic matter from water, and radioactive decay. The study of how these reactions proceed is called *reaction kinetics*. The *rate of reaction, r*, is used to describe the rate of formation or disappearance of a compound. Reactions that take place in a single phase (that is, liquid, gas, or solid) are called *homogeneous* reactions. Those that occur at surfaces between phases are called *heterogeneous*. For each type of reaction, the rate may be defined as follows:

For homogeneous reactions

$$r = \frac{\text{moles or milligrams}}{\text{(unit volume) (unit time)}} \tag{3-42}$$

For heterogeneous reactions

$$r = \frac{\text{moles or milligrams}}{\text{(unit surface) (unit time)}} \tag{3-43}$$

Production of a compound results in a positive sign for the reaction rate $(+r)$, while disappearance of a substance yields a negative sign $(-r)$. Reaction rates are a function of temperature, pressure, and the concentration of reactants. For a stoichiometric reaction of the form:

$$a\text{A} + b\text{B} \rightarrow c\text{C}$$

where a, b, and c are the proportionality coefficients for the reactants A, B, and C, the change in concentration of compound A is equal to the reaction rate equation for compound A:

$$\frac{d[\text{A}]}{dt} = r_\text{A} = -k\,[\text{A}]^\alpha\,[\text{B}]^\beta = k[\text{C}]^\gamma \tag{3-44}$$

where [A], [B], and [C] are the concentrations of the reactants, and α, β, and γ are empirically determined exponents. The proportionality term, k, is called the *reaction rate constant*. It is often not a constant but, rather, is dependent on the temperature and pressure. Since A and B are disappearing, the sign of the reaction rate equation is negative. It is positive for C because C is being formed.

The *order of reaction* is defined as the sum of the exponents in the reaction rate equation. The exponents may be either integers or fractions. Some sample reaction orders are shown in Table 3-4.

TABLE 3-4
Example reaction orders

Reaction order	Rate Equation
Zero	$r_A = -k$
First	$r_A = -k[\text{A}]$
Second	$r_A = -k[\text{A}]^2$
Second	$r_A = -k[\text{A}][\text{B}]$

TABLE 3-5
Plotting procedure to determine order of reaction by method of integration for plug flow reactor and for a batch reactor[a]

Order	Rate equation	Integrated equation	Linear plot	Slope	Intercept
0	$\frac{d[A]}{dt} = -k$	$[A] - [A_0] = -kt$	$[A]$ vs. t	$-k$	$[A_0]$
1	$\frac{d[A]}{dt} = -k[A]$	$\ln \frac{[A]}{[A_0]} = -kt$	$\ln [A]$ vs. t	$-k$	$\ln [A_0]$
2	$\frac{d[A]}{dt} = -k[A]^2$	$\frac{1}{[A]} - \frac{1}{[A_0]} = kt$	$\frac{1}{[A]}$ vs. t	k	$\frac{1}{[A_0]}$

[a] *Source:* J. G. Henry and G. W. Heinke, *Environmental Science and Engineering,* Englewood Cliffs, NJ: Prentice Hall, 1989, p. 201.

For elementary reactions where the stoichiometric equation represents both the mass balance and the molecular scale process, the coefficients of proportionality (a, b, c) are equivalent to the exponents in the reaction rate equation:

$$r_A = -k[A]^a[B]^b \tag{3-45}$$

The overall reaction rate, r, and the individual reaction rates are related:

$$r = \frac{r_A}{a} = \frac{r_B}{b} = \frac{r_C}{c} \tag{3-46}$$

The reaction rate constant, k, may be determined experimentally by obtaining data on the concentrations of the reactants as a function of time and plotting on a suitable graph. The form of the graph is determined from the result of integration of the equations in Table 3-4. The integrated forms and the appropriate graphical forms are shown in Table 3-5.

Gas transfer. An important example of time-dependent reactions is the mass transfer (dissolution or volatilization) of gas from water. In 1924 Lewis and Whitman[7] postulated a two-film theory to describe the mass transfer of gases. According to their theory, the boundary between the gas phase and the liquid phase (also called the *interface*) is composed of two distinct films that serve as a barrier between the bulk phases (Figure 3-5). For a molecule of gas to go into solution, it must pass through the bulk of the gas, the gas film, the liquid film, and into the bulk of the liquid (Figure 3-5*a*). To leave the liquid, the gas molecule must follow the reverse course (Figure 3-5*b*). The driving force causing the gas to move, and hence the mass transfer, is the concentration gradient: $C_s - C$. C_s is the saturation concentration of the gas in the liquid, and C is the actual concentration. When C_s is greater than C, the gas will go into solution. When C is greater than C_s, the gas will desorb.

[7] W. K. Lewis and W. G. Whitman, "Principles of Gas Absorption," *Ind. Eng. Chem.*, vol. 16, p. 1215, 1924.

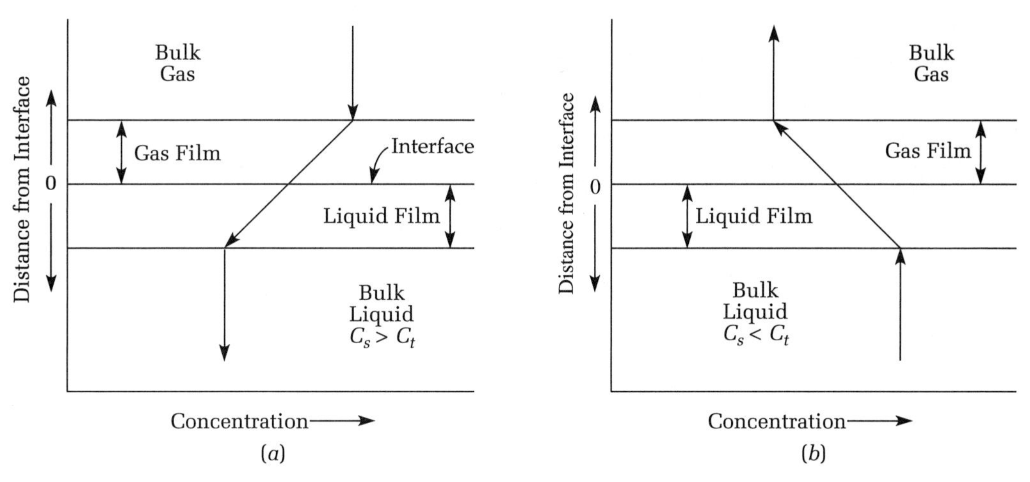

FIGURE 3-5
Two-film model of the interface between gas and liquid: (*a*) absorption mode and (*b*) desorption mode.

The relationship between the equilibrium concentration of gas dissolved in solution and the partial pressure of the gas is defined by *Henry's law:*

$$P = \mathrm{H}C_{\mathrm{equil}} \tag{3-47}$$

where P = partial pressure of gas in equilibrium, kPa
H = Henry's law constant, kPa · m^3/g
C_{equil} = equilibrium concentration of gas dissolved in liquid, g/m^3

For a given temperature and pressure, the equilibrium concentration is a constant and is called the *saturation concentration.*

The partial pressure of a gas in air is the product of its mole fraction and the air pressure. For example, the atmosphere contains about 21 percent oxygen by volume, so the partial pressure would be 0.21 times the atmospheric pressure. Another useful expression of Henry's law is written as:

$$\mathrm{X} = K_H P_{\mathrm{g}} \tag{3-48}$$

where X = mole fraction of the gas dissolved in the liquid at equilibrium
K_H = Henry's law constant, atm^{-1}
P_{g} = partial pressure of gas in equilibrium with liquid, atm

The mole fraction of one gas in water may be calculated as

$$\mathrm{X}_1 = \frac{[\mathrm{gas}_1]}{[\mathrm{H_2O}] + [\mathrm{gas}_1] + [\mathrm{gas}_2] + \ldots} \simeq \frac{[\mathrm{gas}_1]}{[\mathrm{H_2O}]} \tag{3-49}$$

where $[\mathrm{gas}_1]$ = the concentration of dissolved gas in moles/L

Since the molar concentration of H_2O = (1,000 g/L)(18 g/mole) = 55.56 mol/L, Equation 3-48 may be rewritten as:

$$[\mathrm{gas}] = 55.56 K_H P_{\mathrm{g}}$$

The Henry's law coefficient varies both with the temperature and the concentration of other dissolved substances. Henry's law constants are given in Appendix A.

The rate of mass transfer can be described by the following equation:

$$\frac{dC}{dt} = k_a(C_s - C) \tag{3-50}$$

where k_a = rate constant or mass transfer coefficient, s^{-1}.

The difference between the saturation concentration and the actual concentration $(C_s - C)$ is called the *deficit*. Since the saturation concentration is a constant for a constant temperature and pressure, this is a first-order reaction.

Example 3-9. A falling raindrop initially has no dissolved oxygen. The saturation concentration for the drop is 9.20 mg/L. If, after falling for two seconds, the droplet has an oxygen concentration of 3.20 mg/L, how long must the droplet fall (from the start of the fall) to achieve a concentration of 8.20 mg/L?

Solution. We begin by calculating the deficit after two seconds, and that at a concentration of 8.20 mg/L:

$$\text{Deficit at 2 sec} = 9.20 - 3.20 = 6.00 \text{ mg/L}$$

$$\text{Deficit at } t \text{ sec} = 9.20 - 8.20 = 1.00 \text{ mg/L}$$

Now using the integrated form of the first-order rate equation from Table 3-5, noting that the rate of change is proportional to deficit and, hence, $[A] = (C_s - C)$ and that $[A_0] = (9.20 - 0.00)$,

$$\ln \frac{6.00}{9.20} = -k\,(2.00 \text{ s})$$

$$k = 0.2137 \text{ s}^{-1}$$

With this value of k, we can calculate a value for t:

$$\ln \frac{(9.20 - 8.20)}{9.20} = -(0.2137)\,(t)$$

$$t = 10.4 \text{ s}$$

Water Quality[8]

Precipitation in the form of rain, hail, or sleet contains very few impurities. It may contain trace amounts of mineral matter, gases, and other substances as it forms and falls through the earth's atmosphere. The precipitation, however, has virtually no bacterial content.

---

[8]Portions of this discussion are from the following source: U.S. Department of Health, Education and Welfare, *Manual of Individual Water Supply Systems* (PHS Publication No. 24), Washington, DC: U.S. Government Printing Office, 1962.

Once precipitation reaches the earth's surface, many opportunities are presented for the introduction of mineral and organic substances, microorganisms, and other forms of pollution (contamination).[9] When water runs over or through the ground surface, it may pick up particles of soil. This is noticeable in the water as cloudiness or turbidity. It also picks up particles of organic matter and bacteria. As surface water seeps downward into the soil and through the underlying material to the water table, most of the suspended particles are filtered out. This natural filtration may be partially effective in removing bacteria and other particulate materials. However, the chemical characteristics of the water may change and vary widely when it comes in contact with mineral deposits. As surface water seeps down to the water table, it dissolves some of the minerals contained in the soil and rocks. Groundwater, therefore, often contains more dissolved minerals than surface water.

The following four categories are used to describe drinking water quality:

1. Physical: Physical characteristics relate to the quality of water for domestic use and are usually associated with the appearance of water, its color or turbidity, temperature, and, in particular, taste and odor.
2. Chemical: Chemical characteristics of waters are sometimes evidenced by their observed reactions, such as the comparative performance of hard and soft waters in laundering. Most often, differences are not visible.
3. Microbiological: Microbiological agents are very important in their relation to public health and may also be significant in modifying the physical and chemical characteristics of water.
4. Radiological: Radiological factors must be considered in areas where there is a possibility that the water may have come in contact with radioactive substances. The radioactivity of the water is of public health concern in these cases.

Consequently, in the development of a water supply system, it is necessary to examine carefully all the factors that might adversely affect the intended use of a water supply source.

Physical Characteristics

Turbidity. The presence of suspended material such as clay, silt, finely divided organic material, plankton, and other particulate material in water is known as *turbidity.* The unit of measure is a Turbidity Unit (TU) or Nephlometric Turbidity Unit (NTU). It is determined by reference to a chemical mixture that produces a reproducible refraction of light. Turbidities in excess of 5 TU are easily detectable in a glass of water and are usually objectionable for aesthetic reasons.

---

[9]Pollution as used in this text means the presence in water of any foreign substances (organic, inorganic, radiological, or biological) that tend to lower its quality to such a point that it constitutes a health hazard or impairs the usefulness of the water.

Clay or other inert suspended particles in drinking water may not adversely affect health, but water containing such particles may require treatment to make it suitable for its intended use. Following a rainfall, variations in the groundwater turbidity may be considered an indication of surface or other introduced pollution.

Color. Dissolved organic material from decaying vegetation and certain inorganic matter cause color in water. Occasionally, excessive blooms of algae or the growth of aquatic microorganisms may also impart color. While color itself is not usually objectionable from the standpoint of health, its presence is aesthetically objectionable and suggests that the water needs appropriate treatment.

Taste and odor. Taste and odor in water can be caused by foreign matter such as organic compounds, inorganic salts, or dissolved gases. These materials may come from domestic, agricultural, or natural sources. Drinking water should be free from any objectionable taste or odor at point of use.

Temperature. The most desirable drinking waters are consistently cool and do not have temperature fluctuations of more than a few degrees. Groundwater and surface water from mountainous areas generally meet these criteria. Most individuals find that water having a temperature between 10°–15° C is most palatable.

Chemical Characteristics

Chloride. Most waters contain some chloride. The amount present can be caused by the leaching of marine sedimentary deposits or by pollution from sea water, brine, or industrial or domestic wastes. Chloride concentrations in excess of about 250 mg/L usually produce a noticeable taste in drinking water. Domestic water should contain less than 100 mg/L of chloride. In some areas, it may be necessary to use water with a chloride content in excess of 100 mg/L. In these cases, all of the other criteria for water purity must be met.

Fluorides. In some areas, water sources contain natural fluorides. Where the concentrations approach optimum levels, beneficial health effects have been observed. In such areas, the incidence of dental caries has been found to be below the levels observed in areas without natural fluorides. The optimum fluoride level for a given area depends upon air temperature, since temperature greatly influences the amount of water people drink. Excessive fluorides in drinking water supplies may produce fluorosis (mottling) of teeth, which increases as the optimum fluoride level is exceeded.[10] State or local health departments should be consulted for their recommendations.

[10]Mottled teeth are characterized by black spots or streaks and may become brittle when exposed to large amounts of fluoride.

Iron. Small amounts of iron frequently are present in water because of the large amount of iron in the geologic materials. The presence of iron in water is considered objectionable because it imparts a brownish color to laundered goods and affects the taste of beverages such as tea and coffee.

Lead. Exposure of the body to lead, however brief, can be seriously damaging to health. Prolonged exposure to relatively small quantities may result in serious illness or death. Lead taken into the body in quantities in excess of certain relatively low "normal" limits is a cumulative poison.

Manganese. Manganese imparts a brownish color to water and to cloth that is washed in it. It flavors coffee and tea with a medicinal taste.

Sodium. The presence of sodium in water can affect persons suffering from heart, kidney, or circulatory ailments. When a strict sodium-free diet is recommended, any water should be regarded with suspicion. Home water softeners may be of particular concern because they add large quantities of sodium to the water. (See the discussion of ion-exchange softening for an explanation of the chemistry and operation of softeners.)

Sulfate. Waters containing high concentrations of sulfate, caused by the leaching of natural deposits of magnesium sulfate (Epsom salts) or sodium sulfate (Glauber's salt), may be undesirable because of their laxative effects.

Zinc. Zinc is found in some natural waters, particularly in areas where zinc ore deposits have been mined. Zinc is not considered detrimental to health, but it will impart an undesirable taste to drinking water.

Toxic inorganic substances. *Nitrates* (NO_3), cyanides (CN), and *heavy metals* constitute the major classes of inorganic substances of health concern. Methemoglobinemia (infant cyanosis or "blue baby syndrome") has occurred in infants who have been given water or fed formula prepared with water having high concentrations of nitrate. CN ties up the hemoglobin sites that bind oxygen to red blood cells. This results in oxygen deprivation. A characteristic symptom is that the patient has a blue skin color. This condition is called cyanosis. CN causes chronic effects on the thyroid and central nervous system. The toxic heavy metals include arsenic (As), barium (Ba), cadmium (Cd), chromium (Cr), lead (Pb), mercury (Hg), selenium (Se), and silver (Ag). The heavy metals have a wide range of effects. They may be acute poisons (As and Cr^{6+} for example), or they may produce chronic disease (Pb, Cd, and Hg for example).

Toxic organic substances. There are over 120 toxic organic compounds listed on the U.S. Environmental Protection Agency's Priority Pollutant List (Table 1-3). These include pesticides, insecticides, and solvents. Like the inorganic substances, their effects may be acute or chronic.

Microbiological Characteristics

Water for drinking and cooking purposes must be made free from disease-producing organisms (*pathogens*). These organisms include viruses, bacteria, protozoa, and helminths (worms).

Some organisms which cause disease in people originate with the fecal discharges of infected individuals. Others are from the fecal discharge of animals.

Unfortunately, the specific disease-producing organisms present in water are not easily identified. The techniques for comprehensive bacteriological examination are complex and time-consuming. It has been necessary to develop tests that indicate the relative degree of contamination in terms of an easily defined quantity. The most widely used test estimates the number of microorganisms of the *coliform group*. This grouping includes two genera: *Escherichia coli* and *Aerobacter aerogenes*. The name of the group is derived from the word *colon*. While *E. coli* are common inhabitants of the intestinal tract, *Aerobacter* are common in the soil, on leaves, and on grain; on occasion they cause urinary tract infections. The test for these microorganisms, called the *Total Coliform Test,* was selected for the following reasons:

1. The coliform group of organisms normally inhabits the intestinal tracts of humans and other mammals. Thus, the presence of coliforms is an indication of fecal contamination of the water.
2. Even in acutely ill individuals, the number of coliform organisms excreted in the feces outnumber the disease-producing organisms by several orders of magnitude. The large numbers of coliforms make them easier to culture than disease-producing organisms.
3. The coliform group of organisms survives in natural waters for relatively long periods of time, but does not reproduce effectively in this environment. Thus, the presence of coliforms in water implies fecal contamination rather than growth of the organism because of favorable environmental conditions. These organisms also survive better in water than most of the bacterial pathogens. This means that the absence of coliforms is a reasonably safe indicator that pathogens are not present.
4. The coliform group of organisms is relatively easy to culture. Thus, laboratory technicians can perform the test without expensive equipment.

Current research indicates that testing for *Escherichia coli* specifically may be warranted. Some agencies prefer the examination for *E. coli* as a better indicator of biological contamination than total coliforms.

Radiological Characteristics

The development and use of atomic energy as a power source and the mining of radioactive materials, as well as naturally occurring radioactive materials, have made it necessary to establish limiting concentrations for the intake into the body of radioactive substances, including drinking water.

The effects of human exposure to radiation or radioactive materials are harmful, and any unnecessary exposure should be avoided. Humans have always been exposed to natural radiation from water, food, and air. The amount of radiation to which the individual is normally exposed varies with the amount of background radioactivity. Water with high radioactivity is not normal and is confined in great degree to areas where nuclear industries are situated.

Water Quality Standards[11]

President Ford signed the National Safe Drinking Water Act (SDWA) into law on December 16, 1974. The Environmental Protection Agency (EPA) was directed to establish *maximum contaminant levels* (MCLs) for public water systems to prevent the occurrence of any known or anticipated adverse health effects with an adequate margin of safety. EPA defined a public water system to be any system that provides piped water for human consumption, if such a system has at least 15 service connections or regularly serves an average of at least 25 individuals daily at least 60 days out of the year. This definition includes private businesses, such as service stations, restaurants, motels, and others that serve more than 25 persons per day for greater than 60 days out of the year.

From 1975 through 1985, the EPA regulated 23 contaminants in drinking water supplied by public water systems. These regulations are known as interim primary drinking water regulations (IPDWRs). In June of 1986, the SDWA was amended. The amendments required EPA to set *maximum contaminant level goals* (MCLGs) and MCLs for 83 specific substances. This list included 22 of the IPDWRs (all except trihalomethanes). The amendments also required EPA to regulate 25 additional contaminants every three years beginning in January, 1991 and continuing for an indefinite period of time.

As of February, 1994, regulations had not been established for 7 of the 83 contaminants: arsenic, sulfate, and five radionuclides. Arsenic and three of the radionuclides (radium, alpha emitters, and beta-photon emitters) were among the 23 contaminants that had been regulated prior to 1986. The IPDWRs remain in effect for these contaminants until superseded by the new regulations.

Table 3-6 lists each regulated contaminant and summarizes its adverse health effects. Some of these contaminant levels are being considered for revision. The notation "TT" in the table means that a treatment technique is specified rather than

[11]The following references provide a thorough discussion of the development of the standards:
F.W. Pontius, "D-DBP Rule to Set Tight Standards," *Journal of the American Water Works Association, 85,* Nov., 1993, pp. 22–30, 100.
J. Auerbach, "Costs and Benefits of Current SDWA Regulations," *Journal of the American Water Works Association, 86,* Feb., 1994, pp. 69–78.
F.W. Pontius and J.A. Roberson, "The Current Regulatory Agenda: An Update," *Journal of the American Water Works Association, 86,* Feb., 1994, pp. 54–63.
F.W. Pontius, "An Update of the Federal Drinking Water Regs," *Journal of the American Water Works Association, 87,* Feb., 1995, pp. 48–58.

TABLE 3-6
Standards and potential health effects of the contaminants regulated under the SDWA

Contaminant	Maximum Contaminant Level Goal mg/L	Maximum Contaminant Level mg/L	BAT	Potential Health Effects
Organics				
Acrylamide	Zero	TT	PAP	Cancer, nervous system effects
Alachor	Zero	0.002	GAC	Cancer
Aldicarb	0.001	0.003	GAC	Nervous system effects
Aldicarb sulfone	0.001	0.002	GAC	Nervous system effects
Aldicarb sulfoxide	0.001	0.004	GAC	Nervous system effects
Atrazine	0.003	0.003	GAC	Liver, kidney, lung, cardiovascular effects; possible carcinogen
Benzene	Zero	0.005	GAC, PTA	Cancer
Benzo(a)pyrene	Zero	0.0002	GAC	Cancer
Bromodichloromethane	Zero	NA	EC	Cancer
Bromoform	Zero	NA	EC	Cancer
Carbofuran	0.04	0.04	GAC	Nervous system, reproductive system effects
Carbon tetrachloride	Zero	0.005	GAC, PTA	Cancer
Chloral hydrate	0.04	TT	EC	Cancer ?
Chlordane	Zero	0.002	GAC	Cancer
Chloroform	Zero	NA	EC	Cancer
2,4-D	0.07	0.07	GAC	Liver, kidney effects
Dalapon	0.2	0.2	GAC	Kidney, liver effects
Di(2-ethylhexyl)adipate	0.5	0.5	GAC, PTA	Reproductive effects
Di(2-ethylhexyl)phthalate	Zero	0.006	GAC	Cancer
Dibromochloromethane	0.06	NA	EC	Cancer
Dibromochloropropane (DBCP)	Zero	0.0002	GAC, PTA	Cancer
Dichloroacetic acid	Zero	NA	EC	Cancer ?
p-Dichlorobenzene	0.075	0.075	GAC, PTA	Kidney effects, possible carcinogen
o-Dichlorobenzene	0.6	0.6	GAC, PTA	Kiver, kidney, blood cells effects
1,2-Dichloroethane	Zero	0.005	GAC, PTA	Cancer
1,1-Dichloroethylene	0.007	0.007	GAC, PTA	Liver, kidney effects, possible carcinogen

TABLE 3-6
Standards and potential health effects of the contaminants regulated under the SDWA (*continued*)

Contaminant	Maximum Contaminant Level Goal mg/L	Maximum Contaminant Level mg/L	BAT	Potential Health Effects
Organics				
cis-1,2-Dichloroethylene	0.07	0.07	GAC, PTA	Liver, kidney, nervous system, circulatory effects
trans-1,2-Dichloroethylene	0.1	0.1	GAC, PTA	Liver, kidney, nervous system, circulatory effects
Dichloromethane (methylene chloride)	Zero	0.005	PTA	Cancer
1,2-Dichloropropane	Zero	0.005	GAC, PTA	Cancer
Dinoseb	0.007	0.007	GAC	Thyroid, reproductive effects
Diquat	0.02	0.02	GAC	Ocular, liver, kidney effects
Endothall	0.1	0.1	GAC	Liver, kidney, gastrointestinal effects
Endrin	0.002	0.002	GAC	Liver, kidney, heart effects
Epichlorohydrin	Zero	TT	PAP	Cancer
Ethylbenzene	0.7	0.7	GAC, PTA	Liver, kidney, nervous system effects
Ethylene dibromide (EDB)	Zero	0.00005	GAC, PTA	Cancer
Glyphosate	0.7	0.7	OX	Liver, kidney effects
Haloacetic acids (sum of 5; HAA5)[1]		0.030	EC + GAC	Cancer ?
Heptachlor	Zero	0.0004	GAC	Cancer
Heptachlor epoxide	Zero	0.0002	GAC	Cancer
Hexachlorobenzene	Zero	0.001	GAC	Cancer
Hexachlorocyclopentadiene	0.05	0.05	GAC, PTA	Kidney, stomach effects
Lindane	0.0002	0.0002	GAC	Liver, kidney, nervous system, immune system, circulatory system effects
Methoxychlor	0.04	0.04	GAC	Development, liver, kidney, nervous system effects
Monochlorobenzene	0.1	0.1	GAC, PTA	Cancer
Oxamyl (vydate)	0.2	0.2	GAC	Kidney effects
Pentachlorophenol	Zero	0.001	GAC	Cancer
Picloram	0.5	0.5	GAC	Kidney, liver effects
Polychlorinated biphenyls (PCBs)	Zero	0.0005	GAC	Cancer

TABLE 3-6
Standards and potential health effects of the contaminants regulated under the SDWA (*continued*)

Contaminant	Maximum Contaminant Level Goal mg/L	Maximum Contaminant Level mg/L	BAT	Potential Health Effects
Simazine	0.004	0.004	GAC	Body weight and blood effects, possible carcinogen
Styrene	0.1	0.1	GAC, PTA	Liver, nervous system effects, possible carcinogen
2,3,7,8-TCDD (dioxin)	Zero	5×10^{-8}	GAC	Cancer
Tetrachloroethylene	Zero	0.005	GAC, PTA	Cancer
Toluene	1	1	GAC, PTA	Liver, kidney, nervous system, circulatory system effects
Toxaphene	Zero	0.005	GAC	Cancer
2,4,5-TP (silvex)	0.05	0.05	GAC	Liver, kidney effects
Trichloroacetic acid	0.3	NA	EC	Cancer ?
1,2,4-Trichlorobenzene	0.07	0.07	GAC, PTA	Liver, kidney effects
1,1,1-Trichloroethane	0.2	0.2	GAC, PTA	Liver, nervous system effects
1,1,2-Trichloroethane	0.003	0.005	GAC, PTA	Kidney, liver effects, possible carcinogen
Trichloroethylene	Zero	0.005	GAC, PTA	Cancer
Trihalomethanes (sum of 4; TTHMs)[2]	NA	0.040	EC + GAC	Cancer
Vinyl chloride	Zero	0.002	PTA	Cancer
Xylenes (total)	10	10	GAC, PTA	Liver, kidney, nervous system effects
Inorganics				
Antimony	0.005	0.006	C-F,[3] RO	Decreased longevity, blood effects
Arsenic	NA	0.05	NA	Dermal, nervous system effects
Asbestos (Fibers/L > 10μm)	7 million (fibers/L)	7 million (fibers/L)	C-F,[3] DF, DEF	Possible carcinogen by ingestion
Barium	2	2	IX, RO, LS[3]	Blood pressure effects
Beryllium	Zero	0.001	IX, RO, C-F[3] LS,[3] AA, IX	Bone, lung effects, cancer
Bromate	Zero	0.010	DC	
Cadmium	0.005	0.005	C-F,[3] LS,[3] IX RO	Kidney effects

TABLE 3-6
Standards and potential health effects of the contaminants regulated under the SDWA (*continued*)

Contaminant	Maximum Contaminant Level Goal mg/L	Maximum Contaminant Level mg/L	BAT	Potential Health Effects
Chlorite	0.08	1.0	DC	
Chromium (total)	0.1	0.1	C-F[3], LS[3], (Cr III), IX, RO	Liver, kidney, circulatory system effects
Copper	1.3	TT	CC, SWT	Gastrointestinal effects
Cyanide	0.2	0.2	IX, RO, Cl_2	Thyroid, central nervous system effects
Fluoride	4	4	AA, RO	Skeletal fluorosis
Lead	Zero	TT	CC, PE, SWT LSLR	Cancer, kidney, central and peripheral nervous system effects
Mercury	0.002	0.002	C-F[3] (influent $\leq$ 10 μg/L), LS[3], GAC, RO, (influent $\leq$ 10 μg/L)	Kidney, central nervous system effects
Nickel	0.1	0.1	LS[3], IX, RO	Liver effects
Nitrate (as N)	10	10	IX, RO, ED	Methemoglobinemia (blue baby syndrome)
Nitrite (as N)		1	IX, RO	Methemoglobinemia (blue baby syndrome)
Nitrate + nitrite (both as N)	10	10	IX, RO	
Selenium	0.05	0.05	C-F[3] (Se IV), LS[3] AA, RO, ED	Nervous system effects
Sulfate	500	500	RO, IX, ED	
Thallium	0.0005	0.002	IX, AA	Liver, kidney, brain, intestine effects
Radionuclides				
Beta particle and photon emitters	Zero	4 mrem	C-F, IX, RO	Cancer
Alpha emitters	Zero	15 pCi/L	C-F, RO	Cancer

TABLE 3-6
Standards and potential health effects of the contaminants regulated under the SDWA (*continued*)

Contaminant	Maximum Contaminant Level Goal mg/L	Maximum Contaminant Level mg/L	BAT	Potential Health Effects
Radium-226 + radium-228		5 pCi/L		Cancer
Radium-226	Zero	20 pCi/L	LS[3], IX, RO	Cancer
Radium-228	Zero	20 pCi/L	LS[3], IX, RO	Cancer
Radon	Zero	300 pCi/L	AR	Cancer
Uranium	Zero	20 μg/L	C-F[3], LS[3], AX	Cancer
Microbials				
Cryptosporidium	Zero	TT	C-F, SSF, DEF, DF, D	Gastroenteric disease
E. coli	Zero	[5]	D	Gastroenteric disease
Fecal coliforms	Zero	[5]	D	Gastroenteric disease
Giardia lamblia	Zero	TT	C-F, SSF, DEF, D	Gastroenteric disease
Heterotrophic bacteria		TT	C-F, SSF, DEF, DF, D	
Legionella	Zero	TT	C-F, SSF, DEF DF, D	Pneumonialike effects
Total coliforms	Zero	[4]	D	Indicator of gastroenteric infections
Turbidity		PS	C-F, SSF, DEF, DF, D	Interferes with disinfection, indicator of filtration performance
Viruses	Zero	TT	C-F, SSF, DEF, DF, D	Gastroenteric disease, respiratory disease, and other diseases (e.g., hepatitis, myocarditis)

AA-activated alumina, AD-alternative disinfectants, AR-aeration, AX-anion exchange, CC-corrosion control, C-F-coagulation and filtration, Cl_2-chlorination, D-disinfection, DC-disinfection system control, DEF-diatomaceous earth filtration, DF-direct filtration, EC-enhanced coagulation, ED-electrodialysis, GAC-granular activated carbon, IX-ion exchange, LS-lime softening, LSLR-lead service line replacement, NA-not applicable, OX-oxidation, PAP-polymer addition practices, PE-public educatiuon, PR-precursor removal, PS-performance standard, PTA-packed-tower aeration, RO-reverse osmosis, SPC-stop prechlorination, SSF-slow sand filtration, SWT-source water treatment, TT-treatment technique.

1. Sum of the concentrations of mono-, di-, and trichloroacetic acids and mono- and dibromoacetic acids.

2. Sum of the concentrations of bromodichloromethane, dibromonochloromethane, bromoform, and chloroform.

3. Coagulation-filtration and lime-softening are not BAT for small systems for variance unless treatment is already installed.

4. No more than 5 percent of the samples per month may be positive. For systems collecting fewer than 40 samples per month, no more than 1 sample per month may be positive.

5. If a repeat total coliform sample is fecal coliform- or E. coli-positive, the system is in violation of the MCL for total coliforms. The system is also in violation of the MCL for total coliforms if a routine sample is fecal coliform- or E. coli-positive and is followed by a total coliform-positive repeat sample.

a contaminant level. The treatment techniques are specific processes that are used to treat the water. Some examples include coagulation and filtration, lime softening, and ion exchange. These will be discussed in the following sections.

Lead and copper. In June 1988, EPA issued proposed regulations to define MCLs and MCLGs for lead and copper, as well as to establish a monitoring program and a treatment technique for both. The MCLG proposed for lead is zero; for copper, 1.3 mg/L. The MCL action levels, applicable to water entering the distribution system, are 0.005 mg/L for lead and 1.3 mg/L for copper.

Compliance with the regulations is also based on the quality of the water at the consumer's tap. Monitoring is required by means of collection of first-draw samples at residences. The number of samples required to be collected will range from 10 per year to 50 per quarter, depending on the size of the water system.

The SDWA amendments forbid the use of pipe, solder, or flux that is not lead-free in the installation or repair of any public water system or in any plumbing system providing water for human consumption. This does not, however, apply to leaded joints necessary for the repair of cast iron pipes.

Disinfectants and disinfectant by-products (D-DBPs). The disinfectants used to destroy pathogens in water and the by-products of the reaction of these disinfectants with organic materials in the water are of potential health concern. One class of DBPs has been regulated since 1979. This class is known as trihalomethanes (THMs). THMs are formed when a water containing an organic precursor is chlorinated (*precursor* means forerunner). In this case it means an organic compound capable of reacting to produce a THM. The precursors are natural organic substances formed from the decay of vegetative matter, such as leaves, and aquatic organisms. THMs are of concern because they are known or potential carcinogens. The four THMs that were regulated in the 1979 rules are: chloroform ($CHCl_3$), bromodichloromethane ($CHBrCl_2$), dibromochloromethane ($CHBr_2Cl$), and bromoform ($CHBr_3$). Of these four, chloroform appears most frequently and is found in the highest concentrations.

The D-DBP rule was developed through a negotiated rule-making process (called *reg-neg*), in which individuals representing major interest groups concerned with the rule (for example, public-water-system owners, state and local government officials, and environmental groups) publicly work with the EPA representatives to reach a consensus on the contents of the proposed rule.

Maximum residual disinfectant level goals (MRDLGs) and maximum residual disinfectant levels (MRDLs) were established for chlorine, chloramine, and chlorine dioxide (Table 3-7). Because ozone reacts too quickly to be detected in the distribution system, no limits on ozone were set.

The MCLGs and MCLs for the disinfection byproducts are listed in Table 3-8. In addition to regulating individual compounds, the D-DBP rule set levels for two groups of compounds: HAA5 and TTHMs. These groupings were made to recognize the potential cumulative effect of several compounds. HAA5 is the sum of five haloacetic acids (monochloroacetic acid, dichloroacetic acid, trichloroacetic acid, monobromoacetic acid, and dibromoacetic acid). TTHMs (total trihalomethanes) is the sum of the concentrations of chloroform ($CHCl_3$), bromodichloromethane ($CHBrCl_2$), dibromochloromethane ($CHBr_2Cl$), and bromoform ($CHBr_3$).

TABLE 3-7
Maximum residual disinfectant goals (MRDLGs) and maximum residual disinfectant levels (MRDLs)

Disinfectant residual	MRDLGs mg/L	MRDL mg/L
Chlorine (free)	4	4.0
Chloramines (as total chlorine)	4	4.0
Chlorine dioxide	0.08	0.8

The D-DBP rule is quite complex. In addition to the regulatory levels shown in the tables, credits are given for chlorine dioxide oxidation (known as C x T credit) and levels are established for precursors. The amount of precursor to be removed is a function of the water source, the alkalinity of the water, the pH, and the amount of *total organic carbon* (TOC) present.

The D-DBP rule is scheduled to be implemented in stages. Stage 1 of the rule will be promulgated in November 1998. Stage 2 is to be implemented after data are gathered. Requirements set under Stage 1 will be reconsidered in a reg-neg process anticipated for the development of the Stage 2 requirements. These are to be promulgated in May 2002.

Surface water treatment rule (SWTR). The Surface Water Treatment Rule (SWTR) sets forth primary drinking water regulations requiring treatment of surface water supplies or groundwater supplies under the direct influence of surface water. The regulations require a specific treatment technique—filtration and/or

TABLE 3-8
Maximum contaminant level goals (MCLGs) and maximum contaminant levels (MCLs) for disinfection by-products (DBPs)

Contaminant	MCLG mg/L	Stage 1 MCL mg/L	Stage 2 MCL mg/L
Bromate	Zero	0.010	
Bromodichloromethane	Zero		
Bromoform	Zero		
Chloral hydrate	0.005		
Chlorite	0.3	1.0	
Chloroform	Zero		
Dibromochloromethane	0.06		
Dichloroacetic acid	Zero		
Trichloroacetic acid	0.10		
HAA5		0.060	0.030
TTHMs		0.080	0.040

disinfection—in lieu of establishing maximum contaminant levels (MCLs) for turbidity, *Cryptosporidium, Giardia lamblia,* viruses, *Legionella,* and heterotrophic bacteria, as well as many other pathogenic organisms that are removed by these treatment techniques. The regulations also establish a maximum contaminant level goal (MCLG) of zero for *Giardia lamblia,* viruses, and *Legionella.* No MCLG is established for heterotrophic plate count or turbidity.

Turbidity limits. Treatment by conventional or direct filtration must achieve a turbidity level of less than 0.5 NTU in at least 95 percent of the samples taken each month. Individual states may increase the 0.5 limit to 1 NTU if they determine that the overall treatment achieves the disinfection requirements. Those systems using slow sand filtration must achieve a turbidity level of less than 5 NTU at all times and not more than 1 NTU in more than 5 percent of the samples taken each month. The 1 NTU limit may be increased by the state up to 5 NTU if it determines that there is no significant interference with disinfection. Other filtration technologies may be used if they meet the turbidity requirements set for slow sand filtration, provided they achieve the disinfection requirements and are approved by the state.

Turbidity measurements must be performed on representative samples of the system's filtered water every four hours or by continuous monitoring. For any system using slow sand filtration or a filtration treatment other than conventional treatment, direct filtration, or diatomaceous earth filtration, the state may reduce the monitoring requirements to once per day.

Disinfection requirements. Filtered water supplies must achieve the same disinfection as required for unfiltered systems (that is, 99.9 or 99.99% removal, also known as 3-log and 4-log removal or inactivation, for *Giardia lamblia* and viruses) through a combination of filtration and application of a disinfectant.

Total coliform. On June 19, 1989, the EPA promulgated the revised National Primary Drinking Water Regulations for total coliforms, including fecal coliforms and *E. coli.* These regulations apply to all public water systems.

The regulations establish a maximum contaminant level (MCL) for coliforms based on the presence or absence of coliforms. Larger systems that are required to collect at least 40 samples per month cannot obtain coliform-positive results in more than 5 percent of the samples collected each month to stay in compliance with the MCL. Smaller systems that collect fewer than 40 samples per month cannot have coliform-positive results in more than one sample per month.

The EPA will accept any one of the five analytical methods noted below for the determination of total coliforms:

Multiple-tube fermentation technique (MTF)
Membrane filter technique (MF)
Minimal media ONPG-MUG test (colilert system) (MMO-MUG)
Presence-absence coliform test (P-A)
Colisure technique

Regardless of the method used, the standard sample volume required for total coliform testing is 100 mL.

A public water system must report a violation of the total coliform regulations to the state no later than the end of the next business day. In addition to this, the system must make public notification according to the general public notification requirements of the Safe Drinking Water Act, but with special wording prescribed by the total coliform regulations.

Secondary maximum contaminant levels (SMCLs). The National Safe Drinking Water Act also provided for the establishment of an additional set of standards to prescribe maximum limits for those contaminants that tend to make water disagreeable to use, but that do not have any particular adverse public health effect. These secondary maximum contaminant levels are the advisable maximum level of a contaminant in any public water supply system. The levels are shown in Table 3-9.

AWWA goals. The primary and secondary maximum contaminant levels are the maximum allowed (or recommended) values of the various contaminants. However, a reasonable goal may be much lower than the MCLs themselves. The American Water Works Association (AWWA) has issued its own set of goals to which its members try to adhere. These goals are shown in Table 3-10.

Water Classification and Treatment Systems

Water classification by source. Potable water is most conveniently classified as to its source, that is, groundwater or surface water. Generally, groundwater is uncontaminated but may contain aesthetically or economically undesirable impurities. Surface water must be considered to be contaminated with bacteria, viruses, or in-

TABLE 3-9
Secondary maximum contaminant levels

Contaminant	SMCL (mg/L)[a]
Chloride	250
Color	15 color units
Copper	1
Corrosivity	Noncorrosive
Foaming agents	0.5
Hydrogen sulfide	0.05
Iron	0.3
Manganese	0.05
Odor	3 threshold odor number
pH	6.5–8.5 units
Sulfate	250
Total dissolved solids (TDS)	500
Zinc	5

[a] All quantities are mg/L except those for which units are given.

TABLE 3-10
American Water Works Association water quality goals

Contaminant	Goal (mg/L)[a]
Turbidity	< 0.1 TU
Color	< 3 color units
Odor	None
Taste	None objectionable
Aluminum	< 0.05
Copper	< 0.2
Iron	< 0.05
Manganese	< 0.01
Total dissolved solids (TDS)	200.0
Zinc	< 1.0
Hardness	80.0

[a] All quantities are mg/L except those for which units are given.

organic substances which could present a health hazard. Surface water may also have aesthetically unpleasing characteristics for a potable water. Table 3-11 shows a comparison between groundwater and surface water.

Groundwater is further classified as to its source—deep or shallow wells. Municipal water quality factors of safety, temperature, appearance, taste and odor, and chemical balance are most easily satisfied by a deep well source. High concentrations of calcium, iron, manganese, and magnesium typify well waters. Some supplies contain hydrogen sulfide, while others may have excessive concentrations of chloride, sulfate, or carbonate.

Shallow wells are recharged by a nearby surface watercourse. They may have qualities similar to the deep wells, or they may take on the characteristics of the surface recharge water. A sand aquifer between the shallow well supply and the surface watercourse may act as an effective filter for removal of organic matter and as a heat exchanger for buffering temperature changes. To predict water quality from shallow wells, careful studies of the aquifer and nature of recharge water are necessary.

TABLE 3-11
General characteristics of groundwater and surface water

Ground	Surface
Constant composition	Varying composition
High mineralization	Low mineralization
Little turbidity	High turbidity
Low or no color	Color
Bacteriologically safe	Microorganisms present
No dissolved oxygen	Dissolved oxygen
High hardness	Low hardness
H_2S, Fe, Mn	Tastes and odors
	Possible chemical toxicity

TABLE 3-12
Raw water quality as a function of water source

Source	Turbidity (TU)	Color (Pt-Co units)	Average alum dose (mg/L)
Reservoir	11	18	16
Lake	16	28	22
River	26	44	29

Source: D. A. Cornwell and J. A. Susan, "Characterization of Acid Treated Alum Sludges." *Journal of the American Water Works Association,* October 1979.

Surface water supplies are classified as to whether they come from a lake, reservoir, or river. A comparison of the three is shown in Table 3-12. Generally, a river has the lowest water quality and a reservoir the highest. Water quality in rivers depends upon the character of the watershed. River quality is largely influenced by pollution (or lack thereof) from municipalities, industries, and agricultural practices. The characteristics of a river can be highly variable. During rains or periods of runoff, turbidity may increase substantially. Many rivers will show an increase in color and taste and in odor-producing compounds. In warm months, algal blooms frequently cause taste and odor problems.

Reservoir and lake sources have much less day-to-day variation than rivers. Additionally, the quiescent conditions will reduce both the turbidity and, on occasion, the color. As in rivers, summer algal blooms can create taste and odor problems.

Treatment systems. Treatment plants can be classified as simple disinfection, filter plants, or softening plants. Plants employing simple chlorination have a high water quality source and chlorinate to ensure that the water reaching customers contains safe bacteria levels. Generally, a filtration plant is used to treat surface water and a softening plant to treat groundwater.

In a filtration plant, rapid mixing, flocculation, sedimentation, filtration, and disinfection are employed to remove color, turbidity, taste and odors, and bacteria. Additional operations may include bar racks or coarse screens if floating debris and fish are a problem. Figure 3-6 shows a typical flow diagram of a filtration plant. The

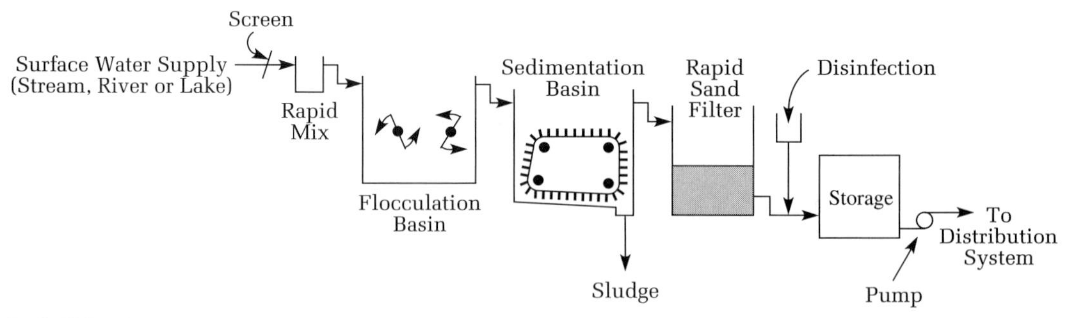

FIGURE 3-6
Flow diagram of a conventional surface water treatment plant ("filtration plant").

raw (untreated) surface water enters the plant via low-lift pumps. Usually screening has taken place prior to pumping. During mixing, chemicals called coagulants are added and rapidly dispersed through the water. The chemical reacts with the desired impurities and forms precipitates (*flocs*) that are slowly brought into contact with one another during flocculation. The objective of flocculation is to allow the flocs to collide and "grow" to a settleable size. The particles are removed by gravity (sedimentation). This is done to minimize the amount of solids that are applied to the filters. For treatment works with a high-quality raw water, it may be possible to omit sedimentation and perhaps flocculation. This modification is called direct filtration. Filtration is the final polishing (removal) of particles. During filtration the water is passed through sand or similar media to screen out the fine particles that will not settle. Disinfection is the addition of chemicals (usually chlorine) to kill or reduce the number of pathogenic organisms. Disinfection of the raw water is neither economical nor efficient. The color and turbidity consume the disinfectant thus requiring the use of excessive amounts of chemical. In addition, the presence of turbidity may shield the pathogens from the action of the disinfectant and thereby prevent efficient destruction. Storage may be provided at the plant or located within the community to meet peak demands and to allow the plant to operate on a uniform schedule. The high-lift pumps provide sufficient pressure to convey the water to its ultimate destination. The precipitated chemicals, original turbidity, and suspended material are removed from the sedimentation basins and from the filters. This *sludge* must be disposed of properly.

Softening plants utilize the same unit operations as filtration plants, but use different chemicals. The primary function of a softening plant is to remove hardness (calcium and magnesium). In a softening plant (a typical flow diagram is shown in Figure 3-7), the design considerations of the various facilities are different than those in filtration. Also the chemical doses are much higher in softening, and the corresponding sludge production is greater.

During rapid mix, chemicals are added to react with and precipitate the hardness. Precipitation occurs in the reaction basin. The other unit operations are the same as in a filtration plant except for the additional recarbonation step employed in softening to adjust the final pH.

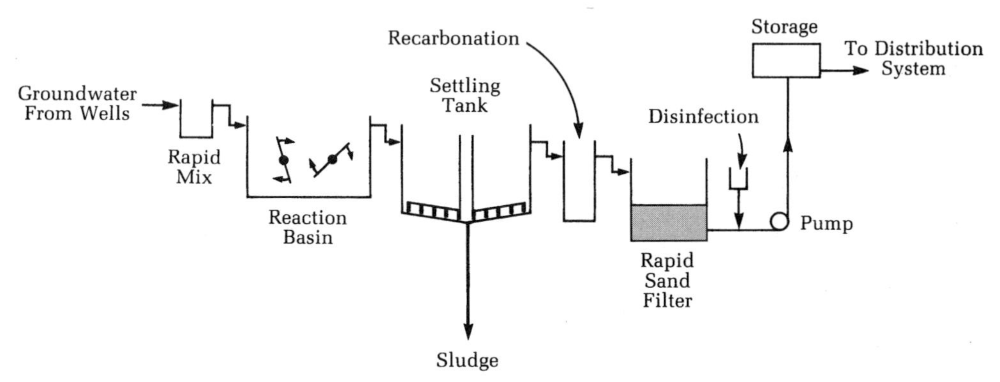

FIGURE 3-7
Flow diagram of water softening plant.

In the next two sections of this chapter, we discuss coagulation chemistry and softening chemistry, respectively. The subsequent sections describe the physical processes themselves. These are applicable to both filtration and softening plants.

3-2 COAGULATION

Surface waters must be treated to remove turbidity, color, and bacteria. When sand filtration was developed around 1885, it became immediately apparent that filtration alone would not produce a clear water. Experience has demonstrated that direct filtration is largely ineffective in removing bacteria, viruses, soil particles, and color.

The object of coagulation (and subsequently flocculation) is to turn the small particles of color, turbidity, and bacteria into larger flocs, either as precipitates or suspended particles. These flocs are then conditioned so that they will be readily removed in subsequent processes. Technically, coagulation applies to the removal of colloidal particles. However, the term has been applied more loosely to removal of dissolved ions, which is actually precipitation. Coagulation in this chapter will refer to colloid removal only. We define *coagulation* as a method to alter the colloids so that they will be able to approach and adhere to each other to form larger floc particles.

Colloid Stability

Before discussing colloid removal, we should understand why the colloids are suspended in solution and can't be removed by sedimentation or filtration. Very simply, the particles in the colloid range are too small to settle in a reasonable time period, and too small to be trapped in the pores of a filter. For colloids to remain stable they must remain small. Most colloids are stable because they possess a negative charge that repels other colloidal particles before they collide with one another.[12] The colloids are continually involved in *Brownian movement,* which is merely random movement. Charges on colloids are measured by placing DC electrodes in a colloidal dispersion. The particles migrate to the pole of opposite charge at a rate proportional to the potential gradient. Generally, the larger the surface charge, the more stable the suspension.

Colloid Destabilization

Since colloids are stable because of their surface charge, in order to destabilize the particles, we must neutralize this charge. Such neutralization can take place by the addition of an ion of opposite charge to the colloid. Since most colloids found in water are negatively charged, the addition of sodium ions (Na^+) should reduce the

[12]Some colloids are stabilized by their affinity for water, but these types of colloids are of less importance.

charge. Figure 3-8 shows such an effect. The plot shows surface charge as a function of distance from the colloid for no-salt (NaCl) addition, low-salt addition, and high-salt addition. As we would have predicted, the higher the concentration of sodium we add, the lower the charge, and therefore the lower the repelling forces around the colloid. If, instead of adding a monovalent ion such as sodium, we add a divalent or trivalent ion, the charge is reduced even faster, as shown in Figure 3-9. In fact, it was found by Schulze and Hardy that one mole of a trivalent ion can reduce the charge as much as 30 to 50 moles of a divalent ion and as much as 1,500 to 2,500 moles of a monovalent ion (often referred to as the *Schulze-Hardy rule*).

Coagulation

The purpose of coagulation is to alter the colloids so that they can adhere to each other. During coagulation a positive ion is added to water to reduce the surface charge to the point where the colloids are not repelled from each other. A *coagulant* is the substance (chemical) that is added to the water to accomplish coagulation. There are three key properties of a coagulant:

1. Trivalent cation. As indicated in the last section, the colloids most commonly found in natural waters are negatively charged, hence a cation is required to neutralize the charge. A trivalent cation is the most efficient cation.
2. Nontoxic. This requirement is obvious for the production of a safe water.
3. Insoluble in the neutral pH range. The coagulant that is added must precipitate out of solution so that high concentrations of the ion are not left in the water. Such precipitation greatly assists the colloid removal process.

The two most commonly used coagulants are aluminum (Al^{3+}) and ferric iron (Fe^{3+}). Both meet the above three requirements, and their reactions are outlined here.

Aluminum. Aluminum can be purchased as either dry or liquid alum [$Al_2(SO_4)_3 \cdot 14H_2O$]. Commercial alum has an average molecular weight of 594. Liquid alum is sold as approximately 48.8 percent alum (8.3% Al_2O_3) and 51.2 percent water. If it is sold as a more concentrated solution, there can be problems with crystallization of the alum during shipment and storage. A 48.8 percent alum solution has a crystallization point of $-15.6°$ C. A 50.7 percent alum solution will crystallize at $+18.3°$ C. Dry alum costs about 50 percent more than an equivalent amount of liquid alum so that only users of very small amounts of alum purchase dry alum.

When alum is added to a water containing alkalinity, the following reaction occurs:

$$Al_2(SO_4)_3 \cdot 14H_2O + 6HCO_3^- \rightleftharpoons 2Al(OH)_3(s) + 6CO_2 + 14H_2O + 3SO_4^{2-} \quad (3\text{-}51)$$

such that each mole of alum added uses six moles of alkalinity and produces six moles of carbon dioxide. The above reaction shifts the carbonate equilibrium and

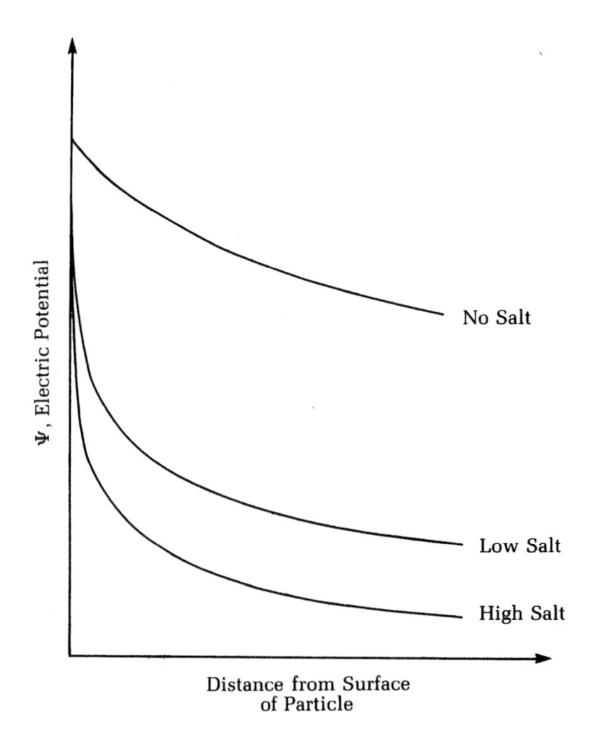

FIGURE 3-8
Effect of salt on electric potential.

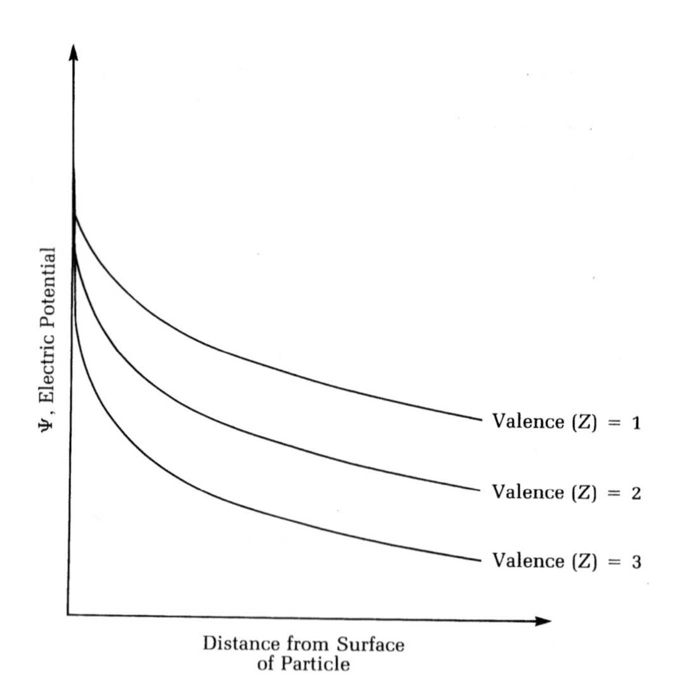

FIGURE 3-9
Effect of valence on electric potential.

decreases the pH. However, as long as sufficient alkalinity is present and $CO_2(g)$ is allowed to evolve, the pH is not drastically reduced and is generally not an operational problem. When sufficient alkalinity is not present to neutralize the sulfuric acid production, the pH may be greatly reduced:

$$Al_2(SO_4)_3 \cdot 14H_2O \rightleftharpoons 2Al(OH)_3(s) + 3H_2SO_4 + 8H_2O \qquad (3\text{-}52)$$

If the second reaction occurs, lime or sodium carbonate may be added to neutralize the acid.

Two important factors in coagulant addition are pH and dose. The optimum dose and pH must be determined from laboratory tests. The optimal pH range for alum is approximately 5.5 to 6.5 with adequate coagulation possible between pH 5 to pH 8 under some conditions.

An important aspect of coagulation is that the aluminum ion does not really exist as Al^{3+} and the final product is more complex than $Al(OH)_3$. When the alum is added to the water, it immediately dissociates, resulting in the release of an aluminum ion surrounded by six water molecules. The aluminum ion immediately starts reacting with the water, forming large $Al \cdot OH \cdot H_2O$ complexes. Some have suggested that it forms $[Al_8(OH)_{20} \cdot 28H_2O]^{4+}$ as the product that actually coagulates. Regardless of the actual species produced, the complex is a very large precipitate that removes many of the colloids by enmeshment as it falls through the water. This precipitate is referred to as a *floc*. Floc formation is one of the important properties of a coagulant for efficient colloid removal.

Example 3-10. One of the most common methods to evaluate coagulation efficiency is to conduct jar tests. Jar tests are performed in an apparatus such as shown in Figure 3-10. Six beakers are filled with water and then each is mixed and flocculated uniformly by a gang stirrer. A test is often conducted by first dosing each jar with the same alum dose and varying the pH in each jar. The test can then be repeated by holding the pH constant and varying the coagulant dose.

Two sets of such jar tests were conducted on a raw water containing 15 TU and an HCO_3^- alkalinity concentration of 50 mg/L expressed as $CaCO_3$. Given the data below, find the optimal pH, coagulant dose, and the theoretical amount of alkalinity that would be consumed at the optimal dose.

Jar test I

	1	2	3	4	5	6
pH	5.0	5.5	6.0	6.5	7.0	7.5
Alum dose (mg/L)	10	10	10	10	10	10
Settled turbidity (TU)	11	7	5.5	5.7	8	13

Jar test II

pH	6.0	6.0	6.0	6.0	6.0	6.0
Alum dose (mg/L)	5	7	10	12	15	20
Settled turbidity (TU)	14	9.5	5	4.5	6	13

(*a*)

(*b*)

FIGURE 3-10
Jar tests: (*a*) Phipps and Bird jar tester; (*b*) left beaker: lower coagulant dose; right beaker: higher coagulant dose.

Solution. The results of the two jar tests are plotted in Figure 3-11. The optimal pH was chosen as 6.25 and the optimal alum dose was about 12.5 mg/L. The experimenter would probably try to repeat the test using a pH of 6.25 and varying the alum dose between 10 and 15 to pinpoint the optimal conditions.

The amount of alkalinity that will be consumed is found by using Equation 3-51, which shows us that one mole of alum consumes six moles of HCO_3^-. With the molecular weight of alum equal to 594, the moles of alum added per liter is found by using Equation 3-7:

$$\frac{12.5 \times 10^{-3} \text{ g/L}}{594 \text{ g/mole}} = 2.1 \times 10^{-5} \text{ moles/L}$$

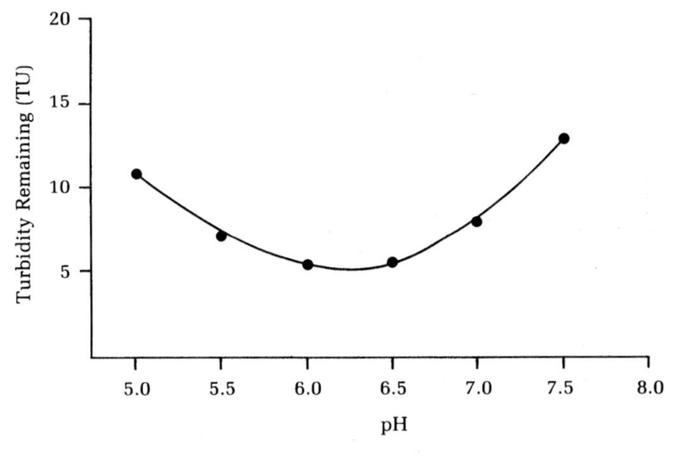

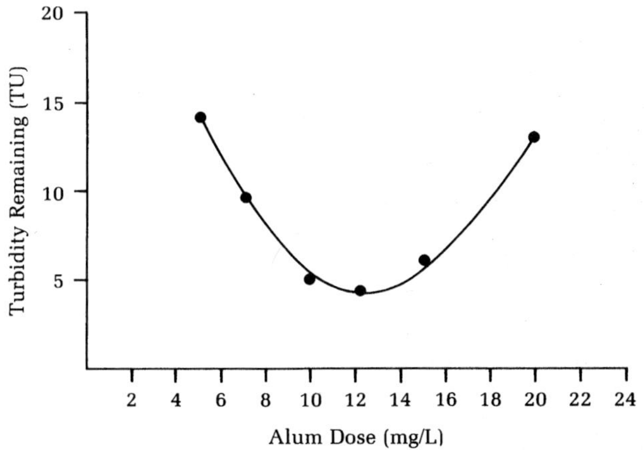

FIGURE 3-11
Results of jar test.

which will consume

$$6(2.1 \times 10^{-5}) = 1.26 \times 10^{-4}\ M\ HCO_3^-$$

The molecular weight of HCO_3^- is 61, so

$$(1.26 \times 10^{-4} \text{ moles/L})(61 \text{ g/mole})(10^3 \text{ mg/g}) = 7.7 \text{ mg/L } HCO_3^-$$

are consumed, which can be expressed as $CaCO_3$ by using Equation 3-40:

$$7.7\,\frac{50}{61} = 6.31 \text{ mg/L } HCO_3^- \text{ as } CaCO_3$$

Iron. Iron can be purchased as either the sulfate salt ($Fe_2(SO_4)_3 \cdot xH_2O$) or the chloride salt ($FeCl_3 \cdot xH_2O$). It is available in various forms, and the individual supplier should be consulted for the specifics of the product. Dry and liquid forms are available. The properties of iron with respect to forming large complexes, dose, and pH curves are similar to those of alum. An example of the reaction of $FeCl_3$ in the presence of alkalinity is

$$FeCl_3 + 3HCO_3^- \rightleftharpoons Fe(OH)_3(s) + 3CO_2 + 3Cl^- \quad (3\text{-}53)$$

and without alkalinity

$$FeCl_3 + 3H_2O \rightleftharpoons Fe(OH)_3(s) + 3HCl \quad (3\text{-}54)$$

forming hydrochloric acid which in turn lowers the pH. Ferric salts generally have a wider pH range for effective coagulation than aluminum, that is, pH ranges from 4 to 9.

Coagulant aids. The four basic types of coagulant aids are pH adjusters, activated silica, clay, and polymers. Acids and alkalies are both used to adjust the pH of the water into the optimal range for coagulation. The acid most commonly used for lowering the pH is sulfuric acid. Either lime [$Ca(OH)_2$] or soda ash (Na_2CO_3) are used to raise the pH.

When activated silica is added to water, it produces a stable solution that has a negative surface charge. The activated silica can unite with the positively charged aluminum or with iron flocs, resulting in a larger, denser floc that settles faster and enhances enmeshment. The addition of activated silica is especially useful for treating highly colored, low-turbidity waters because it adds weight to the floc. However, activation of silica does require proper equipment and close operational control, and many plants are hesitant to use it.

Clays can act much like activated silica in that they have a slight negative charge and can add weight to the flocs. Clays are also most useful for colored, low-turbidity waters.

Polymers can have a negative charge (*anionic*), positive charge (*cationic*), positive and negative charge (*polyamphotype*), or no charge (*nonionic*). Polymers are long-chained carbon compounds of high molecular weight that have many active sites. The active sites adhere to flocs, joining them together and producing a larger, tougher floc that settles better. This process is called *interparticle bridging.* The type of polymer, dose, and point of addition must be determined for each water, and requirements may change within a plant on a seasonal, or even daily, basis.

3-3 SOFTENING

Hardness. The term hardness is used to characterize a water that does not lather well, causes a scum in the bath tub, and leaves hard, white, crusty deposits (scale) on coffee pots, tea kettles, and hot water heaters. The failure to lather well and the formation of scum on bath tubs is the result of the reactions of calcium and magnesium with the soap. For example:

$$Ca^{2+} + (Soap)^- \rightleftharpoons Ca(Soap)_2(s) \quad (3\text{-}55)$$

As a result of this complexation reaction, soap cannot interact with the dirt on clothing, and the calcium-soap complex itself forms undesirable precipitates.

TABLE 3-13
Hard water classification

Hardness range (mg/L $CaCO_3$)	Description
0–75	Soft
75–100	Moderately hard
100–300	Hard
>300	Very hard

Hardness is defined as the sum of all polyvalent cations (in consistent units). The common units of expression are mg/L as $CaCO_3$ or meq/L. Qualitative terms used to describe hardness are listed in Table 3-13. Because many people object to water containing hardness greater than 150 mg/L as $CaCO_3$, suppliers of public water have considered it a benefit to *soften* the water, that is, to remove some of the hardness. A common water treatment goal is to provide water with a hardness in the range of 75 to 120 mg/L as $CaCO_3$.

Although all polyvalent cations contribute to hardness, the predominant contributors are calcium and magnesium. Thus, our focus for the remainder of this discussion will be on calcium and magnesium.

The natural process by which water becomes hard is shown schematically in Figure 3-12. As rainwater enters the topsoil, the respiration of microorganisms increases the CO_2 content of the water. As shown in Equation 3-35, the CO_2 reacts with the water to form H_2CO_3. Limestone, which is made up of solid $CaCO_3$ and

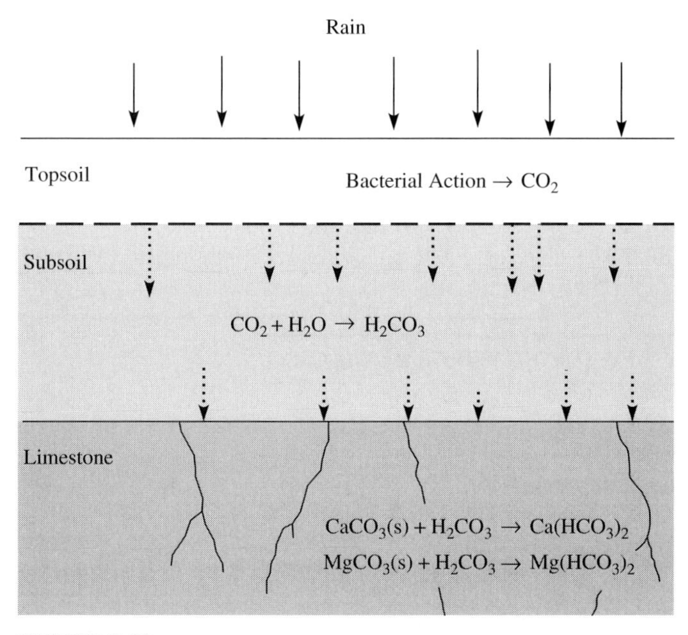

FIGURE 3-12
Natural process by which water is made hard.

$MgCO_3$, reacts with the carbonic acid to form calcium bicarbonate [$Ca(HCO_3)_2$] and magnesium bicarbonate [$Mg(HCO_3)_2$]. While $CaCO_3$ and $MgCO_3$ are both insoluble in water, the bicarbonates are quite soluble. Gypsum ($CaSO_4$) and $MgSO_4$ may also go into solution to contribute to the hardness.

Because calcium and magnesium predominate, it is often convenient in performing softening calculations to define the *total hardness* (TH) of a water as the sum of these elements

$$TH = Ca^{2+} + Mg^{2+} \tag{3-56}$$

where the concentrations of each element are in consistent units (mg/L as $CaCO_3$ or meq/L). Total hardness is often broken down into two components: (1) that associated with the HCO_3^- anion (called *carbonate hardness* and abbreviated CH), and (2) that associated with other anions (called *noncarbonate hardness* and abbreviated NCH).[13] Total hardness, then, may also be defined as

$$TH = CH + NCH \tag{3-57}$$

Carbonate hardness is defined as the amount of hardness equal to the total hardness or the total alkalinity, whichever is less. Carbonate hardness is often called temporary hardness because heating the water removes it. When the pH is less than 8.3, HCO_3^- is the dominant form of alkalinity, and total alkalinity is nominally taken to be equal to the concentration of HCO_3^-.

Noncarbonate hardness is defined as the total hardness in excess of the alkalinity. If the alkalinity is equal to or greater than the total hardness, then there is no noncarbonate hardness. Noncarbonate hardness is called permanent hardness because it is not removed when water is heated.

Bar charts of water composition are often useful in understanding the process of softening. By convention, the bar chart is constructed with cations in the upper bar and anions in the lower bar. In the upper bar, calcium is placed first and magnesium second. Other cations follow without any specified order. The lower bar is constructed with bicarbonate placed first. Other anions follow without any specified order. Construction of a bar chart is illustrated in Example 3-11.

Example 3-11. Given the following analysis of a groundwater, construct a bar chart of the constituents, expressed as $CaCO_3$.

Ion	mg/L as ion	EW $CaCO_3$/EW ion	mg/L as $CaCO_3$
Ca^{2+}	103	2.50	258
Mg^{2+}	5.5	4.12	23
Na^+	16	2.18	35
HCO_3^-	255	0.82	209
SO_4^{2-}	49	1.04	51
Cl^-	37	1.41	52

[13]Note that this does not imply that the compounds exist as compounds in solution. They are dissociated.

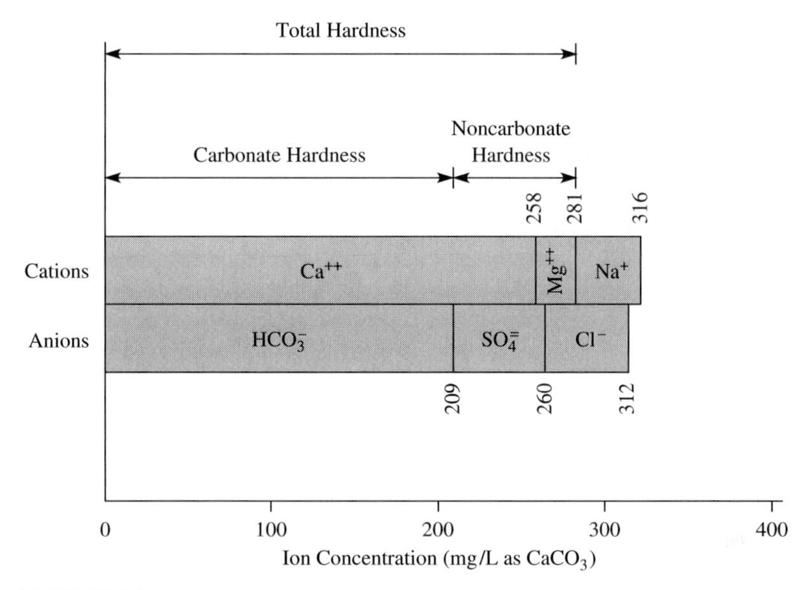

FIGURE 3-13
Bar graph of groundwater constituents.

Solution. The concentrations of the ions have been converted to $CaCO_3$ equivalents. The results are plotted in Figure 3-13.

The cations total 316 mg/L as $CaCO_3$, of which 281 mg/L as $CaCO_3$ is hardness. The anions total 312 mg/L as $CaCO_3$, of which the carbonate hardness is 209 mg/L as $CaCO_3$. There is a discrepancy between the cation and anion totals because there are other ions that were not analyzed. If a complete analysis were conducted, and no analytical error occurred, the equivalents of cations would equal exactly the equivalents of anions. Typically, a complete analysis may vary $\pm 5\%$ because of analytical errors.

The relationships between total hardness, carbonate hardness, and noncarbonate hardness are illustrated in Figure 3-14. In Figure 3-14a, the total hardness is 250 mg/L as $CaCO_3$, the carbonate hardness is equal to the alkalinity ($HCO_3^- =$ 200 mg/L as $CaCO_3$), and the noncarbonate hardness is equal to the difference between the total hardness and the carbonate hardness ($NCH = TH - CH = 250 - 200 = 50$ mg/L as $CaCO_3$). In Figure 3-14b, the total hardness is again 250 mg/L as $CaCO_3$. However, since the alkalinity (HCO_3^-) is greater than the total hardness, and since the carbonate hardness cannot be greater than the total hardness (see Equation 3-57), the carbonate hardness is equal to the total hardness, that is, 250 mg/L as $CaCO_3$. With the carbonate hardness equal to the noncarbonate hardness, then all of the hardness is carbonate hardness and there is no noncarbonate hardness. Note that in both cases it may be assumed that the pH is less than 8.3 because HCO_3^- is the only form of alkalinity present.

Example 3-12. A water has an alkalinity of 200 mg/L as $CaCO_3$. The Ca^{2+} concentration is 160 mg/L as the ion, and the Mg^{2+} concentration is 40 mg/L as the ion. The pH is 8.1. Find the total, carbonate, and noncarbonate hardness.

Solution. The molecular weights of calcium and magnesium are 40 and 24 respectively. Since each has a valence of 2^+, the corresponding equivalent weights are 20 and 12. Using Equation 3-40 to convert mg/L as the ion to mg/L as $CaCO_3$ and adding the two ions as shown in Equation 3-56, the total hardness is

$$TH = 160 \text{ mg/L}\left(\frac{50 \text{ mg/meq}}{20 \text{ mg/meq}}\right) + 40 \text{ mg/L}\left(\frac{50 \text{ mg/meq}}{12 \text{ mg/meq}}\right) = 567 \text{ mg/L as } CaCO_3$$

where 50 is the equivalent weight of $CaCO_3$.

By definition, the carbonate hardness is the lesser of the total hardness or the alkalinity. Since, in this case, the alkalinity is less than the total hardness, the carbonate hardness (CH) is equal to 200 mg/L as $CaCO_3$. The noncarbonate hardness is equal to the difference

$$NCH = TH - CH = 567 - 200 = 367 \text{ mg/L as } CaCO_3$$

Note that we can only add and subtract concentrations of Ca^{2+} and Mg^{2+} if they are in equivalent units, for example, moles/L or milliequivalents/L or mg/L as $CaCO_3$.

Softening can be accomplished by either the lime-soda process or by ion exchange. Both methods are discussed in the following sections.

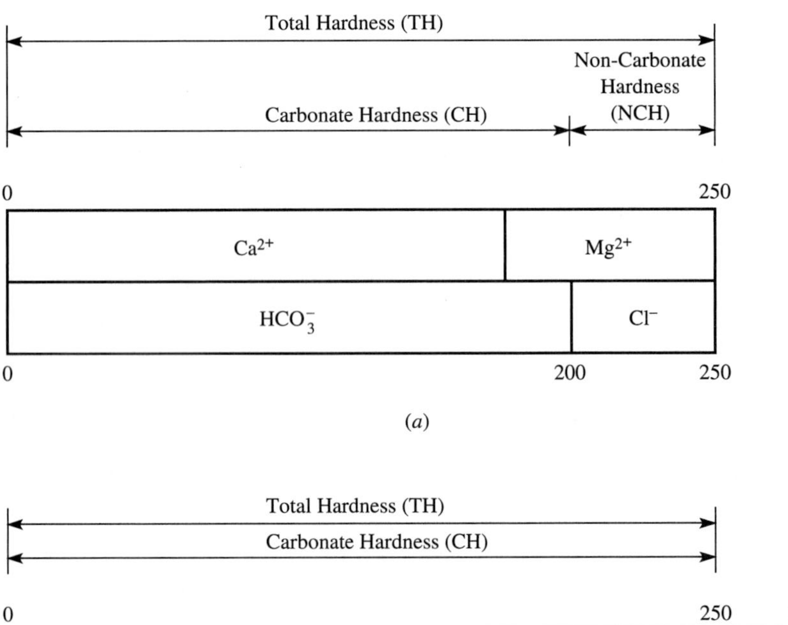

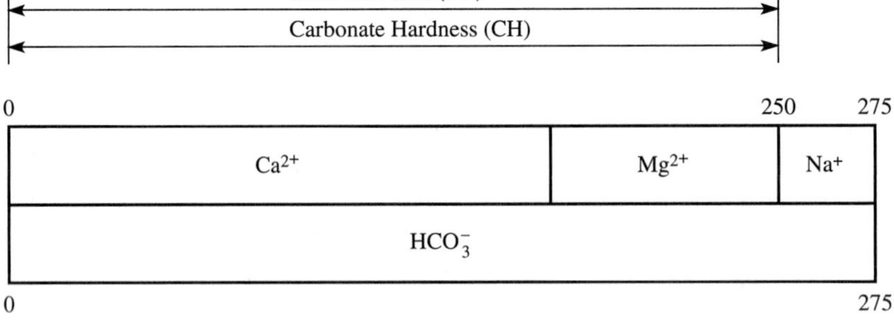

FIGURE 3-14
Relationships between total hardness, carbonate hardness, and noncarbonate hardness.

Lime-Soda Softening

In lime-soda softening it is possible to calculate the chemical doses necessary to remove hardness. Hardness precipitation is based on the following two solubility reactions:

$$Ca^{2+} + CO_3^{2-} \rightleftharpoons CaCO_3(s) \tag{3-58}$$

and

$$Mg^{2+} + 2OH^- \rightleftharpoons Mg(OH)_2(s) \tag{3-59}$$

The objective is to precipitate the calcium as $CaCO_3$ and the magnesium as $Mg(OH)_2$. In order to precipitate $CaCO_3$, the pH of the water must be raised to about 10.3. To precipitate magnesium, the pH must be raised to about 11. If there is not sufficient naturally occurring bicarbonate alkalinity (HCO_3^-) for the $CaCO_3(s)$ precipitate to form (that is, there is noncarbonate hardness), we must add CO_3^{2-} to the water. Magnesium is more expensive to remove than calcium, so we leave as much Mg^{2+} in the water as possible. It is more expensive to remove noncarbonate hardness than carbonate hardness because we must add another chemical to provide the CO_3^{2-}. Therefore, we leave as much noncarbonate hardness in the water as possible.

Softening chemistry. The chemical processes used to soften water are a direct application of the law of mass action. We increase the concentration of CO_3^{2-} and/or OH^- by the addition of chemicals, and drive the reactions given in Equations 3-58 and 3-59 to the right. Insofar as possible, we convert the naturally occurring bicarbonate alkalinity (HCO_3^-) to carbonate (CO_3^{2-}) by the addition of hydroxyl ions (OH^-). Hydroxyl ions cause the carbonate buffer system (Equation 3-35) to shift to the right and, thus, provide the carbonate for the precipitation reaction (Equation 3-58).

The common source of hydroxyl ions is calcium hydroxide [$Ca(OH)_2$]. Many water treatment plants find it more economical to buy quicklime (CaO), commonly called lime, than hydrated lime [$Ca(OH)_2$]. The quicklime is converted to hydrated lime at the water treatment plant by mixing CaO and water to produce a slurry of $Ca(OH)_2$, which is fed to the water for softening. The conversion process is called *slaking:*

$$CaO + H_2O \rightleftharpoons Ca(OH)_2 + \text{heat} \tag{3-60}$$

The reaction is exothermic. It yields almost 1 MJ per gram mole of lime. Because of this high heat release, the reaction must be controlled carefully. All safety precautions for handling a strong base should be observed. Because the chemical is purchased as lime, it is common to speak of chemical additions as addition of "lime," when in fact we mean calcium hydroxide. When carbonate ions must be supplied, the most common chemical chosen is sodium carbonate (Na_2CO_3). Sodium carbonate is commonly referred to as *soda ash* or *soda.*

Softening reactions. The softening reactions are regulated by controlling the pH. First, any free acids are neutralized. Then pH is raised to precipitate the $CaCO_3$; if necessary, the pH is raised further to remove $Mg(OH)_2$. Finally, if necessary, CO_3^{2-} is added to precipitate the noncarbonate hardness.

Six important softening reactions are discussed below. In each case, the chemical that has been added to the water is printed in bold type. Remember that (s) designates the solid form, and hence indicates that the substance has been removed from the water. The following reactions are presented sequentially, although in reality they occur simultaneously.

1. Neutralization of carbonic acid (H_2CO_3).

In order to raise the pH, we must first neutralize any free acids that may be present in the water. CO_2 is the principal acid present in unpolluted, naturally occurring water.[14] You should note that no hardness is removed in this step.

$$CO_2 + \mathbf{Ca(OH)_2} \rightleftharpoons CaCO_3(s) + H_2O \tag{3-61}$$

2. Precipitation of carbonate hardness due to calcium.

As we mentioned previously, we must raise the pH to about 10.3 to precipitate calcium carbonate. To achieve this pH we must convert all of the bicarbonate to carbonate. The carbonate then serves as the common ion for the precipitation reaction.

$$Ca^{2+} + 2HCO_3^- + \mathbf{Ca(OH)_2} \rightleftharpoons 2CaCO_3(s) + 2H_2O \tag{3-62}$$

3. Precipitation of carbonate hardness due to magnesium.

If we need to remove carbonate hardness that results from the presence of magnesium, we must add more lime to achieve a pH of about 11. The reaction may be considered to occur in two stages. The first stage occurs when we convert all of the bicarbonate in step 2 above.

$$Mg^{2+} + 2HCO_3^- + \mathbf{Ca(OH)_2} \rightleftharpoons MgCO_3 + CaCO_3(s) + 2H_2O \tag{3-63}$$

Note that the hardness of the water did not change because $MgCO_3$ is soluble. With the addition of more lime the hardness due to magnesium is removed.

$$Mg^{2+} + CO_3^{2-} + \mathbf{Ca(OH)_2} \rightleftharpoons Mg(OH)_2(s) + CaCO_3(s) \tag{3-64}$$

4. Removal of noncarbonate hardness due to calcium.

If we need to remove noncarbonate hardness due to calcium, no further increase in pH is required. Instead we must provide additional carbonate in the form of soda ash.

$$Ca^{2+} + \mathbf{Na_2CO_3} \rightleftharpoons CaCO_3(s) + 2Na^+ \tag{3-65}$$

---

[14] CO_2 and H_2CO_3 in water are essentially the same:

$$CO_2 + H_2O \rightleftharpoons H_2CO_3$$

Thus, the number of reaction units (n) for CO_2 is two.

5. Removal of noncarbonate hardness due to magnesium.

If we need to remove noncarbonate hardness due to magnesium, we will have to add both lime and soda. The lime provides the hydroxyl ion for precipitation of the magnesium.

$$Mg^{2+} + \mathbf{Ca(OH)_2} \rightleftharpoons Mg(OH)_2(s) + Ca^{2+} \qquad (3\text{-}66)$$

Note that although the magnesium is removed, there is no change in the hardness because the calcium is still in solution. To remove the calcium we must add soda.

$$Ca^{2+} + \mathbf{Na_2CO_3} \rightleftharpoons CaCO_3(s) + 2Na^{+} \qquad (3\text{-}67)$$

Note that this is the same reaction as the one to remove noncarbonate hardness due to calcium.

These reactions are summarized in Figure 3-15.

Process limitations and empirical considerations. Lime-soda softening cannot produce a water completely free of hardness because of the solubility of $CaCO_3$ and $Mg(OH)_2$, the physical limitations of mixing and contact, and the lack of sufficient time for the reactions to go to completion. Thus, the minimum calcium hardness that can be achieved is about 30 mg/L as $CaCO_3$, and the minimum magnesium hardness is about 10 mg/L as $CaCO_3$. Because of the slimy condition that results when soap is used with a water that is too soft, we normally strive for a final total hardness on the order of 75 to 120 mg/L as $CaCO_3$.

In order to achieve reasonable removal of hardness in a reasonable time period, an excess of $Ca(OH)_2$ beyond the stoichiometric amount usually is provided. Based on our empirical experience, a minimum excess of 20 mg/L of $Ca(OH)_2$ expressed as $CaCO_3$ must be provided.

Neutralization of Carbonic Acid

$$CO_2 + \mathbf{Ca(OH)_2} \rightleftharpoons CaCO_3(s) + H_2O$$

Precipitation of Carbonate Hardness

$$Ca^{2+} + 2HCO_3^- + \mathbf{Ca(OH)_2} \rightleftharpoons 2CaCO_3(s) + 2H_2O$$

$$Mg^{2+} + 2HCO_3^- + \mathbf{Ca(OH)_2} \rightleftharpoons MgCO_3 + CaCO_3(s) + 2H_2O$$

$$MgCO_3 + \mathbf{Ca(OH)_2} = Mg(OH)_2(s) + CaCO_3(s)$$

Precipitation of Noncarbonate Hardness Due to Calcium

$$Ca^{2+} + \mathbf{Na_2CO_3} \rightleftharpoons CaCO_3(s) + 2Na^{+}$$

Precipitation of Noncarbonate Hardness Due to Magnesium

$$Mg^{2+} + \mathbf{Ca(OH)_2} \rightleftharpoons Mg(OH)_2(s) + Ca^{2+}$$

$$Ca^{2+} + \mathbf{Na_2CO_3} \rightleftharpoons CaCO_3(s) + 2Na^{+}$$

FIGURE 3-15

Summary of softening reactions. (*Note:* The chemical added is printed in bold type. The precipitate is designated by (s). The arrow indicates where a compound formed in one reaction is used in another reaction.)

Magnesium in excess of about 40 mg/L as $CaCO_3$ forms scales on heat exchange elements in hot water heaters. Because of the expense of removing magnesium, we normally remove only that magnesium which is in excess of 40 mg/L as $CaCO_3$. For magnesium removals less than 20 mg/L as $CaCO_3$, the basic excess of lime mentioned above is sufficient to ensure good results. For magnesium removals between 20 and 40 mg/L as $CaCO_3$, we must add an excess of lime equal to the magnesium to be removed. For magnesium removals greater than 40 mg/L as $CaCO_3$, the excess lime we need to add is 40 mg/L as $CaCO_3$. Addition of excess lime in amounts greater than 40 mg/L as $CaCO_3$ does not appreciably improve the reaction kinetics.

The chemical additions (as $CaCO_3$) to soften water may be summarized as follows:

Step	Chemical addition[a]	Reason
Carbonate hardness		
1.	Lime = CO_2	Destroy H_2CO_3
2.	Lime = HCO_3^-	Raise pH; convert HCO_3^- to CO_3^{2-}
3.	Lime = Mg^{2+} to be removed	Raise pH; precipitate $Mg(OH)_2$
4.	Lime = required excess	Drive reaction
Noncarbonate hardness		
5.	Soda = noncarbonate hardness to be removed	Provide CO_3^{2-}

[a] The terms "Lime =" and "Soda =" refer to mg/L of $Ca(OH)_2$ and Na_2CO_3 as $CaCO_3$ equal to mg/L of ion (or gas in the case of CO_2) as $CaCO_3$.

These steps are diagrammed in the flow chart shown in Figure 3-16. The next three examples illustrate the technique.

Example 3-13. From the water analysis presented below, determine the amount of lime and soda (in mg/L as $CaCO_3$) necessary to soften the water to 80.00 mg/L hardness as $CaCO_3$.

Water Composition (mg/L)

Ca^{2+}:	95.20	CO_2:	19.36	HCO_3^-:	241.46
Mg^{2+}:	13.44			SO_4^{2-}:	53.77
Na^+:	25.76			Cl^-:	67.81

Solution. We begin by converting the elements and compounds to $CaCO_3$ equivalents.

Ion	mg/L as ion	EW $CaCO_3$/EW ion	mg/L as $CaCO_3$
Ca^{2+}	95.20	2.50	238.00
Mg^{2+}	13.44	4.12	55.37
Na^+	25.76	2.18	56.16
HCO_3^-	241.46	0.820	198.00
SO_4^{2-}	53.77	1.04	55.92
Cl^-	67.81	1.41	95.61
CO_2	19.36	2.28	44.14

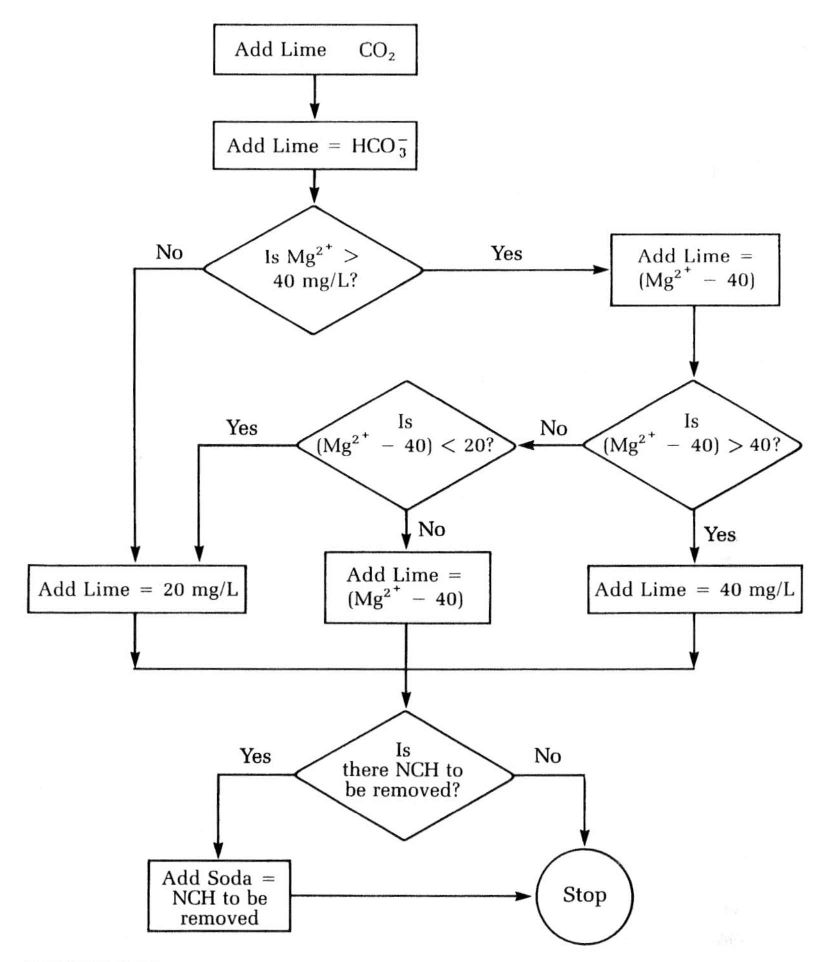

FIGURE 3-16
Flow diagram for solving softening problems. (*Note:* All additions are "as $CaCO_3$." NCH means noncarbonate hardness.)

The resulting bar chart would appear as shown below.

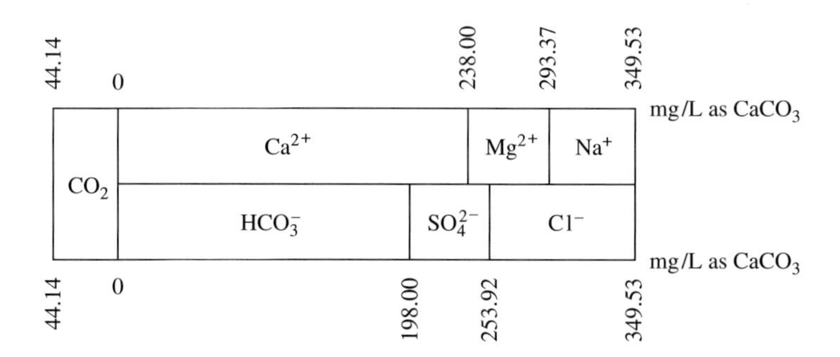

From the bar chart we note the following: CO_2 does not contribute to the hardness; the total hardness (TH) = 293.37 mg/L as $CaCO_3$; the carbonate hardness (CH) = 198.00 mg/L as $CaCO_3$; and, finally, the noncarbonate hardness (NCH) is equal to TH − CH = 95.37 mg/L as $CaCO_3$.

Using Figure 3-16 to guide our logic, we determine the lime dose as follows:

Step	Dose (mg/L as $CaCO_3$)
Lime = CO_2	44.14
Lime = HCO_3^-	198.00
Lime = $Mg^{2+} - 40 = 55.37 - 40$	15.37
Lime = excess	20.00
	277.51

The amount of lime to add is 277.51 mg/L as $CaCO_3$. The excess chosen was the minimum since ($Mg^{2+} - 40$) was less than 20.

Now we must determine if any NCH need be removed. The amount of NCH that can be left (NCH_f) is equal to the final hardness desired (80.00 mg/L) minus the CH left due to solubility and other factors (40.00 mg/L):

$$NCH_f = 80.00 - 40.00 = 40.00 \text{ mg/L}$$

Thus, 40.00 mg/L may be left. The NCH that must be removed (NCH_R) is the initial NCH_i (95.37 mg/L) minus the NCH_f:

$$NCH_R = NCH_i - NCH_f$$

$$NCH_R = 95.37 - 40.00 = 55.37 \text{ mg/L}$$

Thus, the amount of soda to be added is 55.37 mg/L as $CaCO_3$. The fact that this number is equal to the Mg^{2+} concentration is a coincidence of the numbers chosen.

Example 3-14. From the water analysis presented below, determine the amount of lime and soda (in mg/L as $CaCO_3$) necessary to soften the water to 90.00 mg/L as $CaCO_3$.

Water Composition (mg/L as $CaCO_3$)

Ca^{2+}:	149.2	CO_2:	29.3	HCO_3^-:	185.0
Mg^{2+}:	65.8			SO_4^{2-}:	29.8
Na^+:	17.4			Cl^-:	17.6

Solution. The bar chart for this water may be plotted directly as shown below.

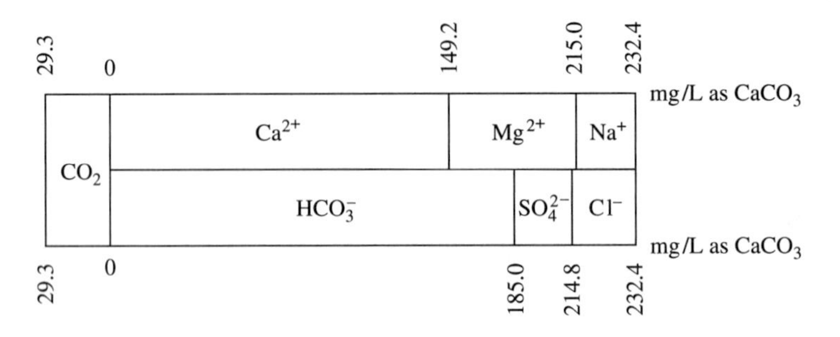

From the bar chart we note the following:

$$\text{TH} = 215.0 \text{ mg/L as } CaCO_3$$
$$\text{CH} = 185.0 \text{ mg/L as } CaCO_3$$
$$\text{NCH} = 30.0 \text{ mg/L as } CaCO_3$$

Following the logic of Figure 3-16, we calculate the lime dose as follows:

Step	Dose (mg/L as $CaCO_3$)
Lime = CO_2	29.3
Lime = HCO_3^-	185.0
Lime = $Mg^{2+} - 40 = 65.8 - 40 =$	25.8
Lime = excess	25.8
	265.9

The amount of lime to add is 265.9 mg/L as $CaCO_3$. The excess chosen was equal to the difference between the Mg^{2+} concentration and 40 since that difference was between 20 and 40, that is, $Mg^{2+} - 40 = 25.8$.

The amount of NCH_R is calculated as in Example 3-13:

$$NCH_f = 90.00 - 40.00 = 50.00$$

$$NCH_R = 30.00 - 50.00 = -20.00$$

Since NCH_R is a negative number, there is no need to remove NCH and, therefore, *no soda ash is required.*

Example 3-15. Given the following water, determine the amount (mg/L) of 90 percent purity CaO and 97 percent purity Na_2CO_3 that must be purchased to treat the water to a final hardness of 85 mg/L; 120 mg/L.

Ion	(mg/L as $CaCO_3$)
CO_2	21
HCO_3^-	209
Ca^{2+}	183
Mg^{2+}	97

Solution. First find the total hardness (TH), carbonate hardness (CH), and noncarbonate hardness (NCH):

$$\text{TH} = Ca^{2+} + Mg^{2+} = 183 + 97 = 280 \text{ mg/L}$$

$$\text{CH} = HCO_3^- = 209 \text{ mg/L}$$

$$\text{NCH} = \text{TH} - \text{CH} = 71 \text{ mg/L}$$

Use Figure 3-16 to find the lime dose as $CaCO_3$, assuming that we will leave 40 mg/L Mg^{2+} in the water:

$$Ca(OH)_2 = [21 + 209 + (97 - 40) + 40] = 327 \text{ mg/L as } CaCO_3$$

Since one mole of CaO equals one mole of $Ca(OH)_2$, we find 327 mg/L of CaO as $CaCO_3$. The molecular weight of CaO is 56 (equivalent weight = 28), and correcting for 90 percent purity:

$$CaO = 327\left(\frac{28}{50}\right)\left(\frac{1}{.9}\right) = 203 \text{ mg/L as CaO}$$

The amount of NCH that can be left in solution is equal to the final hardness desired (85 mg/L) minus the CH left behind due to solubility, inefficient mixing, etc. (40 mg/L), and is equal to 85 − 40 = 45 mg/L.

Therefore, the NCH_R is the initial NCH (71 mg/L) minus the NCH which can be left (45 mg/L) and is 71 − 45 = 26 mg/L. From Figure 3-16:

$$Na_2CO_3 = 26 \text{ mg/L as } CaCO_3$$

The equivalent weight of soda ash is 53 and the purity 97 percent:

$$Na_2CO_3 = 26\left(\frac{53}{50}\right)\left(\frac{1}{.97}\right) = 28 \text{ mg/L as } Na_2CO_3$$

If the final hardness desired is 120 mg/L, then the allowable final NCH is 120 − 40 = 80 mg/L, which is greater than the initial NCH of 71, so no soda ash is necessary. The final hardness would be about 40 mg/L (carbonate hardness solubility) plus 71 mg/L (noncarbonate hardness) or 111 mg/L.

More Advanced Concepts in Lime-Soda Softening

Estimating CO_2 concentration. CO_2 is of importance in two instances in softening. In the first instance, it consumes lime that otherwise could be used to remove Ca and Mg. When the concentration of CO_2 exceeds 10 mg/L as CO_2 (22.7 mg/L as $CaCO_3$ or 0.45 meq/L), the economics of removal of CO_2 by aeration (stripping) are favored over removal by lime neutralization. In the second instance, CO_2 is used to neutralize the high pH of the effluent from the softening process. These reactions are an application of the concepts of the carbonate buffer system discussed in Section 3-1.

The concentration of CO_2 may be estimated by using the equilibrium expressions for the dissociation of water and carbonic acid with the definition of alkalinity (Equation 3-36). The pH and alkalinity of the water must be determined to make the estimate. Example 3-16 illustrates a simple case where one of the forms of alkalinity predominates. Problems 3-26 through 3-30 illustrate more complex cases.

Example 3-16. What is the estimated CO_2 concentration of a water with a pH of 7.65 and a total alkalinity of 310 mg/L as $CaCO_3$?

Solution. When the raw water pH is less than 8.5, we can assume that the alkalinity is predominately HCO_3^-. Thus, we can ignore the dissociation of bicarbonate to form carbonate.

With this assumption, the procedure to solve the problem is

a. Calculate the $[H^+]$ from the pH.

b. Calculate the $[HCO_3^-]$ from the alkalinity.

c. Solve the first equilibrium expression of the carbonic acid dissociation for $[H_2CO_3]$.

d. Use the assumption that $[CO_2] = [H_2CO_3]$ to estimate the CO_2 concentration.

Following this approach, the $[H^+]$ concentration is

$$[H^+] = 10^{-7.65} = 2.24 \times 10^{-8} \text{ moles/L}$$

The $[HCO_3^-]$ concentration is

$$[HCO_3^-] = 310 \text{ mg/L}\left(\frac{61 \text{ mg/meq}}{50 \text{ mg/meq}}\right)\left(\frac{1}{(61 \text{ g/mole})(10^3 \text{ mg/g})}\right)$$

$$= 6.20 \times 10^{-3} \text{ moles/L}$$

Since the alkalinity is reported as mg/L as $CaCO_3$, it must be converted to mg/L as the species using Equation 3-40 before the molar concentration may be calculated. The ratio 61/50 is the ratio of the equivalent weight of HCO_3^- to the equivalent weight of $CaCO_3$.

The equilibrium expression for the dissociation of carbonic acid is written in the form of Equation 3-33 using the reaction and pK_a given in Table 3-3.

$$K_a = \frac{[H^+][HCO_3^-]}{[H_2CO_3]}$$

where $K_a = 10^{-6.35} = 4.47 \times 10^{-7}$

Solving for $[H_2CO_3]$

$$[H_2CO_3] = \frac{(2.24 \times 10^{-8} \text{ moles/L})(6.20 \times 10^{-3} \text{ moles/L})}{4.47 \times 10^{-7}}$$

$$[H_2CO_3] = 3.11 \times 10^{-4} \text{ moles/L}$$

We may assume that all the CO_2 in water forms carbonic acid. Thus, the estimated CO_2 concentration is

$$[CO_2] = 3.11 \times 10^{-4} \text{ moles/L}$$

In other units for comparison and calculation:

$$CO_2 = (3.11 \times 10^{-4} \text{ moles/L})(44 \times 10^3 \text{ mg/mole}) = 13.7 \text{ mg/L as } CO_2$$

$$CO_2 = (13.7 \text{ mg/L as } CO_2)\left(\frac{50 \text{ mg/meq}}{22 \text{ mg/meq}}\right) = 31.14 \text{ or } 31.1 \text{ mg/L as } CaCO_3$$

The equivalent weight of CO_2 is taken as 22 because it effectively behaves as carbonic acid (H_2CO_3) and thus $n = 2$.

Softening to practical limits. In Example 3-14, the NCH_R was found to be a negative number. This implies that the water will be softened to a hardness less than the desired value. One way to overcome this difficulty is to treat a portion of the water to the practical limits and then blend the treated water with the raw water to achieve the desired hardness. Unlike the flowchart method used in Examples 3-13 through 3-15, calculations of chemical additions to soften to the practical limits of softening (that is 0.60 meq/L or 30 mg/L as $CaCO_3$ of Ca and 0.20 meq/L or 10 mg/L as $CaCO_3$ of $Mg(OH)_2$), do not take into account the desired final hardness. Stoichiometric amounts of lime and soda are added to remove all of the Ca and Mg. Example 3-17 illustrates the technique using both mg/L as $CaCO_3$ and milliequivalents/L as units of measure.

Example 3-17. Determine the chemical dosages for softening the following water to the practical solubility limits.

Constituent	mg/L	EW	EW $CaCO_3$/EW ion	mg/L as $CaCO_3$	mEq/L
CO_2	9.6	22.0	2.28	21.9	0.44
Ca^{2+}	95.2	20.0	2.50	238.0	4.76
Mg^{2+}	13.5	12.2	4.12	55.6	1.11
Na^+	25.8	23.0	2.18	56.2	1.12
Alkalinity				198	3.96
Cl^-	67.8	35.5	1.41	95.6	1.91
SO_4^{2-}	76.0	48.0	1.04	76.0	1.58

Bar chart of raw water in mg/L as $CaCO_3$:

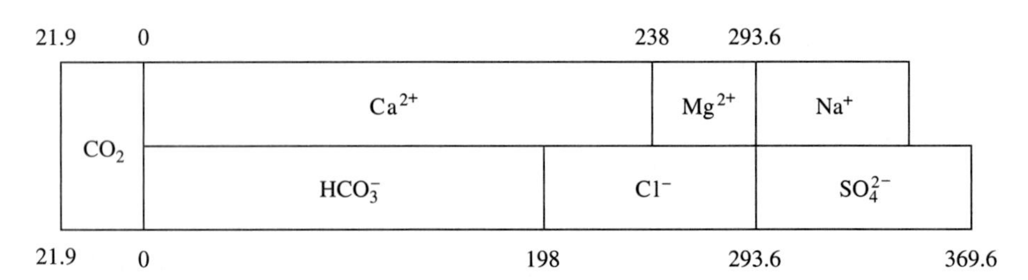

Solution. To soften to the theoretical solubility limits, lime and soda must be added as shown below.

Addition = to:	Lime mg/L as $CaCO_3$	Lime mEq/L	Soda mg/L as $CaCO_3$	Soda mEq/L
CO_2	21.9	0.44		
HCO_3^-	198.0	3.96		
Ca-HCO_3^-			40	0.80
Mg^{2+}	55.6	1.11	55.6	1.11
	275.5	5.51	95.6	1.91

Since the difference Mg − 40 = 15.6 mg/L as $CaCO_3$, the minimum excess lime of 20 mg/L as $CaCO_3$ is selected. The total lime addition is 295.5 mg/L as $CaCO_3$ or 165.5

mg/L as CaO. The soda addition is 95.6 mg/L as $CaCO_3$ or

$$95.6 \text{ mg/L as } CaCO_3(53/50) = 101.3 \text{ mg/L as } Na_2CO_3$$

Note that (53/50) is the equivalent weight of Na_2CO_3/equivalent weight of $CaCO_3$.
Reaction with CO_2:

$$CO_2 + Ca(OH)_2 \longrightarrow CaCO_3(s) + H_2O$$

Bar chart after removal of CO_2:

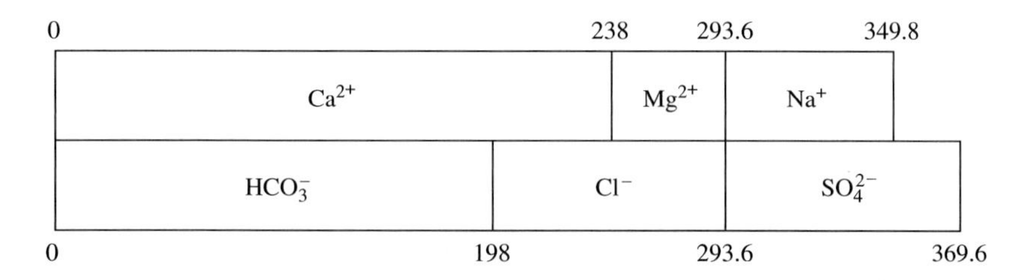

Reaction with HCO_3^-

$$Ca^{2+} + 2HCO_3^- + Ca(OH)_2 \longrightarrow 2CaCO_3(s) + 2H_2O$$

Bar chart after reaction with HCO_3^-:

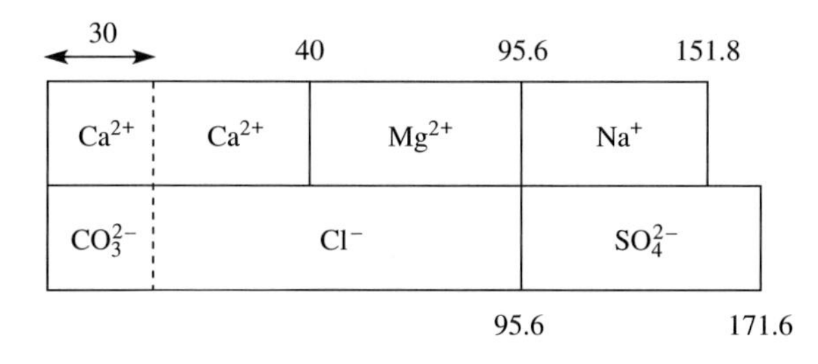

The 30 mg/L as $CaCO_3$ to the left of the dashed line is the calcium carbonate that remains because of solubility product limitations.

Reaction with calcium and soda:

$$Ca^{2+} + Na_2CO_3 \longrightarrow CaCO_3(s) + 2Na^+$$

Bar chart after reaction with calcium and soda:

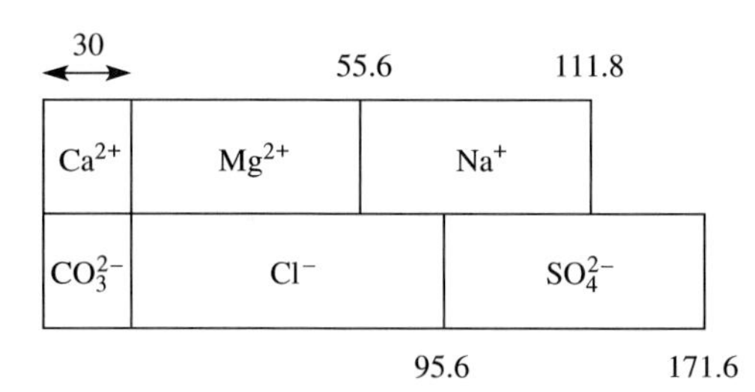

Reactions with magnesium, lime, and soda:

$$Mg^{2+} + Ca(OH)_2 \longrightarrow Mg(OH)_2(s) + Ca^{2+}$$

$$Ca^{2+} + Na_2CO_3 \longrightarrow CaCO_3(s) + 2Na^+$$

Bar chart of finished water:

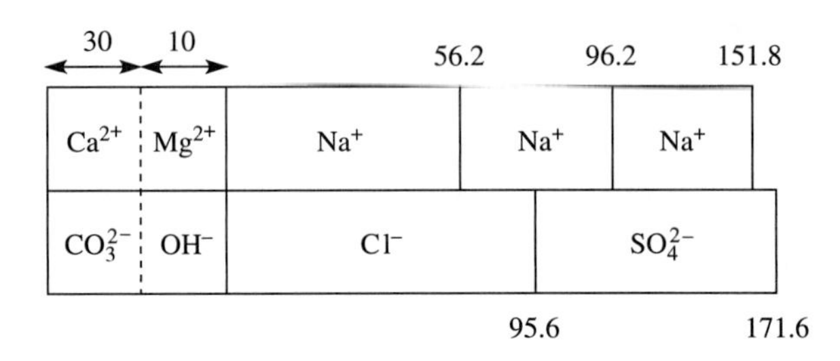

Total hardness of finished water is 30 mg/L as $CaCO_3$ + 10 mg/L as $CaCO_3$ = 40 mg/L as $CaCO_3$.

Split treatment. As shown in Figure 3-17, in split treatment a portion of the raw water is bypassed around the softening reaction tank and the settling tank. This serves several functions. First, it allows the water to be tailored to yield a product water that has 0.80 meq/L or 40 mg/L as $CaCO_3$ of magnesium (or any other value above the solubility limit). Second, it allows for reduction in capital cost of tankage because the entire flow does not need to be treated. Third, it minimizes operating costs for chemicals by treating only a fraction of the flow. Fourth, it uses the natural alkalinity of the water to lower the pH of the product water and assist in stabilization.

The fractional amount of the split is calculated as

$$X = \frac{Mg_f - Mg_i}{Mg_r - Mg_i} \tag{3-68}$$

where Mg_f = final magnesium concentration, mg/L as $CaCO_3$
Mg_i = magnesium concentration from first stage, mg/L as $CaCO_3$
Mg_r = raw water magnesium concentration, mg/L as $CaCO_3$

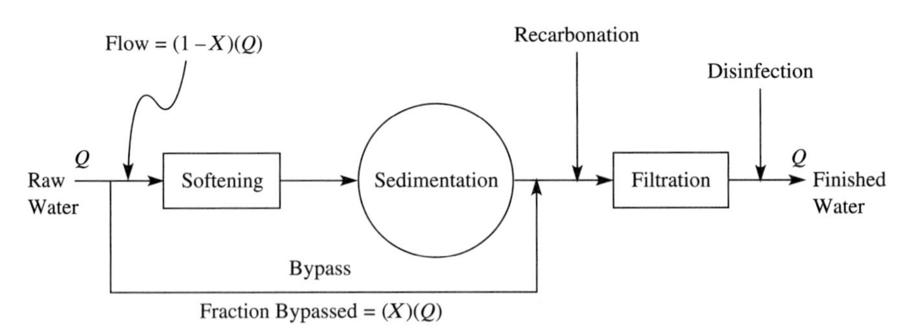

FIGURE 3-17
Split-flow treatment scheme.

The first stage is operated to soften the water to the practical limits of softening. Thus, the value for Mg_i is commonly taken to be 10 mg/L as $CaCO_3$. Since the desired concentration of Mg is nominally set at 40 mg/L as $CaCO_3$ as noted previously, Mg_f is commonly taken as 40 mg/L as $CaCO_3$.

Example 3-18. Determine the chemical dosages for split treatment softening of the following water. The finished water criteria is a maximum magnesium hardness of 40 mg/L as $CaCO_3$ and a total hardness in the range 80 to 120 mg/L as $CaCO_3$.

Constituent	mg/L	EW	EW $CaCO_3$/EW ion	mg/L as $CaCO_3$	mEq/L
CO_2	11.0	22.0	2.28	25.0	0.50
Ca^{2+}	95.2	20.0	2.50	238	4.76
Mg^{2+}	22.0	12.2	4.12	90.6	1.80
Na^+	25.8	23.0	2.18	56.2	1.12
Alkalinity				198	3.96
Cl^-	67.8	35.5	1.41	95.6	1.91
SO_4^{2-}	76.0	48.0	1.04	76.0	1.58

Solution. In the first stage the water is softened to the theoretical solubility limits; lime and soda must be added as shown below.

Addition = to:	Lime mg/L as $CaCO_3$	Lime mEq/L	Soda mg/L as $CaCO_3$	Soda mEq/L
CO_2	25.0	0.50		
HCO_3^-	198.0	3.96		
Ca-HCO_3^-			40	0.80
Mg^{2+}	90.6	1.80	90.6	1.80
	313.6	6.26	130.6	2.60

The split is calculated in terms of mg/L as $CaCO_3$:

$$X = (40 - 10)/(90.6 - 10) = 0.372$$

The fraction of water passing through the first stage is then $1 - 0.372 = 0.628$. The total hardness of the water after passing through the first stage is the theoretical solubility limit, that is, 40 mg/L as $CaCO_3$. Since the total hardness in the raw water is $238 + 90.6 = 328.6$ mg/L as $CaCO_3$, the mixture of the treated and bypass water has a hardness of:

$$(0.372)(328.6) + (0.628)(40) = 147.4 \text{ mg/L as } CaCO_3$$

This is above the acceptable range of 80–120 mg/L as $CaCO_3$, so further treatment is required. Since the split is designed to yield the required 40 mg/L as $CaCO_3$ of magnesium, more calcium must be removed. Removal of the calcium equivalent to the bicarbonate will leave 40 mg/L as $CaCO_3$ of calcium hardness plus the 40 mg/L as $CaCO_3$ of magnesium hardness for a total of 80 mg/L as $CaCO_3$. The additions are as follows.

Constituent	Lime mg/L as $CaCO_3$	Lime mEq/L
CO_2	25.0	0.50
HCO_3^-	198.0	3.96
	223.0	4.46

Addition of lime = to CO_2 and HCO_3^- (even in second stage) is necessary to get pH high enough.

The total chemical additions are in proportion to the flows:

$$\text{Lime} = 0.628(313.6) + 0.372(223) = 280 \text{ mg/L as } CaCO_3$$

$$\text{Soda} = 0.628(130.6) + 0.372(0.0) = 82 \text{ mg/L as } CaCO_3$$

Cases. The selection of chemicals and their dosage depends on the raw water composition and the desired final water composition. If we use a Mg concentration of 40 mg/L as $CaCO_3$ as a product water criterion, then six cases illustrate the dosage schemes. Three of the cases occur when the Mg concentration is less than 40 mg/L as $CaCO_3$ (Figure 3-18*a*, *b*, and *c*) and three cases occur when Mg is greater than

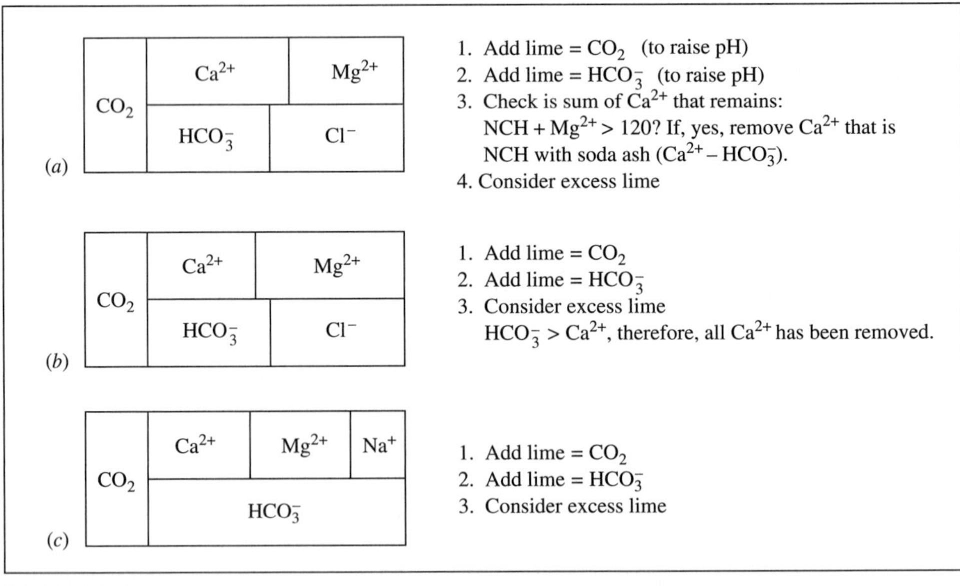

FIGURE 3-18
Dosage schemes when Mg^{2+} concentration is less than 40 mg/L as $CaCO_3$ and no split treatment is required. Note that no Mg^{2+} is removed and that reactions deal with CO_2 and Ca^{2+} only.

40 mg/L as $CaCO_3$ (Figure 3-19*a*, *b*, and *c*). In the cases illustrated in Figure 3-18, no split treatment is required. Conversely, the cases illustrated in Figure 3-19 are for the first stage of a split treatment flow scheme (softening to the practical limits). In the cases illustrated in Figure 3-19, the hardness of the mixture of the treated and raw water must be checked to see if an acceptable hardness has been achieved. As mentioned previously, this is generally below 120 mg/L as $CaCO_3$. If the hardness after blending is above this concentration, then further softening in a second stage is required (Figure 3-20). Since the design of the split is to achieve a desired Mg concentration of 40 mg/L as $CaCO_3$, no further Mg removal is required. Only treatment of the Ca is required.

Ion-Exchange Softening

Ion exchange can be defined as the reversible interchange of ions between a solid and a liquid phase in which there is no permanent change in the structure of the solid. Typically, in water softening by ion exchange, the water containing the hardness is passed through a column containing the ion-exchange material. The hardness in the water exchanges with an ion from the ion-exchange material. Generally, the ion exchanged with the hardness is sodium, as illustrated in Equation 3-69:

$$Ca(HCO_3)_2 + 2NaR \rightleftharpoons CaR_2 + 2NaHCO_3 \tag{3-69}$$

where R represents the solid ion-exchange material. By the above reaction, calcium (or magnesium) has been removed from the water and replaced by an equivalent amount of sodium, that is, two sodium ions for each cation. The alkalinity remains unchanged. The exchange results in essentially 100 percent removal of the hardness from the water until the exchange capacity of the ion-exchange material is reached, as shown in Figure 3-21. When the ion-exchange material becomes saturated, no hardness will be removed. At this point *breakthrough* is said to have occurred because the hardness passes through the bed. At this point the column is taken out of service, and the ion-exchange material is regenerated. That is, the hardness is removed from the material by passing water containing a large amount of Na^+ through the column. The mass action of having so much Na^+ in the water causes the hardness of the ion-exchange material to enter the water and exchange with the sodium:

$$CaR_2 + 2NaCl \rightleftharpoons 2NaR + CaCl_2 \tag{3-70}$$

The ion-exchange material can now be used to remove more hardness. The $CaCl_2$ is a waste stream that must be disposed of.

There are some large water treatment plants that utilize ion-exchange softening, but the most common application is for residential water softeners. The ion-exchange material can either be naturally occurring clays, called *zeolites,* or synthetically made resins. There are several manufacturers of synthetic resins. The resins or zeolites are characterized by the amount of hardness that they will remove per volume of resin material and by the amount of salt required to regenerate the resin. The synthetically produced resins have a much higher exchange capacity and require less salt for regeneration. However, they also cost much more. People who work in the water softening industry often work in units of grains of hardness per gallon of water (gr/gal). It is useful to remember that 1 gr/gal equals 17.1 mg/L.

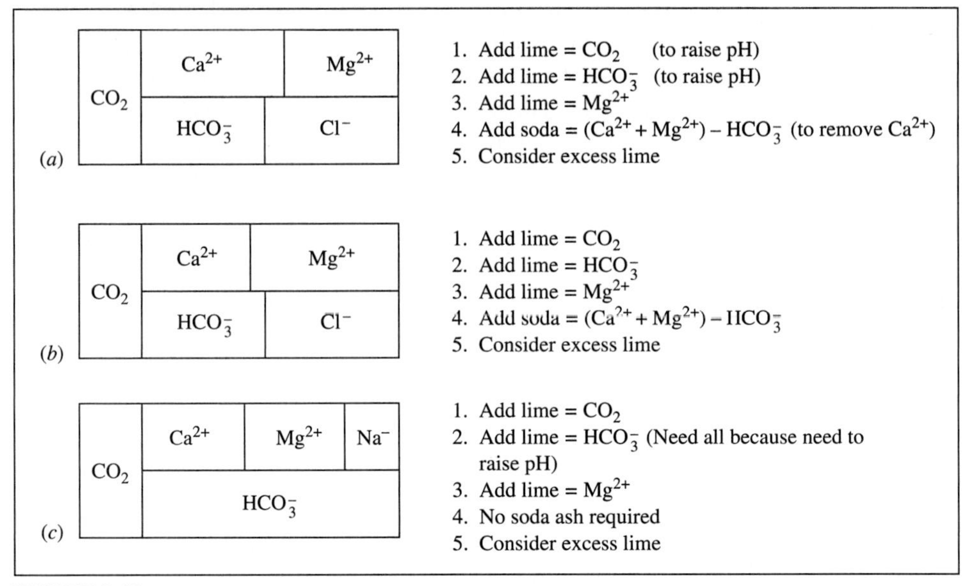

FIGURE 3-19
Cases when Mg^{2+} concentration is greater than 40 mg/L as $CaCO_3$ and split treatment is required. Note that these cases illustrate softening to the practical limits in the first stage of the split-flow scheme.

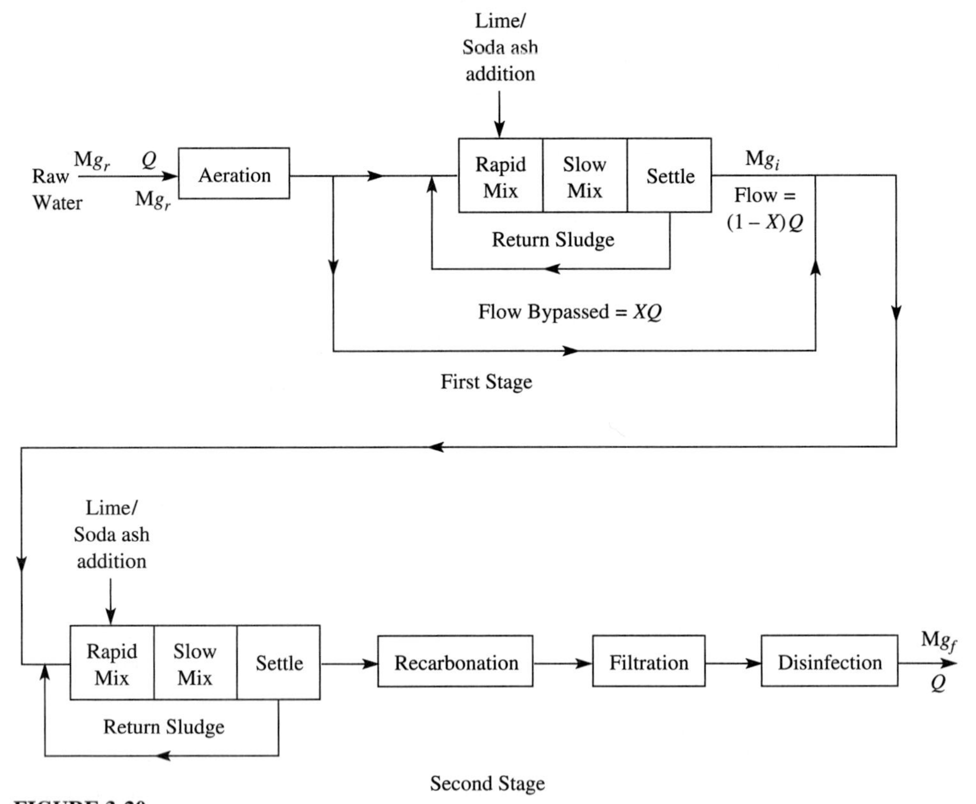

FIGURE 3-20
Flow diagram for a two stage split-treatment lime–soda ash softening plant. [Adapted from J. L. Cleasby and J. H. Dillingham, "Rational Aspects of Split Treatment," *Proc. Am. Soc. Civil Engrs., J. San. Eng. Div.* 92 (SA2) (1966): 1–7.]

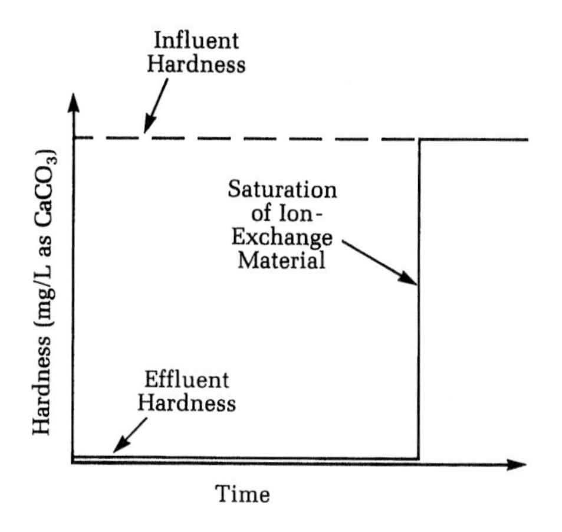

FIGURE 3-21
Hardness removal in ion-exchange column.

Since the resin removes virtually 100 percent of the hardness, it is necessary to bypass a portion of the water and then blend in order to obtain the desired final hardness.

$$\%\text{Bypass} = (100)\,\frac{\text{Hardness}_{\text{desired}}}{\text{Hardness}_{\text{initial}}} \tag{3-71}$$

Example 3-19. A home water softener has 0.1 m^3 of ion-exchange resin with an exchange capacity of 57 kg/m^3. The occupants use 2,000 L of water per day. If the water contains 280.0 mg/L of hardness as $CaCO_3$ and it is desired to soften it to 85 mg/L as $CaCO_3$, how much should be bypassed? What is the time between regeneration cycles?

Solution. The percentage of water to be bypassed is found using Equation 3-71:

$$\%\text{Bypass} = (100)\,\frac{85}{280} = 30.36 \text{ or } 30\%$$

The length of time between regeneration cycles is determined from the exchange capacity of the ion-exchange material (media). This is also called the "time to breakthrough," that is, the time to saturate the exchange material. If 30 percent of the water is being bypassed, then 70 percent of the water is being treated and the hardness loading rate is

$$\text{Loading rate} = (0.7)(280 \text{ mg/L})(2{,}000 \text{ L/d}) = 392{,}000 \text{ mg/d}$$

Since the bed capacity is 57 kg/m^3 and the bed contains 0.1 m^3 of ion-exchange media, the breakthrough time is approximately

$$\text{Breakthrough time} = \frac{(57 \text{ kg/m}^3)(0.1 \text{ m}^3)}{(392{,}000 \text{ mg/d})(10^{-6} \text{ kg/mg})} = 14.5 \text{ d}$$

3-4 REACTORS

The tanks used to perform physical, chemical, and biochemical reactions are called reactors. The reactors are classified based on their flow characteristics and their mixing conditions. *Batch reactors* are of the fill-and-draw type: Materials are added to the tank, mixed for sufficient time to allow the reaction to occur, and then drained. Although the reactor is well mixed and the contents are uniform at any instant in time, the composition within the tank changes with time as the reaction proceeds. A batch reaction is unsteady. *Flow reactors* have a continuous type of operation: Material flows into, through, and out of the reactor at all times. Flow reactors may be further classified by mixing conditions. The contents of a *completely stirred tank reactor* (CSTR) [also called a completely mixed flow reactor (CMF)] ideally are uniform throughout the tank. The composition of the effluent is the same as the composition in the tank. If the mass input rate into the tank remains constant, the composition of the effluent remains constant. In *plug-flow reactors,* fluid particles pass through the tank in sequence. Those that enter first leave first. In the ideal case, it is assumed that there is no mixing in the lateral direction. Although composition varies along the length of the tank, so long as the flow conditions remain steady, the composition of the effluent remains constant. Real continuous-flow reactors generally fall somewhere in between these two ideal conditions.

For time-dependent reactions (see Section 3-1), the time that a fluid particle remains in the reactor obviously affects the degree to which the reaction goes to completion. In ideal reactors the time in the reactor (*detention time* or *retention time*) is defined as:

$$t_o = \frac{\forall}{Q} \tag{3-72}$$

where t_o = theoretical detention time, s
$\forall$ = volume of fluid in basin, m^3
Q = flow rate into basin, m^3/s

Real reactors do not behave as ideal reactors because of density differences due to temperature or other causes, short circuiting because of uneven inlet or outlet conditions, and local turbulence or dead spots in the tank corners. The detention time in real tanks is generally less than the theoretical detention time calculated from Equation 3-72.

Reactor design. The mass-balance approach may be used to design ideal reactors and to approximate the design of real reactors. Recalling the general mass-balance equation (Equation 1-5):

$$\text{Accumulation rate} = \text{Input rate} - \text{Output rate} \pm \text{Transformation rate}$$

For a batch reactor, the input and output are by definition zero, so the equation reduces to

$$\text{Accumulation rate} = \text{Transformation rate}$$

With a constant-volume reactor transforming compound A, the differential form may be written

$$\forall\frac{d[\text{A}]}{dt} = -r(\forall) \tag{3-73}$$

TABLE 3-14
Example solutions to mass-balance reactions for CSTRs at steady state

Reaction order	Integrated equation
0	$kt = [A_0] - [A]$
1	$kt = \frac{[A_0]}{[A]} - 1$
2	$kt = \frac{1}{[A]}\left(\frac{[A_0]}{[A]} - 1\right)$

Since the volume is constant it may be eliminated by dividing through by $\forall$. The equation may be solved by integrating from $[A] = [A_0]$ at time $t = 0$ to $[A] = [A]$ at time $= t$. The solution for zero-, first-, and second-order reactions is shown in Table 3-5.

Solution of the mass balance for plug-flow reactors (Table 3-5) results in exactly the same equations as those for batch reactors, with the constraint that t is defined by the detention time of a fluid particle in the reactor (Equation 3-72). For CSTRs, accumulation is by definition zero, and the mass-balance equation is

$$0 = Q_{in}\,[A_0] - Q_{out}\,[A] - r_A(\forall) \tag{3-74}$$

The solutions for different reaction orders are shown in Table 3-14. Again, t is defined as the detention time.

3-5 MIXING AND FLOCCULATION

Clearly, if the chemical reactions in coagulating and softening a water are going to take place, the chemical must be mixed with the water. In this section we will begin to look at the physical methods necessary to accomplish the chemical processes of coagulation and softening.

Mixing, or rapid mixing as it is called, is the process whereby the chemicals are quickly and uniformly dispersed in the water. Ideally, the chemicals would be instantaneously dispersed throughout the water. During coagulation and softening the chemical reactions that take place in rapid mixing form precipitates. Either aluminum hydroxide or iron hydroxide form during coagulation, while calcium carbonate and magnesium hydroxide form during softening. The precipitates formed in these processes must be brought into contact with one another so that they can agglomerate and form larger particles, called *flocs*. This contacting process is called *flocculation* and is accomplished by slow, gentle mixing.

In the treatment of water and wastewater the degree of mixing is measured by the velocity gradient, G. The velocity gradient is best thought of as the amount of shear taking place; that is, the higher the G value, the more violent the mixing. The velocity gradient is a function of the power input into a unit volume of water. It is given by

$$G = \sqrt{\frac{P}{\mu \forall}} \tag{3-75}$$

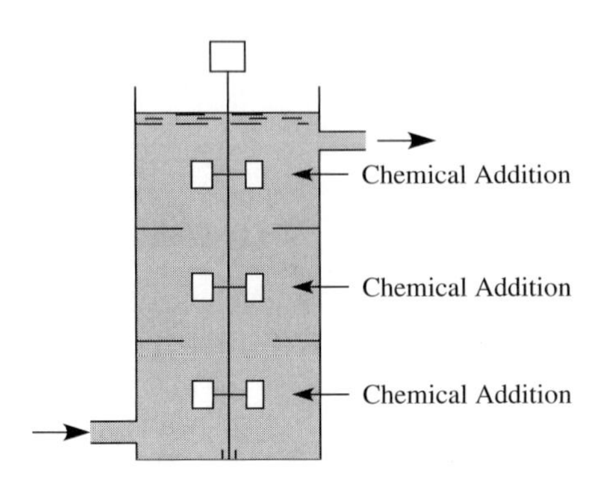

FIGURE 3-22
Rapid mix tank.

where G = velocity gradient, s^{-1}
P = power input, W
$\forall$ = volume of water in mixing tank, m^3
μ = dynamic viscosity, Pa · s

From literature, experience, laboratory, or pilot plant work it is possible to select a G value for a particular application. The total number of particle collisions is proportional to Gt_o, where t_o is the detention time in the basin as given by Equation 3-72.

Rapid Mix

Rapid mixing is probably the most important physical operation affecting coagulant dose efficiency. The chemical reaction in coagulation is completed in less than 0.1 s; therefore, it is imperative that mixing be as instantaneous and complete as possible. Rapid mixing can be accomplished within a tank utilizing a vertical shaft mixer (Figure 3-22), within a pipe using an in-line blender (Figure 3-23), or in a pipe using a

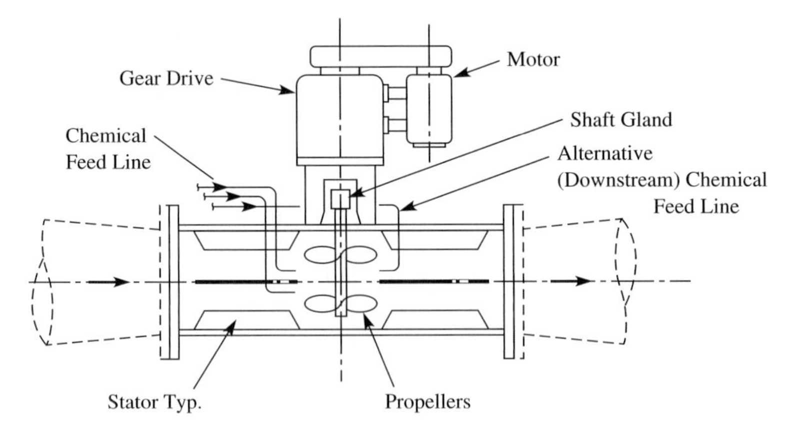

FIGURE 3-23
Typical in-line blender. (*Source:* American Water Works Association, *Water Treatment Plant Design,* 2nd ed., New York: McGraw-Hill, 1990.)

FIGURE 3-24
STATIC MIXER: A succession of reversing, flow-twisting, and flow-splitting elements provides positive dispersion proportional to number of elements. (*Source: Chemical Engineering,* March 22, 1971.)

static mixer (Figure 3-24). Other methods, such as Parshall flumes, hydraulic jumps, baffled channels, or air mixing may also be used.

The selection of G and Gt_o values for coagulation is dependent on the mixing device, the chemicals selected, and the anticipated reactions. Coagulation occurs predominately by two mechanisms: adsorption of the soluble hydrolysis species on the colloid and destabilization or sweep coagulation where the colloid is trapped in the hydroxide precipitate. The reactions in adsorption-destabilization are extremely fast and occur within 1 second. Sweep coagulation is slower and occurs in the range of 1 to 7 seconds.[15] Jar test data may be used to identify whether adsorption-destabilization or sweep coagulation is predominant. If charge reversal is apparent from the dose-turbidity curve (see, for example, Figure 3-11), then adsorption-destabilization is the predominant mechanism. If the dose-turbidity curve does not show charge reversal (that is, the curve is relatively flat at higher doses), then the predominant mechanism is sweep coagulation. G values in the range of 3,000 to 5,000 s^{-1} and detention times on the order of 0.5 s are recommended for adsorption-desorption reactions. These values are most commonly achieved in an in-line blender. For sweep coagulation, detention times of 1 to 10 s and G values in the range of 600 to 1,000 s^{-1} are recommended.[16]

Softening. For dissolution of $CaO/Ca(OH)_2$ mixtures for softening, detention times on the order of 5 to 10 minutes may be required. G values to disperse and maintain particles in suspension may be on the order of 700 s^{-1}. In-line blenders are not used to mix softening reagents.

Rapid-mix tanks. The volume of a rapid-mix tank seldom exceeds 8 m^3 because of mixing equipment and geometry constraints. The mixing equipment consists of an electric motor, gear-type speed reducer, and either a turbine or axial-flow impeller as shown in Figure 3-25. The turbine impeller provides more turbulence and is preferred for rapid mixing. The tanks should be horizontally baffled into at least

[15] A. Amirtharajah, "Design of Rapid Mix Units," in *Water Treatment Plant Design for the Practicing Engineer,* edited by R. L. Sanks, Ann Arbor Science, Ann Arbor, MI, p. 132, 1978.

[16] A. Amirtharajah, *ibid.,* pp. 141 and 143.

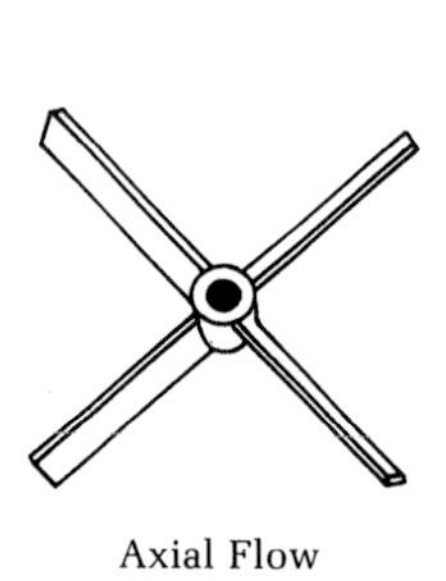

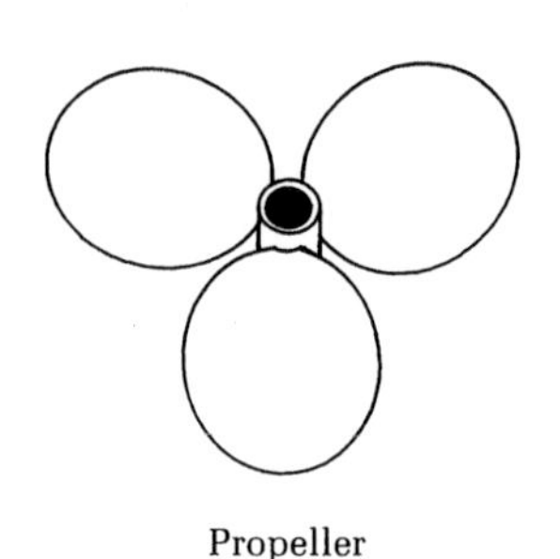

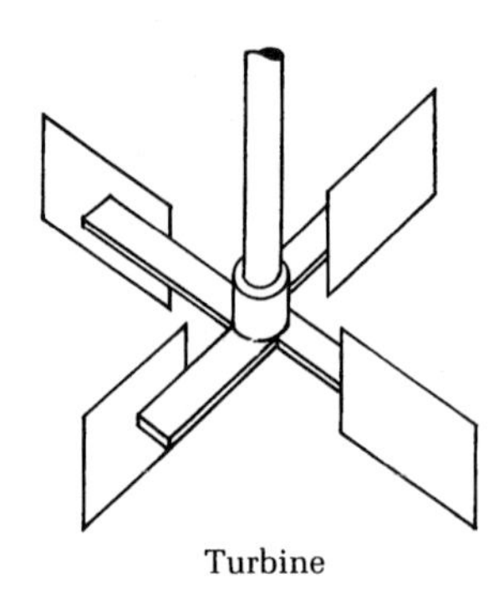

FIGURE 3-25
Basic impeller styles.

two and preferably three compartments in order to provide sufficient residence time. They are also baffled vertically to minimize vortexing. Chemicals should be added below the impeller, the point of the most mixing. Some common rules-of-thumb[17,18] are that the design liquid depth be 0.5 to 1.1 times the basin diameter or width; that the impeller diameter be between 0.30 and 0.50 times the tank diameter or width; and that the vertical baffles extend into the tank about 10 percent of the tank width or diameter. Although they may be larger, impellers normally do not exceed 1 m in diameter. The liquid depth may be increased to between 1.1 and 1.6 times the tank diameter if dual impellers on the shaft are employed. When dual impellers are employed on gear-driven mixers, they are spaced approximately two impeller diameters apart. We normally assume an efficiency of transfer of motor power to water power of 0.8 for a single impeller.

Flocculation

While rapid mix is the most important physical factor affecting coagulant efficiency, flocculation is the most important factor affecting particle-removal efficiency. The objective of flocculation is to bring the particles into contact so that they will collide, stick together, and grow to a size that will readily settle. Enough mixing must be provided to bring the floc into contact and to keep the floc from settling in the flocculation basin. Too much mixing will shear the floc particles so that the floc is small and finely dispersed. Therefore, the velocity gradient must be controlled within a relatively narrow range. Flexibility should also be built into the flocculator so that the plant operator can vary the G value by a factor of two to three. The heavier the floc and the higher the suspended solids concentration, the more mixing is required to keep the floc in suspension. This is reflected in Table 3-15. Softening floc is heavier than coagulation floc and therefore requires a higher G value. An increase in the floc

---

[17]T. D. Reynolds, *Unit Operations and Processes in Environmental Engineering,* Boston, MA: PWS-Kent Publishing, pp. 33–54, 1982.

[18]Personal communication, JWI, Inc., Holland, MI, May 1990.

TABLE 3-15
Gt_o values for flocculation

Type	$G(s^{-1})$	Gt_o (unitless)
Low-turbidity, color removal coagulation	20–70	60,000 to 200,000
High-turbidity, solids removal coagulation	30–80	36,000 to 96,000
Softening, 10% solids	130–200	200,000 to 250,000
Softening, 39% solids	150–300	390,000 to 400,000

concentration (as measured by the suspended solids concentration) also increases the required G. With water temperatures of approximately 20°C, modern plants provide about 20 minutes of flocculation time (t_0) at plant capacity. With lower temperatures, the detention time is increased. At 15°C the detention time is increased by 7 percent, at 10°C it is increased 15 percent, and at 5°C it is increased 25 percent.

Flocculation is normally accomplished with an axial-flow impeller (Figure 3-25), a paddle flocculator (Figure 3-26), or a baffled chamber (Figure 3-27). Axial-flow impellers have been recommended over the other types of flocculators because they impart a nearly constant G throughout the tank.[19] The flocculator basin should be divided into at least three compartments. The velocity gradient is tapered so that the G values decrease from the first compartment to the last and that the average of the compartments is the design value selected from Table 3-15. Some common rules-of-thumb for axial-flow impellers are that the diameter of the impeller is between 0.2 and 0.5 times the width of the chamber and that the maximum impeller diameter is about 3 m.

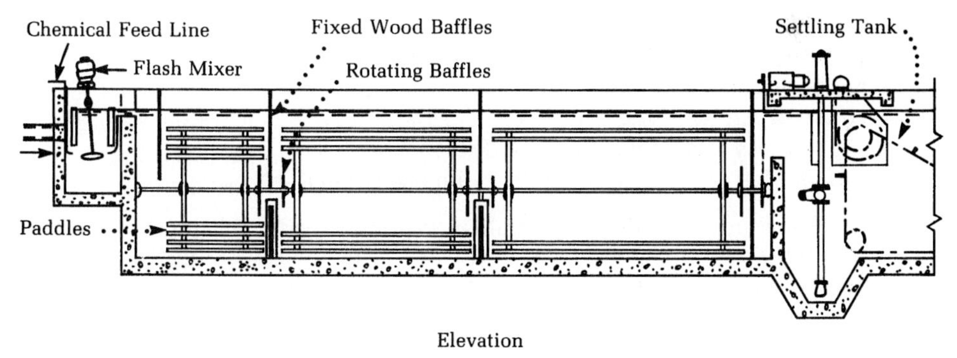

FIGURE 3-26
Paddle flocculator. (Courtesy of Envirex, Inc., a Rexnord Company.)

[19]H. E. Hudson, *Water Clarification Processes, Practical Design and Evaluation,* New York: Van Nostrand Reinhold, pp. 115–117, 1981.

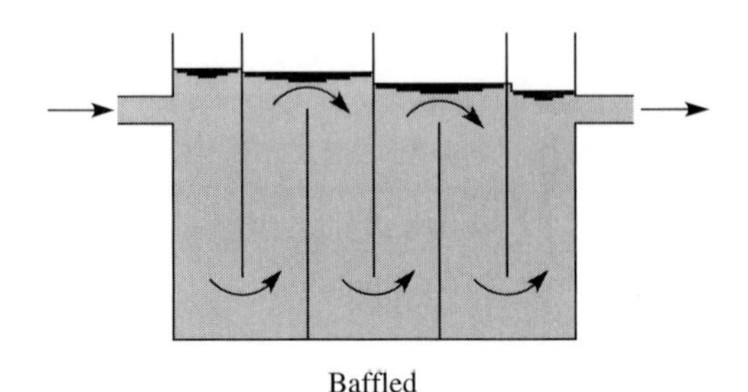

FIGURE 3-27
Baffled chamber flocculator.

Power Requirements

In the design of mixing equipment for rapid-mix and flocculation tanks, the power imparted to the liquid in a baffled tank by an impeller may be described by the following equation for fully turbulent flow developed by Rushton[20]:

$$P = K_T(n)^3(D_i)^5\ \rho \tag{3-76}$$

where P = power, W
K_T = impeller constant
n = rotational speed, revolutions/s
D_i = impeller diameter, m
ρ = density of liquid, kg/m^3

In an unbaffled tank the power imparted may be as low as one-sixth of that predicted by this equation. Values for the impeller constant may be obtained from Table 3-16.

The power imparted by a paddle mixer (Figure 3-26) is a function of the drag force on the paddles

$$P = \frac{C_D A \rho (v_p)^3}{2} \tag{3-77}$$

where P = power imparted, W
C_D = coefficient of drag of paddle
ρ = density of fluid, kg/m^3
A = cross-sectional area of paddles, m^2
v_p = relative velocity of paddles with respect to fluid, m/s

It has been found that the peripheral velocity of the paddle blades should range from 0.1 to 1.0 m/s and that the relative velocity of the paddles to the fluid should be 0.6 to 0.75 times the paddle-tip speed. The drag coefficient (C_D) varies with the length-to-width ratio (for example: for L:W of 5, C_D = 1.20, for L:W of 20, C_D = 1.50, and for L:W of infinity, C_D = 1.90). It is also recommended that the total paddle-blade area on a horizontal shaft not exceed 15 to 20 percent of the total basin cross-sectional area to avoid excessive rotational flow.

[20]J. H. Rushton, "Mixing of Liquids in Chemical Processing," *Industrial & Engineering Chemistry,* v. 44, no. 12, p. 2931, 1952.

TABLE 3-16
Values of the impeller constant K_T

Type of impeller	K_T
Propeller, pitch of 1, 3 blades	0.32
Propeller, pitch of 2, 3 blades	1.00
Turbine, 6 flat blades, vaned disc	6.30
Turbine, 6 curved blades	4.80
Fan turbine, 6 blades at 45°	1.65
Shrouded turbine, 6 curved blades	1.08
Shrouded turbine, with stator, no baffles	1.12

Note: Constant assumes baffled tanks having four baffles at the tank wall with a width equal to 10 percent of the tank diameter.
Source: J. H. Rushton, "Mixing of Liquids in Chemical Processing," *Industrial & Engineering Chemistry,* v. 44, no. 12, p. 2931, 1952.

For pneumatic mixing, the power imparted is given by

$$P = KQ_a \ln\left(\frac{h + 10.33}{10.33}\right) \tag{3-78}$$

where P = power imparted, W
K = constant = 1.689
Q_a = air flow rate at atmospheric pressure, m^3/min
h = air pressure at the point of discharge, m

The power imparted by static-mixing devices may be computed as

$$P = \gamma Q h \tag{3-79}$$

where P = power imparted, kW
γ = specific weight of fluid, kN/m^3
Q = flow rate, m^3/s
h = head loss through the mixer, m

The specific weight of water is equal to the product of the density and the acceleration due to gravity ($\gamma = \rho g$). At normal temperatures, the specific weight of water is taken to be 9.81 kN/m^3.

Upflow solids-contact. Mixing, flocculation, and clarification may be conducted in a single tank such as that in Figure 3-28. The influent raw water and chemicals are mixed in the center cone-like structure. The solids flow down under the cone (sometimes called a "skirt"). As the water flows upward, the solids settle to form a *sludge blanket.* This design is called an upflow solids-contact basin. The main advantage of this unit is its reduced size. The units are best suited to treat a feed water that has a relatively constant quality. It is often favored for softening because the water quality from wells is relatively constant and the sludge blanket provides a further opportunity to drive the precipitation reactions to completion.

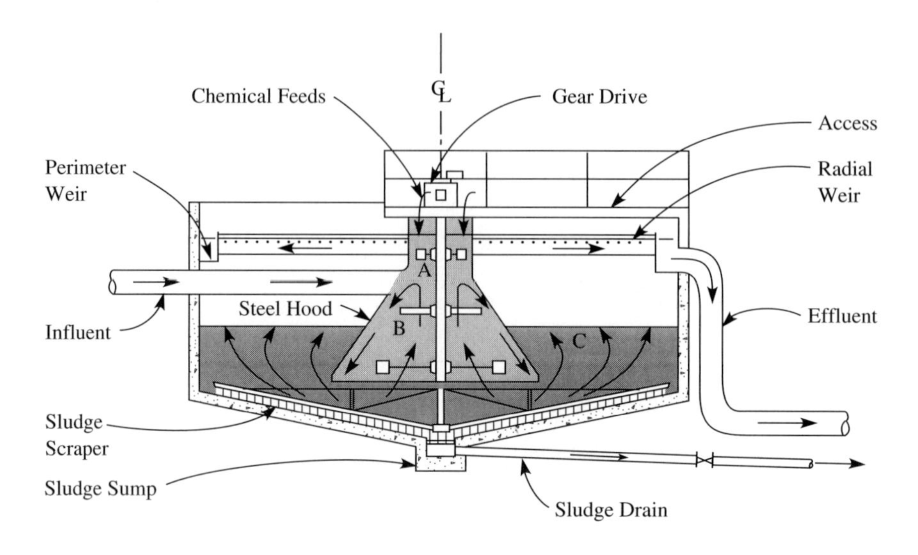

FIGURE 3-28
Typical upflow solids-contact unit. (*Source:* Adapted from American Water Works Association *Water Treatment Plant Design,* 2nd ed., New York: McGraw-Hill, 1990.)

Example 3-20. The city of Redpherric is planning for the installation of a water treatment plant to remove iron. A low-turbidity iron coagulation plant has been proposed with the following design parameters:

$Q = 2\text{m}^3/\text{s}$

Rapid mix $t_o = 10$ s

Rapid mix $G = 1{,}000\ \text{s}^{-1}$

Floc $t_o = 20$ min

Floc $G = 30\ \text{s}^{-1}$

Water temperature $= 18°\text{C}$

Design the rapid-mix and flocculation basins and size the mixing equipment.

Solution

a. Rapid-mix design

The volume of the rapid-mix tank by Equation 3-72 is

$$\forall = (Q)(t_o) = (2\ \text{m}^3/\text{s})(10\ \text{s}) = 20\ \text{m}^3$$

This greatly exceeds our guideline of 8 m^3 as the maximum volume for a rapid-mix tank. To meet this guideline we will need to provide tanks in parallel. Since we are also constrained by the availability of mixers, we need to assess the limitations imposed by the power available from standard mixers.

As an example, the following mixers are available.

Model	Rotational speeds, rpm	Power, kW	Model	Rotational speeds, rpm	Power, kW
JTQ25	30, 45	0.18	JTQ300	110, 175	2.24
JTQ50	30, 45	0.37	JTQ500	110, 175	3.74
JTQ75	45, 70	0.56	JTQ750	110, 175	5.59
JTQ100	45, 110	0.75	JTQ1000	110, 175	7.46
JTQ150	45, 110	1.12	JTQ1500	110, 175	11.19
JTQ200	70, 110	1.50			

rpm = revolutions per minute
JTQ-F models have variable speeds from 1–45 rpm

The largest available mixer can achieve a water power of

$$(11.19 \text{ kW})(0.8) = 8.95 \text{ kW}$$

The factor of 0.8 is the assumed efficiency of transfer of motor power to water power. The required volume for mixing can be found by using Equation 3-75. Given a G of 1,000 s^{-1}, a water power of 8.95 kW, and finding the viscosity at 18°C from Appendix A as 1.053×10^{-3} Pa · s, we find

$$\forall = \frac{P}{G^2\mu} = \frac{8.95 \times 10^3 \text{ W}}{(1{,}000 \text{ s}^{-1})^2 (1.053 \times 10^{-3} \text{ Pa} \cdot \text{s})}$$

$$= 8.50 \text{ m}^3$$

This means that to supply our 20 m^3 volume requirement we will need

$$\frac{20 \text{ m}^3}{8.50 \text{ m}^3/\text{tank}} = 2.35 \text{ or 3 rapid mix tanks.}$$

With three tanks, the volume per tank will be 6.67 m^3.

Assuming a rotational speed of 110 rpm (1.83 rps) for the JTQ1500 and a turbine with six flat blades (vaned disc), we can estimate the impeller diameter using Equation 3-76:

$$D_i = \left(\frac{(P)}{(K_T)(n)^3(\rho)}\right)^{1/5} = \left(\frac{(8.95 \times 10^3 \text{ W})}{(6.30)(1.83 \text{ rps})^3(1{,}000 \text{ kg/m}^3)}\right)^{1/5}$$

$$= (0.23)^{1/5} = 0.75 \text{ m}$$

Using a ratio of impeller diameter to tank diameter of 0.33, the tank diameter would be

$$\frac{0.75 \text{ m}}{0.33} = 2.27 \text{ m}$$

The surface area of the tank would be

$$\frac{\pi(2.27)^2}{4} = 4.05 \text{ m}^2$$

With three 6.67 m^3 tanks, the depth of the tank would be

$$\frac{6.67 \text{ m}^3}{4.05 \text{ m}^2} = 1.65 \text{ m}$$

Checking the liquid-depth to tank-diameter ratio, we find

$$\frac{1.65 \text{ m}}{2.27 \text{ m}} = 0.73$$

This is within our guideline of 0.5 to 1.1.

b. Flocculator design

The volume for flocculation is

$$\begin{aligned} \forall &= (Q)(t_o) \\ &= (2 \text{ m}^3/\text{s})(20 \text{ min})(60 \text{ s/min}) = 2{,}400 \text{ m}^3 \end{aligned}$$

Since we have designed three rapid mix tanks, it is logical to divide this tank volume in a similar fashion, that is, provide three flocculation tanks each with a volume of

$$\frac{2{,}400 \text{ m}^3}{3} = 800 \text{ m}^3$$

Each of these tanks will be subdivided into three compartments to achieve tapered flocculation. The design G for the compartments will be 40 s^{-1}, 30 s^{-1}, and 20 s^{-1} to yield an average G of 30 s^{-1}.

Following the same approach we used for the rapid mix system, we calculate the maximum volume that can be mixed:

$$\begin{aligned} \forall &= \frac{P}{G^2\mu} = \frac{8.95 \times 10^3 \text{ W}}{(30 \text{ s}^{-1})^2(1.053 \times 10^{-3} \text{ Pa} \cdot \text{s})} \\ &= 9{,}444 \text{ m}^3 \end{aligned}$$

Obviously, at this G value, the mixer power will not be limiting. Using the 800 m^3 tank and dividing it into three 267 m^3 compartments, the power required for the second compartment would be

$$\begin{aligned} P &= G^2\mu\forall = (30 \text{ s}^{-1})^2(1.053 \times 10^{-3} \text{ Pa} \cdot \text{s})(267 \text{ m}^3) \\ &= 253 \text{ W} = 0.253 \text{ kW} \end{aligned}$$

With 80 percent efficiency, the motor power should be

$$\frac{0.253 \text{ kW}}{0.8} = 0.32 \text{ kW}$$

A variable speed JTQ50-F at 0.37 kW would be selected for a trial calculation.

Assuming a depth of liquid in the flocculator compartment of 4 m and a square tank arrangement, the dimensions of the tank compartment would be

$$\frac{267 \text{ m}^3}{4 \text{ m}} = 66.75 \text{ m}^2$$

and

$$(66.75 \text{ m}^2)^{1/2} = 8.17 \text{ m on each side.}$$

Using a diameter-to-width ratio of 0.3, the impeller diameter would be

$$(8.17 \text{ m})(0.3) = 2.45 \text{ m}$$

Using a three-blade propeller with a pitch of 2, the required rotational speed would be

$$n^3 = \frac{(P)}{(K_T)(D_i)^5(\rho)} = \frac{(370\ \text{W})(0.8)}{(1.00)(2.45\ \text{m})^5(1{,}000\ \text{kg/m})}$$

$$n = (0.0034)^{1/3} = 0.150\ \text{rps}$$

$$n = (0.150\ \text{rps})(60\ \text{s/min}) = 9\ \text{rpm}$$

This is within the variable speed range of 1 to 45 rpm for the JTQ-F.

To obtain the proper G for the other compartments, the same motor may be used and the power input may be altered by adjusting the rotational speed.

3-6 SEDIMENTATION

Overview

Particles that will settle within a reasonable period of time can be removed in a sedimentation basin (also called a clarifier). Sedimentation basins are usually rectangular or circular with either a radial or upward water flow pattern. Regardless of the type of basin, the design can be divided into four zones: inlet, settling, outlet, and sludge storage. While our intent is to present the concepts of sedimentation and to design a sedimentation tank, a brief discussion of all four zones is helpful in understanding the sizing of the settling zone. A schematic showing the four zones is shown in Figure 3-29.

The purpose of the inlet zone is to evenly distribute the flow and suspended particles across the cross section of the settling zone.[21] The inlet zone consists of a series of inlet pipes and baffles placed about 1 m into the tank and extending the full depth of the tank. Following the baffle system, the flow takes on a flow pattern determined by the inlet structure. At some point the flow pattern is evenly distributed, and the water velocity slowed to the design velocity of the sedimentation zone. At that point the inlet zone ends and the settling zone begins. With a well-designed inlet baffle system, the inlet zone extends approximately 1.5 m down the length of the tank. Proper inlet zone design may well be the most important aspect of removal efficiency.

With improper design, the inlet velocities may never subside to the settling-zone design velocity. Typical design numbers are usually conservative enough that an inlet zone length does not have to be added to the length calculated for the settling zone. In an accurate design, the inlet and settling zones are each designed separately and their lengths added together.

The configuration and depth of the sludge storage zone depends upon the method of cleaning, the frequency of cleaning, and the quantity of sludge estimated to be produced. All these variables can be evaluated and a sludge storage zone

[21]The cross section is the area through which the flow moves. For example, in Figure 3-28*a* the cross section is the settling zone width × depth and in 3-28*b* it is the bottom circular area.

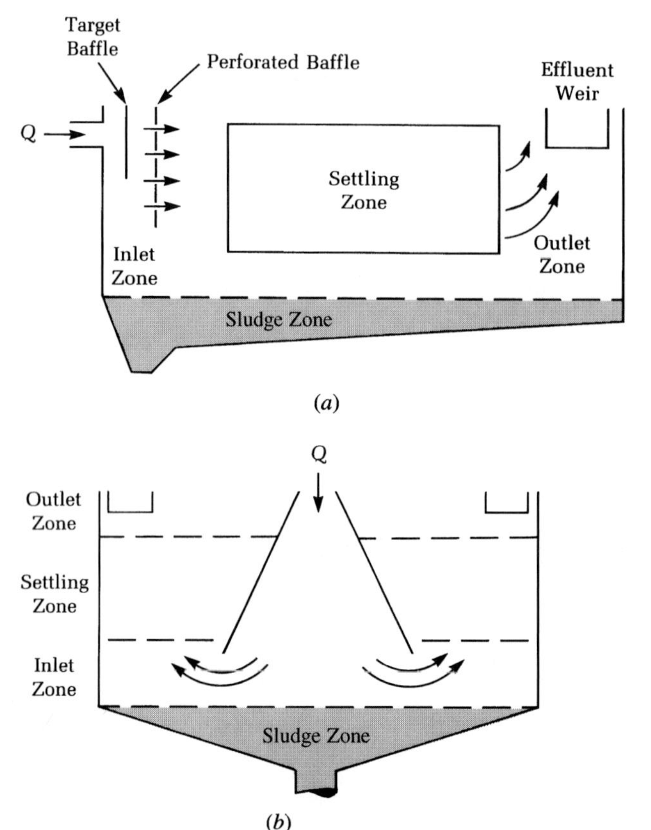

FIGURE 3-29
Zones of sedimentation: (*a*) horizontal flow clarifier; (*b*) upflow clarifier.

designed. In lieu of these design details, some general guidelines can be presented. With a well-flocculated solid and good inlet design, over 75 percent of the solids may settle in the first fifth of the tank. For coagulant floc, Hudson recommends a sludge storage depth of about 0.3 m near the outlet and 2 m or more near the inlet.[22]

If the tank is long enough, storage depth can be provided by bottom slope; if not, a sludge hopper is necessary at the inlet end. Mechanically-cleaned basins may be equipped with a bottom scraper, such as shown in Figure 3-30. The sludge is continuously scraped to a hopper where it is pumped out. For mechanically cleaned basins, a 1 percent slope toward the sludge withdrawal point is used. A sludge hopper is designed with sides sloping with a vertical to horizontal ratio of 1.2:1 to 2:1.

The outlet zone is designed so as to remove the settled water from the basin without carrying away any of the floc particles. A fundamental property of water is that the velocity of flowing water is proportional to the flow rate divided by the area through which the water flows, that is,

$$v = \frac{Q}{A_c} \tag{3-80}$$

[22]H. E. Hudson, *Water Clarification Processes, Practical Design and Evaluation*, 1981.

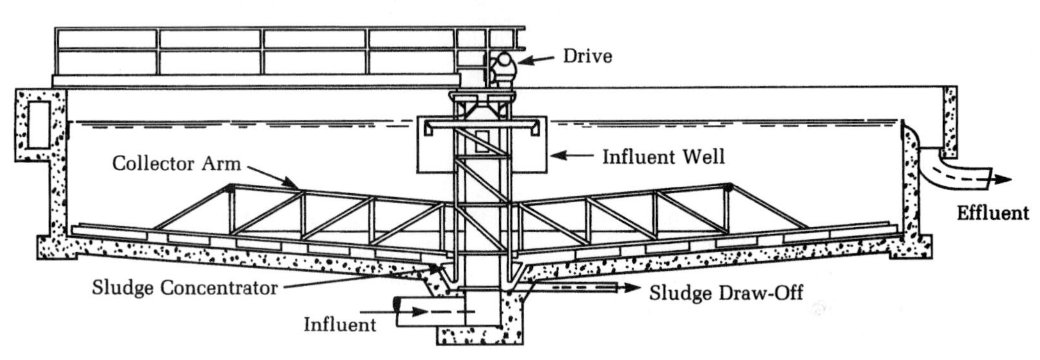

FIGURE 3-30
Photograph and schematic diagram of sludge collector for circular sedimentation basins. [*Source:* Walker Process Equipment, Inc., Division of Chicago Bridge and Iron Company (Bulletin Number 9-W-65, 1973) Aurora, Illinois.]

where v = water velocity, m/s
Q = water flow, m^3/s
A_c = cross-sectional area, m^2

Within the sedimentation tank, the flow is going through a very large area (basin depth times width); consequently, the velocity is slow. To remove the water from the basin quickly, it is desirable to direct the water into a pipe or small channel for easy transport, which will produce a significantly higher velocity. If a pipe were to be placed at the end of the sedimentation basin, all the water would "rush" to the pipe. This rushing water would create high velocity profiles in the basin, which would tend to raise the settled floc from the basin and into the effluent water. This phenomenon of washing out the floc is called *scouring,* and one way to create scouring is with an improper outlet design. Rather than put a pipe at the end of the sedimentation basin, it is desirable to first put a series of troughs, called *weirs,* which provide a large area for the water to flow through and minimize the velocity in the sedimentation tank near the outlet zone. The weirs then feed into a central channel or pipe for transport of the settled water. Figure 3-31 shows various weir arrangements. The length of weir required is a function of the type of solids. The heavier the solids, the harder it is to scour them, and the higher the allowable outlet velocity. Therefore, heavier particles require a shorter length of weir than do light particles. Each state generally has a set of standards which must be followed, but Table 3-17 shows typical design values for weir loadings. The units for weir overflow rates are $m^3/d \cdot m$, which is water flow (m^3/d) per unit length of weir (m).

In a rectangular basin, the weirs should cover at least one-third, and preferably up to one-half, of the basin length. Spacing may be as large as 5 to 6 m on-centers but is frequently on the order of one-half this distance.

Example 3-21. The town of Urbana has a low-turbidity raw water and is designing its overflow weir at a loading rate of 150 $m^3/d \cdot m$. If its plant flow rate is 0.5 m^3/s, how many linear meters of weir are required?

Solution

$$\frac{(0.5\ \text{m}^3/\text{s})(86{,}400\ \text{s/d})}{150\ \text{m}^3/\text{d} \cdot \text{m}} = 288\ \text{m}$$

Sedimentation Concepts

There are two important terms to understand in sedimentation zone design. The first is the particle (floc) *settling velocity,* v_s. The second is the velocity at which the tank is designed to operate, called the *overflow rate,* v_o. The easiest way to understand these two concepts is to view an upward-flow sedimentation tank as shown in Figure 3-32. In this design, the particles fall downward and the water rises vertically. The rate at which the particle is settling downward is the particle-settling velocity, and the velocity of the liquid rising is the overflow rate. Obviously, if a particle is to be

TABLE 3-17
Typical weir overflow rates

Type of floc	Weir overflow rate ($m^3/d \cdot m$)
Light alum floc (low-turbidity water)	143–179
Heavier alum floc (higher-turbidity water)	179–268
Heavy floc from lime softening	268–322

Source: Walker Process Equipment, Inc., Division of Chicago Bridge and Iron Company, *Walker Process Circular Clarifiers* (Bulletin 9-W-65), Aurora, Illinois, 1973.

(*a*)

(*b*)

FIGURE 3-31
Weir arrangements: (*a*) rectangular; (*b*) circular. [*Source:* Walker Process Equipment, Inc., Division of Chicago Bridge and Iron Company (Bulletin Numbers 9-W-65, 1973, and 1600-S-107, 1972) Aurora, Illinois.]

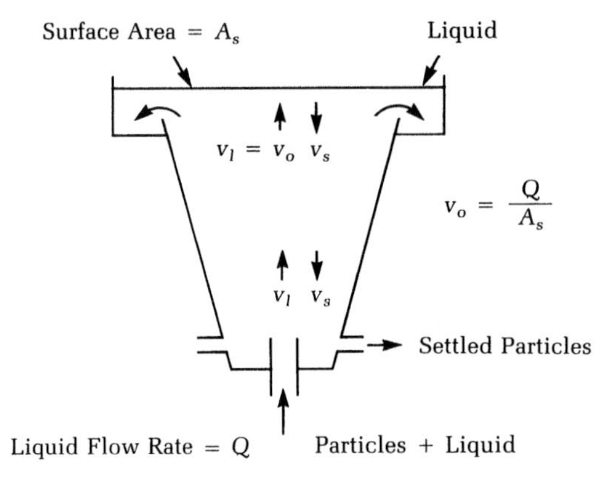

FIGURE 3-32
Settling in an upflow clarifier. (Legend: v_l = velocity of liquid; v_s = terminal settling velocity of particle.)

removed from the bottom of the clarifier and not go out in the settled water, then the particle-settling velocity must be greater than the liquid-rise velocity ($v_s > v_o$). If v_s is greater than v_o, one would expect 100 percent particle removal, and if v_s is less than v_o, one would expect 0 percent removal. In design, the procedure would be to determine the particle-settling velocity and set the overflow rate at some lower value. Often v_o is set at 50 to 70 percent of v_s for an *upflow clarifier.*

Let us now consider why the liquid-rise velocity is called an overflow rate and what its units are. The term *overflow rate* is used since the water is flowing over the top of the tank into the weir system. It is sometimes referred to as the *surface loading* rate because it has units of $m^3/d \cdot m^2$. The units are flow of water (m^3/d) being applied to a m^2 of surface area. This can be thought of as the amount of water that goes through each m^2 of tank surface area per day, which is similar to a loading rate. Recall from Equation 3-80 that the velocity of flow is equal to the flow rate divided by the area through which it flows. Hence an overflow rate is the same as a liquid velocity:

$$v_o = \frac{\text{Volume/Time}}{\text{Surface Area}} = \frac{\text{(Depth)(Surface area)}}{\text{(Time)(Surface area)}} = \frac{\text{Depth}}{\text{Time}} = \text{Liquid velocity} \tag{3-81}$$

$$v_o = \frac{\forall/t_o}{A_s} = \frac{(h)(A_s)}{(t_o)(A_s)} = \frac{h}{t_o}$$

It can be seen from the above discussion that particle removal is independent of the depth of the sedimentation tank. As long as v_s is greater than v_o, the particles will settle downward and be removed from the bottom of the tank regardless of the depth. Sedimentation zones vary from a depth of a few centimeters to a depth of 6 m or greater.[23]

We can show that particle removal in a horizontal sedimentation tank is likewise dependent only upon the overflow rate. An ideal horizontal sedimentation tank

[23]Tube settlers are designed with a very shallow settling zone. Their use is beyond the presentation of this text.

is based upon three assumptions:[24]

1. Particles and velocity vectors are evenly distributed across the tank cross section. This is the function of the inlet zone.
2. The liquid moves as an ideal slug down the length of the tank.
3. Any particle hitting the bottom of the tank is removed.

Using Figure 3-33*a* to illustrate the concept, let us consider a particle which is released at point "A." In order to be removed from the water it must have a settling

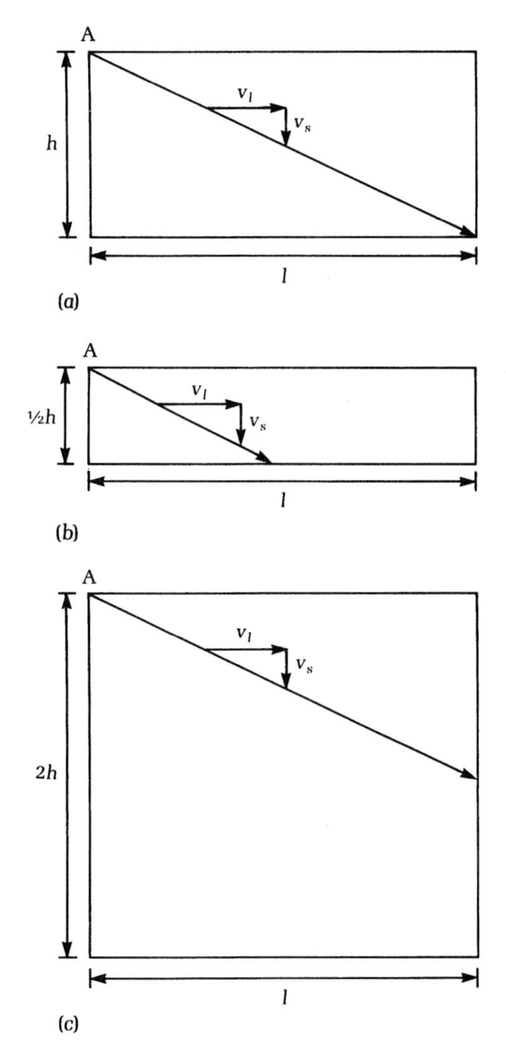

FIGURE 3-33
Ideal horizontal sedimentation tank.

[24] A. Hazen, "On Sedimentation," *Transactions of the American Society of Civil Engineers*, vol. 53, p. 45, 1904; and T. R. Camp, "Sedimentation and Design of Settling Tanks," *Transactions of the American Society of Civil Engineers*, vol. 111, p. 895, 1946.

velocity great enough so that it reaches the bottom of the tank during the detention time (t_o) of the water in the tank. We may say its settling velocity must equal the depth of the tank divided by the detention time, that is,

$$v_s = \frac{h}{t_o} \tag{3-82}$$

Now we can also show that the settling velocity of the particle must be equal to or greater than the overflow rate of the tank in order to be removed. Using the definition of detention time from Equation 3-72 and substituting into Equation 3-82:

$$v_s = \frac{h}{(\forall/Q)} = \frac{hQ}{\forall} \tag{3-83}$$

But we know that the tank volume is described by the product of the height, length, and width so we can rewrite this as:

$$v_s = \frac{hQ}{l \times w \times h} = \frac{Q}{l \times w} \tag{3-84}$$

And the product ($l \times w$) is the surface area (A_s) so

$$v_s = \frac{Q}{A_s} \tag{3-85}$$

which is the overflow rate (v_o). This implies that the removal of a horizontal clarifier is independent of depth! This is indeed strange and runs counter to our intuitive feeling of how the sedimentation tank should work. Why is this so? Figure 3-33*b* should help clear up this apparent ambiguity. What we see is that particles with v_s greater than or equal to v_o are removed in tanks having a depth equal to one-half the depth of Figure 3-33*a*. If the depth were greater, particles having settling velocities equal to v_o would not be completely removed (Figure 3-33*c*). But some removal would take place since those particles entering the tank at lower depths would have the correct trajectory to reach the bottom. This leads us to the concept of partial removal.

In a horizontal sedimentation tank, unlike an upflow clarifier, some percentage of the particles with a v_s less than v_o will be removed. For example, consider particles having a settling velocity of 0.5 v_o entering uniformly into the settling zone. Figure 3-34 shows that 50 percent of these particles (those below half the depth of the tank)

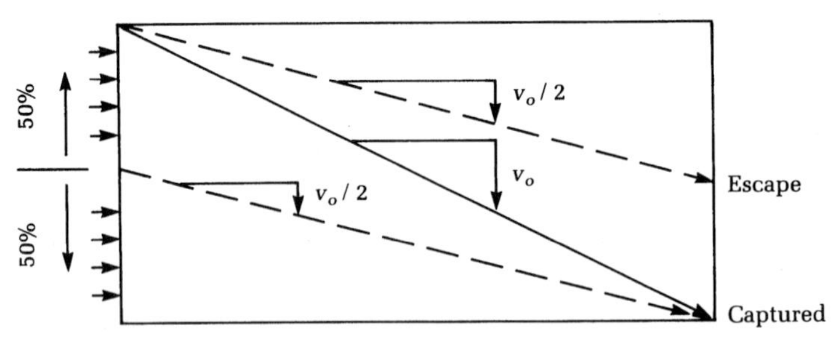

FIGURE 3-34
Partial solids removal in ideal sedimentation tank.

will be removed. That is, they will hit the bottom of the tank before being carried out because they only have to settle one-half the tank depth. Likewise, one-fourth of the particles having a settling velocity of 0.25 v_o will be removed. The percentage of particles removed, P, with a settling velocity of v_s in a sedimentation tank designed with an overflow rate of v_o is

$$P = 100 \frac{v_s}{v_o} \tag{3-86}$$

Example 3-22. The town of San Jose has an existing horizontal-flow sedimentation tank with an overflow rate of 17 $m^3/d \cdot m^2$, and it wishes to remove particles that have settling velocities of 0.1 mm/s, 0.2 mm/s, and 1 mm/s. What percentage of removal should be expected for each particle in an ideal sedimentation tank?

Solution

a. $v_s = 0.1$ mm/s

First we need to convert the overflow rate to compatible units:

$$17 \frac{\text{m}^3}{\text{d} \cdot \text{m}^2} = 17 \frac{\text{m}}{\text{d}} \frac{(1{,}000 \text{ mm/m})}{(86{,}400 \text{ s/d})} = 0.2 \text{ mm/s}$$

Since $v_s < v_o$ for a v_s of 0.1 mm/s, some fraction of the particles will be removed, as given by Equation 3-86.

$$P = 100 \frac{(0.1)}{(0.2)} = 50\%$$

b. $v_s = 0.2$ mm/s

These particles have $v_s = v_o$, and ideally will be 100 percent removed.

c. $v_s = 1$ mm/s

These particles have $v_s > v_o$, and 100 percent of the particles should be easily removed.

Determination of v_s

In design of an ideal sedimentation tank, one first determines the settling velocity (v_s) of the particle to be removed and then sets the overflow rate (v_o) at some value less than or equal to v_s.

Determination of the particle-settling velocity is different for different types of particles. Settling properties of particles are often categorized into one of three classes:

Type I sedimentation. Type I sedimentation is characterized by particles that settle discretely at a constant settling velocity. They settle as individual particles and do not flocculate or stick to other particles during settling. Examples of these particles are sand and grit material. Generally speaking, the only application of Type I settling is during pre-sedimentation for sand removal prior to coagulation in a potable water plant, in settling of sand particles during cleaning of rapid sand filters, and in grit chambers (see Section 5-5).

Type II sedimentation. Type II sedimentation is characterized by particles that flocculate during sedimentation. Since they flocculate, their size is constantly changing; therefore, the settling velocity is changing. Generally the settling velocity is increasing. These types of particles occur in alum or iron coagulation, in primary sedimentation (see Section 5-6), and in settling tanks in trickling filtration (see Section 5-7).

Type III or zone sedimentation. In zone sedimentation the particles are at a high concentration (greater than 1,000 mg/L) such that the particles tend to settle as a mass, and a distinct clear zone and sludge zone are present. Zone settling occurs in lime-softening sedimentation, activated-sludge sedimentation, and sludge thickeners (see Section 5-11).

Determination of v_o

There are five ways to determine effective particle-settling velocities and consequently to determine overflow rates.

Calculation. In the case of Type I sedimentation, the particle-settling velocity can be calculated and the basin designed to remove a specific size particle. In 1687, Sir Isaac Newton showed that a particle falling in a quiescent fluid accelerates until the frictional resistance, or drag, on the particle is equal to the gravitational force of the particle (Figure 3-35).[25] The three forces are defined as follows:

$$F_G = (\rho_s)g\forall_p \tag{3-87}$$

$$F_B = (\rho)g\forall_p \tag{3-88}$$

$$F_D = C_D A_p(\rho)\frac{v^2}{2} \tag{3-89}$$

where F_G = gravitational force
F_B = buoyancy force
F_D = drag force
ρ_s = density of particle, kg/m^3
ρ = density of fluid, kg/m^3
g = acceleration due to gravity, m/s^2
$\forall_p$ = volume of particle, m^3
C_D = drag coefficient
A_p = cross sectional area of particle, m^2
v = velocity of particle, m/s

The driving force for acceleration of the particle is the difference between the gravitational and buoyant force:

$$F_G - F_B = (\rho_s - \rho)\, g\forall_p \tag{3-90}$$

[25] Isaac Newton, *Philosophiae Naturalis Principia Mathematica*, 1687.

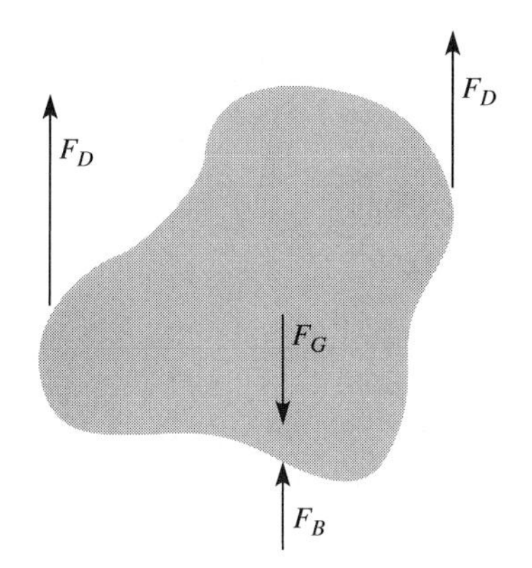

FIGURE 3-35
Forces acting on a free-falling particle in a fluid (F_D = drag force; F_G = gravitational force; F_B = buoyancy force).

When the drag force is equal to the driving force, the particle velocity reaches a constant value called the *terminal settling velocity* (v_s).

$$F_G - F_B = F_D \tag{3-91}$$

$$(\rho_s - \rho)\, g\Psi_p = C_D A_p\, (\rho)\, \frac{v_s^2}{2} \tag{3-92}$$

For spherical particles with a diameter $= d$,

$$\frac{\Psi_p}{A_p} = \frac{4/3\,(\pi)\,(d/2)^3}{(\pi)\,(d/2)^2} = \frac{2}{3}\,d \tag{3-93}$$

Using 3-93 to solve for the terminal settling velocity:

$$v_s = \left[\frac{4\, g(\rho_s - \rho)\, d}{3\, C_D\, \rho}\right]^{1/2} \tag{3-94}$$

The drag coefficient takes on different values depending on the flow regime surrounding the particle. The flow regime may be characterized qualitatively as laminar, turbulent, or transitional. In laminar flow, the fluid moves in layers, or laminas, one layer gliding smoothly over adjacent layers with only molecular interchange of momentum. In turbulent flow, the fluid motion is very erratic with violent transverse interchange of momentum. Osborne Reynolds[26] developed a quantitative means of describing the different flow regimes using a dimensionless ratio that is called the *Reynolds number*. For spheres moving through a liquid this number is defined as

$$\mathbf{R} = \frac{(d)\, v_s}{\nu} \tag{3-95}$$

[26] O. Reynolds, "An Experimental Investigation of the Circumstances Which Determine Whether the Motion of Water Shall Be Direct or Sinuous, and of the Laws of Resistance in Parallel Channels," *Transactions of the Royal Society of London*, vol. 174, 1883.

where $\mathbf{R}$ = Reynolds number
d = diameter of sphere, m
v_s = velocity of sphere, m/s
ν = kinematic viscosity, $m^2/s = \mu/\rho$
ρ = density of fluid, kg/m^3
μ = dynamic viscosity, $Pa \cdot s$

Thomas Camp[27] developed empirical data relating the drag coefficient to Reynolds number (Figure 3-36). For eddying resistance for spheres at high Reynolds numbers ($\mathbf{R} > 10^4$), C_D has a value of about 0.4. For viscous resistance at low Reynolds numbers ($\mathbf{R} < 0.5$) for spheres:

$$C_D = \frac{24}{\mathbf{R}} \tag{3-96}$$

For the transition region of $\mathbf{R}$ between 0.5 and 10^4, the drag coefficient for spheres may be approximated by the following:

$$C_D = \frac{24}{\mathbf{R}} + \frac{3}{\mathbf{R}^{1/2}} + 0.34 \tag{3-97}$$

Sir George Gabriel Stokes showed that, for spherical particles falling under laminar (quiescent) conditions, Equation 3-94 reduces to the following:[28]

$$v_s = \frac{g(\rho_s - \rho)d^2}{18\mu} \tag{3-98}$$

where μ = dynamic viscosity, $Pa \cdot s$
18 = a constant

Equation 3-98 is called *Stokes' law.* Dynamic viscosity (also called absolute viscosity) is a function of the water temperature. A table of dynamic viscosities is given in Appendix A. Fair, Geyer, and Okun recommend that v_0 be set at 0.33 to 0.7 times v_s depending upon the efficiency desired.[29]

Flocculant sedimentation lab or pilot data. There is no adequate mathematical relationship that can be used to describe Type II settling. The Stokes equation cannot be used because the flocculating particles are continually changing in size and shape, and when water is entrapped in the floc, in specific gravity. Laboratory tests with settling columns are used to develop design data.

A settling column is filled with the suspension to be analyzed. The suspension is allowed to settle. Samples are withdrawn from the sample ports at selected time

[27] T. R. Camp, "Sedimentation and the Design of Settling Tanks," *Transactions of the American Society of Civil Engineers*, vol. 111, p. 897, 1946.

[28] G. G. Stokes, *Transactions of the Cambridge Philosophical Society*, vol. 8, p. 287, 1845.

[29] G. M. Fair, J. C. Geyer, and D. A. Okun, *Water and Wastewater Engineering*, vol. 2, New York: Wiley, 1968.

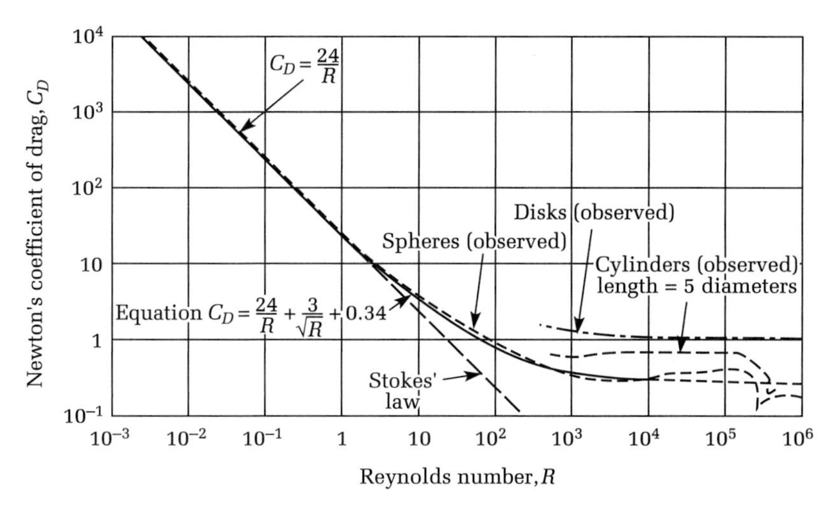

FIGURE 3-36
Newton's coefficient of drag as a function of Reynolds number. (*Source:* T. R. Camp, "Sedimentation and the Design of Settling Tanks," *Transactions of the American Society of Civil Engineers,* vol. 111, p. 897, 1946.)

intervals. The concentration of suspended solids is determined for each sample and the percent removal is calculated:

$$R\% = 1 - \frac{C_t}{C_o}(100\%) \tag{3-99}$$

where $R\%$ = percent removal at one depth and time, %
C_t = concentration at time, t, and given depth, mg/L
C_o = initial concentration, mg/L

Percent removal versus depth is then plotted as shown in Figure 3-37. The circled numbers are the calculated percentages. Interpolations are made between these plotted points to construct curves of equal concentration at reasonable percentages, that is, 5- or 10-percent increments. Each intersection point of an isoconcentration line and the bottom of the column defines an overflow rate (v_o):

$$v_o = \frac{H}{t_i} \tag{3-100}$$

where H = height of column, m
t_i = time defined by intersection of isoconcentration line and bottom of column (x-axis) where the subscript, i, refers to the first, second, third, etc., intersection points, d

A vertical line is drawn from t_i to intersect all the isoconcentration lines crossing the t_i time. The midpoints between isoconcentration lines define heights H_1, H_2, H_3, etc. used to calculate the fraction of solids removed. For each time, t_i, defined by the intersection of the isoconcentration line and the bottom of the column (x-axis), you can construct a vertical line and calculate the fraction of solids removed:

$$R_{T_a} = R_a + \frac{H_1}{H}(R_b - R_a) + \frac{H_2}{H}(R_c - R_b) + \cdots \tag{3-101}$$

where R_{T_a} = total fraction removed for settling time, t_a
R_a, R_b, R_c = isoconcentration fractions a, b, c, etc.

The series of overflow rates and removal fractions are used to plot curves of suspended solids removal versus detention time and suspended solids removal versus overflow rate that can be used to size the settling tank. Eckenfelder[30] recommends that scale-up factors of 0.65 for overflow rate and 1.75 for detention time be used to design the tank.

Example 3-23. The city of Urbana is planning to install a new water treatment plant. Design a settling tank to remove 65 percent of the influent suspended solids from their design flow of 0.5 m^3/s. A batch-settling test using a 2.0 m column and coagulated water from their existing plant yielded the following data:

Percent removal as a function of time and depth

	Sampling Time, min						
Depth, m	**5**	**10**	**20**	**40**	**60**	**90**	**120**
0.5	41	50	60	67	72	73	76
1.0	19	33	45	58	62	70	74
2.0	15	31	38	54	59	63	71

Solution. The plot is shown in Figure 3-37.

Calculate the overflow rate for each intersection point. For example, for the 50 percent line,

$$v_o = \frac{2.0 \text{ m}}{(35 \text{ min})}(1{,}440 \text{ min/d}) = 82.3 \text{ m/d}$$

The corresponding removal fraction is

$$\begin{aligned} R_{T50} &= 50 + \frac{1.5}{2.0}(55 - 50) + \frac{0.85}{2.0}(60 - 55) \\ &\quad + \frac{0.60}{2.0}(65 - 60) + \frac{0.40}{2.0}(70 - 65) \\ &\quad + \frac{0.20}{2.0}(75 - 70) + \frac{0.05}{2.0}(100 - 75) \\ &= 59.5 \text{ or } 60\% \end{aligned}$$

---

[30]W. W. Eckenfelder, *Principles of Water Quality Management*, Boston: CBI Publishing, 1980.

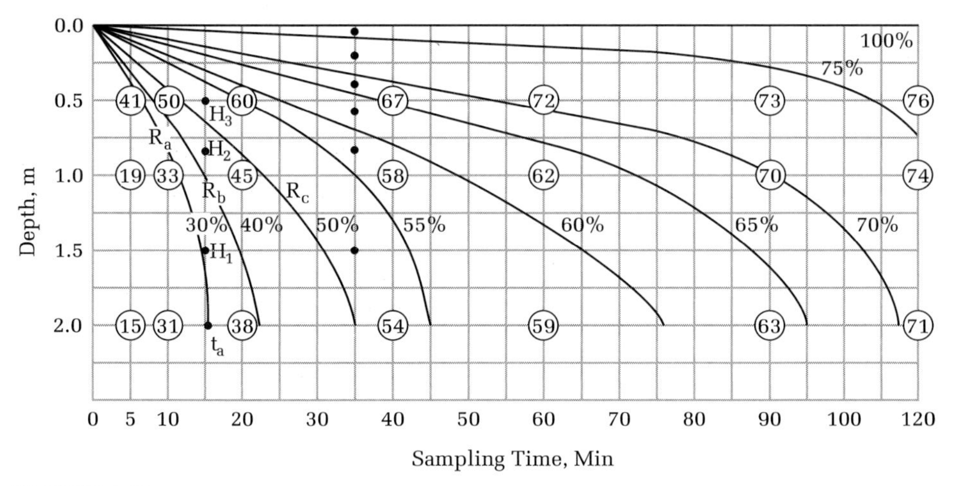

FIGURE 3-37
Isoconcentration lines for Type II settling test using a 2 m deep column.

The corresponding detention time is taken from the intersection of the isoconcentration line and the x-axis used to define the overflow rate, that is, 35 minutes for the 50 percent line.

This calculation is repeated for each isoconcentration line that intersects the x-axis except the last ones for which data are too sparse, that is, 30, 40, 50, 55, 60, and 65 percent, but not 70 or 75 percent.

Two graphs are then constructed (see Figures 3-38 and 3-39). From these graphs the bench-scale detention time and overflow rate for 65 percent removal are found to be 54 minutes and 50 m/d.

Applying the scale-up factors:

$$t_o = (54 \text{ min})(1.75) = 94.5 \text{ or } 95 \text{ min}$$

$$v_o = (50 \text{ m/d})(0.65) = 32.5 \text{ m/d}$$

Zone sedimentation lab data. For zone sedimentation, values can also be obtained from the lab. The design overflow is again set at about 0.5 to 0.7 times the lab value.

Jar test data. A technique has been developed to determine settling velocities of coagulant flocs from jar test data.[31]

Experience. Typical design numbers exist for all types of sedimentation basins. These numbers can be used in lieu of laboratory or pilot work. The applicability of the typical numbers to particles in different situations is unknown. For this reason, the typical numbers are often quite conservative. These conservative numbers also correct for ineffective inlet or outlet zone design. A design engineer with sufficient

[31] H. E. Hudson, *Water Clarification Processes, Practical Design and Evaluation*, 1981.

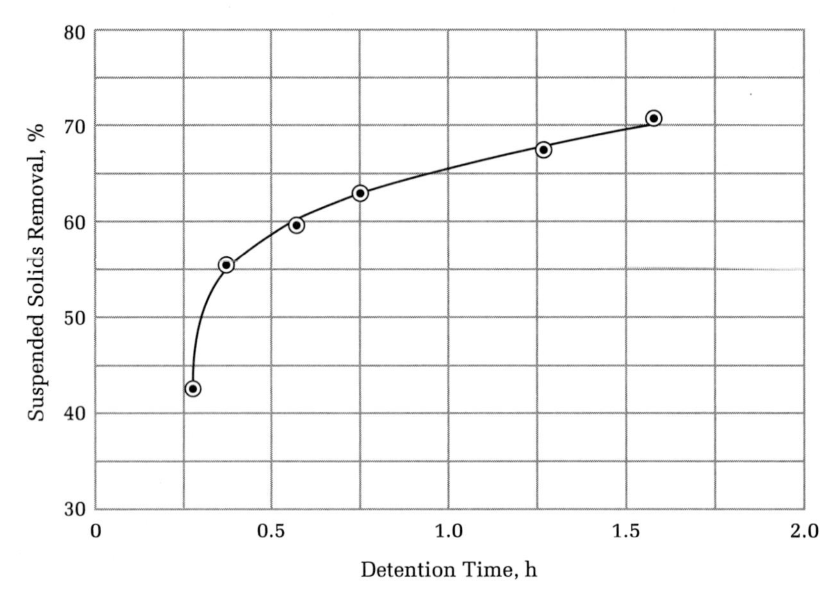

FIGURE 3-38
Suspended solids removal versus detention time.

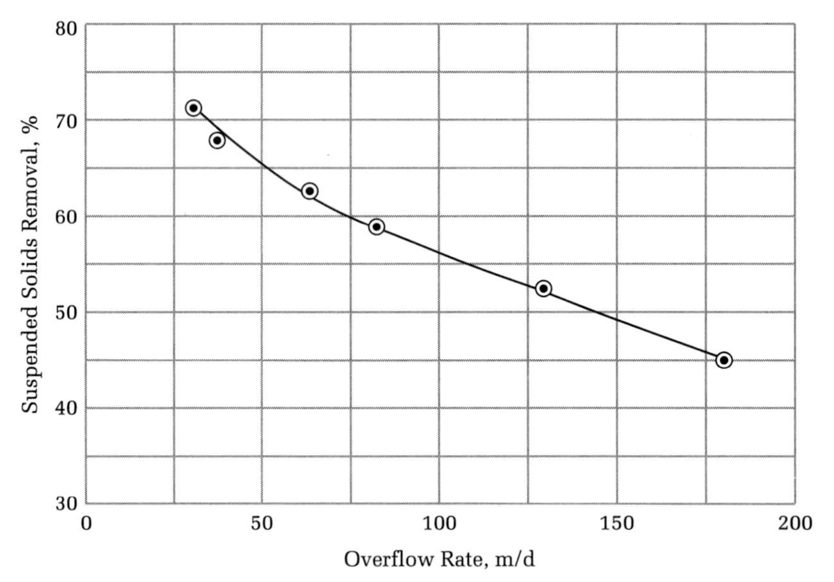

FIGURE 3-39
Suspended solids removal versus overflow rate.

TABLE 3-18
Typical sedimentation tank overflow rates

Application	Long rectangular and circular $m^3d/ \cdot m^2$	Upflow solids-contact $m^3/d \cdot m^2$
Alum or iron coagulation		
Turbidity removal	40	50
Color removal	30	35
High algae	20	
Lime softening		
Low magnesium	70	130
High magnesium	57	105

Source: American Water Works Association, *Water Treatment Plant Design,* 2nd ed., New York: McGraw-Hill, 1990.

time and funds can generally save his or her client capital costs by designing a good inlet system and conducting lab tests for proper sedimentation zone design. However, clients are not always willing to expend such funds and the engineer has little choice other than to select conservative design numbers. Table 3-18 shows some overflow rates for potable water treatment.

Typical detention times for waters coagulated with alum or iron salts are on the order of 2 to 8 hours. In lime-soda softening plants, the detention times range from 4 to 8 hours.[32]

Example 3-24. Determine the surface area of a settling tank for the city of Urbana's 0.5 m^3/s design flow using the design overflow rate found in Example 3-23. Compare this surface area with that which results from assuming a typical overflow rate of 20 $m^3/d \cdot m^2$. Find the depth of the clarifier for the overflow rate and detention time found in Example 3-23.

Solution

a. Find the surface area.

First change the flow rate to compatible units:

$$\left(\frac{0.5\ m^3}{s}\right)\left(\frac{86{,}400\ s}{d}\right) = 43{,}200\ m^3/d$$

Using the overflow rate from Example 3-23, the surface area is

$$A_s = \frac{43{,}200\ m^3/d}{32.5\ m^3/d \cdot m^2} = 1{,}329.23\ m^2 \text{ or } 1{,}330\ m^2$$

---

[32]T. D. Reynolds, *Unit Operations and Processes in Environmental Engineering,* Boston, MA: PWS-Kent, 1996, p. 256.

Using the conservative value

$$A_s = \frac{43{,}200 \text{ m}^3/\text{d}}{20 \text{ m}^3/\text{d} \cdot \text{m}^2} = 2{,}160 \text{ m}^2$$

Obviously, the use of conservative data would, in this case, result in a 60 percent overdesign of the tank area.

Common length-to-width ratios for settling are between 2:1 and 5:1, and lengths seldom exceed 100 m. A minimum of two tanks is always provided.

Assuming two tanks, each with a width of 12 m, a total surface area of 1,330 m^2 would imply a tank length of

$$\text{Length} = \frac{1{,}330 \text{ m}^2}{(2 \text{ tanks})(12 \text{ m wide})} = 55.4 \text{ m or } 55 \text{ m}$$

This meets our length-to-width ratio of 5:1.

b. Find the tank depth.

First find the total tank volume from Equation 3-72 using a detention time of 95 minutes from Example 3-23:

$$\forall = (0.5 \text{ m}^3/\text{s})(95 \text{ min})(60 \text{ s/min}) = 2{,}850 \text{ m}^3$$

This would be divided into two tanks as noted above. The depth is found as the total tank volume divided by the total surface area:

$$\text{Depth} = \frac{2{,}850 \text{ m}^3}{1{,}330 \text{ m}^2} = 2.1428 \text{ m or } 2 \text{ m}$$

This depth would not include the sludge storage zone.

The final design would then be two tanks, each having the following dimensions: 12 m wide × 55 m long × 2 m deep plus sludge storage depth.

3-7 FILTRATION

The water leaving the sedimentation tank still contains floc particles. The settled water turbidity is generally in the range from 1 to 10 TU with a typical value being 3 TU. In order to reduce this turbidity to 0.3 TU, a filtration process is normally used. Water filtration is a process for separating suspended or colloidal impurities from water by passage through a porous medium, usually a bed of sand or other medium. Water fills the pores (open spaces) between the sand particles, and the impurities are left behind, either clogged in the open spaces or attached to the sand itself.

There are several methods of classifying filters. One way is to classify them according to the type of medium used, such as sand, coal (called anthracite), dual media (coal plus sand), or mixed media (coal, sand, and garnet). Another common way to classify the filters is by allowable loading rate. *Loading rate* is the flow rate of water applied per unit area of the filter. It is the velocity of the water approaching the face of the filter:

$$v_a = \frac{Q}{A_s} \tag{3-102}$$

where v_a = face velocity, m/d
= loading rate, $m^3/d \cdot m^2$
Q = flow rate onto filter surface, m^3/d
A_s = surface area of filter, m^2

Based on loading rate, the filters are described as being slow sand filters, rapid sand filters, or high-rate sand filters.

Slow sand filters were first introduced in the 1800s. The water is applied to the sand at a loading rate of 2.9 to 7.6 $m^3/d \cdot m^2$. As the suspended or colloidal material is applied to the sand, the particles begin to collect in the top 75 mm and to clog the pore spaces. As the pores become clogged, water will no longer pass through the sand. At this point the top layer of sand is scraped off, cleaned, and replaced. Slow sand filters require large areas of land and are operator intensive.

In the early 1900s there was a need to install filtration systems in large numbers in order to prevent epidemics. Rapid sand filters were developed to meet this need. These filters have graded (layered) sand within the bed. The sand grain size distribution is selected to optimize the passage of water while minimizing the passage of particulate matter.

Rapid sand filters are cleaned in place by forcing water backwards through the sand. This operation is called *backwashing*. The washwater flow rate is such that the sand is expanded and the filtered particles are removed from the bed. After backwashing, the sand settles back into place. The largest particles settle first, resulting in a fine sand layer on top and a coarse sand layer on the bottom. Rapid sand filters are the most common type of filter used in water treatment today.

Traditionally, rapid sand filters have been designed to operate at a loading rate of 120 $m^3/d \cdot m^2$. Experiments conducted at the Chicago water treatment plant[33] have demonstrated that satisfactory water quality can be obtained with rates as high as 235 $m^3/d \cdot m^2$. Normally, a minimum of two filters are constructed to ensure redundancy. For larger plants (>0.5 m^3/s), a minimum of four filters is suggested.[34] The surface area of the filter tank (often called a filter box) is generally restricted in size to about 100 m^2.

In the wartime era of the early 1940s, dual-media filters were developed. They are designed to utilize more of the filter depth for particle removal. In a rapid sand filter, the finest sand is on the top; hence, the smallest pore spaces are also on the top. Therefore, most of the particles will clog in the top layer of the filter. In order to use more of the filter depth for particle removal, it is necessary to have the large particles on top of the small particles. This was accomplished by placing a layer of coarse coal on top of a layer of fine sand. Coal has a lower specific gravity than sand, so, after backwash, it settles slower than the sand and ends up on top. Dual-media filters are operated up to loading rates of 300 $m^3/d \cdot m^2$.

---

[33]American Water Works Association, *Water Quality and Treatment*, New York: McGraw-Hill, p. 259, 1971.

[34]J. M. Montgomery, *Water Treatment Principles and Design*, New York: Wiley, p. 535, 1985.

In the mid 1980s, deep-bed monomedia filters came into use. The filters are designed to achieve higher loading rates while at the same time producing lower finished water turbidities. The filters typically consist of 1.0-mm to 1.5-mm diameter anthracite about 1.5 m to 2.5 m deep. They operate at loading rates up to 800 $m^3/d \cdot m^2$.

Example 3-25. As part of their proposed new treatment plant, Urbana is going to install rapid sand filters after their sedimentation tanks. The design loading rate to the filter is 200 $m^3/d \cdot m^2$. How much filter surface area should be provided for their design flow rate of 0.5 m^3/s? If the surface area per filter box is to be limited to 50 m^2, how many filter boxes are required?

Solution. The surface area required is the flow rate divided by the loading rate:

$$A_s = \frac{Q}{V_a} = \frac{(0.5 \text{ m}^3/\text{s})(86{,}400 \text{ s/d})}{200 \text{ m}^3/\text{d} \cdot \text{m}^2} = 216 \text{ m}^2$$

If the maximum surface area of any one tank is 50 m^2, then the number of filters required is

$$\text{Number} = \frac{216 \text{ m}^2}{50 \text{ m}^2} = 4.32$$

Since we cannot build 0.32 filter, we need to round to an integer. Normally, we build an even number of filters to make construction easier and to reduce costs. In this case we would propose to build four filters and check to see that the design loading does not exceed our guideline values. With four filters the loading would be

$$v_a = \frac{Q}{A_s} = \frac{(0.5 \text{ m}^3/\text{s})(86{,}400 \text{ s/d})}{(4 \text{ filters})(50 \text{ m}^2/\text{filter})} = 216 \text{ m/d}$$

This is less than the 235 m/d recommended maximum loading rate and would be acceptable except that many states require that the filter capacity be sufficient to handle the design flow rate with one filter out of service. Therefore, we must check the loading with three filters in service

$$v_a = \frac{(0.5 \text{ m}^3/\text{s})(86{,}400 \text{ s/d})}{(3 \text{ filters})(50 \text{ m}^2/\text{filter})} = 288\text{m/d}$$

This exceeds the recommended maximum loading rate. If the 50 m^2/filter cannot be altered, another filter box is required. Since the filter box may be as large as 100 m^2/filter, we would expect that four slightly larger filters would be constructed to meet the required loading with one filter out of service.

Figure 3-40 shows a cutaway of a rapid sand filter. The bottom of the filter consists of a support media and collection system. The support media is designed to keep the sand in the filter and prevent it from leaving with the filtered water. Layers of graded gravel (large on bottom, small on top) traditionally have been used for the support. A perforated pipe is one method of collecting the filtered water.

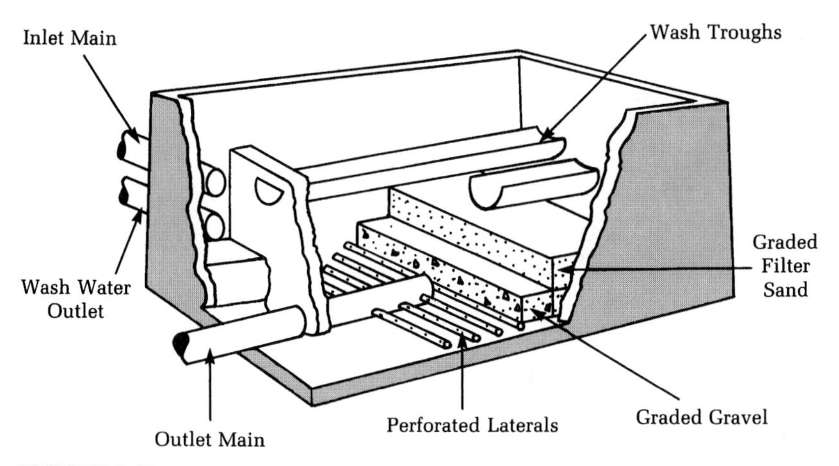

FIGURE 3-40
Typical cross section of a rapid sand filter. (*Source:* American Water Works Association, *Water Treatment Plant Design,* 1969.)

On top of the support media is a layer of graded sand. The sand depth varies between 0.5 and 0.75 m. If a dual media filter is used, the sand is about 0.3 m thick and the coal about 0.45 m thick. Approximately 0.7 m to 1 m above the top of the sand are the washwater troughs. The washwater troughs collect the backwash water used to clean the filter. The troughs are placed high enough above the sand layer so that sand will not be carried out with the backwash water. Generally a total depth of 1.8 m to 3 m is allowed above the sand layer for water to build up above the filter. This depth of water provides sufficient pressure to force the water through the sand during filtration.

Figure 3-41 shows a slightly simplified version of a rapid sand filter. Water from the settling basins enters the filter and seeps through the sand and gravel bed, through a false floor, and out into a clear well that acts as a storage tank for finished water. During filtration, valves "A" and "C" are open.

During filtration the filter bed will become more and more clogged. As the filter clogs, the water level will rise above the sand as it becomes harder to force water through the bed. Eventually, the water level will rise to the point that the filter bed must be cleaned. This point is called *terminal head loss.* When this occurs, the operator turns off valves "A" and "C." This stops the supply of water from the sedimentation tank and prevents any more water from entering the clear well. The operator then opens valves "E" and "B." This allows a large flow of washwater (clean water stored in an elevated tank or pumped from a clear well) to enter below the filter bed. This rush of water forces the sand bed to expand and sets individual sand particles in motion. By rubbing against each other, the light colloidal particles that were trapped in the pore spaces are released and escape with the washwater. The washwater is a waste stream that must be treated. After a few minutes the washwater is shut off and filtration resumed.

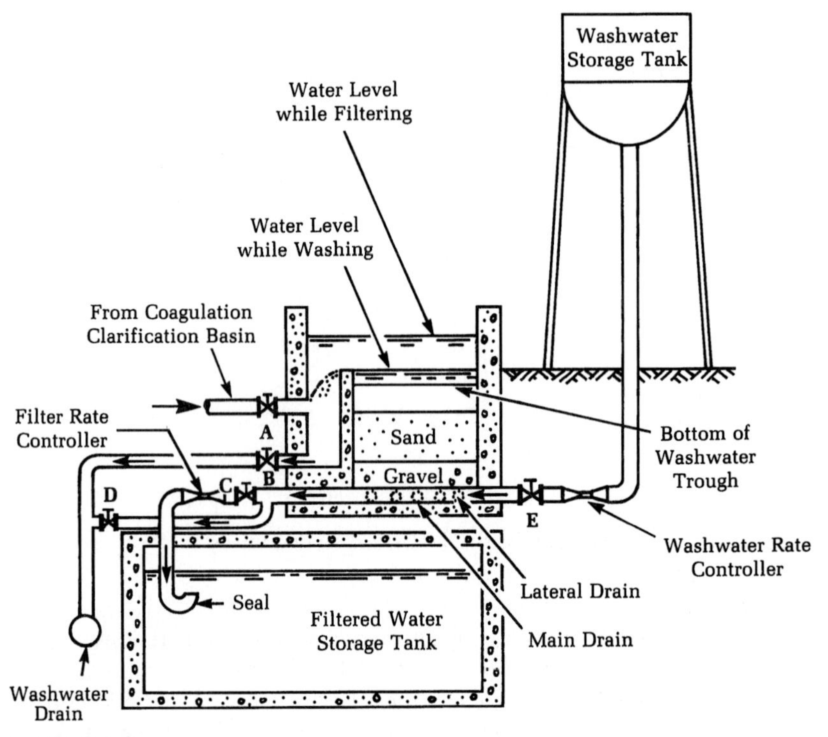

FIGURE 3-41
Operation of a rapid sand filter. (*Source:* E. W. Steel and T. J. McGhee, *Water Supply and Sewerage,* New York: McGraw-Hill, 1979. Reprinted by permission.)

Grain Size Characteristics[35]

The size distribution or variation of a sample of granular material is determined by sieving the sample through a series of standard sieves (screens). One such standard series is the U.S. Standard Sieve Series. The U.S. Standard Sieve Series (Table 3-19) is based on a sieve opening of 1 mm. Sieves in the "fine series" stand successively in the ratio of $(2)^{1/4}$ to one another, the largest opening in this series being 5.66 mm and the smallest 0.037 mm. All material that passes through the smallest sieve opening in the series is caught in a pan that acts as the terminus of the series.

The grain size analysis begins by placing the sieve screens in ascending order with the largest opening on top and the smallest opening on the bottom. A sand sample is placed on the top sieve and the stack is shaken for a prescribed amount of time. At the end of the shaking period, the mass of material retained on each sieve is determined. The cumulative mass is recorded and converted into percentages by mass

[35]Excerpted from G. M. Fair and J. C. Geyer, *Water Supply and Wastewater Disposal,* New York: Wiley, pp. 664–670, 1954.

TABLE 3-19
U.S. Standard Sieve Series

Sieve designation number	Size of opening (mm)	Sieve designation number	Size of opening (mm)
200	0.074	20	0.84
140	0.105	(18)	(1.00)
100	0.149	16	1.19
70	0.210	12	1.68
50	0.297	8	2.38
40	0.42	6	3.36
30	0.59	4	4.76

Source: Excerpted from G. A. Fair and J. C. Geyer, *Water Supply and Wastewater Disposal,* New York: Wiley, pp. 664-670, 1954.

equal to or less than the size of separation of the overlying sieve. Then the cumulative frequency distribution is plotted. For many natural granular materials, this curve approaches geometric normality. Logarithmic-probability paper, therefore, assures an almost straight-line plot which facilitates interpolation. The geometric mean (X_g) and geometric standard deviation (S_g) are useful parameters of central tendency and variation. Their magnitudes may be determined from the plot. The parameters most commonly used, however, are the *effective size, E,* or 10-percentile, P_{10}, and the *uniformity coefficient, U,* or ratio of the 60-percentile to the 10-percentile, P_{60}/P_{10}. Use of the 10-percentile was suggested by Allen Hazen because he had observed that resistance to the passage of water offered by a bed of sand within which the grains are distributed homogeneously remains almost the same, irrespective of size variation (up to a uniformity coefficient of about 5.0), provided that the 10-percentile remains unchanged.[36] Use of the ratio of the 60-percentile to the 10-percentile as a measure of uniformity was suggested by Hazen because this ratio covered the range in size of half the sand.[37] On the basis of logarithmic normality, the probability integral establishes the following relations between the effective size and uniformity coefficient and the geometric mean size and geometric standard deviation:

$$E = P_{10} = (X_g)(S_g)^{-1.282} \tag{3-103}$$

$$U = P_{60}/P_{10} = (S_g)^{1.535} \tag{3-104}$$

Experience has suggested that, for silica sand, the effective size should be in the range of 0.35 to 0.55 mm with a maximum of about 1.0 mm. The uniformity coefficient should range between 1.3 and 1.7. Smaller effective sizes result in a product

[36] A. Hazen, *Annual Report of the Massachusetts State Board of Health*, 1892.

[37] It would be more logical to speak of this ratio as a coefficient of nonuniformity because the coefficient increases in magnitude as the nonuniformity increases.

water that is lower in turbidity, but they also result in higher pressure losses in the filter and shorter operating cycles between cleanings.

Example 3-26. For the size frequencies by weight and by count of the sample of sand listed below, find the effective size, E, and uniformity coefficient, U.

U. S. Standard Sieve No.	Analysis of stock sand (Cumulative mass % passing)
140	0.2
100	0.9
70	4.0
50	9.9
40	21.8
30	39.4
20	59.8
16	74.4
12	91.5
8	96.8
6	99.0

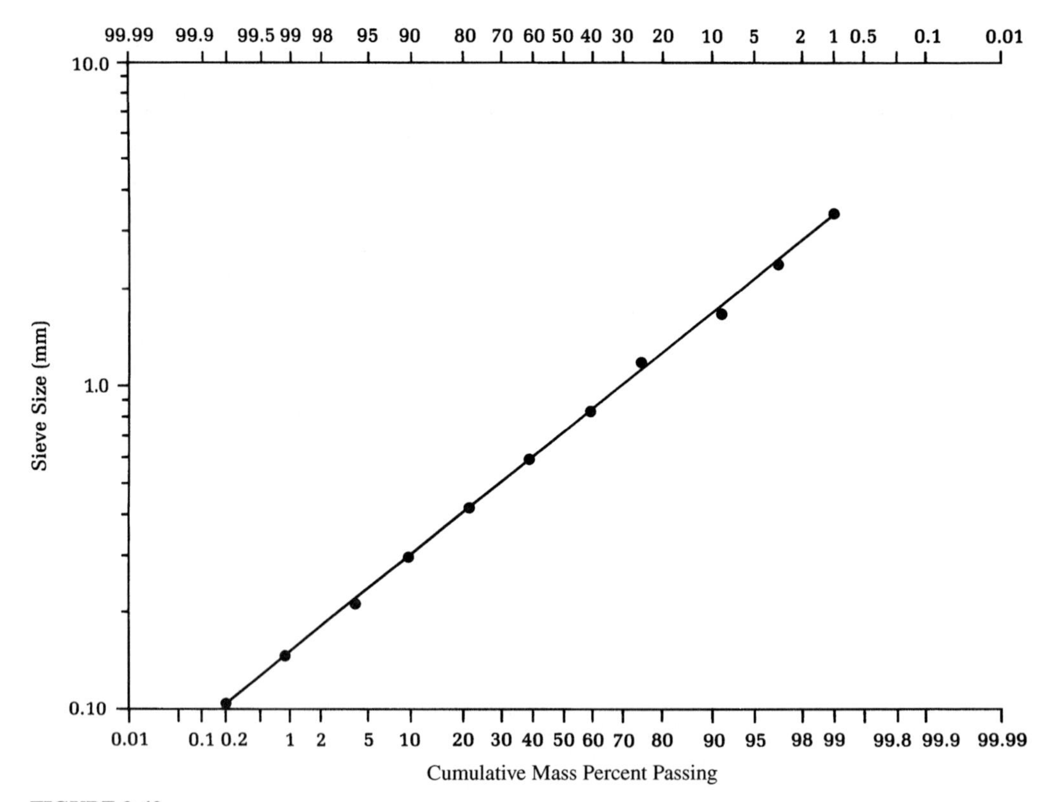

FIGURE 3-42
Grain size analysis of run-of-bank sand.

Solution. First we must plot the data on log-probability paper as shown in Figure 3-42. From this plot we then find the effective size:

$$E = P_{10} = 0.30 \text{ mm}$$

The uniformity coefficient is then

$$U = \frac{P_{60}}{P_{10}} = \frac{0.85}{0.30} = 2.8$$

Sand excavated from a natural deposit is called *run-of-bank* sand. Run-of-bank sand may be too coarse, too fine, or too nonuniform for use in filters. Within economical limits, proper sizing and uniformity are obtained by screening out coarse components and washing out fine components. In rapid sand filters, the removal of "fines" may be accomplished by stratifying the bed through backwashing and then scraping off the layer that includes the unwanted sand.

Filter Hydraulics

The loss of pressure (commonly termed *head loss*) through a clean stratified-sand filter with uniform porosity was described by Rose[38,39] in the following form:

$$h_L = \frac{1.067\,(v_a)^2(D)}{(\phi)(g)(\epsilon)^4} \sum_{i=1}^{n} \frac{(C_D)(f)}{d} \tag{3-105}$$

where h_L = frictional head loss through the filter, m
v_a = approach velocity, m/s
D = depth of filter sand, m
C_D = drag coefficient
f = mass fraction of sand particles of diameter d
d = diameter of sand grains, m
ϕ = shape factor
g = acceleration due to gravity, m/s^2
ϵ = porosity

The drag coefficient is defined in Equations 3-96 and 3-97. The Reynolds number used to calculate the drag coefficient is multiplied by the shape factor to account for nonspherical sand grains. The summation term may be calculated using the size distribution of the sand particles found from a sieve analysis. Although the Rose equation is limited to clean filter beds, it does provide an opportunity to examine the initial stages of filtration and the effects of sand grain size distribution on head

---

[38]H. E. Rose, "On the Resistance Coefficient-Reynolds Number Relationship of Fluid Flow through a Bed of Granular Material," *Proceedings of the Institute of Mechanical Engineers*, vol. 153, p. 493, 1945.

[39]Other headloss equations include those by Carmen-Kozeny, Fair-Hatch, and Hazen. These equations are summarized in Metcalf & Eddy, Inc. *Wastewater Engineering: Treatment, Disposal, and Reuse,* 3rd edition, New York: McGraw-Hill, p. 269, 1991.

loss. As the filter clogs, the head loss will increase so that the calculated results are the minimum expected head losses. Initial head losses in excess of 0.6 m imply either that the loading rate is too high or that the sand has too large a proportion of fine grain sizes. The design of the filter must account for the additional losses that occur as the filter runs. Thus, the filter box must be at least as deep as the highest design head loss. As mentioned above, this is about 3 m maximum.

The hydraulic head loss that occurs during backwashing is calculated to determine the placement of the backwash troughs above the filter bed. The trough bottom should be at least 0.15 m above the expanded bed to prevent loss of filter material. Fair and Geyer[40] developed the following relationship to predict the depth of expanded bed:

$$D_e = (1 - \epsilon)(D) \sum_{i=1}^{n} \frac{f}{(1 - \epsilon_e)} \tag{3-106}$$

where D_e = depth of the expanded bed, m
ϵ = porosity of the bed
ϵ_e = porosity of expanded bed
f = mass fraction of sand with expanded porosity

The porosity of the expanded bed may be calculated for a given particle by

$$\epsilon_e = \left(\frac{v_b}{v_s}\right)^{0.22} \tag{3-107}$$

where v_b = velocity of backwash, m/s
v_s = settling velocity, m/s

Strictly speaking, this form of the expanded bed porosity is applicable only for laminar conditions. Since the conditions during backwash are turbulent, a more representative model equation is that given by Richardson and Zaki[41]:

$$\epsilon_e = \left(\frac{v_b}{v_s}\right)^{0.2247\mathbf{R}^{0.1}} \tag{3-108}$$

where the Reynolds number is defined as

$$\mathbf{R} = \frac{v_s \, d_{60\%}}{\nu}$$

where $d_{60\%}$ = 60-percentile diameter, m

A more sophisticated model developed by Dharmarajah and Cleasby[42] is also available.

[40] G. M. Fair and J. C. Geyer, *Water Supply and Wastewater Disposal*, p. 678, 1954.

[41] J. F. Richardson and W. N. Zaki, "Sedimentation and Fluidization," Part I. *Transactions of the Institute of Chemical Engineers,* (Brit.), **32,** pp. 35–53, 1954.

[42] A. H. Dharmarajah and J. L. Cleasby, "Predicting the Expansion Behavior of Filter Media," *Journal of the American Water Works Association,* pp. 66–76, Dec., 1986.

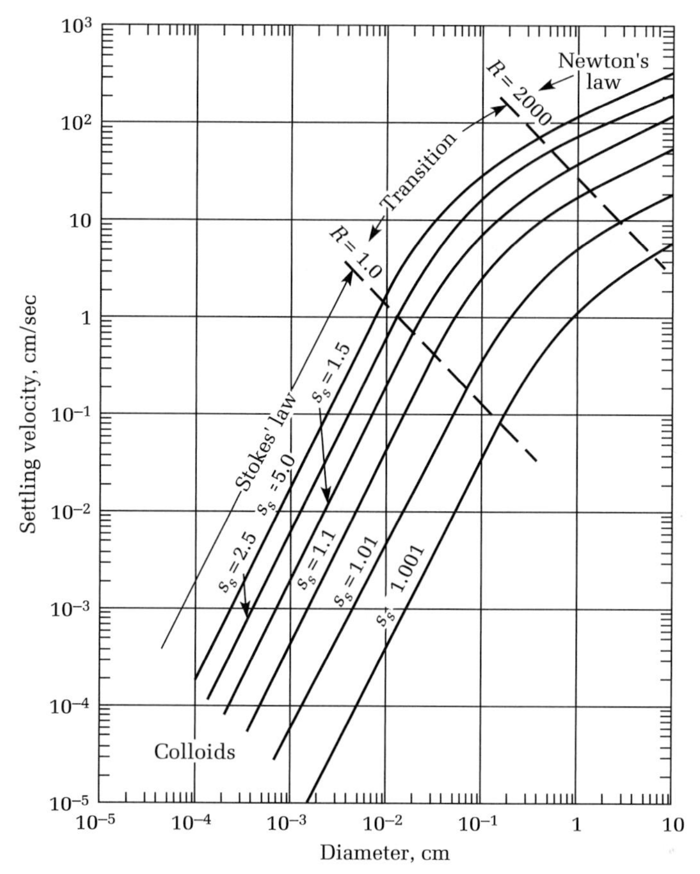

FIGURE 3-43
Settling and rising velocities of discrete spherical particles in quiescent water at 10°C. For other temperatures, multiply the Stokes values by $\nu/(1.31 \times 10^{-2})$, where ν is the kinematic viscosity at the stated temperature. (*Source:* G. M. Fair, J. C. Geyer, and D. A. Orun, *Elements of Water Supply and Wastewater Disposal,* 2nd Edition, New York: Wiley, p. 371, 1971.)

The determination of D_e is not straightforward. From Equation 3-107, it is obvious that the expanded bed porosity is a function of the settling velocity. The particle settling velocity is determined by Equation 3-94. To solve Equation 3-94 you must calculate the drag coefficient (C_D). The drag coefficient is a function of the Reynolds number which, in turn, is a function of the settling velocity. Thus, you need the settling velocity to find the settling velocity! To resolve this dilemma, you must begin with an estimated settling velocity. Knowing the sand grain diameter and specific gravity, you can use Figure 3-43 to obtain a first estimate for the settling velocity to use in calculating the Reynolds number.

Backwash rates normally vary between 880 m/d and 1,200 m/d. However, the limiting factor in choosing a backwash rate is the terminal settling velocity of the smallest sand grains that are to be retained in the filter. Since the filter backwashing process is effectively an upflow clarifier, the backwash rate becomes the overflow rate that determines whether a particle is retained in the filter or is washed out through the backwash trough.

Example 3-27. Estimate the clean filter head loss in Urbana's proposed new sand filter (Example 3-25) using the sand described in Example 3-26 and determine if it is reasonable. Use the following assumptions: specific gravity of sand is 2.65, the shape factor is 0.82, the bed porosity is 0.45, the water temperature is 10°C, and the depth of sand is 0.5 m.

Solution. The computations are shown in the table below.

Sieve No.	% Retain	d(m)	R	C_D	$\frac{(C_D)(f)}{d}$
8–12	5.3	.002	3.1370	9.684551	256.64
12–16	17.1	.00142	2.2272	13.12587	1,580.7
16–20	14.6	.001	1.5685	18.03689	2,633.4
20–30	20.4	.000714	1.1199	24.60549	7,030.1
30–40	17.6	.000505	.79208	34.01075	11,853
40–50	11.9	.000357	.55995	47.21035	15,737
50–70	5.9	.000252	.39526	60.72009	14,216
70–100	3.1	.000178	.27919	85.96328	14,971
100–140	.7	.000126	.19763	121.4402	6,746.7
				Total $(C_D)(f)/d$ =	75,025

In the first two columns, the grain size distribution from Example 3-26 is rearranged to show the fraction retained between sieves. The third column is the geometric mean diameter of the sand grain computed from the upper and lower sieve size. The fourth column is the Reynolds number computed from Equation 3-95 with the correction for nonspherical sand grains. For the first row, using the loading rate from Example 3-25,

$$\mathbf{R} = \frac{\phi\,(d)\,v_a}{\nu} = \frac{(.82)(.002\text{ m})(0.0025\text{ m/s})}{1.307 \times 10^{-6}\text{ m}^2\text{/s}} = 3.137$$

The filtration velocity is simply the conversion of the filtration rate to compatible units:

$$v_a = \frac{216\text{ m}^3\text{/d} \cdot \text{m}^2}{86{,}400\text{ s/d}} = 0.0025\text{ m/s}$$

The kinematic viscosity is determined from Appendix A using the water temperature of 10°C. The factor of 10^{-6} is to convert from μm^2/s to m^2/s.

The drag coefficient is calculated in column 5 using either Equation 3-96 or 3-97, depending on the Reynolds number. For the first row,

$$C_D = \frac{24}{\mathbf{R}} + \frac{3}{\mathbf{R}^{1/2}} + 0.34$$

$$= 7.6507 + 1.6938 + 0.34 = 9.6846$$

The final column is the product of the fractional mass retained and the drag coefficient divided by the diameter. For the first row,

$$\frac{(C_D)(f)}{d} = \frac{(9.6846)(0.053)}{0.002} = 256.64$$

The last column is summed and the head loss calculated using Equation 3-105:

$$h_L = \frac{1.067\,(0.0025)^2\,(0.5)}{(0.82)(9.8)(0.45)^4}(7.5025 \times 10^4)$$

$$= (1.0119 \times 10^{-5})(7.5025 \times 10^4) = 0.76 \text{ m}$$

This initial head loss exceeds the guideline of 0.6 m. Either the filtration rate should be lowered or the fraction of fines should be reduced.

Example 3-28. Determine the height that the backwash troughs must be placed above the filter bed for the sand filter being designed for Urbana. Assume that Equation 3-107 applies.

Solution. To begin, we must select a backwash rate. Assuming that we wish to retain the finest sand grains used in building the filter, the backwash rate must not wash out particles with a diameter of 0.000126 m (0.0126 cm). Using Figure 3-43, we find that, for a 0.0126 cm particle with a specific gravity of 2.65, the terminal settling velocity is approximately 1 cm/s (864 m/d). Thus, our backwash rate may not exceed 864 m/d rather than the nominal minimum of 880 m/d.

The computations are shown in the table below.

Estimated v_s**(m/s)**	**R**	C_D	**Calculated** v_s**(m/s)**	ϵ_e	$\frac{f}{1-\epsilon_e}$
.30	376.435	.558380	.2778839	.4812	.10216
.20	178.179	.699442	.2092095	.5122	.35058
.15	94.1086	.904272	.1544058	.5476	.32275
.10	44.7957	1.32400	.1078248	.5927	.50080
.07	22.1783	2.05917	.0727132	.6463	.49762
.05	11.1989	3.37953	.0477221	.7091	.40902
.03	4.74308	6.77751	.0283125	.7954	.28831
.02	2.23351	13.0928	.0171201	.8884	.27788
0.015	1.18577	23.3350	.0107893	.9834	.42232
				$\sum f/(1-\epsilon_e) =$	3.1715

The estimated settling velocities in the first column were found from Figure 3-43. The Reynolds number was then computed with this estimated velocity. For the first row:

$$\mathbf{R} = \frac{\phi\,(d)\,v_s}{\nu} = \frac{(.82)(.002 \text{ m})(0.30 \text{ m/s})}{1.307 \times 10^{-6} \text{ m}^2/\text{s}} = 376.435$$

You should note that the shape factor, sand particle diameter, and viscosity are all the same as in Example 3-27. The drag coefficient (C_D) is calculated in the same fashion as Example 3-27. The settling velocity is calculated using Equation 3-94 assuming the density of water is 1,000 kg/m^3. For the first row:

$$v_s = \left[\frac{(4)(9.8 \text{ m/s}^2)(2{,}650 \text{ kg/m}^3 - 1{,}000 \text{ kg/m}^3)(0.002 \text{ m})}{(3)(0.55838)(1{,}000 \text{ kg/m}^3)}\right]^{1/2}$$

$$= 0.2778839 \text{ m/s}$$

The density of the sand grain is simply the product of the specific gravity (from Example 3-27) and the density of water:

$$\rho_s = (2.65)(1{,}000 \text{ kg/m}^3) = 2{,}650 \text{ kg/m}^3$$

The expanded bed porosity (column 5) is calculated from Equation 3-107. For the first row,

$$\epsilon_e = \left(\frac{v_b}{v_s}\right)^{0.22} = \left(\frac{0.01 \text{ m/s}}{0.2778839 \text{ m/s}}\right)^{0.22} = 0.4812$$

The first row of the last column is then

$$\frac{f}{1 - \epsilon_e} = \frac{0.053}{1 - 0.4812} = 0.10216$$

where 0.053 is taken from the first row in Example 3-27 and is the mass fraction of sand having a geometric mean diameter of 0.002 m, that is, between sieve numbers 8 and 12.

Using Equation 3-106 with a porosity of 0.45 and an undisturbed bed depth of 0.5 m from Example 3-27, the depth of the expanded bed is then

$$D_e = (1 - .45)(0.5 \text{ m})(3.1715) = 0.87 \text{ m}$$

Allowing for a safety margin of 0.15 m as mentioned above, the bottom of the backwash trough should be placed

$$(0.87 - 0.5) \text{ m} + 0.15 \text{ m} = 0.52 \text{ m or } 0.5 \text{ m}$$

above the top of the sand surface. (Note: $(0.87 - 0.5) = D_e - D$.)

3-8 DISINFECTION

Disinfection is used in water treatment to reduce pathogens (disease-producing microorganisms) to an acceptable level. Disinfection is not the same as sterilization. Sterilization implies the destruction of all living organisms. Drinking water need not be sterile.

Three categories of human enteric pathogens are normally of consequence: bacteria, viruses, and amebic cysts. Purposeful disinfection must be capable of destroying all three.

To be of practical service, such water disinfectants must possess the following properties:

1. They must destroy the kinds and numbers of pathogens that may be introduced into water within a practicable period of time over an expected range in water temperature.

2. They must meet possible fluctuations in composition, concentration, and condition of the waters or wastewaters to be treated.
3. They must be neither toxic to humans and domestic animals nor unpalatable or otherwise objectionable in required concentrations.
4. They must be dispensable at reasonable cost and safe and easy to store, transport, handle, and apply.
5. Their strength or concentration in the treated water must be determined easily, quickly, and (preferably) automatically.
6. They must persist within disinfected water in a sufficient concentration to provide reasonable residual protection against its possible recontamination before use, or—because this is not a normally attainable property—the disappearance of residuals must be a warning that recontamination may have taken place.

Disinfection Kinetics

Under ideal conditions, when an exposed microorganism contains a single site vulnerable to a single unit of disinfectant, the rate of die-off follows *Chick's law,*[43] which states that the number of organisms destroyed in a unit time is proportional to the number of organisms remaining:

$$-\frac{dN}{dt} = kN \tag{3-109}$$

This is a first-order reaction. Under real conditions the rate of kill may depart significantly from Chick's law. Increased rates of kill may occur because of a time lag in the disinfectant reaching vital centers in the cell. Decreased rates of kill may occur because of declining concentrations of disinfectant in solution or poor distribution of organisms and disinfectant.

Chlorine Reactions in Water

Chlorine is the most common disinfecting chemical used. The term *chlorination* is often used synonymously with disinfection. Chlorine may be used as the element (Cl_2), as sodium hypochlorite (NaOCl), or as calcium hypochlorite [$Ca(OCl)_2$].

When chlorine is added to water, a mixture of hypochlorous acid (HOCl) and hydrochloric acid (HCl) is formed:

$$Cl_2(g) + H_2O \rightleftharpoons HOCl + H^+ + Cl^- \tag{3-110}$$

This reaction is pH dependent and essentially complete within a very few milliseconds. In dilute solution and at pH levels above 1.0, the equilibrium is displaced to the right and very little Cl_2 exists in solution. Hypochlorous acid is a weak acid and dissociates poorly at levels of pH below about 6. Between pH 6.0 and 8.5 there occurs

[43] H. Chick, "Investigation of the Law of Disinfection," *Journal of Hygiene,* vol. 8, p. 92, 1908.

a very sharp change from undissociated HOCl to almost complete dissociation:

$$HOCl \rightleftharpoons H^+ + OCl^- \tag{3-111}$$

$$pK = 7.537 \text{ at } 25°C$$

Thus, chlorine exists predominantly as HOCl at pH levels between 4.0 and 6.0. Below pH 1.0, depending on the chloride concentration, the HOCl reverts back to Cl_2 via Equation 3-110. At 20°C, above about pH 7.5, and at 0°C, above about pH 7.8, hypochlorite ions (OCl^-) predominate. Hypochlorite ions exist almost exclusively at levels of pH around 9 and above. Chlorine existing in the form of HOCl and/or OCl^- is defined as free available chlorine.

Hypochlorite salts dissociate in water to yield hypochlorite ions:

$$NaOCl \rightleftharpoons Na^+ + OCl^- \tag{3-112}$$

$$Ca(OCl)_2 \rightleftharpoons Ca^{2+} + 2OCl^- \tag{3-113}$$

The hypochlorite ions establish equilibrium with hydrogen ions (in accord with Equation 3-111), again depending on pH. Thus, the same active chlorine species and equilibrium are established in water regardless of whether elemental chlorine or hypochlorites are used. The significant difference is in the resultant pH and its influence on the relative amounts of HOCl and OCl^- existing at equilibrium. Elemental chlorine tends to decrease pH; each mg/L of chlorine added reduces the alkalinity by up to 1.4 mg/L as $CaCO_3$. Hypochlorites, on the other hand, always contain excess alkali to enhance their stability and tend to raise the pH somewhat. We seek to maintain the design pH within a range of 6.5 to 7.5 to optimize disinfecting action.

Chlorine-Disinfecting Action

Chlorine disinfection involves a very complex series of events and is influenced by the kind and extent of reactions with chlorine-reactive materials (including nitrogen), temperature, pH, the viability of test organisms, and numerous other factors. Such factors greatly complicate attempts to determine the precise mode of action of chlorine on bacteria and other microorganisms. Over the years, several theories have been advanced. One early theory held that chlorine reacts directly with water to produce nascent oxygen; another held that the action of chlorine is due to complete oxidative destruction of organisms. These theories were nullified because small concentrations of hypochlorous acid were observed to destroy bacteria whereas other oxidants (such as hydrogen peroxide or potassium permanganate) failed to do likewise. A later theory suggested that chlorine reacts with protein and amino acids of cells to alter and ultimately destroy cell protoplasm. Currently it is considered that the bactericidal action of chlorine is physiochemical, but among yet-unanswered questions are those pertaining to phenomena such as variations in resistance of bacteria, spores, cysts, and viruses, and the appearance of mutants.

Microorganism kill by disinfectants is assumed to follow the CT concept, that is, the product of disinfectant concentration (C) and time (T) yields a constant. CT is widely used in the SWTR as a criteria for cyst and virus disinfection. CT is an empirical expression for defining the nature of biological inactivation where:

$$CT = 0.9847C^{0.1758}\text{pH}^{2.7519}\text{temp}^{-0.1467} \quad (3\text{-}114)$$

where C = disinfectant concentration
T = contact time between the microorganism and the disinfectant
pH = $-\log[H^+]$
temp = temperature, °C

The relationship shown in Equation 3-114 means that the combination of concentration and time (CT) required to produce a 3-log reduction in *Giardia* cysts by free chlorine can be estimated if the free chlorine concentration, pH, and water temperature are known.

Table 3-20 shows examples of the required CT times for free chlorine under the SWTR. EPA used empirical data and a safety factor to develop the table. Generally, 3-log reduction of *Giardia* is required.

Chlorine/Ammonia Reactions

The reactions of chlorine with ammonia are of great significance in water chlorination processes. When chlorine is added to water that contains natural or added ammonia (ammonium ion exists in equilibrium with ammonia and hydrogen ions), the ammonia reacts with HOCl to form various *chloramines* which, like HOCl, retain the oxidizing power of the chlorine. The reactions between chlorine and ammonia may be represented as follows:[44]

$$NH_3 + HOCl \rightleftharpoons NH_2Cl + H_2O \quad (3\text{-}115)$$
Monochloramine

$$NH_2Cl + HOCl \rightleftharpoons NHCl_2 + H_2O \quad (3\text{-}116)$$
Dichloramine

$$NHCl_2 + HOCl \rightleftharpoons NCl_3 + H_2O \quad (3\text{-}117)$$
Trichloramine or nitrogen trichloride

The distribution of the reaction products is governed by the rates of formation of monochloramine and dichloramine, which are dependent upon pH, temperature, time, and initial Cl_2:NH_3 ratio. In general, high Cl_2:NH_3 ratios, low temperatures, and low pH levels favor dichloramine formation.

Chlorine also reacts with organic nitrogenous materials, such as proteins and amino acids, to form organic chloramine complexes. Chlorine that exists in water in chemical combination with ammonia or organic nitrogen compounds is defined as *combined available chlorine.*

The oxidizing capacity of free chlorine solutions varies with pH because of variations in the resultant HOCl:OCl^- ratios. This applies also in the case of

[44] American Water Works Association, *Water Quality and Treatment,* p. 188, 1971.

TABLE 3-20
CT values (in mg/L · min) for inactivation of *Giardia* cysts by free chlorine at 10°C

Chlorine concentration	pH = 6.0						pH = 7.0						pH = 8.0						pH = 9.0					
	Log inactivations						Log inactivations						Log inactivations						Log inactivations					
mg/L	0.5	1.0	1.5	2.0	2.5	3.0	0.5	1.0	1.5	2.0	2.5	3.0	0.5	1.0	1.5	2.0	2.5	3.0	0.5	1.0	1.5	2.0	2.5	3.0
≤0.4	11	23	34	46	57	69	18	33	33	70	82	105	25	51	76	101	126	152	35	70	105	140	175	210
0.6	12	25	37	49	62	74	19	38	56	75	94	113	27	54	81	109	136	163	38	75	113	150	188	225
0.8	13	26	39	52	65	78	20	40	59	79	99	119	29	57	86	114	143	171	40	79	119	153	198	237
1.0	13	27	40	54	67	81	21	41	62	82	103	123	30	59	89	119	149	178	41	82	123	164	205	247
1.2	14	28	42	56	69	83	21	42	64	85	106	127	31	61	93	126	161	194	42	85	127	170	212	255
1.4	14	29	43	57	71	86	22	44	65	87	109	131	32	63	95	126	158	189	44	87	131	174	218	262
1.6	15	29	44	58	73	88	22	45	67	89	112	134	32	65	97	129	161	194	45	89	134	179	223	268
1.8	15	30	45	60	75	90	23	46	68	91	114	137	33	66	99	132	165	198	46	91	137	182	228	273
2.0	15	30	46	61	76	91	23	46	70	93	116	139	34	67	101	134	168	201	46	93	139	186	232	278
2.2	15	31	46	62	77	93	24	47	71	95	118	142	34	68	102	137	171	205	47	94	142	189	236	283
2.4	16	31	47	63	79	94	24	48	72	96	120	144	35	69	104	139	173	208	47	94	142	189	236	283
2.6	16	32	48	64	80	96	24	49	73	97	122	146	35	70	105	141	176	211	49	97	146	194	243	292
2.8	16	32	48	65	81	97	25	49	74	99	123	148	36	71	107	142	178	214	49	98	148	197	246	295
3.0	16	33	49	65	82	98	25	50	75	100	125	150	36	72	108	144	180	216	50	100	150	199	249	299

Source: U.S. Environmental Protection Agency, *Guidance Manual for Compliance with Filtration and Disinfection Requirements for Public Water Systems Using Surface Water Sources,* Criteria and Standards Division, Office of Drinking Water (U.S.E.P.A. NTIS Publication No. PB 90-148016), Washington, DC: U.S. Government Printing Office, October, 1979.

chloramine solutions as a result of varying $NHCl_2$:NH_2Cl ratios, and where monochloramine predominates at high pH levels. The disinfecting ability of the chloramines is much lower than that of free available chlorine, indicating that the ammonia chloramines also are less reactive than free available chlorine.

Practices of Water Chlorination

Evolution. Early water chlorination practices (variously termed "plain chlorination," "simple chlorination," and "marginal chlorination") were applied for the purpose of disinfection. Chlorine-ammonia treatment was soon thereafter introduced to limit the development of objectionable tastes and odors often associated with marginal chlorine disinfection. Subsequently, "super-chlorination" was developed for the additional purpose of destroying objectionable taste- and odor-producing substances often associated with chlorine-containing organic materials. The introduction of "breakpoint chlorination" and the recognition that chlorine residuals can exist in two distinct forms established contemporary water chlorination as one of two types: combined residual chlorination or free residual chlorination.

Combined residual chlorination. Combined residual chlorination practice involves the application of chlorine to water in order to produce, with natural or added ammonia, a combined available chlorine residual, and to maintain that residual through part or all of a water-treatment plant or distribution system. Combined available chlorine forms have lower oxidation potentials than free available chlorine forms and, therefore, are less effective as oxidants. Moreover, they are also less effective disinfectants. In fact, about 25 times as much combined available residual chlorine as free available residual chlorine is necessary to obtain equivalent bacterial kills (*S. typhosa*) under the same conditions of pH, temperature, and contact time. And about 100 times longer contact is required to obtain equivalent bacterial kills under the same conditions and for equal amounts of combined and free available chlorine residuals.

When a combined available chlorine residual is desired, the characteristics of the water will determine how it can be accomplished:

1. If the water contains sufficient ammonia to produce with added chlorine a combined available chlorine residual of the desired magnitude, the application of chlorine alone suffices.
2. If the water contains too little or no ammonia, the addition of both chlorine and ammonia is required.
3. If the water has an existing free available chlorine residual, the addition of ammonia will convert the residual to combined available residual chlorine. A combined available chlorine residual should contain little or no free available chlorine.

The practice of combined residual chlorination is especially applicable after filtration (posttreatment) for controlling certain algae and bacterial growths and for providing and maintaining a stable residual throughout the system to the point of consumer use.

Although combined chlorine residual is not a good disinfectant, it has an advantage over free chlorine residual in that it is reduced more slowly and, therefore, persists for a longer time in the distribution system. Thus, it is useful as an indicator of major contamination. Water plant personnel routinely monitor the chlorine level in the distribution system. The presence of available chlorine (either combined or free) indicates that no major contamination has occurred. If major contamination does take place, the combined chlorine residual will be depleted, albeit at a slow rate. This depletion serves as a warning that contamination may have taken place.

Because of its relatively poor disinfecting power, combined residual chlorination is often preceded by free residual chlorination to ensure the production of potable water.

Free residual chlorination. Free residual chlorination practice involves the application of chlorine to water to produce, either directly or through the destruction of ammonia, a free available chlorine residual and to maintain that residual through part or all of a water treatment plant or distribution system. Free available chlorine forms have higher oxidation potentials than combined available chlorine forms and therefore are more effective as oxidants. Moreover, as already noted, they are also the most effective disinfectants.

When free available chlorine residual is desired, the characteristics of the water will determine how it can be accomplished:

1. If the water contains no ammonia (or other nitrogenous materials), the application of chlorine will yield a free residual.
2. If the water does contain ammonia that results in the formation of a combined available chlorine residual, it must be destroyed by applying an excess of chlorine.

With molar Cl_2:NH_3 (as N), concentrations up to 1:1 (5:1 mass basis) monochloramine and dichloramine will be formed. The relative amounts of each depend on pH and other factors. Chloramine residuals generally reach a maximum at equimolar concentrations of chlorine and ammonia. Further increases in the Cl_2:NH_3 ratio result in the oxidation of ammonia and reduction of chlorine. These oxidation/reduction reactions are essentially complete when two moles of chlorine have been added for each mole of ammonia present. Sufficient time must be provided to allow the reaction to go to completion. Chloramine residuals decline to a minimum value, the breakpoint, when the molar Cl_2:NH_3 ratio is about 2:1. At this point, oxidation/reduction reactions are essentially complete. Further addition of chlorine produces free residual chlorine as illustrated in Figure 3-44.

Chlorine Dioxide

Another very strong oxidant is chlorine dioxide. Chlorine dioxide (ClO_2) is formed on-site by combining chlorine and sodium chlorite. Chlorine dioxide is often used as a primary disinfectant, inactivating the bacteria and cysts, followed by the use of chloramine as a distribution system disinfectant. Chlorine dioxide does not maintain

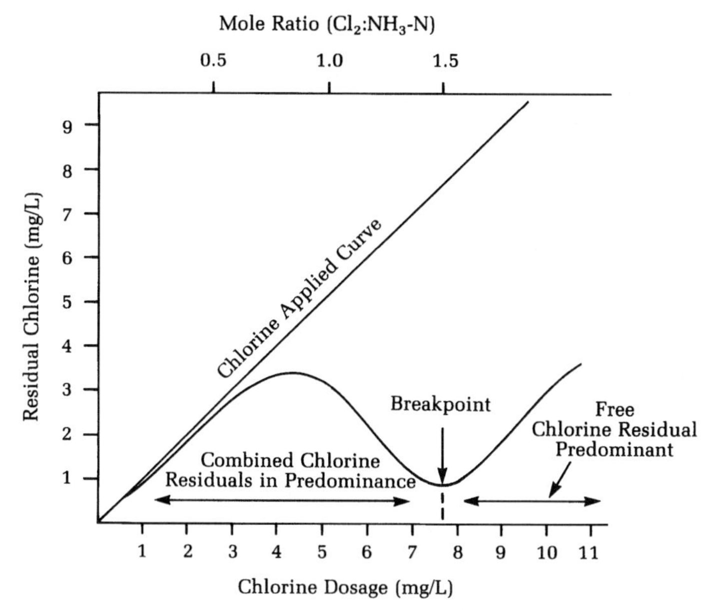

FIGURE 3-44
Breakpoint chlorination. (*Source:* American Water Works Association, *Water Treatment Plant Design,* 1969.)

a residual long enough to be useful as a distribution-system disinfectant. The advantage of chlorine dioxide over chlorine is that chlorine dioxide does not react with precursors to form THMs.

When chlorine dioxide reacts in water, it forms two by-products, chlorite and, to some extent, chlorate. These compounds may be associated with a human health risk, and therefore many state regulatory agencies limit the dose of chlorine dioxide to 1.0 mg/L. In many cases this may not be a sufficient dose to provide good disinfection. Chlorine dioxide use has also been associated with tastes and odors in some communities. The health concerns, taste and odors, and relatively high cost have tended to limit the use of chlorine dioxide. However, many utilities have successfully used chlorine dioxide as a primary disinfectant.

Ozonation

Ozone is a pungent-smelling, unstable gas. It is a form of oxygen in which three atoms of oxygen are combined to form the molecule O_3. Because of its instability, it is generated at the point of use. The ozone-generating apparatus commonly is a discharge electrode. To reduce corrosion of the generating apparatus, air is passed through a drying process and then into the ozone generator. The generator consists of two plates or a wire and tube with an electric potential of 15,000 to 20,000 volts. The oxygen in the air is dissociated by the impact of electrons from the discharge electrode. The atomic oxygen then combines with atmospheric oxygen to form ozone in the following reaction:

$$O + O_2 \rightarrow O_3 \tag{3-118}$$

TABLE 3-21
CT values for 99.99 percent *Giardia* cyst inactivation

	Temperature					
	0.5°C	5°C	10°C	15°C	20°C	25°C
Chlorine dioxide	81	54	40	27	21	14
Ozone	4.5	3	2.5	2	1.5	1
Chloramines	3800	2200	1850	1500	1100	750

Source: Guidance Manual for Compliance with Filtration and Disinfection Requirements for Public Drinking Water Systems Using Surface Water Sources, October 1989.

Approximately 0.5 to 1.0 percent by volume of the air exiting from the apparatus will be ozone. The resulting ozone-air mixture is then diffused into the water that is to be disinfected.

Ozone is widely used in drinking water treatment in Europe and is continuing to gain popularity in the United States. It is a powerful oxidant, more powerful even than hypochlorous acid. It has been reported to be more effective than chlorine in destroying viruses and cysts. Table 3-21 shows the CT values for ozone, chlorine dioxide, and chloramine for 3-log inactivation (99.99 percent removal) of *Giardia* cysts. Table 3-21 can be compared to the values for chlorine in Table 3-20 to note how strong an oxidant ozone is. (Also the weak disinfection ability of chloramines is obvious.)

In addition to being a strong oxidant, ozone has the advantage of not forming THMs or any of the chlorinated DBPs. As with chlorine dioxide, ozone will not persist in the water, decaying back to oxygen in minutes. Therefore, a typical flow schematic would be to add ozone either to the raw water or between the sedimentation basins and filter for primary disinfection, followed by chloramine addition after the filters as the distribution system disinfectant.

Ultraviolet Radiation

Ultraviolet (UV) light is that part of the electromagnetic spectrum that falls in the range 0.2 to 0.39 μm. It is that part of the spectrum that gives us suntans. The UV method of disinfection involves exposure of a thin layer of water to light from a mercury vapor arc lamp that emits UV in the range of 0.2 to 0.29 μm. It has been found that submergence of the UV bulb encased in a quartz tube is superior to overhead irradiation. The depth of light penetration still limits the liquid film thickness around each lamp to about 50 to 80 mm. Multiple lamps are used to provide greater coverage. The major factor in achieving good microorganism kill is the ability of the UV light to pass through the water to get to the target organism. Thus, the lamps must be kept free of slime and precipitates, and the water must be free of turbidity.

UV performs very well against both bacteria and viruses. Its major disadvantages are that it leaves no residual protection for the distribution system and it is very expensive.

Advanced Oxidation Processes (AOPs)

AOPs are combinations of disinfectants designed to produce hydroxyl radicals (OH·). Hydroxyl radicals are highly reactive nonselective oxidants able to decompose many organic compounds. Most noteworthy of the AOP processes is ozone plus hydrogen peroxide.

3-9 ADSORPTION

Adsorption is a mass transfer process wherein a substance is transferred from the liquid phase to the surface of a solid where it is bound by chemical or physical forces.

Generally, in water treatment, the adsorbent (solid) is activated carbon, either granular (GAC) or powdered (PAC). PAC is fed to the raw water in a slurry and is generally used to remove taste- and odor-causing substances or to provide some removal of synthetic organic chemicals (SOCs). GAC is added to the existing filter system by replacing the anthracite with GAC, or an additional contactor is built and is placed in the flow scheme after primary filtration. The design of the GAC contactor is very similar to a filter box, although deeper.

At present, the applications of adsorption in water treatment in the United States are predominately for taste and odor removal. However, adsorption is increasingly being considered for removal of SOCs, VOCs, and naturally occurring organic matter, such as THM precursors and DBPs.

Biologically derived earthy-musty odors in water supplies are a widespread problem. Their occurrence interval and concentration vary greatly from season to season and is often unpredictable. As mentioned, one of the most popular methods for removing these compounds is the addition of PAC to the raw water. The dose is generally less than 10 mg/L. The advantage of PAC is that the capital equipment is relatively inexpensive and it can be used on an as-needed basis. The disadvantage is that the adsorption is often incomplete. Sometimes even doses of 50 mg/L are not sufficient.

As an alternative for taste and odor control, many plants have replaced the anthracite in the filters with GAC. The GAC will last from one to three years and then must be replaced. It is very effective in removing many taste and odor compounds.

Concern about SOCs in drinking water has motivated interest in adsorption as a treatment process for removal of toxic and potentially carcinogenic compounds present in minute, but significant, quantities. Few other processes can remove SOCs to the required low levels. Generally, GAC is used for SOC removal either as a filter media replacement or as a separate contactor. The data for how long the GAC will last for any given SOC are somewhat limited and must be evaluated on a case-by-case basis. If the GAC is designed to remove SOCs from a periodic "spill" into the source water, then filter media replacement may be adequate because the GAC is not being used every day and is acting as a barrier. However, if the GAC is to be used continuously for SOC removal, then a separate contactor may be warranted.

GAC has been proposed to be used to remove naturally occurring organic matter that would, in turn, reduce the formation of DBPs, particularly THMs. Testing has shown that GAC will remove these organics. It must operate in a separate contactor since the depth of a conventional filter is inadequate. The GAC will typically last 90 to 120 days until it loses its adsorptive capacity. Because of its short life, the GAC would need to be regenerated on site or be destroyed by burning in a high-temperature furnace. This is clearly an expensive proposition.

GAC has also been considered for removal of THMs. However, the capacity is very low and the carbon may only last up to 30 days. GAC is not considered practical for THM removal.

3-10 WATER PLANT WASTE MANAGEMENT

The precipitated chemicals and other materials removed from raw water to make it potable and palatable are termed *sludge*. Satisfactory treatment and disposal of water treatment plant sludge can be the single most complex and costly operation in the plant.

The sludges withdrawn from coagulation and softening plants are composed largely of water. As much as 98 percent of the sludge mass may be water. Thus, for example, 20 kg of solid chemical precipitate is accompanied by 980 kg of water. Assuming equal densities for the precipitate and water (a bad assumption at best), approximately 1 m^3 of sludge is produced for each 20 kg of chemicals added to the water. For even a small plant (say 0.05 m^3/s) this might mean up to 800 m^3/y of sludge—a substantial volume to say the least!

Water treatment plants and the wastes they produce can be broadly divided into four general categories. First are those treatment plants that coagulate, filter, and oxidize a surface water for removal of turbidity, color, bacteria, algae, some organic compounds, and often iron and/or manganese. These plants generally use alum or iron salts for coagulation and produce two waste streams. The majority of the waste produced from these plants is sedimentation basin (or clarifier) sludge and filter backwash wastes. The second type of treatment plants are those that practice softening for the removal of calcium and magnesium by the addition of lime, sodium hydroxide, and/or soda ash. These plants produce clarifier basin sludges and filter backwash wastes. On occasion, plants practice both coagulation and softening. Softening plant wastes can also contain trace inorganics such as radium that could affect their proper handling. The third type of plants are those that are designed to specifically remove trace inorganics such as nitrate, fluoride, radium, arsenic, etc. These plants use processes such as ion exchange, reverse osmosis, or adsorption. They produce liquid wastes or solid wastes such as spent adsorption material. The fourth category of treatment plants are those that produce air-phase wastes during the stripping of volatile compounds. The major types of treatment plant wastes produced are shown in Table 3-22. Since 95 percent of the wastes produced are coagulants or softening sludge, they will be stressed in this section.

Hydrolyzing metal salts or synthetic organic polymers are added in the water treatment process to coagulate suspended and dissolved contaminants and yield relatively clean water suitable for filtration. Most of these coagulants and the impurities they remove settle to the bottom of the settling basin where they become part of the sludge. These sludges are referred to as alum, iron, or polymeric sludge according

TABLE 3-22
Major water treatment plant wastes

Solid/Liquid Wastes
1. Alum sludges
2. Iron sludges
3. Polymeric sludges
4. Softening sludges
5. Backwash wastes
6. Spent GAC or discharge from carbon systems
7. Slow sand filter wastes
8. Wastes from iron and manganese removal plants
9. Spent precoat filter media

Liquid-Phase Wastes
10. Ion-exchange regenerant brine
11. Waste regenerant from activated alumina
12. Reverse osmosis waste streams

Gas-Phase Wastes
13. Air stripping off-gases

to which primary coagulant is used. These wastes account for approximately 70 percent of the water plant waste generated. The sludges produced in treatment plants where water softening is practiced using lime or lime and soda ash account for an additional 25 percent of the industry's waste production. It is therefore apparent that most of the waste generation involves water treatment plants using coagulation or softening processes.

The most logical sludge management program attempts to use the following approach in disposing of the sludge:

1. Minimization of sludge generation
2. Chemical recovery of precipitates
3. Sludge treatment to reduce volume
4. Ultimate disposal in an environmentally safe manner

With a short digression to identify the sources of water plant sludges and their production rates, we have organized the following discussion along these lines.

Sludge Production and Characteristics

In water treatment plants, sludge is most commonly produced in the following treatment processes: presedimentation, sedimentation, and filtration (filter backwash).

Presedimentation. When surface waters are withdrawn from watercourses that contain a large quantity of suspended materials, presedimentation prior to coagulation may be practiced. The purpose of this is to reduce the accumulation of solids in subsequent units. The settled material generally consists of fine sand, silt, clays, and organic decomposition products.

Softening sedimentation basin. The residues from softening by precipitation with lime [$Ca(OH)_2$] and soda ash (Na_2CO_3) will vary from a nearly pure chemical to a highly variable mixture. The softening process discussed in Section 3-3 produces a sludge containing primarily $CaCO_3$ and $Mg(OH)_2$.

Theoretically, each mg/L of calcium hardness removed produces 1 mg/L of $CaCO_3$ sludge; each mg/L of magnesium hardness removed produces 0.6 mg/L of sludge; and each mg/L of lime added produces 1 mg/L of sludge. The theoretical sludge production can be calculated as:

$$M_s = 86.40\, Q(\mathrm{Ca_R} + 0.58\, \mathrm{Mg_R} + \mathrm{L_A}) \tag{3-119}$$

where M_s = dry sludge produced, kg/d
Q = plant flow, m^3/s
$\mathrm{Ca_R}$ = calcium hardness removed, mg/L as $CaCO_3$
$\mathrm{Mg_R}$ = magnesium hardness removed, mg/L as $CaCO_3$
$\mathrm{L_A}$ = lime added, mg/L as $CaCO_3$

When surface waters are softened, this equation is not valid. There will be additional sludge from coagulation of suspended materials and precipitation of metal coagulants. The solids content of lime-softening sludge in the sedimentation basin ranges between 2 and 15 percent. A nominal value of 10 percent solids is often used.

Coagulation sedimentation basin. Aluminum or iron salts are generally used to accomplish coagulation. (The chemistry of the two salts was discussed in Section 3-2.) The pH range of 6 to 8 is where most water treatment plants effect the coagulation process. In this range the insoluble aluminum hydroxide complex of $Al(H_2O)_3(OH)_3$ probably predominates. This species results in the production of 0.44 kg of chemical sludge for each kg of alum added. Any suspended solids present in the water will produce an equal amount of sludge. The amount of sludge produced per turbidity unit is not as obvious; however, in many waters a correlation does exist. Carbon, polymers, and clay will produce about 1 kg of sludge per kg of chemical addition. The sludge production for alum coagulation may then be approximated by:

$$M_s = 86.40\, Q(0.44\mathrm{A} + \mathrm{SS} + \mathrm{M}) \tag{3-120}$$

where M_s = dry sludge produced, kg/d
Q = plant flow, m^3/s
A = alum dose, mg/L
SS = suspended solids in raw water, mg/L
M = miscellaneous chemical additions such as clay, polymer, and carbon, mg/L

Alum sludge leaving the sedimentation basin usually has a suspended solids content in the range of 0.5 to 2 percent. It is often less than 1 percent. Twenty to 40 percent of the solids are organic; the remainder are inorganic or silts. The pH of alum sludge is normally in the 5.5 to 7.5 range. Alum sludge from sedimentation basins may include large numbers of microorganisms, but it generally does not exhibit an unpleasant odor. The sludge flow rate is often in the range of 0.3 to 1 percent of the treatment plant flow.

Filter backwash. All water treatment plants that practice filtration produce a large volume of washwater containing a low suspended solids concentration. The volume of backwash water is usually 2 to 3 percent of the treatment plant flow. The solids in backwash water resemble those found in sedimentation units. Since filters can support biological growth, the filter backwash may contain a larger fraction of organic solids than do the solids from the sedimentation basins.

Mass balance analysis. Clarifier sludge production can be estimated by a mass balance analysis of the sedimentation basin. Since there is no reaction taking place, the mass balance equation reduces to the form:

$$\text{Accumulation Rate} = \text{Input Rate} - \text{Output Rate} \tag{3-121}$$

The input rate of solids may be estimated from Equation 3-119 or 3-120. To estimate the mass flow (output rate) of solids leaving the clarifier through the weir, you must have an estimate of the concentration of solids and the flow rate. The mass flow out through the weir is then

$$\text{Weir output rate} = (\text{Concentration, mg/L})(\text{Flow Rate, m}^3/\text{s}) = \text{mg/s}$$

Example 3-29. A coagulation treatment plant with a flow of 0.5 m^3/s is dosing alum at 23.0 mg/L. No other chemicals are being added. The raw-water suspended-solids concentration is 37.0 mg/L. The effluent suspended-solids concentration is measured at 12.0 mg/L. The sludge solids content is 1.00 percent and the specific gravity of the sludge solids is 3.01. What volume of sludge must be disposed of each day?

Solution. The mass-balance diagram for the sedimentation basin is

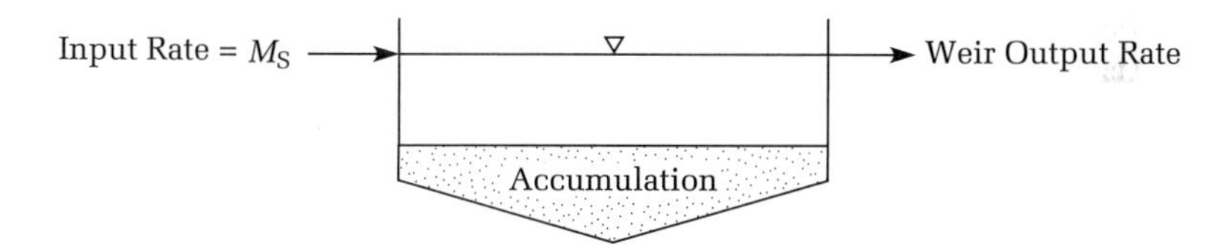

The mass of solids (sludge) flowing into the clarifier is estimated from Equation 3-120:

$$M_s = 86.40(0.50\ \text{m}^3/\text{s})(0.44(23.0\ \text{mg/L}) + 37.0\ \text{mg/L} + 0)$$
$$= 2{,}035.58\ \text{kg/d}$$

Recognizing that g/m^3 = mg/L, the mass of solids leaving the weir is

$$\text{Weir output rate} = (12.0\ \text{g/m}^3)(0.50\ \text{m}^3/\text{s})(86{,}400\ \text{s/d})(10^{-3}\ \text{kg/g})$$
$$= 518.4\ \text{kg/d}$$

The accumulation is then

$$\text{Accumulation} = 2{,}035.58 - 518.4 = 1{,}517.18 \text{ or } 1{,}517\ \text{kg/d}$$

Because this is a dry mass and the sludge has only 1.00 percent solids, we must account for the volume of water in estimating the volume to be removed each day. Using Equation 3-4, we can solve for the mass of water (W_o):

$$1.00 = \frac{1{,}517}{1{,}517 + W_o}(100)$$

$$W_o = 150{,}183 \text{ kg/d}$$

Now we use the definition of density (mass per unit volume) to find the volume of sludge and water:

$$\text{Volume} = \frac{\text{Mass}}{\text{Density}}$$

$$\begin{aligned} V_T &= \text{volume of solids} + \text{volume of water} \\ &= \frac{1{,}517 \text{ kg/d}}{(3.01)(1{,}000 \text{ kg/m}^3)} + \frac{150{,}183 \text{ kg/d}}{1{,}000 \text{ kg/m}^3} \\ &= 0.50 + 150.18 = 150.7 \text{ m}^3\text{/d or } 150 \text{ m}^3\text{/d} \end{aligned}$$

The specific gravity of the solids is 3.01 and the density of water is 1,000 kg/m^3.

Obviously, the solids account for only a small fraction of the total volume. This is why sludge dewatering is an important part of the water treatment process.

Minimization of Sludge Generation

Minimizing sludge generation can have an advantageous effect on the requirements and economics of handling, treating, and disposing of water treatment plant sludges. Minimization also results in the conservation of raw materials, energy, and labor.

There are three methods to minimize the quantity of metal hydroxide precipitates in the sludge:

1. Changing the water treatment process to direct filtration
2. Substituting other coagulants and in particular using polymers that are more effective at lower dosage
3. Conserving chemicals by determining optimum dosage at frequent intervals as raw water characteristics change

Sludge Treatment

The treatment of solid/liquid wastes produced in water treatment processes involves the separation of the water from the solid constituents to the degree necessary for the selected disposal method. Therefore, the required degree of treatment is a direct function of the ultimate disposal method.

There are several sludge treatment methodologies which have been practiced in the water industry. Figure 3-45 shows the most common sludge-handling options available, listed by general categories of thickening, dewatering, and disposal. In

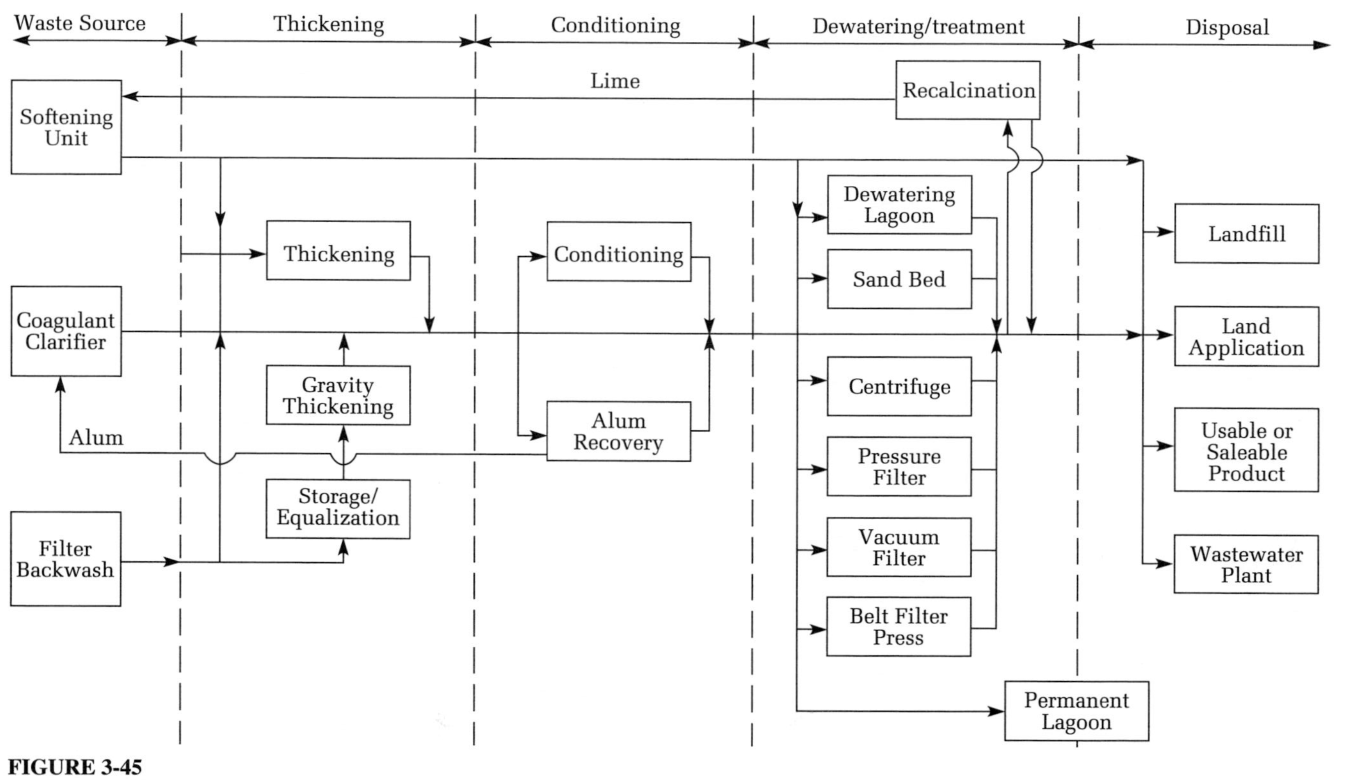

FIGURE 3-45
Sludge handling options.

TABLE 3-23
Range of cake solid concentrations obtainable

	Lime sludge, %	Coagulation sludge, %
Gravity thickening	15–30	3–4
Basket centrifuge		10–15
Scroll centrifuge	55–65	10–20
Belt filter press		10–15
Vacuum filter	45–65	n/a
Pressure filter	55–70	30–45
Sand drying beds	50	20–25
Storage lagoons	50–60	7–15

choosing a combination of the possible treatment process trains, it is probably best to first identify the available disposal options and their requirements for a final cake solids concentration. Most landfill applications will require a "handleable" sludge and this may limit the type of dewatering devices which are acceptable. Methods and costs of transportation may affect the decision "how dry is dry enough." The criteria should not be to simply reach a given solids concentration but rather to reach a solids concentration of desired properties for the handling, transport, and disposal options available.

Table 3-23 shows a generalized range of results which have been obtained for final solids concentrations from different dewatering devices for coagulant and lime sludges.

To give you an appreciation of these solids concentrations, a sludge cake with 35 percent solids would have the consistency of butter, while a 15 percent sludge would have a consistency much like rubber cement.

After removal of the sludge from the clarifier or sedimentation basin, the first treatment step is usually thickening. Thickening assists the performance of any subsequent treatment, gets rid of much water quickly, and helps to equalize flows to the subsequent treatment device. An approximation for determining sludge volume reduction via thickening is given by:

$$\frac{\forall_2}{\forall_1} = \frac{P_1}{P_2} \tag{3-122}$$

where $\forall_1$ = volume of sludge before thickening, m^3
$\forall_2$ = volume of sludge after thickening, m^3
P_1 = percent of solids concentration of sludge before thickening
P_2 = percent of solids concentration after thickening

Thickening is usually accomplished by using circular settling basins similar to a clarifier (Figure 3-46). Thickeners can be designed based on pilot evaluations or using data obtained from similar plants. Lime sludges are typically loaded at 100 to 200 $kg/m^2/d$, and coagulant sludge loading rates are about 15 to 25 $kg/m^2/d$.

Following thickening of the sludge, dewatering can take place by either mechanical or nonmechanical means. In nonmechanical devices, sludge is spread out with the free water draining and the remaining water evaporating. Sometimes the

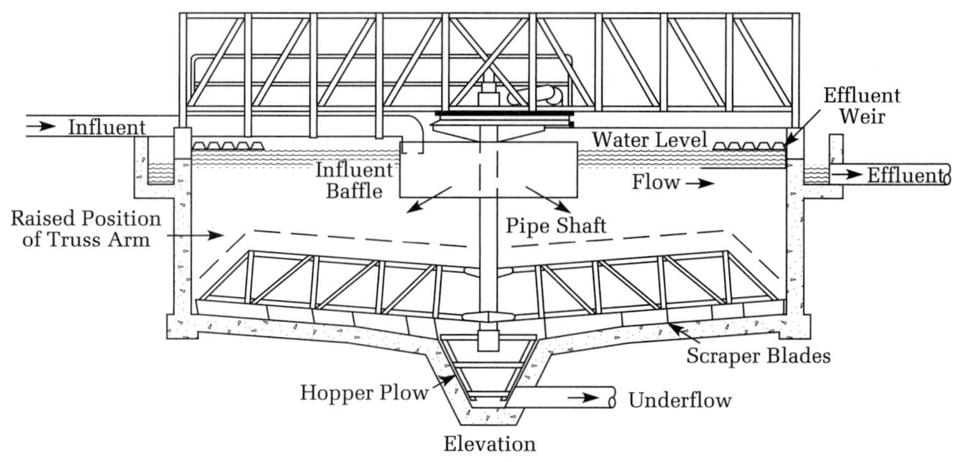

FIGURE 3-46
Continuous-flow gravity thickener. (Courtesy of Link Belt.)

amount of free water available to drain is enhanced by natural freeze-thaw cycles. In mechanical dewatering, some type of device is used to force the water out of the sludge.

We begin our discussion with the nonmechanical methods and follow with the mechanical methods.

Lagoons. A lagoon is essentially a large hole that is dug out for the sludge to flow into. Lagoons can be constructed as either storage lagoons or dewatering lagoons. Storage lagoons are designed to store and collect the solids for some predetermined amount of time. They will generally have decant capabilities but no underdrain system. Storage lagoons should be equipped with sealed bottoms to protect the groundwater. Once the storage lagoon is full or decant can no longer meet discharge limitations, it must be abandoned or cleaned. To facilitate drying, the standing water may be removed by pumping, leaving a wet sludge. Coagulant sludges can only be expected to reach a 7- to 10-percent solids concentration in storage lagoons. The remaining solids must be either cleaned out wet or allowed to evaporate. Depending upon the depth of the wet solids, evaporation can take years. The top layers will often form a crust, preventing evaporation of the bottom layers of sludge.

The primary difference between a dewatering lagoon and a storage lagoon is that a dewatering lagoon has a sand and underdrain-bottom, similar to a drying bed. Dewatering lagoons can be designed to achieve a dewatered sludge cake. The advantage of a dewatering lagoon over a drying bed is that storage is built into the system to assist in meeting peak solids production or to assist in handling sludge during wet weather. The disadvantage of bottom sand layers compared to conventional drying beds is that the bottom sand layers can plug up or "blind" with multiple loadings, thereby increasing the required surface area. Polymer treatment can be useful in preventing this sand blinding.

Storage lagoons, which are generally earthen basins, have no size limitations but have been designed in areas from 2,000 to 60,000 m^2, ranging in depth from 2 to 10 or more meters. Storage and dewatering lagoons may be equipped with inlet structures designed to dissipate the velocity of the incoming sludge. This minimizes turbulence in the lagoons and helps prevent carryover of solids in the decant. The lagoon outlet structure is designed to skim the settled supernatant and is sometimes provided with flash boards to vary the draw-off depths. Any design of a storage lagoon must consider how the sludge will be ultimately removed unless the site is to be abandoned.

The basis for design of dewatering lagoons is essentially the same as that for sand drying beds. The difference is that the applied depth is high and the number of applications per year is greatly reduced.

Sand-drying beds. Sand-drying beds operate on the simple principle of spreading the sludge out and letting it dry. As much water as possible is removed by drainage or decant and the rest of the water must evaporate until the desired final solids concentration is reached. Sand-drying beds have been built simply by cleaning an area of land, dumping the sludge, and hoping something happens. At the other end of the spectrum sophisticated automated drying systems have been built.

Drying beds may be roughly categorized as follows:

1. Conventional rectangular beds with side walls and a layer of sand on gravel with underdrain piping to carry away the liquid. They are built either with or without provisions for mechanical removal of the dried sludge and with or without a roof or a greenhouse-type covering (see Figure 3-47).
2. Paved rectangular drying beds with a center sand drainage strip with or without heating pipes buried in the paved section and with or without covering to prevent incursion of rain.
3. "Wedge-water" drying beds that include a wedge wire septum incorporating provision for an initial flood with a thin layer of water, followed by introduction of liquid sludge on top of the water layer, controlled formation of cake, and provision for mechanical cleaning.
4. Rectangular vacuum-assisted drying beds with provision for application of vacuum to assist gravity drainage.

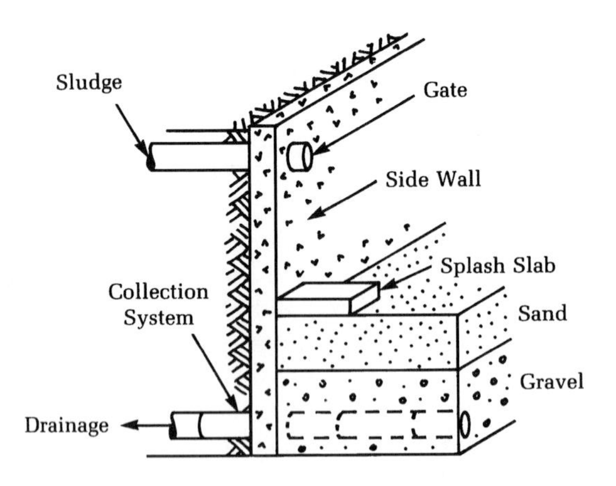

FIGURE 3-47
Typical sludge drying bed construction. (*Source:* U.S. Environmental Protection Agency, *Process Design Manual, Sludge Treatment and Disposal,* Washington, DC: U.S. Government Printing Office, 1979.)

The dewatering of sludge on sand beds is accomplished by two major factors: drainage and evaporation. The removal of water from sludge by drainage is a two-step process. First, the water is drained from the sludge, into the sand, and out the underdrains. This process may last a few days until the sand is clogged with fine particles or until all the free water has drained away. Further drainage by decanting can occur once a supernatant layer has formed (if beds are provided with a means of removing surface water). Decanting for removal of rain water can also be particularly important with sludges that do not crack.

The water that does not drain or is not decanted must evaporate. Obviously climate plays a role here. Phoenix would be a more efficient area for a sand bed than Seattle! Much of the lower Midwest, for example, would have an annual evaporation rate of about 0.75 m, so typical annual loadings to a sand bed in that area would be 100 $kg/m^2/y$. This may be applied and cleaned in about ten cycles during the year.

The filtrate from the sand-drying beds can be either recycled, treated, or discharged to a watercourse depending on its quality. Laboratory testing of the filtrate should be performed in conjunction with sand-drying bed pilot testing before a decision is made as to what is to be done with it.

Current United States practice is to make drying beds rectangular with dimensions of 4 to 20 m by 15 to 50 m with vertical side walls. Usually 100 to 230 mm of sand is placed over 200 to 460 mm of graded gravel or stone. The sand is usually 0.3 to 1.2 mm in effective diameter and has a uniformity coefficient less than 5.0. Gravel is normally graded from 3 to 25 mm in effective diameter. Underdrain piping has normally been of vitrified clay, but plastic pipe is also becoming acceptable. The pipes should be no less than 100 mm in diameter, should be spaced 2.2 to 6 m apart, and should have a minimum slope of 1 percent. When the cost of labor is high, newly constructed beds are designed for mechanical sludge removal.

Freeze treatment. Dewatering sludge by either of the nonmechanical methods may be enhanced by physical conditioning of the sludge through alternate natural freezing and thawing cycles. The freeze-thaw process dehydrates the sludge particles by freezing the water that is closely associated with the particles. The dewatering process takes place in two stages. The first stage reduces sludge volume by selectively freezing the water molecules. Next, the solids are dehydrated when they become frozen. When thawed, the solid mass forms granular-shaped particles. This coarse material readily settles and retains its new size and shape. This residue sludge dewaters rapidly and makes suitable landfill material.

The supernatant liquid from this process can be decanted, leaving the solids to dewater by natural drainage and evaporation. Pilot-scale systems can be utilized to evaluate this method's effectiveness and to establish design parameters. Elimination of rain and snow from the dewatering system by the provision of a roof will enhance the process considerably.

The potential advantages of a freeze-thaw system are

1. Insensitivity to variations in sludge quality
2. No conditioning required
3. Minimum operator attention

4. Natural process in cold climates (winter)
5. Solids cake more acceptable at landfills
6. Sludge easily worked with conventional equipment

Several natural freeze-thaw installations are located in New York state. At the alum coagulation plant of the Metropolitan Water Board of Oswego County, filter backwash is discharged to lagoons that act as decant basins. Thickened sludge is pumped from the lagoons to special freeze-thaw basins in layers about 450 mm thick. The sludge has never been deeper than 300 mm during freezing because of additional water losses. The 300-mm sludge layer reduces to about 75 mm of dried material after freeze-thaw.

Centrifuging. A centrifuge uses centrifugal force to speed up the separation of sludge particles from the liquid. In a typical unit (Figure 3-48), sludge is pumped into a horizontal, cylindrical bowl, rotating at 800 to 2,000 rpm. Polymers used for sludge conditioning also are injected into the centrifuge. The solids are spun to the outside of the bowl where they are scraped out by a screw conveyor. The liquid, or *centrate,* is returned to the treatment plant. Two types of centrifuges are currently used for sludge dewatering: the solid bowl and the basket bowl. For dewatering water-treatment-plant sludges, the solid bowl has proven to be more successful than the basket bowl. Centrifuges are very sensitive to changes in the concentration or composition of the sludge, as well as to the amount of polymer applied.

Because of its calcium carbonate content, lime softening-sludge dewaters with relative ease in a centrifuge. A cake dryness of 15 to 17 percent is about the best that

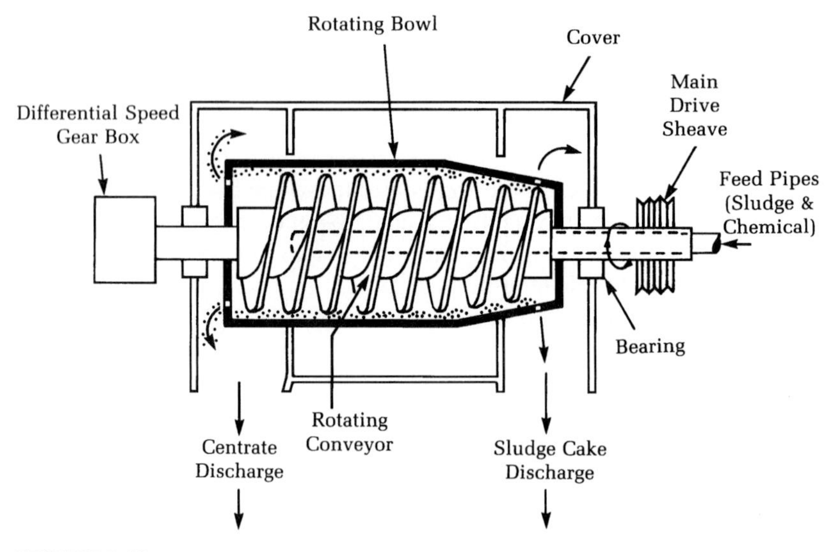

FIGURE 3-48
Solid bowl centrifuge. (Courtesy of Bird Machine Company.)

can be expected for a centrifuged alum sludge. A solids content of 50 percent or higher can be achieved with a lime sludge.

Vacuum filtration. A vacuum filter consists of a cylindrical drum covered with a filtering material or fabric, which rotates partially submerged in a vat of conditioned sludge (Figure 3-49). A vacuum is applied inside the drum to extract water, leaving the solids, or *filter cake,* on the filter medium. As the drum completes its rotational cycles, a blade scrapes the filter cake from the filter and the cycle begins again. Two basic types of rotary-drum vacuum filters are used in water treatment: the *traveling medium* and the *precoat medium* filters. The traveling medium filter is made of fabric or stainless steel coils. This filter is continuously removed from the drum, allowing it to be washed from both sides without diluting the sludge in the sludge vat. The precoat medium filter is coated with 50 to 75 mm of inert material, which is shaved off in 0.1 mm increments as the drum moves.

Continuous belt filter press (CBFP). The belt filter press operates on the principle that bending a sludge cake contained between two filter belts around a roll introduces shear and compressive forces in the cake, allowing water to work its way to the surface and out of the cake, thereby reducing the cake moisture content. The device employs double moving belts to continuously dewater sludges through one or more stages of dewatering (Figure 3-50). Typically the CBFP includes the following stages

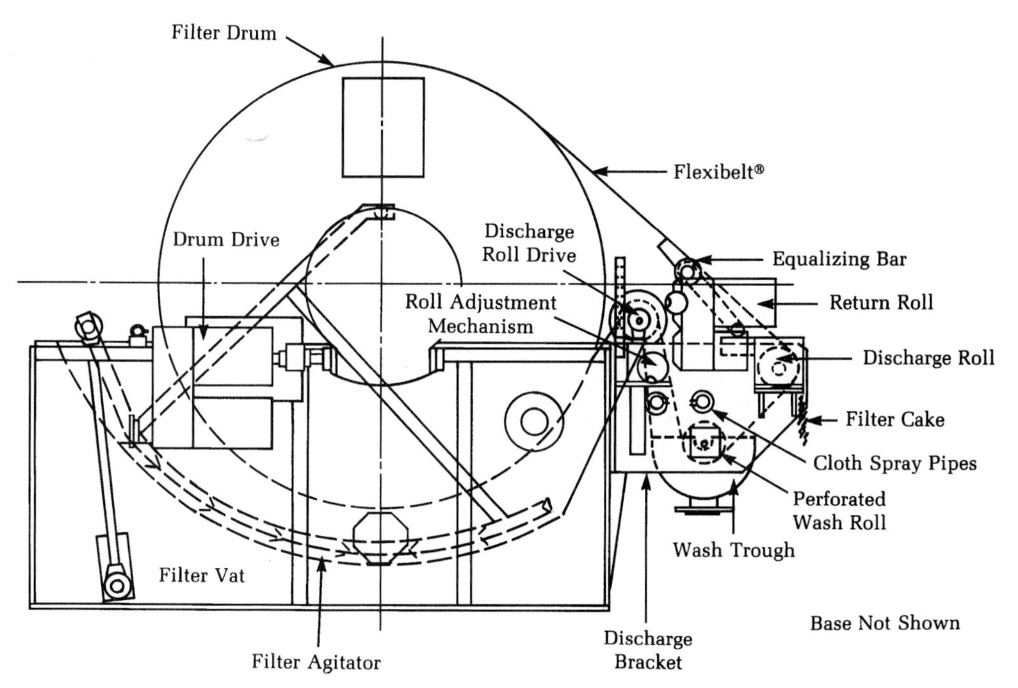

FIGURE 3-49
Vacuum filter. (Courtesy of Komline-Sanderson Engineering Corporation.)

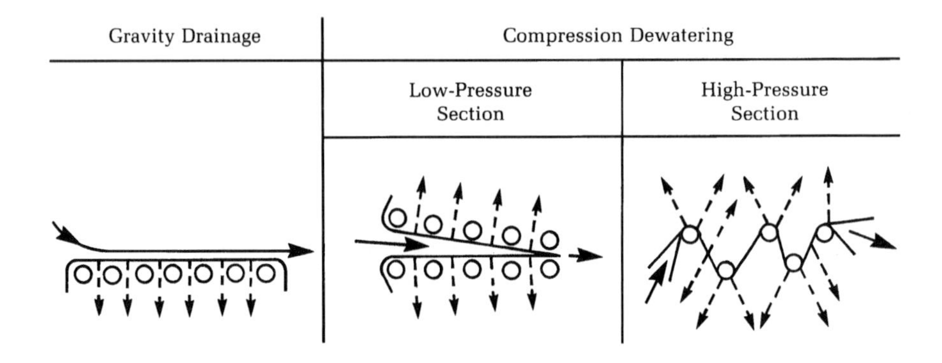

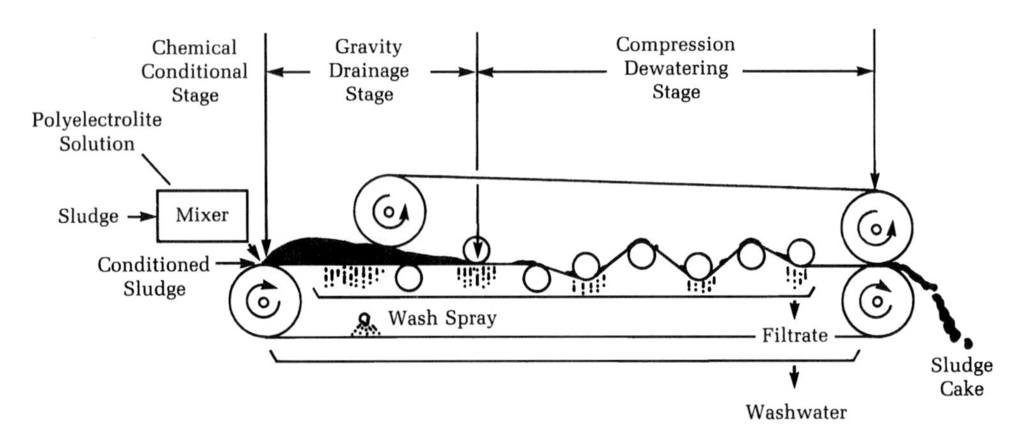

FIGURE 3-50
Continuous belt filter press. (*Source:* U.S. Environmental Protection Agency, *Process Design Manual, Sludge Treatment and Disposal,* 1979.)

of treatment:

1. A reactor/conditioner to remove free-draining water
2. A low pressure zone of belts with the top belt being solid and the bottom belt being a sieve; here further water removal occurs and a sludge mat having significant dimensional stability is formed
3. A high pressure zone of belts with a serpentine or sinusoidal configuration to add shear to the pressure dewatering mechanisms

Plate pressure filters. The basic component of a plate filter is a series of recessed vertical plates. Each plate is covered with cloth to support and contain the sludge cake. The plates are mounted in a frame consisting of two head supports connected by two horizontal parallel bars. A schematic cross section is illustrated in Figure 3-51. Conditioned sludge is pumped into the pressure filter and passes through feed holes

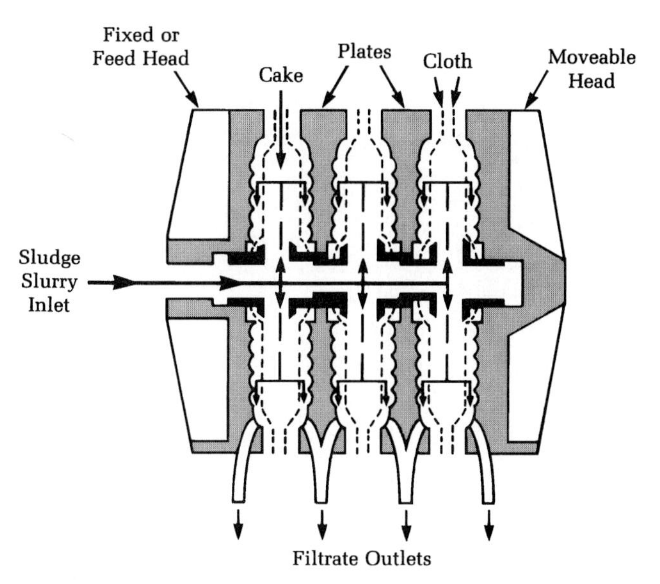

FIGURE 3-51
Schematic cross section of a fixed volume recessed plate filter assembly. (*Source:* U.S. Environmental Protection Agency, *Process Design Manual, Sludge Treatment and Disposal,* 1979.)

in the filter plates along the length of the filter and into the recessed chambers. As the sludge cake forms and builds up in the chamber, the pressure gradually increases to a point where further sludge injection would be counterproductive. At this time the injection ceases.

A typical pressure filtration cycle begins with the closing of the press to the position shown on Figure 3-51. Sludge is fed for a 20- to 30-minute period until the press is effectively full of cake. The pressure at this point is generally the designed maximum (700 to 1,700 kPa) and is maintained for 1- to 4-hours, during which more filtrate is removed and the desired cake solids content is achieved. The filter is then mechanically opened, and the dewatered cake is dropped from the chambers onto a conveyor belt for removal. Cake breakers are usually required to break up the rigid cake into conveyable form. Because recessed-plate pressure filters operate at high pressures and because many units use lime for conditioning, the cloths require routine washing with high-pressure water, as well as periodic washing with acid.

The Erie County Water Authority's Sturgeon Point Plant in Erie County, New York, and the Monroe County Water Authority's Shoremont Plant in Rochester, New York, serve as typical examples of the application of filter presses. Both plants include gravity settling and chemical conditioning of the sludge followed by mechanical dewatering via pressure plate filtration. A typical process flow diagram of the sludge treatment system used at both plants is shown in Figure 3-52. The Sturgeon Point Plant includes a 1.6-m diameter press, while the Monroe County facility includes a 2 m by 2 m press. Under actual operating conditions, the dewatered sludge

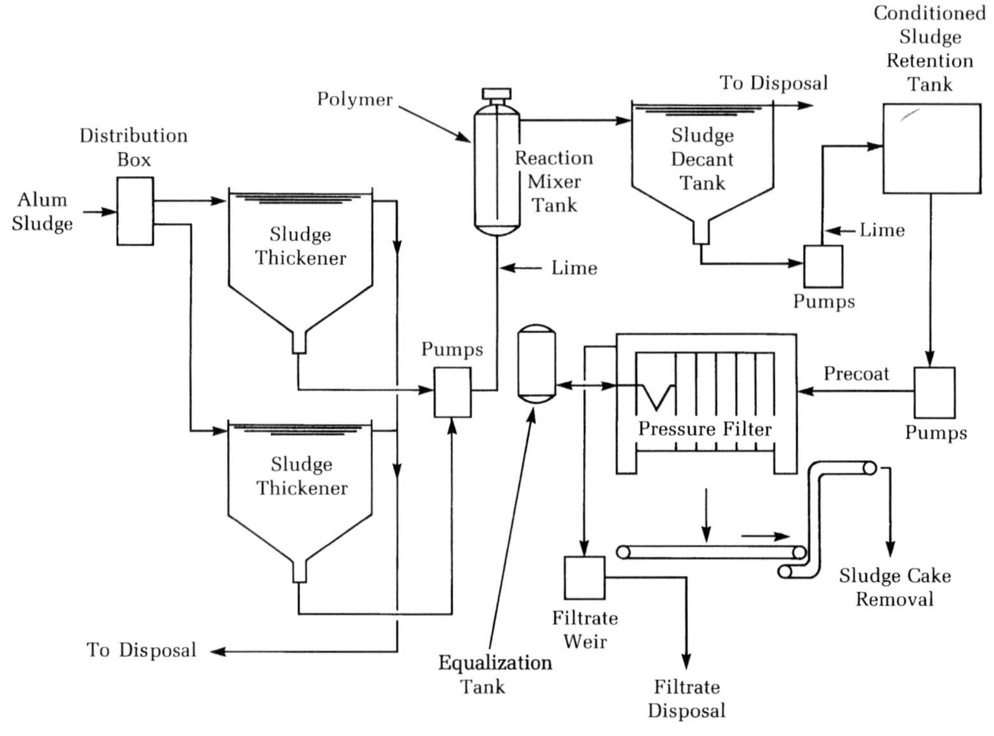

FIGURE 3-52
Pressure plate filter process for alum sludge treatment.

cakes produced at Sturgeon Point and at the Shoremont Plant contain between 45 and 50 percent dry solids by mass. Of the dry solids, as much as 30 percent may be conditioning chemical solids and/or fly ash. Thus, the corrected dry solids achieved is about 35 percent. This is a significantly better product than what the other mechanical methods produce.

Ultimate Disposal

After all possible sludge treatment has been accomplished, a residual sludge remains, which must be sent to ultimate disposal. Of the many theoretical alternatives for ultimate disposal, only three are of practical interest:

1. Co-disposal with sewage sludge
2. Landfilling
3. Land spreading

Of these three methods, only the last two are true ultimate disposal techniques. The first is a subterfuge to pass the problem to the wastewater plant operator.

3-11 CHAPTER REVIEW

When you have completed studying this chapter you should be able to do the following without the aid of your textbook or notes:

1. Define potable and palatable and explain why we must provide a water that is both potable and palatable.
2. Distinguish between dissolved substances, suspended solids, and colloidal substances based on their size and the mechanism by which they can be removed from water.
3. Define and calculate quantities of a given substance in water in percent by weight, parts per million (ppm), and milligrams per liter (mg/L), and convert from one unit of measure to the others.
4. Define alkalinity in terms of all the chemical species found, that is, Equation 3-36.
5. Define buffer.
6. Explain the effect of various chemical additions to the carbonate buffer system. Your explanation should include the effect on the displacement of the reaction (left or right), effect on CO_2 (into or out of solution), and effect on pH (increase, decrease, or no change).
7. List the four categories of water quality for drinking water.
8. List the four categories of physical characteristics.
9. Select the appropriate category of chemical standard for a given constituent, for example, zinc—esthetics, iron—esthetics/economics, nitrates—toxicity.
10. Define pathogen, SDWA, MCL, VOC, SOC, DBP, THM, and SWTR.
11. Identify the microorganism group used as an indicator of fecal contamination of water and explain why it was selected.
12. Sketch a water softening plant and a filtration plant, labeling all of the parts and explaining their functions.
13. Define the Schulze-Hardy rule and use it to explain the effectiveness of ions of differing valence in coagulation (Figures 3-8 and 3-9).
14. Explain the significance of alkalinity in coagulation.
15. Differentiate between coagulation and flocculation.
16. Write the reaction chemistry of alum and ferric chloride when alkalinity is present and when no alkalinity is present.
17. Explain the effect of pH on alum and ferric chloride solubility.
18. Explain how to conduct a jar test to obtain an optimum coagulant dose.
19. List the four basic types of coagulant aids; explain how each aid works and when it should be employed.
20. Define hardness in terms of the chemical constituents that cause it and in terms of the results as seen by the users of hard water.
21. Using diagrams and chemical reactions, explain how water becomes hard.
22. Given the total hardness and alkalinity, calculate the carbonate hardness and noncarbonate hardness.

23. Write the general equations for softening by ion exchange and by chemical precipitation.

24. Explain the significance of alkalinity in lime-soda softening.

25. Calculate the theoretical detention time or volume of tank if you are given the flow rate and the volume or detention time.

26. Explain how an upward-flow sedimentation tank (upflow clarifier) works, using a vector arrow diagram of a settling particle.

27. Define overflow rate in terms of liquid flow and settling-basin geometry, and state its units.

28. Using a vector arrow diagram, show how, in a horizontal-flow clarifier, a particle with a settling velocity that is less than the overflow rate may be captured (see Figure 3-34).

29. Calculate the percent of particles retained in a settling basin given the overflow rate, settling velocity, and basin flow scheme (that is, horizontal flow or upward flow). (Problems 3-66 and 3-67)

30. Explain the difference between Type I, Type II, and Type III sedimentation.

31. Compare slow sand filters, rapid sand filters, and dual media filters with respect to operating procedures and loading rates.

32. Explain how a rapid sand filter is cleaned.

33. Sketch and label a rapid sand filter identifying the following pertinent features: inlet main, outlet main, washwater outlet, collection laterals, support media (graded gravel), graded filter sand, and backwash troughs.

34. Define effective size and uniformity coefficient and explain their use in designing a rapid sand filter.

35. Explain why a disinfectant that has a residual is preferable to one that does not.

36. Write the equations for the dissolution of chlorine gas in water and the subsequent dissociation of hypochlorous acid.

37. Explain the difference between free available chlorine and combined available chlorine, and state which is the more effective disinfectant.

38. Sketch a breakpoint chlorination curve and label the axes, breakpoint, and regions of predominantly combined and predominantly free residual.

39. Define the terms thickening, conditioning, and dewatering.

40. List and describe three methods of nonmechanical dewatering of sludge.

41. List and describe four mechanical methods of dewatering of sludge.

With the aid of the text you should be able to do the following:

1. Calculate the gram equivalent weight of a chemical or compound.

2. Calculate the molarity, normality, and concentration of a given chemical compound in milligrams per liter (mg/L) and convert from one unit of measure to the others.

3. Calculate the equilibrium concentration of a compound when it is in equilibrium with its precipitate.

4. Calculate the pH of a solution containing a strong or weak acid alone (neglecting the dissociation of water).
5. Convert a concentration of a compound to calcium carbonate equivalents.
6. Calculate the reaction rate constant (k) from a set of experimental data.
7. Calculate consumption of alkalinity for a given dose of alum or ferric chloride.
8. Calculate the production of CO_2, SO_4^{2-}, or acid for a given dose of alum or ferric chloride.
9. Calculate the amount of lime (as $CaCO_3$) that must be added if insufficient alkalinity is present in order to neutralize the acid produced by the addition of alum or ferric chloride to a given water.
10. Estimate the amount of lime and soda ash required to soften water of a stated composition.
11. Calculate the fraction of the "split" for a lime-soda softening system or an ion-exchange softening system.
12. Size a rapid-mix and flocculation basin for a given type of water treatment plant and determine the required power input (Problems 3-55 and 3-56).
13. Design a mixer system for either a rapid-mix or flocculation basin.
14. Size a sedimentation basin and estimate the required weir length. (Problem 3-70)
15. Perform a grain size analysis and determine the effective size and uniformity coefficient.
16. Size a rapid sand filter (Problems 3-75 and 3-76) and determine the clean-sand head loss and the depth of the expanded bed during backwash (Problems 3-81 and 3-82).

3-12 PROBLEMS

3-1. Show that a density of 1 g/mL is the same as a density of 1,000 kg/m^3. (*Hint:* Some useful conversions are listed inside the back cover of this book.)

3-2. Show that a 4.50 percent by weight mixture contains 45.0 kg of substance in a cubic meter of water (that is, 4.50% = 45.0 kg/m^3).

3-3. Show that 1 mg/L = 1 g/m^3.

3-4. In a now antiquated and, we hope, soon to be forgotten system of measurements, it was common to consider water flows in terms of millions of gallons per day (MGD). Determine the number of MGD equivalent to the following flows in m^3/s: 0.0438; 0.05; 0.438; 0.5; 4.38; and 5.

Record both your calculated answer and the answer rounded to include only significant figures.

3-5. Calculate the molarity and normality of the following:

a. 200.0 mg/L HCl
b. 150.0 mg/L H_2SO_4
c. 100.0 mg/L $Ca(HCO_3)_2$
d. 70.0 mg/L H_3PO_4

Answers:

Molarity (M)	Normality (N)
a. 0.005485	0.005485
b. 0.001529	0.003059
c. 0.0006168	0.001234
d. 0.000714	0.00214

3-6. Calculate the molarity and normality of the following:

a. 80 μg/L HNO_3
b. 135 μg/L $CaCO_3$
c. 10 μg/L $Cr(OH)_3$
d. 1000 μg/L $Ca(OH)_2$

3-7. Calculate the mg/L of the following:

a. 0.01000 N Ca^{2+}
b. 1.000 M HCO_3^-
c. 0.02000 N H_2SO_4
d. 0.02000 M SO_4^{2-}

3-8. Calculate the μg/L of the following:

a. 0.0500 N H_2CO_3
b. 0.0010 M $CHCl_3$
c. 0.0300 N $Ca(OH)_2$
d. 0.0080 M CO_3

3-9. How many mg/L of magnesium ion will remain in solution in water which is 0.001000 M in hydroxyl ion and at 25°C?

Answer: 0.4423 mg/L

3-10. The Pherric, New Mexico, groundwater contains 1.800 mg/L of iron as Fe^{3+}. What pH is required to precipitate all but 0.30 mg/L of the iron at 25°C?

3-11. Given a saturated solution of calcium carbonate, how many moles of calcium ion will remain in solution after the addition of 3.16×10^{-4} M of Na_2CO_3 at 25°C?

3-12. The solubility product of calcium fluoride (CaF_2) is 3×10^{-11}. Will a fluoride concentration of 1.0 mg/L be soluble in a water containing 200 mg/L of calcium?

3-13. What amount of NaOH (a strong base), in mg, would be required to neutralize the acid in Example 3-6 (see Section 3-1)?

Answer: 81.568 or 81.6 mg

3-14. The pH of a finished water from a softening process is 10.74. What amount of 0.02000 N sulfuric acid, in milliliters, is required to neutralize 1.000 liter of the finished water?

3-15. How many milliliters of 0.02000 N hydrochloric acid would be required to perform the neutralization in Problem 3-14?

3-16. Calculate the pH of a water at 25°C that contains 0.6580 mg/L of carbonic acid. Assume that $[H^+] = [HCO^-]$ at equilibrium and neglect the dissociation of water.

Answer: pH = 5.66

3-17. What is the pH of a water at 25°C that contains 0.5000 mg/L of hypochlorous acid? Assume equilibrium has been achieved. Neglect the dissociation of water. Although it may not be justified based on the data available to you, report the answer to two decimal places.

3-18. If the pH in Problem 3-17 is adjusted to 7.00, what would the OCl^- concentration in mg/L be at 25°C?

3-19. Convert the following from mg/L as the ion to mg/L as $CaCO_3$:
a. 83.00 mg/L Ca^{2+}
b. 27.00 mg/L Mg^{2+}
c. 48.00 mg/L CO_2 (*Hint:* See footnote 14)
d. 220.00 mg/L HCO_3^-
e. 15.00 mg/L CO_3^{2-}

Answers:

$$\begin{aligned} Ca &= 207.25 \text{ or } 207.3 \text{ mg/L as } CaCO_3 \\ Mg &= 111.20 \text{ or } 111.2 \text{ mg/L as } CaCO_3 \\ CO_2 &= 109.18 \text{ or } 109.2 \text{ mg/L as } CaCO_3 \\ HCO_3 &= 180.41 \text{ or } 180.4 \text{ mg/L as } CaCO_3 \\ CO_3 &= 25.02 \text{ or } 25.0 \text{ mg/L } CaCO_3 \end{aligned}$$

3-20. Convert the following from mg/L as the ion or compound to mg/L as $CaCO_3$:
a. 200.00 mg/L HCl
b. 280.00 mg/L CaO
c. 123.45 mg/L Na_2CO_3
d. 85.05 mg/L $Ca(HCO_3)_2$
e. 19.90 mg/L Na

3-21. Convert the following from mg/L as $CaCO_3$ to mg/L as the ion or compound:
a. 100.00 mg/L SO_4^{2-}
b. 30.00 mg/L HCO_3^-
c. 150.00 mg/L Ca^{2+}
d. 10.00 mg/L H_2CO_3
e. 150.00 mg/L Na^+

Answers:

$$\begin{aligned} SO_4^{2-} &= 95.98 \text{ or } 96.0 \text{ mg/L} \\ HCO_3^- &= 36.58 \text{ or } 36.6 \text{ mg/L} \\ Ca^{2+} &= 60.07 \text{ or } 60.1 \text{ mg/L} \\ H_2CO_3 &= 6.198 \text{ or } 6.20 \text{ mg/L} \\ Na^+ &= 68.91 \text{ mg/L} \end{aligned}$$

3-22. Convert the following from mg/L as $CaCO_3$ to mg/L as the ion or compound:
a. 10.00 mg/L CO_2
b. 13.50 mg/L $Ca(OH)_2$
c. 481.00 mg/L H_3PO_4
d. 81.00 mg/L H_2PO_4
e. 40.00 mg/L Cl^-

3-23. What is the "exact" alkalinity of a water that contains 0.6580 mg/L of bicarbonate, as the ion, at a pH of 5.66? No carbonate is present.

Answer: 0.4302 mg/L as $CaCO_3$

3-24. Calculate the "approximate" alkalinity (in mg/L as $CaCO_3$) of a water containing 120 mg/L of bicarbonate ion and 15.00 mg/L of carbonate ion.

3-25. Calculate the "exact" alkalinity of the water in Problem 3-24 if the pH is 9.43.

3-26. Using Equations 3-28, 3-36, 3-38, and 3-39, derive two equations which allow calculation of the bicarbonate and carbonate alkalinities in mg/L as $CaCO_3$ from measurements of the total alkalinity (A) and the pH.

Answers: (in mg/L as $CaCO_3$)

$$HCO_3^- = \frac{50{,}000\left\{\left(\dfrac{A}{50{,}000}\right) + [H^+] - \left(\dfrac{K_w}{[H^+]}\right)\right\}}{1 + \left(\dfrac{2\,K_2}{[H^+]}\right)}$$

$$CO_3^{2-} = \left(\frac{2\,K_2}{[H^+]}\right)(HCO_3^-)$$

where
A = total alkalinity, mg/L as $CaCO_3$
K_2 = second dissociation constant of carbonic acid
= 4.68×10^{-11} at 25°C
K_w = ionization constant of water
= 1×10^{-14} at 25°C
(HCO_3^-) = bicarbonate alkalinity in mg/L as $CaCO_3$
CO_3^{2-} = carbonate alkalinity in mg/L as $CaCO_3$

3-27. Using the solution to Problem 3-26, calculate the bicarbonate and carbonate alkalinities, in mg/L as $CaCO_3$, of a water having a total alkalinity of 233.0 mg/L as $CaCO_3$ and a pH of 10.47.

3-28. If a water has a carbonate alkalinity of 120.00 mg/L as the ion and a pH of 10.30, what is the bicarbonate alkalinity in mg/L as the ion?

Answer: HCO_3^- = 130.686 or 130.7 mg/L

3-29. What is the pH of a water that contains 120.00 mg/L of bicarbonate ion and 15.00 mg/L of carbonate ion?

3-30. Calculate the alkalinity of the water in Problem 3-28 using the equations in Problem 3-26 and the "exact" method of Example 3-8 (see Section 3-1).

3-31. The following mineral analysis was reported for a water sample taken from Well No. 1 at the Eastwood Manor Subdivision near McHenry, Illinois.[45]

Well No. 1, Lab No. 02694, November 9, 1971

(Note: All reported as "mg/L as the ion" unless stated otherwise.)

Iron	0.2	Silica (SiO_2)	20.0
Manganese	0.0	Fluoride	0.35
Ammonium	0.5	Boron	0.1
Sodium	4.7	Nitrate	0.0
Potassium	0.9	Chloride	4.5
Calcium	67.2	Sulfate	29.0
Magnesium	40.0	Alkalinity	284.0 as $CaCO_3$
Barium	0.5	pH (as recorded)	7.6 units

[45] D. M. Woller and E. W. Sanderson, *Public Water Supplies in McHenry County* (Illinois State Water Survey, Publication No. 60–19), Urbana, Illinois, 1976.

Determine the total, carbonate, and noncarbonate hardness in mg/L as $CaCO_3$ using the predominant polyvalent cation definition in Section 3-3.

Answers:

$$TH = 332.8 \text{ mg/L as } CaCO_3$$
$$CH = 284.0 \text{ mg/L as } CaCO_3$$
$$NCH = 48.8 \text{ mg/L as } CaCO_3$$

3-32. Calculate the total, carbonate, and noncarbonate hardness in Problem 3-31 using all of the polyvalent cations. What is the percent error in using only the predominant cations?

3-33. The following mineral analysis was reported for a water sample taken from Well No. 1 at Magnolia, Illinois.[46] Determine the total, carbonate, and noncarbonate hardness in mg/L as $CaCO_3$ using the predominant polyvalent cation definition of hardness.

Well No. 1, Lab No. B109535, April 23, 1973

(Note: All reported as "mg/L as the ion" unless stated otherwise.)

Iron	0.42	Zinc	0.01
Manganese	0.04	Silica (SiO_2)	20.0
Ammonium	11.0	Fluoride	0.3
Sodium	78.0	Boron	0.3
Potassium	2.6	Nitrate	0.0
Calcium	78.0	Chloride	9.0
Magnesium	32.0	Sulfate	0.0
Barium	0.5	Alkalinity	494.0 as $CaCO_3$
Copper	0.01	pH (as recorded)	7.7 units

3-34. The following mineral analysis was reported for Michigan State University well water.[47] Determine the total, carbonate, and noncarbonate hardness in mg/L as $CaCO_3$. *Note:* All units are mg/L unless otherwise stated.

Fluoride	1.1	Silica (SiO_2)	13.4
Chloride	4.0	Bicarbonate	318.0
Nitrate	0.0	Sulfate	52.0
Sodium	14.0	Iron	0.5
Potassium	1.6	Manganese	0.07
Calcium	96.8	Zinc	0.27
Magnesium	30.4	Barium	0.2

3-35. The following data were obtained for an irreversible elementary reaction. Plot the data, determine the order of the reaction (zero, first, or second) and the rate constant (K).

[46] D. M. Woller and E. W. Sanderson, *Public Groundwater Supplies in Putnam County* (Illinois State Water Survey, Publication No. 60–15), Urbana, Illinois, 1976.

[47] Michigan Department of Public Health, *Annual Data Summary,* 1979.

Time, min	Reactant "A" Concentration, mmoles/L
0	2.80
1	2.43
2	2.12
5	1.39
10	0.69
20	0.17

3-36. Repeat Problem 3-35 for the following data.

Time, min	Reactant "A" Concentration, mmoles/L
0	48.0
1	6.22
2	3.32
3	2.27
5	1.39
10	0.704

3-37. Shown below are the results of water quality analyses of the Thames River in London. If the water is treated with 60.00 mg/L of alum to remove turbidity, how much alkalinity will remain? Ignore side reactions with phosphorus and assume all the alkalinity is HCO_3^-.

Thames River, London

Constituent	Expressed as	Milligrams per liter
Total hardness	$CaCO_3$	260.0
Calcium hardness	$CaCO_3$	235.0
Magnesium hardness	$CaCO_3$	25.0
Total iron	Fe	1.8
Copper	Cu	0.05
Chromium	Cr	0.01
Total alkalinity	$CaCO_3$	130.0
Chloride	Cl	52.0
Phosphate (total)	PO_4	1.0
Silica	SiO_2	14.0
Suspended solids	—	43.0
Total solids	—	495.0
pH[a]	—	7.4

[a] Not in mg/L

Answer: Alkalinity remaining = 99.65 or 100 mg/L as $CaCO_3$.

3-38. Shown below are the results of water quality analyses of the Mississippi River at Baton Rouge, Louisiana. If the water is treated with 30.00 mg/L of ferric chloride for turbidity coagulation, how much alkalinity will remain? Ignore the side reactions with phosphorus and assume all the alkalinity is HCO_3^-.

Mississippi River, Baton Rouge, Louisiana

Constituent	Expressed as	Milligrams per liter
Total hardness	$CaCO_3$	164.0
Calcium hardness	$CaCO_3$	108.0
Magnesium hardness	$CaCO_3$	56.0
Total iron	Fe	0.9
Copper	Cu	0.01
Chromium	Cr	0.03
Total alkalinity	$CaCO_3$	136.0
Chloride	Cl	32.0
Phosphate (total)	PO_4	3.0
Silica	SiO_2	10.0
Suspended solids	TSS	29.9
Turbidity[a]	NTU	12.0
pH[a]	—	7.6

[a] Not in mg/L

3-39. Shown below are the results of water quality analyses of the Crater Lake at Mount Mazama, Oregon. If the water is treated with 40.00 mg/L of alum for turbidity coagulation, how much alkalinity will remain? Assume all the alkalinity is HCO_3^-.

Crater Lake, Mount Mazama, Oregon

Constituent	Expressed as	Milligrams per liter
Total hardness	$CaCO_3$	28.0
Calcium hardness	$CaCO_3$	19.0
Magnesium hardness	$CaCO_3$	9.0
Total iron	Fe	0.02
Sodium	Na	11.0
Total alkalinity	$CaCO_3$	29.5
Chloride	Cl	12.0
Sulfate	SO_4	12.0
Silica	SiO_2	18.0
Total dissolved solids	—	83.0
pH[a]	—	7.2

[a] Not in mg/L

3-40. Prepare a bar chart of the water described in Problem 3-31. (*Note:* Valences may be found in Appendix A.) Because all of the constituents were not analyzed, an ion balance is not achieved.

3-41. Prepare a bar chart of the water described in Problem 3-33. Because all of the constituents were not analyzed, an ion balance is not achieved.

3-42. Prepare a bar chart of the water described in Problem 3-34. Because all of the constituents were not analyzed, an ion balance is not achieved.

3-43. Determine the lime and soda ash dose, in mg/L as $CaCO_3$, to soften the following water to a final hardness of 80.0 mg/L as $CaCO_3$. The ion concentrations reported below are all mg/L as $CaCO_3$.

$$Ca^{2+} = 120.0$$
$$Mg^{2+} = 30.0$$
$$HCO_3^- = 70.0$$
$$CO_2 = 10.0$$

3-44. Determine the lime and soda ash dose, in mg/L as CaO and Na_2CO_3, to soften the following water to a final hardness of 80.0 mg/L as $CaCO_3$. The ion concentrations reported below are all mg/L as $CaCO_3$. Assume lime is 90 percent pure and soda ash is 97 percent pure.

$$Ca^{2+} = 210.0$$
$$Mg^{2+} = 23.0$$
$$HCO_3^- = 165.0$$
$$CO_2 = 5.0$$

3-45. Determine the lime and soda ash dose, in mg/L as $CaCO_3$, to soften the water described in Problem 3-31 to a final hardness of 100.0 mg/L as $CaCO_3$.

3-46. Determine the lime and soda ash dose, in mg/L as $CaCO_3$, to soften the following water to a final hardness of 70.0 mg/L as $CaCO_3$. The ion concentrations reported below are all as $CaCO_3$.

$$Ca^{2+} = 220.0$$
$$Mg^{2+} = 75.0$$
$$HCO_3^- = 265.0$$
$$CO_2 = 17.0$$

3-47. Determine the lime and soda ash dose, in mg/L as CaO and Na_2CO_3, to soften the following water to a final hardness of 80.0 mg/L as $CaCO_3$. The ion concentrations reported below are all as $CaCO_3$. Assume lime is 93 percent pure and soda ash is 95 percent pure.

$$Ca^{2+} = 137.0$$
$$Mg^{2+} = 56.0$$
$$HCO_3^- = 128.0$$
$$CO_2 = 7.0$$

3-48. Determine the lime and soda ash dose, in mg/L as $CaCO_3$, to soften the water described in Problem 3-37 to a final hardness of 90.0 mg/L as $CaCO_3$.

3-49. Determine the lime and soda ash dose, in mg/L as $CaCO_3$, to soften the following water to a final hardness of 80.0 mg/L as $CaCO_3$. If the price of lime, purchased as CaO, is \$100.00 per megagram (Mg), and the price of soda ash, purchased as Na_2CO_3 is \$200.00 per Mg, what is the annual chemical cost of treating 0.500 m^3/s of this water?

Assume the lime and soda ash are 100 percent pure. The ion concentrations reported below are all mg/L as $CaCO_3$.

$$Ca^{2+} = 200.0$$
$$Mg^{2+} = 100.0$$
$$HCO_3^- = 150.0$$
$$CO_2 = 22.0$$

Answers:

Lime = 272.00 mg/L as $CaCO_3$

Soda = 110.00 mg/L as $CaCO_3$

Total Annual Cost = \$607,703.25 or \$608,000.00

3-50. Determine the lime and soda ash dose, in mg/L as $CaCO_3$, to soften the following water to a final hardness of 120.0 mg/L as $CaCO_3$. If the price of lime, purchased as CaO, is \$61.70 per megagram (Mg), and the price of soda ash, purchased as Na_2CO_3, is \$172.50 per Mg, what is the annual chemical cost of treating 1.35 m^3/s of this water? Assume the lime is 87 percent pure and the soda ash is 97 percent pure. The ion concentrations reported below are all mg/L as $CaCO_3$.

$$Ca^{2+} = 293.0$$
$$Mg^{2+} = 55.0$$
$$HCO_3^- = 301.0$$
$$CO_2 = 3.0$$

3-51. Determine the lime and soda ash dose, in mg/L as $CaCO_3$, to soften the Hardin, Illinois, water to a final hardness of 95.00 mg/L as $CaCO_3$.[48] Using the price and purity information supplied in Problem 3-49, determine the annual chemical cost of treating 0.150 m^3/s of this water.

Well No. 2, Hardin, Illinois

Iron	0.10	Silica (SiO_2)	21.6
Manganese	0.64	Fluoride	0.3
Ammonium	0.0	Boron	0.38
Sodium	21.8	Nitrate	8.4
Potassium	3.0	Chloride	32.0
Calcium	102.0	Sulfate	65.0
Magnesium	45.2	Alkalinity	344.0 as $CaCO_3$
Copper	0.01	pH (as recorded)	7.2 units
Zinc	0.13		

Note: All analyses reported as "mg/L as the ion" except as noted.

3-52. Determine the lime and soda ash dose, in mg/L as $CaCO_3$, to soften the following water to a final hardness of 90.0 mg/L as $CaCO_3$. If the price of lime, purchased as CaO, is \$61.70 per megagram (Mg), and the price of soda ash, purchased as Na_2CO_3,

[48]D. M. Woller, *Public Groundwater Supplies in Calhoun County* (Illinois State Water Survey, Publication No. 60-16), Urbana, Illinois, 1975.

is $172.50 per Mg, what is the annual chemical cost of treating 0.050 m^3/s of this water? Assume the lime is 90 percent pure and the soda ash is 97 percent pure. The ion concentrations reported below are all mg/L as $CaCO_3$.

Ca^{2+} = 137.0
Mg^{2+} = 40.0
HCO_3^- = 197.0
CO_2 = 9.0

3-53. Design a split treatment softening process (flow scheme/split, chemical dose in mg/L as $CaCO_3$) for the following water. Compounds are given in mg/L as the ion stated unless otherwise specified:

CO_2	42.7	HCO_3^-	344.0 mg/L as $CaCO_3$
Ca^{2+}	102.0	SO_4^{2-}	65.0
Mg^{2+}	45.2	Cl^-	32.0
Na^+	21.8		

3-54. Given the following water (all in meq/L), design a process to soften the water (flow scheme/split; amount of lime and/or soda required in mg/L as CaO and Na_2CO_3 respectively) and find the final hardness.

CO_2	0.40	Mg^{2+}	1.12
Ca^{2+}	2.16	HCO_3^-	2.72

3-55. What is the volume required for a rapid-mix basin that is to be used to treat 0.05 m^3/s of water if the detention time is 60 seconds?
Answer: 3 m^3

3-56. Two parallel flocculation basins are to be used to treat a water flow of 0.150 m^3/s. If the design detention time is 20 minutes, what is the volume of each tank?

3-57. Determine the power input required for the tank designed in Problem 3-55 if the water temperature is 20°C and the velocity gradient is 700 s^{-1}.
Answer: 1473 W or 1.5 kW

3-58. The flocculation tanks in Problem 3-56 were designed for an average velocity gradient of 36 s^{-1} and a water temperature of 17°C. What power input is required?

3-59. What power input is required for Problem 3-57 if the water temperature falls to 10°C?

3-60. What power input is required for Problem 3-58 if the water temperature falls to 10°C?

3-61. The town of Eau Gaullie has requested proposals for a new coagulation water treatment plant. The design flow for the plant is 0.1065 m^3/s. The average annual water temperature is 19°C. The following design assumptions for a rapid-mix tank have been made:

Number of tanks = 2
Tank configuration: circular with liquid depth = diameter
Detention time = 40 s
Velocity gradient = 800 s^{-1}
Mixer: Select from Example 3-20 in Section 3-5.
Impeller type: turbine, 6 flat blades, vaned disc

Design the rapid-mix system by providing the following:

1. Water power input in kW
2. Tank dimensions in m

3. Mixer model number
4. Diameter of the impeller in m
5. Rotational speed of impeller in rpm

Answers:

P = 1.4 kW

Diameter = 1.39 m; depth = 1.39 m

Mixer = JTQ300

Impeller diameter = 0.41 m

Rotational speed = 175 rpm

3-62. Laramie is planning for a new softening plant. The design flow is 0.168 m^3/s. The average water temperature is 5°C. The following design assumptions for a rapid mix tank have been made:

Tank configuration: square plan with depth = 1.25 times width

Detention time = 60 s

Velocity gradient = 700 s^{-1}

Mixer: Select from Example 3-20 in Section 3-5.

Design the rapid mix system by providing the following:

1. Number of tanks
2. Water power input in kW
3. Tank dimensions in m
4. Mixer model number
5. Diameter of the impeller in m
6. Rotational speed of impeller in rpm
7. Impeller type

3-63. Continuing the preparation of the proposal for the Eau Gaullie treatment plant (Problem 3-61), design the flocculation tank by providing the following for the first two compartments only:

1. Water power input in kW
2. Tank dimensions in m
3. Mixer model number (See Example 3-20 in Section 3-5.)
4. Diameter of the impeller in m
5. Rotational speed of impeller in rpm

Use the following assumptions:

1. Number of tanks = number of rapid mix tanks
2. Tapered G in three compartments: 90 s^{-1}, 60 s^{-1}, and 30 s^{-1}
3. Gt_o = 120,000
4. Length = width = depth
5. Impeller type: propeller, pitch of 2, 3 blades

Answers: For first compartment only

P = 0.29 kW

L = W = D = 3.29 m

Mixer = JTQ50

Impeller diameter = 0.93 m

Rotational speed = 45 rpm

Note: Because of the low power input requirement in the third compartment, there is no suitable mixer available from the list provided in Example 3-20.

3-64. Continuing the preparation of the proposal for the Laramie treatment plant (Problem 3-62), design the flocculation tank by providing the following for the first two compartments only:

1. Water power input in kW
2. Tank dimensions in m
3. Mixer model number (See Example 3-20 in Section 3-5.)
4. Diameter of the impeller in m
5. Rotational speed of impeller in rpm

Use the following assumptions:

1. Number of tanks = number of rapid mix tanks
2. Tapered G in three compartments: 90 s^{-1}, 60 s^{-1}, and 30 s^{-1}
3. Gt_o = 120,000
4. Length = width = depth
5. Impeller type: propeller, pitch of 1, 3 blades

3-65. Complete Example 3-20 in Section 3-5 by computing the rotational speed of the impellers in compartments 1 and 3.

3-66. If the settling velocity of a particle is 0.70 cm/s and the overflow rate of a horizontal clarifier is 0.80 cm/s, what percent of the particles are retained in the clarifier?

Answer: 88 percent

3-67. If the settling velocity of a particle is 2.80 mm/s and the overflow rate of an upflow clarifier is 0.560 cm/s, what percent of the particles are retained in the clarifier?

3-68. If the flowrate of the original plant in Problem 3-66 is increased from 0.150 m^3/s to 0.200 m^3/s, what percent removal of particles would be expected?

3-69. If the flowrate of the original plant in Problem 3-67 is doubled, what percent removal of particles would be expected?

3-70. If a 1.0 m^3/s flow water treatment plant uses ten sedimentation basins with an overflow rate of 15 $m^3/d \cdot m^2$, what should be the surface area (m^2) of each tank?

Answer: 576.0 m^2

3-71. Assuming a conservative value for an overflow rate, determine the surface area (in m^2) of each of two sedimentation tanks that together must handle a flow of 0.05162 m^3/s of lime softening floc.

3-72. Repeat Problem 3-71 for an alum or iron floc.

3-73. Determine the detention time and overflow rate for a settling tank that will reduce the influent suspended solids concentration from 33.0 mg/L to 15.0 mg/L. The following batch settling column data are available. The data given are percent removals at the sample times and depths shown.

	Depths, m				
Time, min	**0.5**	**1.5**	**2.5**	**3.5**	**4.5**
10	50	32	20	18	15
20	75	45	35	30	25
40	85	65	48	43	40
55	90	75	60	50	46
85	95	87	75	65	60
95	95	88	80	70	63

3-74. The following test data were gathered to design a settling tank. The initial suspended solids concentration for the test was 20.0 mg/L. Determine the detention time and overflow rate that will yield 60 percent removal of suspended solids. The data given are suspended solids concentrations in mg/L.

	Time, min					
Depth, m	**10**	**20**	**35**	**50**	**70**	**85**
0.5	14.0	10.0	7.0	6.2	5.0	4.0
1.0	15.0	13.0	10.6	8.2	7.0	6.0
1.5	15.4	14.2	12.0	10.0	7.8	7.0
2.0	16.0	14.6	12.6	11.0	9.0	8.0
2.5	17.0	15.0	13.0	11.4	10.0	8.8

3-75. For a flow of 0.8 m^3/s, how many rapid sand filter boxes of dimensions 10 m × 20 m are needed for a loading rate of 110 $m^3/d \cdot m^2$?

Answer: 4 filters (rounding to nearest even number)

3-76. If a dual media filter with a loading rate of 300 $m^3/d \cdot m^2$ were built instead of the standard filter in Problem 3-75, how many filter boxes would be required?

3-77. The water flow meter at the Westwood water plant is on the blink. The plant superintendent tells you the four dual media filters (each 5.00 m × 10.0 m) are loaded at a velocity of 280 m/d. What is the flow rate through the filters in m^3/s?

3-78. The Orono Sand and Gravel Company has made a bid to supply sand for Eau Gaullie's new sand filter. The request for bids stipulated that the sand have an effective size in the range 0.35 to 0.55 mm and a uniformity coefficient in the range 1.3 to 1.7. Orono supplied the following sieve analysis as evidence that their sand will meet the specifications. Perform a grain size analysis (log-probability plot) and determine whether or not the sand meets the specifications.

U.S. Standard Sieve No.	Mass Percent Retained
8	0.0
12	0.01
16	0.39
20	5.70
30	25.90
40	44.00
50	20.20
70	3.70
100	0.10

3-79. The Lexington Sand and Gravel Company has made a bid to supply sand for Laramie's new sand filter. The request for bids stipulated that the sand have an effective size in the range 0.35 to 0.55 mm and a uniformity coefficient in the range 1.3 to 1.7. Lexington supplied the following sieve analysis (sample size 500.00 g) as evidence that their sand will meet the specifications. Perform a grain size analysis (log-probability plot) and determine whether or not the sand meets the specifications.

U.S. Standard Sieve No.	Mass Retained, g
12	0.000
16	2.000
20	65.500
30	272.500
40	151.000
50	8.925
70	0.075

3-80. Rework Example 3-26 (Section 3-7) with the 70, 100, and 140 sieve fractions removed. Assume original sample contained 100 g.

3-81. The rapid sand filter being designed for Eau Gaullie has the characteristics and sieve analysis shown below. Determine the head loss for the clean filter bed in a stratified condition.

Depth = 0.60 m

Filter loading = 120 $m^3/d \cdot m^2$

Sand specific gravity = 2.50

Shape factor = 1.00

Stratified bed porosity = 0.42

Water temperature = 19°C

U.S. Standard Sieve No.	Mass Percent Retained
8–12	0.01
12–16	0.39
16–20	5.70
20–30	25.90
30–40	44.00
40–50	20.20
50–70	3.70
70–100	0.10

3-82. Determine the height of the expanded bed for the sand used in Problem 3-81 if the backwash rate is 1,000 m/d. Assume Equation 3-107 applies.

3-83. The rapid sand filter being designed for Laramie has the characteristics shown below. Determine the head loss for the clean filter bed in a stratified condition.

Depth = 0.75 m

Filter loading = 230 $m^3/d \cdot m^2$

Sand specific gravity = 2.80

Shape factor = 0.91

Stratified bed porosity = 0.50

Water temperature = 5°C

Sand analysis

U.S. Standard Sieve No.	Mass Percent Retained
8–12	0.00
12–16	0.40
16–20	13.10
20–30	54.50
30–40	30.20
40–50	1.785
50–70	0.015

3-84. Determine the maximum backwash rate and the height of the backwash troughs above the sand used in Problem 3-83. Assume Equation 3-107 applies.

3-85. As noted in Example 3-27 in Section 3-7, the head loss was too high. Rework the example without the 100–140 sieve fraction to see how much this would improve the head-loss characteristics.

3-86. What effect does removing the 100–140 sieve fraction have on the depth of the expanded bed in Example 3-28 (Section 3-7)?

3-13 DISCUSSION QUESTIONS

3-1. Would you expect a carbonated beverage to have a pH above, below, or equal to 7.0? Explain why.

3-2. Explain the word "turbidity" in terms that the mayor of a community could understand.

3-3. Which of the chemicals added to treat a surface water aids in making the water palatable?

3-4. Microorganisms play a role in the formation of hardness in groundwater. True or false? Explain.

3-5. If there is no bicarbonate present in a well water that is to be softened to remove magnesium, which chemicals must you add?

3-6. Use a scale drawing to sketch a vector diagram of a horizontal-flow sedimentation tank that shows how 25 percent of the particles with a settling velocity one-quarter that of the overflow rate will be removed.

3-7. In the U.S., chlorine is preferred as a disinfectant over ozone because it has a residual. Why is the presence of a residual important?

3-8. A new water softening plant is being designed for Lubbock, Texas. The climate is dry and land is readily available at a reasonable cost. What methods of sludge dewatering would be most appropriate? Explain your reasoning.

3-14 ADDITIONAL READING

Books

American Water Works Association, *Water Treatment,* New York: McGraw-Hill, 1995.

American Water Works Association, *Water Treatment Plant Design,* 2nd ed., New York: McGraw-Hill, 1990.

L. D. Benefield, J. F. Judkins, Jr., and B. L. Weand, *Process Chemistry for Water and Wastewater Treatment,* Englewood Cliffs, NJ: Prentice Hall, 1982.

J. W. Clark, W. Viessman, and M. J. Hammer, *Water Supply and Pollution Control,* New York: Harper & Row, 1977.

R. L. Droste, *Theory and Practice of Water and Wastewater Treatment,* New York: Wiley, 1997.

J. G. Henry and G. W. Heinke, *Environmental Science and Engineering,* Englewood Cliffs, NJ: Prentice Hall, 1989.

G. Kiely, *Environmental Engineering,* London: McGraw-Hill, 1997.

G. M. Masters, *Introduction to Environmental Engineering and Science,* Englewood Cliffs, NJ: Prentice Hall, 1991.

J. M. Montgomery, Consulting Engineers, Inc., *Water Treatment Principles and Design,* New York: Wiley, 1985.

H. S. Peavy, D. R. Rowe, and G. Tchobanoglous, *Environmental Engineering,* New York: McGraw-Hill, 1979.

B. T. Ray, *Environmental Engineering,* Boston, MA: PWS, 1995.

T. D. Reynolds and P. A. Richards, *Unit Operations and Processes in Environmental Engineering,* 2nd ed., Boston, MA: PWS, 1996.

A. P. Sincero and G. A. Sincero, *Environmental Engineering, A Design Approach,* Upper Saddle River, NJ: Prentice Hall, 1996.

E. W. Steel and T. J. McGhee, *Water Supply and Sewerage,* New York: McGraw-Hill, 1979.

G. Tchobanoglous and E. D. Schroeder, *Water Quality,* Reading, MA: Addison Wesley, 1985.

P. A. Vesilind, *Introduction to Environmental Engineering,* Boston, MA: PWS, 1997.

Journals

Journal of the American Water Works Association

Water Research

CHAPTER 4

WATER QUALITY MANAGEMENT

4-1 INTRODUCTION

4-2 WATER POLLUTANTS AND THEIR SOURCES

4-3 WATER QUALITY MANAGEMENT IN RIVERS
- Effect of Oxygen-Demanding Wastes on Rivers
- Biochemical Oxygen Demand
- Graphical Determination of BOD Constants
- Laboratory Measurement of Biochemical Oxygen Demand
- Additional Notes on Biochemical Oxygen Demand
- Nitrogen Oxidation
- DO Sag Curve
- Effect of Nutrients on Water Quality in Rivers

4-4 WATER QUALITY MANAGEMENT IN LAKES
- Stratification and Turnover
- Biological Zones
- Lake Productivity
- Eutrophication
- Algal Growth Requirements
- The Limiting Nutrient
- Control of Phosphorus in Lakes
- Acidification of Lakes

4-5 CHAPTER REVIEW

4-6 PROBLEMS

4-7 DISCUSSION QUESTIONS

4-8 ADDITIONAL READING

4-1 INTRODUCTION

The uses we make of water in lakes, rivers, ponds, and streams is greatly influenced by the quality of the water found in them. Activities such as fishing, swimming, boating, shipping, and waste disposal have very different requirements for water quality. Water of a particularly high quality is needed for potable water supplies. In many parts of the world, the introduction of pollutants from human activity has seriously degraded water quality even to the extent of turning pristine trout streams into foul open sewers with few life forms and fewer beneficial uses.

Water quality management is concerned with the control of pollution from human activity so that the water is not degraded to the point that it is no longer suitable for intended uses. The lone frontier family, settled on the banks of the Ohio River, did not significantly degrade water quality in that mighty river even though it threw all its wastes into the river. The city of Cincinnati, however, could not discharge its untreated wastes into the Ohio River without disastrous consequences. Thus, water quality management is also the science of knowing how much is too much for a particular water body.

To know how much waste can be tolerated (the technical term is *assimilated*) by a water body, you must know the type of pollutants discharged and the manner in which they affect water quality. You must also know how water quality is affected by natural factors such as the mineral heritage of the watershed, the geometry of the terrain, and the climate of the region. A small, tumbling mountain brook will have a very different assimilative capacity than a sluggish, meandering lowland river, and lakes are different from moving waters.

Originally, the intent of water quality management was to protect the intended uses of a water body while using water as an economic means of waste disposal within the constraints of its assimilative capacity. In 1972, the Congress of the United States established that it was in the national interest to "restore and maintain the chemical, physical, and biological integrity of the nation's waters." In addition to making the water safe to drink, the Congress also established a goal of "water quality which provides for the protection and propagation of fish, shellfish, and wildlife, and provides for recreation in and on the water." By understanding the impact of pollutants on water quality, the environmental engineer can properly design the treatment facilities to remove these pollutants to acceptable levels.

This chapter deals first with the major types of pollutants and their sources. In the remainder of the chapter, water quality management in rivers and in lakes is discussed, placing the emphasis on the categories of pollutants found in domestic wastewaters. For both rivers and lakes, the natural factors affecting water quality will be discussed as the basis for understanding the impact of human activities on water quality.

This chapter was written by John A. Eastman, Ph.D., of Lockwood, Jones and Beals, Inc., Kettering, OH.

4-2 WATER POLLUTANTS AND THEIR SOURCES

The wide range of pollutants discharged to surface waters can be grouped into broad classes, as shown in Table 4-1. Domestic sewage and industrial wastes are called *point sources* because they are generally collected by a network of pipes or channels and conveyed to a single point of discharge into the receiving water. Domestic sewage consists of wastes from homes, schools, office buildings, and stores. The term municipal sewage is used to mean domestic sewage into which industrial wastes are also discharged. In general, point source pollution can be reduced or eliminated through waste minimization and proper wastewater treatment prior to discharge to a natural water body.

Non-point sources. Urban and agricultural runoff are characterized by multiple discharge points. These are called non-point sources. Often the polluted water flows over the surface of the land or along natural drainage channels to the nearest water body. Even when urban or agricultural runoff waters are collected in pipes or channels, they are generally transported the shortest possible distance for discharge, so that wastewater treatment at each outlet is not economically feasible. Much of the non-point source pollution occurs during rainstorms or spring snowmelt resulting in large flow rates that make treatment even more difficult. Reduction of agricultural non-point source pollution generally requires changes in land use practices and improved education. Non-point pollution from urban storm water and, in particular, storm water collected in *combined sewers* that carry both storm water and municipal sewage may require major engineering work to correct. The original design of combined sewers provided a flow structure that diverted excess storm water mixed with raw sewage (above the design capacity of the wastewater treatment plant) directly to the nearest river or stream. The elimination of *combined sewer overflow* (CSO) may involve not only provision of separate storm and sanitary sewers but also the provision of storm water retention basins and expanded treatment facilities to treat the storm water. This is particularly complex and expensive because the combined

TABLE 4-1
Major pollutant categories and principal sources of pollutants

	Point sources		Non-point sources	
Pollutant category	**Domestic sewage**	**Industrial wastes**	**Agricultural runoff**	**Urban runoff**
Oxygen-demanding material	X	X	X	X
Nutrients	X	X	X	X
Pathogens	X	X	X	X
Suspended solids/sediments	X	X	X	X
Salts		X	X	X
Toxic metals		X		X
Toxic organic chemicals		X	X	
Heat		X		

sewers frequently occur in the oldest, most developed portions of the community. Thus, paved streets, utilities, and commercial activities will be disrupted. The installation of combined sewers is now prohibited in the United States.

Oxygen-demanding material. Anything that can be oxidized in the receiving water with the consumption of dissolved molecular oxygen is termed oxygen-demanding material. This material is usually biodegradable organic matter but also includes certain inorganic compounds. The consumption of *dissolved oxygen,* DO (pronounced "dee oh"), poses a threat to fish and other higher forms of aquatic life that must have oxygen to live. The critical level of DO varies greatly among species. For example, brook trout may require about 7.5 mg/L of DO, while carp may survive at 3 mg/L. As a rule, the most desirable commercial and game fish require high levels of dissolved oxygen. Oxygen-demanding materials in domestic sewage come primarily from human waste and food residue. Particularly noteworthy among the many industries which produce oxygen-demanding wastes are the food processing and paper industries. Almost any naturally occurring organic matter, such as animal droppings, crop residues, or leaves, which get into the water from non-point sources, contribute to the depletion of DO.

Nutrients. Nitrogen and phosphorus, two nutrients of primary concern, are considered pollutants because they are too much of a good thing. All living things require these nutrients for growth. Thus, they must be present in rivers and lakes to support the natural food chain.[1] Problems arise when nutrient levels become excessive and the food web is grossly disturbed, which causes some organisms to proliferate at the expense of others. As will be discussed in a later section, excessive nutrients often lead to large growths of algae, which in turn become oxygen-demanding material when they die and settle to the bottom. Some major sources of nutrients are phosphorus-based detergents, fertilizers, and food-processing wastes.

Pathogenic organisms. Microorganisms found in wastewater include bacteria, viruses, and protozoa excreted by diseased persons or animals. When discharged into surface waters, they make the water unfit for drinking (that is, nonpotable). If the concentration of pathogens is sufficiently high, the water may also be unsafe for swimming and fishing. Certain shellfish can be toxic because they concentrate pathogenic organisms in their tissues, making the toxicity levels in the shellfish much greater than the levels in the surrounding water.

Suspended solids. Organic and inorganic particles that are carried by the wastewater into a receiving water are termed *suspended solids* (SS). When the speed of the water is reduced by flowing into a pool or a lake, many of these particles settle to the bottom as sediment. In common usage, the word sediment also includes eroded

---

[1]In simplistic terms, a food chain is the collection of interrelated organisms in which the lower levels are the "eatees" and the upper levels are the "eaters."

soil particles which are being carried by water even if they have not yet settled. Colloidal particles, which do not settle readily, cause the turbidity found in many surface waters. Organic suspended solids may also exert an oxygen demand. Inorganic suspended solids are discharged by some industries but result mostly from soil erosion, which is particularly bad in areas of logging, strip mining, and construction activity. As excessive sediment loads are deposited into lakes and reservoirs, the usefulness of the water is reduced. Even in rapidly moving mountain streams, sediment from mining and logging operations has destroyed many living places (*ecological habitats*) for aquatic organisms. For example, salmon eggs can only develop and hatch in stream beds of loose gravel. As the pores between the pebbles are filled with sediment, the eggs suffocate and the salmon population is reduced.

Salts. Although most people associate salty water with oceans and salt lakes, all water contains some salt. These salts are often measured by evaporation of a filtered water sample. The salts and other things that don't evaporate are called *total dissolved solids* (TDS). A problem arises when the salt concentration in normally fresh water increases to the point where the natural population of plants and animals is threatened or the water is no longer useful for public water supplies or irrigation. High concentrations of salts are discharged by many industries, and the use of salt on roads during the winter causes high salt levels in urban runoff, especially during the spring snowmelt. Of particular concern in arid regions, where water is used extensively for irrigation, is that the water picks up salts every time it passes through the soil on its way back to the river. In addition, evapotranspiration causes the salts to be further concentrated. Thus, the salt concentration continuously increases as the water moves downstream. If the concentration gets too high, crop damage or soil poisoning can result.

Toxic metals and toxic organic compounds. Agricultural runoff often contains pesticides and herbicides that have been used on crops. Urban runoff is a major source of zinc in many water bodies. The zinc comes from tire wear. Many industrial wastewaters contain either toxic metals or toxic organic substances. If discharged in large quantities, many of these materials can render a body of water nearly useless for long periods of time. The lower James River in Virginia has been reduced to use only as a shipping channel because of a large industrial discharge of highly toxic and persistent organic compounds. Many toxic compounds are concentrated in the food chain, making fish and shellfish unsafe for human consumption. Thus, even small quantities of toxic compounds in the water can be incompatible with the natural ecosystem and many human uses.

Heat. Although heat is not often recognized as a pollutant, those in the electric power industry are well aware of the problems of disposing of waste heat. Also, waters released by many industrial processes are much hotter than the receiving waters. In some environments an increase of water temperature can be beneficial. For example, production of clams and oysters can be increased in some areas by warming the water. On the other hand, increases in water temperature can have negative impacts. Many important commercial and game fish, such as salmon and trout, live only in cool water. In some instances the discharge of heated water from a power plant can completely block salmon migration. Higher temperatures also increase the rate of oxygen depletion in areas where oxygen-demanding wastes are present.

4-3 WATER QUALITY MANAGEMENT IN RIVERS

The objective of water quality management is simple to state: to control the discharge of pollutants so that water quality is not degraded to an unacceptable extent below the natural background level. However, controlling waste discharges must be a quantitative endeavor. We must be able to measure the pollutants, predict the impact of the pollutant on water quality, determine the background water quality which would be present without human intervention, and decide the levels acceptable for intended uses of the water.

To most people, the tumbling mountain brook, crystal clear and icy cold, fed by the melting snow, and safe to drink is the epitome of high water quality. Certainly a stream in that condition is a treasure, but we cannot expect the Mississippi River to have the same water quality. It never did and never will. Yet both need proper management if the water is to remain usable for intended purposes. The mountain brook may serve as the spawning ground for desirable fish and must be protected from heat and sediment as well as chemical pollution. The Mississippi, however, is already warmed from hundreds of kilometers of exposure to the sun and carries the sediment from thousands of square kilometers of land. But even the Mississippi can be damaged by organic matter and toxic chemicals. Fish do live there and the river is used as a water supply for millions of people.

The impact of pollution on a river depends both on the nature of the pollutant and the unique characteristics of the individual river.[2] Some of the most important characteristics include the volume and speed of water flowing in the river, the river's depth, the type of bottom, and the surrounding vegetation. Other factors include the climate of the region, the mineral heritage of the watershed, land use patterns, and the types of aquatic life in the river. Water quality management for a particular river must consider all these factors. Thus, some rivers are highly susceptible to pollutants such as sediment, salt, and heat, while other rivers can tolerate large inputs of these pollutants without much damage.

Some pollutants, particularly oxygen-demanding wastes and nutrients, are so common and have such a profound impact on almost all types of rivers that they deserve special emphasis. This is not to say that they are always the most significant pollutants in any one river, but rather that no other pollutant category has as much overall effect on our nation's rivers. For these reasons, the next sections of this chapter will be devoted to a more detailed look at how oxygen-demanding material and nutrients affect water quality in rivers.

---

[2]Here we will use the word "river" to include streams, brooks, creeks, and any other channel of flowing, fresh water.

Effect of Oxygen-Demanding Wastes on Rivers

The introduction of oxygen-demanding material, either organic or inorganic, into a river causes depletion of the dissolved oxygen in the water. This poses a threat to fish and other higher forms of aquatic life if the concentration of oxygen falls below a critical point. To predict the extent of oxygen depletion, it is necessary to know how much waste is being discharged and how much oxygen will be required to degrade the waste. However, because oxygen is continuously being replenished from the atmosphere and from photosynthesis by algae and aquatic plants, as well as being consumed by organisms, the concentration of oxygen in the river is determined by the relative rates of these competing processes. Organic oxygen-demanding materials are commonly measured by determining the amount of oxygen consumed during degradation in a manner approximating degradation in natural waters. This section begins by considering the factors affecting oxygen consumption during the degradation of organic matter, then moves on to inorganic nitrogen oxidation. Finally, the equations for predicting dissolved oxygen concentrations in rivers from degradation of organic matter are developed and discussed.

Biochemical Oxygen Demand

The amount of oxygen required to oxidize a substance to carbon dioxide and water may be calculated by stoichiometry if the chemical composition of the substance is known. This amount of oxygen is known as the *theoretical oxygen demand* (ThOD).

Example 4-1. Compute the ThOD of 108.75 mg/L of glucose ($C_6H_{12}O_6$).

Solution. We begin by writing a balanced equation for the reaction.

$$C_6H_{12}O_6 + 6O_2 \rightleftharpoons 6CO_2 + 6H_2O$$

Next, compute the gram molecular weights of the reactants using the table on the inside cover of the book.

glucose	*oxygen*
6C = 72	(6)(2)O = 192
12H = 12	
6O = 96	
180	

Thus, it takes 192 g of oxygen to oxidize 180 g of glucose to CO_2 and H_2O.

The ThOD of 108.75 mg/L of glucose is

$$(108.75 \text{ mg/L of glucose})\left(\frac{192 \text{ g } O_2}{180 \text{ g glucose}}\right) = 116 \text{ mg/L } O_2$$

In contrast to the ThOD, the *chemical oxygen demand,* COD (pronounced "see oh dee"), is a measured quantity that does not depend on knowledge of the chemical composition of the substances in the water. In the COD test, a strong chemical oxidizing agent (chromic acid) is mixed with a water sample and then boiled. The difference between the amount of oxidizing agent at the beginning of the test and that remaining at the end of the test is used to calculate the COD.

If the oxidation of an organic compound is carried out by microorganisms using the organic matter as a food source, the oxygen consumed is known as *biochemical oxygen demand,* or BOD (pronounced "bee oh dee"). The actual BOD is less than the ThOD due to the incorporation of some of the carbon into new bacterial cells. The test is a bioassay that utilizes microorganisms in conditions similar to those in natural water to measure indirectly the amount of biodegradable organic matter present. Bioassay means to measure by biological means. A water sample is inoculated with bacteria that consume the biodegradable organic matter to obtain energy for their life processes. Because the organisms also utilize oxygen in the process of consuming the waste, the process is called *aerobic* decomposition. This oxygen consumption is easily measured. The greater the amount of organic matter present, the greater the amount of oxygen utilized. The BOD test is an indirect measurement of organic matter because we actually measure only the change in dissolved oxygen concentration caused by the microorganisms as they degrade the organic matter. Although not all organic matter is biodegradable and the actual test procedures lack precision, the BOD test is still the most widely used method of measuring organic matter because of the direct conceptual relationship between BOD and oxygen depletion in receiving waters.

Only under rare circumstances will the ThOD, COD, and BOD be equal. If the chemical composition of all of the substances in the water is known and they are capable of being completely oxidized both chemically and biologically, then the three measures of oxygen demand will be the same.

When a water sample containing degradable organic matter is placed in a closed container and inoculated with bacteria, the oxygen consumption typically follows the pattern shown in Figure 4-1. During the first few days the rate of oxygen

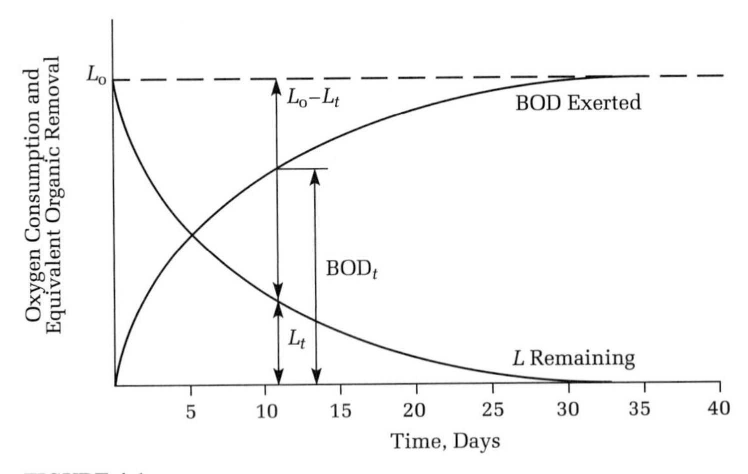

FIGURE 4-1
BOD and oxygen-equivalent relationships.

depletion is rapid because of the high concentration of organic matter present. As the concentration of organic matter decreases, so does the rate of oxygen consumption. During the last part of the BOD curve, oxygen consumption is mostly associated with the decay of the bacteria that grew during the early part of the test. It is generally assumed that the rate at which oxygen is consumed is directly proportional to the concentration of degradable organic matter remaining at any time. As a result, the BOD curve in Figure 4-1 can be described mathematically as a first-order reaction. Using our definition of reaction rate and reaction order from Chapter 3, this may be expressed as:

$$\frac{dL_t}{dt} = -r_A \tag{4-1}$$

where L_t = oxygen equivalent of the organics remaining at time t, mg/L
$-r_A = -kL_t$
k = reaction rate constant, d^{-1}

Rearranging Equation 4-1 and integrating yields:

$$\frac{dL_t}{L_t} = -k\,dt$$

$$\int_{L_o}^{L} \frac{dL_t}{L_t} = -k\int_0^t dt$$

$$\ln\frac{L_t}{L_o} = -kt$$

or

$$L_t = L_o\, e^{-kt} \tag{4-2}$$

where

$$L_o = \text{oxygen equivalent of organics at time} = 0$$

Rather than L_t, our interest is in the amount of oxygen used in the consumption of the organics (BOD_t). From Figure 4-1, it is obvious that BOD_t is the difference between the initial value of L_o and L_t, so

$$\begin{aligned} BOD_t &= L_o - L_t \\ &= L_o - L_o\, e^{-kt} \\ &= L_o(1 - e^{-kt}) \end{aligned} \tag{4-3}$$

L_o is often referred to as the *ultimate BOD,* that is, the maximum oxygen consumption possible when the waste has been completely degraded. Equation 4-3 is called the BOD rate equation and is often written in base 10:

$$BOD_t = L_o(1 - 10^{-Kt}) \tag{4-4}$$

Note that lower case k is used for the reaction rate constant in base e and that capital K is used for the constant in base 10. They are related: $k = 2.303(K)$.

Example 4-2. If the BOD_3 of a waste is 75 mg/L and the K is 0.150 day^{-1}, what is the ultimate BOD?

Solution. Note that the rate constant is given in base 10 (K versus k), and substitute the given values into Equation 4-4 and solve for L_o:

$$75 = L_o(1 - 10^{-(.150)(3)}) = 0.645\ L_o$$

or

$$L_o = \frac{75}{0.645} = 116 \text{ mg/L}$$

In base e,

$$k = 2.303(K) = 0.345, \text{ and}$$

$$75 = L_o(1 - e^{-(.345)(3)}) = 0.645\ L_o$$

so

$$L_o = 116 \text{ mg/L}$$

You should note that the ultimate BOD (L_0) is defined as the maximum BOD exerted by the waste. It is denoted by the horizontal line in Figure 4-1. Since BOD_t approaches L_0 asymptotically, it is difficult to assign an exact time to achieve ultimate BOD. Indeed, based on Equation 4-1, it is achieved only in the limit as t approaches infinity. However, from a practical point of view, we can observe that when the BOD curve is approximately horizontal, the ultimate BOD has been achieved. In Figure 4-1, we would take this to be at about 35 days. In computations, we use a rule of thumb that if BOD_t and L_o agree when rounded to three significant figures, then the time to reach ultimate BOD has been achieved. Given the vagaries of the BOD test, there are occasions when rounding to two significant figures would not be unrealistic.

While the ultimate BOD best expresses the concentration of degradable organic matter, it does not, by itself, indicate how rapidly oxygen will be depleted in a receiving water. Oxygen depletion is related to both the ultimate BOD and the BOD rate constant (k). While the ultimate BOD increases in direct proportion to the concentration of degradable organic matter, the numerical value of the rate constant is dependent on the following:

1. The nature of the waste
2. The ability of the organisms in the system to utilize the waste
3. The temperature

Nature of the waste. There are literally thousands of naturally occurring organic compounds, not all of which can be degraded with equal ease. Simple sugars and starches are rapidly degraded and will therefore have a very large BOD rate constant. Cellulose (for example, toilet paper) degrades much more slowly, and hair

TABLE 4-2
Typical values for the BOD rate constant

Sample	K (20°C) (day^{-1})	k (20°C) (day^{-1})
Raw sewage	0.15–0.30	0.35–0.70
Well-treated sewage	0.05–0.10	0.12–0.23
Polluted river water	0.05–0.10	0.12–0.23

and fingernails are almost undegradable in the BOD test or in normal wastewater treatment. Other compounds are intermediate between these extremes. The BOD rate constant for a complex waste depends very much on the relative proportions of the various components. A summary of typical BOD rate constants is shown in Table 4-2. The lower rate constants for treated sewage compared to raw sewage result from the fact that easily degradable organics are more completely removed than less readily degradable organics during wastewater treatment.

Ability of organisms to utilize waste. Any given microorganism is limited in its ability to utilize organic compounds. As a consequence, many organic compounds can be degraded by only a small group of microorganisms. In a natural environment receiving a continuous discharge of organic waste, that population of organisms which can most efficiently utilize this waste will predominate. However, the culture used to inoculate the BOD test may contain only a very small number of organisms that can degrade the particular organic compounds in the waste. This problem is especially common when analyzing industrial wastes. The result is that the BOD rate constant would be lower in the laboratory test than in the natural water. This is an undesirable outcome. The BOD test should therefore be conducted with organisms which have been acclimated to the waste so that the rate constant determined in the laboratory can be compared to that in the river.[3]

Temperature. Most biological processes speed up as the temperature increases and slow down as the temperature drops. Because oxygen utilization is caused by the metabolism of microorganisms, the rate of utilization is similarly affected by temperature. Ideally, the BOD rate constant should be experimentally determined for the temperature of the receiving water. There are two difficulties with this ideal. Often the receiving-water temperature changes throughout the year, so a large number of tests would be required to define k. An additional difficulty is the task of comparing data from various locations having different temperatures. Laboratory testing is therefore done at a standard temperature of 20°C, and the BOD rate constant is adjusted to the receiving-water temperature using the following expression:

$$k_T = k_{20}(\theta)^{T-20} \tag{4-5}$$

[3]The word "acclimated" means that the organisms have had time to adapt their metabolisms to the waste or that organisms that can utilize the waste have been given the chance to predominate in the culture.

where T = temperature of interest, °C
k_T = BOD rate constant at the temperature of interest, day^{-1}
k_{20} = BOD rate constant determined at 20°C, day^{-1}
θ = temperature coefficient.[4] This has a value of 1.135 for temperatures between 4 and 20°C and 1.056 for temperatures between 20 and 30°C.

Example 4-3. A waste is being discharged into a river that has a temperature of 10°C. What fraction of the maximum oxygen consumption has occurred in four days if the BOD rate constant determined in the laboratory under standard conditions is 0.115 day^{-1} (base *e*)?

Solution. Determine the BOD rate constant for the waste at the river temperature using Equation 4-5:

$$k_{10°\text{C}} = 0.115(1.135)^{10-20}$$
$$= 0.032 \text{ day}^{-1}$$

Use this value of k in Equation 4-3 to find the fraction of maximum oxygen consumption occurring in four days:

$$\frac{\text{BOD}_4}{L_o} = [1 - e^{-(.032)(4)}]$$
$$= 0.12$$

Graphical Determination of BOD Constants

A variety of methods may be used to determine k and L_o from an experimental set of BOD data. The simplest and least accurate method is to plot BOD versus time. This results in a hyperbolic first-order curve of the form shown in Figure 4-1. The ultimate BOD is estimated from the asymptote of the curve. The rate equation is used to solve for k. It is often difficult to fit an accurate hyperbola to data that are frequently scattered. Methods that linearize the data are preferred. The usual graphical methods for first-order reactions cannot be used because the semilog plot requires knowledge of the initial concentration which, in this case, is one of the constants we are trying to determine, that is, L_o! One simple method around this impasse is called Thomas' Graphical Method.[5] The method relies on the similarity of the series expansion of the following two functions:

$$F_1 = 1 - e^{-kt} \tag{4-6}$$

[4] G. J. Schroepfer, M. L. Robins, and R. H. Susag, "Research Program on the Mississippi River in the Vicinity of Minneapolis and St. Paul," *Advances in Water Pollution Research,* vol. 1, Part 1, p. 145, 1964.

[5] H. A. Thomas, "Graphical Determination of B.O.D. Curve Constants," *Water and Sewage Works,* pp. 123–124, 1950.

and

$$F_2 = (kt)(1 + (1/6)\,kt)^{-3} \tag{4-7}$$

The series expansion of these functions yields:

$$F_1 = (kt)\left[1 - 0.5(kt) + \frac{1}{6}(kt)^2 - \frac{1}{24}(kt)^3 + \ldots\right] \tag{4-8}$$

$$F_2 = (kt)\left[1 - 0.5(kt) + \frac{1}{6}(kt)^2 - \frac{1}{21.9}(kt)^3 + \ldots\right] \tag{4-9}$$

The first two terms are identical and the third differs only slightly. Replacing Equation 4-8 by 4-9 in the BOD rate equation results in the following approximate equation:

$$\text{BOD}_t = L_o(kt)[1 + (1/6)kt]^{-3} \tag{4-10}$$

By rearranging terms and taking the cube root of both sides, Equation 4-10 can be transformed to:

$$\left(\frac{t}{\text{BOD}_t}\right)^{1/3} = \frac{1}{(kL_o)^{1/3}} + \frac{(k)^{2/3}}{6(L_o)^{1/3}}(t) \tag{4-11}$$

A plot of $(t/\text{BOD}_t)^{1/3}$ versus t is linear (Figure 4-2). The intercept is defined as:

$$A = (kL_o)^{-1/3} \tag{4-12}$$

The slope is defined by:

$$B = \frac{(k)^{2/3}}{6(L_o)^{1/3}} \tag{4-13}$$

Solving for $L_o^{1/3}$ in Equation 4-12, substituting into Equation 4-13, and solving for k yields:

$$k = 6(B/A) \tag{4-14}$$

Likewise, substituting Equation 4-14 into 4-12 and solving for L yields:

$$L_o = \frac{1}{6(A)^2(B)} \tag{4-15}$$

The procedure for determining the BOD constants by this method is as follows:

1. From the experimental results of BOD for various values of t, calculate $(t/\text{BOD}_t)^{1/3}$ for each day.
2. Plot $(t/\text{BOD}_t)^{1/3}$ versus t on arithmetic graph paper and draw the line of best fit by eye.
3. Determine the intercept (A) and slope (B) from the plot.
4. Calculate k and L_o from Equations 4-14 and 4-15.

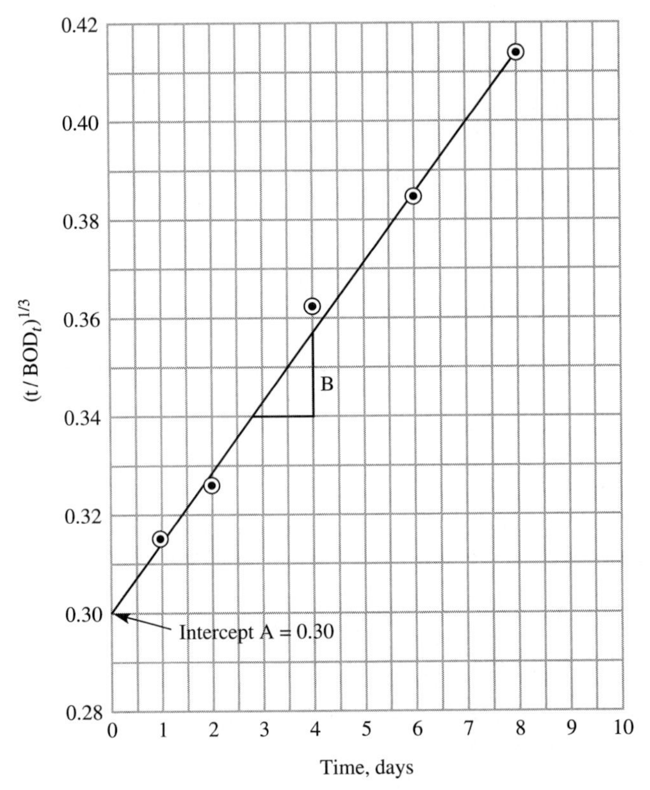

FIGURE 4-2
Plot of $(t/BOD_t)^{1/3}$ versus t for Thomas' Graphical Method.

Example 4-4. The following data were obtained from an experiment to determine the BOD rate constant and ultimate BOD for an untreated wastewater:

Day	0	1	2	4	6	8
BOD, mg/L	0	32	57	84	106	111

Solution. First calculate values of $(t/BOD_t)^{1/3}$

Day	0	1	2	4	6	8
$(t/BOD_t)^{1/3}$	—	0.315	0.327	0.362	0.384	0.416

The plot of $(t/BOD_t)^{1/3}$ versus time for these data is shown in Figure 4-2. From the figure, $A = 0.30$ and

$$B = \frac{(0.416 - 0.300)}{(8 - 0)} = 0.0145$$

Substituting in Equations 4-14 and 4-15:

$$k = \frac{6(0.0145)}{0.30} = 0.29\ \mathrm{d}^{-1}$$

$$L_o = \frac{1}{6(0.30)^2(0.0145)} = 128\ \mathrm{mg/L}$$

Laboratory Measurement of Biochemical Oxygen Demand

In order to have as much consistency as possible, it is important to standardize testing procedures when measuring BOD. In the paragraphs that follow, the standard BOD test is outlined with emphasis placed on the reason for each step rather than the details. The detailed procedures can be found in *Standard Methods for the Examination of Water and Wastewater*,[6] which is the authoritative reference of testing procedures in the water pollution control field.

Step 1. A special 300 mL BOD bottle (Figure 4-3) is completely filled with a sample of water that has been appropriately diluted and inoculated with microorganisms.

FIGURE 4-3
BOD bottles. The bottle on the left is shown with the cap removed to illustrate the shape of the glass stoppers. The point on the end of the stopper is to ensure that no air is trapped in the bottle. The bottle in the center is shown with the stopper in place. Water is placed in the small cup formed by the lip. This acts as a seal to further exclude air. The bottle on the right is shown with plastic wrap over the stopper. This is to prevent evaporation of the water seal.

[6]American Public Health Association, *Standard Methods for the Examination of Water and Wastewater,* 19th ed., Washington, DC, 1995.

The bottle is then stoppered to exclude air bubbles. Samples require dilution because the only oxygen available to the organisms is dissolved in the water. The most oxygen that can dissolve is about 9 mg/L, so the BOD of the diluted sample should be between 2 and 6 mg/L. Samples are diluted with a special dilution water that contains all of the trace elements required for bacterial metabolism so that degradation of the organic matter is not limited by lack of bacterial growth. The dilution water also contains an inoculum of microorganisms so that all samples tested on a given day contain approximately the same type and number of microorganisms.

The ratio of undiluted to diluted sample is called the sample size, usually expressed as a percentage, while the inverse relationship is called the dilution factor. Mathematically, these are

$$\text{Sample size (\%)} = \frac{\text{vol. of undiluted sample}}{\text{vol. of diluted sample}} \times 100 \tag{4-16}$$

$$\text{Dilution factor} = \frac{\text{vol. of diluted sample}}{\text{vol. of undiluted sample}} = \frac{100}{\text{sample size (\%)}} \tag{4-17}$$

The appropriate sample size to use can be determined by dividing 4 mg/L (the midpoint of the desired range of diluted BOD) by the estimated BOD concentration in the sample being tested. A convenient volume of undiluted sample is then chosen to approximate to this sample size.

Example 4-5. The BOD of a wastewater sample is estimated to be 180 mg/L. What volume of undiluted sample should be added to a 300 mL bottle? Also, what are the sample size and dilution factor using this volume? Assume that 4 mg/L BOD can be consumed in the BOD bottle.

Solution. Estimate the sample size needed:

$$\text{Sample size} = \frac{4}{180} \times 100 = 2.22\%$$

Estimate the volume of undiluted sample needed since the volume of diluted sample is 300 mL:

$$\text{Vol. of undiluted sample} = 0.0222 \times 300 \text{ mL} = 6.66 \text{ mL}$$

Therefore a convenient sample volume would be 7.00 mL.

Compute the actual sample size and dilution factor:

$$\text{Sample size} = \frac{7.00 \text{ mL}}{300 \text{ mL}} \times 100 = 2.33\%$$

$$\text{Dilution factor} = \frac{300 \text{ mL}}{7.00 \text{ mL}} = 42.9$$

Step 2. Blank samples containing only the inoculated dilution water are also placed in BOD bottles and stoppered. Blanks are required to estimate the amount of oxygen consumed by the added inoculum in the absence of the sample.

Step 3. The stoppered BOD bottles containing diluted samples and blanks are incubated in the dark at 20°C for the desired number of days. For most purposes, a standard time of five days is used. To determine the ultimate BOD and the BOD rate constant, additional times are used. The samples are incubated in the dark to prevent photosynthesis from adding oxygen to the water and invalidating the oxygen consumption results. As mentioned earlier, the BOD test is conducted at a standard temperature of 20°C so that the effect of temperature on the BOD rate constant is eliminated and results from different laboratories can be compared.

Step 4. After the desired number of days has elapsed, the samples and blanks are removed from the incubator and the dissolved oxygen concentration in each bottle is measured. The BOD of the undiluted sample is then calculated using the following equation:

$$\text{BOD}_t = (\text{DO}_{b,t} - \text{DO}_{s,t}) \times \text{dilution factor} \tag{4-18}$$

where $\text{DO}_{b,t}$ = dissolved oxygen concentration in blank after t days of incubation, mg/L

$\text{DO}_{s,t}$ = dissolved oxygen concentration in sample after t days of incubation, mg/L

Example 4-6. What is the BOD_5 of the wastewater sample of Example 4-5 if the DO values for the blank and diluted sample after five days are 8.7 and 4.2 mg/L, respectively?

Solution. Substitute the appropriate values into Equation 4-18:

$$\text{BOD}_5 = (8.7 - 4.2) \times 42.9 = 193 \text{ or } 190 \text{ mg/L}$$

Additional Notes on Biochemical Oxygen Demand

Although the five-day BOD has been chosen as the standard value for most wastewater analysis and for regulatory purposes, ultimate BOD is actually a better indicator of total waste strength. For any one type of waste having a defined BOD rate constant, the ratio between ultimate BOD and BOD_5 is constant so that BOD_5 indicates relative waste strength. For different types of wastes having the same BOD_5, the ultimate BOD is the same only if, by chance, the BOD rate constants are the same. This is illustrated in Figure 4-4 for a municipal wastewater having a $K = 0.15 \text{ day}^{-1}$ and an industrial wastewater having a $K = 0.05 \text{ day}^{-1}$. Although both wastewaters have a BOD_5 of 200 mg/L, the industrial wastewater has a much higher ultimate BOD and can be expected to have a greater impact on dissolved oxygen in a river. For the industrial wastewater, a smaller fraction of the BOD was exerted in the first five days due to the lower rate constant.

Proper interpretation of BOD_5 values can also be illustrated in another way. Consider a sample of polluted river water for which the following values were determined using standard laboratory techniques: $\text{BOD}_5 = 50$ mg/L, and $K = 0.115 \text{ day}^{-1}$. The ultimate BOD calculated from Equation 4-4 is, therefore, 68 mg/L.

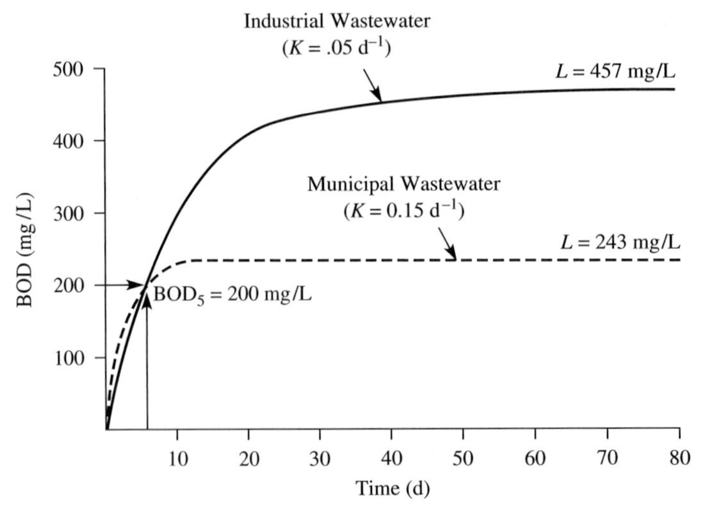

FIGURE 4-4
The effect of K on ultimate BOD for two wastewaters having the same BOD_5.

However, because the river temperature is 10°C, the K value in the river is only 0.032 day^{-1} (see Example 4-3). As shown graphically in Figure 4-5, the laboratory value of BOD_5 seriously overestimates the actual oxygen consumption in the river. Again, a smaller fraction of the BOD is exerted in five days when the BOD rate constant is lower.

The five-day BOD was chosen as the standard value for most purposes because the test was devised by sanitary engineers in England, where rivers have travel times to the sea of less than five days, so there was no need to consider oxygen demand at longer times. Since there is no other time which is any more rational than five days, this value has become firmly entrenched.

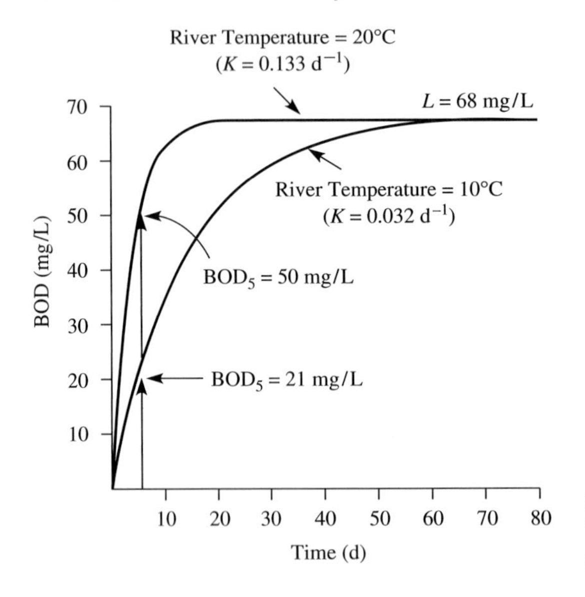

FIGURE 4-5
The effect of K on BOD_5, when the ultimate BOD is constant.

Nitrogen Oxidation

Up to this point an unstated assumption has been that only the carbon in organic matter is oxidized. Actually many organic compounds, such as proteins, also contain nitrogen that can be oxidized with the consumption of molecular oxygen. However, because the mechanisms and rates of nitrogen oxidation are distinctly different from those of carbon oxidation, the two processes must be considered separately. Logically, oxygen consumption due to oxidation of carbon is called *carbonaceous BOD* (CBOD), while that due to nitrogen oxidation is called *nitrogenous BOD* (NBOD).

The organisms that oxidize the carbon in organic compounds to obtain energy cannot oxidize the nitrogen in these compounds. Instead, the nitrogen is released into the surrounding water as ammonia (NH_3). At normal pH values, this ammonia is actually in the form of the ammonium cation (NH_4^+). The ammonia released from organic compounds, plus that from other sources such as industrial wastes and agricultural runoff (that is, fertilizers), is oxidized to nitrate (NO_3^-) by a special group of nitrifying bacteria as their source of energy in a process called *nitrification.* The overall reaction for ammonia oxidation is

$$NH_4^+ + 2O_2 \xrightarrow{\text{microorganisms}} NO_3^- + H_2O + 2H^+ \qquad (4\text{-}19)$$

From this reaction the theoretical NBOD can be calculated as follows:

$$\text{NBOD} = \frac{\text{grams of oxygen used}}{\text{grams of nitrogen oxidized}} = \frac{4 \times 16}{14} = 4.57 \text{ g } O_2/\text{g N}$$

The actual nitrogenous BOD is slightly less than the theoretical value due to the incorporation of some of the nitrogen into new bacterial cells, but the difference is only a few percent.

Example 4-7. Compute the theoretical NBOD of a wastewater containing 30 mg/L of ammonia as nitrogen. (We often say "ammonia nitrogen" and write the expression as NH_3-N.) If the wastewater analysis was reported as 30 mg/L of ammonia (NH_3), what would the theoretical NBOD be?

Solution. In the first part of the problem, the amount of ammonia was reported as NH_3-N. Therefore, we can use the theoretical relationship developed from Equation 4-19.

$$\text{Theo. NBOD} = (30 \text{ mg N/L})(4.57 \text{ mg } O_2/\text{mg N}) = 137 \text{ mg } O_2/\text{L}$$

To answer the second question, we must convert mg/L of ammonia to NH_3-N by multiplying by the ratio of gram molecular weights of N to NH_3.

$$(30 \text{ mg } NH_3/\text{L})\left(\frac{14 \text{ g N}}{17 \text{ g } NH_3}\right) = 24.7 \text{ mg N/L}$$

Now we may use the relationship developed from Equation 4-19.

$$\text{Theo. NBOD} = (24.7 \text{ mg N/L})(4.57 \text{ mg } O_2/\text{mg N}) = 113 \text{ mg } O_2/\text{L}$$

The rate at which the NBOD is exerted depends heavily on the number of nitrifying organisms present. In untreated sewage, there are few of these organisms, while in a well-treated effluent, the concentration is high. When samples of untreated and treated sewage are subjected to the BOD test, oxygen consumption follows the pattern shown in Figure 4-6. In the case of untreated sewage, the NBOD is exerted after much of the CBOD has been exerted. The lag is due to the time it takes for the nitrifying bacteria to reach a sufficient population for the amount of NBOD exertion to be significant compared with that of the CBOD. In the case of the treated sewage, a higher population of nitrifying organisms in the sample reduces the lag time. Once nitrification begins, however, the NBOD can be described by Equation 4-1 with a BOD rate constant comparable to that for the CBOD of a well-treated effluent ($K = 0.04$ to $0.10\ d^{-1}$). Because the lag before the nitrogenous BOD is highly variable, BOD_5 values are often difficult to interpret. When measurement of only carbonaceous BOD is desired, chemical inhibitors are added to stop the nitrification process. The rate constant for nitrification is also affected by temperature and can be adjusted using Equation 4-2.

DO Sag Curve

The concentration of dissolved oxygen in a river is an indicator of the general health of the river. All rivers have some capacity for self-purification. As long as the discharge of oxygen-demanding wastes is well within the self-purification capacity, the DO level will remain high and a diverse population of plants and animals, including game fish, can be found. As the amount of waste increases, the self-purification capacity can be exceeded, causing detrimental changes in plant and animal life. The stream loses its ability to cleanse itself and the DO level decreases. When the DO drops below about 4 to 5 mg/L, most game fish will have been driven out. If the DO is completely removed, fish and other higher animals are killed or driven out and extremely noxious conditions result. The water becomes blackish and foul smelling

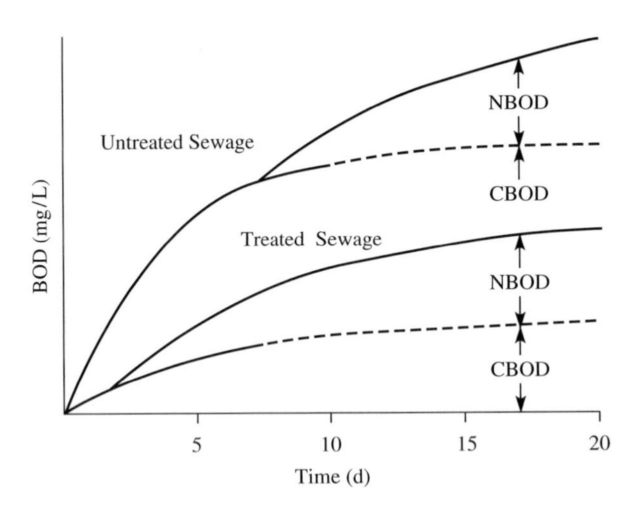

FIGURE 4-6
BOD curves showing both carbonaceous and nitrogenous BOD.

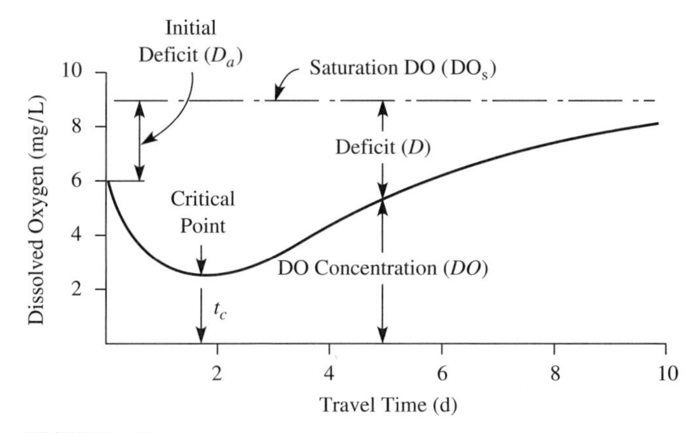

FIGURE 4-7
Typical DO sag curve.

as the sewage and dead animal life decompose under *anaerobic* conditions (that is, without oxygen).

One of the major tools of water quality management in rivers is the ability to assess the capability of a stream to absorb a waste load. This is done by determining the profile of DO concentration downstream from a waste discharge. This profile is called the DO sag curve (see Figure 4-7) because the DO concentration dips as oxygen-demanding materials are oxidized and then rises again further downstream as the oxygen is replenished from the atmosphere. As depicted in Figure 4-8, the biota of the stream are often a reflection of the dissolved oxygen conditions in the stream.

To develop a mathematical expression for the DO sag curve, the sources of oxygen and the factors affecting oxygen depletion must be identified and quantified. The only significant sources of oxygen are reaeration from the atmosphere and photosynthesis of aquatic plants. Oxygen depletion is caused by a larger range of factors, the most important being the BOD, both carbonaceous and nitrogenous, of the waste discharge, and the BOD already in the river upstream of the waste discharge. The second most important factor is that the DO in the waste discharge is usually less than that in the river. Thus, the DO at the river is lowered as soon as the waste is added, even before any BOD is exerted. Other factors affecting dissolved oxygen depletion include non-point source pollution, the respiration of organisms living in the sediments (*benthic demand*), and the respiration of aquatic plants. Following the classical approach, the DO sag equation will be developed by considering only initial DO reduction, carbonaceous BOD, and reaeration from the atmosphere. Subsequently, the equation will be expanded to include the nitrogenous BOD. Finally, the other factors affecting DO levels will be discussed qualitatively; a quantitative discussion is beyond the scope of this book.

Mass-balance approach. Simplified mass balances help us understand and solve the DO sag curve problem. Three conservative (those without chemical reaction) mass balances may be used to account for initial mixing of the waste stream and the

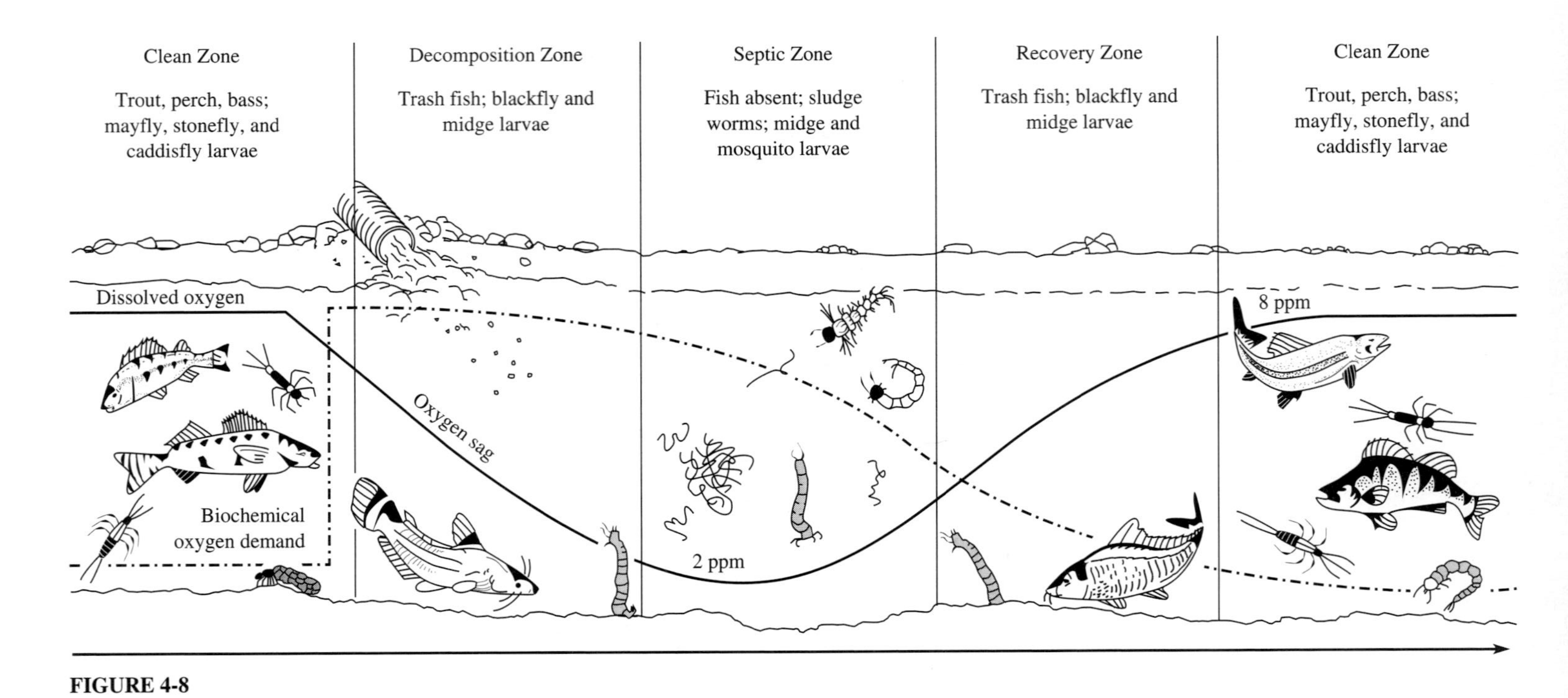

FIGURE 4-8
Oxygen sag downstream of an organic source. *Source:* U.S. EPA.

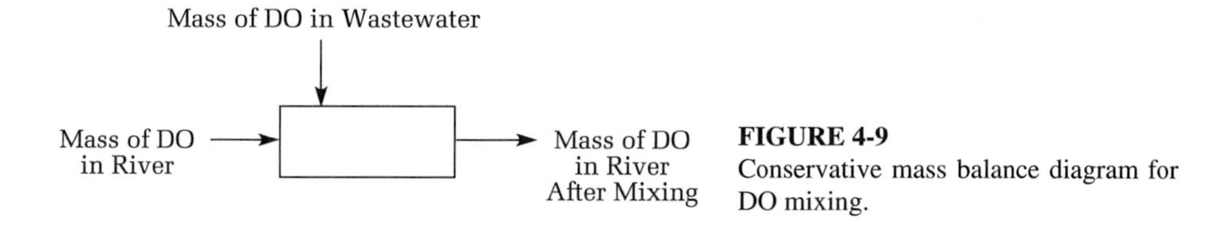

FIGURE 4-9
Conservative mass balance diagram for DO mixing.

river. DO, carbonaceous BOD, and temperature all change as the result of mixing of the waste stream and the river. Once these are accounted for, the DO sag curve may be viewed as a nonconservative mass balance, that is, one with reactions.

The conservative mass balance diagram for oxygen (mixing only) is shown in Figure 4-9. The product of the water flow and the DO concentration yields a mass of oxygen per unit of time:

$$\text{Mass of DO in wastewater} = Q_w\text{DO}_w \tag{4-20}$$

$$\text{Mass of DO in river} = Q_r\text{DO}_r \tag{4-21}$$

where Q_w = volumetric flow rate of wastewater, m^3/s
Q_r = volumetric flow rate of the river, m^3/s
DO_w = dissolved oxygen concentration in the wastewater, g/m^3
DO_r = dissolved oxygen concentration in the river, g/m^3

The mass of DO in the river after mixing equals the sum of the mass flows:

$$\text{Mass of DO after mixing} = Q_w\text{DO}_w + Q_r\text{DO}_r \tag{4-22}$$

In a similar fashion for ultimate BOD:

$$\text{Mass of BOD after mixing} = Q_wL_w + Q_rL_r \tag{4-23}$$

where L_w = ultimate BOD of the wastewater, mg/L
L_r = ultimate BOD of the river, mg/L

The concentrations of DO and BOD after mixing are the respective masses per unit time divided by the total flow rate (that is, the sum of the wastewater and river flows):

$$\text{DO} = \frac{Q_w\text{DO}_w + Q_r\text{DO}_r}{Q_w + Q_r} \tag{4-24}$$

$$\text{L}_a = \frac{Q_wL_w + Q_rL_r}{Q_w + Q_r} \tag{4-25}$$

where L_a = initial ultimate BOD after mixing.

Example 4-8. The town of State College discharges 17,360 m^3/d of treated wastewater into the Bald Eagle Creek. The treated wastewater has a BOD_5 of 12 mg/L and a k of 0.12 d^{-1} at 20°C. Bald Eagle Creek has a flow rate of 0.43 m^3/s and an ultimate BOD of 5.0 mg/L. The DO of the river is 6.5 mg/L and the DO of the wastewater is 1.0 mg/L. Compute the DO and initial ultimate BOD after mixing.

Solution. The DO after mixing is given by Equation 4-24. To use this equation we must convert the wastewater flow to compatible units, that is, m^3/s.

$$Q_w = \frac{(17{,}360 \text{ m}^3/\text{d})}{(86{,}400 \text{ s/d})} = 0.20 \text{ m}^3/\text{s}$$

The DO after mixing is then

$$\text{DO} = \frac{(0.20 \text{ m}^3/\text{s})(1.0 \text{ mg/L}) + (0.43 \text{ m}^3/\text{s})(6.5 \text{ mg/L})}{0.20 \text{ m}^3/\text{s} + 0.43 \text{ m}^3/\text{s}} = 4.75 \text{ mg/L}$$

Before we can determine the initial ultimate BOD after mixing, we must first determine the ultimate BOD of the wastewater. Solving Equation 4-3 for L_o:

$$L_o = \frac{\text{BOD}_5}{(1 - e^{-kt})} = \frac{12 \text{ mg/L}}{(1 - e^{-(0.12)(5)})} = \frac{12}{(1 - .55)} = 26.6 \text{ mg/L}$$

Note that we used the subscript of 5 days in BOD_5 to determine the value of t in the equation. Now setting $L_w = L_o$, we can determine the initial ultimate BOD after mixing using Equation 4-25:

$$L_a = \frac{(0.20 \text{ m}^3/\text{s})(26.6 \text{ mg/L}) + (0.43 \text{ m}^3/\text{s})(5.0 \text{ mg/L})}{0.20 \text{ m}^3/\text{s} + 0.43 \text{ m}^3/\text{s}} = 11.86 \text{ or } 12 \text{ mg/L}$$

For temperature, we must consider a heat balance rather than a mass balance. This is an application of a fundamental principle of physics:

$$\text{Loss of heat by hot bodies} = \text{gain of heat by cold bodies} \tag{4-26}$$

The change in *enthalpy* or "heat content" of a mass of a substance may be defined by the following equation:

$$H = mC_p\Delta T \tag{4-27}$$

where H = change in enthalpy, J
m = mass of substance, g
C_p = specific heat at constant pressure, $\text{J/g} \cdot \text{K}$
ΔT = change in temperature, K

The specific heat of water varies slightly with temperature. For natural waters, a value of 4.19 will be a satisfactory approximation. Using our fundamental heat loss = heat gain equation, we may write

$$(m_w)(4.19)\Delta T_w = (m_r)(4.19)\Delta T_r \tag{4-28}$$

The temperature after mixing is found by solving this equation for the final temperature by recognizing that ΔT on each side of the equation is the difference between the final river temperature (T_f) and the starting temperature of the wastewater and the river water, respectively:

$$T_f = \frac{Q_wT_w + Q_rT_r}{Q_w + Q_r} \tag{4-29}$$

Oxygen deficit. The DO sag equation has been developed using oxygen deficit rather than dissolved oxygen concentration, to make it easier to solve the integral equation that results from the mathematical description of the mass balance. The oxygen deficit is the amount by which the actual dissolved oxygen concentration is less than the saturation value with respect to oxygen in the air:

$$D = \mathrm{DO}_s - \mathrm{DO} \tag{4-30}$$

where D = oxygen deficit, mg/L
DO_s = saturation concentration of dissolved oxygen, mg/L
DO = actual concentration of dissolved oxygen, mg/L

The saturation value of dissolved oxygen is heavily dependent on water temperature—it decreases as the temperature increases. Values of DO_s for fresh water are given in Table A-2 of Appendix A.

Initial deficit. The beginning of the DO sag curve is at the point where a waste discharge mixes with the river. The initial deficit is calculated as the difference between saturated DO and the concentration of the DO after mixing (Equation 4-24):

$$D_a = \mathrm{DO}_s - \frac{Q_w \mathrm{DO}_w + Q_r \mathrm{DO}_r}{Q_w + Q_r} \tag{4-31}$$

where D_a = initial deficit after river and waste have mixed, mg/L
DO_s = saturation concentration of dissolved oxygen at the temperature of the river after mixing, mg/L

Values of the saturation concentration of dissolved oxygen in fresh water may be found in Appendix A.

Example 4-9. Calculate the initial deficit of the Bald Eagle Creek after mixing with the wastewater from the town of State College (see Example 4-8 for data). The stream temperature is 10°C and the wastewater temperature is 10°C.

Solution. With the stream temperature, the saturation value of dissolved oxygen (DO_s) can be determined from the table in Appendix A. At 10°C, DO_s = 11.33 mg/L. Since we calculated the concentration of DO after mixing as 4.75 mg/L in Example 4-8, the initial deficit after mixing is

$$\mathrm{D}_a = 11.33 \text{ mg/L} - 4.75 \text{ mg/L} = 6.58 \text{ mg/L}$$

Because wastewater commonly has a higher temperature than river water, especially during the winter, the river temperature downstream of the discharge is usually higher than that upstream. Since we are interested in downstream conditions, it is important to use the downstream temperature when determining the saturation concentration of dissolved oxygen.

DO sag equation. A mass balance diagram of DO in a small *reach* (stretch) of river is shown in Figure 4-10*a*. This is a comprehensive mass balance that accounts

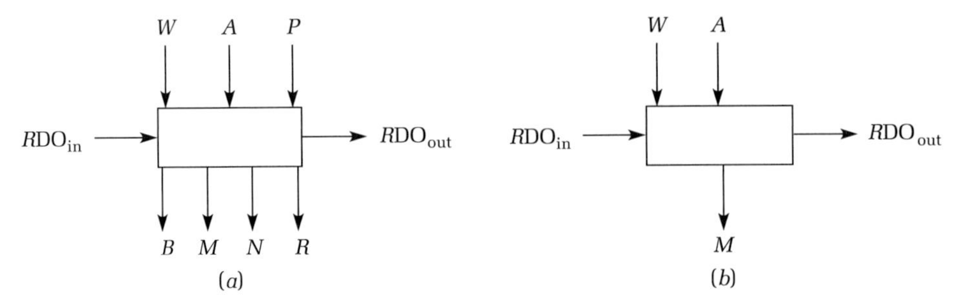

Legend

RDO_{in}, RDO_{out} = mass of DO flowing in and out of reach
W = mass of DO in wastewater flowing into reach
A = mass of DO entering from atmosphere
P = mass of DO entering from algae photosynthetic oxygen production
B = mass of DO consumed by benthic demand
M = mass of DO removed by microbial degradation of carbonaceous BOD
N = mass of DO removed by microbial degradation of nitrogenous BOD
R = mass of DO consumed by algal respiration

FIGURE 4-10
Mass balance diagram of DO in a small reach (*a*) and simplified mass balance for Streeter-Phelps model (*b*).

for all of the inputs and outputs. As mentioned above, we are going to limit our development to the classical Streeter-Phelps model.[7] The simplified mass balance diagram is shown in Figure 4-10*b*. The mass balance equation is then:

$$RDO_{in} + W + A - M - RDO_{out} = 0 \tag{4-32}$$

where RDO_{in} = mass of DO in river flowing into reach
W = mass of DO in wastewater flowing into reach
A = mass of DO added from atmosphere
M = mass of DO removed by microbial degradation of carbonaceous BOD
RDO_{out} = mass of DO in river flowing out of reach

In Equation 4-32 we can account for $RDO_{in} + W$. Our goal is to find RDO_{out} in terms of mass per unit volume (mg/L). This leaves A and M to be accounted for before we can solve the mass balance equation.

The rate at which DO disappears from the stream as a result of microbial action (M) is exactly equal to rate of increase in the deficit. With the assumption that the saturation value for DO remains constant $[d(DO_s)/dt = 0]$, differentiation of Equation 4-30 yields:

$$\frac{d(DO)}{dt} + \frac{dD}{dt} = 0$$

[7]H. W. Streeter and E. B. Phelps, *A Study of the Pollution and Natural Purification of the Ohio River*, U.S. Public Health Service Bulletin No. 146, 1925.

and

$$\frac{d(\text{DO})}{dt} = -\frac{dD}{dt} \tag{4-33}$$

The rate at which DO disappears coincides with the rate that BOD is degraded, so

$$\frac{d(\text{DO})}{dt} = -\frac{dD}{dt} = -\frac{d(\text{BOD})}{dt} \tag{4-34}$$

Remembering that BOD_t was defined as

$$\text{BOD}_t = L_o - L_t$$

and noting that L_o is a constant, we may say that the change in BOD with time is

$$\frac{d(\text{BOD})}{dt} = -\frac{dL_t}{dt} \tag{4-35}$$

This leads us to see that the rate of change in deficit at time t due to BOD is a first-order reaction proportional to the oxygen equivalent of the organics remaining:

$$\frac{dD}{dt} = kL_t \tag{4-36}$$

The rate constant, k, is called the *deoxygenation rate constant* and is designated k_d.

The rate of oxygen mass transfer into solution from the air (A) has been shown to be a first-order reaction proportional to the difference between the saturation value and the actual concentration:

$$\frac{d(\text{DO})}{dt} = k(\text{DO}_s - \text{DO}) \tag{4-37}$$

From Equations 4-30 and 4-33 we can see that

$$\frac{dD}{dt} = -kD \tag{4-38}$$

The rate constant is called the *reaeration rate constant,* k_r.

From Equations 4-36 and 4-38 we can see that the oxygen deficit is a function of the competition between oxygen utilization and reaeration from the atmosphere:

$$\frac{dD}{dt} = k_d L - k_r D \tag{4-39}$$

where $\frac{dD}{dt}$ = the change in oxygen deficit (D) per unit of time, mg/L · d

k_d = deoxygenation rate constant, d^{-1}

L = ultimate BOD of river water, mg/L

k_r = reaeration rate constant, d^{-1}

D = oxygen deficit in river water, mg/L

By integrating Equation 4-39, and using the boundary conditions (at $t = 0, D = D_a$ and $L = L_a$, and at $t = t, D = D$ and $L = L$), the DO sag equation is obtained:

$$D = \frac{k_d L_a}{k_r - k_d}(e^{-k_d t} - e^{-k_r t}) + D_a(e^{-k_r t}) \tag{4-40}$$

where D = oxygen deficit in river water after exertion of BOD for time, t, mg/L
L_a = initial ultimate BOD after river and wastewater have mixed (Equation 4-25), mg/L
k_d = deoxygenation rate constant, d^{-1}
k_r = reaeration rate constant, d^{-1}
t = time of travel of wastewater discharge downstream, d
D_a = initial deficit after river and wastewater have mixed (Equation 4-31), mg/L

When $k_r = k_d$, Equation 4-40 reduces to:

$$D = (k_d t L_a + D_a)(e^{-k_d t}) \tag{4-41}$$

where the terms are as previously defined.

Deoxygenation rate constant. The deoxygenation rate constant differs from the BOD rate constant because there are physical and biological differences between a river and a BOD bottle. In general, BOD is exerted more rapidly in a river because of turbulent mixing, larger numbers of "seed" organisms, and BOD removal by organisms on the stream bed as well as by those suspended in the water. While k rarely has a value greater than 0.7 day^{-1}, k_d may be as large as 7 day^{-1} for shallow, rapidly flowing streams. However, for deep, slowly moving rivers, the value of k_d is very close to that for k.

Bosko has developed a method of estimating k_d from k using characteristics of the stream:[8]

$$k_d = k + \frac{v}{H}\eta \tag{4-42}$$

where k_d = deoxygenation rate constant at 20°C, d^{-1}
v = average speed of stream flow, m/s
k = BOD rate constant determined in laboratory at 20°C, d^{-1}
H = average depth of stream, m
η = bed-activity coefficient

The bed-activity coefficient may vary from 0.1 for stagnant or deep water to 0.6 or more for rapidly flowing streams. Note that the bed-activity coefficient includes a

[8] K. Bosko, "An Explanation of the Difference Between the Rate of BOD Progression Under Laboratory and Stream Conditions," *Advances in Water Pollution Research,* Proceedings of the Third International Conference, Munich, p. 43, 1966.

conversion factor to make the second term dimensionally correct. After determining k_d from Equation 4-42, it should be corrected for temperature using Equation 4-5 if the stream temperature is not 20°C.

Example 4-10. Determine the deoxygenation rate constant for the reach of Bald Eagle Creek (Examples 4-8 and 4-9) below the wastewater outfall (discharge pipe). The average speed of the stream flow in the creek is 0.03 m/s. The depth is 5.0 m and the bed-activity coefficient is 0.35.

Solution. From Example 4-8, the value of k is 0.12 d^{-1}. Using Equation 4-42, the deoxygenation rate constant at 20°C is

$$k_d = 0.12 \text{ d}^{-1} + \frac{0.03 \text{ m/s}}{5.0 \text{ m}}(0.35) = 0.1221 \text{ or } 0.12 \text{ d}^{-1}$$

Note that the units are not consistent. As we have noted before, empirical expressions, such as that in Equation 4-42, may have implicit conversion factors. Thus, you must be careful to use the same units as those used by the author of the equation.

We also note that the deoxygenation rate constant of 0.1221 d^{-1} is at 20°C. In Example 4-9, we noted that the stream temperature was 10°C. Thus, we must correct the estimated k_d using Equation 4-5.

$$\begin{aligned} k_d \text{ at } 10°\text{C} &= (0.1221 \text{ d}^{-1})(1.135)^{10-20} = (0.1221)(0.2819) \\ &= 0.03442 \text{ or } 0.034 \text{ d}^{-1} \end{aligned}$$

Reaeration. The value of k_r depends on the degree of turbulent mixing, which is related to stream velocity, and on the amount of water surface exposed to the atmosphere compared to the volume of water in the river. A narrow, deep river will have a much lower k_r than a wide, shallow river. The reaeration rate constant can be estimated from the following equation:[9]

$$k_r = \frac{3.9v^{0.5}}{H^{1.5}} \tag{4-43}$$

where k_r = reaeration rate constant at 20°C, day^{-1}
v = average stream velocity, m/s
H = average depth, m

Note that the factor of 3.9 includes a conversion factor to make the equation dimensionally correct. The reaeration rate constant is also affected by temperature and can be adjusted to the river temperature using Equation 4-5 but with a temperature coefficient (θ) of 1.024. For various streams, k_r can range from 0.05 to greater than 18 day^{-1}.

[9] D. J. O'Connor and W. E. Dobbins, "Mechanism of Reaeration in Natural Streams," *American Society of Civil Engineers Transactions,* vol. 153, p. 641, 1958.

To relate travel time to a physical distance downstream, one must also know the average stream velocity. Once D has been found at any point downstream, the DO can be found from Equation 4-30. Note that it is physically impossible for the DO to be less than zero. If the deficit calculated from Equation 4-40 is greater than the saturation DO, then all the oxygen was depleted at some earlier time and the DO is zero. If the result of your calculations yields a negative DO, report it as zero because it cannot be less than zero!

The lowest point on the DO sag curve, which is called the *critical point,* is of major interest since it indicates the worst conditions in the river. The time to the critical point (t_c) can be found by differentiating Equation 4-40, setting it equal to zero, and solving for t using base e values for k_r and k_d:

$$t_c = \frac{1}{k_r - k_d} \ln \left[\frac{k_r}{k_d} \left(1 - D_a \frac{k_r - k_d}{k_d L_a} \right) \right] \tag{4-44}$$

or when $k_r = k_d$:

$$t_c = \frac{1}{k_d} \left(1 - \frac{D_a}{L_a} \right) \tag{4-45}$$

The critical deficit (D_c) is then found by using this critical time in Equation 4-40.

In some instances there may not be a sag in the DO downstream. The lowest DO may occur in the mixing zone. In these instances Equation 4-44 will not give a useful value.

Example 4-11. Determine the DO concentration at a point 5 km downstream from the State College discharge into the Bald Eagle Creek (Examples 4-8, 4-9, 4-10). Also determine the critical DO and the distance downstream at which it occurs.

Solution. All of the appropriate data are provided in the three previous examples. With the exceptions of the travel time, t, and the reaeration rate, the values needed for Equations 4-40 and 4-44 have been computed in Examples 4-8, 4-9, and 4-10. The first step then is to calculate k_r.

$$k_r \text{ at } 20°\text{C} = \frac{(3.9)(0.03 \text{ m/s})^{0.5}}{(5.0 \text{ m})^{1.5}} = 0.0604 \text{ d}^{-1}$$

Since this is at 20°C and the stream temperature is at 10°C, Equation 4-5 must be used to correct for the temperature difference.

$$k_r \text{ at } 10°\text{C} = (0.0604 \text{ d}^{-1})(1.024)^{10-20} = (0.0604)(0.7889) = 0.04766 \text{ d}^{-1}$$

Note that the temperature coefficient is the one noted in the text above rather than the ones reported with Equation 4-5.

The travel time t is computed from the distance downstream and the speed of the stream:

$$t = \frac{(5 \text{ km})(1,000 \text{ m/km})}{(0.03 \text{ m/s})(86,400 \text{ s/d})} = 1.929 \text{ d}$$

Although it is not warranted by the significant figures in the computation, we have elected to keep four significant figures because of the computational effects of truncating the value.

The deficit is estimated using Equation 4-40.

$$D = \frac{(0.03442)(11.86)}{0.04766 - 0.03442}[e^{-(0.03442)(1.929)} - e^{-(0.04766)(1.929)}] + 6.58[e^{-(0.04766)(1.929)}]$$

$$D = (30.83)(0.9358 - 0.9122) + 6.58(0.9122) = 6.7299 \text{ or } 6.73 \text{ mg/L}$$

and the dissolved oxygen is

$$\text{DO} = 11.33 - 6.73 = 4.60 \text{ mg/L}$$

The critical time is computed using Equation 4-44:

$$t_c = \frac{1}{0.04766 - 0.03442} \ln\left\{\frac{0.04766}{0.03442}\left[1 - 6.58\frac{0.04766 - 0.03442}{(0.03442)(11.86)}\right]\right\}$$

$$t_c = 6.45 \text{ d}$$

Using t_c for the time in Equation 4-40, calculate the critical deficit as

$$D_c = \frac{(0.03442)(11.86)}{0.04766 - 0.03442}[e^{-(0.03442)(6.45)} - e^{-(0.04766)(6.45)}] + 6.58[e^{-(0.04766)(6.45)}]$$

$$D_c = 6.85 \text{ mg/L}$$

and the critical DO is

$$\text{DO}_c = 11.33 - 6.85 = 4.48 \text{ mg/L}$$

The critical DO occurs downstream at a distance of

$$(6.45\text{d})(86{,}400 \text{ s/d})(0.03 \text{ m/s})\left(\frac{1}{1{,}000 \text{ m/km}}\right) = 16.7 \text{ km}$$

from the wastewater discharge point. (Note that 0.03 m/s is the speed of the stream.)

Management strategy. The beginning point for water quality management in rivers using the DO sag curve is to determine the minimum DO concentration that will protect the aquatic life in the stream. This value, called the DO standard, is generally set to protect the most sensitive species that exist or could exist in the particular river. For a known waste discharge and a known set of river characteristics, the DO sag equation can be solved to find the DO at the critical point. If this value is higher than the standard, the stream can adequately assimilate the waste. If the DO at the critical point is less than the standard, then additional waste treatment is needed. Usually, the environmental engineer has control over just two parameters, L_a and D_a. By increasing the efficiency of the existing treatment processes or by adding additional treatment steps, the ultimate BOD of the waste discharge can be reduced, thereby reducing L_a. Often a relatively inexpensive method for improving stream quality is to reduce D_a by adding oxygen to the wastewater to bring it close to saturation prior to discharge. To determine whether a proposed improvement will be adequate, the new values for L_a and D_a are used to determine whether the DO standard will be violated at the critical point. Under unusual conditions, the engineer may artificially aerate the river with mechanical systems to increase the DO.

When using the DO sag curve to determine the adequacy of wastewater treatment, it is important to use the river conditions that will cause the lowest DO concentration. Usually these conditions occur in the late summer when river flows are low and temperatures are high. A frequently used criterion is the "10-year, 7-day low flow," which is the recurrence interval of the average low flow for a 7-day period estimated using the partial duration series technique (Chapter 2). Low river flows reduce the dilution of the waste entering the river, causing higher values for L_a and D_a. The value of k_r is usually reduced by low river flows because of reduced velocities. In addition, higher temperatures increase k_d more than k_r and also decrease DO saturation, thus making the critical point more severe.

Example 4-12. The Pitts Canning Company is considering opening a new plant at one of two possible locations: the Green River and its twin, the White River. Among the decisions to be made are what effect the plant discharge will have on each river and which river would be impacted less. Effluent data from the Pitts A Plant and the Pitts B Plant are considered to be representative of the potential discharge characteristics. In addition, measurements from each river at summer low-flow conditions are available.

Effluent parameter	A plant	B plant
Flow, m^3/s	0.0500	0.0500
Ultimate BOD at 25°C, kg/d	129.60	129.60
DO, mg/L	0.900	0.900
Temperature, °C	25.0	25.0
K at 20°C, d^{-1}	0.0500	0.0300

River parameter	Green River	White River
Flow, m^3/s	0.500	0.500
Ultimate BOD at 25°C, mg/L	19.00	19.00
DO, mg/L	5.85	5.85
Temperature, °C	25.0	25.0
Speed, m/s	0.100	0.200
Average depth, m	4.00	4.00
Bed-activity coefficient	0.200	0.200

There are four combinations to be evaluated:

A–Green　　B–Green
A–White　　B–White

Solution. This problem will be worked in base 10. Note that the BOD rate constant (K) is given in base 10. Also note that for the purpose of explaining the calculations, the number of significant figures given for the data is greater than can probably be measured. The only difference in the combinations is the change in deoxygenation and reaeration coefficients. Thus, we need calculate only one value of L_a and one value of D_a.

We begin by converting the mass flow of ultimate BOD (kg/d) to a concentration (mg/L). Following our general approach for calculating concentration from mass flow, we divide the mass discharge (kg/d) by the flow of the water carrying the waste (Q_w, Q_r, or the sum $Q_w + Q_r$):

$$\frac{\text{Mass of ultimate BOD discharged (kg/d)}}{\text{Flow of water-carrying waste (m}^3\text{/s)}}$$

The mass discharge units are then converted to mg/d and the water flow to L/d so that the days cancel.

$$\frac{(\text{kg/d}) \times (1 \times 10^6 \text{ mg/kg})}{(\text{m}^3/\text{s}) \times (86,400 \text{ s/d})(1 \times 10^3 \text{ L/m}^3)}$$

For either Plant A or B:

$$\begin{aligned} L_w &= \frac{(129.60 \text{ kg/d})(1 \times 10^6 \text{ mg/kg})}{(0.0500 \text{ m}^3/\text{s})(86,400 \text{ s/d})(1 \times 10^3 \text{ L/m}^3)} \\ &= \frac{129.60 \times 10^6 \text{ mg}}{4.320 \times 10^6 \text{ L}} \\ &= 30.00 \text{ mg/L} \end{aligned}$$

Now we can compute the mixed BOD using Equation 4-25.

$$\begin{aligned} L_a &= \frac{(0.0500)(30.00) + (0.500)(19.00)}{0.0500 + 0.500} \\ &= 20.0 \text{ mg/L} \end{aligned}$$

From Table A-2 of Appendix A, we find that the DO saturation at 25°C is 8.38 mg/L. Then using Equation 4-31 we determine the initial deficit:

$$\begin{aligned} D_a &= 8.38 - \frac{(0.0500)(0.900) + (0.500)(5.85)}{0.0500 + 0.500} \\ &= 8.38 - 5.4 \\ &= 2.98 \text{ mg/L} \end{aligned}$$

For the combination of the A Plant discharging to the Green River, the deoxygenation coefficient and reaeration coefficient are calculated using Equation 4-42 and Equation 4-43, in base 10:

$$\begin{aligned} K_d &= 0.0500 + \frac{0.100 \times 0.200}{2.3 \times 4.00} \\ &= 0.05217 \text{ d}^{-1} \text{ at } 20°\text{C} \end{aligned}$$

and

$$\begin{aligned} K_r &= \frac{1.7(0.100)^{0.5}}{(4.00)^{1.5}} \\ &= 0.067198 \text{ d}^{-1} \text{ at } 20°\text{C} \end{aligned}$$

(Note: The factor 2.3 in the K_d equation and the 1.7 in the K_r equation are the conversions in base 10. Hence, they differ from the equations in the text.)

Since the temperature of the river is 25°C and the wastewater effluent temperature is also 25°C, we do not have to calculate a temperature after mixing. However, we will have to adjust K_d and K_r to 25°C. For K_d, we use Equation 4-5 with a value of θ of 1.056.

$$\begin{aligned} K_d &= 0.05217(1.056)^{25-20} \\ &= 0.068513 \text{ or } 0.0685 \text{ d}^{-1} \end{aligned}$$

From the discussion that follows Equation 4-43, we note that $\theta = 1.024$ for reaeration, and thus

$$\begin{aligned} K_r &= 0.067198(1.024)^{25-20} \\ &= 0.075658 \text{ or } 0.0757 \text{ d}^{-1} \end{aligned}$$

Although perhaps not justified by the coefficients, we round to three significant figures because we will want to calculate travel time to two decimal places.

Now we have all the information we need to calculate the time to the critical point. Using Equation 4-44:

$$\begin{aligned} t_c &= \frac{1}{0.0757 - 0.0685} \log \left\{ \frac{0.0757}{0.0685} \left[1 - 3.0 \left(\frac{0.0757 - 0.0685}{0.0685 \times 20.0} \right) \right] \right\} \\ &= 138.89 \log [1.105(1 - 3.0(0.005255))] \\ &= 138.89 \log [1.0876] \\ &= 5.07 \text{ d} \end{aligned}$$

Using this value for t in Equation 4-40 (in base 10 because K values are in base 10), we can calculate the deficit at the critical point:

$$\begin{aligned} D_c &= \frac{(0.0685)(20.0)}{0.0757 - 0.0685}[10^{-(0.0685)(5.07)} - 10^{-(0.0757)(5.07)}] + 2.98[10^{-(0.0757)(5.07)}] \\ &= 191.76\,[(0.4493) - (0.4133)] + 2.98[0.4133] \\ &= 6.903 + 1.232 \\ &= 8.13 \text{ mg/L} \end{aligned}$$

The DO at the critical point is determined by solving Equation 4-30 for DO and substituting the appropriate value for the DO saturation that we obtained earlier from Table A-2 of Appendix A:

$$\begin{aligned} \text{DO} &= \text{DO}_s - D \\ &= 8.38 - 8.13 = 0.25 \text{ mg/L} \end{aligned}$$

Thus, the lowest DO for the A–Green combination is 0.25 mg/L, and it occurs at a travel time of 5.07 days downstream from the A Plant outfall. Since the Green River travels at a speed of 0.100 m/s, this would be

$$\frac{(0.100 \text{ m/s})(5.07 \text{ d})(86{,}400 \text{ s/d})}{1{,}000 \text{ m/km}} = 43.8 \text{ km}$$

downstream.

The results of the other combinations are summarized below:

	A–Green	A–White	B–Green	B–White
K_d	0.0685	0.0714	0.0422	0.0451
K_r	0.0757	0.107	0.0757	0.107
t_c	5.07	3.99	5.93	4.44
D	8.14	6.93	6.27	5.31
DO	0.25	1.47	2.13	3.08

It is obvious that the best combination is the B Plant on the White River.

Using a spreadsheet program, we have generated the deficit values for a series of times for each of the combinations and plotted the results in Figure 4-11. From this figure we can make the following general observations:

1. Increasing the reaeration rate, while holding everything else as it is, reduces the deficit and displaces the critical point upstream.
2. Decreasing the reaeration rate, while holding everything else as it is, increases the deficit and displaces the critical point downstream.
3. Increasing the deoxygenation rate, while holding everything else as it is, increases the deficit and displaces the critical point upstream.
4. Decreasing the deoxygenation rate, while holding everything else as it is, decreases the deficit and displaces the critical point downstream.

Nitrogenous BOD. Up to this point, only carbonaceous BOD has been considered in the DO sag curve. However, in many cases nitrogenous BOD has at least as much impact on dissolved oxygen levels. Modern wastewater treatment plants can routinely produce effluents with $CBOD_5$ of less than 30 mg/L. A typical effluent also contains approximately 30 mg/L of nitrogen, which would mean an NBOD of about 137 mg/L if it were discharged as ammonia (see Example 4-7). Nitrogenous BOD can be incorporated into the DO sag curve by adding an additional term to Equation 4-40:

$$D = \frac{k_d L_a}{k_r - k_d}(e^{-k_d t} - e^{-k_r t}) + D_a(e^{-k_r t}) + \frac{k_n L_n}{k_r - k_n}(e^{-k_n t} - e^{-k_r t}) \qquad (4\text{-}46)$$

where k_n = the nitrogenous deoxygenation coefficient, day^{-1}; L_n = ultimate nitrogenous BOD after waste and river have mixed, mg/L; and the other terms are as previously defined. It is important to note that with the additional term for NBOD, it is not possible to find the critical time using Equation 4-44. Instead, it must be found by a trial and error solution of Equation 4-46.

Other factors affecting DO levels in rivers. The classical DO sag curve assumes that there is only one point-source discharge of waste into the river. In reality, this is rarely the case. Multiple point sources can be handled by dividing the river up into reaches with a point source at the head of each reach. A *reach* is a length of river

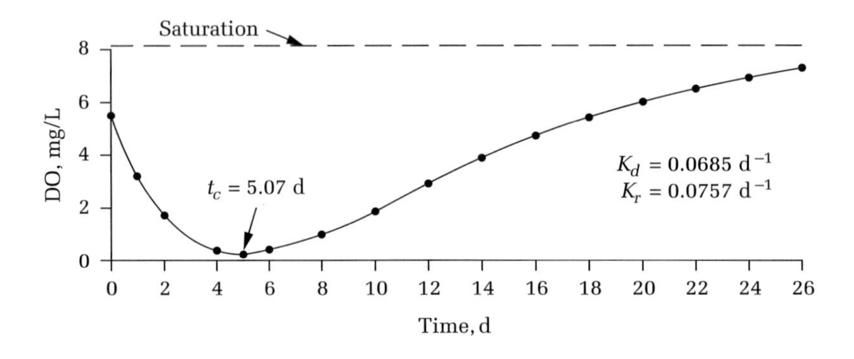

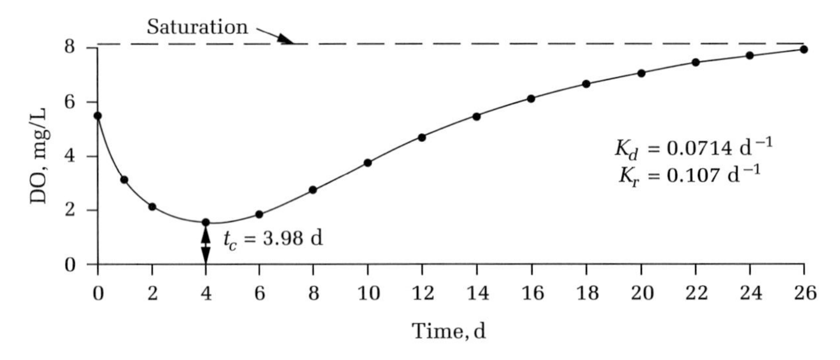

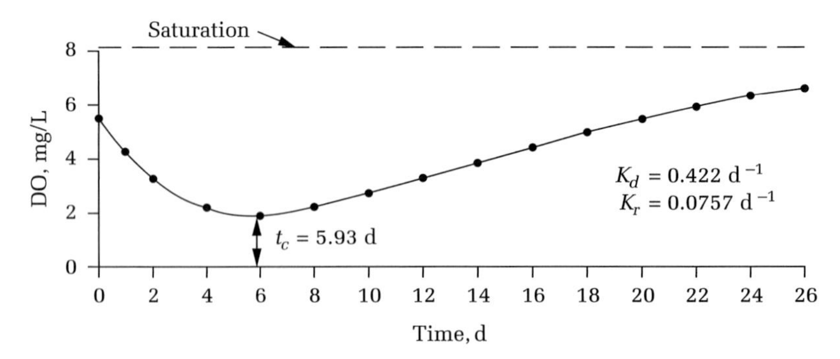

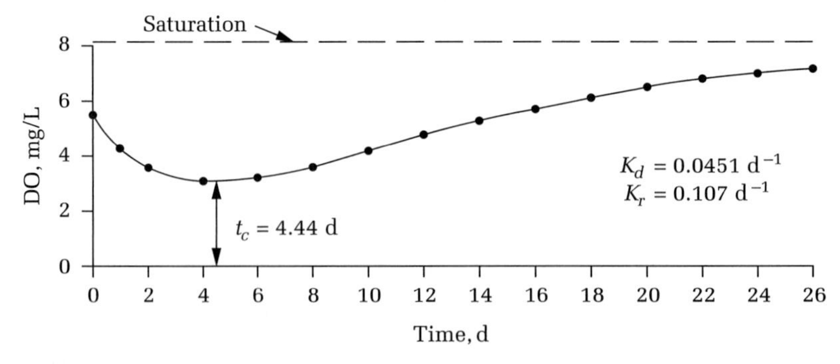

FIGURE 4-11
Effect of K_d and K_r on DO sag curve.

specified by the engineer on the basis of its homogeneity, that is, channel shape, bottom composition, slope, etc. The oxygen deficit and residual BOD can be calculated at the end of each reach. These values are then used to determine new values of D_a and L_a at the beginning of the following reach. Non-point source pollution can also be handled this way if the reaches are made small enough. Non-point source pollution can also be incorporated directly into the DO sag equation for a more sophisticated analysis. Dividing the river into reaches is also necessary whenever the flow regime changes, since the reaeration coefficient would also change. In small rivers, rapids play a major role in maintaining high DO levels. Eliminating rapids by dredging or damming a river can have a severe impact on DO, although DO levels immediately downstream of dams are usually high because of the turbulence of the falling water.

Some rivers contain large deposits of organic matter in the sediments. These can be natural deposits of leaves and dead aquatic plants or can be sludge deposits from wastewaters receiving little or no treatment. In either case, decomposition of this organic matter places an additional burden on the stream's oxygen resources, since the oxygen demand must be supplied from the overlying water. When this benthic demand is significant, compared to the oxygen demand in the water column, it must be included quantitatively in the sag equation.

Aquatic plants can also have a substantial effect on DO levels. During the day, their photosynthetic activities produce oxygen that supplements the reaeration and can even cause oxygen supersaturation. However, plants also consume oxygen for respiration processes. Although there is a net overall production of oxygen, plant respiration can severely lower DO levels during the night. Plant growth is usually highest in the summer when flows are low and temperatures are high, so that large nighttime respiration requirements coincide with the worst cases of oxygen depletion from BOD exertion. In addition, when aquatic plants die and settle to the bottom, they increase the benthic demand. As a general rule, large growths of aquatic plants are detrimental to the maintenance of a consistently high DO level.

Effect of Nutrients on Water Quality in Rivers

Although oxygen-demanding wastes are definitely the most important river pollutants on an overall basis, nutrients can also contribute to deteriorating water quality in rivers by causing excessive plant growth. Nutrients are those elements required by plants for their growth. They include, in order of abundance in plant tissue: carbon, nitrogen, phosphorus, and a variety of trace elements. When there are sufficient quantities of all nutrients available, plant growth is possible. By limiting the availability of any one nutrient, further plant growth is prevented.

Some plant growth is desirable, since plants form the base of the food chain and thus support the animal community. However, excessive plant growth can create a number of undesirable conditions such as thick slime layers on rocks and dense growths of aquatic weeds.

The availability of nutrients is not the only requirement for plant growth. In many rivers, the turbidity caused by eroded soil particles, bacteria, and other factors prevents light from penetrating far into the water, thereby limiting total plant growth in deep water. It is for this reason that slime growths on rocks usually occur in shallow water. Strong water currents also prevent rooted plants from taking hold, and thus limit their growth to quiet backwaters where the currents are weak and the water is shallow enough for light to penetrate.

Effects of nitrogen. There are three reasons why nitrogen is detrimental to a receiving body:

1. In high concentrations, NH_3-N is toxic to fish.
2. NH_3, in low concentrations, and NO_3^- serve as nutrients for excessive growth of algae.
3. The conversion of NH_4^+ to NO_3^- consumes large quantities of dissolved oxygen.

Effects of phosphorus. The major deleterious effect of phosphorus is that it serves as a vital nutrient for the growth of algae. If the phosphorus availability meets the growth demands of the algae, there is an excessive production of algae. When the algae die, they become an oxygen-demanding organic material as bacteria seek to degrade them. This oxygen demand frequently overtaxes the DO supply of the water body and, as a consequence, causes fish to die.

Management strategy. The strategy for managing water quality problems associated with excessive nutrients is based on the sources for each nutrient. Except under rare circumstances, there is plenty of carbon available for plant growth. Plants use carbon dioxide, which is available from the bicarbonate alkalinity of the water and from the bacterial decomposition of organic matter. As carbon dioxide is removed from the water, it is replenished from the atmosphere. Generally, the major source of trace elements is the natural weathering of rock minerals, a process over which the environmental engineer has little control. However, since the acid rain caused by air pollution accelerates the weathering process, air pollution control can help reduce the supply of trace elements. Even when substantial amounts of trace elements are found in wastewater, their removal is difficult. In addition, such small amounts are needed for plant growth that nitrogen or phosphorus is more likely to be the limiting nutrient. Therefore, the practical control of nutrient-caused water-quality problems in streams is based on removal of nitrogen and/or phosphorus from wastewaters before they are discharged.

4-4 WATER QUALITY MANAGEMENT IN LAKES

Oxygen-demanding wastes can also be important lake pollutants, especially when the waste is discharged to a contained area such as a bay. Pathogens are of particular concern near bathing beaches. Again, as with rivers, there are special classes of lakes which are most seriously affected by other pollutants such as toxic chemicals from industrial discharges. However, phosphorus so dominates other pollutants in controlling water quality in the vast majority of lakes that we will give it special emphasis.

A knowledge of lake systems is essential to an understanding of the role of phosphorus in lake pollution. The study of lakes is called *limnology*. This section is essentially a short course in limnology as it relates to phosphorus pollution.

Stratification and Turnover

Nearly all lakes in the temperate zone become stratified during the summer and overturn (turnover) in the fall due to changes in the water temperature that result from the annual cycle of air temperature changes. In addition, lakes in cold climates undergo winter stratification and spring overturn as well. These physical processes, which are described below, occur regardless of the water quality in the lake. Nonetheless, they do help determine the water quality.

During the summer, the surface water of a lake is heated both indirectly by contact with warm air and directly by sunlight. Warm water, being less dense than cool water, remains near the surface until mixed downward by turbulence from wind, waves, boats, and other forces. Because this turbulence extends only a limited distance below the water surface, the result is an upper layer of well-mixed, warm water (the *epilimnion*) floating on the lower water (the *hypolimnion*), which is poorly mixed and cool, as shown in Figure 4-12*a*. Because of good mixing the epilimnion will be *aerobic* (have DO). The hypolimnion will have a lower DO and may become *anaerobic* (devoid of oxygen). The boundary is called the *thermocline* because of the sharp temperature change (and therefore density change) that occurs within a relatively short distance. The thermocline may be defined as a change in temperature with depth that is greater than 1°C/m. You may have experienced the thermocline while swimming in a small lake. As long as you are swimming horizontally, the water is warm, but as soon as you tread water or dive, the water turns cold. You have penetrated the thermocline. The depth of the epilimnion is related to the

(*a*)

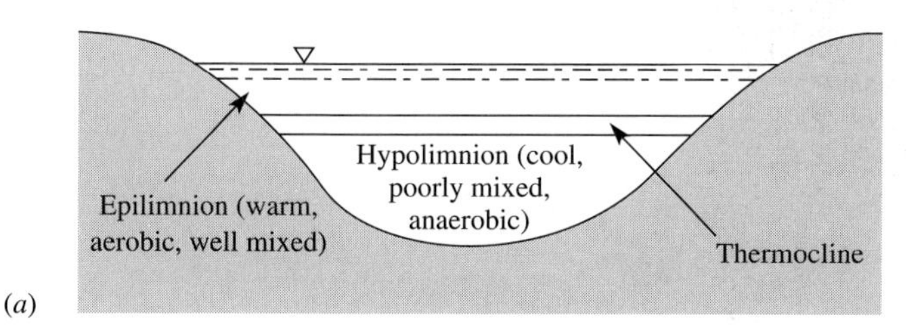

(*b*)

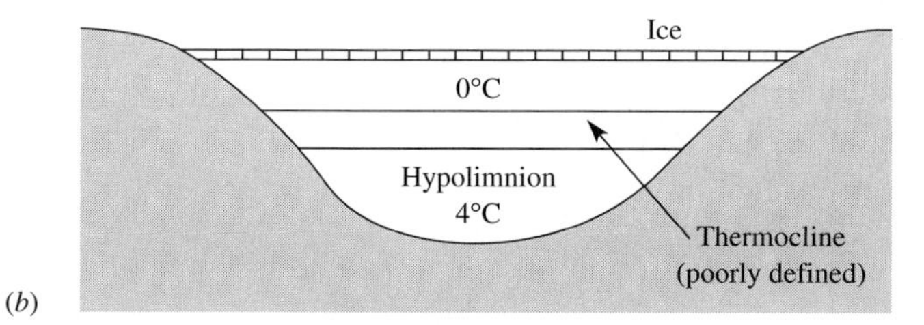

FIGURE 4-12
Stratification of a lake during (*a*) summer and (*b*) winter.

size of the lake. It is as little as one meter in small lakes and as much as 20 meters or more in large lakes. The depth of the epilimnion is also related to storm activity in the spring when stratification is developing. A major storm at the right time will mix warmer water to a substantial depth and thus create a deeper than normal epilimnion. Once formed, lake stratification is very stable. It can be broken only by exceedingly violent storms. In fact, as the summer progresses, the stability increases because the epilimnion continues to warm, while the hypolimnion remains at a fairly constant temperature.

In the fall, as temperatures drop, the epilimnion cools until it is more dense than the hypolimnion. The surface water then sinks, causing overturning. The water of the hypolimnion rises to the surface where it cools and again sinks. The lake thus becomes completely mixed. If the lake is in a cold climate, this process stops when the temperature reaches 4°C, since this is the temperature at which water is most dense. Further cooling or freezing of the surface water results in winter stratification, as shown in Figure 4-12*b*. As the water warms in the spring, it again overturns and becomes completely mixed. Thus, temperate climate lakes have at least one, if not two, cycles of stratification and turnover every year.

Biological Zones

Lakes contain several distinct zones of biological activity, largely determined by the availability of light and oxygen. The most important biological zones, shown in Figure 4-13, are the euphotic, littoral, and benthic zones.

Euphotic zone. The upper layer of water through which sunlight can penetrate is called the *euphotic zone.* All plant growth occurs in this zone. In deep water, algae are the most important plants, while rooted plants grow in shallow water near

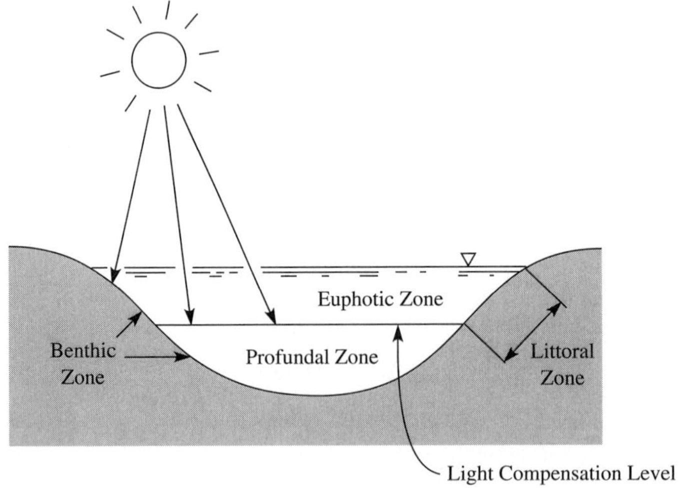

FIGURE 4-13
Biological zones in a lake.

the shore. The depth of the euphotic zone is determined by the amount of turbidity blocking sunlight penetration. In most lakes, the turbidity is due to algal growth, although color and suspended clays may substantially reduce sunlight penetration in some lakes. In the euphotic zone, plants produce more oxygen by photosynthesis than they remove by respiration. Below the euphotic zone lies the *profundal zone.* The transition between the two zones is called the *light compensation level.* The light compensation level corresponds roughly to a depth at which the light intensity is about one percent of unattenuated sunlight. It is important to note that the bottom of the euphotic zone only rarely coincides with the thermocline.

Littoral zone. The shallow water near the shore in which rooted water plants can grow is called the *littoral zone.* The extent of the littoral zone depends on the slope of the lake bottom and the depth of the euphotic zone. The littoral zone cannot extend deeper than the euphotic zone.

Benthic zone. The bottom sediments comprise the *benthic zone.* As organisms living in the overlying water die, they settle to the bottom where they are decomposed by organisms living in the benthic zone. Bacteria are always present. The presence of higher life forms such as worms, insects, and crustaceans depends on the availability of oxygen.

Lake Productivity

The productivity of a lake is a measure of its ability to support a food web. Algae form the base of this food web, supplying food for the higher organisms. A lake's productivity may be determined by measuring the amount of algal growth that can be supported by the available nutrients. Although a more productive lake usually will have a higher fish population, the number of the most desirable fish may decline. In fact, increased productivity generally results in reduced water quality because of undesirable changes that occur as algal growth increases. Because of the important role productivity plays in determining water quality, it forms a basis for classifying lakes.

Oligotrophic lakes. Oligotrophic lakes have a low level of productivity due to a severely limited supply of nutrients to support algal growth. As a result, the water is clear enough that the bottom can be seen at considerable depths. In this case, the euphotic zone often extends into the hypolimnion, which is aerobic. Oligotrophic lakes, therefore, support cold water game fish. Lake Tahoe on the California-Nevada border is a classic example of an oligotrophic lake.

Eutrophic lakes. Eutrophic lakes have a high productivity because of an abundant supply of algal nutrients. The algae cause the water to be highly turbid, so the euphotic zone may extend only partially into the epilimnion. As the algae die, they settle to the lake bottom where they are decomposed by benthic organisms. In a eutrophic lake, this decomposition is sufficient to deplete the hypolimnion of oxygen during summer stratification. Because the hypolimnion is anaerobic during the summer, eutrophic lakes support only warm-water fish. In fact, most cold-water fish are driven out of the lake long before the hypolimnion becomes anaerobic because they generally require dissolved oxygen levels of at least 5 mg/L. Highly eutrophic lakes may also have large mats of floating algae that typically impart unpleasant tastes and odors to the water.

Mesotrophic lakes. Lakes which are intermediate between oligotrophic and eutrophic are called mesotrophic. Although substantial depletion of oxygen may have occurred in the hypolimnion, it remains aerobic.

Senescent lakes. These are very old, shallow lakes which have thick organic sediments and rooted water plants in great abundance. These lakes will eventually become marshes.

Eutrophication

Eutrophication is a natural process in which lakes gradually become shallower and more productive through the introduction and cycling of nutrients. Thus, oligotrophic lakes gradually pass through the mesotrophic, eutrophic, and senescent stages, eventually filling completely. The time for this process to occur depends on the original size of the lake and on the rate at which sediments and nutrients are introduced. In some lakes the eutrophication process is so slow that thousands of years may pass with little change in water quality. Other lakes may have been eutrophic from the day they were formed, if nutrient levels were high at that time.

Cultural eutrophication is caused when human activity speeds the processes naturally occurring by increasing the rate at which sediments and nutrients are added to the lake. Thus, lake pollution can be seen as the intensification of a natural process. This is not to say that eutrophic lakes are necessarily polluted, but that pollution contributes to eutrophication. Water quality management in lakes is primarily concerned with slowing eutrophication to at least the natural rate. To understand the factors involved in eutrophication, it is necessary to understand the factors contributing to algal growth.

Algal Growth Requirements

All algae require macronutrients, such as carbon, nitrogen, and phosphorus, and micronutrients, such as trace elements. For algae to grow, all nutrients must be available. Lack of any one nutrient will limit the total algal population. The availability of each nutrient and its natural cycle are summarized below.

Carbon. Algae obtain their carbon from carbon dioxide dissolved in the water. Since the carbon dioxide is in equilibrium with the bicarbonate buffer system (see Chapter 3), the immediately available carbon is determined by the alkalinity of the water. However, as carbon dioxide is removed from the water, it is replenished from the atmosphere. The atmosphere is, of course, a virtually inexhaustible source of this gas. When algae are either consumed by higher organisms or die and decompose, the organic carbon is oxidized back to carbon dioxide which returns either to the water or to the atmosphere to complete the carbon cycle.

Nitrogen. Nitrogen in lakes is usually in the form of nitrate (NO_3^-) and comes from external sources by way of inflowing streams or groundwater. When taken up for algal growth, the nitrogen is chemically reduced to amino-nitrogen (NH_2^-) and incorporated into organic compounds. When dead algae undergo decomposition, the organic nitrogen is released to the water as ammonia (NH_3). The ammonia is then oxidized back to nitrate by bacteria in the same nitrification process discussed earlier in river systems.

Nitrogen cycles from nitrate to organic nitrogen, to ammonia, and back to nitrate as long as the water remains aerobic. However, in anaerobic sediments, and in the hypolimnion of eutrophic lakes, when algal decomposition has depleted the oxygen supply, nitrate is reduced by anaerobic bacteria to nitrogen gas (N_2) and lost from the system in a process called *denitrification.* Denitrification reduces the average time nitrogen remains in the lake system. The denitrification reaction is

$$2NO_3^- + \text{organic carbon} \rightleftharpoons N_2 + CO_2 + H_2O \qquad (4\text{-}47)$$

Some photosynthetic microorganisms can also fix nitrogen gas from the atmosphere by converting it to organic nitrogen. In lakes the most important nitrogen-fixing microorganisms are photosynthetic bacteria called cyanobacteria, formerly known as blue-green algae because of the pigments they contain. Because of their nitrogen-fixing ability, cyanobacteria have a competitive advantage over green algae when nitrate and ammonium concentrations are low but other nutrients are sufficiently abundant. These cyanobacteria are generally undesirable because of their tendency to aggregate in unsightly floating mats and because they impart unpleasant odor and taste to the water. Cyanobacteria can also produce toxins which kill fish. Fortunately, these nuisance organisms are not prevalent unless the supply of soluble fixed nitrogen is reduced to low levels.

Phosphorus. Phosphorus in lakes originates from external sources and is taken up by algae in the inorganic form (PO_4^{3-}) and incorporated into organic compounds. During algal decomposition, phosphorus is returned to the inorganic form. The release of phosphorus from dead algal cells is so rapid that only a little of it leaves the epilimnion with the settling algal cells. However, little by little, phosphorus is transferred to the sediments, some of it in undecomposed organic matter; some of it in precipitates of iron, aluminum, and calcium; and some bound to clay particles. To a large extent, the permanent removal of phosphorus from the overlying waters to the sediments depends on the amount of iron, aluminum, calcium, and clay entering the lake along with phosphorus.

Trace elements. The quantities of trace elements required to support algal growth are so small that most fresh waters have sufficient amounts for a substantial algal population.

The Limiting Nutrient

In 1840, Justin Liebig formulated the idea that "growth of a plant is dependent on the amount of foodstuff that is presented to it in minimum quantity." This is now known as *Liebig's law of the minimum.* As applied to algae, it means that algal growth will be limited by the nutrient that is least available. Of all the nutrients, only phosphorus is not readily available from the atmosphere or the natural water supply. For this reason, phosphorus is deemed the *limiting nutrient* in lakes. The amount of phosphorus controls the quantity of algal growth and therefore the productivity of lakes. This can be seen from Figure 4-14 in which the concentration of *chlorophyll a* is plotted against phosphorus concentration. Chlorophyll *a*, one of the green pigments involved in photosynthesis, is found in all algae, so it is used to distinguish the amount of algae in the water from other organic solids such as bacteria. It has been estimated that

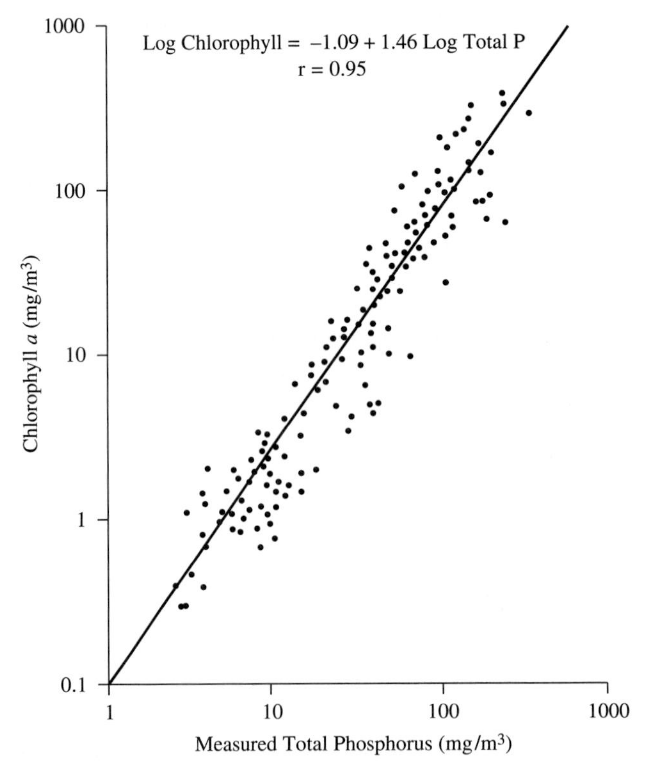

FIGURE 4-14
Relationship between summer levels of chlorophyll *a* and measured total phosphorus concentration for 143 lakes. (*Source:* J. R. Jones and R. W. Bachmann, "Prediction of Phosphorus and Chlorophyll Levels in Lakes," *Journal of the Water Pollution Control Federation,* vol. 48, p. 2176, 1976.)

the phosphorus concentration should be below 0.010 to 0.015 mg/L to limit algae blooms.[10]

Control of Phosphorus in Lakes

Since phosphorus is usually the limiting nutrient, control of cultural eutrophication must be accomplished by reducing the input of phosphorus to the lake. Once the input is reduced, the phosphorus concentration will gradually fall as phosphorus is buried in the sediment or flushed from the lake. Other strategies for reversing or slowing the eutrophication process, such as precipitating phosphorus with additions of aluminum (alum) or removing phosphorus-rich sediments by dredging, have been proposed. However, if the input of phosphorus is not also curtailed, the eutrophication process will continue. Thus, dredging or precipitation alone can result only in temporary improvement in water quality. In conjunction with reduced phosphorus inputs, these measures can help speed up the removal of phosphorus already in the lake system. Of course, the need to speed the recovery process must be weighed against the potential damage from inundating shoreline areas with sludge and stirring up toxic compounds buried in the sediment.

To be able to reduce phosphorus inputs, it is necessary to know the sources of phosphorus and the potential for their reduction. The natural source of phosphorus is the weathering of rock. Phosphorus released from the rock can enter the water directly, but more commonly it is taken up by plants and enters the water in the form of dead plant matter. It is exceedingly difficult to reduce the natural inputs of phosphorus. If these sources are large, the lake is generally naturally eutrophic. For many lakes the principal sources of phosphorus are the result of human activity. The most important sources are municipal and industrial wastewaters, seepage from septic tanks, and agricultural runoff that carries phosphorus fertilizers into the water.

Municipal and industrial wastewaters. All municipal sewage contains phosphorus from human excrement. Many industrial wastes are high in this nutrient. In these cases, the only effective way of reducing phosphorus is through advanced waste treatment processes, which are discussed in Chapter 5. Municipal wastewaters also contain large quantities of phosphorus from detergents containing polyphosphate, which is a chain of phosphate ions (usually three) linked together. The polyphosphate binds with hardness in water to make the detergent a more effective cleaning agent. By the 1970s, phosphorus loading from detergents was approximately twice that from human excrement. Phosphorus from detergents can be removed by advanced waste treatment, but phosphorus can also be removed from detergents so that it never enters the wastewater to begin with. As a direct application of the waste minimization philosophy, several states have passed laws banning phosphate detergents as a rapid method of reducing phosphorus inputs to lakes.

---

[10] R. A. Vollenweider, "Input-output Models with Special Reference to the Phosphorus Loading Concept in Limnology," *Schweiz. Z. Hydro., 37,* pp. 53–83, 1975.

Septic tank seepage. The shores of many lakes are dotted with homes and summer cottages, each with its own septic tank and tile field for waste disposal. As treated wastewater moves through the soil toward the lake, phosphorus is adsorbed by soil particles, especially clay. Thus, during the early life of the tile field, very little phosphorus gets to the lake. However, with time, the capacity of the soil to adsorb phosphorus is exceeded and any additional phosphorus will pass on into the lake, contributing to eutrophication. The time it takes for phosphorus to break through to the lake depends on the type of soil, the distance to the lake, the amount of wastewater generated, and the concentration of phosphorus in that wastewater. To prevent phosphorus from reaching the lake, it is necessary to put the tile field far enough from the lake that the adsorption capacity of the soil is not exceeded. If this is not possible, it may be necessary to replace the septic tanks and tile fields with a sewer to collect the wastewater and transport it to a treatment facility.

Agricultural runoff. Because phosphorus is a plant nutrient, it is an important ingredient in fertilizers. As rain water washes off fertilized fields, some of the phosphorus is carried into streams and then into lakes. Most of the phosphorus not taken up by growing plants is bound to soil particles. Bound phosphorus is carried into streams and lakes through soil erosion. Waste minimization can be applied to the control of phosphorus loading to lakes from agricultural fertilization by encouraging farmers to fertilize more often with smaller amounts and to take effective action to stop soil erosion.

Acidification of Lakes

Pure rainwater is slightly acid. As we discussed in Chapter 3, CO_2 dissolves in water to form carbonic acid (H_2CO_3). The equilibrium concentration of H_2CO_3 results in a rainwater pH of approximately 5.6. Thus, acid rain is usually defined to be precipitation with a pH less than 5.6. The northeastern U.S. and Canada frequently record rainwater pH values between 4 and 5 (Figure 4-15). These low pH values have been attributed to emissions of sulfur and nitrogen oxides from the combustion of fossil fuels (see Chapter 6).

Fish, and in particular trout and Atlantic salmon, are very sensitive to low pH levels. Most are severely stressed if the pH drops below 5.5, and few are able to survive if the pH falls below 5.0. If the pH falls below 4.0, cricket frogs and spring peepers experience mortalities in excess of 85 percent.

High aluminum concentrations are often the trigger which kills fish. Aluminum is abundant in soil but it is normally bound up in the soil minerals. At normal pH values aluminum rarely occurs in solution. Acidification of the water releases highly toxic Al^{3+} to the water.

Most lakes are buffered by the carbonate buffer system (see Chapter 3). To the extent that the buffer capacity of the lake is not exceeded, the pH of the lake will not be appreciably affected by acid rain. If there is a source of carbonate to replace that consumed by the acid rain, the buffering capacity can be quite large. Calcareous soils are those containing large quantities of calcium carbonate ($CaCO_3$). As shown in

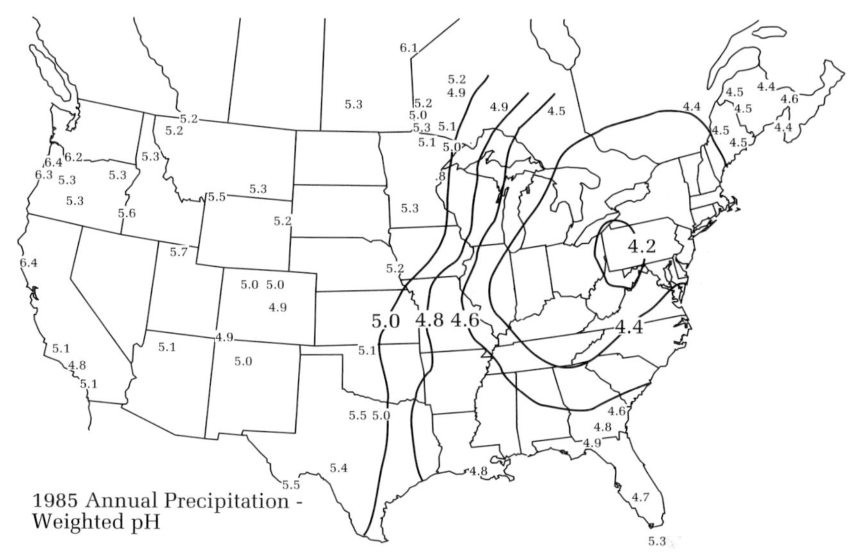

FIGURE 4-15
1985 contour and selected station map of pH in rain over the United States. (*Source:* J. L. Kulp, "NAPAP's Interim Assessment," *Proceedings of the Air Pollution Control Association: Acidic Precipitation,* January 1988.)

Figure 3-12, carbonic acid releases bicarbonate into solution. H^+ from acid rain will also release bicarbonate. Thus, lakes formed in calcareous soils tend to be resistant to acidification.

Other factors that affect the susceptibility of a lake to acidification are the permeability and depth of the soil, the bedrock, the slope and size of the watershed, and the type of vegetation. Thin, impermeable soils provide little time for contact between the soil and the precipitation. This reduces the potential for the soil to buffer the acid precipitation. Likewise, small watersheds with steep slopes reduce the time for buffering to occur. Deciduous foliage tends to decrease acidity. Coniferous foliage tends to yield runoff that is more acid than the precipitation itself. Granite bedrock offers little potential to buffer acid rain. Galloway and Cowling[11] used bedrock geology to predict areas where lakes are potentially most sensitive to acid rain (Figure 4-16). You may note that the predicted areas of sensitivity are also those subjected to very acid precipitation.

The control of lake acidification is related to the control of atmospheric emissions of sulfur and nitrogen oxides. The role of air pollution in acid deposition is discussed in more detail in Chapter 6.

---

[11] J.N. Galloway and E.B. Cowling, "The Effects of Precipitation on Aquatic and Terrestrial Ecosystems: A Proposed Precipitation Chemistry Network," *Journal of the Air Pollution Control Association,* 28, pp. 229–235, 1978.

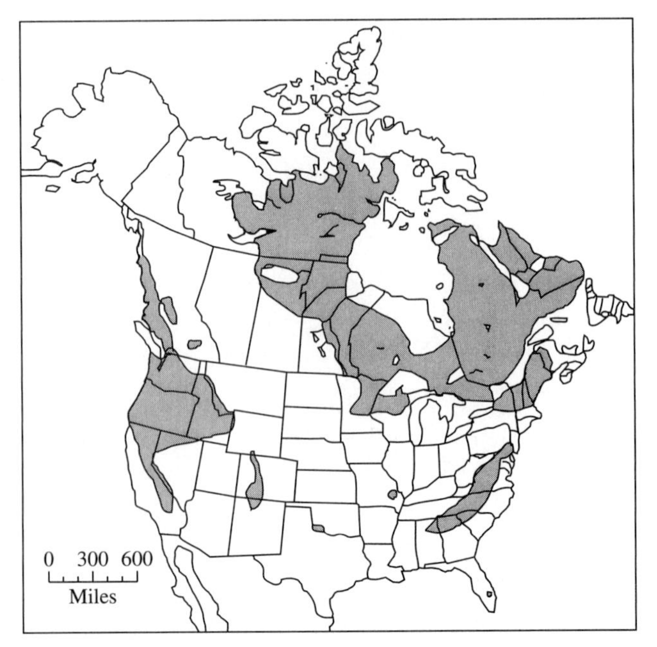

FIGURE 4-16
Regions in North America containing lakes sensitive to acidification by acid precipitation. The shaded areas have igneous or metamorphic bedrock geology; the unshaded areas have calcereous or sedimentary bedrock geology. Regions having low alkalinity lakes are concurrent with regions of igneous and metamorphic bedrock geology.

4-5 CHAPTER REVIEW

When you have completed studying this chapter, you should be able to do the following without the aid of your text or notes:

1. List the major pollutant categories (there are four) that are produced by each of the four principal sources of wastewater.
2. List the two nutrients of primary concern with respect to a receiving body of water.
3. Define biochemical oxygen demand (BOD).
4. Explain the procedure for determining BOD and specify the nominal values of temperature and time used in the test.
5. List three reasons why the BOD rate constant may vary.
6. Sketch a graph showing the effect of varying rate constant on five-day BOD if the ultimate BOD is the same, and the effect on ultimate BOD if the five-day BOD is the same.

7. Utilizing Equation 4-19 in your answer, explain what causes nitrogenous BOD.
8. Sketch a series of curves that show the deoxygenation, reaeration, and DO sag in a river. Show the effect of a change in the deoxygenation or reaeration rate on the location of the critical point and the magnitude of the DO deficit.
9. List three reasons why ammonia nitrogen is detrimental to a receiving body of water and its inhabitants.
10. Sketch and compare the epilimnion and hypolimnion with respect to the following: location in a lake, temperature, and oxygen abundance (that is, DO).
11. Describe the process of stratification and turnover in lakes.
12. Explain what determines the euphotic zone of a lake and what significance this has for biological growth.
13. Given a description of a lake that includes the productivity, clarity, and oxygen levels, classify it as oligotrophic, mesotrophic, eutrophic, or senescent.
14. Explain the process of eutrophication.
15. State Liebig's law of the minimum.
16. Name the most common "limiting nutrient" in lakes and explain why it is a limiting nutrient.
17. List three sources of phosphorus that must be controlled to reduce cultural eutrophication of lakes.
18. Explain why the pH of pure rainwater is about 5.6.
19. Define acid rain.
20. Explain why acid rain is of concern.
21. Explain the role of calcareous soils in protecting lakes from acidification.
22. Other than rainwater pH, list six variables that determine the extent of lake acidification and explain how increasing or decreasing the value of each might be expected to change the extent of acidification.

With the aid of this text you should be able to do the following:

1. Calculate the BOD_5, given the sample size and oxygen consumption, or calculate the sample size, given the allowable oxygen consumption and estimated BOD_5.
2. Calculate the ultimate BOD (L_o), given the BOD exerted (BOD_t) in time t and rate constant, or calculate the rate constant, k, given L_o and BOD_t.
3. Calculate a new k for a temperature other than 20°C, given a value at T°C.
4. Calculate the BOD rate constant (k) and ultimate BOD (L_o) from experimental data of BOD versus time.
5. Calculate the oxygen deficit, D, in a length of stream (reach), given the required input data.
6. Calculate the critical oxygen deficit, D_c, at the DO sag point (minimum).

4-6 PROBLEMS

4-1. Glutamic acid ($C_5H_9O_4N$) is used as one of the reagents for a standard to check the BOD test. Determine the theoretical oxygen demand of 150 mg/L of glutamic acid. Assume the following reactions apply:

$$C_5H_9O_4N + 4.5O_2 \rightleftharpoons 5CO_2 + 3H_2O + NH_3$$

$$NH_3 + 2O_2 \rightleftharpoons NO_3^- + H^+ + H_2O$$

4-2. Bacterial cells have been represented by the chemical formula $C_5H_7NO_2$. Compute the theoretical oxygen demand assuming the following reactions apply:

$$C_5H_7NO_2 + 5O_2 \rightleftharpoons 5CO_2 + 2H_2O + NH_3$$

$$NH_3 + 2O_2 \rightleftharpoons NO_3^- + H^+ + H_2O$$

4-3. If the BOD_5 of a waste is 220.0 mg/L and the ultimate BOD is 320.0 mg/L, what is the rate constant (base 10)?

Answer: $K = 0.101\ d^{-1}$.

4-4. If the BOD of a municipal wastewater at the end of seven days is 60.0 mg/L and the ultimate BOD is 85.0 mg/L, what is the rate constant (base 10)?

4-5. Convert the rate constant found in Problem 4-3 to base *e*.

Answer: $k = 0.233\ d^{-1}$.

4-6. Convert the rate constant found in Problem 4-4 to base *e*.

4-7. Assuming that the data in Problem 4-3 were taken at 20°C, compute the rate constant at a temperature of 15°C.

Answer: $K = 0.0536\ d^{-1}$.

4-8. Assuming that the data in Problem 4-4 were taken at 25°C, compute the rate constant at 16°C.

4-9. What is the BOD_5 of a waste that yields an oxygen consumption of 2.00 mg/L from a 1.00% sample?

Answer: $BOD_5 = 200$ mg/L.

4-10. What sample size (in percent) is required for a BOD_5 of 350.0 mg/L if the oxygen consumed is to be limited to 6.00 mg/L?

4-11. If the ultimate BOD of two wastes having K values of 0.0800 d^{-1} and 0.120 d^{-1} is 280.0 mg/L, what would be the five-day BOD for each?

Answer: For $K = 0.08\ d^{-1}$, $BOD_5 = 169$ mg/L; for $K = 0.12\ d^{-1}$, $BOD_5 =$ 210 mg/L.

4-12. If the BOD_5 of two wastes having K values of 0.0800 d^{-1} and 0.120 d^{-1} is 280.0 mg/L, what would be the ultimate BOD for each?

4-13. Plot the BOD curves that would result for the data given in Problem 4-11. At approximately what day (± 5.0 d) does the ultimate BOD occur? Check your answer using Equation 4-4.

Answer: Ultimate BOD occurs at $t = 34.5$ d.

4-14. Plot the BOD curves that would result for the data given in Problem 4-12. At approximately what day (±5.0 d) would the ultimate BOD occur for each waste? Check your answer using Equation 4-4.

4-15. Using the Thomas Method, calculate the BOD rate constant in base e from the following data:

Day	BOD, mg/L
2	86
5	169
10	236
20	273
35	279.55

4-16. Using the Thomas Method, calculate the BOD rate constant in base e from the following data:

Day	BOD, mg/L
2	119
5	210
10	262
20	279
35	279.98

4-17. Using the data from Problem 4-1, calculate the NBOD of glutamic acid.

4-18. Using the data from Problem 4-2, calculate the NBOD of bacterial cells.

4-19. Derive an expression for the final temperature (T_s) of the mixture of wastewater flow (Q_w) at temperature (T_w) and river flow (Q_r) at temperature (T_r). Assume that the specific heat and density of the wastewater and the river are the same.

4-20. A tannery with a wastewater flow of 0.011 m^3/s and a BOD_5 of 590 mg/L discharges into the Cattaraugus Creek. The creek has a 10-year, 7-day low flow of 1.7 m^3/s. Upstream of the tannery, the BOD_5 of the creek is 0.6 mg/L. The BOD rate constants (k) are 0.115 d^{-1} for the tannery and 3.7 d^{-1} for the creek. Calculate the initial ultimate BOD after mixing.

4-21. A short distance downstream from the tannery in Problem 4-20, a glue factory and a municipal wastewater treatment plant also discharge into Cattaraugus Creek. The wastewater flows and ultimate BODs for these discharges are listed below. Determine the initial ultimate BOD after mixing of the creek and the three wastewater discharges.

Source	Flow, m^3/s	Ultimate BOD, mg/L
Glue factory	0.13	255
Municipal WWTP	0.02	75

4-22. Compute the deoxygenation rate constant and reaeration rate constant (base *e*) for the following wastewater and stream conditions.

Source	k, d^{-1}	Temp., °C	H, m	v, m/s	η
Wastewater	0.20	20			
Stream		20	1.0	0.5	0.4

4-23. During flood stage, the stream conditions in Problem 4-22 change as shown below. Determine the values of k_d and k_r for flood conditions.

Source	k, d^{-1}	Temp., °C	H, m	v, m/s	η
Wastewater	0.20	20			
Stream		20	4.0	2.5	0.6

4-24. The initial ultimate BOD after mixing of the Noir River is 50 mg/L. The DO in the river after the wastewater and river have mixed is at saturation. The river temperature is 10°C. At 10°C, the deoxygenation rate constant (k_d) is 0.30 d^{-1} and the reaeration rate constant (k_r) is 0.30 d^{-1}. Determine the critical point (t_c) and the critical DO.

4-25. Repeat Problem 4-24 assuming the river temperature rises to 15°C so that k_d and k_r change.

4-26. The discharge from a sugar beet plant causes the DO at the critical point to fall to 4.0 mg/L. The stream has a negligible BOD and the initial deficit after the river and wastewater have mixed is zero. What DO will result if the concentration of the waste (L_w) is reduced by 50 percent? Assume that the flows remain the same and that the saturation value of DO is 10.83 mg/L in both cases.

4-27. The Watertown town council has asked that you determine whether or not the discharge of the town's wastewater into the Green River will reduce the DO below the state standard of 5.00 mg/L at Smithville, 5.79 km downstream, or at any other point downstream. The pertinent data are as follows:

Parameter	Watertown wastewater	Green River
Flow, m^3/s	0.280	0.877
Ultimate BOD at 28°C, mg/L	6.44	7.00
DO, mg/L	1.00	6.00
K_d at 28°C, d^{-1}	N/A	0.199
K_r at 28°C, d^{-1}	N/A	0.370
Speed, m/s	N/A	0.650
Temperature, °C	28°C	28°C

Answers: DO at Smithville = 4.75 mg/L

Critical DO = 4.72 mg/L at t_c = 0.3149 d.

4-28. If the Green River temperature in Problem 4-27 decreases to 12°C, will the discharge from Watertown reduce the DO in the river below 5.00 mg/L at a distance of 5.79 km downstream? *Note:* You must calculate a new temperature of the mixed river water and wastewater, then correct K_d and K_r of the river only.

4-29. The town of Avepitaeonmi has filed a complaint with the state Department of Natural Resources (DNR) that the City of Watapitae is restricting its use of the Wash River because of the discharge of raw sewage. The DNR water quality criterion for the Wash River is 5.00 mg/L of DO. Avepitaeonmi is 15.55 km downstream from Watapitae. What is the DO at Avepitaeonmi? What is the critical DO and where (at what distance) downstream does it occur? Is the assimilative capacity of the river restricted? The following data pertain to the 7-year, 10-day low flow at Watapitae.

Parameter	Watapitae wastewater	Wash River
Flow, m^3/s	0.1507	1.08
BOD_5 at 16°C, mg/L	128.00	N/A
BOD_u at 16°C, mg/L	N/A	11.40
DO, mg/L	1.00	7.95
Temperature, °C	16.0	16.0
k at 20°C	0.4375	N/A
Speed, m/s	N/A	0.390
Depth, m	N/A	2.80
Bed-activity coefficient	N/A	0.200

4-30. Under the provisions of the Clean Water Act, the U.S. Environmental Protection Agency established a requirement that municipalities had to provide secondary treatment of their waste. This was defined to be treatment that resulted in an effluent BOD_5 that did not exceed 30 mg/L. The discharge from Watapitae (Problem 4-29) is clearly in violation of this standard. Given the data in Problem 4-29, rework the problem assuming that Watapitae provides treatment to lower the BOD_5 to 30.00 mg/L.

4-31. If the population and water use of Watapitae (Problems 4-29 and 4-30) are growing at 5 percent per year with a corresponding increase in wastewater flow, how many years' growth may be sustained before secondary treatment becomes inadequate? Assume that the treatment plant continues to maintain an effluent BOD_5 of 30.00 mg/L.

4-32. When ice covers a river, it severely limits the reaeration. There is some compensation for the reduced aeration because of the reduced water temperature. The lower temperature reduces the biological activity and, thus, the deaeration rate and, at the same time, the DO saturation level increases. Assuming a winter condition, rework Problem 4-29 with the reaeration reduced to 0 and the river water temperature at 2°C.

4-33. What combination of BOD reduction and/or wastewater DO increase is required so the Watertown wastewater in Problem 4-27 does not reduce the DO below 5.00 mg/L anywhere along the Green River? Assume that the cost of BOD reduction is three to five times that of increasing the effluent DO. Since the cost of adding extra DO is high, limit the excess above the minimum amount such that the critical DO falls between 5.00 mg/L and 5.25 mg/L.

Answer: Raising the wastewater DO to 2.7 mg/L is the most cost-effective remedy.

4-34. What amount of ultimate BOD, in kg/d, may Watapitae (Problem 4-29) discharge and still allow Avepitaeonmi 1.50 mg/L of DO above the DNR water quality criteria for assimilation of its waste?

4-35. Assuming that the mixed oxygen deficit (D_a) is zero and that the ultimate BOD (L_r) of the Looking Glass River above the wastewater outfall from Carrollville is zero, calculate the amount of ultimate BOD, in kg/d, that can be discharged if the DO must be kept at 4.00 mg/L at a point 8.05 km downstream. The stream deoxygenation rate (K_d) is 1.80 d^{-1} at 12°C, and the reaeration rate (K_r) is 2.20 d^{-1} at 12°C. The river temperature is 12°C. The river flow is 5.95 m^3/s with a speed of 0.300 m/s. The Carrollville wastewater flow is 0.0130 m^3/s.

Answer: $Q_w L_w = 1.14 \times 10^4$ kg/d of ultimate BOD.

4-36. Calculate the DO at a point 1.609 km downstream from a waste discharge point for the following conditions. Report answers to two decimal places. Rate constants are already temperature adjusted.

Parameters	Stream
k_d	1.911 d^{-1}
k_r	4.49 d^{-1}
Flow	2.4 m^3/s
Speed	0.100 m/s
D_a (after mixing)	0.00
Temperature, °C	17.00
BOD_L (after mixing)	1100.00 kg/d

4-37. Assume that the Carrollville wastewater (Problem 4-35) also contains 3.0 mg/L of ammonia nitrogen with a stream deoxygenation rate of 0.900 d^{-1} at 12°C. What is the amount of ultimate carbonaceous BOD, in kg/d, that Carrollville can discharge and still meet the DO level of 4.00 mg/L at a point 8.05 km downstream? Assume also that the theoretical amount of oxygen will ultimately be consumed in the nitrification process.

4-7 DISCUSSION QUESTIONS

4-1. Students in a graduate-level environmental engineering laboratory took samples of the influent (raw sewage) and effluent (treated sewage) of a municipal wastewater treatment plant. They used these samples to determine the BOD rate constant (k). Would you expect the rate constants to be the same or different? If different, which would be higher and why?

4-2. If it were your job to set standards for a water body and you had a choice of either BOD_5 or ultimate BOD, which would you choose and why?

4-3. A summer intern has turned in his log book for temperature measurements for a limnology survey. He was told to take the measurements in the air 1 m above the lake, 1 m deep in the lake, and at a depth of 10 m. He turned in the following results but did not record which temperatures were taken where. If the measurements were made at noon in July in Missouri, what is your best guess as to the location of the measurements (i.e. air, 1-m deep, 10-m deep)? The recorded values were: 33°C, 18°C, and 21°C.

4-4. If the critical point in a DO sag curve is found to be 18 km downstream from the discharge point of untreated wastewater, would you expect the critical point to move upstream (toward the discharge point), downstream, or remain in the same place, if the wastewater is treated?

4-5. You have been assigned to conduct an environmental study of a remote lake in Canada. Aerial photos and a ground-level survey reveal no anthropogenic waste sources are contributing to the lake. When you investigate the lake, you find a highly turbid lake with abundant mats of floating algae and a hypolimnion DO of 1.0 mg/L. What productivity class would you assign to this lake? Explain your reasoning.

4-6. The lakes in Illinois, Indiana, western Kentucky, the lower peninsula of Michigan, and Ohio do not appear to be subject to acidification even though the rainwater pH is 4.4. Based on your knowledge (or what you can discover by research) of the topography, vegetation, and bedrock, explain why the lakes in these areas are not acidic.

4-8 ADDITIONAL READING

S. D. Morton, *Water Pollution Causes and Cures,* Madison, Wisconsin: Mimir Publications, 1976.
N. L. Nemerow, *Scientific Stream Pollution Analysis,* New York: McGraw-Hill, 1974.
G. Tchobanoglous and E. D. Schroeder, *Water Quality,* Reading, MA: Addison Wesley, 1985.

CHAPTER 5

WASTEWATER TREATMENT

5-1 WASTEWATER MICROBIOLOGY
Role of Microorganisms
Classification of Microorganisms
Some Microbes of Interest in Wastewater Treatment
Bacterial Biochemistry
Decomposition of Waste
Population Dynamics
5-2 CHARACTERISTICS OF DOMESTIC WASTEWATER
Physical Characteristics of Domestic Wastewater
Chemical Characteristics of Domestic Wastewater
Characteristics of Industrial Wastewater
5-3 ON-SITE DISPOSAL SYSTEMS
Without Water Carriage
With Water Carriage
5-4 MUNICIPAL WASTEWATER TREATMENT SYSTEMS
Pretreatment of Industrial Wastes
5-5 UNIT OPERATIONS OF PRETREATMENT
Bar Racks
Grit Chambers
Comminutors
Equalization
5-6 PRIMARY TREATMENT
5-7 UNIT PROCESSES OF SECONDARY TREATMENT
Overview

Trickling Filters
Activated Sludge
Oxidation Ponds
Rotating Biological Contactors (RBCs)
5-8 DISINFECTION
5-9 ADVANCED WASTEWATER TREATMENT
Filtration
Carbon Adsorption
Phosphorus Removal
Nitrogen Control
5-10 LAND TREATMENT
Slow Rate
Overland Flow
Rapid Infiltration
5-11 SLUDGE TREATMENT
Sources and Characteristics of Various Sludges
Solids Computations
Thickening
Stabilization
Sludge Conditioning
Sludge Dewatering
Reduction
5-12 SLUDGE DISPOSAL
Ultimate Disposal
Land Spreading
Landfilling
Dedicated Land Disposal (DLD)
Utilization
Sludge Disposal Regulations
5-13 CHAPTER REVIEW
5-14 PROBLEMS
5-15 DISCUSSION QUESTIONS
5-16 ADDITIONAL READING

5-1 WASTEWATER MICROBIOLOGY

Role of Microorganisms

The stabilization of organic matter is accomplished biologically using a variety of microorganisms. The microorganisms convert the colloidal and dissolved carbonaceous organic matter into various gases and into protoplasm. Because protoplasm has a specific gravity slightly greater than that of water, it can be removed from the treated liquid by gravity settling.

It is important to note that unless the protoplasm produced from the organic matter is removed from the solution, complete treatment will not be accomplished because the protoplasm, which itself is organic, will be measured as BOD in the effluent. If the protoplasm is not removed, the only treatment that will be achieved is that associated with the bacterial conversion of a portion of the organic matter originally present to various gaseous end products.[1]

Classification of Microorganisms

By kingdoms. Microorganisms are organized into five broad groups based on their structural and functional differences. The groups are called *kingdoms*. The five kingdoms are *animals, plants, protista, fungi,* and *bacteria.* Representative examples and characteristics of differentiation are shown in Figure 5-1.

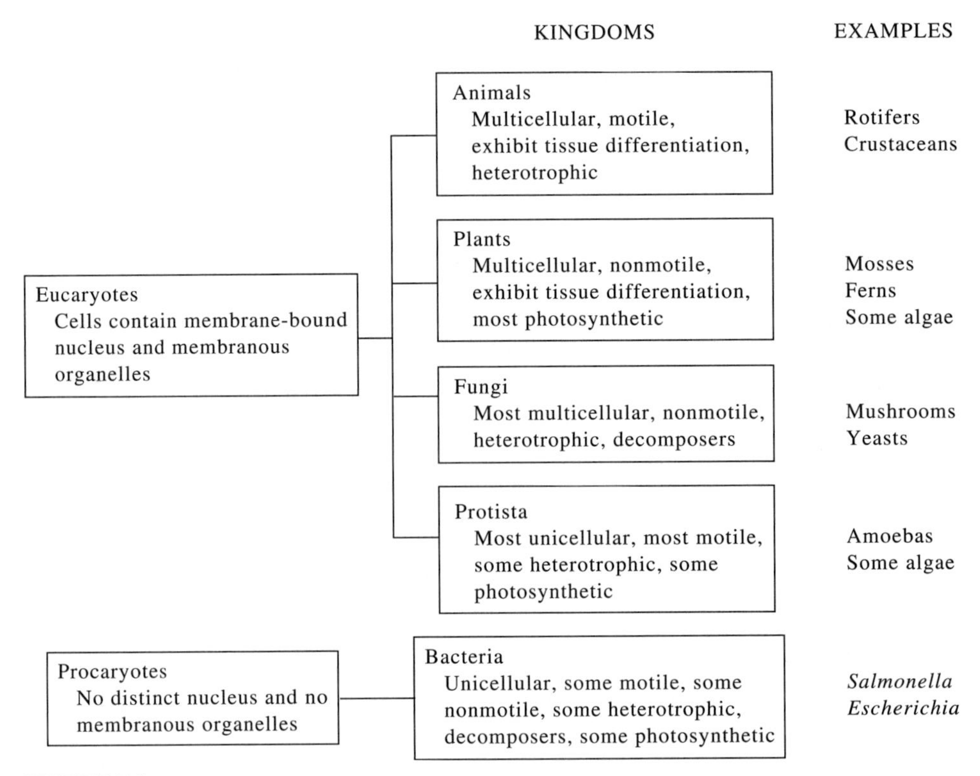

FIGURE 5-1
Classification of microorganisms by kingdom.

---

[1] Metcalf & Eddy, Inc., and G. Tchobanoglous, *Wastewater Engineering: Treatment, Disposal, Reuse,* New York: McGraw-Hill, p. 395, 1979. Reprinted by permission.

By energy and carbon source. The relationship between the source of carbon and the source of energy for the microorganism is important. Carbon is the basic building block for cell synthesis. A source of energy must be obtained from outside the cell to enable synthesis to proceed. Our goal in wastewater treatment is to convert both the carbon and the energy in the wastewater in the cells of microorganisms, which we can remove from the water by settling. Therefore, we wish to encourage the growth of organisms that use organic material for both their carbon and energy source.

If the microorganism uses organic material as a supply of carbon, it is called *heterotrophic. Autotrophs* require only CO_2 to supply their carbon needs.

Organisms that rely only on the sun for energy are called *phototrophs. Chemotrophs* extract energy from organic or inorganic oxidation/reduction reactions. *Organotrophs* use organic materials, while *lithotrophs* oxidize inorganic compounds.[2]

By their relationship to oxygen. Bacteria also are classified by their ability or inability to utilize oxygen as a terminal electron acceptor[3] in oxidation/reduction reactions. *Obligate aerobes* are microorganisms that must have oxygen as the terminal electron acceptor. When wastewater contains oxygen and can support obligate aerobes, it is called *aerobic.*

Obligate anaerobes are microorganisms that cannot survive in the presence of oxygen. They cannot use oxygen as a terminal electron acceptor. Wastewater that is devoid of oxygen is called *anaerobic. Facultative anaerobes* can use oxygen as the terminal electron acceptor and, under certain conditions, they can also grow in the absence of oxygen.

Under *anoxic* conditions, a group of facultative anaerobes called *denitrifiers* utilizes nitrites (NO_2^-) and nitrates (NO_3^-) as the terminal electron acceptor. Nitrate nitrogen is converted to nitrogen gas in the absence of oxygen. This process is called *anoxic denitrification.*

By their preferred temperature regime. Each species of bacteria reproduces best within a limited range of temperatures. Four temperature ranges are used to classify

---

[2]J. E. Bailey and D. F. Ollis, *Biochemical Engineering Fundamentals*, New York: McGraw-Hill, p. 222, 1977.

[3]An organic substrate is not directly oxidized to carbon dioxide and water in a single chemical step because there is no energy-conserving mechanism that could trap so much energy. Thus, biological oxidation occurs in small steps. Oxidation requires the transfer of an electron from the substance being oxidized to some acceptor molecule that will subsequently be reduced. In most biological systems, each step in the oxidation process involves the removal of two electrons and the simultaneous loss of two protons (H^+). The combination of the two losses is equivalent to the molecule having lost two hydrogen atoms. The reaction is often referred to as *dehydrogenation*. The electrons and protons are not released into the cell, but are transferred to an acceptor molecule. The acceptor molecule will not accept the protons until it has accepted the electrons and thus it is referred to as an electron acceptor. Since the net result of accepting an electron and proton is the same as accepting a hydrogen atom, such acceptors are also called hydrogen acceptors. (C. P. L. Grady and H. C. Lim, *Biological Wastewater Treatment, Theory and Applications,* New York: Marcel Dekker, 1980.)

bacteria. Those that grow best at temperatures below 20°C are called *psychrophiles. Mesophiles* grow best at temperatures between 25 and 40°C. Between 45 and 60°C, the *thermophiles* grow best. Above 60°C, *stenothermophiles* grow best. The growth range of *facultative thermophiles* extends from the thermophilic range into the mesophilic range. These ranges are qualitative and somewhat subjective. You will note the gaps between 20 and 25°C and between 40 and 45°C. Don't make the mistake of saying that an organism that grows well at 20.5°C is a mesophile. The rules just aren't that hard and fast. Bacteria will grow over a range of temperatures and will survive at a very large range of temperatures. For example, *Escherichia coli,* classified as mesophiles, will grow at temperatures between 20 and 50°C and will reproduce, albeit very slowly, at temperatures down to 0°C. If frozen rapidly, they and many other microorganisms can be stored for years with no significant death rate.

Some Microbes of Interest in Wastewater Treatment

Bacteria. The highest population of microorganisms in a wastewater treatment plant will belong to the bacteria. They are single-celled organisms which use soluble food. Conditions in the treatment plant are adjusted so that chemoheterotrophs predominate. No particular species is selected as "the best."

Fungi. Fungi are multicellular, nonphotosynthetic, heterotrophic organisms. Fungi are obligate aerobes that reproduce by a variety of methods including fission, budding, and spore formation. Their cells require only half as much nitrogen as bacteria so that in a nitrogen-deficient wastewater, they predominate over the bacteria.[4]

Algae. This group of microorganisms are photoautotrophs and may be either unicellular or multicellular. Because of the chlorophyll contained in most species, they produce oxygen through photosynthesis. In the presence of sunlight, the photosynthetic production of oxygen is greater than the amount used in respiration. At night they use up oxygen in respiration. If the daylight hours exceed the night hours by a reasonable amount, there is a net production of oxygen.

Protozoa. Protozoa are single-celled organisms that can reproduce by *binary fission* (dividing in two). Most are aerobic chemoheterotrophs, and they often consume bacteria. They are desirable in wastewater effluents because they act as polishers in consuming the bacteria.

Rotifers and crustaceans. Both rotifers and crustaceans are animals—aerobic, multicellular chemoheterotrophs. The rotifer derives its name from the apparent

[4]Ross E. McKinney, *Microbiology for Sanitary Engineers,* New York: McGraw-Hill, p. 40, 1962.

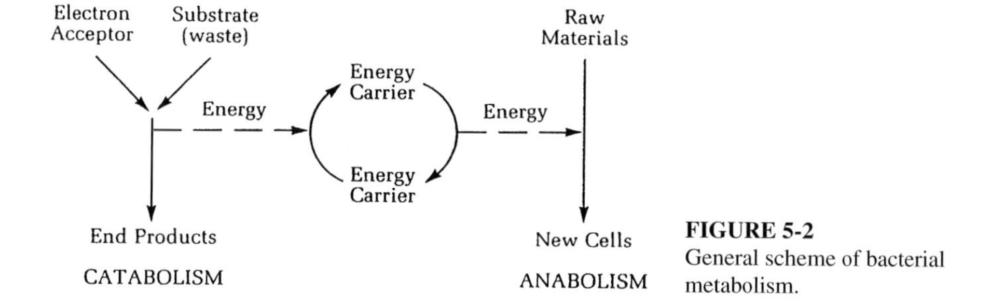

FIGURE 5-2
General scheme of bacterial metabolism.

rotating motion of two sets of cilia on its head. The cilia provide mobility and a mechanism for catching food. Rotifers consume bacteria and small particles of organic matter.

Crustaceans, a group that includes shrimp, lobsters, and barnacles, are characterized by their shell structure. They are a source of food for fish and are not found in wastewater treatment systems to any extent except in underloaded lagoons. Their presence is indicative of a high level of dissolved oxygen and a very low level of organic matter.

Bacterial Biochemistry

Metabolism. The general term that describes all of the chemical activities performed by a cell is *metabolism.* This in turn is divided into two parts: catabolism and anabolism. *Catabolism* includes all the biochemical processes by which a substrate is degraded to end products with the release of energy.[5] In wastewater treatment, the substrate is oxidized. The oxidation process releases energy that is transferred to an energy carrier which stores it for future use by the bacterium (Figure 5-2). Some chemical compounds released by catabolism are used by the bacterial cell for its life functions.

Anabolism includes all the biochemical processes by which the bacterium synthesizes new chemical compounds needed by the cells to live and reproduce. The synthesis process is driven by the energy that was stored in the energy carrier.

Decomposition of Waste

The type of electron acceptor available for catabolism determines the type of decomposition (that is, aerobic, anoxic, or anaerobic) used by a mixed culture of microorganisms. Each type of decomposition has peculiar characteristics which affect its use in waste treatment.

---

[5]Substrate is food. For our application, "food" is the organic material from the human digestive tract and other biodegradable wastes.

Aerobic decomposition. From our discussion of bacterial metabolism you will recall that molecular oxygen (O_2) must be present as the terminal electron acceptor for decomposition to proceed by aerobic oxidation. As in natural water bodies, the oxygen is measured as DO. When oxygen is present, it is the only terminal electron acceptor used. Hence, the chemical end products of decomposition are primarily carbon dioxide, water, and new cell material (Table 5-1). Odiferous gaseous end products are kept to a minimum. In healthy natural water systems, aerobic decomposition is the principal means of self-purification.

A wider spectrum of organic material can be oxidized aerobically than by any other type of decomposition. This fact, coupled with the fact that the final end products are oxidized to a very low energy level, results in a more stable end product (that is, one that can be disposed of without damage to the environment and without creating a nuisance condition) than can be achieved by the other oxidation systems.

Because of the large amount of energy released in aerobic oxidation, most aerobic organisms are capable of high growth rates. Consequently, there is a relatively large production of new cells in comparison with the other oxidation systems. This means that more biological sludge is generated in aerobic oxidation than in the other oxidation systems.

Aerobic decomposition is the method of choice for large quantities of dilute wastewater (BOD_5 less than 500 mg/L) because decomposition is rapid, efficient, and has a low odor potential. For high-strength wastewater (BOD_5 is greater than 1,000 mg/L), aerobic decomposition is not suitable because of the difficulty in supplying enough oxygen and because of the large amount of biological sludge produced. In small communities and in special industrial applications where aerated lagoons (see Section 5-7) are used, wastewaters with BOD_5 up to 3,000 mg/L may be treated satisfactorily by aerobic decomposition.

Anoxic decomposition. Some microorganisms can use nitrate (NO_3^-) as the terminal electron acceptor in the absence of molecular oxygen. Oxidation by this route is called denitrification.

The end products from denitrification are nitrogen gas, carbon dioxide, water, and new cell material. The amount of energy made available to the cell during denitrification is about the same as that made available during aerobic decomposition. As a consequence, the rate of production of new cells, although not as high as in aerobic decomposition, is relatively high.

Denitrification is of importance in wastewater treatment where nitrogen must be removed to protect the receiving body. In this case, a special treatment step is added to the conventional process for removal of carbonaceous material. Denitrification will be discussed in detail later.

One other important aspect of denitrification is in relation to final clarification of the treated wastewater. If the environment of the final clarifier becomes anoxic, the formation of nitrogen gas will cause large globs of sludge to float to the surface and escape from the treatment plant into the receiving water. Thus, it is necessary to ensure that anoxic conditions do not develop in the final clarifier.

TABLE 5-1
Waste decomposition end products

Substrates	Representative end products		
	Aerobic decomposition	**Anoxic decomposition**	**Anaerobic decomposition**
Proteins and other organic nitrogen compounds	Amino acids Ammonia → nitrites → nitrates Alcohols, Organic acids } → $CO_2 + H_2O$	Amino acids Nitrates → nitrites → N_2 Alcohols, Organic acids } → $CO_2 + H_2O$	Amino acids Ammonia Hydrogen sulfide Methane Carbon dioxide Alcohols Organic acids
Carbohydrates	Alcohols, Fatty acids } → $CO_2 + H_2O$	Alcohols, Fatty acids } → $CO_2 + H_2O$	Carbon dioxide Alcohols Fatty acids Methane
Fats and related substances	Fatty acids + glycerol Alcohols, Lower fatty acids } → $CO_2 + H_2O$	Fatty acids + glycerol Alcohols, Lower fatty acids } → $CO_2 + H_2O$	Fatty acids + glycerol Carbon dioxide Alcohols Lower fatty acids Methane

Source: After Pelczar and Reid, *Microbiology,* New York: McGraw-Hill, 1958.

Anaerobic decomposition. In order to achieve anaerobic decomposition, molecular oxygen and nitrate must not be present as terminal electron acceptors. Sulfate (SO_4^{2-}), carbon dioxide, and organic compounds that can be reduced serve as terminal electron acceptors. The reduction of sulfate results in the production of hydrogen sulfide (H_2S) and a group of equally odiferous organic sulfur compounds called *mercaptans.*

The anaerobic decomposition (fermentation) of organic matter generally is considered to be a two-step process. In the first step, complex organic compounds are fermented to low-molecular-weight fatty acids (volatile acids). In the second step, the organic acids are converted to methane. Carbon dioxide serves as the electron acceptor.

Anaerobic decomposition yields carbon dioxide, methane, and water as the major end products. Additional end products include ammonia, hydrogen sulfide, and mercaptans. As a consequence of these last three compounds, anaerobic decomposition is characterized by an unbelievably horrid stench!

Because only small amounts of energy are released during anaerobic oxidation, the amount of cell production is low. Thus, sludge production is low. We make use of this fact in wastewater treatment by using anaerobic decomposition to stabilize sludges produced during aerobic and anoxic decomposition.

Direct anaerobic decomposition of wastewater generally is not feasible for dilute waste.[6] The optimum growth temperature for the anaerobic bacteria is at the upper end of the mesophilic range. Thus, to get reasonable biodegradation, we must elevate the temperature of the culture. For dilute wastewater, this is not practical. For concentrated wastes (BOD_5 greater than 1,000 mg/L), anaerobic digestion is quite appropriate.

Population Dynamics

Bacterial growth requirements. In the discussion of the behavior of bacterial cultures which follows, there is the inherent assumption that all the requirements for growth are initially present. Since these requirements are fairly extensive and stringent, it is worth taking a moment to recapitulate them. The following list summarizes the major requirements that must be satisfied:

1. A terminal electron acceptor
2. Macronutrients
   a. Carbon to build cells
   b. Nitrogen to build cells
   c. Phosphorus for ATP (energy carrier) and DNA

[6] Some researchers are exploring the use of anaerobic systems for treatment of dilute wastes, especially groundwater contaminated with hazardous waste.

3. Micronutrients
   a. Trace metals
   b. Vitamins are required by some bacteria
4. Appropriate environment
   a. Moisture
   b. Temperature
   c. pH

Growth in pure cultures. As an illustration, let us examine a hypothetical situation in which 1,400 bacteria of a single species are introduced into a synthetic liquid medium. Initially nothing appears to happen. The bacteria must adjust to their new environment and begin to synthesize new protoplasm. On a plot of bacterial growth versus time (Figure 5-3), this phase of growth is called the *lag phase.*

At the end of the lag phase the bacteria begin to divide. Since all of the organisms do not divide at the same time, there is a gradual increase in population. This phase is labeled *accelerated growth* on the growth plot.

At the end of the accelerated growth phase, the population of organisms is large enough and the differences in generation time are small enough that the cells appear to divide at a regular rate. Since reproduction is by binary fission (each cell divides producing two new cells), the increase in population follows in geometric progression: 1 $\rightarrow$ 2 $\rightarrow$ 4 $\rightarrow$ 8 $\rightarrow$ 16 $\rightarrow$ 32, and so forth. The population of bacteria

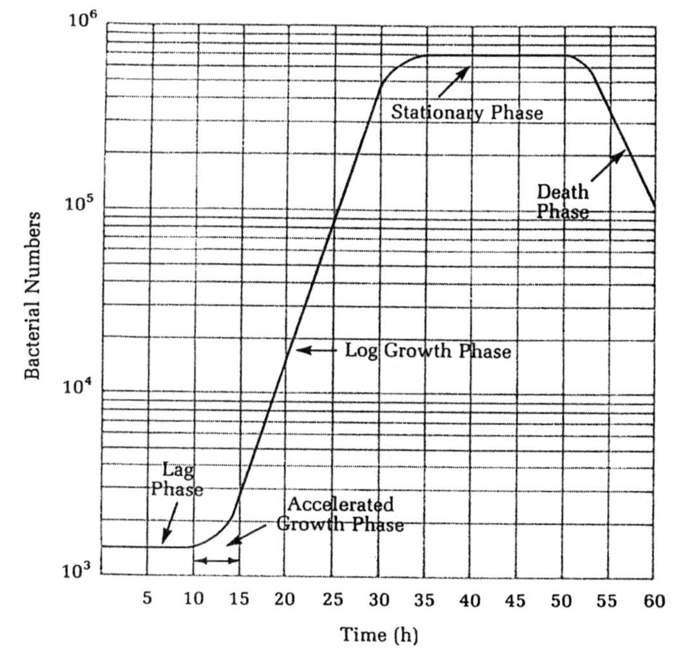

FIGURE 5-3
Bacterial growth in a pure culture: the "log-growth curve."

(P) after the nth generation is given by the following expression:

$$P = P_0(2)^n \tag{5-1}$$

where P_0 is the initial population at the end of the accelerated growth phase. If we take the log of both sides of Equation 5-1, we obtain the following:

$$\log P = \log P_0 + n \log 2 \tag{5-2}$$

This means that if we plot bacterial population on a logarithmic scale, this phase of growth would plot as a straight line of slope n and intercept P_0 at t_0 equal to the end of the accelerated growth phase. Thus, this phase of growth is called the *log growth* or *exponential growth phase.*

The log growth phase tapers off as the substrate becomes exhausted or as toxic by-products build up. Thus, at some point the population becomes constant either as a result of cessation of fission or a balance in death and reproduction rates. This is depicted by the *stationary phase* on the growth curve.

Following the stationary phase, the bacteria begin to die faster than they reproduce. This death phase is due to a variety of causes that are basically an extension of those which lead to the stationary phase.

Growth in mixed cultures. In wastewater treatment, as in nature, pure cultures of microorganisms do not exist. Rather, a mixture of species compete and survive within the limits set by the environment. *Population dynamics* is the term used to describe the time-varying success of the various species in competition. It is expressed quantitatively in terms of relative mass of microorganisms.[7]

The prime factor governing the dynamics of the various microbial populations is the competition for food. The second most important factor is the predator-prey relationship.

The relative success of a pair of species competing for the same substrate is a function of the ability of the species to metabolize the substrate. The more successful species will be the one that metabolizes the substrate more completely. In so doing, it will obtain more energy for synthesis and consequently will achieve a greater mass.

Because of their relatively smaller size and, thus, larger surface area per unit mass, which allows a more rapid uptake of substrate, bacteria will predominate over fungi. For the same reason, the fungi predominate over the protozoa.

When the supply of soluble organic substrate becomes exhausted, the bacterial population is less successful in reproduction and the predator populations increase. In a closed system with an initial inoculum of mixed microorganisms and substrate, the populations will cycle as the bacteria give way to higher level organisms which in turn die for lack of food and are then decomposed by a different set of bacteria (Figure 5-4). In an open system, such as a wastewater treatment plant or a river,

[7]If each individual organism of species A has, on the average, twice the mass at maturity as each individual organism of species B, and both compete equally, we would expect that both would have the same total biomass, but that there would be twice as many of species B as there would be of A.

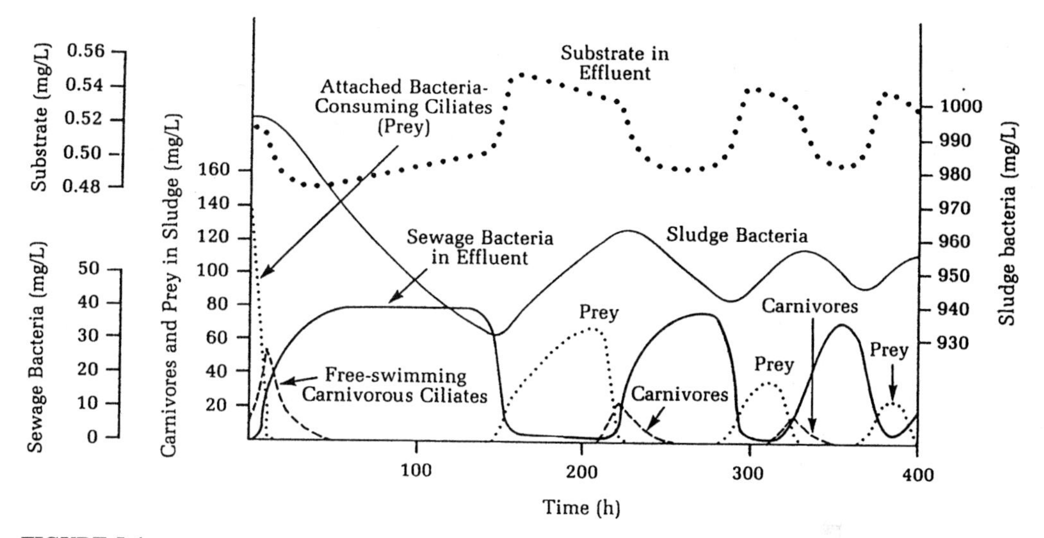

FIGURE 5-4
Population dynamics in a closed system. (*Source:* Curds, "A Theoretical Study of Factors Influencing the Microbial Population Dynamics of the Activated Sludge Process-I." *Water Resources,* vol. 7, p. 1269, 1973.)

with a continuous inflow of new substrate, the predominant populations will change through the length of the plant (Figure 5-5). This condition is known as *dynamic equilibrium.* It is a highly sensitive state, and changes in influent characteristics must be regulated closely to maintain the proper balance of the various populations.

The Monod equation. For the large numbers and mixed cultures of microorganisms found in waste treatment systems, it is convenient to measure biomass rather

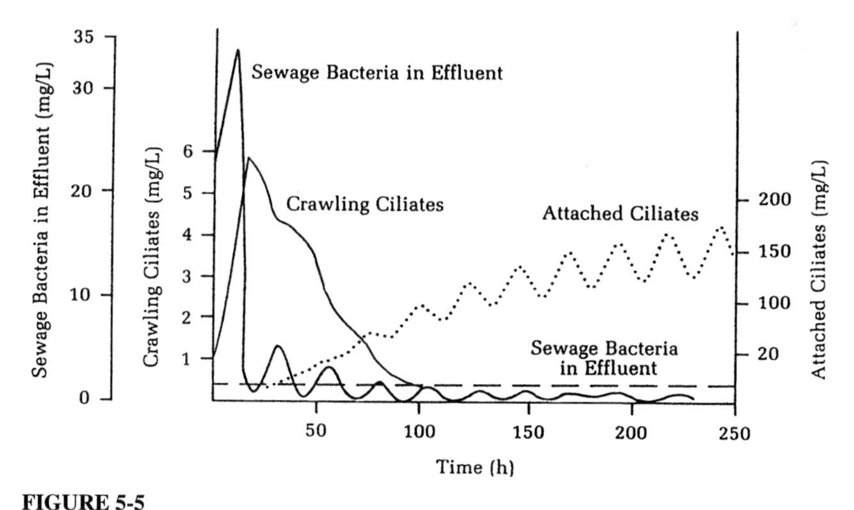

FIGURE 5-5
Population dynamics in an open system. (*Source:* Curds, "A Theoretical Study of Factors Influencing the Microbial Population Dynamics of the Activated Sludge Process-I." *Water Resources,* vol. 7, p. 1269, 1973.)

than numbers of organisms.[8] In the log-growth phase, the rate expression for biomass increase is

$$\frac{dX}{dt} = \mu X \tag{5-3}$$

where $\frac{dX}{dt}$ = growth rate of the biomass, mg/L · t
μ = growth rate constant, t^{-1}
X = concentration of biomass, mg/L

Because of the difficulty of direct measurement of μ in mixed cultures, Monod[9] developed a model equation that assumes that the rate of food utilization, and therefore the rate of biomass production, is limited by the rate of enzyme reactions involving the food compound that is in shortest supply relative to its need. The Monod equation is

$$\mu = \frac{\mu_m S}{K_s + S} \tag{5-4}$$

where μ_m = maximum growth rate constant, t^{-1}
S = concentration of limiting food in solution, mg/L
K_s = half saturation constant, mg/L
= concentration of limiting food when $\mu = 0.5\mu_m$

The growth rate of biomass follows a hyperbolic function as shown in Figure 5-6.

Two limiting cases are of interest in the application of Equation 5-4 to wastewater treatment systems. In those cases where there is an excess of the limiting food, then $S \gg K_s$ and the growth rate constant, μ, is approximately equal to μ_m. Equation 5-3 then becomes first-order in biomass. At the other extreme, when $S \ll K_s$, the system is food-limited and the growth rate becomes zero-order with respect to biomass, that is, it is independent of the biomass.

Equation 5-4 assumes only growth of microorganisms and does not take into account natural die-off. It is generally assumed that the death or decay of the microbial mass is a first-order expression in biomass and hence Equations 5-3 and 5-4 are expanded to

$$\frac{dX}{dt} = \frac{\mu_m S X}{K_s + S} - k_d X \tag{5-5}$$

where k_d = endogenous decay rate constant, t^{-1}.

[8]Frequently, this is done by measuring suspended solids or volatile suspended solids (those that burn at 550 ± 50°C). When the wastewater contains only soluble organic matter, the volatile suspended solids test is reasonably representative. The presence of organic particles (which is often the case in municipal wastewater) confuses the issue completely.

[9]J. Monod, "The Growth of Bacterial Cultures," *Annual Review of Microbiology*, vol. 3, 1949.

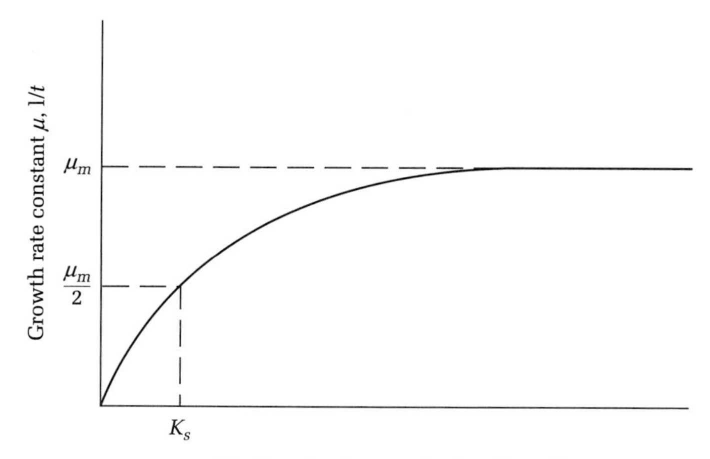

FIGURE 5-6
Monod growth rate constant as a function of limiting food concentration.

If all of the food in the system were converted to biomass, the rate of food utilization (dS/dt) would equal the rate of biomass production. Because of the inefficiency of the conversion process, the rate of food utilization will be greater than the rate of biomass utilization, so

$$-\frac{dS}{dt} = \frac{1}{Y}\frac{dX}{dt} \tag{5-6}$$

where Y = decimal fraction of food mass converted to biomass
= yield coefficient, $\frac{\text{mg/L biomass}}{\text{mg/L food utilized}}$

Combining Equations 5-3, 5-4, and 5-6,

$$-\frac{dS}{dt} = \frac{1}{Y}\frac{\mu_m SX}{K_s + S} \tag{5-7}$$

Equations 5-5 and 5-7 are a fundamental part of the development of the design equations for wastewater treatment processes.

5-2 CHARACTERISTICS OF DOMESTIC WASTEWATER

Physical Characteristics of Domestic Wastewater

Fresh, aerobic, domestic wastewater has been said to have the odor of kerosene or freshly turned earth. Aged, septic sewage is considerably more offensive to the olfactory nerves. The characteristic rotten-egg odor of hydrogen sulfide and the mercaptans is indicative of septic sewage. Fresh sewage is typically gray in color. Septic sewage is black.

Wastewater temperatures normally range between 10 and 20°C. In general, the temperature of the wastewater will be higher than that of the water supply. This is because of the addition of warm water from households and heating within the plumbing system of the structure.

One cubic meter of wastewater weighs approximately 1,000,000 grams. It will contain about 500 grams of solids. One-half of the solids will be dissolved solids such as calcium, sodium, and soluble organic compounds. The remaining 250 grams will be insoluble. The insoluble fraction consists of about 125 grams of material that will settle out of the liquid fraction in 30 minutes under quiescent conditions. The remaining 125 grams will remain in suspension for a very long time. The result is that wastewater is highly turbid.

Chemical Characteristics of Domestic Wastewater

Because the number of chemical compounds found in wastewater is almost limitless, we normally restrict our consideration to a few general classes of compounds. These classes often are better known by the name of the test used to measure them than by what is included in the class. The biochemical oxygen demand (BOD_5) test, which we discussed in Chapter 4, is a case in point. Another closely related test is the *chemical oxygen demand* (COD) test.

The COD test is used to determine the oxygen equivalent of the organic matter that can be oxidized by a strong chemical oxidizing agent (potassium dichromate) in an acid medium. The COD of a waste, in general, will be greater than the BOD_5 because more compounds can be oxidized chemically than can be oxidized biologically, and because BOD_5 does not equal ultimate BOD.

The COD test can be conducted in about three hours. If it can be correlated with BOD_5, it can be used to aid in the operation and control of the wastewater treatment plant (WWTP).

Total Kjeldahl nitrogen (TKN) is a measure of the total organic and ammonia nitrogen in the wastewater.[10] TKN gives a measure of the availability of nitrogen for building cells, as well as the potential nitrogenous oxygen demand that will have to be satisfied.

Phosphorus may appear in many forms in wastewater. Among the forms found are the orthophosphates, polyphosphates, and organic phosphate. For our purpose, we will lump all of these together under the heading "Total Phosphorus (as P)."

Three typical compositions of untreated domestic wastewater are summarized in Table 5-2. The pH for all of these wastes will be in the range of 6.5 to 8.5, with a majority being slightly on the alkaline side of 7.0.

---

[10] Pronounced "kell dall" after J. Kjeldahl, who developed the test in 1883.

TABLE 5-2
Typical composition of untreated domestic wastewater

Constituent	Weak	Medium	Strong
	(all mg/L except settleable solids)		
Alkalinity (as $CaCo_3$)[a]	50	100	200
BOD_5 (as O_2)	100	200	300
Chloride[a]	30	50	100
COD (as O_2)	250	500	1,000
Suspended solids (SS)	100	200	350
Settleable solids, mL/L	5	10	20
Total dissolved solids (TDS)	200	500	1,000
Total Kjeldahl nitrogen (TKN) (as N)	20	40	80
Total organic carbon (TOC) (as C)	75	150	300
Total phosphorus (as P)	5	10	20

[a]To be added to amount in domestic water supply. Chloride is exclusive of contribution from water-softener backwash.

Characteristics of Industrial Wastewater

Industrial processes generate a wide variety of wastewater pollutants. The characteristics and levels of pollutants vary significantly from industry to industry. The Environmental Protection Agency has grouped the pollutants into three categories: conventional pollutants, nonconventional pollutants, and priority pollutants. The conventional and nonconventional pollutants are listed in Table 5-3. The priority pollutants are listed in Table 1-3.

Because of the wide variety of industries and levels of pollutants, we can only present a snapshot view of the characteristics. A sampling of a few industries for two conventional pollutants is shown in Table 5-4.

A similar sampling for nonconventional pollutants is shown in Table 5-5.

TABLE 5-3
EPA's conventional and nonconventional pollutant categories

Conventional	*Nonconventional*
Biochemical oxygen demand (BOD_5)	Ammonia (as N)
Total suspended solids (TSS)	Chromium VI (hexavalent)
Oil and grease	Chemical oxygen demand (COD)
Oil (animal, vegetable)	COD/BOD_7
Oil (mineral)	Fluoride
pH	Manganese
	Nitrate (as N)
	Organic nitrogen (as N)
	Pesticide active ingredients (PAI)
	Phenols, total
	Phosphorus, total (as P)
	Total organic carbon (TOC)

Sources: 40 CFR parts 413.02, 464.02, 467.02, and 469.12

TABLE 5-4
Examples of industrial wastewater concentrations for BOD_5 and suspended solids

Industry	BOD_5, mg/L	Suspended solids, mg/L
Ammunition	50–300	70–1,700
Fermentation	4,500	10,000
Slaughterhouse (cattle)	400–2,500	400–1,000
Pulp and paper (kraft)	100–350	75–300
Tannery	700–7,000	4,000–20,000

TABLE 5-5
Examples of industrial wastewater concentrations for nonconventional pollutants

Industry	Pollutant	Concentration, mg/L
Coke by-product (steel mill)	Ammonia (as N)	200
	Organic nitrogen (as N)	100
	Phenol	2,000
Metal plating	Chromium VI	3–550
Nylon polymer	COD	23,000
	TOC	8,800
Plywood-plant glue waste	COD	2,000
	Phenol	200–2,000
	Phosphorus (as PO_4)	9–15

5-3 ON-SITE DISPOSAL SYSTEMS

Without Water Carriage

The pit privy. Although most modern environmental engineering texts would skip this subject, the mere existence of 10,000 of these or their modern equivalent in the United States is just too much for us to ignore. Furthermore, the facts of the matter are that junior engineers are the most likely candidates for designing, erecting, operating, dismantling, and closing the beasts.

Figure 5-7 provides most of the information you will ever want to know about the construction of an outhouse. The slab is usually poured over flat ground on top of roofing paper. The riser hole is formed using 12-gauge galvanized iron. Once the slab has set, it is lifted into place over the pit. The concrete is a 1:2:3 mix, that is, one part Portland cement, two parts sand, and three parts gravel less than 25 mm in diameter.

The principle of operation of the pit privy is that the liquid materials percolate into the soil through the cribbing and the solids "dry out." A pit of the dimensions shown in Figure 5-7 should last a family of four about ten years. Rainwater is to be prevented from entering the pit. A cup of kerosene at weekly intervals discourages mosquito breeding, and odors can be reduced by the use of a cup of hydrated

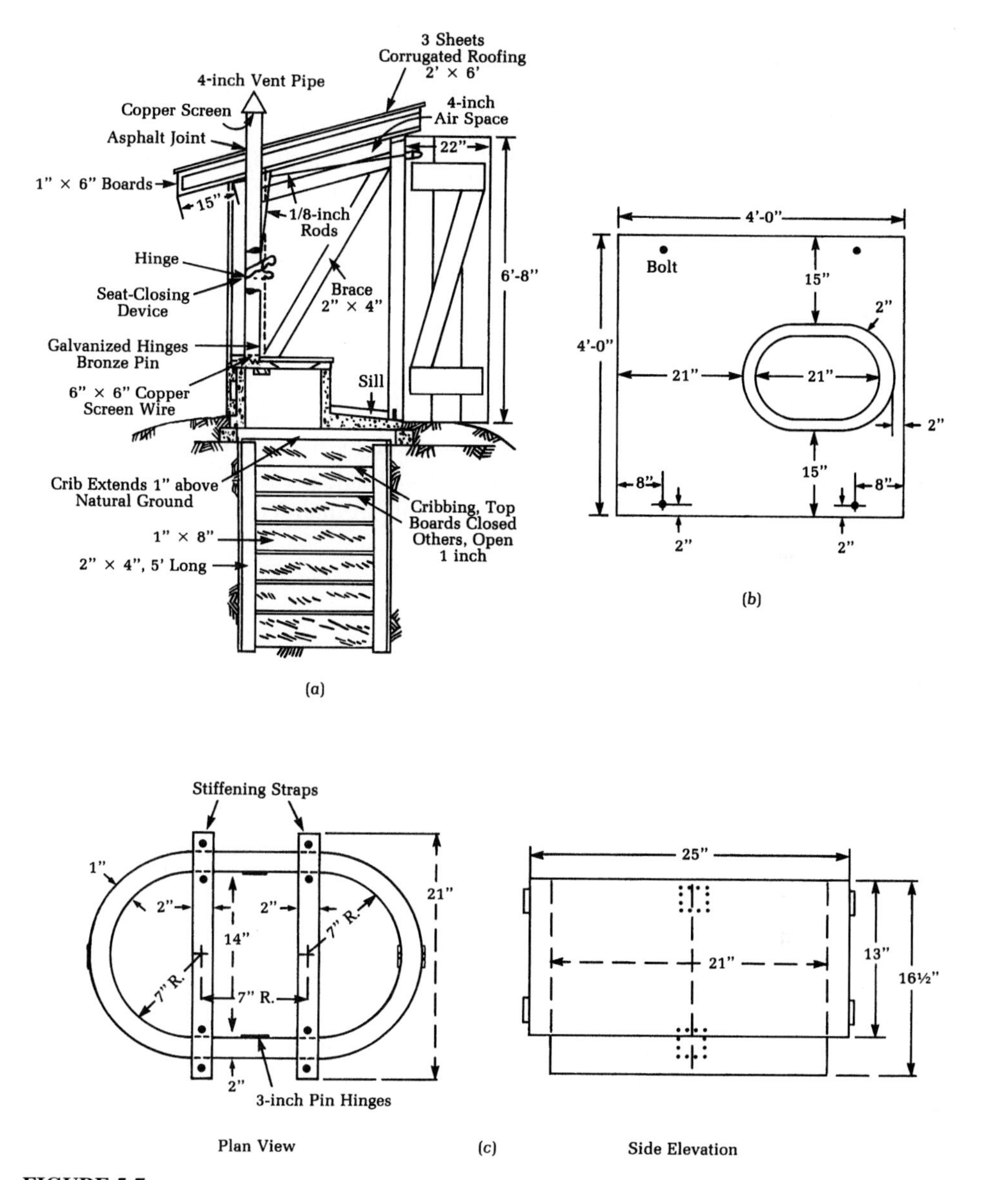

FIGURE 5-7
Construction details of the pit privy: (*a*) cross section; (*b*) plan of concrete slab; and (*c*) details of riser form. (*Source:* Ehlers and Steel, *Municipal and Rural Sanitation,* New York: McGraw-Hill, 1943. Reprinted by permission.)

lime. Unfortunately, the lime also slows the decomposition of paper, so its use is not encouraged. Disinfectants should never be used.

The vault toilet. This is the modern version of the pit privy. Its construction is the same as that of the pit privy with the exception that the pit is formed as a watertight vault. A special truck (fondly called a "honey wagon") is used to pump out the vault at regular intervals. Because of the liquefying action of the bacteria and incipient anaerobic decomposition, vault toilets are much more odiferous than the old pit privies. Many masking agents (perfumes) and disinfectants are available to mitigate the stench. Unfortunately, most of them have unpleasant odors themselves. If electricity is at hand, an ozone generator, set to vent into the gas space above the waste, will perform near-miracles in odor reduction.

The chemical toilet. The airplane toilet, the coach-bus toilet, and the self-contained toilets of recreation vehicles are all versions of the chemical toilet. The essence of the system is a strong disinfectant chemical used to carry the waste to a holding tank and render it inoffensive until it can be pumped from the holding tank. While these vehicular systems are quite effective, the chemical must be selected with an eye toward its impact on the treatment system which ultimately must receive it. The chemical toilet has not found wide acceptance in permanent installations. This is due to the cost of the chemical and to the impracticality of maintenance.

With Water Carriage

Septic tanks and tile fields. A typical septic tank and tile field arrangement for a residential dwelling is illustrated in Figure 5-8. The septic tank and tile field are a unit. Neither part will function as intended without the other.

The main function of the septic tank is to remove large particles and grease which would otherwise clog the tile field. Heavy solids settle to the bottom where they undergo anaerobic decomposition. Grease floats to the surface and is trapped. It is only slightly decomposed.

Since the septic tank is not heated, little reduction in BOD_5 occurs. Rather, the solid organic material which settles out is liquefied. It then passes to the tile field. Since not all of the solid material can be liquefied, the tank must be pumped at periodic intervals. The time interval between pumping depends on the amount of use and the objects which find their way to the tank. Toilet paper is easily degraded; however, plastic-lined disposable diapers cannot be degraded within a reasonable time. A family of four with young children can expect to have their septic tank pumped every two years. A household of two may not have to have its septic tank pumped in five or ten years of use. Grease accumulation is often the major factor in determining the frequency of cleaning.

In the past, the volume of the septic tank has been a function of the number of bedrooms in the dwelling. Current practice suggests that a 24-hour hydraulic detention time at design flow be used. In any case, the tank should not be less than 4.0 m^3 in volume.

In the tile field, the waste flows out of the joints between the tiles and through the gravel layer. The gravel serves to trap some of the solids that escaped from the septic tank. It also provides a storage area for holding the liquid while it seeps into the soil. Bacteria on the gravel degrade some of the trapped particulate matter. Bacteria in the soil aerobically degrade the liquefied organic material. The treated water percolates into the groundwater system.

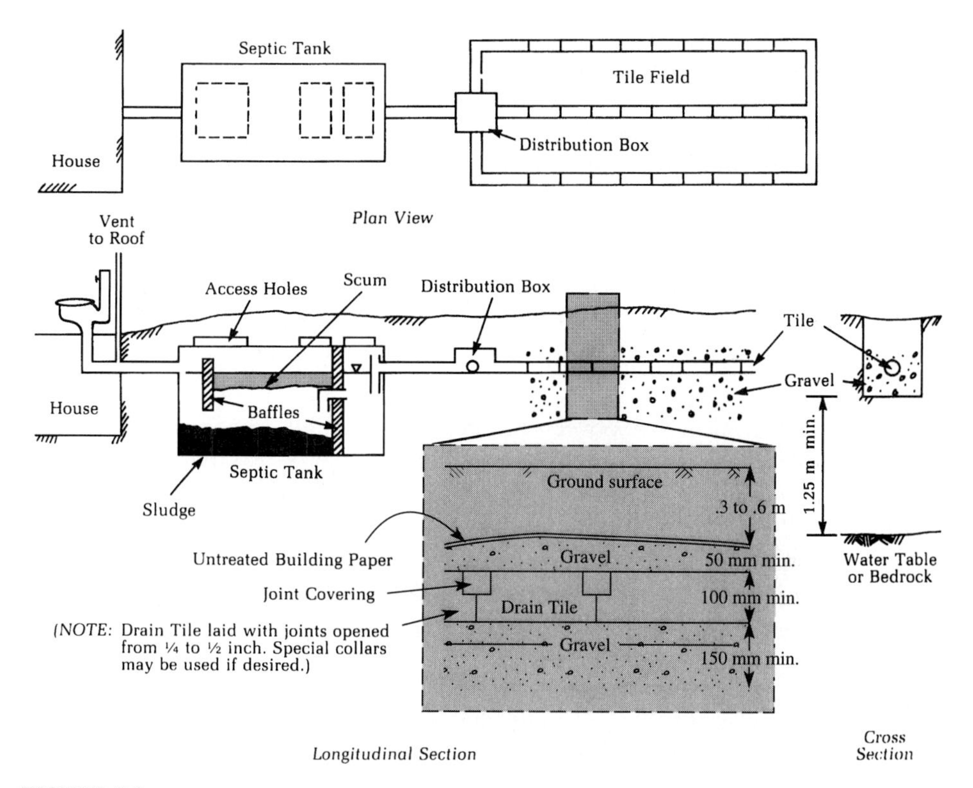

FIGURE 5-8
Schematic layout of a septic tank and tile field.

A septic tank and tile field can be used only when soil conditions are favorable. One method used to determine whether or not a tile system may be installed is the *soil percolation test,* better known as the *perc test* (or sometimes *perk*). In simple terms, the test is performed by digging a hole of prescribed size, filling it with water, and measuring the rate at which the water percolates into the soil. An alternative, and preferred, method for determining the suitability of the soil is to dig a trench in the area proposed for the tile field and visually inspect it. The inspector looks for unsuitable soil (clay, for example) and the presence of mottled (discolored) soil. Mottled soil indicates that the groundwater table has, at some point in time, risen to a level which would interfere with the operation of the tile field and, perhaps more important, bring the groundwater into direct contact with sewage. The information in Table 5-6 is then used to determine the size of the tile field.

Further limitations on the use of a septic-tank tile-field system usually include the following:

1. The tile field must be located more than 30 m from any well, surface water, footing drain, or storm drain.
2. The tile field must be located at least 3 m from any property line.

TABLE 5-6
Maximum acceptable application rates for tile fields

Soil texture and structure	Percolation rate mm/h	Percolation rate min/mm	Maximum acceptable application rate (m^3 of vol/m^2 area)
Coarse and medium sand	≥ 150	<0.40	0.04
Fine and loamy sand	75–150	0.40–0.80	0.03
Sandy loam	50–75	0.80–1.20	0.02
Loam and sandy clay	35–50	1.20–1.71	0.01
Loams	<35	>1.71	Not permitted
Clays, silts, muck, peat, marl	$\ll 35$	$\gg 1.71$	Not permitted

3. The minimum distance between the bottom of the absorption trench and the groundwater table or any impermeable layer must not be less than 1.25 m.
4. The earth cover placed over the absorption tile must not be less than 0.3 m nor more than 0.6 m deep.
5. A clean aggregate graded between 12 and 36 mm must be placed around the tile pipe. It must be a minimum of 50 mm above the pipe and 150 mm below the pipe, with a total depth of not less than 300 mm.

Most states limit septic tank/tile field installation to facilities producing less than 40 m^3/d of wastewater. This limits their use to single family residences, small apartments, freeway rest areas, parks, and isolated commercial establishments.

Example 5-1. John and Mary Jones are considering the purchase of a plot of land on which to build a retirement home. Based on their water bills for the past five years, their average daily water consumption is about 0.4 m^3. What size septic tank and tile field should they expect to put on the lot if it perks at 1.00 min/mm?

Solution. If the septic tank must provide a detention time of 24 h, then its volume should be

$$\forall = 0.4 \text{ m}^3/\text{d} \times 1 \text{ d} = 0.4 \text{ m}^3$$

However, the minimum recommended volume is 4.0 m^3. Good septic tank design practice calls for length to width (l/w) ratios greater than 2 to 1 and a minimum liquid depth of about 1.2 m. Using these criteria and a 4.0 m^3 volume, the liquid surface area would be

$$A_s = \frac{4.0}{1.2} = 3.33 \text{ m}^2$$

If we choose a width of 1.15 m and a length of 3 m, we will have a well-sized tank of 4.14 m^3 and a l/w ratio of 2.61 to 1.

From Table 5-6 we find that a perk rate of 1.00 min/mm will allow an application rate of 0.02 m^3/m^2 of trench. The bottom area of trench should then be about

$$A = \frac{0.4 \text{ m}^3}{0.02 \text{ m}^3/\text{m}^2} = 20.0 \text{ m}^2$$

One trench 1.0 m wide and 20.0 m long would meet the requirements; however, our preference is to use a 0.6 m trench width and three trenches about 12 m in length.

Barriered-landscape water-renovation system (BLWRS). In the summer of 1969, Dr. A. Earl Erickson demonstrated the efficacy of utilizing a BLWRS (pronounced "blowers," like "flowers") to denitrify water containing 100 mg/L of nitrate. Subsequently, he and his associates demonstrated that the BLWRS could be used to renovate both dairy cow and swine feedlot wastewater (Table 5-7).[11] The system is, of course, equally applicable to domestic wastewater.

The BLWRS consists of a mound of soil underlain by an impervious water barrier (Figures 5-9*a* and 5-9*b*). As the renovated water passes beyond the edge of the barrier, it may be collected in drains or be allowed to recharge the aquifer. The mound is constructed of a fine sand. The dimensions of the BLWRS depend on the soil texture and expected wastewater application rates (Table 5-8). A 0.15 m layer of topsoil is used to cover the sand. A water-hardy grass (quack grass or volunteer weed cover) must be established on the surface and banks to maintain the soil's permeability and stability. A carbon source is installed to penetrate the anoxic zone that forms along the barrier. The carbon source is a mixture of one part corn and 100 parts peat.

TABLE 5-7
BLWRS wastewater renovation efficiencies

	Average influent concentration (mg/L)	Average effluent concentration (mg/L)	Efficiency (%)
Swine waste[a]			
BOD_5	1,131	18.9	98.3
P	18	0.02	99.9
SS	3,000	NIL	~100.0
TKN	937	187.4	80.0
Dairy waste[b]			
BOD_5	1,637.0	18.9	98.8
P	38.5	0.23	99.4
SS	4,400.0	NIL	~100.0
TKN	917.0	27.5	97.0

[a] Average application rate of 15 mm/d for 503 d.
[b] Average application rate of 8.8 mm/d for 450 d.
Source: See Erickson, et al., footnote 11.

[11] A. E. Erickson, B. G. Ellis, J. M. Tiedje, A. R. Wolcott, C. M. Hausen, F. R. Peabody. E. C. Miller, and J. W. Thomas, *Soil Modification for Denitrification and Phosphate Reduction of Feedlot Waste* (Environmental Protection Agency Report No. EPA-660/2-74-057), Washington, DC: U.S. Government Printing Office, 1974.

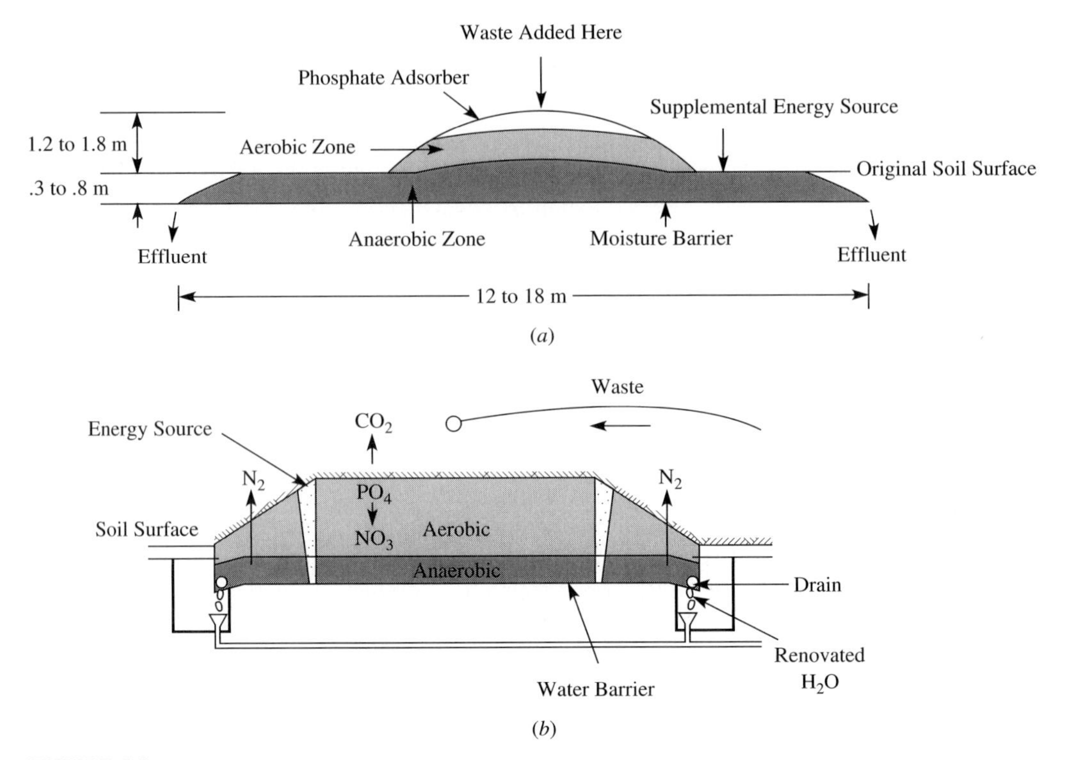

FIGURE 5-9
(*a*) Common dimensions of barriered-landscape water-renovation system (BLWRS); (*b*) water chemistry change in a BLWRS.

The wastewater is spread on the top of the mound by a sprinkler. As the wastewater percolates down, the organic particles are filtered out and remain on the surface. The particles are oxidized by soil microorganisms. The soluble organics and other ions move into the aerobic soil zone. Most of the soluble organic matter is oxidized by bacteria in the highly active aerobic soil. The phosphate ions are held on the clay fraction of the soil and sand bed. (Iron slag and/or limestone can be used to enhance the phosphorus adsorption capacity.) The ammonium ions are held on the soil until they are nitrified to nitrate. The downward movement of the nitrified water is stopped by the barrier. The water then is forced to move laterally through the anoxic layer. Denitrification occurs as the waste passes out of the carbon source.

The BLWRS must be operated in a cyclic fashion to allow the soil microorganisms time to degrade the waste and to maintain aerobic conditions in the soil. Application rates between 9 and 18 mm of wastewater per day may be used provided that the BLWRS is "rested" for one-third of the time. The physical conditions of the soil govern the application rates. The soils used in the original research with their respective application rates are characterized in Table 5-8. Ponding on the surface indicates excessive application rates.

TABLE 5-8
Example BLWRS soil characteristics and application rates

Application rate	Percent sand					Total percent sand	Percent silt	Percent clay
	Very coarse	Coarse	Medium	Fine	Very fine			
9 mm/d								
Surface	4.8	7.6	18.8	38.2	10.3	79.7	12.9	7.4
Subsurface	1.0	4.2	17.1	54.4	15.4	92.1	5.5	2.4
18 mm/d								
Surface	2.1	4.3	8.8	23.3	14.4	52.9	35.1	12.0
Subsurface	0.9	0.6	2.8	30.8	45.5	80.6	17.3	2.1

Notes:
1. Soil characteristics are U.S. Department of Agriculture Soil Textural Classes.
2. Surface refers to depth from 0.0 to 0.15 m; subsurface refers to depth from 0.15 m to 1.80 m, that is, the bottom of the BLWRS.

Source: See Erickson, et al., footnote 11.

5-4 MUNICIPAL WASTEWATER TREATMENT SYSTEMS

This discussion of municipal wastewater treatment systems follows the U.S. Environmental Protection Agency publication, *Environmental Pollution Control Alternatives: Municipal Wastewater.*[12]

The alternatives for municipal wastewater treatment fall into three major categories (Figure 5-10): (1) primary treatment, (2) secondary treatment, and (3) advanced treatment. It is commonly assumed that each of the "degrees of treatment" noted in Figure 5-10 includes the previous steps. For example, primary treatment is assumed to include the pretreatment processes: bar rack, grit chamber, and equalization basin. Likewise, secondary treatment is assumed to include all the processes of primary treatment: bar rack, grit chamber, equalization basin, and primary settling tank.

The purpose of pretreatment is to provide protection to the wastewater treatment plant (WWTP) equipment that follows. In some older municipal plants the equalization step may not be included.

The major goal of primary treatment is to remove from wastewater those pollutants that will either settle or float. Primary treatment will typically remove about 60 percent of the suspended solids in raw sewage and 35 percent of the BOD_5. Soluble pollutants are not removed. At one time, this was the only treatment used by many cities. Now federal law requires that municipalities provide secondary treatment. Although primary treatment alone is no longer acceptable, it is still frequently used as the first treatment step in a secondary-treatment system. The major goal of

[12]U.S. Environmental Protection Agency, *Environmental Pollution Control Alternatives: Municipal Wastewater* (Environmental Protection Agency Technology Transfer Publication No. EPA-625/5-76-012), Washington, DC: U.S. Government Printing Office, 1976.

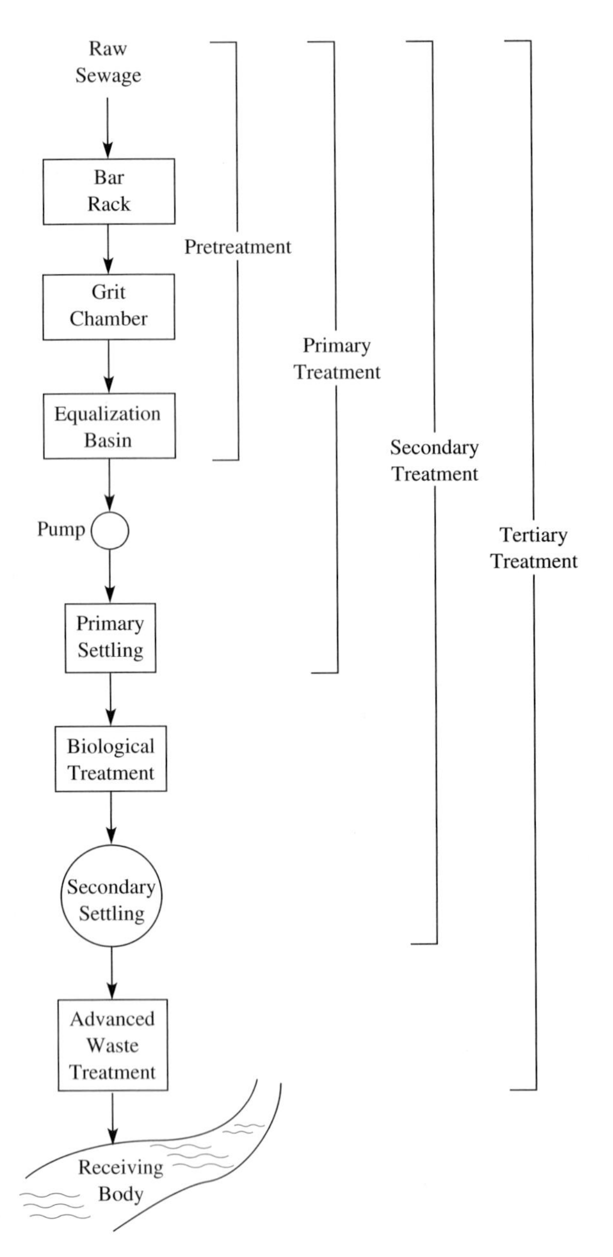

FIGURE 5-10
Degrees of treatment.

secondary treatment is to remove the soluble BOD_5 that escapes the primary process and to provide added removal of suspended solids. Secondary treatment is typically achieved by using biological processes. These provide the same biological reactions that would occur in the receiving water if it had adequate capacity to assimilate the wastewater. The secondary treatment processes are designed to speed up these natural processes so that the breakdown of the degradable organic pollutants can be achieved in relatively short time periods. Although secondary treatment may remove more than 85 percent of the BOD_5 and suspended solids, it does not remove significant amounts of nitrogen, phosphorus, or heavy metals, nor does it completely remove pathogenic bacteria and viruses.

In cases where secondary levels of treatment are not adequate, additional treatment processes are applied to the secondary effluent to provide advanced wastewater treatment (AWT). These processes may involve chemical treatment and filtration of the wastewater—much like adding a typical water treatment plant to the tail end of a secondary plant—or they may involve applying the secondary effluent to the land in carefully designed irrigation systems where the pollutants are removed by a soil-crop system. Some of these processes can remove as much as 99 percent of the BOD_5, phosphorus, suspended solids and bacteria, and 95 percent of the nitrogen. They can produce a sparkling clean, colorless, odorless effluent indistinguishable in appearance from a high-quality drinking water. Although these processes and land treatment systems are often applied to secondary effluent for advanced treatment, they have also been used in place of conventional secondary treatment processes.

Most of the impurities removed from the wastewater do not simply vanish. Some organics are broken down into harmless carbon dioxide and water. Most of the impurities are removed from the wastewater as a solid, that is, sludge. Because most of the impurities removed from the wastewater are present in the sludge, sludge handling and disposal must be carried out carefully to achieve satisfactory pollution control.

Pretreatment of Industrial Wastes

In municipalities, industrial wastewaters can pose serious hazards because wastewater collection and treatment systems have not been designed to treat them. The wastes can damage sewers and interfere with the operation of treatment plants. They may pass through the WWTP untreated or they may concentrate in the sludge, rendering it a hazardous waste.

The Clean Water Act gives the Environmental Protection Agency (EPA) the authority to establish and enforce pretreatment standards for discharge of industrial wastewaters into municipal treatment systems. Specific objectives of the pretreatment program are:

- To prevent the introduction to the WWTPs of pollutants that will interfere with the operation of a WWTP, including interference with its use or disposal of municipal sludge.
- To prevent the introduction to WWTPs of pollutants that will pass through the treatment works or otherwise be incompatible with such works.
- To improve opportunities to recycle and reclaim municipal and industrial wastewaters and sludge.

EPA has established "prohibited discharge standards" (40 CFR 403.5) that apply to all nondomestic discharges to the WWTP and "categorical pretreatment standards" that are applicable to specific industries (40 CFR 405–471). Congress assigned the primary responsibility for enforcing these standards to local WWTPs.

In the General Pretreatment Regulations, industrial users (IU) are prohibited from introducing the following into a WWTP:

1. Pollutants that create a fire or explosion hazard in the municipal WWTP, including, but not limited to, waste streams with a closed-cup flashpoint of less than or equal to 60°C using the test methods specified in 40 CFR 261.21.
2. Pollutants that will cause corrosive structural damage to the municipal WWTP (but in no case discharges with a pH lower than 5.0) unless the WWTP is specifically designed to accommodate such discharges.
3. Solid or viscous pollutants in amounts that will cause obstruction to the flow in the WWTP resulting in interference.
4. Any pollutant, including oxygen-demanding pollutants (such as BOD), released in a discharge at a flow rate and/or concentration that will cause interference with the WWTP.
5. Heat in amounts that will inhibit biological activity in the WWTP resulting in interference, but in no case heat in such quantities that the temperature at the WWTP exceeds 40°C unless the approval authority, upon request of the POTW, approves alternative temperature limits.
6. Petroleum oil, nonbiodegradable cutting oil, or products of mineral oil origin in amounts that will cause interference or will pass through.
7. Pollutants that result in the presence of toxic gases, vapors, or fumes within the POTW in a quantity that may cause acute worker health and safety problems.
8. Any trucked or hauled pollutants, except at discharge points designated by the POTW.

5-5 UNIT OPERATIONS OF PRETREATMENT

Several devices and structures are placed upstream of the primary treatment operation to provide protection to the wastewater treatment plant (WWTP) equipment. These devices and structures are classified as pretreatment because they have little effect in reducing BOD_5. In industrial WWTPs where only soluble compounds are present, bar racks and grit chambers may be absent. Equalization is frequently required in industrial WWTPs.

Bar Racks

Typically, the first device encountered by the wastewater entering the plant is a bar rack (Figure 5-11). The primary purpose of the rack is to remove large objects that would damage or foul pumps, valves, and other mechanical equipment. Rags, logs, and other objects that find their way into the sewer are removed from the wastewater on the racks. In modern WWTPs, the racks are cleaned mechanically. The solid material is stored in a hopper and removed to a sanitary landfill at regular intervals.

FIGURE 5-11
Bar rack. (Courtesy of BIF Sanitrol, a unit of General Signal.)

Bar racks (or bar screens) may be categorized as trash racks, manually cleaned racks, and mechanically cleaned racks. Trash racks are those with large openings, 40 to 150 mm, that are designed to prevent very large objects such as logs from entering the plant. These are normally followed by racks with smaller openings. Manually cleaned racks have openings that range from 25 to 50 mm. Channel approach velocities are designed to be in the range of 0.3 to 0.6 m/s. As mentioned above, manually cleaned racks are not frequently employed. They do find application in bypass channels that are infrequently used. Mechanically cleaned racks have openings ranging from 5 to 40 mm. Maximum channel approach velocities range from 0.6 to 1.2 m/s. Minimum velocities of 0.3 to 0.6 m/s are necessary to prevent grit accumulation. Regardless of the type of rack, two channels with racks are provided to allow one to be taken out of service for cleaning and repair.

Grit Chambers

Inert dense material, such as sand, broken glass, silt, and pebbles, is called grit. If these materials are not removed from the wastewater, they abrade pumps and other mechanical devices, causing undue wear. In addition, they have a tendency to settle in corners and bends, reducing flow capacity and, ultimately, clogging pipes and channels.

There are three basic types of grit-removal devices: velocity controlled, aerated, and constant-level short-term sedimentation basins. We will discuss only the first two, since they are the most common.

Velocity controlled. This type of grit chamber, also known as a *horizontal-flow* grit chamber, can be analyzed by means of the classical laws of sedimentation for discrete, nonflocculating particles (Type I sedimentation).

Stokes' law (See Section 3-6) may be used for the analysis and design of horizontal-flow grit chambers if the horizontal liquid velocity is maintained at about 0.3 m/s. Liquid velocity control is achieved by placing a specially designed weir at the end of the channel. A minimum of two channels must be employed so that one can be out of service without shutting down the treatment plant. Cleaning may be either by mechanical devices or by hand. Mechanical cleaning is favored for plants having average flows over 0.04 m^3/s. Theoretical detention times are set at about one minute for average flows. Washing facilities are normally provided to remove organic material from the grit.

Example 5-2. Will a grit particle with a radius of 0.10 mm and a specific gravity of 2.65 be collected in a horizontal grit chamber that is 13.5 m in length if the average grit-chamber flow is 0.15 m^3/s, the width of the chamber is 0.56 m, and the horizontal velocity is 0.25 m/s? The wastewater temperature is 22°C.

Solution. Before we can calculate the terminal settling velocity of the particle, we must gather some information from Table A-1 in Appendix A. At a wastewater temperature of 22°C, we find the water density to be 997.774 kg/m^3. We will use 1,000 kg/m^3 as a sufficiently close approximation. Since the particle radius is given to only two significant figures, this approximation is reasonable. From the same table, we find the viscosity to be 0.995 mPa · s. As noted in the footnote, we must multiply this by 10^{-3} to obtain the viscosity in units of Pa · s. Using a particle diameter of 0.20×10^{-3} m, we can calculate the terminal settling velocity using Equation 3-98.

$$v_s = \frac{g(\rho_s - \rho)d^2}{18\mu}$$

$$v_s = \frac{9.80(2{,}650 - 1000)(0.20 \times 10^{-3})^2\ (\mathrm{m/s^2})(\mathrm{kg/m^3})(\mathrm{m^2})}{18(0.000995)(\mathrm{kg \cdot m/m^2 \cdot s^2})(\mathrm{s})}$$

$$v_s = 3.61 \times 10^{-2}\ \mathrm{m/s\ or\ about\ 36\ mm/s}$$

Note that the product of the specific gravity of the particle (2.65) and the density of water is the density of the particle (ρ_s). The Reynolds number for this settling velocity (3.61×10^{-2} m/s) and particle size is 7.54. This is within the laminar range and Stokes' law is valid.

With a flow of 0.15 m^3/s and a horizontal velocity of 0.25 m/s, the cross-sectional area of flow may be estimated to be

$$A_c = \frac{0.15}{0.25} = 0.60\ \mathrm{m^2}$$

The depth of flow is then estimated by dividing the cross-sectional area by the width of the channel.

$$h = \frac{0.60}{0.56} = 1.07\ \mathrm{m}$$

If the grit particle in question enters the grit chamber at the liquid surface, it will take h/v_s seconds to reach the bottom.

$$t = \frac{1.07}{0.0361} = 29.6 \text{ s}$$

Since the chamber is 13.5 m in length and the horizontal velocity is 0.25 m/s, the liquid remains in the chamber.

$$t = \frac{13.5 \text{ m}}{0.25 \text{ m/s}} = 54 \text{ s}$$

Thus, the particle will be captured in the grit chamber.

Aerated grit chambers. The spiral roll of the aerated grit chamber liquid "drives" the grit into a hopper which is located under the air diffuser assembly (Figure 5-12). The shearing action of the air bubbles is supposed to strip the inert grit of much of the organic material that adheres to its surface.

Aerated grit chamber performance is a function of the roll velocity and detention time. The roll velocity is controlled by adjusting the air feed rate. Nominal air flow values are in the range of 0.15 to 0.45 cubic meters per minute of air per meter of tank length ($m^3/min \cdot m$). Liquid detention times are usually set to be about three minutes at maximum flow. Length-to-width ratios range from 2.5:1 to 5:1 with depths on the order of 2 to 5 m.

Grit accumulation in the chamber varies greatly, depending on whether the sewer system is a combined type or a separate type, and on the efficiency of the chamber. For combined systems, 90 m^3 of grit per million cubic meters of sewage

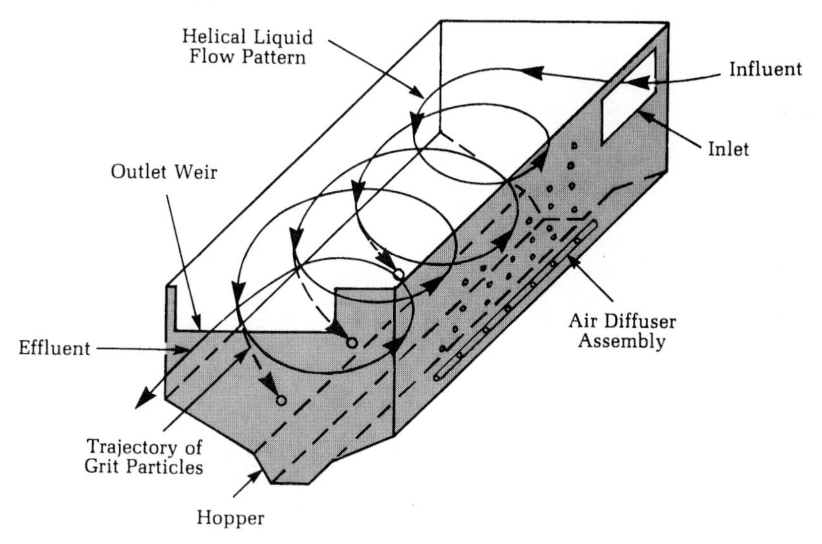

FIGURE 5-12
Aerated grit chamber. (*Source:* Metcalf & Eddy, Inc., and G. Tchobangolous, *Wastewater Engineering: Treatment, Disposal, Reuse,* New York: McGraw-Hill, 1979. Reprinted by permission.)

($m^3/10^6\ m^3$) is not uncommon. In separate systems you might expect something less than 30 $m^3/10^6\ m^3$. Normally the grit is buried in a sanitary landfill.

Comminutors

Devices that are used to macerate wastewater solid (rags, paper, plastic, and other materials) by revolving cutting bars are called *comminutors* (Figure 5-13). These devices are placed downstream of the grit chambers to protect the cutting bars from abrasion. They are used as a replacement for the downstream bar rack but must be installed with a hand-cleaned rack in parallel in case they fail.

Equalization

Flow equalization is not a treatment process *per se,* but a technique that can be used to improve the effectiveness of both secondary and advanced wastewater treatment

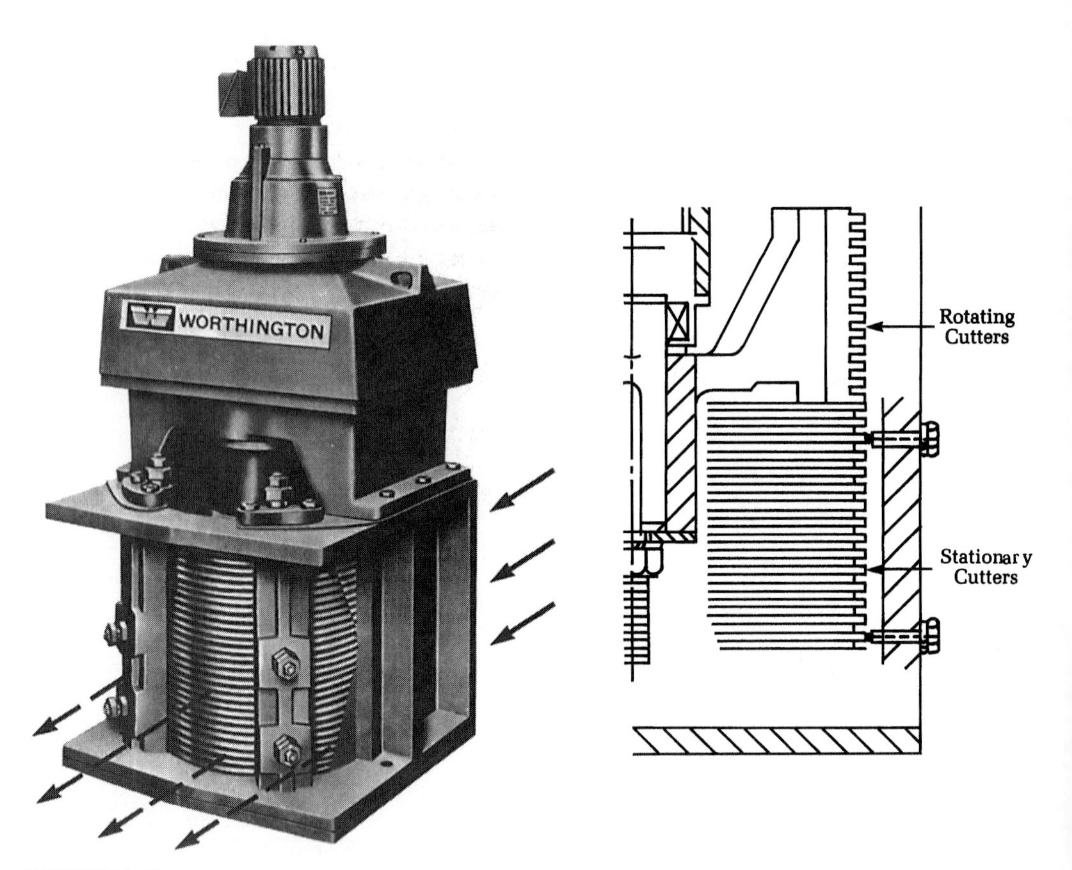

FIGURE 5-13
Drawing and schematic diagram of comminutor. (Courtesy of Worthington Pump, Inc.)

processes.[13] Wastewater does not flow into a municipal wastewater treatment plant at a constant rate (see Figure 1-4); the flow rate varies from hour to hour, reflecting the living habits of the area served. In most towns, the pattern of daily activities sets the pattern of sewage flow and strength. Above-average sewage flows and strength occur in mid-morning. The constantly changing amount and strength of wastewater to be treated makes efficient process operation difficult. Also, many treatment units must be designed for the maximum flow conditions encountered, which actually results in their being oversized for average conditions. The purpose of flow equalization is to dampen these variations so that the wastewater can be treated at a nearly constant flow rate. Flow equalization can significantly improve the performance of an existing plant and increase its useful capacity. In new plants, flow equalization can reduce the size and cost of the treatment units.

Flow equalization is usually achieved by constructing large basins that collect and store the wastewater flow and from which the wastewater is pumped to the treatment plant at a constant rate. These basins are normally located near the head end of the treatment works, preferably downstream of pretreatment facilities such as bar screens, comminutors, and grit chambers. Adequate aeration and mixing must be provided to prevent odors and solids deposition. The required volume of an equalization basin is estimated from a mass balance of the flow into the treatment plant with the average flow the plant is designed to treat. The theoretical basis is the same as that used to size reservoirs (see Section 2-4).

Example 5-3. Design an equalization basin for the following cyclic flow pattern. Provide a 25 percent excess capacity for equipment, unexpected flow variations, and solids accumulation. Evaluate the impact of equalization on the mass loading of BOD_5.

Time, h	Flow, m^3/s	BOD_5, mg/L	Time, h	Flow, m^3/s	BOD_5, mg/L
0000	0.0481	110	1200	0.0718	160
0100	0.0359	81	1300	0.0744	150
0200	0.0226	53	1400	0.0750	140
0300	0.0187	35	1500	0.0781	135
0400	0.0187	32	1600	0.0806	130
0500	0.0198	40	1700	0.0843	120
0600	0.0226	66	1800	0.0854	125
0700	0.0359	92	1900	0.0806	150
0800	0.0509	125	2000	0.0781	200
0900	0.0631	140	2100	0.0670	215
1000	0.0670	150	2200	0.0583	170
1100	0.0682	155	2300	0.0526	130

Solution. Because of the repetitive and tabular nature of the calculations, a computer spreadsheet is ideal for this problem. The spreadsheet solution is easy to verify if the calculations are set up with judicious selection of the initial value. If the initial value

---

[13]U.S. Environmental Protection Agency, *Environmental Pollution Control Alternatives: Municipal Wastewater,* pp. 52–53.

is the first flow rate greater than the average after the sequence of nighttime low flows, then the last row of the computation should result in a storage value of zero.

The first step then is to calculate the average flow. In this case it is 0.05657 m^3/s. Next, the flows are arranged in order beginning with the time and flow that first exceeds the average. In this case it is at 0900 h with a flow of .0631 m^3/s. The tabular arrangement is shown on the following page. An explanation of the calculations for each column follows.

The third column converts the flows to volumes using the time interval between flow measurements:

$$\forall = (0.0631 \text{ m}^3/\text{s})(1 \text{ h})(3600 \text{ s/h}) = 227.16 \text{ m}^3$$

The fourth column is the average volume that leaves the equalization basin.

$$\forall = (0.05657 \text{ m}^3/\text{s})(1 \text{ h})(3600 \text{ s/h}) = 203.655 \text{ m}^3$$

The fifth column is the difference between the inflow volume and the outflow volume.

$$dS = \forall_{\text{in}} - \forall_{\text{out}} = 227.16 \text{ m}^3 - 203.655 \text{ m}^3 = 23.505 \text{ m}^3$$

The sixth column is the cumulative sum of the difference between the inflow and outflow. For the second time interval, it is

$$\text{Storage} = \sum dS = 37.55 \text{ m}^3 + 23.51 \text{ m}^3 = 61.06 \text{ m}^3$$

Note that the last value for the cumulative storage is 0.12 m^3. It is not zero because of round-off truncation in the computations. At this point the equalization basin is empty and ready to begin the next day's cycle.

The required volume for the equalization basin is the maximum cumulative storage. With the requirement for 25 percent excess, the volume would then be

$$\text{Storage volume} = (863.74 \text{ m}^3)(1.25) = 1{,}079.68 \text{ or } 1{,}080 \text{ m}^3$$

The mass of BOD_5 into the equalization basin is the product of the inflow (Q), the concentration of BOD_5 (S_o), and the integration time (Δt):

$$M_{\text{BOD-in}} = (Q)(S_o)(\Delta t)$$

The mass of BOD_5 out of the equalization basin is the product of the average outflow (Q_{avg}), the average concentration (S_{avg}) in the basin, and the integration time (Δt):

$$M_{\text{BOD-out}} = (Q_{\text{avg}})(S_{\text{avg}})(\Delta t)$$

The average concentration is determined as

$$S_{\text{avg}} = \frac{(\forall_{\text{i}})(S_{\text{o}}) + (\forall_{\text{s}})(S_{\text{prev}})}{\forall_{\text{i}} + \forall_{\text{s}}}$$

where $\forall_i$ = volume of inflow during time interval Δt, m^3

S_o = average BOD_5 concentration during time interval Δt, g/m^3

$\forall_s$ = volume of wastewater in the basin at the end of the previous time interval, m^3

S_{prev} = concentration of BOD_5 in the basin at the end of the previous time interval

= (previous S_{avg}), g/m^3

Time	Flow, m^3/s	Vol_{in}, m^3	Vol_{out}, m^3	dS, m^3	$\sum dS$, m^3	BOD_5, mg/L	$M_{BOD\text{-}in}$, kg	S, mg/L	$M_{BOD\text{-}out}$, kg
0900	0.0631	227.16	203.65	23.51	23.51	140	31.80	140.00	28.51
1000	0.067	241.2	203.65	37.55	61.06	150	36.18	149.11	30.37
1100	0.0682	245.52	203.65	41.87	102.93	155	38.06	153.83	31.33
1200	0.0718	258.48	203.65	54.83	157.76	160	41.36	158.24	32.23
1300	0.0744	267.84	203.65	64.19	221.95	150	40.18	153.06	31.17
1400	0.075	270	203.65	66.35	288.3	140	37.80	145.89	29.71
1500	0.0781	281.16	203.65	77.51	365.81	135	37.96	140.51	28.62
1600	0.0806	290.16	203.65	86.51	452.32	130	37.72	135.86	27.67
1700	0.0843	303.48	203.65	99.83	552.15	120	36.42	129.49	26.37
1800	0.0854	307.44	203.65	103.79	655.94	125	38.43	127.89	26.04
1900	0.0806	290.16	203.65	86.51	742.45	150	43.52	134.67	27.43
2000	0.0781	281.16	203.65	77.51	819.96	200	56.23	152.61	31.08
2100	0.067	241.2	203.65	37.55	857.51	215	51.86	166.79	33.97
2200	0.0583	209.88	203.65	6.23	863.74	170	35.68	167.42	34.10
2300	0.0526	189.36	203.65	−14.29	849.45	130	24.62	160.69	32.73
0000	0.0481	173.16	203.65	−30.49	818.96	110	19.05	152.11	30.98
0100	0.0359	129.24	203.65	−74.41	744.55	81	10.47	142.42	29.00
0200	0.0226	81.36	203.65	−122.29	622.26	53	4.31	133.61	27.21
0300	0.0187	67.32	203.65	−136.33	485.93	35	2.36	123.98	25.25
0400	0.0187	67.32	203.65	−136.33	349.6	32	2.15	112.79	22.97
0500	0.0198	71.28	203.65	−132.37	217.23	40	2.85	100.46	20.46
0600	0.0226	81.36	203.65	−122.29	94.94	66	5.37	91.07	18.55
0700	0.0359	129.24	203.65	−74.41	20.53	92	11.89	91.61	18.66
0800	0.0509	183.24	203.65	−20.41	0.12	125	22.91	121.64	24.77

Noting that 1 mg/L = 1 g/m^3, the first row (the 0900 h time) computations are

$$M_{\text{BOD-in}} = (0.0631 \text{ m}^3/\text{s})(140 \text{ g/m}^3)(1 \text{ h})(3{,}600 \text{ s/h})(10^{-3} \text{ kg/g})$$
$$= 31.8 \text{ kg}$$

$$S_{avg} = \frac{(227.16 \text{ m}^3)(140 \text{ g/m}^3) + 0}{227.16 \text{ m}^3 + 0}$$
$$= 140 \text{ mg/L}$$

$$M_{\text{BOD-out}} = (0.05657 \text{ m}^3/\text{s})(140 \text{ g/m}^3)(1 \text{ h})(3{,}600 \text{ s/h})(10^{-3} \text{ kg/g})$$
$$= 28.5 \text{ kg}$$

Note that the zero values in the computation of S_{avg} are valid only at startup of an empty basin. Also note that in this case $M_{\text{BOD-in}}$ and $M_{\text{BOD-out}}$ differ only because of the difference in flow rates. For the second row (1000 h), the computations are

$$M_{\text{BOD-in}} = (0.0670 \text{m}^3/\text{s})(150 \text{ g/m}^3)(1 \text{ h})(3{,}600 \text{ s/h})(10^{-3} \text{ kg/g})$$
$$= 36.2 \text{ kg}$$

$$S_{avg} = \frac{(241.20 \text{ m}^3)(150 \text{ g/m}^3) + (23.51 \text{ m}^3)(140 \text{ g/m}^3)}{241.20 \text{ m}^3 + 23.51 \text{ m}^3}$$
$$= 149.11 \text{ mg/L}$$

$$M_{\text{BOD-out}} = (0.05657 \text{ m}^3/\text{s})(149.11 \text{ g/m}^3)(1 \text{ h})(3{,}600 \text{ s/h})(10^{-3} \text{ kg/g})$$
$$= 30.37 \text{ kg}$$

Note that $\forall_s$ is the volume of wastewater in the basin at the end of the previous time interval. Therefore, it equals the accumulated dS. The concentration of BOD_5 (S_{prev}) is the average concentration at the end of previous interval (S_{avg}) and *not* the influent concentration for the previous interval (S_o).

For the third row (1100 h), the concentration of BOD_5 is

$$S_{avg} = \frac{(245.52 \text{ m}^3)(155 \text{ g/m}^3) + (61.06 \text{ m}^3)(149.11 \text{ g/m}^3)}{245.52 \text{ m}^3 + 61.06 \text{ m}^3}$$
$$= 153.83 \text{ mg/L}$$

5-6 PRIMARY TREATMENT

With the screening completed and the grit removed, the wastewater still contains light organic suspended solids, some of which can be removed from the sewage by gravity in a sedimentation tank. These tanks can be round or rectangular. The mass of settled solids is called *raw sludge*. The sludge is removed from the sedimentation tank by mechanical scrapers and pumps (Figure 5-14). Floating materials, such as grease and oil, rise to the surface of the sedimentation tank, where they are collected by a surface skimming system and removed from the tank for further processing.

Primary sedimentation basins (*primary tanks*) are characterized by Type II flocculant settling. The Stokes equation cannot be used because the flocculating particles are continually changing in size, shape, and, when water is entrapped in the floc, specific gravity. There is no adequate mathematical relationship that can be used to describe Type II settling. Laboratory tests with settling columns are used to develop design data (see Chap. 3).

Rectangular tanks with common-wall construction are frequently chosen because they are advantageous for sites with space constraints. Typically, these tanks

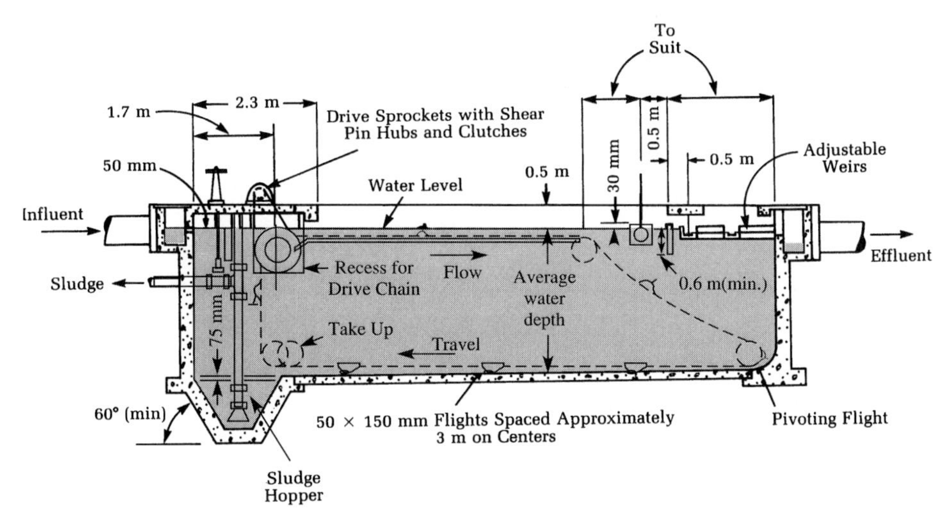

FIGURE 5-14
Primary settling tank.

range from 15 to 100 m in length and 3 to 24 m in width. Common length-to-width ratios for the design of new facilities range from 3:1 to 5:1. Existing plants have length-to-width ratios ranging from 1.5:1 to 15:1. The width is often controlled by the availability of sludge collection equipment. Side water depths range from 2 to 5 m. Typically the depth is about 3.5 m.

Circular tanks have diameters from 3 to 90 m. Side water depths range from 2.4 to 5 m.

As in water treatment clarifier design, overflow rate is the controlling parameter for the design of primary settling tanks. At average flow, overflow rates typically range from 25 to 60 $^3/m^2 \cdot d$ (or 25 to 60 m/d). When waste-activated sludge is returned to the primary tank, a lower range of overflow rates is chosen (25 to 35 m/d). Under peak flow conditions, overflow rates may be in the range of 80 to 120 m/d.

Hydraulic detention time in the sedimentation basin ranges from 1.5 to 2.5 hours under average flow conditions. A 2.0-hour detention time is typical.

The Great Lakes–Upper Mississippi River Board of State Sanitary Engineers (GLUMRB) recommends that *weir loading* (hydraulic flow over the effluent weir) rates not exceed 120 m^3/d of flow per m of weir length ($m^3/d \cdot m$) for plants with average flows less than 0.04 m^3/s. For larger flows, the recommended rate is 190 $m^3/d \cdot m$.[14] If the side water depths exceed 3.5 m, the weir loading rates have little effect on performance.

Two different approaches have been used to place the weirs. Some designers believe in the “long” approach and place the weirs to cover 33 to 50 percent of

[14]Great Lakes–Upper Mississippi River Board of State Sanitary Engineers, *Recommended Standards for Sewage Works,* 1978 edition, Albany, NY: Health Education Service, Inc., pp. 60–63, 1976.

the length of the tank. Those of the "short school" assume the weir length is less important and place it across the width of the end of the tank as shown in Figure 5-14. The spacing may vary from 2.5 to 6 m between weirs.

As mentioned previously, approximately 50 to 60 percent of the raw sewage suspended solids and as much as 30 to 35 percent of the raw sewage BOD_5 may be removed in the primary tank.

Example 5-4. Evaluate the following primary tank design with respect to detention time, overflow rate, and weir loading.

Design data:

Flow = 0.150 m^3/s
Influent SS = 280 mg/L
Sludge concentration = 6.0%
Efficiency = 60%
Length = 40.0 m (effective)
Width = 10.0 m
Liquid depth = 2.0 m
Weir length = 75.0 m

The detention time is simply the volume of the tank divided by the flow:

$$t = \frac{\forall}{Q} = \frac{40.0 \times 10.0 \times 2.0}{0.150}$$
$$= 5333.33 \text{ s or } 1.5 \text{ h}$$

This is a reasonable detention time.

The overflow rate is the flow divided by the surface area:

$$v_o = \frac{0.150}{40.0 \times 10.0}$$
$$= 3.75 \times 10^{-4} \text{ m/s} \times 86{,}400 \text{ s/d} = 32 \text{ m/d}$$

This is an acceptable overflow rate.

The weir loading is calculated in the same fashion:

$$WL = \frac{0.150}{75.0}$$
$$= 0.0020 \text{ m}^3/\text{s} \cdot \text{m} \times 86{,}400 \text{ s/d} = 172.8 \text{ or } 173 \text{ m}^3/\text{d} \cdot \text{m}$$

This is an acceptable weir loading.

5-7 UNIT PROCESSES OF SECONDARY TREATMENT

Overview

The major purpose of secondary treatment is to remove the soluble BOD that escapes primary treatment and to provide further removal of suspended solids. The basic ingredients needed for conventional aerobic secondary biologic treatment are the availability of many microorganisms, good contact between these organisms and the organic material, the availability of oxygen, and the maintenance of other favorable environmental conditions (for example, favorable temperature and sufficient time for the organisms to work). A variety of approaches have been used in the past to meet these basic needs. The most common approaches are (1) trickling filters, (2) activated sludge, and (3) oxidation ponds (or lagoons).

A process that does not fit precisely into either the trickling filter or the activated sludge category but does employ principles common to both is the *rotating biological contactor* (RBC).

Trickling Filters

A trickling filter consists of a bed of coarse material, such as stones, slats, or plastic materials (media), over which wastewater is applied. Trickling filters have been a popular biologic treatment process.[15] The most widely used design for many years was simply a bed of stones from 1 to 3 m deep through which the wastewater passed. The wastewater is typically distributed over the surface of the rocks by a rotating arm (Figure 5-15). Rock filter diameters may range up to 60 m.

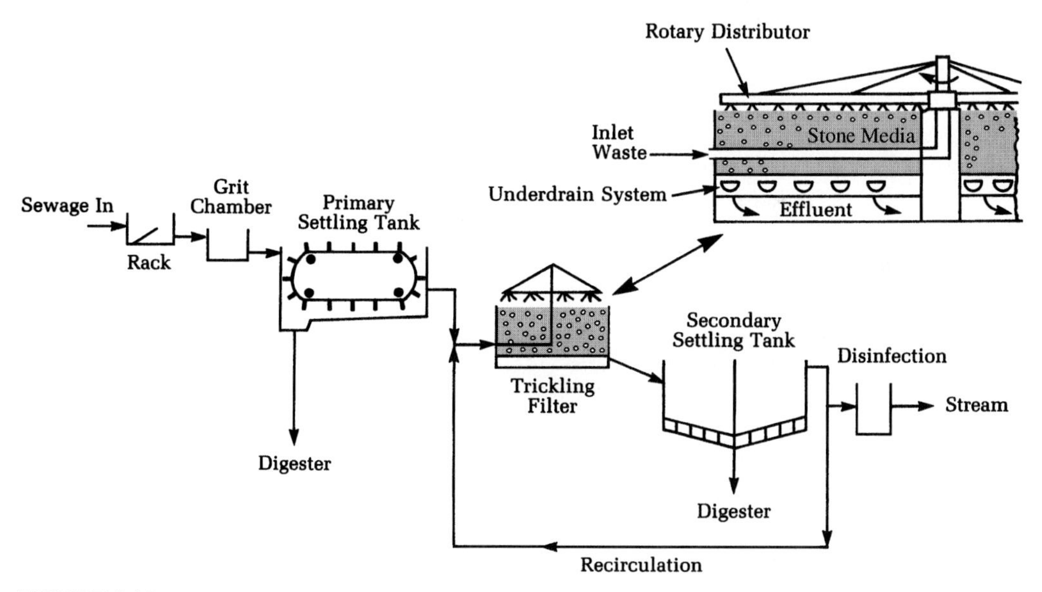

FIGURE 5-15
Trickling-filter plant with enlargement of trickling filter.

[15]U.S. Environmental Protection Agency, *Environmental Pollution Control Alternatives, Municipal Wastewater,* pp. 9–12.

As the wastewater trickles through the bed, a microbial growth establishes itself on the surface of the stone or packing in a fixed film. The wastewater passes over the stationary microbial population, providing contact between the microorganisms and the organics.

Trickling filters are not primarily a filtering or straining process as the name implies. The rocks in a rock filter are 25 to 100 mm in diameter, and hence have openings too large to strain out solids. They are a means of providing large amounts of surface area where the microorganisms cling and grow in a slime on the rocks as they feed on the organic matter.

Excess growths of microorganisms wash from the rock media and would cause undesirably high levels of suspended solids in the plant effluent if not removed. Thus, the flow from the filter is passed through a sedimentation basin to allow these solids to settle out. This sedimentation basin is referred to as a *secondary clarifier,* or *final clarifier,* to differentiate it from the sedimentation basin used for primary settling.

Although rock trickling filters have performed well for years, they have certain limitations. Under high organic loadings, the slime growths can be so prolific that they plug the void spaces between the rocks, causing flooding and failure of the system. Also, the volume of void spaces is limited in a rock filter, which restricts the circulation of air and the amount of oxygen available for the microbes. This limitation, in turn, restricts the amount of wastewater that can be processed.

To overcome these limitations, other materials have become popular for filling the trickling filter. These materials include modules of corrugated plastic sheets (Figure 5-16) and plastic rings. These media offer larger surface areas for slime growths (typically 90 square meters of surface area per cubic meter of bulk volume, as compared to 40 to 60 square meters per cubic meter for 75 mm rocks) and greatly increase void ratios for increased air flow. The materials are also much lighter than rock (by a factor of about 30), so that the trickling filters can be much taller without facing structural problems. While rock in filters is usually not more than 3 m deep, synthetic

FIGURE 5-16
Plastic media trickling filter. (Courtesy of Dow Chemical Company.)

TABLE 5-9
Comparison of different types of trickling filters[a]

Design characteristics	Low or standard rate	Intermediate rate	High rate (stone media)	Super rate (plastic media)	Roughing
	Trickling filter classification				
Hydraulic loading, m/d	1 to 4	4 to 10	10 to 40	15 to 90[b]	60 to 180[b]
Organic loading, kg $BOD_5/d \cdot m^3$	0.08 to 0.32	0.24 to 0.48	0.32 to 1.0	0.32 to 1.0	Above 1.0
Recirculation ratio	0	0 to 1	1 to 3	0 to 1	1 to 4
Filter flies	Many	Varies	Few	Few	Few
Sloughing	Intermittent	Varies	Continuous	Continuous	Continuous
Depth, m	1.5 to 3	1.5 to 2.5	1 to 2	Up to 12	1 to 6
BOD_5 removal, %	80 to 85	50 to 70	65 to 80	65 to 85	40 to 65
Effluent quality	Well nitrified	Some nitrification	Nitrites	Limited nitrification	No nitrification

[a]Adapted from Joint Committee of the American Society of Civil Engineers and the Water Pollution Federation, *Wastewater Treatment Plant Design* (ASCE Manuals and Reports on Engineering Practice No. 36, WPCF Manual of Practice No. 8). Lancaster, PA: Lancaster Press, Inc., p. 285, 1977.
[b]Not including recirculation.

media depths may reach 12 m, thus reducing the overall space requirements for the trickling-filter portion of the treatment plant.

Trickling filters are classified according to the applied hydraulic and organic load. The hydraulic load may be expressed as cubic meters of wastewater applied per day per square meter of bulk filter surface area ($m^3/d \cdot m^3$) or, preferably, as the depth of water applied per unit of time (mm/s or m/d). Organic loading is expressed as kilograms of BOD_5 per day per cubic meter of bulk filter volume ($kg/d \cdot m^3$). Common hydraulic and organic loadings for the the various filter classifications are summarized in Table 5-9.

An important element in trickling filter design is the provision for return of a portion of the effluent to flow through the filter. This practice is called *recirculation.* The ratio of the returned flow to the incoming flow is called the *recirculation ratio* (r). Recirculation is practiced in stone filters for the following reasons:

1. To increase contact efficiency by bringing the waste into contact more than once with active biological material.
2. To dampen variations in loadings over a 24-hour period. The strength of the recirculated flow lags behind that of the incoming wastewater. Thus, recirculation dilutes strong influent and supplements weak influent.
3. To raise the DO of the influent.
4. To improve distribution over the surface, thus reducing the tendency to clog and also reduce filter flies.
5. To prevent the biological slimes from drying out and dying during nighttime periods when flows may be too low to keep the filter wet continuously.

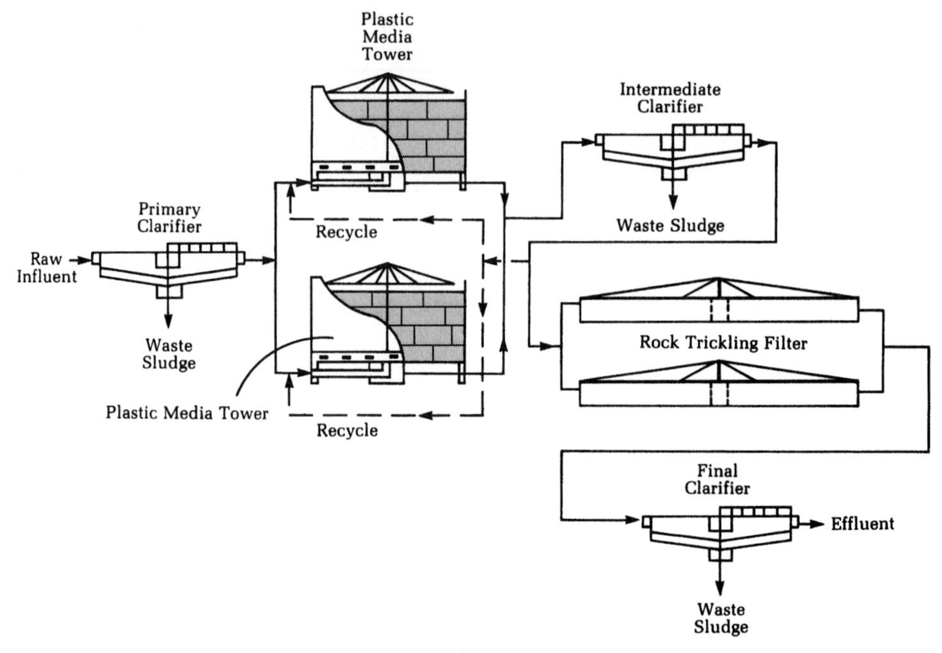

FIGURE 5-17
Two-stage trickling-filter plant. (Courtesy of Dow Chemical Company.)

Recirculation may or may not improve treatment efficiency. The more dilute the incoming wastewater, the less likely it is that recirculation will improve efficiency.

Recirculation is practiced for plastic media to provide the desired wetting rate to keep the microorganisms alive. Generally, increasing the hydraulic loading above the minimum wetting rate does not increase BOD_5 removal. The minimum wetting rate normally falls in the range of 25 to 60 m/d.

Two-stage trickling filters (Figure 5-17) provide a means of improving the performance of filters. The second stage acts as a polishing step for the effluent from the primary stage by providing additional contact time between the waste and the microorganisms. Both stages may use the same media or each stage may have different media as shown in Figure 5-17. The designer will select the types of media and their arrangement based on the desired treatment efficiencies and an economic analysis of the alternatives.

Design formulas. Numerous investigators have attempted to correlate operating data with the bulk design parameters of trickling filters. Rather than attempt a comprehensive review of these formulations, we have selected the National Research Council (NRC) equations[16] and Schulze's equation[17] as illustrations. A thorough

[16]National Research Council, "Sewage Treatment at Military Installations," *Sewage Works Journal,* 18, p. 787, 1946.

[17]K. L. Schulze, "Load and Efficiency of Trickling Filters," *Journal of Water Pollution Control Federation,* 32, p. 245, 1960.

review of several of the more important equations is given in the Water Environment Federation's publication on wastewater treatment plant design.[18]

During World War II, the NRC made an extensive study of the operation of trickling filters serving military installations. From this study, empirical equations were developed to predict the efficiency of the filters based on the BOD load, the volume of the filter media, and the recirculation. For a single-stage filter or the first stage of a two-stage filter, the efficiency is

$$E_1 = \frac{1}{1 + 4.12\left(\dfrac{QC_{\text{in}}}{\forall F}\right)^{0.5}} \tag{5-8}$$

where E_1 = fraction of BOD_5 removal for first stage at 20°C, including recirculation and sedimentation
Q = wastewater flow rate, m^3/s
C_{in} = influent BOD_5, mg/L
$\forall$ = volume of filter media, m^3
F = recirculation factor

The recirculation factor is

$$F = \frac{1 + R}{(1 + 0.1R)^2} \tag{5-9}$$

where R = recirculation ratio $= Q_r/Q$
Q_r = recirculation flow rate, m^3/s
Q = wastewater flow rate, m^3/s

The recirculation factor represents the average number of passes of the raw wastewater BOD through the filter. The factor $0.1R$ is to account for the empirical observation that the biodegradability of the organic matter decreases as the number of passes increases. For the second stage filter, the efficiency is

$$E_2 = \frac{1}{1 + \dfrac{4.12}{1 - E_1}\left(\dfrac{QC_e}{\forall F}\right)^{0.5}} \tag{5-10}$$

where E_2 = fraction of BOD_5 removal for second stage filter at 20°C, including recirculation and sedimentation, %
E_1 = fraction of BOD_5 removed in first stage
C_e = effluent BOD_5 from first stage, mg/L

[18] Joint Task Force of the Water Environment Federation and the American Society of Civil Engineers, *Design of Municipal Wastewater Treatment Plants Vol. I,* Manual of Practice No. 8, Alexandria, VA: pp. 705–714, 1992.

The effect of temperature on the efficiency may be estimated from the following equation:

$$E_T = E_{20}\,\theta^{(T-20)} \tag{5-11}$$

where a value of 1.035 is used for θ.

Some care should be used in applying the NRC equations. Military wastewater during this period (World War II) had a higher strength than domestic wastewater today. The filter media was rock. Clarifiers associated with the trickling filters were shallower and carried higher hydraulic loads than current practice would permit. The second stage filter is assumed to be preceded by an intermediate settling tank (see Figure 5-17).

Example 5-5. Using the NRC equations, determine the BOD_5 of the effluent from a single-stage, low-rate trickling filter that has a filter volume of 1,443 m^3, a hydraulic loading of 1,900 m^3/d, and a recirculation factor of 2.78. The influent BOD_5 is 150 mg/L.

Solution. To use the NRC equation, the hydraulic loading must first be converted to the correct units.

$$Q = (1{,}900\ \text{m}^3/\text{d})\left(\frac{1}{86{,}400\ \text{s/d}}\right) = 0.022\ \text{m}^3/\text{s}$$

The efficiency of a single-stage filter is

$$E_1 = \frac{1}{1 + 4.12\left(\dfrac{(0.022)(150)}{(1{,}443)(2.78)}\right)^{0.5}} = 0.8943$$

The concentration of BOD_5 in the effluent is then

$$C_e = (1 - 0.8943)(150) = 15.8\ \text{mg/L}$$

Schulze[17] proposed that the time of wastewater contact with the biological mass in the filter is directly proportional to the depth of the filter and inversely proportional to the hydraulic loading rate:

$$t = \frac{CD}{(Q/A)^n} \tag{5-12}$$

where t = contact time, d
C = mean active film per unit volume
D = filter depth, m
Q = hydraulic loading, m^3/d
A = filter area over which wastewater is applied, m^2
n = empirical constant based on filter media

The mean active film per unit volume may be approximated by

$$C \simeq \frac{1}{D^m} \tag{5-13}$$

where m is an empirical constant that is an indicator of biological slime distribution. It is normally assumed that the distribution is uniform and that $m = 0$. Thus, C is 1.0.

Schulze combined his relationship with Velz's[19] first-order equation for BOD removal

$$\frac{S_t}{S_o} = \exp\left[-\frac{KD}{(Q/A)^n}\right] \tag{5-14}$$

where K is an empirical rate constant with the units of

$$\frac{(\text{m/d})^n}{\text{m}}$$

The temperature correction for K may be computed with Equation 5-11 if K_T is substituted for E_T and K_{20} is substituted for E_{20}.

Example 5-6. Determine the BOD_5 of the effluent from a low-rate trickling filter that has a diameter of 35.0 m and a depth of 1.5 m if the hydraulic loading is 1,900 m^3/d and the influent BOD_5 is 150.0 mg/L. Assume the rate constant is 2.3 $(m/d)^n/m$ and $n = 0.67$.

Solution. We begin by computing the area of the filter.

$$A = \frac{\pi(35.0)^2}{4}$$

$$= 962.11 \text{ m}^2$$

This area is then used to compute the loading rate.

$$\frac{Q}{A} = \frac{1{,}900 \text{ m}^3/\text{d}}{962.11 \text{ m}^2}$$

$$= 1.97 \text{ m}^3/\text{d} \cdot \text{m}^2$$

Now we can compute the effluent BOD using Equation 5-14.

$$S_t = (150)\exp\left[\frac{-(2.3)(1.5)}{(1.97)^{0.67}}\right]$$

$$= 16.8 \text{ mg/L}$$

---

[19]C. J. Velz, "A Basic Law for the Performance of Biological Filters," *Sewage Works Journal,* 20, p. 607, 1948.

Activated Sludge

The activated sludge process is a biological wastewater treatment technique in which a mixture of wastewater and biological sludge (microorganisms) is agitated and aerated. The biological solids are subsequently separated from the treated wastewater and returned to the aeration process as needed.

The activated sludge process derives its name from the biological mass formed when air is continuously injected into the wastewater. In this process, microorganisms are mixed thoroughly with the organics under conditions that stimulate their growth through use of the organics as food. As the microorganisms grow and are mixed by the agitation of the air, the individual organisms clump together (flocculate) to form an active mass of microbes (biologic floc) called *activated sludge.*

In practice, wastewater flows continuously into an aeration tank (Figure 5-18) where air is injected to mix the activated sludge with the wastewater and to supply the oxygen needed for the organisms to break down the organics. The mixture of activated sludge and wastewater in the aeration tank is called *mixed liquor.* The mixed liquor flows from the aeration tank to a secondary clarifier where the activated sludge is settled out. Most of the settled sludge is returned to the aeration tank (and hence is called *return sludge*) to maintain the high population of microbes that permits rapid breakdown of the organics. Because more activated sludge is produced than is desirable in the process, some of the return sludge is diverted or wasted to the sludge handling system for treatment and disposal. In conventional activated sludge systems, the wastewater is typically aerated for six to eight hours in long, rectangular aeration basins. About 8 m^3 of air is provided for each m^3 of wastewater treated. Sufficient air is provided to keep the sludge in suspension (Figure 5-19). The air is injected near the bottom of the aeration tank through a system of diffusers (Figure 5-20). The volume of sludge returned to the aeration basin is typically 20 to 30 percent of the wastewater flow.

The activated sludge process is controlled by wasting a portion of the microorganisms each day in order to maintain the proper amount of microorganisms to

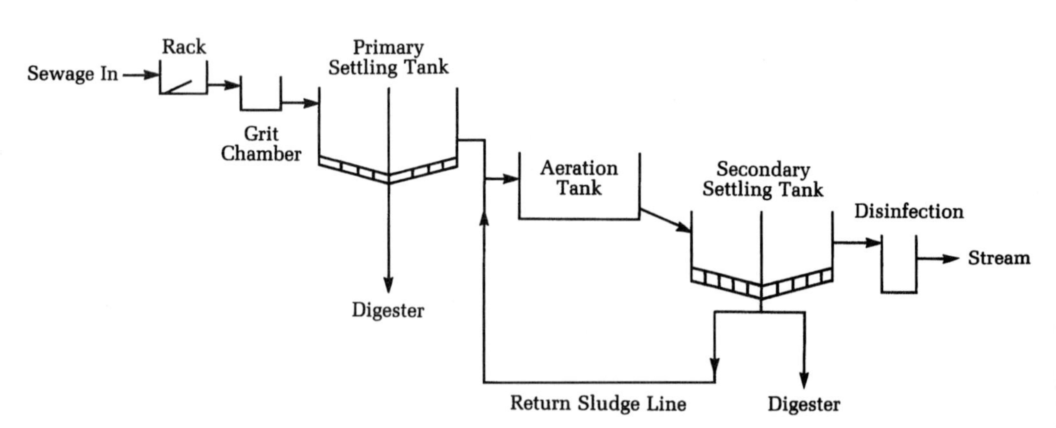

FIGURE 5-18
Conventional activated sludge plant.

FIGURE 5-19
Activated sludge aeration tank under air. (*Note:* White foam is from detergent.)

efficiently degrade the BOD_5. "Wasting" means that a portion of the microorganisms is discarded from the process. The discarded microorganisms are called *waste activated sludge* (WAS). A balance is then achieved between growth of new organisms and their removal by wasting. If too much sludge is wasted, the concentration of microorganisms in the mixed liquor will become too low for effective treatment. If too little sludge is wasted, a large concentration of microorganisms will accumulate and, ultimately, overflow the secondary tank and flow into the receiving stream.

The *mean cell residence time* θ_c, also called *solids retention time* (SRT) or *sludge age,* is defined as the average amount of time that microorganisms are kept in the system.

Many modifications of the conventional activated sludge process have been developed to address specific treatment problems. A brief description of these is given in Table 5-10. We have selected the completely mixed and conventional plug-flow processes for further discussion.

Completely mixed activated sludge process. The design formulas for the completely mixed activated sludge process are a mass-balance application of the equations used to describe the kinetics of microbial growth. A mass-balance diagram for the completely mixed system (CSTR) is shown in Figure 5-21. The mass-balance equations are written for the system boundary shown by the dashed line. Two mass balances are required to define the design of the reactor: one for biomass and one for food (soluble BOD_5).

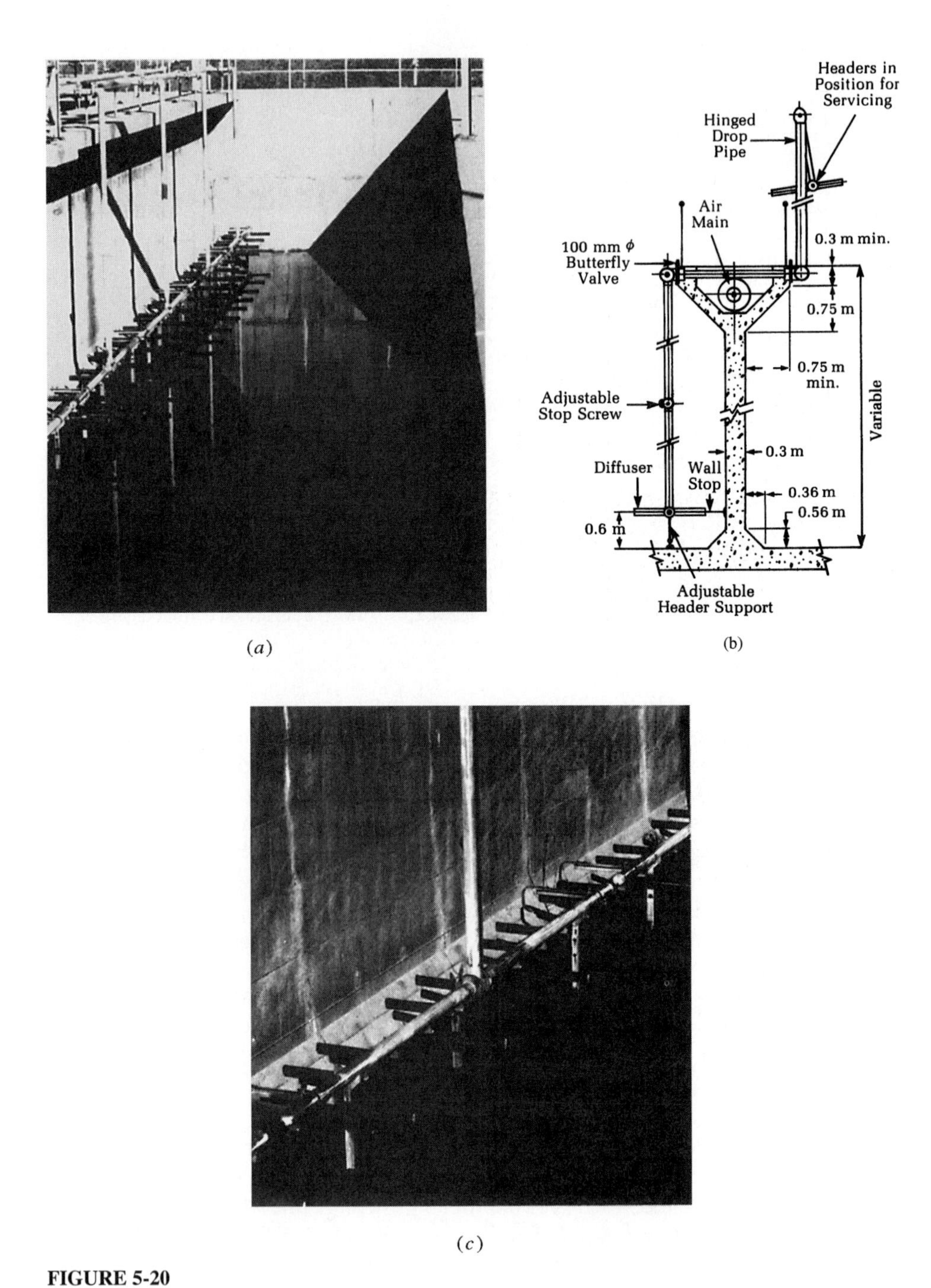

FIGURE 5-20
(*a*) Diffuser system for activated sludge tank; (*b*) cross section through tank illustrating hinged drop pipe; (*c*) close-up of a set of diffusers.

TABLE 5-10
Description of activated sludge processes and process modifications[a]

Process or process modification	Description
Conventional plug-flow	Settled wastewater and recycled activated sludge enter the head end of the aeration tank and are mixed by diffused-air or mechanical aeration. Air application is generally uniform throughout tank length. During the aeration period, adsorption, flocculation, and oxidation of organic matter occur. Activated sludge solids are separated in a secondary settling tank.
Complete-mix	Process is an application of the flow regime of a continuous-flow stirred-tank reactor. Settled wastewater and recycled activated sludge are introduced typically at several points in the aeration tank. The organic load on the aeration tank and the oxygen demand are uniform throughout the tank length.
Tapered aeration	Tapered aeration is a modification of the conventional plug-flow process. Varying aeration rates are applied over the tank length depending on the oxygen demand. Greater amounts of air are supplied to the head end of the aeration tank, and the amounts diminish as the mixed liquor approaches the effluent end. Tapered aeration is usually achieved by using different spacing of the air diffusers over the tank length.
Step-feed aeration	Step feed is a modification of the conventional plug-flow process in which the settled wastewater is introduced at several points in the aeration tank to equalize the F/M ratio, thus lowering peak oxygen demand. Generally three or more parallel channels are used. Flexibility of operation is one of the important features of this process.
Modified aeration	Modified aeration is similar to the conventional plug-flow process except that shorter aeration times and higher F/M ratios are used. BOD removal efficiency is lower than other activated sludge processes.
Contact stabilization	Contact stabilization uses two separate tanks or compartments for the treatment of the wastewater and stabilization of the activated sludge. The stabilized activated sludge is mixed with the influent (either raw or settled) wastewater in a contact tank. The mixed liquor is settled in a secondary settling tank and return sludge is aerated separately in a reaeration basin to stabilize the organic matter. Aeration volume requirements are typically 50 percent less than conventional plug flow.
Extended aeration	Extended aeration process is similar to the conventional plug-flow process except that it operates in the endogenous respiration phase of the growth curve, which requires a low organic loading and long aeration time. Process is used extensively for prefabricated package plants for small communities.
High-rate aeration	High-rate aeration is a process modification in which high MLSS concentrations are combined with high volumetric loadings. This combination allows high F/M ratios and long mean cell-residence times with relatively short hydraulic detention times. Adequate mixing is very important.

(continued)

TABLE 5-10
(continued)

Process or process modification	Description
Kraus process	Kraus process is a variation of the step aeration process used to treat wastewater with low nitrogen levels. Digester supernatant is added as a nutrient source to a portion of the return sludge in a separate aeration tank designed to nitrify. The resulting mixed liquor is then added to the main plug-flow aeration system.
High-purity oxygen	High-purity oxygen is used instead of air in the activated sludge process. Oxygen is difused into covered aeration tanks and is recirculated. A portion of the gas is wasted to reduce the concentration of carbon dioxide. pH adjustment may also be required. The amount of oxygen added is about four times greater than the amount that can be added by conventional aeration systems.
Oxidation ditch	The oxidation ditch consists of a ring- or oval-shaped channel and is equipped with mechanical aeration devices. Screened wastewater enters the ditch, is aerated, and circulates at about 0.25 to 0.35 m/s. Oxidation ditches typically operate in an extended aeration mode with long detention and solids retention times. Secondary sedimentation tanks are used for most applications.
Sequencing batch reactor	The sequencing batch reactor is a fill-and-draw type reactor system involving a single complete-mix reactor in which all steps of the activated sludge process occur. Mixed liquor remains in the reactor during all cycles, thereby eliminating the need for separate secondary sedimentation tanks.
Deep-shaft reactor	The deep vertical-shaft reactor is a form of the activated sludge process. A vertical shaft about 120 to 150 m deep replaces the primary clarifiers and aeration basin. The shaft is lined with a steel shell and fitted with a concentric pipe to form an annular reactor. Mixed liquor and air are forced down the center of the shaft and allowed to rise upward through the annulus.
Single-stage nitrification	In single-stage nitrification, both BOD and ammonia reduction occur in a single biological stage. Reactor configurations can be either a series of complete-mix reactors or plug flow.
Separate stage nitrification	In separate stage nitrification, a separate reactor is used for nitrification, operating on a feed waste from a preceding biological treatment unit. The advantage of this system is that operation can be optimized to conform to the nitrification needs.

[a]*Source:* Metcalf & Eddy, *Wastewater Engineering: Treatment, Disposal and Reuse,* New York: McGraw-Hill, 1991.

Under steady-state conditions, the mass balance for biomass may be written as:

$$\text{Biomass in influent} + \text{Biomass accumulated} = \text{Biomass in effluent} + \text{Biomass wasted} \tag{5-15}$$

The biomass in the influent is the product of the concentration of microorganisms in the influent (X_o) and the flow rate of wastewater (Q). The concentration of microorganisms in the influent (X_o) is measured as suspended solids (mg/L). The biomass that accumulates in the aeration tank is the product of the volume of the

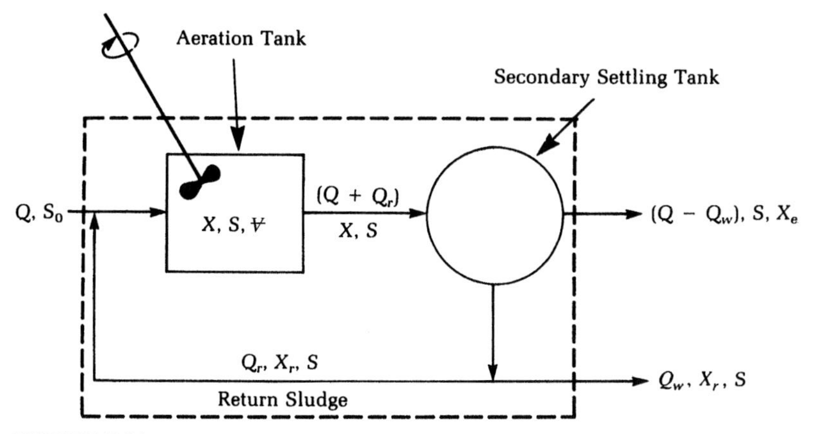

FIGURE 5-21
Completely mixed biological reactor with solids recycle.

tank ($\forall$) and the Monod expression for growth of microbial mass (Equation 5-5)

$$(\forall)\left(\frac{\mu_m SX}{K_s + S} - k_d X\right) \tag{5-16}$$

The biomass in the effluent is the product of flow rate of treated wastewater leaving the plant ($Q - Q_w$) and the concentration of microorganisms that does not settle in the secondary clarifier (X_e). The flow rate of wastewater leaving the plant does not equal the flow rate into the plant because some of the microorganisms must be wasted. The flow rate of wasting (Q_w) is deducted from the flow exiting the plant.

The biomass that is wasted is the product of concentration of microorganisms in the WAS flow (X_r) and the WAS flow rate (Q_r). The narrative mass-balance equation may be rewritten as

$$QX_o + (\forall)\left(\frac{\mu_m SX}{K_s + S} - k_d X\right) = (Q - Q_w)X_e + Q_w X_r \tag{5-17}$$

The variable are summarized as follows:

Q = wastewater flow rate into the aeration tank, m^3/d
X_o = microorganism concentration (volatile suspended solids or VSS)[20] entering aeration tank, mg/L
$\forall$ = volume of aeration tank, m^3
μ_m = maximum growth rate constant, d^{-1}

[20]Suspended solids means that the material will be retained on a filter, unlike dissolved solids such as NaCl. The amount of the suspended solids that volatilizes at 500±50°C is taken to be a measure of active biomass concentration. The presence of nonliving organic particles in the influent wastewater will cause some error (usually small) in the use of volatile suspended solids as a measure of biomass.

S = soluble BOD_5 in aeration tank and effluent, mg/L
X = microorganism concentration (mixed-liquor volatile suspended solids or MLVSS)[21] in the aeration tank, mg/L
K_s = half velocity constant
= soluble BOD_5 concentration at one-half the maximum growth rate, mg/L
k_d = decay rate of microorganisms, d^{-1}
Q_w = flow rate of liquid containing microorganisms to be wasted, m^3/d
X_e = microorganism concentration (VSS) in effluent from secondary settling tank, mg/L
X_r = microorganism concentration (VSS) in sludge being wasted, mg/L

At steady-state, the mass-balance equation for food (soluble BOD_5) may be written

$$\text{Food in influent} + \text{Food consumed} = \text{Food in effluent} + \text{Food in WAS} \tag{5-18}$$

The food in the influent is the product of the concentration of soluble BOD_5 in the influent (S_o) and the flow rate of wastewater (Q). The food that is consumed in the aeration tank is the product of the volume of the tank ($\forall$) and the expression for rate of food utilization (Equation 5-7)

$$(\forall)\left(\frac{\mu_m SX}{Y(K_s + S)}\right) \tag{5-19}$$

The food in the effluent is the product of flow rate of treated wastewater leaving the plant ($Q - Q_w$) and the concentration of soluble BOD_5 in the effluent (S). The concentration of soluble BOD_5 in the effluent (S) is the same as that in the aeration tank because we have assumed that the aeration tank is completely mixed. Since the BOD_5 is soluble, the secondary settling tank will not change the concentration. Thus, the effluent concentration from the secondary settling tank is the same as the influent concentration.

The food in the waste activated sludge flow is the product of the concentration of soluble BOD_5 in the influent (S) and the WAS flow rate (Q_r). The narrative mass-balance equation for steady-state conditions may be rewritten as

$$QS_o - (\forall)\left(\frac{\mu_m SX}{Y(K_s + S)}\right) = (Q - Q_w)S + Q_w S \tag{5-20}$$

where Y = yield coefficient (see Equation 5-6).

[21]Mixed-liquor volatile suspended solids is a measure of the active biological mass in the aeration tank. The term "mixed liquor" implies a mixture of activated sludge and wastewater. The phrase "volatile suspended solids" has the same meaning as in footnote 20.

To develop working design equations we make the following assumptions:

1. The influent and effluent biomass concentrations are negligible compared to that in the reactor.
2. The influent food (S_o) is immediately diluted to the reactor concentration in accordance with the definition of a CSTR.
3. All reactions occur in the CSTR.

From the first assumption we may eliminate the following terms from Equation 5-17: QX_o, and $(Q-Q_w)X_e$ because X_o, and X_e, are negligible compared to X. Equation 5-17 may be simplified to

$$(\forall)\left(\frac{\mu_m SX}{K_s + S} - k_d X\right) = +Q_w X_r \tag{5-21}$$

For convenience, we may rearrange Equation 5-21 in terms of the Monod equation

$$\left(\frac{\mu_m S}{K_s + S}\right) = \frac{Q_w X_r}{\forall X} + k_d \tag{5-22}$$

Equation 5-20 may also be rearranged in terms of the Monod equation

$$\left(\frac{\mu_m S}{K_s + S}\right) = \frac{Q}{\forall}\frac{Y}{X}(S_o - S) \tag{5-23}$$

Noting that the left side of Equations 5-22 and 5-23 are the same, we set the right-hand side of these equations equal and rearrange to give:

$$\frac{Q_w X_r}{\forall X} = \frac{Q}{\forall}\frac{Y}{X}(S_o - S) - k_d \tag{5-24}$$

Two parts of this equation have physical significance in the design of a completely mixed activated sludge system. The inverse of $Q/\forall$ is the *hydraulic detention time* (θ) of the reactor:

$$\frac{\forall}{Q} = \theta \tag{5-25}$$

The inverse of the left side of Equation 5-24 defines the mean cell-residence time (θ_c):

$$\frac{\forall X}{Q_w X_r} = \theta_c \tag{5-26}$$

The mean cell-residence time expressed in Equation 5-26 must be modified if the effluent biomass concentration is not negligible. Equation 5-27 accounts for effluent losses of biomass in calculating θ_c.

$$\theta_c = \frac{\forall X}{Q_w X_r + (Q - Q_w)(X_e)} \tag{5-27}$$

From Equation 5-22, it can be seen that once θ_c is selected, the concentration of soluble BOD_5 in the effluent (S) is fixed:

$$S = \frac{K_s(1 + k_d\theta_c)}{\theta_c(\mu_m - k_d) - 1} \tag{5-28}$$

Typical values of the microbial growth constants are given in Table 5-11. Note that the concentration of soluble BOD_5 leaving the system (S) is affected only by the mean cell-residence time and not by the amount of BOD_5 entering the aeration tank or by the hydraulic detention time. It is also important to reemphasize that S is the soluble BOD_5 and not the total BOD_5. Some fraction of the suspended solids that do not settle in the secondary settling tank also contributes to the BOD_5 load to the receiving body. To achieve a desired effluent quality both the soluble and insoluble fractions of BOD_5 must be considered. Thus, to use Equation 5-28 to achieve a desired effluent quality (S) by solving for θ_c, some estimate of the BOD_5 of the suspended solids must be made first. This value is then subtracted from the total allowable BOD_5 in the effluent to find the allowable S:

$$S = \text{Total BOD}_5 \text{ allowed} - \text{BOD}_5 \text{ in suspended solids} \tag{5-29}$$

From Equation 5-24, it is also evident that the concentration of microorganisms in the aeration tank is a function of the mean cell-residence time, hydraulic detention time, and difference between the influent and effluent concentrations:

$$X = \frac{\theta_c(Y)(S_o - S)}{\theta(1 + k_d\theta_c)} \tag{5-30}$$

Plug-flow with recycle. A plug-flow reactor is one in which the fluid particles pass through the tank in sequence (see Section 3-4 for further discussion). Although it is difficult to achieve true plug flow, many long, narrow aeration tanks may be better approximated by plug flow than by completely mixed models. A kinetic model of

TABLE 5-11
Values of growth constants for domestic wastewater[a]

Parameter	Basis	Value[b] Range	Typical
K_s	mg/L BOD_5	25–100	60
k_d	d^{-1}	0.025–0.075	0.06
μ_m	d^{-1}	2–10	5
Y	mg VSS/mg BOD_5	0.4–0.8	0.6

[a]Source: Metcalf & Eddy, *Wastewater Engineering: Treatment, Disposal, and Reuse,* 3rd ed., New York: McGraw-Hill, p. 394.
[b]Values are for 20°C.

a plug-flow system is difficult to develop from basic mass-balance equations. With two simplifying assumptions, Lawrence and McCarty[22] have developed a useful equation. The assumptions are:

1. The concentration of microorganisms in the influent to the aeration tank is approximately the same as that in the effluent from the aeration tank. This assumption applies if θ_c/θ is greater than 5.
2. The rate of soluble BOD_5 utilization as the waste passes through the aeration tank is given by

$$r_u = -\frac{\mu_m S X_{avg}}{K_s + S} \tag{5-31}$$

where X_{avg} is the average concentration of microoorganisms in the aeration tank. The design equation is

$$\frac{1}{\theta_c} = \frac{Y\mu_m(S_o - S)}{(S_o - S) + (1 + \alpha)K_s \ln(S_i/S)} - k_d \tag{5-32}$$

where α = recycle ratio, Q_r/Q
ln = logarithm to base e
S_i = influent concentration to aeration tank after dilution with recycle flow, mg/L
$= \dfrac{S_o + \alpha S}{1 + \alpha}$

Other terms are the same as those defined previously.

Example 5-7. The town of Gatesville has been directed to upgrade its primary WWTP to a secondary plant that can meet an effluent standard of 30.0 mg/L BOD_5 and 30.0 mg/L suspended solids (SS). They have selected a completely mixed activated sludge system.

Assuming that the BOD_5 of the SS may be estimated as equal to 63 percent of the SS concentration, estimate the required volume of the aeration tank. The following data are available from the existing primary plant.

Existing plant effluent characteristics
Flow = 0.150 m^3/s
BOD_5 = 84.0 mg/L

Assume the following values for the growth constants: K_s = 100 mg/L BOD_5; μ_m = 2.5 d^{-1}; k_d = 0.050 d^{-1}; Y = 0.50 mg VSS/mg BOD_5 removed.

[22] A. W. Lawrence and P. L. McCarty, "A Unified Basis for Biological Treatment Design and Operation," *Journal of the Sanitary Engineering Division,* American Society of Civil Engineers, 96, No. SA3, 1970.

Solution. Assuming that the secondary clarifier can produce an effluent with only 30.0 mg/L SS, we can estimate the allowable soluble BOD_5 in the effluent using the 63-percent assumption from above and Equation 5-29.

$$S = BOD_5 \text{ allowed} - BOD_5 \text{ in suspended solids}$$

$$S = 30.0 - (0.630)(30.0) = 11.1 \text{ mg/L}$$

The mean cell-residence time can be estimated with Equation 5-28 and the assumed values for the growth constants.

$$11.1 = \frac{100.0(1 + (0.050)\theta_c)}{\theta_c(2.5 - 0.050) - 1}$$

Solving for θ_c

$$(11.1)(2.45\theta_c - 1) = 100.0 + 5.00\theta_c$$

$$27.20\theta_c - 11.1 = 100.0 + 5.00\theta_c$$

$$\theta_c = \frac{111.1}{22.2} = 5.00 \text{ or } 5.0 \text{ d}$$

If we assume a value of 2,000 mg/L for the MLVSS, we can solve Equation 5-30 for the hydraulic detention time.

$$2{,}000 = \frac{5.00(0.50)(84.0 - 11.1)}{\theta(1 + (0.050)(5.00))}$$

$$\theta = \frac{2.50(72.9)}{2{,}000(1.25)}$$

$$\theta = 0.073 \text{ d or } 1.8 \text{ h}$$

The volume of the aeration tank is then estimated using Equation 5-25:

$$1.8 = \frac{\forall}{0.150 \times 3{,}600 \text{ s/h}}$$

$$\forall = 972 \text{ m}^3 \text{ or } 970 \text{ m}^3$$

Another commonly used parameter in the activated sludge process is the food to microorganism ratio (F/M), which is defined as:

$$\frac{\text{F}}{\text{M}} = \frac{QS_o}{\forall X} \tag{5-33}$$

The units of the F/M ratio are

$$\frac{\text{mg BOD}_5\text{/d}}{\text{mg MLVSS}} = \frac{\text{mg}}{\text{mg} \cdot \text{d}}$$

The F/M ratio is controlled by wasting part of the microbial mass, thereby reducing the MLVSS. A high rate of wasting causes a high F/M ratio. A high F/M yields organisms that are saturated with food. The result is that efficiency of treatment is poor. A low rate of wasting causes a low F/M ratio, which yields organisms that are starved. This results in more complete degradation of the waste.

A long θ_c (low F/M) is not always used, however, because of certain trade-offs that must be considered. A long θ_c means a larger and more costly aeration tank. It also means a higher requirement for oxygen and, thus, higher power costs. Problems with poor sludge "settleability" in the final clarifier may be encountered if θ_c is too long. However, because the waste is more completely degraded to final end products and less of the waste is converted into microbial cells when the microorganisms are starved at a low F/M, there is less sludge to handle.

Because both the F/M ratio and the cell-detention time are controlled by wasting of organisms, they are interrelated. A high F/M corresponds to a short θ_c and a low F/M corresponds to a long θ_c. F/M values typically range from 0.1 to 1.0 mg/mg·d for the various modifications of the activated sludge process.

Example 5-8. Two "fill and draw," batch-operated sludge tanks are operated as follows:

Tank A is settled once each day, and half the liquid is removed with care not to disturb the sludge that settles to the bottom. This liquid is replaced with fresh settled sewage. A plot of MLVSS concentration versus time takes the shape shown below.

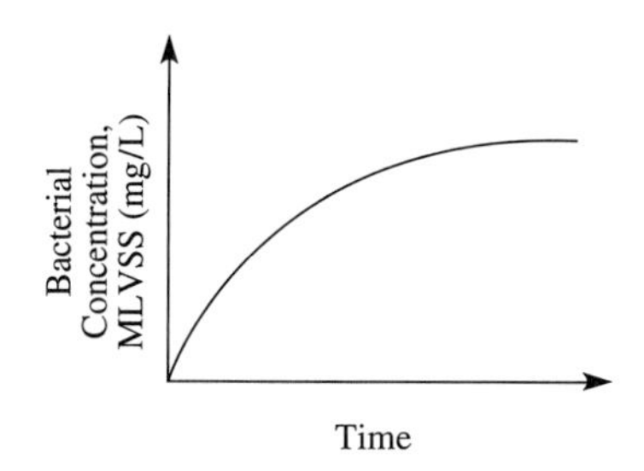

Tank B is not settled. Rather, once each day, half the mixed liquor is removed while the tank is being violently agitated. The liquid is replaced with fresh settled sewage. A plot of MLVSS concentration versus time is shown below.

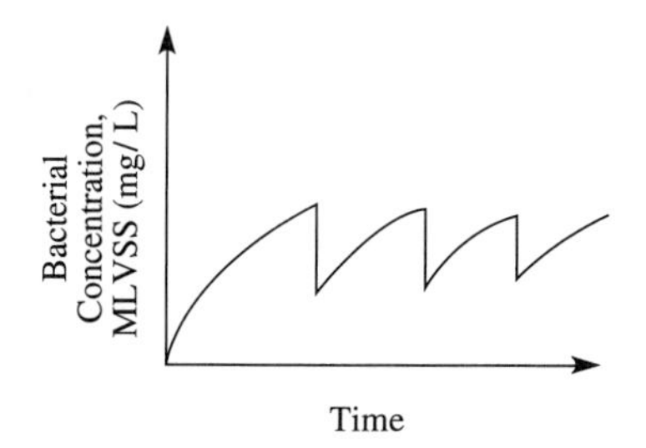

A comparison of the operating characteristics of the two systems is shown in the following table.

Parameter	Tank A	Tank B
F/M	Low	High
θ_c	Long	Short
Sludge	None	Much
Oxygen required	High	Low
Power	High	Low

The optimum choice is somewhere between these extremes. A balance must be struck between the cost of sludge disposal and the cost of power to provide oxygen (air).

Example 5-9. Compute the F/M ratio for the new activated-sludge plant at Gatesville (Example 5-7).

Solution. Using the data from Example 5-7 and Equation 5-33:

$$\frac{\mathrm{F}}{\mathrm{M}} = \frac{(0.150\ \mathrm{m^3/s})(84.0\ \mathrm{mg/L})(86{,}400\ \mathrm{s/d})}{(970\ \mathrm{m^3})(2{,}000\ \mathrm{mg/L})}$$

$$= 0.56\ \mathrm{mg/\ mg \cdot d}$$

This is well within the typical range of F/M ratios.

Sludge return. The purpose of sludge return is to maintain a sufficient concentration of activated sludge in the reactor basin. One method used to control the rate of sludge return to the reactor basin is based on the empirical measurement known as the *sludge volume index* (SVI).

SVI is determined from a standard laboratory test.[23] The procedure involves measuring the MLSS and sludge settleability. A one-liter sample of mixed liquor is obtained from the aeration tank at the discharge end. The sludge settleability is measured by filling a standard one-liter graduated cylinder to the 1.0 liter mark, allowing undisturbed settling for 30 minutes, and then reading the volume occupied by the settled sludge. The MLSS is determined by filtering, drying, and weighing a second portion of the mixed liquor. The SVI, which is defined as the volume in milliliters occupied by 1 g of activated sludge after the aerated liquor has settled 30 min, is calculated as follows:

$$\mathrm{SVI} = \frac{\mathrm{SV}}{\mathrm{MLSS}} \times 1{,}000\ \mathrm{mg/g} \tag{5-34}$$

where SVI = sludge volume index, mL/g
SV = volume of settled solids in one-liter graduated cylinder after 30 min settling, mL/L
MLSS = mixed liquor suspended solids, mg/L

[23] American Public Health Association, *Standard Methods for the Examination of Waste and Wastewater*, 18th ed., Washington, DC: pp. 2–66, 1992.

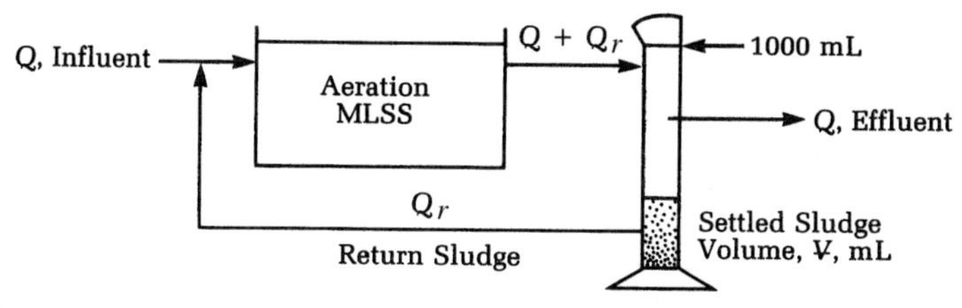

FIGURE 5-22
Hypothetical relationship between settled sludge volume from SVI test and return sludge flow. (*Source:* M. J. Hammer, *Water and Wastewater Technology, SI Version,* New York: Wiley, 1977. Reprinted by permission.)

Conceptually, SVI can be related to the quantity and solids concentration in the secondary settling tank as we have depicted in Figure 5-22. In the following discussion and mathematical relationships, the secondary tank is assumed to respond identically to the graduated cylinder used in the SVI test. This assumption is extraordinary to say the least. In fact, Vesilind has shown that for MLSS concentrations of less than 5,000 mg/L, the settling rates are 10 to 20 percent greater than might be expected in a final clarifier.[24] Nonetheless, environmental engineers have developed a large body of empirical data based on it.

The SVI can be used as an indication of the settling characteristics of the sludge, thereby impacting on return rates and MLSS. Typical values of SVI for activated sludge plants operating with an MLSS concentration of 2,000 to 3,500 mg/L range from 80 to 150 mL/g. As the sludge concentration is increased to the 3,000 to 5,000 mg/L range, there is a higher solids loading on the settling basin, and as a consequence, a lower SVI or larger settling basin is required to avoid the loss of solids caused by "washout" or hydraulic displacement.

The SVI is a key factor in the system design. Indirectly, it limits the reactor basin MLSS concentration and, in turn, the MLVSS that can be achieved, because it controls the settling tank underflow concentration. Thus, for a given SVI and return sludge rate, the maximum MLSS and MLVSS are fixed within narrow limits.

Most activated sludge plants are designed to permit variable sludge return from 10 to 100 percent of the raw waste flow. This range of return sludge flow gives the operator reasonable flexibility to adjust the MLSS to the desired concentration. In general, the return sludge ratio should be limited to or below 100 percent. This is particularly true if the SVI is higher than 150 mL/g and there is no provision for additional floor area in the final clarification step.

Without operating data, the Joint Task Force suggests that MLSS be limited to 5,000 mg/L (lower at temperatures of less than 20°C), even though the SVIs may be very low.[25] Design values over 5,000 mg/L generally will lead to inordinately low

---

[24] P. Aarne Vesilind, "Discussion of Evaluation of Sludge Thickening Theories," *Journal of the Sanitary Engineering Division,* American Society of Civil Engineers, vol. 94, p. 185, 1968.

[25] Joint Task Force of the Water Environment Federation and the American Society of Civil Engineers, *Design of Municipal Wastewater Treatment Plants Vol. I,* Manual of Practice No. 8, Alexandria, VA, 1992.

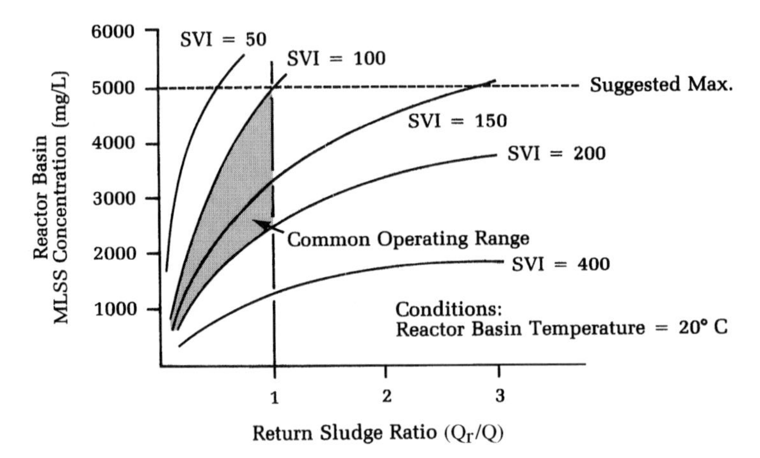

FIGURE 5-23
Design MLSS versus SVI and return sludge ratio. (*Source:* Joint Task Force of the Water Environment Federation and the American Society of Civil Engineers, *Design of Municipal Wastewater Treatment Plants Vol. I,* Manual of Practice No. 8, Chapters 1–12, Alexandria, VA, 1992. Reprinted by permission.)

detention times that are more subject to washout unless surge control is planned. Design MLSS values should not be any higher than needed, since the final settling basin operations become critical at high MLSS levels.

The mixed liquor concentration as a function of the SVI and the return sludge ratio (Q_r/Q) is shown in Figure 5-23. The return sludge pumping rate may be determined from a mass balance around the settling tank in Figure 5-21. Assuming that the amount of sludge in the secondary settling tank remains constant (steady-state conditions) and that the effluent suspended solids (X_e) are negligible, the mass balance is

$$\text{accumulation} = \text{inflow} - \text{outflow} \tag{5-35}$$

$$0 = (Q + Q_r)(X') - (Q_r X_r' + Q_w X_r') \tag{5-36}$$

where Q = wastewater flow rate, m^3/d
Q_r = return sludge flow rate, m^3/d
X' = mixed liquor suspended solids (MLSS), g/m^3
X_r' = maximum return sludge concentration, g/m^3
Q_w = sludge wasting flow rate, m^3/d

Solving for the return sludge flow rate:

$$Q_r = \frac{QX' - Q_w X_r'}{X_r' - X'} \tag{5-37}$$

Frequently, the assumption that the effluent suspended solids are negligible is not valid. If the effluent suspended solids are significant, the mass balance may then be expressed as

$$0 = (Q + Q_r)(X') - (Q_r X_r' + Q_w X_r' + (Q - Q_w)X_e) \tag{5-38}$$

Solving for the return sludge flow rate:

$$Q_r = \frac{QX' - Q_w X_r' - (Q - Q_w)X_e}{X_r' - X'} \tag{5-39}$$

Note that X_r' and X' include both the volatile and inert fractions. Thus, they differ from X_r and X by a constant factor. With the volume of the tank and the mean cell-residence time, the sludge wasting flow rate can be determined with Equation 5-26 if the maximum return sludge concentration (X_r') can be determined. The maximum return sludge concentration is related to the SVI as follows:

$$X_r' = \frac{1{,}000 \text{ mg/g}(1{,}000 \text{ mL/L})}{\text{SVI}} = \frac{10^6}{\text{SVI}} \text{ mg/L} \tag{5-40}$$

Figure 5-23 has been constructed on the basis of rapid sludge removal and uses the concentration achieved in the 30-minute settling test as the settling basin underflow concentration. Practice has shown this to be a relatively valid approach.

The maximum achievable underflow concentration is also a function of temperature. Temperature affects zone settling velocity, as well as the SVI. In cold weather, the SVI increases because of poor settling. The Joint Task Force's recommended mixed-liquor design concentration as a function of the minimum design reactor-basin temperature for several SVI values is shown in Figure 5-24. (The SVI is taken at the temperature of the reactor basin contents.)

Example 5-10. In the continuing saga of the Gatesville plant expansion, we now wish to consider the question of the return sludge design. Based on the aeration tank design (Example 5-7) and an informed, reliable source, we have the following data:

Design data:

Flow = 0.150 m^3/s

MLVSS(X) = 2,000 mg/L

MLSS(X') = 1.43 (MLVSS)

Effluent suspended solids = 30 mg/L

Wastewater temperature = 18.0°C

Solution. We begin by computing the anticipated concentration of the MLSS.

MLSS = 1.43(2,000)

MLSS = 2,860 mg/L

We can't really predict SVI but Figure 5-24 gives us a reasonable range to assume a value. Alternatively, we could assume a return sludge concentration. Using Figure 5-24, we select an SVI of 175 based on the calculated MLSS and the reactor basin temperature.

Now, using Equation 5-40, we can determine the return sludge concentration.

$$X_r' = \frac{10^6}{175}$$

$$X_r' = 5{,}714.29 \text{ or } 5{,}700 \text{ mg/L}$$

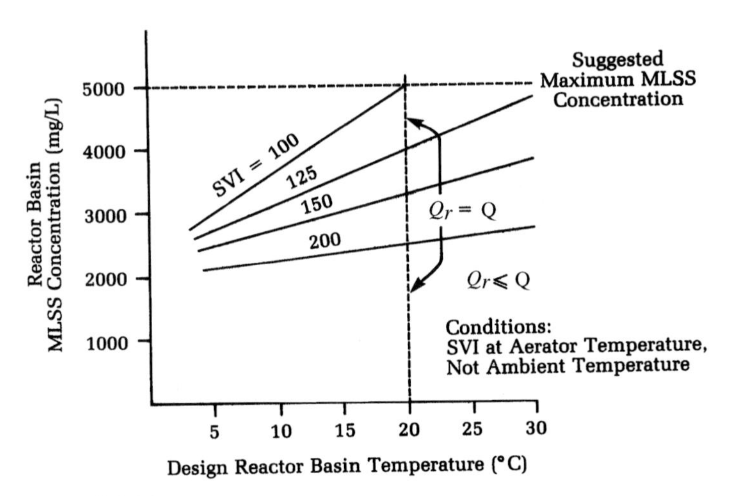

FIGURE 5-24
Recommended maximum MLSS design versus temperature and SVI. (*Source:* Joint Task Force of the Water Environment Federation and the American Society of Civil Engineers, *Design of Municipal Wastewater Treatment Plants Vol. I,* Manual of Practice No. 8, Chapters 1–12, Alexandria, VA, 1992. Reprinted by permission.)

The return sludge flow rate may be computed using Equations 5-26 and 5-37. Solving Equation 5-26 for the sludge wasting flow rate using the data from Example 5-7 and noting that $X_r = X_r'/1.43 = 3{,}986$ mg/L:

$$Q_w = \frac{\forall X}{\theta_c X_r} = \frac{(970\ \text{m}^3)(2{,}000\ \text{mg/L})}{(5\ \text{d})(3{,}986\ \text{mg/L})} = 97.3\ \text{m}^3/\text{d}$$

Converting Q_w to m^3/s:

$$\frac{(97.3\ \text{m}^3/\text{d})}{(86{,}400\ \text{s/d})} = 0.0011\ \text{m}^3/\text{s}$$

Noting that 1 mg/L $=$ 1 g/m^3, if we ignore the effluent suspended solids, the estimated return sludge flow rate is

$$Q_r = \frac{(0.150\ \text{m}^3/\text{s})(2{,}860\ \text{g/m}^3) - (0.0011\ \text{m}^3/\text{s})(5{,}714\ \text{g/m}^3)}{5{,}714\ \text{g/m}^3 - 2{,}860\ \text{g/m}^3}$$

$$= 0.148 \text{ or } 0.15\ \text{m}^3/\text{s}$$

If the effluent suspended solids are not neglected, the estimated return sludge flow rate is

$$Q_r = \frac{(0.150\ \text{m}^3/\text{s})(2{,}860\ \text{g/m}^3) - (0.0011\ \text{m}^3/\text{s})(5{,}714\ \text{g/m}^3) - (0.150 - 0.0011)(30\ \text{g/m}^3)}{5{,}714\ \text{g/m}^3 - 2{,}860\ \text{g/m}^3}$$

$$= 0.146 \text{ or } 0.15\ \text{m}^3/\text{s}$$

We can check this result using Figure 5-23. Although it appears high on first consideration, it is a valid result.

Sludge production. The activated sludge process removes substrate, which exerts an oxygen demand by converting the food into new cell material and degrading this cell material while generating energy. This cell material ultimately becomes sludge, which must be disposed of. Despite the problems in doing so, researchers have attempted to develop enough basic information on sludge production to permit a reliable design basis. Heukelekian and Sawyer both reported that a net yield of 0.5 kg MLVSS/kg BOD_5 removed could be expected for a completely soluble organic substrate.[26] Most researchers agree that, depending on the inert solids in the system and the SRT, 0.40 to 0.60 kg MLVSS/kg BOD_5 removed will normally be observed.

The amount of sludge that must be wasted each day is the difference between the amount of increase in sludge mass and the suspended solids (SS) lost in the effluent:

$$\text{Mass to be wasted} = \text{increase in MLSS} - \text{SS lost in effluent} \quad (5\text{-}41)$$

The net activated sludge produced each day is determined by:

$$Y_{obs} = \frac{Y}{1 + k_d\theta_c} \quad (5\text{-}42)$$

and

$$P_x = Y_{obs}Q(S_o - S)(10^{-3}\ \text{kg/g}) \quad (5\text{-}43)$$

where P_x = net waste activated sludge produced each day in terms of VSS, kg/d
Y_{obs} = observed yield, kg MLVSS/kg BOD_5 removed

Other terms are as defined previously.

The increase in MLSS may be estimated by assuming that VSS is some fraction of MLSS. It is generally assumed that VSS is 60 to 80 percent of MLVSS. Thus, the increase in MLSS in Equation 5-43 may be estimated by dividing P_x by a factor of 0.6 to 0.8 (or multiplying by 1.25 to 1.667). The mass of suspended solids lost in the effluent is the product of the flow rate $(Q - Q_w)$ and the suspended solids concentration (X_e).

Example 5-11. Estimate the mass of sludge to be wasted each day from the new activated sludge plant at Gatesville (Examples 5-7 and 5-10).

Solution. Using the data from Example 5-7, calculate Y_{obs}:

$$Y_{obs} = \frac{0.50\ \text{kg VSS/kg BOD}_5\ \text{removed}}{1 + [(0.050\ \text{d}^{-1})(5\ \text{d})]}$$

$$= 0.40\ \text{kg VSS/kg BOD}_5\ \text{removed}$$

---

[26]H. Heukelekian, H. Orford, and R. Manganelli, "Factors Affecting the Quantity of Sludge Production in the Activated Sludge Process," *Sewage & Industrial Wastes,* vol. 23, p. 8, 1951; and C. N. Sawyer, "Bacterial Nutrition and Synthesis," *Biological Treatment of Sewage and Industrial Wastes,* vol. I, p. 3, 1956.

The net waste activated sludge produced each day is

$$P_x = (0.40)(0.150 \text{ m}^3/\text{s})(84.0 \text{ g/m}^3 - 11.1 \text{ g/m}^3)(86{,}400 \text{ s/d})(10^{-3} \text{ kg/g})$$

$$= 377.9 \text{ kg/d of VSS}$$

The total mass produced includes inert materials. Using the relationship between MLSS and MLVSS in Example 5-10,

$$\text{Increase in MLSS} = (1.43)(377.9 \text{ kg/d}) = 540.4 \text{ kg/d}$$

The mass of solids (both volatile and inert) lost in the effluent is

$$(Q - Q_w)(X_e) = (0.150 \text{ m}^3/\text{s} - 0.0011 \text{ m}^3/\text{s})(30 \text{ g/m}^3)(86{,}400 \text{ s/d})(10^{-3} \text{ kg/g})$$

$$= 385.9 \text{ kg/d}$$

The mass to be wasted is then

$$\text{Mass to be wasted} = 540.4 - 385.9 = 154.5 \text{ kg/d}$$

Note that this mass is calculated as dry solids. Because the sludge is mostly water, the actual mass will be considerably larger. This is discussed further in Section 5-11.

Oxygen demand. Oxygen is used in those reactions required to degrade the substrate to produce the high-energy compounds required for cell synthesis and for respiration. For long SRT systems, the oxygen needed for cell maintenance can be of the same order of magnitude as substrate metabolism. A minimum residual of 0.5 to 2 mg/L DO is usually maintained in the reactor basin to prevent oxygen deficiencies from limiting the rate of substrate removal.

An estimate of the oxygen requirements may be made from the BOD_5 of the waste and amount of activated sludge wasted each day. If we assume all of the BOD_5 is converted to end products, the total oxygen demand can be computed by converting BOD_5 to BOD_L. Since a portion of waste is converted to new cells that are wasted, the BOD_L of the wasted cells must be subtracted from the total oxygen demand. An approximation of the oxygen demand of the wasted cells may be made by assuming cell oxidation can be described by the following reaction:

$$\underbrace{C_5H_7NO_2}_{\text{cells}} + 5O_2 \rightleftharpoons 5CO_2 + 2H_2O + NH_3 + \text{energy} \qquad (5\text{-}44)$$

The ratio of gram molecular weights is

$$\frac{5(32)}{113} = 1.42$$

Thus the oxygen demand of the waste activated sludge may be estimated as 1.42 (P_x).

The mass of oxygen required may be estimated as:

$$M_{O_2} = \frac{Q(S_o - S)(10^{-3} \text{ kg/g})}{f} - 1.42\,(P_x) \qquad (5\text{-}45)$$

where Q = wastewater flow rate into the aertion tank, m^3/d
S_o = influent soluble BOD_5, mg/L
S = effluent soluble BOD_5, mg/L
f = conversion factor for converting BOD_5 to ultimate BOD_L
P_x = waste activated sludge produced (see Equation 5-43)

The volume of air to be supplied must take into account the percent of air that is oxygen and the transfer efficiency of the dissolution of oxygen into the wastewater.

Example 5-12. Estimate the volume of air to be supplied (m^3/d) for the new activated sludge plant at Gatesville (Examples 5-7 and 5-10). Assume that BOD_5 is 68 percent of the ultimate BOD and that the oxygen transfer efficiency is 8 percent.

Solution. Using the data from Examples 5-7 and 5-11

$$M_{O_2} = \frac{(0.150 \text{ m}^3/\text{s})(84.0 \text{ g/m}^3 - 11.1 \text{ g/m}^3)(86{,}400 \text{ s/d})(10^{-3} \text{ kg/g})}{0.68}$$

$$- 1.42(377.9 \text{ kg/d of VSS})$$

$$= 1{,}389.4 - 536.6 = 852.8 \text{ kg/d of oxygen}$$

From Table A-5 in Appendix A, air has a density of 1.185 kg/m^3 at standard conditions. By mass, air contains about 23.2 percent oxygen. At 100 percent transfer efficiency, the volume of air required is

$$\frac{852.8 \text{ kg/d}}{(1.185 \text{ kg/m}^3)(0.232)} = 3{,}101.99 \text{ or } 3{,}100 \text{ m}^3/\text{d}$$

At 8 percent transfer efficiency

$$\frac{3{,}101.99 \text{ m}^3/\text{d}}{0.08} = 38{,}774.9 \text{ or } 38{,}000 \text{ m}^3/\text{d}$$

Process design considerations. The SRT (i.e., θ_c) selected for design is a function of the degree of treatment required. A high SRT (or older sludge age) results in a higher quantity of solids being carried in the system and a higher degree of treatment being obtained. A long SRT also results in the production of less waste sludge.

SRT values for design of carbonaceous BOD_5 removal as a function of the minimum temperature at which the reactor basin will be operated are depicted in Figure 5-25. The SRT values given are those for normal domestic wastewater. It is expected that the soluble BOD_5 in the effluent from the aeration system will be 4 to 8 mg/L.

If industrial wastes are discharged to the municipal system, several additional concerns must be addressed. Municipal wastewater generally contains sufficient nitrogen and phosphorus to support biological growth. The presence of large volumes of industrial wastewater that is deficient in either of these nutrients will result in poor removal efficiencies. Addition of supplemental nitrogen and phosphorus may

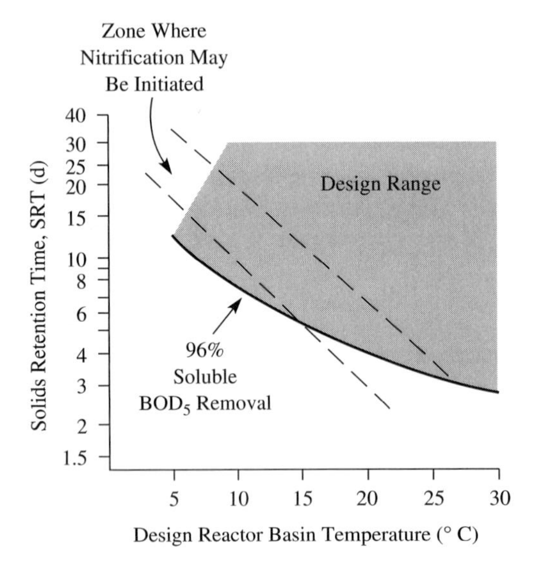

FIGURE 5-25
Design SRT for carbonaceous BOD_5 removal. (*Source:* Joint Task Force of the Water Environment Federation and the American Society of Civil Engineers, *Design of Municipal Wastewater Treatment Plants Vol. I,* Manual of Practice No. 8, Chapters 1–12, Alexandria, VA, 1992. Reprinted by permission.)

be required. The ratio of nitrogen to BOD_5 should be 1:32. The ratio of phosphorus to BOD_5 should be 1:150.

Although toxic metals and organics may be at low enough levels that they do not interfere with the operation of the plant, two other untoward effects may result if they are not excluded in a pretreatment program. Volatile organics may be stripped from solution into the atmosphere in the aeration tank. Thus, the WWTP may become a source of air pollution. The toxic metals may precipitate into the waste sludges. Thus, otherwise nonhazardous sludges may be rendered hazardous.

Oil and grease that pass through the primary treatment system will form grease balls on the surface of the aeration tank. The microorganisms cannot degrade this material because it is not in the water where they can physically come in contact with it. Special consideration should be given to the surface skimming equipment in the secondary clarifier to handle the grease balls.

Secondary clarifier design considerations. Although the secondary settling tank (Figure 5-26) is an integral part of both the trickling filter and the activated sludge process, environmental engineers have focused particular attention on the secondary clarifier used after the activated sludge process. A secondary clarifier is important because of the high solids loading and fluffy nature of the activated sludge biological floc. Also, it is highly desirable that sludge recycle be well thickened.

Secondary settling tanks for activated sludge are generally characterized as having Type III settling. Some authors would argue that Types I and II also occur.

The following guidance has been excerpted from the Joint Task Force of the Water Pollution Control Federation and the American Society of Civil Engineers.[27]

[27] Joint Task Force of the Water Environment Federation and the American Society of Civil Engineers, *Design of Municipal Wastewater Treatment Plants Vol. I,* Manual of Practice No. 8, Alexandria, VA, 1992.

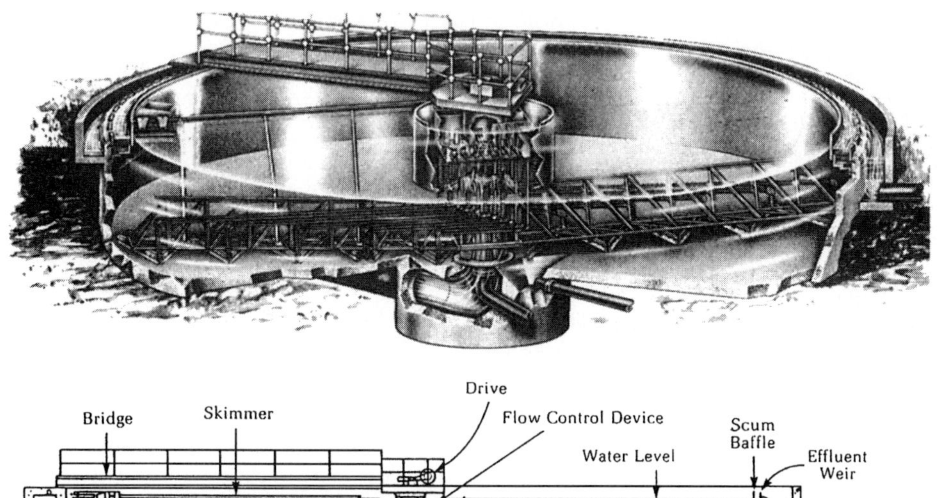

FIGURE 5-26
Rendering and cross-sectional diagram of secondary settling tank. (Courtesy of FMC Corporation.)

The design factors discussed here are the result of the experiences of investigators, plant superintendents, and equipment manufacturers. The criteria primarily apply to circular (or square) center-fed basins, which comprise the majority of activated sludge secondary settling units designed in the last 25 years.

An overflow rate between 20 and 34 m/d for the average flow in a conventional process can be expected to result in good separation of liquid and SS. The design engineer also must check the peak hydraulic rates that will be imposed on the settling basin.

Suggested secondary settling tank side water depths (SWD) and solids loading rates are shown in Table 5-12 and Figure 5-27, respectively.

The GLUMRB has set maximum recommended weir loadings for secondary settling basins at 125 to 250 m^3/d per m of weir length (m^3/d · m). This criterion is based on effluent quality of operating units. It appears that the settling basin design may have had much to do with this limitation. Also, most of the observations apply to rectangular settling basins.

One of the continuing difficulties with the design of secondary settling tanks is the prediction of effluent suspended solids concentrations as a function of common design and operating parameters. Little theoretical work has been conducted, and empirical correlations have been less than satisfactory.[28]

---

[28]Rex Chainbelt, Inc., *A Mathematical Model of a Final Clarifier* (Environmental Protection Agency Report No. 17090 FJW 02/72), Washington, DC: U.S. Government Printing Office, 1972.

TABLE 5-12
Final settling basin side water depth

	Side water depth, m	
Tank diameter, m	**Minimum**	**Recommended**
<12	3.0	3.4
12 to 20	3.4	3.7
20 to 30	3.7	4.0
30 to 42	4.0	4.3
>42	4.3	4.6

Source: Joint Task Force of the Water Environment Federation and the American Society of Civil Engineers, *Design of Municipal Wastewater Treatment Plants Vol. I,* Manual of Practice No. 8, Chapters 1–12, Alexandria, VA, 1992.

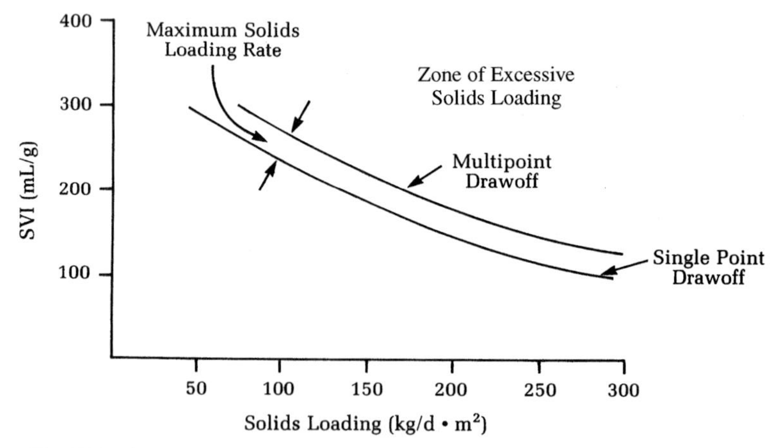

FIGURE 5-27
Design solids loading versus SVI. (*Note:* Rapid sludge removal design assumes that there will be no inventory in the settling tank. *Source:* Joint Task Force of the Water Environment Federation and the American Society of Civil Engineers, *Design of Municipal Wastewater Treatment Plants Vol. I,* Manual of Practice No. 8, Chapters 1–12, Alexandria, VA, 1992. Reprinted by permission.

Example 5-13. The secondary clarifier for the Gatesville plant (Examples 5-7, and 5-9–5-12) must be able to handle an MLSS load of 2,860 mg/L. The flow to the secondary clarifier is 0.300 m^3/s, of which one-half is contributed by the return flow. Determine the tank diameter, depth, and weir length.

Solution. Utilizing an average overflow rate of 33 m/d, we determine the diameter of the secondary tank as follows:

First, compute the surface area required:

$$A_s = \frac{0.150 \text{ m}^3/\text{s} \times 86{,}400 \text{ s/d}}{33 \text{ m/d}}$$

$$= 392.73 \text{ m}^2$$

Note that since only one-half of the flow leaves through the surface of the tank (the remainder leaves through the bottom as Q_r), only one-half of the hydraulic load is used to compute the overflow rate.

The diameter of the tank is then

$$392.73 = \frac{\pi D^2}{4}$$

$$D = (500.04)^{1/2} = 22.36 \text{ or } 22 \text{ m}$$

From Table 5-12 we select an SWD of 4.0 m.

Now we must check the solids loading. Using the equality 1 mg/L = 1 g/m^3:

$$\text{SL} = \frac{2{,}860 \text{ g/m}^3 \times 0.300 \text{ m}^3\text{/s}}{\dfrac{\pi(22 \text{ m})^2}{4}}$$

$$= \frac{858.00 \text{ g/s}}{380.13 \text{ m}^2} \times 10^{-3} \text{ kg/g} \times 86{,}400 \text{ s/d}$$

$$= 195 \text{ kg/d} \cdot \text{m}^2$$

Checking this rate with the maxima shown in Figure 5-27, we find that for an SVI of 175 (from Example 5-10), we have the maximum allowable loading of 200 kg/d $\cdot$ m^2.

The weir loading for a single weir located at the periphery is

$$\text{WL} = \frac{0.150 \times 86{,}400 \text{ s/d}}{\pi(22)}$$

$$= 187.51 \text{ m}^3\text{/d} \cdot \text{m}$$

This is less than the prescribed loading given by GLUMRB and, therefore, is acceptable.

Sludge problems. A *bulking sludge* is one that has poor settling characteristics and poor compactability. There are two principal types of sludge bulking. The first is caused by the growth of filamentous organisms, and the second is caused by water trapped in the bacterial floc, thus reducing the density of the agglomerate and resulting in poor settling.

Filamentous bacteria have been blamed for much of the bulking problem in activated sludge. Although filamentous organisms are effective in removing organic matter, they have poor floc-forming and settling characteristics. Bulking may also be caused by a number of other factors, including long, slow-moving collection-system transport; low available ammonia nitrogen when the organic load is high; low pH, which could favor acid-favoring fungi; and the lack of macronutrients, which stimulates predomination of the filamentous actinomycetes over the normal floc-forming bacteria. The lack of nitrogen also favors slime-producing bacteria, which have a low specific gravity, even though they are not filamentous. The multicellular fungi cannot compete with the bacteria normally, but can compete under specific environmental conditions, such as low pH, low nitrogen, low oxygen, and high carbohydrates. As the pH decreases below 6.0, the fungi are less affected than the bacteria and tend to predominate. As the nitrogen concentrations drop below a BOD_5 : N ratio of 20:1, the fungi, which have a lower protein level than the bacteria, are able to produce normal protoplasm, while the bacteria produce nitrogen-deficient protoplasm.

A sludge that floats to the surface after apparently good settling is called a *rising sludge.* Rising sludge results from denitrification, that is, reduction of nitrates and nitrites to nitrogen gas in the sludge blanket (layer). Much of this gas remains trapped in the sludge blanket, causing globs of sludge to rise to the surface and float over the weirs into the receiving stream.

Rising-sludge problems can be overcome by increasing the rate of return sludge flow (Q_r), by increasing the speed of the sludge-collecting mechanism, by decreasing the mean cell residence time, and, if possible, by decreasing the flow from the aeration tank to the offending tank.

Oxidation Ponds

Treatment ponds have been used to treat wastewater for many years, particularly as wastewater treatment systems for small communities.[29] Many terms have been used to describe the different types of systems employed in wastewater treatment. For example, in recent years, *oxidation pond* has been widely used as a collective term for all types of ponds. Originally, an oxidation pond was a pond that received partially treated wastewater, whereas a pond that received raw wastewater was known as a *sewage lagoon. Waste stabilization pond* has been used as an all-inclusive term that refers to a pond or lagoon used to treat organic waste by biological and physical processes. These processes would commonly be referred to as self-purification if they took place in a stream. To avoid confusion, the classification to be employed in this discussion will be as follows:[30]

1. *Aerobic ponds:* Shallow ponds, less than 1 m in depth, where dissolved oxygen is maintained throughout the entire depth, mainly by the action of photosynthesis.
2. *Facultative ponds:* Ponds 1 to 2.5 m deep, which have an anaerobic lower zone, a facultative middle zone, and an aerobic upper zone maintained by photosynthesis and surface reaeration.
3. *Anaerobic ponds:* Deep ponds that receive high organic loadings such that anaerobic conditions prevail throughout the entire pond depth.

[29]Larry D. Benefield and Clifford W. Randall, *Biological Process Design for Wastewater Treatment,* Upper Saddle River, NJ: Prentice Hall, pp. 322–324, 338–340, 353–354, 1980.

[30]D. H. Caldwell, D. S. Parker, and W. R. Uhte, *Upgrading Lagoons* (Environmental Protection Agency Technology Transfer Seminar Publication), Washington, DC: U.S. Government Printing Office, 1973.

4. *Maturation or tertiary ponds:* Ponds used for polishing effluents from other biological processes. Dissolved oxygen is furnished through photosynthesis and surface reaeration. This type of pond is also known as a *polishing pond.*
5. *Aerated lagoons:* Ponds oxygenated through the action of surface or diffused air aeration.

Aerobic ponds. The aerobic pond is a shallow pond in which light penetrates to the bottom, thereby maintaining active algal photosynthesis throughout the entire system. During the daylight hours, large amounts of oxygen are supplied by the photosynthesis process; during the hours of darkness, wind mixing of the shallow water mass generally provides a high degree of surface reaeration. Stabilization of the organic material entering an aerobic pond is accomplished mainly through the action of aerobic bacteria.

Anaerobic ponds. The magnitude of the organic loading and the availability of dissolved oxygen determine whether the biological activity in a treatment pond will occur under aerobic or anaerobic conditions. A pond may be maintained in an anaerobic condition by applying a BOD_5 load that exceeds oxygen production from photosynthesis. Photosynthesis can be reduced by decreasing the surface area and increasing the depth. Anaerobic ponds become turbid from the presence of reduced metal sulfides. This restricts light penetration to the point that algal growth becomes negligible. Anaerobic treatment of complex wastes involves two distinct stages. In the first stage (known as acid fermentation), complex organic materials are broken down mainly to short-chain acids and alcohols. In the second stage (known as methane fermentation), these materials are converted to gases, primarily methane and carbon dioxide. The proper design of anaerobic ponds must result in environmental conditions favorable to methane fermentation.

Anaerobic ponds are used primarily as a pretreatment process and are particularly suited for the treatment of high-temperature, high-strength wastewaters. However, they have been used successfully to treat municipal wastewaters as well.

Facultative ponds. Of the five general classes of lagoons and ponds, facultative ponds are by far the most common type selected as wastewater treatment systems for small communities. Approximately 25 percent of the municipal wastewater treatment plants in this country are ponds and about 90 percent of these ponds are located in communities of 5,000 people or fewer. Facultative ponds are popular for such treatment situations because long retention times facilitate the management of large fluctuations in wastewater flow and strength with no significant effect on effluent quality. Also capital, operating, and maintenance costs are less than those of other biological systems that provide equivalent treatment.

A schematic representation of a facultative pond operation is given in Figure 5-28. Raw wastewater enters at the center of the pond. Suspended solids contained in the wastewater settle to the pond bottom, where an anaerobic layer develops. Microorganisms occupying this region do not require molecular oxygen as an electron acceptor in energy metabolism, but rather use some other chemical species. Both acid fermentation and methane fermentation occur in the bottom sludge deposits.

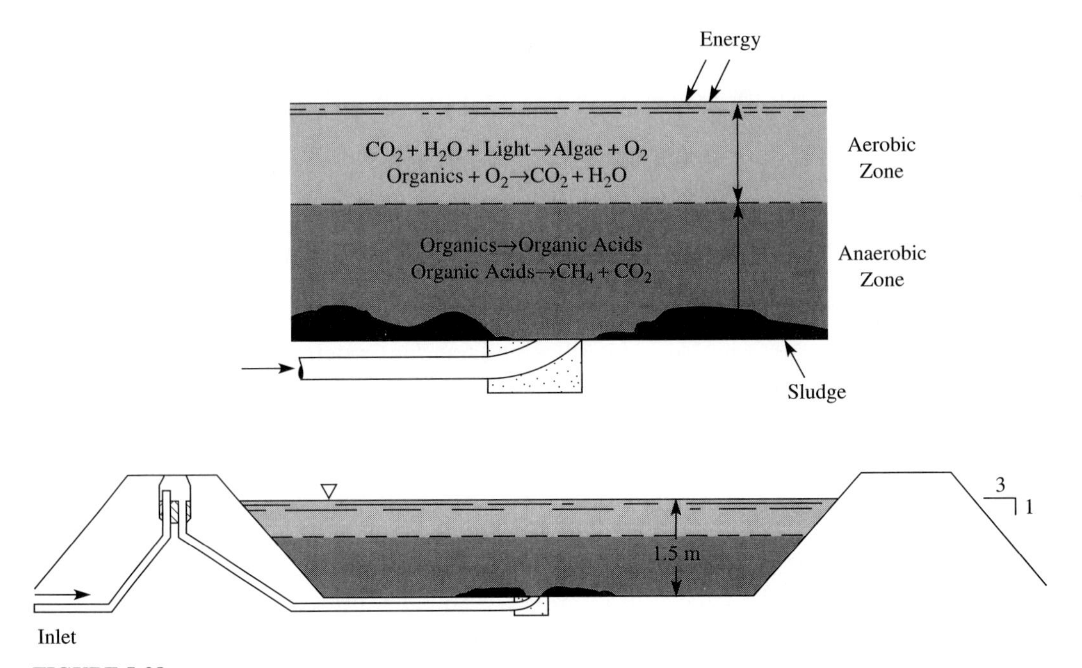

FIGURE 5-28
Schematic diagram of facultative pond relationships.

The facultative zone exists just above the anaerobic zone. This means that molecular oxygen will not be available in the region at all times. Generally, the zone is aerobic during the daylight hours and anaerobic during the hours of darkness.

Above the facultative zone, there exists an aerobic zone that has molecular oxygen present at all times. The oxygen is supplied from two sources. A limited amount is supplied from diffusion across the pond surface. However, the majority is supplied through the action of algal photosynthesis.

Two rules of thumb commonly used in Michigan in evaluating the design of facultative lagoons are as follows:

1. The BOD_5 loading rate should not exceed 22 kg/ha · d on the smallest lagoon cell.
2. The detention time in the lagoon (considering the total volume of all cells but excluding the bottom 0.6 m in the volume calculation) should be six months.

The first criterion is to prevent the pond from becoming anaerobic. The second criterion is to provide enough storage to hold the wastewater during winter months when the receiving stream may be frozen or during the summer when the flow in the stream might be too low to absorb even a small amount of BOD.

Example 5-14. A lagoon having three cells, each 115,000 m^2 in area, a minimum operating depth of 0.6 m, and a maximum operating depth of 1.5 m, receives 1,900 m^3/d of wastewater having an average BOD_5 of 122 mg/L. What is the BOD_5 loading and what is the detention time?

Solution. To compute the BOD loading, we must first compute the mass of BOD_5 entering each day.

$$BOD_5 \text{ mass} = (122 \text{ mg/L})(1{,}900 \text{ m}^3)(1{,}000 \text{ L/m}^3)(1 \times 10^{-6} \text{ mg/kg}) = 231.8 \text{ kg/d}$$

Then, we must convert the area into hectares. Using only one cell,

$$\text{Area} = (115{,}000 \text{ m}^2)(1 \times 10^{-4} \text{ ha/m}^2)$$

$$= 11.5 \text{ ha each}$$

Now we can compute the loading.

$$BOD_5 \text{ loading} = \frac{231.8 \text{ kg/d}}{11.5 \text{ ha}} = 20.2 \text{ kg/ha} \cdot \text{d}$$

This loading rate is acceptable.

The detention time is simply the working volume between the minimum and maximum operating levels divided by the average daily flow.

$$\text{Detention time} = \frac{(115{,}000 \text{ m}^2)(3 \text{ lagoons})(1.5 - 0.6 \text{ m})}{1{,}900 \text{ m}^3/\text{d}}$$

$$= 163.4 \text{ days}$$

This is less than the desired 180 days.

We have ignored the slope of the lagoon walls in this calculation. For large lagoons, this is probably acceptable. In small lagoons, the slope should be considered.

Rotating Biological Contactors (RBCs)

The RBC process consists of a series of closely spaced discs (3 to 3.5 m in diameter) mounted on a horizontal shaft and rotated, while about one-half of their surface area is immersed in wastewater (Figure 5-29). The discs are typically constructed of lightweight plastic. The speed of rotation of the discs is adjustable.

When the process is placed in operation, the microbes in the wastewater begin to adhere to the rotating surfaces and grow there until the entire surface area of the discs is covered with a 1- to 3-mm layer of biological slime. As the discs rotate, they carry a film of wastewater into the air; this wastewater trickles down the surface of the discs, absorbing oxygen. As the discs complete their rotation, the film of water mixes with the reservoir of wastewater, adding to the oxygen in the reservoir and mixing the treated and partially treated wastewater. As the attached microbes pass through the reservoir, they absorb other organics for breakdown. The excess growth of microbes is sheared from the discs as they move through the reservoir. These dislodged organisms are kept in suspension by the moving discs. Thus, the discs serve several purposes:

1. They provide media for the buildup of attached microbial growth.
2. They bring the growth into contact with the wastewater.

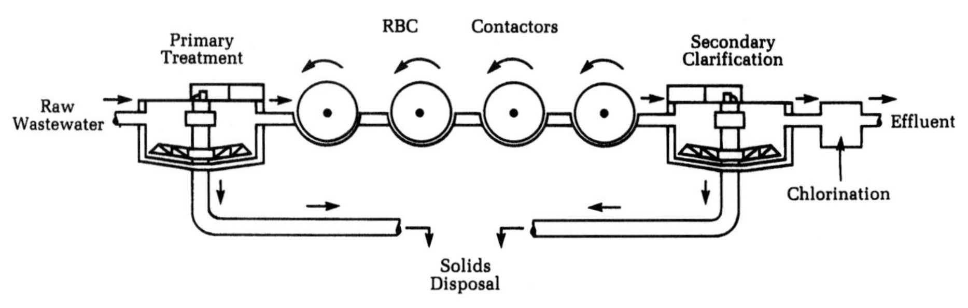

FIGURE 5-29
Photo of RBC and diagram of RBC treatment system. (Courtesy of Environmental Systems Division Geo. A. Hormel & Co.)

3. They aerate the wastewater and the suspended microbial growth in the reservoir.[31]

The attached growths are similar in concept to a trickling filter, except the microbes are passed through the wastewater rather than the wastewater passing over the microbes. Some of the advantages of both the trickling filter and activated sludge processes are realized.

As the treated wastewater flows from the reservoir below the discs, it carries the suspended growths out to a downstream settling basin for removal. The process can achieve secondary effluent quality or better. By placing several sets of discs in series, it is possible to achieve even higher degrees of treatment, including biological conversion of ammonia to nitrates.

---

[31]U.S. Environmental Protection Agency, *Environmental Pollution Control Alternatives: Municipal Wastewater*, pp. 21–24.

5-8 DISINFECTION

The last treatment step in a secondary plant is the addition of a disinfectant to the treated wastewater.[32] The addition of chlorine gas or some other form of chlorine is the process most commonly used for wastewater disinfection in the United States. Chlorine is injected into the wastewater by automated feeding systems. Wastewater then flows into a basin, where it is held for about 15 minutes to allow the chlorine to react with the pathogens.

There is concern that wastewater disinfection may do more harm than good. Early U.S. Environmental Protection Agency rules calling for disinfection to achieve 200 fecal coliforms per 100 mL of wastewater have been modified to a requirement for disinfection only during the summer season when people may come into contact with contaminated water. There were three reasons for this change. The first was that the use of chlorine and, perhaps, ozone causes the formation of organic compounds that are carcinogenic. The second was the finding that the disinfection process was more effective in killing the predators to cysts and viruses than it was in killing the pathogens themselves. The net result was that the pathogens survived longer in the natural environment because there were fewer predators. The third reason was that chlorine is toxic to fish.

5-9 ADVANCED WASTEWATER TREATMENT

Although secondary treatment processes, when coupled with disinfection, may remove over 85 percent of the BOD and suspended solids and nearly all pathogens, only minor removal of some pollutants, such as nitrogen, phosphorus, soluble COD, and heavy metals, is achieved. In some circumstances, these pollutants may be of major concern. In these cases, processes capable of removing pollutants not adequately removed by secondary treatment are used in what is called *tertiary wastewater treatment,* or *advanced wastewater treatment* (AWT). The following sections describe available AWT processes. In addition to solving tough pollution problems, these processes improve the effluent quality to the point that it is adequate for many reuse purposes, and may convert what was originally a wastewater into a valuable resource too good to throw away.

Filtration

Secondary treatment processes, such as the activated-sludge process, are highly efficient for removal of biodegradable colloidal and soluble organics. However, the typical effluent contains a much higher BOD_5 than one would expect from theory. The typical BOD is approximately 20 to 50 mg/L. This is principally because the

[32]U.S. Environmental Protection Agency, *Environmental Pollution Control Alternatives: Municipal Wastewater,* pp. 25–26.

secondary clarifiers are not perfectly efficient at settling out the microorganisms from the biological treatment processes. These organisms contribute both to the suspended solids and to the BOD_5 because the process of biological decay of dead cells exerts an oxygen demand.

By using a filtration process similar to that used in water treatment plants, it is possible to remove the residual suspended solids, including the unsettled microorganisms. Removing the microorganisms also reduces the residual BOD_5. Conventional sand filters identical to those used in water treatment can be used, but they often clog quickly, thus requiring frequent backwashing. To lengthen filter runs and reduce backwashing, it is desirable to have the larger filter grain sizes at the top of the filter. This arrangement allows some of the larger particles of biological floc to be trapped at the surface without plugging the filter. Multimedia filters accomplish this by using low-density coal for the large grain sizes, medium-density sand for intermediate sizes, and high-density garnet for the smallest size filter grains. Thus, during backwashing, the greater density offsets the smaller diameter so that the coal remains on top, the sand remains in the middle, and the garnet remains on the bottom.

Typically, plain filtration can reduce activated sludge effluent suspended solids from 25 to 10 mg/L. Plain filtration is not as effective on trickling filter effluents because trickling filter effluents contain more dispersed growth. However, the use of coagulation and sedimentation followed by filtration can yield suspended solids concentrations that are virtually zero. Typically, filtration can achieve 80 percent suspended solids reduction for activated sludge effluent and 70 percent reduction for trickling filter effluent.

Carbon Adsorption

Even after secondary treatment, coagulation, sedimentation, and filtration, soluble organic materials that are resistant to biological breakdown will persist in the effluent. The persistent materials are often referred to as *refractory organics*. Refractory organics can be detected in the effluent as soluble COD. Secondary effluent COD values are often 30 to 60 mg/L.

The most practical available method for removing refractory organics is by *adsorbing* them on *activated carbon*.[33] *Adsorption* is the accumulation of materials at an *interface*. The interface, in the case of wastewater and activated carbon, is the liquid/solid boundary layer. Organic materials accumulate at the interface because of physical binding of the molecules to the solid surface. Carbon is activated by heating in the absence of oxygen. The activation process results in the formation of many pores within each carbon particle. Since adsorption is a surface phenomenon, the greater the surface area of the carbon, the greater its capacity to hold organic material. The vast areas of the walls within these pores account for most

[33]U.S. Environmental Protection Agency, *Environmental Pollution Control Alternatives: Municipal Wastewater*, pp. 33–35.

of the total surface area of the carbon, which makes it so effective in removing organics.

After the adsorption capacity of the carbon has been exhausted, it can be restored by heating it in a furnace at a temperature sufficiently high to drive off the adsorbed organics. Keeping oxygen at very low levels in the furnace prevents carbon from burning. The organics are passed through an afterburner to prevent air pollution. In small plants where the cost of an on-site regeneration furnace cannot be justified, the spent carbon is shipped to a central regeneration facility for processing.

Phosphorus Removal

All the polyphosphates (molecularly dehydrated phosphates) gradually hydrolyze in aqueous solution and revert to the ortho form (PO_4^{3-}) from which they were derived. Phosphorus is typically found as mono-hydrogen phosphate (HPO_4^{2-}) in wastewater.

The removal of phosphorus to prevent or reduce eutrophication is typically accomplished by chemical precipitation using one of three compounds. The precipitation reactions for each are shown below.

Using ferric chloride:

$$FeCl_3 + HPO_4^{2-} \rightleftharpoons FePO_4\downarrow + H^+ + 3Cl^- \quad (5\text{-}46)$$

Using alum:

$$Al_2(SO_4)_3 + 2HPO_4^{2-} \rightleftharpoons 2AlPO_4\downarrow + 2H^+ + 3SO_4^{2-} \quad (5\text{-}47)$$

Using lime:

$$5Ca(OH)_2 + 3HPO_4^{2-} \rightleftharpoons Ca_5(PO_4)_3OH\downarrow + 3H_2O + 6OH^- \quad (5\text{-}48)$$

You should note that ferric chloride and alum reduce the pH while lime increases it. The effective range of pH for alum and ferric chloride is between 5.5 and 7.0. If there is not enough naturally occurring alkalinity to buffer the system to this range, then lime must be added to counteract the formation of H^+.

The precipitation of phosphorus requires a reaction basin and a settling tank to remove the precipitate. When ferric chloride and alum are used, the chemicals may be added directly to the aeration tank in the activated sludge system. Thus, the aeration tank serves as a reaction basin. The precipitate is then removed in the secondary clarifier. This is not possible with lime since the high pH required to form the precipitate is detrimental to the activated sludge organisms. In some wastewater treatment plants, the $FeCl_3$ (or alum) is added before the wastewater enters the primary sedimentation tank. This improves the efficiency of the primary tank, but may deprive the biological processes of needed nutrients.

Example 5-15. If a wastewater has a soluble orthophosphate concentration of 4.00 mg/L as P, what *theoretical* amount of ferric chloride will be required to remove it completely?

Solution. From Equation 5-46, we see that one mole of ferric chloride is required for each mole of phosphorus to be removed. The pertinent gram molecular weights are as follows:

$$FeCl_3 = 162.21 \text{ g}$$

$$P = 30.97 \text{ g}$$

With a PO_4–P of 4.00 mg/L, the theoretical amount of ferric chloride would be

$$4.00 \times \frac{162.21}{30.97} = 20.95 \text{ or } 21.0 \text{ mg/L}$$

Because of side reactions, solubility product limitations, and day-to-day variations, the actual amount of chemical to be added must be determined by jar tests on the wastewater. You can expect that the actual ferric chloride dose will be 1.5 to 3 times the theoretically calculated amount. Likewise, the actual alum dose will be 1.25 to 2.5 times the theoretical amount.

Nitrogen Control

Nitrogen in any soluble form (NH_3, NH_4^+, NO_2^-, and NO_3^-, but not N_2 gas) is a nutrient and may need to be removed from wastewater to help control algal growth in the receiving body. In addition, nitrogen in the form of ammonia exerts an oxygen demand and can be toxic to fish. Removal of nitrogen can be accomplished either biologically or chemically. The biological process is called *nitrification/denitrification.* The chemical process is called *ammonia stripping.*

Nitrification/denitrification. The natural nitrification process can be forced to occur in the activated-sludge system by maintaining a cell detention time (θ_c) of 15 days in moderate climates and over 20 days in cold climates. The nitrification step is expressed in chemical terms as follows:

$$NH_4^+ + 2O_2 \rightleftharpoons NO_3^- + H_2O + 2H^+ \qquad (5\text{-}49)$$

Of course, bacteria must be present to cause the reaction to occur. This step satisfies the oxygen demand of the ammonium ion. If the nitrogen level is not of concern for the receiving body, the wastewater can be discharged after settling. If nitrogen is of concern, the nitrification step must be followed by anoxic denitrification by bacteria:

$$2NO_3^- + \text{organic matter} \rightarrow N_2 + CO_2 + H_2O \qquad (5\text{-}50)$$

As indicated by the chemical reaction, organic matter is required for denitrification. Organic matter serves as an energy source for the bacteria. The organic matter may be obtained from within or outside the cell. In multistage nitrogen-removal systems, because the concentration of BOD_5 in the flow to the denitrification process is usually quite low, a supplemental organic carbon source is required for rapid denitrification.

(BOD_5 concentration is low because the wastewater previously has undergone carbonaceous BOD removal and the nitrification process.) The organic matter may be either raw, settled sewage or a synthetic material such as methanol (CH_3OH). Raw, settled sewage may adversely affect the effluent quality by increasing the BOD_5 and ammonia content.

Ammonia stripping. Nitrogen in the form of ammonia can be removed chemically from water by raising the pH to convert the ammonium ion into ammonia, which can then be stripped from the water by passing large quantities of air through the water. The process has no effect on nitrate, so the activated sludge process must be operated at a short cell-detention time to prevent nitrification. The ammonia stripping reaction is

$$NH_4^+ + OH^- \rightleftharpoons NH_3 + H_2O \tag{5-51}$$

The hydroxide is usually supplied by adding lime. The lime also reacts with CO_2 in the air and water to form a calcium carbonate scale, which must be removed periodically. Low temperatures cause problems with icing and reduced stripping ability. The reduced stripping ability is caused by the increased solubility of ammonia in cold water.

5-10 LAND TREATMENT

This discussion on land treatment follows two EPA publications: *Environmental Control Alternatives: Municipal Wastewater* and *Land Treatment of Municipal Wastewater Effluents, Design Factors I.*[34]

An alternative to the previously discussed AWT processes for producing an extremely high-quality effluent is offered by an approach called *land treatment.* Land treatment is the application of effluents, usually following secondary treatment, on the land by one of the several available conventional irrigation methods. This approach uses wastewater, and often the nutrients it contains, as a resource rather than considering it as a disposal problem. Treatment is provided by natural processes as the effluent moves through the natural filter provided by soil and plants. Part of the wastewater is lost by evapotranspiration, while the remainder returns to the hydrologic cycle through overland flow or the groundwater system. Most of the groundwater eventually returns, directly or indirectly, to the surface water system.

Land treatment of wastewaters can provide moisture and nutrients necessary for crop growth. In semiarid areas, insufficient moisture for peak crop growth and limited water supplies make this water especially valuable. The primary nutrients (nitrogen, phosphorus, and potassium) are reduced only slightly in conventional secondary treatment processes, so that most of these elements are still present in

---

[34]U.S. Environmental Protection Agency, *Environmental Pollution Control Alternatives: Municipal Wastewater,* pp. 46–51; and C. E. Pound, R. W. Crites, and D. A. Griffes, *Land Treatment of Municipal Wastewater Effluents, Design Factors I* (Environmental Protection Agency Technology Transfer Seminar Publication), Washington, DC: U.S. Government Printing Office, Jan., 1976.

secondary effluent. Soil nutrients that are consumed each year by crop removal and by losses through soil erosion may be replaced by the application of wastewater.

Land application is the oldest method used for treatment and disposal of wastes. Cities have used this method for more than 400 years. Several major cities, including Berlin, Melbourne, and Paris, have used "sewage farms" for at least 60 years for waste treatment and disposal. Approximately 600 communities in the United States reuse municipal wastewater treatment plant effluent in surface irrigation.

Land treatment systems use one of the three basic approaches:

1. Slow rate
2. Overland flow
3. Rapid infiltration

Each method, shown schematically in Figure 5-30, can produce renovated water of different quality, can be adapted to different site conditions, and can satisfy different overall objectives.

Slow Rate

Irrigation, the predominant land application method in use today, involves the application of effluent to the land for treatment and for meeting the growth needs of plants. The applied effluent is treated by physical, chemical, and biological means as it seeps into the soil.[35] Effluent can be applied to crops or vegetation (including forestland) either by sprinkling or by surface techniques, for purposes such as:

1. Avoidance of surface discharge of nutrients
2. Economic return from use of water and nutrients to produce marketable crops
3. Water conservation by exchange when lawns, parks, or golf courses are irrigated
4. Preservation and enlargement of greenbelts and open space

Where water for irrigation is valuable, crops can be irrigated at consumptive use rates (3.5 to 10 mm/d, depending on the crop), and the economic return from the sale of the crop can be balanced against the increased cost of the land and distribution system. On the other hand, where water for irrigation is of little value, hydraulic loadings can be maximized (provided that renovated water quality criteria are met), thereby minimizing system costs. Under high-rate irrigation (10 to 15 mm/d), water-tolerant grasses with high nutrient uptake become the crop of choice.

---

[35]The discussions on land treatment, overland flow, and rapid infiltration follow C. E. Pound, R. W. Crites, and D. A. Griffes, *Land Treatment of Municipal Wastewater Effluents, Design Factors I.*

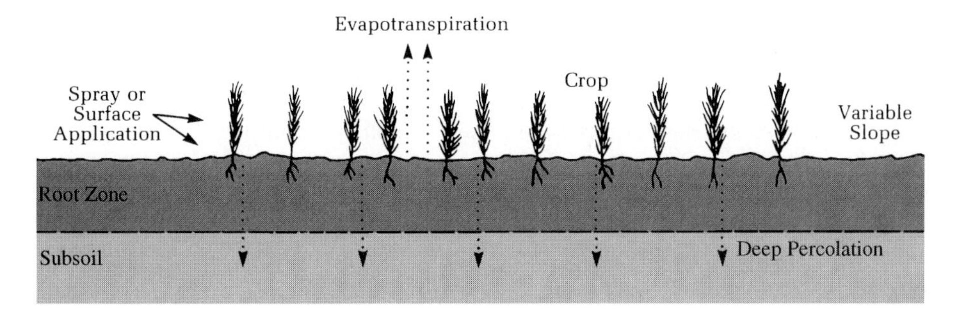

Slow Rate

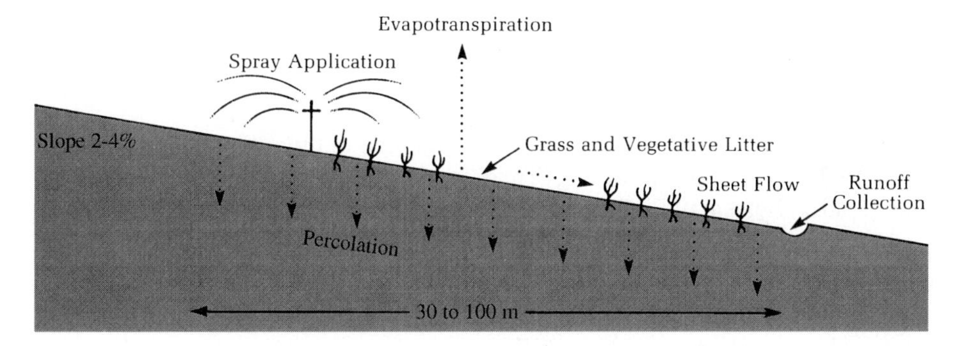

Overland Flow

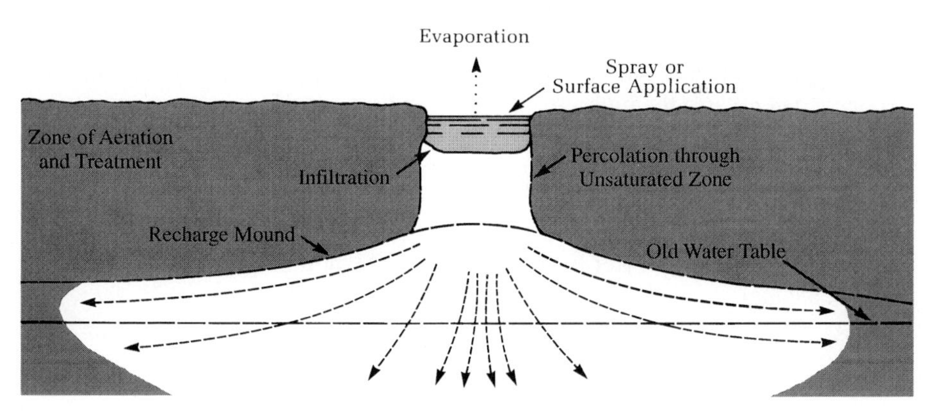

Rapid Infiltration

FIGURE 5-30
Methods of land application. (*Source:* U.S. Environmental Protection Agency, *Land Treatment of Municipal Wastewater, Design Factors I,* Washington, DC: U.S. Government Printing Office, Jan., 1976.)

Overland Flow

Overland flow is essentially a biological treatment process in which wastewater is applied over the upper reaches of sloped terraces and allowed to flow across the vegetated surface to runoff collection ditches. Renovation is accomplished by physical, chemical, and biological means as the wastewater flows in a thin sheet down the relatively impervious slope.

Overland flow can be used as a secondary treatment process where discharge of a nitrified effluent low in BOD is acceptable or as an advanced wastewater treatment process. The latter objective will allow higher rates of application (18 mm/d or more), depending on the degree of advanced wastewater treatment required. Where a surface discharge is prohibited, runoff can be recycled or applied to the land in irrigation or infiltration-percolation systems.

Rapid Infiltration

In infiltration-percolation systems, effluent is applied to the soil at higher rates by spreading in basins or by sprinkling. Treatment occurs as the water passes through the soil matrix. System objectives can include:

1. Groundwater recharge
2. Natural treatment followed by pumped withdrawal or the use of underdrains for recovery
3. Natural treatment where renovated water moves vertically and laterally in the soil and recharges a surface watercourse

Where groundwater quality is being degraded by salinity intrusion, groundwater recharge can reverse the hydraulic gradient and protect the existing groundwater. Where existing groundwater quality is not compatible with expected renovated quality, or where existing water rights control the discharge location, a return of renovated water to surface water can be designed, using pumped withdrawal, underdrains, or natural drainage. At Phoenix, Arizona, for example, the native groundwater quality is poor, and the renovated water is to be withdrawn by pumping, with discharge into an irrigation canal.

5-11 SLUDGE TREATMENT

In the process of purifying the wastewater, another problem is created: sludge. The higher the degree of wastewater treatment, the larger the residue of sludge that must be handled. The exceptions to this rule are where land applications or polishing lagoons are used. Satisfactory treatment and disposal of the sludge can be the single most complex and costly operation in a municipal wastewater treatment system.[36]

[36]U.S. Environmental Protection Agency, *Environmental Pollution Control Alternatives: Municipal Wastewater*, pp. 54–55.

The sludge is made of materials settled from the raw wastewater and of solids generated in the wastewater treatment processes.

The quantities of sludge involved are significant. For primary treatment, they may be 0.25 to 0.35 percent by volume of wastewater treated. When treatment is upgraded to activated sludge, the quantities increase to 1.5 to 2.0 percent of this volume of water treated. Use of chemicals for phosphorus removal can add another 1.0 percent. The sludges withdrawn from the treatment processes are still largely water, as much as 97 percent. Sludge treatment processes, then, are concerned with separating the large amounts of water from the solid residues. The separated water is returned to the wastewater plant for processing.

The basic processes for sludge treatment are as follows:

1. *Thickening:* Separating as much water as possible by gravity or flotation.
2. *Stabilization:* Converting the organic solids to more refractory (inert) forms so that they can be handled or used as soil conditioners without causing a nuisance or health hazard through processes referred to as "digestion." (These are biochemical oxidation processes.)
3. *Conditioning:* Treating the sludge with chemicals or heat so that the water can be readily separated.
4. *Dewatering:* Separating water by subjecting the sludge to vacuum, pressure, or drying.
5. *Reduction:* Converting the solids to a stable form by wet oxidation or incineration. (These are chemical oxidation processes; they decrease the volume of sludge, hence the term reduction.)

Although a large number of alternative combinations of equipment and processes are used for treating sludges, the basic alternatives are fairly limited. The ultimate depository of the materials contained in the sludge must either be land, air, or water. Current policies discourage practices such as ocean dumping of sludge. Air pollution considerations necessitate air pollution control facilities as part of the sludge incineration process.

The following sections discuss the processes commonly used. The basic alternative routes by which these processes may be employed are shown in Figure 5-31.

Sources and Characteristics of Various Sludges

Before we begin the discussion of the various treatment processes, it is worthwhile to recapitulate the sources and nature of the sludges that must be treated.

Grit. The sand, broken glass, nuts, bolts, and other dense material that is collected in the grit chamber is not true sludge in the sense that it is not fluid. However, it still requires disposal. Because grit can be drained of water easily and is relatively stable in terms of biological activity (it is not biodegradable), it is generally trucked directly to a landfill without further treatment.

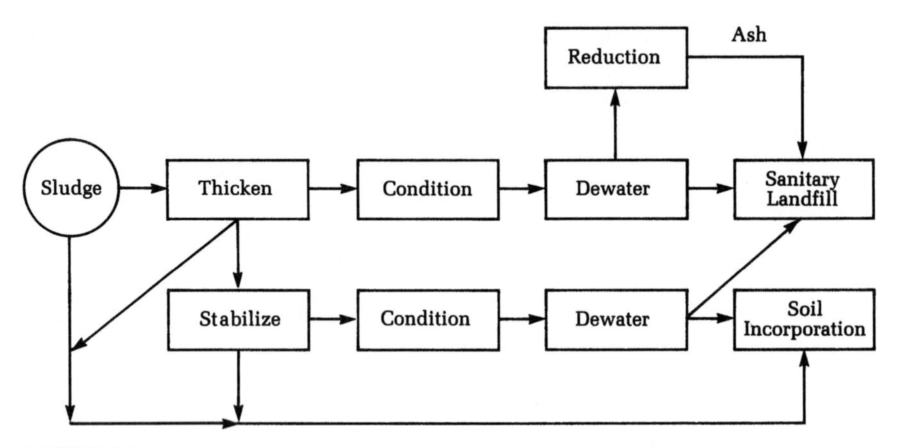

FIGURE 5-31
Basic sludge handling alternatives.

Primary or raw sludge. Sludge from the bottom of the primary clarifiers contains from 3 to 8 percent solids (1 percent solids ≃ 1 g solids/100 mL sludge volume), which is approximately 70 percent organic. This sludge rapidly becomes anaerobic and is highly odiferous.

Secondary sludge. This sludge consists of microorganisms and inert materials that have been wasted from the secondary treatment processes. Thus, the solids are about 90 percent organic. When the supply of air is removed, this sludge also becomes anaerobic, creating noxious conditions if not treated before disposal. The solids content depends on the source. Wasted activated sludge is typically 0.5 to 2 percent solids, while trickling filter sludge contains 2 to 5 percent solids. In some cases, secondary sludges contain large quantities of chemical precipitates because the aeration tank is used as the reaction basin for the addition of chemicals to remove phosphorus.

Tertiary sludges. The characteristics of sludges from the tertiary treatment processes depend on the nature of the process. For example, phosphorus removal results in a chemical sludge that is difficult to handle and treat. When phosphorus removal occurs in the activated sludge process, the chemical sludge is combined with the biological sludge, making the latter more difficult to treat. Nitrogen removal by denitrification results in a biological sludge with properties very similar to those of waste activated sludge.

Solids Computations

Volume–mass relationships. Since most WWTP sludges are primarily water, the volume of the sludge is primarily a function of the water content. Thus, if we know the percent solids and the specific gravity of the solids we can estimate the volume of the sludge. The solid matter in wastewater sludge is composed of fixed (mineral) solids and volatile (organic) solids. The volume of the total mass of solids may be expressed as

$$\forall_{\text{solids}} = \frac{M_s}{S_s \rho} \tag{5-52}$$

where M_s = mass of solids, kg
S_s = specific gravity of solids
ρ = density of water = 1,000 kg/m^3

Since the total mass is composed of fixed and volatile fractions, Equation 5-52 may be rewritten as:

$$\frac{M_s}{S_s \rho} = \frac{M_f}{S_f \rho} + \frac{M_v}{S_v \rho} \tag{5-53}$$

where M_f = mass of fixed solids, kg
M_v = mass of volatile solids, kg
S_f = specific gravity of fixed solids
S_v = specific gravity of volatile solids

The specific gravity of the solids may be expressed in terms of the specific gravities of the fixed and solid fractions by solving Equation 5-53 for S_s:

$$S_s = M_s \left[\frac{S_f S_v}{M_f S_v + M_s S_f} \right] \tag{5-54}$$

The specific gravity of sludge (S_{sl}) may be estimated by recognizing that, in a similar fashion to the fractions of solids, the sludge is composed of solids and water so that

$$\frac{M_{sl}}{S_{sl} \rho} = \frac{M_s}{S_s \rho} + \frac{M_w}{S_w \rho} \tag{5-55}$$

where M_{sl} = mass of sludge, kg
M_w = mass of water, kg
S_{sl} = specific gravity of sludge
S_w = specific gravity of water

It is customary to report solids concentrations as percent solids, where the fraction of solids (P_s) is computed as

$$P_s = \frac{M_s}{M_s + M_w} \tag{5-56}$$

and the fraction of water (P_w) is computed as

$$P_w = \frac{M_w}{M_s + M_w} \tag{5-57}$$

Thus, it is more convenient to solve Equation 5-52 in terms of percent solids. If we divide each term in Equation 5-55 by ($M_s + M_w$) and recognize that $M_{sl} = M_s + M_w$, then Equation 5-55 may be expressed as

$$\frac{1}{S_{sl}\rho} = \frac{P_s}{S_s\rho} + \frac{P_w}{S_w\rho} \tag{5-58}$$

If the specific gravity of water is taken as 1.0000, as it can be without appreciable error, then solving for S_{sl} yields

$$S_{sl} = \frac{S_s}{P_s + (S_s)(P_w)} \tag{5-59}$$

With these expressions in hand, or at least where you can find them, you can calculate the volume of sludge (V_{sl}) with the following equation:

$$V_{sl} = \frac{M_s}{(\rho)(S_{sl})(P_s)} \tag{5-60}$$

Example 5-16. Using the following primary settling-tank data, determine the daily sludge production.

Operating Data:
Flow = $0.150 \text{ m}^3/\text{s}$
Influent SS = 280.0 mg/L = 280.0 g/m^3
Removal efficiency = 59.0%
Sludge concentration = 5.00%
Volatile solids = 60.0%
Specific gravity of volatile solids = 0.990
Fixed solids = 40.0%
Specific gravity of fixed solids = 2.65

Solution. We begin by calculating S_s. We can do this without calculating M_s, M_f, and M_v directly by recognizing that they are proportional to the percent composition.

With

$$M_s = M_f + M_v$$
$$= 0.400 + 0.600 = 1.00$$

Then Equation 5-54 gives the following:

$$S_s = \frac{(2.65)(0.990)}{[(0.990)(0.400)] + [(2.65)(0.600)]}$$
$$= 1.321 \text{ or } 1.32$$

The specific gravity of the sludge is calculated with Equation 5-59:

$$S_{sl} = \frac{1.321}{0.05 + (1.321 \times 0.950)}$$
$$= 1.012 \text{ or } 1.01$$

The mass of the sludge is estimated from the incoming suspended solids concentration and the removal efficiency of the primary tank.

$$M_s = 0.59 \times 280.0 \text{ mg/L} \times 0.15 \text{ m}^3/\text{s} \times 86{,}400 \text{ s/d} \times 10^{-3} \text{ kg/g}$$

$$= 2.14 \times 10^3 \text{ kg/d}$$

The sludge volume is then calculated with Equation 5-60:

$$\forall_{sl} = \frac{2.14 \times 10^3 \text{ kg/d}}{1{,}000 \text{ kg/m}^3 \times 1.012 \times 0.05}$$

$$= 42.29 \text{ or } 42.3 \text{ m}^3/\text{d}$$

Mass balance. Barring black holes[37] and the like, we all understand that the physical, chemical, and biological processes of wastewater treatment neither create nor destroy matter. This fact allows us to employ Equation 1-3 in a new context.

$$\frac{dS}{dt} = M_{\text{in}} - M_{\text{out}} \tag{5-61}$$

where M_{in} and M_{out} refer to the mass of dissolved chemicals, solids, or gas entering and leaving a process or group of processes. If we assume steady-state conditions, then $dS/dt = 0$ and Equation 5-61 reduces to the following:

$$M_{\text{in}} = M_{\text{out}} \tag{5-62}$$

Several interrelated processes are examined together in the flowsheet shown in Figure 5-32. When labeled with mass flows, the flowsheet may be called a *quantitative flow diagram* (QFD). The solids mass balance can be an important aid to a designer in predicting long-term average solids loadings on sludge treatment components. This allows the designer to establish such factors as operating costs and quantities of sludge for ultimate disposal. However, it does not establish the solids loading that each equipment item must be capable of processing. A particular component should be sized to handle the most rigorous loading conditions it is expected to encounter. This loading is usually not determined by applying steady-state models because of storage and plant scheduling considerations. Thus, the rate of solids reaching any particular piece of equipment does not usually rise and fall in direct proportion to the rate of solids arriving at the plant headworks.

The mass balance calculation is carried out in a step-by-step procedure:

1. Draw the flowsheet (as in Figure 5-32).
2. Identify all streams. For example, Stream A contains raw sewage solids plus chemical solids generated by dosing the sewage with chemicals. Let the *mass flow rate* of solids in Stream A be equal to *A* kg per day.

---

[37] We are, of course, referring to the Einsteinian black hole and not the Black Hole of Calcutta.

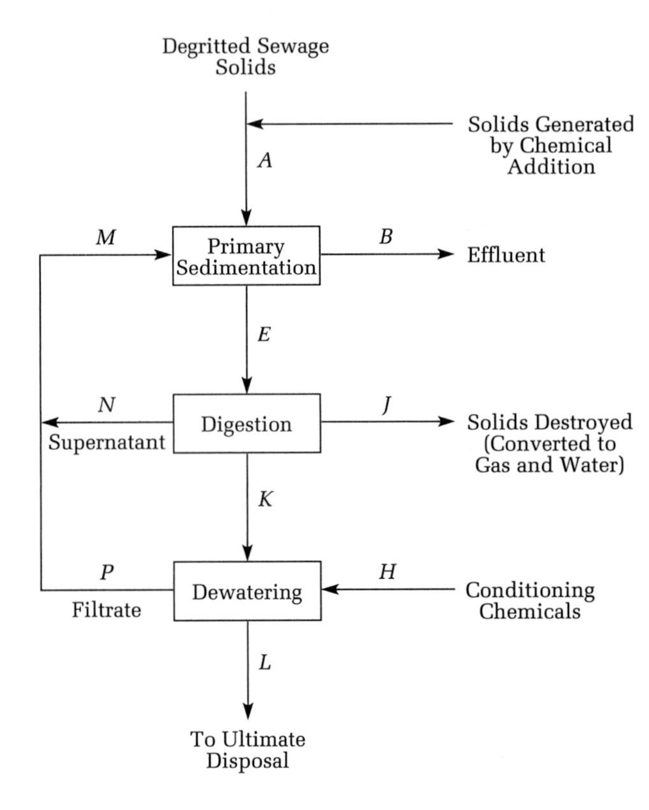

FIGURE 5-32
Primary WWTP flowsheet. [*Source:* U.S. Environmental Protection Agency, *Process Design Manual, Sludge Treatment and Disposal,* Environmental Protection Agency (Publication No. EPA 625/1-79-011), Washington, DC: U.S. Government Printing Office, Sept., 1979.]

3. For each processing unit, identify the relationship of entering and leaving streams to one another in terms of mass. For example, for the primary sedimentation tank, let the ratio of solids in the tank underflow (E) to entering solids ($A + M$) be equal to η_E. η_E is actually an indicator of solids separation efficiency. The general form in which such relationships are expressed is:

$$\eta_i = \frac{\text{mass of solids in stream } i}{\text{mass of solids entering the unit}} \tag{5-63}$$

For example,

$$\eta_p = \frac{P}{K + H}; \eta_j = \frac{J}{E}$$

The processing unit's performance is specified when a value is assigned to η_i.

4. Combine the mass balance relationships so as to reduce them to one equation describing a specific stream in terms of given or known quantities, or ones which can be calculated from a knowledge of the process behavior.

Example 5-17. Using Figure 5-32 and assuming that A, η_E, η_j, η_N, η_P, and η_H are known or can be determined from a knowledge of water chemistry and an understanding of the general solids separation/destruction efficiencies of the processing involved, derive an expression for E, the mass flow out of the primary sedimentation tank.

Solution. The derivation is carried out as follows.

a. Define M by solids balances on streams around the primary sedimentation tank:

$$\eta_E = \frac{E}{A + M} \tag{i}$$

Therefore,

$$M = \frac{E}{\eta_E} - A \tag{ii}$$

b. Define M by balances on recycle streams:

$$M = N + P \tag{iii}$$

$$N = \eta_N E \tag{iv}$$

$$P = \eta_P(H + K) \tag{v}$$

$$H = \eta_H K \tag{vi}$$

Therefore,

$$P = \eta_P(1 + \eta_H)K \tag{vii}$$

$$K + J + N = E \tag{viii}$$

Therefore,

$$K = E - J - N = E - \eta_j E - \eta_N E = E(1 - \eta_j - \eta_N) \tag{ix}$$

and

$$P = \eta_P E(1 - \eta_j - \eta_N)(1 + \eta_H) \tag{x}$$

Therefore,

$$M = E[\eta_N + \eta_P(1 - \eta_j - \eta_N)(1 + \eta_H)] \tag{xi}$$

c. Equate equations (ii) and (xi) to eliminate M:

$$\frac{E}{\eta_E} - A = E[\eta_N + \eta_P(1 - \eta_j - \eta_N)(1 + \eta_H)]$$

$$E = \frac{A}{\dfrac{1}{\eta_E} - \eta_N - \eta_P(1 - \eta_j - \eta_N)(1 + \eta_H)}$$

E is expressed in terms of assumed or known influent solids loadings and solids separation/destruction efficiencies.

Once the equation for E is derived, equations for other streams follow rapidly; in fact, most have already been derived. These are summarized in Table 5-13.

TABLE 5-13
Mass balance equations for Figure 5-32

$$E = \frac{A}{\frac{1}{\eta_E} - \eta_N - \eta_P(1 - \eta_J - \eta_N)(1 + \eta_H)}$$

$$M = \frac{E}{\eta_E} - A$$

$$B = (1 - \eta_E)(A + M)$$

$$J = \eta_J E$$

$$N = \eta_N E$$

$$K = E(1 - \eta_J - \eta_N)$$

$$H = \eta_H K$$

$$P = \eta_P(1 + \eta_H)K$$

$$L = K(1 + \eta_H)(1 - \eta_P)$$

Source: U.S. Environmental Protection Agency, *Process Design Manual, Sludge Treatment and Disposal.*

The example just worked was relatively simple. A more complex system is illustrated in Figure 5-33. Mass balance equations for this system are summarized in Table 5-14. For this flowsheet the following information must be specified:

A = Influent solids

X = Effluent solids, that is, overall suspended solids removal must be specified

$\eta_E, \eta_G, \eta_J, \eta_N, \eta_R,$ and η_T = straightforward assumptions about the degree of solids removal, addition, or destruction

η_D = describes the net solids destruction reduction or the net solids synthesis in the biological system, and must be estimated from yield data. A positive η_D signifies net solids destruction. A negative η_D signifies net solids growth. In this example, 8 percent of the solids entering the biological process are assumed destroyed, that is, converted to gas or liquified.

Note that alternative processing schemes can be evaluated simply by manipulating appropriate variables. For example:

Filtration can be eliminated by setting η_R to zero.

Thickening can be eliminated by setting η_G to zero.

Digestion can be eliminated by setting η_J to zero.

Dewatering can be eliminated by setting η_P to zero.

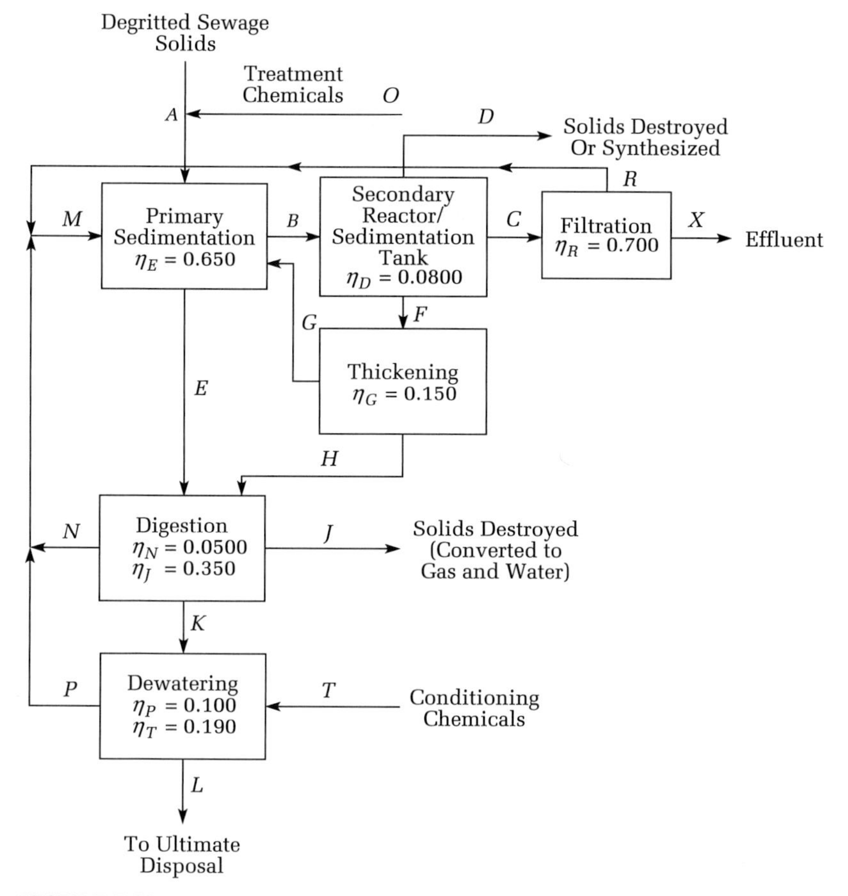

FIGURE 5-33

Flowsheet for a complex WWTP. (*Source:* U.S. Environmental Protection Agency, *Process Design Manual, Sludge Treatment and Disposal.*)

A system without primary sedimentation can be simulated by setting η_E equal to approximately zero, for example, 1×10^{-8}. η_E cannot be set equal to exactly zero, since division by η_E produces indeterminate solutions when computing E.

A set of different mass balance equations must be derived if flow paths between processing units are altered. For example, the equations of Table 5-14 do not describe operations in which the dilute stream from the thickener (Stream G) is returned to the secondary reactor instead of the primary sedimentation tank.

Thickening

Thickening is usually accomplished in one of two ways: the solids are floated to the top of the liquid (*flotation*) or are allowed to settle to the bottom (*gravity thickening*).

TABLE 5-14
Mass balance equations for Figure 5-33

$$E = \frac{A - \left(\frac{X}{1 - \eta_R}\right)(\gamma - \eta_R)}{\frac{1}{\eta_E} - \alpha - \beta(\gamma)}$$

Where $\alpha = \eta_P(1 - \eta_J - \eta_N)(1 + \eta_T) + \eta_N$

$$\beta = \frac{(1 - \eta_E)(1 - \eta_D)}{\eta_E}$$

$$\gamma = \eta_G + \alpha(1 - \eta_G)$$

$$B = \frac{(1 - \eta_E)E}{\eta_E}$$

$$C = \frac{X}{1 - \eta_R}$$

$$D = \eta_D B$$

$$F = \beta E - \frac{X}{1 - \eta_R}$$

$$G = \eta_G F$$

$$H = (1 - \eta_G)F$$

$$J = \eta_J(E + H)$$

$$K = (1 - \eta_J - \eta_N)(E + H)$$

$$L = K(1 + \eta_T)(1 - \eta_P)$$

$$M = \frac{E}{\eta_E} - G - A$$

$$N = \eta_N(E + H)$$

$$P = \eta_P(1 + \eta_T)K$$

$$R = \frac{\eta_R}{1 - \eta_R}X$$

$$T = \eta_T K$$

Source: U.S. Environmental Protection Agency, *Process Design Manual, Sludge Treatment and Disposal.*

The goal is to remove as much water as possible before final dewatering or digestion of the sludge. The processes involved offer a low-cost means of reducing sludge volumes by a factor of two or more. The costs of thickening are usually more than offset by the resulting savings in the size and cost of downstream sludge processing equipment.

Flotation. In the flotation thickening process (Figure 5-34) air is injected into the sludge under pressure (275 to 550 kPa). Under this pressure, a large amount of air can be dissolved in the sludge. The sludge then flows into an open tank where, at atmospheric pressure, much of the air comes out of solution as minute bubbles. The bubbles attach themselves to sludge solids particles and float them to the surface. The sludge forms a layer at the top of the tank; this layer is removed by a skimming

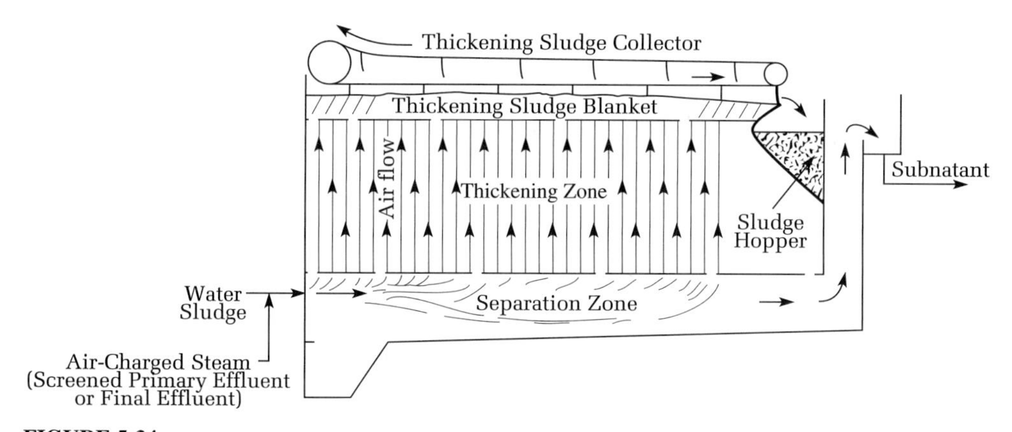

FIGURE 5-34
Air flotation thickener.

mechanism for further processing. The process typically increases the solids content of activated sludge from 0.5–1 percent to 3–6 percent. Flotation is especially effective on activated sludge, which is difficult to thicken by gravity.

Gravity thickening. Gravity thickening is a simple and inexpensive process that has been used widely on primary sludges for many years. It is essentially a sedimentation process similar to that which occurs in all settling tanks. Sludge flows into a tank that is very similar in appearance to the circular clarifiers used in primary and secondary sedimentation (Figure 5-35); the solids are allowed to settle to the bottom where a heavy-duty mechanism scrapes them to a hopper from which they are withdrawn for further processing. The type of sludge being thickened has a major effect on performance. The best results are obtained with purely primary sludges. As the proportion of activated sludge increases, the thickness of settled sludge solids decreases. Purely primary sludges can be thickened from 1–3 percent to 10 percent solids. The current trend is toward using gravity thickening for primary sludges

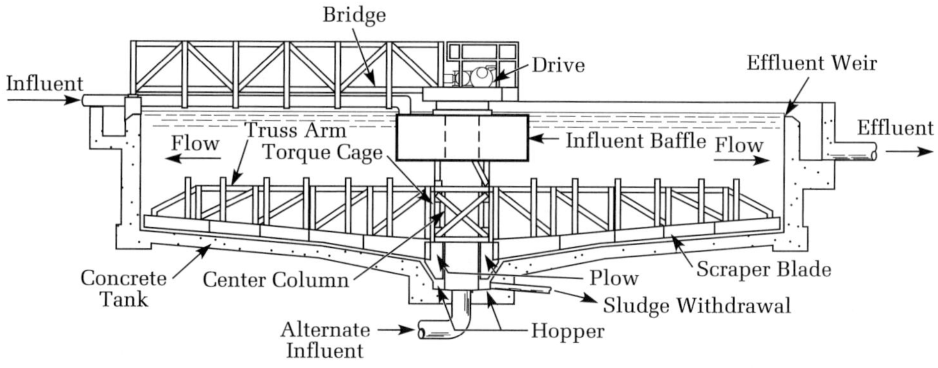

FIGURE 5-35
Gravity thickener.

and flotation thickening for activated sludges, and then blending the thickened sludges for further processing.

Dick[38] has described a graphical procedure for sizing gravity thickeners using a *batch flux curve. Flux* is the term used to describe the rate of settling of solids. It is defined as the mass of solids which pass through a horizontal unit area per unit of time (kg/d · m^2). This may be expressed mathematically as follows:

$$F_s = (C_u)(v) = (C_s)(\text{zone settling velocity}) \quad (5\text{-}64)$$

where F_s = solids flux, kg/m^2 · d
C_s = suspended solids concentration, kg/m^3
C_u = concentration of solids in underflow, that is, sludge withdrawal pipe, kg/m^3
v = underflow velocity, m/d

The sizing procedure begins with a batch settling curve such as that shown in Figure 5-36. Data from the batch settling curve are used to construct a batch flux curve (Figure 5-37). Knowing the desired underflow concentration, a line through the desired concentration and tangent to the batch flux curve is constructed. The extension of this line to the axis of ordinates yields the design flux. From this flux and the inflow solids concentration, the surface area may be determined.

Example 5-18. A gravity thickener is to be designed to thicken the sludge from the primary tank described in Example 5-16. The thickened sludge should have an underflow solids concentration of 10.0 percent. Assume that the sludge yields a batch settling curve such as that shown in Figure 5-36.

Solution. First we must compute the solids flux for several arbitrarily selected suspended solids concentrations.

SS, kg/m^3	v, m/d	F_s, kg/d · m^2	SS, kg/m^3	v, m/d	F_s, kg/d · m^2
100	0.125	12.5	20	5.30	106.
80	0.175	14.0	10	34.0	340.
60	0.30	18.	5	62.0	310.
50	0.44	22.	4	68.0	272.
40	0.78	31.	3	76.0	228.
30	1.70	51.	2	83.0	166.

[38]R. I. Dick, "Thickening," *Advances in Water Quality Improvement—Physical and Chemical Processes,* E. F. Gloyna and W. W. Eckenfelder, eds., Austin, TX: University of Texas Press, p. 358, 1970. The original development of this method was by N. Yoshioka and others. See "Continuous Thickening of Homogenous Flocculated Slurries," *Chemical Engineering, 21,* Tokyo 1957 (in Japanese).

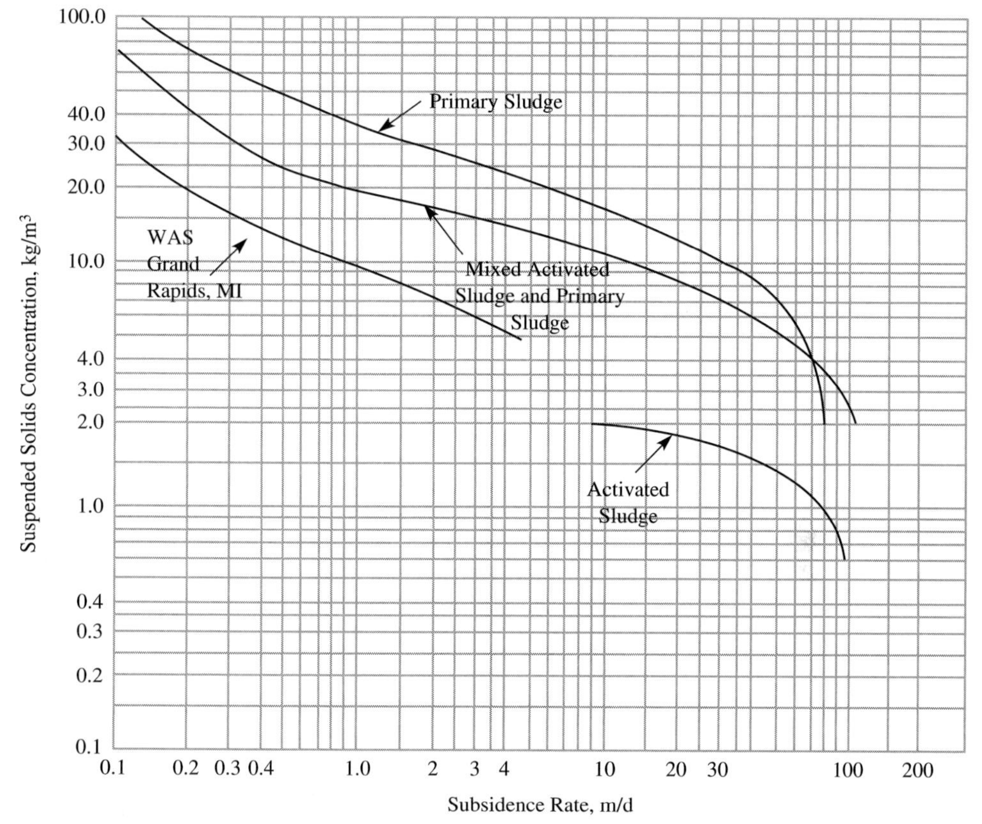

FIGURE 5-36
Batch settling curve.

The data in the first column were selected arbitrarily. The data in the second column were read from Figure 5-36 at the abscissa points noted in the first column. The data in the third column are the products of the first and second column, that is, $100.0 \times 0.125 = 12.5$, $80.0 \times 0.175 = 14.0$, etc.

The percent solids concentration is simply 0.10 times the SS in kg/m^3. Converting the first column to percent and plotting it versus the last column yields the batch flux curve (Figure 5-37).

The tangent line from 10.0 percent yields a solids flux of 43 kg/d $\cdot$ m^2.

From Example 5-16, we find the solids mass loading to be 2.14×10^3 kg/d. Therefore, the required surface area of the thickener is

$$A_s = \frac{2.14 \times 10^3}{43}$$

$$= 49.77 \text{ or } 50 \text{ m}^2$$

Typical gravity-thickener design criteria are summarized in Table 5-15. Wasting to the thickener may or may not be continuous, depending upon the size of the WWTP. Frequently, smaller plants will waste intermittently because of work

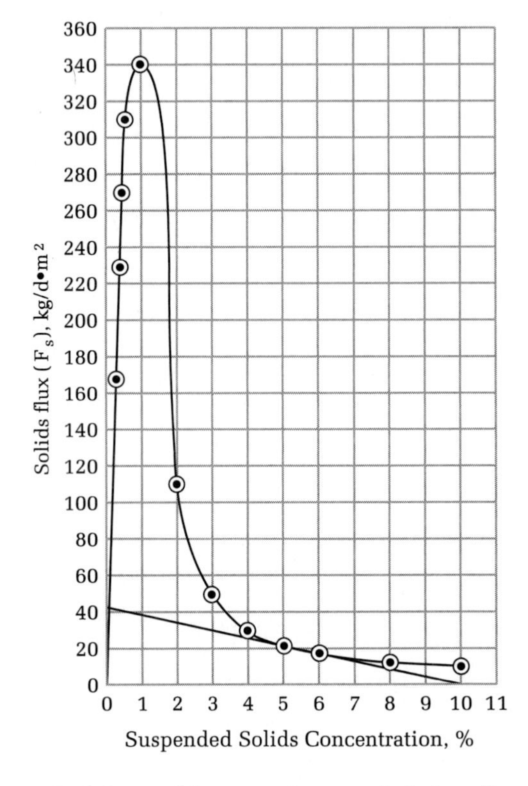

FIGURE 5-37
Batch flux curve.

schedules and lower volumes of sludge. Some examples of thickener performance are listed in Table 5-16. You should note that the supernatant suspended solids levels are quite high. Thus, the supernatant must be returned to the head end of the WWTP.

Example 5-19. One hundred cubic meters per day (100.0 m^3/d) of mixed sludge at 4.0 percent solids is to be thickened to 8.0 percent solids. What is the approximate volume of the sludge after thickening?

Solution. A "4.0 percent sludge" contains 4.0 percent by mass of solids and 96.0 percent by mass of water. Assuming that the specific gravity is not appreciably different from that of water, we can approximate the relationship between volume and percent solids as follows:

$$\frac{V_1}{V_2} = \frac{P_2}{P_1}$$

In this case then, the volume of sludge after thickening would be

$$\frac{100.0}{V_2} = \frac{0.080}{0.040}$$

$$V_2 = 50.0 \text{ m}^3$$

Thus, we can see a substantial reduction in the volume that must be handled by thickening the sludge from 4 to 8 percent solids.

TABLE 5-15
Typical gravity-thickener design criteria

Sludge Source	Influent SS, %	Expected underflow concentration, %	Mass loading kg/h · m^2
Individual sludges			
PS	2–7	5–10	4–6
TF	1–4	3–6	1.5–2.0
RBC	1–3.5	2–5	1.5–2.0
WAS	0.5–1.5	2–3	0.5–1.5
Tertiary sludges			
High CaO	3–4.5	12–15	5–12
Low CaO	3–4.5	10–12	2–6
Fe	0.5–1.5	3–4	0.5–2.0
Combined sludges			
PS + WAS	0.5–4	4–7	1–3.5
PS + TF	2–6	5–9	2–4
PS + RBC	2–6	5–8	2–3
PS + Fe	2	4	1
PS + Low CaO	5	7	4
PS + High CaO	7.5	12	5
PS + (WAS + Fe)	1.5	3	1
PS + (WAS + Al)	0.2–0.4	4.5–6.5	2–3.5
(PS + Fe) + TF	0.4–0.6	6.5–8.5	3–4
(PS + Fe) + WAS	1.8	3.6	1
WAS + TF	.5–2.5	2–4	0.5–1.5

(*Source:* U.S. Environmental Protection Agency, *Process Design Manual, Sludge Treatment and Disposal,* pp. 5–7.)
Legend: PS = primary sedimentation; TF = trickling filter; RBC = rotating biological contactor; WAS = waste activated sludge; High CaO = high lime; Low CaO = low lime; Fe = iron; Al = alum; + = mixture of sludges from processes indicated; ( ) = chemical added to process is within parentheses.

TABLE 5-16
Reported operation results for gravity thickeners

Location	Sludge source	Influent SS, %	Mass loading, kg/h · m^2	Underflow concentration, %	Overflow SS, mg/L
Port Huron, MI	PS + WAS	0.6	1.7	4.7	2,500
Sheboygan, WI	PS + TF	0.3	2.2	8.6	400
	PS + (TF + Al)	0.5	3.6	7.8	2,400
Grand Rapids, MI	WAS	1.2	2.1	5.6	140
Lakewood, OH	PS + (WAS + Al)	0.3	2.9	5.6	1,400

(*Source:* U.S. Environmental Protection Agency, *Process Design Manual, Sludge Treatment and Disposal,* pp. 5–8.)
(*Note:* Values shown are average values only.)

Stabilization

The principal purposes of sludge stabilization are to break down the organic solids biochemically so that they are more stable (less odorous and less putrescible) and more dewaterable, and to reduce the mass of sludge.[39] If the sludge is to be dewatered and burned, stabilization is not used. There are two basic stabilization processes in use. One is carried out in closed tanks devoid of oxygen and is called *anaerobic digestion.* The other approach injects air into the sludge to accomplish *aerobic digestion.*

Aerobic digestion. The aerobic digestion of biological sludges is nothing more than a continuation of the activated sludge process. When a culture of aerobic heterotrophs is placed in an environment containing a source of organic material, the microorganisms remove and utilize most of this material. A fraction of the organic material removed will be used for the synthesis of new biomass. The remaining material will be channeled into energy metabolism and oxidized to carbon dioxide, water, and soluble inert material to provide energy for both synthesis and maintenance (life-support) functions. Once the external source of organic material is exhausted, however, the microorganisms enter into endogenous respiration, where cellular material is oxidized to satisfy the energy of maintenance (that is, energy for life-support requirements). If this condition is continued over an extended period of time, the total quantity of biomass will be considerably reduced. Furthermore, that portion remaining will exist at such a low energy state that it can be considered biologically stable and suitable for disposal in the environment. This forms the basic principle of aerobic digestion.

Aerobic digestion is accomplished by aerating the organic sludges in an open tank resembling an activated sludge aeration tank. Like the activated sludge aeration tank, the aerobic digestor must be followed by a settling tank unless the sludge is to be disposed of on land in liquid form. Unlike the activated sludge process, the effluent (supernatant) from the clarifier is recycled back to the head end of the plant. This is because the supernatant is high in suspended solids (100 to 300 mg/L), BOD_5 (to 500 mg/L), TKN (to 200 mg/L), and total P (to 100 mg/L).

Because the fraction of volatile matter is reduced, the specific gravity of the digested sludge solids will be higher than it was before digestion. Thus, the sludge settles to a more compact mass, and the clarifier underflow concentration can be expected to reach 3 percent. Beyond this, its dewatering properties are terrible.

Anaerobic digestion. The anaerobic treatment of complex wastes involves two distinct stages. In the first stage, complex waste components, including fats, proteins,

[39]The discussion on stabilization follows L. D. Benefield and C. W. Randall, *Biological Process Design for Wastewater Treatment,* pp. 256–264, 461–464, 479; and U.S. Environmental Protection Agency, *Environmental Pollution Control Alternatives: Municipal Wastewater,* pp. 60–61.

and polysaccharides, are hydrolyzed to their component subunits. This is accomplished by a heterogeneous group of facultative and anaerobic bacteria. These bacteria then subject the products of hydrolysis (triglycerides, fatty acids, amino acids, and sugars) to fermentation and other metabolic processes leading to the formation of simple organic compounds. These compounds are mainly short-chain (volatile) acids and alcohols. The first stage is commonly referred to as *acid fermentation.* In this stage, organic material is simply converted to organic acids, alcohols, and new bacterial cells, so that little stabilization of BOD or COD is realized. In the second stage, the end products of the first stage are converted to gases (mainly methane and carbon dioxide) by several different species of strictly anaerobic bacteria. Thus, it is here that true stabilization of the organic material occurs. This stage is generally referred to as *methane fermentation.* The two stages of anaerobic waste treatment are illustrated in Figure 5-38. You must understand that even though the anaerobic process is presented as being sequential in nature, both stages take place simultaneously and synergistically. The primary acids produced during acid fermentation are propionic and acetic. The significance of these acids as precursors for methane formation is illustrated in Figure 5-38.

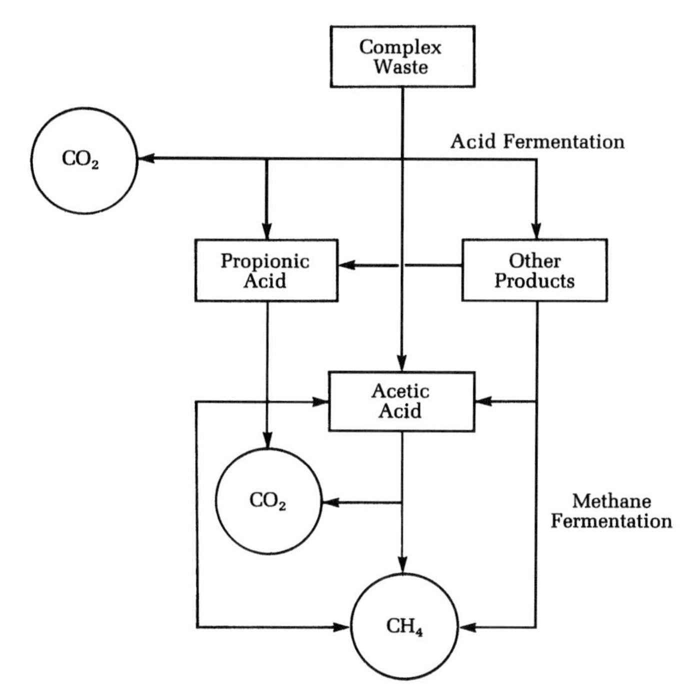

FIGURE 5-38
Pathways in anaerobic digestion. (After McCarty, "Anaerobic Treatment of Soluble Wastes," *Advances in Water Quality Improvement,* E. F. Gloyna and W. W. Eckenfelder, eds., Austin, TX: University of Texas Press, 1968.)

The bacteria responsible for acid fermentation are relatively tolerant to changes in pH and temperature and have a much higher rate of growth than the bacteria responsible for methane fermentation. As a result, methane fermentation is generally assumed to be the rate-controlling step in anaerobic waste treatment processes.

Considering 35°C as the optimum temperature for anaerobic waste treatment, Lawrence proposes that, in the range of 20 to 35°C, the kinetics of methane fermentation of long- and short-chain fatty acids will adequately describe the overall kinetics of anaerobic treatment.[40] Thus, the kinetic equations we presented to describe the completely mixed activated sludge process are equally applicable to the anaerobic process.

There are essentially two types of anaerobic digestion processes used today: the standard-rate process and the high-rate process.

The standard-rate process does not employ sludge mixing, but rather the digester contents are allowed to stratify into zones, as illustrated in Figure 5-39. Sludge feeding and withdrawal are intermittent rather than continuous. The digester is generally heated to increase the rate of fermentation and therefore decrease the required retention time. Retention time ranges between 30 and 60 days for heated digesters. The organic loading rate for a standard-rate digester is between 0.48 and 1.6 kg total volatile solids per m^3 of digester volume per day.

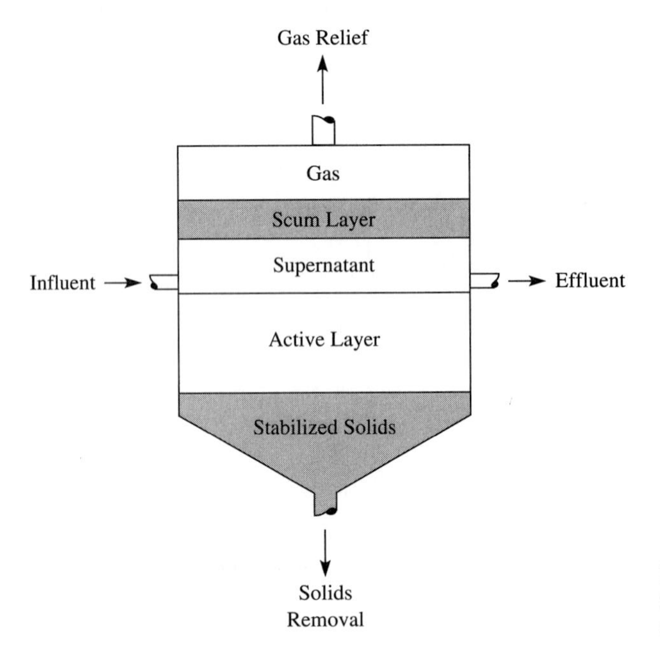

FIGURE 5-39
Schematic of a standard-rate anaerobic digester.

[40] A. W. Lawrence and T. R. Milnes, "Discussion Paper," *Journal of the Sanitary Engineering Division*, American Society of Civil Engineers, vol. 97, p. 121, 1971.

The major disadvantage of the standard-rate process is the large tank volume required because of long retention times, low loading rates, and thick scum-layer formation. Only about one-third of the tank volume is utilized in the digestion process. The remaining two-thirds of the tank volume contains the scum layer, stabilized solids, and the supernatant. Because of this limitation, systems of this type are generally used only at treatment plants having a capacity of 0.04 m^3/s or less.

The high-rate system evolved as a result of continuing efforts to improve the standard-rate unit. In this process, two digesters operating in series separate the functions of fermentation and solids/liquid separation (see Figure 5-40). The contents of the first-stage, high-rate unit are thoroughly mixed and the sludge is heated to increase the rate of fermentation. Because the contents are thoroughly mixed, temperature distribution is more uniform throughout the tank volume. Sludge feeding and withdrawal are continuous or nearly so. The retention time required for the first-stage unit is normally between 10 and 15 days. Organic loading rates vary between 1.6 and 8.0 kg total volatile solids per m^3 of digester per day.

The primary functions of the second-stage digester are solids/liquid separation and residual gas extraction. First-stage digesters may be equipped with fixed or floating covers. Second-stage digester covers are often of the floating type (Figure 5-41). Second-stage units are generally not heated.

The first-stage digester of a high-rate system approximates a completely mixed reactor without solids recycle. Hence, the biological solids retention time (SRT) and the hydraulic retention time are equal for this system. As with the aerobic digesters, the most important operating parameters affecting VSS reduction are solids retention time and digestion temperature.

The BOD remaining at the end of digestion is still quite high. Likewise, the suspended solids may be as high as 12,000 mg/L, while the TKN may be on the order

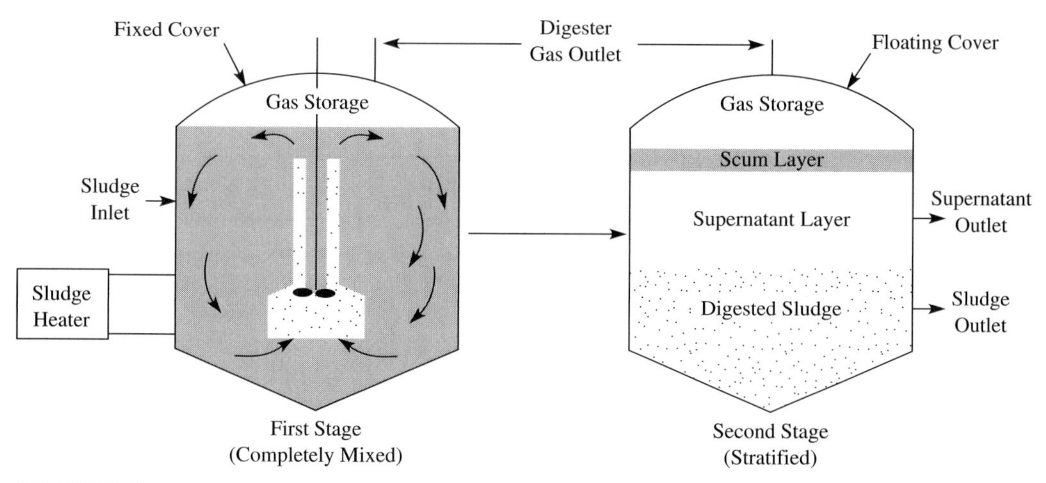

FIGURE 5-40
Schematic of a high-rate anaerobic digester.

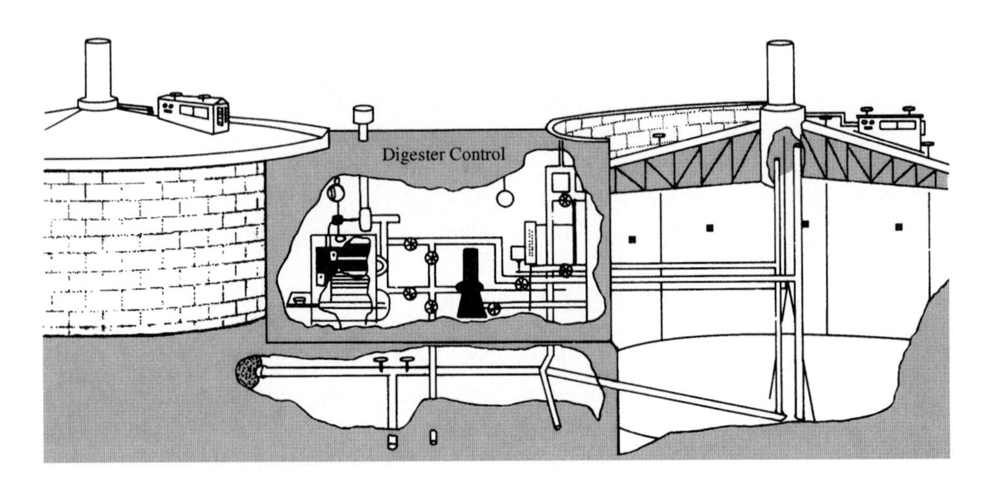

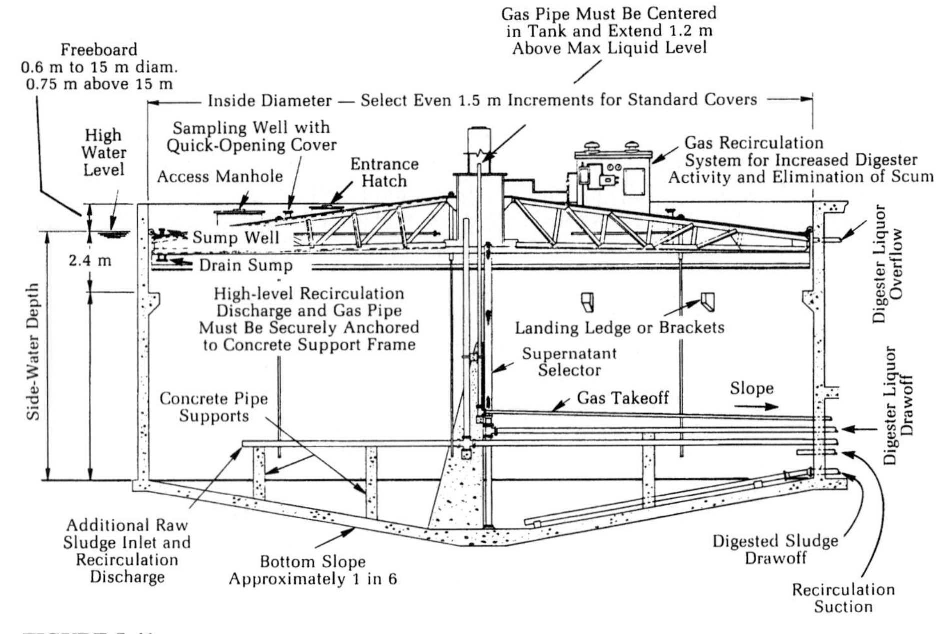

FIGURE 5-41
Phantom view of high-rate anaerobic digester and cross section of detail of floating cover. (Courtesy of Envirex.)

of 1,000 mg/L. Thus, the supernatant from the secondary digester (in the high-rate process) is returned to the head end of the WWTP. The settled sludge is conditioned and dewatered for disposal.

Sludge Conditioning

Chemical conditioning. Several methods of conditioning sludge to facilitate the separation of the liquid and solids are available. One of the most commonly used is the addition of coagulants such as ferric chloride, lime, or organic polymers. Ash from incinerated sludge has also found use as a conditioning agent. As happens when coagulants are added to turbid water, chemical coagulants act to clump the solids together so that they are more easily separated from the water. In recent years, organic polymers have become increasingly popular for sludge conditioning. Polymers are easy to handle, require little storage space, and are very effective. The conditioning chemicals are injected into the sludge just before the dewatering process and are mixed with the sludge.

Heat treatment. Another conditioning approach is to heat the sludge at high temperatures (175 to 230°C) and pressures (1,000 to 2,000 kPa). Under these conditions, much like those of a pressure cooker, water that is bound up in the solids is released, improving the dewatering characteristics of the sludge. Heat treatment has the advantage of producing a sludge that dewaters better than chemically conditioned sludge. The process has the disadvantages of relatively complex operation and maintenance and the creation of highly polluted cooking liquors that when recycled to the treatment plant impose a significant added treatment burden.

Sludge Dewatering

Sludge drying beds. The most popular method of sludge dewatering in the past has been the use of sludge drying beds. These beds are especially popular in small plants because of their simplicity of operation and maintenance. In 1977, two-thirds of all United States wastewater treatment plants utilized drying beds; one-half of all the municipal sludge produced in the United States was dewatered by this method. Although the use of drying beds might be expected in the warmer, sunny regions, they are also used in several large facilities in northern climates.

Operational procedures common to all types of drying beds involve the following steps:

1. Pump 0.20 to 0.30 m of stabilized liquid sludge onto the drying bed surface.
2. Add chemical conditioners continuously, if conditioners are used, by injection into the sludge as it is pumped onto the bed.
3. When the bed is filled to the desired level, allow the sludge to dry to the desired final solids concentration. (This concentration can vary from 18 to 60 percent, depending on several factors, including type of sludge, processing rate needed, and degree of dryness required for lifting. Nominal drying times vary from 10 to 15 d under favorable conditions, to 30 to 60 d under barely acceptable conditions.)
4. Remove the dewatered sludge either mechanically or manually.
5. Repeat the cycle.

Sand drying beds are the oldest, most commonly used type of drying bed. Many design variations are possible, including the layout of drainage piping, thickness and type of gravel and sand layers, and construction materials. Sand drying beds for wastewater sludge are constructed in the same manner as water treatment plant sludge-drying beds. Current U.S. practice was discussed and illustrated in Section 3-9.

Sand drying beds can be built with or without provision for mechanical sludge removal, and with or without a roof. When the cost of labor is high, newly constructed beds are designed for mechanical sludge removal.

Vacuum filtration. A vacuum filter consists of a cylindrical drum covered with a filtering material or fabric, which rotates partially submerged in a vat of conditioned sludge (Figure 5-42). A vacuum is applied inside the drum to extract water, leaving the solids, or filter cake, on the filter medium. As the drum completes its rotational cycle, a blade scrapes the filter cake from the filter and the cycle begins again. In some systems, the filter fabric passes off the drum over small rollers to dislodge the cake. There is a wide variety of filter fabrics, ranging from Dacron to stainless-steel coils, each with its own advantages. The vacuum filter can be applied to digested sludge to produce a sludge cake dry enough (15 to 30 percent solids) to handle and dispose of by burial in a landfill or by application to the land as a relatively dry fertilizer. If the sludge is to be incinerated, it is not stabilized. In this

FIGURE 5-42
Vacuum filters. (Courtesy of Komline-Sanderson Engineering Corporation.)

case, the vacuum filter is applied to the raw sludge to dewater it. The sludge cake is then fed to the furnace to be incinerated.

Continuous belt filter presses (CBFP). The CBFP equipment used in treating wastewater sludges is the same as that used for water treatment plant sludges. This is described and illustrated in Section 3-10.

The CBFP is successful with many normal mixed sludges. Typical dewatering results for digested mixed sludges with initial feed solids of 5 percent give a dewatered cake of 19 percent solids at a rate of 32.8 $kg/h \cdot m^2$. In general, most of the results with these units closely parallel those achieved with rotary vacuum filters. An advantage of CBFPs is that they do not have the sludge pickup problem that sometimes occurs with rotary vacuum filters. Additionally, they have a lower energy consumption.

Reduction

Incineration. If sludge use as a soil conditioner is not practical, or if a site is not available for landfill using dewatered sludge, cities may turn to the alternative of sludge reduction. Incineration completely evaporates the moisture in the sludge and combusts the organic solids to a sterile ash. To minimize the amount of fuel used, the sludge must be dewatered as completely as possible before incineration. The exhaust gas from an incinerator must be treated carefully to avoid air pollution.

5-12 SLUDGE DISPOSAL

Ultimate Disposal

The WWTP process residuals (leftover sludges, either treated or untreated) are the bane of design and operating personnel. Of the five possible disposal sites for residuals, two are feasible and only one is practical. Conceivably, one could ultimately dispose of residues in the following places: in the air, in the ocean, in "outer space," on the land, or in the marketplace. Disposal in the air by burning is in reality not ultimate disposal but only temporary storage until the residue falls to the ground. If you use air pollution control devices, then the residue from these devices must be disposed of. Disposal of sewage sludge at sea by barging is now prohibited in the United States. "Outer space" is not a suitable disposal site. Thus, we are left with land disposal and utilization of the sludge to produce a product.

For ease of discussion, we have divided land disposal into three categories: land spreading, landfilling, and dedicated land disposal. We have grouped all of the utilization ideas under one category.

Land Spreading

The practice of applying WWTP residuals for the purposes of recovering nutrients, water, or reclaiming despoiled land such as strip mine spoils is called *land spreading*. In contrast to the other land disposal techniques, land spreading is land-use intensive.

Application rates are governed by the character of the soil and the ability of the crops or forests on which the sludge is spread to accommodate it.

Landfilling

Sludge landfill can be defined as the planned burial of wastewater solids, including processed sludge, screenings, grit, and ash, at a designated site. The solids are placed into a prepared site or excavated trench and covered with a layer of soil. The soil cover must be deeper than the depth of the plow zone (about 0.20 to 0.25 m). For the most part, landfilling of screenings, grit, and ash is accomplished with methods similar to those used for sludge landfilling.

Dedicated Land Disposal (DLD)

Dedicated land disposal means the application of heavy sludge loadings to some finite land area that has limited public access and has been set aside or dedicated for all time to the disposal of wastewater sludge. Dedicated land disposal does not mean in-place utilization. No crops may be grown. Dedicated sites typically receive liquid sludges. While application of dewatered sludges is possible, it is not common. In addition, disposal of dewatered sludge in landfills is generally more cost-effective.

Utilization

Wastewater solids may sometimes be used beneficially in ways other than as a soil nutrient. Of the several methods worthy of note, composting and co-firing with municipal solid waste are two which have received increasing amounts of interest in the last few years. The recovery of lime and the use of the sludge to form activated carbon have also been in practice to a lesser extent.

Sludge Disposal Regulations

On February 19, 1993, the EPA promulgated risk-based regulations that govern the use or disposal of sewage sludge. These regulations are codified as 40 CFR Part 503 and have become known as the "503 Regulations." The regulations apply to sewage sludge generated from the treatment of domestic sewage that is land-applied, placed on a surface disposal site, or incinerated in an incinerator that accepts only sewage sludge. The regulations do not apply to sludge generated from treatment of industrial process wastes at an industrial facility, hazardous sewage sludge, sewage sludge with polychlorinated biphenyls (PCB) concentrations of 50 mg/L or greater, or drinking water sludge.

Figure 5-43 summarizes the sludge quality requirements for use or disposal. The regulation establishes two levels of sewage sludge quality with respect to heavy-metal concentrations: ceiling concentration limits and pollution concentration limits. To be land-applied, bulk sewage sludge must meet the pollutant ceiling concentration limits *and* cumulative pollutant loading rates (CPLR) *or* the pollutant concentration limits (Table 5-17). Bulk sewage sludge applied to lawns and home gardens must meet the pollutant concentration limits. Sewage sludge sold or given away in bags

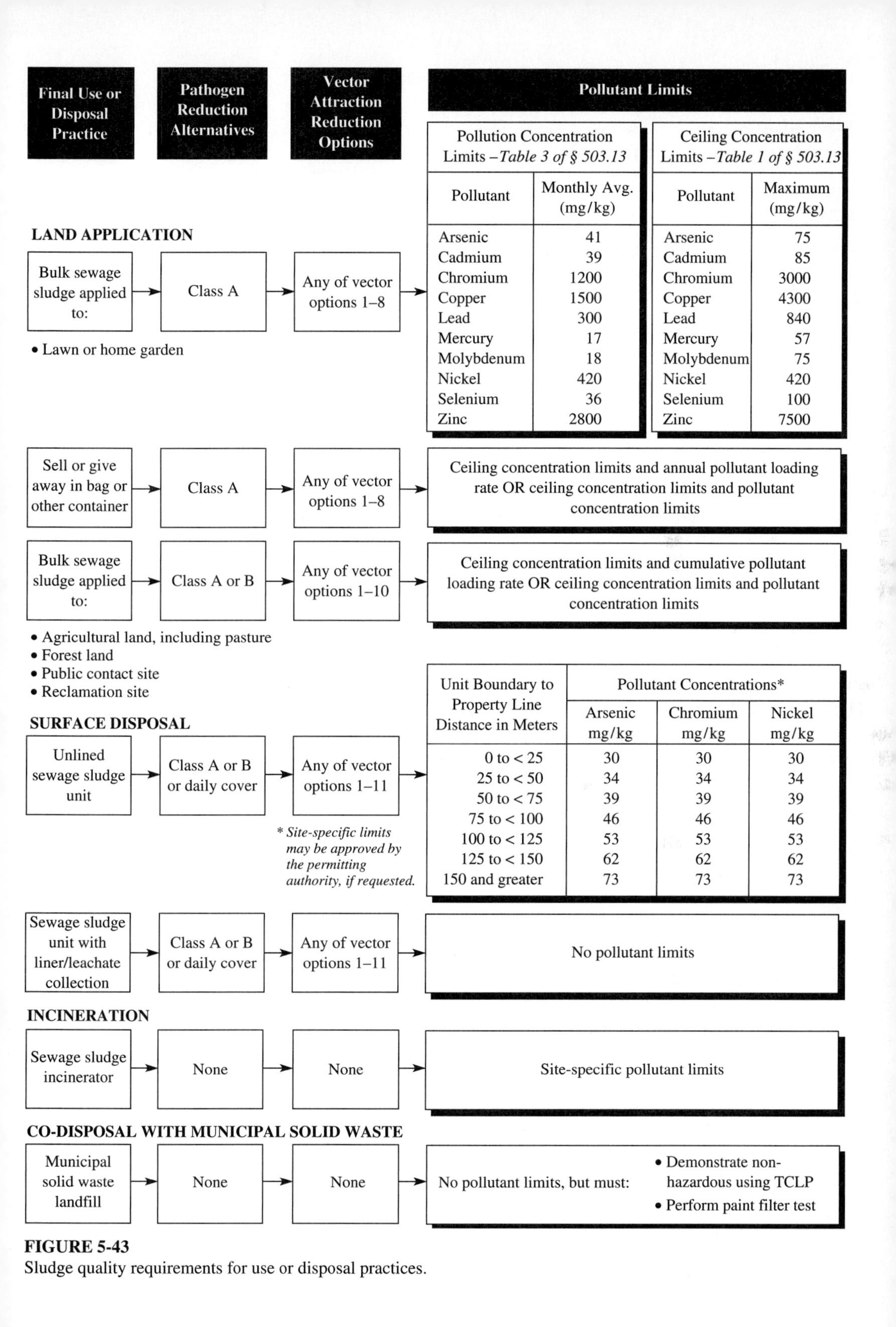

FIGURE 5-43
Sludge quality requirements for use or disposal practices.

TABLE 5-17
Land application limits for heavy metals[a,b]

Pollutant	Ceiling concentration limits, mg/kg	Cumulative pollutant loading rates, kg/ha	Pollutant concentration limits, mg/kg	Annual pollutant loading rates, kg/ha/y
Arsenic	75	41	41	2.0
Cadmium	85	39	39	1.9
Chromium	3,000	3,000	1,200	150
Copper	4,300	1,500	1,500	75
Lead	840	300	300	15
Mercury	57	17	17	0.85
Molybdenum	75	18	18	0.90
Nickel	420	420	420	21
Selenium	100	100	36	5.0
Zinc	7,500	2,800	2,800	140

[a]Source: 40 CFR Part 503.13
[b]Concentrations are on a dry-weight basis

must meet the pollutant concentration limits *or* the annual sewage sludge product application rates that are based on the annual pollutant loading rates.

Two levels of quality for pathogen densities (class A and class B) are defined in the regulation. All class A pathogen reduction alternatives require that either fecal coliform density be less than 1,000 most probable number (MPN) per gram of total solids, or *Salmonella* bacteria be less than 3 MPN per 4 grams of total solids. The class A treatment alternatives include treating the sludge for a specified time and temperature combination, heat-enhanced alkaline stabilization, treatment in a process to further reduce pathogens (PFRP), and use of processes that are proven to reduce virus plaque-forming units and helminth ova to less than 1 per 4 grams of sludge. PFRPs include composting, heat drying, heat treatment, thermophilic aerobic digestion, beta- and gamma-ray irradiation, and pasteurization. The class B pathogen standard is less than 2 million fecal coliforms per gram of sludge or treatment in a process to significantly reduce pathogens (PSRP). The PSRPs include aerobic digestion, air drying, anaerobic digestion, composting, and lime stabilization. Sludges meeting the class A pathogen densities may be land-disposed immediately. Time restrictions are placed on harvesting crops, grazing of animals, and public access to sites on which class B sludge is applied.

Vectors are insects (or other animals) that transmit disease. The organic nature of sludge often attracts vectors after the sludge is land-applied. The 503 regulations provide 11 alternatives to reduce vector attraction. Some of the alternatives are: volatile solids reduction of 38 percent of more, achieving a standard oxygen uptake rate of less than 1.5 mg O_2 per hour per gram of dry solids at 20°C, aerobic treatment at greater than 40°C with an average temperature greater than 45°C for 14 days, alkaline stabilization, sludge drying, surface incorporation, and soil cover.

The 503 regulations are "self-implementing" in that permits are not required to require conformance.

5-13 CHAPTER REVIEW

When you have completed studying this chapter, you should be able to do the following without the aid of your text or notes:

1. For each type of decomposition (aerobic, anoxic, and anaerobic), list the electron acceptor, important end products, and relative advantages and disadvantages as a waste treatment process.
2. List the growth requirements of bacteria and explain why the bacterium needs them.
3. Sketch and label the bacterial growth curve for a pure culture. Define or explain each phase labeled on the curve.
4. List a BOD value for strong, medium, and weak municipal waste.
5. List and describe five on-site alternatives for treating and/or disposing of domestic sewage.
6. Choose the correct on-site treatment/disposal system based on population, land use, and soil conditions.
7. Explain the difference between pretreatment, primary treatment, secondary treatment, and tertiary treatment, and show how they are related.
8. Sketch a graph showing the average variation of daily flow at a municipal wastewater treatment plant (WWTP).
9. Define and explain the purpose of equalization.
10. Sketch, label, and explain the function of the parts of an activated sludge plant and a trickling filter plant.
11. Define θ_c, SRT, and sludge age, and explain their use in regulating the activated sludge process.
12. Explain the purpose of the F/M ratio and define F and M in terms of BOD_5 and mixed liquor volatile suspended solids.
13. Explain the relationship between F/M and θ_c.
14. Explain how cell production is regulated using F/M and/or θ_c.
15. Compare two systems operating at two different F/M ratios.
16. Define SVI and explain its use in the design and operation of an activated sludge plant.
17. Explain the difference between bulking sludge and rising sludge and what circumstances cause each to occur.
18. List and explain the relationship of the five types of oxidation ponds to oxygen.
19. Explain what an RBC is and how it works.
20. Compare the positive and negative effects of disinfection of wastewater effluents.
21. List the four common advanced wastewater treatment (AWT) processes and the pollutants they remove.
22. Explain why removal of residual suspended solids effectively removes residual BOD_5.

23. Describe refractory organics and the method used to remove them.
24. List three chemicals used to remove phosphorus from wastewaters.
25. Explain biological nitrification and denitrification either in words or with an equation.
26. Explain ammonia stripping either in words or with an equation.
27. Describe the three basic approaches to land treatment of wastewater.
28. State the two major purposes of sludge stabilization.
29. Explain the purpose of each of the sludge treatment steps and describe the major processes used.
30. Describe the locations for ultimate disposal of sludges and the treatment steps needed prior to ultimate disposal.

With the aid of this text, you should be able to do the following:

1. Calculate the bacterial population at a time, *t*, given the initial population and the number of generations.
2. Determine the volume of a septic tank and the area of a tile field to treat wastewater from a family or institution, given the proper data.
3. Determine whether or not a grit particle of given diameter and density will be captured in a given velocity-controlled grit chamber, or determine the minimum diameter that will be captured under a given set of conditions.
4. Determine the required volume of an equalization basin to dampen a given periodic flow.
5. Determine the effect of equalization on mass loading of a pollutant.
6. Evaluate or size primary and secondary sedimentation tanks with respect to detention time, overflow rate, solids loading, and weir loading.
7. Use the appropriate trickling filter equation to determine one or more of the following, given the appropriate data: treatment efficiency, filter volume, filter depth, hydraulic loading rate.
8. Estimate the soluble BOD_5 in the effluent from a completely mixed or plug-flow activated sludge plant; determine the mean cell residence time or the hydraulic detention time to achieve a desired degree of treatment; determine the "wasting" flow rate to achieve a desired mean cell residence time or F/M ratio.
9. Calculate the F/M ratio given an influent BOD_5, flow, and detention time, or calculate the volume of the aeration basin given F/M, BOD_5, and flow.
10. Calculate SVI and utilize it to determine return sludge concentration and/or flow rate.
11. Calculate the required mass of sludge to be wasted from an activated sludge process given the appropriate data.
12. Calculate the theoretical mass of oxygen required and the amount of air required to supply it given the appropriate data.
13. Perform a sludge mass balance, given the separation efficiencies and appropriate mass flow rate.

5-14 PROBLEMS

5-1. If the population of microorganisms is 3.0×10^5 at time t_0 and 36 hours later it is 9.0×10^8, how many generations have occurred?

Answer: $n = 11.55$ or 12 generations

5-2. The following data have been gathered in a bacterial growth experiment. At approximately what time did log growth start and terminate? How many generations occurred during log growth?

Time, h	Bacterial count
0	1×10^3
5	1×10^3
10	1.5×10^3
15	5.4×10^3
20	2.0×10^4
25	7.5×10^4
30	2.85×10^5
35	1.05×10^6
40	1.15×10^6
45	1.15×10^6

5-3. Design a septic tank and tile field system for a highway rest area. Use the following assumptions:

a. Average daily traffic = 6,000 vehicles/d

b. % turn in = 10 percent

c. Use rate = 20.0 liters/turn in

maximum use rate = $2.5 \times$ average

d. Terrain = Flat

e. GWT = Average 4.2 m below grade

f. Soil percolation rate: 5 min/cm

5-4. Ginger Snap is planning to expand her Kookie Jar restaurant to a full-size restaurant to be called the Pretzel Bowl. The existing septic tank has a volume of 4.0 m^3 and the existing tile field has a trench area of 100.0 m^2. If the anticipated wastewater production from the Pretzel Bowl is 4000 L/d, will Ms. Snap have to expand either the septic tank or the tile field or both? Assume the soil is a sandy loam.

5-5. If a particle having a 0.0170 cm radius and density of 1.95 g/cm^3 is allowed to fall into quiescent water having a temperature of 4°C, what will be the terminal settling velocity? Assume the density of water = 1,000 kg/m^3.

Answer: 3.82×10^{-2} m/s

5-6. If the terminal settling velocity of a particle falling in quiescent water having a temperature of 15°C is 0.0950 cm/s, what is its diameter? Assume a particle density of 2.05 g/cm^3 and density of water equal to 1,000 kg/m^3.

5-7. A treatment plant being designed for Cynusoidal City requires an equalization basin to even out flow and BOD variations. The average daily flow is 0.400 m^3/s. The following flows and BOD_5 have been found to be typical of the average variation over a day. What size equalization basin, in cubic meters, is required to provide for a uniform outflow equal to the average daily flow? Assume the flows are hourly averages.

Time	Flow, m^3/s	BOD_5, mg/L	Time	Flow, m^3/s	BOD_5, mg/L
0000	0.340	123	1200	0.508	268
0100	0.254	118	1300	0.526	282
0200	0.160	95	1400	0.530	280
0300	0.132	80	1500	0.552	268
0400	0.132	85	1600	0.570	250
0500	0.140	95	1700	0.596	205
0600	0.160	100	1800	0.604	168
0700	0.254	118	1900	0.570	140
0800	0.360	136	2000	0.552	130
0900	0.446	170	2100	0.474	146
1000	0.474	220	2200	0.412	158
1100	0.482	250	2300	0.372	154

Answer: $\forall = 6{,}110 \text{ m}^3$

5-8. What volume equalization basin is required to even outflow and BOD variations shown in the data below? Assume the flows are hourly averages.

Time	Flow, m^3/s	BOD_5, mg/L	Time	Flow, m^3/s	BOD_5, mg/L
0000	0.0875	110	1200	0.135	160
0100	0.0700	81	1300	0.129	150
0200	0.0525	53	1400	0.123	140
0300	0.0414	35	1500	0.111	135
0400	0.0334	32	1600	0.103	130
0500	0.0318	42	1700	0.104	120
0600	0.0382	66	1800	0.105	125
0700	0.0653	92	1900	0.116	150
0800	0.113	125	2000	0.127	200
0900	0.131	140	2100	0.128	215
1000	0.135	150	2200	0.121	170
1100	0.137	155	2300	0.110	130

5-9. Compute and plot the unequalized and the equalized hourly BOD mass loadings to the Cynusoidal City WWTP (Problem 5-7). Using the plot and computations, determine the following ratios for BOD mass loading: peak to average; minimum to average; peak to minimum.

Answers:

	Unequalized	Equalized
P/A	1.97	1.47
M/A	0.14	0.63
P/M	14.05	2.34

5-10. Repeat Problem 5-9 using the data from Problem 5-8.

5-11. Determine the detention time and overflow rate for a primary settling tank that will reduce the influent suspended solids concentration from 330 mg/L to 150 mg/L. The following batch settling column data are available. The data given are percent removals at the sample times and depths shown.

	Depths, m				
Time (min)	0.5	1.5	2.5	3.5	4.5
10	50	32	20	18	15
20	75	45	35	30	25
40	85	65	48	43	40
55	90	75	60	50	46
85	95	87	75	65	60
95	95	88	80	70	63

5-12. The following test data were gathered to design a primary settling tank for a municipal wastewater treatment plant. The initial suspended solids concentration for the test was 200 mg/L. Determine the detention time and overflow rate that will yield 60 percent removal of suspended solids. The data given are suspended solids concentrations in mg/L.

	Time, min					
Depth, m	10	20	35	50	70	85
0.5	140	100	70	62	50	40
1.0	150	130	106	82	70	60
1.5	154	142	120	100	78	70
2.0	160	146	126	110	90	80
2.5	170	150	130	114	100	88

5-13. Using an overflow rate of 26.0 m/d and a detention time of 2.0 h, size a primary sedimentation tank for the average flow at Cynusoidal City (Problem 5-9). What would the overflow rate be for the unequalized maximum flow? Assume 15 sedimentation tanks with length to width ratio of 4.7.

Answers: Tank dimensions = 15 tanks @ 2.17 m deep by 4.34 m by 20.4 m. Maximum overflow rate = 39.3 m/d

5-14. Determine the surface area of a primary settling tank sized to handle a maximum hourly flow of 0.570 m^3/s at an overflow rate of 60.0 m/d. If the effective tank depth is 3.0 m, what is the effective theoretical detention time?

Answers: Surface area = 820.80 or 821 m^2; t_0 = 1.2 h

5-15. If an equalization basin is installed ahead of the primary tank in Problem 5-14, the average flow to the tank is reduced to 0.400 m^3/s. What is the new overflow rate and detention time?

5-16. Envirotech Systems markets a synthetic media for use in the construction of trickling filters. Envirotech uses the following formula to determine BOD removal efficiency:

$$\frac{L_e}{L_i} = \exp\left[-\frac{k\theta D}{Q^n}\right]$$

where L_e = BOD_5 of effluent, mg/L
L_i = BOD_5 of influent, mg/L
k = treatability factor, $\frac{(\text{m/d})^{0.5}}{\text{m}}$
θ = temperature correction factor
= $(1.035)^{T-20}$
T = wastewater temperature, ° C
D = media depth, m
Q = hydraulic loading rate, m/d
n = 0.5

Using the following data for domestic wastewater, determine the treatability factor, k.

Wastewater temperature = 13°C
Hydraulic loading rate = 41.1 m/d

% BOD remaining	Media depth, m
100.0	0.00
80.3	1.00
64.5	2.00
41.6	4.00
17.3	8.00

Answer: $k = 1.79 \frac{(\text{m/d})^{0.5}}{\text{m}}$ at 20°C

5-17. Koon, *et al.*, suggest that recirculation for a synthetic media filter may be considered by the following formula:[41]

[41] J. H. Koon, R. F. Curran, C. E. Adams, and W. W. Eckenfelder, *Evaluation and Upgrading of a Multistage Trickling Filter Facility* (Environmental Protection Agency Publication No. EPA 600/2-76-195), Washington, DC: U.S. Government Printing Office, Dec., 1976.

$$\frac{L_e}{L_i} = \frac{\exp\left[-\frac{kD}{Q^n}\right]}{(1+r) - r\exp\left[-\frac{kD}{Q^n}\right]}$$

where r = recirculation ratio and all other terms are as described in Problem 5-16. Use this equation to determine the efficiency of a 1.8 m deep synthetic media filter loaded at a hydraulic loading rate of 5.00 m/d with a recirculation ratio of 2.00. The wastewater temperature is 16°C and the treatability factor is 1.79 $\frac{(\text{m/d})^{0.5}}{\text{m}}$ at 20°C.

5-18. Determine the concentration of the effluent BOD_5 for the two-stage trickling filter described below. The wastewater temperature is 17°C. Assume the NRC equations apply.

Design flow = 0.0509 m^3/s
Influent BOD_5 (after primary treatment) = 260 mg/L
Diameter of each filter = 24.0 m
Depth of each filter = 1.83 m
Recirculation flow rate for each filter = 0.0594 m^3/s

5-19. Determine the diameter of a single-stage, rock media filter to reduce an applied BOD_5 of 125 mg/L to 25 mg/L. Use a hydraulic loading rate of 14 $m^3/m^2 \cdot d$, a recirculation ratio of 12.0, and a filter depth of 1.83 m. Assume the NRC equations apply.

5-20. Using the assumptions given in Example 5-7, the rule of thumb values for growth constants, and the further assumption that the influent BOD_5 was reduced by 32.0 percent in the primary tank, estimate the liquid volume of an aeration tank required to treat the wastewater in Problem 5-7. Assume an MLVSS of 2,000 mg/L.

Answer: Volume = 4,032 or 4,000 m^3

5-21. Repeat Problem 5-20 using the wastewater in Problem 5-8.

5-22. Using a spreadsheet program you have written, rework Example 5-7 using the following MLVSS concentrations instead of the 2,000 mg/L used in the example: 1,000 mg/L; 1,500 mg/L; 2,500 mg/L; and 3,000 mg/L.

5-23. Using a spreadsheet program you have written, determine the effect of MLVSS concentration on the effluent soluble BOD_5 (S) using the data in Example 5-7. Assume the volume of the aeration tank remains constant at 970 m^3. Use the same MLVSS values used in Problem 5-22.

5-24. If the F/M of a 0.4380 m^3/s activated sludge plant is 0.200 d^{-1}, the influent BOD_5 after primary settling is 150 mg/L, and the MLVSS is 2,200 mg/L, what is the volume of the aeration tank?

Answer: Volume = 1.29×10^4 m^3

5-25. What sludge volume would you expect to find after settling the mixed liquor described in Example 5-10 for 30 minutes in a one-liter graduated cylinder (*magna cum laude*).

Answer: Volume = 500 mL

5-26. What MLVSS and SVI must be achieved to reduce the return sludge flow rate of Example 5-10 from 0.150 m^3/s to 0.0375 m^3/s? (Note that there are several combinations that will be satisfactory.)

5-27. Two activated sludge aeration tanks at Turkey Run, Indiana, are operated in series. Each tank has the following dimensions: 7.0 m wide by 30.0 m long by 4.3 m effective liquid depth. The plant operating parameters are as follows:

Flow = 0.0796 m^3/s
Soluble BOD_5 after primary settling = 130 mg/L
MLVSS = 1,500 mg/L
MLSS = 1.40 (MLVSS)
Settled sludge volume after 30 min = 230.0 mL/L
Aeration tank liquid temperature = 15°C

Determine the following: aeration period, F/M ratio, SVI, solids concentration in the return sludge, and return sludge rate.

Answers: aeration period = 6.3 h; F/M = 0.33; SVI = 110 mL/g; $X_r' = 9{,}130$ mg/L; $Q_r = 0.0238\ m^3/s$

5-28. The 500-bed Lotta Hart Hospital has a small activated sludge plant to treat its wastewater. The average daily hospital discharge is 1,200 liters per day per bed, and the average soluble BOD_5 after primary settling is 500 mg/L. The aeration tank has effective liquid dimensions of 10.0 m wide by 10.0 m long by 4.5 m deep. The plant operating parameters are as follows:

MLVSS = 2,000 mg/L
MLSS = 1.20 (MLVSS)
Settled sludge volume after 30 min = 200 mL/L

Determine the following: aeration period, F/M ratio, SVI, solids concentration in return sludge, and return sludge rate.

5-29. Using the following assumptions, determine the sludge age and cell wastage-flow rate for the Turkey Run WWTP (Problem 5-27).

Assume: SS in the effluent are negligible
Wastage is from the aeration tank
Yield coefficient = 0.40
Bacterial decay rate = 0.040 d^{-1}
Effluent BOD_5 = 5.0 mg/L (soluble)

Answers: $\theta_c = 11.50$ d; $Q_w = 0.00182\ m^3/s$

5-30. Using the following assumptions, determine the solids retention time and the cell wastage flow rate for the Lotta Hart Hospital WWTP (Problem 5-28).

Assume: SS in effluent = 30.0 mg/L
Wastage is from the return sludge line
Yield coefficient = 0.60
Bacterial decay rate = $0.060\ d^{-1}$
Inert fraction of SS = 66.67%
Allowable BOD in effluent = 30.0 mg/L

5-31. The two secondary settling tanks at Turkey Run (Problem 5-27) are 16.0 m in diameter and 4.0 m deep at the side wall. The effluent weir is a single launder set on the tank wall. Evaluate the overflow rate, depth, solids loading, and weir length of this tank for conformance to standard practice.

Answers: v_o = 17.1 m/d $<$ 33 m/d. OK.
SWD $>$ 3.7 m recommended depth. OK.
SL = 46.65 $kg/m^2 \cdot d \ll 253\ kg/m^2 \cdot d$. OK.
WL = 68.4 $m^3/d \cdot m$, which is acceptable.

5-32. The single secondary settling tank at the Lotta Hart Hospital WWTP (Problem 5-28) is 10.0 m in diameter and 3.4 m deep at the side wall. The effluent weir is a single launder set on the tank wall. Evaluate the overflow rate, depth, solids loading, and weir length for conformance to standard practice.

5-33. An oxidation pond having a surface area of 90,000 m^2 is loaded with a waste flow of 500 m^3/d containing 180 kg of BOD_5. The operating depth is from 0.8 to 1.6 m. Using the Michigan rules of thumb, determine whether or not this design is acceptable.

Answers: Loading rate = 20.0 kg/ha · d
Detention time = 180 d
This design is acceptable.

5-34. Determine the required surface area and the loading rate for a facultative oxidation pond to treat a waste flow of 3,800 m^3/d with a BOD_5 of 100.0 mg/L.

5-35. Rework Example 5-15 using alum $[Al_2(SO_4)_3 \cdot 18H_2O]$ to remove the phosphorus.

Answer: 86.1 mg/L of alum

5-36. Rework Example 5-15 using lime (CaO) to remove the phosphorus.

5-37. Prepare a monthly water balance and estimate of the storage volume required (in m^3) for a spray irrigation system being designed for Wheatville, Iowa.

The design population is 1,000 and the design wastewater generation rate is 280.0 Lpcd. Based on a nitrogen balance, the allowable application rate is 27.74 mm/mo. The area available is 40.0 ha. The percolation rate during the spray season is 150 mm/mo. Assume that the runoff is contained and reapplied. Assume "spray season" is when temperature is above 0°C. In fact, spraying can continue to about −4°C but once spraying has stopped, it may not recommence until temperatures exceed +4°C. The following climatological data (from Kansas City) may be used. This is a direct application of the hydrologic balance equation. Equation 2-2 may be rewritten as:

$$\frac{dS}{dt} = \mathrm{P} + \mathrm{WW} - \mathrm{ET} - \mathrm{G} = \mathrm{R}$$

where $\frac{dS}{dt}$ = change in storage, mm/mo
P = precipitation, mm/mo
WW = wastewater application rate, mm/mo
ET = evapotranspiration, mm/mo
G = groundwater infiltration, mm/mo
R = runoff, mm/mo

Climatological data from Kansas City, Missouri

Month	Average temperature (°C)	Evapotranspiration (mm)	Precipitation (mm)
JAN	−0.2	23	36
FEB	2.1	28	32
MAR	6.3	43	63
APR	13.2	79	90
MAY	18.7	112	112
JUN	24.4	155	116
JUL	27.5	203	81
AUG	26.6	198	96
SEP	21.8	152	83
OCT	15.7	114	73
NOV	7.0	64	46
DEC	2.1	25	39

5-38. Prepare a monthly water balance and estimate the storage volume (in m^3) for Flushing Meadows. The design population for Flushing Meadows is 8,880. The average wastewater generation rate is 485.0 Lpcd. The area available is 125.0 ha. The percolation rate is 200 mm/mo during the spray season. Assume that runoff is to be contained and reapplied. Assume also that the following climatological data apply. Assume "spray season" is when temperature is above 0°C. In fact, spraying can continue to about −4°C but once spraying has stopped it may not recommence until temperatures exceed +4°C.

Climatological data from Columbus, Ohio

Month	Average temperature (°C)	Evapotranspiration (mm)	Precipitation (mm)
JAN	−1.2	15	80
FEB	−0.5	20	59
MAR	3.8	28	80
APR	10.4	58	59
MAY	16.4	89	102
JUN	21.9	117	106
JUL	23.8	142	100
AUG	22.9	130	73
SEP	18.8	104	67
OCT	12.3	76	54
NOV	5.2	41	63
DEC	−0.3	15	59

5-39. Determine the daily and annual primary sludge production for a WWTP having the following operating characteristics:

Operating data:
Flow = 0.0500 m^3/s
Influent suspended solids = 155.0 mg/L
Removal efficiency = 53.0%
Volatile solids = 70.0%
Specific gravity of volatile solids = 0.970
Fixed solids = 30.0%
Specific gravity of fixed solids = 2.50
Sludge concentration = 4.50%

Answer: $\forall_{sl} = 7.83\ m^3/d$

5-40. Repeat Problem 5-39 using the following data:

Operating data:
Flow = 2.00 m^3/s
Influent suspended solids = 179.0 mg/L
Removal efficiency = 47.0%
Sludge concentration = 5.20%
Volatile solids = 68.0%
Specific gravity of volatile solids = 0.999
Fixed solids = 32.0%
Specific gravity of fixed solids = 2.50

5-41. Using Figure 5-32, Table 5-13, and the following data, determine B, E, J, K, and L in megagrams per day (Mg/d).

$A = 185.686$ Mg/d
$\eta_E = 0.900$; $\eta_j = 0.250$; $\eta_N = 0.00$; $\eta_P = 0.150$
$\eta_H = 0.190$

Answers: $B = 21.112$ or 21.1 Mg/d
$E = 190.011$ or 190. Mg/d
$J = 47.503$ or 47.5 Mg/d
$K = 142.509$ or 143. Mg/d
$L = 144.147$ or 144. Mg/d

5-42. Rework Problem 5-41 assuming that the digestion solids are not dewatered prior to ultimate disposal, that is, $K = L$.

5-43. The flowsheet for the Doubtful WWTP is shown in Figure P-5-43. Assuming that the appropriate values of η given in Figure 5-33 may be used when needed and that $A = 7.250$ Mg/d, $X = 1.288$ Mg/d, and $N = 0.000$ Mg/d, what is the mass flow (in kg/d) of sludge to be sent to ultimate disposal?

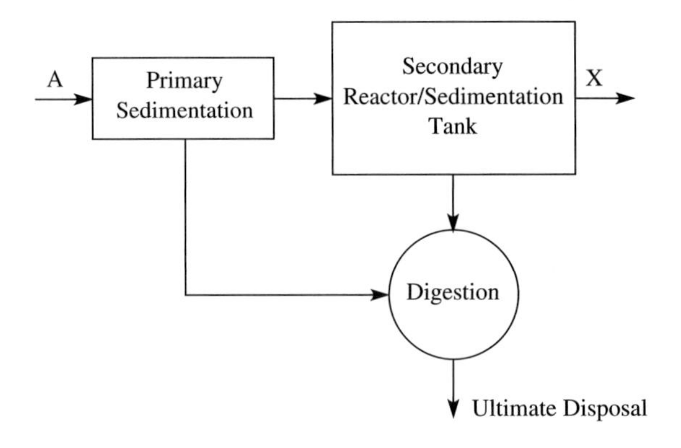

Answer: $L = K = 3.743$ Mg/d or 3,743 kg/d

5-44. Using the following mass flow data from the Doubtful WWTP (Problem 5-43) determine η_E, η_D, η_N, η_J, and η_X.

Mass Flows for Doubtful WWTP in Mg/d:

$A = 7.280$ $\quad J = 4.755$
$B = 7.798$ $\quad K = 6.422$
$D = 0.390$ $\quad N = 9.428$
$E = 8.910$ $\quad X = 0.468$
$F = 6.940$

5-45. Determine the surface area required for the gravity thickeners (assume that no thickener is greater than 30.0 m in diameter) to thicken the waste activated sludge (WAS) at Grand Rapids, Michigan, from 10,600 mg/L to 2.50 percent solids. The waste activated sludge flow is 3,255 m^3/d. Assume that the batch settling curves of Figure 5-36 apply.

Answer: A_s = 2,379.5 or 2,380 m^2 depending on graph reading. Thus, choose four thickeners at 27.5 m diameter.

5-46. Determine the surface area required for the gravity thickeners of Problem 5-45 if 710 m^3/d of primary sludge is mixed with the WAS to form a sludge having 2.00 percent solids. The final sludge is to have a solids concentration of 5.00 percent. The batch settling curve for mixed WAS and PS in Figure 5-36 is assumed to apply.

5-47. The Pomdeterra wastewater treatment plant produces thickened sludge that has a suspended solids concentration of 3.8 percent. They are investigating a filter press that will yield a solids concentration of 24 percent. If they now produce 33 m^3/d of sludge, what annual volume savings will they achieve if they install the press?

5-48. Ottawa's anaerobic digester produces 13 m^3/d of sludge with a suspended solids concentration of 7.8 percent. What volume of sludge must they dispose of each year if their sand drying beds yield a solids concentration of 35 percent?

5-15 DISCUSSION QUESTIONS

1. You are touring the research labs of the environmental engineers at your university. Two biological reactors are in a controlled temperature room that has a temperature of 35°C. Reactor A has a strong odor. Reactor B has virtually no odor. What electron acceptors are being used in each reactor?

2. If the state regulatory agency requires tertiary treatment of a municipal wastewater, what, if any, processes would you expect to find preceding the tertiary process?

3. What is the purpose of recirculation and how does it differ from return sludge?

4. In which of the following cases is the cost of sludge disposal higher?

a. θ_c = 3 days

b. θ_c = 10 days

5. Would an industrial wastewater containing only NH_4 at a pH of 7.00 be denitrified if pure oxygen was bubbled into it? Explain your reasoning.

5-16 ADDITIONAL READING

Books

L. D. Benefield and C. W. Randall, *Biological Process Design for Wastewater Treatment,* Upper Saddle River, NJ: Prentice Hall, 1980.

J. W. Clark, W. Viessman, Jr., and M. J. Hammer, *Water Supply and Pollution Control,* New York: Harper & Row, 1977.

R. L. Droste, *Theory and Practice of Water and Wastewater Treatment,* New York: Wiley, 1997.

C. P. L. Grady and H. C. Lim, *Biological Wastewater Treatment: Theory and Applications,* Marcel Dekker, 1980.
M. J. Hammer, *Water and Wastewater Technology, SI Version,* New York: Wiley, 1977.
J. G. Henry and G. W. Heinke, *Environmental Science and Engineering,* Upper Saddle River, NJ: Prentice Hall, 1989.
G. Kiely, *Environmental Engineering,* London: McGraw-Hill, 1997.
G. M. Masters, *Introduction to Environmental Engineering and Science,* Upper Saddle River, NJ: Prentice Hall, 1991.
Metcalf & Eddy, Inc., and G. Tchobanoglous, *Wastewater Engineering: Treatment, Disposal, Reuse,* New York: McGraw-Hill, 1979.
H. S. Peavy, D. R. Rowe, and G. Tchobanoglous, *Environmental Engineering,* New York: McGraw-Hill, 1985.
B. T. Ray, *Environmental Engineering,* New York: PWS Publishing, 1995.
T. D. Reynolds and P. A. Richards, *Unit Operations and Processes in Environmental Engineering,* 2nd ed., Boston, MA: PWS Publishing, 1996.
A. P. Sincero and G. A. Sincero, *Environmental Engineering: A Design Approach,* Upper Saddle River, NJ: Prentice Hall, 1996.
E. W. Steel and T. J. McGhee, *Water Supply and Sewerage,* New York: McGraw-Hill, 1979.
G. Tchobanoglous and E. D. Schroeder, *Water Quality,* Reading, MA: Addison Wesley, 1985.
P. A. Vesilind, *Introduction to Environmental Engineering*, Boston, MA: PWS Publishing, 1997.
P. A. Vesilind, *Treatment and Disposal of Wastewater Sludges,* Ann Arbor, MI: Ann Arbor Science, 1974.

Journals

Water Environment & Technology (Water Environment Federation)
Water Environment Research (formerly *Journal of the Water Pollution Control Federation*)
Water Environment and Technology
Water Research

CHAPTER 6

AIR POLLUTION

6-1 PHYSICAL AND CHEMICAL FUNDAMENTALS
  Ideal Gas Law
  Dalton's Law of Partial Pressures
  Adiabatic Expansion and Compression
  Units of Measure
6-2 AIR POLLUTION PERSPECTIVE
6-3 AIR POLLUTION STANDARDS
6-4 EFFECTS OF AIR POLLUTANTS
  Effects on Materials
  Effects on Vegetation
  Effects on Health
6-5 ORIGIN AND FATE OF AIR POLLUTANTS
  Carbon Monoxide
  Hazardous Air Pollutants (HAPs)
  Lead
  Nitrogen Dioxide
  Photochemical Oxidants
  Sulfur Oxides
  Particulates
6-6 MICRO AND MACRO AIR POLLUTION
  Indoor Air Pollution
  Acid Rain
  Ozone Depletion
  Greenhouse Effect

6-7 AIR POLLUTION METEOROLOGY
The Atmospheric Engine
Turbulence
Stability
Terrain Effects
6-8 ATMOSPHERIC DISPERSION
Factors Affecting Dispersion of Air Pollutants
Dispersion Modeling
6-9 INDOOR AIR QUALITY MODEL
6-10 AIR POLLUTION CONTROL OF STATIONARY SOURCES
Gaseous Pollutants
Flue Gas Desulfurization (FGD)
Control Technologies for Nitrogen Oxides
Particulate Pollutants
6-11 AIR POLLUTION CONTROL OF MOBILE SOURCES
Engine Fundamentals
Control of Automobile Emissions
6-12 WASTE MINIMIZATION
6-13 CHAPTER REVIEW
6-14 PROBLEMS
6-15 DISCUSSION QUESTIONS
6-16 ADDITIONAL READING

6-1 PHYSICAL AND CHEMICAL FUNDAMENTALS

Ideal Gas Law

Although polluted air may not be "ideal" from the biological point of view, we may treat its behavior with respect to temperature and pressure as if it were ideal. Thus, we assume that at the same temperature and pressure, different kinds of gases have densities proportional to their molecular masses. This may be written as

$$\rho = \frac{1}{R}\frac{PM}{T} \tag{6-1}$$

where ρ = density of gas, kg/m^3
P = absolute pressure, kPa
M = molecular mass, grams/mole
T = absolute temperature, K
R = universal gas constant = 8.3143 J/K · mole

Since density is mass per unit volume, or the number of moles per unit volume, $n/\forall$, the expression may be rewritten in the general form as

$$P\forall = nRT \tag{6-2}$$

where $\forall$ is the volume occupied by n moles of gas. At 273.15 K and 101.325 kPa, one mole of an ideal gas occupies 22.414 L.

Dalton's Law of Partial Pressures

Stack and exhaust sampling measurements are made with instruments calibrated with air. Because combustion products have an entirely different composition than air, the readings must be adjusted ("corrected" in sampling parlance) to reflect this difference. Dalton's law forms the basis for the calculation of the correction factor. Dalton found that the total pressure exerted by a mixture of gases is equal to the sum of the pressures that each type of gas would exert if it alone occupied the container. In mathematical terms,

$$P_t = P_1 + P_2 + P_3 + \dots \tag{6-3}$$

where P_t = total pressure of mixture

P_1, P_2, P_3 = pressure of each gas if it were in container alone, that is, *partial pressure*

Dalton's law also may be written in terms of the ideal gas law:

$$P_t = n_1\frac{RT}{\forall} + n_2\frac{RT}{\forall} + n_3\frac{RT}{\forall} + \dots$$

$$= (n_1 + n_2 + n_3 + \dots)\frac{RT}{\forall}$$

Adiabatic Expansion and Compression

Air pollution meteorology is, in part, a consequence of the thermodynamic processes of the atmosphere. One such process is adiabatic expansion and contraction. An *adiabatic* process is one that takes place with no addition or removal of heat and with sufficient slowness so that the gas can be considered to be in equilibrium at all times.

As an example, let us consider the piston and cylinder in Figure 6-1. The cylinder and piston face are assumed to be perfectly insulated. The gas is at pressure P. A force, F, equal to PA must be applied to the piston to maintain equilibrium. If the force is increased and the volume is compressed, the pressure will increase and work will be done on the gas by the piston. Since no heat enters or leaves the gas, the work will go into increasing the thermal energy of the gas in accordance with the first principle of thermodynamics, that is,

(Heat added to gas) = (increase in thermal energy) + (external work done by or on the gas)

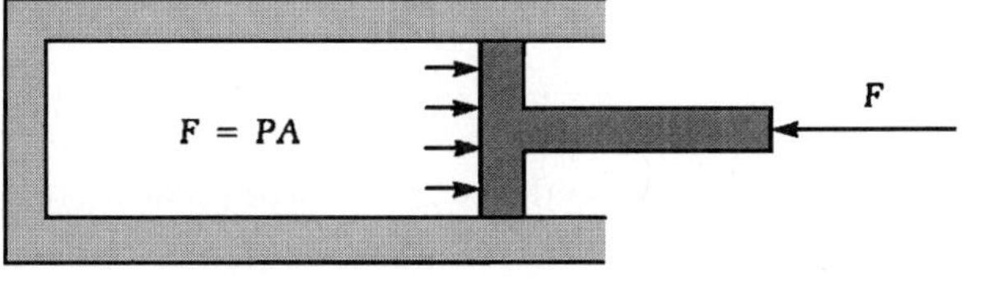

FIGURE 6-1
Work done on gas.

Since the left side of the equation is zero (because it is an adiabatic process), the increase in thermal energy is equal to the work done. The increase in thermal energy is reflected by an increase in the temperature of the gas. If the gas is expanded adiabatically, its temperature will decrease.

Units of Measure

The three basic units of measure used in reporting air pollution data are *micrograms per cubic meter* ($\mu g/m^3$), *parts per million* (ppm), and the *micron* (μ) or, preferably, its equivalent, the *micrometer* (μm). Micrograms per cubic meter and parts per million are measures of concentration. Both $\mu g/m^3$ and ppm are used to indicate the concentration of a gaseous pollutant. However, the concentration of particulate matter may be reported only as $\mu g/m^3$. The μm is used to report particle size.

There is an advantage to the unit ppm that frequently makes it the unit of choice. The advantage results from the fact that ppm is a volume-to-volume ratio. (Note that this is different than ppm in water and wastewater, which is a mass-to-mass ratio.) Changes in temperature and pressure do not change the ratio of the volume of pollutant gas to the volume of air that contains it. Thus, it is possible to compare ppm readings from Denver and Washington, DC, without further conversion.

Converting $\mu g/m^3$ to ppm. The conversion between $\mu g/m^3$ and ppm is based on the fact that at standard conditions (0°C and 101.325 kPa), one mole of an ideal gas occupies 22.414 L. Thus, we may write an equation that converts the mass of the pollutant M_p in grams to its equivalent volume $\forall_p$ in liters at standard temperature and pressure (STP):

$$\forall_p = \frac{M_p}{\text{GMW}} \times 22.414 \text{ L/GM} \tag{6-4}$$

where GMW is the gram molecular weight of the pollutant. For readings made at temperatures and pressures other than standard conditions, the standard volume, 22.414 L/GM, must be corrected. We use the ideal gas law to make the correction:

$$22.414 \text{ L/GM} \times \frac{T_2}{273 \text{ K}} \times \frac{101.325 \text{ kPa}}{P_2} \tag{6-5}$$

where T_2 and P_2 are the absolute temperature and absolute pressure at which the readings were made. Since ppm is a volume ratio, we may write

$$\text{ppm} = \frac{\forall_p}{\forall_a} \tag{6-6}$$

where $\forall_a$ is the volume of air in cubic meters at the temperature and pressure of the reading. We then combine Equations 6-4, 6-5, and 6-6 to form Equation 6-7.

$$\text{ppm} = \frac{\frac{M_p}{\text{GMW}} \times 22.414 \times \frac{T_2}{273 \text{ K}} \times \frac{101.325 \text{ kPa}}{P_2}}{\forall_a \times 1{,}000 \text{ L/m}^3} \tag{6-7}$$

where M_p is in μg. The factors converting μg to g and L to millions of L cancel one another. Unless otherwise stated, it is assumed that $\forall = 1.00 \text{ m}^3$.

Example 6-1. A one-cubic-meter sample of air was found to contain 80 $\mu g/m^3$ of SO_2. The temperature and pressure were 25°C and 103.193 kPa when the air sample was taken. What was the SO_2 concentration in ppm?

Solution. First we must determine the GMW of SO_2. From the chart inside the front cover, we find

$$\text{GMW of } SO_2 = 32.06 + 2(15.9994) = 64.06$$

Next we must convert the temperature to absolute temperature. Thus,

$$25°C + 273\text{ K} = 298\text{ K}$$

Now we may make use of Equation 6-7.

$$\text{ppm} = \frac{\frac{80 \mu g}{64.06} \times 22.414 \times \frac{298}{273} \times \frac{101.325}{103.193}}{1.00\text{ m}^3 \times 1{,}000\text{ L/m}^3} = 0.0300 \text{ ppm of } SO_2$$

Relativity. Before we launch into the esoterics of air pollution, let's take a moment to look at the relationship of a ppm and a μm to something relevant to daily life. Four crystals of common table salt in one cup of granulated sugar is approximately equal to 1 ppm on a volume-to-volume basis. Figure 6-2 should help you visualize the size of a μm. Note that a hair has an average diameter of approximately 80 μm.

6-2 AIR POLLUTION PERSPECTIVE

Air pollution is of public health concern on several scales: micro, meso, and macro. Indoor air pollution results from products used in construction materials, adequacy of general ventilation, and geophysical factors that may result in exposure to naturally occurring radioactive materials. Industrial and mobile sources contribute to meso-scale air pollution that contaminates the ambient air that surrounds us outdoors. Macro-scale impacts include transport of ambient air pollutants over large distances and global impact. Examples of macro-scale impacts include acid rain and ozone pollution. Global impacts of air pollution result from sources that may potentially change the upper atmosphere. Examples include depletion of the ozone layer and global warming. While micro- and macro-scale effects are of concern, our focus will predominately be on meso-scale air pollution.

6-3 AIR POLLUTION STANDARDS

The 1970 Clean Air Act (CAA) required the U.S. Environmental Protection Agency (EPA) to investigate and describe the environmental effects of any air pollutant emitted by stationary or mobile sources that may adversely affect human health or the environment. The EPA used these studies to establish the National Ambient Air Quality Standards (NAAQS). These standards are for the ambient air, that is, the outdoor air that normally surrounds us. EPA calls the pollutants listed in Table 6-1 *criteria pollutants* because they were developed on health-based criteria. The *primary standard* was established to protect human health with an "adequate margin of

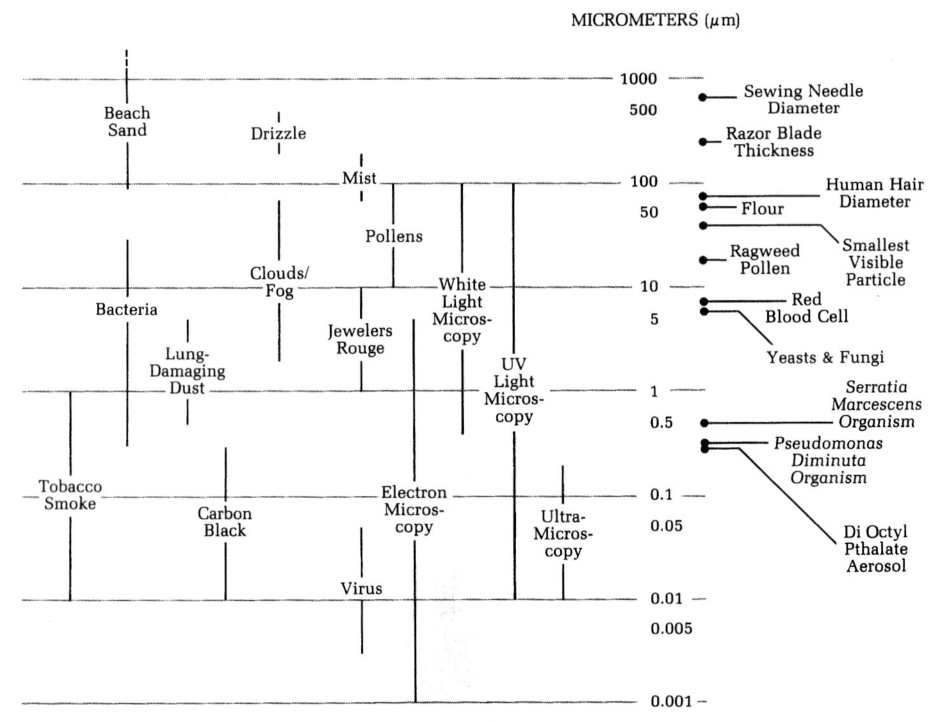

FIGURE 6-2
Relative sizes of small particles. (Courtesy of Gelman Instrument Co.)

safety." The *secondary standards* are intended to prevent environmental and property damage. In 1987, the EPA revised the NAAQS. The standard for hydrocarbons was dropped and the standard for Total Suspended Particulates (TSP) was replaced with a particulate standard based on the mass of particulate matter with an aerodynamic diameter less than or equal to 10 μm. This standard is referred to as the PM_{10} standard.

States are divided into *Air Quality Control Regions (AQRs).* An AQR that has air quality equal to or better than the primary standard is called an *attainment area.* Those areas that do not meet the primary standard are called *nonattainment areas.*

Under the 1970 CAA, the EPA was directed to establish regulations for *hazardous air pollutants (HAPs)* using a risk-based approach. These were called NESHAPs—national emission standards for hazardous air pollutants. Because EPA had difficulty defining "an ample margin of safety" as required by the law, only seven HAPs were regulated between 1970 and 1990: asbestos, arsenic, benzene, beryllium, mercury, vinyl chloride, and radionuclides. The Clean Air Act Amendments of 1990 directed EPA to establish a HAP emissions control program based on technology for 189 chemicals (see Table 1-5 for the list). EPA will establish emission allowances based on *Maximum Achievable Control Technology (MACT)* for 174 categories of industrial sources that potentially emit 9.08 megagrams (Mg) per year of a single HAP or 22.7 Mg per year of a combination of HAPs. A MACT can include process changes, material substitutions, or air pollution control equipment.

TABLE 6-1
National Ambient Air Quality Standards (NAAQS)

Criteria pollutant	Standard type	Concentration $\mu g/m^3$	Concentration ppm	Averaging period or method	Allowable exceedances[a]
CO	Primary	10,000	9	8-hour average	Once per year
	Primary	40,000	35	1-hour average	Once per year
Lead	Primary and secondary	1.5		Maximum arithmetic mean measured over a calendar quarter	—
NO_2	Primary and secondary	100	0.053	Annual arithmetic mean	—
Ozone	Primary and secondary	235	0.12	Maximum hourly average	Once per year
Ozone[b]	Primary and secondary	157	0.08	8-hour average	[c]
Particulate matter (PM_{10})[d]	Primary and secondary	150	—	24-hour average	One day per year
	Primary and secondary	50	—	Annual arithmetic mean	—
($PM_{2.5}$)[b]	Primary and secondary	65	—	24-hour average	one day per year
		15	—	Annual arithmetic mean	—
SO_2	Primary	80	0.03	Annual arithmetic mean	—
	Primary	365	0.14	Maximum 24-hour concentration	Once per year
SO_2	Secondary	1,300	0.5	Maximum 3-hour concentration	Once per year

[a] Allowable exceedances may actually be an average value over a multi-year period.
[b] Proposed by EPA July 1997.
[c] Average fourth highest concentration over 3-year period.
[d] Particulate matter standard applies to particles with an aerodynamic diameter $\leq$ 10 microns.
Source: 40 CFR 50.4–50.12.

6-4 EFFECTS OF AIR POLLUTANTS

Effects on Materials

Mechanisms of deterioration. Five mechanisms of deterioration have been attributed to air pollution: abrasion, deposition and removal, direct chemical attack, indirect chemical attack, and electrochemical corrosion.[1]

Solid particles of large enough size and traveling at high enough speed can cause deterioration by abrasion. With the exception of soil particles in dust storms

[1] The discussions on mechanisms of deterioration and factors that influence deterioration follow John E. Yocom and Roy O. McCaldin, "Effects on Materials and the Economy," *Air Pollution, VI,* 2nd ed., Arthur C. Stern, ed., New York: Academic Press, 1968.

and lead particles from automatic weapons fire, most air pollutant particles either are too small or travel at too slow a speed to be abrasive.

Small liquid and solid particles that settle on exposed surfaces do not cause more than aesthetic deterioration. For certain monuments and buildings, such as the White House, this form of deterioration is in itself quite unacceptable. For most surfaces, it is the cleaning process that causes the damage. Sandblasting of buildings is an obvious case in point. Frequent washing of clothes weakens their fiber, while frequent washing of painted surfaces dulls their finish.

Solubilization and oxidation/reduction reactions typify direct chemical attack. Frequently, water must be present as a medium for these reactions to take place. Sulfur dioxide and SO_3 in the presence of water react with limestone ($CaCO_3$) to form calcium sulfate ($CaSO_4$) and gypsum ($CaSO_4 \cdot 2H_2O$). Both $CaSO_4$ and $CaSO_4 \cdot 2H_2O$ are more soluble in water than $CaCO_3$, and both are leached away when it rains. The tarnishing of silver by H_2S is a classic example of an oxidation/reduction reaction.

Indirect chemical attack occurs when pollutants are absorbed and then react with some component of the absorbent to form a destructive compound. The compound may be destructive because it forms an oxidant, reductant, or solvent. Further, a compound can be destructive by removing an active bond in some lattice structure. Leather becomes brittle after it absorbs SO_2, which reacts to form sulfuric acid because of the presence of minute quantities of iron. The iron acts as a catalyst for the formation of the acid. A similar result has been noted for paper.

Oxidation/reduction reactions cause local chemical and physical differences on metal surfaces. These differences, in turn, result in the formation of microscopic anodes and cathodes. Electrochemical corrosion results from the potential that develops in these microscopic batteries.

Factors that influence deterioration. Moisture, temperature, sunlight, and position of the exposed material are among the more important factors that influence the rate of deterioration.

Moisture, in the form of humidity, is essential for most of the mechanisms of deterioration to occur. Metal corrosion does not appear to occur even at relatively high SO_2 pollution levels until the relative humidity exceeds 60 percent. On the other hand, humidities above 70 to 90 percent will promote corrosion without air pollutants. Rain reduces the effects of pollutant-induced corrosion by dilution and washing away of the pollutant.

Higher air temperatures generally result in higher reaction rates. However, when low air temperatures are accompanied by cooling of surfaces to the point where moisture condenses, then the rates may be accelerated.

In addition to the oxidation effect of its ultraviolet wave lengths, sunlight stimulates air pollution damage by providing the energy for pollutant formation and cyclic reformation. The cracking of rubber and the fading of dyes have been attributed to ozone produced by these photochemical reactions.

The position of the exposed surface influences the rate of deterioration in two ways. First, whether the surface is vertical or horizontal or at some angle affects deposition and wash-off rates. Second, whether the surface is an upper or lower one may alter the rate of damage. When the humidity is sufficiently high, the lower side usually deteriorates faster because rain does not remove the pollutants as efficiently.

Effects on Vegetation

Cell and leaf anatomy. Since the leaf is the primary indicator of the effects of air pollution on plants, we shall define some terms and explain how the leaf functions. A typical plant cell (Figure 6-3) has three main components: the cell wall, the protoplast, and the inclusions. Much like human skin, the cell wall is thin in young plants and gradually thickens with age. Protoplast is the term used to describe the protoplasm of one cell. It consists primarily of water, but it also includes protein, fat, and carbohydrates. The nucleus contains the hereditary material (DNA), which controls the operation of the cell. The protoplasm located outside the nucleus is called cytoplasm. Within the cytoplasm are tiny bodies or plastids. Examples include chloroplasts, leucoplasts, chromoplasts, and mitochondria. Chloroplasts contain the chlorophyll that manufactures the plant's food through photosynthesis. Leucoplasts convert starch into starch grains. Chromoplasts are responsible for the red, yellow, and orange colors of the fruit and flowers.

A cross section through a typical mature leaf (Figure 6-4) reveals three primary tissue systems: the epidermis, the mesophyll, and the vascular bundle (veins). Chloroplasts are usually not present in epidermal cells. The opening in the underside of the leaf is called a stoma. (The plural of stoma is stomata.) The mesophyll, which

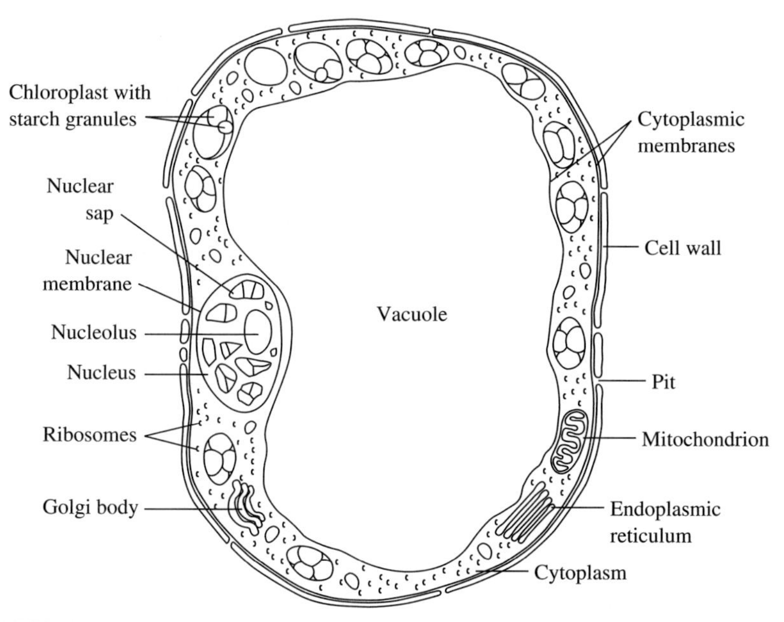

FIGURE 6-3
Typical plant cell. (*Source:* Adapted from H. J. Fuller, *The Plant World,* New York: Henry Holt and Company, 1960.)

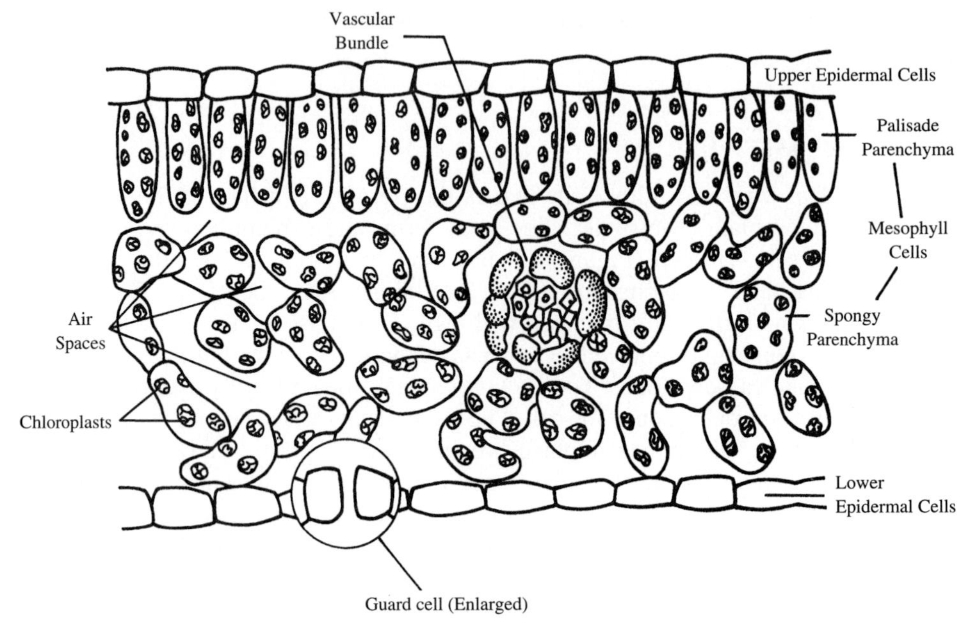

FIGURE 6-4
Cross section of intact leaf. (*Source:* I. Hindawi, *Air Pollution Injury to Vegetation.*)

includes both the palisade parenchyma and the spongy parenchyma, contains chloroplasts. It is the food production center. The vascular bundles carry water, minerals, and food throughout the leaf and to and from the main stem of the plant.

The guard cells regulate the passage of gases and water vapor in and out of the leaf. When it is hot, sunny, and windy, the processes of photosynthesis and respiration are increased. The guard cells open, which allows increased removal of water vapor that otherwise would accumulate because of the increased transport of water and minerals from the roots.

Pollutant damage. Ozone injures the palisade cells.[2] The chloroplasts condense and ultimately the cell walls collapse. This results in the formation of red-brown spots that turn white after a few days. The white spots are called fleck. Ozone injury appears to be the greatest during midday on sunny days. The guard cells are more likely to be open under these conditions and thus allow pollutants to enter the leaf.

Plant growth may be inhibited by continuous exposure to 0.5 ppm of NO_2. Levels of NO_2 in excess of 2.5 ppm for periods of four hours or more are required to produce *necrosis* (surface spotting due to plasmolysis or loss of protoplasm).

[2]I. Hindawi, *Air Pollution Injury to Plants* (U.S. Department of Health, Education and Welfare, National Air Pollution Control Administration Publication No. AP-71), Washington, DC: U.S. Government Printing Office, p. 13, 1971.

Sulfur dioxide injury is also typified by necrosis, but at much lower levels. A concentration of 0.3 ppm for eight hours is sufficient.[3] Lower levels for longer periods of exposure will produce a diffuse *chlorosis* (bleaching).

The net result of air pollutant damage goes beyond the apparent superficial damage to the leaves. A reduction in surface area results in less growth and small fruit. For commercial crops this results in a direct reduction in income for the farmer. For other plants the net result is likely to be an early death.

Fluoride deposition on plants not only causes them damage but may result in a second untoward effect. Grazing animals may accumulate an excess of fluoride that mottles their teeth and ultimately causes them to fall out.

Problems of diagnosis. Various factors make it difficult to diagnose actual air pollution damage. Droughts, insects, diseases, herbicide overdoses, and nutrient deficiencies all can cause injury that resembles air pollution damage. Also, combinations of pollutants that alone cause no damage are known to produce acute effects when combined.[4] This effect is known as *synergism.*

Effects on Health

Susceptible population. It is difficult at best to assess the effects of air pollution on human health. Personal pollution from smoking results in exposure to air pollutant concentrations far higher than the low levels found in the ambient atmosphere. Occupational exposure may also result in pollution doses far above those found outdoors. Tests on rodents and other mammals are difficult to interpret and apply to human anatomy. Tests on human subjects are usually restricted to those who would be expected to survive. This leads us to a question of environmental ethics. If the allowable concentration levels (standards) are based on results from tests on rodents, they would be rather high. If the allowable concentration levels must also protect those with existing cardiorespiratory ailments, they should be lower than those resulting from the observed effects on rodents.

We noted earlier that the air quality standards were established to protect public health with an "adequate margin of safety." In the opinion of the Administrator of the EPA, the standards must protect the most sensitive responders. Thus, as you will note in the following paragraphs, the standards have been set at the lowest level of observed effect. This decision has been attacked by some theorists. They say it would make better economic sense to build more hospitals.[5] However, one also might apply this kind of logic in establishing speed limits for highways, that is, raise the speed limit and build more hospitals, junk yards, and cemeteries!

Anatomy of the respiratory system. The respiratory system is the primary indicator of air pollution effects in humans. The major organs of the respiratory system are

[3]P. J. O'Gara, "Sulfur Dioxide and Fume Problems and Their Solutions," *Industrial Engineering Chemistry,* vol. 14, p. 744, 1922.

[4]I. Hindawi, *Air Pollution Injury to Vegetation.*

[5]Charles H. Connolly, *Air Pollution and Public Health,* New York: Holt, Rinehart & Winston, p. 7, 1972.

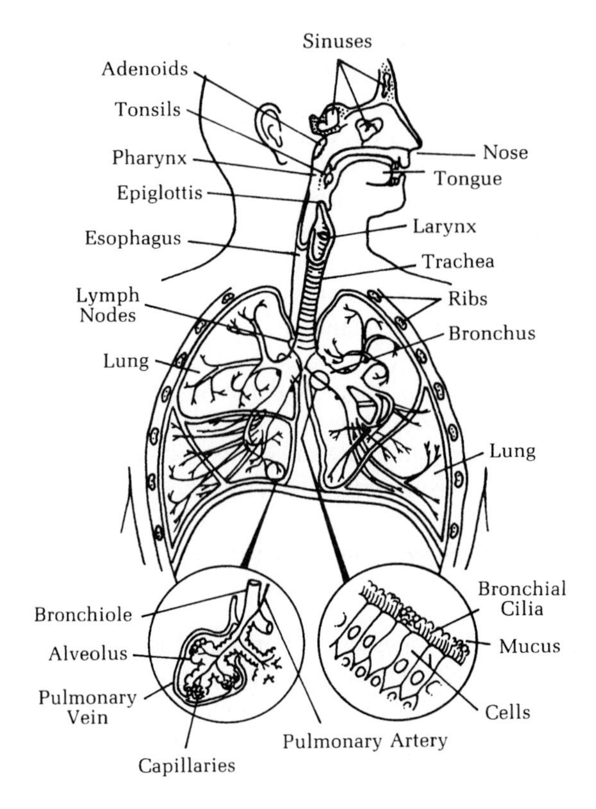

FIGURE 6-5
The respiratory system. (*Source: Air Pollution Primer.* Copyright 1988 by National Tuberculosis and Respiratory Diseases Association. Reprinted by permission.)

the nose, pharynx, larynx, trachea, bronchi, and lungs (Figure 6-5). The nose, pharynx, larynx, and trachea together are called the upper respiratory tract (URT). The primary effects of air pollution on the URT are aggravation of the sense of smell and inactivation of the sweeping motion of cilia, which remove mucus and entrapped particles. The lower respiratory tract (LRT) consists of the branching structures known as bronchi and the lung itself, which is composed of grape-like clusters of sacs called alveoli. The alveoli are approximately 300 μm in diameter. The walls of alveoli are lined with capillaries. Carbon dioxide diffuses through the capillary wall into the alveolus, while oxygen diffuses out of the alveolus into the blood cell. The difference in partial pressure of each of the gases causes it to move from the higher to lower partial pressure.

Inhalation and retention of particles. The degree of penetration of particles into the LRT is primarily a function of the size of the particles and the rate of breathing. Particles greater than 5 to 10 μm are screened out by the hairs in the nose. Sneezing also helps the screening process. Particles in the 1 to 2 μm size range penetrate to the alveoli. These particles are small enough to bypass screening and deposition in the URT, however they are big enough that their terminal settling velocity allows them to deposit where they can do the most damage. Particles that are 0.5 μm in diameter do not have a large enough terminal settling velocity to be removed efficiently. Smaller particles diffuse to the alveolar walls. Refer to Figure 6-2 and note that the size of "Lung Damaging Dust" falls in the critical particle size range.

Chronic respiratory disease. Several long-term diseases of the respiratory system are seriously aggravated by and perhaps may be caused by air pollution. Airway resistance is the narrowing of air passages because of the presence of irritating substances. The result is that breathing becomes difficult. Bronchial asthma is a form of airway resistance that results from an allergy. An asthma "attack" is the result of the narrowing of the bronchioles because of a swelling of the mucous membrane and a thickening of the secretions. The bronchioles return to normal after the attack. Chronic bronchitis is currently defined to be present in a person when excess mucus in the bronchioles results in a cough for three months a year for two consecutive years. Lung infections, tumors, and heart disease must be absent. Pulmonary emphysema is characterized by a breakdown of the alveoli. The small grape-like clusters become a large nonresilient balloon-like structure. The amount of surface area for gas exchange is reduced drastically. Cancer of the bronchus (lung cancer) is characterized by abnormal, disorderly new cell growth originating in the bronchial mucous membrane. The growth closes off the bronchioles. It is usually fatal.

Carbon monoxide (CO). This colorless, odorless gas is lethal to humans within a few minutes at concentrations exceeding 5,000 ppm. CO reacts with hemoglobin in the blood to form carboxyhemoglobin (COHb). Hemoglobin has a greater affinity for CO than it does for oxygen. Thus, the formation of COHb effectively deprives the body of oxygen. At COHb levels of 5 to 10 percent, visual perception, manual dexterity, and ability to learn are impaired. A concentration of 50 ppm of CO for eight hours will result in a COHb level of about 7.5 percent. At COHb levels of 2.5 to 3 percent, people with heart disease are not able to perform certain exercises as well as they might in the absence of COHb. A concentration of 20 ppm of CO for eight hours will result in a COHb level of about 2.8 percent.[6] (We should note here that the average concentration of CO inhaled in cigarette smoke is 200 to 400 ppm!) The sensitive populations are those with heart and circulatory ailments, chronic pulmonary disease, developing fetuses, and those with conditions that cause increased oxygen demand, such as fever.

Hazardous air pollutants (HAPs). Most of the information on the direct human health effects of hazardous air pollutants (also known as air toxics) comes from studies of industrial workers. Exposure to air toxics in the work place is generally much higher than in the ambient air. We know relatively little about the specific effects of the HAPs at the low levels normally found in ambient air.

The HAPs regulated under the NESHAP program were identified as causal agents for a variety of diseases. For example, asbestos, arsenic, benzene, coke oven emissions, and radionuclides may cause cancer. Beryllium primarily causes lung

---

[6]B. G. Ferris, "Health Effects of Exposure to Low Levels of Regulated Air Pollutants," *Journal of the Air Pollution Control Association,* vol. 28, pp. 482–497, 1978.

disease but also affects the liver, spleen, kidneys, and lymph glands. Mercury attacks the brain, kidneys, and bowels. Other potential effects from the HAPs are birth defects and damage to the immune and nervous systems.[7]

Lead (Pb). In contrast to the other major air pollutants, lead is a cumulative poison. A further difference is that it is ingested in food and water, as well as being inhaled. Of that portion taken by ingestion, approximately 5 to 10 percent is absorbed in the body. Between 20 and 50 percent of the inspired portion is absorbed. Those portions that are not absorbed are excreted in the feces and urine. Lead is measured in the urine and blood for diagnostic evidence of lead poisoning.

An early manifestation of acute lead poisoning is a mild anemia (deficiency of red blood cells). Fatigue, irritability, mild headache, and pallor indistinguishable from other causes of anemia occur when the blood level of lead increases to 60 to 120 μg/100 g of whole blood. Blood levels in excess of 80 μg/100 g result in constipation and abdominal cramps. When an acute exposure results in blood levels of lead greater than 120 μg/100 g, acute brain damage (encephalopathy) may result.[8] Such acute exposure results in convulsions, coma, cardiorespiratory arrest, and death. Acute exposures may occur over a period of one to three weeks.

Chronic exposure to lead may result in brain damage characterized by seizures, mental incompetence, and highly active aggressive behavior. Weakness of extensor muscles of the hands and feet and eventual paralysis may also result.

Atmospheric lead occurs as a particulate. The particle size range is between 0.16 and 0.43 μm. Nonsmoking residents of suburban Philadelphia exposed to approximately 1 $\mu g/m^3$ of lead in air have blood levels averaging 11 μg/100 g. Nonsmoking residents of downtown Philadelphia exposed to approximately 2.5 $\mu g/m^3$ of lead have blood levels averaging 20 μg/100 g.[9]

Nitrogen dioxide (NO_2). Exposure to NO_2 concentrations above 5 ppm for 15 minutes results in cough and irritation of the respiratory tract. Continued exposure may produce an abnormal accumulation of fluid in the lung (pulmonary edema). The gas is reddish brown in concentrated form and gives a brownish yellow tint at lower concentrations. At 5 ppm it has a pungent sweetish odor. The average NO_2 concentration in tobacco smoke is approximately 5 ppm. Slight increases in respiratory illness and decrease in pulmonary function have been associated with concentrations of about 0.10 ppm.[10]

---

[7] A. S. Kao, "Formation and Removal Reactions of Hazardous Air Pollutants," *Journal of the Air and Waste Management Association,* 44, pp. 683–696, 1994.

[8] Robert A. Goyer and J. Julian Chilsolm, "Lead," *Metallic Contaminants and Human Health,* Douglas H. K. Lee, ed., New York: Academic Press, pp. 57–95, 1972.

[9] U.S. Department of Health, Education and Welfare, *Survey of Lead in the Atmosphere of Three Urban Communities* (U.S. Public Health Service Publication No. 999-AP-12), Washington, DC: U.S. Government Printing Office, 1965.

[10] The discussion on nitrogen dioxide, photochemical oxidants, sulfur oxides, and total suspended particulates follows B. G. Ferris, "Health Effects of Exposure to Low Levels of Regulated Air Pollutants," *Journal of the Air Pollution Control Association,* pp. 485–491.

Photochemical oxidants. Although the photochemical oxidants include peroxyacetyl nitrate (PAN), acrolein, peroxybenzoyl nitrates (PBzN), aldehydes, and nitrogen oxides, the major oxidant is ozone (O_3). Ozone is commonly used as an indicator of the total amount of oxidant present. Oxidant concentrations above 0.1 ppm result in eye irritation. At a concentration of 0.3 ppm, cough and chest discomfort are increased. Those people who suffer from chronic respiratory disease are particularly susceptible.

PM_{10}. As noted earlier, large particles are not inhaled deeply into the lungs. This is why EPA switched from an air quality standard based on total suspended matter to one based on particles with an aerodynamic diameter less than 10 μm (PM_{10}). Studies in the United States, Brazil, and Germany have related higher levels of particulates to increased risk of respiratory, cardiovascular, and cancer-related deaths, as well as pneumonia, lung function loss, hospital admissions, and asthma.[11]

Recent investigations have pointed toward particle sizes smaller than 2.5 μm as a major contributor to elevated death rates in polluted cities.[12] These epidemiological investigations are based on statistical correlations and at this time have not found wide acceptance in the scientific community. This is primarily because of the lack of a suitable biological mechanism to explain what causes the deaths. Nonetheless, in order to provide an "adequate margin of safety" the EPA Administrator proposed a new standard based on $PM_{2.5}$ in July 1997.

Sulfur oxides (SO_x) and total suspended particulates (TSP). The sulfur oxides include sulfur dioxide (SO_2), sulfur trioxide (SO_3), their acids, and the salts of their acids. Rather than try to separate the effects of SO_2 and SO_3, they are usually treated together. There is speculation that a definite synergism exists whereby fine particulates carry absorbed SO_2 to the LRT. The SO_2 in the absence of particulates would be absorbed in the mucous membranes of the URT.

Patients suffering from chronic bronchitis have shown an increase in respiratory symptoms when the TSP levels exceeded 350 $\mu g/m^3$ and the SO_2 level was above 0.095 ppm. Studies made in Holland at an interval of three years showed that pulmonary function improved as SO_2 and TSP levels dropped from 0.10 ppm and 230 $\mu g/m^3$ to 0.03 ppm and 80 $\mu g/m^3$, respectively.

Air pollution episodes. The word *episode* is used as a refined form of the word *disaster.*[13] Indeed, it was the shock of these disasters that stimulated the first modern legislative action to require control of air pollutants. The characteristics of the three major episodes are summarized in Table 6-2. Careful study of the table will reveal that all of the episodes had some things in common. Comparison of these situations and others where no episode occurred (that is, where the number of dead and ill was

---

[11]T. Reichhardt, "Weighing the Health Risks of Airborne Particulates," *Environmental Science and Technology,* 29, pp. 360A–364A, 1995.

[12]C. A. Pope, et al., *American Journal of Respiratory and Critical Care Medicine,* 151, pp. 669–674, 1995.

[13]In the nuclear power business, they would call it an "incident."

TABLE 6-2
Three major air pollution episodes

	Meuse Valley, 1930 (Oct. 1–5)	Donora, 1948 (Oct. 26–31)	London, 1952 (Dec. 5–9)
Population	No data	12,300	8,000,000
Weather	Anticyclone, inversion, and fog	Anticyclone, inversion, and fog	Anticyclone, inversion, and fog
Topography	River valley	River valley	River plain
Most probable source of pollutants	Industry (including steel and zinc plants)	Industry (including steel and zinc plants)	Household coal-burning
Nature of the illnesses	Chemical irritation of exposed membranous surfaces	Chemical irritation of exposed membranous surfaces	Chemical irritation of exposed membranous surfaces
No. of deaths	63	17	4,000
Time of deaths	Began after second day of episode	Began after second day of episode	Began on first day of episode
Suspected proximate cause of irritation	Sulfur oxides with particulates	Sulfur oxides with particulates	Sulfur oxides with particulates

(*Source:* World Health Organization, *Air Pollution*, 1961, p. 180.)

considerably less) has revealed that four ingredients are essential for an episode. If one ingredient is omitted, fewer people will get sick and only a few people can be expected to die. The crucial ingredients are: (1) a large number of pollution sources, (2) a restricted air volume, (3) failure of officials to recognize that anything is wrong, and (4) the presence of water droplets of the "right" size.[14]

Although a sufficient quantity of any pollutant is lethal by itself, it is generally agreed that some mix is required to achieve the results seen in these episodes. Atmospheric levels of individual pollutants seldom rise to lethal levels without an explosion or transportation accident. However, the proper combination of two or more pollutants will yield untoward symptoms at much lower levels. The sulfur oxides and particulates were the most suspect in the three major episodes.

The meteorology must be such that there is little air movement. Thus, the pollutants cannot be diluted. Although a valley is most conducive to a stagnation effect, the London episode proved that it isn't necessary. The stagnant conditions must persist for several days. Three days appears to be the minimum.

Tragically, each of these hazardous air pollution conditions became lethal because of the failure of city officials to notice anything strange. If they have no measurements of pollution levels or reports from hospitals and morgues, city authorities have no reason to alert the public, shut down factories, or restrict traffic.

[14] J. R. Goldsmith, "Effects of Air Pollution on Human Health," *Air Pollution,* A. C. Stern, ed., New York: Academic Press, pp. 554–557, 1968.

The last and, perhaps, most crucial element is fog.[15] The fog droplets must be of the "right" size, namely, in the 1 to 2 μm diameter range or, perhaps, in the range below 0.5 μm. As mentioned earlier, these particle sizes are most likely to penetrate into the LRT. Pollutants that dissolve into the fog droplet are thus carried deep into the lung and deposited there.

6-5 ORIGIN AND FATE OF AIR POLLUTANTS

Carbon Monoxide

Incomplete oxidation of carbon results in the production of carbon monoxide. The natural anaerobic decomposition of carbonaceous material by soil microorganisms releases approximately 9×10^{15} moles of methane (CH_4) to the atmosphere each year worldwide.[16] The natural formation of CO results from an intermediate step in the oxidation of the methane. The hydroxyl radical ($OH\cdot$) serves as the initial oxidizing agent. It combines with CH_4 to form an alkyl radical:[17]

$$CH_4 + OH\cdot \rightarrow CH_3\cdot + H_2O \qquad (6\text{-}8)$$

This reaction is followed by a complex series of 39 reactions, which we have oversimplified to the following:

$$CH_3\cdot + O_2 + 2(h\nu) \rightarrow CO + H_2 + OH\cdot \qquad (6\text{-}9)$$

This says that $CH_3\cdot$ and O_2 are each zapped by a photon of light energy ($h\nu$). The symbol ν stands for the frequency of the light. The h is Planck's constant $= 6.626 \times 10^{-34}$ J/Hz.

Anthropogenic sources (those associated with the activities of human beings) include motor vehicles, fossil fuel burning for electricity and heat, industrial processes, solid waste disposal, and miscellaneous burning of such things as leaves and brush. Approximately 1×10^{13} moles of CO are released by these sources. Motor vehicles account for more than 60 percent of the emission.

No significant change in the global atmospheric CO level has been observed over the past 20 years. Yet the worldwide anthropogenic contribution of combustion sources has doubled over the same time period. Since there is no apparent change in the atmospheric concentration, a number of mechanisms (sinks) have been proposed

---

[15]The word "smog" is a term coined by Londoners before World War I to describe the combination of smoke and fog that accounted for much of their weather. Los Angeles smog is a misnomer since little smoke and no fog is present. In fact, as we shall see later, Los Angeles smog cannot occur without a lot of sunshine. "Photochemical smog" is the correct term to describe the Los Angeles haze.

[16]John H. Seinfeld, *Air Pollution, Physical and Chemical Fundamentals,* New York, McGraw-Hill, p. 71, 1975.

[17]S. C. Wofsy, J. C. McConnell, and M. B. McElroy, "Atmospheric CH_4, CO and CO_2," *Journal of Geophysical Research,* vol. 67, pp. 4477–4493, 1972.

to account for the missing CO. The two most probable are

1. Reaction with hydroxyl radicals to form carbon dioxide
2. Removal by soil microorganisms

It has been estimated that these two sinks annually consume an amount of CO that just equals the production.[18]

Hazardous Air Pollutants (HAPs)

The EPA has identified 166 categories of major sources and 8 categories of area sources for the HAPs listed in Table 1-5.[19] The source categories represent a wide range of industrial groups: fuel combustion, metal processing, petroleum and natural gas production and refining, surface coating processes, waste treatment and disposal processes, agricultural chemicals production, and polymers and resins production. There are also a number of miscellaneous source categories, such as dry cleaning and electroplating.

In addition to these direct emissions, air toxics can result from chemical formation reactions in the atmosphere. These reactions involve chemicals emitted to the atmosphere that are not listed HAPs and may not be toxic themselves, but can undergo atmospheric transformations to generate HAPs. For organic compounds present in the gas phase, the most important transformation processes involve photolysis and chemical reactions with ozone, hydroxyl radicals ($OH\cdot$), and nitrate radicals.[20] *Photolysis* is the chemical fragmentation or rearrangement of a chemical upon the adsorption of radiation of the appropriate wavelength. Photolysis is only important during the daytime for those chemicals that absorb strongly within the solar radiation spectrum. Otherwise, reaction with $OH\cdot$ or O_3 is likely to predominate. The HAPs most often formed are formaldehyde and acetaldehyde.

The major removal mechanisms appear to be OH abstraction or addition. The reaction products lead to the formation of CO and CO_2. Eighty-nine of the 189 HAPs have atmospheric lifetimes of less than one day.

Lead

Volcanic activity and airborne soil are the primary natural sources of atmospheric lead. Smelters and refining processes, as well as incineration of lead-containing wastes, are major point sources of lead. Approximately 70 to 80 percent of the lead which used to be added to gasoline was discharged to the atmosphere.

---

[18] John H. Seinfeld, *Air Pollution, Physical and Chemical Fundamentals,* p. 71.

[19] Federal Register, "Initial List of Categories of Sources Under Section 112(c)(1) of the Clean Air Act Amendments of 1990," Federal Register, 57, p. 31576, 1992.

[20] A. S. Kao, "Formation and Removal Reactions of Hazardous Air Pollutants," *Journal of the Air and Waste Management Association,* 44, pp. 683–696, 1994.

Submicron lead particles, which are formed by volatilization and subsequent condensation, attach to larger particles or they form nuclei before they are removed from the atmosphere. Once they have attained a size of several microns, they either settle out or are washed out by rain.

Nitrogen Dioxide

Bacterial action in the soil releases nitrous oxide (N_2O) to the atmosphere. In the upper troposphere and stratosphere, atomic oxygen reacts with the nitrous oxide to form nitric oxide (NO).

$$N_2O + O \rightarrow 2NO \tag{6-10}$$

The atomic oxygen results from the dissociation of ozone. The nitric oxide further reacts with ozone to form nitrogen dioxide (NO_2).

$$NO + O_3 \rightarrow NO_2 + O_2 \tag{6-11}$$

The global formation of NO_2 by this process is estimated to be 0.45 Pg annually.[21]

Combustion processes account for 96 percent of the anthropogenic sources of nitrogen oxides. Although nitrogen and oxygen coexist in our atmosphere without reaction, their relationship is much less indifferent at high temperatures and pressures. At temperatures in excess of 1,600 K, they react.

$$N_2 + O_2 \overset{\Delta}{\rightleftharpoons} 2NO \tag{6-12}$$

If the combustion gas is rapidly cooled after the reaction by exhausting it to the atmosphere, the reaction is quenched and NO is the byproduct. The NO in turn reacts with ozone or oxygen to form NO_2. The anthropogenic contribution to global emissions of NO_2 by this route amounted to 0.48 Tg in 1965.[22]

Ultimately, the NO_2 is converted to either NO_2^- or NO_3^- in particulate form. The particulates are then washed out by precipitation. The dissolution of nitrate in a water droplet allows for the formation of nitric acid (HNO_3). This, in part, accounts for "acid" rain found downwind of industrialized areas.

Photochemical Oxidants

Unlike the other pollutants, the photochemical oxidants result entirely from atmospheric reactions and are not directly attributable to either people or nature. Thus, they are called *secondary pollutants.* They are formed through a series of reactions that are initiated by the absorption of a photon by an atom, molecule, free radical, or ion. Ozone is the principal photochemical oxidant. Its formation is usually attributed to the nitrogen dioxide photolytic cycle. Hydrocarbons modify this cycle by reacting with atomic oxygen to form free radicals (highly reactive organic species). The

---

[21] John H. Seinfeld, *Air Pollution, Physical and Chemical Fundamentals,* p. 73.

[22] John H. Seinfeld, *Air Pollution, Physical and Chemical Fundamentals,* p. 73.

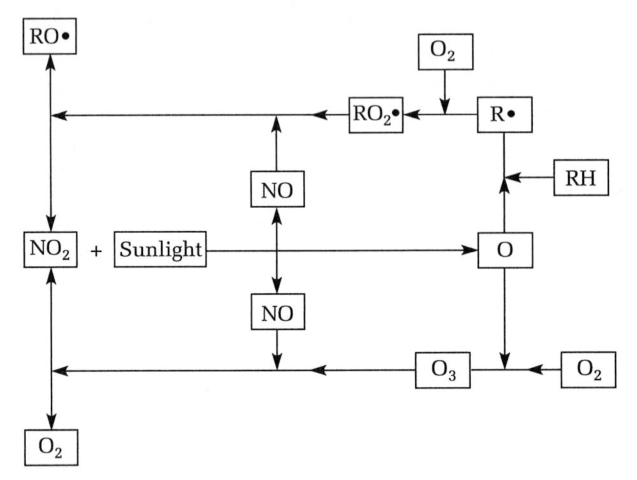

FIGURE 6-6
Interaction of hydrocarbons with atmospheric nitrogen oxide photolytic cycle. [*Source: Air Quality Criteria for Photochemical Oxidants* (U.S. Department of Health, Education and Welfare, National Air Pollution Control Administration Publication No. AP-63), Washington, DC: U.S. Government Printing Office, 1970.]

hydrocarbons, nitrogen oxides, and ozone react and interact to produce more nitrogen dioxide and ozone. This cycle is represented in summary form in Figure 6-6. The whole reaction sequence depends on an abundance of sunshine. A result of these reactions is the photochemical "smog" for which Los Angeles is famous.

Sulfur Oxides

Sulfur oxides may be both primary and secondary pollutants. Power plants, industry, volcanoes, and the oceans emit SO_2, SO_3, and SO_4^{2-} directly as primary pollutants. In addition, biological decay processes and some industrial sources emit H_2S, which is oxidized to form the secondary pollutant SO_2. In terms of sulfur, approximately 0.125 Pg are emitted annually by natural sources. Approximately 45 Tg of sulfur may be attributed to anthropogenic sources each year.[23]

The most important oxidizing reaction for H_2S appears to be one involving ozone:

$$H_2S + O_3 \rightarrow H_2O + SO_2 \tag{6-13}$$

The combustion of fossil fuels containing sulfur yields sulfur dioxide in direct proportion to the sulfur content of the fuel:

$$S + O_2 \rightarrow SO_2 \tag{6-14}$$

[23]John H. Seinfeld, *Air Pollution, Physical and Chemical Fundamentals,* p. 57.

This reaction implies that for every gram of sulfur in the fuel, two grams of SO_2 are emitted to the atmosphere. Because the combustion process is not 100 percent efficient, we generally assume that 5 percent of the sulfur in the fuel ends up in the ash, that is, 1.90 g SO_2 per gram of sulfur in the fuel is emitted.

Example 6-2. An Illinois coal is burned at a rate of 1.00 kg per second. If the analysis of the coal reveals a sulfur content of 3.00 percent, what is the annual rate of emission of SO_2?

Solution. Using the mass-balance approach, we begin by drawing a mass-balance diagram:

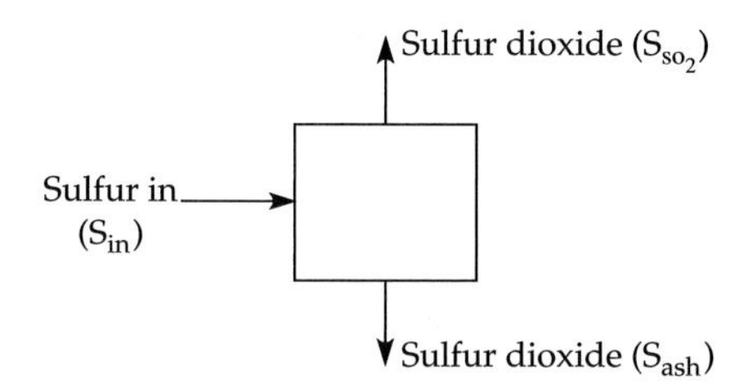

The mass balance equation may be written as

$$S_{in} = S_{ash} + S_{SO_2}$$

From the problem data, the mass of "sulfur in" is

$$S_{in} = 1.00 \text{ kg/s} \times 0.030 = 0.030 \text{ kg/s}$$

In one year,

$$S_{in} = 0.030 \text{ kg/s} \times 86{,}400 \text{ s/d} \times 365 \text{ d/y} = 9.46 \times 10^5 \text{ kg/y}$$

The sulfur in the ash is 5 percent of the input sulfur:

$$S_{ash} = (0.05)(9.46 \times 10^5 \text{ kg/y}) = 4.73 \times 10^4 \text{ kg/y}$$

The amount of sulfur available for conversion to SO_2:

$$S_{SO_2} = S_{in} - S_{ash} = 9.46 \times 10^5 - 4.73 \times 10^4 = 8.99 \times 10^5 \text{ kg/y}$$

The amount of sulfur formed is determined from the proportional weights of the oxidation reaction (Equation 6-14):

$$S + O_2 \rightarrow SO_2$$

$$\text{GMW} = 32 + 32 = 64$$

The amount of sulfur dioxide formed is then 64/32 of the sulfur available for conversion:

$$S_{SO_2} = \frac{64}{32}(8.99 \times 10^5 \text{ kg/y}) = 1.80 \times 10^6 \text{ kg/y}$$

The ultimate fate of most of the SO_2 in the atmosphere is conversion to sulfate salts, which are removed by sedimentation or by washout with precipitation. The conversion to sulfate is by either of two routes: catalytic oxidation or photochemical oxidation. The first process is most effective if water droplets containing Fe^{3+}, Mn^{2+}, or NH_3 are present:

$$2SO_2 + 2H_2O + O_2 \xrightarrow{\text{catalyst}} 2H_2SO_4 \tag{6-15}$$

At low relative humidities, the primary conversion process is photochemical oxidation. The first step is photoexcitation of the SO_2.[24]

$$SO_2 + h\nu \rightarrow \overset{*}{SO_2} \tag{6-16}$$

The excited molecule then readily reacts with O_2 to form SO_3:

$$\overset{*}{SO_2} + O_2 \rightarrow SO_3 + O \tag{6-17}$$

The trioxide is very hygroscopic and consequently is rapidly converted to sulfuric acid:

$$SO_3 + H_2O \rightarrow H_2SO_4 \tag{6-18}$$

This reaction in large part accounts for acid rain (that is, precipitation with a pH value less than 5.6) found in industrialized areas. Normal precipitation has a pH of 5.6 due to the carbonate buffer system.

Particulates

Sea salt, soil dust, volcanic particles, and smoke from forest fires account for 1.404 Pg of particulate emissions each year. Anthropogenic emissions from fossil fuel burning and industrial processes account for emissions of 92 Tg per year.[25] Secondary sources of particulates include the conversion of H_2S, SO_2, NO_x, NH_3, and hydrocarbons. H_2S and SO_2 are converted to sulfates. NO_x and NH_3 are converted to nitrates. The hydrocarbons react to form products that condense to form particles at atmospheric temperatures. Natural sources of secondary pollutants yield about 1.099 Pg annually. Anthropogenic sources yield about 0.204 Pg annually.[26]

Dust particles that are entrained (picked up) by the wind and carried over long distances tend to sort themselves out to the sizes between 0.5 and 50 μm in diameter. Sea salt nuclei have sizes between 0.05 and 0.5 μm. Particles formed as a result of photochemical reactions tend to have very small diameters ($< 0.4\ \mu$m). Smoke and fly ash particles cover a wide range of sizes from 0.05 to 200 μm or more.

---

[24]Photoexcitation is the displacement of an electron from one shell to another, thereby storing energy in the molecule. Photoexcitation is represented in reactions by an asterisk.

[25]John H. Seinfeld, *Air Pollution, Physical and Chemical Fundamentals,* p. 83.

[26]John H. Seinfeld, *Air Pollution, Physical and Chemical Fundamentals,* p. 83.

Particle mass distributions in urban atmospheres generally exhibit two maxima. One is between 0.1 and 1 μm in diameter. The other is between 1 and 30 μm. The smaller fraction is the result of condensation. The coarse fraction consists of fly ash and dust generated by mechanical abrasion.

Small particles are removed from the atmosphere by accretion to water droplets, which grow in size until they are large enough to precipitate. Larger particles are removed by direct washout by falling raindrops.

6-6 MICRO AND MACRO AIR POLLUTION

Air pollution problems may occur on three scales: micro, meso, and macro. Micro-scale problems range from those covering less than a centimeter to those the size of a house or slightly larger. Meso-scale air pollution problems are those of a few hectares up to the size of a city or county. Macro-scale problems extend from counties to states, nations, and in the broadest sense, the globe. Much of the remaining discussion in this chapter is focused on the meso-scale problem. In this section we will address the general micro-scale and macro-scale problems recognized today.

Indoor Air Pollution

People who live in cold climates may spend from 70 to 90 percent of their time indoors. In the last two decades, researchers have become interested in identifying sources, concentrations, and impacts of air pollutants that arise in conventional domestic residences. The startling results indicate that, in certain instances, indoor air may be substantially more polluted than the outdoor air.

Carbon monoxide from improperly operating furnaces has long been a serious concern. In numerous instances, people have died from furnace malfunction. More recently, chronic low levels of CO pollution have been recognized. Gas ranges, ovens, pilot lights, gas and kerosene space heaters, and cigarette smoke all contribute (Table 6-3). Although little or no effort has been exerted to reduce or eliminate the danger from ranges, ovens, etc., the public has come to expect that the recreational habits of smokers should not interfere with the quality of the air others breathe. The results of a general ban on cigarette smoking in one office are shown in Table 6-4. Smokers were allowed to smoke only in the designated lounge area. Period 1 was prior to the implementation of the new policy. It is obvious that the new policy had a positive effect outside of the lounge.

Nitrogen oxide sources are also shown in Table 6-3. NO_2 levels have been found to range from 70 $\mu g/m^3$ in air-conditioned houses with electric ranges to 182 $\mu g/m^3$ in non-air-conditioned houses with gas stoves.[27] The latter value is quite high in comparison to the national ambient air quality limits. SO_2 levels were found to be very low in all houses investigated. On the other hand, respirable particulate

[27] R. Hosein, *et al.*, "The Relationship Between Pollutant Levels in Homes and Potential Sources," *Transactions, Indoor Air Quality in Cold Climates, Hazards and Abatement Measures,* Pittsburgh: Air Pollution Control Association, pp. 250–260, April 1985.

TABLE 6-3
Tested combustion sources and their emission rates

	Range of emission rates,[a] mg/MJ				
Source	**NO**	**NO_2**	**NO_x (as NO_2)**	**CO**	**SO_2**
Range-Top Burner[b]	15–17	9–12	32–37	40–244	—[c]
Range Oven[d]	14–29	7–13	34–53	12–19	—
Pilot Light[e]	4–17	8–12	[f]	40–67	—
Gas Space Heaters[g]	0–15	1–15	1–37	14–64	—
Gas Dryer[h]	8	8	20	69	—
Kerosene Space Heaters[i]	1–13	3–10	5–31	35–64	11–12
Cigarette Smoke[j]	2.78[j]	0.73[j]	[f]	88.43[j]	—

[a]The lowest and highest mean values of emission rates for combustion sources tested in milligrams per mega-Joule (mg/MJ). Note: it takes 4.186 Joules to raise the temperature of 1.0 g of water from 14.5° C to 15.5° C at 100% efficiency.

[b]Three ranges were evaluated. Reported results are for blue flame condition.

[c]Dash (–) means combustion source is not emitting the pollutant.

[d]Three ranges were evaluated. Ovens were operated for several different settings (bake, broil, self-clean cycle, etc.).

[e]One range was evaluated with all three pilot lights, two top pilots, and a bottom pilot.

[f]Emission rates not reported.

[g]Three space heaters including one convective, radiant, and catalytic were tested.

[h]One gas dryer was evaluated.

[i]Two kerosene heaters including a convective and radiant type were tested.

[j]One type of cigarette. Reported emission rates are in mg/cigarette (800 mg tobacco/cigarette).

[*Source:* D. J. Moschandreas, *et al., Characterization of Emissions from Indoor Combustion Sources,* (Gas Research Institute Report No. 85/0075), Chicago, IL, May 1985.]

TABLE 6-4
Mean respirable particulates (RSP), CO and CO_2 levels measured on the test floor

	RSP ($\mu g/m^3$)	**CO (ppm)**	**CO_2 (ppm)**
Period 1			
Floor	26	1.67	624
Lounge	51	1.98	642
Period 2			
Floor	18	1.09	569
Lounge	189	2.40	650

Source: H. K. Lee, *et al.,* "Impact of a New Smoking Policy on Office Air Quality," *Transactions of the Air Pollution Control Association:* Pittsburgh, PA. *Indoor Air Quality in Cold Climates,* April 1985.

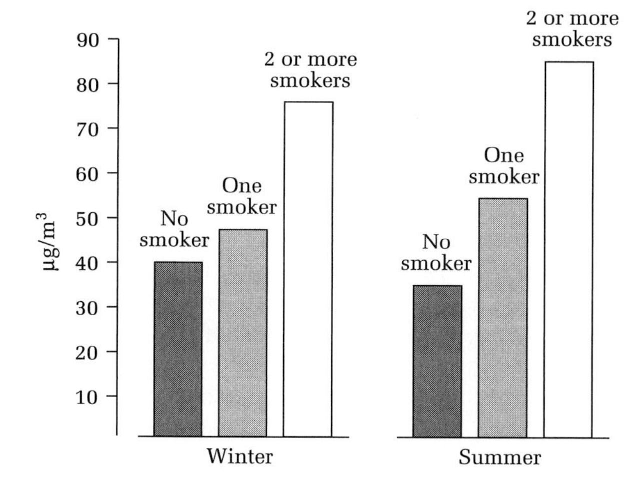

FIGURE 6-7
Respirable suspended particulate levels as a function of smoking. (*Source:* R. Hosein, *et al.*, "The Relationship Between Pollutant Levels in Homes and Potential Sources," *Transactions, Indoor Air Quality in Cold Climates, Hazards and Abatement Measures,* Pittsburgh: Air Pollution Control Association, pp. 250–260, April, 1985.)

matter (RSP) was found to increase with one smoker and to rise dramatically with two (Figure 6-7).

Radon is not regulated as an ambient air pollutant but has been found in dwellings at alarmingly high concentrations. We will address the radon issue in depth in Chapter 10. Suffice it to say at this juncture that radon is a radioactive gas that emanates from natural geologic formations and, in some cases, from construction materials. It is not generated from the activities of the householder as the pollutants discussed above.

Over 300 *volatile organic compounds* (VOCs) have been identified in indoor air. Aldehydes, alkanes, alkenes, ethers, ketones, and polynuclear aromatic hydrocarbons (PAHs) are among them. Although they are not all present all the time, frequently there are several present at the same time. Formaldehyde (CH_2O) has been singled out as one of the more prevalent, as well as one of the more toxic, compounds.[28]

Formaldehyde may not be generated directly by the activity of the homeowner. It is emitted by a variety of consumer products and construction materials including pressed wood products, insulation materials (urea-formaldehyde foam insulation in trailers has been particularly suspect), textiles, and combustion sources. In one study, CH_2O concentrations ranged from 0.0455 ppm to 0.19 ppm.[29] Some mobile homes in Wisconsin had concentrations as high as 0.65 ppm. (For comparison, the American Society of Heating, Refrigeration and Air Conditioning Engineers (ASHRAE) set a guideline concentration of 0.1 ppm.)

---

[28] A. L. Hines, *et al., Indoor Air Quality & Control,* Upper Saddle River, NJ: PTR Prentice Hall, p. 21, 1993.

[29] R. S. Dumont, "The Effect of Mechanical Ventilation on Rn, NO_2, and CH_2O Concentrations in Low-Leakage Houses and a Simple Remedial Measure for Reducing Rn Concentration," *Transactions, Indoor Air Quality in Cold Climates, Hazards and Abatement Measures,* Pittsburgh, PA: Air Pollution Control Association, pp. 90–104, April 1985.

Unlike the other air pollution sources that continue to emit as long as there is anthropogenic activity (or in the case of radon, for geologic time), CH_2O is not regenerated unless new materials are brought into the residence. If the house is ventilated over a period of time, the concentration will drop.

The primary source of heavy metals indoors is from infiltration of outdoor air and soil and dust that is tracked into the building. Arsenic, cadmium, chromium, mercury, lead, and nickel have been measured in indoor air. Lead and mercury may be generated from indoor sources such as paint. Old lead paint is a source of particulate lead as it is abraded or during removal. Mercury vapor is emitted from latex-based paints that contain diphenyl mercury dodecenyl succinate to prevent fungus growth.

Indoor tobacco smoking is of particular concern because of the increasing evidence of the carcinogenic properties of the smoke. While *mainstream smoking* (taking a puff) exposes the smoker to large quantities of toxic compounds, the smoldering cigarette in the ashtray (*sidestream smoke*) also adds a considerable burden to the room environment. Table 6-5 illustrates the emission rates of mainstream and sidestream smoke from a cigarette.

TABLE 6-5
Emission of chemicals from mainstream and sidestream smoke

Chemicals	Mainstream (μg/cigarette)	Sidestream (μg/cigarette)
Gas and Vapor Phase		
Carbon monoxide	1000–20,000	25,000–50,000
Carbon dioxide	20,000–60,000	160,000–480,000
Acetaldehyde	18–1400	40–3100
Hydrogen cyanide	430	110
Methyl chloride	650	1300
Acetone	100–600	250–1500
Ammonia	10–150	980–150,000
Pyridine	9–93	90–930
Acrolein	25–140	55–130
Nitric oxide	10–570	2300
Nitrogen dioxide	0.5–30	625
Formaldehyde	20–90	1300
Dimethylnitrosamine	10–65	520–3300
Nitrosopyrolidine	10–35	270–945
Particulates		
Total suspended particles	36,200	25,800
Nicotine	100–2500	2700–6750
Total phenols	228	603
Pyrene	50–200	180–420
Benzo (a) pyrene	20–40	68–136
Naphthalene	2.8	4.0
Methyl naphthalene	2.2	60
Aniline	0.36	16.8
Nitrosonornicotine	0.1–0.55	0.5–2.5

Source: A. L. Hines, et al., *Indoor Air Quality & Control,* Upper Saddle River, NJ: Prentice Hall, 1993. Reprinted by permission.

It is doubtful that there will be any regulatory effort to reduce the emissions of these pollutants in the near future. Thus the house or apartment dweller has little recourse other than to replace gas appliances, remove or cover formaldehyde sources, and put out the smokers.

Acid Rain

Unpolluted rain is naturally acidic because CO_2 from the atmosphere dissolves to a sufficient extent to form carbonic acid (see Section 3-1). The equilibrium pH for pure rainwater is about 5.6. Measurements taken over North America and Europe have revealed lower pH values. In some cases individual readings as low as 3.0 have been recorded. The average pH in rain weighted by the amount of precipitation over the United States and lower Canada in 1985 is shown in Figure 4-15.

Chemical reactions in the atmosphere convert SO_2, NO_x, and volatile organic compounds (VOCs) to acidic compounds and associated oxidants (Figure 6-8). The primary conversion of SO_2 in the eastern United States is through the aqueous phase reaction with hydrogen peroxide (H_2O_2) in clouds. Nitric acid is formed by the reaction of NO_2 with OH radicals formed photochemically. Ozone is formed and then protected by a series of reactions involving both NO_x and VOCs.

The concern about acid rain relates to potential effects of acidity on aquatic life, damage to crops and forests, and damage to building materials. Lower pH values may affect fish directly by interfering with reproductive cycles or by releasing otherwise insoluble aluminum, which is toxic. Dramatic dieback of trees in Central Europe has stimulated concern that similar results could occur in North America. It is hypothesized that the acid rain leaches calcium and magnesium from the soil (see Figure 3-12). This lowers the molar ratio of calcium to aluminum which, in turn, favors the uptake of aluminum by the fine roots, which ultimately leads to their deterioration.

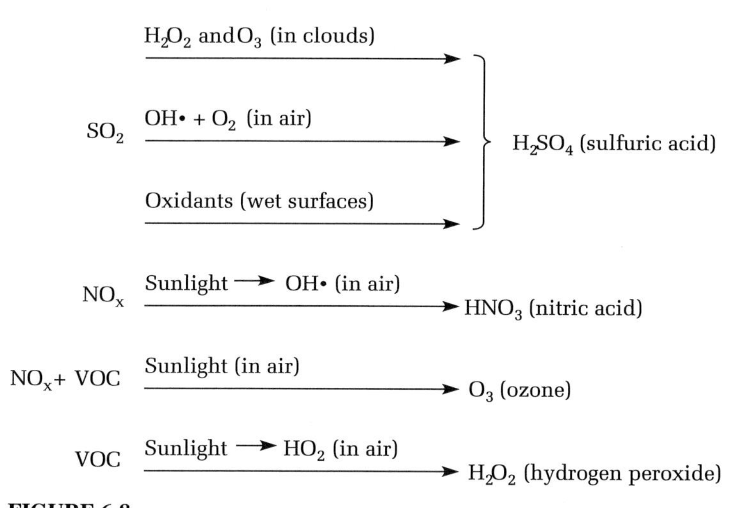

FIGURE 6-8
Acid rain precursors and products.

In 1980, Congress authorized a 10-year study to assess the causes and effects of acidic deposition. This study was titled the National Acid Precipitation Assessment Program (NAPAP). In September 1987, the NAPAP released an interim report that indicated that acidic precipitation appeared to have no measurable and consistent effect on crops, tree seedlings, or human health, and that a small percentage of lakes across the United States were experiencing pH values lower than 5.0.[30] On the other hand, oxidant damage was measurable. The NAPAP reported to Congress in 1992 that there remains no evidence of widespread decline of forest tree species caused by acidic deposition.[31] Eighty-one selected sites in the Northeast and Upper Midwest have been monitored for acidity since the early 1980s. Most of the lakes and streams experienced no measurable changes in acidity. Sulfate levels decreased while nitrate levels increased. The widespread decrease in sulfate concentration parallels the general decrease in national emissions of sulfur dioxide since 1980. However, acidic deposition has been firmly implicated as a causal factor in northeastern high-elevation red spruce decline. The effect is to reduce midwinter cold tolerance by 4° to 10°C. A major new finding was that some watersheds in regions receiving high nitrogen deposition are becoming nitrogen saturated, but there is no direct evidence of the source of the nitrate leaching from these watersheds.

While these conclusions appear to be reassuring, many uncertainties remain. These include questions about the distribution and type of natural and anthropogenic VOCs; relative importance of air pollutants versus natural stresses in causing the decline of forests above the cloud base; probability and percentage of lakes likely to become more acidic at present levels of air pollution emissions over the next few decades; importance of acidic episodes on fish populations; and possible long-term health effects of acidic aerosols.

Approximately 70 percent of the SO_2 emissions in the United States are attributable to electric utilities. In order to decrease the SO_2 emissions, the Congress developed a two-phase control program under Title IV of the 1990 Clean Air Act Amendments. Phase I sets emission allowances for 110 of the largest emitters in the Eastern half of the U.S. Phase II will include smaller utilities. The utilities may buy or sell allowances. Each allowance is equal to about 1 Mg of SO_2 emissions. If a company does not expend its maximum allowance, it may sell it to another company. This program is called a *market-based system.* As a result of this program utility emissions have decreased by 9 million Mg. In May 1993 NAPAP reported a 10 percent decrease in SO_2, a 6 percent decrease in NO_x, and a 17 percent decrease in VOC emissions.

Ozone Depletion

Without ozone, every living thing on the earth's surface would be incinerated. (On the other hand, as we have already noted, ozone can be lethal.) The presence of ozone

---

[30] A. S. Lefohn and S. V. Krupa, "Conference Overview," *Acidic Precipitation, A Technical Amplification of NAPAP's Findings,* Proceedings of an APCA International Conference, Pittsburgh, PA, p. 1, January 1988.

[31] National Acid Precipitation Assessment Program, 1992 Report to Congress, pp. 5–6.

in the upper atmosphere (20 to 40 km and up) provides a barrier to ultraviolet (UV) radiation. The small amounts that do seep through provide you with your summer tan. Too much UV will cause skin cancer. Although oxygen also serves as a barrier to UV radiation, it absorbs only over a narrow band centered at a wavelength of 0.2 μm. The photochemistry of these reactions is shown in Figure 6-9. The M refers to any third body (usually N_2).

In 1974, Molina and Rowland revealed a potential air pollution threat to this protective ozone shield.[32] They hypothesized that chlorofluorocarbons (CF_2Cl_2 and $CFCl_3$—often abbreviated as CFC), which are used as aerosol propellants and refrigerants, react with ozone (Figure 6-10). The frightening aspects of this series of reactions are that the chlorine atom removes ozone from the system, and that the chlorine atom is continually recycled to convert more ozone to oxygen. It has been estimated

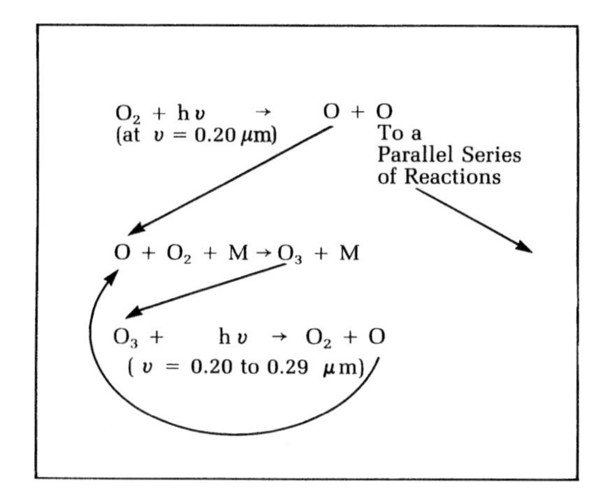

FIGURE 6-9
Photoreactions of ozone.

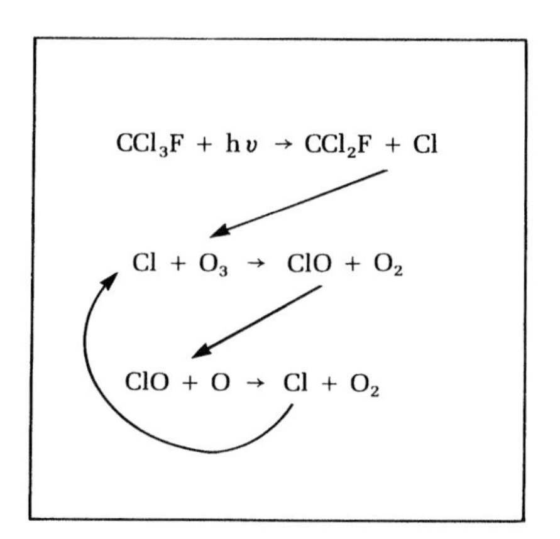

FIGURE 6-10
Ozone destruction by chlorofluoromethane.

[32]M. J. Molina and F. S. Rowland, "Stratospheric Sink for Chlorofluoromethanes: Chlorine Atom Catalysed Destruction of Ozone," *Nature,* vol. 248, pp. 810–812, 1974.

that a 5 percent reduction in ozone could result in nearly a 10 percent increase in skin cancer.[33] Thus, CFCs that are rather inert compounds in the lower atmosphere become a serious air pollution problem at higher elevations.

By 1987, the evidence that CFCs destroy ozone in the stratosphere above Antarctica every spring had become irrefutable. In 1987, the ozone hole was larger than ever. More than half of the total ozone column was wiped out and essentially all ozone disappeared from some regions of the stratosphere.

Research confirmed that the ozone layer, on a worldwide basis, shrunk approximately 2.5 percent in the preceding decade.[34] Initially, it was believed that this phenomenon was peculiar to the geography and climatology of Antarctica and that the warmer northern hemisphere was strongly protected from the processes that lead to massive ozone losses. Studies of the North Pole stratosphere in the winter of 1989 revealed that this is not the case.[35]

In September 1987, the Montreal Protocol on Substances That Deplete the Ozone Layer was developed. The Protocol, which has been ratified by 36 countries and became effective in January 1989, proposed that CFC production first be frozen and then reduced 50 percent by 1998. Yet, under the terms of the Protocol, the chlorine content of the atmosphere would continue to grow because the fully halogenated CFCs have such long lifetimes in the atmosphere. CF_2Cl_2, for example, has a lifetime of 110 years.[36] Eighty countries met at Helsinki, Finland, in the spring of 1989 to assess new information. The delegates gave their unanimous assent to a five-point "Helsinki Declaration":

1. All join the 1985 Vienna Convention for the Protection of the Ozone Layer and the follow-up Montreal Protocol.
2. Phase out production and consumption of ozone-depleting CFCs no later than 2000.
3. Phase out production and consumption as soon as feasible of halons and such chemicals as carbon tetrachloride and methyl chloroform that also contribute to ozone depletion.
4. Commit themselves to accelerated development of environmentally acceptable alternative chemicals and technologies.
5. Make relevant scientific information, research results, and training available to developing countries.[37]

---

[33] Interdepartmental Committee for Atmospheric Sciences, *The Possible Impact of Fluorocarbons and Hydrocarbons on Ozone* (Federal Council for Science and Technology, National Science Foundation Publication No. ICAS 18a-FY 75), Washington, DC: U.S. Government Printing Office, p. 3, May 1975.

[34] P. S. Zurer, "Studies on Ozone Destruction Expand Beyond Antarctic," *C&E News,* pp. 18–25, May 30, 1988.

[35] P. S. Zurer, "Scientists Find Arctic May Face Ozone Hole," *C&E News,* p. 5, February 27, 1989.

[36] M. Reisch and P. S. Zurer, "CFC Production: DuPont Seeks Total Phaseout," *C&E News,* p. 4, April 4, 1988.

[37] D. A. Sullivan, "International Gathering Plans Ways to Safeguard Atmospheric Ozone," *C&E News,* pp. 33–36, June 26, 1989.

The Montreal Protocol was strengthened in 1990 and 1992. The current terms of the treaty ban production of CFCs, carbon tetrachloride, and methyl chloroform as of January 1996. A ban on halon production took effect in January 1995.[38]

A number of alternatives to the fully chlorinated and, hence, more destructive CFCs have been developed. The two groups of compounds that emerged as significant replacements for the CFCs are hydrofluorocarbons (HFCs) and hydrochlorofluorocarbons (HCFCs). In contrast to the CFCs, HFCs and HCFCs contain one or more C-H bonds. This makes them susceptible to attack by OH radicals in the lower atmosphere. Because HFCs do not contain chlorine, they do not have the ozone depletion potential associated with the chlorine cycle shown in Figure 6-10. Although HCFCs contain chlorine, this chlorine is not transported to the stratosphere because OH scavenging in the troposphere is relatively efficient.

The implementation of the Montreal Protocol appears to be working. Worldwide CFC production has dropped by more than 50 percent since 1986. In 1993 alone, it dropped 15 percent.[37] Buildup of CFCs and halons has slowed, and the levels of methyl chloroform have been declining since 1991.[39,40] Because the HCFCs contribute to ozone depletion, albeit at a lower level than CFCs, it is anticipated that the peak chlorine levels will continue to rise until the year 2006 if chemical firms make use of their full allocation of HCFCs.[41]

The picture of the future development or retreat of the ozone hole is blurred by both the weather and volcanic eruptions. A very cold spring and strong winds forming a polar vortex that prevents replenishment of ozone-rich air will result in development of a large ozone hole in the Northern Hemisphere. The eruption of Mount Pinatubo in 1991 contributed sulfuric acid aerosols that cause ozone depletion that adds to the effect of anthropogenic contributions.

Greenhouse Effect

Unlike ozone, carbon dioxide is relatively transparent to shortwave ultraviolet light from the sun. It does, however, absorb and emit longwave radiation at wavelengths typical of the earth and atmosphere. Hence, CO_2 acts much like the glass on a greenhouse: It lets in shortwave radiation from the sun that heats the ground surface, but it restricts the loss of heat by radiation from the ground surface. The more CO_2 in the atmosphere, the more effective it is in restricting the outflow of radiative energy.

Since the first systematic measurements were made at Mauna Loa in Hawaii in 1958, CO_2 levels have risen from 315 ppm to 350 ppm. From analysis of air trapped in ice cores in Greenland and Antarctica, we know that preindustrial levels were about 280 ppm. The ice core records indicate that, over the last 160,000 years, no

[38]P. Zurer, "Scientists Expect Ozone Loss to Peak about 1998," *C&E News,* p. 5, September 12, 1994.

[39]A. Newman, "CFC Phase-out Moving Quickly," *Environmental Science & Technology,* 28, p. 35A, 1994.

[40]P. Zurer, "Global Monitoring Shows Ozone Treaty Is Working," *C&E News,* pp. 7–8, July 17, 1995.

[41]J. Rose, "HCFCs May Slow Ozone Layer Recovery," *Environmental Science & Technology,* 28, p. 111A, 1994.

fluctuations of CO_2 have been larger than 70 ppm.[42] Other gases have now been recognized as contributing to the greenhouse effect. Methane (CH_4), nitrous oxide (N_2O), and CFCs are similar to CO_2 in their radiative behavior. Even though their concentrations are much lower than CO_2, these gases are now estimated to trap about 60 percent as much longwave radiation as CO_2. In recent studies, a drop-off in the rate of methane accumulation has occurred. In 1989–90, atmospheric methane was increasing at a rate of about 10 or 11 parts per billion (ppb) per year. An abrupt decrease to about 2 ppb per year occurred in 1992. This decrease has been attributed to a reduction in natural gas leaks such as those from pipelines.[43]

The bulk of the carbon dioxide that has been added to the atmosphere has been attributed to the combustion of fossil fuel. In the 1980s, massive deforestation has been identified as a possible contribution. Both the burning of timber and the release of carbon from bacterial degradation contribute. Perhaps more important, deforestation removes a mechanism for removing CO_2 from the atmosphere (commonly referred to as a *sink*). In normal respiration, green plants utilize CO_2 much as we use oxygen. This CO_2 is fixed in the biomass by photosynthetic processes. A rapidly growing rain forest can fix between 1 and 2 kg per year of carbon per square meter of ground surface. Cultivated fields, in contrast, fix only about 0.2 to 0.4 kg/m^2—and this amount is recycled by bioconsumption and conversion to CO_2.

Attempts to understand the consequences of global warming are based on mathematical models of the global circulation of the atmosphere and oceans. To date these models have a "good news–bad news" conclusion. Based on a 2° to 6°C rise in global temperature, the following good news is predicted for North America:[44]

1. A decrease in heating costs of up to 25 percent (partly offset by increased air-conditioning cost increases of 10 percent).
2. A longer growing season for crops.
3. Much easier navigation in the Arctic seas.

The bad news:

1. Drier crop conditions in the Midwest and Great Plains, requiring more irrigation.
2. Widespread melting of permanently frozen ground with adverse effects on building technology in Alaska and northern Canada.
3. A rise in sea level from 0.5 to 1.5 m that would result in an increase in the severity of flooding, damage to coastal structures, destruction of wetlands, and saltwater intrusion into drinking water supplies in coastal areas such as Florida.

There is still much speculation about the cause-effect relationship of increases in CO_2 and the possible outcomes. Although the data indicate increases in CO_2 concen-

---

[42] B. Hileman, "Global Warming," *C&E News,* pp. 25–44, March 13, 1989.

[43] *C&E News,* p. 31, Feb. 28, 1994.

[44] F. K. Hare and T. C. Hutchinson, "Human Environmental Disturbances," *Environmental Science and Engineering,* J. G. Henry and G. W. Heinke, eds., Upper Saddle River, NJ: Prentice Hall, pp. 121–122, 1989.

tration, the direct radiative effect of increasing CO_2 alone is not sufficient to explain current trends that show an increase in nighttime temperatures but not an increase in daytime highs.[45] The input sources of greenhouse gases and their sinks are not yet well described. The measurement of temperature on a global basis is not sufficiently uniform in technique and separated from local influence to separate the "noise" of local variability from true trends. Natural changes such as increases in cloud cover may not have been accurately depicted in existing models of climate change.

In general, projections of global warming have been based on assumptions regarding the growth of greenhouse gases. If it is assumed that they will continue to grow exponentially, by the year 2040 the change in atmospheric concentration of greenhouse gases would be the equivalent of doubling of the CO_2 concentration from its preindustrial level. It is this doubling that leads the National Research Council to estimate a temperature rise of 1° to 5°C.[46] In another projection of emissions by the Intergovernmental Panel on Climate Change, global temperatures are expected to rise between 0.8° and 3.5°C by 2100.[47] Obviously, there is still considerable disagreement about the potential for global warming. On the other hand, the consequences of ignoring these trends are sufficiently dramatic that intensive research will continue in the next decade. Even without the risks of climate change, improvements in energy efficiency to reduce CO_2 emissions and to eliminate CFCs are justified. The expectation of damages from climate change provides a rationale for pursuing these programs vigorously.

6-7 AIR POLLUTION METEOROLOGY

The Atmospheric Engine

The atmosphere is somewhat like an engine. It is continually expanding and compressing gases, exchanging heat, and generally raising chaos. The driving energy for this unwieldy machine comes from the sun. The difference in heat input between the equator and the poles provides the initial overall circulation of the earth's atmosphere. The rotation of the earth coupled with the different heat conductivities of the oceans and land produce weather.

Highs and lows. Because air has mass, it also exerts pressure on things under it. Like water, which we intuitively understand to exert greater pressures at greater depths, the atmosphere exerts more pressure at the surface than it does at higher elevations. The highs and lows depicted on weather maps are simply areas of greater and lesser pressure. The elliptical lines shown on more detailed weather maps are lines of constant pressure, or *isobars*. A two-dimensional plot of pressure and distance through a high- or low-pressure system would appear as shown in Figure 6-11.

---

[45] G. Kukla and T. R. Karl, "Nighttime Warming and the Greenhouse Effect," *Environmental Science and Technology,* 27, pp. 1468–1474, 1993.

[46] L. B. Lave and H. Dowlatabadi, "Climate Change: The Effects of Personal Beliefs and Scientific Uncertainty," *Environmental Science and Technology,* 27, pp. 1962–1972, 1993.

[47] *C&E News,* p. 20, August 28, 1995.

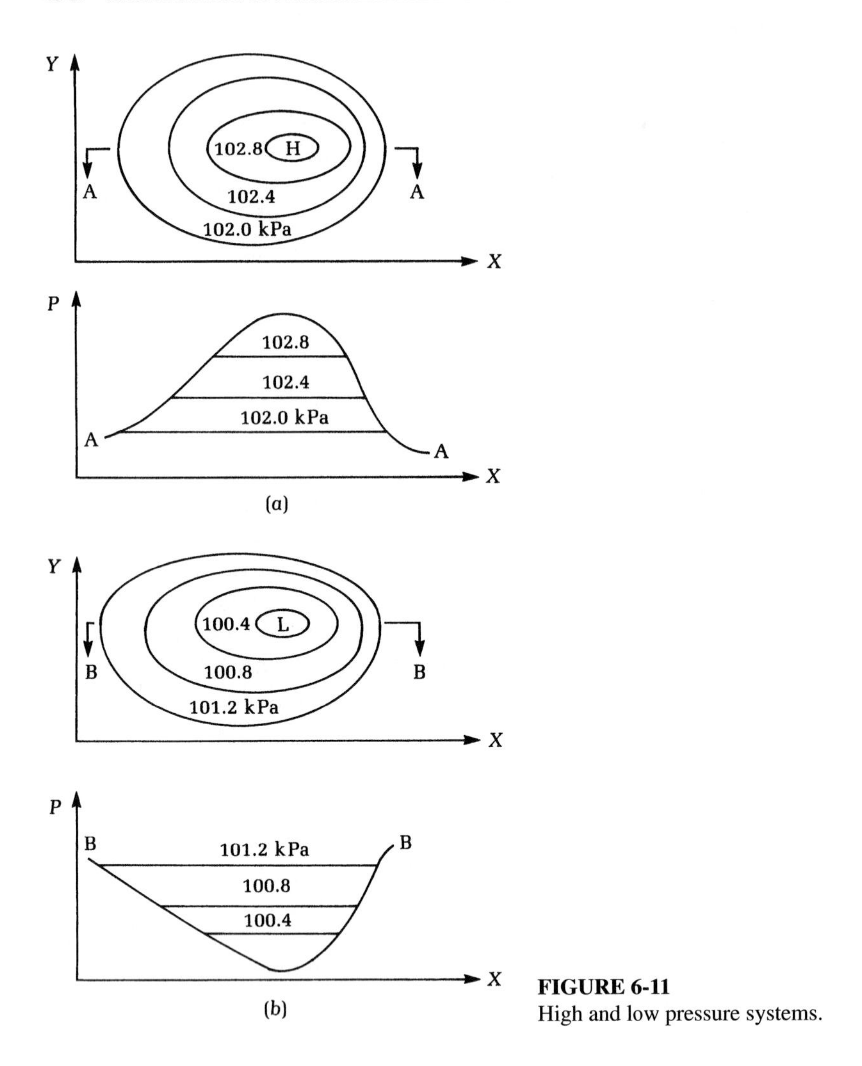

FIGURE 6-11
High and low pressure systems.

The wind flows from the higher pressure areas to the lower pressure areas. On a nonrotating planet, the wind direction would be perpendicular to the isobars (Figure 6-12*a*). However, since the earth rotates, an angular thrust called the Coriolis effect is added to this motion. The resultant wind direction in the northern hemisphere is as shown in Figure 6-12*b*. The technical names given to these systems are *anticyclones* for highs and *cyclones* for lows. Anticyclones are associated with good weather. Cyclones are associated with foul weather. Tornadoes and hurricanes are the foulest of the cyclones.

Wind speed is in part a function of the steepness of the pressure surface. When the isobars are close together, the pressure gradient (slope) is said to be steep and the wind speed relatively high. If the isobars are well spread out, the winds are light or nonexistent.

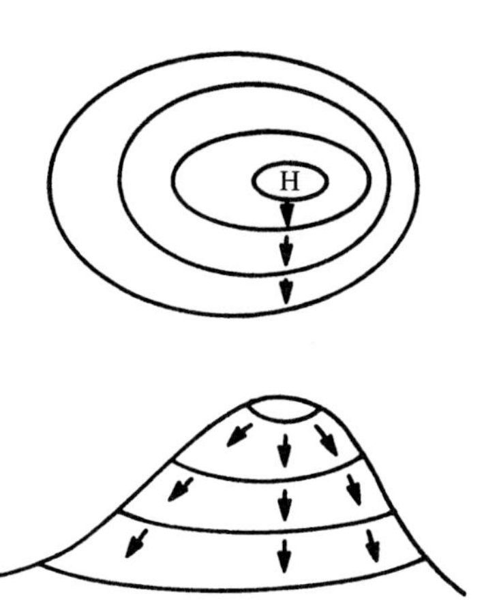

(a) Anticyclone without Coriolis effect

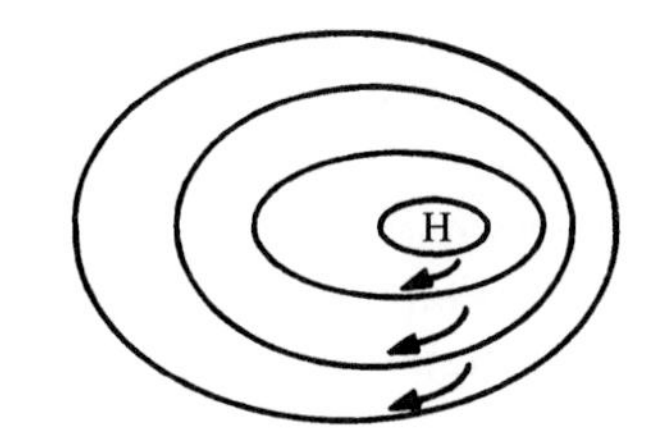

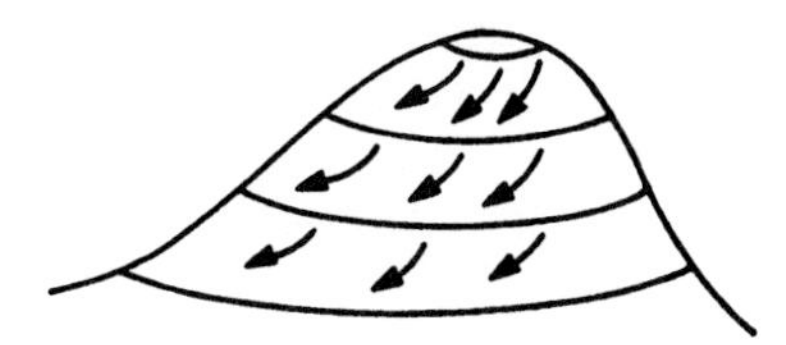

(b) Anticyclone with Coriolis effect

FIGURE 6-12
Wind flow due to pressure gradient.

Turbulence

Mechanical turbulence. In its simplest terms, we may consider turbulence to be the addition of random fluctuations of wind velocity (that is, speed and direction) to the overall average wind velocity. These fluctuations are caused, in part, by the fact that the atmosphere is being sheared. The shearing results from the fact that the wind speed is zero at the ground surface and rises with elevation to near the speed imposed by the pressure gradient. The shearing results in a tumbling, tearing motion as the mass just above the surface falls over the slower moving air at the surface. The swirls thus formed are called *eddies.* These small eddies feed larger ones. As you might expect, the greater the mean wind speed, the greater the mechanical turbulence. The more mechanical turbulence, the easier it is to disperse and spread atmospheric pollutants.

Thermal turbulence. Like all other things in nature, the rather complex interaction that produces mechanical turbulence is confounded and further complicated by a third party. Heating of the ground surface causes turbulence in the same fashion that heating the bottom of a beaker full of water causes turbulence. At some point below boiling, you can see density currents rising off the bottom. Likewise, if the earth's surface is heated strongly and in turn heats the air above it, thermal turbulence will be generated. Indeed, the "thermals" sought by glider pilots and hot air balloonists are these thermal currents rising on what otherwise would be a calm day.

The converse situation can arise during clear nights when the ground radiates its heat away to the cold night sky. The cold ground, in turn, cools the air above it, causing a sinking density current.

Stability

The tendency of the atmosphere to resist or enhance vertical motion is termed *stability.* It is related to both wind speed and the change of air temperature with height (*lapse rate*). For our purpose, we may use the lapse rate alone as an indicator of the stability condition of the atmosphere.

There are three stability categories. When the atmosphere is classified as *unstable,* mechanical turbulence is enhanced by the thermal structure. A *neutral* atmosphere is one in which the thermal structure neither enhances nor resists mechanical turbulence. When the thermal structure inhibits mechanical turbulence, the atmosphere is said to be *stable.* Cyclones are associated with unstable air. Anticyclones are associated with stable air.

Neutral stability. The lapse rate for a neutral atmosphere is defined by the rate of temperature increase (or decrease) experienced by a parcel of air that expands (or contracts) *adiabatically* (without the addition or loss of heat) as it is raised through the atmosphere. This rate of temperature decrease (dT/dz) is called the *dry adiabatic lapse rate.* It is designated by the Greek letter gamma (Γ). It has a value of approximately -1.00°C/100 m. (Note that this is not a slope in the normal sense, that is, it is not dy/dx.) In Figure 6-13*a*, the dry adiabatic lapse rate of a parcel of air is shown as a dashed line and the temperature of the atmosphere (ambient lapse rate) is shown as a solid line. Since the ambient lapse rate is the same as Γ, the atmosphere is said to have a neutral stability.

Unstable atmosphere. If the temperature of the atmosphere falls at a rate greater than Γ (for example, -1.01°C/100 m), the lapse rate is said to be *superadiabatic,* and the atmosphere is unstable. Using Figure 6-13*b,* we can see that this is so. The actual lapse rate is shown by the solid line. If we capture a balloon full of polluted air at elevation A and adiabatically displace it 100 m vertically to elevation B, the temperature of the air inside the balloon will decrease from 21.15° to 20.15°C. At a lapse rate of -1.25°C/100 m, the temperature of the air outside the balloon will decrease from 21.15° to 19.90°C. The air inside the balloon will be warmer than the air outside; this temperature difference gives the balloon buoyancy. It will behave as a hot gas and continue to rise without any further mechanical effort. Thus, mechanical turbulence is enhanced and the atmosphere is unstable. If we adiabatically displace the balloon downward to elevation C, the temperature inside the balloon

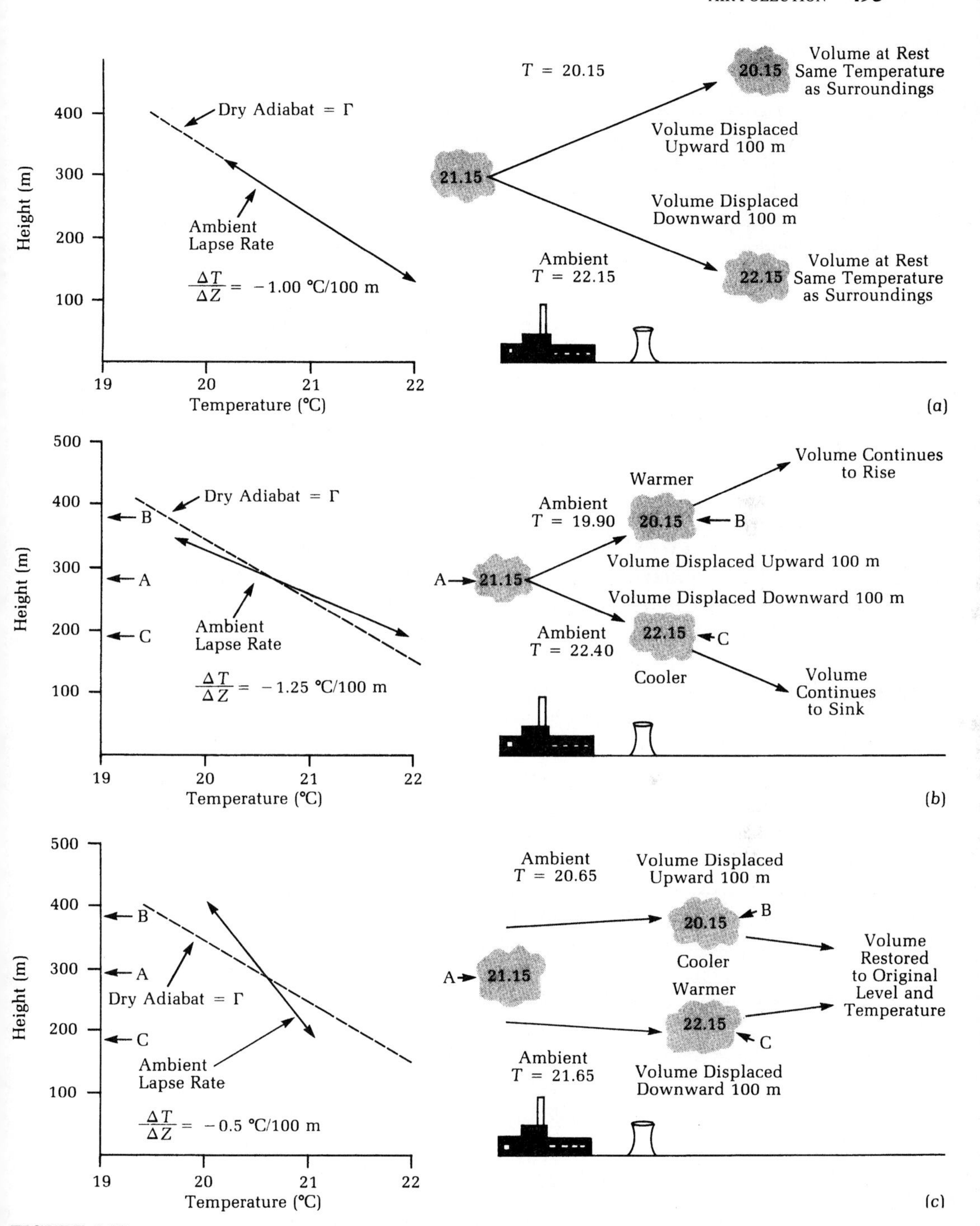

FIGURE 6-13
Lapse rate and displaced air volume. (*Source:* Atomic Energy Commission, *Meteorology and Atomic Energy,* Washington, DC: U.S. Government Printing Office, 1955.)

would rise at the rate of the dry adiabat. Thus, in moving 100 m, the temperature will increase from 21.15° to 22.15°C. The temperature outside the balloon will increase at the superadiabatic lapse rate to 22.40°C. The air in the balloon will be cooler than the ambient air and the balloon will have a tendency to sink. Again, mechanical turbulence (displacement) is enhanced.

Stable atmosphere. If the temperature of the atmosphere falls at a rate less than Γ (for example, −0.99°C/100 m), it is called *subadiabatic,* and the atmosphere is stable. If we again capture a balloon of polluted air at elevation A (Figure 6-13*c*) and adiabatically displace it vertically to elevation B, the temperature of the polluted air will decrease at a rate equal to the dry adiabatic rate. Thus, in moving 100 m, the temperature will decrease from 21.15 to 20.15°C as before. However, since the ambient lapse rate is −0.5°C/100 m, the temperature of the air outside the balloon will have dropped to only 20.65°C. Since the air inside the balloon is cooler than the air outside the balloon, the balloon will have a tendency to sink. Thus, the mechanical displacement (turbulence) is inhibited.

In contrast, if we displace the balloon adiabatically to elevation C, the temperature inside the balloon would increase to 22.15°C, while the ambient temperature would increase to 21.65°C. In this case, the air inside the balloon would be warmer than the ambient air and the balloon would tend to rise. Again, the mechanical displacement would be inhibited.

There are two special cases of subadiabatic lapse rate. When there is no change of temperature with elevation, the lapse rate is called *isothermal.* When the temperature increases with elevation, the lapse rate is called an *inversion.* The inversion is the most severe form of a stable temperature profile. It is often associated with restricted air volumes that cause air pollution episodes.

Example 6-3. Given the following temperature and elevation data, determine the stability of the atmosphere.

Elevation (m)	Temperature (°C)
2.00	14.35
324.00	11.13

Solution. Begin by determining the existing lapse rate:

$$\frac{\Delta T}{\Delta Z} = \frac{T_2 - T_1}{Z_2 - Z_1}$$

$$= \frac{11.13 - 14.35}{324.00 - 2.00} = \frac{-3.22}{322.00}$$

$$= -0.0100°\text{C/m} = -1.00°\text{C/100 m}$$

Now we compare this with Γ and find that they are equal. Thus, the atmospheric stability is neutral.

Plume types. The smoke trail or plume from a tall stack located on flat terrain has been found to exhibit a characteristic shape that is dependent on the stability of the atmosphere. The six classical plumes are shown in Figure 6-14, along with the corresponding temperature profiles. In each case, Γ is given as a broken line to allow

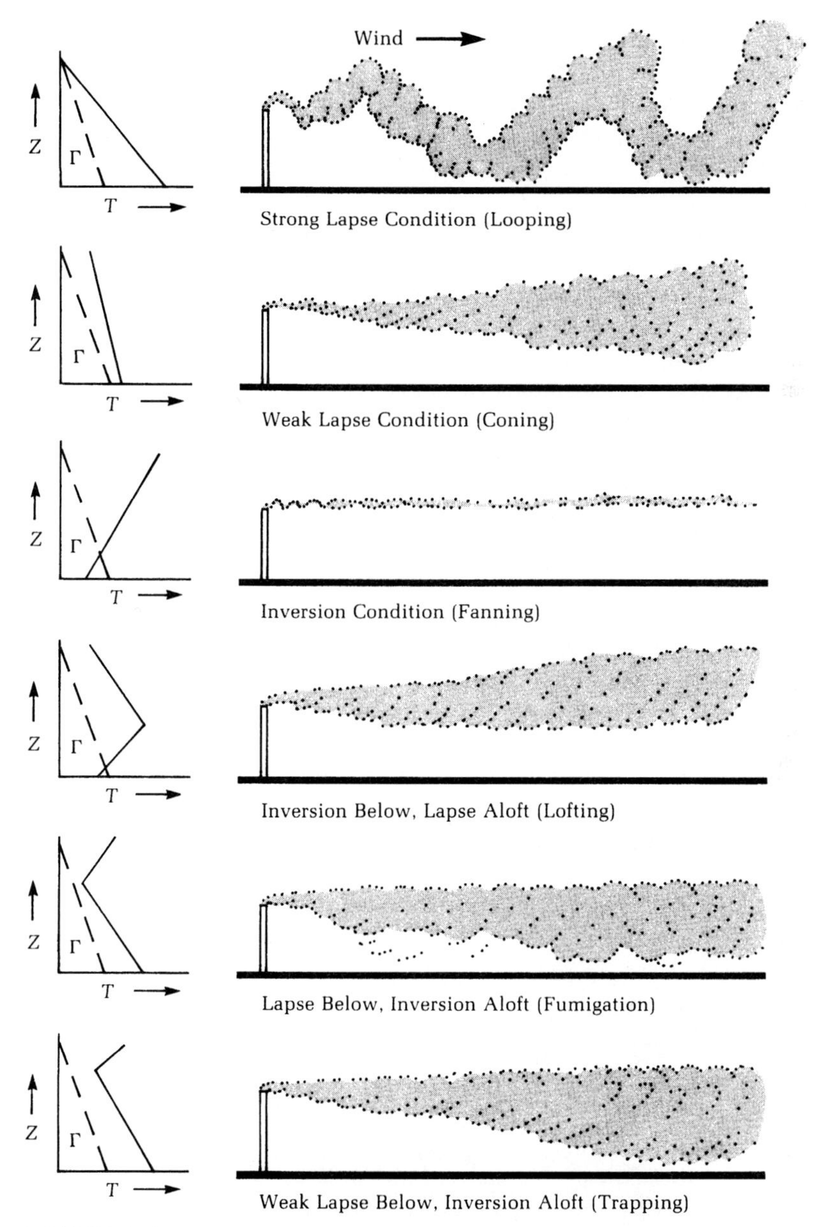

FIGURE 6-14
Six types of plume behavior. (*Source:* P. E. Church, "Dilution of Waste Stack Gases in the Atmosphere," *Industrial Engineering Chemistry,* vol. 41, pp. 3753–3756, 1949.)

comparison with the actual lapse rate, which is given as a solid line. In the bottom three cases, particular attention should be given to the location of the inflection point with respect to the top of the stack.

Terrain Effects

Heat islands. A heat island results from a mass of material, either natural or anthropogenic, that absorbs and reradiates heat at a greater rate than the surrounding area. This causes moderate to strong vertical convection currents above the heat island. The effect is superimposed on the prevailing meteorological conditions. It is nullified by strong winds. Large industrial complexes and small to large cities are examples of places that would have a heat island.

Because of the heat island effect, atmospheric stability will be less over a city than it is over the surrounding countryside. Depending upon the location of the pollutant sources, this can be either good news or bad news. First, the good news: For ground level sources such as automobiles, the bowl of unstable air that forms will allow a greater air volume for dilution of the pollutants. Now the bad news: Under stable conditions, plumes from tall stacks would be carried out over the countryside without increasing ground level pollutant concentrations. Unfortunately, the instability caused by the heat island mixes these plumes to the ground level.

Land/sea breezes. Under a stagnating anticyclone, a strong local circulation pattern may develop across the shoreline of large water bodies. During the night, the land cools more rapidly than the water. The relatively cooler air over the land flows toward the water (a land breeze, Figure 6-15). During the morning the land heats faster than water. The air over the land becomes relatively warm and begins to rise. The rising air is replaced by air from over the water body (a sea or lake breeze, Figure 6-16).

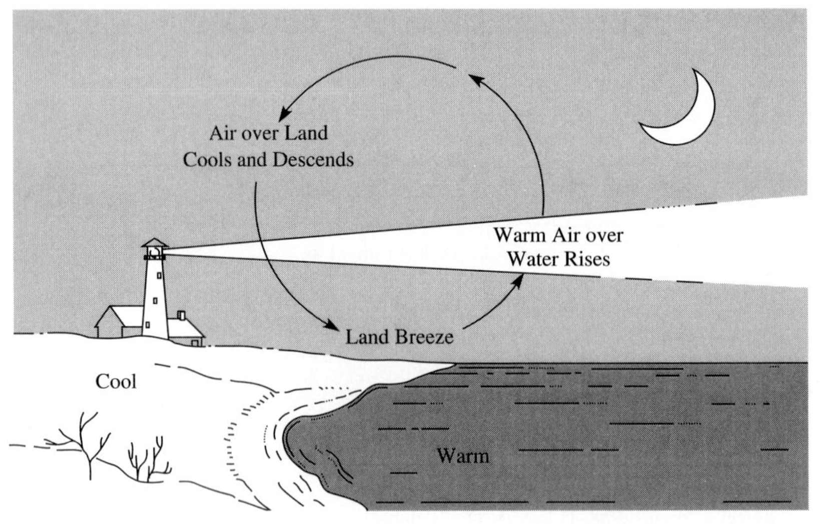

FIGURE 6-15
Land breeze during the night.

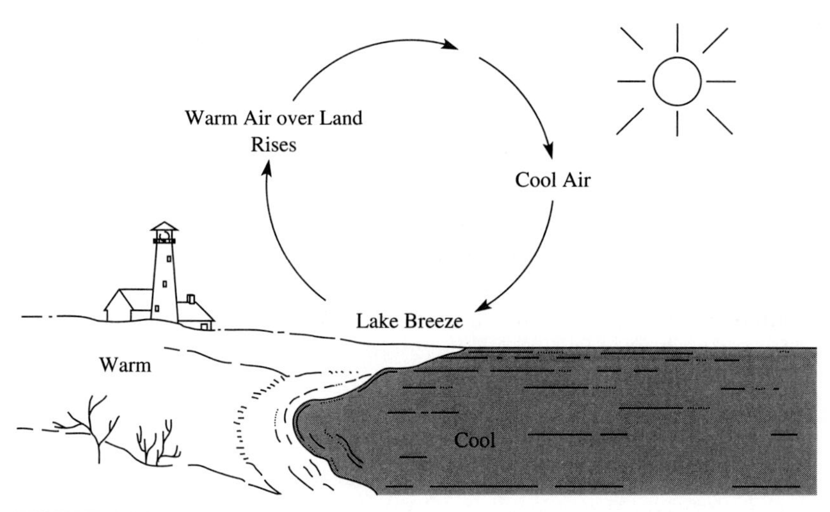

FIGURE 6-16
Lake breeze during the day.

The effect of the lake breeze on stability is to impose a surface-based inversion on the temperature profile. As the air moves from the water over the warm ground, it is heated from below. Thus, for stack plumes originating near the shoreline, the stable lapse rate causes a fanning plume close to the stack (Figure 6-17). The lapse condition grows to the height of the stack as the air moves inland. At some point inland, a fumigation plume results.

Valleys. When the general circulation imposes moderate to strong winds, valleys that are oriented at an acute angle to the wind direction channel the wind. The valley

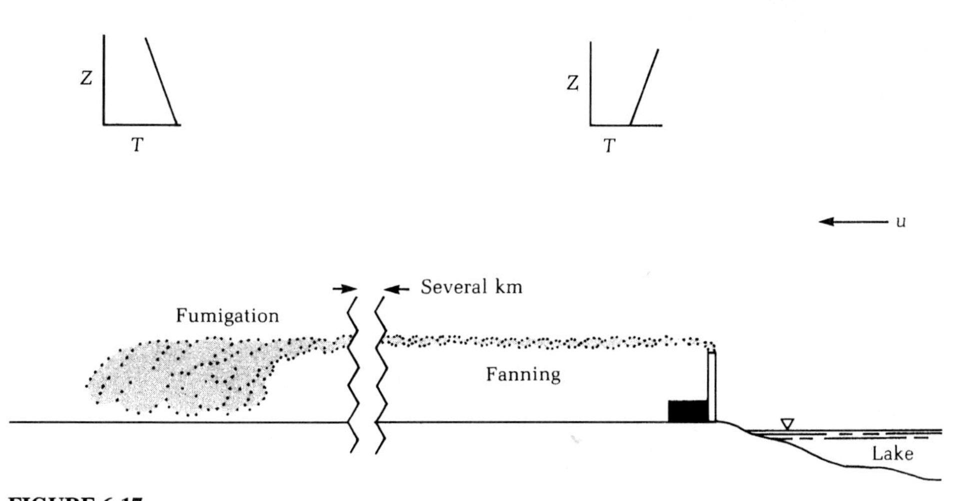

FIGURE 6-17
Effect of lake breeze on plume dispersion.

effectively peels off part of the wind and forces it to follow the direction of the valley floor.

Under a stagnating anticyclone, the valley will set up its own circulation. Warming of the valley walls will cause the valley air to be warmed. It will become more buoyant and flow up the valley. At night the cooling process will cause the wind to flow down the valley.

Valleys oriented in the north-south direction are more susceptible to inversions than level terrain. The valley walls protect the floor from radiative heating by the sun. Yet the walls and floor are free to radiate heat away to the cold night sky. Thus, under weak winds, the ground cannot heat the air rapidly enough during the day to dissipate the inversion that formed during the night.

6-8 ATMOSPHERIC DISPERSION

Factors Affecting Dispersion of Air Pollutants

This discussion follows the training documents of the Texas Air Quality Control Board.

The factors that affect the transport, dilution, and dispersion of air pollutants can generally be categorized in terms of the emission point characteristics, the nature of the pollutant material, meteorological conditions, and effects of terrain and anthropogenic structures. We have discussed all of these except the source conditions. Now we wish to integrate the first and third factors to describe the qualitative aspects of calculating pollutant concentrations. We shall follow this with a simple quantitative model for a point source. More complex models for point sources (in rough terrain, in industrial settings, or for long time periods), area sources, and mobile sources are left for more advanced texts.

Source characteristics. Most industrial effluents are discharged vertically into the open air through a stack or duct. As the contaminated gas stream leaves the discharge point, the plume tends to expand and mix with the ambient air. Horizontal air movement will tend to bend the discharge plume toward the downwind direction. At some point between 300 and 3,000 m downwind, the effluent plume will level off. While the effluent plume is rising, bending, and beginning to move in a horizontal direction, the gaseous effluents are being diluted by the ambient air surrounding the plume. As the contaminated gases are diluted by larger and larger volumes of ambient air, they are eventually dispersed toward the ground.

The plume rise is affected by both the upward inertia of the discharge gas stream and by its buoyancy. The vertical inertia is related to the exit gas velocity and mass. The plume's buoyancy is related to the exit gas mass relative to the surrounding air mass. Increasing the exit velocity or the exit gas temperature will generally increase the plume rise. The plume rise, together with the physical stack height, is called the *effective stack height.*

The additional rise of the plume above the discharge point as the plume bends and levels off is a factor in the resultant downwind ground level concentrations. The higher the plume rises initially, the greater distance there is for diluting the contaminated gases as they expand and mix downward.

For a specific discharge height and a specific set of plume dilution conditions, the ground level concentration is proportional to the amount of contaminant materials discharged from the stack outlet for a specific period of time. Thus, when all other conditions are constant, an increase in the pollutant discharge rate will cause a proportional increase in the downwind ground level concentrations.

Downwind distance. The greater the distance between the point of discharge and a ground level receptor downwind, the greater will be the volume of air available for diluting the contaminant discharge before it reaches the receptor.

Wind speed and direction. The wind direction determines the direction in which the contaminated gas stream will move across local terrain. Wind speed affects the plume rise and the rate of mixing or dilution of the contaminated gases as they leave the discharge point. An increase in wind speed will decrease the plume rise by bending the plume over more rapidly. The decrease in plume rise tends to increase the pollutant's ground level concentration. On the other hand, an increase in wind speed will increase the rate of dilution of the effluent plume, tending to lower the downwind concentrations. Under different conditions, one or the other of the two wind speed effects becomes the predominant effect. These effects, in turn, affect the distance downwind of the source at which the maximum ground level concentration will occur.

Stability. The turbulence of the atmosphere follows no other factor in power of dilution. The more unstable the atmosphere, the greater the diluting power. Inversions that are not ground based, but begin at some height above the stack exit, act as a lid to restrict vertical dilution.

Dispersion Modeling

General considerations and use of models. A dispersion model is a mathematical description of the meteorological transport and dispersion process that is quantified in terms of source and meteorologic parameters during a particular time. The resultant numerical calculations yield estimates of concentrations of the particular pollutant for specific locations and times.

To verify the numerical results of such a model, actual measured concentrations of the particular atmospheric pollutant must be obtained and compared with the calculated values by means of statistical techniques. The meteorological parameters required for use of the models include wind direction, wind speed, and atmospheric stability. In some models, provisions may be made for including lapse rate and vertical mixing height. Most models will require data about the physical stack height, the diameter of the stack at the emission discharge point, the exit gas temperature and velocity, and the mass rate of emission of pollutants.

Models are usually classified as either short-term or climatological models. Short-term models are generally used under the following circumstances: (1) to estimate ambient concentrations where it is impractical to sample, such as over rivers or lakes, or at great distances above the ground; (2) to estimate the required emergency source reductions associated with periods of air stagnations under air pollution episode alert conditions; and (3) to estimate the most probable locations of high, short-term, ground-level concentrations as part of a site selection evaluation for the location of air monitoring equipment.

Climatological models are used to estimate mean concentrations over a long period of time or to estimate mean concentrations that exist at particular times of the day for each season over a long period of time. Long-term models are used as an aid for developing emissions standards. We will be concerned only with short-term models in their most simple application.

Basic point source Gaussian dispersion model. The basic Gaussian diffusion equation assumes that atmospheric stability is uniform throughout the layer into which the contaminated gas stream is discharged. The model assumes that turbulent diffusion is a random activity and hence the dilution of the contaminated gas stream in both the horizontal and vertical direction can be described by the Gaussian or normal equation. The model further assumes that the contaminated gas stream is released into the atmosphere at a distance above ground level that is equal to the physical stack height plus the plume rise. The model assumes that the degree of dilution of the effluent plume is inversely proportional to the wind speed (u). The model also assumes that pollutant material that reaches ground level is totally reflected back into the atmosphere like a beam of light striking a mirror at an angle. Mathematically, this ground reflection is accounted for by assuming a virtual or imaginary source located at a distance of $-H$ with respect to ground level, and emitting an imaginary plume with the same source strength as the real source being modeled. The same general idea can be used to establish other boundary layer conditions for the equations, such as limiting horizontal or vertical mixing.

The model. We have selected the model equation in the form presented by D. B. Turner.[48] It gives the ground level concentration (χ) of pollutant at a point (coordinates x and y) downwind from a stack with an effective height (H) (Figure 6-18). The standard deviation of the plume in the horizontal and vertical directions is designated by s_y and s_z, respectively. The standard deviations are functions of the downward distance from the source and the stability of the atmosphere. The equation is as follows:

$$\chi_{(x,y,0,H)} = \left[\frac{E}{\pi s_y s_z u}\right]\left[\exp\left[-\frac{1}{2}\left(\frac{y}{s_y}\right)^2\right]\right]\left[\exp\left[-\frac{1}{2}\left(\frac{H}{s_z}\right)^2\right]\right] \tag{6-19}$$

[48]D. Bruce Turner, *Workbook of Atmospheric Dispersion Estimates* (U.S. Department of Health, Education and Welfare, Public Health Service, National Center for Air Pollution Control, Publication No. 999-AP-28), Washington, DC: U.S. Government Printing Office, p. 6, 1967. (*Note:* Turner provides guidelines on the accuracy of this model. It is an estimating tool and not a definitive model to be used indiscriminately.)

where $\chi_{(x,y,0,H)}$ = downwind concentration at ground level, g/m^3
E = emission rate of pollutant, g/s
s_y, s_z = plume standard deviations, m
u = wind speed, m/s
$x, y, z,$ and H = distances, m
exp = exponential e such that terms in brackets immediately following are powers of e, that is, $e^{[\,]}$ where e = 2.7182

The value for the effective stack height is the sum of the physical stack height (h) and the plume rise ΔH:

$$H = h + \Delta H \tag{6-20}$$

ΔH may be computed from Holland's formula as follows:[49]

$$\Delta H = \frac{v_s d}{u}\left[1.5 + \left(2.68 \times 10^{-2}(P)\left(\frac{T_s - T_a}{T_s}\right)d\right)\right] \tag{6-21}$$

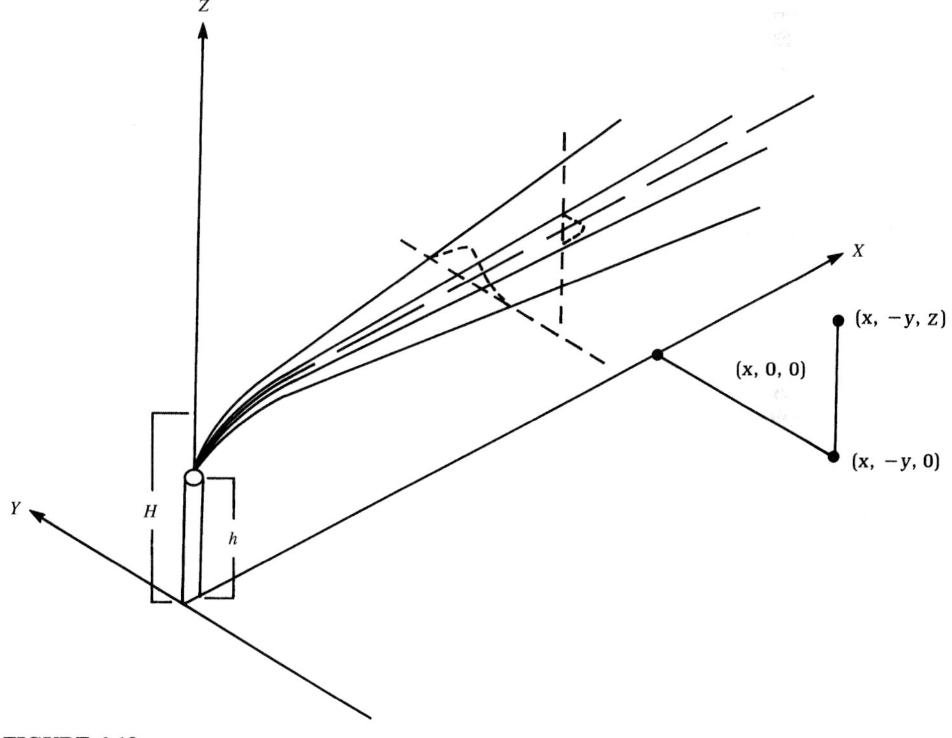

FIGURE 6-18
Plume dispersion coordinate system. [*Source:* D. Bruce Turner, *Workbook of Atmospheric Dispersion Estimates* (U.S. Department of Health, Education and Welfare, Public Health Service, National Center for Air Pollution Control, Publication No. 999-AP-26), Washington, DC: U.S. Government Printing Office, 1967.]

[49] J. Z. Holland, *A Meteorological Survey of the Oak Ridge Area* (U.S. Atomic Energy Commission Report No. ORO-99), Washington, DC: U.S. Government Printing Office, p. 540, 1953.

where v_s = stack velocity, m/s
d = stack diameter, m
u = wind speed, m/s
P = pressure, kPa
T_s = stack temperature, K
T_a = air temperature, K

The values of s_y and s_z depend upon the turbulent structure or stability of the atmosphere. Figures 6-19 and 6-20 provide graphical relationships between the down-

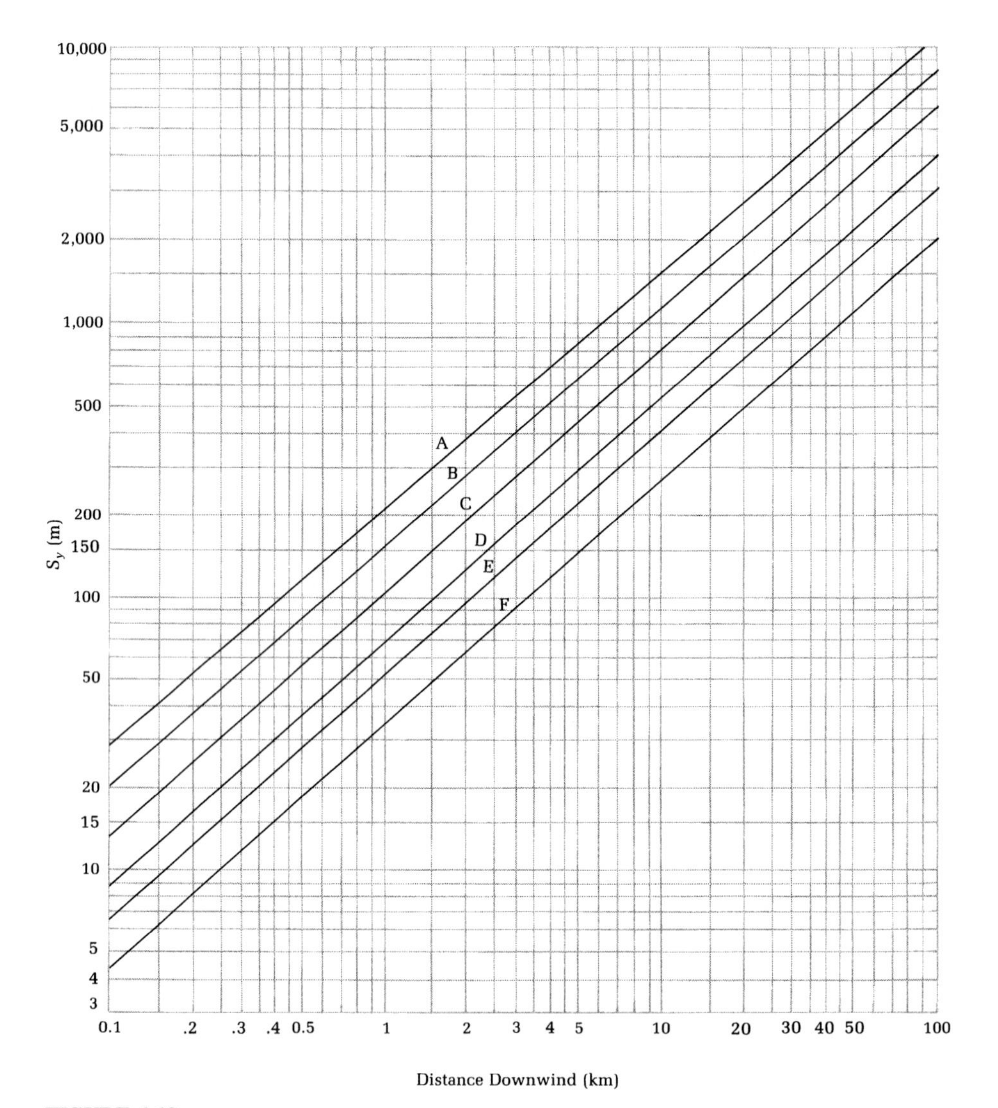

FIGURE 6-19
Horizontal dispersion coefficient. [*Source:* Turner, *Workbook of Atmospheric Dispersion Estimates* (U.S. Department of Health, Education and Welfare, Public Health Service, National Center for Air Pollution Control, Publication No. 999-AP-28), Washington, DC: U.S. Government Printing Office, 1967.]

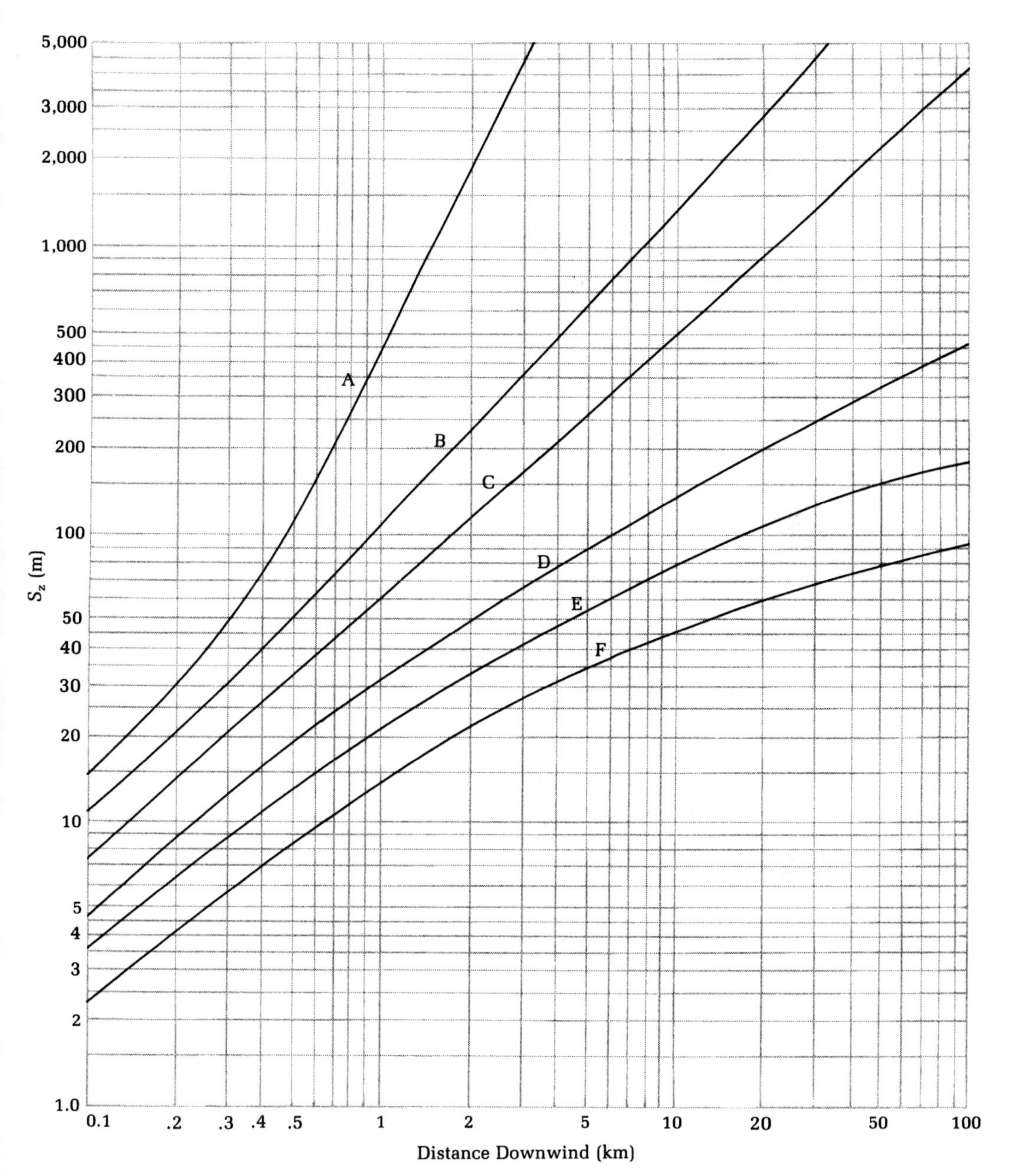

FIGURE 6-20
Vertical dispersion coefficient. (*Source:* Turner, *Workbook of Atmospheric Dispersion Estimates.*)

wind distance x in kilometers and values of s_y and s_z in meters. The curves on the two figures are labeled "A" through "F." The label "A" refers to very unstable atmospheric conditions, "B" to unstable atmospheric conditions, "C" to slightly unstable to neutral conditions, "D" to stable conditions, "E" to stable atmospheric conditions,

TABLE 6-6
Key to stability categories

Surface wind speed (at 10 m) (m/s)	Day[a]			Night[a]	
	Incoming solar radiation				
	Strong	Moderate	Slight	Thinly overcast or ≥ 4/8 Low cloud	≤ 3/8 Cloud
<2	A	A–B	B		
2–3	A–B	B	C	E	F
3–5	B	B–C	C	D	E
5–6	C	C–D	D	D	D
>6	C	D	D	D	D

[a] The neutral class, D, should be assumed for overcast conditions during day or night. Note that "thinly overcast" is not equivalent to "overcast."

Notes: Class A is the most unstable and class F is the most stable class considered here. Night refers to the period from one hour before sunset to one hour after sunrise. Note that the neutral class, D, can be assumed for overcast conditions during day or night, regardless of wind speed.

"Strong" incoming solar radiation corresponds to a solar altitude greater than 60° with clear skies; "slight" insolation corresponds to a solar altitude from 15° to 35° with clear skies. Table 170, Solar Altitude and Azimuth, in the Smithsonian Meteorological Tables, can be used in determining solar radiation. Incoming radiation that would be strong with clear skies can be expected to be reduced to moderate with broken (5/8 to 7/8 cloud cover) middle clouds and to slight with broken low clouds.

Source: D. Bruce Turner, *Workbook of Atmospheric Dispersion Estimates.*

and "F" to very stable atmospheric conditions. Each of these stability parameters represents an averaging time of approximately 3 to 15 min.

Other averaging times may be approximated by multiplying by empirical constants, for example, 0.36 for 24 hours. Turner presented a table and discussion that allows an estimate of stability based on wind speed and the conditions of solar radiation. This is given in Table 6-6.

For computer solutions of the dispersion model, it is convenient to have an algorithm to express the stability class lines in Figures 6-19 and 6-20. D. O. Martin[50]

TABLE 6-7
Values of a, c, d, and f for calculating s_y and s_z

Stability class	a	$x \leq 1$ km c	d	f	$x \geq 1$ km c	d	f
A	213	440.8	1.941	9.27	459.7	2.094	−9.6
B	156	100.6	1.149	3.3	108.2	1.098	2
C	104	61	0.911	0	61	0.911	0
D	68	33.2	0.725	−1.7	44.5	0.516	−13
E	50.5	22.8	0.678	−1.3	55.4	0.305	−34
F	34	14.35	0.74	−0.35	62.6	0.18	−48.6

Source: D. O. Martin.

[50]D. O. Martin, Comment on the Change of Concentration Standard Deviations with Distance, *Journal of the Air Pollution Control Association,* 26, pp. 145–146, 1976.

has developed the following equations that provide an approximate fit.

$$s_y = ax^{0.894} \tag{6-22}$$

$$s_z = cx^d + f \tag{6-23}$$

where the constants a, c, d, and f are defined in Table 6-7. These equations were developed to yield s_y and s_z in meters for downwind distance x in kilometers.

Example 6-4. It has been estimated that the emission of SO_2 from a coal-fired power plant is 1,656.2 g/s. At 3 km downwind on an overcast summer afternoon, what is the centerline concentration of SO_2 if the wind speed is 4.50 m/s? (Note: "centerline" implies $y = 0$.)

Stack parameters:
Height = 120.0 m
Diameter = 1.20 m
Exit velocity = 10.0 m/s
Temperature = 315°C

Atmospheric conditions:
Pressure = 95.0 kPa
Temperature = 25.0°C

Solution. We begin by determining the effective stack height (H).

$$\Delta H = \frac{(10.0)(1.20)}{4.50}\left[1.5 + (2.68 \times 10^{-2}(95.0)\frac{588 - 298}{588}1.20)\right]$$

$$\Delta H = 8.0 \text{ m}$$

$$H = 120.0 + 8.0 = 128.0 \text{ m}$$

Next, we must determine the atmospheric stability class. The footnote to Table 6-6 indicates that the D class should be used for overcast conditions.

From Figures 6-19 and 6-20, we can determine that at 3 km downwind with a D stability the plume standard deviations are as follows:

$$s_y = 190 \text{ m}$$

$$s_z = 65 \text{ m}$$

Thus,

$$\chi = \frac{1{,}656.2}{\pi(190)(65)(4.50)} \exp\left[-\frac{1}{2}\left(\frac{0}{s_y}\right)^2\right] \exp\left[-\frac{1}{2}\left(\frac{128}{65}\right)^2\right]$$

$$= 1.36 \times 10^{-3} \text{ g/m}^3 \text{ or } 1.4 \times 10^{-3} \text{ g/m}^3 \text{ of } SO_2$$

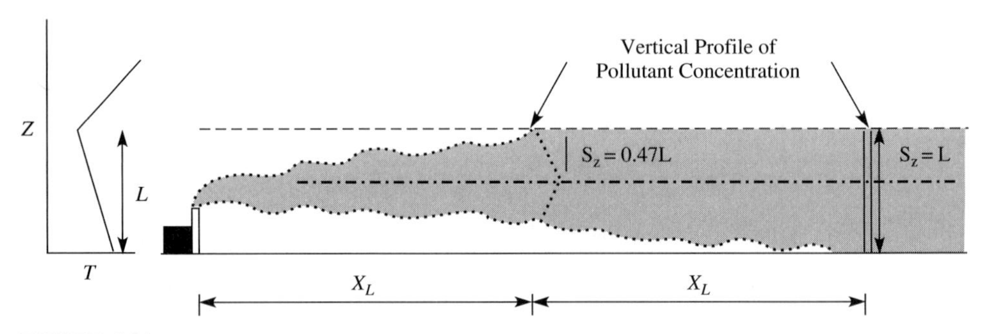

FIGURE 6-21
Effect of elevated inversion on dispersion.

Inversion aloft. When an inversion is present, the basic diffusion equation must be modified to take into account the fact that the plume cannot disperse vertically once it reaches the inversion layer. The plume will begin to mix downward when it reaches the base of the inversion layer (Figure 6-21). The downward mixing will begin at a distance x_L downwind from the stack. The x_L distance is a function of the stability in the layer below the inversion. It has been determined empirically that the vertical standard deviation of the plume can be calculated with the following formula at the distance x_L:

$$s_z = 0.47(L - H) \tag{6-24}$$

where L = height to bottom of inversion layer, m
H = effective stack height, m

When the plume reaches twice the distance to initial contact with the inversion base, the plume is said to be completely mixed throughout the layer below the inversion. Beyond a distance equal to $2x_L$ the centerline concentration of pollutants may be estimated by using the following equation:

$$\chi = \frac{E}{(2\pi)^{1/2} s_y(u)(L)} \tag{6-25}$$

Note that s_y is determined by the stability of the layer below the inversion and the distance to the receptor. We call this the "inversion" or "short form" of the dispersion equation.

Example 6-5. Determine the distance downwind from a stack at which we must switch to the "inversion form" of the dispersion model given the following meteorologic situation:

Effective stack height: 50 m
Inversion base: 350 m
Wind speed: 7.3 m/s
Cloud cover: none
Time: 1130 h
Season: summer

Solution. Determine the stability class using Table 6-6. At > 6 m/s with strong radiation, the stability class is C.

Calculate the value of s_z.

$$s_z = 0.47(350 \text{ m} - 50 \text{ m}) = 141 \text{ m}$$

Using Figure 6-20, find x_L. With $s_z = 141$, draw a horizontal line to stability class C. Drop a vertical line to the "distance downwind." Find $x_L = 2.5$ km.

Therefore, at any distance equal to or greater than 5 km downwind ($2x_L$), use the "inversion form" of the equation (Equation 6-25.)

For distances less than 5 km, we use Equation 6-19 with s_z determined from the distance to the point of interest and the stability. Thus, in no case do we use s_z computed from Equation 6-24 to calculate χ.

6-9 INDOOR AIR QUALITY MODEL

If we envision a house or room in a house or other enclosed space as a simple box (Figure 6-22), then we can construct a simple mass balance model to explore the behavior of the indoor air quality as a function of infiltration of outdoor, indoor sources and sinks, and leakage to the outdoor air. If we assume the contents of the box are well mixed, then

$$\begin{array}{c}\text{Rate of}\\ \text{pollutant}\\ \text{increase}\\ \text{in box}\end{array} = \begin{array}{c}\text{Rate of}\\ \text{pollutant}\\ \text{entering}\\ \text{box from}\\ \text{outdoors}\end{array} + \begin{array}{c}\text{Rate of}\\ \text{pollutant}\\ \text{entering box}\\ \text{from indoor}\\ \text{emissions}\end{array} - \begin{array}{c}\text{Rate of}\\ \text{pollutant}\\ \text{leaving box}\\ \text{by leakage}\\ \text{to outdoors}\end{array} - \begin{array}{c}\text{Rate of}\\ \text{pollutant}\\ \text{leaving box}\\ \text{by decay}\end{array} \qquad (6\text{-}26)$$

or

$$\forall \frac{dC}{dt} = QC_a + E - QC - kC\forall \qquad (6\text{-}27)$$

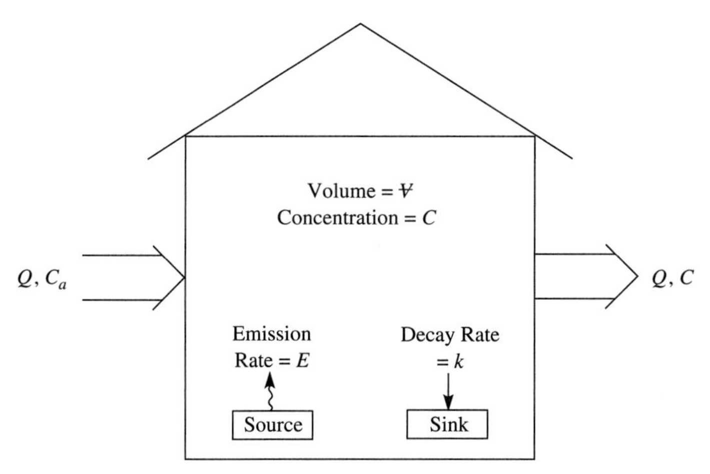

FIGURE 6-22
Mass-balance model for indoor air pollution.

TABLE 6-8
Reaction rate coefficients for selected pollutants

Pollutant	k, s^{-1}
CO	0.0
HCHO	1.11×10^{-4}
NO	0.0
NO_x (as N)	4.17×10^{-5}
Particulates (< 0.5 μm)	1.33×10^{-4}
Radon	2.11×10^{-6}
SO_2	6.39×10^{-5}

Source: G. W. Traynor, *et al., Indoor Air Pollution from Portable Kerosene-Fired Space Heaters, Wood-Burning Stoves, and Wood-Burning Furnaces,* Lawrence Berkeley Laboratory, Report No. LBL-14027, March, 1982.

where $\forall$ = volume of box, m^3
C = concentration of pollutant, g/m^3
Q = rate of infiltration of air into and out of box, m^3/s
C_a = concentration of pollutant in outdoor air, g/m^3
E = emission rate of pollutant into box from indoor source, g/s
k = pollutant reaction rate coefficient, s^{-1}

Reaction rate coefficients for a selected list of pollutants are given in Table 6-8.

The general solution for Equation 6-27 is

$$C_t = \frac{\frac{E}{\forall} + C_a \frac{Q}{\forall}}{\frac{Q}{\forall} + k}\left[1 - \exp\left(-\left(\frac{Q}{\forall} + k\right)t\right)\right] + C_o \exp\left[-\left(\frac{Q}{\forall} + k\right)t\right] \quad (6\text{-}28)$$

The steady-state solution for Equation 6-27 may be found by setting $dC/dt = 0$ and solving for C:

$$C = \frac{QC_a + E}{Q + k\forall} \quad (6\text{-}29)$$

When the pollutant is conservative and does not decay with time or have a significant reactivity, $k = 0$. In the special case when the pollutant is conservative and the ambient concentration is negligible and the initial indoor concentration is zero, Equation 6-27 reduces to:

$$C_t = \frac{E}{Q}\left[1 - \exp\left(-\left(\frac{Q}{\forall}\right)t\right)\right] \quad (6\text{-}30)$$

Example 6-6. An unvented kerosene heater is operated for one hour in an apartment having a volume of 200 m^3. The heater emits SO_2 at a rate of 50 $\mu g/s$. The ambient air concentration (C_a) and the initial indoor air concentration (C_0) of SO_2 are 100 $\mu g/m^3$.

If the rate of ventilation is 50 L/s, and the apartment is assumed to be well mixed, what is the indoor air concentration of SO_2 at the end of one hour?

Solution. The concentration may be determined using the general solution form of the indoor air quality model (Equation 6-28). The decay rate for SO_2 from Table 6-8 is $6.39 \times 10^{-5}\ s^{-1}$ and 50 L/s is equivalent to 0.050 m^3/s.

$$C_t = \frac{\dfrac{50\ \mu g/s}{(200\ m^3)} + 100\ \mu g/m^3 \dfrac{0.050\ m^3/s}{200\ m^3}}{\dfrac{0.050\ m^3/s}{200\ m^3} + 6.39 \times 10^{-5}\ s^{-1}}$$

$$\times \left[1 - \exp\left(-\left(\frac{0.050\ m^3/s}{200\ m^3} + 6.39 \times 10^{-5}\ s^{-1}\right)(3600\ s)\right)\right]$$

$$+ (100\ \mu g/m^3)\exp\left[-\left(\frac{0.050\ m^3/s}{200\ m^3} + 6.39 \times 10^{-5}\ s^{-1}\right)(3600\ s)\right]$$

$$= 876.08(1 - \exp(-1.13)) + 100\exp(-1.13) = 876.08(1 - 0.323) + 100(0.323)$$

$$= 593.09 + 32.3 = 625.39 \text{ or } 630\ \mu g/m^3$$

6-10 AIR POLLUTION CONTROL OF STATIONARY SOURCES

Gaseous Pollutants

Absorption. Control devices based on the principle of absorption attempt to transfer the pollutant from a gas phase to a liquid phase. This is a mass transfer process in which the gas dissolves in the liquid (see Section 3-1). The dissolution may or may not be accompanied by a reaction with an ingredient of the liquid. Mass transfer is a diffusion process wherein the pollutant gas moves from points of higher concentration to points of lower concentration. The removal of the pollutant gas takes place in three steps:

1. Diffusion of the pollutant gas to the surface of the liquid
2. Transfer across the gas/liquid interface (dissolution)
3. Diffusion of the dissolved gas away from the interface into the liquid

Structures such as spray chambers (Figure 6-23) and towers or columns (Figure 6-24) are two classes of devices employed to absorb pollutant gases. In scrubbers, which are a type of spray chamber, liquid droplets are used to absorb the gas. In towers, a thin film of liquid is used as the absorption medium. Regardless of the type of device, the solubility of the pollutant in the liquid must be relatively high. If water is the solute, this generally limits the application to a few inorganic gases such as NH_3, Cl_2, and SO_2. Scrubbers are relatively inefficient absorbers but have the advantage of being able to simultaneously remove particulates. Towers are much more efficient absorbers but they become plugged by particulate matter.

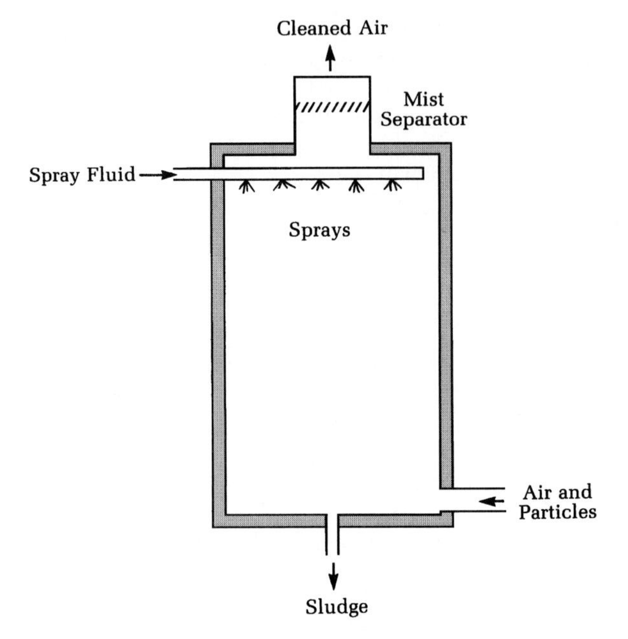

FIGURE 6-23
Spray chamber.

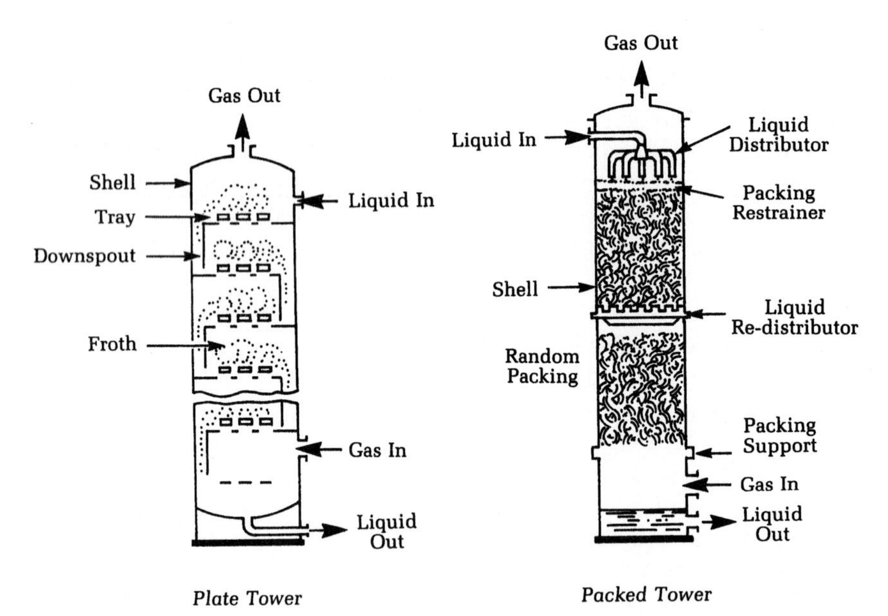

FIGURE 6-24
Absorption systems.

The amount of absorption that can take place for a nonreactive solution is governed by the partial pressure of the pollutant. For dilute solutions, as we have in pollution control systems, the relationship between partial pressure and the concentration of the gas in solution is given by *Henry's law:*

$$P_g = K_H C_{equil} \tag{6-31}$$

where P_g = partial pressure of gas in equilibrium with liquid, kPa
K_H = Henry's law constant, kPa · m^3/g
C_{equil} = concentration of pollutant gas in the liquid phase, g/m^3

Equation 6-31 implies that the partial pressure of the gas must increase as the liquid accumulates more pollutant or else it will come out of solution. Since the liquid is removing pollutant from the gas phase, this means the partial pressure is decreasing as the gas is cleaned. This is just the reverse of what we want to happen. The easiest way to get around this problem is to run the gas and liquid in opposite directions. This is called *countercurrent flow.* In this manner, the high concentration gas is absorbed into a liquid with a high pollutant concentration. The lower concentration gas is absorbed by liquid with no pollutants in it.

A mass-balance diagram of a countercurrent flow absorption column is shown in Figure 6-25. The mass-balance equation is

$$(G_{m1})(y_1) - (G_{m2})(y_2) = (L_{m1})(x_1) - (L_{m2})(x_2) \tag{6-32}$$

where G_{m1}, G_{m2} = total gas flow (air plus pollutant) into and out of the column respectively, kg · mole/h

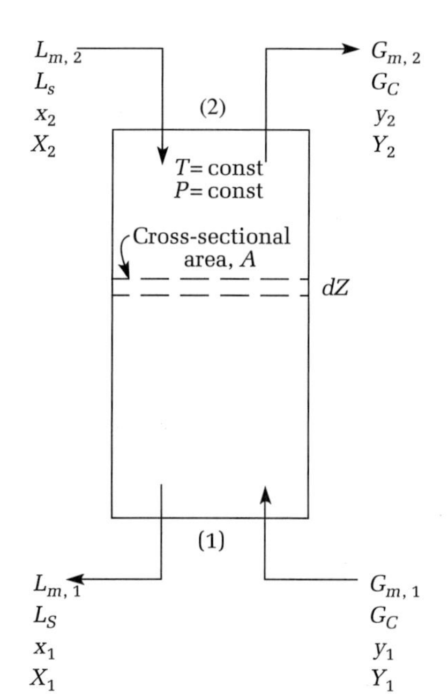

FIGURE 6-25
Notation for a counterflow packed absorption tower.

y_1, y_2 = mole fraction of pollutant in the gas phase at inlet and outlet of column, respectively[51]

L_{m1}, L_{m2} = total liquid flow (solvent plus absorbed pollutant) out of and into the column respectively, kg · mole/h

x_1, x_2 = mole fraction of pollutant in the liquid phase out of and into the column, respectively

Three variables of interest in the design of a packed tower are the gas flow rate, the liquid flow rate, and the height of the tower. As you might expect, the three are related. If we consider a differential height of the absorber, dZ, as shown in Figure 6-25, the total interfacial area open to mass transfer is defined as

$$\text{area for mass transfer} = (a)(A)(dZ) \tag{6-33}$$

where a = area per unit volume of packing
A = cross sectional area of column

As we did in Section 3-1, we may describe the rate of mass transfer of a gas, i, into solution (N_i) by the following differential equation:

$$N_i = \frac{dC}{dt} = K_y(y - y^*) \tag{6-34}$$

where K_y = overall mass transfer coefficient for gas
y, y^* = mole fraction of gaseous pollutant and equilibrium mole fraction, respectively

The rate of transfer of species i then is

$$\text{rate of mass transfer} = (N_i)(A)(a)(dZ) \tag{6-35}$$

This mass is equal to the mass loss from the gas phase as it passes through the differential height dZ:

$$\text{mass loss} = d(G_m y) \tag{6-36}$$

We may expand this expression by defining two new terms: the mass flow rate per unit area G'_m and the mole ratio:

$$Y = \frac{y}{1 - y} \tag{6-37}$$

and noting that

$$G_c Y = G_m y \tag{6-38}$$

---

[51] A mole fraction is defined as follows:

$$y = \frac{P}{P_t}$$

P = partial pressure of gas
P_t = total pressure of gas
$y^* = (P^*/P_t)$

where G_c is the mass flow of the carrier gas without the pollutant. Equating the mass transfer (Equation 6-34) with the mass loss (Equation 6-36) and making substitutions from Equations 6-35, 6-37, and 6-38 yields

$$K_y a(y - y^*)dZ = \frac{G'_m\,dy}{1 - y} \tag{6-39}$$

or

$$dZ = \frac{G'_m\,dy}{K_y a(y - y^*)(1 - y)} \tag{6-40}$$

The overall driving force $(y - y^*)$ at any location in the tower may be written in the form

$$y - y^* = (1 - y^*) - (1 - y) \tag{6-41}$$

It is convenient then to define the log-mean value of $(1 - y^*)$ and $(1 - y)$:

$$(1 - y)_{LM} = \frac{(1 - y^*) - (1 - y)}{\ln[(1 - y^*)/(1 - y)]} \tag{6-42}$$

Multiplying the numerator and denominator of Equation 6-40 by $(1-y)_{LM}$, we obtain

$$dZ = \left(\frac{G'_m}{K_y a(1 - y)_{LM}}\right)\left(\frac{(1 - y)_{LM}dy}{(y - y^*)(1 - y)}\right) \tag{6-43}$$

Although G'_m, $K_y a$, and $(1 - y)_{LM}$ vary along the absorption column, the first term of this equation is reasonably constant. This quantity is called the *overall height of a transfer unit* (H_{og}). As a first approximation to the height of the column, Equation 6-43 may be rewritten as

$$Z = (H_{og})\int_{y_2}^{y_1} \frac{(1 - y)_{LM}\,dy}{(y - y^*)(1 - y)} \tag{6-44}$$

The integral is called the *number of transfer units* (N_{og}). The height of the tower is computed from the following equation:

$$Z_t = (H_{og})(N_{og}) \tag{6-45}$$

For dilute solutions that obey Henry's law, the number of overall gas transfer units may be calculated as follows:[52]

$$N_{og} = \frac{\ln\left[\left(\dfrac{y_1 - mx_2}{y_2 - mx_2}\right)(1 - A) + A\right]}{1 - A} \tag{6-46}$$

where y_1, y_2 = mole fraction of pollutant in the gas phase at inlet and outlet of tower, respectively

[52]Robert E. Treybal, *Mass Transfer Operations,* New York: McGraw-Hill, p. 253, 1988.

m = slope of equilibrium curve defined by Henry's law = y^*/x^* in mole fraction units (m has no units)
x_2 = mole fraction of pollutant in the liquid phase entering the tower
$A = mQ_g/Q_l$
Q_l = liquid flow rate, kg · mole/h · m^2
Q_g = gas flow rate, kg · mole/h · m^2

The height of a single overall mass transfer unit (HTU) may also be expressed as the sum of the gas and liquid HTUs.

$$H_{og} = H_g + AH_l \tag{6-47}$$

where H_g and H_l are complex functions of the flow rate, surface area of the packing, viscosity of the liquid and air, and the diffusivity of the pollutant gas.

Example 6-7. Determine the height of a packed tower that is to reduce NH_3 in air from a concentration of 0.10 kg/m^3 to a concentration of 0.0005 kg/m^3 given the following data:

Column diameter = 3.00 m
Operating temperature = 20.0°C
Operating pressure = 101.325 kPa
H_g = 0.438 m
H_l = 0.250 m
$Q_g = Q_l$ = 10.0 kg/s
Incoming liquid is water free of NH_3

Solution. We begin by converting to mole fractions. NH_3 has a GMW of 17.030. For air we assume a GMW of 28.970 and a density of 1.185 kg/m^3 at 25°C. Since the operating temperature is 20°C, we correct the density of the air:

$$1.185 \times \frac{298}{293} = 1.205 \text{ kg/m}^3$$

Now we compute the mole fractions at the inlet (y_1) and outlet (y_2):

$$y_1 = \frac{\dfrac{0.10 \text{ kg/m}^3}{17.030 \text{ GMW NH}_3}}{\dfrac{1.205 \text{ kg/m}^3}{28.970 \text{ GMW air}}} = \frac{0.005872}{0.04159} = 0.14118$$

In a like manner, y_2 = 0.000706. Since the incoming liquid has no NH_3, the mole fraction is zero, that is, x_2 = 0.0.

The Henry's law constant in mole fraction units must be determined from experimental data. From the *Chemical Engineers' Handbook* we find the following data:[53]

[53]Robert H. Perry and Cecil H. Chilton, eds., *Chemical Engineers' Handbook,* 5th ed., New York: McGraw-Hill, pp. 3–96, 1973.

P_{NH_3} (kPa)	kg NH_3 per 100 kg H_2
15.199	15
9.319	10
4.266	5
1.600	1

If we convert each value to mole fractions and plot x^* versus y^* (the asterisk refers to the steady-state condition), the slope of the line will be m. An example calculation is shown for the first value of x^* and y^*. The total pressure is taken to be 101.325 kPa. The GMW of H_2O is 18.015. For 15 kg NH_3 per 100 kg H_2O:

$$x^* = \frac{\dfrac{15\text{ kg}}{17.030\text{ GMW NH}_3}}{\dfrac{15\text{ kg}}{17.030\text{ GMW NH}_3} + \dfrac{100\text{ kg}}{18.015\text{ GMW H}_2\text{O}}}$$

$$x^* = 0.1369$$

$$y^* = \frac{15.199\text{ kPa}}{101.325\text{ kPa}}$$

$$y^* = 0.1500$$

The value of m is then found by a least squares linear regression fit of a line through the four pairs of x^* and y^* values. The slope of the line is m.

$$m = 1.068$$

The value of A is computed in mole units as follows:

$$A = \frac{1.068\left[\dfrac{10.0\text{ kg/s of air}}{28.97\text{ GMW of air}}\right]}{\dfrac{10.0\text{ kg/s of H}_2\text{O}}{18.015\text{ GMW of H}_2\text{O}}}$$

$$= 0.6641$$

The number of gas transfer units is then

$$N_{og} = \frac{\ln\left[\dfrac{0.14118 - 1.068(0)}{0.000706 - 1.068(0)}(1 - 0.6641) + 0.6641\right]}{1 - 0.6641}$$

$$= 12.5545$$

The height of an individual gas transfer unit is

$$H_{og} = 0.438 + 0.6641(0.250) = 0.6040$$

The height of the tower is then

$$Z_t = (0.6040)(12.5545) = 7.5832$$

Since the limiting concentration data were given to only one significant figure, the answer would be

$$Z_t = 8\text{ m}$$

Before we leave this example, we should look back and see what we have wrought. Since the absorption tower neither creates nor destroys matter, the mass of NH_3 entering and leaving the column must be the same. If we assume isothermal, steady-state conditions (that is, gas and liquid rates in and out are equal), we can solve the mass-balance equation (Equation 6-32) for x_1. After some calculations we find $x_1 = 0.08734$. This is 90,300 mg/L of NH_3. This is a classic example of a multimedia problem. In solving an air pollution problem, we have created a serious water pollution problem. Catch-22!

Adsorption. This is a mass-transfer process in which the gas is bonded to a solid. It is a surface phenomenon. The gas (the *adsorbate*) penetrates into the pores of the solid (the *adsorbent*) but not into the lattice itself. The bond may be physical or chemical. Electrostatic forces hold the pollutant gas when physical bonding is significant. Chemical bonding is by reaction with the surface. Pressure vessels having a fixed bed are used to hold the adsorbent (Figure 6-26). Active carbon (activated charcoal), molecular sieves, silica gel, and activated alumina are the most common adsorbents. Active carbon is manufactured from nut shells (coconuts are great) or coal subjected to heat treatment in a reducing atmosphere. Molecular sieves are dehydrated zeolites (alkali-metal silicates). Sodium silicate is reacted with sulfuric acid to make silica gel. Activated alumina is a porous hydrated aluminum oxide. The common property of these adsorbents is a large "active" surface area per unit volume after treatment. They are very effective for hydrocarbon pollutants. In addition, they can capture H_2S

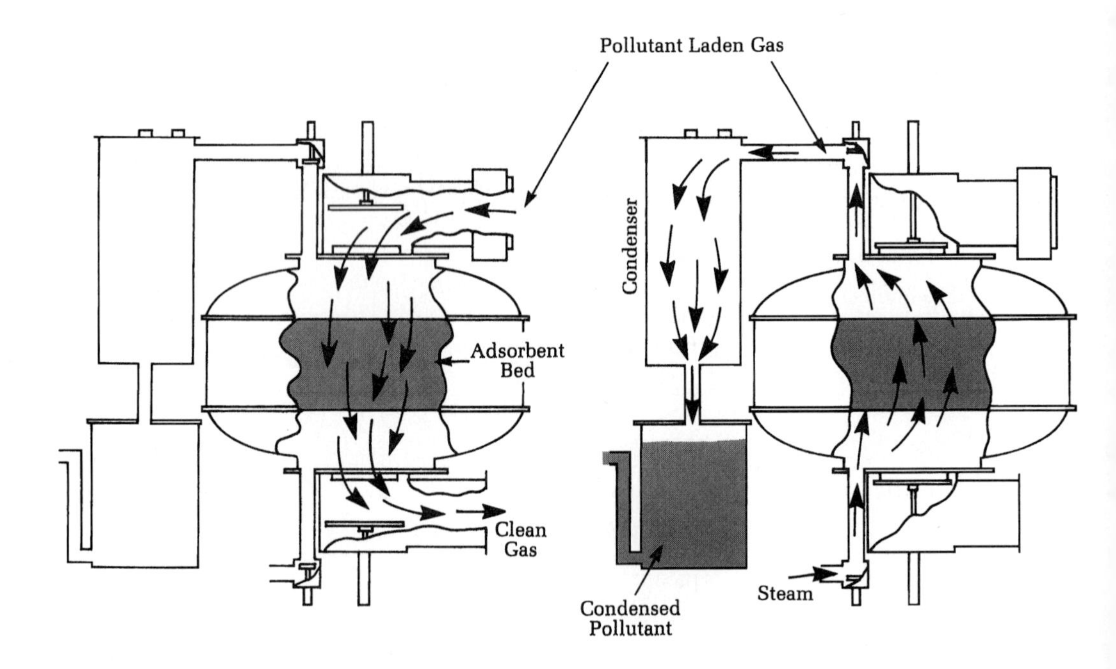

FIGURE 6-26
Adsorption system.

and SO_2. One special form of molecular sieve can also capture NO_2. With the exception of the active carbons, adsorbents have the drawback that they preferentially select water before any of the pollutants. Thus, water must be removed from the gas before it is treated. All of the adsorbents are subject to destruction at moderately high temperatures (150°C for active carbon, 600°C for molecular sieves, 400°C for silica gel, and 500°C for activated alumina). They are very inefficient at these high temperatures. In fact, their activity is regenerated at these temperatures!

The relation between the amount of pollutant adsorbed and the equilibrium pressure at constant temperature is called an *adsorption isotherm.* The equation that best describes this relation for gases is the one derived by Langmuir.[54]

$$W = \frac{aC_g^*}{1 + bC_g^*} \tag{6-48}$$

where W = amount of gas per unit mass of adsorbent, kg/kg
a, b = constants determined by experiment
C_g^* = equilibrium concentration of gaseous pollutant, g/m^3

In the analysis of experimental data, Equation 6-48 is rewritten as follows:

$$\frac{C_g^*}{W} = \frac{1}{a} + \frac{b}{a}C_g^* \tag{6-49}$$

In this arrangement, a plot of (C_g^*/W) versus C_g^* should yield a straight line with a slope of (b/a) and an intercept equal to $(1/a)$.

In contrast to absorption towers where the collected pollutant is continuously removed by flowing liquid, the collected pollutant remains in the adsorption bed. Thus, while the bed has sufficient capacity, no pollutants are emitted. At some point in time, the bed will become saturated with pollutant. As saturation is approached, pollutant will begin to leak out of the bed. This is called *breakthrough.* When the bed capacity is exhausted, the influent and effluent concentration will be equal. A typical breakthrough curve is shown in Figure 6-27. In order to allow for continuous operation, two beds are provided (Figure 6-26). While one is collecting pollutant, the other is being regenerated. The concentrated gas released during regeneration is usually returned to the process as recovered product. The critical factor in the operation of the bed is the length of time it can operate before breakthrough occurs. The time to breakthrough may be calculated from the following:[55]

$$t_B = \frac{Z_t - \delta}{v_f} \tag{6-50}$$

where Z_t = height of bed, m
δ = width of adsorption zone, m
v_f = velocity of adsorption zone as defined by Equation 6-52, m/s

---

[54] A. J. Buonicore and L. Theodore, *Industrial Control Equipment for Gaseous Pollutants, Vol. I,* Cleveland: CRC Press, pp. 149–150, 1975.

[55] M. Crawford, *Air Pollution Control Theory,* New York: McGraw-Hill, p. 516, 1976.

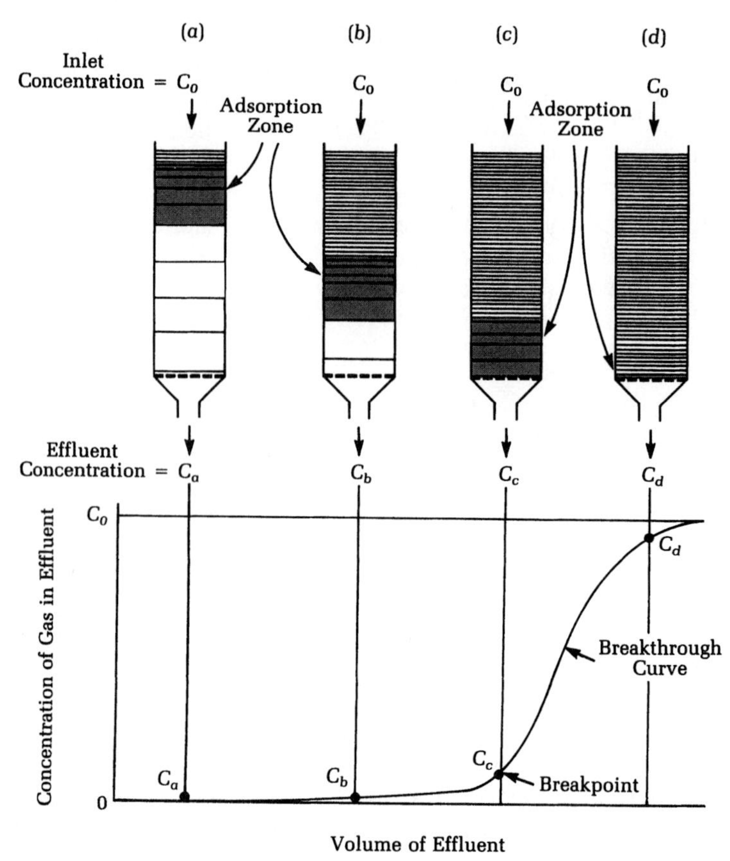

FIGURE 6-27
Adsorption wave and breakthrough curve. (*Source:* Treybal, *Mass Transfer Operations,* New York: McGraw-Hill, 1968. Reprinted by permission.)

The height of the adsorption bed (Z_t) can be determined in the same manner as it was for absorption towers, with a few exceptions. The value of N_{og} must be determined by integration of the following expression:[56]

$$N_{og} = \int_{C_2}^{C} \frac{dC}{C - C_g^*} \tag{6-51}$$

where C_g^* = the equilibrium partial pressure described by Equation 6-48 and C is described by the operating line (Figure 6-28). The H_{og} equation is modified by replacing H_l with H_s. The value of slope m in Equation 6-46 is determined from Equation 6-49.

The width of the adsorption zone is shown in Figure 6-27. It is a function of the shape of the adsorption isotherm.

[56]Robert E. Treybal, *Mass Transfer Operations,* New York: McGraw-Hill, p. 535, 1968.

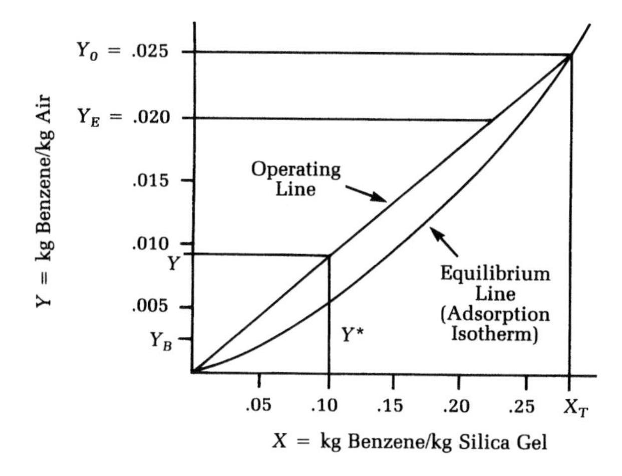

FIGURE 6-28
Equilibrium and operating lines for adsorption of benzene on silica gel. (*Source:* John H. Seinfeld, *Air Pollution,* New York: McGraw-Hill, 1975. Reprinted by permission.)

The velocity of the adsorption zone may be calculated from the properties of the system:

$$v_f = \frac{(Q_g)(1 + bC_g^*)}{a\rho_s\rho_g A_c} \tag{6-52}$$

where ρ_s, ρ_g = density of solid and gas, kg/m^3 (Note that ρ_s is the density of the absorbent "as packed.")
A_c = cross-sectional area of bed, m^2

Example 6-8. Determine the breakthrough time for an adsorption bed that is 0.50 m thick and 10 m^2 in cross section. The operating parameters for the bed are as follows:

Gas flow rate = 1.3 kg/s of air
Gas temperature = 25°C
Gas pressure = 101.325 kPa
Bed density as packed = 420 kg/m^3
Inlet pollutant concentration = 0.0020 kg/m^3
Langmuir parameters: a = 18; b = 124
Width of adsorption zone = 0.03 m

Solution. Using Table A-3 in Appendix A and the gas temperature and pressure, we interpolate to find ρ_g = 1.184 kg/m^3. Then the face velocity of the adsorption wave is

$$v_f = \frac{(1.3 \text{ kg/s})[1 + 124(0.0020 \text{ kg/m}^3)]}{(18)(420 \text{ kg/m}^3)(1.184 \text{ kg/m}^3)(10 \text{ m}^2)}$$
$$= 1.8 \times 10^{-5} \text{ m/s}$$

The breakthrough time is calculated directly from Equation 6-50:

$$t_B = \frac{0.50 \text{ m} - 0.03 \text{ m}}{1.8 \times 10^{-5} \text{ m/s}}$$
$$= 2.6 \times 10^4 \text{ s or } 7.2 \text{ h}$$

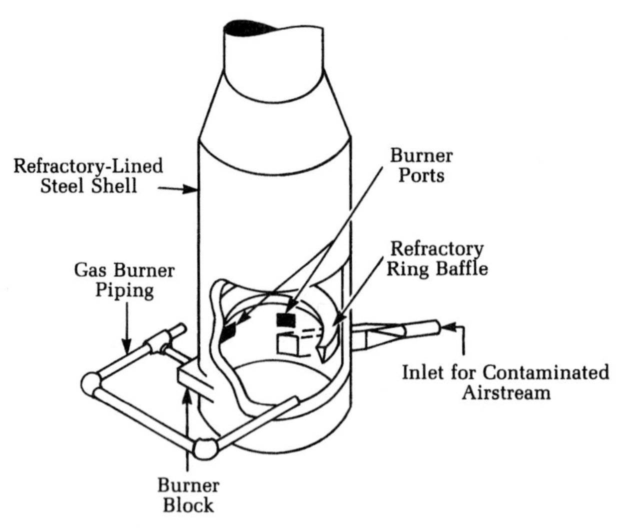

FIGURE 6-29
Direct flame incineration.

Combustion. When the contaminant in the gas stream is oxidizable to an inert gas, combustion is a possible alternative method of control. Typically, CO and hydrocarbons fall into this category. Both direct flame combustion by afterburners (Figure 6-29) and catalytic combustion have been used in commercial applications.

Direct flame incineration is the method of choice if two criteria are satisfied. First, the gas stream must have an energy concentration greater than 3.7 MJ/m^3. At this energy concentration, the gas flame will be self-supporting after ignition. Below this point, supplementary fuel is required. The second requirement is that none of the by-products of combustion be toxic. In some cases the combustion by-product may be more toxic than the original pollutant gas. For example, the combustion of trichloroethylene produces phosgene, which was used as a poison gas in World War I. Direct flame incineration has been successfully applied to varnish-cooking, meat-smokehouse, and paint bake-oven emissions.

Some catalytic materials enable oxidation to be carried out in gases that have an energy content of less than 3.7 MJ/m^3. Conventionally, the catalyst is placed in beds similar to adsorption beds. Frequently, the active catalyst is a platinum or palladium compound. The supporting lattice is usually a ceramic. Aside from expense, a major drawback of the catalysts is their susceptibility to poisoning by sulfur and lead compounds in trace amounts. Catalytic combustion has successfully been applied to printing-press, varnish-cooking, and asphalt-oxidation emissions.

Flue Gas Desulfurization (FGD)

Flue gas desulfurization systems fall into two broad categories: nonregenerative and regenerative. Nonregenerative means that the reagent used to remove the sulfur oxides from the gas stream is used and discarded. Regenerative means that the reagent is recovered and reused. In terms of the number and size of systems installed, nonregenerative systems dominate.

Nonregenerative systems. There are nine commercial nonregenerative systems.[57] All have reaction chemistries based on lime (CaO), caustic soda (NaOH), soda ash (Na_2CO_3), or ammonia (NH_3).

The SO_2 removed in a lime/limestone-based FGD system is converted to sulfite. The overall reactions are generally represented by:[58]

$$SO_2 + CaCO_3 \rightarrow CaSO_3 + CO_2 \tag{6-53}$$

$$SO_2 + Ca(OH)_2 \rightarrow CaSO_3 + H_2O \tag{6-54}$$

when using limestone and lime, respectively. Part of the sulfite is oxidized with the oxygen content in the flue gas to form sulfate:

$$CaSO_3 + \frac{1}{2}O_2 \rightarrow CaSO_4 \tag{6-55}$$

Although the overall reactions are simple, the chemistry is quite complex and not well defined. The choice between lime and limestone, the type of limestone, and method of calcining and slaking can influence the gas-liquid-solid reactions taking place in the absorber.

The principal types of absorbers used in the wet scrubbing systems include venturi scrubber/absorbers, static packed scrubbers, moving-bed absorbers, tray towers, and spray towers.[59]

Spray dryer-based FGD systems consist of one or more spray dryers and a particulate collector.[60] The reagent material is typically a slaked lime slurry or a slurry of lime and recycled material. Although lime is the most common reagent, soda ash has also been used. The reagent is injected in droplet form into the flue gas in the spray dryer. The reagent droplets absorb SO_2 while simultaneously being dried. Ideally, the slurry or solution droplets are completely dried before they impact the wall of the dryer vessel. The flue gas stream becomes more humidified in the process of evaporation of the reagent droplets, but it does not become saturated with water vapor. This is the single most significant difference between spray dryer FGD

[57]S. B. Hance and J. L. Kelly, "Status of Flue Gas Desulfurization Systems," Paper No. 91-157.3, 84th Annual Meeting of the Air and Waste Management Association, 1991.

[58]H. T. Karlsson and H. S. Rosenberg, "Technical Aspects of Lime/Limestone Scrubbers for Coal fired Power Plants, Part I: Process Chemistry and Scrubber Systems," *Journal of the Air Pollution Control Association,* vol. 30 (6), pp. 710–714, 1980.

[59]Black & Veatch Consulting Engineers, *Lime FGD Systems Data Book - Second Edition,* EPRI Publication No. CS-2781, 1983.

[60]Historically, from a mass transfer point of view, spray drying refers to the evaporation of a solvent from an atomized spray. Simultaneous diffusion of a gaseous species into the evaporating droplet is not true spray drying. Nonetheless, many authors have adopted the term "spray drying" as synonymous with dry scrubbing.

and wet scrubber FGD. The humidified gas stream and a significant portion of the particulate matter (fly ash, FGD reaction products, and unreacted reagent) are carried by the flue gas to the particulate collector located downstream of the spray dryer vessel.[61]

Control Technologies for Nitrogen Oxides

Almost all nitrogen oxide (NO_x) air pollution results from combustion processes. They are produced from the oxidation of nitrogen bound in the fuel, from the reaction of molecular oxygen and nitrogen in the combustion air at temperatures above 1,600 K (see Equation 6-12), and from the reaction of nitrogen in the combustion air with hydrocarbon radicals. Control technologies for NO_x are grouped into two categories: those that prevent the formation of NO_x during the combustion process and those that convert the NO_x formed during combustion into nitrogen and oxygen.[62]

Prevention. The processes in this category employ the fact that reduction of the peak flame temperature in the combustion zone reduces NO_x formation. Nine alternatives have been developed to reduce flame temperature: (1) minimizing operating temperatures, (2) fuel switching, (3) low excess air, (4) flue gas recirculation, (5) lean combustion, (6) staged combustion, (7) low NO_x burners, (8) secondary combustion, and (9) water/steam injection.

Routine burner tune-ups and operation with combustion zone temperatures at minimum values reduce the fuel consumption and NO_x formation. Converting to a fuel with a lower nitrogen content or one that burns at a lower temperature will reduce NO_x formation. For example, petroleum coke has a lower nitrogen content and burns with a lower flame temperature than coal. On the other hand, natural gas has no nitrogen content but burns at a relatively high flame temperature and, thus, produces more NO_x than coal.

Low excess air and flue gas recirculation work on the principle that reduced oxygen concentrations lower the peak flame temperatures. In contrast, in lean combustion, additional air is introduced to cool the flame.

In staged combustion and low NO_x burners, initial combustion takes place in a fuel-rich zone that is followed by the injection of air downstream of the primary combustion zone. The downstream combustion is completed under fuel-lean conditions at a lower temperature.

Staged combustion consists of injecting part of the fuel and all of the combustion air into the primary combustion zone. Thermal NO_x production is limited by the low flame temperatures that result from high excess air levels.

---

[61] A. L. Cannell and M. L. Meadows, "Effects of Recent Operating Experience on the Design of Spray Dryer FGD Systems," *Journal of the Air Pollution Control Association,* vol. 35 (7), pp. 782–789, 1985.

[62] A. Prasad, "Air Pollution Control Technologies for Nitrogen Oxides," *The National Environmental Journal,* May/June, pp. 46–50, 1995.

Water/steam injection reduces thermal NO_x emissions by lowering the flame temperature.

Post-combustion. Three processes may be used to convert NO_x to nitrogen gas: selective catalytic reduction (SCR), selective noncatalytic reduction (SNCR), and nonselective catalytic reduction (NSCR).

The SCR process uses a catalyst bed (usually vanadium-titanium, or platinum-based and zeolite) and anhydrous ammonia (NH_3). After the combustion process, ammonia is injected upstream of the catalyst bed. The NO_x reacts with the ammonia in the catalyst bed to form N_2 and water.

In the SNCR process ammonia or urea is injected into the flue gas at an appropriate temperature (870 to 1,090°C). The urea is converted to ammonia, which reacts to reduce the NO_x to N_2 and water.

NSCR uses a three-way catalyst similar to that used in automotive applications. In addition to NO_x control, hydrocarbons and carbon monoxide are converted to CO_2 and water. These systems require a reducing agent similar to CO and hydrocarbons upstream of the catalyst.

Particulate Pollutants

Cyclones. For particle sizes greater than about 10 μm in diameter, the collector of choice is the cyclone (Figure 6-30). This is an inertial collector with no moving parts. The particulate-laden gas is accelerated through a spiral motion, which imparts a centrifugal force to the particles. The particles are hurled out of the spinning gas and impact on the cylinder wall of the cyclone. They then slide to the bottom of

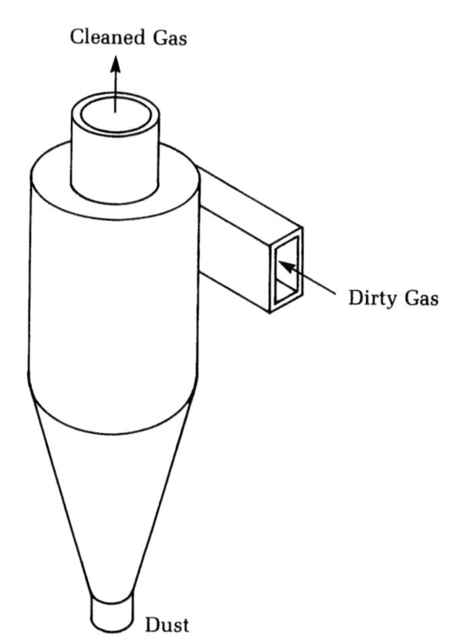

FIGURE 6-30
Reverse flow cyclone. (*Source:* M. Crawford, *Air Pollution Control Theory,* New York: McGraw-Hill, 1976. Reprinted by permission.)

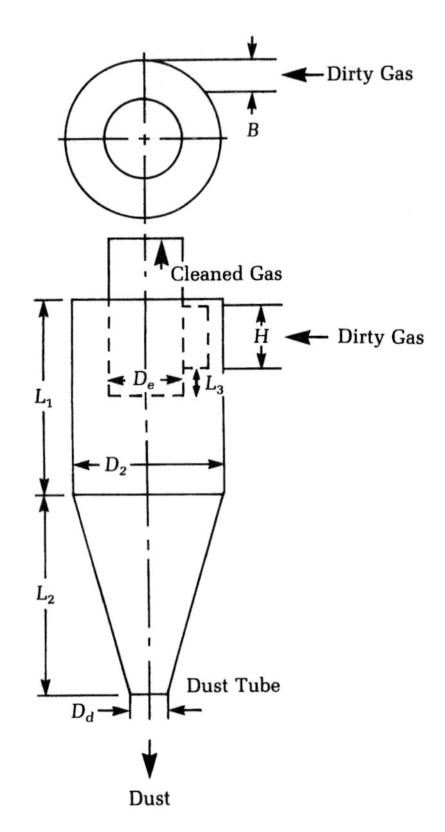

FIGURE 6-31
Standard reverse flow cyclone proportions.
Note: Standard cyclone proportions are as follows:

Length of cylinder, $L_1 = 2D_2$
Length of cone, $L_2 = 2D_2$
Diameter of exit, $D_e = 0.5D_2$
Height of entrance, $H = 0.5D_2$
Width of entrance, $B = 0.25D_2$
Diameter of dust exit, $D_d = 0.25D_2$
Length of exit duct, $L_3 = 0.125D_2$

(*Source:* M. Crawford, *Air Pollution Control Theory,* New York: McGraw-Hill, 1976. Reprinted by permission.)

the cone. Here they are removed through an airtight valving system. The standard single-barrel cyclone will have dimensions proportioned as shown in Figure 6-31.

The efficiency of collection of various particle sizes (η) can be determined from an empirical expression and graph (Figure 6-32) developed by Lapple:[63]

$$d_{0.5} = \left[\frac{9\mu B^2}{\rho_p Q_g}\frac{H}{\theta}\right]^{1/2} \tag{6-56}$$

where $d_{0.5}$ = cut diameter, the particle size for which the collection efficiency is 50 percent
μ = dynamic viscosity of gas, Pa·s
B = width of entrance, m
H = height of entrance, m
ρ_p = particle density, kg/m^{3}s

[63]C. E. Lapple, "Processes Use Many Collection Types," *Chemical Engineer,* vol. 58, pp. 144–151, 1951.

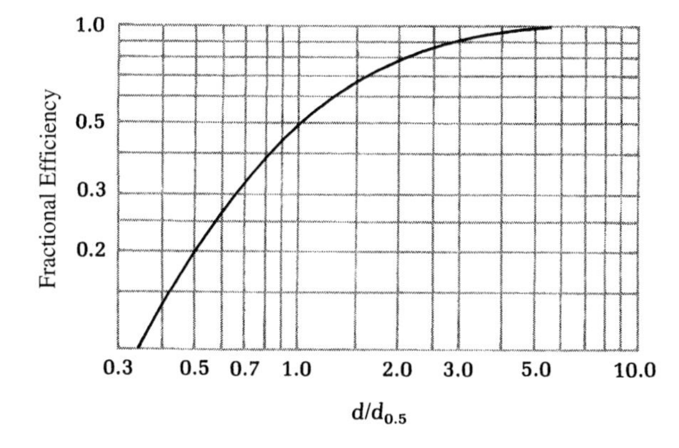

FIGURE 6-32
Empirical cyclone collection efficiency. (*Source:* C. E. Lapple, "Processes Use Many Collection Types," *Chemical Engineer,* vol. 58, p. 147, 1951.)

Q_g = gas flow rate, m^3/s
θ = effective number of turns made in traversing the cyclone as defined in Equation 6-57.

The value of θ may be determined approximately by the following:

$$\theta = \frac{\pi}{H}(2L_1 + L_2) \tag{6-57}$$

where L_1 and L_2 are the length of the cylinder and cone, respectively.

Example 6-9. Determine the efficiency of a "standard" cyclone having the following characteristics for particles 10 μm in diameter with a density of 800 kg/m^3:

Cyclone barrel diameter = 0.50 m
Gas flow rate = 4.0 m^3/s
Gas temperature = 25°C

Solution. From the standard cyclone dimensions we can calculate the following:

$$B = (0.25)(0.50 \text{ m}) = 0.13 \text{ m}$$
$$H = (0.50)(0.50 \text{ m}) = 0.25 \text{ m}$$
$$L_1 = L_2 = (2.00)(0.50 \text{ m}) = 1.0 \text{ m}$$

The number of turns i is then

$$\theta = \frac{\pi}{0.25}[2(1.0) + 1.0]$$
$$= 37.7$$

From the gas temperature and Table A-4 of Appendix A, we find the dynamic viscosity is 18.5 μPa $\cdot$ s. The cut diameter is then

$$d_{0.5} = \left[\frac{9(18.5 \times 10^{-6} \text{ Pa} \cdot \text{s})(0.13 \text{ m})^2(0.25 \text{ m})}{(800 \text{ kg/m}^3)(4.0 \text{ m}^3\text{/s})(37.7)}\right]^{1/2}$$
$$= 2.41 \times 10^{-6} \text{ m} = 2.41 \ \mu\text{m}$$

The ratio of particle sizes is

$$\frac{d}{d_{0.5}} = \frac{10\ \mu\text{m}}{2.41\ \mu\text{m}} = 4.15$$

From Figure 6-32 we find that the collection efficiency is about 95 percent.

As the diameter of the cyclone is reduced, the efficiency of collection is increased. However, the pressure drop also increases. This increases the power requirements for moving the gas through the collector. Since an efficiency increase will result, even if the tangential velocity remains constant, the efficiency may be increased without increasing the power consumption by using multiple cyclones in parallel (*multiclones*).

From the example, you can see that cyclones are quite efficient for particles larger than 10 μm. Conversely, you should note that cyclones are not very efficient for particles 1 μm or less in diameter. Thus, they are employed only for coarse dusts. Some applications include controlling emissions of wood dust, paper fibers, and buffing fibers. Multiclones are frequently used as precleaners for fly-ash control devices in power plants.

Filters. When high efficiency control of particles smaller than 5 μm is desired, a filter may be selected as the control method. Two types are in use: (1) the deep bed filter, and (2) the baghouse (Figure 6-33). The deep bed filter resembles a furnace filter. A packing of fibers is used to intercept particles in the gas stream. For

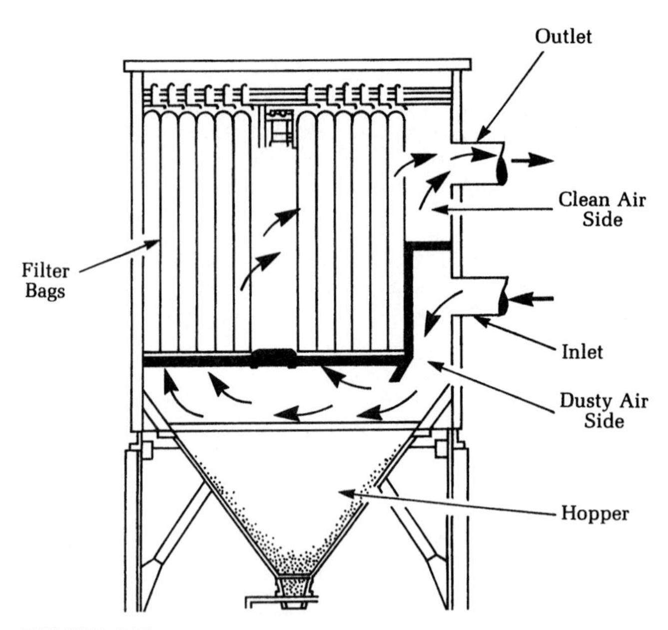

FIGURE 6-33
Baghouse.

relatively clean gases and low volumes, such as air conditioning systems, these are quite effective. For dirty industrial gas with high volumes, the baghouse is preferable.

The fundamental mechanisms of collection by filters include screening or sieving because the particles are larger than the openings between the fibers, interception by the fibers themselves, and electrostatic attraction because of the difference in static charge on the particle and fiber. Once a dust cake begins to form on the fabric, sieving is probably the dominant mechanism.

The bags are made of either natural or synthetic fibers. Synthetic fibers are widely used as filtration fabrics because of their low cost, better temperature- and chemical-resistance characteristics, and small fiber diameter. Bag life varies between one and five years. Two years is considered normal. Bag diameters range from 0.1 to 0.35 m. Their lengths vary between 2 and 10 m. The bags are suspended from the toe and fastened by a collar at the open end. They are arranged in groups in separate compartments.

Reverse-air baghouses operate by directing the dirty gas into the inside of the bag. The particulate matter is collected on the inside of the bag much in the same manner as a vacuum cleaner bag. The bags are cleaned by isolating a compartment and reversing the gas flow. The reverse flow combined with the inward collapse of the bag causes the collected dust cake to fall into the hopper below.

Pulse-jet baghouses are designed with frame structures, called cages, that support the bags. The particulate matter is collected on the outside of the bag. The dust cake is removed by directing a pulsed jet of compressed air into the bag. This causes a sudden expansion of the bag. Dust is removed primarily by inertial forces as the bag reaches maximum expansion.

Baghouses have found a wide variety of applications. Examples include the carbon black industry, cement crushing, feed and grain handling, gypsum, limestone crushing, and sanding machines. Baghouse application to boiler flue gas is finding wide acceptance currently, in contrast to past practice. This is because of the better thermal properties of the bag material. Cotton and wool fiber bags, for example, cannot be used for sustained temperatures above 90–100°C. Glass fiber bags, however, can be used at temperatures up to 260°C. Of all of the particulate control devices, only the filtration method has potential to include the addition of adsorption media to facilitate concurrent removal of gas phase contaminants. Baghouse sizing is based on the ratio of filter area to gas flow rate ($m^3/s \cdot m^2$ of cloth area). Note that this air to cloth ratio is a velocity (m/s). An average value would be about 0.01 $m^3/s \cdot m^2$ for a woven fabric in a conventional baghouse.

Liquid scrubbing. When the particulate matter to be collected is wet, corrosive, or very hot, the fabric filter may not work. Liquid scrubbing might. Typical scrubbing applications include control of emission of talc dust, phosphoric acid mist, foundry cupola dust, and open hearth steel furnace fumes.

Liquid scrubbers vary in complexity. Simple spray chambers are used for relatively coarse particle sizes. For high efficiency removal of fine particles, the combination of a venturi scrubber followed by a cyclone would be selected (Figure 6-34). The underlying principle of operation of the liquid scrubbers is that a differential

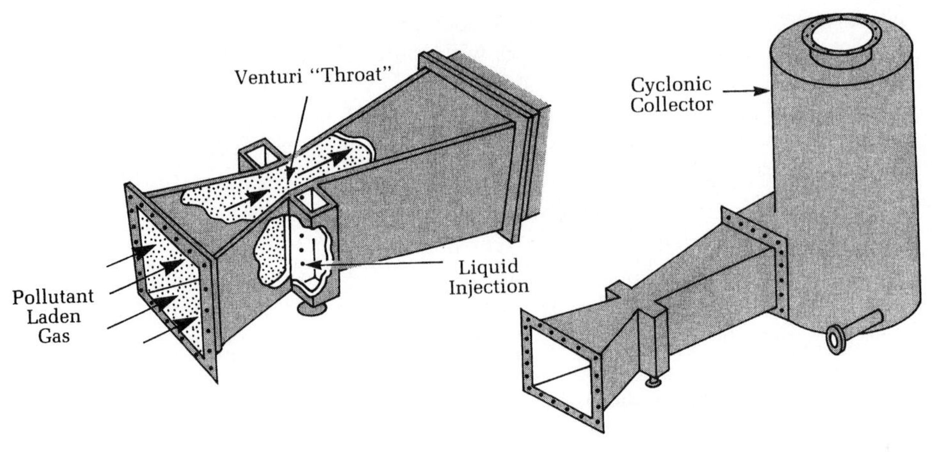

FIGURE 6-34
Venturi scrubber.

velocity between the droplets of collecting liquid and the particulate pollutant allows the particle to impinge onto the droplet. Since the droplet-particle combination is still suspended in the gas stream, an inertial collection device is placed downstream to remove it. Because the droplet enhances the size of the particle, the collection efficiency of the inertial device is higher than it would be for the original particle without the liquid drop.

The most popular collection efficiency equation is that proposed by Johnstone, Field, and Tassler:[64]

$$\eta = 1 - \exp(-\kappa R \sqrt{\psi}) \tag{6-58}$$

where η = efficiency
exp = exponential to base e
κ = correlation coefficient, m^3 of gas/m^3 of liquid
R = liquid flow rate, m^3/m^3 of gas
ψ = inertial impaction parameter defined by Equation 6-59

The inertial impaction parameter (ψ) relates the particle and droplet sizes and relative velocities:

$$\psi = \frac{C\rho_p v_g (d_p)^2}{18 d_d \mu} \tag{6-59}$$

where C = Cunningham correction factor defined by Equation 6-60, unitless
ρ_p = particle density, kg/m^3
v_g = speed of gas at throat, m/s

[64] H. F. Johnstone, R. B. Field, and M. C. Tassler, "Gas Absorption and Aerosol Collection in a Venturi Atomizer," *Industrial Engineering Chemistry,* vol. 46, pp. 1601–1608, 1954.

d_p = diameter of particle, m
d_d = diameter of droplet, m
μ = dynamic viscosity of gas, Pa · s

The Cunningham correction factor accounts for the fact that very small particles do not obey Stokes' settling equation. They tend to "slip" between the gas molecules. Thus, the drag coefficient (C_D) is reduced and the particles fall faster than otherwise would be expected. This is particularly true for particles less than 1 μm in diameter. The Cunningham factor may be approximated with the following equation:[65]

$$C = 1 + \frac{6.21 \times 10^{-4}(T)}{d_p} \tag{6-60}$$

where T = absolute temperature, K
d_p = diameter of particle, μm

Example 6-10. Given the scrubber described below, write an expression for collection efficiency that is a function of particle size. Assume the particles are fly ash with a density of 700 kg/m³ and a minimum size of 10 μm diameter.

Venturi characteristics:

Throat area = 1.00 m²
Gas flow rate = 94.40 m³/s
Gas temperature = 150°C
Liquid flow rate = 0.13 m³/s
Coefficient κ = 200
Droplet diameter = 100 μm

Solution. We begin by determining the value of the Cunningham correction factor for the smallest particle to see if the d_p term in the denominator must be retained.

$$C = 1 + \frac{6.21 \times 10^{-4}(423 \text{ K})}{10\ \mu\text{m}}$$

$$= 1 + 0.0263$$

For this we can see that the term containing d_p will be small for all particles greater than 10 μm and we can use the approximation:

$$C \approx 1$$

Before we can proceed to calculate a value for ψ, we must determine the gas velocity at the throat:

$$v_g = \frac{Q_g}{A_t}$$

[65]H. E. Hesketh, *Fine Particles in Gaseous Media,* Ann Arbor, MI: Ann Arbor Science, p. 19, 1977.

where A_t = cross-sectional area of throat

$$v_g = \frac{94.40 \text{ m}^3/\text{s}}{1.00 \text{ m}^2} = 94.40 \text{ m/s}$$

The dynamic viscosity of the gas is determined from Table A-4 of Appendix A and from the temperature of the gas (150°C). It is 25.2 μPa · s.

Now we can calculate in terms of d_p in μm. Note that $C = 1$ and that 18 is a constant.

$$\psi = \frac{(1)(700 \text{ kg/m}^3)(94.40 \text{ m}^3/\text{s})(1 \times 10^{-12} \ \mu\text{m}^2/\text{m}^2)(d_p)^2}{(18)(100 \times 10^{-6} \text{ m})(25.2 \times 10^{-6} \text{ Pa} \cdot \text{s})}$$

$$= (1.46)(d_p)^2$$

Taking the square root of ψ and computing R as 0.13/94.40, the expression for efficiency as a function of diameter is then

$$\eta = 1 - \exp[-(200)(1.38 \times 10^{-3})(1.21)d_p]$$

$$= 1 - \exp[-0.33(d_p)]$$

Electrostatic precipitation (ESP). High efficiency, dry collection of particles from hot gas streams can be obtained by electrostatic precipitation of the particles. The ESP is usually constructed of alternating plates and wires (Figure 6-35). A large direct current potential (30–75 kV) is established between the plates and wires. This results in the creation of an ion field between the wire and plate (Figure 6-36*a*). As the particle-laden gas stream passes between the wire and the plate, ions attach to the particles, giving them a net negative charge (Figure 6-36*b*). The particles then

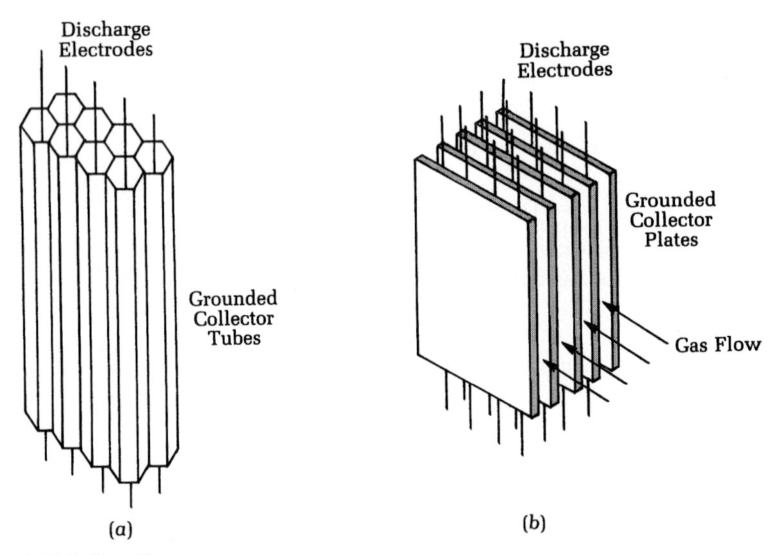

FIGURE 6-35
Electrostatic precipitator with (*a*) wire in tube, (*b*) wire and plate. (*Source:* EPA Training Manual.)

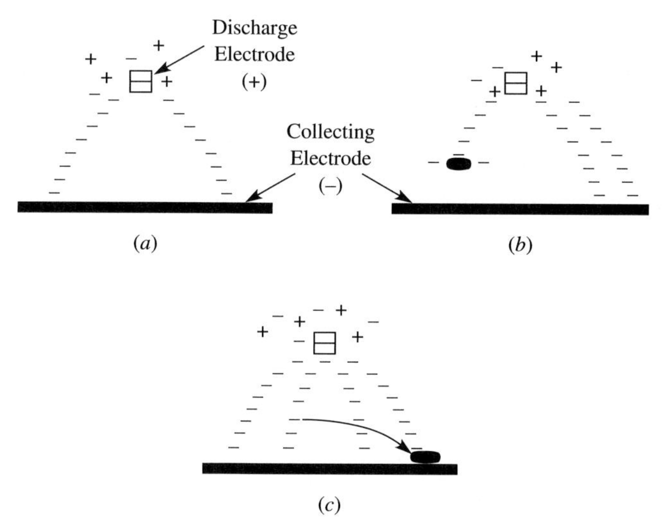

FIGURE 6-36
Particle charging and collection in ESP. (*Source:* EPA Training Manual.)

migrate toward the positively charged plate where they stick (Figure 6-36*c*). The plates are rapped at frequent intervals and the agglomerated sheet of particles falls to a hopper.

Unlike the baghouse, the gas flow between the plates is not stopped during cleaning. The gas velocity through the ESP is kept low (less than 1.5 m/s) to allow particle migration. Thus, the terminal settling velocity of the sheet is sufficient to carry it to the hopper before it exits the precipitator.

The classic ESP efficiency equation is the one proposed by Deutsch.[66]

$$\eta = 1 - \exp\left(-\frac{Aw}{Q_g}\right) \tag{6-61}$$

where A = collection area of plates, m^2
w = migration velocity of particles, m/s
Q_g = gas flow rate, m^3/s

The migration velocity of the particles is a function of the electrostatic force. The migration velocity is described by the following equation:

$$w = \frac{qE_pC}{6\pi r\mu} \tag{6-62}$$

where q = charge, coulombs (C)
E_p = collection field intensity, volts/m
r = particle radius, m

[66]W. Deutsch, *Ann. Phys.* (Leipzig), vol. 68, p. 335, 1922.

μ = dynamic viscosity of gas, Pa · s
C = Cunningham correction factor

Example 6-11. Determine the collection efficiency of the electrostatic precipitator described below for a particle 154 μm in diameter having a drift velocity of 0.184 m/s. What is the effect of reducing the plate spacing to one-half of its current value and doubling the number of plates?

ESP specifications:
Height = 7.32 m
Length = 6.10 m
Number of passages = 5
Plate spacing = 0.28 m
Gas flow rate = 19.73 m^3/s

Solution. First we calculate the area of the plates. For a single plate,

$$A = 7.32 \times 6.10 = 44.65 \text{ m}^2$$

Since there are eight collecting surfaces (two for each plate, 4 plates form 5 passages):

$$A = 44.65 \text{ m}^2 \times 8 = 357.2 \text{ m}^2$$

The efficiency is then calculated in a straightforward manner using Equation 6-61.

$$\eta = 1 - \exp\left[-\frac{(357.2)(0.184)}{19.73}\right]$$
$$= 0.964$$

Therefore the efficiency is 96.4 percent. Now what is the effect of reducing the plate spacing? The spacing enters into the efficiency equation through the calculation of the collection field intensity (E_p). Treating everything else in Equation 6-62 as a constant, we can write the following equation:

$$w = KE_p$$

where E_p is measured in volts per meter. If the distance between the plates is reduced, the collection field intensity is proportionately increased:

$$w = KE_p \frac{0.28}{0.14} = 2KE_p$$

Thus, w increases by a factor of two. In order to maintain the same gas velocity, the number of plates and, hence, the surface area (A) must double. The new efficiency would then be:

$$\eta = 1 - \exp\left[-\frac{(714.4)(0.368)}{19.73}\right]$$
$$= 1 - 0.0000016$$
$$= 0.999998 \text{ or } 1.00$$

Thus, the efficiency would be increased to 100 percent. This plate spacing may not be feasible because of sparkover problems.

One operational problem of ESPs is of particular note. *Fly ash* is a generic term used to describe the particulate matter carried in the effluent gases from furnaces burning fossil fuels. ESPs often are used to collect fly ash. The strongest force holding fly ash to the collection plate is electrostatic and is caused by the flow of current through the fly ash. The fly ash acts like a resistor and, hence, resists the flow of current. This resistance to current flow is called the resistivity of the fly ash. It is measured in units of ohm · cm. If the resistivity is too low (less than 10^4 ohm · cm), not enough charge will be retained to produce a strong force and the particles will not "stick" to the plate. Conversely, and often more importantly, if the resistivity is too high (greater than 10^{10} ohm · cm), there is an insulating effect. The layer of fly ash breaks down locally and a local discharge of current (*back corona*) from the normally passive collection electrode occurs. This discharge lowers the sparkover voltage and produces positive ions that decrease particle charging and, hence, collection efficiency.

The presence of SO_2 in the gas stream reduces the resistivity of the fly ash. This makes particle collection relatively easy. However, the mandate to eliminate SO_2 emissions has frequently been satisfied by switching to low sulfur coal. The result has been increased particulate emissions. This problem can be resolved by adding conditioners such as SO_3 or NH_3 to reduce the resistivity or by building larger precipitators.

Electrostatic precipitators have been used to control air pollution from electric power plants, Portland cement kilns, blast furnace gas, kilns and roasters for metallurgical processes, and mist from acid production facilities.

6-11 AIR POLLUTION CONTROL OF MOBILE SOURCES

Engine Fundamentals

Before we examine some cures for the pollution from the common gasoline auto engine, it may be useful to compare the three familiar types of engines: the gasoline engine, the diesel engine, and the jet engine.

The gasoline engine. Each of the four strokes of the engine is diagrammed in Figure 6-37. In the typical automobile engine with no air pollution controls, a mixture of fuel and air is fed into a cylinder and is compressed and ignited by a spark from the spark plug. The explosive energy of the burning mixture moves the pistons. The pistons' motion is transmitted to the crankshaft that drives the car. The burnt, spent mixture passes out of the engine and out through the tail pipe.

One kg of gasoline can burn completely when mixed with about 15 kg of air. For maximum power, however, the proportion of air to fuel must be less. Most driving takes place at less than the 15-to-1 air-to-fuel ratio. Combustion is incomplete, and substantial amounts of material other than carbon dioxide and water are discharged through the tail pipe. One result of having an inadequate supply of air is the emission of carbon monoxide instead of carbon dioxide. Other by-products are unburned gasoline and hydrocarbons.

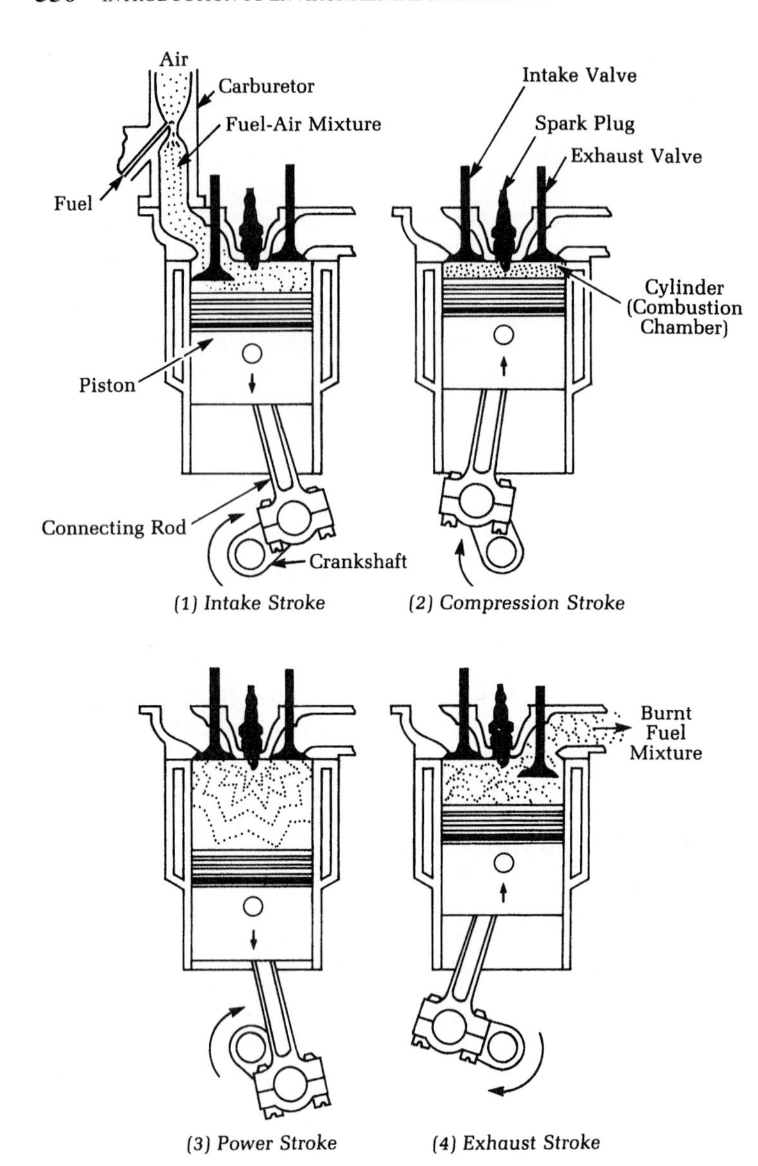

FIGURE 6-37
Combustion in an automobile engine. On the intake stroke (*1*), the piston moves down and a mixture of fuel and air is drawn into the cylinder past the open intake valve. With the compression stroke (*2*), the intake valve closes and the piston moves and compresses the air-fuel mixture. On the power stroke (*3*), a spark from the spark plug ignites the heated, compressed mixture, which begins to burn, expands, and pushes the piston down. For the exhaust stroke (*4*), the exhaust valve opens, the spent, burned mixture exits with its pollutants, and the piston returns to the top of the cylinder. (*Source: Air Pollution Primer,* National Tuberculosis and Respiratory Disease Association, 1969. Reprinted by permission.)

Because of the high temperatures and pressures that exist in the cylinder, copious amounts of NO_x are formed (see Equation 6-12).

The diesel engine. As shown in Figure 6-38, the diesel engine differs from the four-stroke engine in three respects.

First, the air supply is unthrottled; that is, its flow into the engine is unrestricted. Thus, a diesel normally operates at a higher air-to-fuel ratio than does a gasoline engine.

Second, the fuel is injected directly into the combustion chamber, so no carburetor is required. The power output is changed by the rate of fuel injection.

Third, there is no spark ignition system. The air is heated by compression. That is, the air in the engine cylinder is squeezed until it exerts a pressure high enough to raise the air temperature to about 540°C, which is enough to ignite the fuel oil as it is injected into the cylinder.

A well-designed, well-maintained, and properly adjusted diesel engine will emit less CO and hydrocarbons than the four-stroke engine because of the diesel's high air-to-fuel ratio. However, the higher operating temperatures lead to substantially higher NO_x emissions.

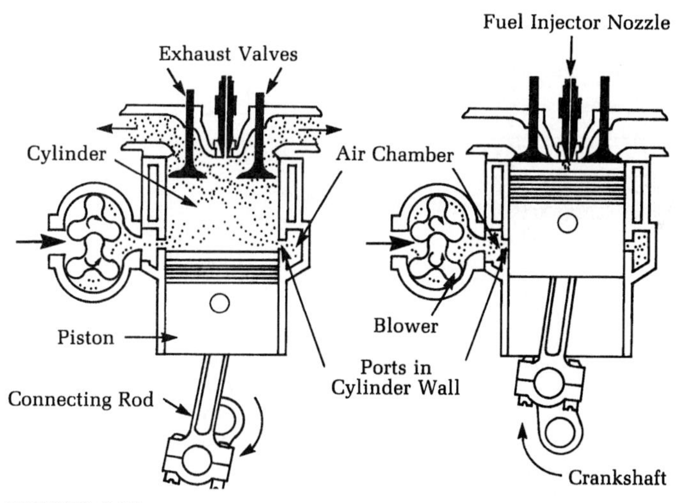

FIGURE 6-38
Combustion in a diesel engine. With the piston at the bottom of the cylinder (shown left), and exhaust valves and ports open, fresh air is forced into the cylinder by the blower, and the used air-fuel mixture—along with any polluting byproducts—left from the previous stroke is forced out. On the second stroke (shown right), the exhaust valves close, the piston rises—shutting off the ports—and compresses the air. When the piston reaches a position near the top of the cylinder, fuel is injected into the now highly compressed, heated air. This heated air ignites the fuel without a spark, and the resulting combustion forces the piston down to its first position. (*Source: Air Pollution Primer,* National Tuberculosis and Respiratory Disease Association, 1969. Reprinted by permission.)

The jet engine. Large commercial aircraft that utilize the thrust of compressed gases for propulsion may contribute significant amounts of particulates and NO_x to urban atmospheres.

As shown in Figure 6-39, air drawn into the front of the engine is compressed and then heated by burning fuel. The expanding gas passes through turbine blades, which drive the compressor. The gas then exits the engine through an exhaust nozzle.

Effect of design and operating variables on emissions. The list of variables that affect internal combustion (automobile) emissions includes the following:[67]

1. air-to-fuel ratio
2. load or power level
3. speed
4. spark timing
5. exhaust back pressure
6. valve overlap
7. intake manifold pressure

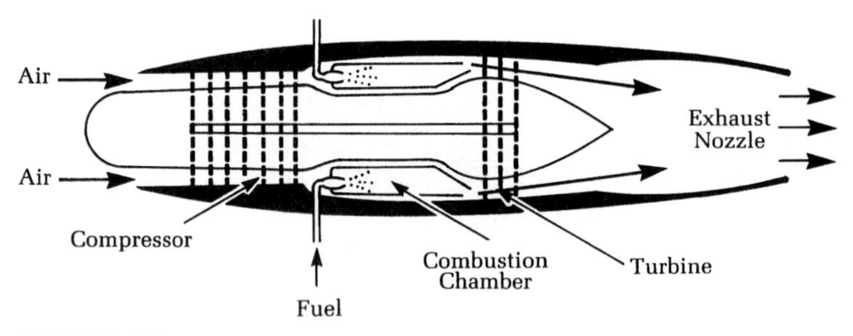

FIGURE 6-39
Combustion in a jet engine. Air enters through the front and goes to a compressor, where it is increasingly compressed and forced into combustion chambers that are arranged in circles around the engine. Fuel is sprayed into the front end of the combustion chamber in a steady stream so that it ignites and burns continuously. The burning air-fuel mixture expands and pushes toward the rear. (On the way, it hits turbine wheel blades and forces them to rotate. This rotation drives the compressor.) As the expanded mixture moves toward the tailpipe, the areaway narrows and the stream of burning air-fuel mixture is compressed into the exceedingly strong jet stream that shoots out of the rear of the plane. (*Source: Air Pollution Primer,* National Tuberculosis and Respiratory Disease Association, 1969. Reprinted by permission.)

---

[67]D. J. Patterson and N. A. Henein, *Emissions from Combustion Engines and Their Control,* Ann Arbor, MI: Ann Arbor Science, p. 143, 1972.

8. combustion chamber deposit buildup
9. surface temperature
10. surface-to-volume ratio
11. combustion chamber design
12. stroke-to-bore ratio
13. displacement per cylinder
14. compression ratio

A discussion of all of these items is beyond the scope of an introductory text such as this. Therefore, we shall restrict ourselves to a few of the variables that serve to illustrate the kinds of problems encountered in trying to design pollution out of an internal combustion engine.

The air-to-fuel ratio (A/F) is fairly easy to regulate. As we noted previously, it has a direct effect on all three emissions. As shown in Figure 6-40*a*, the A/F of 14.6 is the *stoichiometric* mixture for complete combustion.[68] At lower ratios, both CO and HC emissions increase. At higher ratios, to about 15.5, NO_x emissions increase. At very lean mixtures (high A/F ratios), the NO_x emission begins to decrease. Thus, one of the approaches taken to control emissions is to set the A/F ratio at a very lean setting. This is why a cold engine is hard to start. At an A/F greater than 17, the gas mixture will not ignite properly. We would be remiss if we did not show the relationship between A/F and fuel economy and "power" (Figure 6-40*b*).

Increasing the engine speed (not the speed of the vehicle) decreases the emission of hydrocarbons. This occurs because of decreases in the unburned fuel in the cylinder crevices and decreases in the quench gas (the gas above the piston) not reacted in the chamber. NO_x emissions increase until a peak value is reached in A/F.

Retarding the ignition timing decreases the HC emissions as a result of decreasing the unburned fuel from the crevices. NO_x emissions also decrease with increased retarding. Little or no change occurs in CO emissions until the retard becomes excessive, at which point CO emissions go up.

Lowering the compression ratio decreases both the HC and NO_x emissions. It has almost no effect on CO emissions. Of course, lower compression ratios mean a slower response.

Control of Automobile Emissions

Blowby. The flow of air past the moving vehicle is directed through the crankcase in order to rid it of any gas-air mixture that has blown past the pistons, any evaporated lubricating oil, and any escaped exhaust products. The air is drawn in through a vent and emitted through a tube extending from the crankcase at a rate that depends on the speed of the car. About 20 to 40 percent of the car's total hydrocarbon

[68]Note that stoichiometric (stoi-chio-met-ric) means "combined in exactly the proper proportions according to their molecular weight."

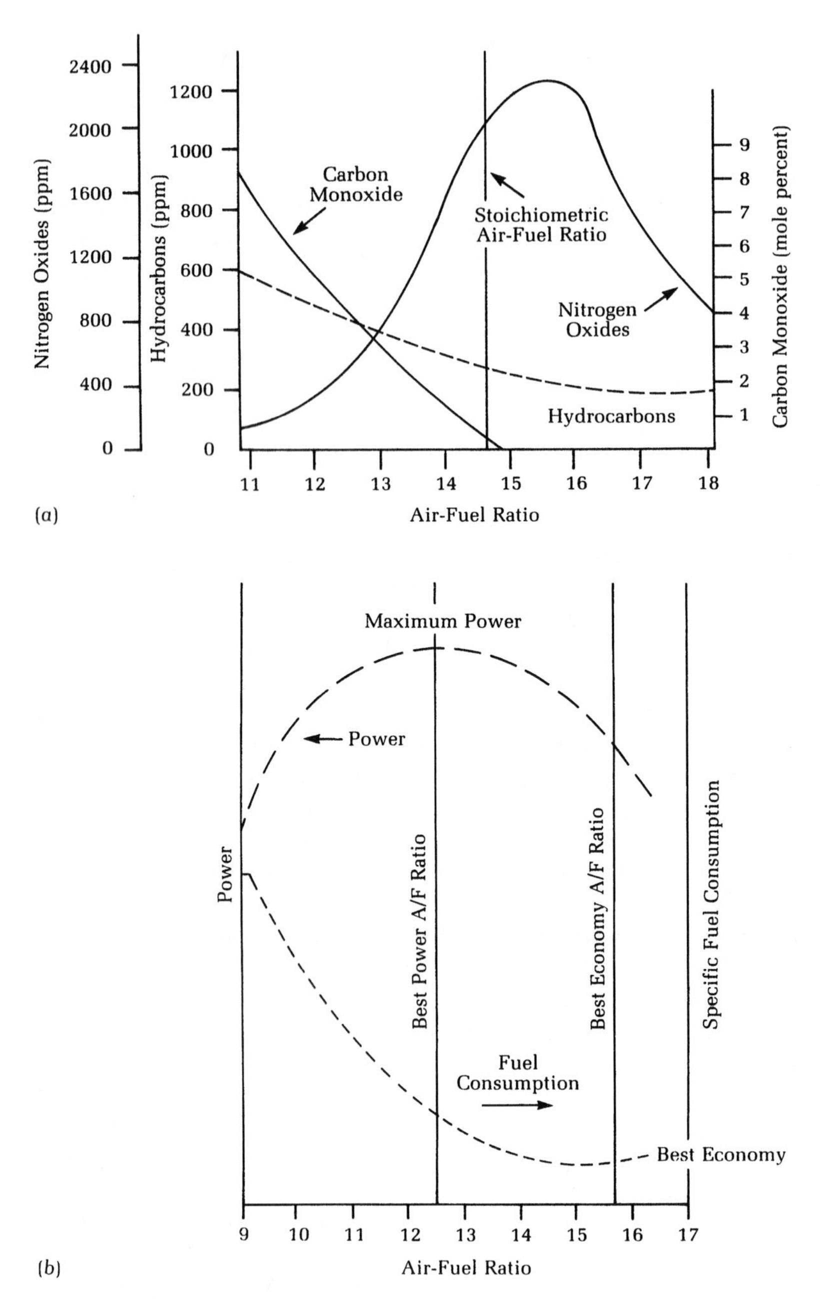

FIGURE 6-40
Effect of air-to-fuel ratio (*a*) on emissions, and (*b*) on power and economy. (*Source:* John H. Seinfeld, *Air Pollution.* Reprinted by permission.)

emissions are sent into the atmosphere from the crankcase. These emissions are called *crankcase blowby.* All vehicles manufactured after 1963 are required to have a positive crankcase ventilation (PCV) valve to eliminate blowby emissions.

Fuel tank evaporation losses. Evaporation of volatile hydrocarbons from the fuel tank is controlled by one of two systems. The simplest system is to place an activated charcoal adsorber in the tank vent line. Thus, as the gasoline expands during warm weather and forces vapor out of the vent, the HC is trapped on the activated carbon.

An alternative system is to vent the tank to the crankcase. With this method, it is more difficult to achieve 100 percent control than with the activated charcoal system.

Carburetor evaporation losses. During engine operation, the hydrocarbon vapors generated in the carburetor are vented internally to the engine intake system. After the engine is shut off, the gasoline in the float bowl continues to evaporate because of the high temperature in the engine compartment. This phenomenon is called *hot soak.* These losses may be controlled by the use of an activated carbon adsorption system (called a canister) or by venting the vapors to the crankcase. The canister system is preferred. Modern fuel injection systems do not have carburetors and thus avoid evaporation losses.

Engine exhaust. The number of techniques for reducing engine exhaust emissions far exceeds the list of engine variables that contribute to the production of emissions. In general, the control strategies can be grouped into three categories: engine modifications, fuel system modifications, and exhaust treatment devices. To some extent all three techniques have been employed. The stringent requirements of the Clean Air Act have led to the general adoption of catalytic converter exhaust treatment systems.

A dual catalyst system consisting of an NO-reducing catalyst followed by an HC/CO-oxidizing catalyst seems to be the best remedy at present. The catalyst materials are either noble metals or base metals deposited on an inert support material.

The major problems with the catalysts are their susceptibility to "poisoning" by lead, phosphorus, and sulfur, and their poor wear characteristics under thermal cycling. The poisoning problem is solved by removing the lead, phosphorus, and sulfur from the fuel.

Another approach being implemented is fuel modification. The use of lead in fuels was completely phased out by January 1996. In addition, diesel fuel refining is being changed so that it will contain less sulfur and emit 20 percent less VOC's. Lowering the gasoline vapor pressure (called the *Reid vapor pressure*) reduces hydrocarbon emissions. *Oxyfuel* is yet another alternative. Oxyfuel is one with more oxygen to allow the fuel to burn more efficiently. Other alternatives include alcohols, liquified petroleum gas, and natural gas.

Inspection/maintenance (I/M) programs. The devices installed by automobile manufacturers are extremely successful in minimizing the pollution from the exhaust and from evaporating fuel. However, as with other aspects of running an automobile, these devices wear out and fail. Since their failure does not inhibit the operation of the automobile, they are not likely to be repaired by the owner. In those areas that have exceeded the NAAQS (nonattainment areas), inspection/maintenance programs have been implemented to ensure that the control devices are in good working order. These programs require periodic checks of the exhaust and, in some instances, the evaporative controls. If the vehicle fails the inspection, the owner is required to provide the required maintenance and have the vehicle reinspected. Failure to pass the inspection may be cause to deny the issuance of license plates or tags.

6-12 WASTE MINIMIZATION

The best and first step in any air pollution control strategy should be to minimize the production of pollutants in the first place. Since a large proportion of air pollutants results from the combustion of fossil fuels, an obvious approach to waste minimization is to conserve energy. Modern technology has yielded more efficient furnaces that improve fuel use, but simple measures such as turning off the lights in unoccupied rooms, turning down the heat at night and, in factories, during weekends and holidays, can have a dramatic impact. Because of the interrelationship between energy consumption and water supply, water conservation also reduces air pollution. In a similar manner, building smaller, lighter automobiles reduces air pollution because less fuel is burned to propel them, but alternatives such as mass transit, walking, and bicycles can contribute significantly to reduced fuel consumption. Alternative sources of energy such as solar, wind, and nuclear also reduce air pollution emissions. (Nuclear power, of course, has a series of pollution problems that may outweigh the benefits of reduced air pollution.)

The chlorofluorocarbon destruction of the ozone layer can only be resolved by waste minimization. Preventing the escape of CFCs from refrigeration systems, the use of alternative propellants for spray cans, and similar measures are the only ones that will be successful, since control devices make no sense. Waste minimization is, in fact, the method of control specified by the Montreal Protocol (see Section 6-6). In a similar fashion, the production of ozone in the lower atmosphere can only be reduced by minimizing the release of precursor hydrocarbons and the production of NO_x. Reduced use of solvents and the substitution of water-based paints for solvent-based paints are examples of methods to reduce hydrocarbon release.

6-13 CHAPTER REVIEW

When you have completed studying this chapter, you should be able to do the following without the aid of your textbook or notes:

1. List the six criteria air pollutants for which the U.S. Environmental Protection Agency has designated National Ambient Air Quality Standards (NAAQS).
2. List and define three units of measure used to report air pollution data (that is, ppm, $\mu g/m^3$, and μm).
3. Explain the difference between ppm in air pollution and ppm in water pollution.
4. Explain the effect of temperature and pressure on readings made in ppm.

5. Explain the influence of moisture, temperature, and sunlight on the severity of air pollution effects on materials.
6. Differentiate between acute and chronic health effects from air pollution.
7. State which particle sizes are more important with respect to alveolar deposition and explain why.
8. Explain why it is difficult to define a causal relationship between air pollution and health effects.
9. List three potential chronic health effects of air pollution.
10. List four common features of air pollution episodes and identify the locations of three "killer" episodes.
11. Discuss the natural and anthropogenic origin of the six criteria air pollutants and identify the likely mechanisms for their removal from the atmosphere.
12. Identify one indoor air pollution source for each of the following pollutants: CH_2O, CO, NO_x, Rn, respirable particulates, and SO_x.
13. Define the term "acid rain" and explain how it occurs.
14. Discuss the photochemistry of ozone in the upper atmosphere using the pertinent chemical reactions. Discuss the hypothesized effect of chlorofluorocarbons on these reactions.
15. Explain the term "greenhouse effect", its hypothesized cause, and why it is being debated, pro and con.
16. Determine the stability (ability to dissipate pollutants) of the atmosphere from vertical temperature readings.
17. Explain why valleys are more susceptible to inversions than is flat terrain.
18. Explain why lake breezes and land breezes occur.
19. Explain how a lake breeze adversely affects the dispersion of pollutants.
20. State the theoretical principle on which each of the following air pollution control devices operates: (*a*) absorption column (either a packed tower or plate tower), (*b*) adsorption column, (*c*) either afterburner or catalytic combustor, (*d*) cyclone, (*e*) baghouse, (*f*) venturi scrubber, and (*g*) electrostatic precipitator.
21. Select the correct air pollution control device for a given pollutant and source.
22. Discuss the pros and cons of FGD and the problem of fly ash resistivity.
23. Explain the difference between prevention and post-combustion techniques for reduction of nitrogen oxide emissions and give one example of each.
24. Graph the relationship between air-to-fuel ratio and emission of CO, HC, and NO_x from automobiles.
25. Explain what "blowby" is and how it is controlled.
26. Explain how evaporative emissions and exhaust emissions are commonly controlled.

With the aid of this text, you should be able to do the following:

1. Solve gas law problems.
2. Convert parts per million (ppm) to micrograms per cubic meter ($\mu g/m^3$) and vice versa.

3. Calculate the amount of SO_2 that will be released from burning coal or fuel oil with a given sulfur content in percent.
4. Calculate the ground level concentration of air pollutants released from a stationary elevated source or the emission rate (Q) for a given ground level concentration.
5. Use the air pollution control equations to analyze the performance and modify the design of absorption and adsorption control devices, cyclones, scrubbers, and ESPs.

6-14 PROBLEMS

6-1. What is the density of oxygen at a temperature of 273.0 K and at a pressure of 98.0 kPa?

Answer: 1.382 kg/m^3

6-2. Determine the density of nitrogen gas at a pressure of 122.8 kPa and a temperature of 298.0 K.

6-3. Show that one mole of any ideal gas will occupy 22.414 L at standard temperature and pressure (STP). (STP is 273.16 K and 101.325 kPa.)

6-4. What volume would one mole of an ideal gas occupy at 25.0°C and 101.325 kPa?

6-5. A sample of air contains 8.583 moles/m^3 of oxygen and 15.93 moles/m^3 of nitrogen at STP. Determine the partial pressures of oxygen and nitrogen in 1.0 m^3 of the air.

Answer: 19.45 kPa; 36.18 kPa

6-6. A 1 m^3 volume tank contains a gas mixture of 18.32 moles of oxygen, 16.40 moles of nitrogen, and 6.15 moles of carbon dioxide. What is the partial pressure of each component in the gas mixture at 25.0°C?

6-7. Calculate the volume occupied by 5.2 kg of carbon dioxide at 152.0 kPa and 315.0 K.

Answer: 2,036 L

6-8. Determine the mass of oxygen contained in a 5.0 m^3 volume under a pressure of 568.0 kPa and at a temperature of 263.0 K.

6-9. A gas mixture at 0°C and 108.26 kPa contains 250 mg/L of H_2S gas. What is the partial pressure exerted by this gas?

Answer: 16.7 kPa/L

6-10. A 28-liter volume of gas at 300.0 K contains 11 g of methane, 1.5 g of nitrogen, and 16 g of carbon dioxide. Determine the partial pressure exerted by each gas.

6-11. Given the gas mixture of Problem 6-10, how many moles of each gas are present in the 28-liter volume?

Answer: 0.688 M of CH_4; 0.054 M of N_2; 0.364 M of CO_2

6-12. The partial pressures of the gases in a 22.414-liter volume of air at STP are: oxygen, 21.224 kPa; nitrogen, 79.119 kPa; argon, 0.946 kPa; and carbon dioxide, 0.036 kPa. Determine the gram-molecular weight of air.

6-13. Convert 80 $\mu g/m^3$ of SO_2 to ppm at 25°C and 101.325 kPa pressure.

Answer: 0.031 ppm

6-14. Convert 0.55 ppm of NO_2 to $\mu g/m^3$ at 290 K and 100.0 kPa pressure.

6-15. Given the following temperature profiles, determine whether the atmosphere is unstable, neutral, or stable. Show all work and explain choices.

a.

Z, M	*T(°C)*
2	−3.05
318	−6.21

b. *Z, M* *T* (°C)
10 5.11
202 3.09
c. *Z, M* *T* (°C)
18 14.03
286 16.71
Answers: (*a*) Neutral; (*b*) Unstable; (*c*) Stable (inversion)

6-16. Determine the atmospheric stability for each of the following temperature profiles. Show all work and explain choices.
a. *Z, m* *T*(°C)
1.5 −4.49
349 0.10
b. *Z, m* *T* (°C)
12 28.05
279 19.67
c. *Z, m* *T* (°C)
8 18.55
339 17.93

6-17. Given the following observations, use the "Key to Stability Categories" (Table 6-6) to determine the stability.
a. Clear winter morning at 9:00 A.M.; wind speed of 5.5 m/s
b. Overcast summer afternoon at 1:30 P.M.; wind speed of 2.8 m/s
c. Clear winter night at 2:00 A.M.; wind speed of 2.8 m/s
d. Summer morning at 11:30 A.M.; wind speed of 4.1 m/s
Answers: (*a*) D; (*b*) D; (*c*) F; (*d*) B

6-18. Determine the atmospheric stability category of the following observations.
a. Clear summer afternoon at 1:00 P.M.; wind speed of 1.6 m/s
b. Overcast summer night at 1:30 A.M.; wind speed of 2.1 m/s
c. Overcast winter morning at 9:30 A.M.; wind speed of 6.6 m/s
d. Thinly overcast winter night at 8:00 P.M.; wind speed of 2.4 m/s

6-19. A power plant in a college town is burning coal on a cold, clear winter morning at 8:00 A.M. with a wind speed of 2.6 m/s and an inversion layer with its base at a height of 697 m. The effective stack height is 30 m. Calculate the distance downwind x_L at which the plume released will reach the inversion layer and begin to mix downward.
Answer: 5.8 km

6-20. A factory releases a plume into the atmosphere on an overcast summer afternoon. At what distance downwind will the plume begin mixing downward if an inversion layer exists at a base height of 414 m and the windspeed is 1.8 m/s? The effective stack height is 45 m.

6-21. Given the same power plant and conditions as were found in Example 6-4, determine the concentration of SO_2 at a point 4 km downwind and 0.2 km perpendicular to the plume centerline ($y = 0.2$ km) if there is an inversion with a base height of 328 m.
Answer: 1.16×10^{-3} g/m^3

6-22. On a clear summer afternoon with a wind speed of 3.20 m/s, the TSP concentration was found to be 1,520 μg/m^3 at a point 2 km downwind and 0.5 km perpendicular to the plume centerline from a coal-fired power plant. Given the following parameters and conditions, determine the TSP emission rate of the power plant:

Stack parameters:
Height = 75.0 m
Diameter = 1.50 m
Exit velocity = 12.0 m/s
Temperature = 322°C

Atmospheric conditions:
Pressure = 100.0 kPa
Temperature = 28.0°C

6-23. Determine the slope of the equilibrium curve defined by Henry's law for HCl gas at 20°C from the following data:

P_{HCl} (kPa)	kg HCl per 100 kg H_2O
0.6533	38.9
0.0871	31.6
0.02733	25.0

Answer: Henry's law is not followed well. By linear regression, $m = 0.120$.

6-24. Find the slope of the equilibrium curve defined by Henry's law for SO_2 gas at 30°C from the following data:

P_{SO_2} (kPa)	kg SO_2 per 100 kg H_2O
10.532	1.0
6.933	0.7
4.800	0.5
2.626	0.3

6-25. Determine the height of a packed tower that is to reduce the concentration in air of H_2S from 0.100 kg/m³ to 0.005 kg/m³ given the following data:

Incoming liquid is water free of H_2S
Operating temperature = 25.0°C
Operating pressure = 101.325 kPa
Henry's law constant, m = 5.522 mole fraction units
H_g = 0.444 m
H_l = 0.325 m
Liquid flow rate = 20.0 kg/s
Gas flow rate = 5.0 kg/s

Answer: 7 m

6-26. The concentration of chlorine gas in air must be reduced from 10.0 mg/m^3 to 2.95 mg/m^3. Determine the height of the packed tower that should be used if the following parameters apply:

Incoming liquid is water free of Cl_2
Operating temperature = 20.0°C
Operating pressure = 101.325 kPa
H_g = 0.662 m
H_l = 0.285 m
Henry's law constant, m = 6.820 mole fraction units
Liquid flow rate = 15.0 kg/s
Gas flow rate = 3.00 kg/s

6-27. Determine the Langmuir constants a and b for the following isotherm data for adsorption of H_2S on a molecular sieve.

P_{H_2S} (kPa)	W (g H_2S/g sieve)
0.840	0.082
1.667	0.1065
2.666	0.118
3.333	0.122

Answer: $a = 18; b = 124$

6-28. Determine the Langmuir constants a and b for the adsorption of benzene on activated carbon, given the following isotherm data.

$P_{C_6H_6}$ (kPa)	W (kg C_6H_6/kg carbon)
0.027	0.129
0.067	0.170
0.133	0.204
0.266	0.240

6-29. Determine the breakthrough time for toluene on an adsorption bed of activated carbon that is 0.75 m thick and 5.0 m^2 in cross section. The operating parameters for the bed are as follows:

Gas flow rate = 1.185 kg/s
Gas temperature = 25°C
Bed density = 450 kg/m^3
Inlet pollutant concentration = 0.00350 kg/m^3
Langmuir parameters: $a = 465; b = 3{,}000$
Width of adsorption zone = 0.045 m

Answer: 17.8 h

6-30. What thickness of molecular sieve adsorption bed is required for the following system to ensure an SO_2 breakthrough time (t_B) of not less than 8.00 h?

Gas flow rate = 2.36 m^3/s of air
Gas temperature = 25.0°C
Gas pressure = 105.0 kPa
Bed density as packed = 390 kg/m^3
Inlet pollutant concentration = 3,000 ppm
Langmuir parameters: $a = 400$; $b = 900$
Width of adsorption zone = 0.028 m
Bed diameter = 3.00 m

6-31. Determine the efficiency of the cyclone in Example 6-9 for particles having a density of 1,000 kg/m^3 and radii of 1.00, 5.00, 10.00, and 25.00 μm.

6-32. Calculate the efficiency of removal of a 2.50 μm diameter particle having a density of 1,250 kg/m^3 for cyclone barrel diameters of 0.25 and 0.50 m. The gas flow rate is 2.80 m^3/s and the gas temperature is 25°C.

6-33. Calculate the overall mass efficiency (η) of the venturi described in Example 6-10 for the following particle size distribution.

Average diameter (μm)	% of total mass
2.5	25
7.5	20
15.0	15
25.0	15
35.0	10
50.0	15

6-34. Calculate the venturi throat area required to achieve 99.0 percent removal of a 1.25 μm radius particle having a density of 1,400 kg/m^3 for the following gas stream and venturi characteristics.

Gas flow rate = 10.0 m^3/s
Gas temperature = 180°C
Liquid flow rate = 0.0100 m^3/s
Coefficient κ = 200
Droplet diameter = 100 μm

6-35. Determine the collection efficiency of an electrostatic precipitator (ESP) tube that is 0.300 m in diameter and is 2.00 m in length for particles that are 1.00 μm in diameter. The flow rate is 0.150 m^3/s, the collection field intensity is 100,000 V/m, the particle charge is 0.300 femtocoulombs (fC), and the gas temperature is 25.0°C.

Answer: = 92.4%

6-36. Rework Problem 6-35 with the gas flow rate reduced to 0.075 m^3/s.

6-15 DISCUSSION QUESTIONS

6-1. A gas sample is collected in a special gas sampling bag that does not react with the pollutants collected but is free to expand and contract. When the sample was collected, the atmospheric pressure was 103.0 kPa. At the time the sample was analyzed the atmospheric pressure was 100.0 kPa. The bag was found to contain 0.020 ppm of SO_2. Would the original concentration of SO_2 be more, less, or the same? Explain.

6-2. Under which of the following conditions would you expect the strongest inversion (largest positive lapse rate) to form?
a. Foggy day in the fall after the leaves have fallen
b. Clear winter night with fresh snow on the ground
c. Clear summer morning just before sunrise
Explain why.

6-3. Cement dust is characterized by very fine particulates. The exhaust gas temperatures from a cement kiln are very hot. Which of the following air pollution control devices would appear to be appropriate? Explain the reasoning for your selection.
a. Venturi scrubber
b. Baghouse
c. Electrostatic precipitator.

6-4. Photochemical oxidants are not directly attributable to either people or natural sources. Why, then, are automobiles singled out as the major cause of the formation of ozone?

6-5. Explain why the $PM_{2.5}$ standard is more appropriate than a "Total Suspended Particulate" for protection of human health.

6-16 ADDITIONAL READING

A. J. Buonicore and W. T. Davis, *Air Pollution Engineering Manual,* Air & Waste Management Association, New York: Van Nostrand Reinhold, 1992.
E. J. Calabrese, *Air Toxics and Risk Assessment,* Chelsea, MI: Lewis Publishers, 1991.
C. D. Cooper and F. C. Alley, *Air Pollution Control: A Design Approach,* Prospect Heights, IL: Waveland Press, 1994.
M. Crawford, *Air Pollution Control Theory,* New York: McGraw-Hill, 1976.
R. C. Flagan, *Fundamentals of Air Pollution Engineering,* Upper Saddle River, NJ: Prentice Hall, 1988.
A. L. Hines, *et al., Indoor Air Quality & Control,* Upper Saddle River, NJ: Prentice Hall, 1993.
H. C. Perkins, *Air Pollution,* New York: McGraw-Hill, 1974.
J. H. Seinfeld, *Air Pollution, Physical and Chemical Fundamentals,* New York: McGraw-Hill, 1975.
A. C. Stern, H. C. Wohlers, R. W. Boubel, and W. P. Lowry, *Fundamentals of Air Pollution,* New York: Academic Press, 1973.
A. C. Stern, ed., *Air Pollution,* Vols. 1, 2, and 3, New York: Academic Press, 1968.
D. B. Turner, *Workbook of Atmospheric Dispersion Estimates,* 2nd ed., Ann Arbor, MI: Lewis Publishers, 1994.
K. Wark and C. F. Warner, *Air Pollution, Its Origin and Control,* 2nd ed., New York: Harper & Row, 1981.

CHAPTER 7

NOISE POLLUTION

7-1 INTRODUCTION
Properties of Sound Waves
Sound Power and Intensity
Levels and the Decibel
Characterization of Noise

7-2 EFFECTS OF NOISE ON PEOPLE
The Hearing Mechanism
Normal Hearing
Hearing Impairment
Damage-Risk Criteria
Speech Interference
Annoyance
Sleep Interference
Effects on Performance
Acoustic Privacy

7-3 RATING SYSTEMS
Goals of a Noise-Rating System
The L_N Concept
The L_{eq} Concept

7-4 COMMUNITY NOISE SOURCES AND CRITERIA
Transportation Noise
Other Internal Combustion Engines
Construction Noise
Zoning and Siting Considerations
Levels to Protect Health and Welfare

7-5 TRANSMISSION OF SOUND OUTDOORS
Inverse Square Law
Radiation Fields of a Sound Source
Directivity
Airborne Transmission
7-6 TRAFFIC NOISE PREDICTION
National Cooperative Highway Research Program 174
L_{eq} Prediction
L_{dn} Prediction
7-7 NOISE CONTROL
Source-Path-Receiver Concept
Control of Noise Source by Design
Noise Control in the Transmission Path
Control of Noise Source by Redress
Protect the Receiver
7-8 CHAPTER REVIEW
7-9 PROBLEMS
7-10 DISCUSSION QUESTIONS
7-11 ADDITIONAL READING

7-1 INTRODUCTION

Noise, commonly defined as unwanted sound, is an environmental phenomenon to which we are exposed before birth and throughout life. Noise can also be considered an environmental pollutant, a waste product generated in conjunction with various anthropogenic activities. Under the latter definition, noise is any sound—independent of loudness—that can produce an undesired physiological or psychological effect in an individual, and that may interfere with the social ends of an individual or group. These social ends include all of our activities—communication, work, rest, recreation, and sleep.

As waste products of our way of life, we produce two general types of pollutants. The general public has become well aware of the first type—the mass residuals associated with air and water pollution—that remain in the environment for extended periods of time. However, only recently has attention been focused on the second general type of pollution, the energy residuals such as the waste heat from manufacturing processes that creates thermal pollution of our streams. Energy in the form of sound waves constitutes yet another kind of energy residual, but, fortunately, one that does not remain in the environment for extended periods of time. The total amount of energy dissipated as sound throughout the earth is not large when compared with other forms of energy; it is only the extraordinary sensitivity of the ear that permits such a relatively small amount of energy to adversely affect us and other biological species.

It has long been known that noise of sufficient intensity and duration can induce temporary or permanent hearing loss, ranging from slight impairment to nearly total deafness. In general, a pattern of exposure to any source of sound that produces high enough levels can result in temporary hearing loss. If the exposure persists over a period of time, this could lead to permanent hearing impairment. Short-term, but frequently serious, effects include interference with speech communication and the perception of other auditory signals, disturbance of sleep and relaxation, annoyance, interference with an individual's ability to perform complicated tasks, and general diminution of the quality of life.

Beginning with the technological expansion of the Industrial Revolution and continuing through a post-World War II acceleration, environmental noise in the United States and other industrialized nations has been gradually and steadily increasing, with more geographic areas becoming exposed to significant levels of noise. Where once noise levels sufficient to induce some degree of hearing loss were confined to factories and occupational situations, noise levels approaching such intensity and duration are today being recorded on city streets and, in some cases, in and around the home.

There are valid reasons why widespread recognition of noise as a significant environmental pollutant and potential hazard or, as a minimum, a detractor from the quality of life, has been slow in coming. In the first place, noise, if defined as unwanted sound, is a subjective experience. What is considered noise by one listener may be considered desirable by another.

Secondly, noise has a short decay time and thus does not remain in the environment for extended periods, as do air and water pollution. By the time the average individual is spurred to action to abate, control, or, at least, complain about sporadic environmental noise, the noise may no longer exist.

Thirdly, the physiological and psychological effects of noise on us are often subtle and insidious, appearing so gradually that it becomes difficult to associate cause with effect. Indeed, to those persons whose hearing may already have been affected by noise, it may not be considered a problem at all.

Further, the typical citizen is proud of this nation's technological progress and is generally happy with the things that technology delivers, such as rapid transportation, labor-saving devices, and new recreational devices. Unfortunately, many technological advances have been associated with increased environmental noise, and large segments of the population have tended to accept the additional noise as part of the price of progress.

The engineering and scientific community has already accumulated considerable knowledge concerning noise, its effects, and its abatement and control. In that regard, noise differs from most other environmental pollutants. Generally, the technology exists to control most indoor and outdoor noise. As a matter of fact, this is one instance in which knowledge of control techniques exceeds the knowledge of biological and physical effects of the pollutant.

Properties of Sound Waves

Sound waves result from the vibration of solid objects or the separation of fluids as they pass over, around, or through holes in solid objects. The vibration and/or separation causes the surrounding air to undergo alternating compression and rarefaction, much in the same manner as a piston vibrating in a tube (Figure 7-1). The

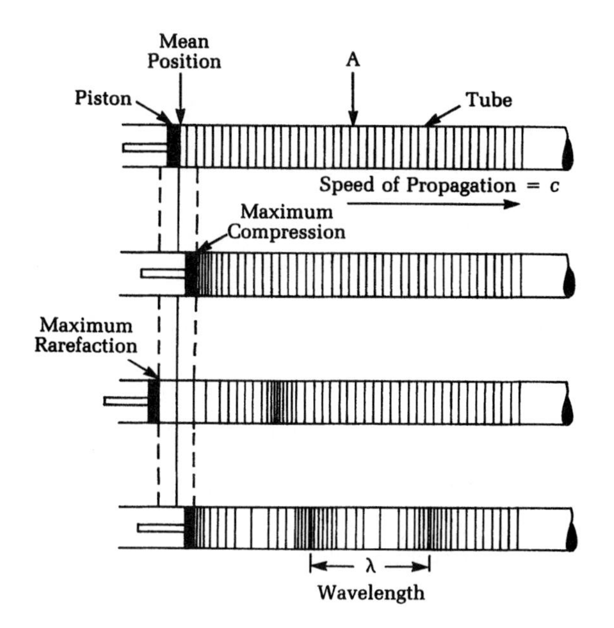

FIGURE 7-1
Alternating compression and rarefaction of air molecules resulting from a vibrating piston.

compression of the air molecules causes a local increase in air density and pressure. Conversely, the rarefaction causes a local decrease in density and pressure. These alternating pressure changes are the sound detected by the human ear.

Let us assume that you could stand at Point A in Figure 7-1. Also let us assume that you have an instrument that will measure the air pressure every 0.000010 seconds and plot the value on a graph. If the piston vibrates at a constant rate, the condensations and rarefactions will move down the tube at a constant speed. That speed is the speed of sound (c). The rise and fall of pressure at point A will follow a cyclic or wave pattern over a "period" of time (Figure 7-2). The wave pattern is called *sinusoidal.* The time between successive peaks or between successive troughs of the oscillation is called the *period (P).* The inverse of this, that is, the number of times a peak arrives in one second of oscillations, is called the *frequency (f).* Period and frequency are then related as follows:

$$P = \frac{1}{f} \tag{7-1}$$

Since the pressure wave moves down the tube at a constant speed, you would find that the distance between equal pressure readings would remain constant. The distance between adjacent crests or troughs of pressure is called the *wavelength* (λ). Wavelength and frequency are then related as follows:

$$\lambda = \frac{c}{f} \tag{7-2}$$

The *amplitude (A)* of the wave is the height of the peak or depth of the trough measured from the zero pressure line (Figure 7-2). From Figure 7-2 we can also note that the average pressure could be zero if an averaging time was selected that

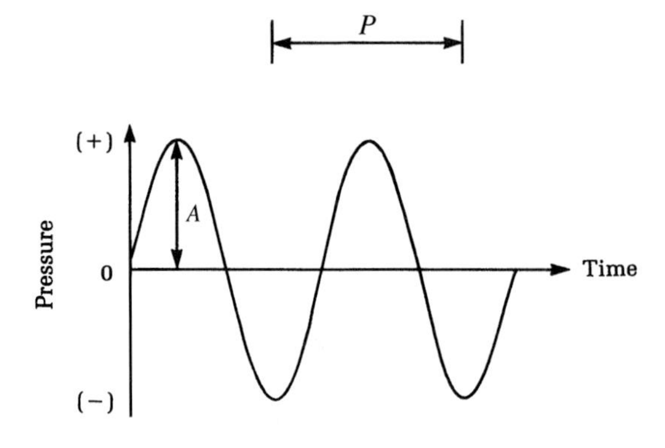

FIGURE 7-2
Sinusoidal wave that results from alternating compression and rarefaction of air molecules. The amplitude is shown as A and the period is P.

corresponded to the period of the wave. This would result regardless of the amplitude! This, of course, is not an acceptable state of affairs. The root mean square (rms) sound pressure (p_{rms}) is used to overcome this difficulty.[1] The rms sound pressure is obtained by squaring the value of the amplitude at each instant in time; summing the squared values; dividing the total by the averaging time; and taking the square root of the total. The equation for rms is

$$p_{rms} = \left(\overline{p^2}\right)^{1/2} = \left[\frac{1}{T}\int_0^T P^2(t)\,dt\right]^{1/2} \tag{7-3}$$

where the overbar refers to the time-weighted average and T is the time period of the measurement.

Sound Power and Intensity

Work is defined as the product of the magnitude of the displacement of a body and the component of force in the direction of the displacement. Thus, traveling waves of sound pressure transmit energy in the direction of propagation of the wave. The rate at which this work is done is defined as the sound power (W).

Sound intensity (I) is defined as the time-weighted average sound power per unit area normal to the direction of propagation of the sound wave. Intensity and power are related as follows:

$$I = \frac{W}{A} \tag{7-4}$$

where A is a unit area perpendicular to the direction of wave motion. Intensity, and hence, sound power, is related to sound pressure in the following manner:

$$I = \frac{(p_{rms})^2}{\rho c} \tag{7-5}$$

[1] Sound pressure = (total atmospheric pressure) − (barometric pressure).

where I = intensity, W/m^2
p_{rms} = root mean square sound pressure, Pa
ρ = density of medium, kg/m^3
c = speed of sound in medium, m/s

Both the density of air and speed of sound are a function of temperature. Given the temperature and pressure, the density of air may be determined from Table A-3 in Appendix A. The speed of sound in air at 101.325 kPa may be determined from the following equation:

$$c = 20.05\sqrt{T} \tag{7-6}$$

where T is the absolute temperature in degrees Kelvin (K) and c is in m/s.

Levels and the Decibel

The sound pressure of the faintest sound that a normal healthy individual can hear is about 0.00002 Pascal. The sound pressure produced by a Saturn rocket at liftoff is greater than 200 Pascal. Even in scientific notation this is an "astronomical" range of numbers.

In order to cope with this problem, a scale based on the logarithm of the ratios of the measured quantities is used. Measurements on this scale are called *levels.* The unit for these types of measurement scales is the *bel,* which was named after Alexander Graham Bell:

$$L' = \log \frac{Q}{Q_o} \tag{7-7}$$

where L' = level, bels
Q = measured quantity
Q_o = reference quantity
log = logarithm in base 10

A bel turns out to be a rather large unit, so for convenience it is divided into 10 subunits called *decibels* (dB). Levels in dB are computed as follows:

$$L = 10 \log \frac{Q}{Q_o} \tag{7-8}$$

The dB does not represent any physical unit. It merely indicates that a logarithmic transformation has been performed.

Sound power level. If the reference quantity (Q_o) is specified, then the dB takes on physical significance. For noise measurements, the reference power level has been established as 10^{-12} watts. Thus, sound power level may be expressed as

$$L_W = 10 \log \frac{W}{10^{-12}} \tag{7-9}$$

Sound power levels computed with Equation 7-9 are reported as dB re: 10^{-12} W.

Sound intensity level. For noise measurements, the reference sound intensity (Equation 7-4) is 10^{-12} W/m^2. Thus, the sound intensity level is given as

$$L_I = 10 \log \frac{I}{10^{-12}} \tag{7-10}$$

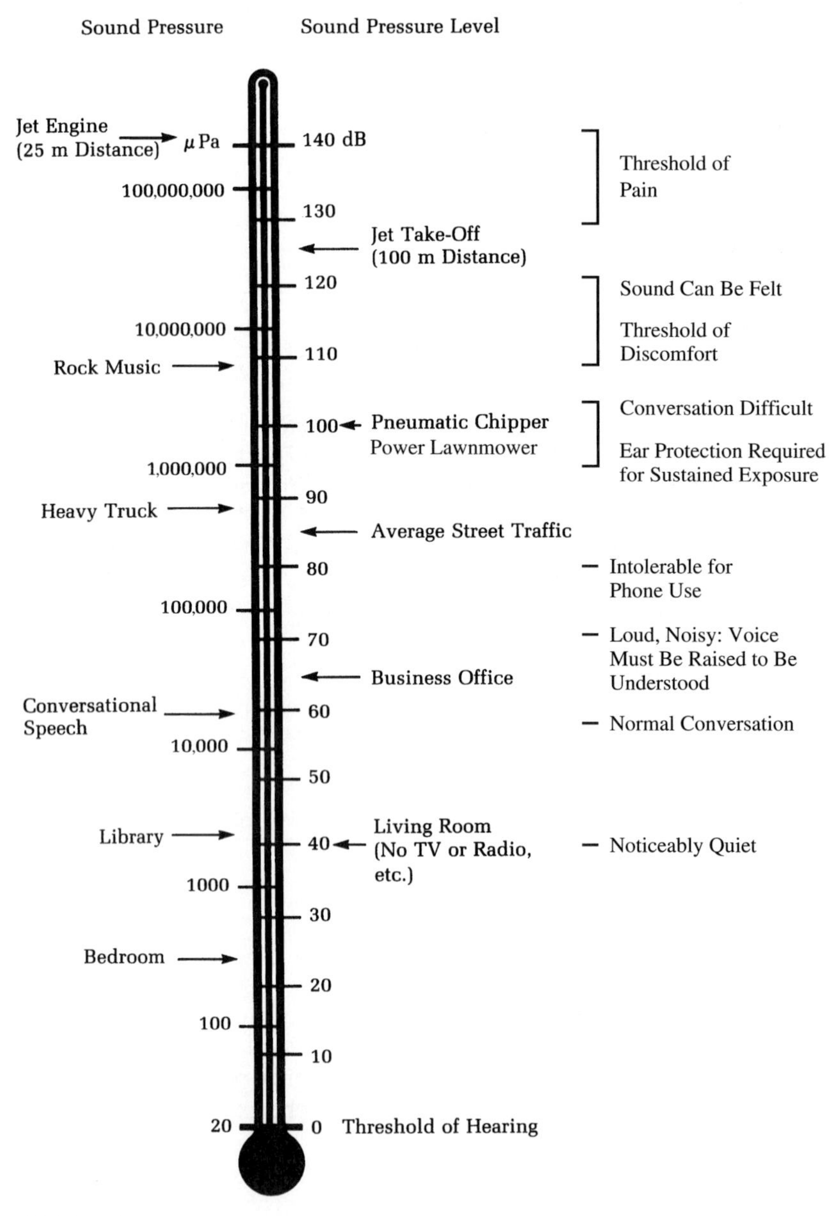

FIGURE 7-3
Relative scale of sound pressure levels.

Sound pressure level. Because sound-measuring instruments measure the root mean square pressure, the sound pressure level is computed as follows:

$$L_P = 10 \log \frac{(p_{rms})^2}{(p_{rms})_o^2} \tag{7-11}$$

which, after extraction of the squaring term, is given as

$$L_P = 20 \log \frac{p_{rms}}{(p_{rms})_o} \tag{7-12}$$

The reference pressure has been established as 20 micropascals (μPa). A scale showing some common sound pressure levels is shown in Figure 7-3.

Combining sound pressure levels. Because of their logarithmic heritage, decibels don't add and subtract the way apples and oranges do. Remember: adding the logarithms of numbers is the same as multiplying them. If you take a 60-decibel noise (re: 20 μPa) and add another 60-decibel noise (re: 20 μPa) to it, you get a 63-decibel noise (re: 20 μPa). If you're strictly an apple-and-orange mathematician, you may take this on faith. For skeptics, this can be demonstrated by converting the dB to sound power level, adding them, and converting back to dB. Figure 7-4 provides a graphical solution for this type of problem. For noise pollution work, results should be reported to the nearest whole number. When there are several levels to be combined, they should be combined two at a time, starting with lower-valued levels and continuing two at a time with each successive pair until one number remains. Henceforth, in this chapter we will assume levels are all "re: 20 μPa" unless stated otherwise.

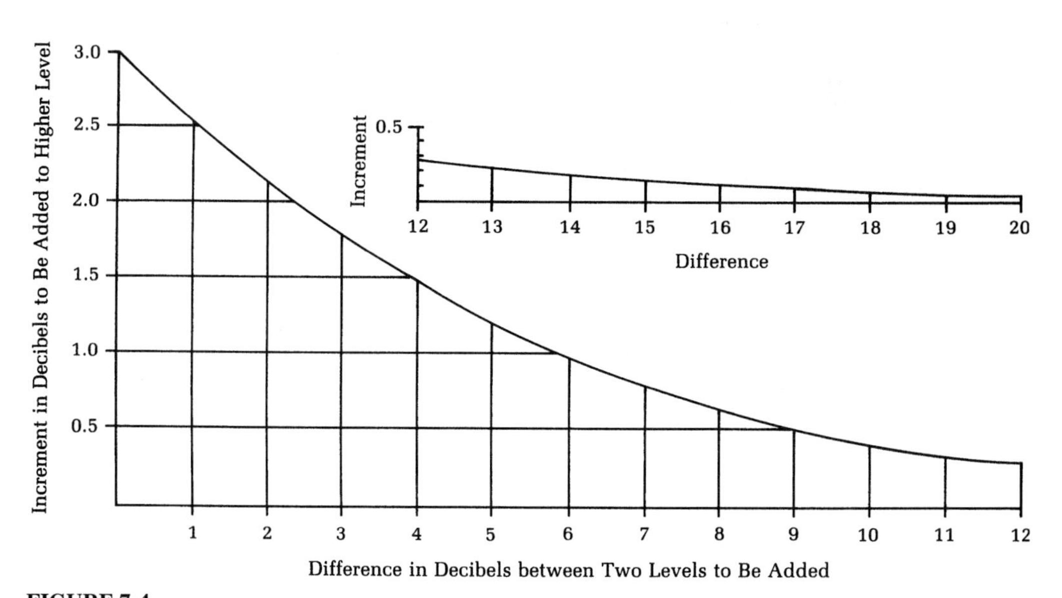

FIGURE 7-4
Graph for solving decibel addition problems.

Example 7-1. What sound power level results from combining the following three levels: 68 dB, 79 dB, and 75 dB?

Solution. We begin by selecting the two lowest levels: 68 dB and 75 dB. The difference between the values is 75 − 68 = 7.00. Using Figure 7-4, draw a vertical line from 7.00 on the abscissa to intersect the curve. A horizontal line from the intersection to the ordinate yields about 0.8 dB. Thus, the combination of 68 dB and 75 dB results in a level of 75.8 dB. This, and the remainder of the computation, is shown diagramatically below.

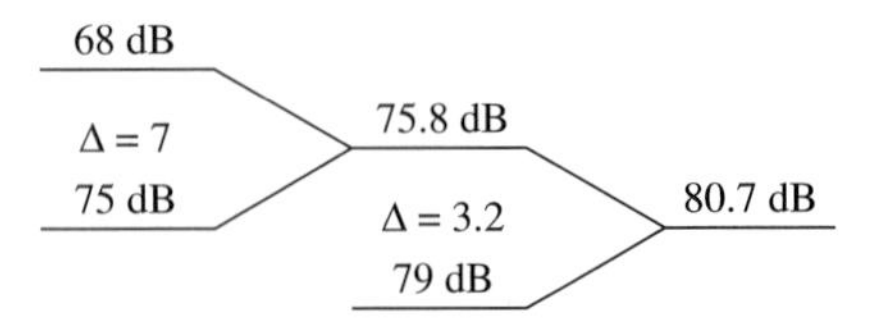

Rounding off to the nearest whole number yields an answer of 81 dB re: 20 μPa.

This problem can also be worked by converting the readings to sound power level, adding them, and converting back to dB.

$$
\begin{aligned}
L_P &= 10 \log \sum 10^{(68/10)} + 10^{(75/10)} + 10^{(79/10)} \\
&= 10 \log(117{,}365{,}173) \\
&= 80.7 \text{ dB}
\end{aligned}
$$

Characterization of Noise

Weighting networks. Because our reasons for measuring noise usually involve people, we are ultimately more interested in the human reaction to sound than in sound as a physical phenomenon. Sound pressure level, for instance, can't be taken at face value as an indication of loudness because the frequency (or pitch) of a sound has quite a bit to do with how loud it sounds. For this and other reasons, it often helps to know something about the frequency of the noise you're measuring. This is where weighting networks come in.[2] They are electronic filtering circuits built into the meter to attenuate certain frequencies. They permit the sound level meter (Figure 7-5) to respond more to some frequencies than to others with a prejudice something like that of the human ear. Writers of the acoustical standards have established three weighting characteristics: A, B, and C. The chief difference among them is that very low frequencies are filtered quite severely by the A network, moderately by the B network, and hardly at all by the C network. Therefore, if the measured sound level of a noise is much higher on C weighting than on A weighting, much

[2] *A Primer of Noise Measurement,* Concord, MA: General Radio Company, 1972. Reprinted by permission.

FIGURE 7-5
Type 2 sound level meter. (Courtesy of Gen-Rad, Inc.)

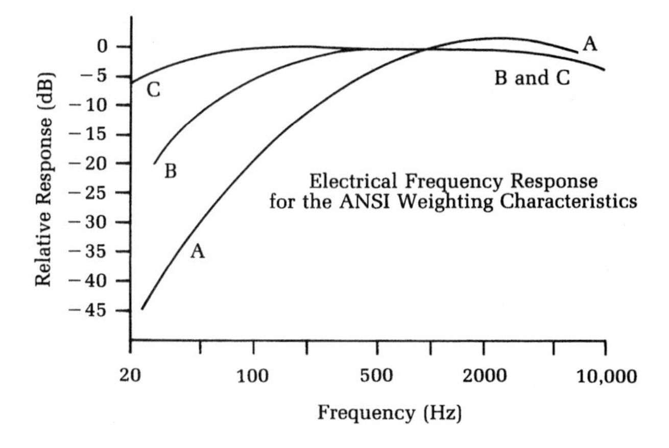

FIGURE 7-6
Response characteristics of the three basic weighting networks.

of the noise is probably of low frequency. If you really want to know the frequency distribution of a noise (and most serious noise measurers do), it is necessary to use a *sound analyzer.* But if you are unable to justify the expense of an analyzer, you can still find out something about the frequency of a noise by shrewd use of the weighting networks of a sound level meter.

Figure 7-6 shows the response characteristics of the three basic networks as prescribed by the American National Standards Institute (ANSI) specification number S1.4–1971. When a weighting network is used, the sound level meter electronically subtracts or adds the number of dB shown at each frequency shown in Table 7-1 from or to the actual sound pressure level at that frequency. It then sums all the resultant numbers by logarithmic addition to give a single reading. Readings taken when a network is in use are said to be "sound levels" rather than "sound pressure levels." The readings taken are designated in decibels in one of the following forms: dB(A); dBa; dBA; dB(B); dBb; dBB; and so on. Tabular notations may refer to L_A, L_B, L_C.

Example 7-2. A new Type 2 sound level meter is to be tested with two pure tone sources that emit 90 dB. The pure tones are at 1,000 Hz and 100 Hz. Estimate the expected readings on the A, B, and C weighting networks.

Solution. From Table 7-1 at 1,000 Hz, we note that the relative response (correction factor) for each of the weighting networks is zero. Thus for the pure tone at 1,000 Hz we would expect the readings on the A, B, and C networks to be 90 dB.

From Table 7-1 at 100 Hz, the relative response for each weighting network differs. For the A network, the meter will subtract 19.1 dB from the actual reading, for the B network, the meter will subtract 5.6 dB from the actual reading, and for the C

TABLE 7-1
Sound level meter network weighting values

Frequency (Hz)	Curve A (dB)	Curve B (dB)	Curve C (dB)
10	−70.4	−38.2	−14.3
12.5	−63.4	−33.2	−11.2
16	−56.7	−28.5	−8.5
20	−50.5	−24.2	−6.2
25	−44.7	−20.4	−4.4
31.5	−39.4	−17.1	−3.0
40	−34.6	−14.2	−2.9
50	−30.2	−11.6	−1.3
63	−26.2	−9.3	−0.8
80	−22.5	−7.4	−0.5
100	−19.1	−5.6	−0.3
125	−16.1	−4.2	−0.2
160	−13.4	−3.0	−0.1
200	−10.9	−2.0	0
250	−8.6	−1.3	0
315	−6.6	−0.8	0
400	−4.8	−0.5	0
500	−3.2	−0.3	0
630	−1.9	−0.1	0
800	−0.8	0	0
1,000	0	0	0
1,250	0.6	0	0
1,600	1.0	0	−0.1
2,000	1.2	−0.1	−0.2
2,500	1.3	−0.2	−0.3
3,150	1.2	−0.4	−0.5
4,000	1.0	−0.7	−0.8
5,000	0.5	−1.2	−1.3
6,300	−0.1	−1.9	−2.0
8,000	−1.1	−2.9	−3.0
10,000	−2.5	−4.3	−4.4
12,500	−4.3	−6.1	−6.2
16,000	−6.6	−8.4	−8.5
20,000	−9.3	−11.1	−11.2

network, the meter will subtract 0.3 dB. Thus, the anticipated readings would be:

A network: 90 − 19.1 = 70.9 or 71 dB(A)
B network: 90 − 5.6 = 84.4 or 84 dB(B)
C network: 90 − 0.3 = 89.7 or 90 dB(C)

Example 7-3. The following sound levels were measured on the A, B, and C weighting networks:

Source 1: 94 dB(A), 95 dB(B), and 96 dB(C)
Source 2: 74 dB(A), 83 dB(B), and 90 dB(C)

Characterize the sources as "low frequency" or "mid/high frequency."

Solution. From Figure 7-6, we can see that readings on the A, B, and C networks will be close together if the source emits noise in the frequency range above about 500 Hz. This range may be classified "mid/high frequency" since we cannot distinguish between "mid" and "high" frequency using a Type 2 sound level meter. Likewise, we can see that below 200 Hz (low frequency), readings on the A, B, and C scale will be substantially different. The readings from the A network will be lower than the readings from the B network, and readings from both the A and B networks will be lower than those from the C network.

Source 1: Note that the sound levels on each of the weighting networks differ by 1 dB. From Figure 7-6, it appears that the sound level will be in the mid/high frequency range.

Source 2: Note that the sound levels on each of the weighting networks differ by several dB and that the reading from the A network is lower than that from the B network and both are below that from the C network. From Figure 7-6, it appears that the sound level will be in the low frequency range.

Octave bands. To completely characterize a noise, it is necessary to break it down into its frequency components or spectra. Normal practice is to consider 8 to 11 octave bands.[3] The standard octave bands and their geometric mean frequencies (center band frequencies) are given in Table 7-2. Octave analysis is performed with a combination precision sound level meter and an octave filter set.

While octave band analysis is frequently satisfactory for community noise control (that is, identifying violators), more refined analysis is required for corrective action and design. One-third octave band analysis provides a slightly more refined picture of the noise source than the full octave band analysis. This improved resolution is usually sufficient for determining corrective action for community noise problems. Narrow band analysis is highly refined and may imply band widths down to 2 Hz. This degree of refinement is only justified in product design and testing or in troubleshooting industrial machine noise and vibration.

[3] An octave is the frequency interval between a given frequency and twice that frequency. For example, given the frequency 22 Hz, the octave band is from 22 to 44 Hz. A second octave band would then be from 44 to 88 Hz.

TABLE 7-2
Octave bands

Octave frequency range (Hz)	Geometric mean frequency (Hz)
22–44	31.5
44–88	63
88–175	125
175–350	250
350–700	500
700–1,400	1,000
1,400–2,800	2,000
2,800–5,600	4,000
5,600–11,200	8,000
11,200–22,400	16,000
22,400–44,800	31,500

Averaging sound pressure levels. Because of the logarithmic nature of the dB, the average value of a collection of sound pressure level measurements cannot be computed in the normal fashion. Instead, the following equation must be used:

$$\overline{L}_p = 20 \log \frac{1}{N} \sum_{j=1}^{N} 10^{(L_j/20)} \qquad (7\text{-}13)$$

where $\overline{L}_p$ = average sound pressure level, dB re: 20 μPa
N = number of measurements
L_j = the jth sound pressure level, dB re: 20 μPa
$j = 1, 2, 3, \ldots N$

This equation is equally applicable to sound levels in dBA. It may also be used to compute average sound power levels if the factors of 20 are replaced with 10s.

Example 7-4. Compute the mean sound level from the following four readings (all dBA): 38, 51, 68, and 78.

Solution. First we compute the sum:

$$\sum_{j=1}^{4} = 10^{(38/20)} + 10^{(51/20)} + 10^{(68/20)} + 10^{(78/20)}$$

$$= 1.09 \times 10^4$$

Now we complete the computation:

$$\overline{L}_p = 20 \log \frac{1.09 \times 10^4}{4}$$

$$= 68.7 \text{ or } 69 \text{ dBA}$$

Straight arithmetic averaging would yield 58.7 or 59 dB.

Types of sounds. Patterns of noise may be qualitatively described by one of the following terms: *steady-state* or *continuous; intermittent;* and *impulse* or *impact.* Continuous noise is an uninterrupted sound level that varies less than 5 dB during the period of observation. An example is the noise from a household fan. Intermittent noise is a continuous noise that persists for more than one second that is interrupted for more than one second. A dentist's drilling would be an example of an intermittent noise. Impulse noise is characterized by a change of sound pressure of 40 dB or more within 0.5 second with a duration of less than one second.[4] The noise from firing a weapon would be an example of an impulsive noise.

Two types of impulse noise generally are recognized. The type A impulse is characterized by a rapid rise to a peak sound pressure level followed by a small negative pressure wave or by decay to the background level (Figure 7-7). The type B impulse is characterized by a damped (oscillatory) decay (Figure 7-8). Where the duration of the type A impulse is simply the duration of the initial peak, the duration

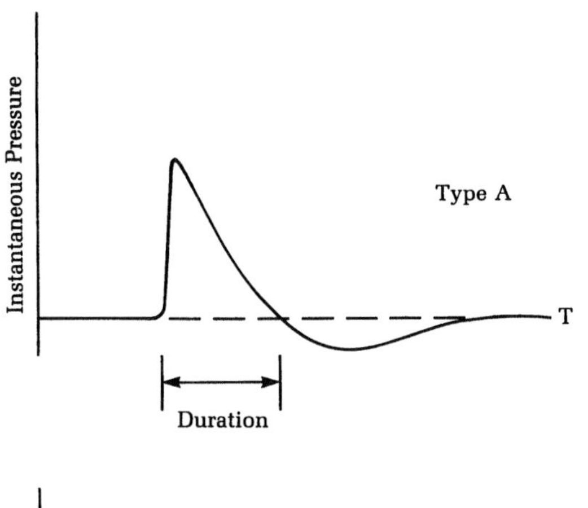

FIGURE 7-7
Type A impulse noise.

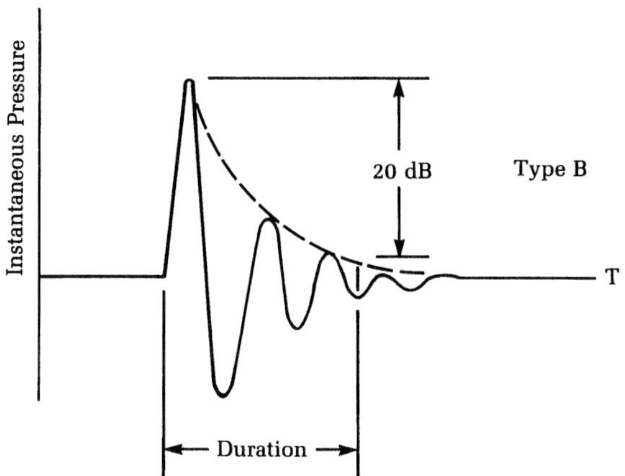

FIGURE 7-8
Type B impulse noise.

[4]The Occupational Safety and Health Administration (OSHA) classifies repetitive events, including impulses, as steady noise if the interval between events is less than 0.5 seconds.

of the type B impulse is the time required for the envelope to decay to 20 dB below the peak. Because of the short duration of the impulse, a special sound-level meter must be employed to measure impulse noise. You should note that the peak sound pressure level is different than the impulse sound level because of the time-averaging used in the latter.

7-2 EFFECTS OF NOISE ON PEOPLE

For the purpose of our discussion, we have classified the effects of noise on people into the following two categories: auditory effects and psychological/sociological effects. Auditory effects include both hearing loss and speech interference. Psychological/sociological effects include annoyance, sleep interference, effects on performance, and acoustical privacy.

The Hearing Mechanism

Before we can discuss hearing loss, it is important to outline the general structure of the ear and how it works.

Anatomically, the ear is separated into three sections: the outer ear, the middle ear, and the inner ear (Figure 7-9). The outer and middle ear serve to convert sound pressure to vibrations. In addition, they perform the protective role of keeping debris and objects from reaching the inner ear. The Eustachian tube extends from the middle ear space to the upper part of the throat behind the soft palate. The tube is normally closed. Contraction of the palate muscles during yawning, chewing, or swallowing opens the tubes. This allows the middle ear to ventilate and equalize pressure. If external air pressure changes rapidly, for example, by a sudden change in elevation, the tube is opened by involuntary swallowing or yawning to equalize the pressure.

The sound transducer mechanism is housed in the middle ear.[5] It consists of the *tympanic membrane* (eardrum) and three *ossicles* (bones) (Figure 7-10). The ossicles are supported by ligaments and may be moved by two muscles or by deflection of the tympanic membrane. The muscle movement is involuntary. Loud sounds cause these muscles to contract. This stiffens and diminishes the movement of the ossicular chain. This presumably offers some protection for the delicate inner ear structure from physical injury. According to J. D. Clemis, "More convincing research is still needed to test the validity of this theory of the function of these muscles." The discussion on the middle ear that follows is excerpted from Clemis.[6]

> The primary function of the middle ear in the hearing process is to transfer sound energy from the outer to the inner ear. As the eardrum vibrates, it transfers its motion to the malleus. Since the bones of the ossicular chain are connected to one another, the

---

[5]A transducer is a device that transmits power from one system to another. In this case, sound power is converted to mechanical displacement, which is later measured and interpreted by the brain.

[6]J. D. Clemis, "Anatomy, Physiology, and Pathology of the Ear," *Industrial Noise and Hearing Conservation,* Julian B. Olishifski and Earl R. Harford, eds., Chicago: National Safety Council, p. 213, 1975.

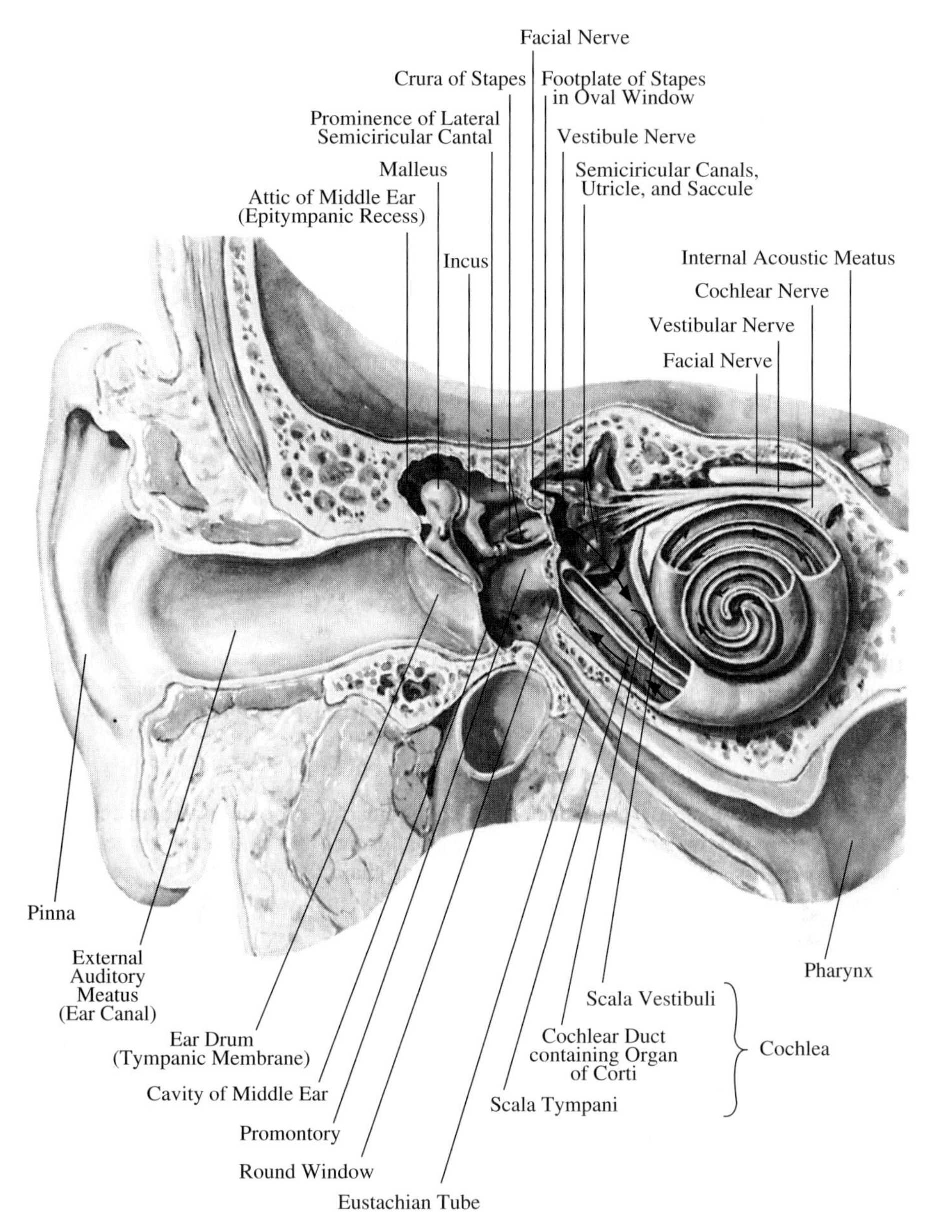

FIGURE 7-9
Anatomical divisions of the ear. (© Copyright 1972 CIBA Pharmaceutical Company, Division of CIBA-GEIGY Corporation. Reproduced, with permission, from *Clinical Symposia,* illustrated by Frank H. Netter, M.D. All rights reserved.)

movements of the malleus are passed on to the incus, and finally to the stapes, which is imbedded in the oval window.

As the stapes moves back and forth in a rocking motion, it passes the vibrations into the inner ear through the oval window. Thus, the mechanical motion of the eardrum is effectively transmitted through the middle ear and into the fluid of the inner ear.

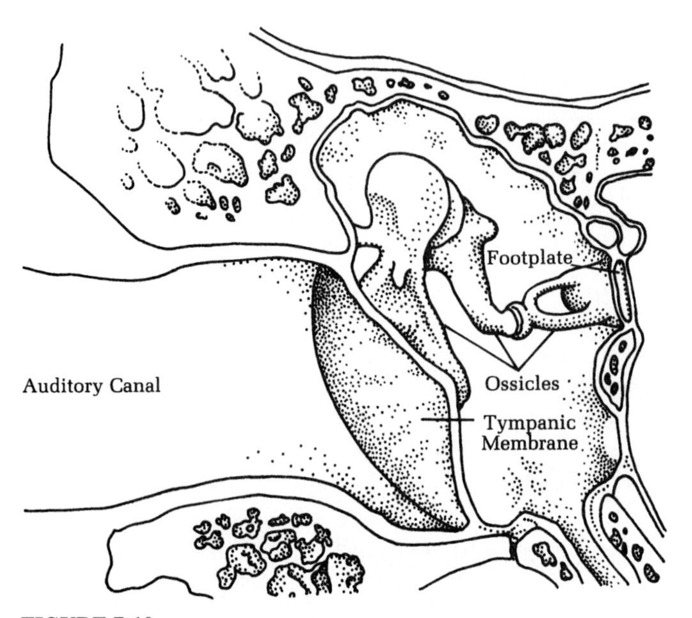

FIGURE 7-10
The sound transducer mechanism housed in the middle ear. (Adapted from an original painting by Frank H. Netter, M.D., for *Clinical Symposia,* copyright by CIBA-GEIGY Corporation.)

> The sound-conducting transducer amplifies sound by two main mechanisms. First, the large surface area of the drum as compared to the small surface area of the base of the stapes (footplate) results in a hydraulic effect. The eardrum has about 25 times as much surface area as the oval window. All of the sound pressure collected on the eardrum is transmitted through the ossicular chain and is concentrated on the much smaller area of the oval window. This produces a significant increase in pressure.
>
> The bones of the ossicular chain are arranged in such a way that they act as a series of levers. The long arms are nearest the eardrum, and the shorter arms are toward the oval window. The fulcrums are located where the individual bones meet. A small pressure on the long arm of the lever produces a much stronger pressure on the shorter arm. Since the longer arm is attached to the eardrum and the shorter arm is attached to the oval window, the ossicular chain acts as an amplifier of sound pressure. The magnification effect of the entire sound-conducting mechanism is about 22-to-1.

The inner ear houses both the balance receptors and the auditory receptors. The auditory receptors are in the *cochlea.* It is a bone shaped like a snail coiled two and one-half times around its own axis (Figure 7-9). A cross section through the cochlea (Figure 7-11) reveals three compartments: the *scala vestibuli;* the *scala media;* and the *scala tympani.* The scala vestibuli and the scala tympani are connected at the apex of the cochlea. They are filled with a fluid called *perilymph,* in which the scala media floats. The hearing organ, the *organ of Corti,* is housed in the scala media. The scala media contains a different fluid, *endolymph,* which bathes the organ of Corti.

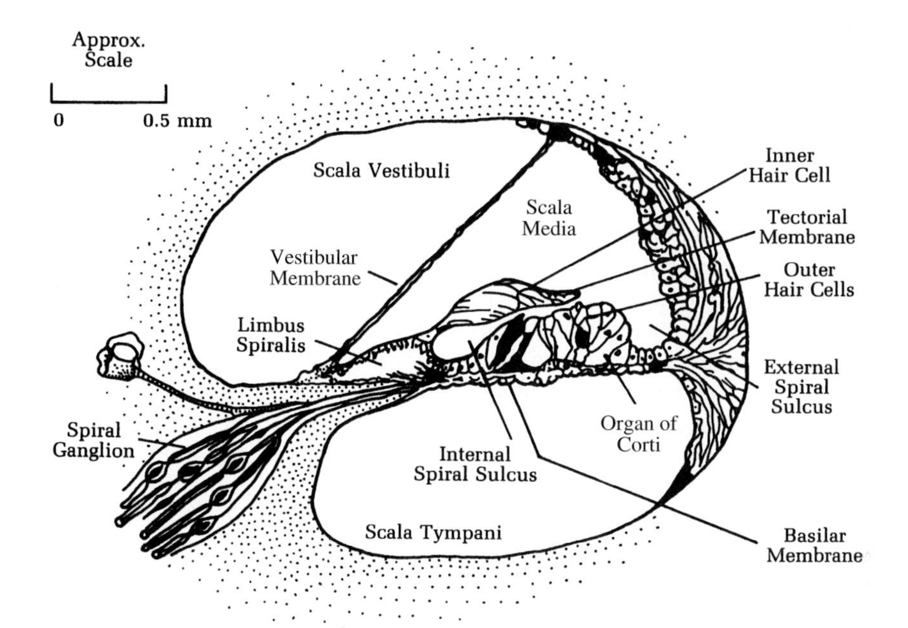

FIGURE 7-11
Cross section through the cochlea.

The scala media is triangular in shape and is about 34 mm in length. As shown in Figure 7-11, there are cells growing up from the *basilar membrane.* They have a tuft of hair at one end and are attached to the hearing nerve at the other end. A gelatinous membrane (*tectorial membrane*) extends over the hair cells and is attached to the *limbus spiralis.* The hair cells are embedded in the tectorial membrane.

Vibration of the oval window by the stapes causes the fluids of the three scalae to develop a wave-like motion. The movement of the basilar membrane and the tectorial membrane in opposite directions causes a shearing motion on the hair cells. The dragging of the hair cells sets up electrical impulses in the auditory nerves, which are transmitted to the brain.

The nerve endings near the oval and round windows are sensitive to high frequencies. Those near the apex of the cochlea are sensitive to low frequencies.

Normal Hearing

Frequency range and sensitivity. The ear of the young, audiometrically healthy, adult male responds to sound waves in the frequency range of 20 to 16,000 Hz. Young children and women often have the capacity to respond to frequencies up to 20,000 Hz. The speech zone lies in the frequency range of 500 to 2,000 Hz. The ear is most sensitive in the frequency range from 2,000 to 5,000 Hz. The smallest perceptible sound pressure in this frequency range is 20 μPa.

A sound pressure of 20 μPa at 1,000 Hz in air corresponds to a 1.0 nm displacement of the air molecules. The thermal motion of the air molecules corresponds to a sound pressure of about 1 μPa. If the ear were much more sensitive, you would hear the air molecules crashing against your ear like waves on the beach!

Loudness. In general, two pure tones having different frequencies but the same sound pressure level will be heard as different loudness levels. Loudness level is a psychoacoustic quantity.

In 1933, Fletcher and Munson conducted a series of experiments to determine the relationship between frequency and loudness.[7] A reference tone and a test tone were presented alternately to the test subjects. They were asked to adjust the sound level of the test tone until it sounded as loud as the reference. The results were plotted as sound pressure level in dB versus the test tone frequency (Figure 7-12). The curves are called the Fletcher-Munson or equal loudness contours. The reference frequency is 1,000 Hz. The curves are labeled in *phons,* which are the sound pressure levels of the 1,000 Hz pure tone in dB. The lowest contour (dashed line) represents the "threshold of hearing." The actual threshold may vary by as much as ±10 dB between individuals with normal hearing.

Audiometry. Hearing tests are conducted with a device known as an *audiometer.* Basically, it consists of a source of pure tones with variable sound pressure level

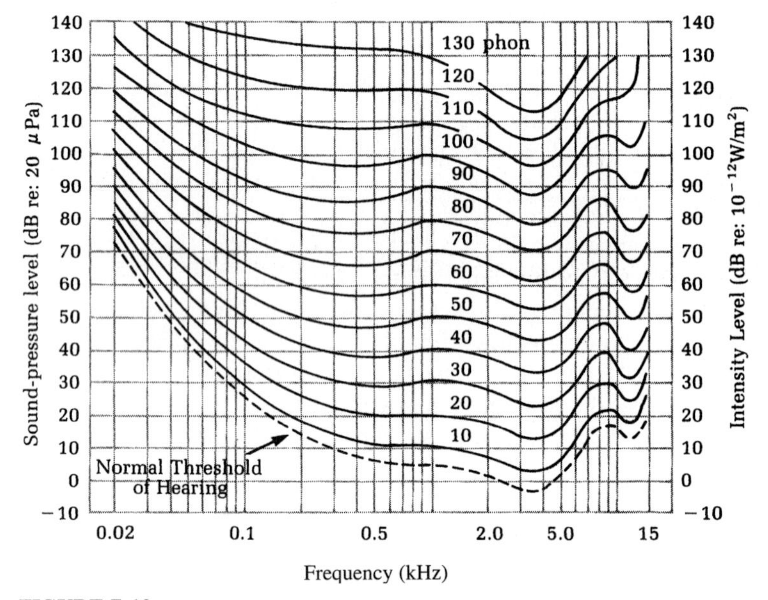

FIGURE 7-12
Fletcher-Munson equal loudness contours. (*Source:* Edward B. Magrab, *Environmental Noise Control,* New York: Wiley, 1975. Reprinted by permission.)

[7] H. Fletcher and W. A. Munson, "Loudness, Its Definition, Measurement and Calculation," *Journal of the Acoustical Society of America,* vol. 5, pp. 82–108, Oct. 1935.

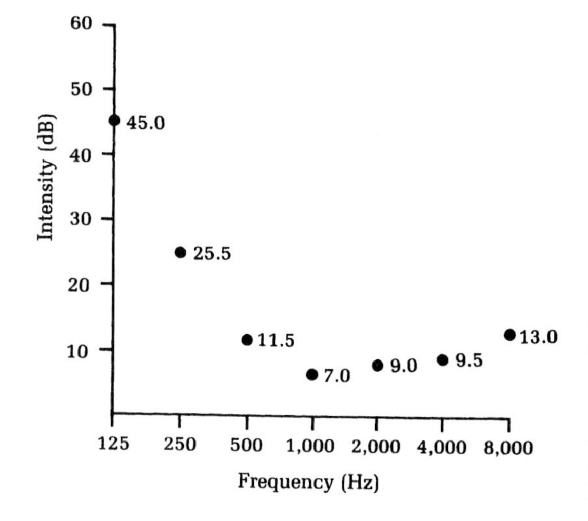

FIGURE 7-13
The ANSI reference values for hearing threshold level. (*Source:* Julian B. Olishifski and Earl R. Harford, eds., *Industrial Noise and Hearing Conservation,* Chicago: National Safety Council, 1975. Reprinted by permission.)

output into a pair of earphones. If the instrument also automatically prepares a graph of the test results (an *audiogram*), then it will include a weighting network called the *hearing threshold level* (HTL) scale.

The HTL scale is one in which the loudness of each pure tone is adjusted by frequency such that "0" dB is the level just audible for the average normal young ear. Two reference standards are in use: ASA–1951 and ANSI–1969. The ANSI reference values are shown in Figure 7-13. Note the similarity to the Fletcher-Munson contours. The initial audiogram prepared for an individual may be referred to as the baseline HTL or simply as the HTL.

The audiogram shown in Figure 7-14 reflects excellent hearing response. The average normal response may vary ±10 dB from the "0" dB value. As noted on the audiogram, this test was conducted with the ANSI–1969 weighting network.

You may have noted that we keep stressing young in our references to normal hearing. This is because there is hearing loss due to the aging process. This type of loss is called *presbycusis.* The average amount of loss as a function of age is shown in Figure 7-15.

Hearing Impairment

Mechanism. With the exception of eardrum rupture from intense explosive noise, the outer and middle ear rarely are damaged by noise. More commonly, hearing loss is a result of neural damage involving injury to the hair cells (Figure 7-16). Two theories are offered to explain noise-induced injury. The first is that excessive shearing forces mechanically damage the hair cells. The second is that intense noise stimulation forces the hair cells into high metabolic activity, which overdrives them to the point of metabolic failure and consequent cell death. Once destroyed, hair cells are not capable of regeneration.

Measurement. Since direct observation of the organ of Corti in persons having potential hearing loss is impossible, injury is inferred from losses in their HTL. The

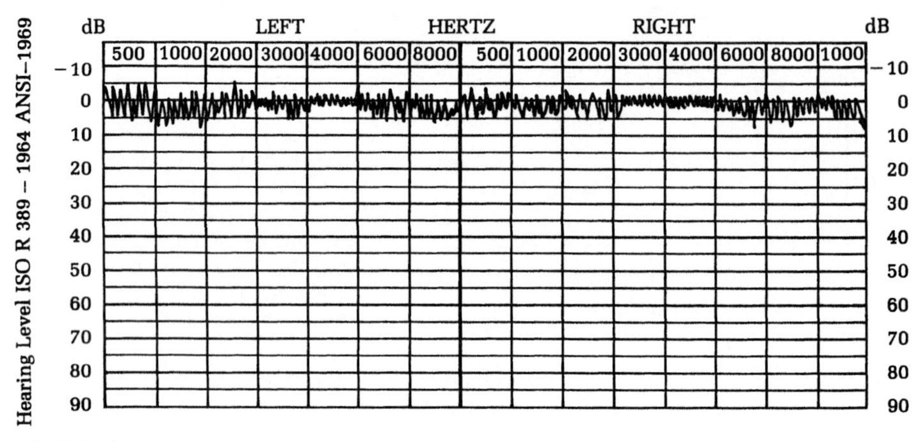

FIGURE 7-14
An audiogram illustrating excellent hearing response.

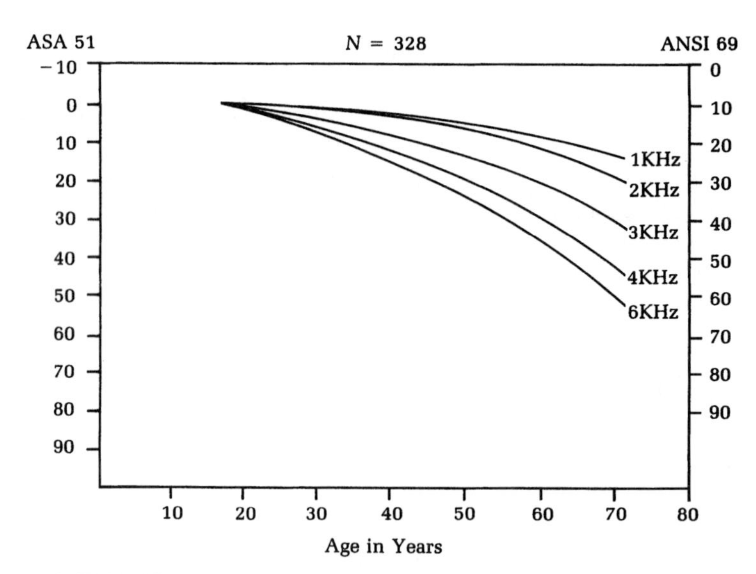

FIGURE 7-15
Hearing loss as a result of presbycusis. (*Source:* Julian B. Olishifski and Earl R. Harford, eds., *Industrial Noise and Hearing Conservation.* Reprinted by permission.)

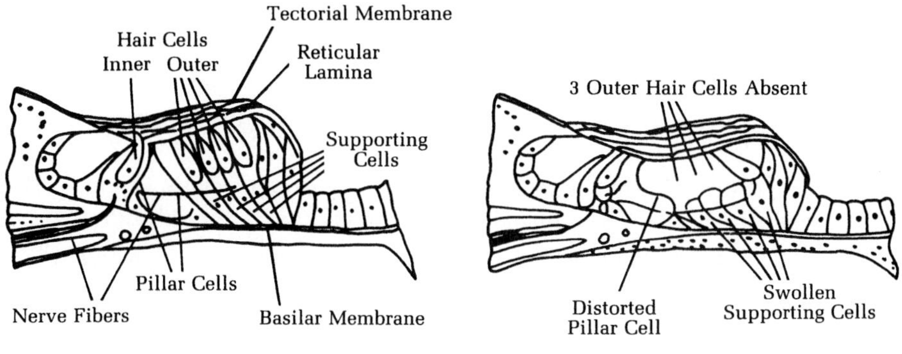

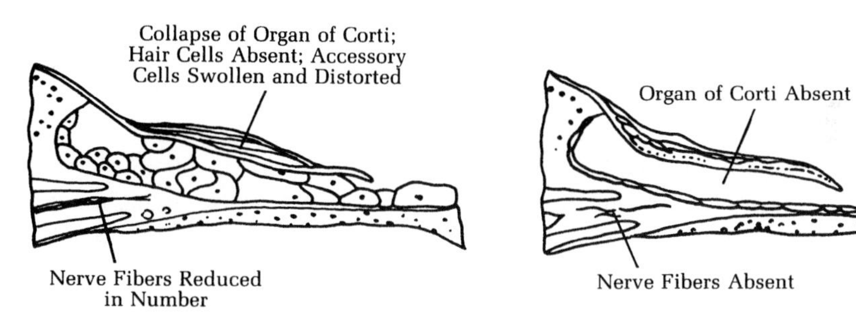

FIGURE 7-16
Various degrees of injury to the hair cells.

increased sound pressure level required to achieve a new HTL is called *threshold shift.* Obviously, any measurement of threshold shift is dependent upon having a baseline audiogram taken before the noise exposure.

Hearing losses may be either temporary or permanent. Noise-induced losses must be separated from other causes of hearing loss such as age (presbycusis), drugs, disease, and blows on the head. Temporary threshold shift (TTS) is distinguished from permanent threshold shift (PTS) by the fact that in TTS removal of the noise overstimulation will result in a gradual return to baseline hearing thresholds.

Factors affecting threshold shift. Important variables in the development of temporary and permanent hearing threshold changes include the following:[8]

1. Sound level: Sound levels must exceed 60 to 80 dBA before the typical person will experience TTS.

[8] U.S. Department of Health, Education and Welfare, National Institute for Occupational Safety and Health, *Criteria for a Recommended Standard: Occupational Exposure to Noise,* Washington, DC: U.S. Government Printing Office, 1972.

2. Frequency distribution of sound: Sounds having most of their energy in the speech frequencies are more potent in causing a threshold shift than are sounds having most of their energy below the speech frequencies.
3. Duration of sound: The longer the sound lasts, the greater the amount of threshold shift.
4. Temporal distribution of sound exposure: The number and length of quiet periods between periods of sound influences the potentiality of threshold shift.
5. Individual differences in tolerance of sound may vary greatly among individuals.
6. Type of sound—steady-state, intermittent, impulse, or impact: The tolerance to peak sound pressure is greatly reduced by increasing the duration of the sound.

Temporary threshold shift (TTS). TTS is often accompanied by a ringing in the ear, muffling of sound, or discomfort of the ears. Most of the TTS occurs during the first two hours of exposure. Recovery to the baseline HTL after TTS begins within the first hour or two after exposure. Most of the recovery that is going to be attained occurs within 16 to 24 hours after exposure.

Permanent threshold shift (PTS). There appears to be a direct relationship between TTS and PTS. Noise levels that do not produce TTS after two to eight hours of exposure will not produce PTS if continued beyond this time. The shape of the TTS audiogram will resemble the shape of the PTS audiogram.

Noise-induced hearing loss generally is first characterized by a sharply localized dip in the HTL curve at the frequencies between 3,000 and 6,000 Hz. This dip commonly occurs at 4,000 Hz (Figure 7-17). This is the *high frequency notch.* The progress from TTS to PTS with continued noise exposure follows a fairly regular pattern. First, the high frequency notch broadens and spreads in both directions. While substantial losses may occur above 3,000 Hz, the individual will not notice any change in hearing. In fact, the individual will not notice any hearing loss until the speech frequencies between 500 and 2,000 Hz average more than a 25 dB increase in HTL on the ANSI—1969 scale. The onset and progress of noise-induced permanent hearing loss is slow and insidious. The exposed individual is unlikely to notice it. Total hearing loss from noise exposure has not been observed.

Acoustic trauma. The outer and middle ear rarely are damaged by intense noise. However, explosive sounds can rupture the tympanic membrane or dislocate the ossicular chain. The permanent hearing loss that results from very brief exposure to a very loud noise is termed *acoustic trauma.*[9] Damage to the outer and middle ear may or may not accompany acoustic trauma. Figure 7-18 is an example of an audiogram that illustrates acoustic trauma.

---

[9]H. Davis, "Effects of High Intensity Noise on Navy Personnel," *U.S. Armed Forces Medical Journal,* vol. 9, pp. 1027–1047, 1958.

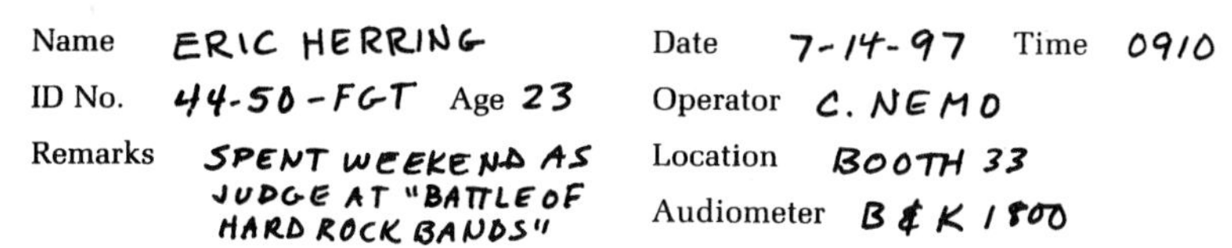

Name ERIC HERRING
Date 7-14-97
Time 0910
ID No. 44-50-FGT
Age 23
Operator C. NEMO
Remarks SPENT WEEKEND AS JUDGE AT "BATTLE OF HARD ROCK BANDS"
Location BOOTH 33
Audiometer B & K 1800

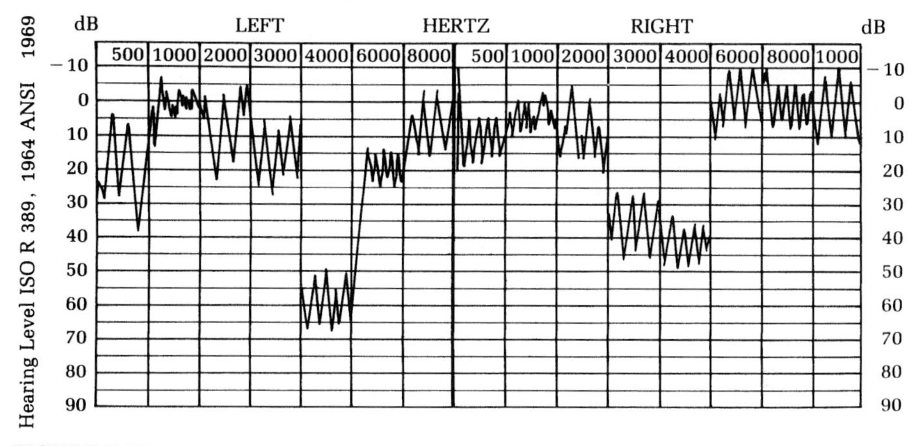

FIGURE 7-17
An audiogram illustrating hearing loss at the high frequency notch.

Protective mechanisms. Although the extent and mechanisms are not clear, it appears that the structures of the middle ear offer some protection to the delicate sensory organs of the inner ear.[10] One mechanism of protection is a change in the mode of vibration of the stapes. As noted earlier, there is some evidence that the muscles of the middle ear contract reflexively in response to loud noise. This contraction results in a reduction in the amplification that this series of levers normally produces. Changes in transmission may be on the order of 5 to 10 dB. However, the reaction time of the muscle/bone structure is on the order of 10 milliseconds. Thus, this protection is not effective against steep acoustic wave fronts that are characteristic of impact or impulsive noise.

Damage-Risk Criteria

A damage-risk criterion specifies the maximum allowable exposure to which a person may be exposed if risk of hearing impairment is to be avoided. The American Academy of Ophthalmology and Otolaryngology has defined hearing impairment as an average HTL in excess of 25 dB (ANSI–1969) at 500, 1,000, and 2,000 Hz. This is called the *low fence.* Total impairment is said to occur when the average HTL exceeds 92 dB. Presbycusis is included in setting the 25 dB ANSI low fence. Two

[10]H. Davis, "The Hearing Mechanism," in *Handbook of Noise Control,* edited by C. M. Harris, New York: McGraw-Hill, pp. 4–6, 1957.

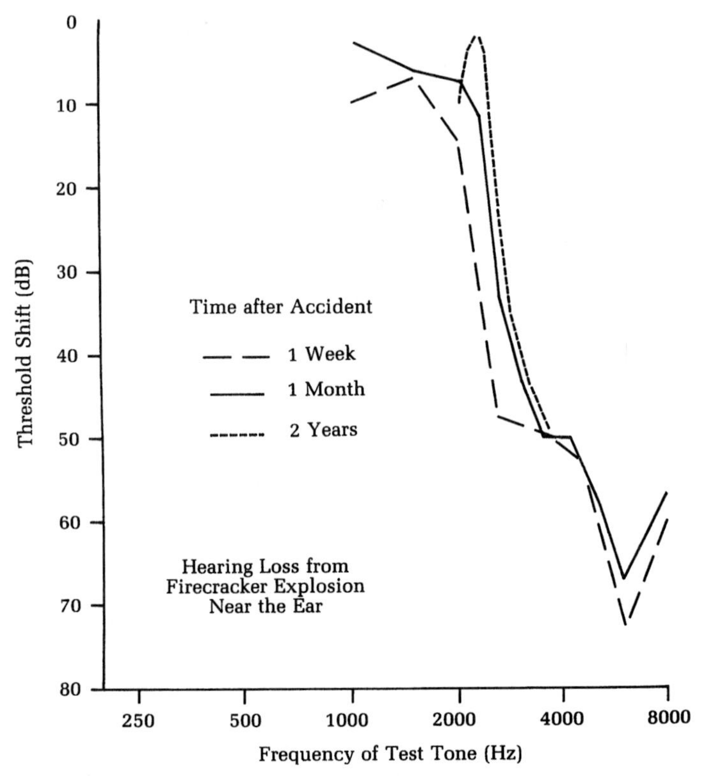

FIGURE 7-18
An example audiogram illustrating acoustic trauma. (*Source:* W. D. Ward and Abram Glorig, "A Case of Firecracker-induced Hearing Loss," *Laryngoscope,* vol. 71, 1961. Copyright by Laryngoscope. Reprinted by permission.)

criteria have been set to provide conditions under which nearly all workers may be repeatedly exposed without adverse effect on their ability to hear and understand normal speech.

Continuous or intermittent exposure. The National Institute for Occupational Safety and Health (NIOSH) has recommended that occupational noise exposure be controlled so that no worker is exposed in excess of the limits defined by line B in Figure 7-19. In addition, NIOSH recommends that new installations be designed to hold noise exposure below the limits defined by line A in Figure 7-19. The Walsh-Healey Act, which was enacted by Congress in 1969 to protect workers, used a damage-risk criterion equivalent to the line A criterion.

Speech Interference

As we all know, noise can interfere with our ability to communicate. Many noises that are not intense enough to cause hearing impairment can interfere with speech communication. The interference, or *masking,* effect is a complicated function of the

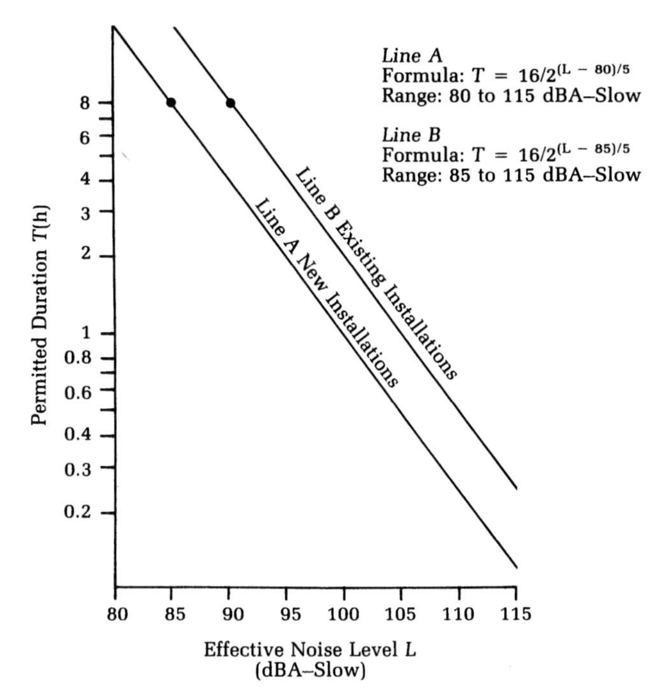

FIGURE 7-19
NIOSH occupational noise exposure limits for continuous or intermittent noise exposure.

distance between the speaker and listener and the frequency components of the spoken words. The Speech Interference Level (SIL) was developed as a measure of the difficulty in communication that could be expected with different background noise levels.[11] It is now more convenient to talk in terms of A-weighted background noise levels and the quality of speech communication (Figure 7-20).

Example 7-5. Consider the problem of a speaker in a quiet zone who wishes to speak to a listener operating a 4.5 Mg (megagram) truck 6.0 m away. The sound level in the truck cab is about 73 dBA.

Solution. Using Figure 7-20, we can see that she is going to have to shout very loudly to be heard. However, if she moved to within about 1.0 m, she would be able to use her "expected" voice level, that is, the unconscious slight rise in voice level that one would normally use in a noisy situation.

It can be seen that at distances not uncommon in living rooms or classrooms (4.5 to 6.0 m), the A-weighted background level must be below about 50 dB for normal conversation.

---

[11] Leo L. Beranek, *Acoustics,* New York: McGraw-Hill, 1954.

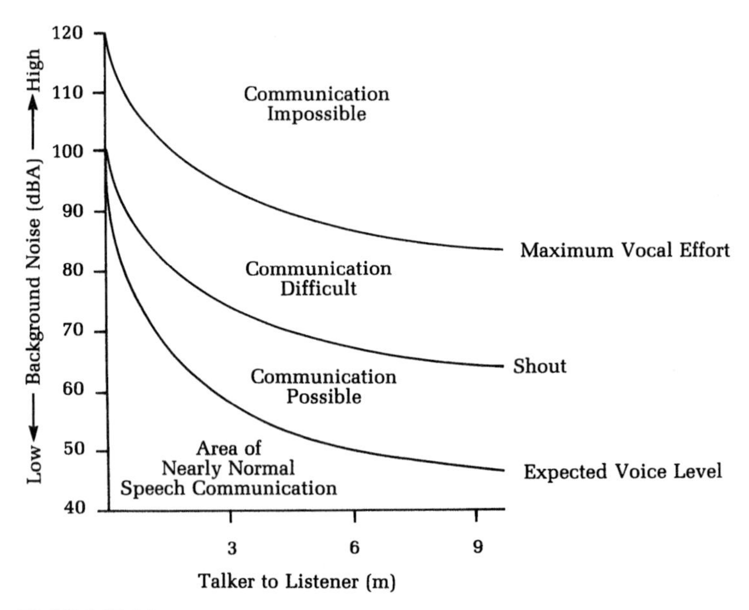

FIGURE 7-20
Quality of speech communication as a function of sound level and distance. [*Source:* James D. Miller, *Effects of Noise on People* (U.S. Environmental Protection Agency Publication No. NTID 300.7), Washington, DC: U.S. Government Printing Office, 1971.]

Annoyance

Annoyance by noise is a response to auditory experience. Annoyance has its base in the unpleasant nature of some sounds, in the activities that are disturbed or disrupted by noise, in the physiological reactions to noise, and in the responses to the meaning of "messages" carried by the noise.[12] For example, a sound heard at night may be more annoying than one heard by day, just as one that fluctuates may be more annoying than one that does not. A sound that resembles another sound that we already dislike and that perhaps threatens us may be especially annoying. A sound that we know is mindlessly inflicted and will not be removed soon may be more annoying than one that is temporarily and regretfully inflicted. A sound, the source of which is visible, may be more annoying than one with an invisible source. A sound that is new may be less annoying. A sound that is locally a political issue may have a particularly high or low annoyance.[13]

The degree of annoyance and whether that annoyance leads to complaints, product rejection, or action against an existing or anticipated noise source depend upon many factors. Some of these factors have been identified, and their

---

[12]J. D. Miller, *Effects of Noise on People* (U.S. Environmental Protection Agency Publication No. NTID 300.7), Washington, DC: U.S. Government Printing Office, p. 93, 1971.

[13]Daryln N. May, *Handbook of Noise Assessment,* New York: Van Nostrand Reinhold, p. 5, 1978.

relative importance has been assessed. Responses to aircraft noise have received the greatest attention. There is less information available concerning responses to other noises, such as those of surface transportation and industry, and those from recreational activities.[14] Many of the noise rating or forecasting systems that are now in existence were developed in an effort to predict annoyance reactions.

Sonic booms. One noise of special interest with respect to annoyance is called *sonic boom* or, more correctly as we shall see, sonic booms.

> The flow of air around an aircraft or other object whose speed exceeds the speed of sound (supersonic) is characterized by the existence of discontinuities in the air known as *shock wave*. These discontinuities result from the sudden encounter of an impenetrable body with air. At subsonic speeds, the air seems to be forewarned; thus, it begins its outward flow before the arrival of the leading edge. At supersonic speeds, however, the air in front of the aircraft is undisturbed, and the sudden impulse at the leading edge creates a region of overpressure (Figure 7-21) where the pressure is higher than atmospheric pressure. This overpressure region travels outward with the speed of sound,

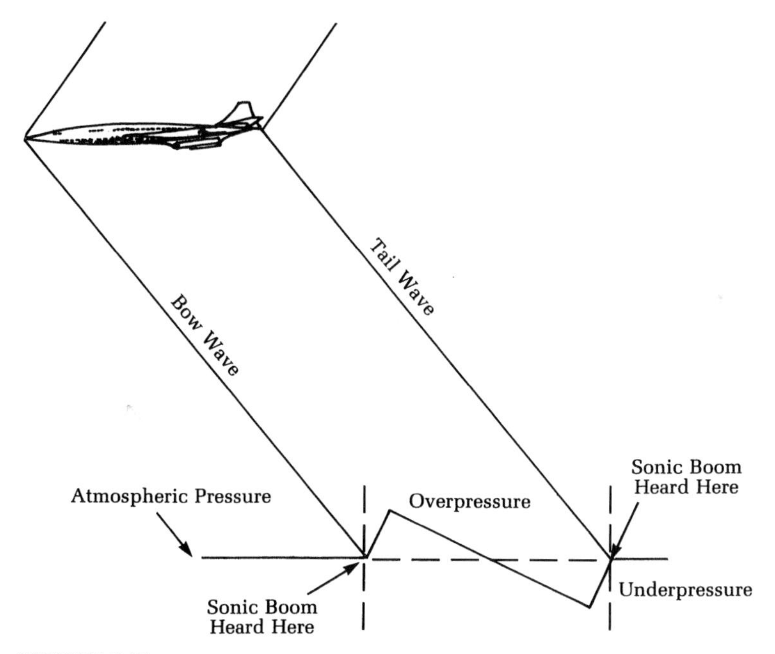

FIGURE 7-21
Sonic booms resulting from bow wave and tail wave set in motion by supersonic flight.

[14]J. D. Miller, *Effects of Noise on People,* p. 93.

> creating a conically shaped shock wave called the *bow wave* that changes the direction of airflow. A second shock wave, the *tail wave*, is produced by the tail of the aircraft and is associated with a region where the pressure is lower than normal. This underpressure discontinuity causes the air behind the aircraft to move sideways.
>
> Major pressure changes are experienced at the ear as the bow and tail shock waves reach an observer. Each of these pressure deviations produces the sensation of an explosive sound.[15]

You should note that the pressure wave and, hence, the sonic boom exist whenever the aircraft is at supersonic speed and not "just when it breaks the sound barrier."

Both the loudness of the noise and the startling effect of the impulse (it makes us "jump") are found to be very annoying. Apparently we can never get used to this kind of noise. Supersonic flight by commercial aircraft is forbidden in the airspace above the United States. Supersonic flight by military aircraft is restricted to sparsely inhabited areas.

Sleep Interference

Sleep interference is a special category of annoyance that has received a great deal of attention and study.[16] Almost all of us have been wakened or kept from falling asleep by loud, strange, frightening, or annoying sounds. It is commonplace to be wakened by an alarm clock or clock radio. But it also appears that one can get used to sounds and sleep through them. Possibly, environmental sounds only disturb sleep when they are unfamiliar. If so, disturbance of sleep would depend only on the frequency of unusual or novel sounds. Everyday experience also suggests that sound can help to induce sleep and, perhaps, to maintain it. The soothing lullaby, the steady hum of a fan, or the rhythmic sound of the surf can serve to induce relaxation. Certain steady sounds can serve as an acoustical shade and mask disturbing transient sounds.

Common anecdotes about sleep disturbance suggest an even greater complexity. A rural person may have difficulty sleeping in a noisy urban area. An urban person may be disturbed by the quiet when sleeping in a rural area. And how is it that a parent may wake to a slight stirring of his or her child, yet sleep through a thunderstorm? These observations all suggest that the relations between exposure to sound and the quality of a night's sleep are complicated.

The effects of relatively brief noises (about three minutes or less) on a person sleeping in a quiet environment have been studied the most thoroughly. Typically, presentations of the sounds are widely spaced throughout a sleep period of five to seven hours. A summary of some of these observations is presented in

---

[15]Richard B. Minnix, "The Nature of Sound," *Noise Control Handbook of Principles and Practices,* D. M. Lipscomb and A. C. Taylor, eds., New York: Van Nostrand Reinhold, pp. 29–30, 1978.

[16]The sections on sleep interference, effects on performance, and acoustic privacy follow J. D. Miller (see footnote 12) pp. 59, 67–69.

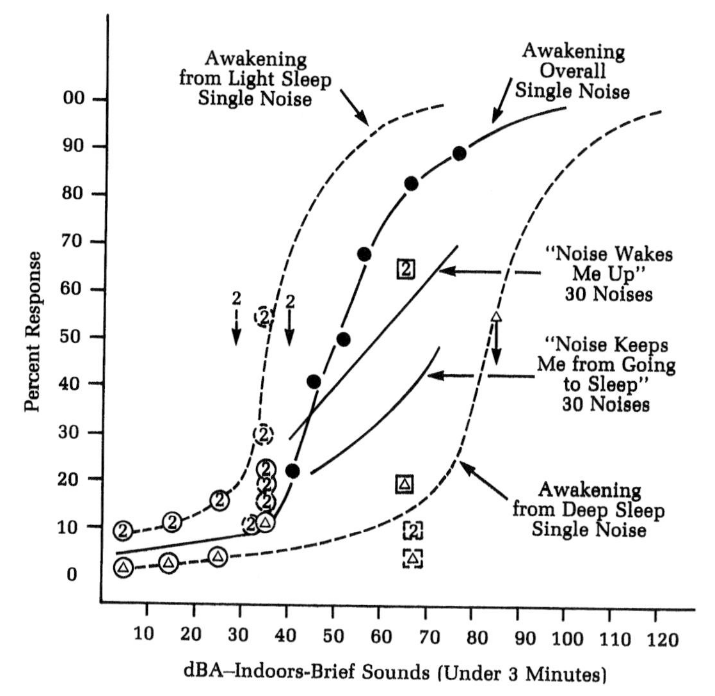

FIGURE 7-22
Effects of brief noise on sleep. [*Source:* J. D. Miller (U.S. Environmental Protection Agency Publication No. NTID 300.7), Washington, DC: U.S. Government Printing Office, 1971.]

Figure 7-22. The dashed lines are hypothetical curves that represent the percent of awakenings under conditions in which the subject is a normally rested young adult male who has been adapted for several nights to the procedures of a quiet sleep laboratory. He has been instructed to press an easily reached button to indicate that he has awakened, and had been moderately motivated to awake and respond to the noise.

While in light sleep, subjects can awake to sounds that are about 30–40 decibels above the level at which they can be detected when subjects are conscious, alert, and attentive. While in deep sleep, the stimulus may have to be 50–80 decibels above the level at which they can be detected by conscious, alert, attentive subjects before they will awaken the sleeping subject.

The solid lines in Figure 7-22 are data from questionnaire studies of persons who live near airports. The percentage of respondents who claim that flyovers wake them or keep them from falling asleep is plotted against the A-weighted sound level of a single flyover. These curves are for the case of approximately 30 flyovers spaced over the normal sleep period of six to eight hours. The filled circles represent the percentage of sleepers that awake to a three-minute sound at each A-weighted sound level (dBA) or lower. This curve is based on data from 350 persons, each tested in his or her own bedroom. These measures were made between 2:00 and 7:00 A.M. It is reasonable to assume that most of the subjects were roused from a light sleep.

Effects on Performance

When a task requires the use of auditory signals, speech or nonspeech, then noise at any intensity level sufficient to mask or interfere with the perception of these signals will interfere with the performance of the task.

Where mental or motor tasks do not involve auditory signals, the effects of noise on their performance have been difficult to assess. Human behavior is complicated, and it has been difficult to discover exactly how different kinds of noises might influence different kinds of people doing different kinds of tasks. Nonetheless, the following general conclusions have emerged. Steady noises without special meaning do not seem to interfere with human performance unless the A-weighted noise level exceeds about 90 decibels. Irregular bursts of noise (intrusive noise) are more disruptive than steady noises. Even when the A-weighted sound levels of irregular bursts are below 90 decibels, they may sometimes interfere with performance of a task. High-frequency components of noise, above about 1,000–2,000 hertz, may produce more interference with performance than low-frequency components of noise. Noise does not seem to influence the overall rate of work, but high levels of noise may increase the variability of the rate of work. There may be "noise pauses" followed by compensating increases in work rate. Noise is more likely to reduce the accuracy of work than to reduce the total quantity of work. Complex tasks are more likely to be adversely influenced by noise than are simple tasks.

Acoustic Privacy

Without opportunity for privacy, either everyone must conform strictly to an elaborate social code or everyone must adopt highly permissive attitudes. Opportunity for privacy avoids the necessity for either extreme. In particular, without opportunity for acoustical privacy, one may experience all of the effects of noise previously described and, in addition, one is constrained because one's own activities may disturb others. Without acoustical privacy, sound, like a faulty telephone exchange, reaches the "wrong number." The result disturbs both the sender and the receiver.

7-3 RATING SYSTEMS

Goals of a Noise-Rating System

An ideal noise-rating system is one that allows measurements by sound level meters or analyzers to be summarized succinctly and yet represent noise exposure in a meaningful way. In our previous discussions on loudness and annoyance, we noted that our response to sound is strongly dependent on the frequency of the sound. Furthermore, we noted that the type of noise (continuous, intermittent, or impulsive) and the time of day that it occurred (night being worse than day) were significant factors in annoyance.

Thus, the ideal system must take frequency into account. It should differentiate between daytime and nighttime noise. And, finally, it must be capable of describing the cumulative noise exposure. A statistical system can satisfy these requirements.

The practical difficulty with a statistical rating system is that it would yield a large set of parameters for each measuring location. A much larger array of numbers would be required to characterize a neighborhood. It is literally impossible for such an array of numbers to be used effectively in enforcement. Thus, there has been a considerable effort to define a single number measure of noise exposure. The following paragraphs describe one of the systems now being used.

The L_N Concept

The parameter L_N is a statistical measure that indicates how frequently a particular sound level is exceeded.[17] If, for example, we write $L_{40} = 72$ dBA, then we know that 72 dB(A) was exceeded for 40 percent of the measuring time. A plot of L_N against N where $N = 1$ percent, 2 percent, 3 percent, and so forth, would look like the cumulative distribution curve shown in Figure 7-23.

Allied to the cumulative distribution curve is the probability distribution curve. A plot of this will show how often the noise levels fall into certain class intervals. In Figure 7-24 we can see that 22 percent of the time the measured noise levels ranged between 70 and 72 dBA; for 17 percent of the time they ranged between 72 and 74 dBA; and so on. The relationship between this picture and the one for L_N is really quite simple. By adding the percentages given in successive class intervals from right to left, we can arrive at a corresponding L_N where N is the sum of the percentages and L is the lower limit of the left-most class interval added, thus, L_{40}:

$$L(2 + 7 + 14 + 17) = 72 \text{ dBA}$$

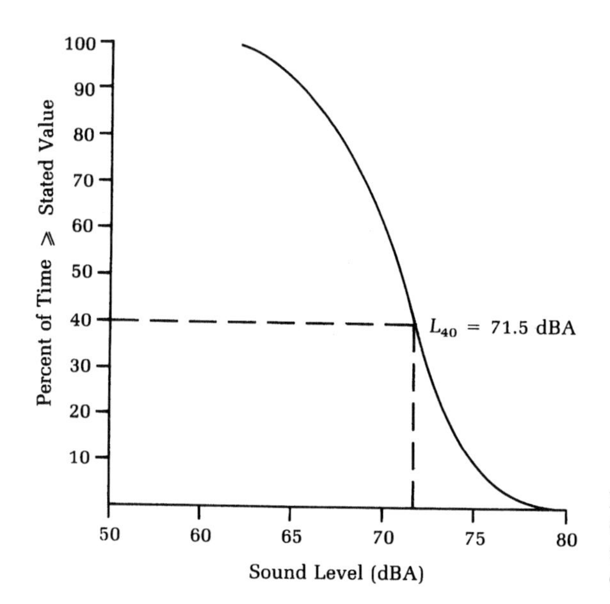

FIGURE 7-23
Cumulative distribution curve. (Courtesy of B & K Instruments, Inc., Cleveland.)

[17]Courtesy of B & K Instruments, Inc., Cleveland.

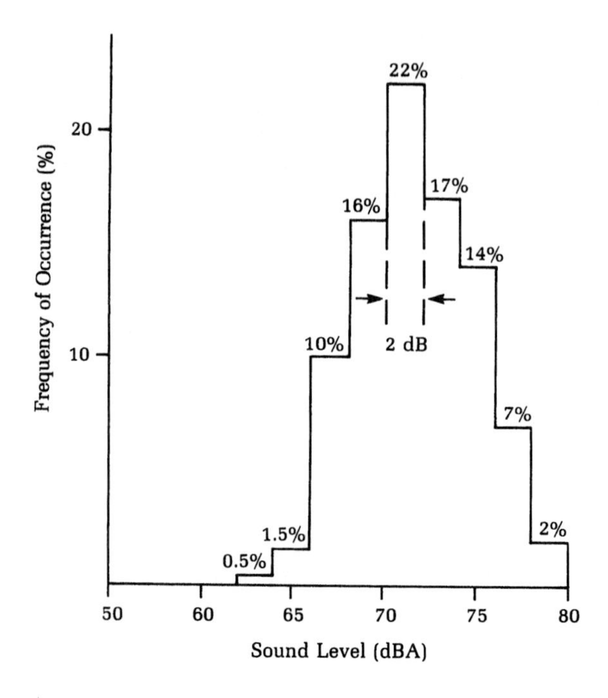

FIGURE 7-24
Probability distribution plot. (Courtesy of B & K Instruments, Inc., Cleveland.)

The L_{eq} Concept

The equivalent continuous equal energy level (L_{eq}) can be applied to any fluctuating noise level.[17] It is that constant noise level that, over a given time, expends the same amount of energy as the fluctuating level over the same time period. It is expressed as follows:

$$L_{eq} = 10 \log \frac{1}{t} \int_{o}^{t} 10^{L(t)/10}\, dt \tag{7-14}$$

where t = the time over which L_{eq} is determined
$L(t)$ = the time varying noise level in dBA

Generally speaking, there is no well-defined relationship between $L(t)$ and time, so a series of discrete samples of $L(t)$ have to be taken. This modifies the expression to:

$$L_{eq} = 10 \log \sum_{i=1}^{i=n} 10^{L_i/10} t_i \tag{7-15}$$

where n = the total number of samples taken
L_i = the noise level in dBA of the ith sample
t_i = fraction of total sample time

Example 7-6. Consider the case where a noise level of 90 dBA exists for five minutes and is followed by a reduced noise level of 60 dBA for 50 minutes. What is the equivalent continuous equal energy level for the 55-minute period? Assume a five-minute sampling interval.

Solution. If the sampling interval is five minutes, then the total number of samples (n) is 11, and the fraction of total sample time (t_i) for each sample is $1/11 = 0.091$. With these preliminary calculations, we may now compute the sum:

$$\sum_{t=1}^{2} = (10^{90/10})(0.091) + (10^{60/10})(0.91)$$

$$= (9.1 \times 10^7) + (9.1 \times 10^5) = 9.19 \times 10^7$$

And finally, we take the log to find

$$L_{eq} = 10\log(9.19 \times 10^7) = 79.6 \text{ or } 80 \text{ dBA}$$

The example calculation is depicted graphically in Figure 7-25. From this you may note that great emphasis is put on occasional high noise levels.

The equivalent noise level was introduced in 1965 in Germany as a rating specifically to evaluate the impact of aircraft noise upon the neighbors of airports.[18] It was almost immediately recognized in Austria as appropriate for evaluating the

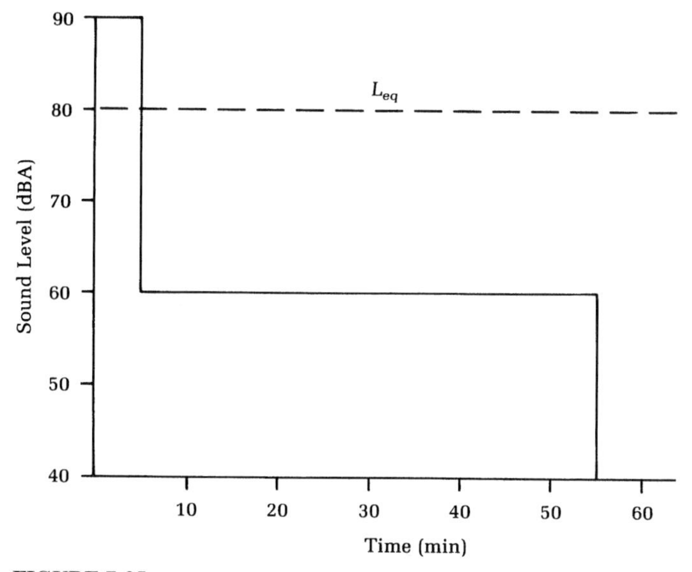

FIGURE 7-25
Graphical illustration of L_{eq} computation given in Example 7-5. (Courtesy of B & K Instruments, Inc., Cleveland.)

---

[18]W. Burck, *et al., Fluglärm,* Gutachten erstatet im Auftrag des Bundesministers fur Gesundheitswesen, Göttingen, 1965.

impact of street traffic noise in dwellings and schoolrooms. It has been embodied in the National Test Standards of Germany for rating the subjective effects of fluctuating noises of all kinds, such as from street and road traffic, rail traffic, canal and river ship traffic, aircraft, industrial operations (including the noise from individual machines), sports stadiums, playgrounds, and the like.

7-4 COMMUNITY NOISE SOURCES AND CRITERIA

It is not our intent to provide a detailed discussion of the noise characteristics of all community noise sources. Likewise, we have not attempted to provide a comprehensive list of noise criteria. Rather, we have selected a few examples to provide you with a feeling for the magnitude and range of the numbers.

Transportation Noise

Aircraft noise. The noise spectra of a wide body fan jet (for example, the Boeing 747) reveal that sound pressure levels are higher on takeoff than during the approach to land. This is typical of all aircraft. With the notable exception of the turbo jets, smaller aircraft have lower sound pressure levels.

The annoyance criteria for aircraft operations are based on extensive field measurements and opinion surveys. The results of annoyance surveys at nine airports in the United States and Great Britain are summarized in Figure 7-26.

Highway vehicle noise. For most automobiles, exhaust noise constitutes the predominant source for normal operation below about 55 km/h (Figure 7-27). Although tire noise is much less of a problem in automobiles than in trucks, it is the dominant noise source at speeds above 80 km/h. While not as noisy as trucks, the total contribution of automobiles to the noise environment is significant because of the very large number in operation.

Diesel trucks are 8 to 10 dB noisier than gasoline-powered ones. At speeds above 80 km/h, tire noise often becomes the dominant noise source on the truck. The "crossbar" tread is the noisiest.

Motorcycle noise is highly dependent on the speed of the vehicle. The primary source of noise is the exhaust. The noise spectra of two-cycle and four-cycle engines are of somewhat different character. The two-cycle engines exhibit more high frequency spectra energy content.

In 1968, Griffiths and Langdon reported on the results of an extensive attitude survey on traffic noise.[19] They correlated their results with the Traffic Noise Index

[19] I. D. Griffiths and F. J. Langdon, "Subjective Response to Road Traffic Noise," *Journal of Sound & Vibration*, vol. 8, pp. 16–32, 1968.

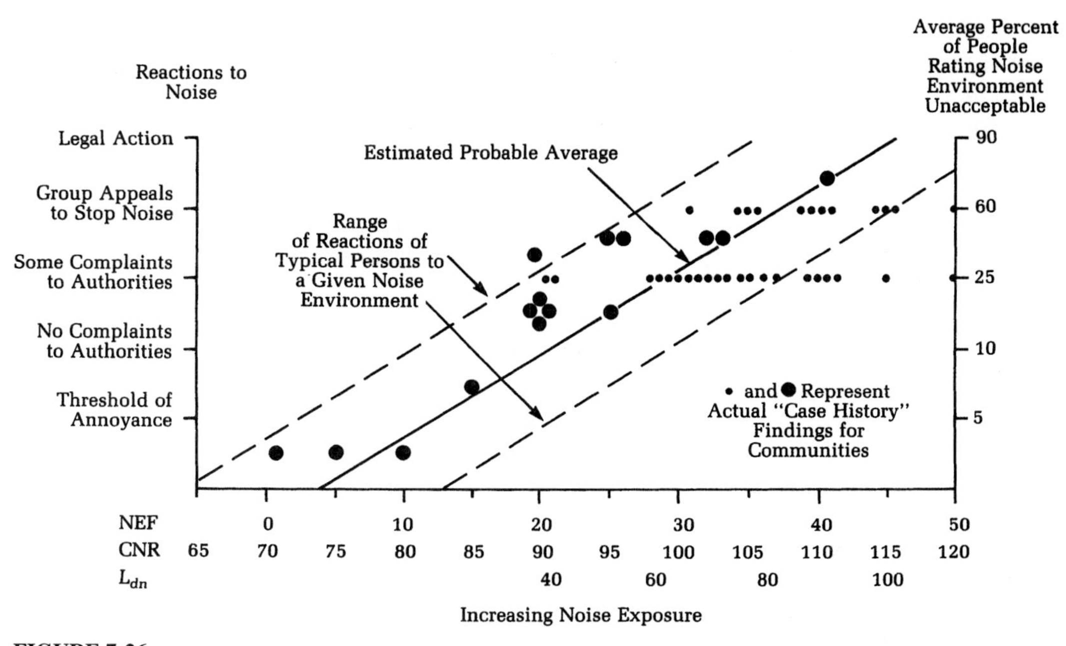

FIGURE 7-26
Relationship between exposure to aircraft noise and annoyance. (*Source:* K. D. Kryter, G. Jansen, D. Parker, H. O. Parrack, G. Thiessen, and H. L. William, *Non-Auditory Effects of Noise.* Report of WG-63, National Academy of Science–National Research Council Committee on Hearing, Bioacoustics, and Biomechanics, Washington, DC: U.S. GPO, 1971.)

rating system (Figure 7-28). The U.S. Federal Highway Administration has developed the standards shown in Table 7-3. The levels are above those that would be expected to yield no problems but are below those of many existing highways.

Other Internal Combustion Engines

Because of their ubiquitous nature and the general interest they stimulate, the combustion engines listed in Table 7-4 are included at this point. "In general, these devices are not significant contributors to average residential noise levels in urban areas. However, the relative annoyance of most of the equipment tends to be high."[20] The eight-hour exposure level is in reference to the equipment operator.

Construction Noise

The range of sound levels found for 19 common types of construction equipment is shown in Figure 7-29. Although the sample was limited, the data appear to be

---

[20]U.S. Environmental Protection Agency, *Transportation Noise and Noise from Equipment Powered by Internal Combustion Engines* (Environmental Protection Agency Publication No. NTID 300.13), Washington, DC: U.S. Government Printing Office, p. 230, 1971.

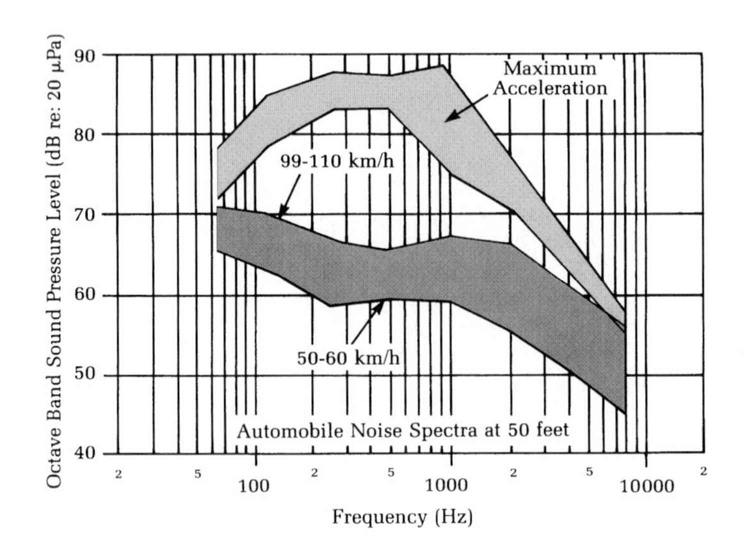

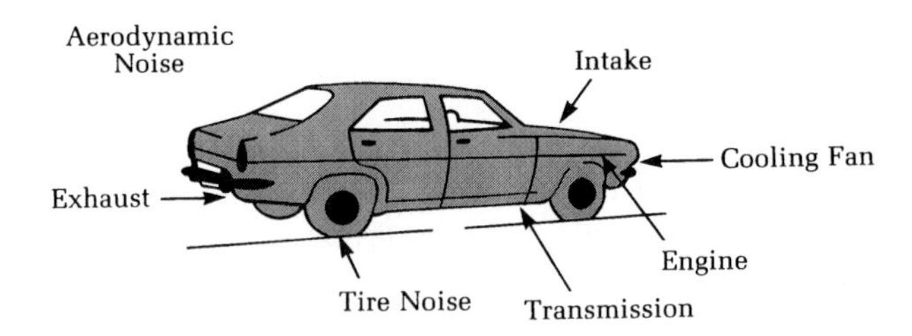

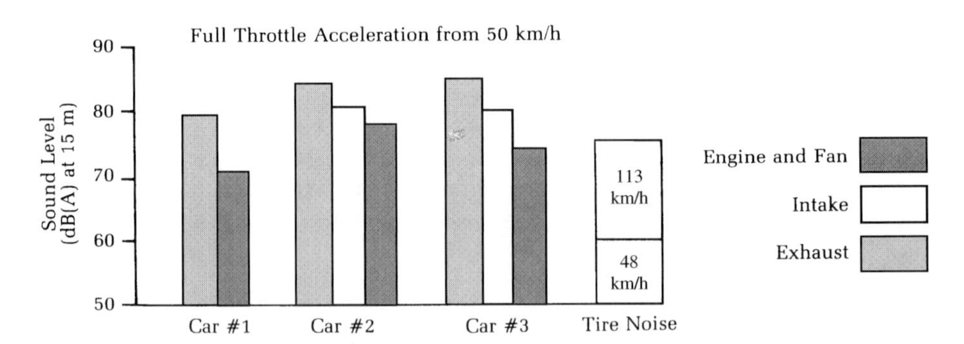

FIGURE 7-27
Typical noise spectra of automobiles. (*Source:* U.S. Environmental Protection Agency, *Transportation Noise,* Washington, DC: U.S. Government Printing Office, 1971.)

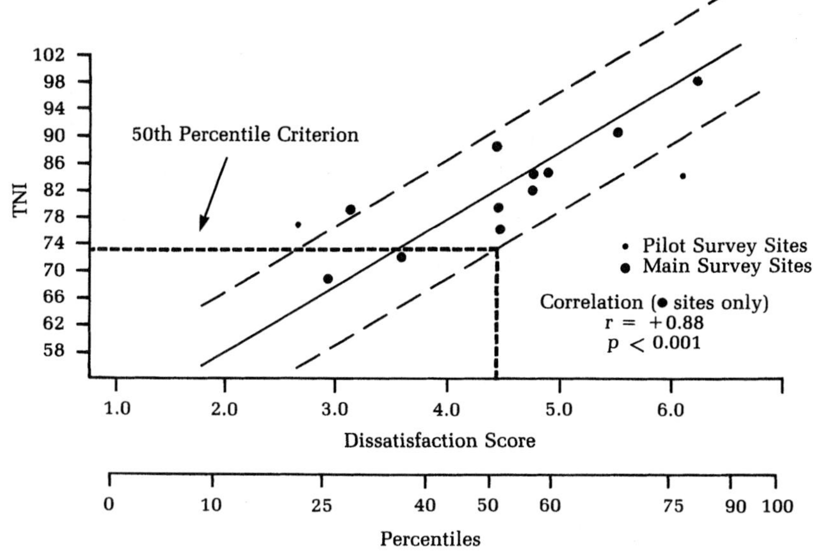

FIGURE 7-28
Annoyance as a function of the Traffic Noise Index (TNI). (*Source*: Alexandre, Barde, Lamure, and Langdon, *Road Traffic Noise,* Copyright 1975 by Applied Science Publishers, Ltd. Reprinted by permission.)

TABLE 7-3
FHA noise standards for new construction

Land use category	Exterior design noise level dBA[a] L_{eq}	Exterior design noise level dBA[a] L_{10}	Description of land use category
A	57	60	Tracts of lands in which serenity and quiet are of extraordinary significance and serve an important public need, and where the preservation of those qualities is essential if the area is to continue to serve its intended purpose. For example, such areas could include amphitheaters, particular parks or portions of parks, or open spaces, which are dedicated or recognized by appropriate local officials for activities requiring special qualities of serenity and quiet.
B	67	70	Residences, motels, hotels, public meeting rooms, schools, churches, libraries, hospitals, picnic areas, recreation areas, playgrounds, active sports areas, and parks.
C	72	75	Developed lands, properties, or activities not included in categories A and B above.
D	Unlimited	Unlimited	Undeveloped lands
E	52 (Interior)	55 (Interior)	Public meeting rooms, schools, churches, libraries, hospitals, and other such public buildings.

[a]Either L_{eq} or L_{10} may be used, but not both. The levels are to be based on a one-hour sample.

TABLE 7-4
Summary of noise characteristics of internal combustion engines

Source	A-weighted noise energy (kW · h/d)[a]	Typical A-weighted noise level at 15.2 m [dB(A)]	8-hr exposure level [db(A)][b]		Typical exposure time (h)
			Average	Maximum	
Lawn mowers	63	74	74	82	1.5
Garden tractors	63	78	N/A	N/A	N/A
Chain saws	40	82	85	95	1
Snow blowers	40	84	61	75	1
Lawn edgers	16	78	67	75	0.5
Model aircraft	12	78	70[c]	79[c]	0.25
Leaf blowers	3.2	76	67	75	0.25
Generators	0.8	71	—	—	—
Tillers	0.4	70	72	80	1

[a]Based on estimates of the total number of units in operation per day.
[b]Equivalent level for evaluation of relative hearing damage risk.
[c]During engine trimming operation.
[*Source:* U.S. Environmental Protection Agency, *Transportation Noise and Noise from Equipment Powered by Internal Combustion Engines* (Environmental Protection Agency Publication No. NTID 300.13), Washington, DC: U.S. Government Printing Office, 1971.]

reasonably accurate. The noise produced by the interaction of the machine and the material on which it acts often contributes greatly to the sound level.

It is difficult, at best, to quantify the annoyance that results from construction noise. The following generalizations appear to hold:

1. Single house construction in suburban communities will generate sporadic complaints if the boundary line eight-hour L_{eq} exceeds 70 dBA.
2. Major excavation and construction in a normal suburban community will generate threats of legal action if the boundary line eight-hour L_{eq} exceeds 85 dBA.

Zoning and Siting Considerations

The U.S. Department of Housing and Urban Development (HUD) set out guideline criteria for noise exposure at residential sites for new construction (Table 7-5). The NEF zones listed in the table are those specified by the Federal Aviation Administration (FAA) for land use compatibility (Table 7-6). These guidelines, and those given above for traffic noise (Table 7-3), if followed in zoning and siting, will minimize annoyance and complaints.

Levels to Protect Health and Welfare

In accordance with the directive from Congress, the U.S. Environmental Protection Agency published noise criteria levels that it deemed necessary to protect the health

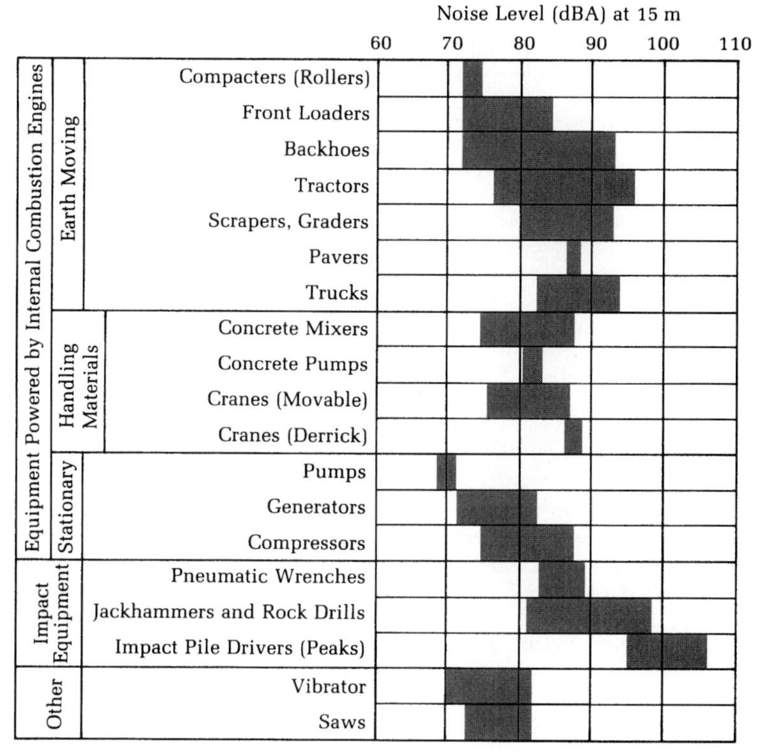

FIGURE 7-29
Range of sound levels from various types of construction equipment (based on limited available data samples). (*Source: Report to the President and Congress on Noise,* 1972.)

TABLE 7-5
HUD noise assessment criteria for new residential construction

General external exposures	Assessment
Exceeds 89 dBA 60 minutes per 24 hours Exceeds 75 dBA 8 hours per 24 hours CNR Zone 3, NEF Zone C (airport environs) $L_{NP} > 88$ dB (NP) (exterior)	Unacceptable
Exceeds 65 dBA 8 hours per 24 hours Loud repetitive sounds on site CNR Zone 2, NEF Zone B (airport environs) L_{NP} 74 to 88 dB (NP) (exterior)	Discretionary: normally unacceptable
Does not exceed 65 dBA more than 8 hours per 24 hours L_{NP} 62 to 74 dB (NP) (exterior)	Discretionary: normally acceptable
Does not exceed 45 dBA more than 30 minutes per 24 hours CNR Zone 1, NEF Zone A (airport environs) $L_{NP} < 62$ dB (NP) (exterior)	Acceptable

TABLE 7-6
Land use compatibility for various NEF values

Land use	Compatibility		
	Zone A (less than 30 NEF)	**Zone B (30–40 NEF)**	**Zone C (above 40 NEF)**
Residential	Yes	(*b*)	No
Hotel, motel, offices, public buildings	Yes	Yes(*c*)	No
Schools, hospitals, churches, indoor theaters, auditoriums	Yes(*c*)	No	No
Commercial, industrial	Yes	Yes	(*c*)
Outdoor amphitheaters, theaters	Yes(*a*),(*c*)	No	No
Outdoor recreational (non-spectator)	Yes	Yes	Yes

Code:

(*a*) A detailed noise analysis should be undertaken by qualified personnel for all indoor music auditoriums and all outdoor theaters.

(*b*) Case history experience indicates that individuals in private residences may complain, perhaps vigorously. Concerted group action is possible. New single-dwelling construction should generally be avoided. For apartment construction, note (*c*) applies.

(*c*) An analysis of building noise reduction requirements should be made and needed noise control features should be included in the design.

and welfare of U.S. citizens (Table 7-7).[21] The EPA maintained that a quiet residential environment is necessary in both urban and rural areas to prevent activity interference and annoyance and to permit the hearing mechanism an opportunity to recuperate if it is exposed to high levels during the day. The L_{dn} of 45 provides a fair margin of safety.

7-5 TRANSMISSION OF SOUND OUTDOORS

Inverse Square Law

If a sphere of radius δ vibrates with a uniform radial expansion and contraction, sound waves radiate uniformly from its surface. If the sphere is placed such that no sound waves are reflected back in the direction of the source, and if the product $\kappa\delta$, where κ is the wave number, is much less than 1, then the sound intensity at any

[21]U.S. Environmental Protection Agency, *Information on Levels of Environmental Noise Requisite to Protect Public Health and Welfare with an Adequate Margin of Safety* (Environmental Protection Agency Publication No. 550/9-74-004), Washington, DC: U.S. Government Printing Office, 1974.

TABLE 7-7
Yearly energy average L_{eq} identified as requisite to protect the public health and welfare with an adequate margin of safety

	Measure	Indoor: Activity interference	Indoor: Hearing loss consideration	To protect against both effects (*b*)	Outdoor: Activity interference	Outdoor: Hearing loss consideration	To protect against both effects (*b*)
Residential with outside space and farm residences	L_{dn}	45		45	55		55
	$L_{eq(24)}$		70			70	
Residential with no outside space	L_{dn}	45		45			
	$L_{eq(24)}$		70				
Commercial	$L_{eq(24)}$	(*a*)	70	70(*c*)	(*a*)	70	70(*c*)
Inside transportation	$L_{eq(24)}$	(*a*)	70	(*a*)			
Industrial	$L_{eq(24)(d)}$	(*a*)	70	70(*c*)	(*a*)	70	70(*c*)
Hospitals	L_{dn}	45		45	55		55
	$L_{eq(24)}$		70			70	
Educational	$L_{eq(24)}$	45		45	55		55
	$L_{eq(24)(d)}$		70			70	
Recreational areas	$L_{eq(24)}$	(*a*)	70	70(*c*)	(*a*)	70	70(*c*)
Farm land and general unpopulated land	$L_{eq(24)}$				(*a*)	70	70(*c*)

Code:

(*a*) Since different types of activities appear to be associated with different levels, identification of a maximum level for activity interference may be difficult except in those circumstances where speech communication is a critical activity.

(*b*) Based on lowest level.

(*c*) Based only on hearing loss.

(*d*) An $L_{eq(8)}$ of 75 dB may be identified in these situations so long as the exposure over the remaining 16 hours per day is low enough to result in negligible contribution to the 24-hour average, that is, no greater than an L_{eq} of 60 dB.

Note: Explanation of identified level for hearing loss: The exposure period that results in hearing loss at the identified level is a period of 40 years.

radial distance r from the sphere is inversely proportional to the square of distance, that is:[22]

$$I = \frac{W}{4\pi r^2} \tag{7-16}$$

where I = sound intensity, watts/m^2
W = sound power of source, watts

This is the *inverse square law.* It explains that portion of the reduction of sound intensity with distance that is due to wave divergence (Figure 7-30). If we measure sound power level (L_w, re: 10^{-12} W) rather than sound power (W), we can rewrite Equation 7-16 in terms of sound pressure level:[23]

$$L_P \cong L_w - 20 \log r - 11 \tag{7-17}$$

where L_P = sound pressure level, dB re: 20 μPa
L_w = sound power level, dB re: 10^{-12} W
r = distance between source and receiver, m
$20 \log r$ = decibel transform = $10 \log r^2$
11 = decibel transform $\cong [10 \log(4\pi) = 10.99]$

The tilde ($\sim$), indicating "approximately," results from the assumptions used above. L_w should be computed for all frequency bands of interest.

Radiation Fields of a Sound Source

The character of the wave radiation from a noise source will vary with distance from the source (Figure 7-31). At locations close to the source, the *near field,* the particle velocity is not in phase with the sound pressure. In this area, L_p fluctuates with distance and does not follow the inverse square law. When the particle velocity and sound pressure are in phase, the location of the sound measurement is said to be in the *far field.* If the sound source is in free space, that is, there are no reflecting surfaces, then measurements in the far field are also *free field measurements.* If the sound source is in a highly reflective space, for example, a room with steel walls, ceiling,

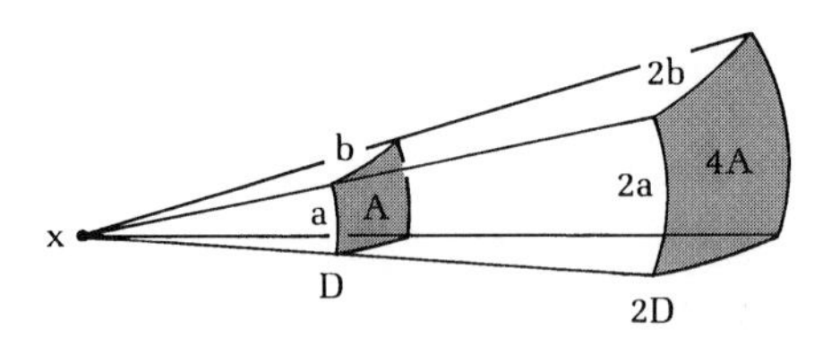

FIGURE 7-30
Illustration of inverse square law.

[22] $\kappa = 2\pi/\lambda$, where λ = wavelength, κ has units of reciprocal length, m^{-1}.

[23] This can be proven by using Equations 7-4, 7-5, 7-9, 7-10, and 7-11, and the assumption that ρc = 400 kg/m^2 · s.

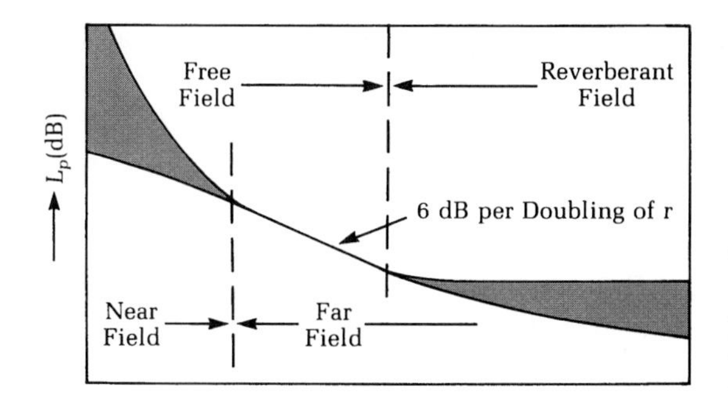

FIGURE 7-31
Variation of sound-pressure level in an enclosure along radius r from a noise source. (*Source:* L. L. Beranek, *Noise and Vibration Control,* New York: McGraw-Hill, 1971. Reprinted by permission.)

and floor, then measurements in the far field are also *reverberant field measurements.* The shaded area in the far field of Figure 7-31 shows that L_p does not follow the inverse square law in the reverberant field.

Directivity

Most real sources do not radiate sound uniformly in all directions. If you were to measure the sound pressure level in a given frequency band at a fixed distance from a real source, you would find different levels for different directions. If you plotted these data in polar coordinates, you would obtain the directivity pattern of the source.

The *directivity factor* is the numerical measure of the directivity of a sound source. In logarithmic form the directivity factor is called the *directivity index.* For a spherical source it is defined as follows:

$$DI_\theta = L_{p\theta} - L_{ps} \tag{7-18a}$$

where $L_{p\theta}$ = sound pressure level measured at distance r' and angle 0° from a directive source radiating power W into an echo-free (*anechoic*) space, dB

L_{ps} = sound pressure level measured at distance r' from a nondirective source radiating power W into anechoic space, dB[24]

For a source located on or near a hard, flat surface, the directivity index takes the following form:

$$DI_\theta = L_{p\theta} - L_{ps} + 3 \tag{7-18b}$$

The 3 dB addition is made because the measurement is made over a hemisphere instead of a sphere. That is, the intensity at a radius, r, is twice as large if a source radiates into a hemisphere rather than the ideal sphere we have used up to this point. Each directivity index is applicable only to the angle at which $L_{p\theta}$ was measured and only for the frequency at which it was measured.

---

[24]This is the same source as the directive source, but acting in the ideal fashion that we assumed in developing the inverse square law.

We assume that the directivity pattern does not change its shape regardless of the distance from the source. This allows us to apply the inverse square law to directive sources simply by adding the directivity index:

$$L_{p\theta} \cong L_w + DI_\theta - 20 \log r - 11 \tag{7-19}$$

You should note that it is not possible to reduce the equation by using the equality given in Equation 7-18*a*. The values of $L_{p\theta}$ are at a distance r, which is different than the r' in Equation 7-18*a*.

Airborne Transmission

Effects of atmospheric conditions. Sound energy is absorbed in quiet isotropic air by molecular excitation and relaxation of oxygen molecules and, at very low temperatures, by heat conduction and viscosity in the air. Molecular excitation is a complex function of the frequency of noise, humidity, and temperature. In general, we may say that as the humidity decreases, sound absorption increases. As the temperature increases to about 10 to 20°C (depending upon the noise frequency), absorption increases. Above 25°C, absorption decreases. Sound absorption is higher at higher frequencies.

The vertical temperature profile greatly alters the propagation paths of sound. If a superadiabatic lapse rate exists, sound rays bend upward and noise shadow zones are formed. If an inversion exists, sound rays are bent back toward the ground. This results in an increase in the sound level. These effects are negligible for short distances but may exceed 10 dB at distances over 800 m.

In a similar fashion, wind speed gradients alter the way noise propagates. Sound traveling with the wind is bent down, while sound traveling against the wind is bent upward. When sound waves are bent down, there is little or no increase in sound levels. But when sound waves are bent upward, there can be a noticeable reduction in sound levels.

Basic point source model. A point source is one for which $\kappa\delta \ll 1$ and for which Equation 7-16 holds. According to Magrab,

> In practice most noise sources cannot be classified as simple point sources. However, the sound field of a complicated sound source will look as if it were a point source if the following two conditions are met: (1) $r/\delta \gg 1$, that is, the distance from the source is large compared to its characteristic dimension, and (2) $\delta/\lambda \ll r/\delta$, that is, the ratio of the size of the source to the wavelength of sound in the medium is small compared to the ratio of the distance from the source to its characteristic dimension. Recall that $r/\delta \gg 1$ from the first condition. A value of $r/\delta > 3$ is a sufficient approximation; therefore, $\delta\lambda \ll 3$.[25]

[25] Edward B. Magrab, *Environmental Noise Control,* New York: Wiley, pp. 4 & 6, 1975.

A directional source having a sound power level L_w will produce a sound pressure level at a receiver, which can be estimated by the following equation:[26]

$$L_{p\theta} = L_w + DI_\theta - 20 \log r - 11 - A_e \tag{7-20}$$

This is the basic point source equation. With the exception of the last term (A_e), it is the modified inverse square law. The A_e term is the *excess attenuation* beyond wave divergence. It is caused by environmental conditions and has units of dB.

The A_e term may be further divided into six terms as follows:[27]

A_{e1} = effect of the difference in value of ρc from 400 mks rayls when the ambient temperature and barometric pressure differ appreciably from values that make $\rho c = 400$, for example, 38.9°C and 101.325 kPa. Units are dB. $\left(1 \text{ rayl} = 1 \frac{\text{N}\cdot\text{s}}{\text{m}^3}\right)$
A_{e2} = attenuation by absorption in the air, dB
A_{e3} = attenuation by rain, sleet, snow, or fog, dB
A_{e4} = attenuation by barriers, dB
A_{e5} = attenuation by grass, shrubbery, and trees, dB
A_{e6} = attenuation and fluctuation owing to wind and temperature gradients, to atmospheric turbulence, and to the characteristics of the ground, dB

The effect of the difference in ρc from 400 mks rayls can be calculated by first computing the change in density (ρ) due to temperature and pressure changes (see "Gas Laws" in Chapter 6). The effect of temperature changes on the speed of sound (c) can be calculated using Equation 7-6. The total attenuation, A_{e1}, is then computed as follows:

$$A_{e1} = 10 \log \frac{\rho c}{400} \tag{7-21}$$

The sign of A_{e1} is +. Thus, positive values of the right side of Equation 7-21 serve to reduce $L_{p\theta}$, while negative values increase it.

Results of laboratory tests of the effects of temperature and humidity on attenuation of sound, A_{e2}, in the frequency range of 125 to 12,500 Hz, for temperatures between −10 and 30°C, and for relative humidities between 10 and 90 percent have been published by the Society of Automotive Engineers.[28] For a temperature of 20°C, the following formula may be used:[29]

$$A_{e2} = 7.4 \times 10^{-8} \frac{f^2 r}{\phi} \tag{7-22}$$

---

[26] Even if the source is nondirective, $DI_\theta = 3$ dB for hemispherical radiation.

[27] L. L. Beranek, *Noise and Vibration Control,* New York: McGraw-Hill, p. 169, 1971.

[28] Society of Automotive Engineers, *Standard Values of Atmospheric Absorption as a Function of Temperature and Humidity for Use in Evaluating Aircraft Flyover Noise* (ARP886), New York: The Society of Automotive Engineers, 1964.

[29] L. Cremer, *Raumakustik,* Vol. II (Germany), S. Herzel Verlag, p. 27, 1961.

where f = geometric mean frequency of band, Hz
r = distance between source and receiver, m
ϕ = relative humidity, %

For other temperatures (20°C ± 10°), an approximate solution may be used:

$$A'_{e2} = \frac{A_{e2}}{1 + (\beta)(\Delta T)(f)} \tag{7-23}$$

where A_{e2} = attenuation at 20°C and ϕ = 50 percent from Equation 7-22, dB
$\beta = 4 \times 10^{-6}$ for T in °C
$\Delta T = T - 20°\text{C}$
T = temperature, °C

The excess attenuation due to rain, mist, fog, hail, sleet, and snow have not been studied extensively. A_{e3} is on the order of 0.5 dB/1,000 m in fog and generally is taken to be zero for conservative estimates.

The attenuation due to barriers (A_{e4}) is a complex function of the path length and the wavelength of the sound. This topic is largely beyond the scope of this text.

The absorption data for grass, shrubbery, and trees (A_{e5}) are not easy to generalize. Attenuations range from 0 to 30 dB/100 m. This analysis, too, is beyond the scope of this text.

The effects of wind and stability (A_{e6}) are treated separately for upwind and downwind receptors. For the downwind case, Figure 7-32 is used to calculate A_{e6}. For the upwind case, one nighttime and two daytime conditions are considered (Table 7-8). The quantity X_0 is the estimated distance from the noise source to the edge of the shadow zone (Figure 7-33). This is where the wind and temperature deflection of the sound waves begins to come into play. Once the value of X_0 is determined, Figure 7-34 can be used to select the attenuation value.

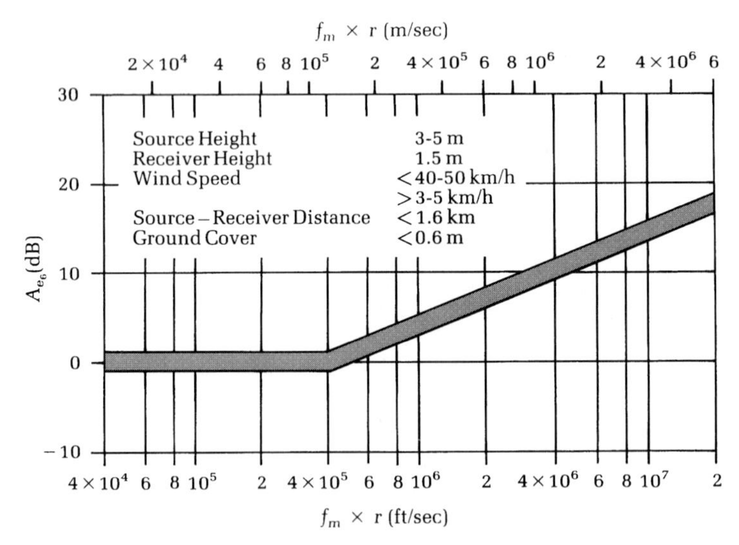

FIGURE 7-32
Downwind excess attenuation, A_{e6}. (*Source:* L. L. Beranek, *Noise and Vibration Control*. Reprinted by permission.)

TABLE 7-8
Estimates of X_0 upwind, 300 to 5,000 Hz, $\phi = 0$

Time		Sky		Temperature profile			Wind speed (m/s)	X_0 (m)
Day	Night	Clear	Overcast	Lapse	Neutral	Inversion		
x		x		x			4–8	75
x			x		x		4–7	120
	x	x				x	1–2	600

Source: F. M. Wiener, "Sound Propagation Outdoors," *Noise Control,* vol. 4, p. 224, 1958.

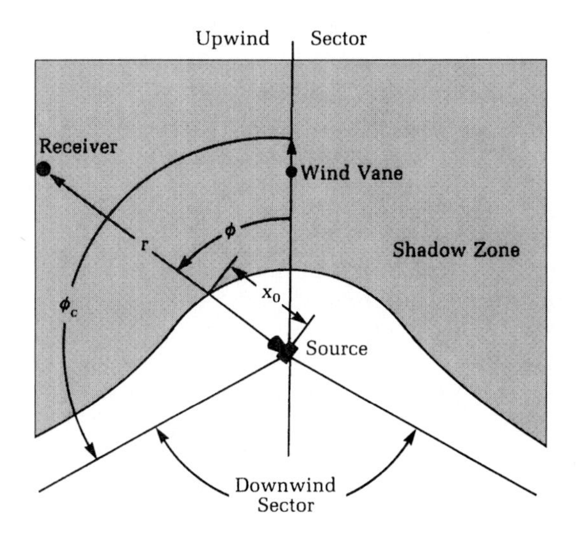

FIGURE 7-33
Definition of the "shadow zone" for upwind excess attenuation.
Code:

x_0 Distance from source to shadow zone
ϕ Angle between wind and sound
ϕ_c Critical angle

(*Source:* L. L. Beranek, *Noise and Vibration Control.* Reprinted by permission.)

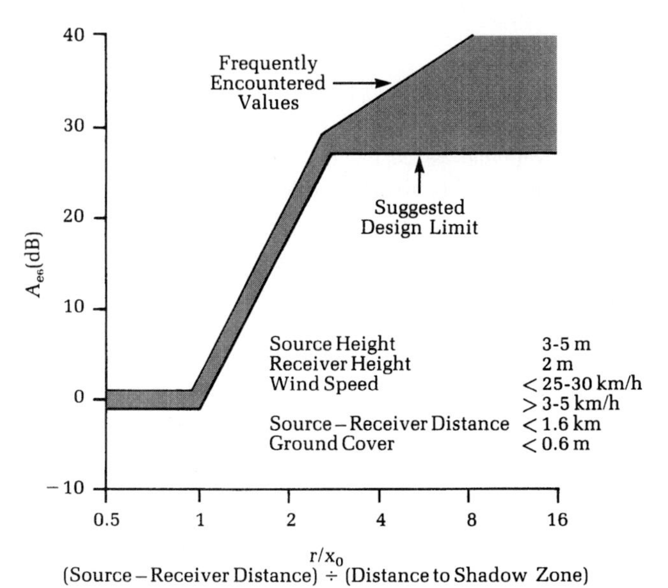

FIGURE 7-34
Upwind excess attenuation. (*Source:* L. L. Beranek, *Noise and Vibration Control.* Reprinted by permission.)

Sound power level unknown. Data on the sound power level (L_w) of many noise sources are not readily available. On the other hand, data on the sound pressure level ($L_{p\theta}$) at some given distance and angle frequently are available.

An alternative equation to Equation 7-20 makes use of existing measurements. It is given as follows:

$$L_{p2} = L_{p1} - 20 \log \frac{r_2}{r_1} - A_e \tag{7-24}$$

where L_{p1} = the measured SPL at angle θ and distance r_1 from source dB
L_{p2} = the desired SPL at angle θ and distance r_2 from source dB
r_1, r_2 = distance from source to measurement L_{p1} and L_{p2}, respectively
A_e = attenuation for the distance $r_2 - r_1$ dB

Example 7-7. The sound pressure level (SPL) measured 50 m from a compressor is 90 dB at 1,000 Hz. Determine the SPL 200 m upwind and downwind on a clear summer afternoon if the wind speed is 5 m/s, the temperature is 20°C, the relative humidity is 50 percent, and the barometric pressure is 100.325 kPa.

Solution. We begin by computing the A_e terms. We assume that the L_{p1} measurement was taken at the same temperature and pressure that the L_{p2} estimate is to be made. Thus, the value of A_{e1} at r_2 is the same as the value at r_1, and the attenuation for the distance $r_2 - r_1$ is zero, that is, $A_{e1} = 0.0$. The attenuation by air absorption is calculated directly from Equation 7-22 with the $r_2 - r_1$ distance being $200 - 50 =$ 150 m.

$$A_{e2} = 7.4 \times 10^{-8} \frac{(1{,}000)^2(150)}{50}$$

$$= 0.22 \text{ dB}$$

We assume that A_{e3}, A_{e4}, and A_{e5} all equal zero for the reasons noted in the text. The attenuation, A_{e6}, for downwind may be determined from Figure 7-32 after we have computed the abscissa quantity. Note that we assume that the 50-m measurement was taken under the same conditions as for the prediction at 200 m.

For r_1:

$$f_m \times r = (1{,}000 \text{ Hz})(50 \text{ m}) = 5.0 \times 10^4 \text{ m/s}$$

and from Figure 7-32 we find A_{e6} is 0 dB. For r_2:

$$f_m \times r = (1{,}000 \text{ Hz})(200 \text{ m}) = 2.00 \times 10^5 \text{ m/s}$$

and $A_{e6} = 2$ dB from Figure 7-32. The resultant value for $r_2 - r_1$ is then

$$A_{e6} = 2 - 0 = 2 \text{ dB}$$

For the upwind case we must start with Table 7-8. From the problem statement, we find it is a clear summer afternoon. From this we can assume a lapse temperature profile, and since 5 m/s is within the limits of the table, $X_0 = 75$ m. At lower wind speeds the shadow zone is removed to quite a large distance and probably can be ignored for all practical purposes. At higher wind speeds the howl of the wind quickly moves X_0 toward the source.

With $X_0 = 75$, we calculate r/X_0 for r_1 and r_2 separately, and take the difference in A_{e6} values.

For r_1:

$$\frac{r}{X_0} = \frac{50}{75} = 0.67$$

and from Figure 7-34, $A_{e6} = 0$. For r_2:

$$\frac{r}{X_0} = \frac{200}{75} = 2.67$$

and $A_{e6} = 25$ dB from Figure 7-34. The resultant value is

$$A_{e6} = 25 - 0 = 25 \text{ dB}$$

The total downwind attenuation is then

$$\sum = 0 + 0.22 + 0 + 0 + 0 + 2$$
$$= 2.22$$

The total upwind attenuation is then

$$\sum = 0 + 0.22 + 0 + 0 + 0 + 25$$
$$= 25.22 \text{ dB}$$

Using Equation 7-24, the downwind SPL is

$$L_{p2} = 90 - 20 \log \frac{200}{50} - 2.22$$
$$= 75.74 \text{ or } 76 \text{ dB at } 1{,}000 \text{ Hz}$$

The upwind SPL is

$$L_{p2} = 90 - 20 \log \frac{200}{50} - 25.22$$
$$= 52.74 \text{ or } 53 \text{ dB at } 1{,}000 \text{ Hz}$$

Obviously, it is much better to be upwind. Note that if you wished to estimate L_{eq}, L_{dn}, L_A, and so on, the value of L_{p2} would have to be calculated at each octave band geometric mean frequency and then summed by decibel addition.

7-6 TRAFFIC NOISE PREDICTION

National Cooperative Highway Research Program 174

The National Cooperative Highway Research Program has developed a series of documents (NCHRP 117, NCHRP 144, and NCHRP 174) that provide design guidance for the prediction and control of highway noise.[30] These documents have been used

[30] B. Andrew Kugler, Daniel E. Commins, and William J. Galloway, *Highway Noise: A Design Guide for Prediction and Control,* National Cooperative Highway Research Program Report 174, 1976.

widely because of their simplicity and relatively high success in making accurate noise predictions. The NCHRP 174 procedure is the last revision in the series. It contains a four-step procedure for the prediction and control of highway noise. We have limited ourselves to the first prediction step, that is, the "short method." The Federal Highway Administration (FHWA) has developed sophisticated computer models to replace the NCHRP 174 manual technique used in this text for illustration purposes. In 1997, the FHWA will release a new version (Traffic Noise Model - TNM, Version 1.0) that will replace a pair of computer programs, Stamina 2.0 and Optima, now in use.

The objective of the "short method" is to obtain a quick and gross (always overpredicting) prediction of the expected noise levels. This is necessary because the prediction of true highway noise levels is a rather complicated subject. In many instances it is desirable to first obtain a rough idea of the potential problem areas before full knowledge of the horizontal and vertical roadway design parameters has been gained. Such is the case, for example, of a location study where a number of alignments must be considered. Also, this first step helps to eliminate areas that do not represent a problem in terms of noise levels, thus simplifying further evaluation.

The "short method" prediction can be performed quickly through use of two *nomographs* and knowledge of a few traffic and roadway parameters.[31] By its design, the "short method" requires many assumptions and approximations and should not be used as a final tool.

The second step (the "complete method") utilizes a microcomputer program to refine the predictions made in the first step. The third step is the selection of a noise control design. The fourth step is to redo the second step and check the design solution. In the following paragraphs we have reproduced the short method as it appears in NCHRP 174, with the addition of clarifying comments and modification to SI units.

Methodology. The flow diagram that illustrates the methodology of the short method is shown in Figure 7-35. The method assumes that the roadway can be approximated by one infinite element with constant traffic parameters and roadway characteristics.

The initial step in using the short method consists of defining an infinite straight-line approximation to the real highway configuration. On-ramps, off-ramps, and interchange ramps are omitted from the short method analysis.

Once the approximate roadway has been chosen, the following parameters must be computed or estimated: (*a*) the traffic parameters, which include the speed and volume of each class of vehicles; (*b*) the propagation characteristics, which describe the location of the receiver relative to the roadway; and (*c*) the roadway-shielding parameters, which describe the shielding provided by the roadway, if any. [Only barriers located within the right-of-way (ROW) may be considered.]

---

[31] A nomograph is a graph that provides the solution to an equation or series of equations containing three or more variables (see Figure 7-39).

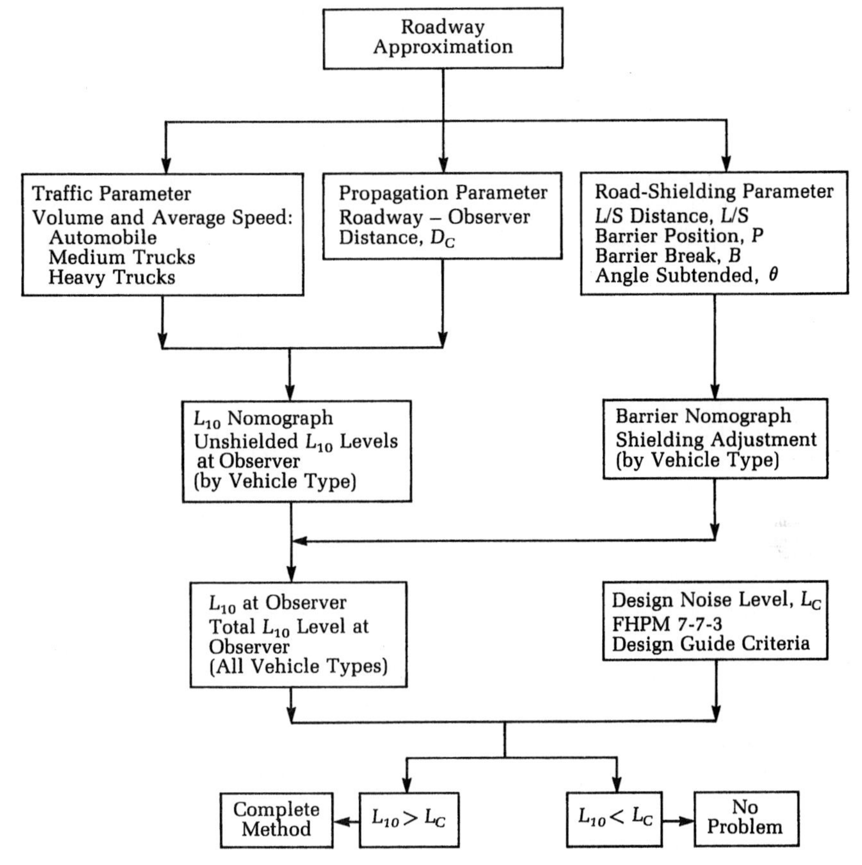

FIGURE 7-35
Flow diagram of methodology for applying NCHRP 174 method for estimating L_{10} from traffic. (*Source: NCHRP 174,* 1976.)

These parameters are used in two operations. First, the traffic and propagation parameters are combined in the L_{10} nomograph to determine, for each type of source, the unshielded L_{10} level at the observer.

The final result is then compared to the criteria level, L_c, at the observer (see, for example, Table 7-3) to define a "no problem" or "potential problem" condition. If a potential problem is identified, the observer location in question should be evaluated using the complete method.

Procedure. The step-by-step procedures necessary to calculate noise levels by the short method are presented in the following numbered paragraphs. In addition to the nomographs, the method uses a noise prediction worksheet to aid the user in the sequential steps. A blank worksheet and larger scale nomographs are included in Appendix B.

1. Observer identification: On a route map of convenient scale, identify all observer locations at which analysis is desired.
2. Roadway approximation: Approximate the roadway alignment by a straight, infinite line. The procedure is as follows: Determine and measure the nearest perpendicular distance, D_c, between the roadway centerline and observer, as shown in Figure 7-36*a*. Enter on line 4 on the noise prediction worksheet (Figure 7-37). Note that the infinite roadway approximation automatically assumes a line perpendicular to the centerline distance, D_c. There is no need to draw this line on the route map for computation reasons. An illustration of this assumption is shown in Figure 7-36*b*. Note that for each observer location, a different roadway approximation might result. Note also that the noise prediction worksheet (NPWS) allows for computations for six different observer locations by entering an observer location identification at the head of each column. Similarly, the NPWS can be used to calculate six different traffic conditions for the same observer location.

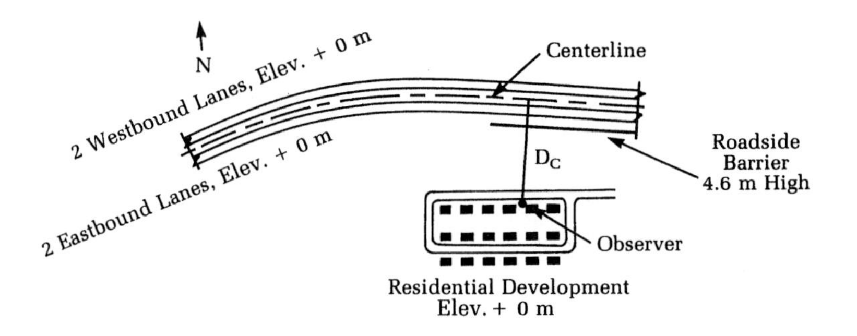

(a) Route Map Showing Observer Location and Observer – Roadway Centerline distance, D_C

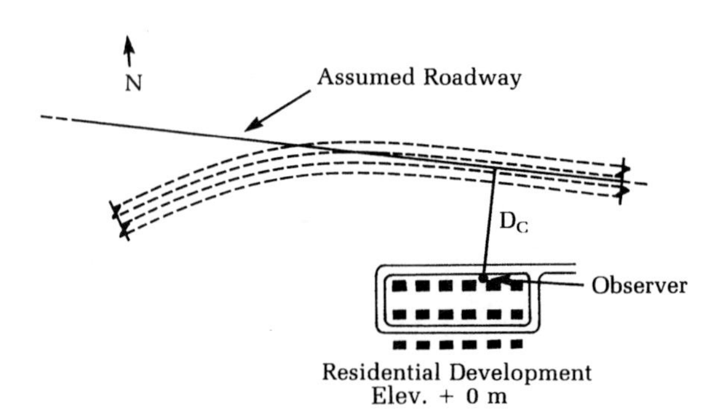

(b) Model of Assumed Roadway Alignment

FIGURE 7-36
(*a*) Roadway and (*b*) roadway approximation. (*Source: NCHRP 174,* 1976.)

Project BRISTOL HWY. Date 13 AUG. 1980 Engineer I. THOMPSON

Step			Ex. 9-10a			Ex. 9-10b														
			A	T_M	T_H	A	T_M	T_H	A	T_M	T_H	A	T_M	T_H	A	T_M	T_H	A	T_M	T_H
1	Traffic	Vehicle Volume, V(Vph)	2000	100	100	2000	100	100												
2		Vehicle Av. Speed, S(km/h)	80.5	80.5	80.5	80.5	80.5	80.5												
3		Combined Veh. Vol.*, V_C(Vph)	3000			3000														
4	Prop.	Observer-Roadway Dist., D_C(m)	60			60														
5	Shielding	Line-of-Sight Dist., L/S(m)	—			60														
6		Barrier Position Dist., P(m)	—			15.2														
7		Break in Barrier., B(m)	—			4.6		2.7												
8		Angle Subtended, θ (deg)				170														
9	Prediction**	Unshield L_{10} Level (dBA)	66	-	68	66	-	68												
10		Shielding Adjust. (dBA)	0		0	13		10												
11		L_{10} at Observer (By Veh. Class)	66	-	68	53	-	58												
12		L_{10} at Observer – Total	70 dBA			59.25 or 59 dBA														

Code:

A = Automobiles, T_M = Medium Trucks, T_H = Heavy Trucks

* Applies only when automobile and medium truck average speeds are equal. $V_C = V_A + (10)V_{T_M}$

** If automobile-medium truck volume V_C is combined, use L_{10} Nomograph prediction only once for these two vehicle classes

FIGURE 7-37
Noise prediction worksheet. (*Source: NCHRP 174*, 1976.)

3. Traffic parameters: Determine the vehicle operating conditions by using the traffic parameters at the roadway point nearest the observer (if these parameters vary along the roadway). The procedure is as follows:

a. Determine the automobile volume (vph) and average speed (km/h) and enter them on lines 1 and 2 under automobiles (A).

b. Determine the medium truck volume (vph) and average speed (km/h) and enter them on lines 1 and 2 under medium trucks (T_M).

c. Determine the heavy truck volume (vph) and average speed (km/h) and enter them on lines 1 and 2 under heavy trucks (T_H).

d. If the automobile and medium truck speeds are the same, multiply the medium truck volume by ten and add to the automobile volume.[32] Enter combined volume, V_c, on line 3 of the NPWS. If the automobile and medium truck volumes are combined, in subsequent operations consider the two vehicle classes as one source. If the speed of the medium trucks differs from that of the automobiles, the volumes are *not* combined *but* the medium truck volume is still multiplied by ten for the determination of the L_{10} value for the reason noted in footnote 32.

4. Roadway-shielding parameters: If the roadway cross section at the nearest point is not at grade (either elevated or depressed), or if a roadside barrier (on the roadway right-of-way) is present, determine the roadway-shielding parameters. If the elevation, depression, or roadside barrier is less than 1.5 m high (compared to the surrounding terrain), disregard it. The procedure is as follows: Determine the barrier parameters and enter on lines 5 through 8 of the NPWS. Use Figure 7-38 for definitions of parameters. The parameters that must be measured are: (*a*) line-of-sight distance, L/S, (*b*) break in line of sight, (*c*) barrier position distance, and (*d*) angle subtended, θ (degree). The angle subtended is measured from the ends of the barrier with respect to the position of the observer. For an observer placed equidistant from the ends of the barrier, the angle may be determined from simple trigonometric principles. Graphical methods may be more appropriate for other configurations. Note that as the barrier length increases, the angle approaches 180 degrees.

5. Unshielded L_{10} level at observer location: Determine the unshielded L_{10} level at the observer location for all three traffic sources (automobiles, medium trucks, and heavy trucks) using the L_{10} nomograph (Figure 7-39). Note that if the automobile and medium truck speeds are equal, these two sources may be evaluated together using the combined volume, V_c, and average speed, S_A or S_M on lines 3 and 2 of NPWS. The procedure is as follows:

a. Automobiles (and medium trucks): Using the vehicle volume, V_A (this corresponds to V_c, the combined auto and medium truck volumes, when the speeds of these two populations are equal), and the average speed, S_A (or S_M), enter the L_{10} nomograph and determine the unshielded L_{10} noise level at the observer. Enter on line 9 of the NPWS.

---

[32]The medium truck volume is multiplied by ten because this traffic noise source behaves similarly to automobile noise, but the overall level is 10 dBA higher than for automobiles.

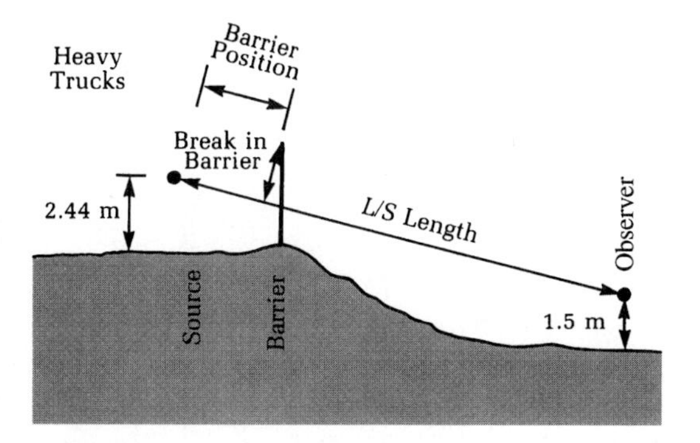

(a) Barrier Parameters for Simple Barrier, Section View

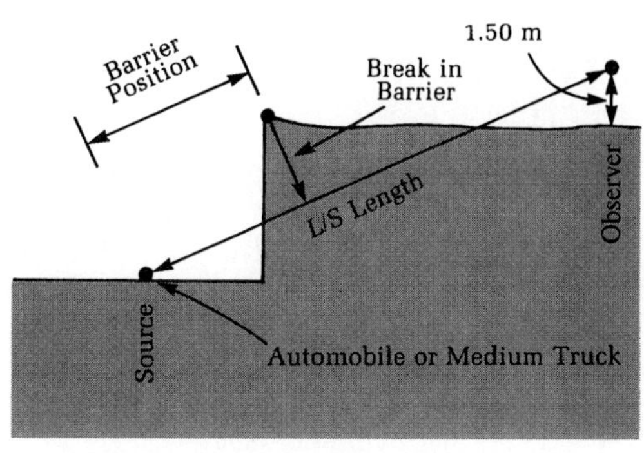

(b) Barrier Parameters for Depressed Roadway, Section View

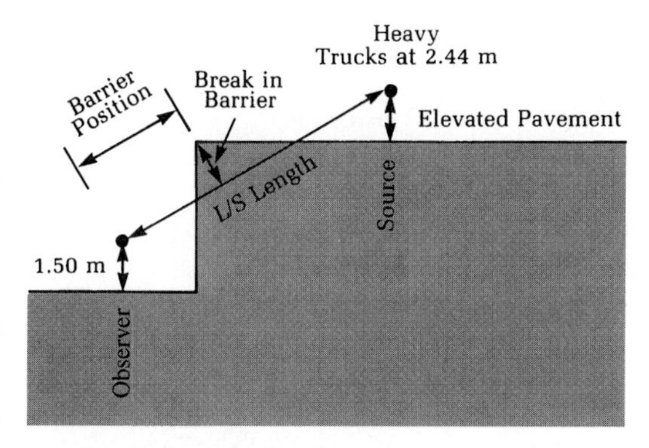

(c) Barrier Parameters for Elevated Roadway, Section View

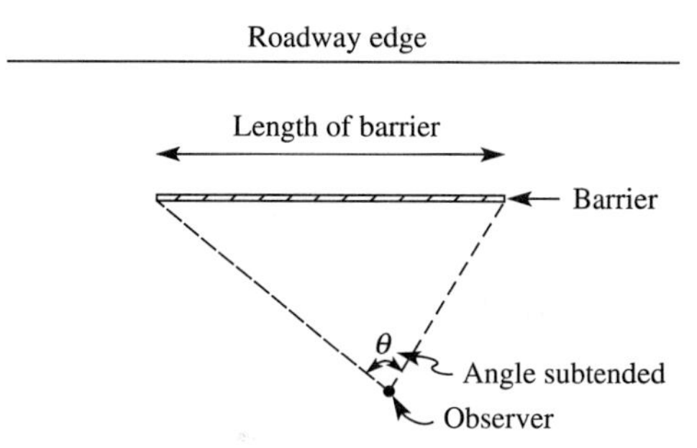

(d) Barrier Parameters, Plan View

FIGURE 7-38
Definitions of barrier parameters. (*Source: NCHRP 174,* 1976.)

b. Medium trucks: Using the vehicle volume, V_M, *multiplied by ten,* and the average speed, S_M, enter the L_{10} nomograph and determine the unshielded L_{10} noise level at the observer. Enter on line 9 of the NPWS. If automobiles and medium trucks were combined in step a, this step should be omitted.

c. Heavy trucks: Using the vehicle volume, V_T, and the average speed, S_T, enter the L_{10} nomograph and determine the unshielded L_{10} noise level at the observer. Enter on line 9 of the NPWS.

6. Shielding adjustment: Determine the noise reduction afforded by the roadway geometry using the barrier nomograph (Figure 7-40) and the roadway parameters listed in the NPWS. This procedure must be performed twice: once for the 0 m source elevation (automobiles and medium trucks) and once for the 2.44 m elevation (heavy trucks).

a. Low sources (0 m): Using the line-of-sight distance, L/S, barrier position distance, P, break in L/S distance, B (for sources at ground level), and the angle

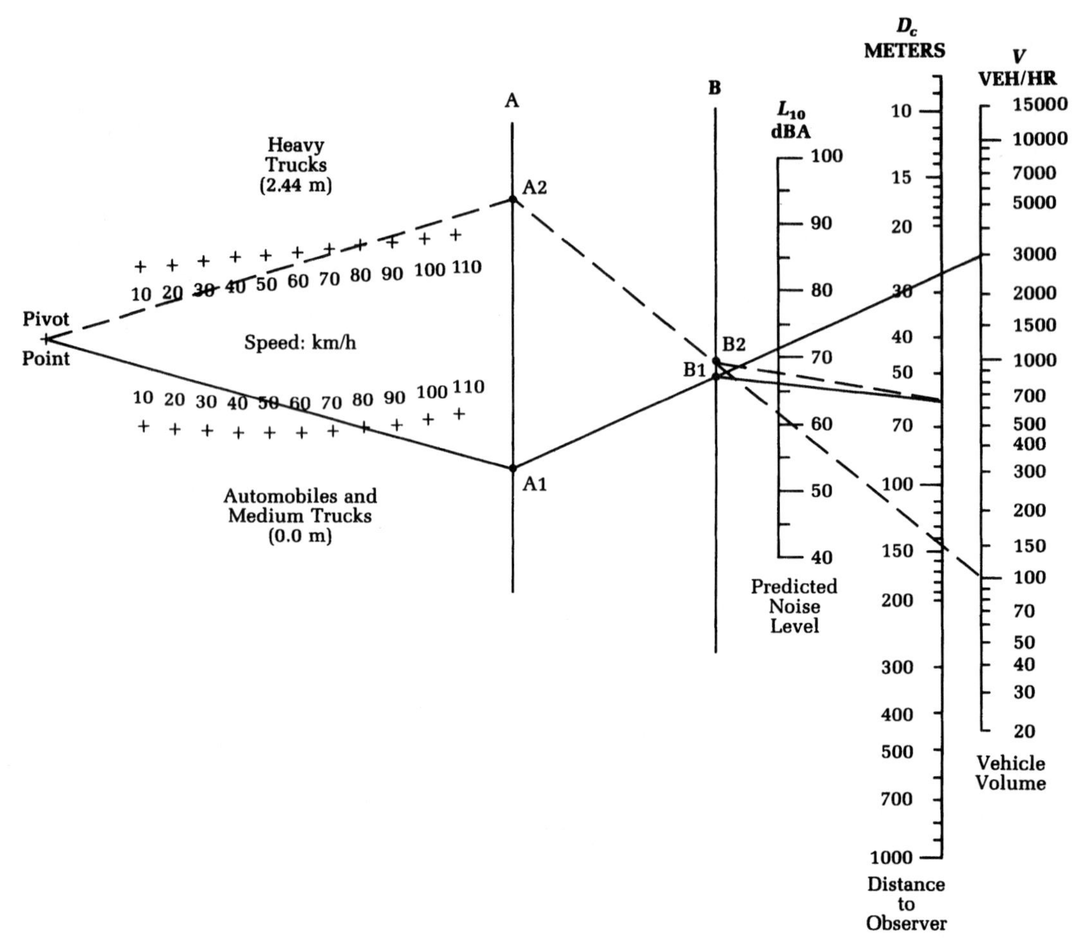

FIGURE 7-39
L_{10} nomograph.

subtended, θ, enter the barrier nomograph and calculate the shielding adjustment. Enter on line 10 of the NPWS under automobile and medium trucks.

b. High sources (2.44 m): Using the line-of-sight distance, L/S, barrier position distance, P, break in L/S distance, B (for 2.44 m sources), and the angle subtended, θ, enter the barrier nomograph and calculate the shielding adjustment. Enter on line 10 of the NPWS under heavy trucks.

7. L_{10} at observer, by vehicle type: Calculate the L_{10} noise level at the observer for each individual source by subtracting the shielding adjustment (line 10) from the unshielded L_{10} level at the observer (line 9), and enter the result in line 11. Note that the shielding adjustment is always negative and can be subtracted algebraically from line 9.

8. Total L_{10} level at observer: Determine the total L_{10} noise level at the observer and enter on line 12. This is done by logarithmically adding (decibel addition) the contributions from automobiles, medium trucks, and heavy trucks computed

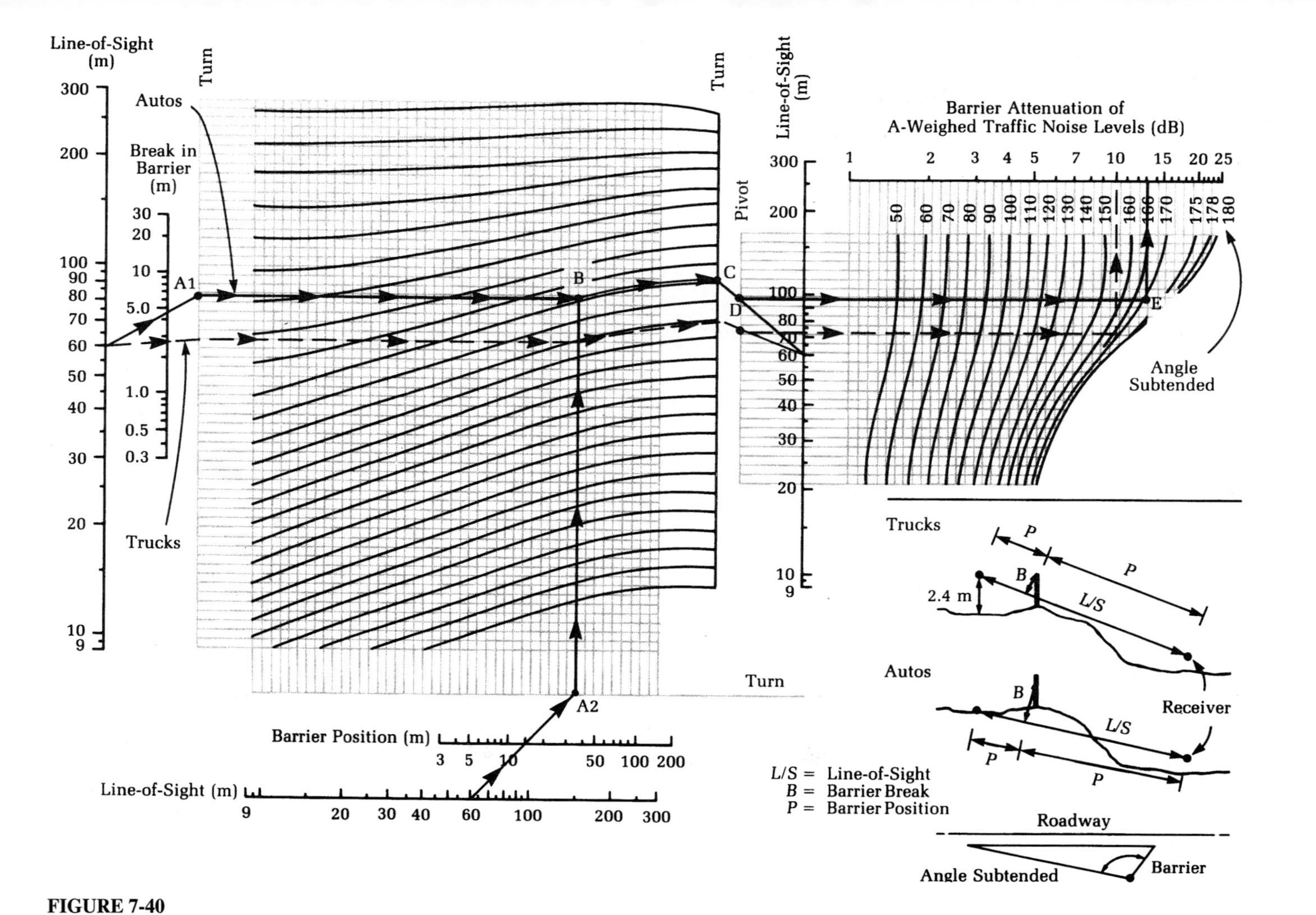

FIGURE 7-40
Barrier nomograph. (*Source: NCHRP 174,* 1976.)

in line 11. Taking two L_{10} levels at a time, find the difference between them and enter the addition scale provided in Figure 7-4 to find the "adjustment." The "adjustment" should be added to the higher of the two L_{10} levels. The operation is then repeated, two levels at a time, until only one level remains. The lowest levels should be added first for maximum accuracy.

Example 7-8. The county road commissioner has requested a noise evaluation of a proposed highway near Bristol. The proposed highway is to be routed such that the centerline of the roadway will be 60 m from a school. Using the following data, determine whether or not the FHA criterion will be met:

Average vehicle speed = 80.5 km/h for all vehicles
Automobiles = 2,000/h
Medium trucks = 100/h
Heavy trucks = 100/h
The terrain is level and no shielding is present.

Solution. According to step 3d, the medium trucks and the cars can be combined by multiplying the medium truck volume by ten, and then combining it with the car volume. The combined volume, V_c, for cars and medium trucks is 2,000 + 10 (100) = 3,000 vph. Using the L_{10} nomograph shown in Figure 7-39, proceed as follows:

1. Draw a straight line from the left pivot point through the 80.5 km/h point on the automobile speed scale. Extend the straight line to turn line A. The intersection is marked A1.
2. Draw a second straight line from the intersection point A1 to the 3,000 vph point on the volume scale on the far right of the figure. The intersection of this straight line with turn line B is marked B1.
3. Draw a third straight line from point B1 to the point on the D_c scale (60 m). The intersection of this third line with the L_{10} scale gives the predicted A-weighted L_{10} level at the observer. For this example, the predicted L_{10} level is 66 dBA. This value is entered on line 9 of the NPWS (Figure 7-37).

Now repeat the procedure for the heavy trucks. It is shown by the dashed line in Figure 7-39. The predicted L_{10} for heavy trucks is 68 dBA.

The combined level of the automobiles and trucks is found by "decibel addition" (Figure 7-4) to be 70 dBA. This just meets the FHA design level for land use category B (Table 7-3). Let us see what effect a barrier will have. We have arbitrarily selected the following characteristics for the barrier:

Height = 5.0 m, which yields a "Break in Barrier" of 4.6 m for automobiles and 2.6 m for heavy trucks
Position = 15.2 m from centerline of roadway
Subtended angle = 170°

Note: The "Break in Barrier" is determined by constructing a scale drawing of the roadway and receiver as shown in Figure 7-38. Because the automobile noise is assumed to originate at an elevation of 0.0 m, the "Break in Barrier" is greater than for the heavy trucks, where the noise is asssumed to originate at an elevation of 2.44 m.

Using the barrier nomograph shown in Figure 7-40, proceed as follows:

1. Starting from the vertical L/S scale on the left: From the 60-m point on the L/S scale, draw a straight line going through the 4.6-m point on the "Break in Barrier" to the turn line. Note that the scale is logarithmic. From the intersection, called A_1, draw a straight horizontal line.
2. Starting from the horizontal L/S scale on the bottom: Draw a straight line through the 60-m point on the L/S scale, and the 15.2-m point on the "Barrier Position" scale to the turn line. From the intersection, called A_2, draw a straight vertical line until it intersects (at B) with the horizontal drawn from A_1.
3. From B, move to the right upward following the nearest curve to turn line C. (If B is on one of the curves, simply follow it; if B is between curves, follow parallel to the nearest curve upward to the right until it intersects with the turn line.)
4. From C, draw a straight line to the line-of-sight (L/S) distance (60 m) point on the vertical L/S scale. It will intersect with the pivot line at D.
5. From D, draw a horizontal line to the right until it intersects the curve corresponding to the subtended angle (170°) at E.
6. Finally, draw a vertical line from E upward until it intersects the barrier attenuation scale. The attenuation can now be read, that is, 13 dBA for automobiles and medium trucks.

The same procedure is followed for heavy trucks using 2.6 m for the "Break in Barrier" because of the higher noise emission elevation from the heavy trucks. Note from Figure 7-38 that this distance is perpendicular to L/S and not perpendicular to the horizontal. The attenuation is about 10 dBA.

The revised overall combined L_{10} would then be 60 dBA. This is very acceptable for Class B land use category. All of the tabulations are summarized in Figure 7-37.

L_{eq} Prediction

At about the same time that the NCHRP 174 report was being finalized, the Ontario Ministry of Transportation and Communications completed development of a predictive equation based on the L_{eq} concept.[33] The empirical equation they developed is as follows:

$$L_{eq} = 42.3 + 10.2 \log(V_c + 6V_t) - 13.9 \log D + 0.13S \qquad (7\text{-}25)$$

where L_{eq} = energy equivalent sound level during one hour, dBA
V_c = volume of automobiles (four tires only), veh/h
V_t = volume of trucks (six or more tires), veh/h
D = distance from edge of pavement to receiver, m
S = average speed of traffic flow during one hour, km/h

[33] J. J. Hajek, "L_{eq} Traffic Noise Prediction Method," *Environmental and Conservation Concerns in Transportation: Energy, Noise and Air Quality* (Transportation Research Record No. 648), Transportation Research Board, National Academy of Sciences, pp. 48–53, 1977.

The simplicity of the equation is an obvious advantage over the NCHRP method. It does have the restriction that it does not account for barriers. A nomograph technique similar to the NCHRP method is available to take barriers into account.

L_{dn} Prediction

As a direct extension of the L_{eq} methodology, the Ontario method was extended to enable the calculation of L_{dn}. The modified model has the following form:

$$L_{dn} = 31.0 + 10.2 \log [\text{AADT} + (\text{T\% AADT/2O})] - 13.9 \log D + 0.13 S \tag{7-26}$$

where L_{dn} = equivalent A-weighted sound level during 24-hour time period with 10 dBA weighting applied to 2200–0700 h, dBA
AADT = annual average daily traffic, veh/d
$\%T$ = average percentage of trucks during a typical day, %

Equation 7-26 has the same advantages and disadvantages as Equation 7-25.

7-7 NOISE CONTROL

Source-Path-Receiver Concept

If you have a noise problem and want to solve it, you have to find out something about what the noise is doing, where it comes from, how it travels, and what can be done about it. A straightforward approach is to examine the problem in terms of its three basic elements: that is, sound arises from a source, travels over a path, and affects a receiver or listener.[34]

The source may be one or any number of mechanical devices that radiate noise or vibratory energy. Such a situation occurs when several appliances or machines are in operation at a given time in a home or office.

The most obvious transmission path by which noise travels is simply a direct line-of-sight air path between the source and the listener. For example, aircraft fly-over noise reaches an observer on the ground by the direct line-of-sight air path. Noise also travels along structural paths. Noise can travel from one point to another via any one path or a combination of several paths. Noise from a washing machine operating in one apartment may be transmitted to another apartment along air passages such as open windows, doorways, corridors, or duct work. Direct physical contact of the washing machine with the floor or walls sets these building components into vibration. This vibration is transmitted structurally throughout the building, causing walls in other areas to vibrate and to radiate noise.

The receiver may be, for example, a single person, a classroom of students, or a suburban community.

---

[34]This discussion in large part was taken from Raymond D. Berendt, Edith L. R. Corliss, and Morris S. Ojalvo, *Quieting: A Practical Guide to Noise Control* (National Bureau of Standards Handbook 119), Washington, DC: U.S. Department of Commerce, pp. 16–41, 1976.

Solution of a given noise problem might require alteration or modification of any or all of these three basic elements:

1. Modifying the source to reduce its noise output
2. Altering or controlling the transmission path and the environment to reduce the noise level reaching the listener
3. Providing the receiver with personal protective equipment

Control of Noise Source by Design

Reduce impact forces. Many machines and items of equipment are designed with parts that strike forcefully against other parts, producing noise. Often, this striking action or impact is essential to the machine's function. A familiar example is the typewriter—its keys must strike the ribbon and paper in order to leave an inked impression. But the force of the key also produces noise as the impact falls on the ribbon, paper, and platen.

Several steps can be taken to reduce noise from impact forces. The particular remedy to be applied will be determined by the nature of the machine in question. Not all of the steps listed below are practical for every machine and for every impact-produced noise. But application of even one suggested measure can often reduce the noise appreciably.

Some of the more obvious design modifications are as follows:

1. Reduce the weight, size, or height of fall of the impacting mass.
2. Cushion the impact by inserting a layer of shock-absorbing material between the impacting surfaces. (For example, insert several sheets of paper in the typewriter behind the top sheet to absorb some of the noise-producing impact of the keys.) In some situations, you could insert a layer of shock-absorbing material behind each of the impacting heads or objects to reduce the transmission of impact energy to other parts of the machine.
3. Whenever practical, one of the impact heads or surfaces should be made of non-metallic material to reduce resonance (ringing) of the heads.
4. Substitute the application of a small impact force over a long time period for a large force over a short period to achieve the same result.
5. Smooth out acceleration of moving parts by applying accelerating forces gradually. Avoid high, jerky acceleration or jerky motion.
6. Minimize overshoot, backlash, and loose play in cams, followers, gears, linkages, and other parts. This can be achieved by reducing the operational speed of the machine, better adjustment, or by using spring-loaded restraints or guides. Machines that are well made, with parts machined to close tolerances, generally produce a minimum of such impact noise.

Reduce speeds and pressures. Reducing the speed of rotating and moving parts in machines and mechanical systems results in smoother operation and lower noise output. Likewise, reducing pressure and flow velocities in air, gas, and liquid circulation systems lessens turbulence, resulting in decreased noise radiation. Some specific suggestions that may be incorporated in design are the following:

1. Fans, impellers, rotors, turbines, and blowers should be operated at the lowest bladetip speeds that will still meet job needs. Use large-diameter, low-speed fans rather than small-diameter, high-speed units for quiet operation. In short, maximize diameter and minimize tip speed.
2. All other factors being equal, centrifugal squirrel-cage type fans are less noisy than vane axial or propeller type fans.
3. In air ventilation systems, a 50 percent reduction in the speed of the air flow may lower the noise output by 10 to 20 dB, or roughly one-quarter to one-half of the original loudness. Air speeds less than 3 m/s measured at a supply or return grille produce a level of noise that usually is unnoticeable in residential or office areas. In a given system, reduction of air speed can be achieved by operating at lower motor or blower speeds, installing a greater number of ventilating grilles, or increasing the cross-sectional area of the existing grilles.

Reduce frictional resistance. Reducing friction between rotating, sliding, or moving parts in mechanical systems frequently results in smoother operation and lower noise output. Similarly, reducing flow resistance in fluid distribution systems results in less noise radiation.

Four of the more important factors that should be checked to reduce frictional resistance in moving parts are the following:

1. Alignment: Proper alignment of all rotating, moving, or contacting parts results in less noise output. Good axial and directional alignment in pulley systems, gear trains, shaft couplings, power transmission systems, and bearing and axle alignment are fundamental requirements for low noise output.
2. Polish: Highly polished and smooth surfaces between sliding, meshing, or contacting parts are required for quiet operation, particularly where bearings, gears, cams, rails, and guides are concerned.
3. Balance: Static and dynamic balancing of rotating parts reduces frictional resistance and vibration, resulting in lower noise output.
4. Eccentricity (out-of-roundness): Off-centering of rotating parts such as pulleys, gears, rotors, and shaft/bearing alignment causes vibration and noise. Likewise, out-of-roundness of wheels, rollers, and gears causes uneven wear, resulting in flat spots that generate vibration and noise.

The key to effective noise control in fluid systems is *streamline flow.* This holds true regardless of whether one is concerned with air flow in ducts or vacuum cleaners, or with water flow in plumbing systems. Streamline flow is simply smooth, nonturbulent, low-friction flow.

The two most important factors that determine whether flow will be streamline or turbulent are the speed of the fluid and the cross-sectional area of the flow path, that is, the pipe or duct diameter. The rule of thumb for quiet operation is to use a low-speed, large-diameter system to meet a specified flow capacity requirement.

However, even such a system can inadvertently generate noise if certain aerodynamic design features are overlooked or ignored. A system designed for quiet operation will employ the following features:

1. Low fluid speed: Low fluid speeds avoid turbulence, which is one of the main causes of noise.
2. Smooth boundary surfaces: Duct or pipe systems with smooth interior walls, edges, and joints generate less turbulence and noise than systems with rough or jagged walls or joints.
3. Simple layout: A well-designed duct or pipe system with a minimum of branches, turns, fittings, and connectors is substantially less noisy than a complicated layout.
4. Long-radius turns: Changes in flow direction should be made gradually and smoothly. It has been suggested that turns should be made with a curve radius equal to about five times the pipe diameter or major cross-sectional dimension of the duct.
5. Flared sections: Flaring of intake and exhaust openings, particularly in a duct system, tends to reduce flow speeds at these locations, often with substantial reductions in noise output.
6. Streamline transition in flow path: Changes in flow path dimensions or cross-sectional areas should be made gradually and smoothly with tapered or flared transition sections to avoid turbulence. A good rule of thumb is to keep the cross-sectional area of the flow path as large and as uniform as possible throughout the system.
7. Remove unnecessary obstacles: The greater the number of obstacles in the flow path, the more tortuous, turbulent, and hence noisier, the flow. All other required and functional devices in the path, such as structural supports, deflectors, and control dampers, should be made as small and as streamlined as possible to smooth out the flow patterns.

Reduce radiating area. Generally speaking, the larger the vibrating part or surface, the greater the noise output. The rule of thumb for quiet machine design is to minimize the effective radiating surface areas of the parts without impairing their operation or structural strength. This can be done by making parts smaller, removing excess material, or by cutting openings, slots, or perforations in the parts. For example, replacing a large, vibrating sheet-metal safety guard on a machine with a guard made of wire mesh or metal webbing might result in a substantial reduction in noise because of the drastic reduction in surface area of the part.

Reduce noise leakage. In many cases, machine cabinets can be made into rather effective soundproof enclosures through simple design changes and the application of some sound-absorbing treatment. Substantial reductions in noise output may be achieved by adopting some of the following recommendations:

1. All unnecessary holes or cracks, particularly at joints, should be caulked.

2. All electrical or plumbing penetrations of the housing or cabinet should be sealed with rubber gaskets or a suitable nonsetting caulk.
3. If practical, all other functional or required openings or ports that radiate noise should be covered with lids or shields edged with soft rubber gaskets to effect an airtight seal.
4. Other openings required for exhaust, cooling, or ventilation purposes should be equipped with mufflers or acoustically lined ducts.
5. Openings should be directed away from the operator and other people.

Isolate and dampen vibrating elements. In all but the simplest machines, the vibrational energy from a specific moving part is transmitted through the machine structure, forcing other component parts and surfaces to vibrate and radiate sound—often with greater intensity than that generated by the originating source itself.

Generally, vibration problems can be considered in two parts. First, we must prevent energy transmission between the source and surfaces that radiate the energy. Second, we must dissipate or attenuate the energy somewhere in the structure. The first part of the problem is solved by *isolation*. The second part is solved by *damping*.

The most effective method of vibration isolation involves the resilient mounting of the vibrating component on the most massive and structurally rigid part of the machine. All attachments or connections to the vibrating part, in the form of pipes, conduits, and shaft couplers, must be made with flexible or resilient connectors or couplers. For example, pipe connections to a pump that is resiliently mounted on the structural frame of a machine should be made of resilient tubing and be mounted as close to the pump as possible. Resilient pipe supports or hangers may also be required to avoid bypassing the isolated system (Figure 7-41).

Damping material or structures are those that have some viscous properties. They tend to bend or distort slightly, thus consuming part of the noise energy in molecular motion. The use of spring mounts on motors and laminated galvanized steel and plastic in air-conditioning ducts are two examples.

When the vibrating noise source is not amenable to isolation, as, for example, in ventilation ducts, cabinet panels, and covers, then damping materials can be used to reduce the noise.

The type of material best suited for a particular vibration problem depends on factors such as size, mass, vibrational frequency, and operational function of the vibrating structure. Generally speaking, the following guidelines should be observed in the selection and use of such materials to maximize vibration damping efficiency:

1. Damping materials should be applied to those sections of a vibrating surface where the most flexing, bending, or motion occurs. These usually are the thinnest sections.
2. For a single layer of damping material, the stiffness and mass of the material should be comparable to that of the vibrating surface to which it is applied. This means that single-layer damping materials should be about two or three times as thick as the vibrating surface to which they are applied.
3. Sandwich materials (*laminates*) made up of metal sheets bonded to mastic (sheet metal viscoelastic composites) are much more effective vibration dampers than

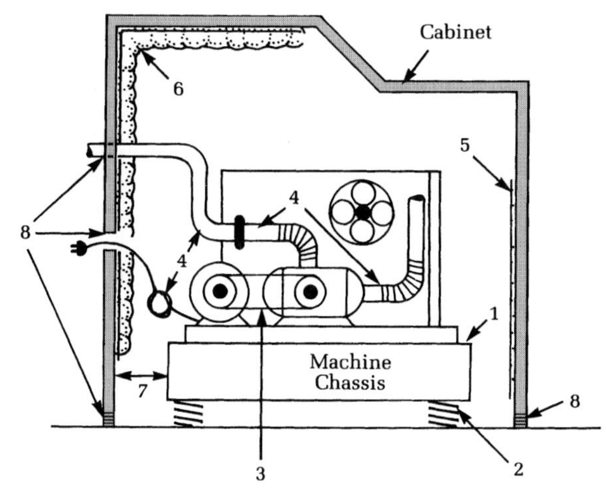

FIGURE 7-41
Examples of vibration isolation.

Code:

1. Motors, pumps, and fans installed on most massive part of the machine
2. Resilient mounts or vibration isolators used for the installation
3. Belt-drive or roller-drive systems used in place of gear trains
4. Flexicle hoses and wiring used instead of rigid piping and stiff wiring
5. Vibration-damping materials applied to surfaces undergoing most vibration
6. Acoustical lining installed to reduce noise buildup inside machine
7. Mechanical contact minimized between the cabinet and the machine chassis
8. Openings at the base and other parts of the cabinet scaled to prevent noise leakage

(*Source: National Bureau of Standards Handbook 119,* 1976.)

single-layer materials; the thickness of the sheet-metal constraining layer and the viscoelastic layer should each be about one-third the thickness of the vibrating surface to which they are applied. Ducts and panels can be purchased already fabricated as laminates.

Provide mufflers/silencers. There is no real distinction between mufflers and silencers. They are often used interchangeably. They are, in effect, acoustical filters and are used when fluid flow noise is to be reduced. The devices can be classified into two fundamental groups: *absorptive mufflers* and *reactive mufflers.* An absorptive muffler is one whose noise reduction is determined mainly by the presence of fibrous or porous materials, which absorb the sound. A reactive muffler is one whose noise reduction is determined mainly by geometry. It is shaped to reflect or expand the sound waves with resultant self-destruction.

Although there are several terms used to describe the performance of mufflers, the most frequently used appears to be *insertion loss* (IL). Insertion loss is the difference between two sound pressure levels that are measured at the same point in space before and after a muffler has been inserted. Since each muffler IL is highly dependent on the manufacturer's selection of materials and configuration, we will not present general IL prediction equations.

Noise Control in the Transmission Path

After you have tried all possible ways of controlling the noise at the source, your next line of defense is to set up devices in the transmission path to block or reduce the flow of sound energy before it reaches your ears. This can be done in several ways: (*a*) absorb the sound along the path, (*b*) deflect the sound in some other direction by placing a reflecting barrier in its path, or (*c*) contain the sound by placing the source inside a sound-insulating box or enclosure.

Selection of the most effective technique will depend upon various factors, such as the size and type of source, intensity and frequency range of the noise, and the nature and type of environment.

Separation. We can make use of the absorptive capacity of the atmosphere, as well as divergence, as a simple, economical method of reducing the noise level. Air absorbs high-frequency sounds more effectively than it absorbs low-frequency sounds. However, if enough distance is available, even low-frequency sounds will be absorbed appreciably.

If you can double your distance from a point source, you will have succeeded in lowering the sound pressure level by 6 dB. It takes about a 10 dB drop to halve the loudness. If you have to contend with a line source such as a railroad train, the noise level drops by only 3 dB for each doubling of distance from the source. The main reason for this lower rate of attenuation is that line sources radiate sound waves that are cylindrical in shape. The surface area of such waves only increases two-fold for each doubling of distance from the source. However, when the distance from the train becomes comparable to its length, the noise level will begin to drop at a rate of 6 dB for each subsequent doubling of distance.

Indoors, the noise level generally drops only from 3 to 5 dB for each doubling of distance in the near vicinity of the source. However, further from the source, reductions of only 1 or 2 dB occur for each doubling of distance due to the reflections of sound off hard walls and ceiling surfaces.

Absorbing materials. Noise, like light, will bounce from one hard surface to another. In noise control work, this is called *reverberation.* If a soft, spongy material is placed on the walls, floors, and ceiling, the reflected sound will be diffused and soaked up (absorbed). Sound-absorbing materials are rated either by their *Sabin absorption coefficients* (α_{SAB}) at 125, 500, 1,000, 2,000, and 4,000 Hz or by a single number rating called the *noise reduction coefficient* (NRC). If a unit area of open window is assumed to transmit all and reflect none of the acoustical energy that reaches it, it is assumed to be 100 percent absorbent. This unit area of totally absorbent surface is called a "sabin."[35] The absorptive properties of acoustical materials are then compared with this standard. The performance is expressed as a fraction or percentage of the sabin (α_{SAB}). The NRC is the average of the α_{SAB}s at 250,

[35] H. J. Sabin, "Notes on Acoustic Impedance Measurement," *Journal of the Acoustical Society of America,* vol. 14, p. 143, 1942.

500, 1,000, and 2,000 Hz rounded to the nearest multiple of 0.05. The NRC has no physical meaning. It is a useful means of comparing similar materials.

Sound-absorbing materials such as acoustical tile, carpets, and drapes placed on ceiling, floor, or wall surfaces can reduce the noise level in most rooms by about 5 to 10 dB for high-frequency sounds, but only by 2 or 3 dB for low-frequency sounds. Unfortunately, such treatment provides no protection to an operator of a noisy machine who is in the midst of the direct noise field. For greatest effectiveness, sound-absorbing materials should be installed as close to the noise source as possible.

If you have a small or limited amount of sound-absorbing material and wish to make the most effective use of it in a noisy room, the best place to put it is in the upper trihedral corners of the room, formed by the ceiling and two walls. Due to the process of reflection, the concentration of sound is greatest in the trihedral corners of a room. Additionally, the upper corner locations also protect the lightweight fragile materials from damage.

Because of their light weight and porous nature, acoustical materials are ineffectual in preventing the transmission of either airborne or structure-borne sound from one room to another. In other words, if you can hear people walking or talking in the room or apartment above, installing acoustical tile on your ceiling will not reduce the noise transmission.

Acoustical lining. Noise transmitted through ducts, pipe chases, or electrical channels can be reduced effectively by lining the inside surfaces of such passageways with sound-absorbing materials. In typical duct installations, noise reductions on the order of 10 dB/m for an acoustical lining 2.5 cm thick are well within reason for high-frequency noise. A comparable degree of noise reduction for the lower frequency sounds is considerably more difficult to achieve because it usually requires at least a doubling of the thickness and/or length of acoustical treatment.

Barriers and panels. Placing barriers, screens, or deflectors in the noise path can be an effective way of reducing noise transmission, provided that the barriers are large enough in size, and depending upon whether the noise is high frequency or low frequency.[36] High-frequency noise is reduced more effectively than low-frequency noise.

The effectiveness of a barrier depends on its location, its height, and its length. Referring to Figure 7-42, we can see that the noise can follow five different paths.

First, the noise follows a direct path to receivers who can see the source well over the top of the barrier. The barrier does not block their line of sight (L/S) and therefore provides no attenuation. No matter how absorptive the barrier is, it cannot pull the sound downward and absorb it.

Second, the noise follows a diffracted path to receivers in the shadow zone of the barrier. The noise that passes just over the top edge of the barrier is diffracted (bent) down into the apparent shadow shown in the figure. The larger the angle of

[36]*NCHRP 174,* pp. 52–53.

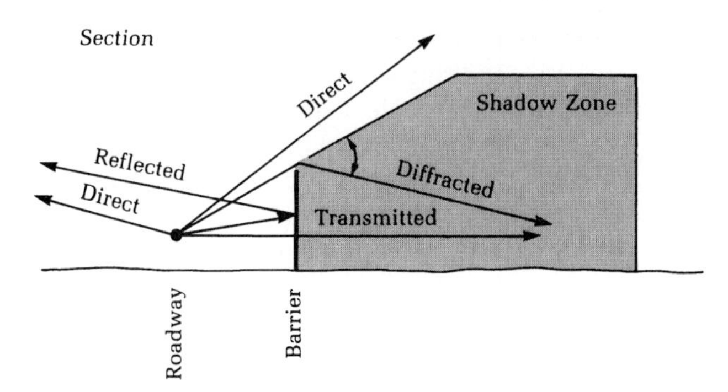

FIGURE 7-42
Noise paths from a source to a receiver. (*Source: NCHRP 174,* 1976.)

diffraction, the more the barrier attenuates the noise in this shadow zone. In other words, less energy is diffracted through large angles than through smaller angles.

Third, in the shadow zone, the noise transmitted directly through the barrier may be significant in some cases. For example, with extremely large angles of diffraction, the diffracted noise may be less than the transmitted noise. In this case, the transmitted noise compromises the performance of the barrier. It can be reduced by constructing a heavier barrier. The allowable amount of transmitted noise depends on the total barrier attenuation desired. More is said about this transmitted noise later.

The fourth path shown in Figure 7-42 is the reflected path. After reflection, the noise is of concern only to a receiver on the opposite side of the source. For this reason, acoustical absorption on the face of the barrier may sometimes be considered to reduce this reflected noise; however, this treatment will not benefit any receivers in the shadow zone. It should be noted that in most practical cases the reflected noise does not play an important role in barrier design. If the source of noise is represented by a line of noise, another short-circuit path is possible. Part of the source may be unshielded by the barrier. For example, the receiver might see the source beyond the ends of the barrier if the barrier is not long enough. This noise from around the ends may compromise, or short-circuit, barrier attenuation. The required barrier length depends on the total net attenuation desired. When 10 to 15 dB attenuation is desired, barriers must, in general, be very long. Therefore, to be effective, barriers must not only break the line of sight to the nearest section of the source, but also to the source far up and down the line.

Of these four paths, the noise diffracted over the barrier into the shadow zone represents the most important parameter from the barrier design point of view. Generally, the determination of barrier attenuation or barrier noise reduction involves only calculation of the amount of energy diffracted into the shadow zone. The procedures presented in the barrier nomograph used to predict highway noise are based on this concept.

Another general principle of barrier noise reduction that is worth reviewing at this point is the relation between noise attenuation expressed in (1) decibels, (2) energy terms, and (3) subjective loudness. Table 7-9 gives these relationships for line sources. As indicated in the loudness column, a barrier attenuation of 3 dB will be barely discerned by the receiver. However, to attain this reduction, 50 percent of the acoustical energy must be removed. To cut the loudness of the source in half,

TABLE 7-9
Relation between sound level reduction, energy, and loudness for line sources

To reduce A-level by dB	Remove portion of energy (%)	Divide loudness by
3	50	1.2
6	75	1.5
10	90	2
20	90	4
30	99.9	8
40	99.99	16

Source: NCHRP 174, 1976.

a reduction of 10 dB is necessary. That is equivalent to eliminating 90 percent of the energy initially directed toward the receiver. As indicated previously, this drastic reduction in energy requires very long and high barriers. In summary, when designing barriers, you can expect the complexity of the design to be about as follows:

Attenuation (dB)	Complexity
5	Simple
10	Attainable
15	Very difficult
20	Nearly impossible

Roadside barriers can be designed using the barrier nomograph in reverse order. A set of typical solutions is summarized in Table 7-10. The noise reduction at 152 m is less than that at 30 m because the barrier does not cast as large a shadow at a distance. The effectiveness of the barrier is reduced for trucks because of the elevated nature of the source.

Transmission loss. When the position of the noise source is very close to the barrier, the diffracted noise is less important than the transmitted noise. If the barrier is in fact a wall panel that is sealed at the edges, the transmitted noise is the only one of concern.

The ratio of the sound energy incident on one surface of a panel to the energy radiated from the opposite surface is called the *sound transmission loss* (TL). The actual energy loss is partially reflected and partially absorbed. Since TL is frequency-dependent, only a complete octave or one-third octave band curve provides a full description of the performance of the barrier.

Enclosures. Sometimes it is much more practical and economical to enclose a noisy machine in a separate room or box than to quiet it by altering its design, operation, or component parts. The walls of the enclosure should be massive and airtight to contain the sound. Absorbent lining on the interior surfaces of the enclosure will reduce the reverberant buildup of noise within it. Structural contact between the noise source

TABLE 7-10
Noise reductions for various highway configurations

Highway configuration[a]		Height or depth	Truck mix	Noise reduction[b] at distance from ROW (dBA)	
Sketch	Description	(m)	(%)	30 m	152 m
	Roadside barriers	6.1	0	13.9	13.3
	7.6 m from edge		5	13.0	12.1
	of shoulders;		10	12.6	11.7
	ROW = 78 m wide		20	12.3	11.3
	Depressed roadway	6.1	0	9.9	11.4
	w/2:1 slopes;		5	8.8	10.3
	ROW = 102 m		10	8.4	9.8
			20	8.1	9.4
	Fill elevated	6.1	0	9.0	6.3
	roadway w/2:1		5	7.6	2.7
	slopes;		10	7.1	1.8
	ROW = 102 m		20	6.7	1.1
	Elevated	7.3	0	9.8	6.0
	structure;		5	9.6	2.4
	ROW = 78 m		10	9.3	1.5
			20	8.8	0.8

[a]Assumes divided 8 lanes with 9.1 m median.
[b]Based on observed 1.5 m above grade.
Source: B. A. Kugler, D. E. Commins, W. J. Galloway, "Highway Noise: Generation and Control," *National Cooperative Highway Research Program Report 173,* 1976.

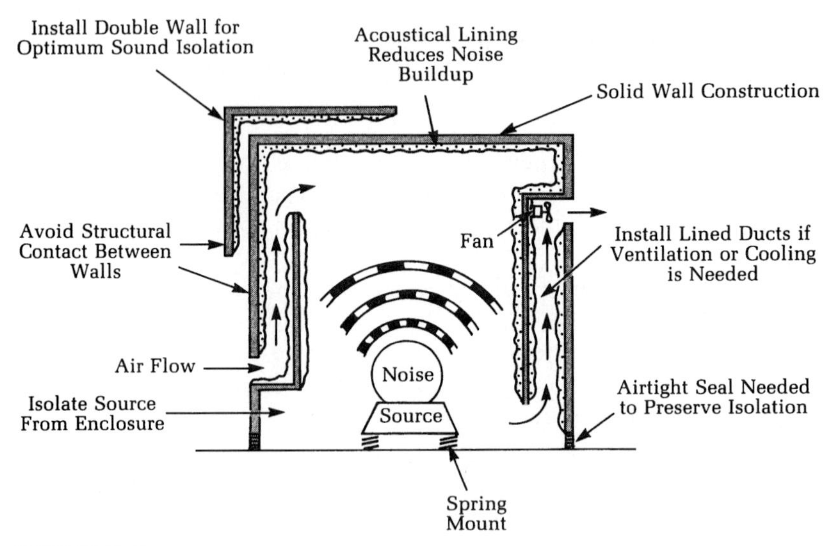

FIGURE 7-43
Enclosures for controlling noise. (*Source: National Bureau of Standards Handbook 119,* 1976.)

and the enclosure must be avoided, so that the source vibration is not transmitted to the enclosure walls, thus short-circuiting the isolation. For maximum effective noise control, all of the techniques illustrated in Figure 7-43 must be employed.

Control of Noise Source by Redress

The best way to solve noise problems is to design them out of the source. However, we are frequently faced with an existing source that, either because of age, abuse, or poor design, is a noise problem. The result is that we must redress, or correct, the problem as it currently exists. The following sections identify some measures that might apply if you are allowed to tinker with the source.

Balance rotating parts. One of the main sources of machinery noise is structural vibration caused by the rotation of poorly balanced parts, such as fans, fly wheels, pulleys, cams, shafts, and so on. Measures used to correct this condition involve the addition of counterweights to the rotating unit or the removal of some weight from the unit. You are probably familiar with noise caused by imbalance in the high-speed spin cycle of washing machines. The imbalance results from clothes not being distributed evenly in the tub. By redistributing the clothes, balance is achieved and the noise ceases. This same principle of balance can be applied to furnace fans and other common sources of such noise.

Reduce frictional resistance. A well-designed machine that has been poorly maintained can become a serious source of noise. General cleaning and lubrication of all rotating, sliding, or meshing parts at contact points should go a long way toward fixing the problem.

Apply damping materials. Since a vibrating body or surface radiates noise, the application of any material that reduces or restrains the vibrational motion of that body will decrease its noise output. Three basic types of redress vibration damping materials are available:

1. Liquid mastics, which are applied with a spray gun and harden into relatively solid materials, the most common being automobile "undercoating"
2. Pads of rubber, felt, plastic foam, leaded vinyls, adhesive tapes, or fibrous blankets, which are glued to the vibrating surface
3. Sheet metal viscoelastic laminates or composites, which are bonded to the vibrating surface

Seal noise leaks. Small holes in an otherwise noise-tight structure can reduce the effectiveness of the noise control measures. As you can see in Figure 7-44, if the designed transmission loss of an acoustical enclosure is 40 dB, an opening that comprises only 0.1 percent of the surface area will reduce the effectiveness of the enclosure by 10 dB.

Perform routine maintenance. We all recognize the noise of a worn muffler. Likewise, studies of automobile tire noise in relation to pavement roughness show that

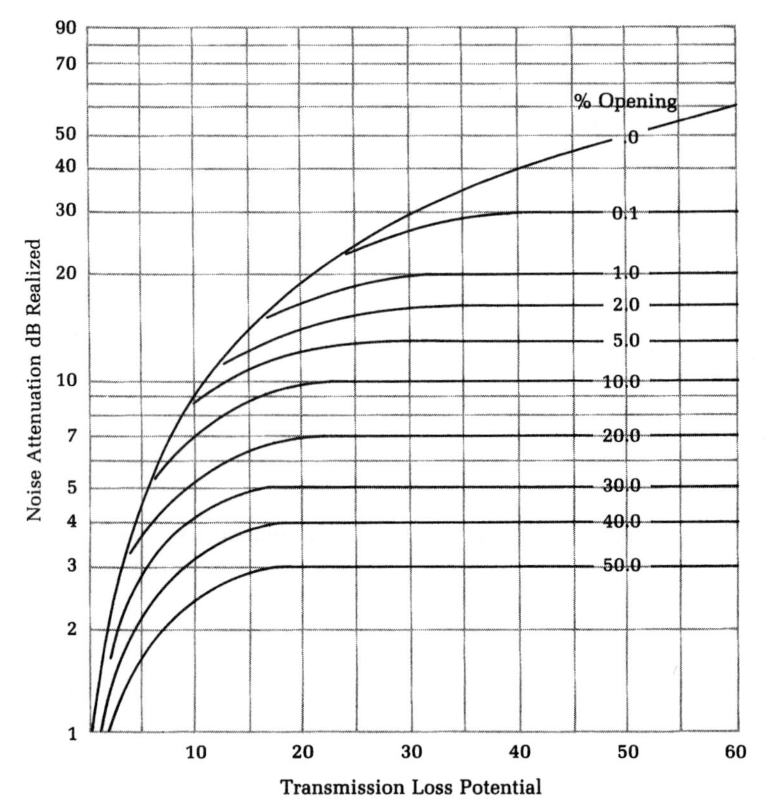

FIGURE 7-44
Transmission loss potential versus transmission loss realized for various opening sizes as a percent of total wall area. (*Source:* Bell, *Fundamentals of Industrial Noise Control,* Trumbull, CT: Harmony Publications, 1973. Reprinted by permission.)

maintenance of the pavement surface is essential to keep noise at minimum levels. Normal road wear can yield noise increases on the order of 6 dBA.

Protect the Receiver

When all else fails. When exposure to intense noise fields is required and none of the measures discussed so far is practical, as, for example, for the operator of a chain saw or pavement breaker, then measures must be taken to protect the receiver. The following two techniques are commonly employed.

Alter work schedule. Limit the amount of continuous exposure to high noise levels. In terms of hearing protection, it is preferable to schedule an intensely noisy operation for a short interval of time each day over a period of several days rather than a continuous eight-hour run for a day or two.

In industrial or construction operations, an intermittent work schedule would benefit not only the operator of the noisy equipment, but also other workers in the vicinity. If an intermittent schedule is not possible, then workers should be given relief time during the day. They should take their relief time at a low-noise-level location, and should be discouraged from trading relief time for dollars, paid vacation, or an "early out" at the end of the day!

Inherently noisy operations, such as street repair, municipal trash collection, factory operation, and aircraft traffic, should be curtailed at night and early morning to avoid disturbing the sleep of the community. Remember: operations between 10 P.M. and 7 A.M. are effectively 10 dBA higher than the measured value.

Ear protection. Molded and pliable earplugs, cup-type protectors, and helmets are commercially available as hearing protectors. Such devices may provide noise reductions ranging from 15 to 35 dB (Figure 7-45). Earplugs are effective only if they are properly fitted by medical personnel. As shown in Figure 7-45, maximum protection can be obtained when both plugs and muffs are employed. Only muffs that have a certification stipulating the attenuation should be used.

These devices should be used only as a last resort, after all other methods have failed to lower the noise level to acceptable limits. Ear protection devices should be used while operating lawn mowers, mulchers, and chippers, and while firing weapons at target ranges. It should be noted that protective ear devices do interfere with speech communication and can be a hazard in some situations where warning calls may be a routine part of the operation (for example, TIMBERRRR!). A modern ear-destructive device is a portable mini-radio/recorder that uses earphones. In this "reverse" muff, high noise levels are directed at the ear without attenuation. If you can hear someone else's radio/recorder, that person is subjecting him- or herself to noise levels in excess of 90–95 dBA!

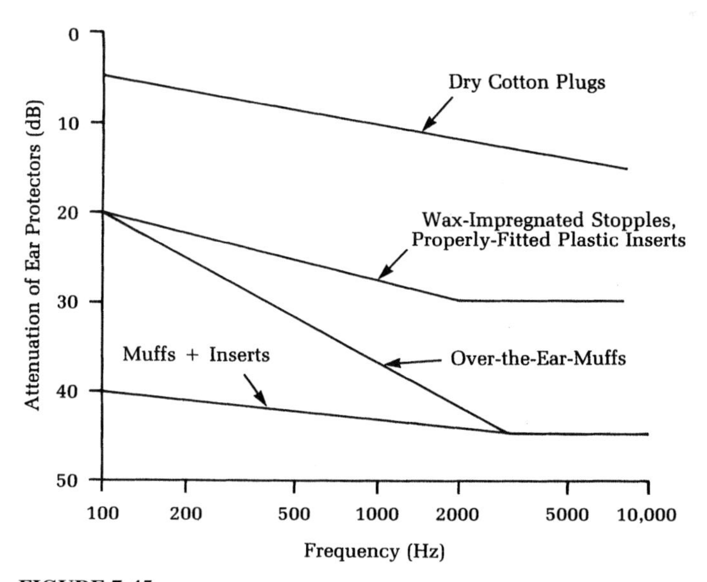

FIGURE 7-45
Attenuation of ear protectors at various frequencies. (*Source: National Bureau of Standards Handbook 119,* 1976.)

7-8 CHAPTER REVIEW

When you have completed studying this chapter, you should be able to do the following without the aid of your textbook or notes:

1. Define frequency, based on a sketch of a harmonic wave you have drawn, and state its units of measure (namely, hertz, Hz).
2. State the basic unit of measure used in measuring sound energy (namely, the decibel) and explain why it is used.
3. Define sound pressure level in mathematical terms, that is,

$$\text{SPL} = 20 \log \frac{P_{rms}}{(P_{rms})_0}$$

4. Explain why a weighting network is used in a sound level meter.
5. List the three common weighting networks and sketch their relative frequency response curves. (Label frequencies, that is, 20, 1,000, and 10,000 Hz; and relative response, that is, 0, −5, −20, and −45 dB, as in Figure 7-6.)
6. Differentiate between a mid/high-frequency noise source and a low-frequency noise source on the basis of A, B, and C scale readings.
7. Explain the purpose of octave band analysis.
8. Differentiate between continuous, intermittent, and impulsive noise.
9. Sketch the curves and label the axes of the two typical types of impulsive noise.
10. Sketch a Fletcher-Munson curve, label the axes, and explain what the curve depicts.
11. Define "phon."
12. Explain the mechanism by which hearing damage occurs.
13. Explain what hearing threshold level (HTL) is.
14. Define presbycusis and explain why it occurs.
15. Distinguish between temporary threshold shift (TTS), permanent threshold shift (PTS), and acoustic trauma with respect to cause of hearing loss, duration of exposure, and potential for recovery.
16. Explain why impulsive noise is more dangerous than steady state noise.
17. Explain the relationship between the allowable duration of noise exposure and the allowable level for hearing protection, that is, damage-risk criteria.
18. List five effects of noise other than hearing damage.
19. List the three basic elements that might require alteration or modification to solve a noise problem.
20. Describe two techniques to protect the receiver when design and/or redress are not practical, that is, when all else fails.

With the aid of this text, you should be able to do the following:

1. Calculate the resultant sound pressure level from a combination of two or more sound pressure levels.
2. Determine the A-, B-, and C-weighted sound levels from octave band readings.

3. Compute the mean sound level from a series of sound level readings.
4. Compute the following noise statistics if you are provided the appropriate data: L_N and/or L_{eq}; L_{dn}.
5. Determine whether or not a noise level will be acceptable given a series of measurements and the criteria listed in Tables 7-3, 7-5, 7-6, 7-7, and/or Figures 7-26 and 7-28.
6. Calculate the sound level at a receptor site after transmission through the atmosphere.
7. Estimate the noise level L_{10} that might be expected for a given roadway configuration and traffic pattern.

7-9 PROBLEMS

7-1. A building located near a road is 6.92 m high. How high is the building in terms of wavelengths of a 50.0 Hz sound? Assume that the speed of sound is 346.12 m/s.

Answer: One wavelength

7-2. Repeat Problem 7-1 for a 500 Hz sound if the temperature is 25.0°C.

7-3. Determine the sum of the following sound levels (all in dB): 68, 82, 76, 68, 74, and 81.

Answer: 85.5 or 86 dB

7-4. A motorcyclist is warming up his racing cycle at a racetrack approximately 200 m from a sound level meter. The meter reading is 56 dBA. What meter reading would you expect if 15 of the motorcyclist's friends join him with motorcycles having exactly the same sound emission characteristics? You may assume that the sources may be treated as ideal point sources located at the same point.

7-5. A law enforcement officer has taken the following readings with her sound level meter. Is the noise source a predominantly low- or middle-frequency emitter? Readings: 80 dBA, 84 dBB, and 90 dBC.

Answer: Predominantly low-frequency

7-6. The following readings have been made outside the open stage door of the opera house: 109 dBA, 110 dBB, and 111 dBC. Is the singer a bass or a soprano? Explain how you arrived at your answer.

7-7. Convert the following octave band measurements to an equivalent A-weighted level.

Band center frequency (Hz)	Band level (dB)
31.5	78
63	76
125	78
250	82
500	81
1,000	80
2,000	80
4,000	73
8,000	65

Answer: 85.5 or 86 dBA

7-8. Using the typical noise spectrum for automobiles traveling at 50 to 60 km/h (Figure 7-27), determine the equivalent A-weighted level using the following octave band geometric mean center frequencies (all in Hz): 63, 125, 250, 500, 1,000, 2,000, 4,000, and 8,000.

7-9. Compute the average sound pressure level of the following readings by simple arithmetic averaging and by logarithmic averaging (Equation 7-13) (all readings in dB): 42, 50, 65, 71, and 47. Does arithmetic averaging underestimate or overestimate the sound pressure level?

Answers:

$$\overline{x} = 55.00 \text{ or } 55 \text{ dB}$$
$$\overline{L}_p = 61.57 \text{ or } 62 \text{ dB}$$

7-10. Repeat Problem 7-9 for the following data (all in dB): 76, 59, 35, 69, and 72.

7-11. The following noise record was obtained in the front yard of a home. Is this a relatively quiet or a relatively noisy neighborhood? Determine the equivalent continuous equal energy level.

Time (h)	Sound level (dBa)
0000–0600	42
0600–0800	45
0800–0900	50
0900–1500	47
1500–1700	50
1700–1800	47
1800–0000	45

Answers: It is a quiet neighborhood. $L_{eq} = 46.2$ or 46 dBA

7-12. A developer has proposed putting a small shopping mall next to a very quiet residential area in Nontroppo, Michigan. Based on the measurements given below, which were taken at a similar size mall in a similar setting, should the developer expect complaints or legal action? Calculate L_{eq}.

Time (h)	Sound level (dBa)
0000–0600	42
0600–0800	55
0800–1000	65
1000–2000	70
2000–2200	68
2200–0000	57

7-13. Two oil-fired boilers for a 600 megawatt (MW) power plant produce an 83 dB noise level at 31.5 Hz, 180.0 m from the induced draft fans. Determine the sound pressure level 488.0 m upwind on an overcast day when the wind speed is 4.50 m/s, the temperature is 20.0°C, the relative humidity is 30.0 percent, and the barometric pressure is 106.0 kPa.

Answer: SPL at 488.0 m = 56 dB at 31.5 Hz

7-14. The sound pressure level from a jet engine test cell is 90 dB at 125 Hz, and at 1,000 Hz at a distance of 400 m. What are the sound pressure levels at 125 and 1,000 Hz, 1,200 m downwind on a clear evening when the wind speed is 1.50 m/s, the temperature is 20.0°C, the relative humidity is 50.0 percent, and the barometric pressure is 103.2 kPa?

7-15. Consider an ideal single lane of road that carries 1,200 vehicles per hour uniformly spaced along the road and determine the following:

a. The average center-to-center spacing of the vehicles for an average traffic speed of 40.0 km/h.

b. The number of vehicles in a one kilometer length of the lane when the average speed is 40.0 km/h.

c. The sound level (dBA) 60.0 m from a one kilometer length of this roadway with automobiles emitting 71 dBA at the edge of an 8.0 m wide roadway. Assume that the autos travel at a speed of 40.0 km/h, that the sound radiates ideally from a hemisphere, and that contributions of less than 0.3 dBA may be ignored.

Answers: a. Average center-to-center spacing = 33.3 m

b. Number of vehicles in a kilometer length = 30 veh/km

c. $L_p = 47.47$ or 48 dBA

7-16. Repeat Problem 7-15 if the vehicle speed is increased to 80.0 km/h and the spacing is decreased or increased appropriately to maintain 1,200 vehicles per hour.

7-17. In preparation for a public hearing on a proposed interstate bypass at Nontroppo, Michigan, the County Road Commission has requested that you prepare an estimate of the potential for violation of FHA noise standards 75 meters from the interstate. The city engineer has supplied sketch maps (see Figure P-7-17*b*, p. 628) and data summary for your use.

Data for I-481 at Pianissimo Avenue

Estimated traffic:

Automobiles: 7,800 per hour at 88.5 km/h

Medium trucks: 520 per hour at 80.5 km/h

Heavy trucks: 650 per hour at 80.5 km/h

Roadway configuration: Depressed

Section length: 857.25 m east and 857.25 m west of center line of Pianissimo Avenue

Assume that the receiver is located on the center line of Pianissimo Avenue 75.00 m from the center line of I-481. Show all work.

Answer: L_{10} at observer = 68.6 or 69 dBA

7-18. Determine the potential for violation of FHA noise standards at the north side of Fermata School. The city engineer has supplied sketch maps (see Figure P-7-17*c*) and data for your use.

Data for I-481 at Fermata School

Estimated traffic: Same as at Pianissimo Avenue

Roadway configuration: At grade

Barrier length: 199.80 m east and 199.80 m west of Fermata School

Assume that the receiver is located just outside of the north side of the school, 123.17 m from the center line of I-481, and is 1.5 m above the ground. Show all work.

7-19. Using the data from Problem 7-17, compute the unattenuated L_{eq} at the receiver for autos only. Assume the edge of the roadway is at the "toe" of the road cut.

Answer: L_{eq} = 70 dBA

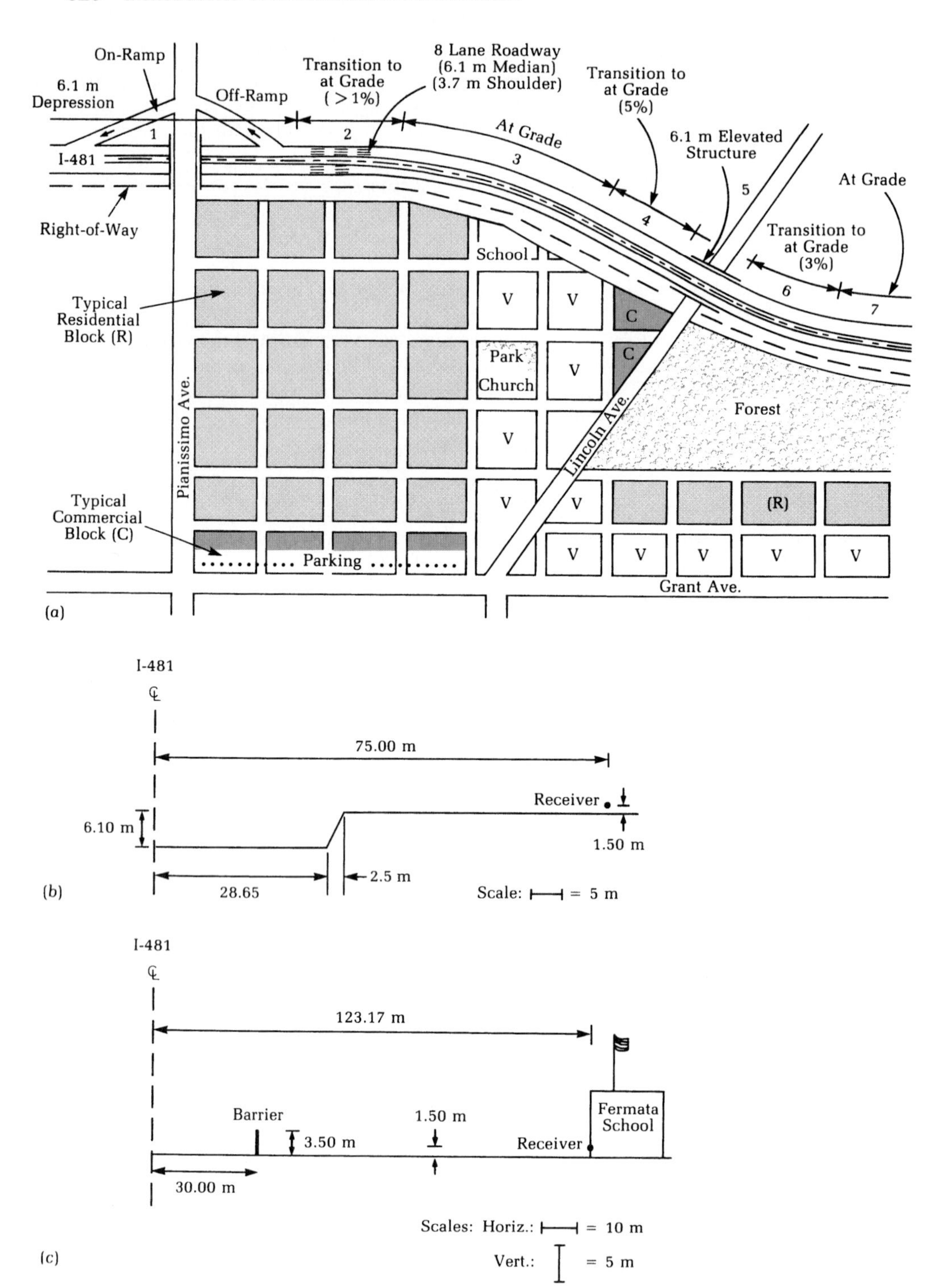

FIGURE P-7-17
Sketch maps for proposed bypass: (*a*) plan view, (*b*) cross section along Pianissimo Avenue, (*c*) cross section at Fermata School. (See Problems 7-17 and 7-18.)

7-20. Rework Problem 7-19 using the data from Problem 7-18. Assume the edge of the roadway is at the barrier.

7-10 DISCUSSION QUESTIONS

7-1. Classify each of the following noise sources by "type," that is, continuous, intermittent or impulse. (Not all sources fit these three classifications.)

(a) electric saw
(b) air conditioner
(c) alarm clock (bell type)
(d) punch press

7-2. Is the following statement true or false? If it is false, correct it in a nontrivial manner.

"A sonic boom occurs when an aircraft breaks the sound barrier."

7-3. Is the following statement true or false? If it is false, correct it in a nontrivial manner.

"Excessive continuous noise causes hearing damage by breaking the stapes."

7-4. As the safety officer of your company, you have been asked to determine the feasibility of reducing exposure time as a method of reducing hearing damage for the following situation:

The worker is operating a high speed grinder on steel girders for a high rise building. The effective noise level at the operator's ear is 100 dBA. She cannot wear protective ear devices because she must communicate with others.

What amount of exposure time would you set as the limit?

7-5. In Figure 7-43, identify where the following noise-control techniques are applied: isolation and/or damping, reduction in noise leakage, use of absorbing materials, use of acoustical lining, enclosure.

7-11 ADDITIONAL READING

L. H. Bell, *Fundamentals of Industrial Noise Control,* Trumbull, CT: Harmony Publications, 1973.

L. L. Beranek, ed., *Noise and Vibration Control,* New York: McGraw-Hill, 1971.

D. M. Lipscomb and A. C. Taylor, *Noise Control Handbook of Principles and Practices,* New York: Van Nostrand Reinhold, 1978.

E. B. Magrab, *Environmental Noise Control,* New York: Wiley, 1975.

J. B. Olishifski and E. R. Harford, *Industrial Noise and Hearing Conservation,* Chicago: National Safety Council, 1975.

CHAPTER 8

SOLID WASTE MANAGEMENT

8-1 PERSPECTIVE
Magnitude of the Problem
Characteristics of Solid Waste
Solid Waste Management Overview
8-2 COLLECTION
Collection Methods
Collection Estimates
Truck Routing
Crew Integration
8-3 INTERROUTE TRANSFER
Maximum Haul Time
Economical Haul Time
8-4 DISPOSAL BY SANITARY LANDFILL
Site Selection
Site Preparation
Equipment
Operation
Environmental Considerations
Leachate
Landfill Design
Completed Sanitary Landfills
8-5 WASTE TO ENERGY
Heating Value of Waste
Fundamentals of Combustion
Conventional Incineration
Recovering Energy from Waste

8-6 RESOURCE CONSERVATION AND RECOVERY
Background and Perspective
Low Technology RC & R
Medium Technology RC & R
High Technology RC & R
8-7 CHAPTER REVIEW
8-8 PROBLEMS
8-9 DISCUSSION QUESTIONS
8-10 ADDITIONAL READING

8-1 PERSPECTIVE

Solid waste is a generic term used to describe the things we throw away. It includes things we commonly describe as garbage, refuse, and trash. The U.S. Environmental Protection Agency's (EPA) regulatory definition is broader in scope. It includes any discarded item; things destined for reuse, recycle, or reclamation; sludges; and hazardous wastes. The regulatory definition specifically excludes radioactive wastes and *in situ* mining wastes.

We have limited the discussion in this chapter to solid wastes generated from residential and commercial sources. Sludges were discussed in Chapters 3 and 5. Hazardous waste will be discussed in Chapter 9, and radioactive waste will be discussed in Chapter 10.

Magnitude of the Problem

Solid waste disposal creates a problem primarily in highly populated areas. The more concentrated the population, the greater the problem becomes. Various estimates have been made of the quantity of solid waste generated and collected per person per day. In 1990, the EPA estimated that the national average rate of solid waste generated was 1.95 kg/capita · day.[1] On this basis, in 1990, the U.S. produced 178 teragrams (Tg) of solid waste.[2] This is an 8 percent increase over the 1988 estimate of 163 Tg and a 29 percent increase over the 1968 estimate of 127 Tg. The EPA projects that solid waste generation will reach 200 Tg by the year 2000. The EPA estimates that 60 percent of the waste stream comes from residential sources, and the remainder is from commercial sources. Individual cities may vary greatly from these estimates. For example, Los Angeles, California, generates about 3.18 kg/capita · day while the rural community of Wilson, Wisconsin, generates about 1.0 kg/capita · day.

[1]U.S. Environmental Protection Agency, *Characterization of Municipal Solid Waste in the United States: 1992 Update,* EPA/530-R-92-019, Washington, D.C., 1992.

[2]In keeping with correct SI notation, we use teragrams (1×10^{12} grams). One Tg is equivalent to 1×10^{9} kilograms (kg) or 1×10^{6} megagrams (Mg). The megagram is often referred to as the "metric ton."

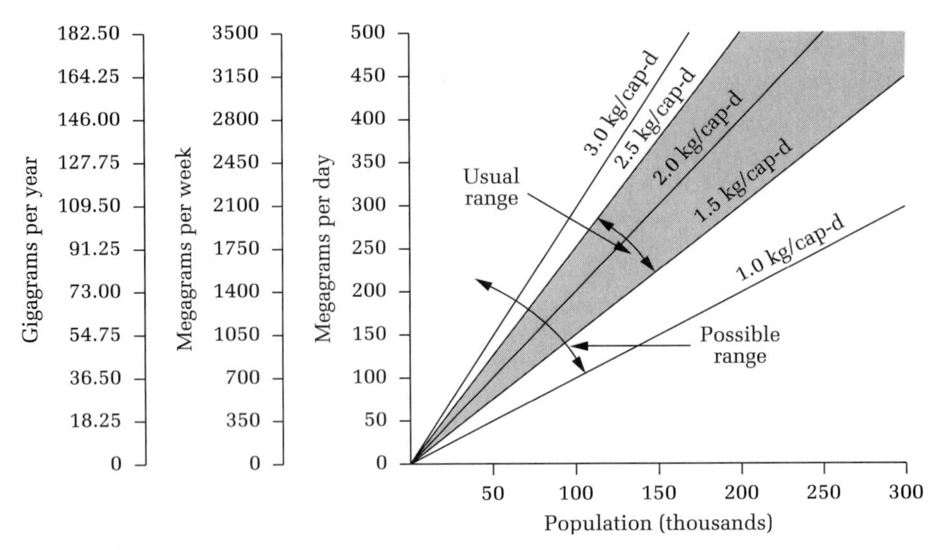

FIGURE 8-1
Solid waste produced: varying per capita figures.

Figure 8-1 shows solid waste production rates. Averages are subject to adjustment depending on many local factors. Studies show there are wide differences in amounts collected by municipalities because of differences in climate, living standards, time of year, education, location, and collection and disposal practices.

Characteristics of Solid Waste

The terms *refuse* and *solid waste* are used more or less synonymously, although the latter term is preferred. The common materials of solid waste can be classified in several different ways. The point of origin is important in some cases, so classification as domestic, institutional, commercial, industrial, street, demolition, or construction may be useful. The nature of the material may be important, so classification can be made on the basis of organic, inorganic, combustible, noncombustible, putrescible, and nonputrescible fractions. One of the most useful classifications is based on the kinds of materials as shown in Table 8-1. Another classification system that is similar to this is the one used by the Incinerator Institute of America (Table 8-2). This is based primarily on the heat content of the waste.

Garbage is the animal and vegetable waste resulting from the handling, preparation, cooking, and serving of food. It is composed largely of putrescible organic matter and moisture; it includes a minimum of free liquids. The term does not include food processing wastes from canneries, slaughterhouses, packing plants, and similar facilities, or large quantities of condemned food products. Garbage originates primarily in home kitchens, stores, markets, restaurants, and other places where food is stored, prepared, or served. Garbage decomposes rapidly, particularly in warm weather, and may quickly produce disagreeable odors. There is some commercial

TABLE 8-1
Refuse materials by kind, composition, and sources

Kind	Composition	Sources
Garbage	Wastes from preparation, cooking, and serving of food; market wastes; wastes from handling, storage, and sale of produce	Households, restaurants, institutions, stores, markets
Rubbish	Combustible: paper, cartons, boxes, barrels, wood, excelsior, tree branches, yard trimmings, wood furniture, bedding, dunnage	
	Noncombustible: metals, tin cans, metal furniture, dirt, glass, crockery, minerals	
Ashes	Residue from fires used for cooking and heating and from on-site incineration	
Street refuse	Sweepings, dirt, leaves, catch basin dirt, contents of litter receptacles	Streets, sidewalks, alleys, vacant lots
Dead animals	Cats, dogs, horses, cows	
Abandoned vehicles	Unwanted cars and trucks left on public property	
Industrial wastes	Food-processing wastes, boiler house cinders, lumber scraps, metal scraps, shavings	Factories, power plants
Demolition wastes	Lumber, pipes, brick, masonry, and other construction materials from razed buildings and other structures	Demolition sites to be used for new buildings, renewal projects, expressways
Construction wastes	Scrap lumber, pipe, other construction materials	New construction, remodeling
Special wastes	Hazardous solids and liquids; explosives, pathological wastes, radioactive materials	Households, hotels, hospitals, institutions, stores, industry
Sewage treatment residue	Solids from coarse screening and from grit chambers; septic tank sludge	Sewage treatment plants, septic tanks

Source: Institute for Solid Wastes, *Municipal Refuse Disposal,* Chicago: American Public Works Association, 1970.

value in garbage as animal food and as a base for commercial feeds. However, this use may be precluded by health considerations.

Rubbish consists of a variety of both combustible and noncombustible solid wastes from homes, stores, and institutions, but does not include garbage. Trash is synonymous with rubbish in some parts of the country, but trash is technically a subcomponent of rubbish. Combustible rubbish (the "trash" component of rubbish)

TABLE 8-2
Incinerator Institute of American waste classification

Classification of wastes to be incinerated								
Classification of Wastes								
Type	Description	Principal components	Approximate composition % by weight	Moisture content %	Incombustible solids %	MJ heat value/kg of refuse as fired	MJ of aux. fuel per kg of waste to be included in combustion calculations	Recommended min. MJ burner input per kg waste
[a]0	Trash	Highly combustible waste, paper, wood, cardboard cartons, including up to 10% treated papers, plastic or rubber scraps; commercial and industrial sources	Trash 100%	10%	5%	19.8	0	0
[a]1	Rubbish	Combustible waste, paper, cartons, rags, wood scraps, combustible floor sweepings; domestic, commercial, and industrial sources	Rubbish 80% Garbage 20%	25%	10%	15.1	0	0
[a]2	Refuse	Rubbish and garbage; residential sources	Rubbish 50% Garbage 50%	50%	7%	10.0	0	3.5
[a]3	Garbage	Animal and vegetable wastes; restaurants, hotels, markets; institutional, commercial, and club sources	Garbage 65% Rubbish 35%	70%	5%	5.8	3.5	7.0
4	Animal solids and organic wastes	Carcasses, organs, solid organic wastes; hospital, laboratory, abattoirs, animal pounds, and similar sources	100% Animal and human tissue	85%	5%	2.3	7.0	18.6 (11.6 Primary) (7.0 Secondary)
5	Gaseous, liquid, or semi-liquid wastes	Industrial process wastes	Variable	Dependent on predominant components	Variable according to wastes survey	Variable according to wastes survey	Variable according to wastes survey	Variable according to wastes survey
6	Semi-solid and solid wastes	Combustibles requiring hearth, retort, or grate burning equipment	Variable	Dependent on predominant components	Variable according to wastes survey	Variable according to wastes survey	Variable according to wastes survey	Variable according to wastes survey

[a]The above figures on moisture content, ash, and MJ as fired have been determined by analysis of many samples. They are recommended for use in computing heat release, burning rate, velocity, and other details of incinerator designs. Any design based on these calculations can accomodate minor variations.

Source: Incinerator Institute of America, *I.I.A. Standards,* 1968.

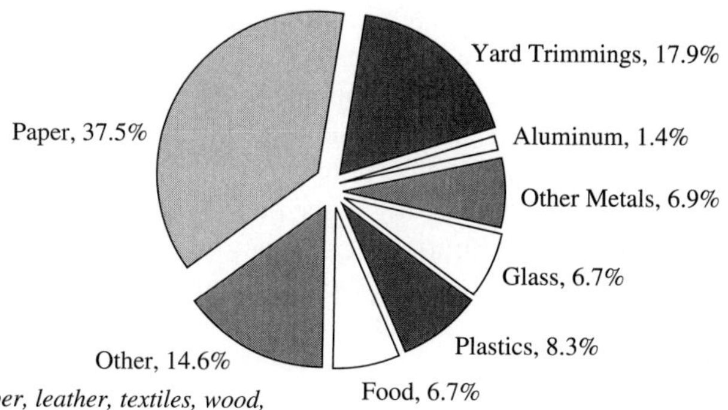

FIGURE 8-2
Materials generated in municipal solid waste (percent by mass), 1990. (*Source:* U.S. Environmental Protection Agency, "Characterization of Municipal Solid Waste in the United States," 1992 update.)

consists of paper, rags, cartons, boxes, wood, furniture, tree branches, yard trimmings, and so on. Some cities have separate designations for yard wastes. Combustible rubbish is not putrescible and may be stored for long periods of time. Noncombustible rubbish is material that cannot be burned at ordinary incinerator temperatures of 700 to 1,100°C. It is the inorganic portion of refuse, such as tin cans, heavy metals, glass, ashes, and so on.

The average municipal solid waste composition in the United States in 1990 is shown in Figure 8-2.

The density of loose combustible refuse is approximately 115 kg/m^3, while the density of collected solid waste is 235 to 300 kg/m^3.

Solid Waste Management Overview

The first objective of solid waste management is to remove discarded materials from inhabited places in a timely manner to prevent the spread of disease, to minimize the likelihood of fires, and to reduce aesthetic insults arising from putrifying organic matter. The second objective, which is equally important, is to dispose of the discarded materials in a manner that is environmentally acceptable.

Policy making. Solid waste system policy making is primarily a function of the public sector rather than the private sector. The goal of a private firm is to minimize a well-defined cost function or to maximize profits. These are generally not the only, or even the primary, constraints of the public sector. The public objective function is more vague and difficult to express formally.

Constraints on the public sector, especially those of a political or a social nature, are difficult to measure, and criteria of effectiveness may not exist in units that can be quantified. Criteria of effectiveness against which public efficiency might be measured include such things as the frequency of collection, types of waste collected, location from which waste is collected, method of disposal, location of disposal site, environmental acceptability of disposal system, and the level of satisfaction of the customers. Public receptivity of a solid waste management system also depends on even less quantifiable parameters, which we group under the term *institutional factors.* Institutional factors include such things as political feasibility of the system, legislative constraints, and administrative simplicity.

Additional constraints on decision making in the public sector are environmental factors and resource conservation. Environmental factors are most important in the areas of waste storage and disposal because these functions represent prolonged exposure of wastes to the environment. Resource conservation is beginning to be considered seriously by local governments as we become increasingly conscious of the limits of our natural resources.

Decisions in solid waste management policy formulation must be made in four basic areas: collection, transport, processing, and disposal. The flowchart in Figure 8-3 illustrates the decisions that must be made from the point of generation to the ultimate disposal of residential solid waste.

In designing a solid waste collection system, one of the first decisions to be made is where the waste will be picked up: the curb or the backyard. This is an important decision because it affects many other collection variables, including choice of storage containers, crew size, and the selection of collection trucks.

Another key decision is frequency of collection. Both point of collection and frequency of collection should be evaluated in terms of their impact on collection costs. Since collection costs generally account for 70 to 85 percent of total solid waste management costs, and labor represents 60 to 75 percent of collection costs, increases in the productivity of collection personnel can dramatically reduce overall costs.

Systems with once-a-week curbside collection help maximize labor productivity and result in significantly lower costs than systems with more frequent collection and/or backyard pickup. The main reason many communities retain twice-a-week backyard service is that the citizens demand this convenience and are willing to pay for it. In warmer regions of the country, twice-a-week service may be deemed essential to prevent gross odors and to break the fly-breeding cycle. The egg-larvae-adult cycle is about 4–5 days.

The choice of solid waste storage containers must be evaluated in terms of both environmental effects and costs. From the environmental standpoint, some storage containers can present health and safety problems to the collectors, as well as to the general public. Therefore, the decision facing a community is which storage system is both environmentally sound and most economical, given the collection system characteristics. For example, paper and plastic bags are superior to many other containers from a health and esthetic standpoint and can increase productivity when used in conjunction with curbside collection. However, with backyard collection systems, bags have little effect on productivity.

Another factor to be considered in examining storage alternatives is home separation of various materials for recovery. The collection of materials for recovery is a growing practice that many cities are implementing. The technique of greatest interest to municipal decision makers is home separation and collection by either the regular collection truck equipped with special bins or by separate trucks.

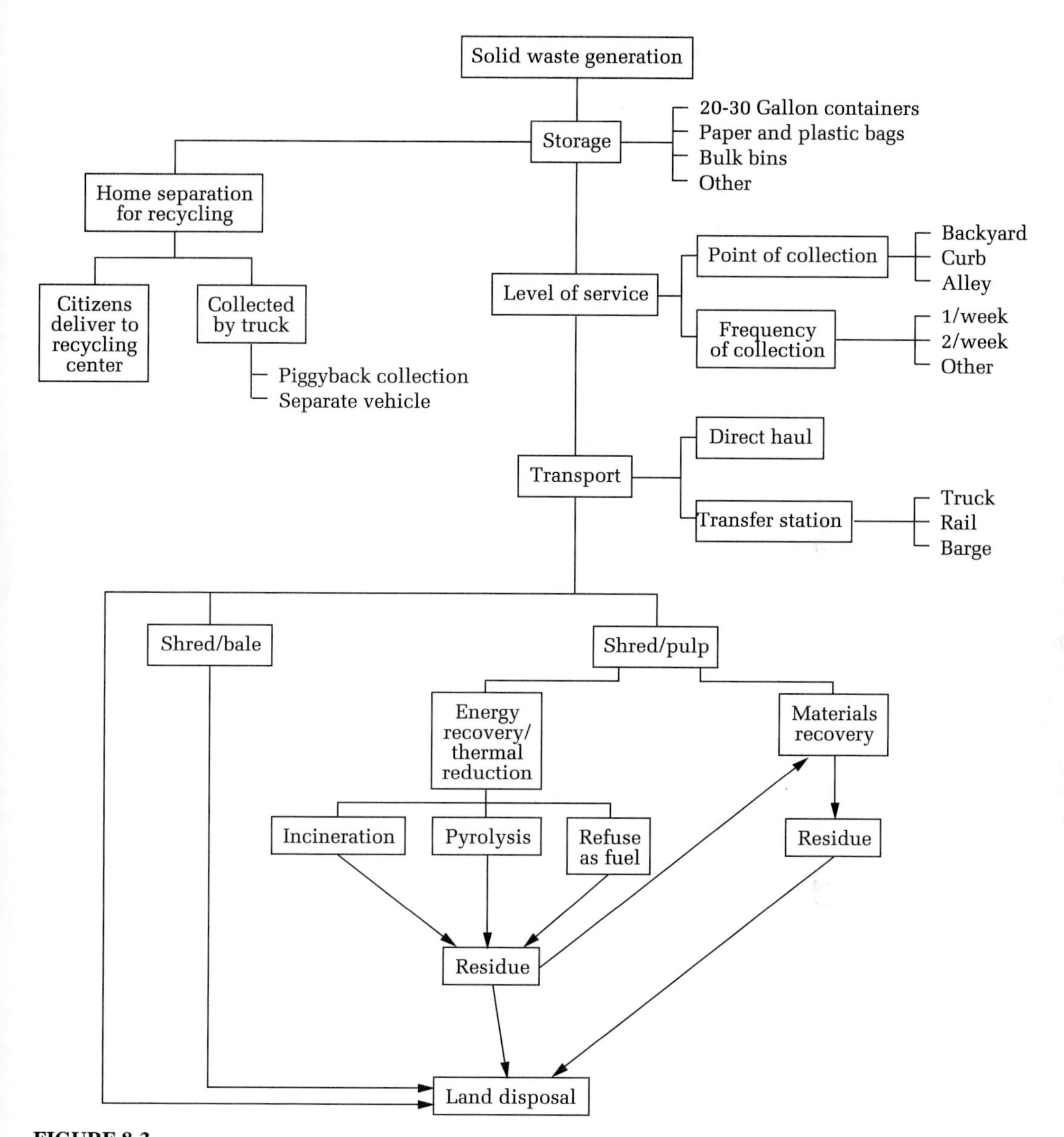

FIGURE 8-3
Solid waste management decision alternatives. (*Source:* U.S. Environmental Protection Agency, *Decision Makers' Guide to Solid Waste Management,* Washington, D.C.: U.S. Government Printing Office, 1974.)

The primary factor to consider in implementing a separate collection system is whether the benefits of recovery outweigh the costs involved. The economic viability of separate collection depends primarily on the local market price for the material and the degree of participation by the citizens. If these factors are positive, it may be possible to implement a recovery system with no increase, and possibly a savings, in collection operating costs; often no additional capital expenditure is required.

The distance between the disposal site and the center of the city will determine the advisability of including a *transfer station* in the transport system.[3] In addition to distance traveled to the disposal site, the time required for the transport is a key factor, especially in traffic-congested large cities.

The tradeoffs involved in transfer station operations are the capital and operating costs of the transfer station as compared to the cost (mostly labor) of having route collection vehicles travel excessive distances to the disposal site. These tradeoffs can be computed to find the point at which transfer becomes economically advantageous.

The sheer quantities of solid waste to be disposed of daily makes the problem of what to do with the waste, once it has been collected, among the most difficult problems confronting community officials. A crisis situation can develop very quickly, for example, in the case of an incinerator or land disposal site forced to shut down because of failure to meet newly passed environmental regulations. Alternatively, a crisis can build gradually over a period of time if needed new facilities are not properly planned for and put into service.

There are three basic alternatives for disposal. Some have subalternatives. The major alternatives are: (1) direct disposal of unprocessed waste in a sanitary landfill, (2) processing of waste followed by land disposal, and (3) processing of waste to recover resources (materials and/or energy) with subsequent disposal of the residues. Most municipal solid waste is landfilled, but the amount landfilled declined from 73 percent in 1988 to 67 percent in 1990. Sixteen percent of the waste was incinerated in 1990. EPA projects an increase in waste incineration and a decrease in landfilling through the 1990s.[4]

Direct haul to a sanitary landfill (with or without transfer and long haul) is usually the cheapest disposal alternative in terms of both operating and capital costs. In 1994, it was estimated that about 5,000 landfills were in operation. Approximately one-third of these were expected to close as a result of regulatory restrictions. Municipalities own 75 percent of the sites.[5] With rising *tipping fees* (the cost to dump solid waste at a disposal facility), a surplus of disposal capacity has replaced the late 1980s predictions of lack of landfill space.

With the second alternative, processing prior to land disposal, the primary objective is to reduce the volume of wastes. Such volume reduction has definite advantages since it reduces hauling costs and ultimate disposal cost, both of which are, to some extent, a function of waste volume. However, the capital and operating cost to achieve this volume reduction are significant and must be balanced against the savings achieved.

An additional consideration is the environmental benefit that might be derived from the volume reduction process. In some cases, shredding and baling may reduce the chances for water pollution from leachate. This alternative is more

---

[3] A transfer station is a place where trucks dump their loads into a larger vehicle where it is compacted. By combining loads, the cost per Mg · km for transport to the landfill is reduced.

[4] U.S. Environmental Protection Agency, *Characterization of Municipal Solid Waste in the United States: 1992 Update,* EPA 530-R-92-019, Washington, D.C., 1992.

[5] B. Wolpin, "Go Figure," *World Wastes,* V. 37, No. 10, p. 4, October, 1994.

conserving of land than sanitary landfilling of unprocessed wastes, but by itself provides no opportunity for material or energy recovery.

The third category of disposal alternatives includes those processes that recover energy or materials from solid waste and leave only a residue for ultimate land disposal. There are significant capital and operating costs associated with all these energy and/or materials recovery systems. However, if markets are available, both energy and materials can be sold to reduce the net costs of recovery.

While resource recovery techniques may be more costly than other disposal alternatives, they do achieve the goal of resource conservation, and the residuals of the processes require much less space for land disposal than unprocessed wastes.

Affecting all four major functions are basic decisions regarding how the solid waste system will be managed and operated. This includes how the system will be financed, which level of government will administer it, and whether a public agency or private firm will operate the collection, transport, processing, and disposal functions. The criteria most relevant for making these decisions are the institutional factors of political feasibility and legislative constraints.

Integrated solid waste management (ISWM). The selection of a combination of techniques, technologies, and management programs to achieve waste management objectives is called *integrated solid waste management* (ISWM). This approach has made major strides in the last decade. The EPA proposed a hierarchy of actions to implement ISWM: source reduction, recycling, waste combustion, and landfilling.[6] Tchobanoglous, Theisen, and Vigil suggest that the term *combustion* is too limiting and that *waste transformation,* which includes such things as separate collection of yard waste and composting, is more appropriate.[7] The most obvious effect of the integrated approach is to reduce the size of the incineration facility. This reduces the capital cost of the incineration facility. Although the energy output is also reduced, the waste that remains has a higher energy content so that the reduction in energy output is less than the reduction in plant size. Recycling also reduces waste elements that can damage the boilers and removes those components that slag in the furnace and foul it.[8]

8-2 COLLECTION

The solid waste collection policies of a city begin with decisions made by elected representatives about whether collection is to be made by: (1) city employees (municipal collection), (2) private firms that contract with city government (contract collection), or (3) private firms that contract with private residents (private collection).

[6]U.S. Environmental Protection Agency, *Decision-Makers Guide to Solid Waste Management,* EPA/530-SW89-072, Washington, D.C., November, 1989.

[7]G. Tchobanoglous, H. Theisen, and S. Vigil, *Integrated Solid Waste Management,* New York: McGraw-Hill, Inc., p. 15, 1993.

[8]J. Shortsleeve and R. Roche, "Analyzing the Integrated Approach," *Waste Age,* pp. 92–94, March, 1990.

The trend since 1964 has been to shift away from exclusive municipal collection and toward a combined system.

Elected officials may also determine what type of solid wastes are to be collected and from whom. In some municipalities broad classes of solid wastes (such as rubbish) are not accepted for collection. In others, certain materials (such as tires, grass trimmings, furniture, or dead animals) may be excluded. Hazardous wastes are generally excluded from regular collections because of disposal and collection dangers. The nature of the service may be governed by limitations of disposal facilities or by the opinion of the legislative body as to what service should be performed. A city may collect garbage only or it may collect everything but garbage. Almost all municipal systems collect residential waste, but only about one-third collect industrial waste.

The final decision concerning collection, which is made by the elected officials, is the frequency of collection. The proper frequency for the most satisfactory and economical service is governed by the amount of solid waste that must be collected and by climate, cost, and public requests. For the collection of solid waste that contains garbage, the maximum period should not be greater than

1. The normal time for the accumulation of the amount that can be placed in containers of reasonable size;
2. The time it takes for fresh garbage to putrefy and emit foul odors under average storage conditions;
3. The length of the fly-breeding cycle, which, during the hot summer months, is less than seven days.

In the last three decades the prevailing frequency of collection has changed from twice a week pickup to once a week. The increased use of once per week service is due to two factors. First, unit costs are reduced when frequency is cut from twice to once per week. Second, the increased percentage of paper and decreased percentage of garbage in the solid waste permit longer periods of acceptable storage.

Once policy has been set, the actual method of collection is determined by engineers or managers. Major considerations include how the solid waste will be collected, how the crews will be managed, and how the trucks will be routed.

Collection Methods

The first decision to be made is how the solid waste container will get from the residence to the collection vehicle. The three basic methods are: (1) *curbside* or alley pickup, (2) *set-out, set-back collection,* and (3) *backyard pickup* or the tote barrel method.

The quickest and most economical point of collection is from curbs or alleys using standard containers. It is the most common type of collection used. It costs only about one-half as much as backyard collection. Usually the city designates what type of containers are to be used. The crews simply empty the containers into the collection vehicles. Whenever possible the crews collect from both sides of the street at the same time. Municipal ordinances or administrative regulations usually specify when the containers must be placed at the curb or in the alley for pickup and also how long they may remain after pickup. Common limits are out by 7 A.M. and back by 7 P.M. When solid wastes are loaded from curbs or alleys, work progresses rapidly. A typical crew consists of a driver and two collectors. Some crews still have three or even four collectors, but the trend is toward fewer collectors. Recent studies indicate that small crews are more efficient than larger ones, since labor costs are a major element of the total cost. Aside from the cost advantage of this method, it also eliminates the need for the collectors to enter private property, and the amount of service given each homeowner is relatively uniform. However, many citizens dislike having to set their solid wastes out at certain times and object to the unsightly appearance on the streets. Some surveys have shown that many homeowners would prefer to pay more in order to receive backyard service.

The set-out, set-back method eliminates most of the disadvantages of the curb method, but it does require the collector to enter private property. This method consists of the following operations: (1) the set-out crew carries the full containers from the residential storage location to the curb or alley before the collection vehicle arrives, (2) the collection crew loads the refuse in the same manner as the curb method, and (3) the set-back crew returns the empty cans. Any of the crew may be required to do more than one step or the homeowner may be required to do one of the steps. This method has not been shown to be more economical or advantageous than the backyard method.

Backyard pickup is usually accomplished by the use of tote barrels. In this method, the collector enters the resident's property, dumps the container into a tote barrel, carries it to the truck, and dumps it. The collector may collect refuse from more than one house before returning to the truck to dump. The primary advantage of this system is in the convenience to the homeowner. The major disadvantage is the high cost. Many homeowners object to having the collectors enter their private property.

Cost analyses have revealed that 70 to 85 percent of the cost of solid waste collection and disposal can be attributed to the collection phase. For this reason, it would seem that a great deal of municipal effort should be directed to studying collection alternatives to determine the most efficient system. However, many analyses begin their studies assuming that waste loads are already collected and waiting for disposal. There are two major reasons why the collection system is not studied more often. First, the collection system is a complex and expensive system to analyze. The primary reasons for this are that it involves people, equipment, and levels of service, plus the possibility of numerous variations in secondary factors such as collection methodology; quantity, nature, and the method of storage of refuse; location of pickup point; equipment type and characteristics of operation; road factors; service density; route topography; climatic factors; and human factors. Human factors would include morale, incentive, fatigue, and other variables that influence the time required to complete a given task. Secondly, most cities are already collecting refuse in some manner, and the cliche "leave well enough alone" often prevails. It is generally on the disposal system that the public is placing pressure for improvement, rather than the collection system.

Most changes in collection systems will require a great deal of investigation and testing. Even if the change is an obvious one, often "proof" of some sort is needed to convince the elected officials. The most important thing to realize about the solid waste collection system is that it is too big, complex, and vital to allow actual experimentation except on a very small scale. Coupled with this are all the other problems peculiar to studying large-scale public systems. A relevant data base is probably nonexistent. The political implications of control of the system and cost distribution may override an otherwise practical solution. A large investment will have already been made in the existing system and the designer is not allowed the luxury of starting at the beginning, but must start with a system that may be founded on a pyramid of errors.

Many techniques have been devised to evaluate and subsequently optimize collection systems. Most of them require a tremendous amount of data collection and analysis, are based on oversimplistic assumptions, or both. One data collection system has been developed that is relatively easy to use and allows very detailed analysis of the collection process.[9] In this procedure, a data collector follows the crew through its collection schedule and records appropriate data on a "crew-machine" chart, such as the one in Figure 8-4. Crew-machine charts show how much time is spent on each task. In this case a record is made of how long the crew members spend on each task. A stopwatch is used, which runs continuously. As a crew member leaves the truck or begins an operation the time is noted. Upon completion of the task the time is recorded and the next task begun. A crew member must always be doing some task so that the whole day is accounted for.

For meaningful results the tasks must be standardized. Typical tasks are

1. *Walk:* This is time spent by the workers in walking along the road from the truck to the next collection point. It is used only in the curb or alley collection method.
2. *Drive:* Time spent driving the truck from collection point to collection point.
3. *Trash:* Time spent placing yard wastes from the curb into the truck. Some crews accomplish this by use of a pitchfork.
4. *Ride:* Time the workers spend riding on the truck between collection points.
5. *Lead:* This is unique to some backyard collection procedures. It is when the worker leaves the truck with a tub and collects refuse from several houses. The collector stops ahead on the route and waits for the truck to catch up.
6. *Collect:* Time spent collecting from houses in the traditional way in backyard pickup. It includes the time from when the collector leaves the truck, walks to the backyard, dumps the cans into a tub, returns to the truck, and dumps the tub. The collector may collect from more than one house before returning.
7. *Dump:* Time spent dumping cans or bags from the curb into the truck.
8. *Delay:* Time spent on any operation other than these listed.

The above tasks can be modified to suit a particular system. After the data are collected, they are converted to histogram format, which shows the percentage of time spent on a given task.

---

[9] D. A. Cornwell, "Digital Simulation of Solid Waste Collection Systems," Master's thesis, University of Florida, 1972.

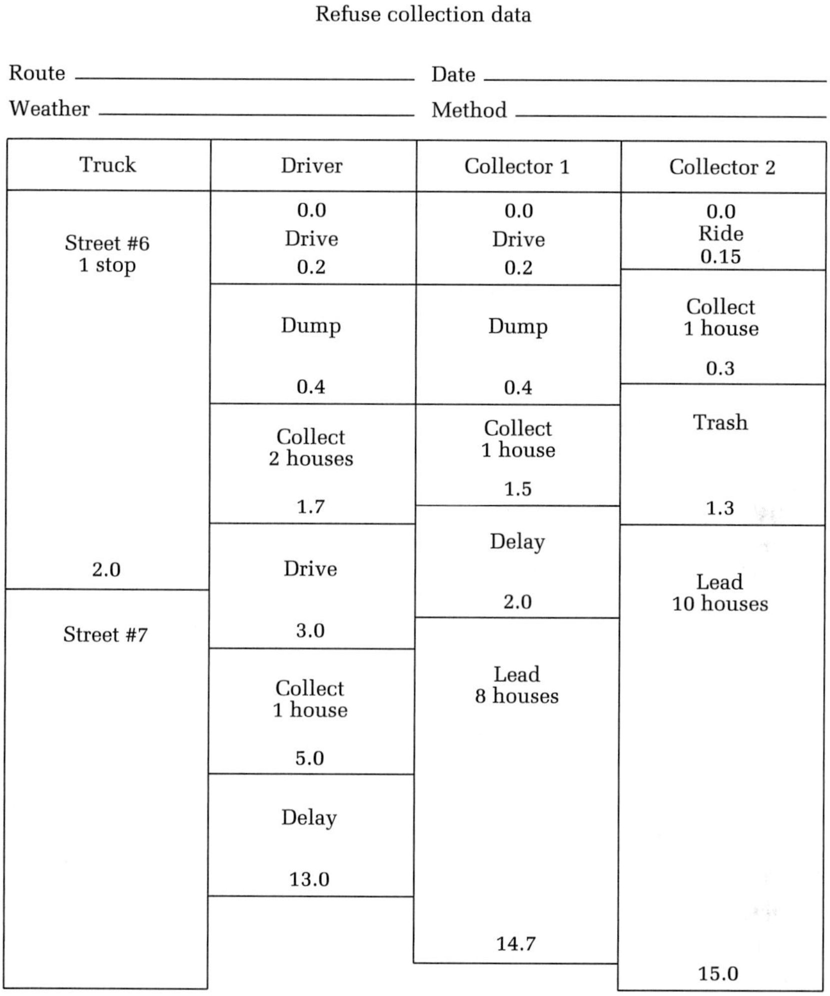

Refuse collection data

Route ______________________ Date ______________________

Weather ______________________ Method ______________________

Truck	Driver	Collector 1	Collector 2
Street #6 1 stop 2.0	0.0 Drive 0.2	0.0 Drive 0.2	0.0 Ride 0.15
Street #7	Dump 0.4	Dump 0.4	Collect 1 house 0.3
	Collect 2 houses 1.7	Collect 1 house 1.5	Trash 1.3
	Drive 3.0	Delay 2.0	Lead 10 houses 15.0
	Collect 1 house 5.0	Lead 8 houses 14.7	
	Delay 13.0		

FIGURE 8-4
Sample data collection form. (*Note:* "Crew-Machine" chart time is recorded in minutes from a start time of 0.0.)

Example 8-1. The procedure as described above was followed for a city in Florida. The city was utilizing a backyard collection method with a crew consisting of one driver and two collectors. City officials were interested in ways to reduce costs and were considering changing to curbside collection. The data were first gathered by following the crew members while they collected refuse by their normal procedure. The data were recorded on crew-machine charts and were then changed to a histogram format. Figure 8-5 shows the resulting plots of task frequency for the driver and each of the collectors.

Solution. Looking first at the driver operations, it can be seen that he only spent about 25 percent of his time actually driving. He spent almost an equal amount of his time

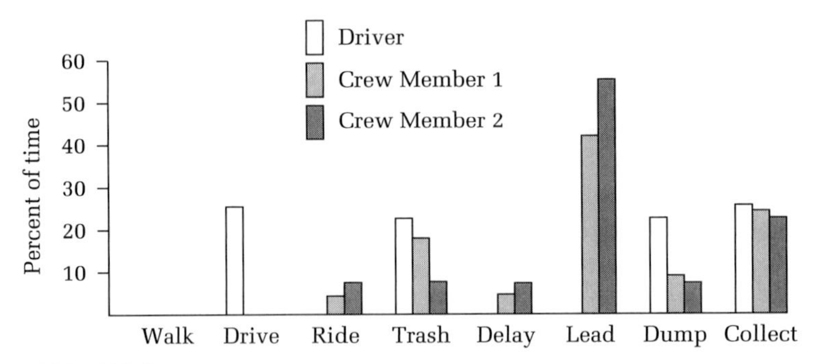

FIGURE 8-5
Operation frequency for crew on backyard method.

collecting refuse from the backyards (collect), dumping cans and bags that had been placed on the curb into the truck (dump), and in dumping yard wastes (trash). The city required that grass and tree clippings be placed in the street next to the curb. The collectors used pitchforks to dump the wastes. The mean time spent on trash collection was 2.05 minutes (standard deviation of 2.57 minutes). Occasionally as much as 30 minutes was required for one trash collection.

The first collector spent 18 percent of her time on trash, 24 percent on collect, and 42 percent on lead. This is considerable deviation from theoretical backyard collection where it would be expected that she would spend almost all her time collecting refuse from houses. This particular crew had adopted a much less regimented collection technique by spending much of their time on their own in lead operation. The average time spent on each lead was 13 minutes, but with a very high standard deviation of 20 minutes. It was not uncommon to not see one of the collectors for an hour. When her tub was full, she would put it on the curb and then carry the homeowner's cans to the street as in the set-out, set-back method.

The second collector spent even more of his time in lead (54 percent). Note that the delay time for all the crew members was very small. Considering the Florida heat, this is quite remarkable. The small delay time is probably attributable to the fact that after the crew members completed the route they were able to go home. This incentive plan also accounts for why the driver spent so much of his time assisting the crew members. His job description did not require him to do any collecting. In planning the routes the city managers did not expect this and therefore the route was often completed two to four hours early.

The collection procedure was altered and new data gathered. Homeowners were given bags and requested to put their solid waste on the curb the day of collection. All yard wastes, except large branches and the like, would be put in bags. Figure 8-6 shows the new distribution of task frequencies. Under the curbside collection method, the driver spent almost all his time (71 percent) in the truck cab, either driving or waiting for the bags to be dumped by the collectors. The collectors spent 21 percent and 35 percent, respectively, of their time riding on the truck between collection points. Most of the ride time was spent with one collector standing on the truck while the other collector was dumping bags from the curb. This indicates low efficiency.

The amount of physical labor required of the workers was much less using this collection procedure. This was a definite plus for curbside collection. In the backyard method the collectors spent about 70 percent of their time walking (lead plus collect), most of the time carrying solid waste. With the curbside method only 20 percent of their

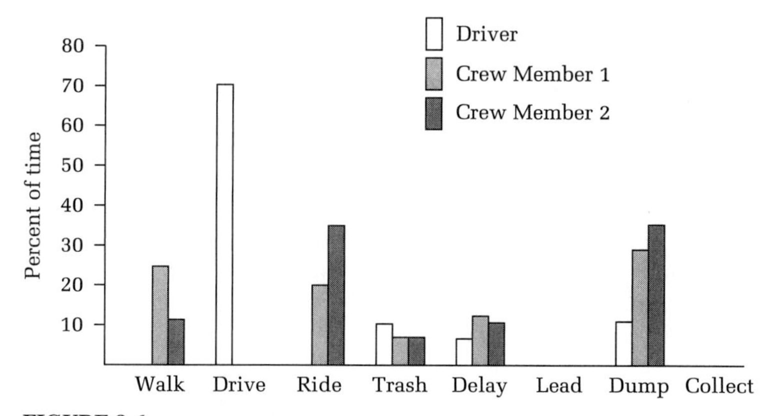

FIGURE 8-6
Operation frequency for crew on curbside collection.

time was spent walking. They carried no solid waste while walking. By placing yard wastes in bags, the trash time dropped from about 3.6 crew-hours to 1.5 crew-hours. Thus, a considerable savings was obtained by incorporating a relatively minor change.

As can be seen, a large amount of information can be gained simply from proper data collection and analysis. This particular study went on to develop a simulation program to study changes in crew size and collection procedure, but this is not always necessary.

Collection Estimates

Often, it is desirable to calculate "quick and dirty" estimates of such things as crew size, desired truck capacity, and labor and capital costs. Simple formulas have been developed that enable such calculations. The formulas are based on crude averages regarding collection times, and they make broad assumptions. An example of a not-always-justifiable assumption is that if one collector can collect a house in one minute, then two can do it in one-half minute. Several such equations follow.

Estimating truck capacity. Given that you are able to estimate a large number of factors, the following equation will allow you to estimate the volume of solid waste a truck must be able to carry.

$$\forall_T = \frac{\forall_p}{rt_p}\left[\frac{H}{N_d} - \frac{2x}{s} - 2t_d - t_u - \frac{B}{N_d}\right] \tag{8-1}$$

where $\forall_T$ = volume of solid waste carried per trip by truck at a mean density, D_T, m^3

$\forall_p$ = volume of solid waste per pickup location or stop, m^3/stop

r = compaction ratio

t_p = mean time per collection stop plus the mean time to reach the next stop, h

H = length of working day,[10] h
N_d = number of trips to the disposal site per day
x = one-way distance to disposal site, km
s = average haul speed to and from disposal site, km/h
t_d = one-way delay time, h/trip
t_u = unloading time at disposal site, h/trip
B = off route time per day, h

The factor of two in Equation 8-1 accounts for travel both to and from the disposal site. The average haul speed is a function of the total round trip distance to the disposal site (Figure 8-7). As noted in the definitions, the volume carried presumes a mean density, D_T. This is the density that results after the waste has been compacted in the truck. The compaction ratio (r) is the ratio of the density after compaction to that before compaction. Typical densities "as discarded" are given for several solid waste components in Table 8-3. If, for example, paper waste was compacted to a density of 163.4 kg/m^3, the compaction ratio would be two to one. Compactor trucks can achieve densities ranging from 300 to 600 kg/m^3.

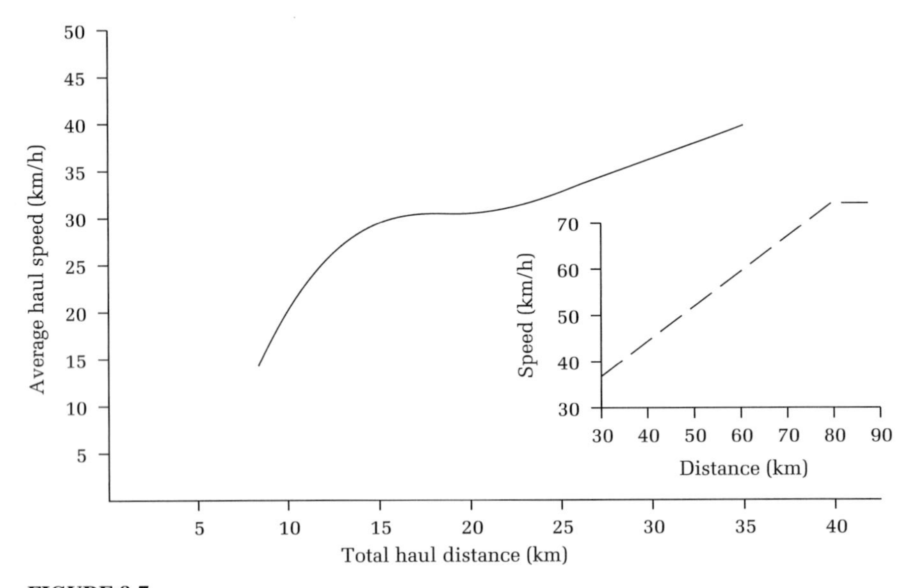

FIGURE 8-7
Effect of haul distance on average haul speed. [Adapted from *An Analysis of Refuse Collection and Sanitary Landfill Disposal* (University of California Technical Bulletin No. 8), 1952.]

[10] We should note that it is standard practice to allow two fifteen-minute breaks during the day. Since the crew is paid for this, the number of hours in the workday (H) are unchanged. However, some allowance must be made for it. Hence the off route time (B) is included in the equation.

TABLE 8-3
Typical properties of uncompacted solid waste as discarded in Davis, California

Component	Mass (kg)	Density (kg/m^3)	Volume (m^3)
Food wastes	4.3	288	0.0149
Paper	19.6	81.7	0.240
Cardboard[a]	2.95	99.3	0.0297
Plastics	0.82	64	0.013
Textiles	0.091	64	0.0014
Rubber	—	128	—
Leather	0.68	160	0.0043
Garden trimmings	6.5	104	0.063
Wood	1.59	240	0.00663
Glass	3.4	194	0.018
Tin cans	2.36	88.1	0.0268
Nonferrous metals	0.68	160	0.0043
Ferrous metals	1.95	320	0.00609
Dirt, ashes, brick	0.50	480	0.0010
Total	45.4		0.429

[a] Cardboard partially compressed by hand before being placed in container.
Source: G. Tchobanoglous, H. Theisen, R. Eliassen, *Solid Wastes Engineering Principles and Management Issues,* New York: McGraw-Hill, p. 135, 1977.

A value for t_p can be estimated from empirical data.[11] The data may be approximated by linear equations of the following form:

$$t'_p = t_{b_p} + a(C_n) + b(PRH) \tag{8-2}$$

where t'_p = mean time per collection stop plus mean time to reach next stop, min/stop

t_{b_p} = mean time between collection stops, min/stop

a, b = coefficients of regression fit to data points

C_n = mean number of containers at each pickup location

PRH = rear of house pickup locations, %

To convert t'_p to t_p, we must divide by 60 min/h.

The number of pickup locations that can be handled by a given crew is simply the available time after haul divided by the mean pickup time:

$$N_p = \frac{\frac{H}{N_d} - \frac{2x}{s} - 2_{t_d} - t_u - \frac{B}{N_d}}{t_p} \tag{8-3}$$

where N_p = number of pickup locations per load

[11] *An Analysis of Refuse Collection and Sanitary Landfill Disposal* (University of California Technical Bulletin No. 8), p. 22, 1952; and R. A. Stone, *A Study of Solid Waste Collection Systems: Comparing One Man with Multi-man Crews* (U.S. Department of Health, Education and Welfare, RPT SW-9C). Washington, DC: U.S. Government Printing Office, pp. 76, 96–98, 1969.

Example 8-2. The solid waste collection vehicle of Watapitae, Michigan is about to expire, and city officials are in need of advice on the size of truck they should purchase. The compactor trucks available from a local supplier are rated to achieve a density (D_T) of 400 kg/m^3 and a dump time of 6.0 minutes. In order to ensure once-a-week pickup the truck must service 250 locations per day. The disposal site is 6.4 km away from the collection route. From past experience, a delay time of 13 minutes can be expected. The data given in Table 8-3 have been found to be typical for the entire city. Each stop typically has three cans containing 4 kg each. About 10 percent of the stops are backyard pickups. Assume that two trips per day will be made to the disposal site. Also assume that the crew size will be two and that the empirical equation of Tchobanoglous, Theisen, and Eliassen for a two-person crew applies.[12] That equation is given as follows:

$$t'_p = 0.72 + 0.18(C_n) + 0.014\,(PRH)$$

$$t'_p = 0.72 + 0.54 + 0.14 = 1.40 \text{ min/stop}$$

$$t_p = \frac{1.40 \text{ min}}{60 \text{ min/h}} = 0.0233 \text{ h}$$

Solution. Using Table 8-3 we determine the mean density of the uncompacted solid waste to be

$$D_u = \frac{\text{Total Mass}}{\text{Total Volume}} = \frac{45.4 \text{ kg}}{0.429 \text{ m}^3} = 105.83 \text{ or } 106 \text{ kg/m}^3$$

The volume per pickup is then

$$\forall_p = \frac{(3 \text{ cans})(4 \text{ kg/can})}{106 \text{ kg/m}^3} = 0.11 \text{ m}^3$$

The compaction ratio is determined from the densities:

$$r = \frac{D_T}{D_u} = \frac{400 \text{ kg/m}^3}{106 \text{ kg/m}^3} = 3.77$$

The average haul speed is determined from Figure 8-7. Since the graph is for total haul distance, we enter with (2)(6.4) = 12.8 km and determine that s = 27 km/h. All of the other required data were given; thus, we can now use Equation 8-1. The factor of 60 is to convert minutes to hours. For two 15-minute breaks, B = 0.50.

$$\forall_t = \frac{0.11}{(3.77)(0.0233)}\left[\frac{8}{2} - \frac{(2)(6.4)}{27} - 2\frac{13 \text{ min}}{60 \text{ min/h}} - \frac{6 \text{ min}}{60 \text{ min/h}} - \frac{0.50}{2}\right]$$

$$= (1.25)(2.74) = 3.43 \text{ m}^3$$

The number of stops that can be handled is given by Equation 8-3.

$$N_p = \frac{2.74}{0.0233} = 117.60 \text{ or } 118 \text{ pickups per load}$$

[12] G. Tchobanoglous, H. Theisen, and R. Eliassen, *Solid Wastes Engineering Principles and Management Issues,* New York: McGraw-Hill, p. 135, 1977.

The smallest compactor truck available is one that will hold 4.0 m^3. Obviously, this will be satisfactory. However, the crew will not be able to reach the required 250 stops per day. Thus, some other alternative must be considered. One would be to extend the workday by 30 minutes.

Estimating costs. Most of the decisions involved in the collection of solid waste are based on economic considerations rather than technical ones. The costs are considered on the basis of a unit mass of solid waste to facilitate comparison between different size vehicles, crews, and the like. Furthermore, truck costs are considered separately from labor costs.

Truck costs include depreciation of the initial capital investment plus the *operating and maintenance (O & M)* costs.[13]

The following equation may be used to estimate the annual cost per Mg:[14]

$$A_T = \frac{1{,}000(F)}{\forall_T D_T N_T Y}\left[1 + \frac{i(Y+1)}{2}\right] + \frac{1{,}000(X_t)(OM)}{\forall_T D_T} \tag{8-4}$$

where A_T = annual truck cost, \$/Mg
F = initial (first) cost of truck, \$
D_T = mean density of solid waste in truck, kg/m^3
N_T = number of trips per year
Y = useful life of truck, y
i = interest rate on capital
X_t = distance per trip, pickup plus haul, km
OM = operating and maintenance cost, \$/km

The factor of 1,000 is to convert kg to Mg.

Labor costs consist of direct wages plus some overhead costs for such things as supervision, secretarial support, phone, utilities, insurance, and fringe benefits. Equation 8-5 can be used to estimate the annual labor cost per Mg:

$$A_L = \frac{1{,}000(CS)(W)(H)}{\forall_T D_T N_d}[1 + (OH)] \tag{8-5}$$

where A_L = annual labor cost, \$/Mg
CS = average crew size
W = average hourly wage rate, \$/h
OH = overhead as a fraction of wages

Again, the factor of 1,000 is to convert kg to Mg.

---

[13]Government-operated collection systems, by the nature of their operation, do not actually depreciate purchases. First of all, they get no tax credit for doing so and, secondly, they do not save or put aside money in a bank and therefore cannot draw interest. In spite of all this, good engineering economics demands that capital costs be depreciated in order to allow valid comparisons between alternatives.

[14]*An Analysis of Refuse Collection and Sanitary Landfill Disposal* (University of California Technical Bulletin No. 8), p. 61, 1952.

Example 8-3. Estimate the customer service charge for the situation of Example 8-2. The initial truck cost of a 4.0 m^3 compactor truck is $104,000, and the average O & M cost over the five-year life of the truck is expected to be $5.50/km. The interest rate is 8.25 percent. The average route length is 6.3 km. The average hourly wage rate is $13.50 per hour with time and a half for overtime. The overhead rate is 125 percent of the hourly wage rate.

Solution. Assuming a five-day work week and ignoring holidays, the number of trips per year would be

$$N_t = N_d(5)(52) = 2(5)(52) = 520$$

Since the average route length is 6.3 km and the average haul distance from Example 8-2 is 2(6.4) = 12.8 km, then

$$X_t = 6.3 + 12.8 = 19.1 \text{ km}$$

For the extended workday proposed at the end of Example 8-2, the volume of solid waste per trip would be

$$V_T = (1.25)(2.74 + 1/2(0.5)) = 3.74 \text{ m}^3$$

The factor of one-half times the extra half hour was selected because we assumed the time to be equally divided between each of the two trips. Note that we do not use the actual volume of the truck, which is somewhat larger than V_T. (The truck size is the nearest standard size.) Now we may compute the annualized truck cost.

$$A_T = \frac{1{,}000(104{,}000)}{(3.74)(400)(520)(5)}\left[1 + \frac{0.0825(5+1)}{2}\right] + \frac{1{,}000(19.1)(5.50)}{(3.74)(400)}$$

$$= (26.74)(1.25) + 70.22 = \$103.65/\text{Mg}$$

Since we have planned for an extra half hour of work each workday, we must adjust the hourly wage rate accordingly before we can use Equation 8-5. The adjustment is simply a determination of the weighted average rate.

$$\text{W} = \frac{(\text{reg. shift hours})(\text{wage}) + (\text{overtime hours})(\text{OT rate})(\text{wage})}{\text{total hours}}$$

$$= \frac{8(13.50) + 0.5(1.5)(13.50)}{8.5} = \$13.90/\text{h}$$

Now we may apply Equation 8-5 directly.

$$A_L = \frac{(1{,}000)(2)(13.90)(8.5)}{(3.74)(400)(2)}[1 + 1.25] = \$177.70/\text{Mg}$$

The total annual cost is then

$$A_{tot} = \$103.65 + \$177.70 = \$281.35/\text{Mg}$$

From Example 8-2, we know that each service stop averages three cans per week at 4 kg per can. Thus, each service stop contributes 3(4)(52) = 624 kg or 0.624 Mg per year. The annual cost per service stop should be ($281.35/Mg)(0.624 Mg) = $175.56. For 52 pickups per year, this is an average cost of about $3.38 per week (that is, $175.56/52).

Truck Routing

The routing of trucks may follow one of four methods. The first possibility is the daily route method. In this method the crew has a definite route that must be finished before going home. When the route is finished the crew can leave, but if necessary, they must work overtime to finish the route. This is the simplest method and the most common. The advantages of this method are as follows:

1. The homeowner knows when the refuse will be picked up.
2. The route sizes can be adjusted for the load to maximize crew and truck utilization.
3. The crew likes the method because it provides an incentive to get done early.

The disadvantages include:

1. If the route is not finished, the crew will work overtime, which will increase the expense.
2. The crew may have a tendency to become careless as they try to finish the job sooner.
3. Frequently the result is underutilization of the crew and equipment due to the increased incentive of the crew.
4. A breakdown seriously affects operations.
5. It is hard to plan routes if the load is variable, because of the disposal of yard wastes and the like.

The next method is the large route method. In this scheme the crew has enough work to last the entire week. The route must be completed in one week. The crew is left on its own to decide when to pick up the route. Usually some time off at the end of the week is the goal of the crew. This method is only good for backyard pickup since the residents don't know when pickup will be. The same advantages and disadvantages apply to this method as to the daily route method.

In the single load method, the routes are planned to get a full truck load. Each crew is assigned as many loads as it can collect per day. The biggest advantage of this method is that it can minimize travel time. The method must consider size of crew, capacity of truck, length of travel, refuse generated, and similar variables. Other advantages include:

1. A full day's work can be provided for maximum utilization of the crew and equipment.
2. It can be used for any type of pickup.

The major disadvantage is that it is hard to predict the number of homes that can be serviced before the truck is filled.

The last method is the definite working day method. As its name implies, the crew works for its assigned number of hours and quits. This method predominates in areas where unions are strong. With this method, the crew and the equipment get maximum utilization. Regularity is sacrificed with this method, and residents have little idea when pickup will occur.

Having determined the method by which the trucks will be managed, it is still necessary to find the actual route the truck will follow through the city. The purpose of routing and districting is to subdivide the community into units that will permit collection crews to work efficiently. No matter what the size of the community, it can be divided into districts, with each district constituting one day's work for the crew. The route is the detailed path of travel for the collection vehicle. The size of each route depends upon various factors as discussed earlier. The Office of Solid Waste Management Programs of the U.S. Environmental Protection Agency has developed a simple, noncomputerized "heuristic" (rule-of-thumb) approach to routing based on logical principles. The goal is to minimize deadheading, delay, and left turns. This method relies on developing, recognizing, and using certain patterns that repeat themselves in every municipality. Routing skills can be quickly acquired by applying the rules and developing experience. The following rules are taken from an EPA publication.[15]

1. Routes should not be fragmented or overlapped. Each route should be compact, consisting of street segments clustered in the same geographical area.
2. Total collection plus haul times should be reasonably constant for each route in the community (equalized workloads).
3. The collection route should be started as close to the garage or motor pool as possible, taking into account heavily traveled and one-way streets. (See rules 4 and 5.)
4. Heavily traveled streets should not be collected during rush hours.
5. In the case of one-way streets, it is best to start the route near the upper end of the street, working down it through the looping process.
6. Services on dead end streets can be considered as services on the street segment that they intersect, since they can only be collected by passing down that street segment. To keep left turns at a minimum, collect the dead end streets when they are to the right of the truck. They must be collected by walking down, backing down, or making a U-turn.
7. When practical, service stops on steep hills should be collected on both sides of the street while the vehicle is moving downhill for safety, ease, speed of collection, wear on vehicle, and conservation of gas and oil.
8. Higher elevations should be at the start of the route.
9. For collection from one side of the street at a time, it is generally best to route with many clockwise turns around blocks. (Authors' note: Heuristic rules 8 and

---

[15] K. A. Shuster and D. A. Schur, *Heuristic Routing for Solid Waste Collection Vehicles* (U.S. Environmental Protection Agency Publication No. SW-113), Washington, DC: U.S. Government Printing Office, 1974.

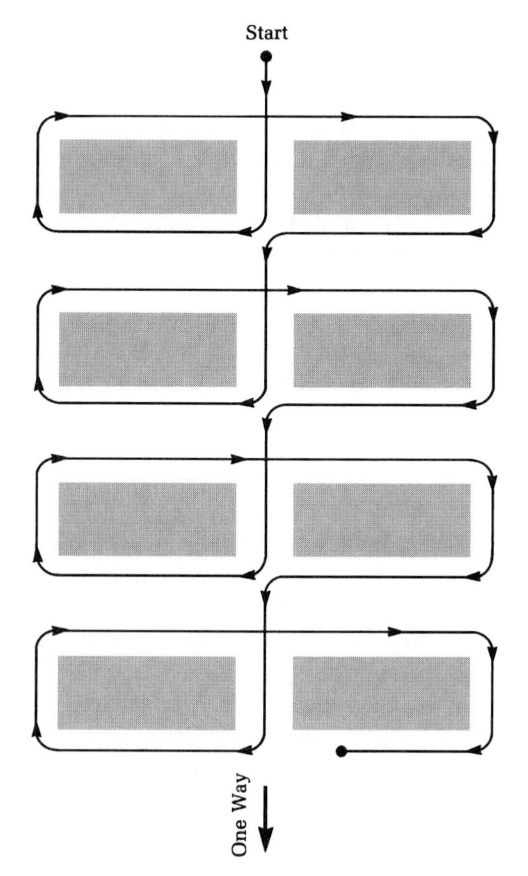

FIGURE 8-8
Arrows show heuristic routing pattern developed for a north-south, one-way street combined with east-west, two-way streets. If both sides of the one-way street cannot be collected in one pass, it is necessary to loop back to the upper end and make a straight pass down the other side. [*Source:* K. A. Shuster and D. A. Schur, *Heuristic Routing for Solid Waste Collection Vehicles* (U.S. Environmental Protection Agency Publication No. SW-113), Washington, DC: U.S. Government Printing Office, 1974.]

9 emphasize the development of a series of clockwise loops in order to minimize left turns, which generally are more difficult and time-consuming than right turns. Especially for right-hand-drive vehicles, right turns are safer.)

10. For collection from both sides of the street at the same time, it is generally best to route with long, straight paths across the grid before looping clockwise.

11. For certain block configurations within the route, specific routing patterns should be applied.

See Figure 8-8 for an example of the heuristic routing procedure.

Crew Integration

Another area of consideration is the integration of several crews. There are four ways of managing crews; usually some combination of the four is employed by any given city.

The swing crew method utilizes an extra crew as standby for heavy pickups, breakdown, or illness. Many times this crew will not report until noon to begin its day.

Crew sizes may be varied because of heavy loads, rain, different route sizes, and other factors. This is referred to as the variable crew method.

With the interroute relay method, when a crew member finishes one job, he or she is put on another route that needs additional help. This method requires more administration to operate, but results in better utilization of personnel and helps ensure that all routes will be completed during the day. Some form of this method has found wide acceptance with good results. Management must be sure that the work load is being balanced fairly and that a faster worker doesn't have to carry the load for others.

The last possibility is the reservoir route method. In this method, the crews work around a central core. When they have finished the route, the crews go to the core and begin picking up there. The core is usually an every day pickup, such as a park or a downtown area.

8-3 INTERROUTE TRANSFER

It is not always economical, or even possible, to haul the solid waste directly to the disposal site in the collection vehicle. In these cases, the solid waste is transferred from several collection vehicles to a larger vehicle, which then carries it to the disposal site. The larger vehicle (*transfer vehicle*) may be a tractor-trailer, railroad car, or barge. A special facility, called a *transfer station,* must be constructed to permit this exchange in a rapid and sanitary fashion.

Among the more important considerations in planning and designing a transfer station are location, type of station, sanitation, access, and accessories such as weighing scales and fences. The use of a transfer station may also provide for present or future resource recovery facilities.

Maximum Haul Time

As in estimating collection times, it is possible to use average values to evaluate tradeoffs in transfer station effectiveness. One such method is to compute the travel time available to the crew to travel to the disposal site and still collect the appointed route. This can be done by rearranging Equation 8-3:

$$T_H = \frac{H}{N_d} - t_p N_p - 2t_d - t_u - \frac{B}{N_d} \tag{8-6}$$

where T_H = maximum available haul time, h

If the maximum available haul time is less than the round trip distance divided by the average route speed ($2x/s$), then you have a problem. Up to a point, changes in t_d, t_u, B, and/or H may alleviate the situation.[16]

[16] Alternatively, you could provide a police escort for the collection vehicle. This might allow you to increase the average haul speed sufficiently to reduce $2x/s$ below T_H.

Economical Haul Time

The travel time in and of itself is not usually the prime consideration. Cost is usually the prime consideration. Costs are saved when a transfer operation is used because

1. The nonproductive time of collectors is reduced, since they no longer ride to and from the disposal site. It may be possible to reduce the number of collection crews needed because of increased productive collection time.
2. Any reduction in mileage traveled by the collection trucks results in a savings in operating costs.
3. The maintenance requirements for collection trucks can be reduced when these vehicles are no longer required to drive into the landfill site. Much of the damage to suspensions, drive trains, and tires occurs at landfills.
4. The capital cost of collection equipment may be reduced; since the trucks will be traveling only on improved roads, lighter duty, less expensive models can be used.[17]

In order to compare "direct haul" with "transfer" costs, the costs are computed on the basis of \$/Mg · km or, preferably, \$/Mg · min. The time-based comparison is preferred because the average haul speed of the collection vehicle will often be greater than that of the transfer vehicle. Since it is time, not distance, that costs money, this gives a fairer comparison. In addition to the travel cost of operating the transfer vehicle, there are fixed costs for the construction and operation of the transfer station and for maneuvering and unloading the transfer vehicle. Figure 8-9 may be used to estimate the cost of the transfer station.

Example 8-4. The disposal site for Watapitae will be closed in two years because of the lack of capacity. An alternative disposal site will be available when the present site is closed. It will be a county-wide regional system that will be 32.5 km from the collection route. Using the data from Examples 8-2 and 8-3 and the following assumptions, determine the maximum haul time for the collection vehicle and the cost for collection vehicle and transfer vehicle haul: $N_d = 1$, $B = 0.50$ h, and the amortized capital cost and operating cost for the transfer station is approximately \$37/Mg.

Solution. First we must determine whether or not the collection vehicle has the time to get to the disposal site while still making all of its pickups.

$$T_H = \frac{8.5}{1} - (0.0233)(250) - 2\frac{13}{60} - \frac{6}{60} - \frac{0.5}{1}$$

$$= 1.64 \text{ h or } 98.5 \text{ min}$$

We now note that the round trip distance is two times the distance from the collection route. The average haul speed can be determined from Figure 8-7. The average haul speed is 64 km/h. Thus, we find the round trip travel time to the regional

---

[17]K. Anderson, *et al.*, *Decision-Makers Guide in Solid Waste Management* (U.S. Environmental Protection Agency Publication No. SW-500), Washington, DC: U.S. Government Printing Office, p. 70, 1976.

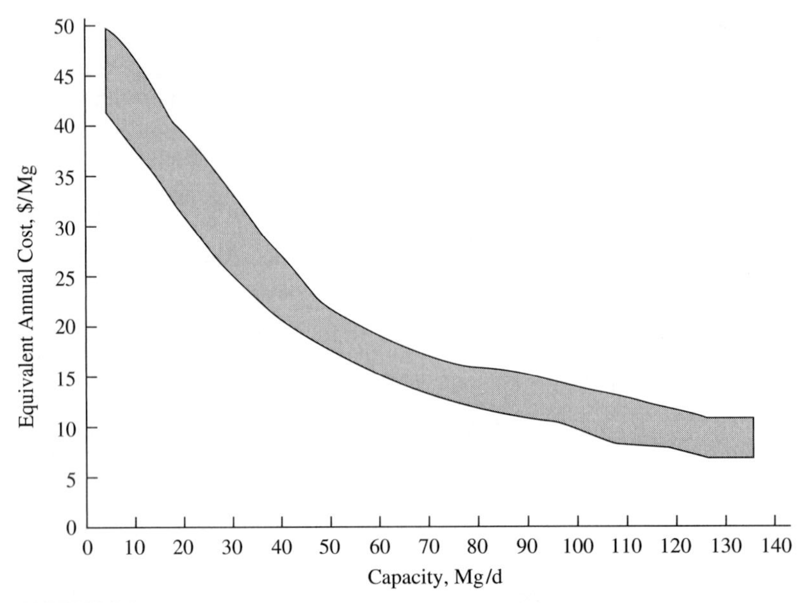

FIGURE 8-9
Transfer station equivalent annual cost as a function of capacity. Costs adjusted to 1996. (*Source:* A. J. Zuena, "Snapshot of Small Transfer Station Costs," *Waste Age,* 1987. Reprinted by permission.)

facility to be

$$\frac{2(32.5\text{ km})}{64\text{ km/h}} = 1.02\text{ h or }61\text{ min}$$

The collection vehicle can make it to the disposal site. However, since we have reduced the number of trips to the disposal site, we must either provide an additional vehicle of the same size or replace the existing one with one that is twice as large. Since the existing crew size can handle the 250 pickups per day, the more logical choice would seem to be to choose the larger vehicle. (This is especially true since the existing one is about to expire.) Let us assume the new vehicle will have a capacity of 10.0 m^3.

Now let us examine the comparative haul costs. First we will look at the collection vehicle. We will take the annual cost for a new vehicle exclusive of O & M to be \$29,851. Assuming eight hours of operation per day for five days a week for 52 weeks per year, the annual cost per minute of operation is

$$\frac{\$29{,}851/\text{y}}{(8\text{ h/d})(60\text{ min/h})(5\text{ d/w})(52\text{ wk/y})} = \$0.2392/\text{min}$$

With the effective wage rate of \$13.90 per hour from Example 8-3, the cost of wages and 125 percent overhead is

$$\frac{(\$13.90 \times 2.25)}{60\text{ min/h}} = \$0.5213/\text{min}$$

per worker or \$1.0425/min for the crew. The operating cost will be about \$5.50 per kilometer. For travel to the disposal site, the cost per minute would be

$$\frac{(\$5.50/\text{km})(32.5\text{ km})(2)}{61} = \$5.8607/\text{min}$$

The factor of two is for the round trip to the disposal site. The total haul cost per trip would be

$$61[(\$0.2392) + (\$1.0425) + (\$5.8607)] = \$435.69$$

The mass of solid waste hauled per trip is

$$(\forall_T)(D_T) = \text{mass}$$

$$(7.48 \text{ m}^3)(400 \text{ kg/m}^3) = 2{,}992 \text{ kg or } 3.0 \text{ Mg}$$

Note that the volume is twice that of a single trip (Example 8-3), but is considerably less than the capacity of the new vehicle. The unit cost of the haul would then be

$$\frac{\$435.69}{3.0 \text{ Mg}} = 145.23 \text{ or } \$145/\text{Mg}$$

Now let us look at the transfer vehicle. Assume that a tractor-trailer rig having a capacity of 46 m^3 has an annual cost exclusive of O & M of \$37,601. The cost per minute is then

$$\frac{\$37{,}601}{(8 \text{ h/d})(60 \text{ min/h})(5 \text{ d/wk})(52 \text{ wk/y})} = \$0.3013/\text{min}$$

Since the tractor-trailer rig requires an operator with higher skill, the wage rate will be higher. Using a rate of \$19.85 per hour and an overhead rate of 125 percent of wages, the cost per minute is

$$\frac{(\$19.85 \times 2.25)}{60 \text{ min/h}} = \$0.7444/\text{min}$$

In contrast to the collection vehicle, the crew is comprised of only the operator. Thus, the crew cost is \$0.7444/min.

The operating cost will be about \$6.50 per kilometer. The time for the rig to travel to the disposal site will be about 25 percent more than the collection vehicle. The travel cost would then be

$$\frac{(\$6.50)(32.5)(2)}{61 \times 1.25} = \$5.541/\text{min}$$

The total haul cost per trip would be

$$(1.25)(61)[(\$0.3013) + (\$0.7444) + (\$5.541)] = \$502.23$$

Since the capacity of the rig is four times that of the collection vehicle, the mass hauled per trip is

$$4(3.0) = 12 \text{ Mg}$$

The unit cost of the haul, including the cost of building and operating the transfer station (approximately \$37/Mg), would be

$$\frac{\$502.23}{12} + \$37 = 78.83 \text{ or } \$79/\text{Mg}$$

Obviously, consideration should be given to the construction and operation of a transfer station as an alternative to direct haul.

8-4 DISPOSAL BY SANITARY LANDFILL

A sanitary landfill is defined as a land disposal site employing an engineered method of disposing of solid wastes on land in a manner that minimizes environmental hazards by spreading the solid wastes to the smallest practical volume, and applying and compacting cover material at the end of each day.

Site Selection

Site location is perhaps the most difficult obstacle to overcome in the development of a sanitary landfill. Opposition by local citizens eliminates many potential sites. In choosing a location for a landfill, consideration should be given to the following variables:

1. Public opposition
2. Proximity of major roadways
3. Speed limits
4. Load limits on roadways
5. Bridge capacities
6. Underpass limitations
7. Traffic patterns and congestion
8. Haul distance (in time)
9. Detours
10. Hydrology
11. Availability of cover material
12. Climate (for example, floods, mud slides, snow)
13. Zoning requirements
14. Buffer areas around the site (for example, high trees on the site perimeter)
15. Historic buildings, endangered species, wetlands, and similar environmental factors.

In October of 1991, under Subtitle D of the Resource Conservation and Recovery Act (RCRA), the EPA promulgated new federal regulations for landfills.[18,19] These included siting criteria that include restrictions on distances from airports, flood plains, and fault areas, as well as limitations on construction in wetlands, seismic impact areas, and other areas of unstable geology such as landslide areas and those susceptible to sink holes. Other restrictions may apply. For example, a landfill

---

[18] *Code of Federal Regulations,* 40 CFR 257 and 258, and "Solid Waste Disposal Facility Criteria: Final Rule," *Federal Register,* 9 OCT 1991.

[19] Because municipal landfills are regulated under Subtitle D of RCRA, they are often referred to as "Subtitle D landfills."

should be more than:

30 m from streams,
160 m from drinking water wells,
65 m from houses, schools, and parks, and
3,000 m from airport runways.

Site Preparation

The plans and specifications for a sanitary landfill should require that certain steps be carried out before operations begin. These steps include grading the site area, constructing access roads and fences, and installing signs, utilities, and operating facilities.

On-site access roads should be of all-weather construction and wide enough to permit two-way truck travel (7.3 m). Grades should not exceed equipment limitations. For loaded vehicles, most uphill grades should be less than 7 percent, and downhill grades should be less than 10 percent.

All sanitary landfill sites should have electric, water, and sanitary services. Remote sites may have to use acceptable substitutes, for example, portable chemical toilets, trucked-in drinking water, and electric generators. Water should be available for drinking, fire-fighting, dust control, and sanitation. Telephone or radio communications are desirable.

A small sanitary landfill operation will usually require only a small building for storing hand tools and equipment parts and a shelter with sanitary facilities. A single building may serve both purposes. Buildings may be temporary and preferably movable.

Equipment

The size, type, and amount of equipment required at a sanitary landfill depends on the size and method of operation, quantities and time of solid waste deliveries, and, to a degree, the experience and preference of the designer and equipment operators. Another factor to be considered is the availability and dependability of service from the equipment.

The most common equipment used on sanitary landfills is the crawler or rubber-tired tractor (Figure 8-10). The tractor can be used with a dozer blade, trash blade, or a front-end loader. A tractor is versatile and can perform a variety of operations: spreading, compacting, covering, trenching, and even hauling the cover material. The decision on whether to select a rubber-tired or a crawler-type tractor, and a dozer blade, trash blade, or front-end loader must be based on the conditions at each individual site (see Table 8-4).

The crawler dozer is excellent for grading and can be economically used for dozing solid waste or soil over distances up to 100 m. The larger trash or landfill blade can be used in lieu of a straight dozer blade, thereby increasing the volume of solid waste that can be dozed. The crawler loader has the capability to lift materials off the ground for carrying. It is an excellent excavator, well suited for trench operations.

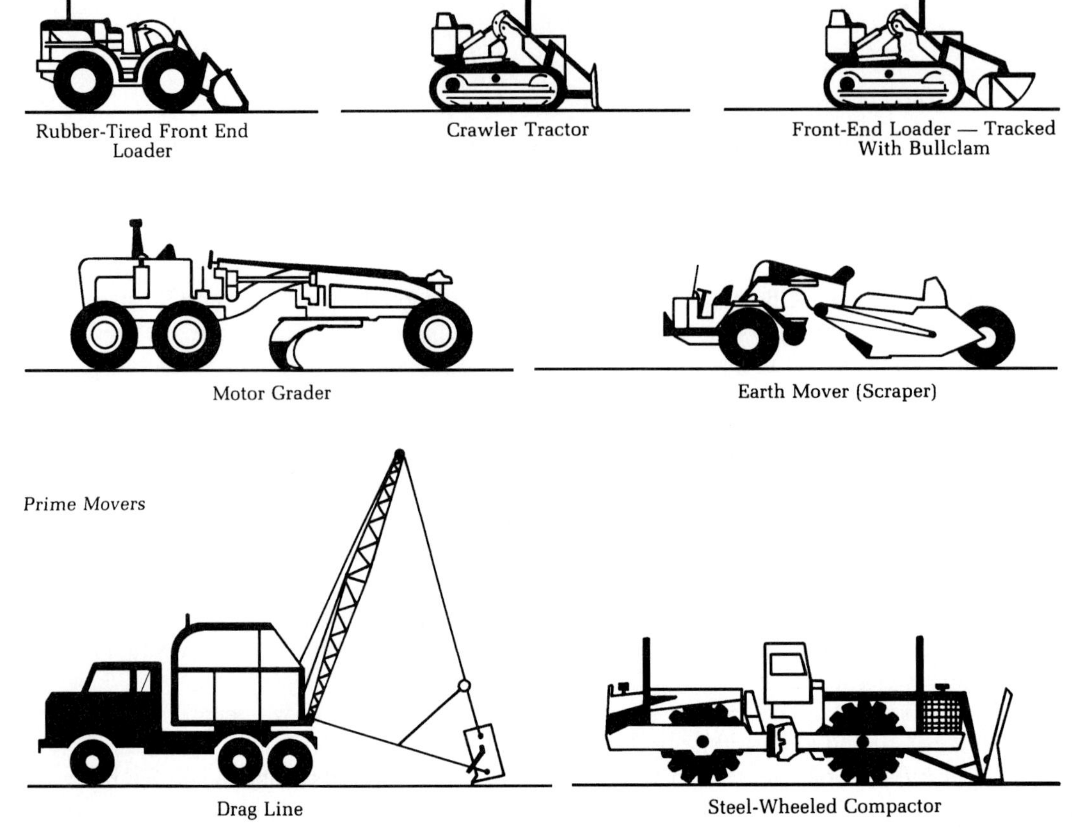

FIGURE 8-10
Sanitary landfill equipment.

Rubber-tired machines are generally faster than crawler machines. Because their loads are concentrated more, rubber-tired machines have less flotation and traction than crawler machines. Rubber-tired machines can be economically operated at distances of up to 200 m.

Steel-wheeled compactors are finding increased application at sanitary landfills. In basic design, compactors are similar to rubber-tired tractors. The unique feature of compactors is the design of their wheels, which are steel and equipped with teeth or lugs of varying shape and configuration. This design is employed to impart greater crushing and demolition forces to the solid waste. Use of compactors should be restricted to solid waste, because their design does not lend them to application of a smooth layer of compacted cover material. Thus, compactors are best used in conjunction with tracked or rubber-tired machines that can be used for cover material application.

Other equipment used at sanitary landfills are scrapers, water wagons, draglines, and graders. This type of equipment is normally found only at large sanitary landfills where specialized equipment increases the overall efficiency.

TABLE 8-4
Performance characteristics of landfill equipment[a]

Equipment	Spreading	Compacting	Excavating	Spreading	Compacting	Hauling	Density of compacted solid waste (kg/m^3)
Crawler dozer	E	G	E	E	G	NA	750
Crawler loader	G	G	E	G	G	NA	—
Rubber-tired dozer	E	G	F	G	G	NA	733
Rubber-tired loader	G	G	F	G	G	NA	—
Steel-wheeled compactor	E	E	P	G	E	NA	809
Scraper	NA	NA	G	E	NA	E	NA
Dragline	NA	NA	E	F	NA	NA	NA

[a]*Basis of evaluation:* Easily workable soil and cover material haul distance greater than 300 m.

Rating key: E, excellent; G, good; F, fair; P, poor; NA, not applicable.

Note: Density of "well-compacted" solid waste resulting from four passes over each square meter. Density measured after daily soil cover emplaced but not including soil in volume and weight measurements.

Source: Data from R. Stone and E. T. Conrad, "Landfill Compaction Equipment Efficiency," *Public Works,* pp. 111–113 and 160, May 1969.

Equipment size depends on the size of the operation. Small sanitary landfills for communities of 15,000 or less, or sanitary landfills handling 50 Mg of solid wastes per day or less, can operate successfully with one tractor in the 20 to 30 Mg range. Heavier equipment in the 30 to 45 Mg range, or larger, can handle more waste and achieve better compaction. Heavy equipment is recommended for sanitary landfill sites serving more than 15,000 people or handling more than 50 Mg per day. Sanitary landfills serving 50,000 people or less or handling no more than about 150 Mg of solid waste per day normally can manage well with one piece of heavy equipment (30 to 45 Mg range).

Operation

Although various titles are used to describe the operating methods employed at sanitary landfills, only two basic techniques are involved. They are termed the *area method* (Figure 8-11) and the *trench method* (Figure 8-12). At many sites, both methods are used, either simultaneously or sequentially.

In the area method, the solid waste is deposited on the surface, compacted, then covered with a layer of compacted soil at the end of the working day. Use of the area method is seldom restricted by topography; flat or rolling terrain, canyons, and other types of depressions are all acceptable. The cover material may come from on or off site.

The trench method is used on level or gently sloping land where the water table is low. In this method a trench is excavated; the solid waste is placed in it and compacted; and the soil that was taken from the trench is then laid on the waste and compacted. The advantage of the trench method is that cover material is readily

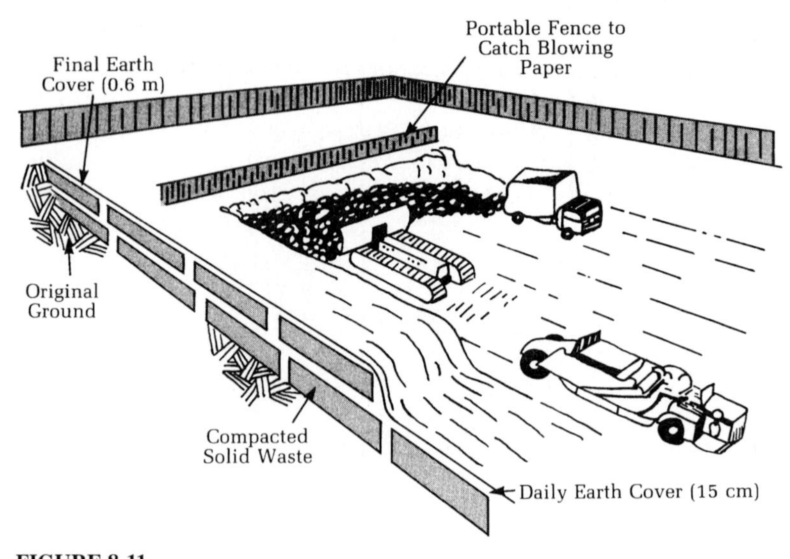

FIGURE 8-11
The area method.

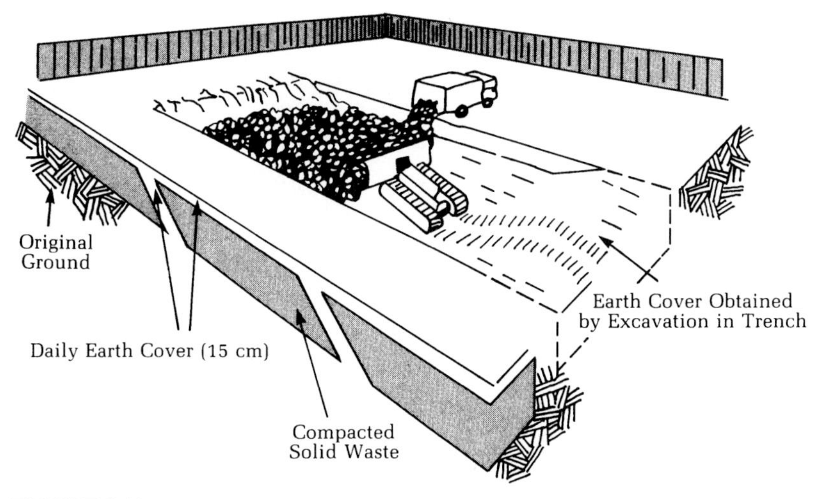

FIGURE 8-12
The trench method.

available as a result of trench excavation. Stockpiles can be created by excavating long trenches, or the material can be dug up daily. The depth depends on the location of the groundwater and/or the character of the soil. Trenches should be at least twice as wide as the compacting equipment so that the treads or wheels can compact all the material on the working area.

A sanitary landfill does not need to be operated by using only the area or trench method. Combinations of the two are possible. The methods used can be varied according to the constraints of the particular site.

A profile view of a typical landfill is shown in Figure 8-13. The waste and the daily cover placed in a landfill during one operational period form a *cell*. The operational period is usually one day. The waste is dumped by the collection and transfer vehicles onto the working *face*. It is spread in 0.4 to 0.6 m layers and compacted by driving a crawler tractor or other compaction equipment over it. At the end of each day *cover* material is placed over the cell. The cover material may be native soil or other approved materials. Recommended depths of cover for various exposure periods are given in Table 8-5. The dimensions of a cell are determined by the amount of waste and the operational period.

A *lift* may refer to the placement of a layer of waste or the completion of the horizontal active area of the landfill. In Figure 8-13 a lift is shown as the completion of the active area of the landfill. An extra layer of intermediate cover may be provided if the lift is exposed for long periods. The active area may be up to 300 m in length and width. The side slopes typically range from 1.5:1 to 2:1. Trenches vary in length from 30 to 300 m with widths of 5 to 15 m. The trench depth may be 3 to 9 m.[20]

[20]G. Tchobanoglous, H. Theisen, and S. Vigil, *Integrated Solid Waste Management,* New York: McGraw-Hill, p. 374, 1993.

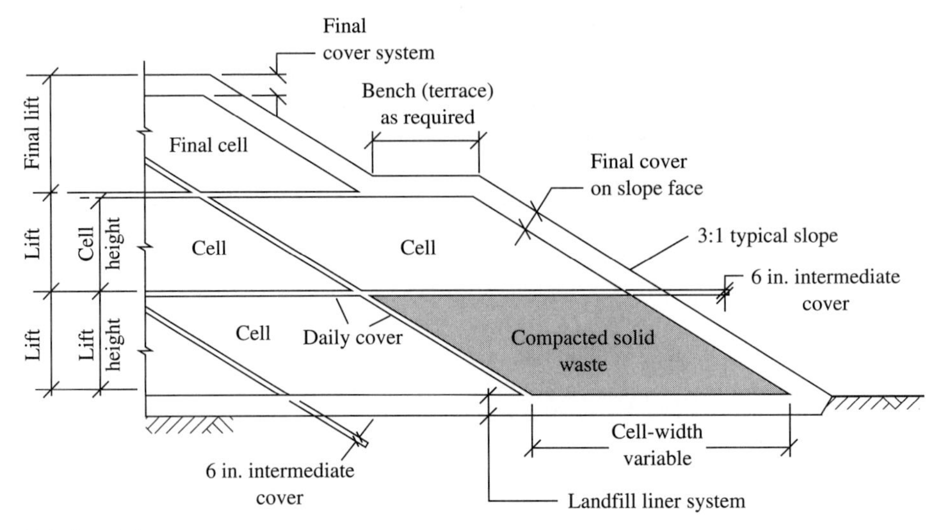

FIGURE 8-13
Sectional view through a sanitary landfill. (*Source:* G. Tchobanoglous, H. Theisen, and S. Vigil, *Integrated Solid Waste Management,* New York: McGraw-Hill, 1993. Reprinted by permission.)

TABLE 8-5
Recommended depths of cover

Type of cover	Minimum depth (m)	Exposure time (d)
Daily	0.15	< 7
Intermediate	0.30	7 to 365
Final	0.60	> 365

Benches are used where the height of the landfill exceeds 15 to 20 m. They are used to maintain the slope stability of the landfill, for the placement of surface water drainage channels, and for the location of landfill gas collection piping.

Final cover is applied to the entire landfill site after all landfilling operations are complete. A modern final cover will contain several different layers of material to perform different functions. These are discussed more fully in the landfill design section of this chapter.

Additional considerations in the operation of the landfill are those required by the 1991 Subtitle D regulations promulgated by EPA. These require exclusion of hazardous waste, use of cover materials, disease vector control, explosive gas control, air quality measurements, access control, runoff and run-on controls, surface water and liquids restrictions, and groundwater monitoring, as well as record keeping.[21]

[21] *Code of Federal Regulations,* 40 CFR 257 and 258, and "Solid Waste Disposal Facility Criteria: Final Rule," *Federal Register,* 9 OCT 1991.

Environmental Considerations

Vectors (carriers of disease) and water and air pollution should not be a problem in a properly operated and maintained landfill. Good compaction of the waste, daily covering of the solid waste with good compaction of the cover, and good housekeeping are musts for control of flies, rodents, and fires.

Burning, which may cause air pollution, is never permitted at a sanitary landfill. If accidental fires should occur, they should be extinguished immediately using soil, water, or chemicals. Odors can be controlled by covering the wastes quickly and carefully, and by sealing any cracks that may develop in the cover.

Landfill gases. The principal gaseous products emitted from a landfill (methane and carbon dioxide) are the result of microbial decomposition. Typical concentrations of landfill gases and their characteristics are summarized in Table 8-6. During the early life of the landfill, the predominant gas is carbon dioxide. As the landfill matures, the gas is composed almost equally of carbon dioxide and methane. Because the methane is explosive, its movement must be controlled. The heat content of this landfill gas mixture (16,000 to 20,000 kJ/m^3), although not as substantial as methane alone (37,000 kJ/m^3), has sufficient economic value that many landfills have been tapped with wells to collect it.

Because of their toxicity, trace gas emissions from landfills are of concern. More than 150 compounds have been measured at various landfills. Many of these may be classified as volatile organic compounds (VOCs). The occurrence of significant VOC concentrations is often associated with older landfills that previously accepted industrial and commercial wastes containing these compounds. The concentrations of 10 compounds measured in landfill gases from several California sites are shown in Table 8-7.

TABLE 8-6
Typical constituents found in MSW landfill gas

Component	Percent (dry volume basis)
Methane	45–60
Carbon dioxide	40–60
Nitrogen	2–5
Oxygen	0.1–1.0
Sulfides, disulfides, mercaptans, etc.	0–1.0
Ammonia	0.1–1.0
Hydrogen	0–0.2
Carbon monoxide	0–0.2
Trace constituents	0.01–0.06
Characteristic	**Value**
Temperature, °C	35–50
Specific gravity	1.02–1.05
Moisture content	Saturated
High heating value, kJ/m^3	16,000–20,000

Source: G. Tchobanoglous, H. Theisen and S. Vigil, *Integrated Solid Waste Management,* New York: McGraw-Hill, 1993. Reprinted by permission.

TABLE 8-7
Concentrations of specified air contaminants measured in landfill gases (in parts per billion)

	Landfill Site						
Compound	**Yolo Co.**	**City of Sacramento**	**Yuba Co.**	**El Dorado Co.**	**L.A.-Pacific (Ukiah)**	**City Of Clovis**	**City Of Willits**
Vinyl chloride	6,900	1,850	4,690	2,200	<2	66,000	7.5
Benzene	1,860	289	963	328	<2	895	<18
Ethylene dibromide	1,270	<10	<50	<1	<1	<1	<0.5
Ethylene dichloride	nr	nr	nr	<20	0.2	<20	4
Methylene chloride	1,400	54	4,500	12,900	<1	41,000	<1
Perchloroethylene	5,150	92	140	233	<0.2	2,850	8.1
Carbon tetrachloride	13	<5	<7	<5	<0.2	<5	<0.2
1,1,1-TCA[1]	1,180	6.8	<60	3,270	0.52	113	0.8
TCE[2]	1,200	470	65	900	<0.6	895	8
Chloroform	350	<10	<5	120	<0.8	1,200	<0.8
Methane	nr	nr	nr	nr	0.11%	17%	0.14%
Carbon dioxide	nr	nr	nr	nr	0.12%	24%	<0.1%
Oxygen	nr	nr	nr	nr	nr	10%	21%

nr: Not reported by operator

[1] 1,1,1-TCA: 1,1,1-trichloroethane, methyl chloroform

[2] TCE: Trichloroethene, trichloroethylene

Source: "The Landfill Gas Testing Program: A Report to the Legislature," State of California Air Resources Board, Stationary Source Division, June, 1988.

Leachate

Liquid that passes through the landfill and that has extracted dissolved and suspended matter from it is called *leachate.* The liquid enters the landfill from external sources such as rainfall, surface drainage, groundwater, and the liquid in and produced from the decomposition of the waste.

Leachate quantity. The amount of leachate generated from a landfill site may be estimated using a hydrologic mass balance for the landfill. Those portions of the global hydrologic cycle (see Chapter 2) that typically apply to a landfill site include precipitation, surface runoff, evaporation, transpiration (when the landfill cover is completed), infiltration, and storage. Precipitation may be estimated in the conventional fashion from climatological records. Surface runoff or run-on may be estimated using the rational formula (Equation 2-12). Evaporation and transpiration are often lumped together as *evapotranspiration.* It may be estimated from regional data such as that provided by the U.S. Geologic Service.[22] Infiltration (and exfiltration) may be estimated using Darcy's law (Equation 2-23). Until the landfill becomes saturated, some of the water infiltration will be stored in both the cover material and the waste. The quantity of water that can be held against the pull of gravity is referred to as *field capacity* (Figure 8-14). Theoretically, when the landfill reaches its field capacity, leachate will begin to be produced. Then, the potential quantity of leachate is the amount of moisture within the landfill in excess of the field capacity. In reality, leachate will begin to be produced almost immediately because of channeling in the waste. The following equation may be used to estimate the field capacity of the waste:[23]

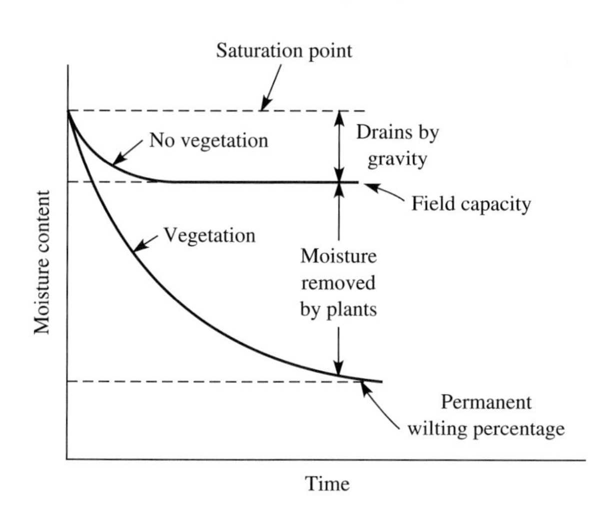

FIGURE 8-14
Moisture relationships in soil. (*Source:* J. T. Pfeffer, *Solid Waste Management Engineering,* Upper Saddle River, NJ: Prentice Hall, 1992. Reprinted by permission.)

[22] Water Resources Information Service, *Water Atlas,* U.S. Geologic Service, Washington, D.C.

[23] G. Tchobanoglous, H. Theisen, and S. Vigil, *Integrated Solid Waste Management,* New York: McGraw-Hill, p. 424, 1993.

$$FC = 0.6 - 0.55\left(\frac{2.205W}{10{,}000 + 2.205W}\right) \tag{8-7}$$

where FC = field capacity (fraction of water in the waste based on dry weight of the waste)
W = overburden mass of waste calculated at midheight of the lift in question, kg

The EPA and the Waterways Experiment Station of the U.S. Army Corps of Engineers developed a microcomputer model of the hydrologic balance called the Hydrologic Evaluation of Landfill Performance (HELP).[24] The program contains extensive data on the characteristics of various soil types, precipitation patterns, and evapotranspiration-temperature relationships as well as the algorithms to perform a routing of the moisture flow through the landfill.

Leachate composition. Solid wastes placed in a sanitary landfill may undergo a number of biological, chemical, and physical changes. Aerobic and anaerobic decomposition of the organic matter results in both gaseous and liquid end products. Some materials are chemically oxidized. Some solids are dissolved in water percolating through the fill. A range of leachate compositions is listed in Table 8-8. The VOCs in the landfill gas often contribute to contamination of groundwater because they dissolve in the leachate as it passes through the landfill. Henry's law (see Chapter 3) may be used to estimate the VOC concentrations that might occur in the leachate. Because of the differential heads (slope of the piezometric surface), the water containing dissolved substances moves into the groundwater system. The result is gross pollution of the groundwater.

Landfill Design

The design of the landfill has many components including site preparation, buildings, monitoring wells, size, liners, leachate collection system, final cover, and gas collection system. In the following discussion we will limit ourselves to introductory consideration of the design of the size of the landfill, the selection of a liner system, the design of a leachate collection system, and a discussion of the final cover system.

Volume required. To estimate the volume required for a landfill, it is necessary to know the amount of refuse being produced and the density of the in-place, compacted refuse. The volume of refuse differs markedly from one city to another because of local conditions.

---

[24]P. R. Schroeder, *et al., The Hydrologic Evaluation of Landfill Performance (HELP) Model Documentation, User's Guide,* EPA/530/SW-84-009, U.S. Environmental Protection Agency, Washington, D.C., 1984.

TABLE 8-8
Typical data of the composition of leachate from new and mature landfills

Constituent	Value, mg/L: New landfill (less than 2 years) Range	Value, mg/L: New landfill (less than 2 years) Typical	Value, mg/L: Mature landfill (greater than 10 years)
BOD_5 (5-day biochemical oxygen demand)	2,000–30,000	10,000	100–200
TOC (total organic carbon	1,500–20,000	6,000	80–160
COD (chemical oxygen demand)	3,000–60,000	18,000	100–500
Total suspended solids	200–2,000	500	100–400
Organic nitrogen	10–800	200	80–120
Ammonia nitrogen	10–800	200	20–40
Nitrate	5–40	25	5–10
Total phosphorus	5–100	30	5–10
Ortho phosphorus	4–80	20	4–8
Alkalinity as $CaCO_3$	1,000–10,000	3,000	200–1,000
pH (no units)	4.5–7.5	6	6.6–7.5
Total hardness as $CaCO_3$	300–10,000	3,500	200–500
Calcium	200–3,000	1,000	100–400
Magnesium	50–1,500	250	50–200
Potassium	200–1,000	300	50–400
Sodium	200–2,500	500	100–200
Chloride	200–3,000	500	100–400
Sulfate	50–1,000	300	20–50
Total iron	50–1,200	60	20–200

Source: G. Tchobanoglous, H. Theisen, and S. Vigil, *Integrated Solid Waste Management,* New York: McGraw-Hill, 1993. Reprinted by permission.

Salvato recommends a formula of the following form for estimating the annual volume required:[25]

$$\forall_{LF} = \frac{PEC}{D_c} \tag{8-8}$$

where $\forall_{LF}$ = volume of landfill, m^3
P = population
E = ratio of cover (soil) to compacted fill
$= \dfrac{\forall_{sw} + \forall_c}{\forall_{sw}}$
$\forall_{sw}$ = volume of solid waste, m^3
$\forall_c$ = volume of cover, m^3
C = average mass of solid waste collected per capita per year, kg/person
D_c = density of compacted fill, kg/m3

The density of the compacted fill is somewhat dependent on the equipment used at the landfill site and the moisture content of the waste. Compacted solid waste

[25] J. A. Salvato, *Environmental Engineering and Sanitation,* New York: Wiley, p. 427, 1972.

TABLE 8-9
Typical compaction ratios[a]

Component	Poorly compacted	Normal compaction	Well-compacted
Food wastes	2.0	2.8	3.0
Paper	2.5	5.0	6.7
Cardboard	2.5	4.0	5.8
Plastics	5.0	6.7	10.0
Textiles	2.5	5.8	6.7
Rubber, leather, wood	2.5	3.3	3.3
Garden trimmings	2.0	4.0	5.0
Glass	1.1	1.7	2.5
Nonferrous metal	3.3	5.6	6.7
Ferrous metal	1.7	2.9	3.3
Ashes, masonry	1.0	1.2	1.3

[a]The ratio of the density after compaction to that as discarded, that is, before pickup by collection vehicle.

Source: G. Tchobanoglous, H. Theisen, and R. Eliassen, *Solid Wastes Engineering Principles and Management Issues,* New York: McGraw-Hill, p. 347, 1977.

densities vary from 300 to 700 kg/m^3. Nominal values are generally in the range of 475 to 600 kg/m^3. The compaction ratios given in Table 8-9 may be used for estimating the density of the compacted fill.

Example 8-5. How much landfill space does Watapitae require for 20 years of operation? Assume that the village will use a cell height of 2.4 m and that it will follow normal practice and use 0.15 m of soil for daily cover; 0.3 m to complete the cell; and a final cover of 0.6 m for every stack of three cells. Assume that compaction will be "normal."

Solution. Although we do not know the population or per capita waste generation rate, we can estimate the mass generated per year from other data. From Example 8-2 we know that 1,250 service stops must be collected each week. From Example 8-3 we know that each service stop contributes an average of 0.624 Mg per year. Then the annual mass generation rate is

$$\text{Mass} = (1{,}250 \text{ stops}) \times (0.624 \text{ Mg/y stop}) = 780 \text{ Mg/y}$$

This is equivalent to the product $(P)(C)$ in Equation 8-8.

In Example 8-2 we determined that the mean density of the uncompacted solid waste was 106 kg/m^3. Using the fractional mass composition of the waste as given in Table 8-3 and the "normal" compaction ratios in Table 8-9, we can determine the weighted compaction ratio by multiplying the fractional mass by the compaction ratio (Table 8-10).

With a compaction ratio of 4.18, the density of the compacted fill is estimated to be

$$D_c = (106 \text{ kg/m}^3) \times (4.18) = 443 \text{ kg/m}^3 \text{ or } 0.443 \text{ Mg/m}^3$$

TABLE 8-10
Weighted compaction ratios for Example 8-5

Component	Mass fraction	Weighted compaction ratio
Food wastes	0.0947	0.27
Paper	0.4317	2.16
Cardboard	0.0650	0.26
Plastics	0.0181	0.12
Textiles	0.0020	0.01
Rubber	—	—
Leather	0.0150	0.05
Garden trimmings	0.1432	0.57
Wood	0.0350	0.12
Glass	0.0749	0.12
Tin cans	0.0520	0.29
Nonferrous metals	0.0150	0.08
Ferrous metals	0.0430	0.12
Dirt, ashes, brick	0.0110	0.01
Total	1.0006	4.18

Note that this implies that waste dumped at the face of the fill in a 1.25-m layer would have to be compressed to a depth of 0.3 m, that is,

$$\left(\frac{1}{4.18}\right)(1.25\text{ m})$$

Before we can estimate E, we must determine the daily volume of solid waste and the area over which it will be spread. For a five-day week, the daily volume is determined as follows:

$$V = \frac{780\text{ Mg/y}}{0.443\text{ Mg/m}^3} \times \frac{1}{52\text{ wk/y}} \times \frac{1}{5\text{ d/wk}} = 6.77\text{ m}^3\text{/d}$$

If this is spread in a 0.3-m layer, then the area would be

$$\frac{6.77\text{ m}^3}{0.3\text{ m}} = 22.57\text{ m}^2\text{/d}$$

This is equivalent to a square 4.75 m on each side. This seems reasonable for a small community.

If 0.15 m of soil is used as cover each day, then 0.45 m will be placed each day and it will take

$$\frac{2.4\text{ m} - 0.15\text{ m}}{0.45\text{ m/day}} = 5.00\text{ days}$$

to complete the cell. (The 0.15 m is the addition to daily cover to complete the cell with 0.3 m of cover.) At this rate we will complete a stack of three cells every three weeks (15 working days).

The soil volume separating a stack of three cells will be about

$$0.3\text{ m thick} \times 2.4\text{ m high} \times 4.75\text{ m long} \times 3\text{ cells} = 10.26\text{ m}^3$$

To account for two sides of the cell, this number needs to be multiplied by two.

$$10.26 \text{ m}^3 \times 2 = 20.52 \text{ m}^3$$

If we ignore this volume, E can be calculated as

$$E = \frac{0.3 + [0.15 + 0.03 + 0.02]}{0.3} = 1.67$$

The terms in the brackets account for the daily cover of 0.15 m; the cell cover of an additional 0.15 m each five days or 0.03 m per day; and the final stack cover of an additional 0.3 m to the three-cell cover each 15 days or 0.02 m per day.

If we do not ignore the soil separating the cells, then the soil volume per stack of three cells as shown in Figure 8-15 is calculated as follows:

$$(3 \text{ cells/stack})(5 \text{ lifts/cell})(22.57 \text{ m}^2)(0.15 \text{ m}) = 50.78 \text{ m}^3$$

plus the 0.15 m of additional soil to bring the weekly cell cover to 0.30 m is

$$(3 \text{ cells/stack})(22.57 \text{ m}^2)(0.15 \text{ m}) = 10.16 \text{ m}^3$$

plus the additional 0.3 m to bring the final cover to 0.6 m,

$$(22.57 \text{ m}^2)(0.3 \text{ m}) = 6.77 \text{ m}^3$$

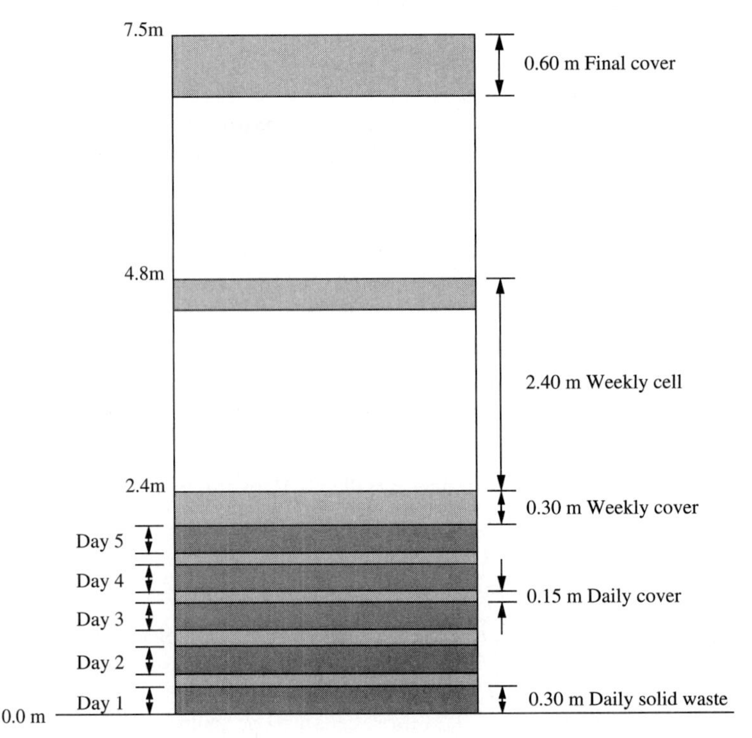

FIGURE 8-15
Schematic diagram of sanitary landfill stack of three cells (Example 8-5).

The total soil volume, including the 20.52 m^3 for the sides of the stack, is

$$50.78 + 10.16 + 6.77 + 20.52 = 88.23 \text{ m}^3$$

The value for $\forall_{sw}$ would then be

$$\forall_{sw} = (6.77 \text{ m}^3/\text{d})(15 \text{ d/stack}) = 101.55 \text{ m}^3/\text{stack}$$

The value for E would then be

$$E = \frac{101.55 + 88.23}{101.55} = 1.87$$

Thus, for this landfill, the separation wall will increase the volume by about 12 percent. This is not insignificant!

The estimated volume requirement for 20 years would be

$$\forall_{LF} = \frac{(780 \text{ Mg/y})(1.87)}{0.443 \text{ Mg/m}^3} \times 20 \text{ y} = 6.59 \times 10^4 \text{ m}^3$$

Since the average landfill depth will be three 2.4 m cells plus an additional 0.3 m final cover, the area will be

$$A_{LF} = \frac{6.59 \times 10^4}{(3)(2.4) + 0.3} = 8.78 \times 10^3 \text{ m}^2$$

An area approximately 100 m on a side would do very nicely.

Liner selection. In order to prevent groundwater contamination, strict leachate control measures are required. Under the 1991 Subtitle D rules promulgated by EPA, new landfills must be lined in a specific manner or meet maximum contaminant levels for the groundwater at the landfill boundary. The specified liner system includes a synthetic membrane (*geomembrane*) at least 30 mils (0.76 mm) thick supported by a compacted soil liner at least 0.6 m thick. The soil liner must have a hydraulic conductivity of no more than 1×10^{-7} cm/s. Flexible membrane liners consisting of high-density polyethylene (HDPE) must be at least 60 mils thick.[26] A schematic of the EPA specified liner system is shown in Figure 8-16.

Several geomembrane materials are available. Some examples include polyvinyl chloride (PVC), high-density polyethylene (HDPE), chlorinated polyethylene (CPE), and ethylene propylene diene monomer (EPDM). Designers show a strong preference for PVC and especially for HDPE. Although the geomembranes are highly impermeable (hydraulic conductivities are often less than 1×10^{-12} cm/s), they can be easily damaged or improperly installed. Damage may occur during construction by construction equipment, by failure due to tensile stress generated by the overburden, tearing as a result of differential settling of the supporting soil, puncture from sharp objects in the overburden, puncture from coarse aggregate in the supporting soil, and tearing by landfill equipment during operation. Installation

---

[26] *Code of Federal Regulations,* 40 CFR 257 and 258, and "Solid Waste Disposal Facility Criteria: Final Rule," *Federal Register,* 9 OCT 1991.

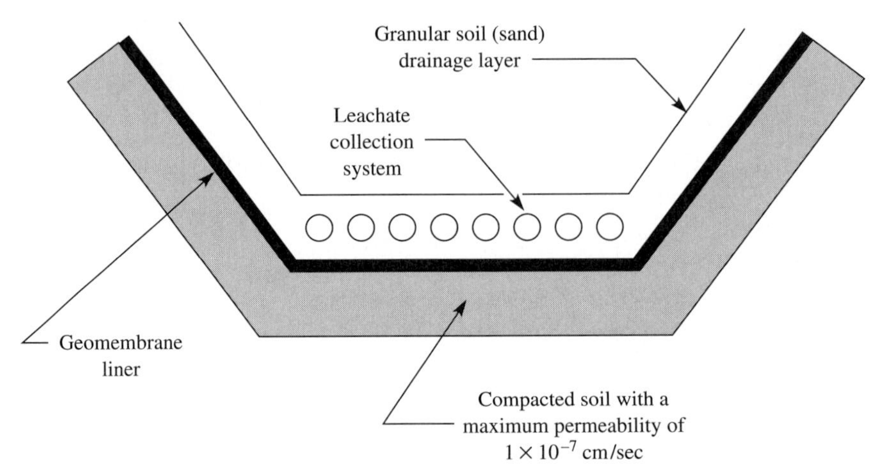

FIGURE 8-16
A composite liner and leachate collection system.

errors primarily occur during seaming when two pieces of geomembrane must be attached or when piping must pass through the liner. A liner placed with adequate quality control should have less than 3 to 5 defects per hectare.

The soil layer under the geomembrane acts as a foundation for the geomembrane and as a backup for control of leachate flow to the groundwater. Compacted clay generally meets the requirement for a hydraulic conductivity of less than 1×10^{-7} cm/s. In addition to having a low permeability, it should be: free of sharp objects greater than 1 cm in diameter, graded evenly without pockets or hillocks, compacted to prevent differential settlement, and free of cracks.

Leachate collection. Under the 1991 Subtitle D rules promulgated by EPA, the leachate collection system must be designed so that the depth of leachate above the liner does not exceed 0.3 m. The leachate collection system is designed by sloping the floor of the landfill to a grid of underdrain pipes[27] that are placed above the geomembrane. A 0.3-m-deep layer of granular material (for example, sand) with a high hydraulic conductivity (EPA recommends greater than 1×10^{-2} cm/s) is placed over the geomembrane to conduct the leachate to the underdrains. In addition to carrying the leachate, this layer also protects the geomembrane from mechanical damage from equipment and solid waste. In some instances a geonet (a synthetic matrix that resembles a miniature chain link fence), with a geofabric (an open-weave cloth) protective layer to keep out the sand, is placed under the sand and above the geomembrane to increase the flow of leachate to the pipe system.

Several different methods for estimating the steady-state maximum leachate depth have been proposed. EPA has proposed the following formula (refer to

[27]Underdrain pipes are perforated pipes designed to collect the leachate.

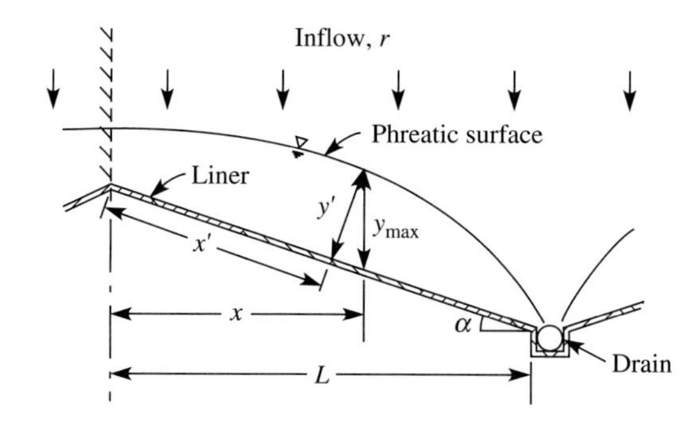

FIGURE 8-17
Geometry and symbols for calculating Y_{max}. (*Source:* B. M. McEnroe, "Maximum Saturated Depth over Landfill Liner," *Journal Environmental Engineering Division,* vol. 119, no. 2, p. 263, 1993. Reprinted by permission.)

Figure 8-17 for an explanation of the notation):[28]

$$y_{max} = L\left(\frac{r}{K}\right)\left[\frac{KS^2}{r} + 1 - \frac{KS}{r}\left(S^2 + \frac{r}{K}\right)^{0.5}\right] \tag{8-9}$$

where y_{max} = maximum saturated depth, m
L = drainage distance, measured horizontal, m
r = vertical flow rate per unit horizontal area, $m^3/s \cdot m^2$
K = hydraulic conductivity of drainage layer, m/s
S = slope of liner ($= \tan\alpha$)

This formula may overestimate the value of y_{max} where the underdrain system has free drainage, that is, it is not undersized or clogged. Since this is commonly the case, McEnroe has proposed the following equations as a better approximation:[29]

$$Y_{max} = (R - RS + R^2S^2)^{0.5}\left[\frac{(1 - A - 2R)(1 + A - 2RS)}{(1 + A - 2R)(1 - A - 2RS)}\right]^{0.5A} \tag{8-10}$$

for $R < 1/4$;

$$Y_{max} = \frac{R(1 - 2RS)}{1 - 2R}\exp\left[\frac{2R(S - 1)}{(1 - 2RS)(1 - 2R)}\right] \tag{8-11}$$

for $R = 1/4$;
and

$$Y_{max} = (R - RS + R^2S^2)^{0.5}\exp\left[\frac{1}{B}\tan^{-1}\left(\frac{2RS - 1}{B}\right) - \frac{1}{B}\tan^{-1}\left(\frac{2R - 1}{B}\right)\right] \tag{8-12}$$

for $R > 1/4$;

[28]U.S. Environmental Protection Agency, *Requirements for Hazardous Waste Landfill Design, Construction and Closure,* EPA/625/4-89/022, p. 89, 1989.

[29]B. M. McEnroe, "Maximum Saturated Depth Over Landfill Liner," *Journal of Environmental Engineering,* American Society of Civil Engineers, V. 119, pp. 262–270, 1993.

where $Y_{max} = y_{max}/(L \tan \alpha)$
$R = r/(K \sin^2 \alpha)$
S = slope of liner (= $\tan \alpha$)
$A = (1 - 4R)^{0.5}$
$B = (4R - 1)^{0.5}$

The collected leachate must be treated because of the high concentration of pollutants it contains. In some instances on-site treatment is provided. This frequently is a biological treatment system. In other cases, the leachate may be pumped to a municipal treatment plant. In some recent designs, the leachate is recirculated through the landfilled waste. This provides moisture for the microbial population and accelerates the stabilization process. It also promotes the production of methane and provides some treatment for the biodegradable fraction of the constituents in the leachate.

Final cover. The major function of the final cover is to prevent moisture from entering the finished landfill. If no moisture enters, then at some point in time the leachate production will reach minimal proportions and the chance of groundwater contamination will be minimized.

Modern final cover design consists of a surface layer, biotic barrier, drainage layer, hydraulic barrier, foundation layer, and gas control. The surface layer is to provide suitable soil for plants to grow. This minimizes erosion. A soil depth of about 0.3 m is appropriate for grass. The biotic barrier is to prevent the roots of the plants from penetrating the hydraulic barrier. At this time, there does not seem to be a suitable material for this barrier. The drainage layer serves the same function here as in the leachate collection system—that is, it provides an easy flow path to a grid of perforated pipes. This collection piping system is subject to differential settling and may fail because of this settling. Some designers do not recommend installing it as they prefer to use the funds to develop a thicker hydraulic barrier. The hydraulic barrier serves the same function as the liner in that it prevents movement of water into the landfill. The EPA recommends a composite liner consisting of a geomembrane and a low hydraulic conductivity soil that also serves as the foundation for the geomembrane. This soil also protects the geomembrane from the rough aggregate in the gas control layer. The gas control layer is constructed of coarse gravel that acts as a vent to carry the gases to the surface. If the gas is to be collected for its energy value, a series of gas recovery wells is installed. A negative pressure is placed on these wells to draw the gas into the system.

Completed Sanitary Landfills

Completed landfills generally require maintenance because of uneven settling. Maintenance consists primarily of regrading the surface to maintain good drainage and filling in small depressions to prevent ponding and possible subsequent groundwater pollution. The final soil cover should be about 0.6 m deep.

Completed landfills have been used for recreational purposes such as parks, playgrounds, or golf courses. Parking and storage areas or botanical gardens are other final uses. Because of the characteristic uneven settling and gas evolution from landfills, construction of buildings on completed landfills should be avoided.

On occasion, one-story buildings and runways for light aircraft might be constructed. In such cases, it is important to avoid concentrated foundation loading, which can result in uneven settling and cracking of the structure. The designer must provide the means for the gas to dissipate into the atmosphere and not into the structure.

8-5 WASTE TO ENERGY

Utilization of the organic fraction of solid waste for fuel, while simultaneously reducing the volume, may be an important part of an integrated waste management plan. Since 1984, 97 waste-to-energy plants have become operational. There are 159 new plants in the planning stage.[30,31]

Heating Value of Waste

The heating value of waste is measured in kilojoules per kilogram (kJ/kg), and is determined experimentally using a bomb calorimeter. A dry sample is placed in a chamber and burned. The heat released at a constant temperature of 25°C is calculated from a heat balance. Because the combustion chamber is maintained at 25°C, combustion water produced in the oxidation reaction remains in the liquid state. This condition produces the maximum heat release and is defined as the *higher heating value* (HHV).

In actual combustion processes, the temperature of the combustion gas remains above 100°C until the gas is discharged into the atmosphere. Consequently, the water from actual combustion processes is always in the vapor state. The heating value for actual combustion is termed the *lower heating value* (LHV). The following equation gives the relationship between HHV and LHV:

$$\text{LHV} = \text{HHV} - [(\Delta H_v)(9\ \text{H})] \tag{8-13}$$

where ΔH_v = heat of vaporization of water
= 2,420 kJ/kg
H = hydrogen content of combusted material

The factor of 9 results because one gram mole of hydrogen will produce 9 gram moles of water (that is, 18/2). Note that this water is only that resulting from the combustion reaction. If the waste is wet, the free water must also be evaporated. The energy required to evaporate this water may be substantial. This results in a very inefficient combustion process from the point of view of energy recovery. The ash content also reduces the energy yield because it reduces the proportion of dry organic matter per kilogram of fuel and because it retains some heat when it is removed from the furnace.

---

[30] B. Wolpin, "Go Figure," *World Wastes,* V. 37, No. 10, p. 4, October, 1994.

[31] B. Shanoff, "WTE Report Focuses on Socioeconomics," *World Wastes,* V. 37, No. 5, p. 14, May, 1994.

Fundamentals of Combustion

Combustion is a chemical reaction where the elements in the fuel are oxidized. In waste-to-energy (WTE) plants, the fuel is, of course, the solid waste. The major oxidizable elements in the fuel are carbon and hydrogen. To a lesser extent sulfur and nitrogen are also present. With complete oxidation, carbon is oxidized to carbon dioxide, hydrogen to water, and sulfur to sulfur dioxide. Some fraction of the nitrogen may be oxidized to nitrogen oxides.

The combustion reactions are a function of oxygen, time, temperature, and turbulence (O, T, T, T). There must be a sufficient excess of oxygen to drive the reaction to completion in a short period of time. The oxygen is most frequently supplied by forcing air into the combustion chamber. Over 100 percent excess air may be provided to ensure a sufficient excess. Sufficient time must be provided for the combustion reactions to proceed. The amount of time is a function of the combustion temperature and the turbulence in the combustion chamber. Some minimum temperature must be exceeded to initiate the combustion reaction (that is, to ignite the waste). Higher temperatures also yield higher quantities of nitrogen oxide emissions, so there is a tradeoff in destroying the solid waste and forming air pollutants. Mixing of the combustion air and the combustion gases is essential for completion of the reaction.

As the solid waste enters the combustion chamber and its temperature increases, volatile materials are driven off as gases. Rising temperatures cause the organic components to thermally "crack" and form gases. When the volatile compounds are driven off, fixed carbon remains. When the temperature reaches the ignition temperature of carbon (700°C), it is ignited. To achieve destruction of all the combustible material (*burn out*), it is necessary to achieve 700°C throughout the bed of waste and ash.[32]

The flame zone is that area where the hot volatilized gases mix with oxygen. This reaction is very rapid. It goes to completion within 1 or 2 seconds if there is sufficient excess air and turbulence.

The evolution of solid waste combustion has led to higher temperatures both to destroy toxic compounds and to increase the opportunity to utilize the waste as an energy source by producing steam.

Conventional Incineration

The basic arrangement of the conventional incinerator is shown in Figure 8-18. Although the solid waste may have some heat value, it is normally quite wet and is not *autogenous* (self-sustaining in combustion) until it is dried. Conventionally, auxiliary fuel is provided for the initial drying stages. Because of the large amount of particulate matter generated in the combustion process, some form of air pollution control device is required. Normally, electrostatic precipitators are chosen. Bulk

[32] J. T. Pfeffer, *Solid Waste Management Engineering,* Upper Saddle River, NJ: Prentice Hall, p. 172, 1992.

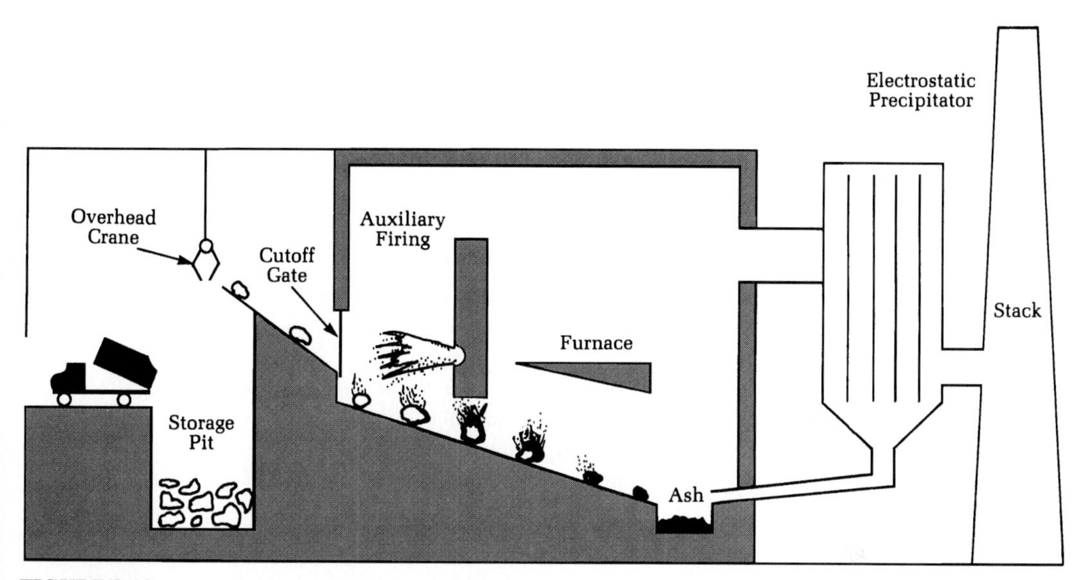

FIGURE 8-18
Schematic of a conventional traveling grate incinerator.

volume reduction in incinerators is about 90 percent. Thus, about 10 percent of the material still must be carried to a landfill.

Recovering Energy from Waste

In order to utilize the heat value of solid waste, most modern combustion devices are designed to recover the energy. The concept is more than 100 years old. The first refuse-to-electricity system was built in Hamburg, Germany, in 1896. In 1903, the first of several solid waste-fired electricity generating plants in the United States was installed in New York City.

There are now more than 160 WTE plants operating in the United States[30] They burn solid waste in a specially designed incinerator furnace jacketed with water-filled tubes to recover the heat as steam. The steam may be used directly for heating or to produce electricity.

Many states require public utilities to buy the electricity produced at these plants. With efficient heat recovery and electric generators, WTE plants can produce about 600 kWh per ton of waste.

Refuse-derived fuel (RDF). Refuse-derived fuel is the combustible portion of solid waste that has been separated from the noncombustible portion through processes such as shredding, screening, and air classifying.[33] By processing municipal solid

[33]The discussions on refuse-derived fuel, modular incinerators, and water wall combustion follows T. D. Vence and D. L. Powers, "Resource Recovery Systems, Part l, Technological Comparison," *Solid Waste Management/Resource Recovery Journal,* pp. 26–28, 32, 34, 72, 92, 93, May, 1980.

waste (MSW), refuse-derived fuel containing 12 to 16 MJ/kg can be produced from between 55 and 85 percent of the refuse received. This system is also called a supplemental fuel system because the combustible fraction is typically marketed as a fuel to outside users (utilities or industries) as a supplement to coal or other solid fuels in their existing boilers.

In a typical system, MSW is fed into a trommel or rotating screen to remove glass and dirt, and the remaining fraction is conveyed to a shredder for size reduction. Shredded wastes may then pass through an air classifier to separate the "light fraction" (plastics, paper, wood, textiles, food wastes, and smaller amounts of light metals) from the "heavy fraction" (metals, aluminum, and small amounts of glass and ceramics).

The light fraction, after being routed through a magnetic system to remove ferrous metals, is ready for fuel use. The heavy fraction is conveyed to another magnetic removal system for recovery of ferrous metals. Aluminum may also be recovered. The remaining glass, ceramics, and other nonmagnetic materials from the heavy fraction are then sent to the landfill.

The first full-scale plant to prepare RDF has been in operation in Ames, Iowa, since 1975. Subsequently, other plants using similar technology have been designed and constructed. Figure 8-19 shows the process flow diagram for the Southeastern Virginia Public Service Authority's RDF plant.

Although there are a number of RDF production systems operating or starting up, they are still developmental in terms of process, equipment, and application. Data still are being gathered for prediction of performance and maintenance requirements.

Modular incinerators. These units are available in various sizes. Their modularity enables them to be coupled with similar units to process available tonnage.

Most modular incinerators that produce energy incorporate a controlled air principle, use unprocessed MSW, and require a small amount of auxiliary fuel for startup. The waste is fed into a primary chamber where it is burned in the absence of sufficient oxygen for complete combustion. The resulting combustible gas passes through a second chamber, where excess air is injected, completing combustion. Auxiliary fuel may also be required in minimal quantities to maintain proper combustion temperatures.

After most of the particulate matter burns off, the hot effluent passes through a waste heat boiler to produce steam. The ash is water-quenched and disposed of at a landfill. The steam can be used directly or can be converted to electricity with the addition of a turbine generator.

The newer waste-to-energy plants are not without their problems. Serious concern has been raised about emissions of dioxins that result from the combustion process. Two approaches are used to reduce the dioxin emission. Because the dioxin is formed as a combustion by-product from chlorinated plastics, it can be minimized by reducing the plastic in the feed stream. The second approach is to utilize sophisticated air pollution control equipment.

A second problem is associated with the ash from the combustion process. There are two categories of ash generated: fly ash from the air pollution control equipment and bottom ash from the furnace. Fly ash is of greater concern because

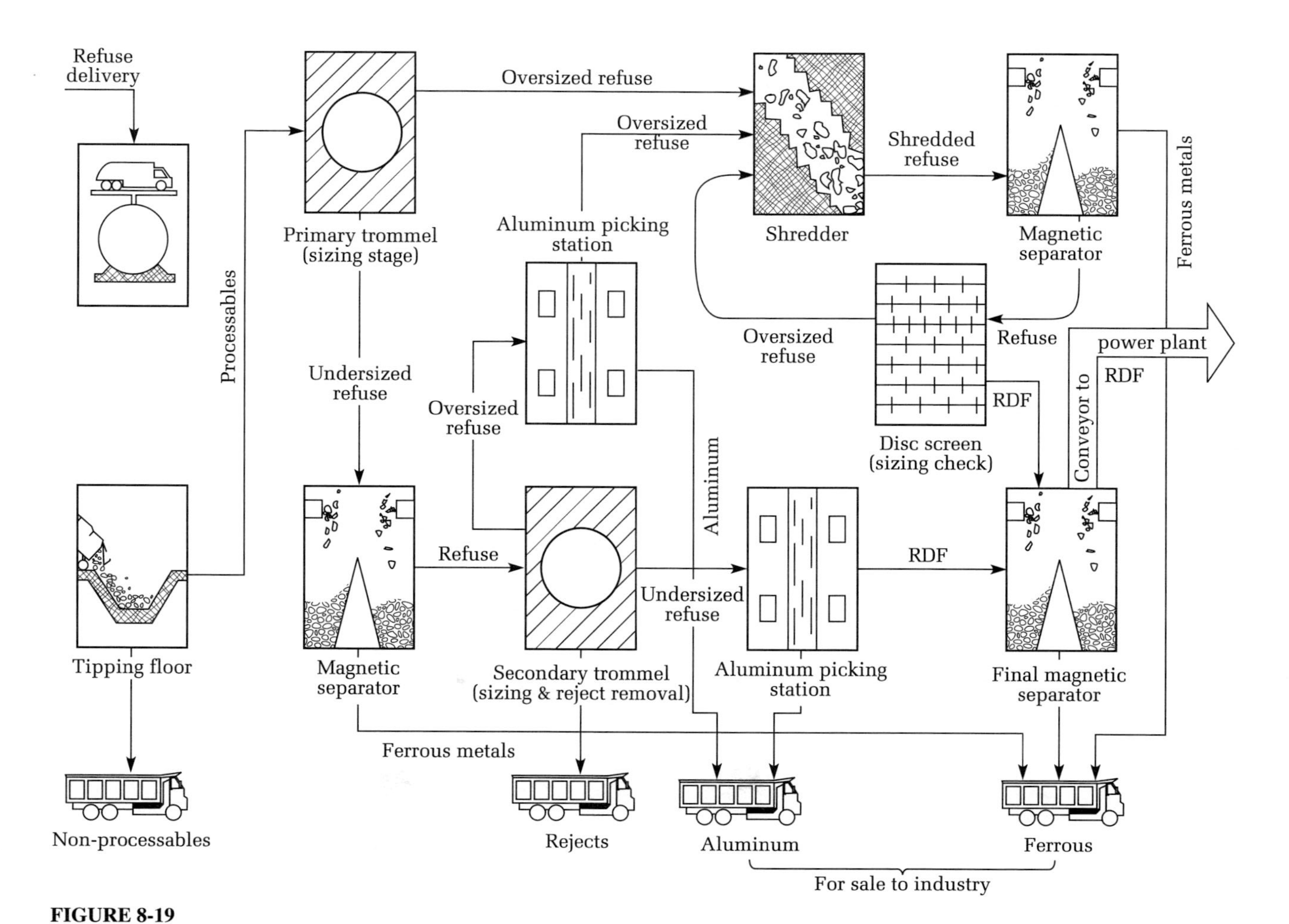

FIGURE 8-19
Southeastern Virginia Public Service Authority's refuse-derived fuel (RDF) plant.

the metals are adsorbed on particulates and are easily leached with water. When fly ash is mixed with bottom ash, the leachability of the metals is reduced. In 1994, the Supreme Court ruled that ash from municipal incinerators is not excluded from being considered as a hazardous waste.[34] It must be tested before it can be landfilled and must be treated if it fails the tests.

8-6 RESOURCE CONSERVATION AND RECOVERY

Background and Perspective

The earth's prime mineral deposits are limited. As high-quality ores are depleted, lower-grade ores must be used. Lower-grade ores require proportionately greater amounts of energy and capital investment to extract. In a broad economic context, we should view with concern the long-term reasonableness of a market-accounting system that applies only current development costs to our use of depletable, non-renewable natural resources such as aluminum, copper, iron, and petroleum. High rates of solid waste production imply high rates of virgin raw material extraction. In the United States, blatant mispricing—including the "depletion allowance" on minerals and unreasonably low rail rate fares on ores in contrast to scrap—is in no small way responsible for this state of affairs. Furthermore, our high-waste, low-recycle lifestyle is inherently wasteful of a bountiful endowment of natural resources.

Our renewable resources, primarily timber, are also under siege. Our prepackaged society, in combination with a wanton lack of care in our forests, has strained nature's capacity for growth and replenishment. Europe, India, and Japan have long been faced with a want of timber. We in the United States should learn from their predicaments.

The prevention of waste generation (resource conservation) and the productive use of waste material (resource recovery) represent means of alleviating some of the problems of solid waste management. At one time in our history, resource recovery played an important role in our industrial production. Until the mid-twentieth century, salvage (recovery and recycling) from household wastes was an increasingly important source of materials. In the five years preceding 1939, recycled copper, lead, aluminum, and paper supplied 44, 39, 28, and 30 percent, respectively, of the total raw materials shipments to fabricators in the United States.[35] Ultimately, it became more economical to process virgin materials than to use recovered materials.

In principle, processable municipal solid waste could provide 95 percent and 73 percent of our nation's needs in glass and paper, respectively. However, in the categories of metals, plastics, and wood, only a fraction of the annual needs of the country is recoverable from municipal solid waste (MSW). Much of the paper and glass that is recoverable in principle could, in practice, never be reclaimed. The

---

[34] *City of Chicago et al. v. Environmental Defense Fund, et al., No. 92-1639,* May, 1994.

[35] National Center for Resource Recovery, Inc., *Resource Recovery from Municipal Solid Waste,* Lexington, MA: Lexington Books, 1974.

Environmental Protection Agency, in 1974, estimated the actual potential recovery, expressed as a percentage of the U.S. consumption, to be as follows: iron, 6.7; aluminum, 8.4; copper, 4.7; lead, 2.8; tin, 18.9; and paper, 14.0 percent. In terms of energy production, if all the MSW collected in the United States were converted to energy, it would supply 3 to 5 percent of the nation's need. This would be enough to supply all of our residential and commercial lighting requirements.[36] Although some communities might be able to generate 7 to 10 percent or more of their need, nationally the actual potential is probably on the order of 0.5 percent.

In light of these facts, it is not difficult to see why "recovery for recovery's sake" (that is, profit) or "energy recovery for energy conservation" (that is, cheap fuel) receive scant attention except in times of national emergency or failure of the local sanitary landfill to accommodate the waste being generated. The fact that, under our current market-accounting system, resource conservation and resource recovery, except in locally favorable situations, costs more than landfilling is a further disincentive to conservation of our natural resources. Why, then, should any governing body even consider resource conservation and resource recovery? The answer, simply stated, is "to protect our environment." In the same context that we have attempted to control air and water pollution, we have an obligation to leave our descendents something better than the accumulated litter of our squandering habits. That there will be some cost should be expected. That the cost should be minimized is sound engineering practice as well as common sense. That changes in the resource depletion allowance and rail rate structure are in order is obvious. As further incentive, EPA has set as a national goal that 25 percent of all solid waste should be recycled. Many states are moving to enact this goal as a requirement.

Forty states and the District of Columbia have enacted over 338 laws on recycling ranging from purchasing preferences to comprehensive recycling goals.[37] Over 6,000 curbside recycling programs, 3,000 composting programs, and 200 municipal recycling facilities are in operation.[38] The EPA has estimated the increases in recovery for recyclables as shown in Figure 8-20. It estimates that overall recovery increased from 13 percent in 1988 to 17 percent in 1990.[39] The recyclable market continues to fluctuate dramatically. For example, the price of old newsprint fell from \$50/Mg in 1988 to less than \$10/Mg in 1993.[40] It rose to over \$100/Mg in 1995.[41]

---

[36]T. D. Vence and D. L. Powers, "Resource Recovery Systems, Part l: Technological Comparison," *Solid Wastes Management/Resource Recovery Journal,* p. 26, May 1980.

[37]M. Roberts, "State Laws for 1993 Reflect Recycling Trends," *World Wastes,* V. 37, No. 3, p. 6, March 1994.

[38]B. Wolpin, "Go Figure," *World Wastes,* V. 37, No. 10, p. 4, October 1994.

[39]U.S. Environmental Protection Agency, *Characterization of Municipal Solid Waste in the United States: 1992 Update,* EPA/530-R-92-019, Washington, D.C., 1992.

[40]M. J. Rogoff and J. F. Williams, "Marketing Efforts Aim to Close Loop," *World Wastes,* V. 38, No. 5, p. 28, May 1995.

[41]S. Paul, "Reaching Equilibrium in Recyclables Markets," *World Wastes,* V. 38, No. 8, p. 52, August 1995.

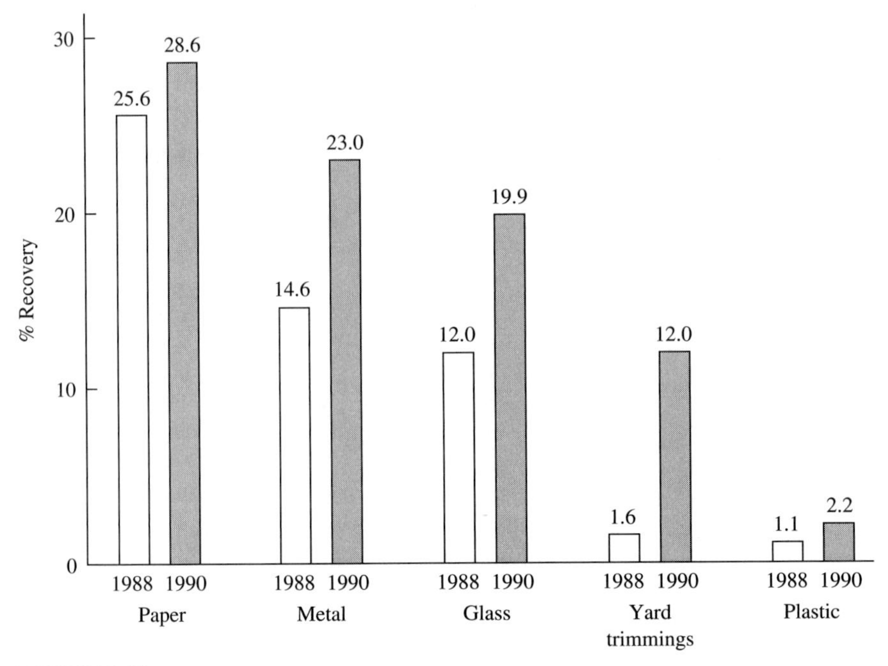

FIGURE 8-20
Recovery rates for some MSW components by mass. (*Source:* U.S. Environmental Protection Agency, *Reusable News,* Fall, 1992.)

The remainder of our discussion will be devoted to the technical details of several of the more promising resource conservation and recovery (RC & R) techniques. We have divided these into three broad categories entitled low technology, medium technology, and high technology. These categories refer to increasing degrees of sophistication in terms of implementation, equipment, and capital investment. No municipal government should be enticed into any one of these schemes with the hope of making money. The best that can be hoped for is defraying the additional costs over conventional landfilling and extending the life of the landfill by some modest amount. In some cases, even these modest goals may not be achieved.

Low Technology RC & R

Returnable beverage containers. The substitution of reusable products for single-use "disposable" products is a workable means of conserving natural resources. Legislation requiring mandatory refunds and/or deposits on both returnable and non-returnable beverage containers has been and will continue to be hotly contested by the beverage and beverage container industries. The arguments of these groups are largely without merit. In those states that have enacted mandatory refund and/or deposit legislation (California, Connecticut, Delaware, Maine, Massachusetts, Michigan, New York, Oregon, and Vermont), there is strong evidence to show that the claims of lost jobs and lost business are either unfounded or are offset by increases in jobs and business in other sectors. After almost ten years' experience in these states, we find the programs to be a great success. Between 90 and 95 percent of the bottles are returned and between 80 and 85 percent of the cans are returned. In Oregon, a reduction in total roadside litter of 39 percent by item count and 47 percent by volume was reported after the second year of implementation of its law. Furthermore, for glass containers there is a significant energy savings in that a glass bottle reused ten times consumes less than one-third of the energy of a single-use container. Average reuse cycles vary from 10 to 20 times per container.

Recycling. The reprocessing of wastes to recover an original raw material was formerly called *salvage* and is now called *recycling.* At its lowest and most appropriate technological level, the materials are separated at the source by the consumer (*source-separation*). This is the most appropriate level because it requires the minimum expenditure of energy. With stringent goals for recycling, municipalities are looking at detailed recycling options.

Generally, the recycling options available to a municipality for residential use include:

curbside collection
drop-off centers
material processing facility
material transfer stations
leaf/yard waste compost
bulky waste collection and processing
tire recovery

The primary method of recycling in the United States today is curbside collection. This has the advantage of being easier on the resident than having to drive to a recycling center. There are two basic types of curbside collection for recycling. In the first, the homeowner is given a number of bins or bags. The homeowner separates the refuse as it is used, placing it in the appropriate bin. On collection day the container is placed on the curb. The primary disadvantage of supplying home storage containers is the cost, which can represent a significant investment. A second method of curbside recycling is to provide the homeowner with only one bin, into which is placed all the recyclable materials. Curbside personnel then separate material as it is being picked up, placing each type of material into a separate compartment in the vehicle.

A second alternative is a drop-off center. Because recycling is a community-specific operation, a drop-off system must be designed around and in consideration of conditions particular to the area of involvement. To evaluate and select the most appropriate drop-off system, we must consider critical factors such as location, materials handled, population, number of centers, operation, and public information. When drop-offs are used to supplement curbside programs, fewer and smaller drop-off sites may be required. When drop-off sites are the only, or primary, recycling system in a community, the system must provide for increased capacity. Careful planning to accommodate traffic flow, as well as storage and collection of materials, must be part of the siting activity.

The convenience of a drop-off center will directly affect the amount of citizen participation. Strategically locating a drop-off center in an area of high traffic flow, where the center is highly visible, will encourage a greater level of participation. Even rural areas with widely scattered populations provide good locations for drop-offs. Rural homeowners have certain common travel patterns that bring them to a few locations at regular intervals—to a grocery store, church, or post office, to name a few. Figure 8-21 shows an example of a drive-through material recycling center.

A third major type of recycling is a materials recovery facility. In this case the recyclable material is taken by the municipality to a central facility where the material is separated via mechanical and labor-intensive means. Figure 8-22 shows an example layout of a separation facility and Figure 8-23 shows a mass balance of what can be expected at such a facility.

Medium Technology RC & R

Product design. Simple changes in product configuration or packaging can result in conservation of resources. Two examples will suffice to illustrate the concept. In the mid-1970s several newspapers (for example, *Los Angeles Times, Washington Post,* and *New York Times*) switched from a traditional eight-column format to a new six-column format for news and nine-column format for advertising. This shift resulted in a 5 percent reduction in the amount of newsprint consumed. A large retail grocery store found that it could eliminate the custom of double bagging groceries by using a slightly heavier-weight bag with a reinforced bottom. This resulted in a 30 percent savings in the amount of fiber consumed.

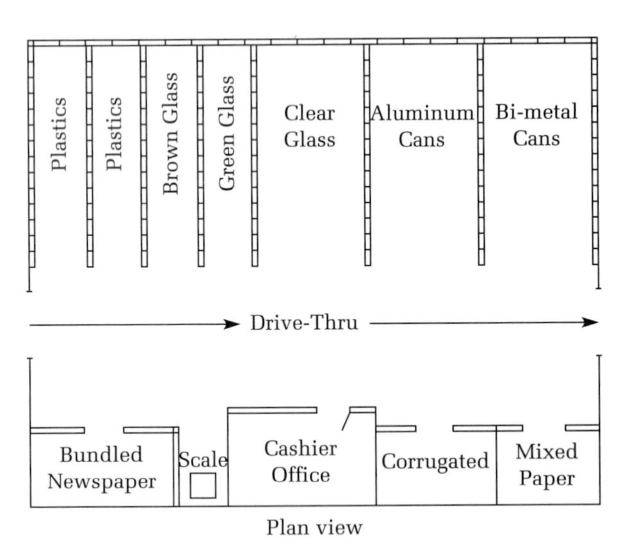

FIGURE 8-21
Enclosed drive-through drop-off center.

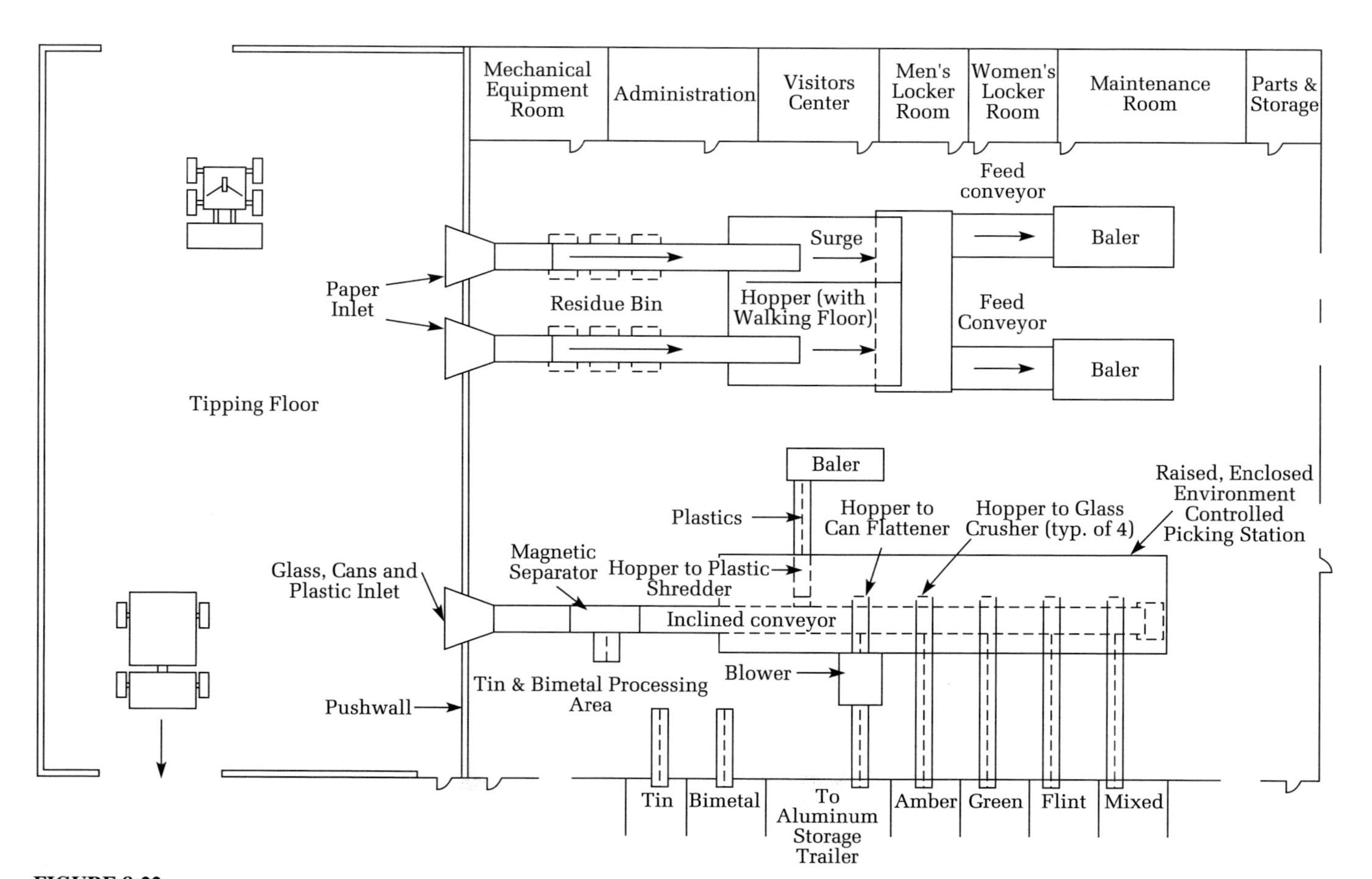

FIGURE 8-22
Material processing conceptual floor plan.

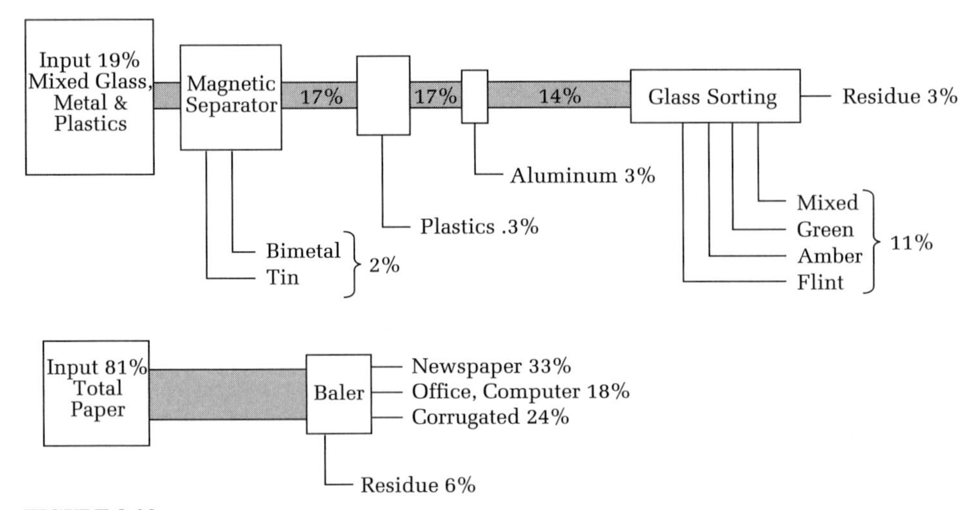

FIGURE 8-23
Material recovery facility process mass flow.

These kinds of changes are generally beyond the scope of the environmental engineer. However, their use can be encouraged, and purchases can be made that support those who use environmentally conservative packages and products.

Shredding and separation. As a first step in a medium technology system or as an add-on to a landfill volume enhancement program, some materials may be reclaimed at a central processing point. The most likely candidates for recycling are paper, nonferrous metals (for example, aluminum), and ferrous metals. Paper generally is removed by hand as the MSW passes along on a conveyor belt.[42] After passing through a shredder, ferrous metals can be removed using a magnetic separator. In large communities, where more than 1,000 Mg/wk of MSW is collected, some consideration may be given to the separation and shredding of auto and truck tires. Asphaltic concrete plants may be able to use the shredded tires in their raw material feedstock. Since tires are troublesome at landfills (because no matter how deep they are buried, they often pop up to the surface), their recovery as a resource is doubly beneficial.

Composting. Compost is a humus-like material that results from the aerobic biological stabilization of the organic materials in solid waste. The most effective composting occurs when the waste stream is free of inorganic materials. Frequently, this makes source-separated yard waste ideal. For the biological process to be effective,

[42]Depending upon the economy, hand sorting may be a losing proposition. An average worker can pick about 2.0 Mg of newspaper in an eight-hour day. At a wage of $5.50/h, a day's wages amount to $44.00, exclusive of overhead and fringe benefits. Using an overhead rate of 100 percent, the cost of sorting is $44.00/Mg. If the price for No. 6 newsprint (a grade of paper) is $22/Mg as it was in 1994, this is a loss of $22/Mg before transportation costs are deducted. Of course, in 1995, when the price was $116/Mg, it was a winning proposition.

the following conditions must be met:[43]

1. Particle size must be small (< 5 cm).
2. Aerobic conditions must be maintained by turning the compost pile or forcing air through it.
3. Adequate, but not excessive, moisture must be present (50 to 60 percent).
4. An adequate population of acclimated microorganisms must be present.
5. The carbon-to-nitrogen ratio must be in the range of 20–25 to 1.

The biodegradation process is exothermic and a well-operating compost will have a temperature between 55 and 60°C during the period of active degradation. These temperatures are effective in destroying pathogens. The processing cycle for composting is about 20 to 25 days with active degradation taking place over a 10- to 15-day period. One of the major drawbacks of composting is odors. Maintenance of aerobic conditions and a proper cure time minimize odor problems.

Compost is useful as a soil conditioner. In this role compost will: (1) improve soil structure, (2) increase moisture-holding capacity, (3) reduce leaching of soluble nitrogen, and (4) increase the buffer capacity of the soil. It should be emphasized that compost is not a valuable fertilizer. It contains only 1 percent or less of the major nutrients, such as nitrogen, phosphorus, and potash.

Composting is one of the fastest-growing aspects of ISWM. The driving force is legislation enacted to extend the life of landfills by removing yard waste from the waste stream. According to the EPA, recovery by composting was negligible in 1988. By 1990, EPA estimated that 2 percent of the nation's solid waste was being composted. The 1995 estimate was that 7 percent of the solid waste was being composted. In 1994, over 3,000 composting facilities were operating in the United States. Sludge composting facilities numbered over 180, and municipal solid waste composting was being practiced by 21 cities.[44]

Methane recovery. In 1980, 23 landfill sites were serving as a source of methane.[45] Methane is produced in sanitary landfills as a result of anaerobic decomposition of the organic fraction of the waste. In addition to gas extraction wells and a collection system, some gas processing equipment is employed. The minimum processing consists of dehydration, gas cooling, and, perhaps, removal of heavy hydrocarbons. The gas produced is a low-Joule gas having heating value of 18.6 MJ/m^3. (MJ/m^3 = megaJoule per cubic meter.) In high-Joule processing systems, carbon dioxide and some hydrocarbons are removed to yield essentially pure methane. The resulting gas

[43]G. Tchobanoglous, H. Theisen, and S. Vigil, *Integrated Solid Waste Management,* New York: McGraw-Hill, pp. 686–695, 1993.

[44]R. B. Monk, "Digging in the Dirt, Unearthing Potential," *World Wastes,* V. 37, No. 4, pp. cs1–cs14, April, 1994.

[45]R. P. Stearns, "Landfill Methane: 23 Sites are Developing Recovery Programs," *Solid Wastes Management/Resource Recovery Journal,* pp. 56–59, June 1980.

is of pipeline quality and has a heating value of approximately 37.3 MJ/m^3. The anticipated quantity of landfill gas (LFG) varies between 0.6 and 8.7 liters per kilogram of solid waste present per year (L/kg · y). The average production rate is 5 L/kg · y.

Although landfill sites as small as 11 ha have yielded substantial quantities of recoverable methane, the capital investment and complexity of the gas processing equipment will limit this technique to the larger sites (>65 ha). Otherwise, the technology is readily available and can make use of a resource that otherwise would dissipate into the atmosphere.

High Technology RC & R

In the mid-1970s, under the auspices of the U.S. Environmental Protection Agency and with federal financing, several innovative high technologies for resource recovery were examined. At the end of the decade, a few workable systems and a large number of unworkable systems were identified.

Since the successful high technology systems depend, to a large measure, on the recovery of energy for their success, we will consider the worth of solid waste as a fuel. As illustrated in Table 8-11, MSW is not a very good fuel. On the other hand, its cost of $0.00/Mg may seem quite attractive. This is especially so when the price of anthracite coal may be $50/Mg and the price of No. 2 fuel oil is $250/Mg. Unfortunately, solid waste, as a fuel, has a hidden cost. Unless the physical characteristics are upgraded by removing metals and glass and by reducing the particle size, MSW cannot be burned in conventional coal-fired power plants. The alternative is the

TABLE 8-11
Net heating value of various materials

Material	Net heating value (MJ/kg)
Charcoal	26.3
Coal, anthracite	25.8
Coal, bituminous (hi volatile B)	28.5
Fuel oil, no. 2 (home heating)	45.5
Fuel oil, no. 6 (bunker C)	42.5
Garbage	4.2
Gasoline (regular, 84 octane)	48.1
Methane[a]	55.5
Municipal solid waste (MSW)	10.5
Natural gas[a]	53.0
Newsprint	18.6
Refuse derived fuel (RDF)	18.3
Rubber	25.6
Sewage gas[a]	21.3 to 26.6
Sewage sludge (dry solids)	23.3
Trash	19.8
Wood, oak	13.3 to 19.3
Wood, pine	14.9 to 22.3

[a]Densities taken as follows (all in kg/m^3): CH_4 = 0.680; natural gas = 0.756; sewage gas = 1.05.

construction of a special power plant that can handle the MSW as it is received. In either case, some cost is imposed.

From our experiences of the 1970s, it appears that if a high technology resource recovery facility is to be successful, it must meet the following criteria:[46]

1. High technology resource recovery can only be economical in large metropolitan areas where landfill sites are unavailable or are very expensive, above $25/Mg, or in geographic locations where the water table makes safe landfilling impossible, as, for example, the city of New Orleans and its surrounding suburbs.
2. There must be an adequate refuse supply committed to the facility (a minimum of 1.8 Gg/d is needed). In general, this implies a population of 250,000 or more.
3. A customer must be obtained for the steam or the power generated by the plant and must be located close by. Firm contracts must be obtained for both the refuse supply and the sale of energy.
4. If the customer is totally dependent on the energy supplied by the facility, the combustion facility must be designed with the capacity to burn fossil fuel when refuse is unavailable or when the plant cannot process the raw refuse due to malfunctions of the processing equipment.
5. The logistics of delivering refuse to the resource recovery facility should be planned long in advance. It may be necessary to establish transfer stations and storage locations that will operate in conjunction with the resource recovery plant.
6. Systems that can dispose of both municipal refuse and sewage sludge will have economic advantages over systems that dispose of refuse only. With the ban of ocean dumping now in effect, local sewage districts are being forced to spend astronomical amounts of money to incinerate sludge. A co-disposal plant should reduce both the refuse and sludge disposal costs. In order to be economically competitive, sewage sludge must be dewatered to the maximum practical extent. A number of co-disposal plants are now in operation in Europe. Except for large installations, there will not be sufficient excess energy to warrant exporting it.

Many of the high technology systems have, as a common starting point, the medium technology materials recovery systems as their first process steps. These were discussed in a previous section.

8-7 CHAPTER REVIEW

When you have completed studying this chapter you should be able to do the following without the aid of your textbooks or notes:

1. State the average mass of solid waste produced per capita per day in the United States in 1990.

[46] A. Serper, "Resource Recovery Field Stands Poised Between Problems, Solutions," *Solid Wastes Management/Resource Recovery Journal,* p. 86, May 1980.

2. Differentiate between garbage, rubbish, refuse, and trash, based on their composition and source.
3. Compare the advantages and disadvantages of public and private solid waste collection systems.
4. List the three pickup methods (backyard, set-out/set-back, and curbside) and explain the advantages and disadvantages of each.
5. Explain the function and use of the crew-machine chart in analyzing solid waste collection.
6. Compare the advantages and disadvantages of the four methods of collection truck routing.
7. Explain the four methods of integrating several crews.
8. Explain what a transfer station is and what purpose it serves.
9. List and discuss the factors pertinent to the selection of a landfill site.
10. Describe the two methods of constructing a sanitary landfill.
11. Explain the purpose of daily cover in a sanitary landfill and state the minimum desirable depth of daily cover.
12. Define leachate and explain why it occurs.
13. Sketch a sanitary landfill that includes proper cover and a leachate collection system.
14. Define or explain the following terms: WTE, autogenous, HHV, LHV, RDF, source-separation.
15. Explain the relationship between oxygen, time, temperature, and turbulence in establishing efficient combustion reactions.
16. Explain the effect of source-separation on the heating value of solid waste and on the potential for hazardous air pollution emissions.
17. List two highly feasible methods of resource conservation and/or recovery in low technology and medium technology RC & R.
18. Describe and explain, in a basic manner, each of the two methods listed in number 17 above such that the average citizen could understand the method.

With the aid of this text you should be able to do the following:

1. Determine the volume and mass of solid waste from various establishments.
2. Determine the required volume capacity of a solid waste collection truck, or conversely, determine the number of stops possible for a given truck volume, or the allowable mean time per collection.
3. Estimate the annual truck and labor cost for solid waste collection and the cost per service stop.
4. Lay out a truck route using the heuristic routing technique.
5. Determine the necessity and/or advisability of constructing a transfer station.
6. Estimate the volume and area requirements for a landfill.
7. Compute the LHV given the HHV and the chemical formula for a compound to be burned.

8-8 PROBLEMS

8-1. The student population of a high school is 881. The school has 30 standard classrooms. Assuming a five-day school week with solid waste pickups on Wednesday and Friday before school starts in the morning, determine the size of storage container ("Dempster Dumpster") required. Assume waste is generated at a rate of 0.11 kg/cap · d plus 3.6 kg per room and that the density of uncompacted solid waste is 120.0 kg/m^3. Standard container sizes are as follows (all in m^3): 1.5, 2.3, 3.0, and 4.6.

Answer: Select one 1.5 m^3 and one 4.6 m^3 container.

8-2. The Bailey Stone Works employs six people. Assuming that the density of uncompacted waste is 480 kg/m^3, determine the annual volume of solid waste produced by the stone works assuming a waste generation rate of 1 kg/cap · d.

8-3. Professor Dexter has made measurements of her household solid waste, shown in the table below. If the container volume is 0.0757 m^3, what is the average density of the solid waste produced in her household? Assume that the mass of each empty container is 3.63 kg.

Date	Can no.	Gross mass[a] (kg)
March 18	1	7.26
	2	7.72
March 25	1	10.89
	2	7.26
	3	8.17
April 8	1	6.35
	2	8.17
	3	8.62

[a]Container plus solid waste.

Answer: Average density = 58.4 kg/m^3

8-4. Phi Systems is considering bidding on a solid waste management contract to collect all of the residential solid waste generated by Midden, Illinois (population 44,000). The average solid waste generation rate is 1.17 kg/cap · d and the average uncompacted density is 144.7 kg/m^3. The request for bids specifies that each residence must have a minimum of two pickups per week (maximum of 4 days between pickups) and that there will be no rear-of-house pickups. Using the following assumptions, determine how many trucks of what size Phi Systems should plan on using.

Assumptions for Midden, Illinois:

Average residential occupancy = 4/residence
Average number of cans per stop = 3/wk at 0.0757 m^3/can
Side loader compactor truck with a crew of one
Truck compactor density rating = 475 kg/m^3
Truck dump time = 7.50 min
Delay time = 20.0 min
Distance to disposal site = 24.0 km
Number of trips to disposal site = 2/d
Time between pickup stops = 18.00 s

Dump time (regression coefficient a) = 12.60 s/can

Standard side-loading compactor truck capacities (all in m^3): 9.0, 10.0, 12.0, 14.0, 15.0, 18.0, and 19.0

Answer: Should have 12 trucks of 9.0 m^3 capacity.

8-5. The city of Darren, California (population 361,564) has requested your assistance in evaluating its solid waste collection system. Determine the mean time per collection stop plus the mean time to reach the next stop, the number of pickup locations per load, and the minimum number of trucks the city must own.

Darren, California, Collection Data:

Average truck capacity = 18.0 m^3

Average observed compaction ratio = 3.97

Crew size = 3

Number of pickups = 1/wk (no rear-of-house service)

Average number of cans per stop = 2.53/wk at 0.1136 m^3/can

Average uncompacted density = 100.76 kg/m^3

Average transport time to disposal site including delays and dumping = 1.00 h/trip

Average number of trips to disposal site = 2/d

Rest breaks = 2 at 15.0 min

Average maintenance downtime = 24.0 min/d

Average work day = 8.00 h

Average percent of trucks out of service for major repairs = 15.0%

8-6. Rework Example 8-4 assuming no rear-of-yard pickup and only one trip per day to the disposal site.

8-7. Rework Problem 8-5 using a time between pickup stops of 28.20 s and a dump time (regression coefficient a) of 12.80 s/can for a side-loading truck and a crew of one. Assume that the truck size remains the same but the number of trips to the disposal site is reduced to one per day.

8-8. Mr. Midas, owner and manager of Phi Systems, would like to make a 20 percent profit (before taxes) on the Midden collection system work (Problem 8-4). Using the data provided by Mr. Midas, shown in the table below, determine the annual cost per megagram (Mg) and the average weekly charge to each household in order for Mr. Midas to make a 20 percent profit before taxes.

Labor costs for Midden

Employee title	Number	Wage rate, $/h
Route supervisor[a]	1	$22.00
Secretary/bookkeeper[a]	1	14.20
Mechanic[b]	1	15.58
Driver/collector	12	14.00
General laborer[a]	2	5.50

[a]Paid by overhead.

[b]Mechanic is included in O & M cost.

Average work week = 40.0 h/wk, 5 d/wk

Overhead rate = 101.38% of total driver/collector wages

Truck Data:

Size = 9.0 m^3

Capital cost = $106,628

O & M cost = $5.50/km

Anticipated life of truck = 5 y

Interest rate = 11.50%

Average annual distance = 16,412 km

8-9. In your continuing examination of the solid waste management system for Darren (Problems 8-5 and 8-7), determine the annual cost per megagram and the average weekly charge per household for a system using a crew of three and for a system using a crew of one. Use $\forall_p$ and r from 8-5, t'_p, from 8-7, and the data shown in the table below.

Labor costs for Darren

Employee title	Number	Wage rate, $/h
Director[a]	1	$38.95
Secretary[a]	1	10.00
Bookkeeper[a]	1	17.13
Route supervisors[a]	4	24.93
Senior mechanic[b]	1	28.04
Mechanic[b]	3	18.69
Crew-of-three		
Driver (1/truck)	[c]	11.25
Collector (2/truck)	[c]	11.00
Crew-of-one		
Driver/collector	[c]	11.50
General laborers[a]	4	5.50

[a]Paid by overhead.

[b]Included in O & M cost.

[c]Dependent upon number and type of trucks required as determined in Problems 8-5 and 8-7.

Average work week = 40 h/wk, 5 d/wk

Overhead rate = 67.64% of total crew wages for crew of three; 75.04% of total crew wages for crew of one.

Truck Data:

Capital cost of 27.0 m^3 truck = $125,000

O & M for 27.0 m^3 truck = $4.60/km

Capital cost of 18.0 m^3 truck = $99,000

O & M cost for 18.0 m^3 truck = $4.55/km

Truck compactor density rating = 400 kg/m^3

Anticipated life = 5 y

Interest rate = 8.25%

Average annual distance = 11,797 km

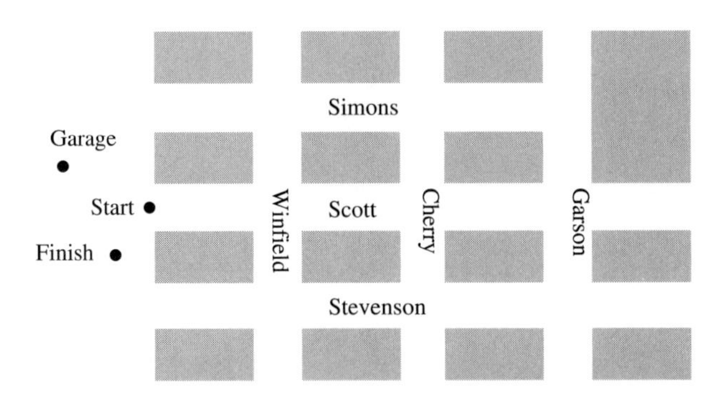

FIGURE P-8-10
Sketch Map No. 1, Redbud, Idaho.

8-10. Using the rules for heuristic routing, plan a collection route for the section of Redbud, Idaho, shown in Figure P-8-10. Assume that all streets are two-way and that the pattern is bounded by two-way streets on all four sides. Also assume that collection is on one side of the street at a time.

Answer: The solution has no dead distance and two left turns. Both left turns occur at the intersection of Simons and Garson.

8-11. Rework Problem 8-10 for the section of Mundy, Minnesota, shown in Figure P-8-11. All of the streets are two-way and the pattern is bounded by two-way streets on all four sides. Collection is from one side of the street at a time.

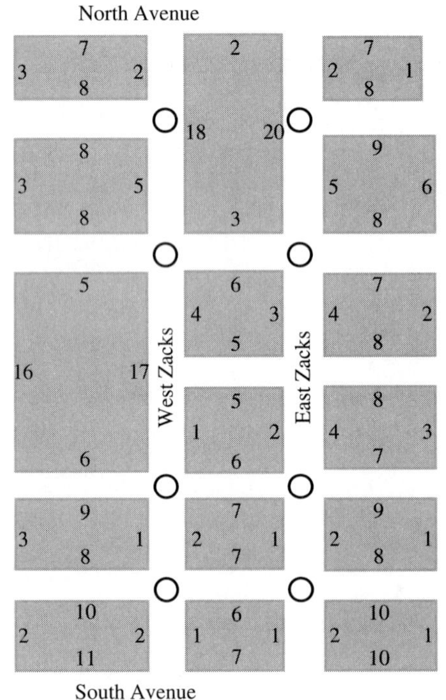

FIGURE P-8-11
Sketch Map No. 2, Mundy, Minnesota. The numbers refer to the number of stops in a block. The circles denote traffic signals.

FIGURE P-8-13
Sketch Map No. 3, Travail, Arkansas.

8-12. Rework Problem 8-11 but assuming that West Zacks is one-way "going north" and that East Zacks is one-way "going south."

Answer: The solution has three dead distances and 14 left turns. The dead distances are in the middle block of North and South Avenues. The left turns occur at the traffic signals.

8-13. Using the rules for heuristic routing, plan a collection route for the section of Travail, Arkansas, shown in Figure P-8-13. Assume collection is on one side of the street at a time and that all undesignated streets are two-way.

8-14. Divide the Olson, North Carolina, collection area shown in Figure P-8-14 into two approximately equal collection routes with starting points at A(1) and A(2). The difference in the number of stops for each route should not exceed 25. Assuming that both sides of the street can be collected in one pass, lay out the collection route that begins at A(2).

Answer: $N_p = 500$

8-15. Repeat Problem 8-14 for the Masters, Mississippi, collection area shown in Figure P-8-15 for the collection route that begins at A(1).

8-16. Write an equation for the relationship between the cost per megagram (y-axis) for hauling solid waste to a disposal site and the round trip time it takes to travel to the disposal site (x-axis) for a crew of one in a 9.0 m^3 compactor truck (see Problems 8-4 and 8-8 for data).

8-17. Repeat Problem 8-16 using a crew of three and a crew of one with 18 m^3 compactor trucks. (See Problems 8-5, 8-7, and 8-9 for data.)

8-18. The town of Trooper, Wyoming, (population 8,500) is closing its open dump and will transport its solid waste (9.53 Mg/wk) to the regional landfill at Tuppance Junction. If the one-way distance to the disposal site is 64.0 km and the crew size is one (Problem 8-16), should Trooper consider using a transfer station? Assume that the capital cost of the transfer station is \$19,000 amortized at 8.5 percent over five years, and that the annual cost of owning and operating the transfer vehicle and station is \$15,000.

Answer: No.

8-19. Calamity, Georgia, (population 35,000) generates 48,800 m^3 of solid waste at a density of 425.0 kg/m^3 each year. Four crews of three each are now averaging 1.08 h/d in haul time to the disposal site. Using the data given below, and the crew-of-three cost curves developed in Problem 8-17, determine whether or not a transfer station should be considered based on an economical analysis.

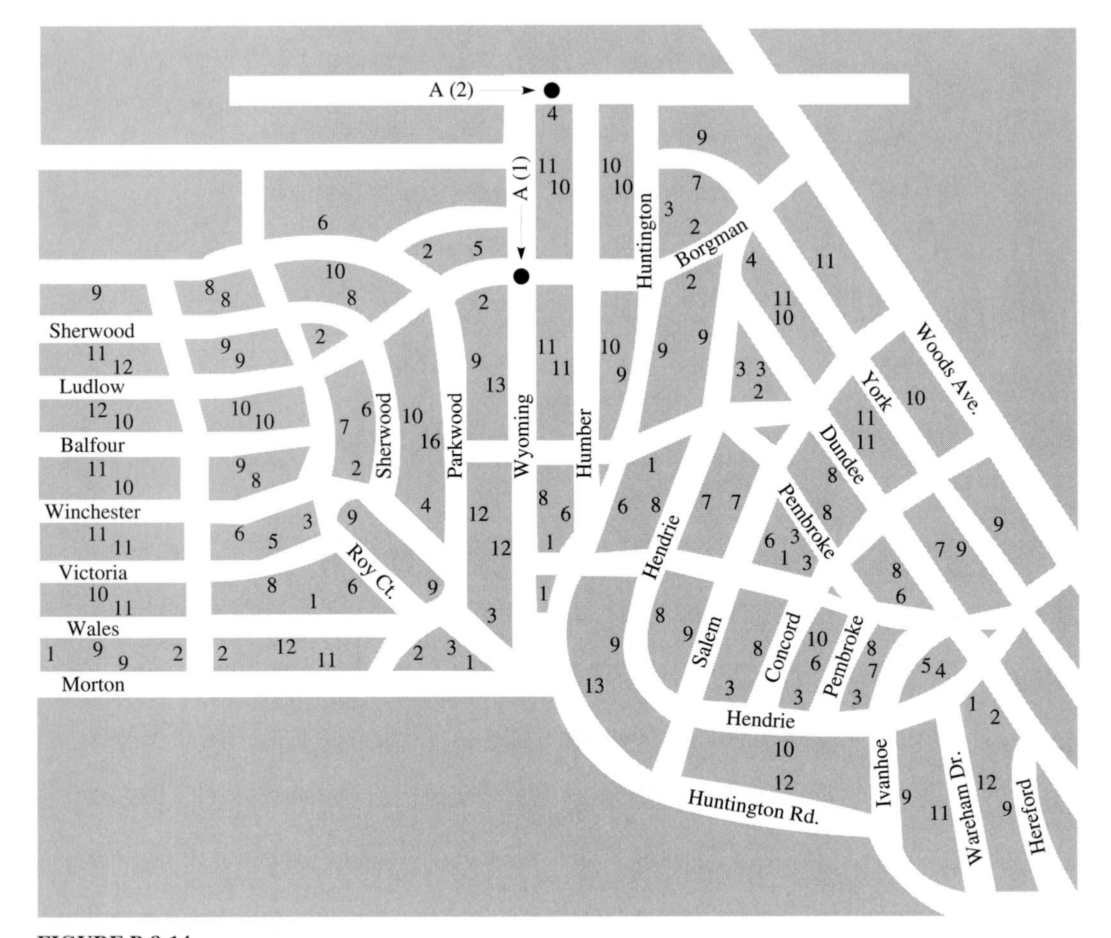

FIGURE P-8-14
Sketch Map No. 4, Olson, North Carolina. The numbers refer to the number of stops in a block.

Transfer Station Data:

Capital cost = \$750,000 at 9.25% over eight years

Transfer vehicle amortization = \$35,980/y

Operator cost (including overhead) = \$20,960/y

O & M = \$4.88/km

Round trip travel and dump time to disposal site = 1.35 h

Number of trips = 5/d

Distance to disposal site = 46.7 km

Savings from Transfer Station:

Number of collection crews will be reduced to two.

Average daily round trip haul time to transfer station for each vehicle = 20 min.

8-20. Estimate the area and volume of landfill to handle the solid waste from Midden (Problem 8-4) for 20 years. The Science Club at Midden High School has furnished the

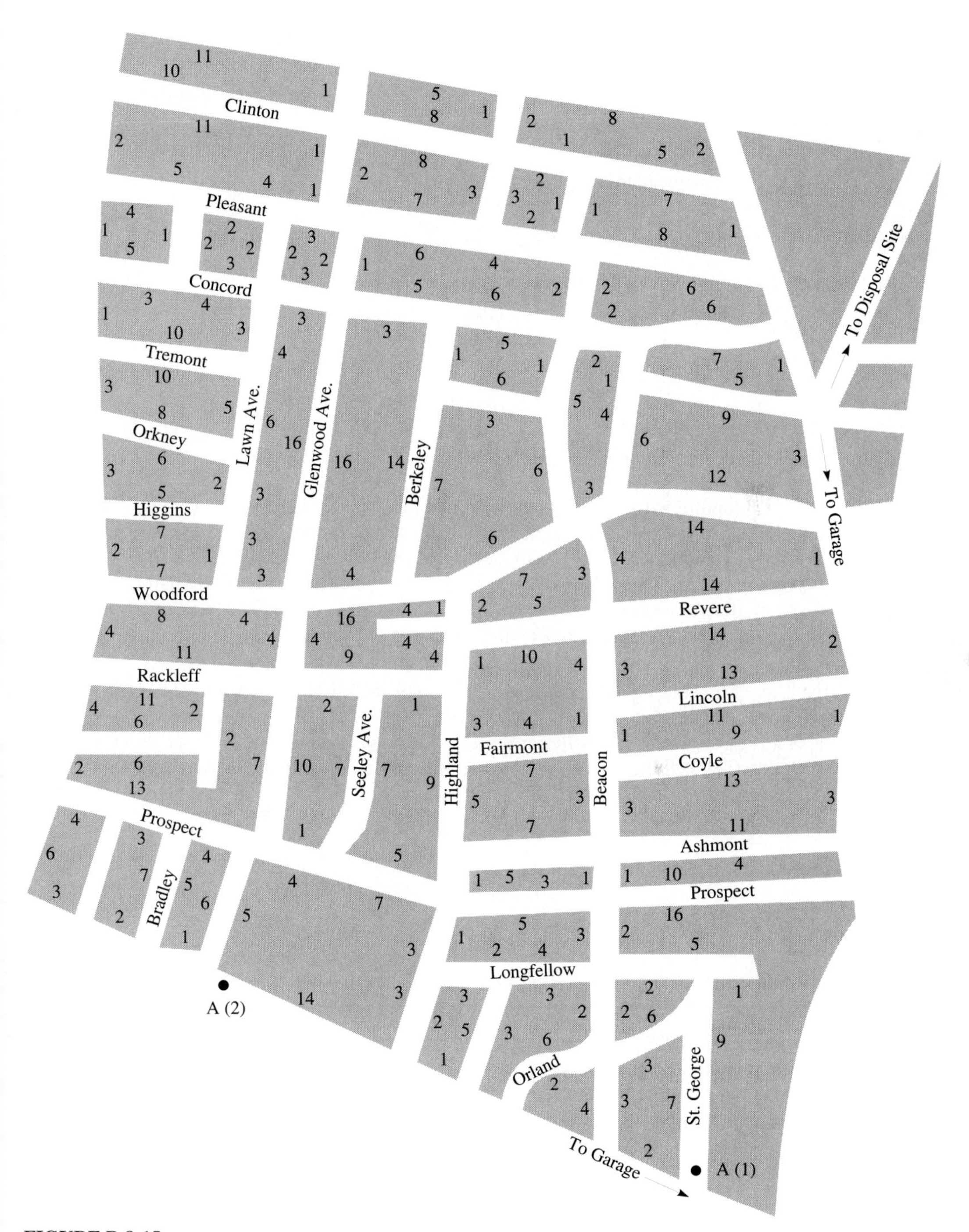

FIGURE P-8-15
Sketch Map No. 5, Masters, Mississippi. The numbers refer to the number of stops in a block.

following data based on a 12-month survey. (One sample having a mass of 1.000 Mg was taken at the existing landfill during normal off-loading operations one day each month.) Assume a cell height of 2.40 m and that the recommended depths of cover will be used and that compaction will be normal.

Characterization of Midden solid waste

Component	Mass fraction
Food waste	0.0926
Paper	0.4954
Plastics, rubber, leather	0.0438
Textiles	0.0379
Metals	0.0741
Glass	0.1668
Miscellaneous	0.0894
Total	1.0000

8-21. A sanitary landfill is being designed to handle solid waste generated by Binford, Vermont, at a rate of 50 Mg/d. It is expected that the waste will be delivered by compactor truck on a 5 d/week basis. The density as spread is 122 kg/m^3. It will be spread in 0.50 m layers and compacted to 0.25 m. Assuming three such lifts per day and a daily cover of 0.15 m, determine the following: (a) annual volume of landfill consumed in m^3, and (b) daily horizontal area covered by the solid waste. Ignore the soil volume between stacks.

8-22. The City of Darren (Problem 8-5) is considering instituting a recycling program in which the residents presort the solid waste into four components: (1) mixed waste, (2) paper, (3) glass, and (4) metallics. From a research study report, we find that the mean time per collection stop plus the mean time to reach the next stop (t_p) can be estimated from the following equation:[47]

$$t_p = 22.6 + 3.80R + 5.50S$$

where t_p = mean collection time, s
R = number of units of mixed waste per stop
S = sum of the number of units of separated paper, glass, and metallics per stop

Assuming that $S = 3.00$ and $R = 1.53$, rework Problems 8-5 and 8-9 to determine what savings in disposal cost is needed to offset the additional cost of collection for a crew of three.

8-23. Rework Example 8-5 assuming that 50 percent of the paper and 80 percent of the glass and metal are separated at the source and recycled.

8-24. The higher heating value for cellulose ($C_6H_{10}O_5$) is 32,600 kJ/kg. Compute the lower heating value.

[47]Richard Tichenor, "Designing a Vehicle to Collect Source-Separated Recyclables," *Compost Science/Land Utilization,* Vol. 21 (1), pp. 36–41, January/February 1980.

8-25. The higher heating value for methane (CH_4) is 888,500 kJ/kg. Compute the lower heating value.

8-9 DISCUSSION QUESTIONS

8-1. What is the effect of crew size on the mean time per collection stop (t'_p)? How does the container location affect the mean time per collection stop?

8-2. Under what conditions would you recommend consideration of a transfer station?

8-3. Which of the following soil types would be suitable for (a) composite liner, (b) drainage layer, (c) gas venting:
(1) Gravel (>2.5 cm diameter)
(2) Glacial till
(3) Clay ($K = 1 \times 10^{-9}$ cm/s)
(4) Clay ($K = 1 \times 10^{-6}$ cm/s)
(5) Sand ($K = 0.1$ cm/s)
(6) Sand ($K = 0.001$ cm/s)

8-4. A WTE plant is being proposed as part of an ISWM plan. The proponents of the WTE argue that recycling is not necessary and will have no effect on the performance of the plant. Do you agree or disagree? Explain.

8-5. Although the market value of compost is negligible, many communities have implemented yard waste composting systems. Explain why.

8-10 ADDITIONAL READING

Books

D. J. Hagerty, J. L. Pavoni, and J. E. Heer, Jr., *Solid Waste Management,* New York: Van Nostrand Reinhold, 1973.

J. T. Pfeffer, *Solid Waste Management Engineering,* Englewood Cliffs, NJ: Prentice Hall, 1992.

C. R. Rhyner, *et al., Waste Management and Resource Recovery,* Boca Raton, FL: CRC Lewis Publishers, 1995.

J. A. Salvato, *Environmental Engineering and Sanitation,* New York: Wiley, 1972.

G. Tchobanoglous, H. Theisen, and R. Eliassen, *Solid Wastes, Engineering Principles and Management Issues,* New York: McGraw-Hill, 1993.

G. Tchobanoglous, H. Theisen, and S. Vigil, *Integrated Solid Waste Management,* New York: McGraw-Hill, Inc., 1993.

P. A. Vesilind and A. E. Rimer, *Unit Operations in Resource Recovery Engineering,* Englewood Cliffs, NJ, Prentice Hall, 1981.

Journals

Waste Age
World Wastes

CHAPTER 9

HAZARDOUS WASTE MANAGEMENT

9-1 THE HAZARD
Dioxins and PCBs
9-2 RISK
Risk Perception
Risk Assessment
Risk Management
9-3 DEFINITION AND CLASSIFICATION OF HAZARDOUS WASTE
EPA's Hazardous Waste Designation System
Ignitability
Corrosivity
Reactivity
Toxicity
9-4 RCRA AND HSWA
Congressional Action on Hazardous Waste
Cradle-to-Grave Concept
Generator Requirements
Transporter Regulations
Treatment, Storage, and Disposal Requirements
Underground Storage Tanks (UST)
9-5 CERCLA AND SARA
The Superfund Law
The National Priority List (NPL)
The Hazard Ranking System (HRS)
The National Contingency Plan (NCP)

Liability
Superfund Amendments and Reauthorization Act (SARA)
9-6 HAZARDOUS WASTE MANAGEMENT
Waste Minimization
Waste Exchange
Recycling
9-7 TREATMENT TECHNOLOGIES
Biological Treatment
Chemical Treatment
Physical/Chemical Treatment
Incineration
Stabilization/Solidification
9-8 LAND DISPOSAL
Deep Well Injection
Land Treatment
The Secure Landfill
9-9 GROUNDWATER CONTAMINATION AND REMEDIATION
The Process of Contamination
EPA's Groundwater Remediation Procedure
Mitigation and Treatment
9-10 CHAPTER REVIEW
9-11 PROBLEMS
9-12 DISCUSSION QUESTIONS
9-13 ADDITIONAL READING

9-1 THE HAZARD

A hazardous waste, in short, is any waste or combination of wastes that poses a substantial danger, now or in the future, to human, plant, or animal life, and which therefore cannot be handled or disposed of without special precautions. The following examples illustrate the potential problems that may arise when special precautions are not taken in the disposal of hazardous waste.

1. Judy Piatt hired Russell Bliss to spray oil around her stables at Moscow Mills, Missouri, to control the dust. A few days later, hundreds of birds fell to the ground and died. Within the next three-and-a-half years, 20 of her cats went bald and died. Sixty-two of her horses died. Bliss's oil was waste from a defunct hexachlorophene plant that had paid him to dispose of it. That same waste oil was used to settle the dust in the streets of Times Beach, Missouri. It contained dioxin as a contaminant. It was 1971 and virtually no one knew that this waste oil was a hazardous waste.
2. The lagoon was once a licensed disposal site for toxic wastes. The owner was Berlin & Farro Liquid Incineration, Inc. At the bottom of the mysterious blue liquid in the lagoon were some barrels. They were thought to contain hydrochloric

acid. The blue liquid in the lagoon was a cyanide waste. The combination of the two chemicals would result in a lethal cloud of cyanide gas. The citizens of nearby Swartz Creek, Michigan, were evacuated while the State Department of Natural Resources oversaw the cleanup.

These actual case histories epitomize our concern with hazardous wastes. That which is common practice today may be the seed of disaster for tomorrow. What is considered good disposal practice may become a nightmare if the operators are not responsible and/or are not good enough business people to make money within the rules, and are therefore tempted to stray beyond them.

Dioxins and PCBs

We would like to elaborate on two particular hazardous wastes that have achieved national prominence: dioxins and PCBs. Because of their newsworthiness, we provide you with a brief summary of what these compounds are, where they come from, and their environmental impact.

Dioxins are found as over twenty different isomers of a basic chlorodioxin structure (Figure 9-1). The most common form, 2,3,7,8-tetrachlorodibenzo-p-dioxin

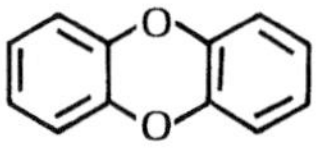

Unsubstituted Dioxin

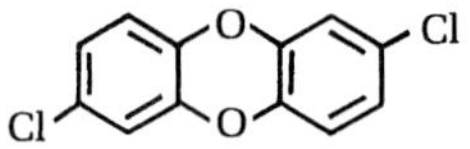

2, 7-DCDD

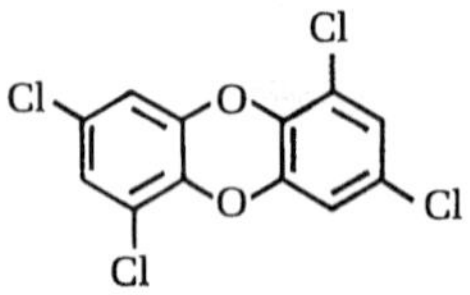

1, 3, 6, 8-TCDD

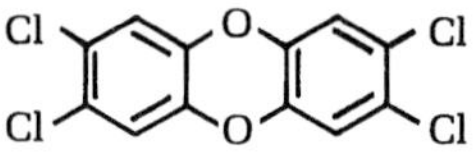

2, 3, 7, 8-TCDD

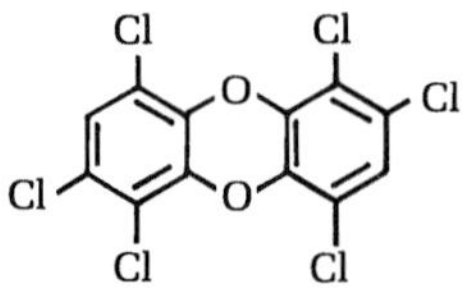

1, 2, 4, 6, 7, 9-HEXA-CDD

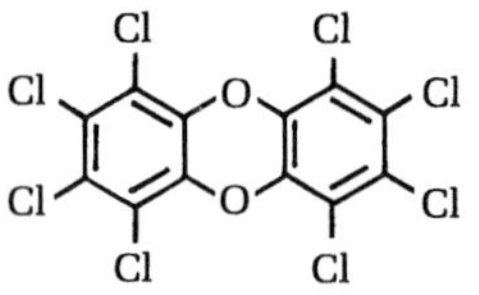

OCDD

FIGURE 9-1
Some examples of dioxins.

(TCDD), has become recognized as probably the most poisonous of all synthetic chemicals. Dioxins are a contaminant by-product that may be thermally generated during the manufacture or burning of chlorophenols; pesticides such as 2,4,5-T; Agent Orange, a defoliant made of a 50/50 mix of 2,4-D and 2,4,5-T; algae-controlling herbicides; insecticides; and preservatives. Dioxins are not manufactured for any commercial purpose. They occur only as a contaminant by-product. To date no dioxin has been found to be formed naturally in the environment. Widespread TCDD contamination has been reported in particulate matter from commercial and domestic combustion processes. Additional background dioxin contamination (0.1 to 10 parts per million, ppm) may persist and bioaccumulate following the field application of herbicides.

TCDD is a crystalline solid at room temperature. It is only slightly soluble in water (0.2 to 0.6 parts per billion, ppb). TCDD is considered to be a highly stable compound. It is thermally degraded at temperatures over 700°C. It is photochemically degraded under ultraviolet light in the presence of a hydrogen-donating solvent such as a solution of olive oil in cyclohexanone.

TCDD contamination was found at ppm levels in 2,4,5-T and 2,4-D used for weed control in the United States and as a defoliant in Vietnam; in wastes at the Love Canal disposal sites; in orthochlorophenol crude spill residues in the Sturgeon, Missouri, train derailment; and in fallout from an explosion at a chlorophenol manufacturing plant spill in Seveso, Italy. It is at this last site that engineers and scientists were challenged to develop environmentally safe control strategies.

The environmental health effects of dioxin in people are not well documented. However, alleged birth defects in newborns in South Vietnam caused researchers to begin animal toxicological investigations. TCDD is known to cause severe skin disorders, such as chloracne. In test animals it is a carcinogen, teratogen, mutagen, and embryo-toxin, and is known to affect immune responses in mammals. It is considered persistent, and it bioaccumulates in aquatic organisms. At this date (1996) no deaths have been directly correlated with low-level TCDD exposure. Nor have epidemiological findings shown any increased incidence of carcinogenesis, teratogenesis, mutagenesis or newborn defects, miscarriages, or similar adverse health effects in people. In 1994, the U.S. Environmental Protection Agency (EPA) released a report compiled by more than 100 scientists, including many not affiliated with EPA, that presents evidence that dioxin, even in trace amounts, may cause adverse human health effects.[1] EPA believes that dioxins are a probable carcinogen and may cause a wide range of other effects including disruption of regulatory hormones, reproductive and immune system disorders, and abnormal fetal development. Levels of dioxins in the environment were negligible until about 1930, peaked about 1970, and have been declining since then. Concentrations of dioxins in human lipid tissue have declined since 1980.

The term PCB (polychlorinated biphenyls) refers to a class of organic chemicals produced by the chlorination of a biphenyl molecule. It is composed of ten

[1]B. Hileman, "EPA Reassesses Dioxins," *C&E News*, p. 6, September 19, 1994.

possible forms and, theoretically, more than 200 isomers. These forms arise from a specified number of chlorine substitutions on the biphenyl molecule and correspond to the chemical nomenclatures monochlorobiphenyl, dichlorobiphenyl, trichlorobiphenyl, and so on. Several isomers for each PCB molecule are possible, the number depending on available substitution sites on each biphenyl portion (2–6, 2′–6′) of the molecule. However, not all possible isomers are likely to be formed during the manufacturing processes. In general, the most common ones are those that have either an equal number of chlorine atoms on both rings or a difference of only one chlorine atom between rings. Some examples are shown in Figure 9-2.

Commercial PCB mixtures were manufactured under a variety of trade names. The chlorine content of any product varied from 18 to 79 percent, depending on the extent of chlorination during the manufacturing process or on the amount of isomeric mixing engaged in by individual producers. Each company had a specific system for identifying the chlorine content of its product. For example, Aroclor 1248, 1254, and 1260 indicate 48 percent, 54 percent, and 60 percent chlorine, respectively; Clophen A60, Phenochlor DP6, and Kaneclor 600 designate that these products contain mixtures of hexachlorobiphenyls.

The only important U.S. producer of PCBs was Monsanto Industrial Chemicals Co., which had plants at Anniston, Alabama, where production of PCBs ended in 1970; and Sauget, Illinois, where production ceased in 1977. Sold under Monsanto's registered trademark of Aroclors, mixtures of PCBs had been used originally as a coolant/dielectric for transformers and capacitors, as heat transfer fluids, and as protective coatings for woods when low flammability was essential or desirable. Producers and users alike, apparently unaware of any potential hazards from expo-

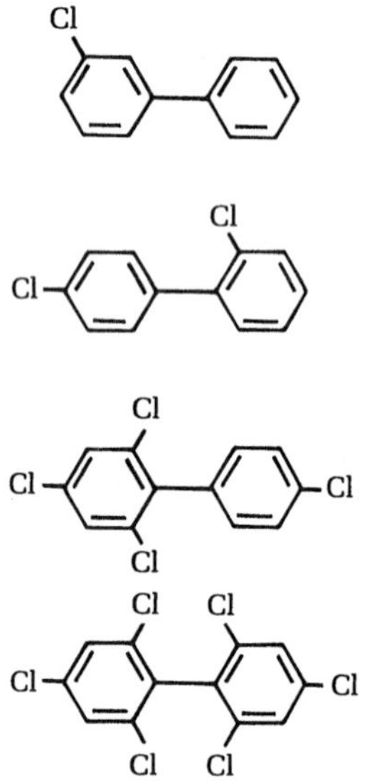

Cl

3-chlorobiphenyl

Cl

Cl

2,4′-dichlorobiphenyl

Cl

Cl

Cl

Cl

2,4,4′,6-tetrachlorobiphenyl

Cl Cl

Cl

Cl

Cl Cl

2,2′,4,4′,6,6′-hexachlorobiphenyl

FIGURE 9-2
Molecular structure and names of a few selected polychlorinated biphenyls.

sure to PCBs, initially operated in accordance with earlier results of toxicity tests that indicated no effects.[2] The expansion of open-ended applications between 1930 and 1960, incorporating PCBs into such commodities as paints, inks, dedusting agents, and pesticides, led to the widespread dissemination of which we are now aware. By 1937, toxic effects were noted in occupationally exposed workers, and threshold limit values were imposed at manufacturing sites.

The general pattern of release of PCBs to the environment changed significantly during the early 1970s. Until then, essentially no restrictions were imposed either on the use or on the disposal of PCBs. After evidence became available in 1969 and 1970 that chronic exposure could result in hazards to human health and the environment, Monsanto voluntarily banned sales of PCBs, and the release rate from industrial use was reduced through stringent control measures. However, significant reservoirs of mobile PCBs (those available for transport among environmental media and biota) still exist along with even larger, currently immobile reservoirs. The latter include those materials containing PCBs that are still in service and those deposited in landfills and dumps. The major factor affecting future release of PCBs from these sources will be government regulations controlling storage and disposal of the chemical.

9-2 RISK

The concepts of risk and hazard are inextricably intertwined. *Hazard* implies a probability of adverse effects in a particular situation. *Risk* is a measure of the probability. In some instances the measure is subjective, or *perceived risk.* Scientists and engineers use models to calculate an estimated risk. In some instances actual data may be used to estimate the risk. In the last two decades an attempt has been made to bring more rigor to the risk estimation process. Today this process is called *quantitative risk assessment,* or more simply *risk assessment.* The use of the results of a risk assessment to make policy decisions is called *risk management.*

Risk Perception[3]

There is an old political saying: "Perception is reality." This is no less true for environmental concerns than it is for politics. People respond to the hazards they perceive. If their perceptions are faulty, risk management efforts to improve environmental protection may be misdirected.

Some risks are well quantified. For example, the frequency and severity of automobile accidents are well documented. In contrast, other hazardous activities such as the use of alcohol and tobacco are more difficult to document. Their assessment requires complex epidemiological studies.

---

[2]Penning, C. H., "Physical Characteristics and Commercial Possibility of Chlorinated Diphenyl," *Industrial & Engineering Chemistry,* Vol. 22, pp. 1180–1183, 1930.

[3]This discussion follows that of P. Slovic, B. Fischoff, and S. Lichtenstein, "Rating the Risks," *Environment,* vol. 21, no. 3, pp. 14–20 and 36–39, 1979.

When lay people (and some experts for that matter) are asked to evaluate risk, they seldom have ready access to the statistics. In most cases, they rely on inferences based on their experience. People are likely to judge an event as likely or frequent if instances of it are easy to imagine or recall. Also, it is evident that acceptable risk is inversely related to the number of people participating in the activity. In addition, recent events such as a disaster can seriously distort risk judgments.

Table 9-1 and Figure 9-3 illustrate different perceptions of risk. Four different groups were asked to rate thirty activities and technologies according to the present

TABLE 9-1
Ordering of perceived risk for 30 activities and technologies[a]

	Group 1: LOWV	Group 2: College Students	Group 3: Active Club Members	Group 4: Experts
Nuclear power	1	1	8	20
Motor vehicles	2	5	3	1
Handguns	3	2	1	4
Smoking	4	3	4	2
Motorcycles	5	6	2	6
Alcoholic beverages	6	7	5	3
General (private) aviation	7	15	11	12
Police work	8	8	7	17
Pesticides	9	4	15	8
Surgery	10	11	9	5
Fire fighting	11	10	6	18
Large construction	12	14	13	13
Hunting	13	18	10	23
Spray cans	14	13	23	26
Mountain climbing	15	22	12	29
Bicycles	16	24	14	15
Commercial aviation	17	16	18	16
Electric power	18	19	19	9
Swimming	19	30	17	10
Contraceptives	20	9	22	11
Skiing	21	25	16	30
X-rays	22	17	24	7
High school & college football	23	26	21	27
Railroads	24	23	20	19
Food preservatives	25	12	28	14
Food coloring	26	20	30	21
Power mowers	27	28	25	28
Prescription antibiotics	28	21	26	24
Home appliances	29	27	27	22
Vaccinations	30	29	29	25

[a] The ordering is based on the geometric mean risk ratings within each group. Rank 1 represents the most risky activity or technology.

Source: P. Slovic, B. Fischoff and S. Lichtenstein, "Rating Risk," *Environment, 21,* pp. 14–20, 1979. Reprinted by permission.

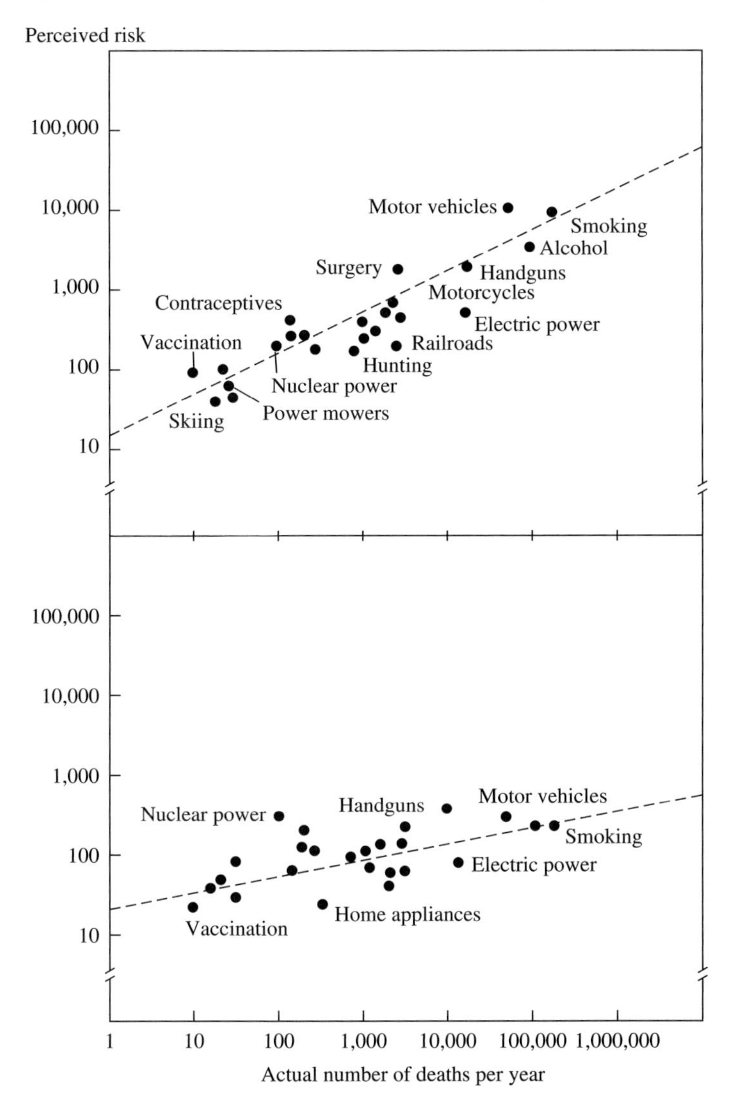

FIGURE 9-3
Judgments of perceived risk for experts (top) and lay people (bottom) plotted against the best technical estimates of annual fatalities for 25 technologies and activities. Each point represents the average responses of the participants. The dashed lines are the straight lines that best fit the points. The experts' risk judgments are seen to be more closely associated with annual fatality rates than are the lay judgments. (*Source:* P. Slovic, B. Fischoff, and S. Lichtenstein, "Rating Risks," *Environment, 21,* pp. 14–20, 1979. Reprinted by permission.)

risk of death from each. Three of the groups were from Eugene, Oregon. They included 30 college students, 40 members of the League of Women Voters (LOWV), and 25 business and professional members of the "Active Club." The fourth group was composed of 15 people selected from across the United States because of their professional involvement in risk assessment. Table 9-1 shows how the various groups ranked the risk of the activities or technologies. The same groups were asked to estimate the mean fatality for the same group of activities and technologies given the fact that the annual death toll from motor vehicle accidents in the U.S. was 50,000. The results are plotted in Figure 9-3. The dashed line is the line of best fit to the results. If the dashed line was at 45 degrees, the estimate would be perfect. The steeper slope of the line for the experts' risk judgments shows that they are more closely associated with the actual annual fatality rates than those of the lay groups.

Putting risk perception in perspective, we can calculate the risk of death from some familiar causes. To begin, we recognize that we will all die at some time. So, as a trivial example, the lifetime risk of death from all causes is 100%, or 1.0. In the late 1980s, there were about 2.1 million deaths per year. Of these, about 460,000 were cancer related. Without considering age factors, the risk of dying from cancer in a lifetime was about

$$\frac{460{,}000}{2.1 \times 10^6} = 0.22$$

The annual risk (assuming a 70-year life expectancy and again neglecting age factors) is about

$$\frac{0.22}{70} = 0.003$$

For comparison, Table 9-2 summarizes the risk of dying from some of the common causes of death.

In developing standards for environmental protection, the EPA often selects a lifetime risk in the range of 10^{-7} to 10^{-4} as acceptable. Table 9-3 shows a comparison of other activities that, based on statistical evidence, yield a risk of 10^{-6}.

TABLE 9-2
Annual risk of death from selected common human activities[a]

	Number of deaths in representative year	Individual risk/year
Coal mining		
Accident	180	1.3×10^{-3} or 1/770
Black lung disease	1,135	8×10^{-3} or 1/125
Fire fighting	—	8×10^{-4} or 1/1,250
Motor vehicle	46,000	2.2×10^{-4} or 1/4,500
Truck driving	400	10^{-4} or 1/10,000
Falls	16,339	7.7×10^{-5} or 1/13,000
Football (averaged over participants		4.10^{-5} or 1/25,000
Home accidents	25,000	1.2×10^{-5} or 1/83,000
Bicycling (assuming one person per bicycle)	1,000	10^{-5} or 1/100,000
Air travel: one transcontinental trip/year		2×10^{-6} or 1/500,000

[a] Selected from Hutt, *Food, Drugs, Cosmetic Law J.*, vol. 33, pp. 558–559, 1978.
Source: J. Rodricks and M. R. Taylor, "Application of Risk Assessment to Good Safety Decision Making," *Regulatory Toxicology and Pharmacology,* V. 3, pp. 275–284, 1983. Reprinted by permission.

TABLE 9-3
Risks which increase chance of death by 0.000001[a]

Smoking 1.4 cigarettes	Cancer, heart disease
Drinking 1/2 liter of wine	Cirrhosis of the liver
Spending 1 hour in a coal mine	Black lung disease
Spending 3 hours in a coal mine	Accident
Living 2 days in New York or Boston	Air pollution
Travelling 6 minutes by canoe	Accident
Travelling 10 miles by bicycle	Accident
Travelling 300 miles by car	Accident
Flying 1,000 miles by jet	Accident
Flying 6,000 miles by jet	Cancer caused by cosmic radiation
Living 2 months in Denver on vacation from N. Y.	Cancer caused by cosmic radiation
Living 2 months in average stone or brick building	Cancer caused by natural radioactivity
One chest X-ray taken in a good hospital	Cancer caused by radiation
Living two months with a cigarette smoker	Cancer, heart disease
Eating 40 tablespoons of peanut butter	Liver cancer caused by aflatoxin B
Drinking Miami drinking water for a year	Cancer caused by chloroform
Living 5 years at a site boundary of a typical nuclear power plant in the open	Cancer caused by radiation
Drinking 1,000 24 oz. soft drinks from banned plastic bottles	Cancer from acrylonitrile monomer
Living 20 years near PVC plant	Cancer caused by vinyl chloride (1976 standard)
Living 150 years within 20 miles of a nuclear power plant	Cancer caused by radiation
Eating 100 charcoal broiled steaks	Cancer from benzopyrene
Risk of accident by living within 5 miles of a nuclear reactor for 50 years	Cancer caused by radiation

[a] (1 part in 1 million)
Source: R. Wilson, "Analyzing the Daily Risks of Life," *Technology Review, 81,* pp. 41–46, 1979. Reprinted by permission.

Of course, if the risk of dying in one year is increased, the risk of dying from another cause in a later year is decreased. Since accidents often occur early in life, a typical accident may shorten life by 30 years. In contrast, diseases, such as cancer, cause death later in life, and life is shortened by about 15 years. Therefore, a risk of 10^{-6} shortens life on the average of 30×10^{-6} years, or 15 minutes for an accident. The same risk for a fatal illness shortens life by about 8 minutes. It has been noted that smoking a cigarette takes ten minutes and shortens life by five minutes.[4]

Risk Assessment

In 1989, the EPA adopted a formal process for conducting a baseline risk assessment.[5] This process includes data collection and evaluation, toxicity assessment, exposure assessment, and risk characterization. Risk assessment is considered to be site-specific. Each step is described briefly below.

Data collection and evaluation. Data collection and evaluation includes gathering and analyzing site-specific data relevant to human health concerns for the purpose of identifying substances of major interest.This step includes gathering background and site information as well as the preliminary identification of potential human exposure through sampling, and development of a sample collection strategy.

When collecting background information, it is important to identify the following:

1. Possible contaminants on the site;
2. Concentrations of the contaminants in key sources and media of interest, characteristics of sources, and information related to the chemical's release potential; and
3. Characteristics of the environmental setting that could affect the fate, transport, and persistence of the contaminants.

The review of the available site information determines basic site characteristics such as groundwater movement or soil characteristics. With these data, it is possible to initially identify potential exposure pathways and exposure points important for assessing exposure. A conceptual model of pathways and exposure points can be formed from the background data and site information. This conceptual model can then be used to help refine data needs.

Toxicity assessment. Toxicity assessment is the process of determining the relationship between the exposure to a contaminant and the increased likelihood of the occurrence or severity of adverse effects to people. This procedure includes hazard identification and dose-response evaluation. Hazard identification determines whether exposure to a contaminant causes increased adverse effects towards humans and to what level of severity. Dose-response evaluation uses quantitative information on the dose of the contaminant and relates it to the incidence of adverse health in an

---

[4] R. Wilson, "Analyzing the Daily Risks of Life," *Technology Review,* vol. 81, no. 4, pp. 41–46, 1979.

[5] *Risk Assessment Guidance for Superfund Volume 1, Human Health Evaluation Manual (Part A),* United States Environmental Protection Agency, EPA/540/1-89/002, 1989.

exposed population. Toxicity values can be determined from this quantitative relationship and used in the risk characterization step to estimate different occurrences of adverse health effects based on various exposure levels.

The single factor that determines the degree of harmfulness of a compound is the dose of that compound.[6] *Dose* is defined as the mass of chemical received by the animal or exposed individual. Dose is usually expressed in units of milligrams per kilogram of body mass (mg/kg). Some authors use parts per million (ppm) instead of mg/kg. Where the dose is administered over time, the units may be mg/kg · d. It should be noted that dose differs from the concentration of the compound in the medium (air, water, or soil) to which the animal or individual is exposed.

For toxicologists to establish the "degree of harmfulness" of a compound, they must be able to observe a quantitative effect. The ultimate effect manifested is death of the organism. Much more subtle effects may also be observed. Effects on body weight, blood chemistry, and enzyme inhibition or induction are examples of *graded responses.* Mortality and tumor formation are examples of *quantal* (all-or-nothing) responses. If a dose is sufficient to alter a biological mechanism, a harmful consequence will result. The experimental determination of the range of changes in a biologic mechanism to a range of doses is the basis of the dose-response relationship.

The statistical variability of organism response to dose is commonly expressed as a cumulative-frequency distribution known as a dose-response curve. Figure 9-4

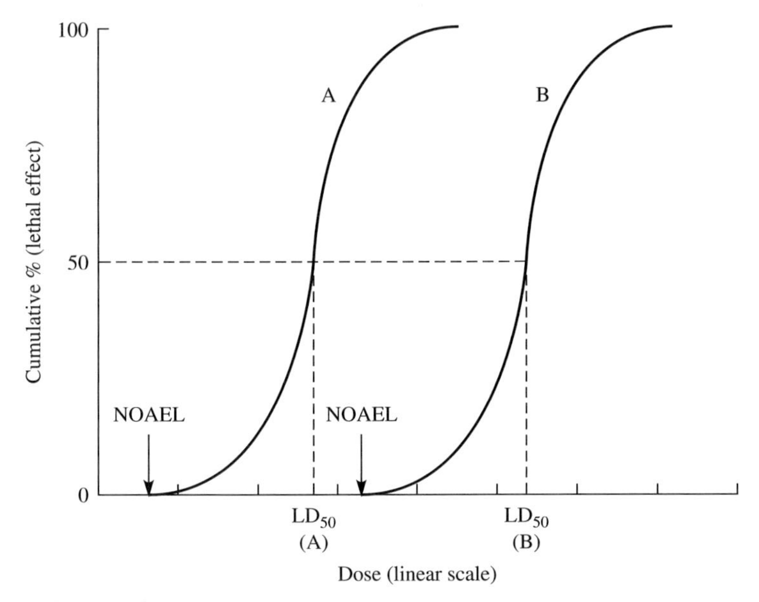

FIGURE 9-4
Hypothetical dose-response curves for two chemical agents (A and B) administered to a uniform population of biologic specimens. (*Source:* T. A. Loomis, *Essentials of Toxicology,* New York: Lea & Febiger, 1968. Reprinted by permission.)

---

[6]T. A. Loomis, *Essentials of Toxicology,* Philadelphia: Lea & Febiger, p. 2, 1978.

illustrates the method by which a common toxicological measure, namely the LD_{50}, or lethal dose for 50 percent of the animals, is obtained. The assumption inherent in the plot of the dose-response curve is that the test population variability follows a Gaussian distribution and, hence, that the dose-response curve has the statistical properties of a Gaussian cumulative-frequency curve.

Toxicity is a relative term. That is, there is no absolute scale for establishing toxicity; one may only specify that one chemical is more or less toxic than another. Comparison of different chemicals is uninformative unless the organism or biologic mechanism is the same and the quantitative effect used for comparison is the same. Figure 9-4 serves to illustrate how a toxicity scale might be developed. Of the two curves in the figure, the LD_{50} for compound B is greater than that for compound A. Thus, for the test animal represented by the graph, compound A is more toxic than compound B as measured by lethality. There are many difficulties in establishing toxicity relationships. Species respond differently to toxicants so that the LD_{50} for a mouse may be very different than that for a human. The shape (slope) of the dose-response curve may differ for different compounds so that a high LD_{50} may be associated with a low "no observed adverse effect level" (NOAEL) and vice versa.

The nature of a statistically obtained value, such as the LD_{50}, tends to obscure a fundamental concept of toxicology: that there is no fixed dose that can be relied on to produce a given biologic effect in every member of a population. In Figure 9-4, the mean value for each test group is plotted. If, in addition, the extremes of the data are plotted as in Figure 9-5, it is apparent that the response of individual

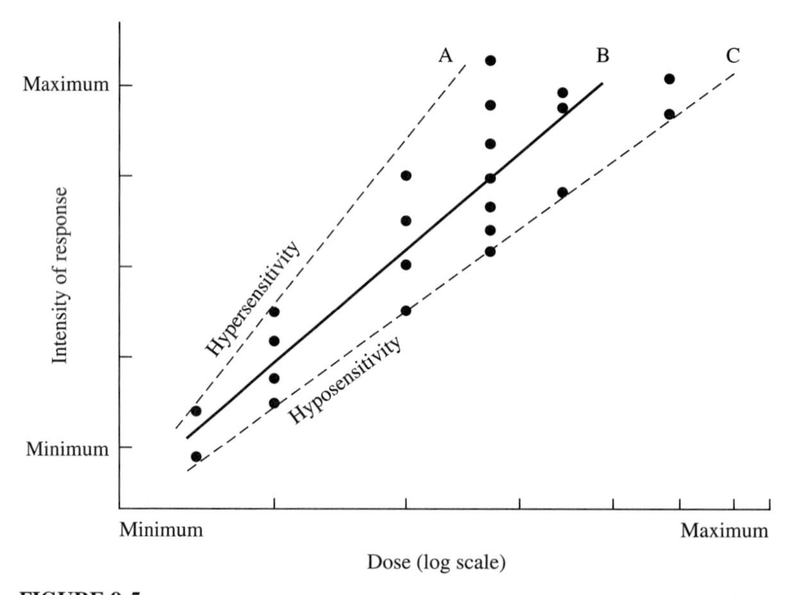

FIGURE 9-5
Hypothetical dose-response relationships for a drug administered to a uniform population of animals. Each point represents response of a single animal. (*Source:* T. A. Loomis, *Essentials of Toxicology,* Philadelphia: Lea & Febiger, 1968. Reprinted by permission.)

members of the population may vary widely from the mean. This implies not only that single point comparisons, such as the LD_{50}, may be misleading, but that even knowing the slope of the average dose-response curve may not be sufficient if one wishes to protect hypersensitive individuals.

Organ toxicity is frequently classified as an acute or subacute effect. Carcinogenesis, teratogenesis, reproductive toxicity, and mutagenesis have been classified as chronic effects.[7] It is self-evident that an organ may exhibit acute, subacute, and chronic effects and that this system of classification is not well bounded.

Virtually all of the data used in hazard identification and, in particular, hazard quantification, is derived from animal studies. Aside from the difficulty of extrapolating from one species to another, the testing of animals to estimate low-dose response is difficult. Example 9-1 illustrates the problem.

Example 9-1. An experiment was developed to ascertain whether a compound has a 5 percent probability of causing a tumor. The same dose of the compound was administered to 10 groups of 100 test animals. A control group of 100 animals was, with the exception of the test compound, exposed to the same environmental conditions for the same period of time. The following results were obtained:

Group	No. of Tumors
A	6
B	4
C	10
D	1
E	2
F	9
G	5
H	1
I	4
J	7

No tumors were detected in the controls (not likely in reality).

Solution. The average number of excess tumors is 4.9 percent. These results tend to confirm that the probability of causing a tumor is 5 percent.

If, instead of using 1,000 animals (10 groups $\times$ 100 animals), only 100 animals were used, it is fairly evident from the data that, statistically speaking, some very anomalous results might be achieved. That is, we might find a risk from 1 percent to 10 percent.

Note that a 5 percent risk (probability of 0.05) is very high in comparison to the EPA's objective of achieving an environmental contaminant risk of 10^{-7} to 10^{-4}.

Animal studies are only capable of detecting risks on the order of 1 percent. To extrapolate the data taken from animals exposed to high doses to humans who

[7] A glossary of these toxicology terms is given in Table 9-4.

TABLE 9-4
Glossary of toxicological terms

Acute toxicity	An adverse effect that has a rapid onset, short course, and pronounced symptoms
Cancer	An abnormal growth process in which cells begin a phase of uncontrolled growth and spread
Carcinogen	A cancer-producing substance
Carcinomas	Cancers of epithelial tissues. Lung cancer and skin cancer are examples of carcinomas.
Chronic toxicity	An adverse effect that frequently takes a long time to run its course and initial onset of symptoms may go undetected
Genotoxic	Toxic to the genetic material (DNA)
Initiator	A chemical that starts the change in a cell that irreversibly converts the cell into a cancerous or precancerous state. Needs to have a promoter to develop cancer.
Leukemias	Cancers of white blood cells and the tissue from which they are derived
Lymphomas	Cancers of the lymphatic system. An example is Hodgkin's disease.
Metastasis	Process of spreading/migration of cancer cells throughout the body
Mutagenesis	Mutagens cause changes in the genetic materal of cells. The mutations may occur either in somatic (body) cells or germ (reproductive) cells.
Neoplasm	A new growth. Usually an abnormally fast-growing tissue.
Oncogenic	Causing cancers to form
Promoter	A chemical that increases the incidence to a previous carcinogen exposure
Reproductive toxicity	Decreases in fertility, increases in miscarriages, and fetal or embryonic toxicity as manifested in reduced birth weight or size
Sarcoma	Cancer of mesodermal tissue such as fat and muscle
Subacute toxicity	Subacute toxicity is measured using daily dosing during the first 10 percent of the organism's normal life expectancy and checking for effects throughout the normal lifetime.
Teratogenesis	Production of a birth defect in the offspring after maternal or paternal exposure

will be exposed to doses several orders of magnitude lower, toxicologists employ mathematical models.

One of the most controversial aspects of toxicological assessment is the method chosen to extrapolate the carcinogenic dose-response curve from the high doses actually administered to test animals to the low doses that humans actually experience in the environment. The conservative worst-case assessment is that one event capable of altering DNA will lead to tumor formation. This is called the *one-hit hypothesis.* From this hypothesis, it is assumed that there is no threshold dose below which the risk is zero, so that for carcinogens, there is no NOAEL and the dose-response curve passes through the origin.

Many models have been proposed for extrapolation to low doses. The selection of an appropriate model is more a policy decision than a scientific one since there are no data to confirm or refute any model. The *one-hit model* is frequently used:

$$P(d) = 1 - \exp(-q_0 - q_1 d) \tag{9-1}$$

where $P(d)$ = lifetime risk (probability) of cancer
d = dose
q_0 and q_1 = parameter to fit data

This model corresponds to the simplest mechanistic model of carcinogenesis, namely that a single chemical hit will induce a tumor.

The background rate of cancer incidence, $P(0)$, may be represented by expanding the exponential as

$$\exp(x) = 1 + x + \frac{x^2}{2!} + \cdots + \frac{x^n}{n!} \tag{9-2}$$

For small values of x, this expansion is approximately

$$\exp(x) \simeq 1 + x \tag{9-3}$$

Assuming the background rate for cancer is small, then

$$P(0) = 1 - \exp(-q_0) \simeq 1 - [1 + (-q_0)] = q_0 \tag{9-4}$$

This implies that q_0 corresponds to the background cancer incidence. For small dose rates, the one-hit model can then be expressed as

$$P(d) \simeq 1 - [1 - (q_0 + q_1 d)] = q_0 + q_1 d = P(0) + q_1 d \tag{9-5}$$

For low doses, the additional cancer risk above the background level may be estimated as

$$A(d) = P(d) - P(0) = (P(0) + q_1 d) - P(0) \tag{9-6}$$

or

$$A(d) = q_1 d \tag{9-7}$$

This model, therefore, assumes that the excess lifetime probability of cancer is linearly related to dose.

Some authors prefer a model that is based on an assumption that tumors are formed as a result of a sequence of biological events. This model is called the *multistage model:*

$$P(d) = 1 - \exp[-(q_0 + q_1 d + q_2 d^2 + \cdots + q_n d^n)] \tag{9-8}$$

where q_i values are selected to fit the data. The one-hit model is a special case of the multistage model.

EPA has selected a modification of the multistage model for toxicological assessment. It is called the *linearized multistage model.* This model assumes that we can extrapolate from high doses to low doses with a straight line. At low doses, the slope of the dose-response curve is represented by a *slope factor* (SF). It has units of *risk per unit dose* or $risk(mg/kg \cdot d)^{-1}$.

The EPA maintains a toxicological data base called IRIS (*Integrated Risk Information System*) that provides background information on potential carcinogens. IRIS includes suggested values for the slope factor. A list of slope factors for several compounds is shown in Table 9-5.

In contrast to the carcinogens, it is assumed that for noncarcinogens there is a dose below which there is no adverse effect; that is, there is an NOAEL. The EPA has estimated the acceptable daily intake, or *reference dose* (RfD), that is likely to be without appreciable risk. The RfD is obtained by dividing the NOAEL by safety factors to account for the transfer from animals to humans, sensitivity, and other uncertainties in developing the data. A list of several compounds and their RfD values is given in Table 9-6.

Exposure assessment. The objective of this step is to estimate the magnitude of exposure to chemicals of potential concern. The magnitude of exposure is based on

TABLE 9-5
Slope factors for potential carcinogens[a]

Chemical	CPS_o kg · d/mg	CPS_i kg · d/mg
Arsenic	1.5	15.1
Benzene	0.029	0.029
Benzo(a)pyrene	7.3	N/A
Cadmium	N/A	6.3
Carbon tetrachloride	0.13	0.0525
Chloroform	0.0061	0.08
Chromium VI	N/A	42.0
DDT	0.34	0.34
1,1-Dichloroethylene	0.6	0.175
Dieldrin	16.0	16.1
Heptachlor	4.5	4.55
Hexachloroethane	0.014	0.014
Methylene chloride	0.0075	0.00164
Polychlorinated biphenyls	7.7	N/A
2,3,7,8-TCDD[b]	1.5×10^5	1.16×10^5
Tetrachloroethylene[b]	0.052	0.002
Trichloroethylene[c]	W	0.006
Vinyl chloride[b]	1.9	N/A

CPS_o = cancer potency slope, oral; CPS_i = cancer potency slope, inhalation; W = withdrawn from IRIS

[a] Values are frequently updated. Refer to IRIS and HEAST for current data.

[b] From Health Effects Assessment Summary Tables (HEAST), 1994

[c] From EPA-NCEA Regional Support—provisional value

Source: With exceptions noted above: U.S. Environmenal Protection Agency, IRIS database, October 1996.

TABLE 9-6
RfDs for chronic noncarcinogenic effects for selected chemicals[a]

Chemical	Oral RfD, mg/kg · d
Acetone	0.1
Barium	0.05
Cadmium	0.0005
Chloroform	0.01
Cyanide	0.02
1,1-Dichloroethylene	0.009
Hydrogen cyanide	0.02
Methyline chloride	0.06
Pentachlorophenol	0.03
Phenol	0.6
PCB	
Aroclor 1016	7.0×10^{-6}
Aroclor 1254	2.0×10^{-6}
Silver	0.003
Tetrachloroethylene	0.01
Toluene	0.2
1,2,4-Trichlorobenzene	0.02
Xylenes	2.0

[a] Values are frequently updated. Refer to IRIS for current data.
Source: U.S. Environmental Protection Agency IRIS database, 1996

chemical intake and pathways of exposure. The most important route (or pathway) of exposure may not always be clearly established. Arbitrarily eliminating one or more routes of exposure is not scientifically sound. The more reasonable approach is to consider an individual's potential contact with all contaminated media through all possible routes of entry. These are summarized in Table 9-7.

The evaluation of all major sources of exposure is known as total exposure assessment.[8] After reviewing the available data, it may be possible to decrease or increase the level of concern for a particular route of entry to the body. Elimination of a pathway of entry can be justified if:

1. The exposure from a particular pathway is less than that of exposure through another pathway involving the same media at the same exposure point;
2. The magnitude of exposure from the pathway is low; or
3. The probability of exposure is low and incidental risk is not high.

There are two methods of quantifying exposure: point estimate methods and probabilistic methods. The EPA utilizes the point estimate procedure by estimating

[8] J. P. Butler, A. Greenberg, P. J. Lioy, G. B. Post, J. M. Waldman, "Assessment of Carcinogenic Risk from Personal Exposure to Benzo(a)pyrene in the Total Human Environmental Exposure Study (THEES)," *Journal of the Air & Waste Management Association,* July 1993, vol. 43, p. 970–977.

TABLE 9-7
Potential contaminated media and corresponding routes of exposure

Media	Routes of potential exposure
Groundwater	Ingestion, dermal contact, inhalation during showering
Surface water	Ingestion, dermal contact, inhalation during showering
Sediment	Ingestion, dermal contact
Air	Inhalation of airborne (vapor phase) chemicals (indoor and outdoor)
	Inhalation of particulates (indoor and outdoor)
Soil/dust	Incidental ingestion, dermal contact
Food	Ingestion

the *reasonable maximum exposure* (RME). Because this method results in very conservative estimates, some scientists believe probabilistic methods are more realistic.[9] We will limit our consideration to the EPA point estimate technique.

RME is defined as the highest exposure that is reasonably expected to occur and is intended to be a conservative estimate of exposure within the range of possible exposures. Two steps are involved in estimating RME: first, exposure concentrations are predicted using a transport model such as the Gaussian plume model for atmospheric dispersion (Section 6-8), then pathway-specific intakes are calculated using these exposure concentration estimates. The following equation is a generic intake equation:

$$\text{CDI} = \text{C}\left[\frac{(\text{CR})(\text{EDF})}{\text{BW}}\right]\left(\frac{1}{\text{AT}}\right) \tag{9-9}$$

where CDI = Chronic daily intake, (mg/kg body weight · day)
C = chemical concentration, contacted over the exposure period (e.g., mg/L water)
CR = contact rate, the amount of contaminated medium contacted per unit time or event (e.g., L/day)
EFD = exposure frequency and duration, describes how long and how often exposure occurs. Often calculated using two terms (EF and ED):
EF = exposure frequency (days/year)
ED = exposure duration (years)
BW = body weight, the average body weight over the exposure period (kg)
AT = averaging time, period over which exposure is averaged (days)

For each different media and corresponding route of exposure, it is important to note that additional variables are used to estimate intake. For example, when calculating intake for the inhalation of airborne chemicals, an inhalation rate and exposure time are required. Specific equations for media and routes of exposure are given in Table 9-8. Standard values for use in the intake equations are shown in Table 9-9.

[9] B. Finley, and D. Paustenbach, "The Benefits of Probabilistic Exposure Assessment: Three Case Studies Involving Contaminated Air, Water, and Soil." *Risk Analysis,* vol. 14, no. 1, pp. 53–73, 1994.

TABLE 9-8
Residential exposure equations for various pathways[a]

Ingestion in drinking water

$$CDI = \frac{(CW)(IR)(EF)(ED)}{(BW)(AT)} \quad (9\text{-}10)$$

Ingestion while swimming

$$CDI = \frac{(CW)(CR)(ET)(EF)(ED)}{(BW)(AT)} \quad (9\text{-}11)$$

Dermal contact with water

$$AD = \frac{(CW)(SA)(PC)(ET)(EF)(ED)(CF)}{(BW)(AT)} \quad (9\text{-}12)$$

Ingestion of chemicals in soil

$$CDI = \frac{(CS)(IR)(CF)(FI)(EF)(ED)}{(BW)(AT)} \quad (9\text{-}13)$$

Dermal contact with soil

$$AD = \frac{(CS)(CF)(SA)(AF)(ABS)(EF)(ED)}{(BW)(AT)} \quad (9\text{-}14)$$

Inhalation of airborn (vapor phase) chemicals

$$CDI = \frac{(CA)(IR)(ET)(EF)(ED)}{(BW)(AT)} \quad (9\text{-}15)$$

Ingestion of contaminated fruits, vegetables, fish and shellfish

$$CDI = \frac{(CF)(IR)(FI)(EF)(ED)}{(BW)(AT)} \quad (9\text{-}16)$$

where ABS = absorption factor for soil contaminant, unitless
AD = absorbed dose, mg/kg · d
AF = soil-to-skin adherence factor, mg/cm^2
AT = averaging time, d
BW = body weight, kg
CA = contaminant concentration in air, mg/m^3
CDI = chronic daily intake, mg/kg · d
CF = volumetric conversion factor for water = 1 L/1,000 cm^3
= conversion factor for soil = 10^{-6} kg/mg
CR = contact rate, L/h
CS = chemical concentration in soil, mg/kg
CW = chemical concentration in water, mg/L
ED = exposure duration, y
EF = exposure frequency, d/y or events/y
ET = exposure time, h/d or h/event
FI = fraction ingested, unitless
IR = ingestion rate, L/d or mg soil/d or kg/meal
= inhalation rate, m^3/h
PC = chemical-specific dermal permeability constant, cm/h
SA = skin surface area available for contact, cm^2

[a] *Source: Risk Assessment Guidance for Superfund Volume I, Human Health Evaluation Manual (Part A),* United States Environmental Protection Agency, EPA/540/1-89/002, 1989.

TABLE 9-9
EPA recommended values for estimating intake[a]

Parameter	Standard value
Average body weight, adult	70 kg
Average body weight, child[b]	
0–1.5 y	10 kg
1.5–5 y	14 kg
5–12 y	26 kg
Amount of water ingested daily, adult	2 L
Amount of water ingested daily, child	1 L
Amount of air breathed daily, adult	20 m^3
Amount of air breathed daily, child	5 m^3
Amount of fish consumed daily, adult	6.5 g/d
Contact rate, swimming	50 mL/h
Skin surface available, adult male	1.94 m^2
Skin surface available, adult female	1.69 m^2
Skin surface available, child	
3–6 y (avg for male and female)	0.720 m^2
6–9 y (avg for male and female)	0.925 ^{2}
9–12 y (avg for male and female)	1.16 m^2
12–15 y (avg for male and female)	1.49 m^2
15–18 y (female)	1.60 m^2
15–18 y (male)	1.75 m^2
Soil ingestion rate, children 1 to 6 yrs	200 mg/d
Soil ingestion rate, persons > 6 yrs	100 mg/d
Skin adherence factor, potting soil to hands	1.45 mg/cm^2
Skin adherence factor, kaolin clay to hands	2.77 mg/cm^2
Exposure duration	
Lifetime	70 y
At one residence, 90th percentile	30 y
National median	5 y
Averaging time	(ED)(365 d/y)
Exposure frequency (EF)	
Swimming	7 d/y
Eating fish and shell fish	48 d/y
Exposure time (ET)	
Shower, 90th percentile	12 min
Shower, 50th percentile	7 min

[a] *Source: Risk Assessment Guidance for Superfund Volume I Human Health Evaluation Manual (Part A),* United States Environmental Protection Agency, EPA/540/1-89/002. 1989.
[b] *Source:* R. Copeland, *et al,* "Use of Probabilistic Methods to Understand the Conservativism in California's Approach to Assessing Health Risks Posed by Air Contaminants, *Journal of Air & Waste Management Association,* vol 44, pp. 1399–1413, 1994.

Example 9-2. Estimate the chronic daily intake of benzene from exposure to a city water supply that contains a benzene concentration equal to the drinking water standard. Assume the exposed individual is an adult male who consumes water at the adult rate for 70 years, that he is an avid swimmer and swims in a local pool (supplied with city water) 3 days a week for 30 minutes and has been doing so since he was 30 years old. He takes a long shower every day. Assume that the average air concentration of benzene during the shower is 5 μg/m^3.[10] From the literature, it is estimated that the dermal uptake from water is 0.0020 m^3/m$^2 \cdot$h (This is PC in Table 9-8. PC also has units of m/h or cm/h.) and that direct dermal absorption during showering is no more than 1 percent of the available benzene because most of the water does not stay in contact with skin long enough.[11]

Solution. From Table 9-8, we note that five routes of exposure are possible from the drinking water medium: (1) ingestion, dermal contact while (2) showering and (3) swimming, (4) inhalation of vapor while showering, and (5) ingestion while swimming. The allowable drinking water concentration is determined to be 0.005 mg/L from Chapter 3, Table 3-6.

We begin by calculating the CDI for ingestion (Equation 9-10):

$$\text{CDI} = \frac{(0.005 \text{ mg/L})(2.0 \text{ L/d})(365 \text{ d/y})(70 \text{ y})}{(70 \text{ kg})(70 \text{ y})(365 \text{ d/y})}$$

$$= 1.43 \times 10^{-4} \text{ mg/kg} \cdot \text{d}$$

The ingestion rate (IR) and body weight (BW) were selected from Table 9-9.

Equation 9-12 may be used to estimate absorbed dose while showering:

$$\text{AD} = \frac{(0.005 \text{ mg/L})(1.94 \text{ m}^2)(0.0020 \text{ m/h})(0.20 \text{ h/event})}{(70 \text{ kg})(70 \text{ y})}$$

$$\times \frac{(1 \text{ event/d})(365 \text{ d/y})(70 \text{ y})(10^3 \text{ L/m}^3)}{(365 \text{ d/y})}$$

$$= 5.54 \times 10^{-5} \text{ mg/kg} \cdot \text{d}$$

But only about 1 percent of this amount is available for adsorption in a shower because of the limited contact time, so the actual adsorbed dose by dermal contact is

$$\text{AD} = (0.01)(5.54 \times 10^{-5} \text{ mg/kg} \cdot \text{d}) = 5.54 \times 10^{-7} \text{ mg/kg} \cdot \text{d}$$

The surface area (SA) and exposure time were obtained from Table 9-9. The permeability constant was given in the problem statement. The exposure time is estimated by converting a long shower of 12 minutes to hours (12/60 = 0.2).

The adsorbed dose for swimming is calculated in the same fashion:

$$\text{AD} = \frac{(0.005 \text{ mg/L})(1.94 \text{ m}^2)(0.0020 \text{ m/h})(0.5 \text{ h/event})}{(70 \text{ kg})(70 \text{ y})}$$

$$\times \frac{(3 \text{ events/w})(52 \text{ w/y})(40 \text{ y})(10^3 \text{ L/m}^3)}{(365 \text{ d/y})}$$

$$= 3.38 \times 10^{-5} \text{ mg/kg} \cdot \text{d}$$

[10] T. E. McKone, "Human Exposure to Volatile Organic Compounds in Household Tap Water: The Indoor Inhalation Pathway," *Environmental Science & Technology,* vol. 21, no. 12, pp. 1194–1201, 1987.

[11] J. L Byard, "Hazard Assessment of 1,1,1-Trichloroethane in Groundwater," in *The Risk Assessment of Environmental Hazards,* edited by D. J. Paustenbach, New York: John Wiley & Sons, pp. 331–344, 1989.

In this case, since there is virtually total body immersion for the entire contact period and since there is virtually an unlimited supply of water for contact, there is no reduction for availability. The value of ET is computed from the swimming time (30 minutes = 0.5 h/event). The exposure frequency is computed from the number of swimming events per week and the number of weeks in a year. The exposure duration (ED) is calculated from the lifetime and beginning time of swimming = 70 y − 30 y = 40 y.

The inhalation rate from showering is estimated from Equation 9-15:

$$\text{CDI} = \frac{(5\ \mu\text{g/m}^3)(10^{-3}\ \text{mg}/\mu\text{g})(0.833\ \text{m}^3/\text{h})(0.20\ \text{h/event})(1\ \text{event/d})(365\ \text{d/y})(70\ \text{y})}{(70\ \text{kg})(70\ \text{y})(365\ \text{d/y})}$$

$$= 1.19 \times 10^{-5}\ \text{mg/kg} \cdot \text{d}$$

The inhalation rate (IR) is taken from Table 9-9 and converted to an hourly basis.

For ingestion while swimming, we apply Equation 9-11:

$$\text{CDI} = \frac{(0.005\ \text{mg/L})(50\ \text{mL/h})(10^{-3}\ \text{L/mL})(0.5\ \text{h/event})(3\ \text{events/w})(52\ \text{w/y})(40\ \text{y})}{(70\ \text{kg})(70\ \text{y})(365\ \text{d/y})}$$

$$= 4.36 \times 10^{-7}\ \text{mg/kg} \cdot \text{d}$$

The contact rate (CR) was determined from Table 9-9. Other values were obtained in the same fashion as those for dermal contact while swimming.

The total exposure would be estimated as:

$$\text{CDI}_\text{T} = 1.43 \times 10^{-4} + 5.54 \times 10^{-7} + 3.38 \times 10^{-5} + 1.19 \times 10^{-5} + 4.36 \times 10^{-7}$$

$$= 1.90 \times 10^{-4}\ \text{mg/kg} \cdot \text{d}$$

From these calculations, it becomes readily apparent that, in this case, drinking the water dominates the intake of benzene.

Risk characterization. In the risk characterization step, all data collected from exposure and toxicity assessments are reviewed to corroborate qualitative and quantitative conclusions about risk. The risk for each media source and route of entry is calculated. This includes the evaluation of compounding effects due to the presence of more than one chemical contaminant and the combination of risk across all routes of entry.

For low-dose cancer risk (risk below 0.01), the quantitative risk assessment for a single compound by a single route is calculated as:

$$\text{Risk} = (\text{Intake})(\text{Slope Factor}) \tag{9-17}$$

where intake is calculated from one of the equations in Table 9-8 or a similar relationship. The slope factor is obtained from IRIS (see, for example, Table 9-5). For high carcinogenic risk levels (risk above 0.01), the one-hit equation is used:

$$\text{Risk} = 1 - \exp[-(\text{Intake})(\text{Slope Factor})] \tag{9-18}$$

The measure used to describe the potential for noncarcinogenic toxicity to occur in an individual is not expressed as a probability. Instead, EPA uses the noncancer hazard quotient, or hazard index (HI):

$$\text{HI} = \frac{\text{Intake}}{\text{RfD}} \tag{9-19}$$

These ratios are not to be interpreted as statistical probabilities. A ratio of 0.001 does *not* mean that there is a one in one thousand chance of an effect occurring. If the HI exceeds unity, there may be concern for potential noncancer effects. As a rule, the greater the value above unity, the greater the level of concern.

To account for multiple substances in one pathway, EPA sums the risks for each constituent:

$$\text{Risk}_{\text{T}} = \sum \text{Risk}_{\text{i}} \tag{9-20}$$

For multiple pathways

$$\text{Total Exposure Risk} = \sum \text{Risk}_{\text{ij}} \tag{9-21}$$

where i = the compounds and j = pathways.

In a like manner, the hazard index for multiple substances and pathways is estimated as

$$\text{Hazard Index}_{\text{T}} = \sum HI_{\text{ij}} \tag{9-22}$$

In EPA's guidance documents, they recommend segregation of the hazard index into chronic, subchronic, and short-term exposure.

Example 9-3. Using the results from Example 9-2, estimate the risk from exposure to drinking water containing the MCL for benzene.

Solution. Equation 9-21 in the form

$$\text{Total Exposure Risk} = \sum \text{Risk}_{\text{j}}$$

may be used to estimate the risk. Since the problem is only to consider one compound, namely benzene, i = 1 and others do not need to be considered. Since the total exposure from Example 9-2 included each of the routes of concern for drinking water, that is all j's, the final sum may be used to compute risk. The slope factor is obtained from Table 9-5. The risk is

$$\text{Risk} = (1.90 \times 10^{-4}\ \text{mg/kg} \cdot \text{d})(2.9 \times 10^{-2}\ (\text{mg/kg} \cdot \text{d})^{-1})$$
$$= 5.5 \times 10^{-6}$$

This is the total lifetime risk (70 years) for benzene in drinking water at the MCL. Another way of viewing this is to estimate the number of people that might develop cancer. For example, in a population of 2 million,

$$(2 \times 10^{6})(5.5 \times 10^{-6}) = 11 \text{ people might develop cancer.}$$

This risk falls within the EPA guidelines of 10^{-4} to 10^{-7} risk. It, of course, does not account for all sources of benzene by all routes. None the less, the risk, compared to some other risks in daily life, appears to be quite small.

Risk Management

Though some might wish it, it is clear that establishment of zero risk cannot be achieved. There are risks in all societal decisions from driving a car to drinking water with benzene at the MCL concentration. Even banning the production of chemicals, as was done for PCBs, for example, does not remove those that already permeate our environment. Risk management is performed to decide the magnitude of risk that is tolerable in specific circumstances.[12] This is a policy decision that weighs the results of the risk assessment against costs and benefits as well as the public acceptance. The risk manager recognizes that if a very high certainty in avoiding risk (that is, a very low risk, for example, 10^{-7}) is required, the costs in achieving low concentrations of the contaminant are likely to be high.

Unfortunately, there is very little guidance that can be provided to the risk manager. We know that people are willing to accept a higher risk for things that they expose themselves to voluntarily than for involuntary exposures, and, hence, insist on lower levels of risk, regardless of cost, for involuntary exposure. We also know that people are willing to accept risk if it approaches that for disease, that is, a fatality rate of 10^{-6} people per person-hour of exposure.[13]

9-3 DEFINITION AND CLASSIFICATION OF HAZARDOUS WASTE

There are two ways a waste material is found to be hazardous:[14] (1) by its presence on the EPA-developed lists, or (2) by evidence that the waste exhibits ignitable, corrosive, reactive, or toxic characteristics.

EPA's Hazardous Waste Designation System

The list of hazardous wastes includes spent halogenated and nonhalogenated solvents; electroplating baths; wastewater treatment sludges from many individual production processes; and heavy ends, light ends, bottom tars, and side-cuts from various distillation processes.

Some commercial chemical products are also listed as being hazardous wastes when discarded. These include "acutely hazardous" wastes such as arsenic acid, cyanides, and many pesticides, as well as "toxic" wastes such as benzene, toluene, and phenols.

EPA has designated five hazardous waste categories. Each hazardous waste is given an EPA Hazardous Waste Number. This is often referred to as the *Hazardous Waste Code.* Each of the five categories may be identified by the prefix letter assigned by EPA. The five categories may be described as follows:

1. Specific types of wastes from nonspecific sources; examples include halogenated solvents, nonhalogenated solvents, electroplating sludges, and cyanide solutions

---

[12]*Risk Assessment in the Federal Government: Managing the Process,* National Research Council, Washington, D.C., National Academy Press, pp. 18–19, 1983.

[13]C. Starr, "Social Benefit Versus Technological Risk," *Science,* vol. 165, pp. 1232–1238, September 1969.

[14]*Code of Federal Regulations,* 40 CFR 260, July 1, 1989. *Note:* CFR means Code of Federal Regulations. The prefix number refers to the volume. Each agency (EPA, DOT, FAA, NRC) has a volume prefix. The suffix number is the paragraph number that spells out the regulation.

from plating batches. (There are 28 listings in this category. See Appendix C.) These wastes have a waste code prefix letter F.

2. Specific types of wastes from specific sources; examples include oven residue from the production of chrome oxide green pigments and brine purification muds from the mercury cell process in chlorine production where separated, prepurified brine is not used. (There are 111 listings in this category. See Appendix C.) These wastes have a waste code prefix letter K.
3. Any commercial chemical product or intermediate, off-specification product, or residue that has been identified as an acute hazardous waste. Examples include potassium silver cyanide, toxaphene, and arsenic oxide. (There are approximately 203 listings in this category.) These wastes have a waste code prefix letter P.
4. Any commercial chemical product or intermediate, off-specification product, or residue that has been identified as hazardous waste. Examples include xylene, DDT, and carbon tetrachloride. (There are approximately 450 listings in this category.) These wastes have a waste code prefix letter U.
5. Characteristic wastes, which are wastes not specifically identified elsewhere, that exhibit properties of ignitability, corrosivity, reactivity, or toxicity. These wastes have a waste code prefix letter D.

The wastes that appear on one of the lists specified in items one through four are called *listed wastes.* Those wastes that are declared hazardous because of their general properties are called *characteristic wastes.* The characteristics of ignitability, corrosivity, and reactivity may be referred to as *ICR*. The toxicity characteristic may be referred to as *TC*.

Ignitability

A solid waste is said to exhibit the characteristic of ignitability if a representative sample of the waste has any of the following properties:

1. It is a liquid, other than an aqueous solution containing less than 24 percent alcohol by volume, and has a flash point less than 60°C.[15]
2. It is not a liquid and is capable, under standard temperature and pressure, of causing fire through friction, absorption of moisture, or spontaneous chemical changes; and, when ignited, burns so vigorously and persistently that it creates a hazard.
3. It is an ignitable, compressed gas.
4. It is an oxidizer.

[15] Although it would seem to be a contradiction in terms, that is, calling a solid waste a liquid, Congress has done what was once only the province of the gods. In Section 1004 (27) of the Resource Conservation and Recovery Act of 1976, they saw fit to violate the laws of physics and make all of the physical states (liquids, gases, and solids) one and the same, that is, solid waste. By their definition, almost any discarded material is solid waste.

A solid waste that exhibits the characteristic of ignitability is given an EPA Hazardous Waste Number of D001.

Corrosivity

A solid waste is said to exhibit the characteristic of corrosivity if a representative sample of the waste has either of the following properties:

1. It is aqueous and has a pH less than or equal to 2 or greater than or equal to 12.5.
2. It is a liquid that corrodes steel at a rate greater than 6.35 mm per year at a test temperature of 55°C.

A solid waste that exhibits the characteristic of corrosivity is given an EPA Hazardous Waste Number of D002.

Reactivity

A solid waste is said to exhibit the characteristic of reactivity if a representative sample of the waste has any of the following properties:

1. It is normally unstable and readily undergoes violent change without detonating.
2. It reacts violently with water.
3. It forms potentially explosive mixtures with water.
4. When mixed with water, it generates toxic gases, vapors, or fumes in a quantity sufficient to present a danger to human health or the environment.
5. It is a cyanide or sulfide-bearing waste that, when exposed to pH between 2 and 12.5, can generate toxic gases, vapors, or fumes in a quantity sufficient to present a danger to human health or the environment.
6. It is capable of detonation or explosive reaction if it is subjected to a strong initiating source or if heated under confinement.
7. It is readily capable of detonation or explosive decomposition or reaction at standard temperature and pressure.
8. It is a forbidden explosive, as defined in Department of Transportation regulations (49 CFR 173.51, 173.53, and 173.88).

A solid waste that exhibits the characteristic of reactivity is given an EPA Hazardous Waste Number of D003.

Toxicity

A solid waste is said to exhibit the characteristic of extraction procedure (EP) toxicity if, using the test methods described in Appendix II of the *Federal Register* (55 FR 11863 and 55 FR 26986),[16] the extract from a representative sample of the waste

[16]The *Federal Register* is published each day that the U.S. Government is in business. It contains proposals for new regulations and promulgation of regulations along with the rationale for the regulations. Once each year the regulations (without the explanations) are codified in the *Code of Federal Regulations*. The notation 55 FR 11863 means that it is the fifty-fifth volume of the Federal Register and

TABLE 9-10
Toxicity characteristic constituents and regulatory levels

EPA HW No.[a]	Constituent	Regulatory level (mg/L)
D004	Arsenic	5.0
D005	Barium	100.0
D018	Benzene	0.5
D006	Cadmium	1.0
D019	Carbon tetrachloride	0.5
D020	Chlordane	0.03
D021	Chlorobenzene	100.0
D022	Chloroform	6.0
D007	Chromium	5.0
D023	o-Cresol	200.0[b]
D024	m-Cresol	200.0[b]
D025	p-Cresol	200.0[b]
D026	Cresol	200.0[b]
D016	2,4-D	10.0
D027	1,4-Dichlorobenzene	7.5
D028	1,2-Dichloroethane	0.5
D029	1,1-Dichloroethylene	0.7
D030	2,4-Dinitrotoluene	0.13[c]
D012	Endrin	0.02
D031	Heptachlor (and its epoxide)	0.008
D032	Hexachlorobenzene	0.13[c]
D033	Hexachloro-1,3-butadiene	0.5
D034	Hexachloroethane	3.0
D008	Lead	5.0
D013	Lindane	0.4
D009	Mercury	0.2
D014	Methoxychlor	10.0
D035	Methyl ethyl ketone	200.0
D036	Nitrobenzene	2.0
D037	Pentachlorophenol	100.0
D038	Pyridine	5.0[c]
D010	Selenium	1.0
D011	Silver	5.0
D039	Tetrachloroethylene	0.7
D015	Toxaphene	0.5
D040	Trichloroethylene	0.5
D041	2,4,5-Trichlorophenol	400.0
D042	2,4,6-Trichlorophenol	2.0
D017	2,4,5-TP (Silvex)	1.0
D043	Vinyl chloride	0.2

[a]Hazardous waste number.
[b]If o-, m-, and p-cresol concentrations cannot be differentiated, the total cresol (D026) concentration is used. The regulatory level for total cresol is 200 mg/L.
[c]Quantitation limit is greater than the calculated regulatory level. The quantitation limit therefore becomes the regulatory level.

---

that the rule is found on page 11,863. When searching for the FR citation it is often handy to have the date on which it was published as it is easier to search by date than page number. 55 FR 26986 was published on June 29, 1990. This procedure was codified in the CFR as 40 CFR 261, Appendix II, Method 1311, Toxicity Characteristic Leaching Procedure (TCLP).

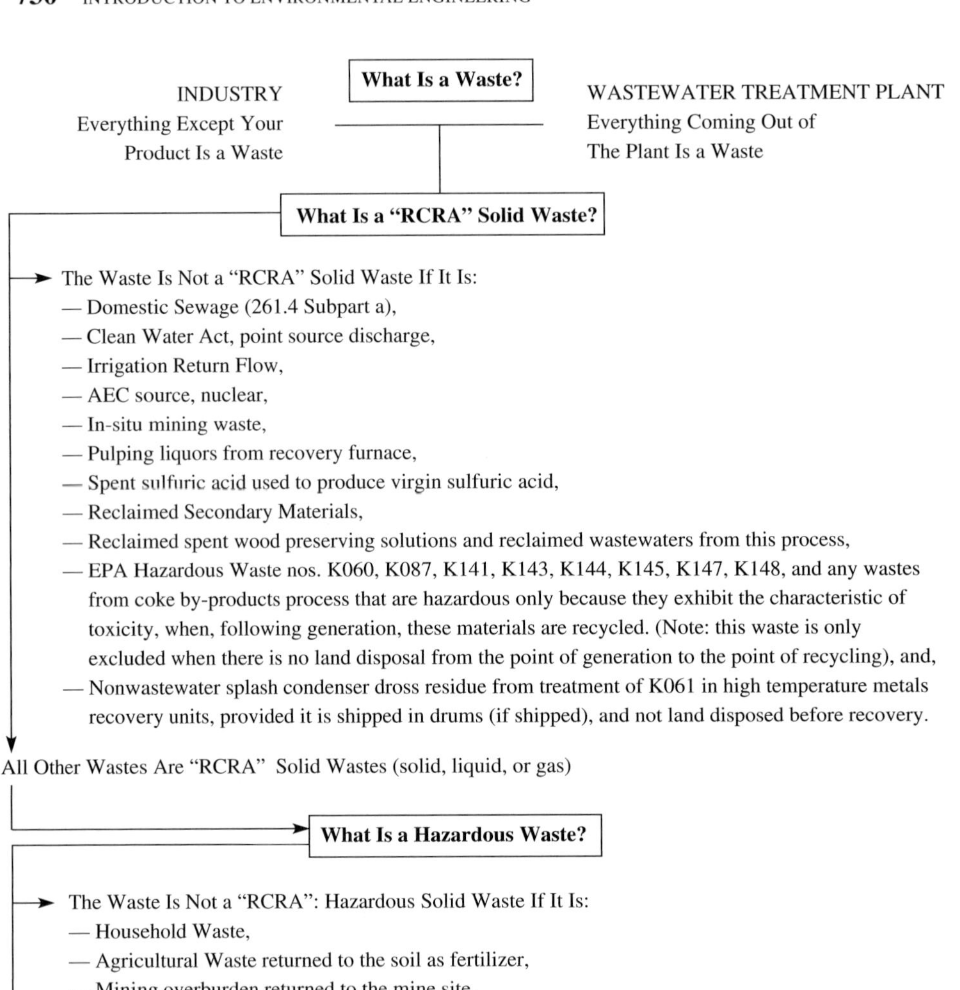

What Is a Waste?
INDUSTRY
Everything Except Your Product Is a Waste
WASTEWATER TREATMENT PLANT
Everything Coming Out of The Plant Is a Waste
What Is a "RCRA" Solid Waste?
The Waste Is Not a "RCRA" Solid Waste If It Is:
— Domestic Sewage (261.4 Subpart a),
— Clean Water Act, point source discharge,
— Irrigation Return Flow,
— AEC source, nuclear,
— In-situ mining waste,
— Pulping liquors from recovery furnace,
— Spent sulfuric acid used to produce virgin sulfuric acid,
— Reclaimed Secondary Materials,
— Reclaimed spent wood preserving solutions and reclaimed wastewaters from this process,
— EPA Hazardous Waste nos. K060, K087, K141, K143, K144, K145, K147, K148, and any wastes from coke by-products process that are hazardous only because they exhibit the characteristic of toxicity, when, following generation, these materials are recycled. (Note: this waste is only excluded when there is no land disposal from the point of generation to the point of recycling), and,
— Nonwastewater splash condenser dross residue from treatment of K061 in high temperature metals recovery units, provided it is shipped in drums (if shipped), and not land disposed before recovery.
All Other Wastes Are "RCRA" Solid Wastes (solid, liquid, or gas)
What Is a Hazardous Waste?
The Waste Is Not a "RCRA": Hazardous Solid Waste If It Is:
— Household Waste,
— Agricultural Waste returned to the soil as fertilizer,
— Mining overburden returned to the mine site,
— Fly ash, scrubber sludge,
— Waste associated with the production of crude oil, gas, or geothermal energy,
— Wastes which fail the test for toxicity characteristic because trivalent chromium (Cr^{3+}) is found, or wastes which are listed in subpart D due to the presence of chromium,
— Waste from extraction, benefication, and processing of ores and minerals,
— Cement kiln dust waste,
— Arsenic treated wood,
— Petroleum-contaminated media and debris,
— Injected groundwater which is hazardous due to the toxicity characteristic,
— Used chlorofluorocarbon refrigerants,
— Non-terne plated used oil filters, and
— Used oil re-refining distillation bottoms (that are used as feedstock to manufacture asphalt products).
The Waste Is Not a "RCRA" Hazardous Solid Waste Unless
— It is listed in Part 261, Subpart D of "RCRA,"
— It is a mixture including a listed (above) substance, and
— It exhibits any of the four specific hazardous waste characteristics.

FIGURE 9-6
Flow scheme for determining if waste is hazardous.

The Waste Is Not a "RCRA" Hazardous Solid Waste If:
— It has been excluded from the lists in Subpart D by petition (delisted).

All Other Wastes Are "RCRA" Hazardous Solid Wastes

What Wastes Are Subject to Regulation?

The "RCRA" Hazardous Solid Waste Is Currently Not Subject to Subtitle C Regulations If:
— The total combined "RCRA" hazardous waste generated at the site is less than 100 kg/mo.,
— It is intended to be legitimately reclaimed or reused, (261.6). However it is subject to RCRA reporting requirements regarding storage and transportion if it is a sludge or contains a Part 261 listed substance.
The "RCRA" Hazardous Solid Waste Is Temporarily Exempt from Certain Regulations If:
— It is a hazardous waste which is generated in a product or raw material storage tank, a product or raw material transport vehicle or vessel, a product or raw material pipeline, or in a manufacturing process unit or an associated non-waste-treatment-manufacturing unit.

All Other "RCRA" Hazardous Solid Wastes Are Subject to Subtitle C of RCRA Regulation with Respect to Disposal, Transport, and Storage.

Requirements for Recyclable Materials

Hazardous Wastes that are not subject to requirements for generator, transporters, and storage facilities:
Regulated under Subparts C through H (261.6):
— Recyclable materials used in a manner constituting disposal,
— Hazardous wastes burned for energy recovery in boilers and industrial furnaces,
— Recyclable materials from which precious metals are reclaimed,
— Spent lead-acid batteries that are being reclaimed.
Not subject to regulation or to the notification requirements of RCRA:
— Industrial ethyl alcohol that is reclaimed,
— Used batteries returned to a battery manufacturer for regeneration,
— Scrap metal,
— Fuels produced from the refining of oil bearing hazardous wastes,
— Oil reclaimed from hazardous waste resulting from normal petroleum refining, production, and transportation practices,
— Hazardous waste fuel produced from oil-bearing hazardous wastes,
— Petroleum coke produced form petroleum refinery hazardous wastes,
Used oil that is recycled and is also a hazardous waste solely because it exhibits a hazardous characteristic is not subject to the requirements of Parts 260–268 of this chapter, but is regulated under Part 279 of this chapter.

FIGURE 9-6
(*continued*)

contains any of the contaminants listed in Table 9-10 at a concentration equal to or greater than the respective value given in the table.

Figure 9-6 shows a generalized flow scheme for determining if a waste is hazardous according to EPA definitions. Of particular importance in the scheme are those things that are not included in the RCRA regulations. For example, domestic sewage, certain nuclear material, household wastes, including toxic and hazardous materials, and small quantities (less than 100 kg/mo) are excluded from the RCRA regulations. This does not mean that these wastes are not regulated at all. In fact they are regulated under other statutes and, thus, do not need to be regulated under RCRA.

Four important but controversial parts of the definition of a hazardous waste are the mixture rule, "contained in" policy, "derived from" rule, and waste-code carry through principle.

The *mixture rule* prevents dilution of waste for the purpose of escaping RCRA Subtitle C regulation. Under 40 CFR 261.3(a)(2), mixtures of a listed hazardous waste and other solid wastes become hazardous wastes. In certain instances for characteristic wastes and in those cases where the listed waste is not to be land disposed, dilution is permitted. The *dilution rules* are summarized in 56 FR 3875, 31 JAN 1991. When the dilution rules apply, the mixture of a hazardous waste with the diluent does not cause the diluent to become hazardous and may render the hazardous waste nonhazardous.

A corollary to the mixture rule is the "*contained in*" policy. Under this policy, media such as soil and water are treated as hazardous wastes if they "contain" listed hazardous wastes.

Any solid waste generated from the treatment, storage, or disposal of a hazardous waste, including any sludge, spill residue, ash, emission-control dust, or leachate (but not precipitation run-off) is a hazardous waste (40 CFR 261.3(c)). This is known as the "*derived from*" rule.

A corollary to the derived from and mixture rules is the "*waste-code carry through*" principle. The principle states that a solid waste derived from a hazardous waste or a mixture of hazardous and nonhazardous waste contains all of the same waste codes as the original waste (53 FR 31138, 31148).

Because of a legal suit and Court order regarding the mixture rule, contained in policy, derived from rule, and waste-code carry through principle, EPA developed and proposed the *Hazardous Waste Identification Rule* (HWIR) (60 FR 66344, 12 DEC 1995). The proposed HWIR establishes exit levels for low-risk solid wastes that are designated as hazardous because they are listed, or have been mixed with, derived from, or contain a listed hazardous waste. If the constituents of the waste are below the exit levels, the waste may be disposed of as a Subtitle D waste.

RCRA provides a petition mechanism (40 CFR 260.20 and 260.22) for excluding a waste from nonspecific sources and at a particular generating facility. Those wastes that successfully pass the petition process are *delisted.* The list of delisted wastes appears in Appendix IX of 40 CFR 261.

Some waste streams do not come under the purview of RCRA but are, nonetheless, considered hazardous. These special wastes include, for example, polychlorinated biphenyls (PCBs) and asbestos. PCBs and asbestos are regulated under the Toxic Substances Control Act (abbreviated TSCA and pronounced "tas-kah").

9-4 RCRA AND HSWA

Congressional Actions on Hazardous Waste

In 1976 Congress passed the Resource Conservation and Recovery Act (abbreviated RCRA and pronounced "rick-rah") directing the U.S. Environmental Protection

Agency to establish hazardous waste regulations. RCRA was amended in 1984 by the Hazardous and Solid Waste Amendments (abbreviated HSWA and pronounced "hiss-wah"). RCRA and HSWA were enacted to regulate the generation and disposal of hazardous wastes. These acts did not address abandoned or closed waste disposal sites or spills. The Comprehensive Environmental Response, Compensation, and Liability Act (abbreviated by CERCLA and pronounced "sir-klah"), commonly referred to as "Superfund," was enacted in 1980 to address these problems. SARA, the Superfund Amendments and Reauthorization Act of 1986, extended the provisions of CERCLA. In the following sections, we shall attempt to tell you about the who, what, where, and how of RCRA, HSWA, CERCLA, and SARA.

Cradle-to-Grave Concept

The EPA's cradle-to-grave hazardous waste management system is an attempt to track hazardous waste from its generation point (the "cradle") to its ultimate disposal point (the "grave"). The system requires generators to attach a manifest (itemized list describing the contents) form to their hazardous waste shipments. This procedure is designed to ensure that wastes are directed to, and actually reach, a permitted disposal site.

Generator Requirements[17]

Generators of hazardous waste are the first link in the cradle-to-grave chain of hazardous waste management established under RCRA. Generators of more than 100 kilograms of hazardous waste or 1 kilogram of acutely hazardous waste per month must (with a few exceptions) comply with all of the generator regulations.

The regulatory requirements for hazardous waste generators include:

1. Obtaining an EPA ID number
2. Handling of hazardous waste before transport
3. Manifesting of hazardous waste
4. Recordkeeping and reporting

EPA assigns each generator a unique identification number. Without this number the generator is barred from treating, storing, disposing of, transporting, or offering for transportation any hazardous waste. Furthermore, the generator is forbidden from offering the hazardous waste to any transporter, or treatment, storage, or disposal (TSD) facility that does not also have an EPA ID number.

Pre-transport regulations are designed to ensure safe transportation of a hazardous waste from origin to ultimate disposal. In developing these regulations, EPA

[17]U.S. Environmental Protection Agency, *RCRA Orientation Manual* (U.S. Environmental Protection Agency Office of Solid Waste, Publication No. EPA/530-SW-86-001), Washington, DC: U.S. Government Printing Office, January 1986.

adopted those used by the Department of Transportation (DOT) for transporting hazardous wastes (49 CFR Parts 172, 173, 178, and 179). These DOT regulations require:

1. Proper packaging to prevent leakage of hazardous waste during both normal transport conditions and in potentially dangerous situations, such as when a drum falls out of a truck.
2. Identification of the characteristics and dangers associated with the wastes being transported through labeling, marking, and placarding of the packaged waste.

These pre-transport regulations only apply to generators shipping waste off site.

In addition to adopting the DOT regulations outlined above, EPA also developed pre-transport regulations that cover the accumulation of waste prior to transport. A generator may accumulate hazardous waste on site for 90 days or less as long as the following requirements are met:

1. *Proper storage:* The waste is properly stored in containers or tanks marked with the words "Hazardous Waste" and the date on which accumulation began.
2. *Emergency plan:* There is a contingency plan and emergency procedures for use in an emergency.
3. *Personnel training:* Facility personnel are trained in the proper handling of hazardous waste.

The 90-day period allows a generator to collect enough waste to make transportation more cost effective, that is, instead of paying to haul several small shipments of waste, the generator can accumulate waste until there is enough for one big shipment.

If the generator accumulates hazardous waste on site for more than 90 days, it is considered an operator of a storage facility and must comply with requirements for such facilities. Under temporary, unforeseen, and uncontrollable circumstances, the 90-day period may be extended for up to 30 days by the EPA Regional Administrator on a case-by-case basis.

There is an exception to this 90-day accumulation period that applies to generators of between 100 and 1,000 kg/mo of hazardous waste who ship their waste off site. People who fall in this category are called *small quantity generators* (SQG). HSWA required, and the EPA developed, regulations that allow such generators to accumulate waste for 180 days (or 270 days if the waste must be shipped over 200 miles) before they are considered the operator of a storage facility.

The Uniform Hazardous Waste Manifest (the *manifest*) is the key to cradle-to-grave waste management (see Figure 9-7). Through the use of a manifest, generators can track the movement of hazardous waste from the point of generation to the point of ultimate treatment, storage, or disposal.

HSWA requires that each manifest certify that the generator has in place a program to reduce the volume and toxicity of the waste to the degree that is economically practicable, as determined by the generator, and that the treatment, storage, or disposal method chosen by the generator is the best practicable method currently available that minimizes the risk to human health and the environment.

Please print or type. *(Form designed for use on elite (12-pitch) typewriter.)* *Form Approved. OMB No. 2000-0404. Expires 7-31-86*

UNIFORM HAZARDOUS WASTE MANIFEST

1. Generator's US EPA ID No. | Manifest Document No.

2. Page 1 of

Information in the shaded areas is not required by Federal law.

3. Generator's Name and Mailing Address

A. State Manifest Document Number

B. State Generator's ID

4. Generator's Phone ( )

5. Transporter 1 Company Name

6. US EPA ID Number

C. State Transporter's ID

D. Transporter's Phone

7. Transporter 2 Company Name

8. US EPA ID Number

E. State Transporter's ID

F. Transporter's Phone

9. Designated Facility Name and Site Address

10. US EPA ID Number

G. State Facility's ID

H. Facility's Phone

11. US DOT Description *(Including Proper Shipping Name, Hazard Class, and ID Number)*	12. Containers		13. Total Quantity	14. Unit Wt/Vol	I. Waste No.
	No.	Type			
a.					
b.					
c.					
d.					

J. Additional Descriptions for Materials Listed Above

K. Handling Codes for Wastes Listed Above

15. Special Handling Instructions and Additional Information

16. GENERATOR'S CERTIFICATION: I hereby declare that the contents of this consignment are fully and accurately described above by proper shipping name and are classed, packed, marked, and labeled, and are in all respects in proper condition for transport by highway according to applicable international and national government regulations.

Unless I am a small quantity generator who has been exempted by statute or regulation from the duty to make a waste minimization certification under Section 3002(b) of RCRA, I also certify that I have a program in place to reduce the volume and toxicity of waste generated to the degree I have determined to be economically practicable and I have selected the method of treatment, storage, or disposal currently available to me which minimizes the present and future threat to human health and the environment.

Printed/Typed Name | Signature | *Month Day Year*

17. Transporter 1 Acknowledgement of Receipt of Materials

Printed/Typed Name | Signature | *Month Day Year*

18. Transporter 2 Acknowledgement of Receipt of Materials

Printed/Typed Name | Signature | *Month Day Year*

19. Discrepancy Indication Space

20. Facility Owner or Operator: Certification of receipt of hazardous materials covered by this manifest except as noted in Item 19.

Printed/Typed Name | Signature | *Month Day Year*

GENERATOR / TRANSPORTER / FACILITY

EPA Form 8700-22 (Rev. 4-85) Previous edition is obsolete.

FIGURE 9-7
Uniform hazardous waste manifest.

It is especially important for the generators to prepare the manifest properly since they are responsible for the hazardous waste they produce and its ultimate disposition.

The manifest is part of a controlled tracking system. Each time the waste is transferred, that is, from a transporter to the designated facility or from a transporter to another transporter, the manifest must be signed to acknowledge receipt of the waste. A copy of the manifest is retained by each link in the transportation chain. Once the waste is delivered to the designated facility, the owner or operator of the facility must send a copy of the manifest back to the generator. This system ensures that the generator has documentation that the hazardous waste has made it to its ultimate destination.

If 35 days pass from the date on which the waste was accepted by the initial transporter and the generator has not received a copy of the manifest from the designated facility, the generator must contact the transporter and/or the designated facility to determine the whereabouts of the waste. If 45 days pass and the manifest still has not been received, the generator must submit an exception report.

The recordkeeping and reporting requirements for generators provide EPA and states with a method for tracking the quantities of waste generated and the movement of hazardous wastes. There are three primary recordkeeping and reporting requirements: biennial reporting; exception reporting; and three-year retention of reports, manifests, and test records.

Biennial reporting. Generators who transport hazardous waste off site must submit a biennial report to the Regional Administrator by March 1 of each even-numbered year. The report details the generator's activities during the previous calendar year including:

1. EPA identification number and name of each transporter used throughout the year
2. EPA identification number, name, and address of each off-site treatment, storage, or disposal facility to which waste was sent during the year
3. Quantities and nature of each hazardous waste generated

HSWA expanded the scope of the biennial report to include a description of efforts taken to reduce the volume and toxicity of the wastes generated and changes in volume or toxicity that were actually achieved, as compared with those achieved in previous years. In other words, the generators must describe what they are doing to minimize waste generation.

Generators who treat, store, or dispose of their hazardous waste on site also must submit a biennial report that contains a description of the type and quantity of hazardous waste the facility handled during the year, and the method(s) of treatment, storage, or disposal used.

Exception reports. In addition to the biennial report, generators who transport waste off site must submit an exception report to the Regional Administrator if they do not receive a copy of the manifest signed and dated by the owner or operator of the designated facility within 45 days from the date on which the initial transporter accepted the waste. The exception report must describe the efforts taken to locate the waste and the results of these efforts.

Three-year retention of reports, manifests, and test records. Generators must keep a copy of each biennial report and any exception reports for a period of at least three years from the date the report was submitted. Generators are also required to keep a copy of all manifests for three years or until they receive a copy of the manifest signed and dated from the owner or operator of the designated facility. The manifest from the facility must then be kept for at least three years from the date on which the hazardous waste was accepted by the initial transporter. Finally, the records of the waste analyses and determinations undertaken by generators must be kept for at least three years from the date that the waste was last sent to an on-site or off-site TSD. The periods of retention can be extended automatically during the course of any unresolved enforcement action regarding the regulated activity or as requested by the Administrator.

Transporter Regulations[18]

Transporters of hazardous waste are the critical link between the generator and the ultimate off-site treatment, storage, or disposal of hazardous waste. The transporter regulations were developed jointly by EPA and the DOT to avoid contradictory requirements coming from the two agencies. Although the regulations are integrated, they are not contained under the same act. A transporter must comply with the regulations under 49 CFR 171-179 (The Hazardous Materials Transportation Act) as well as those under 40 CFR Part 263 (Subtitle C of RCRA).

A transporter under RCRA is defined as any person engaged in the off-site transportation of hazardous waste within the United States, if such transportation requires a manifest under 40 CFR 262. This definition covers transport by air, highway, or water. The transporter regulations do not apply to either the on-site transportation of hazardous waste by generators who have their own TSDs or the TSDs transporting wastes within a facility.

Under certain circumstances a transporter of hazardous waste may be subject to regulatory requirements other than those contained in 40 CFR 263. Once a transporter accepts hazardous waste from a generator or another transporter, he or she can store it at a transfer station for up to 10 days without being subject to any new regulations. However, if the storage time exceeds 10 days, the transporter is considered to be operating a storage facility and must comply with the regulations for such a facility. In addition, transporters who bring hazardous waste into the United States or mix hazardous wastes of different DOT shipping descriptions by placing them in

[18]U.S. Environmental Protection Agency, *RCRA Orientation Manual,* January 1986.

the same container are classified as generators and must comply with the regulations applicable to generators.

A transporter is subject to a number of regulations, including ones for obtaining an EPA ID number, complying with the manifest system, and dealing with hazardous waste discharges.

The transporter is required to deliver the entire quantity of waste that is accepted (from either the generator or another transporter) to the designated facility listed on the manifest. If the waste cannot be delivered as the manifest directs, the transporter must inform the generator and receive further instructions, such as to return the waste or take it to another facility. Before handing the waste over to a TSD, the transporter must have the TSD facility operator sign and date the manifest. One copy of the manifest remains at the TSD facility while the other stays with the transporter. The transporter must retain a copy of the manifest for three years from the date the hazardous waste was accepted by the initial transporter.

Even if generators and transporters of hazardous waste comply with all appropriate regulations, transporting hazardous waste can still be dangerous. There is always the possibility that an accident will occur. To deal with this possibility, the regulations require transporters to take immediate action to protect health and the environment if a release occurs by notifying local authorities and/or diking off the discharge area.

The regulations also give officials special authority to deal with transportation accidents. Specifically, if a federal, state, or local official, with appropriate authority, determines that the immediate removal of the waste is necessary to protect human health or the environment, the official can authorize waste removal by a transporter who lacks an EPA ID and without the use of a manifest.

Treatment, Storage, and Disposal Requirements[19]

Treatment, storage, and disposal facilities (TSDs) are the last link in the cradle-to-grave hazardous waste management system. All TSDs handling hazardous waste must obtain an operating permit and abide by the treatment, storage, and disposal regulations. The TSD regulations establish performance standards that owners and operators must apply to minimize the release of hazardous waste into the environment.

A TSD facility may perform one or more of the following functions:

1. *Treatment:* Any method, technique, or process, including neutralization, designed to change the physical, chemical, or biological character or composition of any hazardous waste so as to neutralize it or render it nonhazardous or less hazardous; to recover it; make it safer to transport, store, or dispose of; or make it amenable for recovery, storage, or volume reduction.

[19]U.S. Environmental Protection Agency, *RCRA Orientation Manual,* January 1986.

2. *Storage:* The holding of hazardous waste for a temporary period, at the end of which the hazardous waste is treated, disposed, or stored elsewhere.
3. *Disposal:* The discharge, deposit, injection, dumping, spilling, leaking, or placing of any solid waste or hazardous waste into or on any land or water so that any constituent thereof may enter the environment or be emitted into the air or discharged into any waters, including ground waters.

The act establishes two categories of TSDs based on a facility's status regarding a permit. The first category is made up of "interim status" facilities, that is, ones that have not obtained a permit. Although the act specifies that only permitted facilities may operate, Congress, in developing this requirement, recognized that it would take many years for EPA to issue all permits. Therefore, it established interim status. This allows owners and operators of facilities in existence on November 19, 1980 (or brought under regulation due to an amendment), who meet certain conditions, to continue operating as if they have a permit until their permit is issued or their application is denied. The second category is made up of facilities that have a permit.

Under Section 3004(a) of the act, EPA was required to develop regulations for all TSDs. Although only one set was required, EPA developed two sets of regulations, one for interim status TSDs, the other for permitted TSDs. The reason for doing this is that during the process of developing TSD regulations, EPA decided that owners and operators of facilities in interim status should meet only a portion of the requirements permitted facilities must meet.

The interim status standards, found in 40 CFR 265, are primarily "good housekeeping practices" that owners and operators must follow to properly manage hazardous wastes during the interim status period. The permit standards found in 40 CFR 264 on the other hand, are "design and operating" criteria that writers of permits include in facility-specific permits. Both interim status and permit standards consist of administrative/nontechnical requirements and technical requirements.

The administrative/nontechnical requirements are nearly identical for interim status and permit standards. The technical requirements applied to interim status and permitted facilities are significantly different.

The purpose of the administrative/nontechnical requirements is to ensure that owners and operators of TSDs establish the necessary procedures and plans to operate a facility properly and to handle any emergencies or accidents. They cover the subject areas shown below:

Subpart	Subject
A	Who is subject to the regulations
B	General facility standards
	Waste analysis, security, inspections, training
	Ignitable, reactive, or incompatible wastes
	Location standards (permitted facilities)
C	Preparedness and prevention
D	Contingency plans and emergency procedures
E	Manifest system, recordkeeping, and reporting

The objective of the interim status technical requirements is to minimize the potential for threats resulting from hazardous waste treatment, storage, and disposal at existing facilities waiting to receive an operating permit. There are two groups of interim status requirements: general standards that apply to several types of facilities and specific standards that apply to a waste management method.

The general standards cover three areas:

1. Groundwater monitoring requirements
2. Closure, postclosure requirements
3. Financial requirements

Groundwater monitoring is only required of owners or operators of a surface impoundment, landfill, land treatment facility, or some waste piles used to manage hazardous waste. The purpose of these requirements is to assess the impact of a facility on the groundwater beneath it. Monitoring must be conducted for the life of the facility except at land disposal facilities, which must continue monitoring for up to 30 years after the facility has closed.

The groundwater monitoring program outlined in the regulations requires a monitoring system of four wells to be installed: one upgradient from the waste management unit and three downgradient. The downgradient wells must be placed so as to intercept any waste migrating from the unit, should such a release occur. The upgradient wells must provide data on groundwater that is not influenced by waste coming from the waste management unit (called background data). If the wells are properly located, comparison of data from upgradient and downgradient wells should indicate if contamination is occurring.

Once the wells have been installed, the owner or operator monitors them for one year to establish background concentrations for selected chemicals. These data form the basis for all future data comparisons. There are three sets of parameters for which background concentrations are established: drinking water parameters, groundwater quality parameters, and groundwater contamination parameters.

Closure is the period when wastes are no longer accepted, during which owners or operators of TSD facilities complete treatment, storage, and disposal operations, apply final covers to or cap landfills, and dispose of or decontaminate equipment, structures, and soil. Postclosure, which applies only to disposal facilities, is the 30-year period after closure during which owners or operators of disposal facilities conduct monitoring and maintenance activities to preserve and look after the integrity of the disposal system.

Financial requirements were established to ensure that funds are available to pay for closing a facility, for rendering post-closure care at disposal facilities, and to compensate third parties for bodily injury and property damage caused by sudden and non-sudden accidents related to the facility's operation (states and federal governments are exempted from abiding by these requirements). There are two kinds of financial requirements: financial assurance for closure/postclosure and liability coverage for injury and property damage.

The first step owners and operators must take in meeting the financial assurance requirements is to prepare written cost estimates for closing their facilities. If postclosure care is required, a cost estimate for providing this care must also be prepared. These cost estimates must reflect the actual cost of conducting all the activities outlined in the closure and postclosure plans and are adjusted annually for inflation. The cost estimate for closure is based on the point in the facility's operating life when closure would be the most expensive. Cost estimates for postclosure monitoring and maintenance are based on projected costs for the entire post-closure period.

Following the preparation of the cost estimates, the owner or operator must demonstrate to EPA the ability to pay the estimated amounts. This is known as financial assurance. Six mechanisms for guaranteeing financial assurance to assure EPA that funds are available for closure and postclosure activities are described in the regulations. All six mechanisms are adjusted annually for inflation, or more frequently if cost estimates change. The six mechanisms are trust fund, surety bond, letter of credit, closure/postclosure insurance, corporate guarantee for closure, and financial test.

An owner or operator is financially responsible or liable for bodily injury and property damage to third parties caused by a sudden accidental occurrence or a non-sudden accidental occurrence due to operations at a facility. Sudden occurrences are usually due to an accident, such as an explosion or fire. Non-sudden occurrences take place over a long period of time and include groundwater and surface water contamination. Separate liability coverage for each of these two types of occurrences must be obtained.

For sudden accidental occurrences, an owner or operator of a TSD must have liability coverage for at least $1 million per occurrence with an annual aggregate of at least $2 million, exclusive of legal defense costs.

Only an owner or operator of a surface impoundment, landfill, land treatment facility, or group of such facilities must maintain liability for non-sudden accidental occurrences. They must maintain at least $3 million per occurrence with an annual aggregate of at least $6 million, exclusive of legal defense costs.

Land ban. The Hazardous and Solid Waste Amendments (HSWA) of 1984 significantly expanded the scope of the Resource Conservation and Recovery Act (RCRA). HSWA was created, in large part, in response to strongly voiced citizen concerns that existing methods of hazardous waste disposal, particularly land disposal, were not safe. Section 3004 of the act sets restrictions on land disposal of specific wastes. This is commonly called the "land ban," or land disposal restrictions (LDR). As specifically required by Section 3004(m), the agency must establish levels or methods of treatment, if any, which substantially reduce the likelihood of migration of hazardous constituents from waste so that short-term and long-term threats to human health and the environment are minimized. Congress established a stringent timetable for development of treatment standards. After the effective date of the promulgated standards, listed and characteristic wastes must be treated to meet the standards before the wastes can be placed in any form of land disposal facility. The only exception is where a special variance is approved based on a showing of no migration of hazardous constituents from the land disposal site for as long as the waste remains hazardous. The last set of Congressionally mandated standards was promulgated in accordance with the timetable on May 8, 1990. EPA has subsequently published revisions to clarify and streamline the standards. The *Universal Treatment Standards* (UTS) are of particular note in this respect.

Prior to 1994, treatment facilities managing hazardous waste often had to meet LDR treatment standards established for many different listed and characteristic wastes. In some cases, a constituent regulated to a given concentration level for one waste was also regulated in another waste at a different concentration level. On 18 SEP 1994, EPA published the UTS to eliminate these differences.[20]

Underground Storage Tanks (UST)[22]

A "UST system"[22] includes an underground storage tank, connected piping, underground ancillary equipment, and containment system, if any. On September 23, 1988, the EPA promulgated the final rules for underground storage tanks.

There are a number of exclusions to the new regulations, including:

Hazardous waste UST systems

Regulated wastewater treatment facilities

Any equipment or machinery that contains regulated substances for operational purposes such as hydraulic lift tanks and electrical equipment tanks

Any UST system of less than 415 liters

Any UST system containing a *de minimis* (negligible) concentration of regulated substances

Any emergency spill or overflow containment system that is expeditiously emptied after use

Some of the technical requirements pertaining to release detection and tank tightness testing are to be phased in over a five-year period for existing UST systems. After the effective date of December 22, 1988, all existing UST systems must fully comply within ten years. The regulations are deferred for certain UST systems, including some wastewater treatment systems, and UST systems with field constructed tanks.

All UST systems installed after December 22, 1988, must have corrosion protection. There are three ways to obtain corrosion protection for tanks: (1) construction

---

[20]59 FR 47980, 18 SEP 1994. Corrections to the UTS were published on 3 JAN 1995, 60 FR 242.

[21]Portions of this discussion were excerpted from J. E. Bair, "Underground Storage Tanks (UST)," *Michigan Environmental Law Journal,* vol. 8, no. 2, p. 2, 1989.

[22]You can imagine the chagrin of regulators and others when the acronym for Leaking Underground Storage Tanks appears on meeting agendas and technical symposia!

of fiberglass-reinforced plastic, (2) steel- and fiberglass-reinforced plastic composite, or (3) a coated steel tank with cathodic protection. Cathodic protection systems must be regularly tested and inspected. All owners and operators must also provide spill and overfill prevention equipment and a certificate of installation to ensure that the methods of installation were in compliance with the regulations.

All existing UST systems must meet the new UST standards, the upgrading requirements, or the closure requirements within ten years, that is, by December 22, 1998. There are certain upgrading requirements that apply to current systems. Upgrading requires that steel tanks be lined and inspected internally at regular intervals thereafter. Cathodic protection may be provided, depending upon the age and integrity of the tank. All existing metal piping must be cathodically protected. Further, all existing UST systems must comply with the new requirements for spill and overfill prevention equipment.

A UST system may be temporarily closed for up to 12 months while meeting certain minimal requirements. If the UST system does not meet the new or upgraded standards, it must be permanently closed after 12 months of being out of service. Permanent closure requires that the UST system be removed or be left in place after removing all regulated substances and cleaning the tank, filling it with an inert substance, and closing it to all future outside access. In addition, owners and operators must perform assessment at the time of closure to ensure that a release has not occurred at the site. If a release has occurred, then corrective action must be taken.

Release (leak) detection must be instituted for all UST systems on a graduated timetable depending on the age of the tank. Several different methods are allowed for petroleum UST systems. However, some systems have specific requirements, for instance, a pressurized delivery system must be equipped with an automatic line leak detector and have an annual line tightness test. All new or upgraded UST systems storing hazardous substances must have secondary containment with interstitial monitoring.

With new systems there is an option of either monthly monitoring or monthly inventory control and tank tightness testing every five years. This choice is only allowed for ten years after installation upgrade. Such monitoring may include automatic tank gauging and the monitoring of vapor, groundwater, and interstitial areas. Similar specific requirements exist for both pressurized and suction piping systems.

The technical requirements of the UST rules are summarized in Figure 9-8.

When release is confirmed, owners and operators must begin corrective action. Immediate corrective action measures include mitigation of safety and fire hazards, removal of saturated soils and floating free product, and an assessment of further corrective action needed. As with any remediation situation, a corrective action plan may be required for long-term cleanups of contaminated soil and groundwater.

The financial responsibility rules generally apply to all owners and operators of petroleum UST systems. Exemptions include government entities whose debts and liabilities are those of a state or the United States, and those owners and operators of UST systems excluded or deferred from compliance with the technical regulations.

Both the owner and operator are liable in the event of noncompliance; either may meet the financial responsibility requirement.

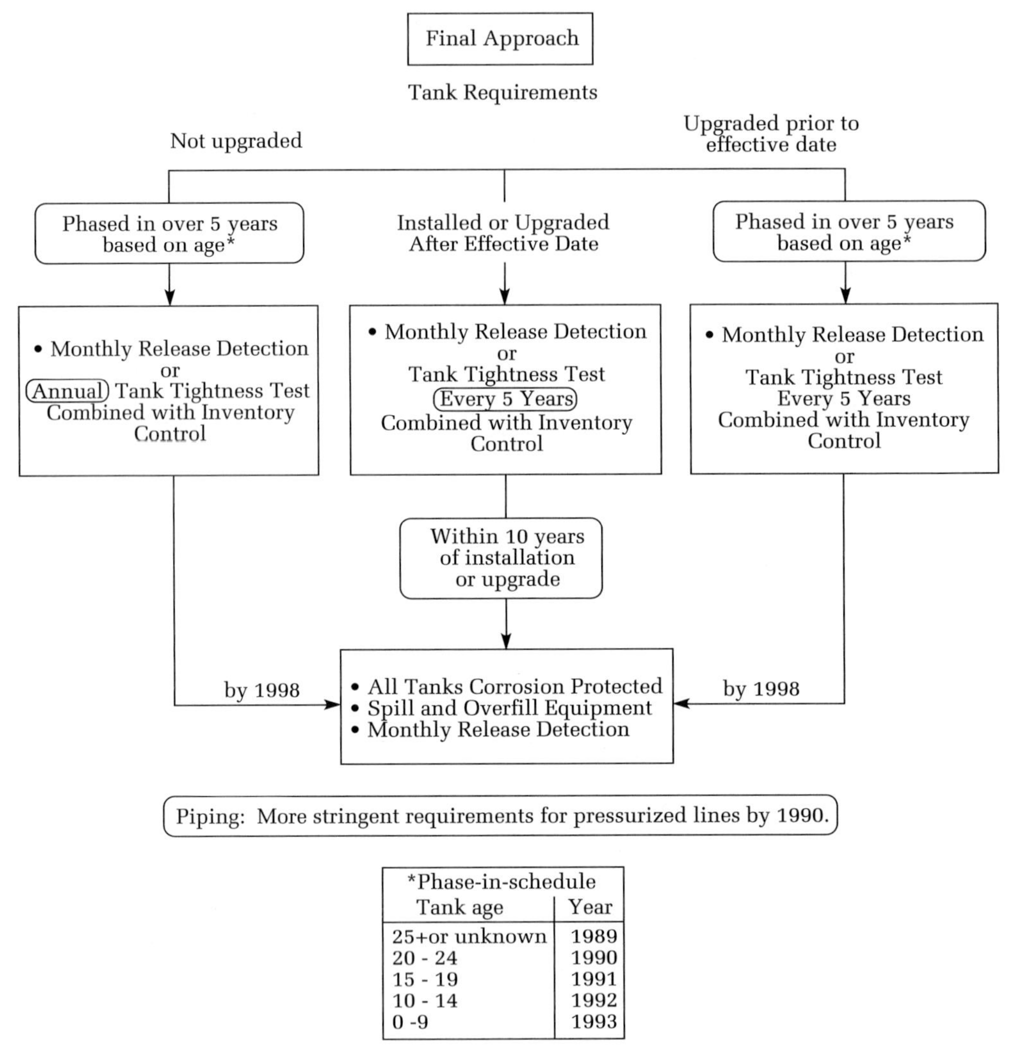

*Phase-in-schedule	
Tank age	Year
25+or unknown	1989
20 - 24	1990
15 - 19	1991
10 - 14	1992
0 -9	1993

FIGURE 9-8
Summary of technical requirements for UST rules.

For purposes of covering costs of corrective action and third-party liability, the following assurance elements apply:

1. $1 million per occurrence: For all owners and operators of petroleum USTs at facilities engaged in petroleum production, refining, or marketing, and owners and operators of USTs with an average monthly throughput of more than 38 cubic meters.
2. $500,000 per occurrence: For owners or operators not engaged in petroleum production, refining, or marketing with an average throughput of 38 cubic meters or less.

All owners or operators must maintain an annual aggregate of $1 million for 1 to 100 individual containment units (tanks not systems) or $2 million for 101 or more tanks.

9-5 CERCLA AND SARA

The Superfund Law

The Comprehensive Environmental Response, Compensation, and Liability Act (CERCLA) of 1980, better known as "Superfund," became law "to provide for liability, compensation, cleanup and emergency response for hazardous substances released into the environment and the cleanup of inactive hazardous waste disposal sites." CERCLA was generally intended to give EPA authority and funds to clean up abandoned waste sites and to respond to emergencies related to hazardous waste. The law provides for both response and enforcement mechanisms. The four major provisions of the law establish:

1. A fund (the "superfund") to pay for investigations and remedies at sites where the responsible people cannot be found or will not voluntarily pay;
2. A priority list of abandoned or inactive hazardous waste sites for cleanup (the National Priority List);
3. The mechanism for action at abandoned or inactive sites (the National Contingency Plan);
4. Liability for those responsible for cleaning up.

Initially the trust fund was supported by taxes on producers and importers of petroleum and 42 basic chemicals. In its first five-year period, Superfund collected about 1.6 billion dollars, with 86 percent of that money coming from industry and the remainder from federal government appropriations. In 1986 the Superfund Amendments and Reauthorization Act (SARA) greatly expanded the money available to remediate Superfund sites. The fund was raised to $8.6 billion for a five-year period by taxing petroleum products ($2.75 billion), business income ($2.5 billion), and chemical feedstocks ($1.4 billion). The remainder is from general revenues.

The National Priority List (NPL)

The NPL serves as a tool for the EPA to use in identifying sites that appear to present a significant risk to public health or the environment and that may merit use of Superfund money. First published in 1982, it is updated three times a year. In 1997 the list contained over 1,200 sites. The first NPL was formulated from notification procedures and existing information sources. Subsequently, a numeric ranking system known as the *Hazard Ranking System* (HRS) was developed. Sites with high HRS scores may be added to the list. Sites on the NPL are eligible for Superfund money. Those with lower scores are not likely to be eligible.

The Hazard Ranking System (HRS)[23,24]

The HRS is a procedure for ranking uncontrolled hazardous waste sites in terms of the potential threat based upon containment of the hazardous substances, route of release, characteristics and amount of the substances, and likely targets. The methodology of the HRS provides a quantitative estimate that represents the relative hazards posed by a site and takes into account the potential for human and environmental exposure to hazardous substances. The HRS score is based on the probability of contamination from four pathways—groundwater, surface water, soil, and air—on the site in question. The groundwater and air migration pathways are evaluated for ingestion and inhalation respectively. The surface water migration and soil exposure pathways are evaluated for multiple intake routes. Surface water is evaluated for (1) drinking water, (2) human food chain, and (3) environmental (contact) exposures. These exposures are evaluated for two separate migration components—overland/flood migration and groundwater to surface water migration. Soil is evaluated for potential exposure to the (1) resident population and (2) nearby population.

Use of the HRS requires considerable information about the site and its surroundings, the hazardous substances present, and the geology of the aquifers and the intervening strata. The factors that most affect an HRS site score are the proximity to a densely populated area or source of drinking water, the quantity of hazardous substances present, and the toxicity of those hazardous substances. The HRS methodology has been criticized for the following reasons:

1. There is a strong bias toward human health effects, with only a slight chance of a site in question receiving a high score if it represents only a threat or hazard to the environment.
2. Because of the human health bias, there is an even stronger bias in favor of highly populated affected areas.
3. The air emission migration route must be documented by an actual release, while groundwater and surface water routes have no such documentation requirement.
4. The scoring for toxicity and persistence of chemicals may be based on site containment, which is not necessarily related to a known or potential release of the toxic chemicals.
5. A high score for one migration route can be more than offset by low scores for the other migration routes.
6. Averaging of the route scores creates a bias against a site that has only one hazard, even though that one hazard may pose extreme threat to human health and the environment.

EPA provides quality assurance and quality control for each HRS score to ensure that these site evaluations are performed on a consistent basis. These HRS scores

---

[23]This discussion follows that of C. A. Wentz, *Hazardous Waste Management,* New York: McGraw-Hill, pp. 392–403, 1989.

[24]40 CFR 300, Appendix A, 1 JUL 1994 and 55 FR 51583, 14 DEC 1990.

range from 0 to 100, with a score of 100 representing the most hazardous site. Occasional exceptions have been made in the HRS priority ranking to meet the CERCLA requirement that a site designated by a state as its top priority be included on the NPL.

The National Contingency Plan (NCP)[25,26]

The National Contingency Plan (NCP) provides detailed direction on the action to be taken at a hazardous waste site, including initial assessment to determine if an emergency or imminent threat exists, emergency response actions, and a method to rank sites (the HRS) and establish priority for future action. When there is sufficient indication that a site poses a potential risk to the environment, a detailed study is required.

The NCP describes the steps to be taken for the detailed evaluation of the risks associated with a site. Such an evaluation is termed a *remedial investigation* (RI). The process of selecting an appropriate remedy is termed the *feasibility study* (FS). The remedial investigation and the feasibility study are often combined into a single measure, known popularly as a remedial investigation/feasibility study (RI/FS). The requirements of the RI/FS are usually outlined in a written work plan, which must be approved by the relevant federal and state agencies before it may be implemented.

A remedial investigation includes the development of detailed plans that address the following items:

1. *Site characterization:* A description of the hydrogeological and geophysical sampling and analytical procedures to be applied in order to discover the nature and extent of the waste materials, the physical characteristics of the site, and any receptors that could be affected by the wastes at the site.
2. *Quality control:* The guidelines to be enforced to ensure that all the data collected from the characterization program are valid and satisfactorily accurate.
3. *Health and safety:* The procedures to be employed to protect the safety of the individuals who will work at the site and perform the site characterization.

The RI activities and subsequent evaluation of the data gathered are termed a *risk assessment* or an endangerment assessment. The remedial investigation report documents the evaluation.

The remedial investigation report serves as a basis for the feasibility study, which evaluates various remedial alternatives. The review criteria include: overall protection of human health and the environment; compliance with applicable or relevant and appropriate regulations; long-term effectiveness; reduction in toxicity,

---

[25]Parts of this discussion were excerpted from O'Brien & Gere Engineers Inc., *Hazardous Waste Site Remediation,* New York: Van Nostrand Reinhold, pp. 11–13, 1988.

[26]40 CFR 300.400, Subpart E, 1 JUL 1994 and 55 FR 8839, 8 MAR 1990.

mobility, or volume; short-term effectiveness; technical and administrative implementability; cost; state acceptance; and community acceptance. All the remedies selected must be capable of reducing the risk at the hazardous waste site to an acceptable level. And, in general, the lowest-cost alternative that achieves this objective is chosen as the course of action. The results of the feasibility study are presented in a written report, called the *record of decision* (ROD). This document serves as a preliminary basis for the design of the selected alternative.

One of the keys to the National Contingency Plan is that it specifies that the degree of cleanup be selected in accordance with several criteria, including the degree of hazard to the "public health, welfare and the environment." Therefore, there is no predetermined level of remediation that can be required or that must be achieved at any site. Rather, the degree of correction is established on a site-by-site basis. What is acceptable in one location may not necessarily be acceptable in another.

Upon completion and approval of the RI/FS, the next step is the preparation of plans and specifications for the selected remedy—the *remedial design* (RD). To complete the process, the actual construction and other activities are undertaken in accordance with the plans and specifications.

Liability

Perhaps the most far reaching provision of CERCLA that has stood the test of the courts was the establishment of *strict, joint, and several liability* for cleanup of an NPL site. Those identified by EPA as *potentially responsible parties* (PRPs) may include generators, present owners, or former owners of facilities or real property where hazardous wastes have been stored, treated, or disposed of, as well as those who accepted hazardous waste for transport and selected the facility. PRPs have strict liability; that is, liability without fault. Neither care nor negligence, neither good nor bad faith, neither knowledge nor ignorance, can be claimed as a defense. Congress correctly predicted that there would be instances where the PRPs would contest their contribution to the problem and would, then, be unwilling to share the costs or the responsibility. The strict liability provision orders that the PRP is liable even if the method of disposal was in accordance with prevailing standards, laws, and practice at the time of disposal. In other words, CERCLA is a "pay now, argue later" statute.[27]

Although the language specific to "joint and several" liability was removed from CERCLA, the courts have interpreted the law as though the language were included. This means that if a PRP contributed any wastes to a site, that PRP can be held accountable for all costs associated with the cleanup. This concept was strongly reaffirmed in SARA. If the PRP refuses to pay, the federal government can sue to recover costs. These actions have been successful. In certain instances where those liable fail, without sufficient cause, to properly provide for cleanup, they may be liable for treble damages!

---

[27]O'Brien & Gere Engineers, Inc., *Hazardous Waste Site Remediation,* p. 11.

Superfund Amendments and Reauthorization Act (SARA)[28]

SARA reaffirmed and strengthened many of the provisions and concepts of the CERCLA program. In SARA, Congress clearly expressed a preference, but not a require ment, for remedies such as incineration or chemical treatment that render a waste nonhazardous rather than transport to another disposal site or simple containment on site.

Another aspect of SARA is that the level of cleanup should achieve compliance with *Applicable or Relevant and Appropriate Requirements* (ARARs). ARARs are environmental standards from programs other than CERCLA and SARA. For example, if a state has a regulation regarding atmospheric emissions from incinerators, then a Superfund cleanup using incineration must meet those *applicable* standards. Furthermore, if a similar standard appears to be *relevant* and *appropriate,* then EPA may elect to apply it. For example, if drums of waste found on an uncontrolled hazardous waste site have contents that appear to have the same constituents as F001-F005 spent solvent, then the UTS standards for RCRA waste may be considered relevant and appropriate even though there is no specific evidence to identify the origin of the waste.

SARA significantly strengthens the requirement to consider damages to natural resources, especially those off site. CERCLA also required such observance, but few sites included this factor in practice. SARA provides a mechanism to include the issue in future investigations and remedies.

Title III. SARA includes a major addition to the provisions of CERCLA, namely Title III—Emergency Planning and Community Right-to-Know. Under the Emergency Planning provisions, facilities must notify the State Emergency Response Commission if they have quantities of extremely hazardous substances that exceed EPA specified *Threshold Planning Quantities* (TPQ). In addition, communities must establish Local Emergency Planning Committees (LEPCs) to develop a chemical emergency response plan. This plan must include identification of regulated facilities, emergency response and notification procedures, training programs, and evacuation plans in case of a chemical release.

If a facility accidentally releases chemicals that are on one of two lists [that is, EPA's Extremely Hazardous Substance list or the CERCLA Section 103(a) list], in regulated quantities (RQ), and the release has the potential for exposure off-site, they must notify the LEPC immediately. The law also requires a report on response actions taken, known or anticipated health risks, and advice on medical attention for exposed individuals.

Perhaps the most revolutionary provision of Title III is the establishment of the *Community's Right-to-Know* amounts of chemicals and their location in facilities in their community. Thus, information about potential hazards from chemicals is available to the public. In addition, each year those facilities that release chemicals above specified threshold amounts must submit a *Toxic Release Inventory* (TRI) on an

---

[28]O'Brien & Gere Engineers Inc., *Hazardous Waste Site Remediation,* pp. 11–16.

EPA-specified form ("Form R"). This inventory includes both accidental and routine releases, as well as off-site shipments of waste. The publication of these data has resulted in strenuous efforts by industry to control their previously unregulated and, hence, uncontrolled emissions because of the public outcry at the large quantities of materials being dumped into their environment.

9-6 HAZARDOUS WASTE MANAGEMENT

A logical priority in managing hazardous waste would be to:

1. Reduce the amount of hazardous wastes generated in the first place.
2. Stimulate "waste exchange." (One factory's hazardous wastes can become another's feedstock; for instance, acid and solvent wastes from some industries can be utilized by others without processing.)
3. Recycle metals, the energy content, and other useful resources contained in hazardous wastes.
4. Detoxify and neutralize liquid hazardous waste streams by chemical and biological treatment.
5. Reduce the volume of waste sludges generated in number four, above, by dewatering.
6. Destroy combustible hazardous wastes in special high-temperature incinerators equipped with proper pollution-control and monitoring systems.
7. Stabilize/solidify sludges and ash from numbers five and six to reduce leachability of metals.
8. Dispose of remaining treated residues in specially designed landfills.

Waste Minimization[29]

The key elements necessary to the success of a waste-minimization program include:

Top-level organizational commitment
Financial resources
Technical resources
Appropriate organization, goals, and strategy

The commitment of senior management is the first element that must be in place. Efforts to establish the other elements can follow. The organizational structure adopted

[29]Portions of this discussion were excerpted from C. H. Fromm, A. Bachrach, and M. S. Callahan, "Overview of Waste Minimization Issues, Approaches and Techniques," in *Transactions of an APCA International Specialty Conference on Performance and Costs of Alternatives to Land Disposal of Hazardous Waste,* E. T. Oppelt, B. L. Blaney, and W. F. Kemner, eds., Pittsburgh, PA: Air Pollution Control Association, pp. 6–20, 1986.

should promote communication and feedback from participants. Often, the best ideas come from line operators who work with the processes day in and day out.

Some firms set quantitative waste-minimization goals. For example, DuPont is currently reviewing a 5 percent per year corporate waste-reduction goal.[30] Other firms are more qualitative in their goal setting.

Waste audit. An important first step in establishing a strategy for waste minimization is to conduct a waste audit. The audit should proceed stepwise:

1. Identify waste streams
2. Identify sources
3. Establish priority of waste streams for waste-minimization activity
4. Screen alternatives
5. Implement
6. Track
7. Evaluate progress

The key question that must be asked at the outset of a waste audit is "why is this waste being generated?" You must first establish the primary cause(s) of waste generation before attempting to find solutions. The audit should be waste stream–oriented in order to produce a list of specific waste-minimization options for additional evaluation or implementation. Once the causes are understood, solution options can be formulated. An efficient materials and waste tracking system that allows computation of mass balances is useful in establishing priorities. Knowing how much material is going in and how much of it is ending up as waste allows you to decide which process and which waste to address first.

Example 9-4. A manufacturing company has, as part of their first audit, gathered the following data. Estimate the potential annual air emissions in kg of VOCs from the company.

Purchasing department records

Material	Purchase Quantity (barrels)
Carbon tetrachloride (CCl_4)	48
Methylene chloride (CH_2Cl_2)	228
Trichloroethylene (C_2HCl_3)	505

---

[30] J. G. Hollod, "Waste Reduction at DuPont," materials distributed at the Government Institute's seminar on waste minimization in Los Angeles, CA, November 6–7, 1986.

Wastewater treatment plant influent

Material	Average Concentration (mg/L)
CCl_4	0.343
CH_2Cl_2	4.04
C_2HCl_3	3.25

(Average flow into treatment plant is 0.076 m^3/s.)

Hazardous waste manifests

Material	Barrels	Concentration (%)
CCl_4	48	80
CH_2Cl_2	228	25
C_2HCl_3	505	80

Unused barrels at end of year

CCl_4	1
CH_2Cl_2	8
C_2HCl_3	13

Solution. The materials balance diagram will be the same for each waste.

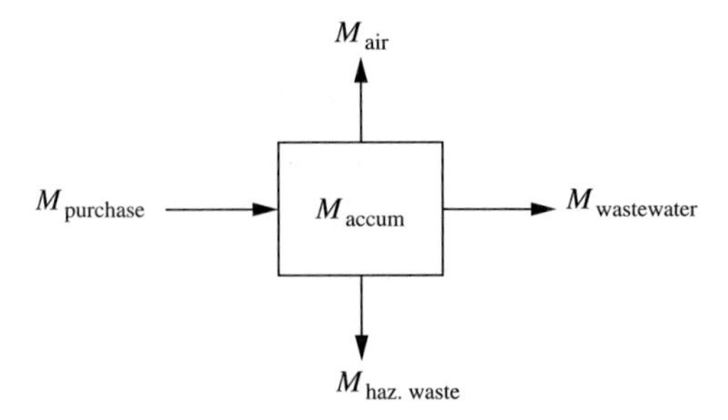

The mass balance equation would be

$$M_{\text{purchase}} = M_{\text{air}} + M_{\text{ww}} + M_{\text{hw}} + M_{\text{accum}}$$

Solving this equation for M_{air} gives us the estimated VOC emission.
First, we calculate the mass purchased. The density of each compound is found in Appendix A.
Mass Purchased

$$M(CCl_4) = (48 \text{ barrels/y})(0.12 \text{ m}^3\text{/barrel})(1{,}594 \text{ kg/m}^3) = 9{,}181.44 \text{ kg/y}$$

$$M(CH_2Cl_2) = (228 \text{ barrels/y})(0.12 \text{ m}^3\text{/barrel})(1{,}326 \text{ kg/m}^3) = 36{,}279.36 \text{ kg/y}$$

$$M(C_2HCl_3) = (505 \text{ barrels/y})(0.12 \text{ m}^3\text{/barrel})(1{,}476 \text{ kg/m}^3) = 89{,}445.60 \text{ kg/y}$$

Now we calculate the mass received at the wastewater treatment plant. (Note that mg/L = g/m^3.)

$$M(CCl_4) = (0.343\ g/m^3)(0.076\ m^3/s)(86{,}400\ s/d)(365\ d/y)(10^{-3}\ kg/g)$$
$$= 822.08\ kg/y$$

$$M(CH_2Cl_2) = (4.04\ g/m^3)(0.076\ m^3/s)(86{,}400)(365)(10^{-3})$$
$$= 9{,}682.81\ kg/y$$

$$M(C_2HCl_3) = (3.25)(0.076)(86{,}400)(365)(10^{-3})$$
$$= 7{,}789.39\ kg/y$$

The mass shipped to the hazardous waste disposal facility is calculated next.

$$M(CCl_4) = (48)(0.12)(1{,}594)(0.80) = 7{,}345.15\ kg/y$$

$$M(CH_2Cl_2) = (228)(0.12)(1{,}326)(0.25) = 9{,}069.84\ kg/y$$

$$M(C_2HCl_3) = (505)(0.12)(1{,}476)(0.80) = 71{,}556.48\ kg/y$$

Accumulated

$$M(CCl_4) = (1)(0.12)(1{,}594) = 191.28\ kg/y$$

$$M(CH_2Cl_2) = (8)(0.12)(1{,}326) = 1{,}272.96\ kg/y$$

$$M(C_2HCL_3) = (13)(0.12)(1{,}476) = 2{,}302.56\ kg/y$$

The estimated air emission for each compound is then

$$M(CCl_4) = 9{,}181.14 - 822.08 - 7{,}345.15 - 191.28$$
$$= 822.63 \text{ or } 820\ kg/y$$

$$M(CH_2Cl_2) = 36{,}279.36 - 9{,}682.81 - 9{,}069.84 - 1{,}272.96$$
$$= 16{,}253.75 \text{ or } 16{,}000\ kg/y$$

$$M(C_2HCL_3) = 89{,}445.60 - 7{,}789.39 - 71{,}556.48 - 2{,}302.56$$
$$= 7{,}797.17 \text{ or } 7{,}800\ kg/y$$

Note that we round to two significant figures because the volume of the barrels is known to only two significant figures. From this analysis, to reduce the mass of air pollutants emitted, the company should attack the methylene chloride source first. We should also point out that simply counting "barrels in" from the purchasing record and "barrels out" from the hazardous waste manifest would give a highly erroneous picture of the environmental impact of this company's emissions. From a waste minimization point of view, it is also apparent that CCl_4 and C_2HCl_3, at 80 percent concentration in barrels going to hazardous waste disposal, are candidates for recycling.

The first four steps of the waste audit allow you to generate a comprehensive set of waste management options following the hierarchy of source reduction first, waste exchange second, recycling third, and treatment last.

The screening of options begins with source control. The source control investigation should focus on (1) changes in input materials, (2) changes in process technology, and (3) changes in the human aspect of production. Input material changes can be classified into three separate elements: purification, substitution, and dilution.

Purification of input materials is performed in order to avoid the introduction of inerts or impurities into the production process. Such an introduction results in waste because the process inventory must be purged in order to prevent the undesirable accumulation of impurities. Examples of purification of feed materials to lower waste generation include the use of deionized rinse water in electroplating or the use of oxygen instead of air in oxychlorination reactors for production of ethylene dichloride.

Substitution is the replacement of a toxic material with one characterized by lower toxicity or higher environmental desirability. Examples include using phosphates in place of dichromates as cooling water corrosion inhibitors or the use of alkaline cleaners in place of chlorinated solvents for degreasing.

Dilution is a minor component of input material changes and is exemplified by use of more dilute plating solutions to minimize dragout.

Technology changes are those made to the physical plant. Examples include process changes, equipment, piping or layout changes, changes to process operational settings, additional automation, energy conservation, and water conservation.

Procedural and/or institutional changes consist of improvements in the ways people affect the production process. Also referred to as "good operating practices" or "good housekeeping," these include operating procedures, loss prevention, waste segregation, and material handling improvements.

Waste Exchange

Waste minimization by consignment of excess unused materials to an independent party for resale to a third party, saves both in waste production and in the cost (environmental and financial) of production from new raw materials. In essence "one person's trash becomes another person's treasure." The difference between a manufacturing by-product, which is costly to treat or dispose, and a usable or salable by-product involves opportunity, knowledge of processes outside the generator's immediate production line, and comparative pricing of virgin material. Waste exchanges serve as information clearinghouses through which the availability and need for various types of materials can be established.

Recycling

Under RCRA and HSWA, EPA has carefully defined recycling to prohibit bogus recyclers that are really TSDs from taking advantage of more lenient rules for recycling. The definition says that a material is *recycled* if it is used, reused, or reclaimed (40 CFR 261.1 (c)(7)). A material is "used or reused" if it is either (1) employed as an ingredient (including its use as an intermediate) to make a product (however, a material will not satisfy this condition if distinct components of the material are recovered as separate end products, as when metals are recovered from metal-containing secondary materials); or (2) employed in a particular function as an effective substitute for a commercial product (40 CFR 261.1 (c)(5)). A material is *reclaimed* if it is processed to recover a useful product or if it is regenerated. Examples include the recovery of lead from spent batteries and the regeneration of spent solvents (40 CFR 261.1 (c)(4)).[31]

Distillation processes can be utilized to recover spent solvent. The principal characteristics that determine the potential for recovery are the boiling points of the various useful constituents and the water content. The more dilute the waste solvent, the less economical it is to recover. Recovered solvents can be reused by the generator or sold for at least a substantial fraction of the cost of virgin material, and the credit for recovered solvent can more than offset the cost of recovery.

There are several technologies for recovery of metals from metal-plating rinse water. Most are applicable only to waste streams containing a single metal constituent. Examples include ion exchange, electrodialysis, evaporation, and reverse osmosis.

In October 1988 a federal appeals court struck down an EPA policy not to list used oil collected for recycling as a hazardous waste. Prior to that ruling, PCB-contaminated oils, petroleum industry sludges, and leaded tank bottoms were the only oils regulated. The majority of oil and oily wastes generated were not classified as hazardous under EPA regulations. These wastes are amenable either to recovery for use as fuel or to refinement for use as lubricants. Although all waste oil is now deemed hazardous, those oils that were formerly recovered may still be recovered, but the requirements for tracking them are more stringent.

9-7 TREATMENT TECHNOLOGIES

The wastes that remain after the implementation of waste minimization must be detoxified and neutralized. There are a large number of treatment technologies available to accomplish this. Many of these are applications of processes we have discussed in earlier chapters. Examples include: biological oxidation (Chapter 5), chemical precipitation, ion exchange, and oxidation-reduction (Chapter 3), and carbon adsorption (Chapter 6). Here we will discuss these as they apply to hazardous waste treatment, and we will introduce some new technologies.

Biological Treatment[32]

In contrast to naturally occurring compounds, *anthropogenic compounds* (those created by human beings) are relatively resistant to biodegradation. One reason is that the organisms that are naturally present often cannot produce the enzymes necessary to bring about transformation of the original compound to a point at which the resultant intermediates can enter into common metabolic pathways and be completely mineralized.

---

[31] *Waste Minimization Opportunity Assessment Manual,* U.S. Environmental Protection Agency Publication No. EPA/625/7-88/003), Cincinnati: Hazardous Waste Engineering Research Lab., p. 2, July 1988.

[32] This discussion follows that of H. Kobayashi and B. E. Rittman, "Microbial Removal of Hazardous Organic Compounds," *Environmental Science and Technology,* vol. 16, pp. 170A–172A, 1982.

Many environmentally important anthropogenic compounds are halogenated, and halogenation is often implicated as a reason for their persistence. The list of halogenated organic compounds includes pesticides, plasticizers, plastics, solvents, and trihalomethanes. Chlorinated compounds are the best known and most studied because of the highly publicized problems associated with DDT and other pesticides and numerous industrial solvents. Hence, chlorinated compounds serve as the basis for most of the information available on halogenated compounds.

Some of the characteristics that appear to confer persistence to halogenated compounds are the location of the halogen atom, the halide involved, and the extent of halogenation. The first step in biodegradation, then, is sometimes dehalogenation, for which there are several biological mechanisms.

Simple generalizations do not appear to be applicable. For example, until recently, oxidative pathways were mostly believed to be the typical means by which halogenated compounds were dehalogenated. Anaerobic, reductive dehalogenation, either biological or nonbiological, is now recognized as the critical factor in the transformation or biodegradation of certain classes of compounds. Compounds that require reductive dechlorination are common among the pesticides, as well as halogenated one- and two-carbon aliphatic compounds.

Reductive dehalogenation involves the removal of a halogen atom by oxidation-reduction. In essence, the mechanism involves the transfer of electrons from reduced organic substances via microorganisms or a nonliving (abiotic) mediator, such as inorganic ions (for example, Fe^{3+}) and biological products (for example, NAD(P), flavin, flavoproteins, hemoproteins, porphyrins, chlorophyll, cytochromes, and glutathione). The mediators are responsible for accepting electrons from reduced organic substances and transferring them to the halogenated compounds. The major requirements for the process are believed to be available free electrons and direct contact between the donor, mediator, and acceptor of electrons. Significant reductive dechlorination usually occurs only when the oxidation-reduction potential of the environment is 0.35 V and lower; the exact requirements appear to depend upon the compound involved.

Although simple studies using pure cultures of microorganisms and single substrates are valuable, if not essential, for determining biochemical pathways, they cannot always be used to predict biodegradability or transformation in more natural situations. The interactions among environmental factors, such as dissolved oxygen, oxidation-reduction potential, temperature, pH, availability of other compounds, salinity, particulate matter, competing organisms, and concentrations of compounds and organisms, often control the feasibility of biodegradation. The compound's physical or chemical characteristics, such as solubility, volatility, hydrophobicity, and octanol-water partition coefficient, contribute to the compound's availability in solution. Often compounds not soluble in the water are not readily available to organisms for biodegradation. There are some exceptions. For example, DDT, which is only slightly soluble in water, may be degraded by the white rot fungus found on decaying trees. This is because the enzymes involved in the white rot reaction are secreted from the cell.

Simple culture studies are similarly inadequate for predicting the fate of substances in the environment if there are many interactions between different organisms. First, substances that cannot be changed significantly in pure culture studies often will be degraded or transformed under mixed culture conditions. A good example of this type of interaction is *cometabolism,* in which a compound, the nongrowth substrate, is not metabolized as a source of carbon or energy, but is incidentally transformed by organisms using other compounds as growth substrates. The growth substrates provide the energy needed to cometabolize the nongrowth substrates. Second, products of the initial transformation by one organism may subsequently be broken down by a series of different organisms until compounds that can be metabolized by normal metabolic pathways are formed. An example is the degradation of DDT, which is reportedly mineralized directly by only one organism, a fungus; other organisms studied appear to degrade DDT only through cometabolism, resulting in numerous transformation products that subsequently can be used by other organisms. For example, *Hydrogenomonas* can metabolize DDT only as far as p-chlorophenylacetic acid (PCPA), while *Arthrobacter* species can then remove the PCPA.

Table 9-11 demonstrates that members of almost every class of anthropogenic compound can be degraded by some microorganism. The table also illustrates the wide variety of microorganisms that participate in environmentally significant biodegradation.

The use of novel microorganisms for biological treatment of anthropogenic compounds is a new concept. A number of advancements are required before large-scale application is possible, among the most important of which is an improved knowledge of metabolic pathways for the biodegradation of specific compounds by different organisms. The metabolic capabilities of many microorganisms, in particular algae and oligotrophic bacteria, are poorly understood. Such knowledge is necessary if limiting reactions are to be determined and the proper types of organisms selected for specific applications. More information about appropriate types of microorganisms to be selected and maintained in "real-world" treatment systems is needed, especially for the more novel microbial cultures. In order to develop special-purpose organisms by genetic manipulation, major advances in the understanding of the genetic structure of the many different types of organisms in nature are needed.

Conventional biological treatment processes such as activated sludge and trickling filters have been used to treat hazardous wastes. The major modification to the activated sludge processes has been to extend the mean cell residence time from the conventional values of 4 to 15 days to much longer periods of 3 to 6 months. In a similar fashion, trickling filter loading rates are much lower than those employed in municipal treatment systems. One innovation that has been adopted by TSD facilities is the *sequencing batch reactor* (SBR). The SBR is a periodically operated, fill-and-draw reactor.[33] Each reactor in an SBR system has five discrete periods in each cycle: fill, react, settle, draw, and idle. Biological reactions are initiated as the

[33]P. A. Herzbrun, R. L. Irvine, and K. C. Malinowski, "Biological Treatment of Hazardous Waste in Sequencing Batch Reactors," *Journal of the Water Pollution Control Federation,* vol. 57, pp. 1163–1167, 1985.

TABLE 9-11
Examples of anthropogenic compounds and microorganisms that can degrade them

Compound	Organism
Aliphatic (nonhalogenated)	
Acrylonitrile	Mixed culture of yeast mold, protozoan bacteria
Aliphatic (halogenated)	
Trichloroethane, trichloroethylene, methyl chloride, methylene chloride	Marine bacteria, soil bacteria, sewage sludge
Aromatic compounds (nonhalogenated)	
Benzene, 2,6-dinitrotoluene, creosol, phenol	*Pseudomonas* sp., sewage sludge
Aromatic compounds (halogenated)	
1,2-; 2,3-; 1,4-dichlorobenzene, hexachlorobenzene, trichlorobenzene	Sewage sludge
Pentachlorophenol	Soil microbes
Polycyclic aromatics (nonhalogenated)	
Benzo(a)pyrene, naphthalene	*Cunninghamella elegans*
Benzo(a)anthracene	*Pseudomonas*
Polycyclic aromatics (halogenated)	
PCBs	*Pseudomonas, Flavobacterium*
4-Chlorobiphenyl	Fungi
Pesticides	
Toxaphene	*Corynebacterium pyrogenes*
Dieldrin	Anacystic nidulans
DDT	Sewage sludge, soil bacteria
Kepone	Treatment lagoon sludge
Nitrosamines	
Dimethylnitrosamines	*Rhodopseudomonas*
Phthalate esters	Micrococcus 12B

Source: Extracted from Table 1 of H. Kobayashi and B. E. Rittman, "Microbial Removal of Hazardous Organic Compounds," *Environmental Science and Technology,* vol. 16, p. 173A, 1982.

raw wastewater fills the tank. During the fill and react phase, the waste is aerated in the same fashion as an activated sludge unit. After the react phase, the mixed liquor suspended solids (MLSS) are allowed to settle. The treated supernatant is discharged during the draw phase. The idle stage, the time between the draw and fill, may be zero or may be a few days depending on wastewater flow demand. The SBR has a major advantage in that wastes may be tested for completeness of treatment before discharge.

Chemical Treatment

Chemical detoxification is a treatment technology, either employed as the sole treatment procedure or used to reduce the hazard of a particular waste prior to transport, incineration, and burial.

It is important to remember that a chemical procedure cannot magically make a toxic chemical disappear from the *matrix* (wastewater, sludge, etc.) in which it is found, but can only convert it to another form. Thus, it is vital to ensure that the products of a chemical detoxification step are less of a problem than the starting material. It is equally important to remember that the reagents for such a reaction can be hazardous.

The spectrum of chemical methods includes complexation, neutralization, oxidation, precipitation, and reduction. An optimum method would be fast, quantitative, inexpensive, and leave no residual reagent, which itself would be a pollution problem. The following paragraphs describe a few of these techniques.

Neutralization. Solutions are neutralized by a simple application of the law of mass balance to bring about an acceptable pH. Sulfuric or hydrochloric acid is added to basic solutions, while caustic (NaOH) or slaked lime [$Ca(OH)_2$] is added to acidic solutions. Though a waste is hazardous at pH values less than 2 or greater than 12.5, and it would seem that simply bringing the pH into the range 2 to 12.5 would be adequate, good treatment practice requires that final pH values be in the range 6 to 8 to protect natural biota.

Oxidation. The cyanide molecule is destroyed by oxidation. Chlorine is the oxidizing agent most frequently used. Oxidation must be conducted under alkaline conditions to avoid the generation of hydrogen cyanide gas. Hence, this process is often referred to as *alkaline chlorination.* In chlorine oxidation, the reaction is carried out in two steps:

$$\mathrm{NaCN + 2NaOH + Cl_2 \rightleftharpoons NaCNO + 2NaCl + H_2O} \qquad (9\text{-}23)$$

$$\mathrm{2NaCNO + 5NaOH + 3Cl_2 \rightleftharpoons 6NaCl + CO_2 + N_2 + NaHCO_3 + 2H_2O} \qquad (9\text{-}24)$$

In the first step, the pH is maintained above 10 and the reaction proceeds in a matter of minutes. In this step, great care must be taken to maintain relatively high pH values, because at lower pHs there is a potential for the evolution of highly toxic hydrogen cyanide gas. The second reaction step proceeds most rapidly around a pH of 8, but it is not as rapid as the first step. Higher pH values may be selected for the second step to reduce chemical consumption in the following precipitation steps. This increases the reaction time. Often the second reaction is not carried out because the CNO is considered non-toxic by current regulations.

Ozone also may be used as the oxidizing agent. Ozone has a higher redox potential than chlorine, thus there is a higher driving force toward the oxidized state. When ozone is used, the pH considerations are similar to those discussed for chlorine. Ozone cannot be purchased. It must be made on site as part of the process.

This technology can be applied to a wide range of cyanide wastes: copper, zinc, and brass plating solutions; cyanide from cyanide salt heating baths; and passivating solutions. The process has been practiced on an industrial scale since the early 1940s. For extremely high cyanide concentrations (>1 percent), oxidation may not be desirable. Cyanide complexes of metals, particularly iron and to some extent nickel, cannot be decomposed easily by cyanide oxidation techniques.

Electrolytic oxidation of cyanide is carried out by anodic electrolysis at high temperatures. The theoretical basis of the process is that cyanide reacts with oxygen in solution in the presence of an electric potential to produce carbon dioxide and nitrogen gas. Normally, the destruction is carried out in a closed cell. Two electrodes are suspended in the solution and a DC current is applied to drive the reaction. The bath temperature must be maintained in the range of 50 to 95°C.

This technology is used for the destruction of cyanide in concentrated spent stripping solutions; in plating solutions for copper, zinc, and brass; in alkaline descalers; and in passivating solutions. It has been more successful for wastes containing high concentrations of cyanide (50,000 to 100,000 mg/L), but it has also been successfully used for concentrations as low as 500 mg/L.

Chemical oxidation methods for organic compounds in wastewater have received extensive study. In general they apply only to dilute solutions and often are considered expensive in comparison to the biological methods. Some examples include wet air oxidation, hydrogen peroxide, permanganate, chlorine dioxide, chlorine, and ozone oxidation. Of these, wet air oxidation and ozonation have shown promise as a pretreatment step for biological processes.

Wet air oxidation, also known as the Zimmerman process, operates on the principle that most organic compounds can be oxidized by oxygen given sufficient temperature and pressure. Wet air oxidation may be described as the aqueous phase oxidation of dissolved or suspended organic particles at temperatures of 175 to 325°C and sufficiently high pressure to prevent excessive evaporation. Air is bubbled through the liquid. The process is fuel efficient; once the oxidation reaction has started, it is usually self-sustaining. As this method is not limited by reagent cost, it is potentially the most widely applicable of all chemical oxidation methods. The method has been shown to be of use in destroying a wide range of organic compounds, including some pesticides. Although wet oxidation can provide acceptable levels of destruction for many hazardous compounds, it generally is not as complete as incineration. In many instances, the addition of metal salt catalysts can increase the destruction efficiency or allow the process to be run at lower temperature and/or pressure.

Precipitation. Metals are often removed from plating rinse waters by precipitation. This is a direct application of the solubility product principle (see Section 3-1). By raising the pH with lime or caustic, the solubility of the metal is reduced (Figure 9-9) and the metal hydroxide precipitates. Optimum removal is achieved by selecting the optimum pH as shown in Figure 9-9. Though there is an optimum for each metal, in many cases, the metals are mixed and the lowest value for an individual metal may not be achievable for the mixture.

Example 9-5. A metal plating firm is installing a precipitation system to remove zinc. They plan to use a pH meter to control the feed of hydroxide solution to the mixing tank. What pH should the controller be set at to achieve a zinc effluent concentration of 0.80 mg/L? The K_{sp} of $Zn(OH)_2$ is 7.68×10^{-17}.

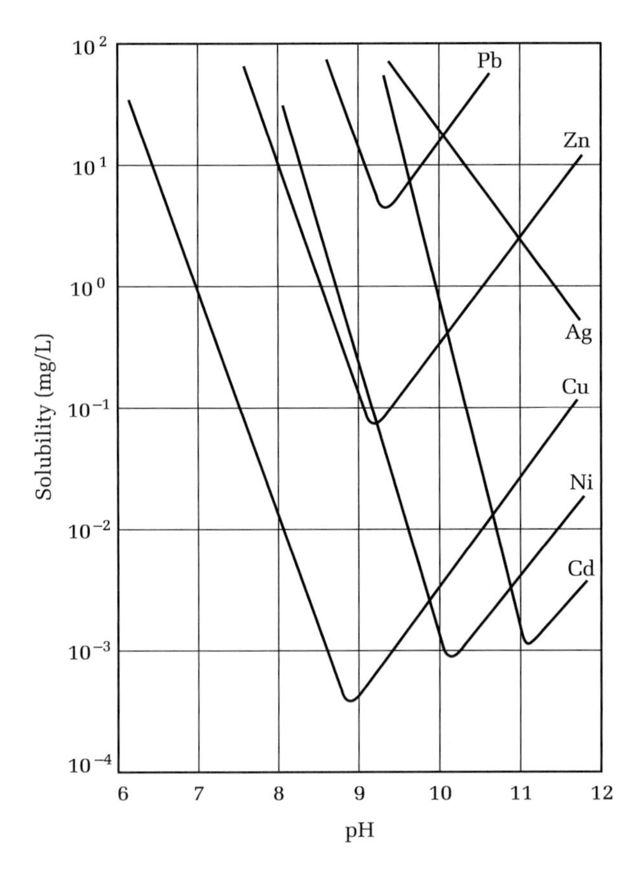

FIGURE 9-9
Solubilities of metal hydroxides as a function of pH. (*Source: Development Document for Effluent Limitations: Guideline and Standards for the Metal Finishing Point Category* (U.S. Environmental Protection Agency Publication No. EPA/440/1-83-091), Washington, DC: U.S. Government Printing Office, 1981.)

Solution. From Table A-9 in Appendix A we find that the zinc hydroxide reaction is

$$Zn^{2+} + 2OH^- \rightleftharpoons Zn(OH)_2$$

As shown in Section 3-1, we can write the solubility product equation as

$$K_{sp} = [Zn^{2+}][OH^-]^2$$

Since we want the zinc concentration to be no greater than 0.80 mg/L, we calculate the moles per liter of zinc.

$$[Zn^{2+}] = \frac{0.80 \text{ mg/L}}{(65.38 \text{ g/mol})(1{,}000 \text{ mg/g})} = 1.224 \times 10^{-5} \text{ moles/L}$$

Now we solve for the hydroxide concentration.

$$[OH^-]^2 = \frac{7.68 \times 10^{-17}}{1.224 \times 10^{-5}} = 6.275 \times 10^{-12}$$

$$= (6.275 \times 10^{-12})^{\frac{1}{2}} = 2.505 \times 10^{-6}$$

The pOH is

$$\text{pOH} = -\log(2.505 \times 10^{-6}) = 5.601$$

And the pH set point for the controller is

$$\begin{aligned} pH &= 14 - pOH \\ &= 14 - 5.601 = 8.399 \text{ or } 8.4 \end{aligned}$$

Reduction. Although most heavy metals readily precipitate as hydroxides, hexavalent chromium used in plating solutions must be reduced to trivalent chromium before it will precipitate. Reduction is usually done with sulfur dioxide (SO_2) or sodium sulfite ($NaHSO_3$). With SO_2 the reaction is

$$3SO_2 + 2H_2CrO_4 + 3H_2O \rightleftharpoons Cr_2(SO_4)_3 + 5H_2O \tag{9-25}$$

Because the reaction proceeds rapidly at low pH, an acid is added to control the pH between 2 and 3.

Physical/Chemical Treatment

Several treatment processes are used to separate hazardous waste from aqueous solution. The waste is not detoxified but only concentrated for further treatment or recovery.

Carbon adsorption. Adsorption is a mass transfer process in which gas vapors or chemicals in solution are held to a solid by intermolecular forces (for example, hydrogen bonding and van der Waals' interactions). It is a surface phenomenon. Pressure vessels having a fixed bed are used to hold the adsorbent (see Section 6-10). Activated carbon, molecular sieves, silica gel, and activated alumina are the most common adsorbents. The active sites become saturated at some point in time. When the organic material has commercial value, the bed is then regenerated by passing steam through it. The vapor-laden steam is condensed and the organic fraction is separated from the water. If the organic compounds have no commercial value, the carbon may be either incinerated or shipped to the manufacturer for regeneration. Carbon systems for recovery of vapor from degreasers and for polishing wastewater effluents have been in commercial application for over 20 years.

Distillation. The separation of more volatile materials from less volatile materials by a process of vaporization and condensation is called *distillation.* When a liquid mixture of two or more components is brought to the boiling point of the mixture, a vapor phase is created above the liquid phase. If the vapor pressures of the pure components are different (which is usually the case), then the constituent(s) having the higher vapor pressure will be more concentrated in the vapor phase than the constituent(s) having the lower vapor pressure. If the vapor phase is cooled to yield a liquid, a partial separation of the constituents will result. The degree of separation depends on the relative differences in the vapor pressures. The larger the differences, the more efficient the separation. If the difference is large enough, a single separation cycle of vaporization and condensation is sufficient to separate the components. If the difference is not large enough, multiple cycles (stages) are required. Four types of distillation may be used: batch distillation, fractionation, steam stripping, and thin film evaporation.[34]

Both batch distillation and fractionation are well proven technologies for recovery of solvents. Batch distillation is particularly applicable for wastes with high solids concentrations. Fractionation is applicable where multiple constituents must be separated and where the waste contains minimal suspended solids.

When the volatility of the organic compound is relatively high and the concentration relatively low, then some form of stripping may be appropriate. *Air stripping* has been used to purge large quantities of contaminated groundwater of small concentrations of volatile organic matter. The behavior of the process is the inverse of absorption discussed in Section 6-10. Air and contaminated liquid are passed countercurrently through a packed tower. The volatiles evaporate into the air, leaving a clean liquid stream. The contaminated air stream must then be treated to avoid an air pollution problem. Frequently this is accomplished by passing the air through an activated carbon column. The carbon is then incinerated. Air stripping has been used to remove tetrachloroethylene, trichloroethylene, and toluene from water.[35]

The air stripper design equation may be developed in the same fashion as the absorber equation in Chapter 6. It is given here without that development:

$$Z_T = \frac{L}{A} \frac{\ln\left[\dfrac{C_1}{C_2} - \dfrac{\mathrm{LRT}_g}{\mathrm{GH}_c}\left(\dfrac{C_1}{C_2} - 1\right)\right]}{K_L a\left(1 - \dfrac{\mathrm{LRT}_g}{\mathrm{GH}_c}\right)} \tag{9-26}$$

where
- Z_T = packed tower depth, m
- L = water flow, m^3/min
- A = cross-sectional area of tower, m^2
- G = air flow, m^3/min
- H_c = Henry's constant, $atm \cdot m^3/mol$
- R = universal gas constant = 8.206×10^{-5} $atm \cdot m^3/mole \cdot K$
- T_g = temperature of air, K
- C_1, C_2 = influent and effluent organic concentration in the water, mol/m^3
- K_L = overall mass transfer coefficient, $mol/min \cdot m^2 \cdot mol/m^3$
- a = effective interfacial area of packing per unit volume for mass transfer, m^2/m^3

[34] Air stripping, though not strictly a distillation process because the condensation step is omitted, employs the same general principles of volatilization and, hence, is included in the discussion.

[35] R. L. Gross and S. G. TerMaath, "Packed Tower Aeration Strips Trichloroethylene from Groundwater," *Environmental Progress,* vol. 4, pp. 119–124, May 1985; U.S. Environmental Protection Agency, *A Compendium of Technologies Used in the Treatment of Hazardous Waste* (U.S. Environmental Protection Agency Publication No. EPA/625/8-87/014), Cincinnati: Center for Environmental Research Information, September 1987.

Example 9-6. Well 12A at the City of Tacoma, WA, is contaminated with 350 μg/L of 1,1,2,2-tetrachloroethane. The water must be cleaned to the detection limit of 1.0 μg/L. Design a packed tower stripping column to meet this requirement using the following design parameters.

Henry's law constant $= 5.0 \times 10^{-4}$ atm $\cdot$ m^3/mol
$K_L a = 10 \times 10^{-3}$ s^{-1}
Air flow rate $= 13.7$ m^3/s
Liquid flow rate $= 0.044$ m^3/s
Temperature $= 25°$C
Column diameter may not exceed 4.0 m
Column height may not exceed 6.0 m

Solution. The Henry's law constants given in Appendix A are in kPa $\cdot$ m^3/moles. To convert these to atm $\cdot$ m^3/mole, divide by the atmospheric pressure at standard conditions, that is, 101.325 kPa/atm.

The stripper equation is then solved for $Z_T A$, the column volume.

$$Z_T A = (0.044)\frac{\ln\left[\dfrac{350}{1} - \dfrac{(0.044)(8.206 \times 10^{-5})(298)}{(13.7)(5.0 \times 10^{-4})}\left(\dfrac{350}{1} - 1\right)\right]}{10 \times 10^{-3}\left[1 - \dfrac{(0.044)(8.206 \times 10^{-5})(298)}{(13.7)(5.0 \times 10^{-4})}\right]}$$

$$= (0.044)(674.74)$$

$$= 29.69 \text{ m}^3$$

Any number of solutions are now possible within the boundary conditions of 4 m diameter and 6 m height. For example:

Diameter (m)	Tower height (m)
4.00	2.36
3.34	3.39
2.58	5.68

For gases of lower volatility or higher concentration (>100 ppm) *steam stripping* may be employed. The physical arrangement of the process is much like that of an air stripper, except that steam is introduced instead of air. The addition of steam enhances the stripping process by decreasing the solubility of the organic in the aqueous phase and by increasing the vapor pressure. Steam stripping has been used to treat aqueous waste contaminated with chlorinated hydrocarbons, xylenes, acetone, methyl ethyl ketone, methanol, and pentachlorophenol. Concentrations treated range from 100 ppm to 10 percent organic.[36]

[36]U.S. Environmental Protection Agency, *A Compendium of Technologies Used in the Treatment of Hazardous Waste,* September 1987.

Recovery of metals by evaporation is accomplished by boiling off sufficient water from the collected rinse stream to allow the concentrate to be returned to the plating bath. The condensed steam is recycled for use as rinse water. The boil-off rate, or evaporator duty, is set to maintain the water balance of the plating bath. Evaporation is usually performed under vacuum to prevent thermal degradation of additives and to reduce the amount of energy required for evaporation of the water.

There are four types of evaporators: rising film, flash evaporators using waste heat, submerged tube, and atmospheric pressure. Rising film evaporators are built so that the evaporative heating surface is covered by a wastewater film and does not lie in a pool of boiling wastewater. Flash evaporators are of similar configuration, but the plating solution is continuously recirculated through the evaporator along with the wastewater. This allows the use of waste heat in the plating bath to augment the evaporation process. In the submerged tube design, the heating coils are submerged in the wastewater. Atmospheric evaporators do not recover the distillate for reuse and they do not operate under vacuum.

Ion exchange. Metals and ionized organic chemicals can be recovered by ion exchange. Ion exchange chemistry was discussed in Section 3-3. In ion exchange, the waste stream containing the ion to be removed is passed through a bed of resin. The resin is selected to remove either cations or anions. In the exchange process, ions of like charge are removed from the resin surface in exchange for ions in solution. Typically, either hydrogen or sodium is exchanged for cations (metal) in solution. When the bed becomes saturated with the exchanged ion, it is shut down and the resin is regenerated by passing a concentrated solution containing the original ion (hydrogen or sodium) back through the bed. The exchanged pollutant is forced off the bed in a concentrated form that may be recycled. A typical ion exchange column is shown in Figure 9-10. A prefilter is required to remove suspended material that would hydraulically foul the column. It also removes organics and oils that would foul the resin.

As a rule, ion exchange systems are suitable for chemical recovery applications where the rinse water feed has a relatively dilute concentration (<1,000 mg/L) and a relatively low concentration is required for recycle. Ion exchange has been demonstrated commercially for recovery of plating chemicals from acid-copper, acid-zinc, nickel, tin, cobalt, and chromium plating baths.

The breakthrough curves for an ion-exchange column and an adsorption column (Section 6-8) are similar. Thomas[37] proposed a kinetic equation to describe the removal of the contaminant in the column:

$$\ln\left(\frac{C_o}{C} - 1\right) = \frac{(k)(q_o)(M)}{Q} - \frac{(k)(C_o)(\mathcal{V})}{Q} \tag{9-27}$$

[37]H. C. Thomas, "Chromatography: A Problem of Kinetics," *Annals of the New York Academy of Science,* vol. 49, p. 161, 1948.

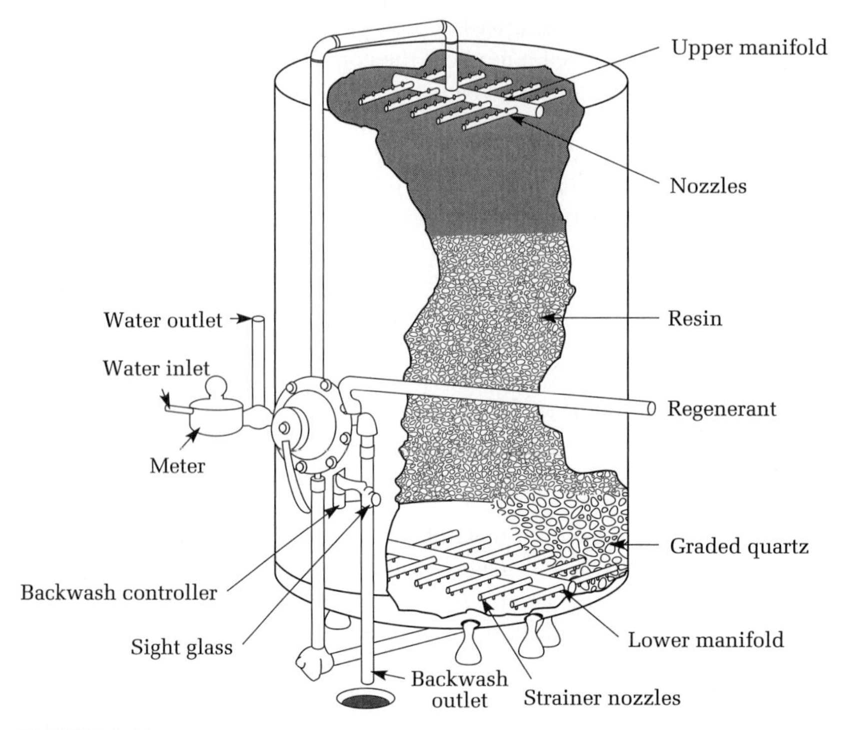

FIGURE 9-10
Typical ion exchange resin column. (*Source:* Kunin, R., "Ion Exchange for the Metal Products Finishers," (3 parts), *Products Finishing,* April-June 1969.)

where C_0 = influent solute concentration, mg/L or milliequivalents/L (meq/L)
C = effluent solute concentration, mg/L or milliequivalents/L (meq/L)
k = rate constant, L/d · equivalent
q_0 = maximum solid phase concentration of exchanged solute, equivalents/kg of resin
M = mass of resin, kg
$\forall$ = volume of solution passed through column, L
Q = flow rate, L/d

This equation is of the form $y = mx + b$

where $y = \ln\left(\frac{C_0}{C} - 1\right)$

$x = \forall$

This allows us to determine the rate constant and the maximum solid phase concentration from a plot of

$$\ln\left(\frac{C_0}{C} - 1\right) \text{ versus } \forall \text{ as shown in Figure 9-11.}$$

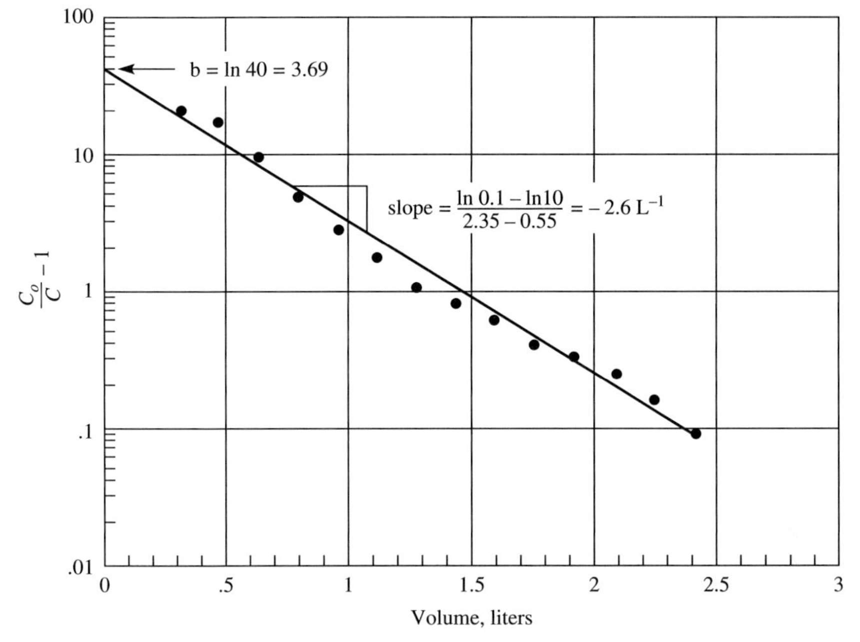

FIGURE 9-11
Plot of breakthrough data to estimate kinetic equation constants. (Note: ordinate scale is logarithm to the base e.)

The slope of the line is equal to

$$\frac{kC_o}{Q}$$

and the intercept is equal to

$$\frac{(k)(q_o)(M)}{Q}$$

Data from a laboratory or pilot scale breakthrough curve are required to obtain the plot. The same flow rate, in terms of bed volumes per unit time, should be used for both the pilot studies and the full scale column.

Example 9-7. An electroplating rinse water containing 49 mg/L of zinc is to be treated by an ion exchange column to meet an allowable effluent concentration of 2.6 mg/L. A laboratory scale column has provided the breakthrough data shown in the first two columns of the table below. The laboratory column data are as follows:

Inside diameter = 1.0 cm

Length = 10.0 cm

Mass of resin (moist basis) = 5.2 g

Water content = 17%

Density of dry resin = 0.65 g/cm^3
Liquid flow rate = 7.87 L/d
Initial concentration of zinc = 49 mg/L

The full scale design must meet the following requirements:

Flow rate = 36,000 L/d
Hours of operation = 8 h/d
Regeneration is to be once every five days

Determine the mass of resin required.

Solution. The laboratory breakthrough data are converted to the form of Equation 9-27 in the following table. The initial concentration of zinc (C_o) is 49 mg/L. The meq/L is determined by first finding the equivalent weight (See Chapter 3) as

$$\frac{\text{GMW}}{\text{n}} = \frac{65.38 \text{ g/mole}}{2 \text{ eq/mole}} = 32.96 \text{ g/eq or mg/meq}$$

and dividing the concentration of zinc by its equivalent weight. The initial concentration (C_o) in meq/L is

$$\frac{49 \text{ mg/L}}{32.69 \text{ mg/meq}} = 1.50 \text{ meq/L}$$

Breakthrough Data

$\forall$, L	C, mg/L	C, meq/L	$\frac{C_o}{C} - 1$
0.32	2.25	0.06826	20.973
0.48	2.74	0.08313	17.044
0.64	4.56	0.13835	9.8421
0.80	8.32	0.25243	4.9423
0.96	12.74	0.38653	2.8807
1.12	17.70	0.53701	1.7932
1.28	23.54	0.71420	1.1003
1.44	27.48	0.83374	0.7991
1.60	30.58	0.92779	0.6167
1.76	35.34	1.07221	0.3990
1.92	37.02	1.12317	0.3355
2.08	39.38	1.19478	0.2555
2.24	42.50	1.28944	0.1632
2.40	45.10	1.36833	0.0962
2.56	44.10	1.33799	0.1211

The plot of these data is shown in Figure 9-11.
From the plot

$$k = (\text{slope})\left(\frac{Q}{C_o}\right) = (2.6 \text{ L}^{-1})\left(\frac{7.87 \text{ L/d}}{1.50 \text{ meq/L}}\right)$$
$$= 13.64 \text{ L/d} \cdot \text{meq}$$

and

$$q_o = \frac{(b)(Q)}{(k)(M)} = \frac{(3.69)(7.87 \text{ L/d})}{(13.64 \text{ L/d} \cdot \text{meq})(4.316 \text{ g})}$$
$$= 0.4933 \text{ meq/g}$$

Note that the mass of the test column resin (M) is corrected for moisture, that is, the dry weight is

$$(5.2 \text{ g})(1 - 0.17) = 4.316 \text{ g}$$

Using these values of k and q_o and reapplying Equation 9-27 we can determine the mass of resin for the full scale column. Since the effluent concentration must not exceed 2.6 mg/L, we can solve the left-hand side of the equation as:

$$\ln\left(\frac{49}{2.6} - 1\right) = 2.882$$

The first term on the right-hand side contains the unknown (M). Using the constants determined above, and the daily flow rate, it may be simplified to

$$\frac{(13.64 \text{ L/d} \cdot \text{meq})(0.4933 \text{ meq/g})(M)}{36{,}000 \text{ L/d}} = 1.87 \times 10^{-4}(M)$$

Using a flow rate of 36,000 L/d and a 5 day operating cycle, the volume to treat ($\forall$) is

$$(36{,}000 \text{ L/d})(5 \text{ d}) = 180{,}000 \text{ L}$$

The second term on the right-hand side of the equation is then

$$\frac{(13.64 \text{ L/d} \cdot \text{meq})(1.50 \text{ meq/L})(180{,}000 \text{ L})}{36{,}000 \text{ L/d}} = 102.30$$

Setting the left-hand side of the equation equal to the right-hand side and solving for M yields

$$2.882 = 1.87 \times 10^{-4}(M) - 102.30$$
$$M = 5.6 \times 10^5 \text{ g or } 560 \text{ kg}$$

In full-scale operation, the resin bed is not allowed to reach saturation because the concentration of the solute will exceed most discharge standards before this occurs. Normal operation then requires either an operating cycle that will allow regeneration of the spent resin during nonworking hours or, in the case of 24 hour, 7 day per week schedules, multiple beds so that one may be taken off-line.

The diameters of ion exchange columns may vary from centimeters to 6 m. Resin bed depths range from 1 to 3 m. Bed height-to-diameter ratios range from 1.5:1 to 1:3. The column shell is designed to allow for 100 percent expansion of the resin bed during backwashing (regeneration). Columns are normally prefabricated and shipped by truck. Column height generally does not exceed 4 m. Multiple columns in series are provided where the design height exceeds 4 m. The maximum column diameter is often controlled by the clearance under bridges passing over the highway.

During ion exchange, the normal flow pattern is downward through the bed. The hydraulic loading may range from 25 to 600 $m^3/d \cdot m^2$. Lower hydraulic loadings result in longer contact periods and better exchange efficiency. Because the surface of the bed acts like a filter, regeneration is often countercurrent, that is, the regenerating solution is pumped into the bottom of the column. This results in a cleansing of the column much like the backwashing of a rapid sand filter cleans it. Regeneration hydraulic loadings range from 60 to 120 $m^3/d \cdot m^2$.

Electrodialysis. The electrodialysis unit uses a membrane to selectively retain or transmit specific molecules. The membranes are thin sheets of ion exchange resin reinforced by a synthetic fiber backing. The construction of the unit is such that anion membranes are alternated with cation membranes in stacks of cells in series (Figure 9-12). An electric potential is applied across the membrane to provide the motive force for ion migration. Cation membranes permit passage of only positively charged ions, while anion membranes permit passage of only negatively charged ions. The

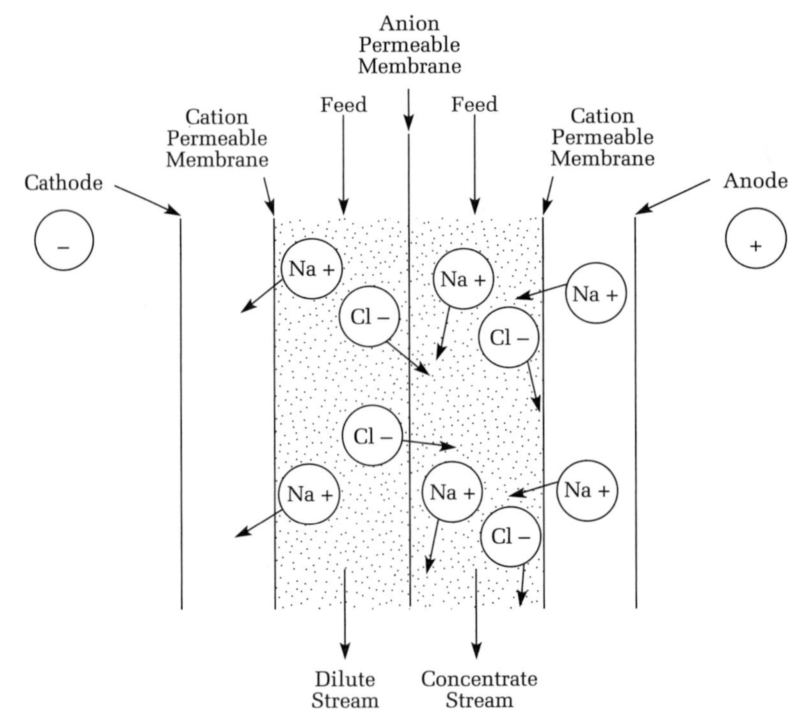

FIGURE 9-12
Electrodialysis. (Cations in the feed water show the same behavior as sodium (Na^+) and anions show the same behavior as chloride (Cl^-). Under the action of an electric field, cation exchange membranes permit passage only of positive ions, while anion exchange membranes permit passage only of negatively charged ions.) (*Source:* Unit Operations for Treatment of Hazardous Industrial Wastes, Noyes Data Corporation, 1978.)

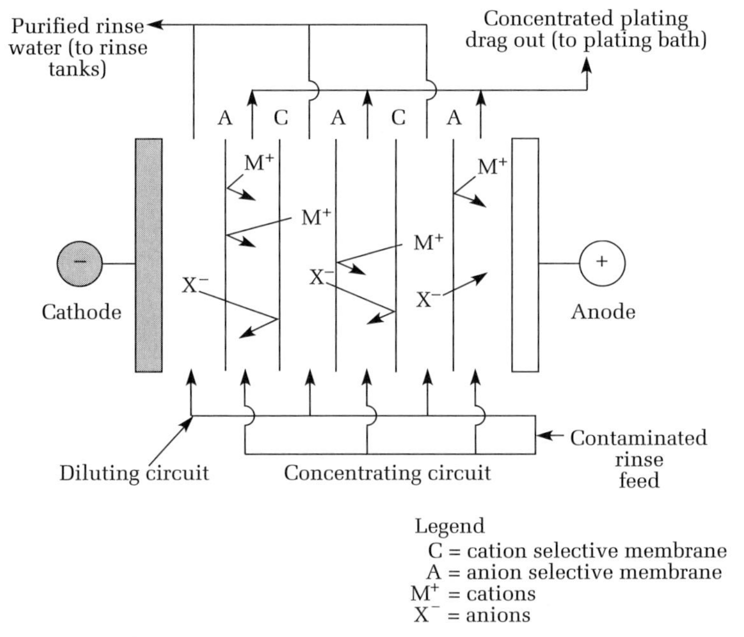

FIGURE 9-13
Electrodialysis unit flow schematic.

flow is directed through the membrane in two hydraulic circuits (Figure 9-13). One circuit is ion-depleted and the other is ion-concentrated. The degree of purification achieved in the dilute circuit is set by the electric potential. The ability to pass the charge is proportional to the concentration of the ionic species in the dilute stream. Because ion migration is proportional to electric potential, the optimum system is a trade-off between energy requirements and degree of contaminant removal.

Electrodialysis has been in commercial operation for more than two decades in the production of potable water from brackish water. It has also been used in de-ashing of sugars, desalting of food products such as whey, and to recover waste developer in the photo processing industry and nickel from a metal-plating rinse water. Typically, electrodialysis can separate a waste stream containing 1,000 to 5,000 mg/L inorganic salts into a dilute stream that contains 100 to 500 mg/L salt and a concentrated stream that contains up to 10,000 mg/L salt.

Reverse osmosis. Osmosis is defined as the spontaneous transport of a solvent from a dilute solution to a concentrated solution across an ideal semipermeable membrane that impedes passage of the solute but allows the solvent to flow. Solvent flow can be reduced by exerting pressure on the solution side of the membrane, as shown in Figure 9-14. If the pressure is increased above the osmotic pressure on the solution side, the flow reverses. Pure solvent will then pass from the solution into the solvent. As applied to metal finishing wastewater, the solute is the metal and the solvent is pure water.

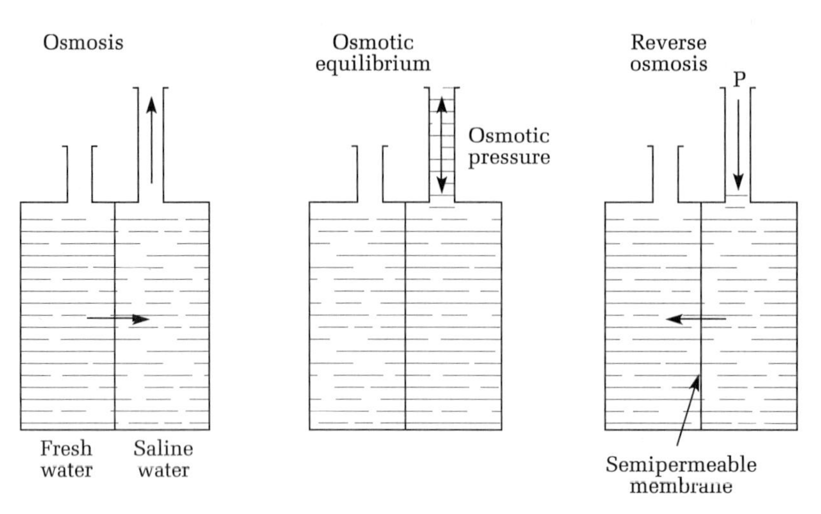

FIGURE 9-14
Direct and reverse osmosis.

Many configurations of the membrane are possible. The driving pressure is on the order of 2,800 to 5,500 kPa. No commercially available membrane polymer has demonstrated tolerance to all extreme chemical factors such as pH, strong oxidizing agents, and aromatic hydrocarbons. However, selected membranes have been demonstrated on nickel, copper, zinc, and chrome baths.

Solvent extraction. Solvent extraction is also called *liquid extraction* and *liquid-liquid extraction.* Contaminants can be removed from a waste stream using liquid-liquid extraction if the wastewater is contacted with a solvent having a greater solubility for the target contaminants than the wastewater. The contaminants will tend to migrate from the wastewater into the solvent. Although predominately a method for separating organic materials, it may also be applied to remove metals if the solvent contains a material that will react with the metal. *Liquid ion exchange* is one kind of these reactions.

In the solvent extraction process, the solvent and the waste stream are mixed to allow mass transfer of the constituent(s) from the waste to the solvent. The solvent, immiscible in water, is then allowed to separate from the water by gravity. The solvent solution containing the extracted contaminant is called the *extract.* The extracted waste stream with the contaminants removed is called the *raffinate.* As in distillation, the separation may need to be done in one or more stages. In general, more stages result in a cleaner raffinate. The degree of complexity of the apparatus varies from simple mixer/settlers to more exotic contacting devices. If the extract is sufficiently enriched, it may be possible to recover useful material. Distillation is often employed to recover the solvent and reusable organic chemicals. For metal recovery, the ion exchange material is regenerated by the addition of an acid or alkali. The process has found wide application in the ore processing industry, in food processing, in pharmaceuticals, and in the petroleum industry.

Incineration

In an incinerator, chemicals are decomposed by oxidation at high temperatures (800°C and greater). The waste, or at least its hazardous components, must be combustible in order to be destroyed. The primary products from combustion of organic wastes are carbon dioxide, water vapor, and inert ash. However, there are a multitude of other products that can be formed.

Products of combustion. The percentages of carbon, hydrogen, oxygen, nitrogen, sulfur, halogens, and phosphorus in the waste, as well as the moisture content, need to be known to determine stoichiometric combustion air requirements and to predict combustion gas flow and composition. Actual incineration conditions generally require excess oxygen to maximize the formation of *products of complete combustion* (POCs) and minimize the formation of *products of incomplete combustion* (PICs).

The incineration of halogenated organics results in the formation of halogenated acids, which require further treatment to ensure environmentally acceptable air emissions from the incineration process. Chlorinated organics are the most common halogenated hydrocarbons found in hazardous waste. The incineration of chlorinated hydrocarbons with excess air results in the formation of carbon dioxide, water, and hydrogen chloride. An example is the following reaction for the incineration of dichlorethane:[38]

$$2C_2H_4Cl_2 + 5O_2 \rightarrow 4CO_2 + 2H_2O + 4HCl \qquad (9\text{-}28)$$

The hydrogen chloride must be removed before the carbon dioxide and steam can be safely exhausted into the atmosphere.

Hazardous waste may contain either organic or inorganic sulfur compounds. When these wastes are incinerated, sulfur dioxide is produced. For example, the destruction of ethyl mercaptan results in the following reaction:

$$2C_2H_5SH + 9O_2 \rightarrow 4CO_2 + 6H_2O + 2SO_2 \qquad (9\text{-}29)$$

The sulfur dioxide produced by the incineration of sulfur-containing wastes must not exceed air quality standards.

Excess air must be provided to ensure complete combustion. However, the amount of the excess can only be determined empirically. For example, a highly volatile, clean, hydrocarbon waste would probably require much less excess air than would a heavy hydrocarbon sludge within a high solids content. Incineration of sludges and solids may require as much as two to three times excess air above stoichiometric equivalents. Too much excess air should be avoided because it increases the fuel required to heat the waste to destruction temperatures, reduces residence time for the hazardous wastes to be oxidized, and increases the volume of air emissions to be handled by the air pollution control equipment.

By-products from the incineration of hazardous wastes may also result from incomplete combustion as well as from the products of combustion. Products of in-

---

[38]C. A. Wentz, *Hazardous Waste Management,* New York: McGraw-Hill, pp. 206–207, 1989.

complete combustion (PICs) include carbon monoxide, hydrocarbons, aldehydes, ketones, amines, organic acids, and polycyclic aromatic hydrocarbons (PAHs). In a well-designed incinerator, these products are insignificant in amount. However, in poorly designed or overloaded incinerators, PICs may pose environmental concerns. Polychlorinated biphenyls, for instance, decompose under such conditions into highly toxic chlorinated dibenzo furans (CDBF). The hazardous material, hexachlorocyclopentadiene (HCCPD), found in many hazardous wastes, is known to decompose into the even more hazardous compound hexachlorobenzene (HCB).[39]

Suspended particulate emissions are also produced during incineration. These include particles of mineral oxides and salts from the mineral constituents in the waste material, as well as fragments of incompletely burned combustibles.

Last, but not least, ash is a product of combustion. The ash is considered a hazardous waste. Metals not volatilized end up in the ash. Unburned organics may also be found in the ash. When organics remain, the ash may simply be incinerated. The metals must be treated prior to land disposal.

Design considerations.[40] The most important factors for proper incinerator design and operation are combustion temperature, combustion gas residence time, and the efficiency of mixing the waste with combustion air and auxiliary fuel.

Chemical and thermal dynamic properties of the waste that are important in determining its time/temperature requirements for destruction are its elemental composition, net heating value, and any special properties (for example, explosive properties) that may interfere with incineration or require special design considerations.

In general, higher heating values are required for solids versus liquids or gases, for higher operating temperatures, and for higher excess air rates if combustion is to be sustained without auxiliary fuel consumption. While sustained combustion (*autogenous combustion*) is possible with heating values as low as 9.3 MJ/kg, in the hazardous waste incineration industry it is common practice to blend wastes (and fuel oil, if necessary) to obtain an overall heating value of 18.6 MJ/kg or greater.

Blending is also used to limit the net chlorine content of chlorinated hazardous waste to a maximum of roughly 30 percent by weight to reduce chlorine concentrations in the combustion gas. The chlorine and, especially, hydrogen chloride that forms from the chlorine, are very corrosive. They oxidize the fire brick in the incinerator which causes it to fail.

Hazardous waste incinerators must be designed to achieve a 99.99 percent destruction and removal efficiency (DRE) of the principal organic hazardous components (POHCs) in the waste. This is commonly referred to as "four 9s DRE"; higher DREs may be referred to as five 9s, six 9s, that is, 99.999 and 99.9999 percent DRE, respectively. Because of the complexity of the wastes being burned, little success has

[39]E. T. Oppelt, "Thermal Destruction Options for Controlling Hazardous Wastes," *Civil Engineering-ASCE,* pp. 72–75, September 1981.

[40]Parts of this discussion follow E. T. Oppelt, "Thermal Destruction Options for Controlling Hazardous Wastes," pp. 72–75, September 1981.

been achieved in predicting the time and temperature requirements for achieving the 99.99 percent DRE. Empirical tests (*trial burns*) are required to demonstrate compliance. Experience has demonstrated that highly halogenated materials are more difficult to destroy than those with low halogen content.

Incinerator types.[41] Two technologies dominate the incineration field: liquid injection and rotary kiln incinerators. Over 90 percent of all incineration facilities use one of these technologies. Of these, more than 90 percent are liquid injection units. Less commonly used incinerators include fluidized beds and starved air/pyrolysis systems.

Horizontal, vertical, and tangential liquid injection units are used. The majority of the incinerators for hazardous wastes inject liquid hazardous waste at 350 to 700 kPa through an atomizing nozzle into the combustion chamber. These liquid incinerators vary in size from 300,000 to 90 million Joules of heat released per second. An auxiliary fuel such as natural gas or fuel oil is often used when the waste is not autogenous. The liquid wastes are atomized into fine droplets as they are injected. A droplet size in the range 40 to 100 μm is obtained with atomizers or nozzles. The droplet volatilizes in the hot gas stream and the gas is oxidized. Efficient destruction of liquid hazardous wastes requires minimizing unevaporated droplets and unreacted vapors.

Residence time, temperature, and turbulence (often referred to as the "three Ts") are optimized to increase destruction efficiencies. Typical residence times are 0.5 to 2 seconds. Incinerator temperatures usually range between 800 and 1600°C. A high degree of turbulence is desirable for achieving effective destruction of the organic chemicals in the waste. Depending on whether the liquid incinerator flow is axial, radial, or tangential, additional fuel burners and separate waste injection nozzles can be arranged to achieve the desired temperature, turbulence, and residence time. Vertical units are less likely to experience ash buildup. Tangential units have a much higher heat release and generally superior mixing.

The rotary kiln is often used in hazardous waste disposal systems because of its versatility in processing solid, liquid, and containerized wastes. Waste is incinerated in a refractory-lined rotary kiln, as shown in Figure 9-15. The shell is mounted at a slight incline from the horizontal plane to facilitate mixing the waste materials with circulating air. Solid wastes and drummed wastes are usually fed by a conveyor system or a ram. Liquids and pumpable sludges are injected through a nozzle. Noncombustible metal and other residues are discharged as ash at the end of the kiln.

Rotary kilns are typically 1.5 to 4 m in diameter and range in length from 3 to 10 m. Rotary kiln incinerators usually have a length-to-diameter ratio (L/D) of between two and eight. Rotational speeds range from 0.5 to 2.5 cm/s, depending on kiln periphery. High L/D ratios, along with slower rotational speeds, are used for wastes requiring longer residence times. The feed end of the kiln has airtight seals to adequately control the initial incineration reactions.

---

[41] Parts of this discussion were excerpted from C. A. Wentz, *Hazardous Waste Management,* pp. 206–207, 1989.

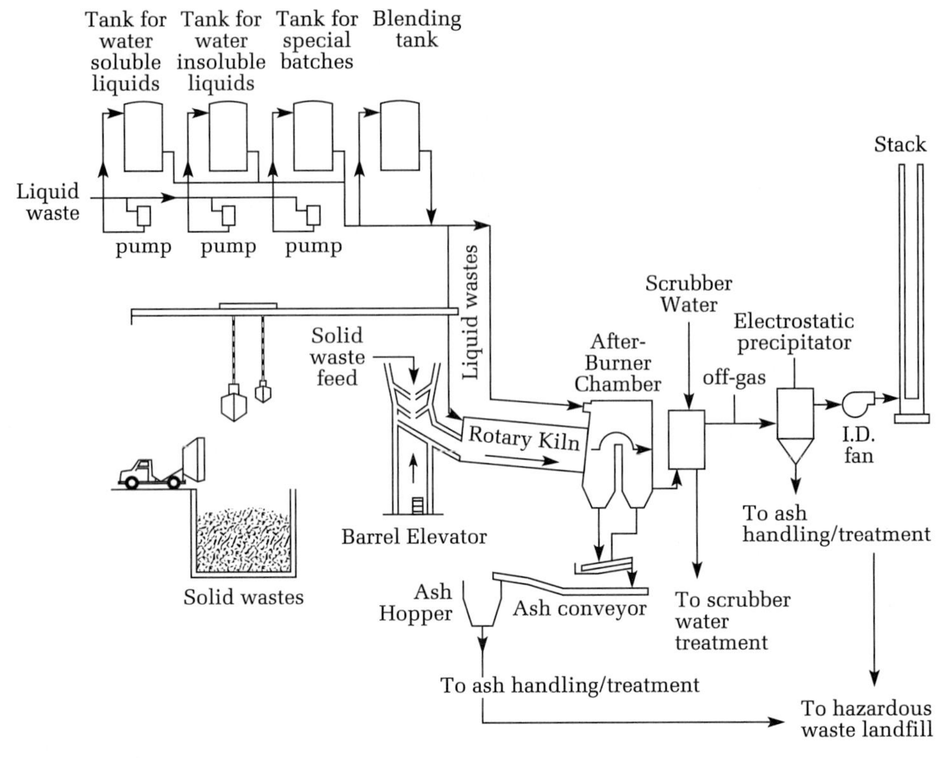

FIGURE 9-15
Rotary kiln incinerator.

Residence times for solid wastes are based on the rotational speed of the kiln and its angle. The residence time to volatilize waste is controlled by the gas velocity. The retention time of solids in the incinerator can be estimated from the following, where the coefficient 0.19 is based on limited experimental data:

$$t_o = \frac{0.19\,L}{NDS} \tag{9-30}$$

where t_o = retention time, min
L = kiln length, m
N = kiln rotational speed, rev/min
D = kiln diameter, m
S = kiln slope, m/m

Rotary kiln systems typically include secondary combustion chambers or afterburners to ensure complete destruction of the hazardous waste. Kiln operating temperatures range from 800 to 1,600°C. Afterburner temperatures range from 1,000 to 1,600°C. Liquid wastes are often injected into the secondary combustion chamber. The volatilized and combusted wastes leave the kiln and enter the secondary chamber, where additional oxygen is available and high heating value liquid wastes or fuel may be introduced. Both the secondary combustion chamber and the kiln are usually equipped with an auxiliary fuel firing system for startup.

Cement kilns are very efficient at destroying hazardous waste. Their long residence times and high operating temperatures exceed the requirements for destruction of most wastes. Hydrochloric acid generated from chlorinated hydrocarbon wastes is neutralized by the lime in the kiln while slightly lowering the alkalinity of the cement products. While cement plants can save energy by incinerating liquid wastes, the expense of obtaining permits and public resistance have inhibited use of this process.

Air pollution control (APC). Typical APC equipment on an incinerator will include an afterburner, liquid scrubber, demister, and fine particulate control device. Afterburners are used to control emission of unburned organic by-products by providing additional combustion volume at an elevated temperature. Scrubbers are used to physically remove particulate matter, acid gases, and residual organics from the combustion gas stream. Metals, of course, are not destroyed in the incineration process. Some are volatilized and then collected in the air pollution control device. The large liquid droplets that escape from the scrubber are captured in a mist collector. The final stage in gas cleaning is to remove the fine particles that remain. Electrostatic precipitators have been used for this step. Scrubber water and residues from other APC devices are still considered hazardous and must be treated before ultimate land disposal.

Permitting of hazardous waste incinerators.[42] The permitting of hazardous waste incinerators is a complex, multifaceted program conducted simultaneously on federal, state, and local levels. Because of the variety of state and local regulations for the handling, transportation, treatment, and disposal of hazardous wastes, as well as those concerning the operation of incinerators, each startup has a unique set of permit requirements.

Generally speaking, hazardous waste incinerators require at least the following permits: federal RCRA, state RCRA, for PCBs—the Toxic Substances and Control Act (TSCA), state and federal wastewater discharge, and state and federal air pollution control. A variety of local permits may also be necessary. Each of these require data substantiating an incinerator's operation at or above performance levels determined by environmental legislation. Each requires a public hearing and discussion of environmental impacts as well.

Hazardous waste incinerators must meet three performance standards:

1. *Principal Organic Hazardous Constituents* (POHC). The DRE for a given POHC is defined as the mass percentage of the POHC removed from the waste. The POHC performance standard requires that the DRE for each POHC *designated*

[42]Portions of this discussion were excerpted from L. Theodore and J. Reynolds, *Introduction to Hazardous Waste Incineration,* New York: Wiley, pp. 76–85, 1987.

in the permit be 99.99 percent or higher. The DRE performance standard implicitly requires sampling and analysis to measure the amounts of the designated POHC(s) in both the waste stream and the stack effluent gas during a trial burn. (The term *designated POHC* is described in more detail later in this section.) The DRE is determined for each designated POHC from a mass balance of the waste introduced into the incinerator and in the stack gas:[43]

$$\mathrm{DRE} = \frac{(W_{\mathrm{in}} - W_{\mathrm{out}})}{W_{\mathrm{in}}} \times 100\% \tag{9-31}$$

where W_{in} = mass feed rate of one POHC in the waste stream
W_{out} = mass emission rate of the same POHC present in exhaust emissions prior to release to the atmosphere

2. *Hydrochloric acid.* An incinerator burning hazardous waste and producing stack emissions of more than 1.8 kg/h of hydrogen chloride (HCl) must control HCl emissions such that the rate of emission is no greater than the larger of either 1.8 kg/h or 1 percent of the HCl in the stack gas prior to entering any pollution control equipment.
3. *Particulates.* Stack emissions of particulate matter are limited to 180 milligrams per dry standard cubic meter (mg/dscm) for the stack gas corrected to 7 percent oxygen. This adjustment is made by calculating a corrected concentration:

$$P_c = P_m \frac{14}{21 - Y} \tag{9-32}$$

where P_c = corrected concentration of particulate, mg/dscm
P_m = measured concentration of particulate, mg/dscm
Y = percent oxygen in the dry flue gas

In this way, a decrease in the particulate concentration due solely to increasing air flow in the stack is not rewarded, and an increase in the particulate concentration due solely to reduction in the air flow in the stack is not penalized. Special rules for this calculation are being developed for oxygen-enriched combustion systems where the oxygen content is greater than the 21 percent found in the atmosphere.

Compliance with these performance standards is documented by a trial burn of the facility's waste streams. As part of the Part B RCRA permit application, a trial burn plan detailing waste analysis, an engineering description of the incinerator, sampling and monitoring procedures, test schedule and protocol, as well as control information, must be developed. After submittal of Part B, EPA reviews the

[43]Note that this is not a mass balance around the incinerator. Hazardous waste that ends up in the scrubber water, APC residue, and ash are not counted. Hence, the oxidation can be very poor and the incinerator can still meet the 99.99 percent rule if the scrubber is efficient and/or the waste ends up in the ash. This is one reason that residues are considered hazardous and must be treated before land disposal.

documents for completeness; if the documentation is incomplete, a notice of deficiency requesting further information is issued; if the documentation is complete, EPA conducts a public hearing. If EPA determines that the design is adequate, a temporary or draft permit is issued. This allows the owner or operator to build the incinerator and initiate the trial burn procedure.

The temporary permit covers four phases of operation. During the first phase, immediately following construction, the unit is operated for *shake-down* purposes to identify possible mechanical deficiencies and to ensure its readiness for the trial burn procedures. This phase of the permit is limited to 720 h of operation using hazardous waste feed. The trial burn is conducted during the second phase. This is the most critical component of the permitting process, since it demonstrates the incinerator's ability to meet the three performance standards. In addition, performance data collected during the trial burn phase are reviewed by the permitting official and become the basis for setting the conditions of the facility permit. These conditions are: (1) allowable waste analysis procedures, (2) allowable waste feed composition (including acceptable variations in the physical or chemical properties of the waste feed), (3) acceptable operating limits for carbon monoxide in the stack, (4) waste feed rate, (5) combustion temperature, (6) combustion gas flow rate, and (7) allowable variations in incinerator design and operating procedures (including a requirement for shutoff of waste feed during startup, shutdown, and at any time when conditions of the permit are violated).

To verify compliance with the POHC performance standard during the trial burn, it is not required that the incinerator DRE for every POHC identified in the waste be measured. The POHCs with the greatest potential for a low DRE, based on the expected difficulty of thermal degradation (incinerability) and the concentration of the POHC in the waste, become the *designated POHCs* for the trial burn. The EPA permit review personnel work with the owners/operators of the incinerator facility in determining which POHCs in a given waste should be designated for sampling and analysis during the trial burn.

If a wide variety of wastes are to be treated, a difficult-to-incinerate POHC at high concentration may be proposed for the trial burn. The substitute POHC is referred to as a *surrogate POHC.* The surrogate POHC does not have to be actually present in the normal waste. It does, however, have to be considered more difficult to incinerate than any POHC found in the waste.

The third phase consists of completing the trial burn and submitting the results. This phase can last several weeks to several months, during which the incinerator is allowed to operate under specified conditions. The data to be reported to regulatory agencies after the burn are: (1) a quantitative analysis of the POHCs in the waste feed, (2) a determination of the concentration of the particulates, POHCs, oxygen, and HCl in the exhaust gas, (3) a quantitative analysis of any scrubber water, ash residues, and other residues to determine the fate of the POHCs, (4) a computation of the DRE for the POHCs, (5) a computation of the HCl removal efficiency if the HCl emission rate exceeds 1.8 kg/h, (6) a computation of particulate emissions, (7) the identification of sources of fugitive emissions and their means of control, (8) a measurement of average, maximum, and minimum temperatures and combustion gas velocities (gas flows), (9) a continuous measurement of carbon monoxide

(CO) in the exhaust gas, and (10) any other information EPA may require to determine compliance.

Provided that performance standards are met in the trial burn, the facility can begin its *fourth* and final phase, which continues through the duration of the permit. In the event that the trial burn results do not demonstrate compliance with standards, the temporary permit must be modified to allow for a second trial burn.

Example 9-8. A test burn waste mixture consisting of three designated POHCs (chlorobenzene, toluene, and xylene) is incinerated at 1,000°C. The waste feed rate and the stack discharge are shown below. The stack gas flow rate is 375.24 dscm/min (dry standard cubic meters per minute). Is the unit in compliance?

Compound	**Inlet (kg/h)**	**Outlet (kg/h)**
Chlorobenzene (C_6H_5Cl)	153	0.010
Toluene (C_7H_8)	432	0.037
Xylene (C_8H_{10})	435	0.070
HCl	—	1.2
Particulates at 7% O_2	—	3.615

Outlet concentrations were measured in the stack after APC equipment.

Solution. We begin by calculating the DRE for each of the POHCs.

$$\text{DRE} = \frac{(W_{\text{in}}) - (W_{\text{out}})}{(W_{\text{in}})} \times 100$$

$$\text{DRE}_{\text{chlorobenzene}} = \frac{153 - 0.010}{153} \times 100 = 99.993\%$$

$$\text{DRE}_{\text{toluene}} = \frac{432 - 0.037}{432} \times 100 = 99.991\%$$

$$\text{DRE}_{\text{xylene}} = \frac{435 - 0.070}{435} \times 100 = 99.984\%$$

The DRE for each designated POHC must be at least 99.99 percent. In this case, the designated POHC xylene fails to meet the standard. The other POHCs exhibit a DRE of greater than 99.99 percent.

Now we check compliance for the HCl emission. The HCl emission may not exceed 1.8 kg/h or 1 percent of the HCl prior to the control equipment, whichever is greater. It is obvious that the 1.2 kg/h emission meets the 1.8 kg/h limit. This would be sufficient to demonstrate compliance, but we will calculate the mass emission rate prior to control for the purpose of comparison. To do this we assume all the chlorine in the feed is converted to HCl. The molar feed rate of chlorobenzene (M_{CB}) is

$$M_{\text{CB}} = \frac{W_{\text{CB}}}{(MW)_{\text{CB}}} = \frac{(153 \text{ kg/h})(1{,}000 \text{ g/kg})}{112.5 \text{ g/mole}}$$

$$= 1{,}360 \text{ mole/h}$$

where M_{CB} = molar flow rate of chlorobenzene
$(MW)_{CB}$ = molecular weight of chlorobenzene

Each molecule of chlorobenzene contains one atom of chlorine. Therefore,

$$M_{HCl} = M_{CB}$$
$$= 1{,}360 \text{ mole/h}$$
$$W_{HCl} = (\text{GMW of HCl})(\text{mole/h})$$
$$= (36.5 \text{ g/mole})(1{,}360 \text{ mole/h})$$
$$= 49{,}640 \text{ g/h or } 49.64 \text{ kg/h}$$

This is the HCl emission prior to control. The emission of 1.2 kg/h is greater than 1 percent of the uncontrolled emission, that is,

$$1\% \text{ of uncontrolled} = (0.01)(49.64)$$
$$= 0.4964 \; kg/h.$$

However, the incinerator passes the HCl limits because the HCl emission is less than 1.8 kg/h.

The particulate concentration was measured at 7 percent O_2 and, therefore, does not need to be corrected. The outlet loading (W_{out}) of the particulates is

$$W_{out} = \frac{(3.615 \text{ kg/h})(10^6 \text{ mg/kg})}{(375.24 \text{ dscm/min})(60 \text{ min/h})}$$
$$= 160 \text{ mg/dscm}$$

This is less than the standard of 180 mg/dscm and is, therefore, in compliance with regard to particulates. However, because the incinerator fails the DRE for xylene, the unit is out of compliance.

Regulations for PCBs.[44] Incineration of PCBs is regulated under the Toxic Substances Control Act (TSCA) rather than RCRA. Thus, some of the permit conditions for incineration of PCBs are different from other RCRA hazardous wastes.

The conditions for incineration of liquid PCBs may be summarized as follows:

1. *Time and temperature.* Either of two conditions must be met. The residence time of the PCBs in the furnace must be 2 seconds at 1200°C ± 100°C with 3 percent excess oxygen in the stack gas or, alternatively, the furnace residence time must be 1.5 seconds at 1600°C ± 100°C with 2 percent excess oxygen in the stack gas.

   The EPA has interpreted these conditions to require a liquid PCB DRE ≥ 99.9999 percent.

[44] Portions of the following discussion were excerpted from C. A. Wentz, *Hazardous Waste Management*, New York: McGraw-Hill, pp. 206–207, 1989.

2. *Combustion efficiency.* The combustion efficiency shall be at least 99.99 percent, computed as follows:

$$\text{Combustion efficiency} = \frac{C_{CO_2}}{C_{CO_2} + C_{CO}} \times 100\% \tag{9-33}$$

where C_{CO_2} = concentration of carbon dioxide in stack gas
C_{CO} = concentration of carbon monoxide in stack gas

3. *Monitoring and controls.* In addition to these permitted limits, owners or operators of incinerators are required to monitor and control the variables that affect performance. The rate and quantity of PCBs fed to the combustion system must be measured and recorded at regular intervals of no longer than 15 minutes. The temperatures of the incineration process must be continuously measured and recorded. The flow of PCBs to the incinerator must stop automatically whenever one of the following occurs: the combustion temperature drops below the temperatures specified, that is, 1200 or 1600°C; when there is a failure of monitoring operations; when the PCB rate and quantity measuring and recording equipment fails; or when excess oxygen falls below the percentage specified. Scrubbers must be used for HCl removal during PCB incineration.

   In addition, a trial burn must be conducted and the following exhaust emissions must be monitored:

   Oxygen (O_2)
   Carbon monoxide (CO)
   Oxides of nitrogen (NO_x)
   Hydrogen chloride (HCl)
   Total chlorinated organic content
   PCBs
   Total particulate matter

   An incinerator used for incinerating nonliquid PCBs, PCB articles, PCB equipment, or PCB containers must comply with the same rules as those for liquid PCBs, and the mass air emissions from the incinerator must be no greater than 0.001 g PCB per kilogram of the PCB introduced into the incinerator, that is, a DRE of 99.9999 percent.

Stabilization/Solidification

Because of their elemental composition, some wastes, such as nickel, cannot be destroyed or detoxified by physical or chemical means. Thus, once they have been separated from aqueous solution and concentrated in ash or sludge, the hazardous constituents must be bound up in stable compounds that meet the LDR restrictions for leachability.

The terminology for this treatment technology has evolved in the last decade. In the early to mid 1980s "chemical fixation," "encapsulation," and "binding" were often used interchangeably with solidification and stabilization. With the promulgation of the LDR restrictions, the EPA established a more precise definition for solidification/stabilization and discouraged the use of the other terms to describe the technology.[45] EPA linked solidification and stabilization because the resultant material from the treatment must be both stable and solid. "Stability" is determined by the degree of resistance of the mixture of the hazardous waste and additive chemical to leaching in the *Toxicity Characteristic Leaching Procedure* (TCLP).[46] In the EPA definition, then, solidification/stabilization refers to chemical treatment processes that chemically reduce the mobility of the hazardous constituent.

Reduced leachability is accomplished by the formation of a lattice structure and/or chemical bonds that bind the hazardous constituent and thereby limit the amount of constituent that can be leached when water or a mild acid solution comes into contact with the waste matrix. There are two principal solidification/stabilization processes: cement based and lime based. The cement or lime additive is mixed with the ash or sludge and water. It is then allowed to cure to form a solid. The correct mix proportions are determined by trial-and-error experiments on waste samples. In both techniques the stabilizing agent may be modified by other additives such as silicates. In general, this technology is applicable to wastes containing metals with little or no organic contamination, oil, or grease.

9-8 LAND DISPOSAL

Deep Well Injection

Deep well injection consists of pumping wastes into geologically secure formations. Pumping of wastes into these formations has been practiced primarily in Louisiana and Texas. In promulgating the final third of the LDR restrictions,[47] the EPA allowed disposal of waste in Class I injection wells for wastes disposed under clean water act regulations.

Land Treatment

Land treatment is sometimes called "land farming" of the waste. In this practice, waste was incorporated with soil material in the manner that fertilizer or manure might be. Microorganisms in the soil degraded the organic fraction of the waste. Under the LDR restrictions, this practice is prohibited.

---

[45]U.S. Environmental Protection Agency, *Best Demonstrated Available Technology (BDAT) Background Document for F006* (U.S. Environmental Protection Agency Publication No. EPA/530-SW-88-0009-I), Washington, DC: U.S. Government Printing Office, May 1988.

[46]*Federal Register,* 55 FR 26986, June 29, 1990.

[47]*Federal Register,* 55 FR 22530, June 1, 1990.

The Secure Landfill

Although far from ideal, the use of land for the disposal of hazardous wastes is a major option for the foreseeable future. Furthermore, we recognize that incinerator ash, scrubber bottoms, and the results of biological, chemical, and physical treatment leave residues of up to 20 percent of the original mass. These residues must be secured in an economical fashion. At this juncture, the secure landfill is the only option.

The basic physical problem with land disposal of hazardous waste stems from the movement of water, originating as precipitation into and through the disposal site. The dissolution of waste material results in contaminants being transported from the waste site to larger regions of the soil zone and, too often, to an underlying aquifer. Problems of groundwater pollution frequently lead to the condemnation of wells and to the contamination of surface water bodies fed by the associated aquifer. In many instances, well contamination is not detected until years after land disposal of waste has begun, owing primarily to the slow movement of the conveying groundwater. For some chemical species, adsorption to soil particles may also act to retard the movement of the contaminant plume.[48]

Water pollution, caused by a hazardous waste facility, may evolve in a variety of ways. Leachate from landfills may drain out of the side of the landfill and appear as surface runoff. Alternatively, it may seep down slowly through the unsaturated zone, if one exists, and enter an underlying aquifer. Breaks in liners of holding ponds or cracks in the bottom of storage tanks also lead to a downward migration of contaminants toward the water table. Sometimes this migration is impeded by a relatively impermeable geologic barrier, such as a clay layer.

Without the institution of remedial measures, buried waste usually acts as a continuing source of pollution. The waste constituents continue to be transported in the subsurface by infiltrating precipitation. Thus, it is generally recommended that sites that handle hazardous wastes be located above a natural barrier, as well as an applied liner. Moreover, the site should be instrumented to continuously monitor the condition of any associated aquifers. If leachate generation is anticipated, there should be a system for the collection and treatment of the leachate.

The technology of the secure landfill may be divided into two phases: siting and construction. The following discussion is drawn primarily from E. F. Wood.[49]

Landfill siting. In siting a hazardous waste landfill, the four main considerations are air quality, groundwater quality, surface water quality, and subsurface migration of gases and leachates. Aside from the sociopolitical aspects, the last three components are the major factors to be considered in siting the landfill.

---

[48] E. F. Wood, R. A. Ferrara, W. G. Gray, and G. F. Pinder, *Groundwater Contamination from Hazardous Waste,* Englewood Cliffs, NJ: Prentice Hall, pp. 2–4, 145–158, 1984.

[49] E. F. Wood, *et al., Groundwater Contamination from Hazardous Waste,* 1984.

Air quality must be considered to prevent adverse effects to the air caused by volatilization, gas generation, gas migration, and wind dispersal of landfilled hazardous wastes. Generally, these can be controlled by proper construction techniques and do not inhibit the siting.

The hydrogeologic siting problem can be divided into four main areas: geology, soil, hydrology, and climate. Bedrock geology determines the structural framework that surfaces as landforms and the structural integrity of the landfill site. Soil is defined as the mantle of weathering between the atmosphere and unweathered rock.

Both groundwater and surface water supplies have the potential to become polluted. The proximity of a site to water supplies and the type of natural materials that occur between a site and a water supply influence contaminant migrations. These changes in topography and hydraulic conductivity of a geological formation influence surface and groundwater flow.

Climate is considered a driving force in contaminant migration, but we may exclude it when considering potential sites within the same region, where climate is unlikely to vary significantly.

Structural integrity of host rock is important in terms of seismic risk zones, dipping, and cleavage. Seismic risk zones are areas where earthquakes have occurred or have a high probability of occurring. They indicate the presence of geologic faults and fractures. The presence of faults and fractures is extremely important because they provide a natural pathway for the flow of contaminants, even in low-permeability and low-porosity rock. Future seismic activity could damage the landfill cells and storage tanks of a land emplacement facility during or after the construction, filling, and closing of the site, unless it is structurally designed to withstand ground motion.

Transport capacity refers to a soil's ability to allow migration of contaminants. Thus, the greater the soil transport capacity, the greater the migration of contaminants, which is undesirable. A soil with low permeability and porosity can lengthen the flow period and act as a natural defense by retarding the movement of contaminants. Glacial outwash plains and deltaic sands are both well-sorted sand and gravel beds with high permeability. Thus, they allow wastes to move faster and further. Clays and silts have lower permeabilities and, thus, inhibit the movement of wastes.

Sorption capacity depends on the organic content, predominant minerals, pH, and soil. Sorption includes both absorption and adsorption of contaminants. Sorption is important in limiting the movement of metals, phosphorus, and organic chemicals. *Cation exchange capacity* (CEC) is a measure of the ability of the soil to trade cations in the soil for those in waste. The higher the CEC, the more metal will be retained. The capacity of soil to retard contaminant migration also depends on the presence of numerous hydrous oxides, particularly iron oxides, and other compounds such as phosphates and carbonates. These compounds precipitate heavy metals out of solution, making them unable to travel further.

The hydrogen-ion concentration (pH) of soil influences the dominant removal mechanism for metal cations. The dominant removal mechanism for metal cations when $pH < 5$ is exchange or adsorption; when $pH > 6$, it is precipitation.

Hydrologic considerations in locating a hazardous waste landfill include distance to the groundwater table, the hydraulic gradient, the proximity of wells, and the proximity to surface waters.

When distance from the surface to the groundwater table is short, contaminant travel time is also short, allowing for little attenuation before pollutants disperse laterally in the saturated zone. It is desirable to have the average distance to the groundwater table large enough so that contaminants may be significantly attenuated. This also facilitates monitoring of the saturated zone. This will permit remedial action to be undertaken, if necessary.

A hydraulic gradient that slopes away from local groundwater supplies is desired. The steeper the hydraulic gradient, the lower the attenuation time and the faster the water movement. Therefore, a moderate hydraulic gradient may be most acceptable.

The distance from the disposal site to water-supply wells and surface waters must be as large as possible to protect them from potential contamination in case the landfill leaks. Furthermore, the proximity to surface waters must take into account the potential for flooding. Site flooding will weaken the structure of a land emplacement facility, causing it to fail and leak wastes. Therefore, it is essential that the facility not be built on a floodplain or area subject to local flooding. The facility should be designed so that it will not be flooded by the 100-year, or lesser, floods.

Landfill construction.[50] A secure landfill means, in essence, that no leachate or other contaminant can escape from the fill and cause adverse impacts on the surface water or groundwater. Leakage from the site is not acceptable during or after operations. Neither is any external or internal displacement, which could be brought about by slumping, sliding, and flooding. Wastes must not be allowed to migrate from the site.

It is nearly impossible to create an impervious burial vault for hazardous wastes and guarantee its integrity forever. Landfill design and operation is regulated to minimize migration of wastes from the site. The current EPA rules (40 CFR 264.300) for hazardous waste landfills require a minimum of (1) two or more liners, (2) a leachate collection system above and between the liners, (3) surface run-on and run-off control to collect and control at least the water volume resulting from a 24-hour, 25-year storm, (4) monitoring wells, and (5) a "cap" (Figure 9-16).

The liner system must include:[51]

1. A top liner designed and constructed of materials (for example, a geomembrane) to prevent migration of hazardous constituents into the liner during the active life and post-closure care period;
2. A composite bottom liner consisting of at least two components. The upper component must be designed and constructed of materials (for example, a geomembrane) to prevent migration of hazardous constituents into the liner during the

[50]This discussion follows that reported by Julian Josephson, "Hazardous Waste Landfill," *Environmental Science and Technology,* vol. 15 (3), p. 250, March 1981.

[51]*Federal Register,* 57 FR 3462, 29 JAN 1992.

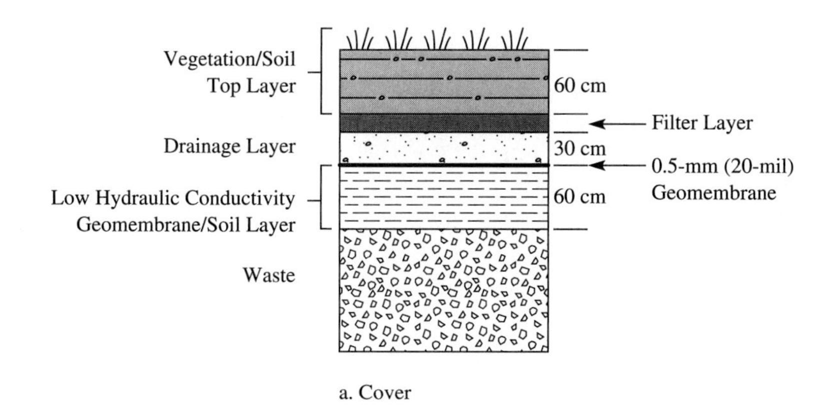

a. Cover

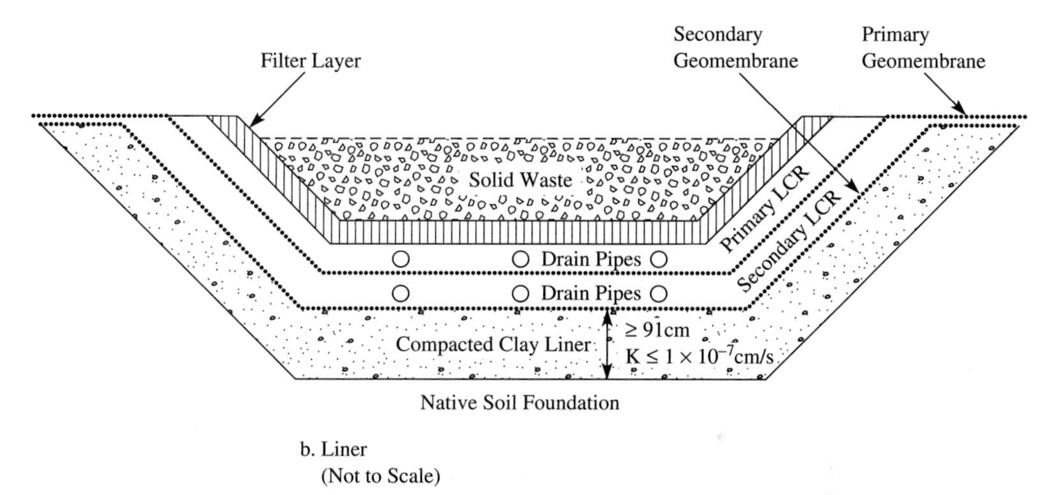

b. Liner
(Not to Scale)

FIGURE 9-16
Minimum technology landfill liner, design and recommended final cover design. (*Source: U.S. EPA Seminar Publications,* "Requirements for Hazardous Waste Landfill Design, Construction and Closure," EPA 625/4-89/022, 1989, and "Design and Construction of RCRA/CERCLA Final Covers," EPA 625/4-91/025, 1991.

active life and post-closure care period. The lower component must be designed and constructed of materials to minimize migration of hazardous constituents if a breach in the upper component were to occur. The lower component must be constructed of at least 91 cm of compacted soil material with a hydraulic conductivity of no more than 1×10^{-7} cm/s.

The leachate collection and removal system (LCR) immediately above the top liner must be designed, constructed, operated, and maintained to collect and remove leachate so that the leachate depth over the liner does not exceed 30 cm. The leachate collection and removal system between the liners and immediately above the bottom liner is also a leak detection system. The leachate collection system must,

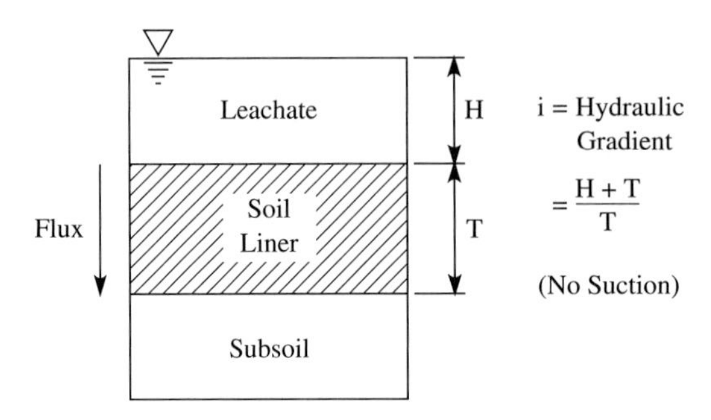

FIGURE 9-17
Definition of hydraulic gradient for landfill liner.

at a minimum, be:

1. Constructed with a bottom slope of one percent or more;
2. Constructed of a granular drainage material with a hydraulic conductivity of 1×10^{-2} cm/s or more and a thickness of 30 cm or more; or be constructed of synthetic or geonet drainage materials with a transmissibility of 3×10^{-5} m^2/s or more;
3. Constructed of sufficient strength to prevent collapse and be designed to prevent clogging.

The design equations for the leachate collection system are the same as those used for a municipal landfill (Section 8-4). The leachate collection system must include pumps of sufficient size to remove the liquids to prevent leachate from backing up into the drainage layer. The leachate must be treated to meet discharge limits. The treated leachate may be discharged into the municipal wastewater treatment system or into a waterway.

The amount of leachate may be estimated using Darcy's law (Equation 2-25). The hydraulic gradient for a liner is defined as shown in Figure 9-17. The flow rate cannot exceed the amount of water available, that is the product of the precipitation rate and the area of the landfill. The travel time of a contaminant through a soil layer may be estimated as the linear length of the flow path (T) divided by the seepage velocity (Equation 2-24).

Example 9-9. How long will it take for leachate to migrate through a 0.9 m clay liner with a hydraulic conductivity of 1×10^{-7} cm/s if the depth of leachate above the clay layer is 30 cm and the porosity of the clay is 55 percent?

Solution. The Darcy velocity is found using Equation 2-19.

$$v = K\left(\frac{dh}{dr}\right)$$

where the hydraulic gradient (dh/dr) is defined as in Figure 9-17:

$$\frac{dh}{dr} = \frac{0.30 \text{ m} + 0.9 \text{ m}}{0.9 \text{ m}} = 1.33$$

The Darcy velocity is then

$$v = (1 \times 10^{-7} \text{ cm/s})(1.33) = 1.33 \times 10^{-7} \text{ cm/s}$$

From Equation 2-24, the seepage velocity is

$$v' = \frac{K(dh/dr)}{\eta} = \frac{1.33 \times 10^{-7} \text{ cm/s}}{0.55} = 2.42 \times 10^{-7} \text{ cm/s}$$

The travel time is then

$$t = \frac{T}{v'} = \frac{(0.9 \text{ m})(100 \text{ cm/m})}{2.42 \times 10^{-7} \text{ cm/s}} = 3.71 \times 10^{8} \text{ s or about 12 years}$$

The site operator must keep careful records of the location and dimensions of each cell and must depict each cell on a map keyed to permanently surveyed vertical and horizontal markers. Records must show the contents of each cell and the approximate location of each hazardous waste type within the cell.

The purpose of groundwater monitoring is to ensure that programs for managing runon, runoff, and leachates are functioning properly so that groundwater remains uncontaminated. If contamination is occurring, early warning can be given and countermeasures taken. The site owner/operator has to place a sufficient number of monitoring wells around the limits of the facility to be able to describe the background (upgradient) and downgradient water quality. The regulations set forth, in detail, how the monitoring wells must be sunk, screened, sealed, sampled, and located, with special emphasis on location of the downgradient wells.

General groundwater quality, especially the suitability of the uppermost aquifer for use as a drinking water source, must meet EPA's primary drinking water standards. The flow rate for each sump must be calculated weekly during the active life and closure period, and monthly during the post-closure care period. If the landfill is leaking to the groundwater, the site operator must file an assessment plan with the EPA that shows how the problem is to be remedied.

9-9 GROUNDWATER CONTAMINATION AND REMEDIATION

The Process of Contamination[52]

Hazardous waste landfills are, of course, not the only source of groundwater contamination. Other sources include municipal landfills, septic tanks, mining and agricultural activities, "midnight dumping," and leaking underground storage tanks. It has been estimated that as many as 5,000 gasoline storage tanks at local service stations are leaking.

---

[52]Portions of this discussion were excerpted from C. A. Wentz, *Hazardous Waste Management,* pp. 277–279, 1989.

In all cases, the threat of contamination to groundwater depends on the specific geologic and hydrologic conditions of the site. The following paragraphs describe general considerations. All conditions may not exist at every site.

Leaking chemicals pass through several different hydrologic zones as they migrate through the soil to the groundwater system. The pore spaces in the unsaturated zone in the top soil layers are occupied by both air and water. Flow in this zone for liquid contaminants is downward by gravity. The upper region of the unsaturated zone is important for pollutant attenuation. Some chemicals are retained by adsorption onto organic material and chemically active soil particles. Some are trapped in the pore spaces and held by surface tension. These adsorbed and trapped chemicals may decompose through abiotic processes such as oxidation, reduction, and hydrolysis, as well as microbial activity, or they may simply remain sorbed onto the particles. Migration of precipitation may leach this sorbed and trapped material and carry it to the underlying aquifer for long periods of time after the source of contamination has been removed.

In the capillary zone just above the saturated zone that marks the groundwater table, spaces between soil particles may be saturated by water rising from the water table by capillary action. Chemicals that are lighter than water will "float" on top of the water table in this zone and move in different directions and rates than dissolved contaminants.

All pore spaces between soil particles below the water table are saturated. Generally, the saturated zone is devoid of oxygen. The lack of dissolved oxygen limits the oxidation of chemicals.

Groundwater flow is laminar, with minimal mixing occurring as the groundwater moves. Dissolved chemicals will flow with groundwater and form distinct plumes. The shape and size of a contaminant plume depends upon the local hydrogeological setting, groundwater flow, the characteristics of the contaminants, and geochemistry. Solubility, adsorption characteristics, and degradation affect mobility. The density of the contaminant is important in determining the shape and movement of the plume. Lighter, less soluble chemicals, like gasoline, will tend to flow on top of the aquifer (Figure 9-18). Water soluble contaminants tend to dissolve in and then flow with groundwater (Figure 9-19). Dense, insoluble contaminants will sink to the bottom (Figure 9-20) of the aquifer. *Volatile organic chemicals* (VOCs) in groundwater are extremely mobile. Polyvalent metal contaminants tend to adsorb onto clays and, hence, are not very mobile.

EPA's Groundwater Remediation Procedure[53]

The federal program for cleanup of contaminated sites follows a procedural sequence as shown in Figure 9-21. Each of these steps is discussed in the following paragraphs.

---

[53]Portions of this discussion were excerpted from C. A. Wentz, *Hazardous Waste Management,* pp. 277–279, 1989.

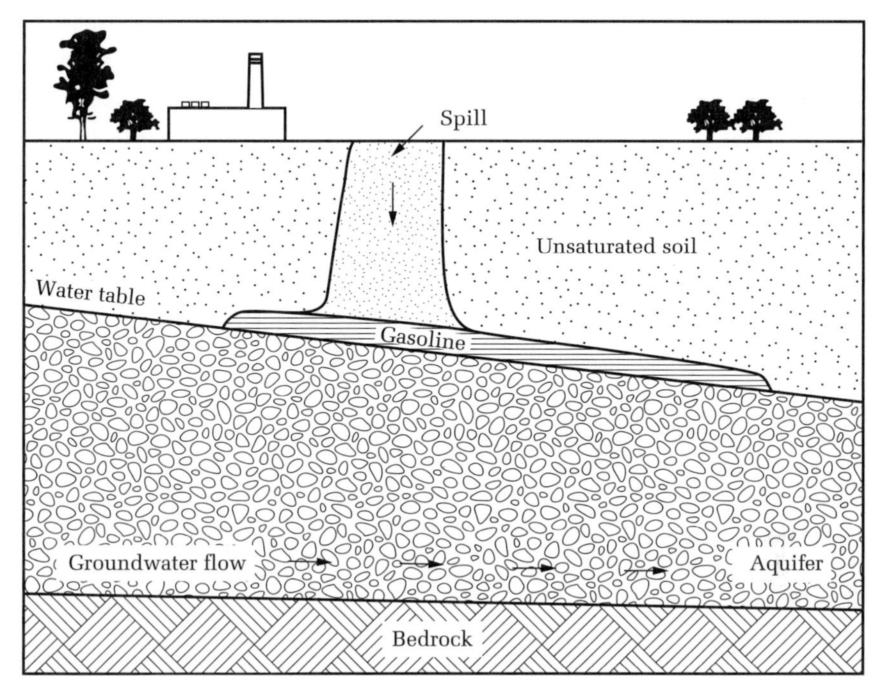

FIGURE 9-18
Immiscible plume less dense than water.

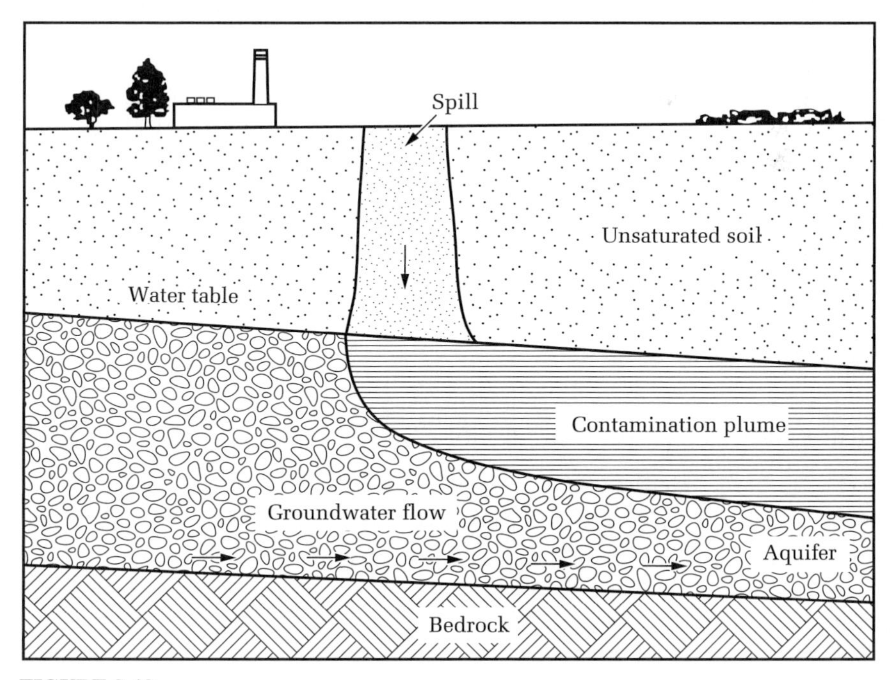

FIGURE 9-19
Dissolved contamination plume.

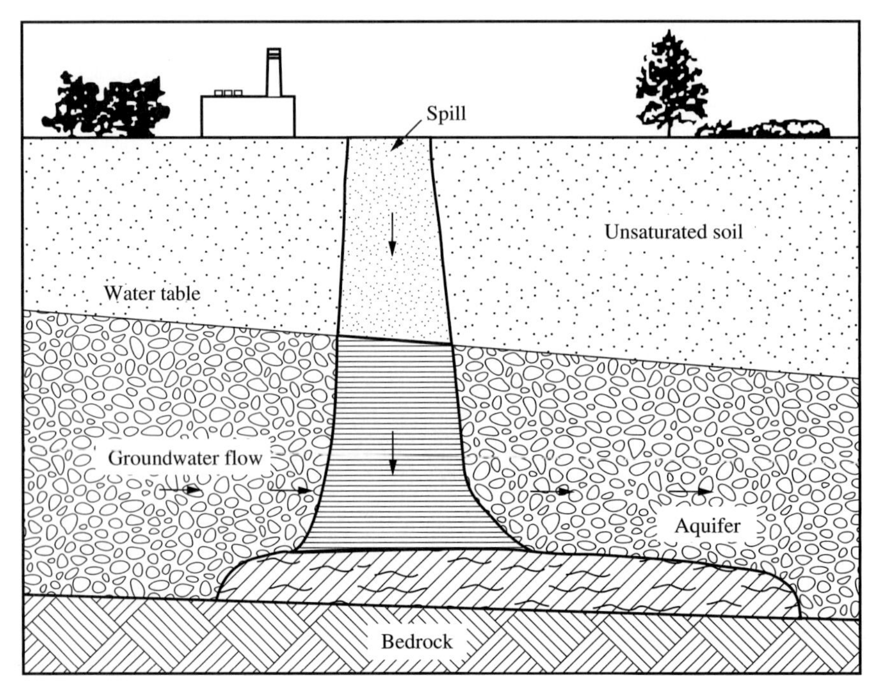

FIGURE 9-20
Immiscible plume denser than water.

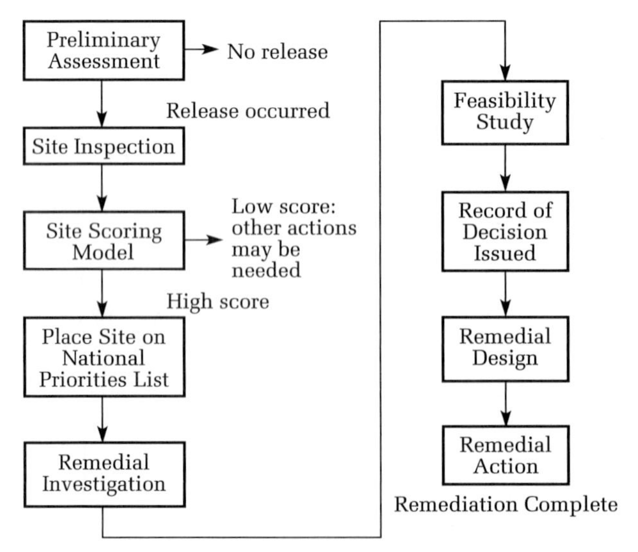

FIGURE 9-21
Steps involved in the Superfund cleanup process.

Preliminary assessment. EPA involvement usually begins with the identification of a potential hazardous waste site. The initial information can come from a variety of sources, including local citizens and officials, state environmental agencies, the site owners themselves, or simply from awareness of potential problems associated with particular industries.

EPA has developed an inventory system called the Comprehensive Environmental Response, Compensation, and Liability Information System (CERCLIS) to document all of the sites in the United States that may be candidates for remedial action. This is a continuing program that identifies sites as information about them becomes available. The growth in the number of CERCLIS sites has been dramatic and is expected to continue for the foreseeable future as additional abandoned and contaminated sites are discovered. As of September 1989, 31,522 sites had been logged into the inventory.[54]

A *preliminary assessment* (PA) is the first step in identifying the potential for contamination from a particular site. The primary objectives of the PA are to determine if there has been a release of contaminant to the environment, if there is immediate danger to persons living or working near the site, and whether a site inspection is necessary. Samples for environmental analysis are generally not taken during the PA. Following the preliminary assessment, EPA or the designated state agency might determine that an immediate threat to residents or employees at the site requires an immediate removal action. Otherwise, on the basis of the preliminary assessment, the site is classified by EPA into one of the three following categories:

1. There is no further action needed, since there is no threat to human health or the environment.
2. Additional information is required to complete the preliminary assessment.
3. Inspection of the site is necessary.

Site inspection. Site inspection requires sampling to determine the types of hazardous substances that are present and to identify the extent of contamination and its migration. The actual site inspection includes preparation of a work plan and an on-site safety plan. The site assessment has three objectives:

1. To determine which releases pose no threat to public health and the environment;
2. To determine if there is any immediate threat to persons living or working near the release;
3. To collect data to determine whether or not a site should be included on the National Priorities List (NPL).

HRS, NPL, RI/FS, and ROD. The next series of steps in the EPA's procedure include calculations to complete the HRS; inclusion on the NPL if the score is suffi-

[54]Office of Technology Assessment, *Coming Clean: Superfund Problems Can Be Solved...*, Congress of the United States, OTA-ITE-433, p. 11, 1989.

ciently high; conduct of a RI/FS; and issuance of an ROD. These steps were discussed in detail in Section 9-5.

Remedial design and remedial action. EPA-funded remedial actions may be taken only at those sites that are on the NPL. This ranking helps ensure that the Superfund dollars are used in the most cost-effective manner and where they will yield the greatest benefit.

Before a remedial action can be taken at a site, a number of questions must be answered. These can be classified as problem definition, design alternatives, and policy.

1. Problem definition questions: What are the contaminants and how much contamination is present? How large is the surface area of the contaminated site? What is the size of the contaminated groundwater plume? Where is the exact location of the plume and in what direction is it moving?
2. Design questions: Based on the alternatives available, what is the best way to clean up the site? How should these alternatives be implemented? What products will be produced during treatment? How long will it take to complete the remediation and what will it cost?
3. Policy questions: What level of protection is adequate? In other words, how clean is clean?

The answers to the first two sets of questions require scientific and engineering background that is supported by extensive sampling of the contaminated site area. The last question cannot be answered objectively; rather it is a subjective and oftentimes political question.

The NCP defines three types of responses for incidents involving hazardous substances. In these responses *removal* is differentiated from *remediation.* Removal is, as its name suggests, the physical relocation of the waste—usually to a secure hazardous waste landfill. Remediation means that the waste is to be treated to make it less toxic and/or less mobile or the site is to be contained to minimize further release. Remediation can take place on site or at a TSD facility. The three types of responses are:

1. *Immediate removal* is a prompt response to prevent immediate and significant harm to human health or the environment. Immediate removals must be completed within six months.
2. *Planned removal* is an expedited removal when some response, not necessarily an emergency response, is required. The same six-month limitation also applies to planned removal.
3. *Remedial response* is intended to achieve a site solution that is a permanent remedy for the particular problem involved.

Immediate removals are done to prevent an emergency involving hazardous substances. These emergencies might include fires; explosions; direct human contact with a hazardous substance; human, animal, or food-chain exposure; or contamination of drinking water sources. An immediate removal involves cleaning up the hazardous site to protect human health and life, containing the hazardous release, and minimizing the potential for damage to the environment. For example, a truck, train, or barge spill could involve an immediate removal determination by EPA to get the spill cleaned up.

Immediate removal responses may include activities such as sample collection and analysis, containment or control of the release, removal of the hazardous substances from the site, provision of alternate water supplies, installation of security fences, evacuation of threatened citizens, or general deterrent of the spread of the hazardous contaminants.

A planned removal involves a hazardous site that does not present an immediate emergency. Under Superfund, EPA may initiate a planned removal if the action will minimize the damage or risk and is consistent with a more effective long-term solution to the problem. Planned removals are carried out by EPA if the responsible party is either unknown or cannot or will not take timely and appropriate action. The state in which the cleanup is located must be willing to match at least 10 percent of the costs of the removal action, as well as agree to nominate the site in question for the National Priority List.

Mitigation and Treatment[55]

Since the spread of contaminants is usually confined to a plume, only localized areas of an aquifer need to be reclaimed and restored. Cleanup of a contaminated aquifer, however, is often troublesome, time-consuming, and costly. The original source of contamination can be eliminated, but the complete restoration of the groundwater is fraught with additional problems, such as defining the site's subsurface environment, locating potential contamination sources, defining potential contaminant transport pathways, determining contaminant extent and concentration, and choosing and implementing an effective remedial process.

Although not a simple task, cleanup is possible. Various methods are being used and have proven successful in certain cases. These efforts have ranged from containment to destruction of the contaminants, either in their original position in the aquifer or by withdrawing the groundwater. Examples of these remedial methods include installing pumping wells to remove contaminated water, building trenches to arrest only the contaminated flow, and stimulating biodegradation of groundwater contaminants.

In many circumstances, the most reasonable and economic remedial approach is to treat the water to attain the necessary quality for a specific use. This treated water may then be used or returned to the aquifer. Certainly, combinations of barriers and treatment methods should be considered. Source control (removal or remediation of the source), physical control, and treatment methods all will have their part in mit-

[55]Portions of this discussion were excerpted from R. D. Griffin, *Principles of Hazardous Materials Management,* Ann Arbor, MI; Lewis Publishers, 1988.

igating groundwater contamination problems. Legal implications may also dictate strategies that may be utilized.

Well systems. A well system serves as an example of a common remedial process for contaminated groundwater. Well systems manipulate the subsurface hydraulic gradient through injection or withdrawal of water. They are designed to control the movement of the groundwater directly and of the subsurface pollutants indirectly. All require installation of wells at selected sites. It is necessary to first conduct a hydrogeologic study to determine the characteristics of the contaminant plume (width, length, depth, and general shape), the hydraulic gradient across the plume, and the hydrogeologic characteristics of the aquifer. There are three major types of well system: well point, deep well, and pressure ridge systems.

Both well point and deep well systems withdraw water. The former utilizes closely spaced, shallow wells, each connected to a main pipe (header) which is connected to a centrally located pump. The well point system is used only for shallow water table aquifers. Deep well systems are used for greater depths and are most often pumped individually. Both well point and deep well systems should be designed so that the radius of influence of the system completely intercepts the contaminant plume.

The principle of pressure ridge systems is the inverse of that of a well system. Clean, uncontaminated water is injected to form an upconing of the water table that acts as a barrier to groundwater flow. The injection well is also called a recharge well. The upconing effect of a recharge well is shown in Figure 9-22.

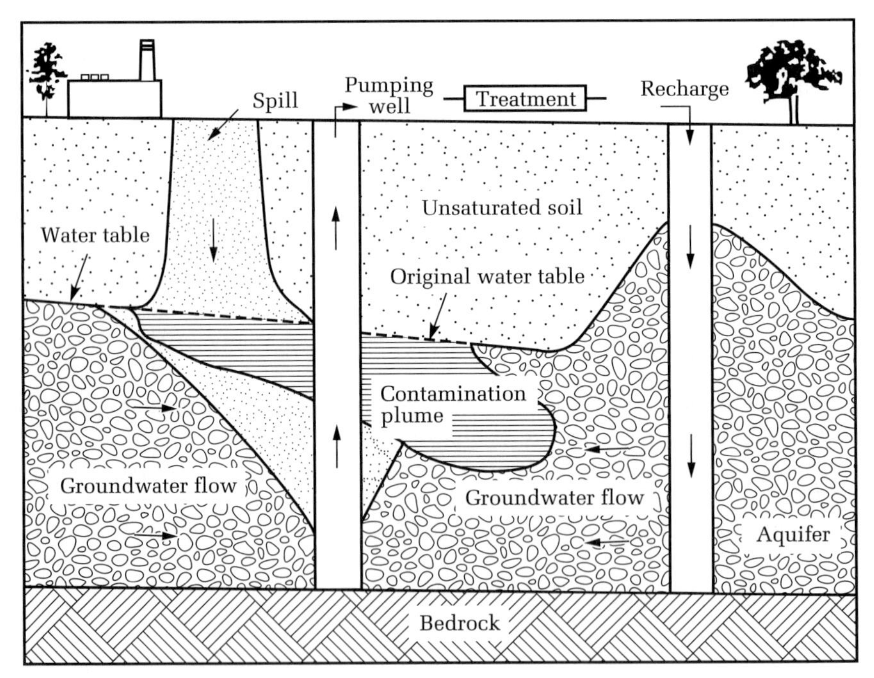

FIGURE 9-22
Effect of a recharge well on groundwater flow patterns.

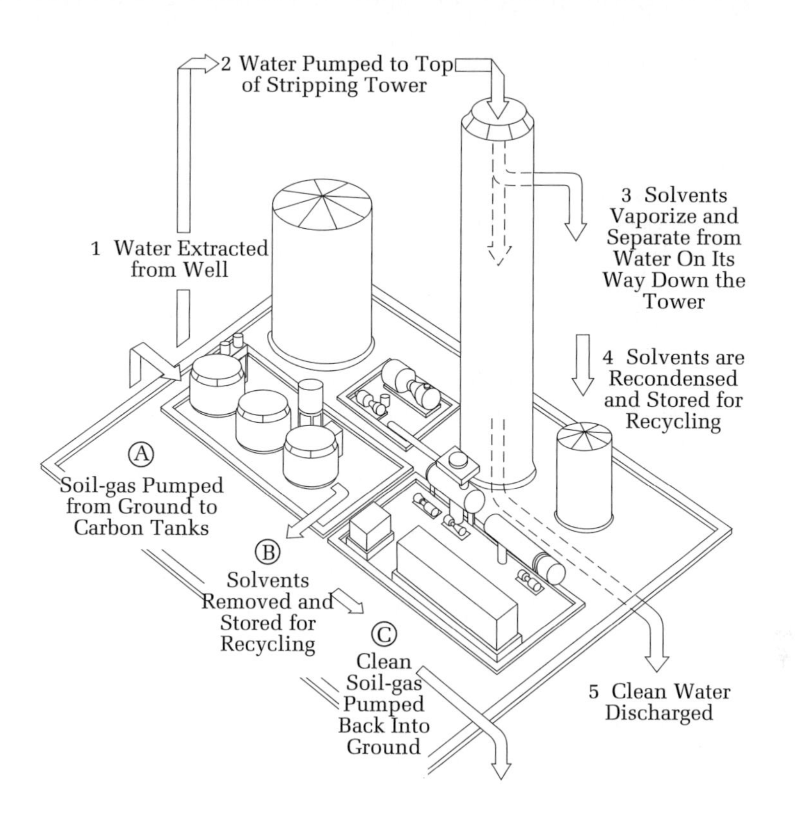

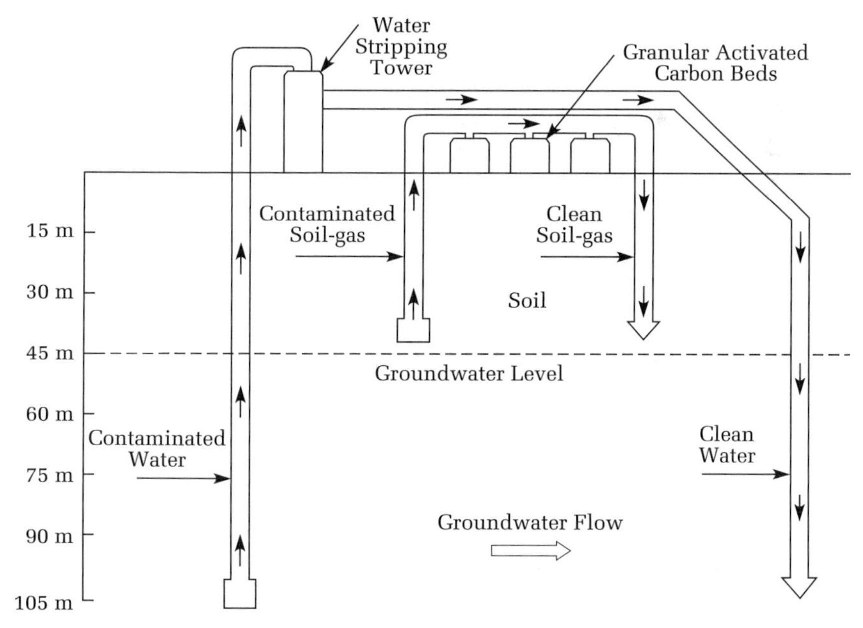

FIGURE 9-23
Lockheed Aeronautical Systems Company's Aqua-Detox groundwater treatment system. (*Source: Hazmat World,* November 1989.)

Non-Aqueous Phase Liquids (NAPLs) such as gasoline are referred to as "product" because their recovery may have some commercial value. When the NAPL floats on the groundwater table, special recovery techniques may be employed to recover it. Product recovery systems to recover NAPL use wells that terminate in the NAPL plume rather than in the aquifer. Because all hydrocarbons are slightly soluble in water, the product recovery system is usually accompanied by a groundwater pumping system to remove and treat the contaminated groundwater.

Treatment technologies. The treatment methods used to clean the contaminated groundwater are fundamentally the same as those discussed in Section 9-7. Each must be tailored to the local situation. An example is a system for removing trichloroethylene (TCE) and perchloroethylene (PCE) developed by Lockheed Aeronautical Systems Company (LASC). It consists of a deep well to remove the water, a steam stripping column with a condenser to remove organic chemicals from the water, a vacuum/pressure system to remove contaminated vapors, an activated carbon system to capture the vapors, and reinjection wells (Figure 9-23). This system treats 0.066 m^3/s of water.

9-10 CHAPTER REVIEW

When you have finished studying this chapter, you should be able to do the following without the aid of your textbooks or notes:

1. Sketch the chemical structure of 2,3,7,8-TCDD.
2. Explain how 2,3,7,8-TCDD occurs and/or when it is found in nature.
3. Sketch the chemical structure of the PCB 2,4′-dichlorobiphenyl.
4. Explain the origin of PCBs.
5. Define and differentiate between risk and hazard.
6. List the four steps in risk assessment and explain what occurs in each step.
7. Define the terms dose, LD_{50}, NOAEL, slope factor, RfD, CDI, IRIS.
8. Explain why it is not possible to establish an absolute scale of toxicity.
9. Explain why an average dose-response curve may not be an appropriate model to develop environmental protection standards.
10. Identify routes of potential exposure for the release of contaminants in multiple media.
11. Explain how risk management differs from risk assessment and the role of risk perception in risk management.
12. Define hazardous waste.
13. List the five ways a waste can be found to be hazardous and briefly explain each.
14. Explain why dioxin and PCB are hazardous wastes.
15. State how long generators may store their waste.
16. Explain what defines a small quantity generator and what "break" the rules give them.
17. Define the abbreviations CFR, FR, RCRA, HSWA, CERCLA, and SARA.

18. Explain the major difference (objective) between RCRA/HSWA and CERCLA/SARA.
19. Define/explain the terms "cradle-to-grave" and manifest system.
20. Explain what "land ban," or LDR, means.
21. Define the abbreviations TSD and UST.
22. Describe the three ways to meet corrosion protection standards for underground storage tanks.
23. List the four major provisions of CERCLA.
24. Define/explain the following abbreviations: NCP, NPL, HRS, RI, FS, ROD, and PRP.
25. Explain why it is important for a site to be placed on the NPL.
26. Explain the concept of "joint and several liability" and the implications to those with wastes found in an abandoned hazardous waste site.
27. List and explain four hazardous waste management techniques.
28. List the objectives of a waste audit.
29. Differentiate between waste minimization, waste exchange, and recycling.
30. List six disposal technologies for hazardous wastes.
31. Explain why seismic risk is important in landfill siting.
32. Explain how permeability, porosity, and sorption capacity of soil limit the migration of hazardous wastes.
33. Explain what hydrologic features are important in siting a landfill.
34. List the minimum EPA requirements for a hazardous waste landfill and sketch a landfill that meets these.
35. Explain the difference between deep well injection and land treatment.
36. Define the following acronyms: PIC, POM, POHC, and DRE, as they apply to incineration.
37. List the most important factors for proper incinerator design and operation.
38. List the two types of incinerators most commonly used for destroying hazardous waste.
39. Explain the terms "designated POHC" and "surrogate" as they apply to a trial burn.
40. Outline the steps in EPA's remediation procedures.
41. Differentiate between "remediation" and "removal" as they pertain to a CERCLA/SARA cleanup.

With the aid of this text, you should be able to do the following:

1. Calculate lifetime risk using the one-hit or multistage model.
2. Calculate chronic daily intake or other variables given the media and values for remaining variables.
3. Perform a risk characterization calculation for carcinogenic and noncarcinogenic threats by multiple contaminants and multiple pathways.

4. Determine whether or not a waste is an EPA hazardous waste based on its composition, source, or characteristics.
5. Perform a mass balance to identify waste sources or waste-minimization opportunities.
6. Write the reactions for oxidation or reduction of chemical contaminants to mineralized form.
7. Perform solubility product calculations to estimate treatment doses for precipitation or the concentration of contaminants that remain in solution.
8. Determine the dimensions of an air stripping column, air or liquid flow rate given the values for remaining variables.
9. Determine the mass of resin and column dimensions for an ion exchange column given laboratory or pilot breakthrough data.
10. Evaluate a chemical feed to an incinerator to determine whether or not the chlorine content is acceptable and design a mix of waste feeds to achieve a desired chlorine feed rate.
11. Evaluate the operating variables for an incinerator to determine regulatory compliance for DRE, HCl emissions, and particulate emissions.
12. Estimate the hydraulic conductivity of a liner material based on laboratory measurements.
13. Estimate the quantity of leachate given the precipitation rate, area, hydraulic gradient, and hydraulic conductivity.
14. Estimate the seepage velocity and travel time of a contaminant through a soil given the hydraulic gradient, hydraulic conductivity, porosity, and length of the flow path.

9-11 PROBLEMS

9-1. Estimate the chronic daily intake of toluene from exposure to a city water supply that contains a toluene concentration equal to the drinking water standard. Assume the exposed individual is an adult female who consumes water at the adult rate for 70 years, that she abhors swimming, and that she takes a long (20 min) bath every day. Assume that the average air concentration of toluene during the bath is 1 $\mu g/m^3$. Assume the dermal uptake from water (PC) is 9.0×10^{-6} m/h and that direct dermal absorption during bathing is no more than 80 percent of the available toluene because she is not completely submerged.

9-2. Estimate the chronic daily intake of 1,1,1-trichloroethane from exposure to a city water supply that contains a 1,1,1-trichloroethane concentration equal to the drinking water standard. Assume the exposed individual is a child who consumes water at the child rate for 5 years, that she swims once a week for 30 min, and that she takes a short (10 min) bath every day. Assume her average age over exposure period is 8. Assume that the average air concentration of 1,1,1-trichloroethane during the bath is 2 $\mu g/m^3$. Assume the dermal uptake from water (PC) is 0.0060 m/h and that direct dermal absorption during bathing is no more than 80 percent of the available 1,1,1-trichloroethane because she is not completely submerged.

9-3. Characterize the risk for a chronic daily exposure by the water pathway (oral) of 0.03 mg/kg · d of toluene, 0.06 mg/kg · d of barium, and 0.3 mg/kg · d of xylenes.

9-4. Characterize the risk for a chronic daily exposure by the water pathway (oral) of 1.43×10^{-4} mg/kg · d of tetrachloroethylene, 1.43×10^{-3} mg/kg · d of arsenic, and 1.43×10^{-4} mg/kg · d of dichloromethane (methylene chloride).

9-5. Determine whether each of the following is a RCRA hazardous waste.

a. Municipal wastewater containing 2.0 mg/L of selenium.

b. Spent pickle liquor from steel finishing operations.

9-6. Determine whether each of the following is a RCRA hazardous waste.

a. Sludge from which an acetic acid extract produced according to EPA procedures in SW 846 yields 5.1 mg/L of silver.

b. An empty pesticide container that a homeowner wishes to discard.

9-7. A dry cleaner accumulates 10 kg per month of carbon tetrachloride (a hazardous waste solvent). To save shipping cost he would like to accumulate six months' worth before he ships it to a TSDF. Can he do this? Explain.

9-8. Using the following data and Figure P-9-8, use the mass balance technique to determine the mass flow rate (kg/d) of organics to the condensate collection tank (sample location 4 in Figure P-9-8).

Sample location	Flow rate, L/min	Total volatile organic	Temperature, °C
1	40.5	5,858 mg/L	25
2	44.8	0.037 mg/L	80
3	57 (vapor)	44.13%	20

The temperature at sample location 3 is 20°C.

Notes:

a. % is volume percent

b. Vapor flow rate is corrected to 1 atm and 20°C.

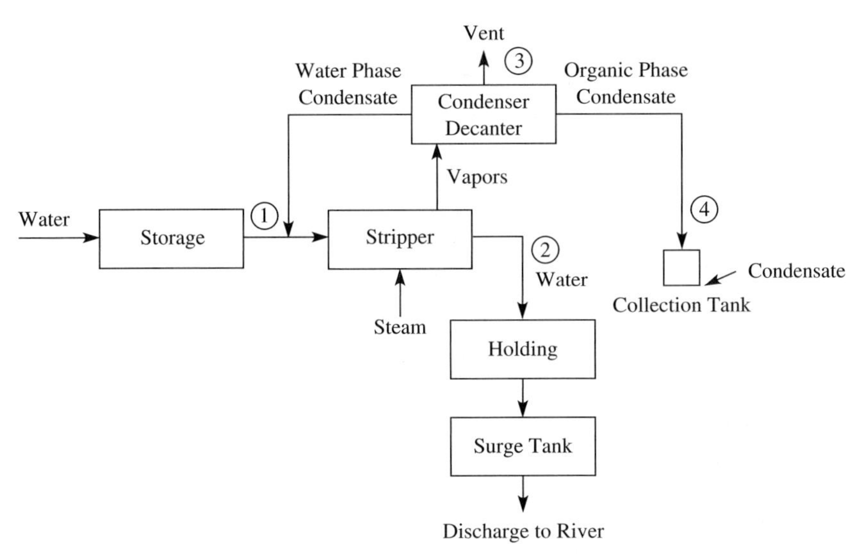

FIGURE P-9-8

c. Liquid organic density may be assumed to be 0.95 kg/L.
d. Assume the molecular weight of the organic vapor is equal to that of methylene chloride.
e. Steam mass flow rate is 252 kg/h at 106°C.

9-9. What is the efficiency of the condenser-decanter in Problem 9-8?

9-10. Given the waste constituent and concentration shown below, determine the quantity (in kg/d) of hydrated lime ($Ca(OH)_2$) required to neutralize the waste. Estimate the total dissolved solids (TDS) after neutralization. Report your answer in mg/L.

Constituent	Concentration, mg/L	Flow, L/min
HCl	100	5

9-11. Given the constituents and concentrations shown below, determine the quantities (in kg/d) of sulfuric acid required to neutralize the waste. Estimate the total dissolved solids (TDS) after neutralization. Report your answer in mg/L.

Constituent	Concentration, mg/L	Flow, L/min
NaOH	250	20

9-12. Write the reaction equation to oxidize sodium cyanide using sodium hypochlorite (NaOCl).

9-13. Write the reaction equation to reduce hexavalent chromium in chromic acid ($H_2Cr_2O_7$) using $NaHSO_3$.

9-14. A metal plating solution contains 20.00 mg/L of copper. Determine the concentration, in moles/L, to which the hydroxide concentration must be raised to precipitate all but 0.500 mg/L of the copper using lime. The K_{sp} of copper hydroxide is 2.00×10^{-19}. Estimate the final pH.

9-15. A metal plating sludge as removed from a clarifier has a solids concentration of 4 percent. If the volume of sludge is 1.0 m^3/d, what volume will result if the sludge is processed in a filter press to a solids concentration of 30 percent? If the pressed sludge is dried to 80 percent solids, what volume will result?

9-16. In Problem 9-15, ferrocyanide is found in the clarifier sludge at a concentration of 400 mg/kg (4 percent solids). Assuming that the ferrocyanide is part of the precipitate and that none escapes from the filter press, what concentration would be expected in filter cake?

9-17. A drinking water supply at Oscoda, Michigan, has been contaminated by trichloroethylene. The average concentration in the water is estimated to be 6,000 μg/L. Using the following design parameters, design a packed-tower stripping column to reduce the water concentration to the State of Michigan discharge limit of 1.5 μg/L. Note that more than one column in series may be required for reasonable tower heights.

Henry's law constant $= 6.74 \times 10^{-3}$ $m^3 \cdot$ atm/mol
$K_L a = 0.720$ min^{-1}
Air flow rate $= 60$ m^3/min

G/L = 18
Temperature = 25°C
Column diameter may not exceed 4.0 m
Column height may not exceed 6.0 m

9-18. Well 13 at Watapitae is contaminated with 440 μg/L of tetrachloroethylene (perchloroethylene). The water must be remediated to achieve a concentration of 0.2 μg/L (the detection limit). Design a packed-tower stripping column to meet this requirement using the following design parameters. Note that more than one column in series may be required for reasonable tower heights.

Henry's law constant = 100×10^{-4} $m^3 \cdot$ atm/mol
$K_L A = 13.5 \times 10^{-3}$ s^{-1}
Air flow rate = 15 m^3/s
Liquid flow rate = 0.22 m^3/s
Temperature = 20°C
Column diameter may not exceed 4.0 m
Column height may not exceed 6.0 m

9-19. An electroplating rinse water containing 55 mg/L of nickel is to be treated by an ion exchange column to meet an allowable effluent concentration of 2.6 mg/L. A laboratory scale column has provided the breakthrough data shown in the table below. The laboratory column data are as follows:

Inside diameter = 1.0 cm
Length = 7.0 cm
Mass of resin (moist basis) = 5.2 g
Water content = 17%
Density of resin = 0.65 g/cm^3
Liquid flow rate = 7.68 L/d
Initial concentration = 55 mg/L

Breakthrough Data

$\forall$, L	C, mg/L	$\forall$, L	C, mg/L	$\forall$, L	C, mg/L	$\forall$, L	C, mg/L
0.160	4.23	0.800	23.62	1.280	39.04	1.760	53.32
0.320	5.14	0.960	29.54	1.440	44.04	1.920	54.14
0.480	10.03	1.120	35.46	1.600	49.54	2.080	53.22
0.640	16.65						

The full scale design must meet the following requirements:

Flow rate = 36,000 L/d
Hours of operation = 8 h/d
Regeneration is to be once every five days

Determine the mass of resin required.

9-20. An electroplating rinse water containing 10 mg/L of silver is to be treated by an ion exchange column to meet an allowable effluent concentration of 0.24 mg/L. A laboratory scale column has provided the breakthrough data shown in the table below. The laboratory column data are as follows:

Inside diameter = 1.0 cm
Length = 14.85 cm
Mass of resin (moist basis) = 7.58 g
Water content = 34%
Density of resin = 0.65 g/cm^3
Liquid flow rate = 4.523 L/d
Initial concentration = 10 mg/L

Breakthrough Data

$\forall$, L	C, mg/L	$\forall$, L	C, mg/L
0.1	0.00	1.1	2.00
0.2	0.00	1.2	3.33
0.3	0.01	1.3	5.00
0.4	0.02	1.4	6.67
0.5	0.04	1.5	8.00
0.6	0.08	1.6	8.89
0.7	0.16	1.7	9.41
0.8	0.31	1.8	9.69
0.9	0.61	1.9	9.84
1.0	1.15	2.0	9.92

The full scale design must meet the following requirements:

Flow rate = 3,600 L/d
Hours of operation = 8 h/d
Regeneration is to be once every five days

Determine the mass of resin required.

9-21. An incinerator operator receives the following shipments of waste for incineration. Can the operator mix these wastes to achieve 30 percent by mass of chlorine in the feed?

Trichloroethylene = 18.9 m^3
1,1,1-Trichloroethane = 5.3 m^3
Toluene = 21.3 m^3
o-Xylene = 4.8 m^3

9-22. An incinerator operator receives the following shipments of waste for incineration. What volume of methanol (CH_3OH) must the operator mix to achieve 30 percent by mass of chlorine in the feed? Assume the density of methanol is 0.7913 g/mL.

Carbon tetrachloride = 1.2 m^3
Hexachlorobenzene = 15.3 m^3
Pentachlorophenol = 25.0 m^3

9-23. A hazardous waste incinerator is being fed methylene chloride at a concentration of 5,858 mg/L in an aqueous stream at a rate of 40.5 L/min. Calculate the mass flow rate of the feed in units of g/min.

9-24. Methylene chloride was measured in the flue gas of a hazardous waste incinerator at a concentration of 211.86 $\mu g/m^3$. If the flow rate of gas from the incinerator was 597.55 m^3/min, what was the mass flow rate of methylene chloride in g/min?

9-25. Assuming that the same incinerator is being evaluated in Problems 9-23 and 9-24, what is the DRE for the incinerator?

9-26. Xylene is fed into an incinerator at a rate of 481 kg/h. If the mass flow rate at the stack is 72.2 g/h, is the unit in compliance with the EPA rules?

9-27. 1,2-Dichlorobenzene is being burned in an incinerator under the following conditions:

Operating temperature = 1,150°C
Feed flow rate = 173.0 L/min
Feed concentration = 13.0 g/L
Residence time = 2.4 s
Oxygen in stack gas = 7.0%
Stack gas flow rate = 6.70 m^3/s at standard conditions
Stack gas concentrations after APC equipment
Dichlorobenzene = 338.8 μg/dscm
HCl = 77.2 mg/dscm
Particulates = 181.6 mg/dscm

Assume all of the chlorine in the feed is converted to HCl. Does the incinerator comply with the EPA rules?

9-28. During a trial burn, an incinerator was fed a mixed feed containing trichloroethylene, 1,1,1-trichloroethane, and toluene in an aqueous solution. Each component accounted for 5.0 percent of the feed solution on a volume basis. The feed rate was 40 L/min. The incinerator was operated at a temperature of 1,200°C. The stack gas flow rate was 9.0 dscm/s with 7 percent oxygen. Assuming that all the chlorine in the feed is converted to HCl, is the unit in compliance with the following emissions measured after the APC equipment?

Trichloroethylene = 170 μg/dscm
1,1,1-Trichloroethane = 353 μg/dscm
Toluene = 28 μg/dscm
HCl = 83.2 mg/dscm
Particulates = 123.4 mg/dscm

9-29. During a trial burn, an incinerator was fed a mixed feed containing hexachlorobenzene (HCB), pentachlorophenol (PCP), and acetone (ACET) in an aqueous solution. Each component accounted for 9.3 percent of the feed solution on a volume basis, that is, HCB = 9.3 percent, PCP = 9.3 percent, and ACET = 9.3 percent. The feed rate was 140 L/min. The incinerator was operated at a temperature of 1,200°C. The stack gas flow rate was 28.32 dscm/s with 14 percent oxygen. Assuming that all the chlorine in the feed is converted to HCl, is the unit in compliance if the following emissions are measured downstream of the APC equipment?

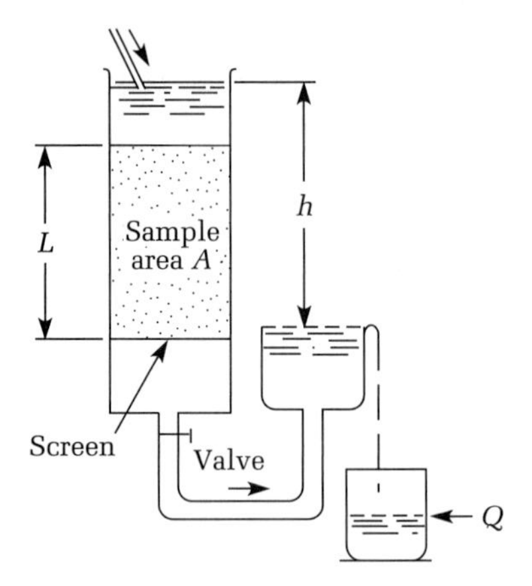

Standard constant head permeater equation:

$$K = \frac{QL}{hAt}$$

where K = hydraulic conductivity
Q = quantity of discharge
L = length of sample
h = hydraulic head
A = cross-sectioned area of sample
t = time

FIGURE P-9-30

Hexachlorobenzene = 170 μg/dscm
Pentachlorophenol = 353 μg/dscm
Acetone = 28 μg/dscm
HCl = 83.2 mg/dscm
Particulates = 123.4 mg/dscm

9-30. A standard permeameter is being considered for testing a clay for a hazardous waste landfill base. If the clay must have a permeability of 10^{-7} cm/s and the dimensions of the permeameter are as shown below, how long will the test take if a minimum of 100.0 milliliter of liquid must be collected for an accurate measurement? See Figure P-9-30 for notation and permeameter equation.

L = 10 cm
n = 1 m
Diameter of sample = 5.0 cm

9-31. A soil sample has been tested to determine permeability using a falling-head permeameter. (See Figure P-9-31.) The following data were recorded:

Diameter of a = 1 mm
Diameter of A = 10 cm
Length, L = 25 cm

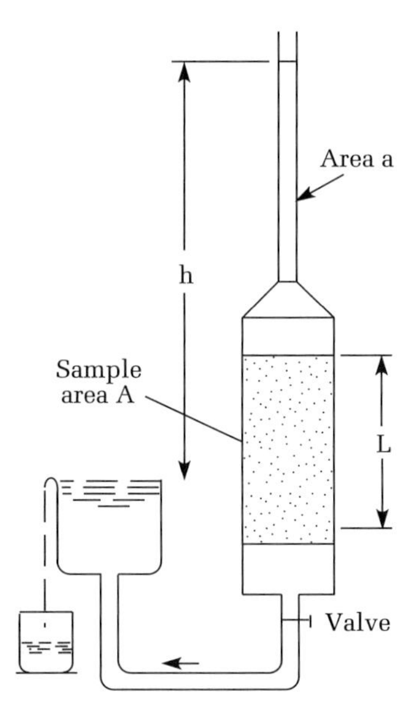

Falling head permeater equation:

$$K = 2.3\frac{a\,L}{A\,t}\log\left(\frac{h_0}{h_1}\right)$$

where K = hydraulic conductivity
a = cross-sectional area of stand pipe
A = cross-sectional area of sample
L = length of sample
t = time
h_0, h_1 = head at beginning of test and at time t, respectively

FIGURE P-9-31

Initial head = 1.0 m

Final head = 25 cm

Duration of test = 14 days

From these data, calculate the permeability of the sample. Assuming the sample is representative of the landfill site, is this a good soil for a hazardous waste landfill base?

9-32. An old hazardous waste landfill was built on a 10 m deep clay liner. An aquifer lies immediately below the clay layer. The clay layer through which the leachate must pass has a hydraulic conductivity of 1×10^{-7} cm/s. If the liquid level (leachate) is 1.0 m deep above the clay layer, how much leachate (in m^3/d) will reach the aquifer when the clay layer becomes saturated? Assume Darcy's law applies.

9-33. The three soil layers described below lie between the bottom of a hazardous waste landfill and the underlying aquifer. The depth of leachate above the top soil layer is 0.3 m. How long will it take (in years) for the leachate to migrate to an aquifer located at the bottom of soil C?

Soil A

Depth = 3.0 m

Hydraulic conductivity = 1.8×10^{-7} cm/s

Porosity = 55%

Soil B

Depth = 10 m

Hydraulic conductivity = 2.2×10^{-5} m/s

Porosity = 25%

Soil C

Depth = 12.0 m

Hydraulic conductivity = 5.3×10^{-5} mm/s

Porosity = 35%

9-34. The practical quantitation limit (PQL) for the solvent trichloroethylene is 5 μg/L. If a barrel (approximately 0.12 m^3) of spent solvent leaked into an aquifer, approximately how many cubic meters of water would be contaminated at the PQL?

9-35. A purge well must be installed at the site of a leaking gasoline storage tank. The depth of the unconfined aquifer is 60.00 m and the permeability is 6.4×10^{-3} m/s. Measurements have shown that the plume does not extend more than 150 m from the center of the leak. At 130 m from the center of the leak the plume is 0.1 in depth. If the purge well is 28 cm in diameter what size pump (m^3/s) is required so that the plume does not migrate any farther? (Note: this is application of well equations in Chapter 2.)

9-12 DISCUSSION QUESTIONS

9-1. It has been stated that based on LD_{50}, 2,3,7,8-TCDD is the most toxic chemical known. Why might this statement be misleading? How would you rephrase the statement to make it more scientifically correct?

9-2. Does the "land ban" actually ban the disposal of hazardous waste on the land? Explain.

9-3. A multimillion dollar company has just learned that one drum out of several hundred found at an abandoned waste disposal site has been identified as theirs. Their attorney explains that they may be potentially responsible for cleanup of all the drums at the site if no other former owners of the drums can be identified. Is this correct? Why or why not?

9-4. Your boss has proposed that your company institute a recycling program to minimize the generation of waste. Is recycling the best first step to investigate in a waste minimization program? If not,what others would you suggest and in what order?

9-5. A metal plater is proposing to treat his waste sludge to recover the nickel from it. Would this be

a. recycling

b. reusing

c. reclaiming

State the correct answer(s) and explain why you made your choice(s).

9-6. It is not necessary to measure every POHC in an incinerator trial burn. True or false? Explain your answer.

9-13 ADDITIONAL READING

Books

W. C. Anderson, ed., *Innovative Site Remediation Technology,* Volumes 1–8, Annapolis, MD: American Academy of Environmental Engineers, 1995.

R. Bellandi, ed., *Hazardous Waste Site Remediation, The Engineer's Perspective,* New York: Van Nostrand Reinhold, 1988.

R. C. Fortuna and D. J. Lennett, *Hazardous Waste Regulations, The New Era,* New York: McGraw-Hill, 1988.

H. M. Freeman, ed., *Hazardous Waste Minimization,* New York: McGraw-Hill, 1990.

H. M. Freeman, ed., *Standard Handbook of Hazardous Waste Treatment and Disposal,* New York: McGraw-Hill, 1989.

R. D. Griffin, *Principles of Hazardous Materials Management,* Ann Arbor, MI: Lewis Publishers, 1988.

C. N. Haas and R. J. Vamos, *Hazardous and Industrial Waste Treatment,* Upper Saddle River, NJ: Prentice Hall, 1995.

S. Hasan, *Geology and Hazardous Waste Management,* Upper Saddle River, NJ: Prentice Hall, 1996.

T. Higgins, *Hazardous Waste Minimization Handbook,* Chelsea, MI: Lewis Publishers, 1989.

M. A. Kamrin, *Toxicology,* Chelsea, MI: Lewis Publishers, 1988.

M. D. LaGrega, P. L. Buckingham, and J. C. Evans, *Hazardous Waste Management,* New York: McGraw-Hill, 1994.

D. J. Paustembach, ed., *The Risk Assessment of Environmental Hazards,* New York: John Wiley, 1989.

J. Quarles, *Federal Regulation of Hazardous Wastes,* Washington, DC: The Environmental Law Institute, 1982.

C. A. Wentz, *Hazardous Waste Management,* New York: McGraw-Hill, 1989.

E. F. Wood, R. A. Ferrara, W. G. Gray, and G. F. Pinder, *Groundwater Contamination from Hazardous Wastes,* Upper Saddle River, NJ: Prentice Hall, 1984.

Journals

Environmental Progress

Environmental Science and Technology

Hazardous Waste and Hazardous Materials

Journal of Air and Waste Management (formerly *Journal of the Air Pollution Control Association*)

CHAPTER 10

IONIZING RADIATION

10-1 FUNDAMENTALS
- Atomic Structure
- Radioactivity and Radiation
- Radioactive Decay
- Radioisotopes
- Fission
- The Production of X-rays
- Radiation Dose

10-2 BIOLOGICAL EFFECTS OF IONIZING RADIATION
- Sequential Pattern of Biological Effects
- Determinants of Biological Effects
- Acute Effects
- Relation of Dose to Type of Acute Radiation Syndrome
- Delayed Effects
- Genetic Effects

10-3 RADIATION STANDARDS

10-4 RADIATION EXPOSURE
- External and Internal Radiation Hazards
- Natural Background
- X-rays
- Radionuclides
- Nuclear Reactor Operations
- Radioactive Wastes

10-5 RADIATION PROTECTION
Reduction of External Radiation Hazards
Reduction of Internal Radiation Hazards
10-6 RADIOACTIVE WASTE
Types of Waste
High-Level Radioactive Waste Management
WIPP
Low-Level Radioactive Waste Management
Long-Term Management and Containment
10-7 CHAPTER REVIEW
10-8 PROBLEMS
10-9 DISCUSSION QUESTIONS

10-1 FUNDAMENTALS[1]

Atomic Structure

We assume that you are familiar with the Bohr model of atomic structure. In this model the atom is described as consisting of a central nucleus surrounded by a number of electrons in closed orbits about the nucleus. The orbital electrons are grouped in shells.

The nucleus itself can be considered as composed of two distinct kinds of particles. These are the protons, which carry a positive unit charge, e^+, and the neutrons, which are uncharged. In a particular atom there are Z electrons, each carrying a charge e^-, orbiting around the nucleus, and a nucleus composed of N neutrons and P protons. The condition of electrical neutrality for the atom as a whole yields $Pe - Ze = 0$, that is, the number of protons in the nucleus is equal to the number of orbital electrons.

The number Z is the atomic charge or atomic number of the atom, while $Z + N$ is the atomic mass number, usually denoted by A. The parameters A and Z completely define a particular atomic species, this being known as a *nuclide*.

The scale on which the masses of nuclides are measured is in terms of the unified atomic mass unit, with the symbol u. This is defined as the unit of mass equal to one-twelfth the mass of an atom of carbon of atomic mass number 12. This gives 1 u as 1.6606×10^{-27} kg. On this scale, the mass of the neutron is 1.0088665 u, the mass of the proton 1.0088925 u, and the mass of the electron 0.0005486 u.

From the definition of the mass scale, giving proton and neutron masses of the order unity, it is clear that the atomic mass number will be a whole number

[1]This discussion follows R. A. Coombe, *An Introduction to Radioactivity for Engineers,* NY: Macmillan/St. Martin's Press, pp. 1–37, 1968.

Ms. Kristin Erickson, Radiation Safety Officer, Office of Radiation, Chemical, and Biological Safety, Michigan State University, contributed to the revisions for the third edition.

approximation to the nuclidic mass in u. For example, a nuclide of magnesium which contains 12 protons and 12 neutrons has $A = 24$ and a nuclidic mass of 23.985045 u. The difference between the nuclidic mass and the atomic mass number is called the *mass excess.*

The chemical properties of the atom, and hence its designation as a particular element, depend upon the number of orbital electrons, that is, on the atomic number Z. Given Z, the element is uniquely defined. As an example, if a given atom has two orbital electrons, it must be helium (assuming that the atom is not ionized or in some similar nonequilibrium state). Similarly an atom with eight electrons must be oxygen.

A particular nuclide is denoted by ${}^{A}_{Z}X$, where X takes the place of the element symbol. But as Z determines the element, Z and X denote the same thing. Thus, the shorthand can be amended to ${}^{A}X$. For example, carbon has six neutrons and six protons. Therefore, this nuclide can be written ${}^{12}C$, or carbon-12.

For each element (determined only by Z) there are several nuclides (determined by Z and A) that have the same Z value but different values of A. These different nuclides of the same element are called isotopes. Hydrogen with $Z = 1$ has three isotopes with atomic mass numbers of 1, 2, and 3. As Z must remain constant at 1, this means that they have 0, 1, and 2 neutrons, respectively. This is illustrated in Figure 10-1. These isotopes all act chemically as hydrogen, but their nuclidic masses are different. The nuclidic mass of ${}^{1}H$ is 1.007825 u, that of ${}^{2}H$ (known as deuterium) is 2.014102 u, and that of ${}^{3}H$ (known as tritium) is 3.016049 u.

The atomic weight of an element is defined as the combined nuclidic masses of all the isotopes, weighted according to their natural relative abundances. It is denoted by A. In the case of hydrogen it follows that the atomic weight is

$$1.007825(0.9844) + 2.014102(0.0156) + 3.016049(0) = 1.00797$$

The masses of the hydrogen isotopes are not obtained by simple addition of neutron masses. For example, the nuclidic mass of ${}^{1}H$ plus a neutron is 2.016490 u, while the mass of deuterium is 2.014102 u. This difference of 0.002388 u is called the *mass defect.* This is because when a proton and a neutron are brought together to form a deuteron (the nucleus of deuterium), energy is released in order to bind them together. Conversely, energy must be supplied to split them apart. This required

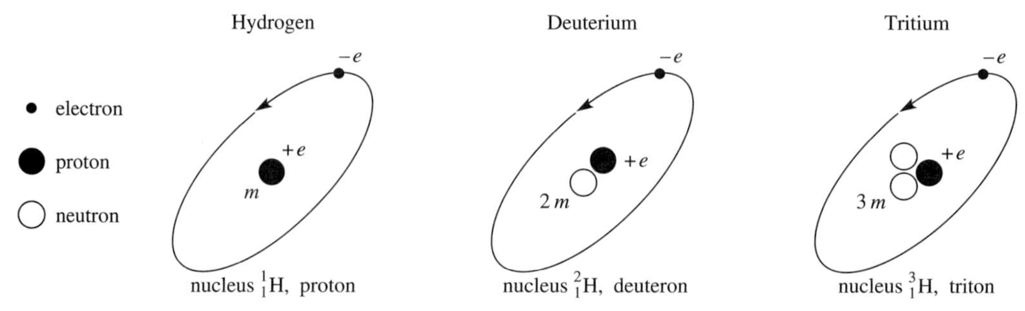

FIGURE 10-1
Three isotopes of hydrogen. (*Source:* R. A. Coombe, *An Introduction to Radioactivity for Engineers,* 1968.)

energy, the *binding energy,* is obtained from Einstein's equation for the conversion of mass into energy,

$$E = \Delta mc^2 \tag{10-1}$$

where here Δm is the mass defect and c is the speed of light.

All energies of emitted radiation and particles, as well as the various atomic and nuclear energy levels, are quoted in terms of the electron volt, eV. This is the energy that would be acquired by an electron in falling through a potential difference of one volt. From this definition the following equivalent units of energy can be established:

$$1 \text{ eV} = 1.602 \times 10^{-12} \text{ erg} = 1.602 \times 10^{-19} \text{ Joule}$$

For nuclear energy levels and radiation energies, the electron volt is usually an inconveniently small unit. The units MeV and KeV are then used for 10^6 eV and 10^3 eV, respectively. Using Equation 10-1, with the information that $c = 2.99793 \times 10^8$ m/s, and 1 u $= 1.6606 \times 10^{-27}$ kg, then the energy equivalent of 1 u is 931.634 MeV. In other words, this means that if an electron of mass 0.0005486 u were completely annihilated, the energy released would be approximately 0.511 MeV.

Radioactivity and Radiation

By definition, isotopes have different ratios of neutrons to protons in the nucleus. Some ratios give rise to unstable conditions. This is usually because the neutron-to-proton ratio is too large. Because of this instability, the nucleus changes its state to attain equilibrium, and in so doing emits either a particle or electromagnetic radiation to carry off the excess energy. This phenomenon of nuclear disintegration is known as *radioactivity,* and an isotope that displays such activity is known as a *radioisotope.*

There are three types of isotopes. Some are stable, others are naturally radioactive, and the third group can be artificially produced and are also radioactive. These artificially produced radioisotopes are the isotopes most used in industrial application.

There are three major types of decay product that carry off the surplus energy when a radioisotope decays: alpha particles, beta particles, and gamma radiation.

Alpha particle emission. Conceptually, the source of the instability of the heavy elements is their size; their nuclei are too large. How can they become smaller? One method would be to eject protons or neutrons. Rather than doing this singly, the heavy-element atoms expel them in "packages" containing two protons and two neutrons. This "package" is called an *alpha particle.* An alpha particle is equivalent to the nucleus of the helium-4 atom, that is, it is simply a body consisting of two protons and two neutrons bound together. Consequently, if an alpha particle is emitted, the nucleus must change to one that has a charge 2e less and a mass approximately 4 u less. The general expression is

$${}^{A}_{Z}X \rightarrow {}^{A-4}_{Z-2}X + {}^{4}_{2}\text{He} \tag{10-2}$$

Atoms that eject the helium "package" are said to decay through emission of alpha radiation. Alpha particle emission occurs mainly in radioisotopes whose atomic number is greater than 82. With increasing atomic number, the occurrence of alpha particle decay increases rapidly, and it is a characteristic of the very heavy elements. It is particularly in evidence in the main decay chains of the natural radioactive isotopes.

You should note that an atom undergoing alpha particle decay changes into a new element. It is a new element because the product nucleus (often called a *daughter*) contains two less protons than the *parent* atom. Through emission of an alpha particle, uranium becomes thorium. Similarly, radium becomes radon.

Beta particle emission. The instability that is the cause of beta particle emission arises from the fact that the neutron-to-proton ratio in the nucleus is too high (there are too many neutrons in the nucleus). To achieve stability, a neutron in the nucleus can decay into a proton and an electron. The proton remains in the nucleus so that the neutron-to-proton ratio is decreased, and the electron is ejected. This ejected electron is known as a *beta particle.* The general expression for the decay is

$$^{A}_{Z}X \longrightarrow {}^{A}_{Z+1}X + \beta^- \qquad (10\text{-}3)$$

You should note that we use the β^- to represent an electron of nuclear origin to differentiate it from electrons from other sources. The negative sign is used with the β if there is any chance of ambiguity, because there also exists a similar particle, called a positron, that carries a positive charge.

Again, as in alpha radiation, emission of a beta particle changes the parent atom into a new element, since the number of protons in the nucleus increases by one. If the daughter product also is radioactive, it will, in turn, emit a beta particle, becoming another new element, etc., until finally a stable neutron-to-proton ratio is reached. Through such a series of changes, for example, the fission-product element krypton becomes rubidium which, in turn, becomes strontium, which finally converts to stable yttrium.

Gamma ray emission. Either alpha or beta particles may be accompanied by gamma radiation. Whereas alpha or beta radiation brings about a change in the size of the nucleus or the number of a particular type of particle therein, the emission of gamma radiation represents only a release of energy. This is the energy that remains in the newly formed nucleus after emission of the alpha or beta particle. Electromagnetic radiation in the form of gamma rays is emitted when a nucleus in an excited state transfers to a more stable state. The nucleus thus retains its original composition, the excess energy being radiated away. If the frequency of the radiation is ν, and the nucleus changes from a state of energy E_1 to a state of energy E_2, then the two energies are related by the equation

$$E_1 - E_2 = \mathrm{h}\nu \qquad (10\text{-}4)$$

where h is Planck's constant, having a value of 6.624×10^{-27} ergs. The energy of the emitted gamma ray is thus $\mathrm{h}\nu$. In equations, the gamma ray is represented by the Greek letter γ.

X-rays. Gamma rays are similar to x-rays. Their difference lies only in their source. Gamma rays originate from a nucleus transferring from one nuclear excited state to another, whereas x-rays originate from electrons transferring from a higher to a lower atomic energy state. As atomic energy levels are in general spaced much closer in terms of energy than nuclear levels, it follows from Equation 10-4 that the frequencies of x-rays are much less than those of gamma rays. As far as industrial applications are concerned, the only difference between them is the penetrating power. Since penetrating power increases with frequency, gamma rays have more penetrating power than x-rays.

Multiple emissions. In the above discussions, only single emission has been considered. In practice, two or more different types of emission are possible, and in a great many cases several particles of the same type but of different energies are emitted. This latter effect is due to the multiplicity of nuclear energy levels both in the original isotope nucleus and in the nucleus formed by particle emission.

Radioactive Decay

Each unstable (radioactive) atom will eventually achieve stability by ejecting an alpha or beta particle. This shift to a more stable form is called *decay.* Each radioactive decay process is characterized by the fact that only a fraction of the unstable nuclei in a given sample will decay in a given time. The probability that a particular nucleus in the sample will decay during a time interval dt is λdt, where λ is the radioactive decay constant. It is defined as the probability that any particular nucleus will disintegrate in unit time.

For a large number of like nuclei together, we make the assumption that λ is independent of the age of a nucleus and is the same for all nuclei. This means that λ is a constant. If N is the number of nuclei present at a time t, then the number of decays occurring in a time dt can be written $\lambda N dt$. As the number of nuclei decreases by dN in this time, we can write

$$dN = -\lambda N dt \tag{10-5}$$

The negative sign denotes that N is decreasing with time. Equation 10-5 shows that the rate of decay is proportional to the number of nuclei present, that is, it is a first-order reaction.

Equation 10-5 can be rearranged and integrated:

$$\int_{N_0}^{N} \frac{dN}{N} = \int_0^t \lambda dt$$

$$\ln \frac{N}{N_0} = -\lambda t$$

or

$$N = N_0 \exp(-\lambda t) \tag{10-6}$$

where N_0 is the number of radioactive nuclei present at time $t = 0$. Equation 10-6 shows that radioactive decay follows an exponential form. In particular, the time taken for a given number of nuclei to decay to half that number, $T_{1/2}$, is obtained from Equation 10-6 as

$$\ln \frac{N_0/2}{N_0} = -\lambda T_{1/2}$$

Solving for $T_{1/2}$ yields

$$T_{1/2} = \frac{\ln 2}{\lambda} = \frac{0.693}{\lambda} \tag{10-7}$$

This equation relates two important parameters of a radioactive species: λ and the half-life, $T_{1/2}$. These quantities are characteristic properties of a particular species. Half-lives of radioisotopes cover an enormous range of values, from microseconds to millions of years. To illustrate this, some values are given in Table 10-1.

Specific activity and the curie. The quantity N is called the *activity* of a sample. The *curie* is the unit used for activity. The abbreviation is Ci. One curie of radioactive material is that quantity of unstable atoms whose frequency of decay is 3.700×10^{10} disintegrations per second. This definition covers all modes of disintegration for both single isotopes and mixtures. The curie is quite a large unit for a lot of purposes. Millicuries (1 mCi $= 10^{-3}$ Ci) or microcuries (1 μCi $= 10^{-6}$ Ci) and even picocuries (1 pCi $= 10^{-12}$ Ci) may be more manageable units to work with.

The *specific activity* of a radioisotope is the activity per gram of the pure radioisotope. The number of atoms of a pure radioisotope in one gram is given by

$$N = \frac{N_A}{A} \tag{10-8}$$

TABLE 10-1
Some radioisotope half-lives

Radioisotope	Half-life
Polonium-212	3.04×10^{-7} s
Carbon-10	19.3 s
Oxygen-15	2.05 min
Carbon-11	20.4 min
Radon-222	3.825 d
Iodine-131	8.06 d
Phosphorus-32	14.3 d
Polonium-210	138.4 d
Calcium-45	165 d
Cobalt-60	5.27 y
Tritium	12.5 y
Strontium-90	28 y
Cesium-137	30 y
Radium-226	1,622 y
Carbon-14	5,570 y
Potassium-40	1.4×10^{9} y

where N_A is Avogadro's number (6.0248×10^{23}) and A is the nuclidic mass. The specific activity S of a particular radioisotope is an intrinsic property of that radioisotope:

$$S = \frac{\lambda N_A}{A} \text{ disintegrations/s} \tag{10-9}$$

Growth of subsidiary products. In the process of decay, a new nuclide is formed, the daughter product. If the daughter product is stable, its concentration will gradually increase as the parent decays. On the other hand, if the daughter product is itself radioactive, the variation in concentrations of parent, daughter, and granddaughter products will be very much dependent upon the relative rates of decay.

There are several cases where a radioactive isotope decays into another nuclide that is itself radioactive. This can continue for a large number of nuclides, resulting in a decay chain. The characteristics of any particular chain depend largely upon the relative decay constants of its various members.

The simplest case is the growth of a radioactive daughter product from the parent atoms. Let us assume we begin with N_1 parent atoms of decay constant λ_1, and N_2 daughter atoms of decay constant λ_2. The rate at which the daughter product is increasing is then the difference between the rate at which it is produced by its parent and the rate at which it decays. This can be written as

$$\frac{dN_2}{dt} = \lambda_1 N_1 - \lambda_2 N_2 \tag{10-10}$$

The rate of production of the daughter is simply the decay rate of the parent.

Using Equation 10-6 with the notation that N_1 is the number of nuclei of the parent and N_{10} is the initial number:

$$N_1 = N_{10} \exp(-\lambda_1 t)$$

Substituting in Equation 10-10,

$$\frac{dN_2}{dt} = N_{10} \exp(-\lambda_1 t) - \lambda_2 N_2 \tag{10-11}$$

Rearranging,

$$\frac{dN_2}{dt} + \lambda_2 N_2 = N_{10} \exp(-\lambda t) \tag{10-12}$$

This equation can readily be solved by multiplying throughout by the factor $e^{\lambda_2 t}$. Thus,

$$\exp(\lambda_2 t)\frac{dN_2}{dt} + \exp(\lambda_2 t)\lambda_2(N_2) = \lambda_1 N_{10} \exp(-\lambda_1 t)\exp(\lambda_2 t) \tag{10-13}$$

and

$$\frac{dN_2 e^{\lambda_2 t}}{dt} = \lambda_1 N_{10} \exp[(\lambda_2 - \lambda_1)t] \tag{10-14}$$

On integration this yields

$$N_2 e^{\lambda_2 t} = \frac{\lambda_1 N_{10}}{\lambda_2 - \lambda_1} \exp\left[(\lambda_2 - \lambda_1)t\right] + C \qquad (10\text{-}15)$$

The integration constant C is determined from the boundary conditions. For this case, at $t = 0$, there was no daughter product present, that is, $N_2 = 0$ at $t = 0$. Using these boundary conditions, Equation 10-15 reduces to

$$N_2 = \frac{\lambda_1 N_{10}}{\lambda_2 - \lambda_1}(e^{-\lambda_1 t} - e^{-\lambda_2 t}) \qquad (10\text{-}16)$$

Characteristics of daughter products. In the derivation of Equation 10-16, it was assumed that N_2 was zero at zero time. Since the daughter nuclide itself decays, then at an infinite time, N_2 will again be zero. Between these two times when $N_2 = 0$, there will be a time, say t', when N_2 will reach a maximum. At this time, the rate of increase will be passing through a turning point, that is, $dN_2/dt = 0$. Using this fact, together with Equation 10-16, it can be shown that

$$t' = \frac{\ln \lambda_2 - \ln \lambda_1}{\lambda_2 - \lambda_1} \qquad (10\text{-}17)$$

Continuous production of parent. The calculations above assumed that at zero time a certain number of parent atoms were present and then decayed. There are many cases of interest where the parent is continuously replenished. Such cases occur for instance in nuclear reactors, where the parent nuclides are continuously being created by neutron bombardment. Another case is the continuous production of carbon-14 by cosmic rays incident on the nuclei present in the upper atmosphere.

End products. Any radioactive decay chain must finally arrive at a nuclide that is stable. The relevant equations can readily be obtained, for any stable nuclide has $\lambda = 0$. For example, consider the case of a radioisotope whose daughter is stable. For this, Equation 10-16 can be used with $\lambda_2 = 0$. Thus,

$$N_2 = N_{10}(1 - e^{-\lambda_1 t}) \qquad (10\text{-}18)$$

Similar modifications can be made to other equations concerned with longer decay chains.

Radioisotopes

Naturally occurring radioisotopes. Most of the 50 naturally occurring radioisotopes are associated with three distinct series, known as the thorium series, the uranium series, and the actinium series. Each one of these series starts with an element of high atomic mass (uranium-238, thorium-232, and uranium-235, respectively) and then decays by a long series of alpha and beta particle emissions to reach a stable nuclide (lead-206, lead-208, and lead-207, respectively). The three chains are associated with the heavy elements, and very few naturally occurring radioisotopes are found with atomic masses less than 82.

The half-lives of the naturally occurring radioisotopes are very long. Presumably they were constituents of the earth at its formation and their activity has not yet died away beyond detection.

Two other important isotopes that occur in the natural environment but that are not strictly naturally occurring are hydrogen-3 (tritium) and carbon-14. These radioisotopes are artificially produced by cosmic rays bombarding the upper atmosphere of the earth. At the present time the quantities of these isotopes are in equilibrium, their production rate by cosmic radiation being balanced by their natural decay rate. Because of this phenomenon these isotopes are of particular use in archaeological dating.

Artificially produced radioisotopes. The artificial production of radioisotopes is mainly carried out either by nuclear reactors or by particle accelerators. The cyclotron is the accelerator in most general use because the required bombarding particle energies are easily obtained and the output is reasonably high. The transmutation of a stable isotope to a radioactive one is effected by bombarding a target nucleus with a suitable projectile, either electromagnetic or a particle, to produce the required isotope from the resultant nuclear reaction.

When an accelerator is used, the bombarding particles are usually either protons, deuterons, or alpha particles. In the nuclear reaction brought about by the bombardment of zinc-64 with energetic deuterons from a cyclotron, the deuteron and zinc-64 nucleus combine to form a new element. The new element has a charge of $30e + e$ and an atomic mass number of $64 + 2$. This compound nucleus is then ^{66}Ga, gallium-66. This intermediate nucleus disintegrates almost immediately by one of several possible modes of decay. If a proton is emitted, for example, the final nucleus must be left with a nuclear charge of 32e and an atomic mass number of 65, so it is ^{65}Zn. This isotope does not occur in nature.

For the production of radioisotopes for industrial application, the most common nuclear reactions used are those from thermal neutrons. A target sample, in a suitable container, is inserted into the core of a reactor and left there for varying amounts of time. In the core of a reactor there is a copious supply of thermal neutrons. These interact with the target nucleus to produce the required radioisotope. This is known as *neutron activation.*

Fission

A *nuclear reactor* is an assembly of fissionable material (such as uranium-235, plutonium-239, or uranium-233) arranged in such a way that a self-sustaining *chain reaction* is maintained. When these nuclei are bombarded with neutrons of the appropriate energy, they split up, or *fission,* into fission fragments and neutrons. For the nuclear reaction to continue, at least one of the neutrons produced must be available to produce another fission instead of escaping from the assembly or being used up in some other nuclear reaction. Thus, there is a minimum (*critical*) mass below which the reaction cannot be self-sustaining. Actual reactors are built with an excess mass to make large amounts of neutrons available. The excess neutron production is controlled by the use of *moderators.* The moderators are made of materials with large neuron-capture cross sections, such as boron, cadmium, or hafnium. These are formed into *control rods* that are moved in and out of the reactor to moderate the excess neutrons.

The fission chain reaction is characterized by an enormous release of heat. This heat must be carried away by an efficient cooling system to prevent mechanical failure of the reactor assembly (*meltdown*) and, ultimately, an uncontrolled fission. The ultimate uncontrolled reaction is, of course, an atomic explosion.

The fission fragments are simply lower mass elements. There are, most commonly, two fission fragments from each nucleus with an energy of the order of 200 MeV shared between them. The uranium nucleus does not split into the same two fragments each time. The breakup is far from symmetrical and can occur in more than 30 different ways. The most commonly produced isotopes are grouped around the mass numbers 95 and 139.

The fragments produced from the fission process have very large neutron-to-proton ratios so that they are highly unstable. Many transitions have to occur before a stable nucleus is finally achieved. These successive decays give rise to a decay chain.

Fission fragments, because of their high mass and very high initial charge, have extremely short ranges in matter. Hence, they are contained within the fuel element when a uranium nucleus fissions. The spent nuclear reactor fuel elements thus provide a very intense radioactive source that presents many problems in the subsequent handling and processing. Fission fragments themselves can sometimes be used as a radioactive source for industrial application.

The Production of X-rays[2]

X-rays were discovered in 1895 by Wilhelm Conrad Roentgen. During the course of some studies, he covered a cathode ray tube with a black cardboard box and observed fluorescence on a screen coated with barium platinocyanide near the tube. After further investigation of this phenomenon, he concluded that the effect was caused by the generation of new invisible rays capable of penetrating opaque materials and producing visible fluorescence in certain chemicals. He called these new invisible rays x-rays. Because of their discoverer, x-rays are also sometimes referred to as Roentgen rays.

As pointed out previously, x-rays are electromagnetic waves and occupy the same portion of the electromagnetic spectrum as gamma rays. Like gamma rays, x-rays can pass through solid material. The mode of interaction of x-rays with matter is the same, as are the biological and photographic effects.

Whereas gamma rays come from within the nucleus of the atom, x-rays are generated outside the nucleus by the interaction of high-speed electrons with the

---

[2]This discussion follows that of the U.S. Public Health Service, *Introduction to Medical X-Ray Protection, Training and Manpower Development Program,* Rockville, MD, 1968.

atom. For this reason, there is a difference in the energy distribution of x- and gamma rays. Gamma rays from any single radionuclide consist only of rays of one or several discrete energies. X-rays consist of a broad continuous spectrum of energies. The continuous spectrum will be discussed in detail later.

The x-ray tube. X-rays are produced whenever a stream of high-speed electrons strikes a substance. This is caused by their sudden stoppage or deflection by atoms within the target material. The x-ray tube (Figure 10-2) is designed to provide the high speed electrons and the interacting material. Essential components of an x-ray tube are (1) a highly evacuated glass envelope containing the cathode and anode; (2) a source of electrons proceeding from a cathode; and (3) a target (or anode) placed in the path of the electron stream.

One early type of x-ray source was the gas tube, which had a cold cathode. The quality and intensity of the radiation were controlled by the gas pressure (about 10^{-4} mm Hg) within the tube. This type of x-ray source utilized the electrons set free through the bombardment of the cathode by positive ions driven by the difference in potential between the target and cathode. The liberated electrons were then accelerated across the tube to the target. The upper limit of operation was about 50 kilovolts.

The development of the hot filament tube by Coolidge in 1913 was a major advance. Most x-ray tubes in use today are of this type. Here, the free electrons are "boiled out" of an incandescent filament within an evacuated tube and given their velocity by accelerating them through an electric field. In the hot filament tube, the quality and intensity of radiation can be controlled independently by simple electrical means. The intensity of radiation is directly proportional to the current and is proportional to the square of the voltage. This allows a much wider range of

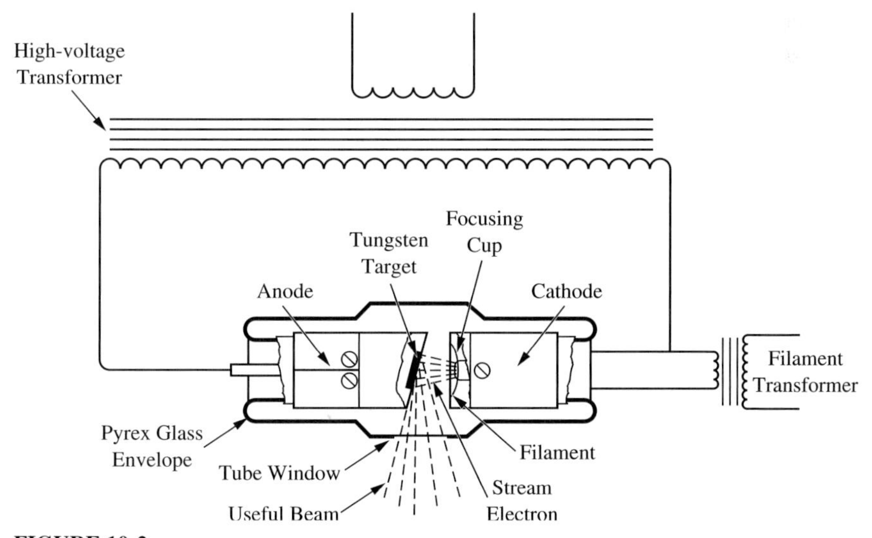

FIGURE 10-2
Typical x-ray tube in self-rectified circuit.

wavelengths and intensities, while the characteristics of the tube remain reasonably constant throughout its useful life.

The high voltages required for x-ray tube operation are best obtained by step-up transformers, whose output is always AC. Since the electrons must flow only from cathode to anode within the tube, some means of rectification is necessary. A self-rectified tube acts as its own rectifier. When an alternating voltage is applied to such a tube, electrons flow only from the cathode to anode as long as the anode remains cool. If the anode becomes hot, the flow of electrons reverses during the second half-cycle and the cathode is damaged. Thus the self-rectified tube is limited to low currents and short periods of operation. The use of "valves" (rectifiers) in the power supply circuit eliminates the inverse voltage on the x-ray tube. Thus, more power can be handled by the x-ray tube, the radiation output is increased, and the time of exposure is shortened.

X-ray production efficiency. On the average, the fraction of the electron energy emitted as electromagnetic radiation increases with the atomic number of the atoms of the target and the velocity of the electrons. This fraction is very small and can be represented by the following empirical equation:

$$F = 1.1 \times 10^{-9} ZV \tag{10-19}$$

where F = fraction of the energy of the electrons converted into x-rays
Z = atomic number of the target
V = energy of the electrons in volts[3]

Typically, less than 1 percent of the electrical power supplied is converted into x-ray energy. The remaining energy (over 99 percent) appears as heat produced at the target (largely through ionization and excitation). As a result, electron bombardment of the target raises it to a high temperature and, if the heat produced is not dissipated fast enough, the target will melt. This heat production is a serious factor in limiting the capacity of an x-ray tube.

A suitable target must have the following characteristics:

1. A high atomic number because efficiency is directly proportional to Z
2. A high melting point because of the high temperatures involved
3. A high thermal conductivity to dissipate the heat
4. Low vapor pressure at high temperatures to prevent target evaporation

The continuous spectrum. When high-speed electrons are stopped by a target, the radiation produced has a continuous distribution of energies (wavelengths). As the fast-moving electrons enter the surface layers of the target, they are abruptly slowed

[3]The electron energy is generally expressed in terms of the voltage applied across the tube.

down by collision with the strong Coulomb field of the nucleus and are diverted from their original direction of motion. Each time the electron suffers an abrupt change of speed, a change in direction, or both, energy in the form of x-rays is radiated. The energy of the x-ray photon emitted depends upon the degree of deceleration that occurs. If the electron is brought to rest in a single collision, the energy of the resulting photon corresponds to the kinetic energy of the electron stopped and will be a maximum. If the electron suffers a less drastic collision, a lower energy photon is produced. Since a variety of types of collisions will be occurring, photons of all energies up to the maximum will be produced. This accounts for the continuous distribution of an x-ray spectrum. The maximum intensity (peak of the curve) occurs at a wavelength about 1.5 times the minimum wavelength. The total intensity of radiation from a given x-ray tube is represented by the area under the spectral curve. The intensity has been found, as might be expected, to be directly proportional to the electron current (number of electrons striking the target).

Radiation Dose[4]

Fundamentally, the harmful consequences of ionizing radiations to a living organism are due to the energy absorbed by the cells and tissues of the organism. This absorbed energy (or dose) produces chemical decomposition of the molecules present in the living cells. The mechanism of the decomposition appears to be related to ionization and excitation interactions between the radiation and atoms within the tissue. The amount of ionization or number of ion pairs produced by ionizing radiations in the cells or tissues provides some measure of the amount of decomposition or physiological damage that might be expected from a given quantity or dose. The ideal basis for radiation dose measurement could be, therefore, the number of ion pairs (or ionizations) taking place within the medium of interest. For certain practical reasons, the medium chosen for defining exposure dose is air.

Exposure dose—the roentgen. The exposure dose of x- or gamma radiation within a specific volume is a measure of the radiation based upon its ability to produce ionization in air. The unit used for expressing the exposure to x- or gamma radiation is the roentgen (R). Its merit lies in the fact that the magnitude of the exposure dose in roentgens can usually be related to the absorbed dose, which is of importance in predicting or quantifying the expected biological effect (or injury) resulting from the radiation.

The *roentgen* is defined as an "exposure dose of x- or gamma radiation such that the associated corpuscular emission per 0.001293 g of air[5] produces, in air, ions carrying one electrostatic unit of quantity of electricity of either sign." Since the

[4]This discussion follows that of the U.S. Public Health Service, *Introduction to Medical X-Ray Protection, Training and Manpower Development Program,* 1968.

[5]One cubic centimeter of air at STP has a mass of 0.001293 g.

ionizing property of radiation provides the basis for several types of detection instruments and methods, such devices may be used to quantify the exposure dose. You should note that this is a unit of exposure dose based upon ionization of air; it is not a unit of ionization, nor is it an absorbed dose in air.

Absorbed dose—the rad. The absorbed dose of any ionizing radiation is the energy imparted to matter by ionizing radiations per unit mass of irradiated material at the place of interest. The unit of absorbed dose is the *rad.* One rad is equivalent to the absorption of 100 ergs/g. It should be emphasized that although the roentgen unit is strictly applicable only to x- or gamma radiation, the rad unit may be used regardless of the type of ionizing radiation or the type of absorbing medium.

A conversion from roentgens to rads can be calculated. Two things must be known: the energy of the incident radiation and the mass absorption coefficient of the absorbing material. For example, let us assume that a dose of 1.0 R of gamma radiation was measured in air. What is the equivalent absorbed dose in air?

To form 1 esu per 0.001293 gram of air, the radiation must produce 1.61×10^{12} ion pairs when absorbed in 1 gram of air. From empirical studies it is known that, on the average, 34 electron volts of energy are transferred (or absorbed) in the process of forming each ion pair in air. Thus, the total energy absorbed is

$$\left(\frac{34\,\text{eV}}{\text{ion pair}}\right)(1.61 \times 10^{12}\ \text{ion pairs/g}) = 5.48 \times 10^{13}\,\text{eV/g}$$

Or, in ergs rather than electron volts:

$$(5.48 \times 10^{13}\ \text{eV/g})(1.602 \times 10^{-12}\ \text{erg/eV}) = 87\,\text{ergs/g}$$

Since 100 ergs/gram is 1 rad, then one roentgen of exposure dose to a specific volume of air at standard conditions results in the absorbed dose of 0.87 rad.

Relative biological effectiveness (quality factor). Although all ionizing radiations are capable of producing similar biological effects, the absorbed dose, measured in rads, which will produce a certain effect may vary appreciably from one type of radiation to another. The difference in behavior, in this connection, is expressed by means of a quantity called the *relative biological effectiveness* (RBE) of the particular radiation. The RBE of a given radiation may be defined as the ratio of the absorbed dose (rads) of gamma radiation (of a specified energy) to the absorbed dose of the given radiation required to produce the same biological effect. Thus, if an absorbed dose of 0.2 rad of slow neutron radiation produces the same biological effect as 1 rad of gamma radiation, the RBE for slow neutrons would be

$$\text{RBE} = \frac{1\ \text{rad}}{0.2\ \text{rad}} = 5$$

The value of the RBE for a particular type of nuclear radiation depends upon several factors, such as the energy of the radiation, the kind and degree of the biological damage, and the nature of the organisms or tissue under consideration.

Tissue weighting factor (W_T). The *tissue weighting factor* is a modifying factor used in dose calculations to correct for the fact that different tissues and organs have varying degrees of radiosensitivity depending on the radioisotope and the chemical form of the radioisotope. Some tissues and organs are very sensitive, while others are not radiosensitive at all. For example, since iodine is easily incorporated in thyroid tissue, the thyroid gland is very sensitive to the radioiodines. The tissue weighting factor (W_T) is, therefore, high for the radioiodines. When the tissue or organ is not radiosensitive, the value of W_T may be very small or zero for that tissue.

The rem. With the concept of the RBE in mind, it is now useful to introduce another unit, known as the *rem,* an abbreviation of "roentgen equivalent man." The rad is a convenient unit for expressing energy absorption, but it does not take into account the biological effect of the particular nuclear radiation absorbed. The rem, however, which is defined by

$$\text{Dose in rems} = \text{RBE} \times \text{dose in rads} \times W_T$$

provides an indication of the extent of biological injury (of a given type) that would result from the absorption of nuclear radiation. Thus, the rem is a unit of biological dose.

10-2 BIOLOGICAL EFFECTS OF IONIZING RADIATION[6]

The fact that ionizing radiation produces biological damage has been known for many years. The first case of human injury was reported in the literature just a few months following Roentgen's original paper in 1895 announcing the discovery of x-rays. As early as 1902, the first case of x-ray induced cancer was reported in the literature.

Early human evidence for harmful effects as a result of exposure to radiation in large amounts existed in the 1920s and 1930s based upon the experience of early radiologists, persons working in the radium industry, and other special occupational groups. The long-term biological significance of smaller, chronic doses of radiation, however, was not widely appreciated until the 1950s, and most of our current knowledge of the biological effects of radiation has been accumulated since World War II.

Sequential Pattern of Biological Effects

The sequence of events following radiation exposure may be classified into three periods: a latent period, a period of demonstrable effect, and a recovery period.

---

[6]This discussion follows that of the U.S. Public Health Service, *Introduction to Medical X-Ray Protection, Training and Manpower Development Program,* 1968.

Latent period. Following the initial radiation event, and often before the first detectable effect occurs, there is a time lag referred to as the latent period. There is a vast time range possible in the latent period. In fact, the biological effects of radiation are arbitrarily divided into short-term, or acute, and long-term, or delayed, effects on this basis. Those effects which appear within a matter of minutes, days, or weeks are called acute effects and those which appear years, decades, and sometimes generations later are called delayed effects.

Demonstrable effects period. During or immediately following the latent period, certain discrete effects can be observed. One of the phenomena that is seen most frequently in growing tissues exposed to radiation is the cessation of mitosis or cell division. This may be temporary or permanent, depending upon the radiation dosage. Other effects observed are chromosome breaks, clumping of chromatin, formation of giant cells or other abnormal mitoses, increased granularity of cytoplasm, changes in staining characteristics, changes in motility or ciliary activity, cytolysis, vacuolization, altered viscosity of protoplasm, and altered permeability of the cell wall. Many of these effects can be duplicated individually with other types of stimuli. However, the entire gamut of effects cannot be reproduced by any single chemical agent.

Recovery period. Following exposure to radiation, recovery can and does take place to a certain extent. This is particularly manifest in the case of the acute effects, that is, those appearing within a matter of days or weeks after exposure. There is, however, a residual damage from which no recovery occurs, and it is this irreparable injury which can give rise to later delayed effects.

Determinants of Biological Effects

The dose-response curve. For any biologically harmful agent, it is useful to correlate the dosage administered with the response or damage produced. "Amount of damage" in the case of radiation might be the frequency of a given abnormality in the cells of an irradiated animal, or the incidence of some chronic disease in an irradiated human population. In plotting these two variables, a dose-response curve is produced. With radiation, an important question has been the nature and shape of this curve. Two possibilities are illustrated in Figures 10-3 and 10-4.

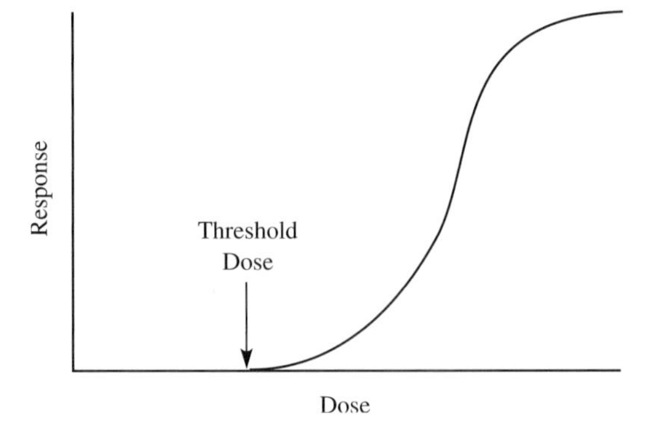

FIGURE 10-3
Dose-response curve depicting "threshold" dose.

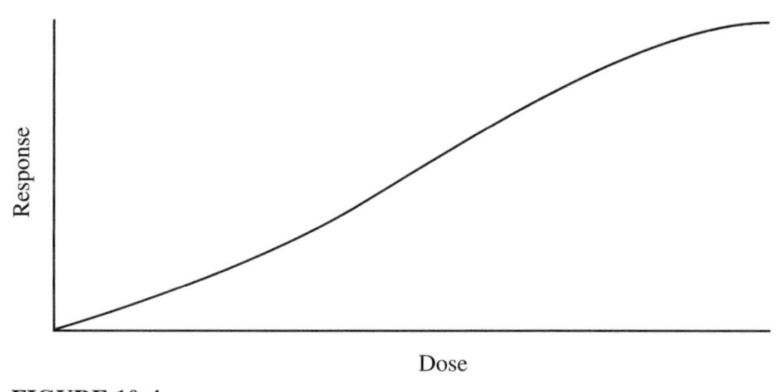

FIGURE 10-4
Dose-response curve depicting "no threshold" dose.

Figure 10-3 is a typical "threshold" curve. The point at which the curve intersects the abscissa is the threshold dose, that is, the dose below which there is no response. If an acute and easily observable radiation effect, such as reddening of the skin, is taken as "response," then this type of curve is applicable. The first evidence of the effect does not occur until a certain minimum dose is reached.

Figure 10-4 represents a linear, or nonthreshold, relationship, in which the curve intersects the abscissa at the origin. Here any dose, no matter how small, involves some degree of response. There is some evidence that the genetic effects of radiation constitute a nonthreshold phenomenon, and one of the underlying (and prudent) assumptions in the establishment of radiation protection guidelines and in radiation control activities in public health programs has been the assumption of a nonthreshold effect. Thus, some degree of risk is assumed when large populations of people are exposed to even very small amounts of radiation. This assumption often makes the establishment of guidelines for acceptable radiation exposure an enormously complex task, since the concept of "acceptable risk" comes into play, in which the benefit to be accrued from a given radiation exposure must be weighed against its hazard.

Rate of absorption. The rate at which the radiation is administered or absorbed is most important in the determination of what effects will occur. Since a considerable degree of recovery occurs from the radiation damage, a given dose will produce less of an effect if divided (thus allowing time for recovery between dose increments) than if it were given in a single exposure.

Area exposed. Generally when an external radiation exposure is referred to without qualification as to the area of the body involved, whole-body irradiation is assumed. The portion of the body irradiated is an important exposure parameter because the larger the area exposed, other factors being equal, the greater the overall damage to the organism. Even partial shielding of the highly radiosensitive blood-forming organs such as the spleen and bone marrow can mitigate the total effect considerably. An example of this phenomenon is in radiation therapy, in which doses that would be lethal if delivered to the whole body are commonly delivered to very limited areas, such as to tumor sites.

Variation in species and individual sensitivity. There is a wide variation in the radiosensitivity of various species. Lethal doses for plants and microorganisms, for example, are usually hundreds of times larger than those for mammals. Even among different species of rodents, it is not unusual for one to demonstrate three or four times the sensitivity of another.

Within the same species, biological variability accounts for a difference in sensitivity among individuals. For this reason the lethal dose for each species is expressed in statistical terms. The LD_{50} for that species, or the dose required to kill 50 percent of the individuals in a large population, is the standard statistical measure. For people, the LD_{50} is estimated to be approximately 450 R.

Variation in cell sensitivity. Within the same individual, a wide variation in susceptibility to radiation damage exists among different types of cells and tissues. In general, cells that are rapidly dividing or have a potential for rapid division are more sensitive than those that do not divide. Further, nondifferentiated cells (that is, primitive, or nonspecialized) are more sensitive than highly specialized cells. Within the same cell families then, the immature forms, which are generally primitive and rapidly dividing, are more radiosensitive than the older, mature cells which have specialized in structure and function and have ceased to divide.

Acute Effects

An acute dose of radiation is one delivered to a large portion of the body during a very short time. If the amount of radiation involved is large enough, acute doses may result in effects which can manifest themselves within a period of hours or days. Here the latent period, or time elapsed between the radiation insult and the onset of effects, is relatively short and grows progressively shorter as the dose level is raised. These short-term radiation effects are comprised of signs and symptoms collectively known as *acute radiation syndrome.*

The stages in acute radiation syndrome may be be described as follows:

1. *Prodrome.* This is the initial phase of the syndrome and is usually characterized by nausea, vomiting, and malaise. It may be considered analogous to the prodrome state in acute viral infections in which the individual is subject to nonspecific systemic reactions.
2. *Latent stage.* During this phase, which may be likened to the incubation period of a viral infection, the subjective symptoms of illness may subside, and the individual may feel well. Changes, however, may be taking place within the blood-forming organs and elsewhere that will subsequently give rise to the next aspect of the syndrome.
3. *Manifest illness stage.* This phase reflects the clinical picture specifically associated with the radiation injury. Among the possible signs and symptoms are fever, infection, hemorrhage, severe diarrhea, prostration, disorientation, and cardiovascular collapse. Which, if any, of the foregoing phenomena are observed in a given individual largely depend on the radiation dose received.
4. *Recovery or death.*

Relation of Dose to Type of Acute Radiation Syndrome

As mentioned above, each kind of cell has a different sensitivity to radiation. At relatively low doses, for example, the most likely cells to be injured are those with greatest sensitivity, such as the immature white blood cells of lymph nodes and bone marrow. At low doses the observable effects during the manifest illness stage would be in these cells. Thus, you would expect to observe fever, infection, and hemorrhage. This is known as the *hematopoietic form* of the acute radiation syndrome.

At higher doses, usually over 600 rads, cells of somewhat lower sensitivity will be injured. Of particular importance are the epithelial cells lining the gastrointestinal tract, for when these are destroyed a vital biological barrier is broken down. As a result, there may be fluid loss, overwhelming infection, and severe diarrhea in the *gastrointestinal form* of the acute radiation syndrome.

In the *cerebral form,* which may result from doses of 10,000 rads or more, the relatively resistant cells of the central nervous system are damaged, and the affected individual undergoes a rapid illness, characterized by disorientation and shock.

Considering the large degree of individual variation that exists in the manifestation of radiation injury, it is difficult to assign a precise dose range to each of the above forms of the syndrome. The following generalizations, however, may serve to provide a rough indication of the kinds of doses involved. At 50 rads or less, ordinary laboratory or clinical methods will show no indications of injury. At 100 rads, most individuals show no symptoms, although a small percentage may show mild blood changes. At 200 rads, most persons show definite signs of injury; this dose level may prove fatal to those individuals most sensitive to the effects of radiation. At 450 rads, the mean lethal dose has been reached, and 50 percent of exposed individuals will succumb. Approximately 600 rads usually marks the threshold of the gastrointestinal form of the acute radiation syndrome, with a very poor prognosis for all individuals involved. A fatal outcome may well be certain at 800 to 1,000 rads.

Delayed Effects

Long-term effects of radiation are those which may manifest themselves years after the original exposure. The latent period, then, is much longer than that associated with acute radiation syndrome. Delayed radiation effects may result from previous acute, high-dose exposures or from chronic, low-level exposures over a period of years.

There is no unique disease associated with the long-term effects of radiation. These effects manifest themselves in human populations simply as a statistical increase in the incidence of certain already existing conditions. Because of the low incidence of these conditions, it is usually necessary to observe large populations of irradiated persons in order to measure these effects. Biostatistical and epidemiologic methods are then used to indicate relationships between exposure and effect. In addition to the large numbers of people needed for human studies of delayed radiation effects, the situation is further complicated by the latent period. In some cases, a radiation-induced increase in a disease may go unrecorded unless the study is continued for many years.

It should also be noted that although it is possible to perform true experiments with animal populations, in which all factors with the exception of radiation exposure are kept identical in study populations, human data is limited to "secondhand" information accrued from populations irradiated for reasons other than radiobiological information. Often a special characteristic of irradiated human populations is the presence of some pre-existing disease, which makes it extremely difficult to draw meaningful conclusions when these groups are compared with nonirradiated ones.

Despite the above difficulties, many epidemiologic investigations of irradiated human beings have provided convincing evidence that ionizing radiation may indeed result in an increased risk of certain diseases long after the initial exposure. This information supplements and corroborates that gained from animal experimentation that demonstrates these same effects.

Among the delayed effects thus far observed have been somatic damage, which may result in an increased incidence of cancer, embryological defects, cataracts, lifespan shortening, and genetic mutations. With proper selection of animal species and strains, and of dose, ionizing radiation may be shown to exert an almost universal carcinogenic action, resulting in tumors in a great variety of organs and tissues. There is human evidence as well that radiation may contribute to the induction of various kinds of neoplastic diseases.

Human evidence. Both empirical observations and epidemiologic studies of irradiated individuals have more or less consistently demonstrated the carcinogenic properties of radiation. Some of these findings are summarized below.

Early in this century, when delayed radiation effects were little recognized, luminous numerals on watches and clocks were painted by hand with fine sable brushes, dipped first in radium-containing paint and then often "tipped" on the lips or tongue. Young women commonly were employed in this occupation. Years later, studies of these individuals who had ingested radium paint have disclosed an increased incidence of bone sarcomas and other malignancies resulting from the burdens of radium which had accumulated in their bones.

Some early medical and dental users of x-rays, largely unaware of the hazards involved, accumulated considerable doses of radiation. As early as the year 1910, there were reports of cancer deaths among physicians, presumably attributable to x-ray exposure. Skin cancer was a notable finding among these early practitioners. Dentists, for example, developed lesions on the fingers with which they repeatedly held dental films in their patients' mouths.

Early in this century, certain large mines in Europe were worked for pitchblende, a uranium ore. Lung cancer was highly prevalent among the miners as a result of the inhalation of large quantities of airborne radioactive materials. It was estimated that the risk of lung cancer in the pitchblende miners was at least 50 percent higher than that of the general population.

One of the strongest supports for the concept that radiation is a leukemogenic agent in people comes from the epidemiologic studies of the survivors of the atomic bombing in Hiroshima. Survivors exposed to radiation above an estimated dose of approximately 100 rem showed a significant increase in the incidence of leukemia.

In addition, leukemia incidence correlated well with the estimated dose (expressed as distance from the detonation), thus strengthening the hypothesis that the excess leukemia cases were indeed attributable to the radiation exposure. There is also some indication of an increase in thyroid cancer among the heavily irradiated survivors.

A pioneering study of children of mothers irradiated during pregnancy purported to show an increased risk of leukemia among young children if they had been irradiated in utero as a result of pelvic x-ray examination of the mother. Mothers of leukemic children were questioned as to their radiation histories during pregnancy with the child in question, and these responses were compared with those of a control group, consisting of mothers of healthy playmates of the leukemic children. Originally this work received much criticism, based partly on the questionnaire technique used to elicit the information concerning radiation history. It was believed that differences in recall between the two groups of mothers might have biased the results. A larger subsequent study designed to correct for the objections to the first one corroborated its essential findings and established the leukemogenic effect on the fetus of prenatal x-rays.

Considering the fact that immature, undifferentiated, and rapidly dividing cells are highly sensitive to radiation, it is not surprising that embryonic and fetal tissues are readily damaged by relatively low doses of radiation. It has been shown in animal experiments that deleterious effects may be produced with doses of only 10 rads delivered to the embryo. There is no reason to doubt that the human embryo is equally susceptible.

The majority of the anomalies produced by prenatal irradiation involve the central nervous system, although the specific type of damage is related to the dose and to the stage of pregnancy during which irradiation takes place. In terms of embryonic death, the very earliest stages of pregnancy, perhaps the first few weeks in human beings, are most radiosensitive. From the standpoint of practical radiation protection, this very early sensitivity is of great significance, because it involves a stage in human embryonic development in which pregnancy may well be unsuspected. For this reason, the International Committee on Radiological Protection has recommended that routine nonemergency diagnostic irradiation involving the pelvic area of women in the childbearing years be limited to the 10-day interval following the onset of menstruation. Such precautions would virtually eliminate the possibility of inadvertently exposing a fertilized egg.

The period from approximately the second through the sixth week of human gestation, when pregnancy could still be unsuspected, is the most sensitive for the production of congenital anomalies in the newborn. During this period, embryonic death is less likely than in the extremely early stage, but the production of morphological defects in the newborn is a major consideration.

During later stages of pregnancy, embryonic tissue is more resistant to gross and easily observable damage. However, functional changes, particularly those involving the central nervous system, may result from such late exposures. These would be difficult to measure or evaluate at birth. They usually involve subtle alterations in such phenomena as learning patterns and development and may have a considerable latent period before they manifest themselves. There is some evidence that the decreasing sensitivity of the fetus to gross radiation damage as pregnancy progresses may not apply for the leukemogenic effects of prenatal irradiation. Another important factor to be considered in evaluating the radiation hazard during late pregnancy is that irradiation may produce true genetic mutations in the immature germ cells of the fetus for which no threshold dose has been established.

Lifespan shortening. In a number of animal experiments, radiation has been demonstrated to have a lifespan shortening effect. The aging process *per se* is complex and largely obscure, and the exact mechanisms involved in aging are as yet uncertain. Irradiated animals in these investigations appear to die of the same diseases as the nonirradiated controls, but they do so at an earlier age. How much of the total effect is due to premature aging and how much to an increased incidence of radiation-induced diseases is still unresolved.

Genetic Effects

Background. The fertilized egg is a single cell resulting from the union of sperm and egg; millions of cell divisions develop it into a complete new organism. The information that produces the characteristics of the new individual is carried in the nucleus of the fertilized egg on rod-shaped structures called chromosomes, arranged in 23 pairs. In each pair, one member is contributed by the mother and the other by the father. With each cell division that the rapidly developing embryonic tissue undergoes, all of this information is faithfully duplicated, so that the nucleus in each new cell of the developing organism contains essentially all of the information. This, of course, includes the germ cells in the new organism which are destined to become sperm or eggs, and thus the information is transmitted from one generation to the next. This hereditary information is often likened to a template or to a code, which is reproduced millions of times over with remarkable accuracy. It is possible to damage the hereditary material in the cell nucleus by means of external influences, and when this is done the garbled or distorted genetic information will be reproduced just as faithfully when the cell divides as was the original message. When this kind of alteration occurs in those cells of the testes or ovaries that will become mature sperm or eggs, it is referred to as *genetic mutation;* if the damaged sperm or egg cell is then utilized in conception, the defect is reproduced in all of the cells of the new organism that results from this conception, including those that will become sperm or eggs, and thus whatever defect resulted from the original mutation can be passed on for many generations.

Most geneticists agree that the great preponderance of genetic mutations are harmful. By virtue of their damaging effects, they can be gradually eliminated from a population by natural means, since individuals afflicted with this damage are less likely to reproduce themselves successfully than are normal individuals. The more severe the defect produced by a given mutation, the more rapidly it will be eliminated and vice versa; mildly damaging mutations may require a great many generations before they gradually disappear.

As a balance to this natural elimination of harmful mutations, fresh ones are constantly occurring. A large number of agents have mutagenic properties, and it is probable that our current knowledge includes just a fraction of these. In addition, mutations can arise within the germ cells of an organism without external insult. Among the various external influences which have been found to be mutagenic are a wide variety of chemicals, certain drugs, and physical factors such as elevated temperatures and ionizing radiation. Natural background radiation probably accounts for a small proportion of naturally occurring mutations. For people, it has been estimated that background radiation probably produces less than 10 percent of these. Anthropogenic radiation, of course, if delivered to the gonads, can also produce mutations over and above those which occur spontaneously. Radiation, it should be noted, is not unique in this respect and is probably one of a number of environmental influences capable of increasing the mutation rate.

Animal evidence. The mutagenic properties of ionizing radiation were first discovered in 1927, using the fruit fly as the experimental animal. Since that time, experiments have been extended to include other species, and many investigations have been carried out on mice. Animal experimentation remains our chief source of information concerning the genetic effects of radiation, and as a result of the intensive experimentation, certain generalizations may be made. Among those of health significance are (1) there is no indication of a threshold dose for the genetic effects of radiation, that is, a dose below which genetic damage does not occur; and (2) the degree of mutational damage that results from radiation exposure seems to be dose-rate dependent, so that a given dose is less effective in producing damage if it is protracted or fractioned over a long period of time.

Human evidence. A major human study on genetic effects was made with the Japanese who survived the atomic bomb in 1945. As the index of a possible increase of the mutation rate, the sex ratio in the offspring of certain irradiated groups (families, for example, in which the mother had been irradiated but the father had not) was observed. Assuming that some of the mutational damage in the mothers would be recessive, lethal, and sex-linked, a shift in the sex ratio among these families might be expected in the direction of fewer male births than in completely nonirradiated groups, and this seemed to be the case in early reports. Later evaluation of more complete data, however, did not bear out the original suggestion of an effect on the sex ratio.

The preconception radiation histories of the parents of leukemic children compared with those of normal children was a part of the subject of another investigation. From the results, it would appear that there is a statistically significant increase in leukemia risk among children whose mothers had received diagnostic x-rays during this period. The effect here is apparently a genetic rather than an embryonic one, since the irradiation occurred prior to the conception of the child.

A somewhat similar study ascertained the radiation exposure histories of the parents of children with Down's Syndrome. Most of this exposure was prior to the conception of the child. A significantly greater number of the mothers of children with Down's Syndrome reported receiving fluoroscopy and x-ray therapy than did mothers of the normal children in the control group.

The findings of these two studies may provide evidence that ionizing radiation is a mutational agent in people. However, the findings should be viewed with some reservations because there could be significant differences to begin with between populations of people requiring x-rays and those not requiring x-rays. These differences alone might account for a slightly higher incidence of leukemia or Down's Syndrome in the offspring of the former group, irrespective of the radiation received. To date, there has been no evidence found of genetic effects in humans from radiation exposure.

10-3 RADIATION STANDARDS

Two population groups are given distinctly different treatment in the establishment of exposure/dose guidelines and rules. Standards are set for those occupationally engaged in work requiring ionizing radiation and for the general public. Although there are many standard-setting bodies, in general, the limits are consistent between groups. The Nuclear Regulatory Commission (NRC) has published guidelines in the Code of Federal Regulations (10 CFR 20) that serve as the standard in the United States. The dose guidelines are in addition to the natural background dose.

The allowable dose for occupational exposure is predicated on the following assumptions: the exposure group is under surveillance and control; it is adult; it is knowledgeable of its work and the associated risks; its exposure is at work, that is, 40 h/wk; and it is in good health. On this basis, no individual is to receive more than 5 rems per year of radiation exposure.

For the population at large, the allowable whole body dose in one calendar year is 0.1 rem. This dose does not include medical and dental doses that, for diagnostic and therapeutic reasons, may far exceed this.

In addition to these dose rules, the NRC has set standards for the discharge of radionuclides into the environment. Table 10-2 is an extract from that list. These concentrations are measured above the existing background concentration and are annual averages. Discharges must be limited such that the amounts shown are not exceeded in ambient air or natural waters. If a mixture of isotopes is released into an unrestricted area, the concentrations shall be limited so that the following relationship exists:

$$\frac{C_A}{\text{MPC}_A} + \frac{C_B}{\text{MPC}_B} + \frac{C_C}{\text{MPC}_C} \leq 1 \tag{10-20}$$

where C_A, C_B, C_C = concentrations of radionuclides A, B, and C, respectively, μCi/mL

$\text{MPC}_A, \text{MPC}_B, \text{MPC}_C$ = maximum permissible concentrations of radionuclides A, B, and C from Table II of Appendix B, Part 20 of the CFR (10 CFR 20)

Radon. Unlike the standards for exposure and releases to the environment, those for radon in indoor air are established by the U.S. Environmental Protection Agency. This is because radon is not the result of anthropogenic activity, but rather occurs naturally. The EPA guidelines suggest that the annual average radon exposure be limited to 4 pCi/L of air.

TABLE 10-2

Selected maximum permissible concentrations of radionuclides in air and water above background (excerpted from Title 10 CFR, Part 20, Appendix B)

		Occupational Values			Effluent Concentrations		Releases to Sewers
Radionuclide	**Class**	**Oral Ingestion ALI (μCi)**	**Inhalation ALI (μCi)**	**Inhalation DAC (μCi)**	**Air (μCi/mL)**	**Water (μCi/mL)**	**Monthly Average Conc. (μCi/mL)**
Barium-131	D, all compounds	3×10^3	8×10^3	3×10^{-6}	1×10^{-8}	4×10^{-5}	4×10^{-4}
Beryllium-7	W, all compounds except those given for Y	4×10^4	2×10^4	9×10^{-6}	3×10^{-8}	6×10^{-4}	6×10^{-3}
	Y, oxides, halides and nitrates	—	2×10^4	8×10^{-6}	3×10^{-8}	—	—
Calcium-45	W, all compounds	2×10^3	8×10^2	4×10^{-7}	1×10^{-9}	2×10^{-5}	2×10^{-4}
Carbon-14	Monoxide	—	2×10^6	7×10^{-4}	2×10^6	—	—
	Dioxide	—	2×10^5	9×10^{-5}	3×10^{-7}	—	—
	Compounds	2×10^3	2×10^3	1×10^{-6}	3×10^{-9}	3×10^{-5}	3×10^{-4}
Cesium-137	D, all compounds	1×10^2	2×10^2	6×10^{-8}	2×10^{-10}	1×10^{-6}	1×10^{-5}
Iodine-131	D all compounds	3×10^1 Thyroid (9×10^1)	5×10^1 Thyroid (2×10^3)	2×10^{-8}	— 2×10^{10}	— 1×10^{-6}	— 1×10^{-5}
Iron-55	D, all compounds except those given for W	9×10^3	2×10^3	8×10^{-7}	3×10^{-9}	1×10^{-4}	1×10^{-3}
	W, oxides, hydorixdes and halides	—	4×10^3	2×10^{-6}	6×10^{-9}	—	—

(continued)

TABLE 10-2
(continued)

		Occupational Values			Effluent Concentrations		Releases to Sewers
Radionuclide	**Class**	**Oral Ingestion ALI (μCi)**	**Inhalation ALI (μCi)**	**Inhalation DAC (μCi)**	**Air (μCi/mL)**	**Water (μCi/mL)**	**Monthly Average Conc. (μCi/mL)**
Phosphorus-32	D, all compounds except those given for W	6×10^2	9×10^2	4×10^{-7}	1×10^{-9}	9×10^{-6}	9×10^{-5}
	W, phosphates of Zn^{2+}, S^{3+}, Mg^{2+}, Fe^{3+}, Bi^{3+} and lanthanides	—	4×10^2	2×10^{-7}	5×10^{-10}	—	—
Radon-222	With daughters removed	—	1×10^4	4×10^{-6}	1×10^{-8}	—	—
	With daughters present	—	1×10^2 (or 4 working level months)	3×10^{-8} (or 0.33 working level)	1×10^{-10}	—	—
Strontium-90	D, all soluble compounds except $SrTiO_3$)	3×10^1 Bone surface 4×10^1	2×10^1 Bone surface 2×10^1	8×10^{-9}	3×10^{-11}	5×10^{-7}	5×10^{-6}
	Y, all insoluble compounds and $SrTiO_3$	—	4	2×10^{-9}	6×10^{-12}	—	—
Zinc-65	Y, all compounds	4×10^2	3×10^2	1×10^{-7}	4×10^{-10}	5×10^{-6}	5×10^{-5}

Note: D, W, and Y are classes denoting the time of retention in the body, days, weeks and years, respectively.
ALI is the Annual Limit of Intake, and DAC is the Derived Air Concentration.

10-4 RADIATION EXPOSURE

External and Internal Radiation Hazards

External radiation hazards result from exposure to sources of ionizing radiation of sufficient energy to penetrate the body and cause harm. Generally speaking, it requires an alpha particle of at least 7.5 MeV to penetrate the 0.07 mm protective layer of the skin. A beta particle requires 70 keV to penetrate the same layer.[7] Unless the sources of alpha and/or beta radiation are quite close to the skin, they pose only a small external radiation hazard. X-rays and gamma rays constitute the most common type of external hazard. When of sufficient energy, both are capable of deep penetration into the body. As a result no radiosensitive organ is beyond the range of their damaging power.

Radioactive materials may gain access to the body by ingestion, by inhalation of air containing radioactive materials, by absorption of a solution of radioactive materials through the skin, and by absorption of radioactive material into the tissue through a cut or break in the skin. The danger of ingesting radioactive materials is not necessarily from swallowing a large amount at one time, but rather from the accumulation of small amounts on the hands, on cigarettes, on foodstuffs, and other objects that bring the material into the mouth.

Any radioactive material that gains entry into the body is an internal hazard. The extent of the hazard depends upon the type of radiation emitted, its energy, the physical and biological half-life of the material, and the radiosensitivity of the organ where the isotope localizes. Alpha and beta emitters are the most dangerous radionuclides from an internal hazard point of view because their specific ionization is very high. Radionuclides with half-lives of intermediate length are the most dangerous, for they combine fairly high activity with life sufficiently long to cause considerable damage. Polonium is an example of a potentially very serious internal hazard. It emits a highly ionizing alpha particle of energy 5.3 MeV and has a half-life of 138 days.

Natural Background

People are exposed to natural radiation from cosmic, terrestrial, and internal sources. Typical gonadal exposures from natural background are summarized in Figure 10-5. Cosmic radiation is that originating outside of our atmosphere. This radiation consists predominately, if not entirely, of protons whose energy spectrum peaks in the range of 1 to 2 GeV. Heavy nuclei are also present. The impact of primary and very high energy secondary cosmic rays produces violent nuclear reactions in which many neutrons, protons, alpha particles, and other fragments are emitted. Most of the neutrons produced by cosmic rays are slowed to thermal energies and, by n, p (neutron/proton)

[7]U.S. Public Health Service, *Radiological Health Handbook* (PHS Publication No. 2016), Rockville, MD, p. 204, 1970.

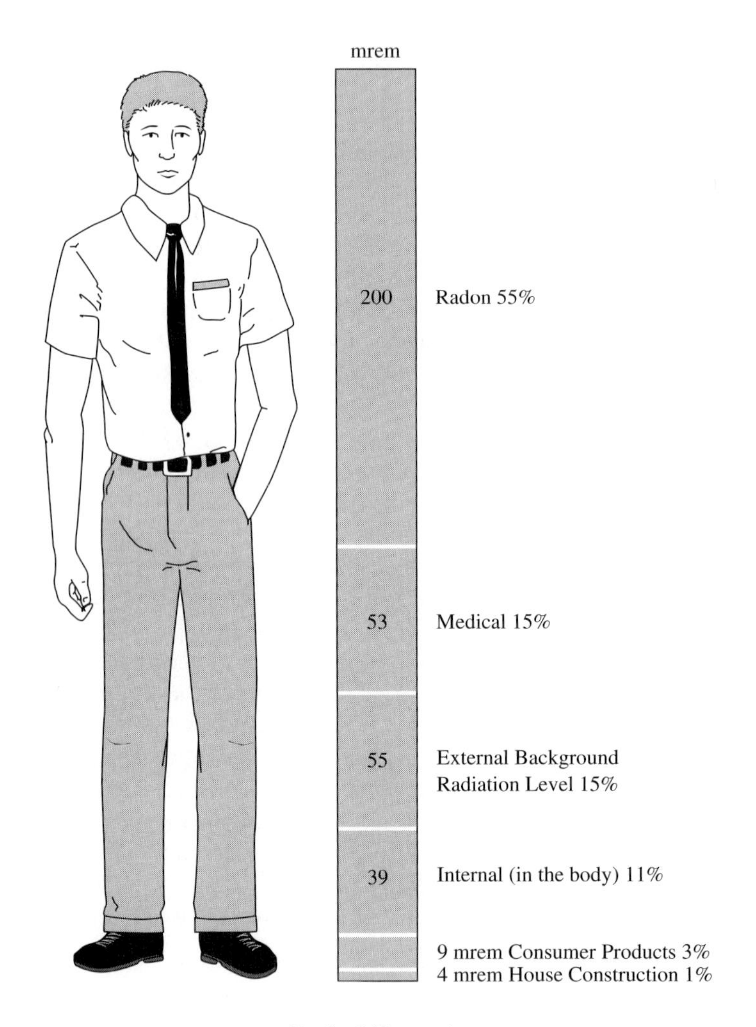

FIGURE 10-5
Average dose per year to person living in the United States. (*Source:* U.S. Department of Energy.)

reaction with ^{14}N, produce ^{14}C. The lifetime of carbon-14 is long enough that it becomes thoroughly mixed with the exchangeable carbon at the earth's surface (carbon dioxide, dissolved bicarbonate in the oceans, living organisms, etc.). Some of the cosmic radiation penetrates to the earth's surface and contributes directly to our whole body dose. Terrestrial radiation exposure comes from the 50 naturally occurring radionuclides found in the earth's crust. Of these, radon has come to have the most significance as a common environmental hazard to the general public.

Radon is the product of the radioactive decay of its parent, radium. Radium is produced from each of the three major series: ^{235}U, ^{238}U, and ^{232}Th. The radon isotopes produced are Rn-222, Rn-220, and Rn-219. These have half-lives of

3.8 d, 55.6 s, and 3.92 s, respectively. Rn-222, because of its longer half-life and the abundance of its parent uranium in geologic materials, is generally more abundant and, hence, is considered the greater environmental hazard. Because the half-life of radium and its parents is so long, the source is essentially undiminished over human time scales.

The hazard of radon does not come from radon itself, but from its radioactive decay products (^{218}Po, ^{214}Pb, ^{214}Bi). The decay products are charged atoms of heavy metals that readily attach themselves to airborne particulates. The main health problem stems from the inhalation of unattached decay products and these particulates. The decay products and particulates become lodged in the lung. As they continue to decay, they release small bursts of energy in the form of alpha, beta, and gamma radiation that damage the lung tissue and could ultimately lead to lung cancer.[8]

Radon is a gas. It is colorless, odorless, and generally chemically inert like other noble gases such as helium, neon, krypton, and argon. It does not sorb, hydrolize, oxidize, or precipitate. Thus, its movement through the ground is not inhibited by chemical interaction with the soil.

Migration of radon occurs by two mechanisms: diffusion as a gas through the pore spaces in the soil and by dissolution and transport in the groundwater. The rate of diffusion or transport is a function of the emanation rate, porosity, structural channels, moisture content, and hydrologic conditions. These migration routes lead to two mechanisms of impact on people. Buildings constructed in areas of high radon emanation may have radon gas penetrate the structure through natural construction openings such as floor drains or joints (see Table 10-3) or through structural failures such as cracks that develop from foundation settlement. In areas where the public water supply is drawn from an aquifer that has radon emanation, shower water may

TABLE 10-3
Radon gas measurements in the floor drains and in the basement air of seven houses

House No.	Radon concentration in floor drain (pCi/L)	Radon concentration in basement air (pCi/L)	Ratio drain/basement
1	169.3	2.51	67.5
2	98.4	2.24	43.9
3	91.4	1.43	63.9
4	413.3	1.87	221
5	255.4	3.95	64.7
6	173.4	3.02	57.4
7	52.1	9.63	5.4
	Average 179.0	3.52	

[8]R. W. Kuennen and R. C. Roth, "Reduction of Radon Working Level by a Room Air Cleaner," presented at the 82nd Annual Meeting of the Air & Waste Management Association, Anaheim, CA, June 1989.

release radon. One rule of thumb is that a radon concentration of 10,000 pCi/L of water, when heated and agitated, will produce about 1 pCi/L of air.[9]

X-rays

X-ray machine use is widespread in industry, medicine, and research. All such uses are potential sources of exposure.

Medical and dental use. In addition to the 300,000 to 400,000 medical-technical personnel that are occupationally exposed to radiation in the use of these machines, a considerable portion of the general population is also exposed. A large portion of the 2,500,000 persons seen daily by physicians have some x-ray diagnostic procedure performed upon them.

Industrial uses. Industrial x-ray devices include radiographic and fluoroscopic units used for the determination of defects in castings, fabricated structures, and welds, and fluoroscopic units used for the detection of foreign material in, for example, airline luggage. Use of these units may result in whole body exposure to the operators and people who are nearby.

Research use. High-voltage x-ray machines are becoming familiar features of research laboratories in universities and similar institutions. Other x-ray equipment used in research includes x-ray diffraction units used for crystal analysis, electron microscopes, and particle accelerators.

Radionuclides

Naturally occurring. Thousands of curies of radium are in use in the medical field. In this use, many individuals besides the patient are potentially exposed to radiation. These include other patients, nurses, technicians, radiologists, and physicians.

Static eliminators, employing polonium or radium as the radioactive source, have been widely used in industry. Typical industries where they may be found are the textile and paper trades, printing, photographic processing, and telephone and telegraph companies.

Artificially produced. Over 6,000 universities, hospitals, and research laboratories in the United States are using radionuclides for medical, biological, industrial, agricultural, and scientific research and for medical diagnosis and therapy. Over a million people in the United States receive radiotherapy treatment each year. Possible exposure from such radionuclides is involved with their preparation, handling, application, and transportation. Exposures, internal and external, might also arise

---

[9] D. M. Murane and J. Spears, "Radon Reduction in New Construction," *Indoor Radon II, Proceedings of the Second APCA International Specialty Conference,* Cherry Hill, NJ, 1987.

through contamination of the environment by wastes originating from the use of these materials.

Nuclear Reactor Operations

Sources of radiation exposure associated with nuclear reactor operations include the reactor itself; its ventilation and cooling wastes; procedures associated with the removal and reprocessing of its "spent" fuel and the resulting fission product wastes; and procedures associated with the mining, milling, and fabrication of new fuels.

Radioactive Wastes

There are three principal sources of radioactive wastes: reactors and chemical processing plants, research facilities, and medical facilities. Regulations for the handling and disposal of radioactive wastes are designed to minimize exposure to the general public, but the regulations obviously provide less protection to those handling the waste.

10-5 RADIATION PROTECTION[10]

The principles discussed here are generally applicable to all types or energies of radiation. However, their application will vary depending upon the type, intensity, and energy of the source. For example, beta particles from radioactive materials require different shielding than high-speed electrons from an accelerator. Ideally, we would like to provide protection that results in a radiation exposure of zero. In actuality, technical and economic limitations force us to compromise so that the risks are small compared to the benefits obtained. The radiation standards set the limit above which the risk is deemed to be too great.

Reduction of External Radiation Hazards

Three fundamental methods are employed to reduce external radiation hazards: distance, shielding, and reduction of exposure time.

Distance. Distance is not only very effective, but also in many instances the most easily applied principle of radiation protection. Beta particles of a single energy have a finite range in air. Sometimes the distance afforded by the use of remote control handling devices will supply complete protection.

The inverse square law for reduction of radiation intensity applies for point sources of x-, gamma, and neutron radiation. The inverse square law states that

---

[10]This discussion follows that of the U.S. Public Health Service, *Introduction to Medical X-Ray Protection,* 1968.

radiation intensity from a point varies inversely as the square of the distance from the source:

$$\frac{I_1}{I_2} = \frac{(R_2)^2}{(R_1)^2} \tag{10-21}$$

where I_1 is the radiation intensity at distance R_1 from the source, and I_2 is the radiation intensity at distance R_2 from the source. Inspection of this formula will show that increasing the distance by a factor of three, for example, reduces the radiation intensity to one-ninth of its value. The inverse square law does not apply to extended sources or to radiation fields from multiple sources.

X-ray tubes act sufficiently like point sources so that reduction calculations by this law are valid. Gamma ray sources whose dimensions are small in comparison to the distances involved may also be considered point sources, as can capsule neutron sources.

Shielding. Shielding is one of the most important methods for radiation protection. It is accomplished by placing some absorbing material between the source and the person to be protected. Radiation is attenuated in the absorbing medium. When so used, "absorption" does not imply an occurrence such as a sponge soaking up water, but rather absorption here refers to the process of transferring the energy of the radiation to the atoms of the material through which the radiation passes. X- and gamma radiation energy is lost by three methods: photoelectric effect, Compton effect, and pair production.

The photoelectric effect is an all-or-none energy loss. The x-ray, or photon, imparts all of its energy to an orbital electron of some atom. This photon, since it consisted only of energy in the first place, simply vanishes. The energy is imparted to the orbital electron in the form of kinetic energy of motion, and this greatly increased energy overcomes the attractive force of the nucleus for the electron and causes the electron to fly from its orbit with considerable velocity. Thus, an ion-pair results. The high-velocity electron (which is called a photoelectron) has sufficient energy to knock other electrons from the orbits of other atoms, and it goes on its way producing secondary ion-pairs until all of its energy is expended.

The Compton effect provides a means of partial energy loss for the incoming x- or gamma ray. Again the ray appears to interact with an orbital electron of some atom, but in the case of Compton interactions, only a part of the energy is transferred to the electron, and the x- or gamma ray "staggers on" in a weakened condition. The high-velocity electron, now referred to as a Compton electron, produces secondary ionization in the same manner as does the photoelectron, and the weakened x-ray continues on until it loses more energy in another Compton interaction or disappears completely via the photoelectric effect. The unfortunate aspect of Compton interaction is that the direction of flight of the weakened x- or gamma ray is different from the direction of the original. In fact, the weakened x- or gamma ray is frequently referred to as a "scattered" photon, and the entire process is known as "Compton scattering." By this mechanism of interaction, the direction of photons in a beam may be randomized, so that scattered radiation may appear around corners and behind shields although at a lesser intensity.

Pair production, the third type of interaction, is much rarer than either the photoelectric or Compton effect. In fact, pair production is impossible unless the x- or gamma ray possesses at least 1 MeV of energy. (Practically speaking, it does not become important until 2 MeV of energy.) Pair production may be thought of as the lifting of an electron from a negative to a positive energy state. The pair is a positron-electron pair that results from the photon ejecting an electron and leaving a "hole"—the positron. If there is any excess energy in the photon above the 1 MeV required to create two electron masses, it is simply shared between the two electrons as kinetic energy of motion, and they fly out of the atom with great velocity. The negative electron behaves in exactly the ordinary way, producing secondary ion-pairs until it loses all of its energy of motion. The positron also produces secondary ionization so long as it is in motion, but when it has lost its energy and slowed almost to a stop, it encounters a free negative electron somewhere in the material. The two are attracted by their opposite charges, and, upon contact, annihilate each other, converting both their masses into pure energy. Thus, two gamma rays of 0.51 MeV arise at the site of the annihilation. The ultimate fate of the annihilation gammas is either photoelectric absorption or Compton scattering followed by photoelectric absorption.

Since the energy of the photon must be greater than 1 MeV for pair production to occur, this process is not a factor in the absorption of x-rays used in dental and medical radiography. The energies of x-rays used in this type of radiography are rarely more than 0.1 MeV.

The predominating mechanism of interaction with the shielding material depends upon the energy of the radiation and the absorbing material. The photoelectric effect is most important at low energies, the Compton effect at intermediate energies, and pair production at high energies. As x- and gamma ray photons travel through an absorber, their decrease in number caused by the above-named absorption processes is governed by the energy of radiation, the specific absorber medium, and the thickness of the absorber traversed. The general attenuation may be expressed as follows:

$$\frac{dI}{dx} = -uI_0 \tag{10-22}$$

where dI = reduction of radiation
I_0 = incident radiation
u = proportionality constant
dx = thickness of absorber traversed

Integrating yields

$$I = I_0 \exp(-ux) \tag{10-23}$$

Using this formula it is easy to calculate the radiation intensity behind a shield of thickness x, or to calculate the thickness of absorber necessary to reduce radiation intensity to a desired level, if the factor u is known. This factor is called the linear absorption coefficient when x is a linear dimension. The value of u depends upon the energy of the radiation and the absorbing medium. The ratio I/I_0 is sometimes called the transmission. Tables and graphs are available which give values of u determined experimentally or that give transmission values for varying thickness or different shielding materials (Figures 10-6 through 10-9).

If the radiation being attenuated does not meet narrow-beam conditions, and/or thick absorbers are involved, the absorption equation becomes

$$I = BI_0 \exp(-ux) \tag{10-24}$$

where B is the buildup factor that takes into account an increasing radiation intensity due to scattered radiation within the absorber.

For alpha and beta emissions from radionuclides (not accelerators), substantial attenuation can be achieved with modest shielding. The amount of shielding required is, of course, a function of the particle energy. For example, a 10 MeV alpha particle

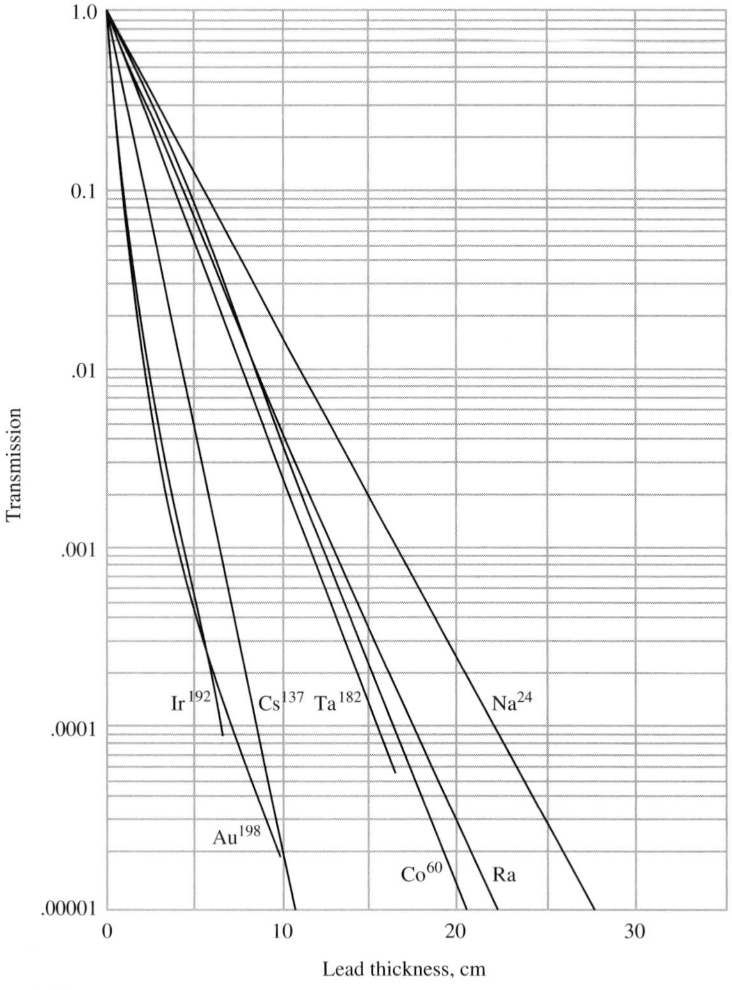

FIGURE 10-6
Transmission through lead of gamma rays from radium; cobalt 60; cesium 137; gold 198; iridium 192; tantalum 182; and sodium 24.

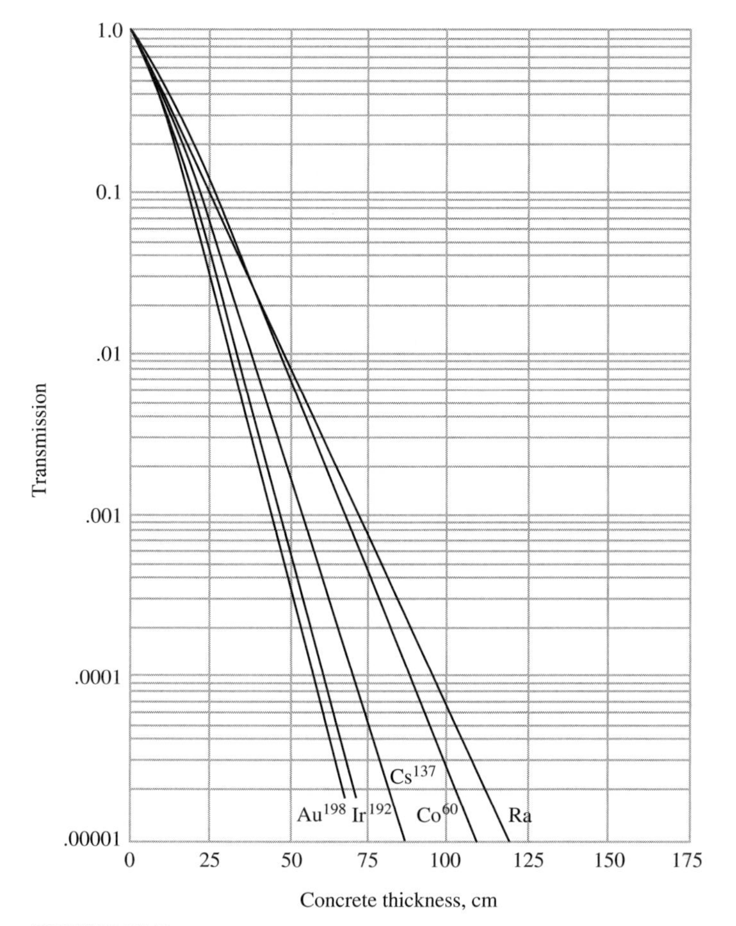

FIGURE 10-7
Transmission through concrete (density 147 lb/ft^3) of gamma rays from radium; cobalt 60, cesium 137, gold 198; iridium 192.

has a range of 1.14 m in air, while a 1 MeV particle has a range of 2.28 cm. Virtually any solid material of any substance can be used to shield alpha particles. Beta particles can also be shielded relatively easily. For example a ^{32}P beta at 1.71 MeV can be attenuated 99.8 percent by 0.25 cm of aluminum. However, materials with high atomic numbers, such as metals, should not be used for high-energy beta shielding due to the production of *Bremsstrahlung radiation* (radiation produced by stopping another kind of radiation). In materials with high atomic numbers, the beta particle is absorbed, but the excess "trapped" energy is released in the form of an x-ray. For this reason, Plexiglas® or Lucite®, typically 6 to 12 mm thick, is often used.

Fast neutrons are poorly absorbed by most materials. Therefore, it is necessary to slow them down for efficient absorption. Since the greatest transfer of energy takes place in collisions between particles of equal mass, hydrogenous materials are most effective for slowing down fast neutrons. Water, paraffin, and concrete are all rich in hydrogen, and thus, important in neutron shielding. Once the neutrons

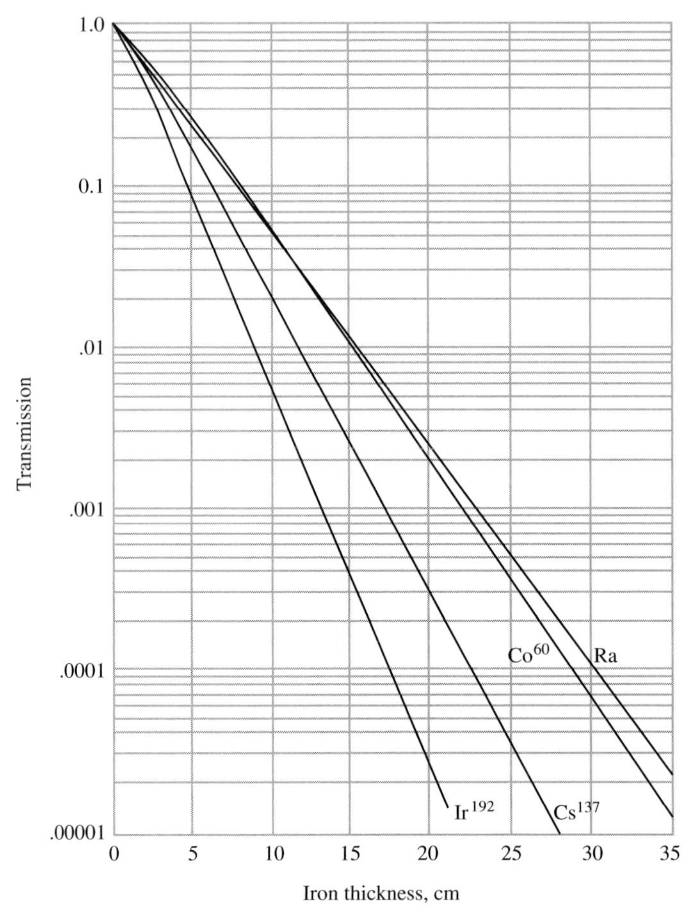

FIGURE 10-8
Transmission through iron of gamma rays from radium; cobalt 60, cesium 137; iridium 192.

have been reduced in energy, they may be absorbed by either boron or cadmium. When a boron atom captures a neutron, it emits an alpha particle, but because of the extremely short range of alpha particles, there is no additional hazard. Neutron capture by cadmium results in the emission of gamma radiation. Lead or a similar gamma absorber must be used as a shield against these gammas. A complete shield for a capsule-type neutron source may consist of, first, a thick layer of paraffin to slow down the neutrons, then a surrounding layer of cadmium to absorb the slow neutrons, and finally, an outer layer of lead to absorb both the gammas produced in the cadmium and those emanating from the capsule.

Some care must be exercised in using shielding to reduce exposure. People outside the "shadow" cast by the shield are not necessarily protected. A wall or partition is not necessarily a safe shield for persons on the other side. Their allowable dose may be less than conceived in the design of the barrier. Radiation can "bounce around corners" because it can be scattered.

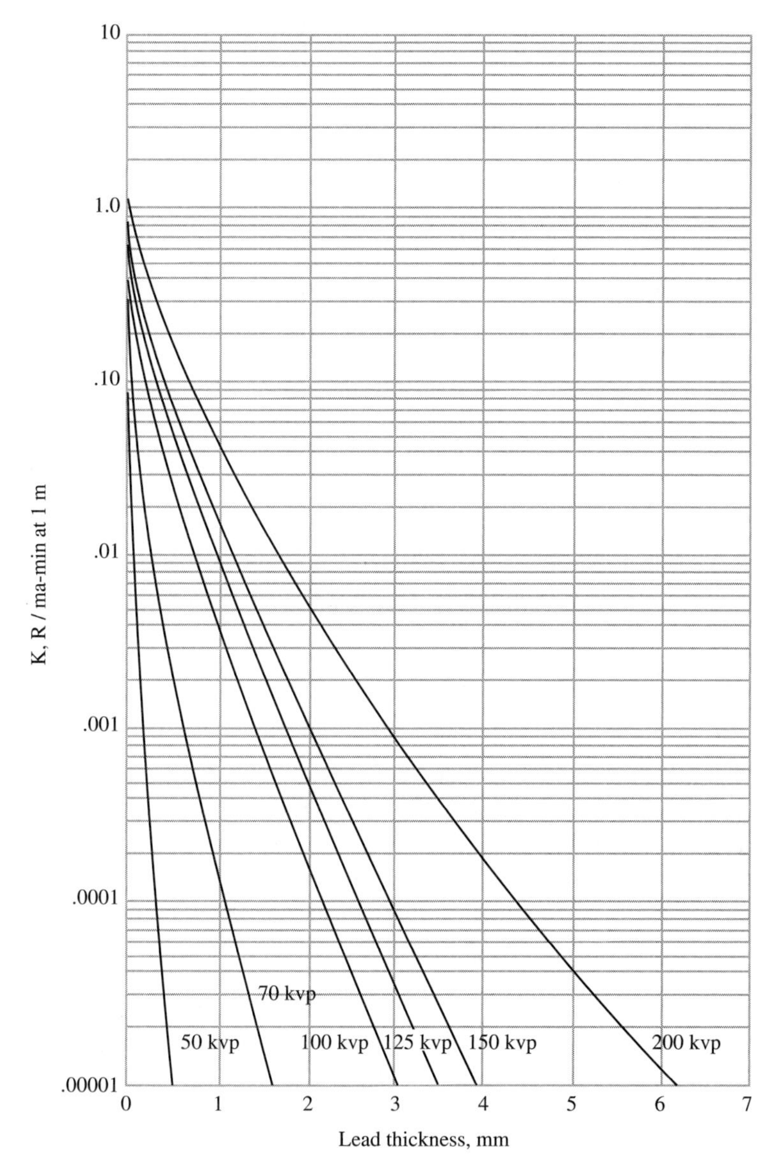

FIGURE 10-9
Transmission through lead of x-rays.

Scattered radiation is present to some extent whenever an absorbing medium is in the path of radiation. The absorber then acts as a new source of radiation. Frequently, room walls, the floor, and other solid objects are near enough to a source of radiation to make scatter appreciable. When a point source is used under these conditions, the inverse square law is no longer completely valid for computing radiation intensity at a distance. Measurement of the radiation is then necessary to determine the potential exposure at any point.

Reduction of exposure time. By limiting the duration of exposure to all radiation sources and by providing ample recuperative time between exposures, the untoward effects of radiation can be minimized. Recognition of the zero threshold theory of damage warrants that exposures, no matter how small, be minimized. The standards established by the NRC are upper bounds to be avoided and not goals to be achieved.

In emergency situations it may occasionally be necessary to work in areas of very high dose rates. This can be done with safety by limiting the total exposure time so that the average permissible value for a day based on the Radiation Protection Guide dose of 0.1 rad per week is not exceeded. This does not imply that a worker should be allowed to extend this practice beyond receiving 0.1 rad in a short period of time, that is, a dose of 0.1 rad one day and no dose for six days would comply with the rule but would be considered excessive. Repetitions of this cycle would be unacceptable. Emergency situations may require that work be done in relays of several people in the same job so that the value of the Radiation Protection Guide is not exceeded by any one person.

Reduction of Internal Radiation Hazards

Occupational. The prevention and control of contamination is the most effective way to reduce internal hazards in the workplace. The use of protective devices and good handling techniques affords a large measure of protection. Dust should be kept to a minimum by elimination of dry sweeping. Laboratory operations should be carried out in a hood. The exhaust air from the hood must be filtered with a high-efficiency filter. The filter must be replaced regularly in an approved manner. Protective clothing should be worn so that normal street clothes do not become contaminated. Respirators should be worn during emergency operations or when dust is generated. Eating must not be permitted in areas where radioactive materials are handled. Proper training in the care and handling of radioactive materials is, perhaps, the most important method for reducing the potential for internal radiation exposure in the workplace.

Radon. Aside from cigarette smoke, the most likely nonoccupational internal radiation hazard is from radon in private dwellings. Since the radon primarily originates in the soil beneath the house, control efforts are aimed at the basement/crawl space.

The EPA suggests two major approaches for new construction: reduction of the pathways for radon entry and reduction of the draft of the house on surrounding and underlying soil. The methods to reduce the pathways for entry are summarized in Figure 10-10. Of particular concern are penetrations into the foundation such as floor drains (Table 10-3) and cracks in the floor. The use of a polyethylene sheet below the slab is particularly effective for controlling leaks that result from slab cracks that develop as the house settles. Because heat in the upper floors tends to rise, creating a draft much like a chimney, the house has a tendency to create a negative pressure on the basement and, hence, "suck in" radon from the soil pore spaces. Figure 10-11 shows some techniques to minimize the draft effect.[11]

[11] D. M. Murane and J. Spears, "Radon Reduction in New Construction," *Indoor Radon II, Proceedings of the Second APCA International Specialty Conference,* Cherry Hill, NJ, pp. 183–194, 1987.

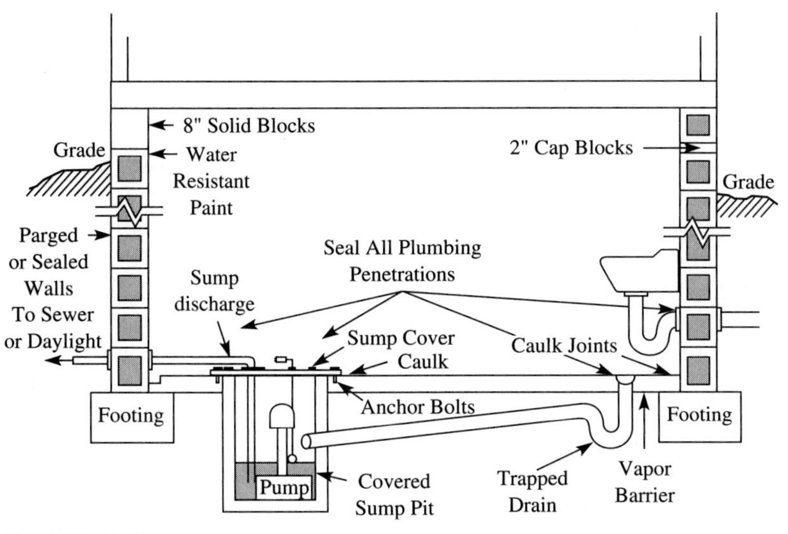

FIGURE 10-10
Methods to reduce pathways for radon entry.

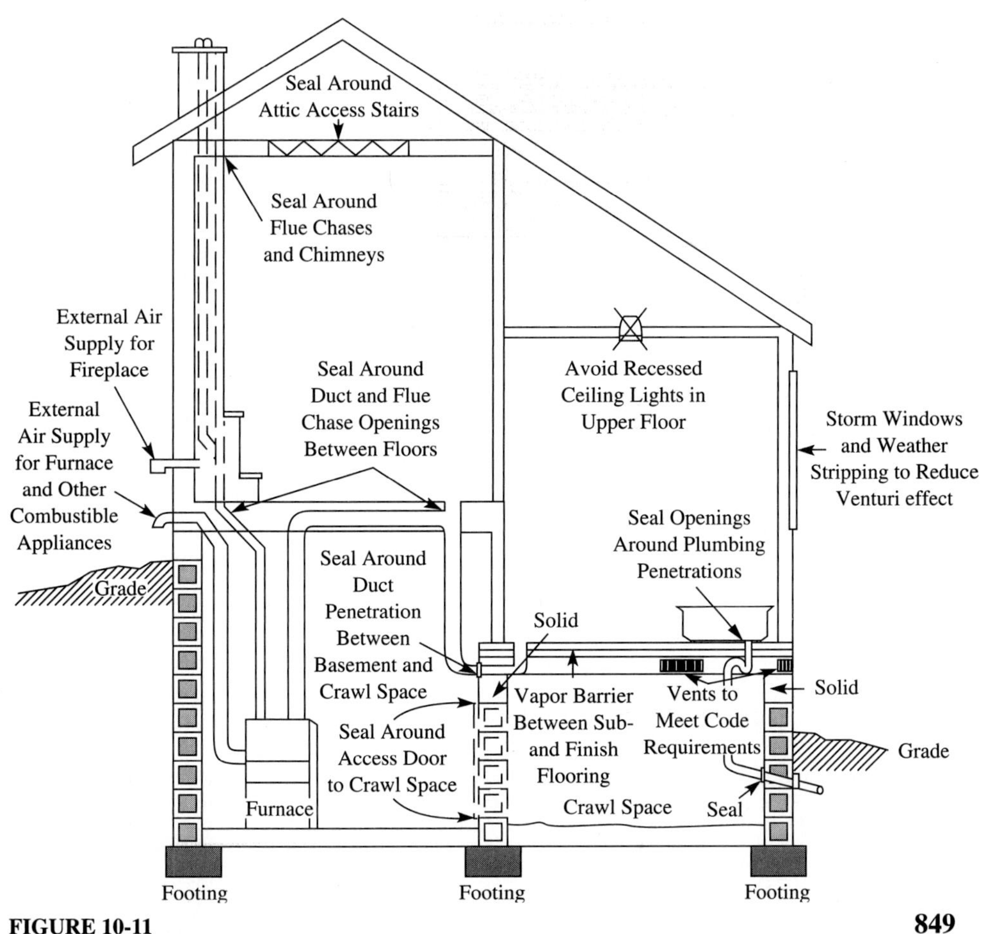

FIGURE 10-11
Methods to reduce the vacuum.

For existing structures, the remedies are more difficult to install, will be expensive, and may not yield satisfactory results. If drain tiles are present around the outside or inside of the perimeter footings, these are ideally located to permit vacuum to be drawn near some of the major soil gas entry routes (the joint between the slab and the foundation wall and the footing region where the radon can enter the voids in the block walls). Other efforts have included drilling holes in the slab itself and creating a vacuum system beneath the whole slab. Several suction points (three to seven) are required for this technique to work.[12] One demonstration project showed that jacking the house off the foundation and sealing the block walls was effective. In addition, a proprietary epoxy coating was applied to the floor and walls (Figure 10-12).[13]

10-6 RADIOACTIVE WASTE

Types of Waste

There is no single scheme that is satisfactory for classifying radioactive waste in a quantitative way. Usage has led us to categorize wastes into "levels." *High-level wastes* are those with activities measured in curies per liter; *intermediate-level wastes* have activities measured in millicuries per liter; *low-level wastes* have activities measured in microcuries per liter. Other classifications skip the intermediate-level wastes and use the terms high-level, *transuranic,* and low-level. The high-level wastes (HLW) are those resulting from reprocessing of spent fuel or the spent fuel itself from nuclear reactors. Transuranic wastes are those containing isotopes above uranium in the periodic table. They are the by-products of fuel assembly, weapons fabrication, and reprocessing. In general their radioactivity is low but they contain long-lived isotopes (those with half-lives greater than 20 years). The bulk of low-level wastes (LLW) has relatively little radioactivity. Most require little or no shielding and may be handled by direct contact.

High-Level Radioactive Waste Management

In 1993, there were about 109 operating reactors in the United States.[14] Roughly 10 m^3 of spent fuel is generated annually from each of these reactors. The construction of the fuel assembly results in considerably less fission product waste. Approximately 0.1 m^3 of the 10 m^3 is fission product waste. Of course, it is evenly distributed throughout the assembly and cannot be easily separated. The management choices

[12] D. B. Henschel and A. G. Scott, "Testing of Indoor Radon Reduction Techniques in Eastern Pennsylvania: An Update," *Indoor Radon II, Proceedings of the Second APCA International Specialty Conference,* Cherry Hill, NJ, pp. 146–159, 1987.

[13] M. T. Ibach and J. H. Gallagher, "Retrofit and Preoccupancy Radon Mitigation Program for Homes," *Indoor Radon II, Proceedings of the Second APCA International Specialty Conference,* Cherry Hill, NJ, pp. 172–182, 1987.

[14] J. Josephson, "Nuclear Power's Dim Future," *Environmental Science and Technology,* vol. 27, no. 6, pp. 1007–1009, 1993.

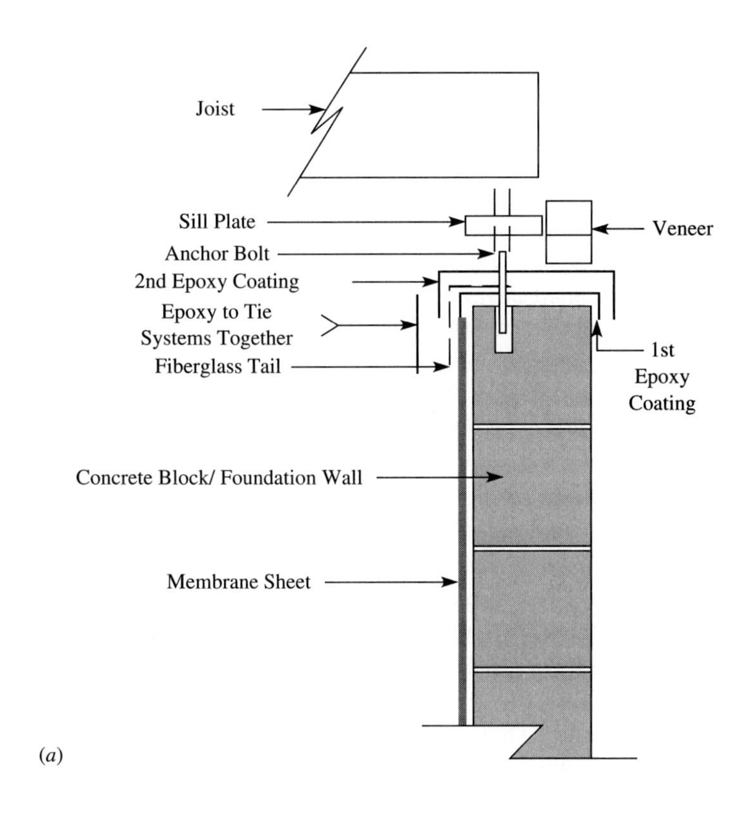

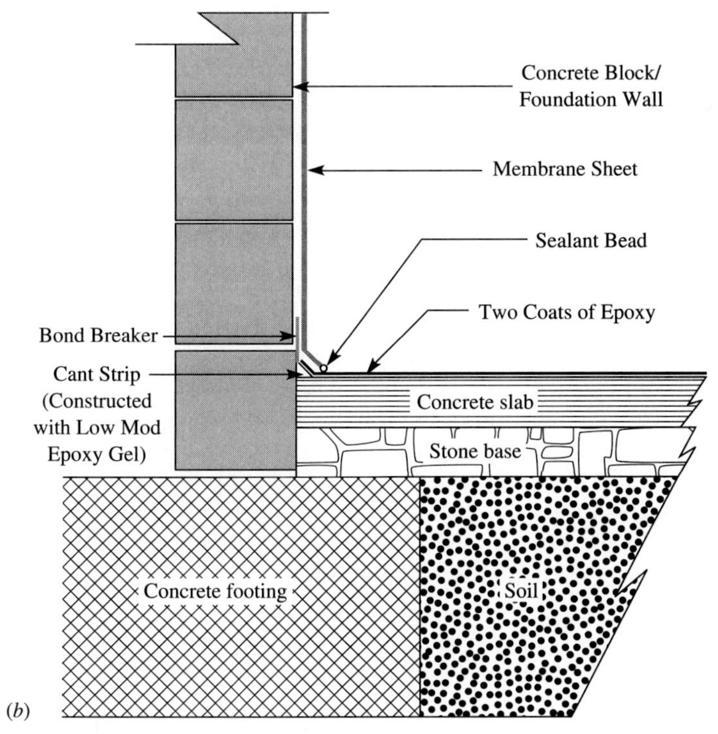

FIGURE 10-12
Interior membrane linings and sealants to prevent radon gas infiltration. (*Source:* Gallagher & Associates, Inc. April 1987.)

are (1) store it indefinitely in the form it was removed from the reactor, (2) reprocess it to extract the fission products and recycle the other materials, or (3) dispose of it by burial or other isolation technique.

Under the Nuclear Waste Policy Act of 1987, Congress has prescribed that a storage facility be constructed that will not become permanent. Designated the Monitored Retrievable Storage Facility, the current and only candidate site is at Yucca Flats, Nevada. The NRC has detailed the rules for the site in the Code of Federal Regulations (10 CFR 60). Some of the important provisions are summarized here:[15]

1. The design and operation of the facility should not pose an unreasonable risk to the health and safety of the public. The radiation dose limit is a small fraction of that due to natural background.
2. A multiple barrier is to be used.
3. A thorough site study must be made. Geologic and hydrologic characteristics of the site must be favorable.
4. The repository must be located where there are no attractive resources, be far from population centers, and be under federal control.
5. High-level wastes are to be retrievable for up to 50 years from the start of operations.
6. The waste package must be designed to take into account all of the possible effects from earthquakes to accidental mishandling.
7. The package is to have a design life of 300 years.
8. Groundwater travel time from the repository to the source of public water is to be at least 1,000 years.
9. The annual release of radionuclides must be less than 1/1000 of a percent of the amount of the radioactivity that is present 1,000 years after the repository is closed.

WIPP

The Waste Isolation Pilot Plant (WIPP) Project was authorized by Congress in 1979. After much political negotiation, the WIPP was authorized as a military transuranic waste facility exempt from licensing by the NRC. Under construction since 1983, the facility consists of 16 km of shafts and tunnels 650 m below ground in southeast New Mexico. The geologic material is a Permian salt basin. It was supposed to begin accepting waste in 1989 but has not done so. Certain risk analyses and compliance with RCRA requirements have delayed its use. The facility was conceived to demonstrate design concepts and to conduct some experiments with military high-level wastes that will later be retrieved. Many design questions about gas generation in the decaying waste and potential contamination scenarios are still being worked out.

[15] R. L. Murray, *Understanding Radioactive Waste,* Richland, WA: Battelle Press, pp. 137–138, 1989.

Low-Level Radioactive Waste Management

Historical perspective. Between 1962 and 1971, six commercial waste disposal sites were licensed. Three were subsequently closed because they failed. The three sites (Maxey Flats, Kentucky; Sheffield, Illinois; and West Valley, New York) all experienced similar problems. They used shallow land burial to dispose of the waste. This was accomplished by excavating a trench about 3 to 6 m deep and placing the drums and other containers (often cardboard boxes) of radionuclides in the trench and covering them with excavated soil. The completed trench was covered with a mound of earth and seeded.

Water seeped through the cover material and animals burrowed through it. The heavy clay sites chosen precisely to limit passage to the groundwater system served as holding ponds for the rainwater and ultimately accelerated the corrosion of the drums. At West Valley, when increased radioactivity called attention to this phenomenon, the trenches were opened and pumped to the nearest stream! Concurrently, it was discovered that the drums were often 30 to 50 percent empty. This, combined with the fact that the backfill material was heavy clay that did not completely fill the void spaces between the drums, allowed significant settlement of the cover material. This enhanced the collection of precipitation that contributed to the corrosion and failure of the drums.

These episodes led to a major rethinking of how we should manage our radioactive wastes. One result was that in 1980, Congress enacted the Low-Level Waste Policy Act. It says that each state is responsible for providing for the availability of capacity either within or outside of the state for disposal of low-level radioactive waste generated within its borders. The law provided for the formation of *compacts* between states to allow a regional approach to management. As of December 1995, the compact organization was as shown in Figure 10-13. The compacts decide what facilities are required and which states will serve as hosts. Although the compacts were supposed to begin accepting waste in 1986, the negotiation process has taken longer than expected and the deadline has been extended to beyond the year 2000. Many compacts have yet to select sites, let alone begin construction. The three currently available sites will soon run out of capacity, so there is some urgency to solve the problem.

Waste minimization.[16] As with all waste problems we have dealt with in this text, the first step in managing low-level radioactive waste is to minimize its production. Since 1980, considerable strides have been made in reducing the volume of LLRW (Figure 10-14). There are a number of procedures that can be effectively employed.

Immediate sorting of solid radioactive waste from nonradioactive waste is an essential initial step in any scheme for the reduction of the volume of that waste and for the recovery of radionuclides from uranium and transuranium waste. It is optimistic to expect much reduction of that volume of waste by sorting out

---

[16]This discussion follows that of the National Research Council, *The Shallow Land Burial of Low-Level Radioactively Contaminated Solid Waste,* Washington, DC: National Academy of Science, 1976.

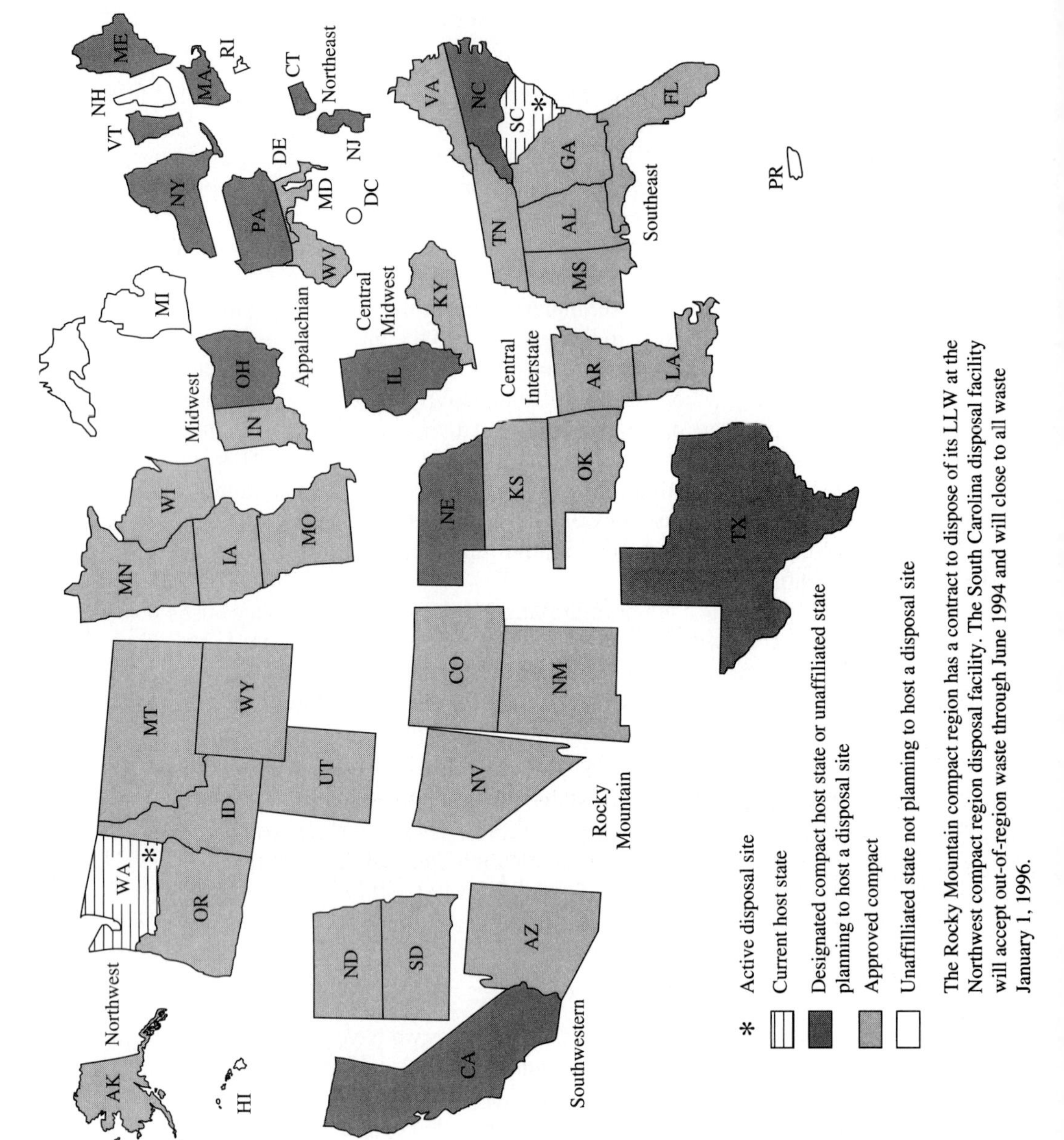

FIGURE 10-13
Low-level radioactive waste compacts (December 1995).

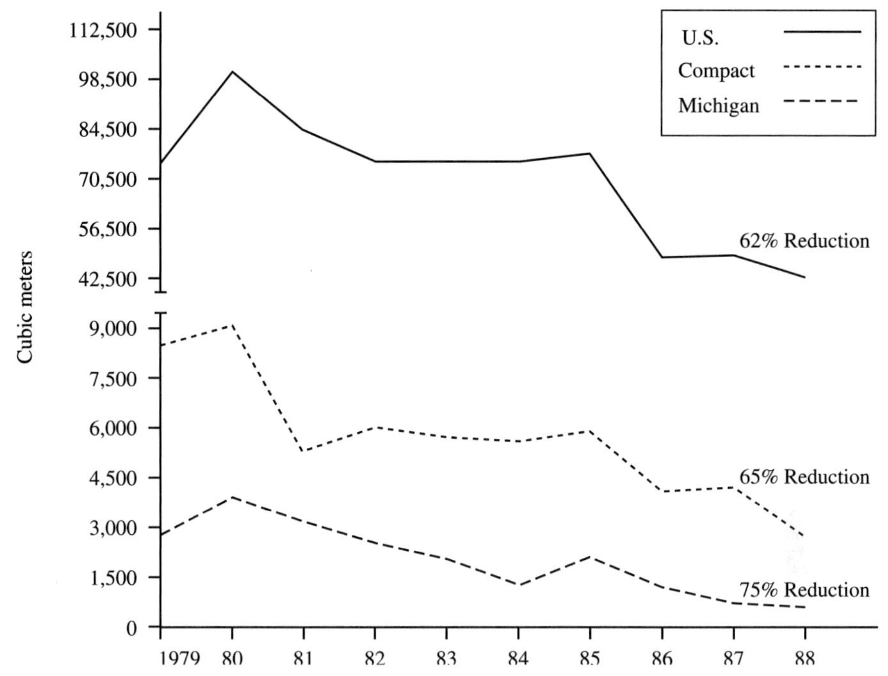

FIGURE 10-14
Reduction in low-level radioactive waste volume. (*Source:* U.S. Department of Energy Figures, 1988.)

uncontaminated waste unless it is done at the point of origin. Training plant personnel to do this work at the point of origin has been reasonably successful. To ask radioactive waste management personnel to do the sorting of an unknown mixture of wastes at a subsequent time and place creates an unacceptable hazard of exposure to radiation by inhalation, injury, or ambient external exposure.

Often material only suspected of being radioactively contaminated is labeled and disposed of as such without necessarily actually being radioactive. Much of the so-called "radioactive" waste fits into such a category merely because of the place where it was generated. The cost of assaying such suspected low-level solid wastes to determine their true radioactive content is such that it is often cheaper to combine suspected waste with known radioactive waste than to separate it. This suspicious but not always radioactive waste takes up burial space unnecessarily. Time, effort, and money are needlessly expended in putting these nonradioactive wastes in the special radioactive waste landfills.

It has been a general practice to assume that all waste is radioactive if it has been generated in a laboratory using radioactive materials or by a radiochemical or similar processing activity. It is termed "Radiation Zone" or "Contaminated Area" waste. Thus, waste that is suitable for disposal in a municipal landfill is mixed with contaminated waste. The burden of proof that the waste is not radioactive is on the person certifying or releasing the waste. Testing of the waste is time-consuming and is often omitted.

Probably the method most likely to succeed in reducing the amount of nonradioactive waste is a careful delineation and reduction of the so-called "Radiation Zone" and "Contamination Areas." It is now common to define such areas rather broadly and to include certain zones and areas from which it should be obvious that the waste would not be radioactive. An example would be the office and administrative areas within a Radiation Zone. Such areas produce much nonradioactive waste that is often included for convenience in the low-level solid radioactive waste from the technical areas. In laboratory situations where nonradioactive wastes are generated alongside the radioactive waste, point source segregation can result in minimal radioactive waste generation.

Separation of combustible or compactible waste at the point of origin both improves waste handling and reduces volume. By sorting, wastes that are not compatible for incineration do not have to be handled at the incinerator. Because the volume reduction in an incinerator is greater than that in a compactor, the more wastes that are capable of incineration that reach the incinerator rather than the compactor, the greater the volume reduction.

Volume reduction by compression. Compression of solid low-level radioactive waste is suitable for about half the waste generated. There are three kinds of compression devices: compactors, balers, and baggers.

Compactors force material into the final storage, shipping, or disposal container. A favorite container is the 0.210 m^3 drum. Some space saving is possible. A variant of the compactor is called the packer. In this device, the material is compressed into a reusable container. At the burial grounds, the compacted material is dumped directly without any effort to retain its compacted form. Space saving is minimal with packer systems.

Balers compress the waste into bales that are wrapped, tied, or banded and then stored, shipped, or disposed of in burial grounds. Considerable space saving is possible with balers.

Baggers compress waste into a predetermined shape that is injected into round or rectangular bags, boxes, or drums before storage, shipment, or disposal. Some space saving is possible with this method of compaction.

These three techniques may be suited to general and sometimes even to unique situations. Unfortunately, such treatment does not reduce the possibility of burning while in storage, and only certain materials are suitable for compaction. These include paper, cloth, rubber, plastics, wood, glass, and small light metal objects. Large, rigid metal objects must be excluded because they are usually relatively incompressible and can damage the container and compressing machinery. Moisture (free or absorbed in large quantities by blotting paper or rags) has to be avoided because of its potential forcible release under high pressure, creating a great hazard to operators. Obviously, corrosive, pyrophoric, and explosive waste must be excluded from such processing, whether it is organic or inorganic.

The compression machinery must be economical, reliable, and easy to operate. Many commercial devices are available, but all must be modified by providing air containment, off-gas ventilation, often filtration, and, if necessary, shielding.

Volume reduction by incineration. Reduction of volumes of solid radioactive waste by incineration has interested managers of low-level radioactive waste, particularly in those parts of the world where land area is at a premium and costs are high. Under these conditions, the advantages of volume reduction are so great that the drawbacks seem only obstacles to be surmounted. In Europe, where land is scarce and more revered, the incineration of solid combustible radioactive waste is a common and apparently satisfactory method of pretreatment before final disposal.

There are certain advantages, such as volume reductions of 80 to 90 percent, reported for selected burnable waste. This may be a high estimate if such factors as residues from off-gas treatment and refractory changes are considered. This would represent a considerable saving in land used for burial, in transportation, and in long-term monitoring. In addition, it would free us from the nagging worry about the possible problem of long-burning subterranean fires. Special attention should be given to the problems of burning organic matter (solvents, ion-exchange resins, etc.) and putrescible biological material (animal cadavers, excreta, etc.). Incineration of radioactive waste must be carried out under controlled conditions to prevent the formation of radioactive aerosols and must comply with both RCRA and NRC rules if the wastes are RCRA wastes as well as being radioactive.

Long-Term Management and Containment

Site selection. One concern in the burial of radioactive waste is that groundwater or infiltrating surface water will leach the waste and mobilize the radioactive materials. The radionuclides would be carried by this water back to the surface as a part of natural groundwater discharge or through a water well. Because of this concern, hydrogeologic and hydrochemical considerations in site selection become paramount.

The types of hydrogeologic and hydrochemical data that may be needed to determine whether or not a site is adequate include:[17]

1. Depth to water table, including perched water tables, if present;
2. Distance to nearest points of groundwater, spring water, or surface water usage (including well and spring inventory, and, particularly, wells available to the public);
3. Ratio of pan evaporation to precipitation minus runoff (by month for a period of at least two years);
4. Water table contour map;
5. Magnitude of annual water table fluctuation;
6. Stratigraphy and structure to base of shallowest confined aquifer;
7. Baseflow data on perennial streams traversing or adjacent to storage site;

[17] S. S. Papadopulos and I. J. Winograd, *Storage of Low-Level Radioactive Wastes in the Ground: Hydrogeological and Hydrochemical Factors* (U.S. Environmental Protection Agency Report 520/3-74-009), Washington, DC: U.S. Government Printing Office, 1974.

8. Chemistry of water in aquifers and aquitards and of leachate from the waste trenches;
9. Laboratory measurements of hydraulic conductivity, effective porosity, and mineralogy of core and grab samples (from trenches) of each lithology in unsaturated and saturated (to base of shallowest confined aquifer) zone—hydraulic conductivity to be measured at different water contents and tensions;
10. Neutron moisture meter measurements of moisture content of unsaturated zone—measurements to be made in specially constructed holes (at least two years' record needed);
11. In situ measurements of soil moisture tension in upper 4.5 to 9 m of unsaturated zone (at least two years' record needed);
12. Three-dimensional distribution of head in all saturated hydrostratigraphic units to base of shallowest confined aquifer;
13. Pumping, bailing, or slug tests to determine transmissivity and storage coefficients;
14. Definition of recharge and discharge areas for unconfined and shallowest confined aquifers;
15. Field measurements of dispersivity coefficients;
16. Laboratory and field determination of the distribution coefficient for movement of critical nuclides through all hydrostratigraphic units; and
17. Rates of denudation and/or slope retreat.

This data is necessary for a complete definition of flow and nuclide transport through both the unsaturated and saturated zones.

It is not possible to immobilize a radioactive contaminant in a burial site for long periods of geologic time (that is, for millions of years) with complete certainty. However, there appear to be hydrogeologic environments in which these contaminants can be kept below the surface and away from people until they have decayed to acceptable levels.

The problem is not merely a matter of ensuring optimum confinement, but also one of ensuring confinement for a minimum but specified time period and/or describing and predicting the performance of these radioactive contaminants in the subsurface until this time period has elapsed. For this reason, burial sites having complex hydrogeology in which such predictions are difficult or impossible are probably not suitable for radioactive waste.

From a geological standpoint, there appear to be two basic approaches to the long-term control of buried radioactive waste. The simplest approach is to prevent water from reaching the waste and thereby to eliminate the possibility of contaminants in the waste being mobilized. In arid climates, where there is little or no infiltration, this appears to be feasible.

In humid climates, where there is infiltration, some sort of engineered container or facilities that would isolate the waste from the water for hundreds of years is necessary. Whether or not such a facility can be designed, constructed, and demonstrated remains to be seen.

The second approach to long-term control involves burying the waste in a hydrogeologic environment that can be demonstrated to be safe despite the fact that radioactive contaminants can and will be mobilized. Demonstrating that such sites are, in fact, safe requires a quantitative evaluation of the factors influencing contaminant movement. Such an evaluation may be quite difficult but appears to be our only option if we wish to bury radioactive waste in humid climates or in climates where infiltration is capable of mobilizing or leaching the buried waste.

It is also important to give attention to the possible biological and microbiological environment of a burial site. Soil microorganisms, earthworms, larger burrowing animals, and the deep taproots of plants seeking water and nourishment (particularly in desert areas) can all be factors in moving components of waste out of a burial place into the biosphere. Some organisms can release organic compounds into the soil that can serve as complexing agents to mobilize otherwise insoluble contaminants. Some organisms can concentrate radionuclides by surprisingly high factors from their environment and so can change both the biochemical availability and the distribution of a radionuclide.

Site selection criteria. Michigan's site selection criteria serve to illustrate the factors that need to be considered in selecting disposal sites.

The first objective is to avoid population centers and conflicts with human activities. Michigan established an isolation distance of 1 km and required that projected population growth must not infringe to the extent that it would interfere with health and safety performance objectives or environmental monitoring.

Areas within 1.6 km of a fault where tectonic movement has occurred within the last 10,000 years are excluded as candidate sites. Likewise excluded are areas where significant earthquake intensity has been measured and flood plains exist. Mass wasting, erosion, and similar geologic processes are to be evaluated for possible damage to the facility.

Areas where groundwater flows from sites more than 30 m in 100 years or where groundwater could reach an aquifer in less than 500 years are excluded. The criteria also exclude areas over sole source aquifers and areas where groundwater discharges to the surface within 1 km. The facility may not be built within 16 km of the Great Lakes.

The criteria specify that the safest transportation net will be used. Highways with low accident rates located away from population centers are favored.

The site must have no complex meteorological characteristics and must avoid resource development conflicts. Likewise, environmentally sensitive areas such as wetlands and shorelands must be avoided. Areas that have formally proposed or approved development plans as of January 1, 1988, are excluded.

These criteria are extremely rigorous. Because of these constraints and the more serious problem of public opposition, no new sites have been finalized in the United States. Some compacts have had severe problems and conflicts that have resulted in the expulsion of one of the states. For example, after being selected as the host state, Michigan failed to identify an acceptable site and was expelled from the Midwest Compact.

A few compacts are proceeding quite well. These compacts are involving the public, community officials, regulators, and generators in joint efforts to identify sites, complete licensing applications, secure contractors, and construct the site. The most successful approach appears to be one of identification of actual candidate sites followed by a volunteer applicant.

The two currently operating sites in the United States are at Hanford, Washington, and Barnwell, South Carolina. These sites are accepting low-level radioactive waste from across the United States. There is a tremendous financial advantage to them in doing so. In November, 1995, the total cost of disposing of a 0.210 m^3 drum was about \$3,000. Many generators have been storing this waste for several years (for example, in Michigan, 55 generators have been storing waste for 5 years) and they are willing to pay these prices because of the lack of space.

Engineered containment structures. As mentioned above, in humid climates, land burial is not acceptable. Michigan, as an example, has passed a statute that prohibits it. The alternative is an engineered structure. Engineered structures for the containment of waste must be designed with the intent to keep water, which can mobilize the contaminants, out of the facility. The Michigan statute specifies that each technology considered fulfills three requirements:

1. Maximum containment until the waste naturally decays to nonhazardous levels;
2. Capability to identify and retrieve wastes if necessary; and
3. Comprehensive monitoring of the facility and its environment.

There are four conceptual designs being considered by the Midwest Compact that would typify the various possible approaches. They are (1) aboveground vault, (2) below-ground vault, (3) aboveground modular concrete canister, and (4) below-ground concrete canister.

An aboveground vault is a large, reinforced concrete structure with access through the top or side walls for placing the waste inside (Figure 10-15). When a cell is filled, the vault will be sealed with a roof of concrete or some other suitable material. It must be designed to withstand earthquakes, tornadoes, floods, and fire.

The below-ground vault is similar to the aboveground vault except that it is located below ground (Figure 10-16). A compacted clay cover serves as part of the seal.

The aboveground concrete canister method consists of placing the low-level radioactive waste in large, precast concrete containers that are then stacked in an engineered structure (Figure 10-17).

The below-ground concrete canister method follows the same principles as the aboveground system, but the canisters are placed in a vault below ground (Figure 10-18).

Monitoring systems. A monitoring system must operate both at permanent burial sites and at storage sites so that surface or air contamination will be detected quickly. Ground and surface water beneath or very near to the burial facilities should be

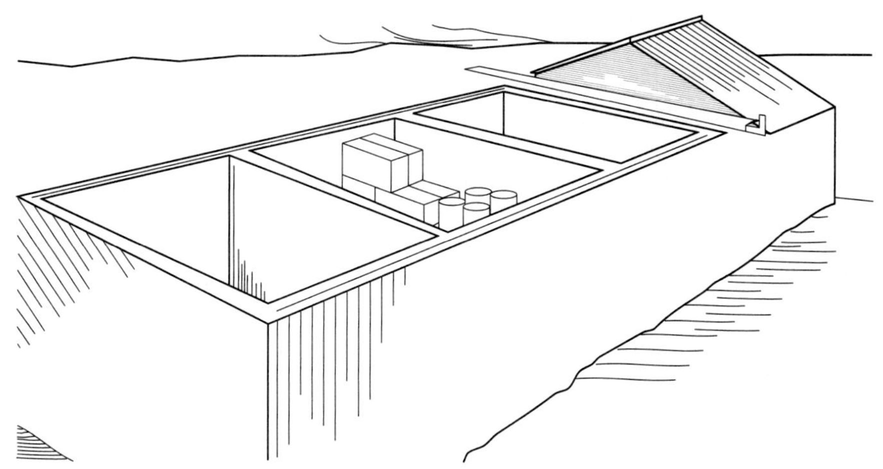

FIGURE 10-15
Aboveground vault for low-level radioactive waste disposal/storage. (*Source:* Midwest Compact.)

FIGURE 10-16
Below-ground vault for low-level radioactive waste disposal/storage. (*Source:* Midwest Compact.)

monitored sufficiently often to give the earliest practical warning of failure of any facilities. "Failure" is defined as significant contamination of the ground or surface water in excess of standards that have been set for the disposal site.

Early detection of contamination is most important. Unlike surface water, groundwater usually moves slowly, and if contaminants move unexpectedly, we must know about it before significant amounts have left the disposal site. Interception of the contaminants is not likely to be simple or prompt if this has not been considered in selection of the site or the design of facilities.

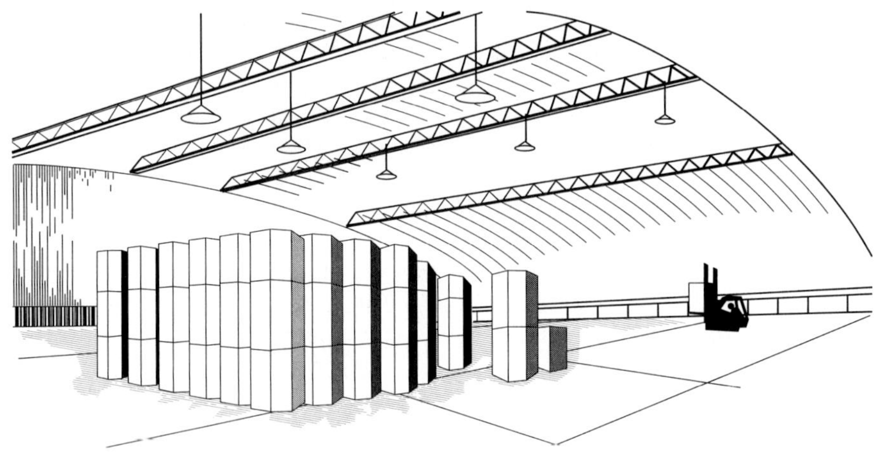

FIGURE 10-17
Aboveground canister storage for low-level radioactive waste. (*Source:* Midwest Compact.)

FIGURE 10-18
Below-ground canister storage for low-level radioactive waste. (*Source:* Midwest Compact.)

Should it be necessary to take remedial measures to eliminate further discharges, the smaller the amount of waste involved, the simpler these measures are likely to be. Early detection of contaminants generally requires that monitoring points be placed as close as possible to the waste.

Air monitoring should be provided around the site. Likewise, monitoring should also include adequate biological and ecological sampling to detect entrance of radionuclides into the local biosphere.

Contingency plan. Contingency plans must be made to cover all foreseeable accidents or failures. They must include plans for corrective action in the event that monitoring shows a hazardous spread of contamination. These plans should include natural disaster precautions as well as more chronic types of failures.

Records management. Duplicate records of the types, quantities, and concentrations of radioactive waste nuclides delivered to a burial site must be made and filed with more than one record bank. Reports on monitoring results and significant incidents, such as spills or unanticipated release of waste, must be filed with more than one record bank. These records should show the real (that is, observed, not calculated) level of contamination of the environment (including the ground area). These records must be in such a form that they will be useful and available for the effective length of time that the waste burial facility will require human attention.

Nonexhumation of radioactive wastes. Exhumation of waste originally buried without any intent of later retrieval is potentially a very hazardous operation. The National Academy of Science recommends that exhumation not be made unless there is a credible reason to believe that a significant radiation hazard could arise from leaving the waste where it is and that the wastes can be exhumed safely.[18] As a corollary to this recommendation, radioactive waste should not be exhumed and put into temporary engineered storage where the material must await a final decision on permanent disposal. Experience has shown that "temporary" storage may in reality be permanent storage because of the political realities in being able to relocate it.

10-7 CHAPTER REVIEW

When you have completed studying this chapter, you should be able to do the following without the aid of your textbook or notes:

1. Explain what an isotope is.
2. Explain why some isotopes are radioactive and others are not.
3. Define/explain how alpha, beta, x-ray, and gamma ray emissions occur and how they differ.
4. Define the unit "curie."
5. Explain the process of fission in a nuclear reactor.
6. Explain how x-rays are produced in an x-ray machine.
7. Define the concept of radiation dose and the units of roentgen, rad, and rem.
8. Explain the concepts of RBE and W_T.
9. List the pattern of biological effects of radiation.
10. Discuss the determinants of biological effects.
11. Discuss the difference between acute and delayed biological effects of radiation.

---

[18]National Research Council, *The Shallow Land Burial of Low-Level Radioactively Contaminated Solid Waste,* Washington, DC: National Academy of Science, p. 14, 1976.

12. List three possible delayed effects of radiation exposure.
13. State the acceptable occupational and nonoccupational dose of radiation as established by the NRC.
14. Explain the difference between internal and external radiation hazard.
15. Select a material and its thickness to protect against alpha or beta radiation.
16. Describe the sources of background radiation.
17. Explain why radon is a hazard and the mechanism by which the hazard is realized.
18. List three fundamental methods of reducing external radiation hazard.
19. Explain how to reduce occupational exposure to internal radiation hazards.
20. Describe how radon enters a house and give some techniques that may be used to inhibit radon entry.
21. List and describe the three types of radioactive waste (IILW, transuranic, and LLW).
22. Describe how each type of radioactive waste is to be disposed of.
23. Discuss waste minimization practice in reducing the volume of LLW.
24. Describe two types of engineered containment structure for LLW.

With the aid of this text, you should be able to do the following:

1. Determine what particles are emitted in a given decay chain.
2. Determine the activity of a radioisotope given the original activity and the time interval.
3. Determine the activity resulting from the growth of a daughter product from a parent radionuclide.
4. Determine the time to achieve maximum activity of a daughter product.
5. Apply the inverse square law to determine radiation intensity.
6. Determine whether or not a combination of radionuclides exceeds the permissible concentrations.
7. Calculate the radiation intensity behind a shielding material or the desired thickness of a shielding material to achieve a reduction of radiation intensity.

10-8 PROBLEMS

10-1. What are the elements ${}^{40}_{18}X$ and ${}^{14}_{7}X$?

10-2. What are the elements ${}^{8}_{4}X$ and ${}^{238}_{92}X$?

10-3. What particles are emitted in each step in the decay chain represented by

$${}^{226}_{88}Ra \longrightarrow {}^{222}_{86}Rn \longrightarrow {}^{218}_{84}Po \longrightarrow {}^{214}_{82}Pb\ ?$$

10-4. What particles are emitted in each step in the decay chain represented by

$${}^{214}_{82}Pb \longrightarrow {}^{214}_{83}Bi \longrightarrow {}^{214}_{84}Po \longrightarrow {}^{210}_{82}Pb \longrightarrow {}^{210}_{83}Bi \longrightarrow {}^{210}_{84}Po \longrightarrow {}^{206}_{82}Pb\ ?$$

10-5. Show that if a positron and electron are annihilated, then an energy of 1.02 MeV is released.

10-6. A laboratory solution containing 0.5 μCi/L of ^{32}P is to be disposed of. How long must the radioisotope be held to meet the allowable discharge activity?

10-7. If, in August 1911, Mme. Curie prepared an international standard containing 20.00 mg of $RaCl_2$, what will the radium content of this standard be in August 2000?

10-8. An accident has contaminated a laboratory with ^{45}Ca. The radiation level is ten times the tolerance level. How long must the room be isolated before the tolerance level is reached?

10-9. What is the mass of a 50 μCi sample of pure ^{131}I?

10-10. By emitting an alpha particle, ^{210}Po decays to ^{206}Pb. If the half-life of ^{210}Po is 138.4 d, what volume of ^{4}He will be produced in one year from 50 Ci of ^{210}Po? Assume the gas is at standard temperature and pressure.

10-11. Calculate and plot the growth curve of ^{222}Rn from an initially pure sample of ^{226}Ra. Assume no ^{222}Rn is present initially.

10-12. When an x-ray unit is operated at 70 kVp and 5 mA, it produces an intensity of D R/min at 1.0 m from the source. What intensity will it produce 2.0 m from the source?

10-13. If the source of x-rays in Problem 10-12 is operated at 15 mA, what intensity will be produced 2.0 m from the source?

10-14. What thickness (in cm) of lead is required to shield a ^{60}Co source so that the transmission is reduced 99.6 percent?

10-15. What is the equivalent thickness (in cm) of concrete to accomplish the same attenuation as the lead in Problem 10-14?

10-16. Determine the proportionality constant (u) for lead when it is used to shield ^{137}Cs.

10-17. Determine the proportionality constant (u) for iron when it is used to shield ^{137}Cs.

10-9 DISCUSSION QUESTIONS

1. Explain why an archaelogical artifact such as wood or bone may be dated by measuring its concentration of carbon-14.

2. Would you expect the tissue weighting factor (W_T) for x-rays to the big toe to be greater than, less than, or the same as that for radioiodine to the thyroid? Explain your choice.

3. What kind of radionuclide emitter (alpha, beta, gamma, or x-ray) is most dangerous from an internal hazard point of view? Explain why.

4. A laboratory worker has requested your advice on a shield for work she is doing with high-energy beta particles. What would you recommend?

5. You have an opportunity to purchase an older home with a basement that is serviced by a floor drain. What measures might you request to limit the migration of radon into the basement?

APPENDIX A

PROPERTIES OF AIR, WATER, AND SELECTED CHEMICALS

TABLE A-1
Physical properties of water at 1 atm

Temperature (°C)	Density, ρ (kg/m^3)	Specific weight, γ (kN/m^3)	Dynamic viscosity, μ (mPa · s)	Kinematic viscosity, ν (μm^2/s)
0	999.842	9.805	1.787	1.787
3.98	1,000.000	9.807	1.567	1.567
5	999.967	9.807	1.519	1.519
10	999.703	9.804	1.307	1.307
12	999.500	9.802	1.235	1.236
15	999.103	9.798	1.139	1.140
17	998.778	9.795	1.081	1.082
18	998.599	9.793	1.053	1.054
19	998.408	9.791	1.027	1.029
20	998.207	9.789	1.002	1.004
21	997.996	9.787	0.998	1.000
22	997.774	9.785	0.955	0.957
23	997.542	9.783	0.932	0.934
24	997.300	9.781	0.911	0.913
25	997.048	9.778	0.890	0.893
26	996.787	9.775	0.870	0.873
27	996.516	9.773	0.851	0.854
28	996.236	9.770	0.833	0.836
29	995.948	9.767	0.815	0.818
30	995.650	9.764	0.798	0.801
35	994.035	9.749	0.719	0.723
40	992.219	9.731	0.653	0.658
45	990.216	9.711	0.596	0.602
50	988.039	9.690	0.547	0.554
60	983.202	9.642	0.466	0.474
70	977.773	9.589	0.404	0.413
80	971.801	9.530	0.355	0.365
90	965.323	9.467	0.315	0.326
100	958.366	9.399	0.282	0.294

Pa · s = (mPa · s) × 10^{-3}
m^2/s = (μm^2/s) × 10^{-6}

TABLE A-2

Saturation values of dissolved oxygen in fresh water exposed to a saturated atmosphere containing 20.9% oxygen under a pressure of 101.325 kPa[a]

Temperature (°C)	Dissolved oxygen (mg/L)	Saturated vapor pressure (kPa)
0	14.62	0.6108
1	14.23	0.6566
2	13.84	0.7055
3	13.48	0.7575
4	13.13	0.8129
5	12.80	0.8719
6	12.48	0.9347
7	12.17	1.0013
8	11.87	1.0722
9	11.59	1.1474
10	11.33	1.2272
11	11.08	1.3119
12	10.83	1.4017
13	10.60	1.4969
14	10.37	1.5977
15	10.15	1.7044
16	9.95	1.8173
17	9.74	1.9367
18	9.54	2.0630
19	9.35	2.1964
20	9.17	2.3373
21	8.99	2.4861
22	8.83	2.6430
23	8.68	2.8086
24	8.53	2.9831
25	8.38	3.1671
26	8.22	3.3608
27	8.07	3.5649
28	7.92	3.7796
29	7.77	4.0055
30	7.63	4.2430
31	7.51	4.4927
32	7.42	4.7551
33	7.28	5.0307
34	7.17	5.3200
35	7.07	5.6236
36	6.96	5.9422
37	6.86	6.2762
38	6.75	6.6264

[a] For other barometric pressures, the solubilities vary approximately in proportion to the ratios of these pressures to the standard pressures.

Source: Calculated by G. C. Whipple and M. C. Whipple from measurements of C. J. J. Fox, *Journal of the American Chemical Society,* vol. 33, p. 362, 1911.

TABLE A-3
Density of dry air in kg/m^3 for °C and absolute pressure in kilopascals

°C	Pressure—kPa										
	110.0	100.0	90.0	80.0	70.0	60.0	50.0	40.0	30.0	20.0	10.0
0	1.4029	1.2754	1.1478	1.0203	0.8928	0.7652	0.6377	0.5102	0.3826	0.2551	0.1275
1	1.3978	1.2707	1.1437	1.0166	0.8895	0.7624	0.6354	0.5083	0.3812	0.2541	0.1271
2	1.3927	1.2661	1.1395	1.0129	0.8863	0.7597	0.6331	0.5064	0.3798	0.2532	0.1266
3	1.3877	1.2615	1.1354	1.0092	0.8831	0.7569	0.6308	0.5046	0.3785	0.2523	0.1262
4	1.3827	1.2570	1.1313	1.0056	0.8799	0.7542	0.6285	0.5028	0.3771	0.2514	0.1257
5	1.3777	1.2525	1.1272	1.0020	0.8767	0.7515	0.6262	0.5010	0.3757	0.2505	0.1252
6	1.3728	1.2480	1.1232	0.9984	0.8736	0.7488	0.6240	0.4992	0.3744	0.2496	0.1248
7	1.3679	1.2435	1.1192	0.9948	0.8705	0.7461	0.6218	0.4974	0.3731	0.2487	0.1244
8	1.3630	1.2391	1.1152	0.9913	0.8674	0.7435	0.6195	0.4956	0.3717	0.2478	0.1239
9	1.3582	1.2347	1.1112	0.9878	0.8643	0.7408	0.6174	0.4939	0.3704	0.2469	0.1235
10	1.3534	1.2303	1.1073	0.9843	0.8612	0.7382	0.6152	0.4921	0.3691	0.2461	0.1230
11	1.3486	1.2260	1.1034	0.9808	0.8582	0.7356	0.6130	0.4904	0.3678	0.2452	0.1226
12	1.3439	1.2217	1.0995	0.9774	0.8552	0.7330	0.6109	0.4887	0.3665	0.2443	0.1222
13	1.3392	1.2174	1.0957	0.9740	0.8522	0.7305	0.6087	0.4870	0.3652	0.2435	0.1217
14	1.3345	1.2132	1.0919	0.9706	0.8492	0.7279	0.6066	0.4853	0.3640	0.2426	0.1213
15	1.3299	1.2090	1.0881	0.9672	0.8463	0.7254	0.6045	0.4836	0.3627	0.2418	0.1209
16	1.3253	1.2048	1.0843	0.9638	0.8434	0.7229	0.6024	0.4819	0.3614	0.2410	0.1205
17	1.3207	1.2007	1.0806	0.9605	0.8405	0.7204	0.6003	0.4803	0.3602	0.2401	0.1201
18	1.3162	1.1965	1.0769	0.9572	0.8376	0.7179	0.5983	0.4786	0.3590	0.2393	0.1197
19	1.3117	1.1924	1.0732	0.9540	0.8347	0.7155	0.5962	0.4770	0.3577	0.2385	0.1192
20	1.3072	1.1884	1.0695	0.9507	0.8319	0.7130	0.5942	0.4753	0.3565	0.2377	0.1188
21	1.3028	1.1843	1.0659	0.9475	0.8290	0.7106	0.5922	0.4737	0.3553	0.2369	0.1184
22	1.2984	1.1803	1.0623	0.9443	0.8262	0.7082	0.5902	0.4721	0.3541	0.2361	0.1180
23	1.2940	1.1763	1.0587	0.9411	0.8234	0.7058	0.5882	0.4705	0.3529	0.2353	0.1176
24	1.2896	1.1724	1.0551	0.9379	0.8207	0.7034	0.5862	0.4690	0.3517	0.2345	0.1172
25	1.2853	1.1684	1.0516	0.9348	0.8179	0.7011	0.5842	0.4674	0.3505	0.2337	0.1168
26	1.2810	1.1645	1.0481	0.9316	0.8152	0.6987	0.5823	0.4658	0.3494	0.2329	0.1165
27	1.2767	1.1607	1.0446	0.9285	0.8125	0.6964	0.5803	0.4643	0.3482	0.2321	0.1161
28	1.2725	1.1568	1.0411	0.9254	0.8098	0.6941	0.5784	0.4627	0.3470	0.2314	0.1157
29	1.2683	1.1530	1.0377	0.9224	0.8071	0.6918	0.5765	0.4612	0.3459	0.2306	0.1153

(continued)

TABLE A-3
(continued)

°C	Pressure—kPa										
	110.0	**100.0**	**90.0**	**80.0**	**70.0**	**60.0**	**50.0**	**40.0**	**30.0**	**20.0**	**10.0**
30	1.2641	1.1492	1.0343	0.9193	0.8044	0.6895	0.5746	0.4597	0.3448	0.2298	0.1149
31	1.2599	1.1454	1.0309	0.9163	0.8018	0.6872	0.5727	0.4582	0.3436	0.2291	0.1145
32	1.2558	1.1416	1.0275	0.9133	0.7991	0.6850	0.5708	0.4567	0.3425	0.2283	0.1142
33	1.2517	1.1379	1.0241	0.9103	0.7965	0.6827	0.5690	0.4552	0.3414	0.2276	0.1138
34	1.2476	1.1342	1.0208	0.9074	0.7939	0.6805	0.5671	0.4537	0.3403	0.2268	0.1134
35	1.2436	1.1305	1.0175	0.9044	0.7914	0.6783	0.5653	0.4522	0.3392	0.2261	0.1131
36	1.2396	1.1269	1.0142	0.9015	0.7888	0.6761	0.5634	0.4507	0.3381	0.2254	0.1127
37	1.2356	1.1232	1.0109	0.8986	0.7863	0.6739	0.5616	0.4493	0.3370	0.2246	0.1123
38	1.2316	1.1196	1.0077	0.8957	0.7837	0.6718	0.5598	0.4479	0.3359	0.2239	0.1120
39	1.2276	1.1160	1.0044	0.8928	0.7812	0.6696	0.5580	0.4464	0.3348	0.2232	0.1116
40	1.2237	1.1125	1.0012	0.8900	0.7787	0.6675	0.5562	0.4450	0.3337	0.2225	0.1112
41	1.2198	1.1089	0.9980	0.8872	0.7763	0.6654	0.5545	0.4436	0.3327	0.2218	0.1109
42	1.2160	1.1054	0.9949	0.8843	0.7738	0.6633	0.5527	0.4422	0.3316	0.2211	0.1105
43	1.2121	1.1019	0.9917	0.8815	0.7713	0.6612	0.5510	0.4408	0.3306	0.2204	0.1102
44	1.2083	1.0984	0.9886	0.8788	0.7689	0.6591	0.5492	0.4394	0.3295	0.2197	0.1098
45	1.2045	1.0950	0.9855	0.8760	0.7665	0.6570	0.5475	0.4380	0.3285	0.2190	0.1095
46	1.2007	1.0916	0.9824	0.8732	0.7641	0.6549	0.5458	0.4366	0.3275	0.2183	0.1092
47	1.1970	1.0882	0.9793	0.8705	0.7617	0.6529	0.5441	0.4353	0.3264	0.2176	0.1088
48	1.1932	1.0848	0.9763	0.8678	0.7593	0.6509	0.5424	0.4339	0.3254	0.2170	0.1085
49	1.1895	1.0814	0.9733	0.8651	0.7570	0.6488	0.5407	0.4326	0.3244	0.2163	0.1081
50	1.1859	1.0780	0.9702	0.8624	0.7546	0.6468	0.5390	0.4312	0.3234	0.2156	0.1078
51	1.1822	1.0747	0.9673	0.8598	0.7523	0.6448	0.5374	0.4299	0.3224	0.2149	0.1075
52	1.1786	1.0714	0.9643	0.8571	0.7500	0.6429	0.5357	0.4286	0.3214	0.2143	0.1071
53	1.1750	1.0681	0.9613	0.8545	0.7477	0.6409	0.5341	0.4273	0.3204	0.2136	0.1068
54	1.1714	1.0649	0.9584	0.8519	0.7454	0.6389	0.5324	0.4259	0.3195	0.2130	0.1065
55	1.1678	1.0616	0.9555	0.8493	0.7431	0.6370	0.5308	0.4247	0.3185	0.2123	0.1062
56	1.1642	1.0584	0.9526	0.8467	0.7409	0.6350	0.5292	0.4234	0.3175	0.2117	0.1058
57	1.1607	1.0552	0.9497	0.8442	0.7386	0.6331	0.5276	0.4221	0.3166	0.2110	0.1055
58	1.1572	1.0520	0.9468	0.8416	0.7364	0.6312	0.5260	0.4208	0.3156	0.2104	0.1052
59	1.1537	1.0488	0.9440	0.8391	0.7342	0.6293	0.5244	0.4195	0.3147	0.2098	0.1049

(continued)

TABLE A-3
(continued)

	Pressure—kPa										
°C	110.0	100.0	90.0	80.0	70.0	60.0	50.0	40.0	30.0	20.0	10.0
60	1.1503	1.0457	0.9411	0.8366	0.7320	0.6274	0.5228	0.4183	0.3137	0.2091	0.1046
61	1.1468	1.0426	0.9383	0.8340	0.7298	0.6255	0.5213	0.4170	0.3128	0.2085	0.1043
62	1.1434	1.0394	0.9355	0.8316	0.7276	0.6237	0.5197	0.4158	0.3118	0.2079	0.1039
63	1.1400	1.0364	0.9327	0.8291	0.7255	0.6218	0.5182	0.4145	0.3109	0.2073	0.1036
64	1.1366	1.0333	0.9300	0.8266	0.7233	0.6200	0.5166	0.4133	0.3100	0.2067	0.1033
65	1.1333	1.0302	0.9272	0.8242	0.7212	0.6181	0.5151	0.4121	0.3091	0.2060	0.1030
66	1.1299	1.0272	0.9245	0.8218	0.7190	0.6163	0.5136	0.4109	0.3082	0.2054	0.1027
67	1.1266	1.0242	0.9218	0.8193	0.7169	0.6145	0.5121	0.4097	0.3073	0.2048	0.1024
68	1.1233	1.0212	0.9191	0.8169	0.7148	0.6127	0.5106	0.4085	0.3064	0.2042	0.1021
69	1.1200	1.0182	0.9164	0.8146	0.7127	0.6109	0.5091	0.4073	0.3055	0.2036	0.1018
70	1.1167	1.0152	0.9137	0.8122	0.7107	0.6091	0.5076	0.4061	0.3046	0.2030	0.1015
71	1.1135	1.0123	0.9110	0.8098	0.7086	0.6074	0.5061	0.4049	0.3037	0.2025	0.1012
72	1.1103	1.0093	0.9084	0.8075	0.7065	0.6056	0.5047	0.4037	0.3028	0.2019	0.1009
73	1.1071	1.0064	0.9058	0.8051	0.7045	0.6039	0.5032	0.4026	0.3019	0.2013	0.1006
74	1.1039	1.0035	0.9032	0.8028	0.7025	0.6021	0.5018	0.4014	0.3011	0.2007	0.1004
75	1.1007	1.0006	0.9006	0.8005	0.7004	0.6004	0.5003	0.4003	0.3002	0.2001	0.1001
76	1.0976	0.9978	0.8980	0.7982	0.6984	0.5987	0.4989	0.3991	0.2993	0.1996	0.0998
77	1.0944	0.9949	0.8954	0.7959	0.6965	0.5970	0.4975	0.3980	0.2985	0.1990	0.0995
78	1.0913	0.9921	0.8929	0.7937	0.6945	0.5953	0.4960	0.3968	0.2976	0.1984	0.0992
79	1.0882	0.9893	0.8904	0.7914	0.6925	0.5936	0.4946	0.3957	0.2968	0.1979	0.0989
80	1.0851	0.9865	0.8878	0.7892	0.6905	0.5919	0.4932	0.3946	0.2959	0.1973	0.0986
81	1.0821	0.9837	0.8853	0.7870	0.6886	0.5902	0.4918	0.3935	0.2951	0.1967	0.0984
82	1.0790	0.9809	0.8828	0.7847	0.6866	0.5886	0.4905	0.3924	0.2943	0.1962	0.0981
83	1.0760	0.9782	0.8804	0.7825	0.6847	0.5869	0.4891	0.3913	0.2935	0.1956	0.0978
84	1.0730	0.9754	0.8779	0.7803	0.6828	0.5853	0.4877	0.3902	0.2926	0.1951	0.0975
85	1.0700	0.9727	0.8754	0.7782	0.6809	0.5836	0.4864	0.3891	0.2918	0.1945	0.0973
86	1.0670	0.9700	0.8730	0.7760	0.6790	0.5820	0.4850	0.3880	0.2910	0.1940	0.0970
87	1.0640	0.9673	0.8706	0.7738	0.6771	0.5804	0.4836	0.3869	0.2902	0.1935	0.0967
88	1.0611	0.9646	0.8682	0.7717	0.6752	0.5788	0.4823	0.3858	0.2894	0.1929	0.0965
89	1.0582	0.9620	0.8658	0.7696	0.6734	0.5772	0.4810	0.3848	0.2886	0.1924	0.0962

(continued)

TABLE A-3
(continued)

	Pressure—kPa										
°C	110.0	100.0	90.0	80.0	70.0	60.0	50.0	40.0	30.0	20.0	10.0
90	1.0552	0.9593	0.8634	0.7674	0.6715	0.5756	0.4797	0.3837	0.2878	0.1919	0.0959
91	1.0523	0.9567	0.8610	0.7653	0.6697	0.5740	0.4783	0.3827	0.2870	0.1913	0.0957
92	1.0495	0.9541	0.8587	0.7632	0.6678	0.5724	0.4770	0.3816	0.2862	0.1908	0.0954
93	1.0466	0.9514	0.8563	0.7612	0.6660	0.5709	0.4757	0.3806	0.2854	0.1903	0.0951
94	1.0437	0.9489	0.8540	0.7591	0.6642	0.5693	0.4744	0.3795	0.2847	0.1898	0.0949
95	1.0409	0.9463	0.8517	0.7570	0.6624	0.5678	0.4731	0.3785	0.2839	0.1893	0.0946
96	1.0381	0.9437	0.8493	0.7550	0.6606	0.5662	0.4719	0.3775	0.2831	0.1887	0.0944
97	1.0353	0.9412	0.8471	0.7529	0.6588	0.5647	0.4706	0.3765	0.2824	0.1882	0.0941
98	1.0325	0.9386	0.8448	0.7509	0.6570	0.5632	0.4693	0.3755	0.2816	0.1877	0.0939
99	1.0297	0.9361	0.8425	0.7489	0.6553	0.5617	0.4681	0.3744	0.2808	0.1872	0.0936
100	1.0270	0.9336	0.8402	0.7469	0.6535	0.5602	0.4668	0.3734	0.2801	0.1867	0.0934

Source: Robert J. List, *Smithsonian Meteorological Tables,* 6th Revised Edition (Publication 4014 of the Smithsonian Institution), 1951.

TABLE A-4

Viscosity of dry air at approximately 100 kPa[a]

Temperature (°C)	Dynamic viscosity (μPa · s)
0	17.1
5	17.4
10	17.7
15	17.9
20	18.2
25	18.5
30	18.7
35	19.0
40	19.3
45	19.5
50	19.8
55	20.1
60	20.3
65	20.6
70	20.9
75	21.1
80	21.4
85	21.7
90	21.9
95	22.2
100	22.5
150	25.2

$\mu = 17.11 + 0.0536\ T + (P/8280)$ where T is in °C and P is in kPa.

TABLE A-5

Properties of air at standard conditions[a]

Molecular weight	M	28.97
Gas constant	R	287 J/kg · K
Specific heat at constant pressure	c_p	1,005 J/kg · K
Specific heat at constant volume	c_v	718 J/kg · K
Density	ρ	1.185 kg/m^3
Dynamic viscosity	μ	1.8515×10^{-5} Pa · s
Kinematic viscosity	ν	1.5624×10^{-5} m^2/s
Thermal conductivity	k	0.0257 W/m · K
Ratio of specific heats, c_p/c_v	k	1.3997
Prandtl number	Pr	0.720

[a] Measured at 101.325 kPa pressure and 298 K temperature.

TABLE A-6
Properties of saturated water at 298 K

Molecular weight	M	18.02
Gas constant	R	461.4 J/kg · K
Specific heat	c	4,181 J/kg · K
Prandtl number	Pr	6.395
Thermal conductivity	k	0.604 W/m · K

TABLE A-7
Frequently used constants

Standard atmospheric pressure	P_{atm}	101.325 kPa
Standard gravitational acceleration	g	9.8067 m/s^2
Universal gas constant	R_u	8,314.3 J/kg · mol · K
Electrical permittivity constant	ϵ_0	8.85×10^{-12} C/V · m
Electron charge	q_e	1.60×10^{-19} C
Boltzmann's constant	k	1.38×10^{-23} J/K

TABLE A-8
Properties of selected organic compounds

Name	Formula	M.W.	Density, g/mL	Vapor pressure, mm Hg	Henry's law constant kPa · m³/mol
Acetone	CH_3COCH_3	58.08	0.79	184	0.01
Benzene	C_6H_6	78.11	0.879	95	0.6
Bromodichloromethane	$CHBrCl_2$	163.8	1.971		0.2
Bromoform	$CHBr_3$	252.75	2.8899	5	0.06
Bromomethane	CH_3Br	94.94	1.6755	1300	0.5
Carbon tetrachloride	CCl_4	153.82	1.594	90	3
Chlorobenzene	C_6H_5Cl	112.56	1.107	12	0.4
Chlorodibromomethane	$CHBr_2Cl$	208.29	2.451	50	0.09
Chloroethane	C_2H_5Cl	64.52	0.8978	700	0.2
Chloroethylene	C_2H_3Cl	62.5	0.912	2550	4
Chloroform	$CHCl_3$	119.39	1.4892	190	0.4
Chloromethane	CH_3Cl	50.49	0.9159	3750	1.0
1,2-Dibromoethane	$C_2H_2Br_2$	187.87	2.18	10	0.06
1,2-Dichlorobenzene	$1,2\text{-}Cl_2\text{-}C_6H_4$	147.01	1.3048	1.5	0.2
1,3-Dichlorobenzene	$1,3\text{-}Cl_2\text{-}C_6H_4$	147.01	1.2884	2	0.4
1,4-Dichlorobenzene	$1,4\text{-}Cl_2\text{-}C_6H_4$	147.01	1.2475	0.7	0.2
1,1-Dichloroethylene	$CH_2{=}CCl_2$	96.94	1.218	500	15
1,2-Dichloroethane	$ClCH_2CH_2Cl$	98.96	1.2351	700	0.1
1,1-Dichloroethane	CH_3CHCl_2	98.96	1.1757	200	0.6
Trans-1,2-Dichloroethylene	$CHCl{=}CHCl$	96.94	1.2565	300	0.6
Dichloromethane	CH_2Cl_2	84.93	1.327	350	0.3
1,2-Dichloropropane	$CH_3CHClCH_2Cl$	112.99	1.1560	50	0.4
Cis-1,3-Dichloropropylene	$ClCH_2CH{=}CHCl$	110.97	1.217	40	0.2
Ethyl benzene	$C_6H_5CH_2CH_3$	106.17	0.8670	9	0.8
Formaldehyde	$HCHO$	30.05	0.815		
Hexachlorobenzene	C_6Cl_6	284.79	1.5691		
Pentachlorophenol	Cl_5C_6OH	266.34	1.978		
Phenol	C_6H_5OH	94.11	1.0576		
1,1,2,2-Tetrachloroethane	$CHCl_2CHCl_2$	167.85	1.5953	5	0.05
Tetrachloroethylene	$CL_2C{=}CCl_2$	165.83	1.6227	15	3
Toluene	$C_6H_5CH_3$	92.14	0.8669	28	0.7
1,1,1-Trichloroethane	CH_3CCl_3	133.41	1.3390	100	3.0
1,1,2-Trichloroethane	$CH_2ClCHCl_2$	133.41	1.4397	25	0.1
Trichloroethylene	$ClCH{=}CCl_2$	131.29	1.476	50	0.9
Vinyl chloride	$CH_2{=}CHCl$	62.50	0.9106	2200	50
o-Xylene	$1,2\text{-}(CH_3)_2C_6H_4$	106.17	0.8802	6	0.5
m-Xylene	$1,3\text{-}(CH_3)_2C_6H_4$	106.17	0.8642	8	0.7
p-Xylene	$1,4\text{-}(CH_3)_2C_6H_4$	106.17	0.8611	8	0.7

Note: Ethene = ethylene; ethyl chloride = chloroethane; ethylene chloride = 1,2-dichloroethane; ethylidene chloride = 1,1-dichloroethane; methyl benzene = toluene; methyl chloride = chloromethane; methyl chloroform = 1,1,1-trichloroethane; methylene chloride = dichloromethane; tetrachloromethane = carbon tetrachloride; tribromomethane = bromoform.

TABLE A-9
Typical solubility product constants

Equilibrium equation	K_{sp} at 25°C
$AgCl \rightleftharpoons Ag^{+} + Cl^{-}$	1.76×10^{-10}
$Al(OH)_3 \rightleftharpoons Al^{3+} + 3OH^{-}$	1.26×10^{-33}
$BaSO_4 \rightleftharpoons Ba^{2+} + SO_4^{2-}$	1.05×10^{-10}
$Cd(OH)_2 \rightleftharpoons Cd^{2+} + 2OH^{-}$	5.33×10^{-15}
$CdS \rightleftharpoons Cd^{2+} + S^{2-}$	1.40×10^{-29}
$CdCO_3 \rightleftharpoons Cd^{2+} + CO_3^{2-}$	6.20×10^{-12}
$CaCO_3 \rightleftharpoons Ca^{2+} + CO_3^{2-}$	4.95×10^{-9}
$CaF_2 \rightleftharpoons Ca^{2+} + 2F^{-}$	1.61×10^{-10}
$Ca(OH)_2 \rightleftharpoons Ca^{2+} + 2OH^{-}$	7.88×10^{-6}
$Ca_3(PO_4)_2 \rightleftharpoons 3Ca^{2+} + 2PO_4^{3-}$	2.02×10^{-33}
$CaSO_4 \rightleftharpoons Ca^{2+} + SO_4^{2}$	3.73×10^{-5}
$Cr(OH)_3 \rightleftharpoons Cr^{3+} + 3OH^{-}$	6.0×10^{-31}
$Cu(OH)_2 \rightleftharpoons Cu^{2+} + 2OH^{-}$	2.0×10^{-19}
$CuS \rightleftharpoons Cu^{2+} + S^{2-}$	1.28×10^{-36}
$Fe(OH)_3 \rightleftharpoons Fe^{3+} + 3OH^{-}$	2.67×10^{-39}
$FeCO_3 \rightleftharpoons Fe^{2+} + CO_3^{2-}$	3.13×10^{-11}
$Fe(OH)_2 \rightleftharpoons Fe^{2+} + 2OH^{-}$	4.79×10^{-17}
$FeS \rightleftharpoons Fe^{2+} + S^{2-}$	1.57×10^{-19}
$PbCO_3 \rightleftharpoons Pb^{2+} + CO_3^{2-}$	1.48×10^{-13}
$Pb(OH)_2 \rightleftharpoons Pb^{2+} + 2OH^{-}$	1.40×10^{-20}
$PbS \rightleftharpoons Pb^{2+} + S^{2-}$	8.81×10^{-29}
$Mg(OH)_2 \rightleftharpoons Mg^{2+} + 2OH^{-}$	1.82×10^{-11}
$MgCO_3 \rightleftharpoons Mg^{2+} + CO_3^{2-}$	1.15×10^{-5}
$MnCO_3 \rightleftharpoons Mn^{2+} + CO_3^{2-}$	2.23×10^{-11}
$Mn(OH)_2 \rightleftharpoons Mn^{2+} + 2OH^{-}$	2.04×10^{-13}
$NiCO_3 \rightleftharpoons Ni^{2+} + CO_3^{2-}$	1.45×10^{-7}
$Ni(OH)_2 \rightleftharpoons Ni^{2+} + 2OH^{-}$	5.54×10^{-16}
$NiS \rightleftharpoons Ni^{2+} + S^{2-}$	1.08×10^{-21}
$SrCO_3 \rightleftharpoons Sr^{2+} + CO_3^{2-}$	5.60×10^{-10}
$Zn(OH)_2 \rightleftharpoons Zn^{2+} + 2OH^{-}$	7.68×10^{-17}
$ZnS \rightleftharpoons Zn^{2+} + S^{2-}$	2.91×10^{-25}

TABLE A-10
Typical valences of elements and compounds in water

Element or compound	Valence
Aluminum	3^+
Ammonium (NH_4)	1^+
Barium	2^+
Boron	3^+
Cadmium	2^+
Calcium	2^+
Carbonate (CO_3)	2^-
Carbon dioxide (CO_2)	[a]
Chloride (*not* chlorine)	1^-
Chromium	$3^+, 6^+$
Copper	2^+
Fluoride (*not* fluorine)	1^-
Hydrogen	1^+
Hydroxide (OH)	1^-
Iron	$2^+, 3^+$
Lead	2^+
Magnesium	2^+
Manganese	2^+
Nickel	2^+
Oxygen	2^-
Nitrogen	$3^+, 5^+, 3^-$
Nitrate (NO_3)	1^-
Nitrite (NO_2)	1^-
Phosphorus	$5^+, 3^-$
Phosphate (PO_4)	3^-
Potassium	1^+
Silver	1^+
Silica	[b]
Silicate (SiO_4)	4^-
Sodium	1^+
Sulfate (SO_4)	2^-
Sulfide (S)	2^-
Zinc	2^+

[a] Carbon dioxide in water is essentially carbonic acid:

$$CO_2 + H_2O \rightleftharpoons H_2CO_3$$

As such, the equivalent weight = GMW/2.

[b] Silica in water is reported as SiO_2. The equivalent weight is equal to the gram molecular weight.

APPENDIX B

NOISE COMPUTATION TABLES AND NOMOGRAPHS

Project ____________________ Date ____________ Engineer ____________

Step			A	T_M	T_H	A	T_M	T_H	A	T_M	T_H	A	T_M	T_H	A	T_M	T_H	A	T_M	T_H
1	Traffic	Vehicle Volume, V(Vph)																		
2		Vehicle Av. Speed, S(km/h)																		
3		Combined Veh. Vol.*, V_C(Vph)																		
4	Prop.	Observer-Roadway Dist., D_C(m)																		
5	Shielding	Line-of-Sight Dist., L/S(m)																		
6		Barrier Position Dist., P(m)																		
7		Break in Barrier, B(m)																		
8		Angle Subtended, θ (deg)																		
9	Prediction**	Unshield L_{10} Level (dBA)																		
10		Shielding Adjust. (dBA)																		
11		L_{10} at Observer (By Veh. Class)																		
12		L_{10} at Observer – Total																		

Code:

A = Automobiles, T_M = Medium Trucks, T_H = Heavy Trucks

* Applies only when automobile and medium truck average speeds are equal. $V_C = V_A + (10)V_{T_M}$

** If automobile-medium truck volume V_C is combined, use L_{10} Nomograph prediction only once for these two vehicle classes

FIGURE B-1
Blank noise prediction worksheet. (*Source: NCHRP 174,* 1976.)

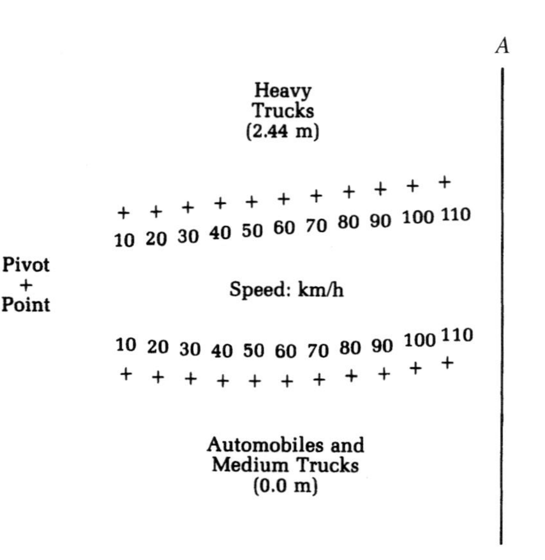

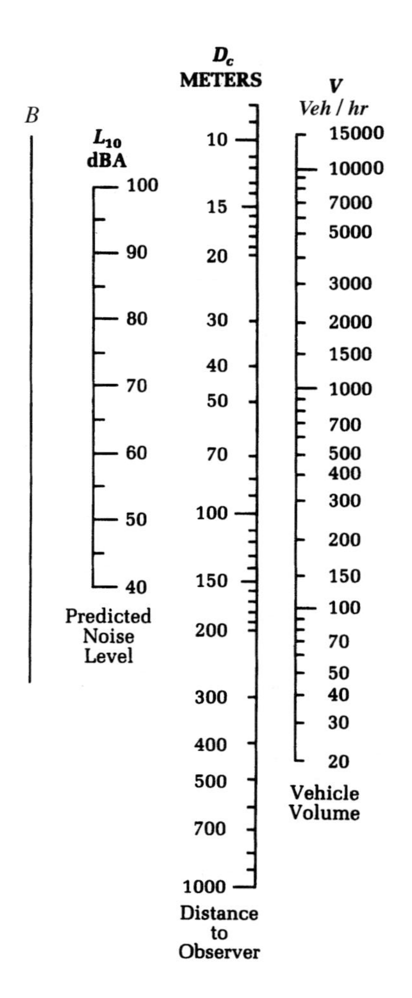

FIGURE B-2
Blank L_{10} nomograph. (*Source: NCHRP 174,* 1976.)

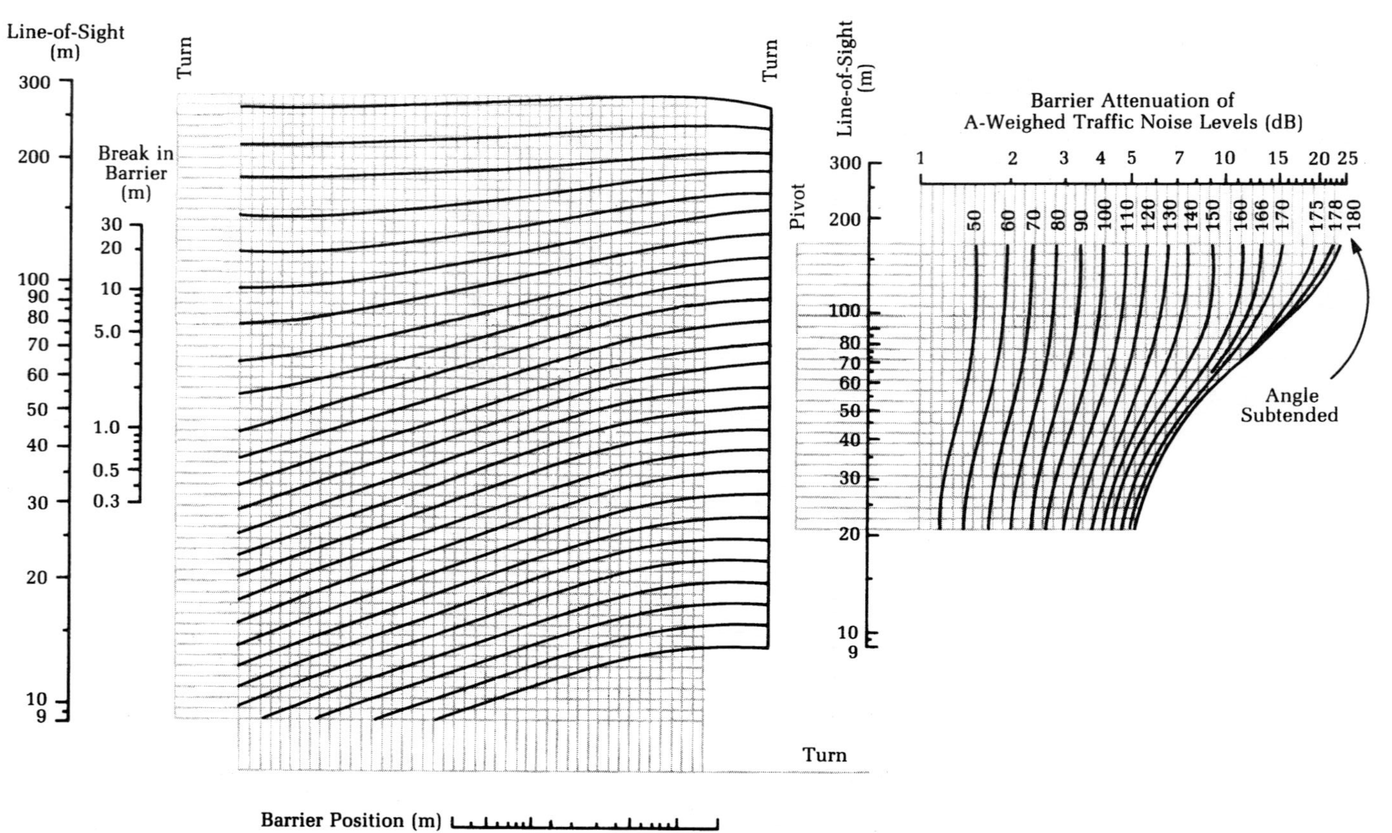

FIGURE B-3
Blank barrier nomograph. (*Source: NCHRP 174,* 1976.)

APPENDIX C

EPA HAZARDOUS WASTE CODES

TABLE C-1
Hazardous wastes from non-specific sources (§261.31)

Industry and EPA hazardous waste No.	Hazardous waste	Hazard code
Generic:		
F001	The following spent halogenated solvents used in degreasing: Tetrachloroethylene, trichloroethylene, methylene chloride, 1,1,1-trichlorethane, carbon tetrachloride, and chlorinated fluorocarbons; all spent solvent mixtures/blends used in degreasing containing, before use, a total of ten percent or more (by volume) of one or more of the above halogenated solvents or those solvents listed in F002, F004, and F005; and still bottoms from the recovery of these spent solvents and spent solvent mixtures.	(T)
F002	The following spent halogenated solvents: Tetrachloroethylene, methylene chloride, trichloroethylene, 1,1,1-trichloroethane, chlorobenzene, 1,1,2-trichloro-1,2,2-trifluoroethane, ortho-dichlorobenzene, trichlorofluoromethane, and 1,1,2-trichloroethane; all spent solvent mixtures/blends containing, before use, a total of ten percent or more (by volume) of one or more of the above halogenated solvents or those listed in F001, F004, or F005; and still bottoms from the recovery of these spent solvents and spent solvent mixtures.	(T)
F003	The following spent non-halogenated solvents: Xylene, acetone, ethyl acetate, ethyl benzene, ethyl ether, methyl isobutyl ketone, n-butyl alcohol, cyclohexanone, and methanol; all spent solvent mixtures/blends containing, before use, only the above spent non-halogenated solvents; and all spent solvent mixtures/blends containing, before use, one or more of the above non-halogenated solvents, and a total of ten percent or more (by volume) of one or more of those solvents listed in F001, F002, F004, and F005; and still bottoms from the recovery of these spent solvents and spent solvent mixtures.	(I)[a]
F004	The following spent non-halogenated solvents: Cresols and cresylic acid, and nitrobenzene; all spent solvent mixtures/blends containing, before use, a total of ten percent or more (by volume) of one or more of the above non-halogenated solvents or those solvents listed in F001, F002, and F005; and still bottoms from the recovery of these spent solvents and spent solvent mixtures.	(T)

[a] (I, T) should be used to specify mixtures containing ignitable and toxic constituents.

(continued)

TABLE C-1
(continued)

Industry and EPA hazardous waste No.	Hazardous waste	Hazard code
F005	The following spent non-halogenated solvents: Toluene, methyl ethyl ketone, carbon disulfide, isobutanol, pyridine, benzene, 2-ethoxyethanol, and 2-nitropropane; all spent solvent mixtures/blends containing, before use, a total of ten percent or more (by volume) of one or more of the above non-halogenated solvents or those solvents listed in F001, F002, or F004; and still bottoms from the recovery of these spent solvents and spent solvent mixtures.	(I, T)
F006	Wastewater treatment sludges from electroplating operations except from the following processes: (1) Sulfuric acid anodizing of aluminum; (2) tin plating on carbon steel; (3) zinc plating (segregated basis) on carbon steel; (4) aluminum or zinc-aluminum plating on carbon steel; (5) cleaning/stripping associated with tin, zinc, and aluminum plating on carbon steel; and (6) chemical etching and milling of aluminum.	(T)
F007	Spent cyanide plating bath solutions from electroplating operations.	(R, T)
F008	Plating bath residues from the bottom of plating baths from electroplating operations where cyanides are used in the process.	(R, T)
F009	Spent stripping and cleaning bath solutions from electroplating operations where cyanides are used in the process.	(R, T)
F010	Quenching bath residues from oil baths from metal heat treating operations where cyanides are used in the process.	(R, T)
F011	Spent cyanide solutions from salt bath pot cleaning from metal heat treating operations.	(R, T)
F012	Quenching wastewater treatment sludges from metal heat treating operations where cyanides are used in the process.	(T)
F019	Wastewater treatment sludges from the chemical conversion coating of aluminum except from zirconium phosphating in aluminum can washing when such phosphating is an exclusive conversion coating process.	(T)
F020	Wastes (except wastewater and spent carbon from hydrogen chloride purification) from the production or manufacturing use (as a reactant, chemical intermediate, or component in a formulating process) of tri- or tetrachlorophenol, or of intermediates used to produce their pesticide derivatives. (This listing does not include wastes from the production of hexachlorophene from highly purified 2,4,5-trichlorophenol.)	(H)
F021	Wastes (except wastewater and spent carbon from hydrogen chloride purification) from the production or manufacturing use (as a reactant, chemical intermediate, or component in a formulating process) of pentachlorophenol, or of intermediates used to produce its derivatives.	(H)
F022	Wastes (except wastewater and spent carbon from hydrogen chloride purification) from the manufacturing use (as a reactant, chemical intermediate, or component in a formulating process) of tetra-, penta-, or hexachlorobenzenes under alkaline conditions.	(H)

[a] (I, T) should be used to specify mixtures containing ignitable and toxic constituents.

(continued)

TABLE C-1
(continued)

Industry and EPA hazardous waste No.	Hazardous waste	Hazard code
F023	Wastes (except wastewater and spent carbon from hydrogen chloride purification) from the production of materials on equipment previously used for the production or manufacturing use (as a reactant, chemical intermediate, or component in a formulating process) of tri- and tetrachlorophenols. (This listing does not include wastes from equipment used only for the production or use of hexachlorophene from highly purified 2,4,5-trichlorophenol.)	(H)
F024	Process wastes, including but not limited to, distillation residues, heavy ends, tars, and reactor clean-out wastes, from the production of certain chlorinated aliphatic hydrocarbons by free radical catalyzed processes. These chlorinated aliphatic hydrocarbons are those having carbon chain lengths ranging from one to and including five, with varying amounts and positions of chlorine substitution. (This listing does not include wastewaters, wastewater treatment sludges, spent catalysts, and wastes listed in §261.31 or §261.32.)	(T)
F025	Condensed light ends, spent filters and filter aids, and spent desiccant wastes from the production of certain chlorinated aliphatic hydrocarbons by free radical catalyzed processes. These chlorinated aliphatic hydrocarbons are those having carbon chain lengths ranging from one to and including five, with varying amounts and positions of chlorine substitution.	(T)
F026	Wastes (except wastewater and spent carbon from hydrogen chloride purification) from the production of materials on equipment previously used for the manufacturing use (as a reactant, chemical intermediate, or component in a formulating process) of tetra-, penta-, or hexachlorobenzene under alkaline conditions.	(H)
F027	Discarded unused formulations containing tri-, tetra-, or pentachlorophenol or discarded unused formulations containing compounds derived from these chlorophenols. (This listing does not include formulations containing hexachlorophene synthesized from prepurified 2,4,5-trichlorophenol as the sole component.)	(H)
F028	Residues resulting from the incineration or thermal treatment of soil contaminated with EPA Hazardous Waste Nos. F020, F021, F022, F023, F026, and F027.	(T)
F032	Wastewaters (except those that have not come into contact with process contaminants), process residuals, preservative drippage, and spent formulations from wood preserving processes generated at plants that currently use or have previously used chlorophenolic formulations (except potentially cross-contaminated wastes that have had the F032 waste code deleted in accordance with 40 CFR 261.35 or potentially cross-contaminated wastes that are otherwise currently regulated as hazardous wastes (i.e., F034 or F035), and where the generator does not resume or initiate use of chlorophenolic formulations). This listing does not include K001 bottom sediment sludge from the treatment of wastewater from wood preserving processes that use creosote and/or pentachlorophenol.	(T)
F034	Wastewaters (except those that have not come into contact with process contaminants), process residuals, preservative drippage, and spent formulations from wood preserving processes generated at plants that use creosote formulations. This listing does not include K001 bottom sediment sludge from the treatment of wastewater from wood preserving processes that use creosote and/or pentachlorophenol.	(T)

[a] (I, T) should be used to specify mixtures containing ignitable and toxic constituents.

(continued)

TABLE C-1
(continued)

Industry and EPA hazardous waste No.	Hazardous waste	Hazard code
F035	Wastewaters (except those that have not come into contact with process contaminants), process residuals, preservative drippage, and spent formulations from wood preserving processes generated at plants that use inorganic preservatives containing arsenic or chromium. This listing does not include K001 bottom sediment sludge from the treatment of wastewater from wood preserving processes that use creosote and/or pentachlorophenol.	(T)
F037	Petroleum refinery primary oil/water/solids separation sludge—Any sludge generated from the gravitational separation of oil/water/solids during the storage or treatment of process wastewaters and oily cooling wastewaters from petroleum refineries. Such sludges include, but are not limited to, those generated in: oil/water/solids separators; tanks and impoundments; ditches and other conveyances; sumps; and stormwater units receiving dry weather flow. Sludge generated in stormwater units that do not receive dry weather flow, sludges generated from non-contact once-through cooling waters segregated for treatment from other process or oily cooling waters, sludges generated in aggressive biological treatment units as defined in §261.31(b)(2) (including sludges and floats generated in one or more additional units after wastewaters have been treated in aggressive biological treatment units) and K051 wastes are not included in this listing.	(T)
F038	Petroleum refinery secondary (emulsified) oil/water/solids separation sludge—Any sludge and/or float generated from the physical and/or chemical separation of oil/water/solids in process wastewaters and oily cooling wastewaters from petroleum refineries. Such wastes include, but are not limited to, all sludges and floats generated in: induced air flotation (IAF) units, tanks and impoundments, and all sludges generated in DAF units. Sludges generated in stormwater units that do not receive dry weather flow, sludges generated from non-contact once-through cooling waters segregated for treatment from other process or oily cooling waters, sludges and floats generated in aggressive biological treatment units as defined in §261.31(b)(2) (including sludges and floats generated in one or more additional units after wastewaters have been treated in aggressive biological treatment units) and F037, K048, and K051 wastes are not included in this listing.	(T)
F039	Leachate (liquids that have percolated through land disposed wastes) resulting from the disposal of more than one restricted waste classified as hazardous under subpart D of this part. (Leachate resulting from the disposal of one or more of the following EPA Hazardous Wastes and no other Hazardous Wastes retains its EPA Hazardous Waste Number(s): F020, F021, F022, F026, F027, and/or F028.)	(T)

[a] (I, T) should be used to specify mixtures containing ignitable and toxic constituents.

TABLE C-2
Hazardous wastes from specific sources (§261.32)

Industry and EPA hazardous waste No.	Hazardous waste	Hazard code
Wood preservation:		
K001	Bottom sediment sludge from the treatment of wastewaters from wood preserving processes that use creosote and/or pentachlorophenol.	(T)
Inorganic pigments:		
K002	Wastewater treatment sludge from the production of chrome yellow and orange pigments.	(T)
K003	Wastewater treatment sludge from the production of molybdate orange pigments.	(T)
K004	Wastewater treatment sludge from the production of zinc yellow pigments.	(T)
K005	Wastewater treatment sludge from the production of chrome green pigments.	(T)
K006	Wastewater treatment sludge from the production of chrome oxide green pigments (anhydrous and hydrated).	(T)
K007	Wastewater treatment sludge from the production of iron blue pigments.	(T)
K008	Oven residue from the production of chrome oxide green pigments.	(T)
Organic chemicals:		
K009	Distillation bottoms from the production of acetaldehyde from ethylene.	(T)
K010	Distillation side cuts from the production of acetaldehyde from ethylene.	(T)
K011	Bottom stream from the wastewater stripper in the production of acrylonitrile.	(R, T)
K013	Bottom stream from the acetonitrile column in the production of acrylonitrile.	(R, T)
K014	Bottoms from the acetonitrile purification column in the production of acrylonitrile.	(T)
K015	Still bottoms from the distillation of benzyl chloride.	(T)
K016	Heavy ends or distillation residues from the production of carbon tetrachloride.	(T)
K017	Heavy ends (still bottoms) from the purification column in the production of epichlorohydrin.	(T)
K018	Heavy ends from the fractionation column in ethyl chloride production.	(T)
K019	Heavy ends from the distillation of ethylene dichloride in ethylene dichloride production.	(T)
K020	Heavy ends from the distillation of vinyl chloride in vinyl chloride monomer production.	(T)
K021	Aqueous spent antimony catalyst waste from fluoromethanes production.	(T)
K022	Distillation bottom tars from the production of phenol/acetone from cumene.	(T)
K023	Distillation light ends from the production of phthalic anhydride from naphthalene.	(T)
K024	Distillation bottoms from the production of phthalic anhydride from naphthalene.	(T)
K025	Distillation bottoms from the production of nitrobenzene by the nitration of benzene.	(T)
K026	Stripping still tails from the production of methyl ethyl pyridines.	(T)
K027	Centrifuge and distillation residues from toluene diisocyanate production.	(R, T)
K028	Spent catalyst from the hydrochlorinator reactor in the production of 1,1,1-trichloroethane.	(T)

(*continued*)

TABLE C-2
(continued)

Industry and EPA hazardous waste No.	Hazardous waste	Hazard code
K029	Waste from the product steam stripper in the production of 1,1,1-trichloroethane.	(T)
K030	Column bottoms or heavy ends from the combined production of trichloroethylene and perchloroethylene.	(T)
K083	Distillation bottoms from aniline production.	(T)
K085	Distillation or fractionation column bottoms from the production of chlorobenzenes.	(T)
K093	Distillation light ends from the production of phthalic anhydride from ortho-xylene.	(T)
K094	Distillation bottoms from the production of phthalic anhydride from ortho-xylene.	(T)
K095	Distillation bottoms from the production of 1,1,1-trichloroethane.	(T)
K096	Heavy ends from the heavy ends column from the production of 1,1,1-trichloroethane.	(T)
K103	Process residues from aniline extraction from the production of aniline.	(T)
K104	Combined wastewater streams generated from nitrobenzene/aniline production.	(T)
K105	Separated aqueous stream from the reactor product washing step in the production of chlorobenzenes.	(T)
K107	Column bottoms from product separation from the production of 1,1-dimethyl-hydrazine (UDMH) from carboxylic acid hydrazines.	(C,T)
K108	Condensed column overheads from product separation and condensed reactor vent gases from the production of 1,1-dimethylhydrazine (UDMH) from carboxylic acid hydrazides.	(I,T)
K109	Spent filter cartridges from product purification from the production of 1,1-dimethylhydrazine (UDMH) from carboxylic acid hydrazides.	(T)
K110	Condensed column overheads from intermediate separation from the production of 1,1-dimethylhydrazine (UDMH) from carboxylic acid hydrazides.	(T)
K111	Product washwaters from the production of dinitrotoluene via nitration of toluene.	(C, T)
K112	Reaction by-product water from the drying column in the production of toluenediamine via hydrogenation of dinitrotoluene.	(T)
K113	Condensed liquid light ends from the purification of toluenediamine in the production of toluenediamine via hydrogenation of dinitrotoluene.	(T)
K114	Vicinals from the purification of toluenediamine in the production of toluenediamine via hydrogenation of dinitrotoluene.	(T)
K115	Heavy ends from the purification of toluenediamine in the production of toluenediamine via hydrogenation of dinitrotoluene.	(T)
K116	Organic condensate from the solvent recovery column in the production of toluene diisocyanate via phosgenation of toluenediamine.	(T)
K117	Wastewater from the reactor vent gas scrubber in the production of ethylene dibromide via bromination of ethene.	(T)
K118	Spent adsorbent solids from purification of ethylene dibromide in the production of ethylene dibromide via bromination of ethene.	(T)
K136	Still bottoms from the purification of ethylene dibromide in the production of ethylene dibromide via bromination of ethene.	(T)

(*continued*)

TABLE C-2
(continued)

Industry and EPA hazardous waste No.	Hazardous waste	Hazard code
K149	Distillation bottoms from the production of alpha- (or methyl-) chlorinated toluenes, ring-chlorinated toluenes, benzoyl chlorides, and compounds with mixtures of these functional groups. (This waste does not include still bottoms from the distillation of benzyl chloride.)	(T)
K150	Organic residuals, excluding spent carbon adsorbent, from the spent chlorine gas and hydrochloric acid recovery processes associated with the production of alpha- (or methyl-) chlorinated toluenes, ring-chlorinated toluenes, benzoyl chlorides, and compounds with mixtures of these functional groups.	(T)
K151	Wastewater treatment sludges, excluding neutralization and biological sludges, generated during the treatment of wastewaters from the production of alpha- (or methyl-) chlorinated toluenes, ring-chlorinated toluenes, benzoyl chlorides, and compounds with mixtures of these functional groups.	(T)
Inorganic chemicals:		
K071	Brine purification muds from the mercury cell process in chlorine production, where separately prepurified brine is not used.	(T)
K073	Chlorinated hydrocarbon waste from the purification step of the diaphragm cell process using graphite anodes in chlorine production.	(T)
K106	Wastewater treatment sludge from the mercury cell process in chlorine production.	(T)
Pesticides:		
K031	By-product salts generated in the production of MSMA and cacodylic acid.	(T)
K032	Wastewater treatment sludge from the production of chlordane.	(T)
K033	Wastewater and scrub water from the chlorination of cyclopentadiene in the production of chlordane.	(T)
K034	Filter solids from the filtration of hexachlorocyclopentadiene in the production of chlordane.	(T)
K035	Wastewater treatment sludges generated in the production of creosote.	(T)
K036	Still bottoms from toluene reclamation distillation in the production of disulfoton.	(T)
K037	Wastewater treatment sludges from the production of disulfoton.	(T)
K038	Wastewater from the washing and stripping of phorate production.	(T)
K039	Filter cake from the filtration of diethylphosphorodithioic acid in the production of phorate.	(T)
K040	Wastewater treatment sludge from the production of phorate.	(T)
K041	Wastewater treatment sludge from the production of toxaphene.	(T)
K042	Heavy ends or distillation residues from the distillation of tetrachlorobenzene in the production of 2,4,5-T.	(T)
K043	2,6-Dichlorophenol waste from the production of 2,4-D.	(T)
K097	Vacuum stripper discharge from the chlordane chlorinator in the production of chlordane.	(T)
K098	Untreated process wastewater from the production of toxaphene.	(T)
K099	Untreated wastewater from the production of 2,4-D.	(T)

(continued)

TABLE C-2
(continued)

Industry and EPA hazardous waste No.	Hazardous waste	Hazard code
K123	Process wastewater (including supernates, filtrates, and washwaters) from the production of ethylenebisdithiocarbamic acid and its salts.	(T)
K124	Reactor vent scrubber water from the production of ethylenebisdithiocarbamic acid and its salts.	(C, T)
K125	Filtration, evaporation, and centrifugation solids from the production of ethylenebisdithiocarbamic acid and its salts.	(T)
K126	Baghouse dust and floor sweepings in milling and packaging operations from the production or formulation of ethylenebisdithiocarbamic acid and its salts.	(T)
K131	Wastewater from the reactor and spent sulfuric acid from the acid dryer from the production of mcthyl bromide.	(C,T)
K132	Spent absorbent and wastewater separator solids from the production of methyl bromide.	(T)
Explosives:		
K044	Wastewater treatment sludges from the manufacturing and processing of explosives.	(R)
K045	Spent carbon from the treatment of wastewater containing explosives.	(R)
K046	Wastewater treatment sludges from the manufacturing, formulation, and loading of lead-based initiating compounds.	(T)
K047	Pink/red water from TNT operations.	(R)
Petroleum refining:		
K048	Dissolved air flotation (DAF) float from the petroleum refining industry.	(T)
K049	Slop oil emulsion solids from the petroleum refining industry.	(T)
K050	Heat exchanger bundle cleaning sludge from the petroleum refining industry.	(T)
K051	API separator sludge from the petroleum refining industry.	(T)
K052	Tank bottoms (leaded) from the petroleum refining industry.	(T)
Iron and steel:		
K061	Emission control dust/sludge from the primary production of steel in electric furnaces.	(T)
K062	Spent pickle liquor generated by steel finishing operations of facilities within the iron and steel industry (SIC Codes 331 and 332).	(C, T)
Primary copper:		
K064	Acid plant blowdown slurry/sludge resulting from the thickening of blowdown slurry from primary copper production.	(T)
Primary lead:		
K065	Surface impoundment solids contained in and dredged from surface impoundments at primary lead smelting facilities.	(T)
Primary zinc:		
K066	Sludge from treatment of process wastewater and/or acid plant blowdown from primary zinc production.	(T)

(*continued*)

TABLE C-2
(*continued*)

Industry and EPA hazardous waste No.	Hazardous waste	Hazard code
Primary aluminum:		
K088	Spent potliners from primary aluminum reduction.	(T)
Ferroalloys:		
K090:	Emission control dust or sludge from ferrochromiumsilicon production.	(T)
K091	Emission control dust or sludge from ferrochromium production.	(T)
Secondary lead:		
K069	Emission control dust/sludge from secondary lead smelting.	(T)
K100	Waste leaching solution from acid leaching of emission control dust/sludge from secondary lead smelting.	(T)
Veterinary pharmaceuticals:		
K084	Wastewater treatment sludges generated during the production of veterinary pharmaceuticals from arsenic or organo-arsenic compounds.	(T)
K101	Distillation tar residues from the distillation of aniline-based compounds in the production of veterinary pharmaceuticals from arsenic or organo-arsenic compounds.	(T)
K102	Residue from the use of activated carbon for decolorization in the production of veterinary pharmaceuticals from arsenic or organo-arsenic compounds.	(T)
Ink formulation:		
K086	Solvent washes and sludges, caustic washes and sludges, or water washes and sludges from cleaning tubs and equipment used in the formulation of ink from pigments, driers, soaps, and stabilizers containing chromium and lead.	(T)
Coking:		
K060	Ammonia still lime sludge from coking operations.	(T)
K087	Decanter tank tar sludge from coking operations.	(T)
K141	Process residues from the recovery of coal tar, including, but not limited to, collecting sump residues from the production of coke from coal or the recovery of coke by-products produced from coal. This listing does not include K087 (decanter tank tar sludges from coking operations).	(T)
K142	Tar storage tank residues from the production of coke from coal or from the recovery of coke by-products produced from coal.	(T)
K143	Process residues from the recovery of light oil, including, but not limited to, those generated in stills, decanters, and wash oil recovery units from the recovery of coke by-products produced from coal.	(T)
K144	Wastewater sump residues from light oil refining, including, but not limited to, intercepting or contamination sump sludges from the recovery of coke by-products produced from coal.	(T)
K145	Residues from naphthalene collection and recovery operations from the recovery of coke by-products produced from coal.	(T)
K147	Tar storage tank residues from coal tar refining.	(T)
K148	Residues from coal tar distillation, including, but not limited to, still bottoms.	(T)

INDEX

Abandoned sites, 29
Abiotic, 756
Absorption:
  and air pollution control, 511–518, 523
  coefficient, Sabin, 616
  defined, 511
  trench for tile fields, 358
Abstraction and runoff, 48
Accelerated growth, log-growth curve, 347
Accelerator, 819
Acceptable risk, radiation, 827
Acid/base reaction, 141, 143–146
Acid fermentation:
  anaerobic digestion, 435
  oxidation ponds, 407
Acidification of lakes, *see* Acid rain
Acid rain, 328–330
  and aluminum, 328, 485
  and carbonate buffer system, 328
  legislation, 21, 26
  and nitrates, 485–486
  and sulfates, 485–486
  and VOCs, 485–486
  and water quality management, 320, 328–330
Acoustic:
  privacy, 580
  trauma, 572
Acoustical lining, 616
Actinium series, 818
Activated alumina, 518
Activated carbon (charcoal), 249, 412, 518
  absorber for fuel tank, 541
Activated silica, 178
Activated sludge, 375, 382–406
  defined, 382
  completely mixed process, 383–390
  growth constants, table of, 390
  F/M ratio, 392–394
  modifications, 385–386
  and nitrogen:BOD ratio, 402
  oxygen demand, 400
  and phosphorus:BOD ratio, 402
  plug-flow, 390–391
  process design considerations, 401
  secondary clarifier, 402
  sludge problems, 405–406
  sludge production, 399
  sludge return, 394
  sequencing batch reactor, 757
  wasting, 382, 383, 392
Activity, specific, 816
Activity coefficients, 150
Acute hazardous waste, 726
Acute radiation syndrome, 828
Adiabatic:
  defined, 461
  lapse rate (Γ), 494
  process, 461
Adsorbate, 518

Adsorbent, 518
Adsorption:
  in air pollution control, 518–522
  defined, 249, 412
  in hazardous waste, 762
  isotherm, 519
  in wastewater treatment, 412–413
  in water treatment, 249–250
Advanced oxidation products (AOPs), 249
Advanced wastewater treatment (AWT), 411–415
Aerated:
  grit chamber, 367–368
  lagoon, 407
Aeration tank, activated sludge, 382–384, 387
Aerobes, obligate, 341
Aerobic:
  decomposition, 290, 343, 344
  defined, 290, 344
  digestion, 434
  lagoons, 407
  lakes, 321
  ponds, 406–407
Afterburner, 518, 776
Agricultural runoff and water quality, 285, 328
Airborne transmission of noise, 590–599
Aircraft noise and annoyance, 28
Air-fuel ration (A/F), 538, 539
Air pollution:
  control of mobile sources, 535–542
  control of stationary sources, 511–535
  and dispersion modeling, 501–509
  effects on:
    health, 469–475
    materials, 465–467
    vegetation, 467–469
  episodes, 473–475
  and incinerators, 678, 680, 777
  meteorology, 474, 491–500
  origin and fate, 475–481
  standards, 463–465
  and water quality management, 328–330
Air Quality Control Regions (AQCRs), 21, 464
Air quality management legislation, 20–26
Air resource management, 13–14, 486
Air stripping, 763–764
Air toxics, *see* Hazardous air pollutants
Airway resistance, 471
Algae:
  growth requirements, 324–327
  in wastewater, 342
Alkaline chlorination, 759
Alkalinity:
  bicarbonate, 149
  carbonate, 149
  and coagulation, 173–175
  defined, 147–149
  and softening, 183
  units, 149
  in wastewater, 353
  and water quality management, 328–329
Alpha particle, 813
  radiation exposure, 837
Alum:
  and coagulation, 173, 175
  and control of phosphorus:
    in lakes, 327
    in wastewater, 413
  sludge, 252
Aluminum:
  and acid rain, 328, 485
  and fish toxicity, 328
  and tree root deterioration, 485
Alveoli, 470
Ambient air quality standards, 21, 22, 465
American Academy of Ophthalmology and Otolaryngology, 573
American National Standards Institute (ANSI), 559, 569
American Society of Civil Engineers (ASCE), 3
American Society for Testing Materials (ASTM), 6
American Water Works Association (AWWA), 168
Ammonia:
  and anaerobic decomposition, 345, 346
  and ESPs, 535
  and flue gas desulfurization, 523
  as NBOD, 301, 320
  and nitrogen oxide control, 525
  as nutrient, 320
  oxidation, 301
  reaction with chlorine, 243, 246
  stripping, 415
  toxicity to fish, 320
Amplitude of a wave, 553

Anabolism, 343
Anaerobes, 341
Anaerobic:
  decomposition, 346, 434–438
  defined, 303, 346
  digestion, 434–438
  lagoons, 406–407
  lakes, 321
  landfills, 665, 689–690
  ponds, 406–407
Angle subtended, noise barrier parameter, 604, 605
Annoyance, noise, 576
Annual series, 77–78
  maxima, 77
  minima, 77
Anoxic decomposition, 341, 344
  and secondary settling tank, 406
Anthropogenic activities, 475, 755, 834
Anticyclone, 492, 493
Aquiclude, 49, 50
Aquifer:
  artesian, 50
  barrier, 106
  confined, 50
  hydraulic properties, 104–106
  non-homogeneous, 106
  porosity, 93
  recharge area, 49, 51, 106
  unconfined, 49
  water table, 49
Aquitard, 50
ARARs for hazardous waste, 749
Assimilation of waste, 284
Atmospheric:
  dispersion, 500–509
  stability, 494–498
    and noise, 594
  stability class, 504–507
    table of, 506
Atomic:
  bomb, 833
  energy:
    legislation, 32–33
    and radioactivity in water, 155, 158–159, 835, 836
  mass unit, 811
  number, 811
  structure, 811
  weight, 812
    table of, *inside front cover*
Atomic Energy Act, 32
Atomic Energy Commission (AEC), 32
ATP and bacterial growth, 346
Attainment area, air pollution, 464
Attenuation of noise, 595
Audiogram, 569
Audiometry, 568–569
Auditory effects of noise, 569–573
Autogenous combustion, 678, 774
Automobile:
  blowby control, 539, 541
  engine combustion, 535–536, 538–539
  noise and annoyance, 584
Autotroph, 341
Average daily water consumption, 9–10, 110
Average haul speed to solid waste disposal site, 646
Avogadro's number, 138, 817

Back corona, ESP, 535
Backwashing filters, 229, 231
Backyard pickup of solid waste, 640, 641
Bacteria:
  biochemistry of, 343
  growth, 347–351
  in wastewater, 342
  in water, 133, 158, 164, 167, 240
Bacterial growth requirements, 346–347
Baffles:
  and flocculators, 205
  and rapid mixing, 203–204
  and settling tanks, 211, 212
Baghouse, 528–529
Bar racks:
  wastewater treatment, 364–365
  water treatment plant, 170
Barrier:
  break, 605
  nomograph, 607
    blank for use, Appendix B
  position distance, 605
Barriered landscape water renovation system (BLWRS), 359–361
Barriers and panels for noise control, 600, 605, 617–619
Base flow, 48
Baseline HTL, 569
Basin, 48
Basket bowl centrifuge, 260

Batch flux curve, 430
Batch reactor, 199–200
Bel, 555
Benthic demand, 303
Benthic zone in lakes, 323
Berlin and Farro Liquid Incineration, 703
Best available treatment (BAT), 17, 18
  for toxic substances, 18
Best conventional treatment (BCT), 17, 18
Best practical treatment (BPT), 17, 18
Beta particle, 813, 814
  radiation exposure, 837
Bicarbonate:
  alkalinity, 149
  buffer system, 147–148
    See also Carbonate, buffer system
  defined, 147
Binary fission of bacteria, 347
Binding energy, 813
Bioassay, 290
Biochemical oxygen demand (BOD), 289–300, 390, 414–415, 434, 437
  bottles, 297
  carbonaceous (CBOD), 301–302
  and COD, 290
  constants, determination of, 294–297
  defined, 290
  digester supernatant, 434, 437
  and microorganisms, 290, 340
  nitrogenous (NBOD), 301–302
  and nitrogen ratio, 401–402
  and phosphorus ratio, 401–402
  rate constant (k), 291–297, 299–300
    table of, 293
  removal efficiency, 359, 361, 377, 379–381, 390, 391, 411
  residual, 412
  and septic tank, 356
  and suspended solids, 390, 411
  and temperature, 293
  test, 297–299
  and treatment, 344, 346, 356, 359, 377–381
  ultimate, 291, 292
  and waste strength, 344, 346, 353, 354
Biodegradable, 290
Biological:
  treatment of hazardous waste, 755–758
  treatment of municipal wastewater, 339, 374–411
  zones in lakes, 322–323
Biomass, 350, 386
Blowby, crankcase, 539, 541
Bottom ash, incinerator, 682
Break in line of sight, noise barrier parameter, 605
Breakpoint chlorination, 245, 247
Breakthrough:
  adsorption, 519, 765
  ion exchange, 197, 199, 765, 767, 768
Bronchi, 470
Bronchial asthma and air pollution, 471
Brownian movement, 172
Bubble policy, 22
Buffer:
  capacity, 149
  capacity of lakes, 328
  carbonate, 147
  defined, 147
Bulking sludge, 405–406
Burn out, 678
Bureau of Mines, 29

Calcium:
  hardness, 178–179
  softening, 184–185
  scale, 178
Calcium carbonate, 179, 183, 185, 328, 466
  alkalinity, expressed in mg/L as, 149
Calcium hydroxide and softening, 183
Calcium oxide and softening, 183, 190
Cancer of bronchus (lung cancer), 471
Capillary fringe, 49
Carbon:
  and adsorption, 412, 762
  and algal growth, 324
  and bacterial growth, 341
Carbon-14, 819, 838
Carbonaceous BOD (CBOD), 301–302
  design SRT, 401–402
Carbonate:
  alkalinity, 148
  buffer system, 147
    acid rain, 328, 466
  defined, 147
Carbonate hardness (CH)
  bar chart of, 181, 182
  defined, 180
  formation of, 179
  precipitation of, 184

Carbon dioxide:
and acid rain, 328, 485
and carbonate buffer system, 147
as carbon source, 341
as electron acceptor, 346
and greenhouse effect, 489–491
and softening, 184, 190–191
Carbonic acid, 145, 147, 191
Carbon monoxide:
emissions and A/F ratio, 539, 540
hazardous waste incineration, 782
health effects, 471
and indoor air pollution, 481
origin and fate of, 475–476
standard, 465
Carbon source for microorganism, 341
Carboxyhemoglobin (COHb), 471
Carburetor evaporation losses, 541
Casing, well, 88–90
Catabolism, 343
Catalytic combustion, 522, 541
Cation exchange capacity (CEC), 785
Cells, waste disposal, 663, 664
Center line distance (D_c), traffic noise, 602
Centrate, 260
Centrifuging sludge, 260
CERCLA, 30, 733, 745–755
CFR, 726
Chain reaction, 819
Chemical oxygen demand (COD), 289–290, 352–354, 412
Chemical toilet, 356
Chemical treatment of hazardous waste, 758–762
Chemical units, 136–139
Chemoheterotrophs, 342
Chemotrophs, 341, 342
Chicks law, 241
Chloramines, 243
Chloride:
in drinking water, 156
in wastewater, 353
Chlorinated:
dibenzo furans (CDBF), 774
wastes, 756
Chlorination and chlorine, 241–246
Chlorine dioxide, 246–247
Chlorofluorocarbons (CFCs), 487–489, 542
Chlorophyll *a*, 326
Chlorosis, 469
Chronic daily intake (CDI), 720–724
Chronic respiratory disease, 471
Cigarettes and indoor air pollution, 484
Civil engineering, 3
Clarifier:
defined, 211
horizontal flow, 212, 216–219
upflow, 212, 215–216
See also Secondary settling tank; Sedimentation basin; Settling tank
Clean Air Act (CAA), 20, 26, 29
amendments (CAAA), 22–26, 28, 486
Titles, 23–26, 486
Clean Water Act (CWA), 17–19, 29
Clear well, 231
Closed system:
hydrologic, 52
microbial, 348
Coagulants, 173–178
Coagulation, 172–178
aids, 178
defined, 172
floc, 175
and pH, 173, 175, 177
sludge, 252
Coarse screens, water treatment, 170
Cochlea, ear, 566
Code of Ethics, 3
Code of Federal Regulations (CFR), 726, 834
Co-disposal of sludge, 264, 691
Co-firing of sludge, 442, 691
Coliform:
group, 158
limits, 164, 167
test, 167
Collection of solid waste:
cost, 636, 649–650
estimates, 645–649
frequency, 636, 640
interroute transfer, 654–657
methods, 636, 640–641
number of stops, 647
truck capacity, 645–649
truck routing, 651–653
Collection works for water, 8
Colloids, 135, 172
Color, 136, 156, 168, 170, 172
Combined available chlorine, 243
Combined residual chlorination, 245
Combined sewers, 12, 285

Combined sewer overflow (CSO), 12–13, 285
Combustion:
and air pollution control, 522
efficiency, 782
fundamentals, 678
and hazardous waste, 774, 775
Cometabolism, 757
Comminutors, 368
Community right-to-know, 749
Community sewage system, 11
Community water system, 88, 132
Compacts, low-level radioactive waste, 853
Compaction ratios for solid waste, 670
Completed sanitary landfill, 676–677
Completely mixed flow reactor (CMF), 200
Completely mixed systems, 38
Completely stirred tank reactor (CSTR), 200, 383
Complete series:
CDF, 81
runoff analysis, 81
yield analysis, 81
Complexed ions, 141, 175
Composite liner, 786
Composting:
of sludge, 442
of solid waste, 683, 688–689
Compound hydrograph, 69–71
Comprehensive Environmental Response, Compensation, and Liability Act (CERCLA), 30–31, 733, 745–755
Comprehensive Environmental Response, Compensation, and Liability Information System (CERCLIS), 793
Cone of depression, well, 91–93
Confining layer, 50
Conservative pollutants, 38
Construction noise, 585, 589
Construction Safety Act, 28
Consumptive use, 10, 12
Continuous belt filter press (CBFP), 261–262
Continuous noise, 563
exposure, 574
Contract collection of solid waste, 639
Control rods, reactor, 820
Conventional pollutants, 353
Coriolis effect, 492, 493
Corrosive hazardous waste, 728
Cosmic rays, 837, 838
Cost-benefit analysis, 13
Countercurrent flow, 513
Cover, landfill, 676
Cradle-to-grave manifest, 733, 735
Crew integration, solid waste collection, 653–654
Criteria air pollutants, 463
Critical deficit (D_c), 312
Critical mass, nuclear, 819
Critical point in DO sag curve, 303, 312
Critical time in DO sag curve (t_c), 312
Crustaceans, 342
CT:
concept, 242–243
values, table of, 244, 248
Cultural eutrophication, 324
Cumulative frequency distribution:
of noise, 581
of sand, 233–234
Cunningham correction factor, 530, 531, 534
Curbside pickup of solid waste, 640
Curbside recycling of solid waste, 682, 685
Curie, 816
Cyanide, 157
Cyanobacteria, 325
Cyclone:
collector, 525–528
meteorologic, 492
Cyclotron, 819

Dalton's law, 461
Damage-risk criteria for noise, 573–574
Damping for noise control, 614
Darcy's law, 97–98
Daughter isotopes, radioactive decay, 814
DDT degradation, 756–757
Death phase, log-growth curve, 347, 348
Decibel (dB):
addition graph, 557
defined, 555
level, 555
Decomposition of waste, 343–346
Dedicated land disposal (DLD), 442
Deep bed filter, 230
Deep well injection, 783

Deficit:
  DO sag curve, 303, 307, 310
  oxygen, 303, 307
Demand, water:
  average, 9, 10, 110
  defined, 8
  peak, 9
Denitrification:
  anoxic decomposition, 325, 341, 344
  BLWRS, 360
  and nitrogen control, 414–415
  and secondary settling, 406
Density:
  air, table of, Appendix A
  of compacted landfill, 670
  defined, 134
  of gas, 460
  organic compounds, table of, Appendix A
  of solid waste, 635, 647
  water, table of, Appendix A
Dental caries and fluoride, 156
Deoxygenation rate constant (k_d), 309, 310–311
Department of Energy (DOE), 32
Department of Transportation (DOT), 728, 733, 737
Department of Transportation Act, 28
Depths of cover, solid waste, 664
Designated POHC, 778
Destruction and removal efficiency (DRE), 774, 777–778
Detention time (t_o):
  activated sludge aeration tank, 389, 414
  anaerobic digester, 437
  defined, 200
  disinfection of wastewater, 411
  flocculation, 205
  grit chambers, 366, 367
  hazardous waste incinerator, 775, 776
  oxidation ponds, 408
  primary settling tank, 373
  rapid mixing, 203
  settling tank, water, 227
Deuterium, 812
Deutsch efficiency equation, for ESP, 533
Dichloramine, 243
Diesel engine, 537
Digester, 382, 436–438
Digestion, stabilization of sludge, 434–438
Dilution factor in BOD analysis, 298
Dioxin, 680, 704–705
Direct filtration, 171
Direct flame incineration, 522
Directivity of noise, 593–594
Direct runoff, 48
  abstractions, 48
Discharge, 62, 63, 65
Disinfection:
  byproducts (DBPs), 165, 248, 249
  drinking water, 170, 171, 240–249
  kinetics, 241
  requirements, 167
  wastewater, 411
  well, 91
Dispersion of air pollutants, 500–509
  coefficients, 504–506
Disposal alternatives for solid waste, 638
Dissociation:
  table of constants, 145
  of weak acids, 145
Dissolved impurities, 135
Dissolved oxygen (DO), 286, 302, 303, 307, 319
  saturation, 303, 307
  table of, Appendix A
Distillation treatment of hazardous waste, 762–765
Distribution:
  system, 8
  works, 8
Divide, 48
DNA and bacterial growth, 346
Domestic:
  sewage, 11, 285
  wastewater, 12
  wastewater characteristics, 352–353
DO sag curve, 302–319
  equation of, 310
  management strategy, 313
  mass balance, 303–306
  point discharge, 317
Dose-response curve, 713, 826
Downwind attenuation of noise, 596
Drag coefficient:
  filter backwash, 237
  filter headloss, 235
  paddle, 206
  settling, 220–222

Drawdown:
aquifer:
confined, 99–100, 101–104
unconfined, 100, 104
well, 92
DRE, 774, 777–778
Dredging:
effect on DO sag curve, 319
effect on phosphorous, 327
Drift velocity, 534
DRH, 63
Drop-off center for solid waste, 685
Drought, 81
Dry adiabatic lapse rate, 494
Dry weather flow, 48
Dual catalyst, 541
Dual media filter, 229
Dupuit equation, 99–100
Duration:
rainfall, 58
unit hydrograph, 66
Dynamic equilibrium of microorganisms, 349
Dynamic viscosity:
air, table of, Appendix A
and cyclone efficiency, 526
and mixing, 201–202
and Reynolds number, 221–222
water, table of, Appendix A

Eardrum, 564–566
Earmuffs, for noise, 623
Earplugs, for noise, 623
Ecological habitat, 287
Effective size of sand grains, 223
Effective stack height, 500, 503
Effluent, 13
BOD, 340, 390
limitations, 19
suspended solids, 390
Einstein's equation, 813
Electrodialysis treatment of hazardous waste, 770–771
Electrolytic oxidation of cyanide, 760
Electron acceptor and decomposition, 341, 343–346
Electron volt (eV), 813
Electrostatic precipitator (ESP), 532–535
Emission offsets, 22
Enclosures for noise control, 619–621
Endogeneous decay, 350
End products of waste decomposition, 344–346
Energy carrier, bacterial metabolism, 343, 346
Energy and noise, 551, 582
Energy Organization Act, 32
Energy Research and Development Administration (ERDA), 32
Energy source, microorganism, 341
Engineering, definition of, 3
Engineer's Council for Professional Development (ECPD), 3
Engine exhaust, 541
Engine fundamentals, 535–538
Enthalpy, 306
Environmental:
ethics, 33–35
legislation, 16–33
systems, 7–16
Environmental engineering, definition of, 4
Environmental Protection Agency (EPA), 16, 18, 20, 22, 30, 159, 165, 353, 442, 463, 464, 465, 588, 590, 591, 658, 664, 674, 683, 689, 718, 719, 725, 726, 745, 834
Epilimnion, 321
Episode, air pollution, 473–475
EP toxicity, 728
Equalization basin, wastewater, 87, 368–372
Equivalent continuous equal energy level (L_{eq}), 582–584, 609–610
Equivalent weight (EW), 138
Escherichia coli, 158, 167, 342
Ethics:
environmental, 33–35
professional, 3
Euphotic zone, 322
Eutrophication, 324
Eutrophic lakes, 323–324
Evaporation, 47, 54–55
Evaporation losses, reservoir, 54–55, 87
Evapotranspiration, 55, 667
Excess attenuation of noise, 595
Exfiltration, 48
Expanded porosity, 236
Exponential growth, log-growth curve, 347, 348

Extract from liquid extraction, 772
Extraction procedure (EP) toxicity, 728

Facultative:
anaerobes, 341
ponds, 406, 407–409
thermophiles, 342
Far field, noise, 592
Feasibility study (FS), 747, 748, 793
Fecal coliform group, 411, 444
Federal Aid Highways Act, 27, 28
Federal Air Pollution Control Act, 20
Federal Aviation Agency (FAA), 75
Federal Highway Administration, 600
Federal Housing Administration (FHA), noise standards, 28, 587
Federal Insecticide, Fungicide and Rodenticide Act (FIFRA), 30
Federal Register (FR), 728
Federal Water Pollution Control Act, 17, 18, 284
national discharge permit system, 17, 19
navigable waters, 18
Fermentation and anaerobic decomposition, 346
Fermentation tube method, 167
Ferric chloride:
as coagulant, 177–178
and phosphorus removal, 413
Ferrous sulfate as coagulant, 177
Field capacity, landfill, 667–668
Filter:
and advanced wastewater treatment, 411–412
backwash sludge, 253
baghouse for air pollution, 528–529
box, 229
cake, 256
continuous belt press, 261, 441
deep-bed, 230
direct, 171
dual media, 229
grain size, 232–235
hydraulics, 235–240
press, 262–264
rapid sand, 229
slow sand, 229
for sludge dewatering, 258–259, 261–264, 439–441
vacuum, 261, 440
washwater, 231
wastewater, 411–412
Filtration plant, 8, 170
Final clarifier,
see Settling tank, secondary
Fission, nuclear, 819
503 regulations, 442–444
Fletcher-Munson equal loudness contours, 568
Floc:
activated sludge, 382
coagulation, 171, 201
defined, 171
softening, 201
Flocculation, 201, 204–211
basin, 170
Gt_o values, 206
particle removal efficiency, 204
Flocculator, 204–205
Flood, 81
Flood data paper, 77
Floodplain and hazardous waste, 786
Flotation thickening of sludge, 427–429
Flow, 52
Flow equalization, 87, 369
Flowing well, artesian, 49, 51
Flow reactor, 200
Flue gas desulfurization (FGD), 522–524
Flue gas recirculation, 524
Fluoride in drinking water, 156, 163
Fluorosis, 156
Flux, batch, 430
Fly ash, 535, 680
Fog and air pollution, 475
Food to microorganism ratio (F/M), 392–394
Formaldehyde and indoor air pollution, 483
Four 9s destruction of hazardous waste, 774
FR, 728
Free available chlorine, 242
Free field, noise, 592
Free residual chlorine, 246
Freeze treatment of sludge, 259–260
Frequency of:
noise, 553, 558–562, 567–574
occurrence, hydrologic, 59
radiation, 814
solid waste collection, 636, 640
Fuel tank evaporation losses, 541

Fungi, 342
Furans, 774

GAC, 249
Gamma radiation, 813, 814
Garbage, 632, 634
Gasoline engine and air pollution, 535–536
Gas transfer, 152–154
Gaussian dispersion, 502–509
Generator, hazardous waste, 733–737
Genetic effects, radiation, 832–833
Geomembrane, 673, 786
Giardia, 167
Global warming, 489–491
GLUMRB, 373
Gonadal dose, 837, 838
Grain per gallon (gr/gal), 197
Grain size analysis of sand, 232–235
Granular activated carbon (GAC), 249
Gravity spring, 50
Gravity thickening, 427, 429–434
Greenhouse effect, 489–491
Grit chamber, 365–368
  sludge, 367, 419
Groundwater, 7, 47, 49, 87–110, 166
  characteristics, 169
  contamination, 110, 789–790
  exfiltration, 48
  and hazardous waste landfills, 740, 784, 786
  and radioactive waste, 857–858
  remediation, 790–798
  supply, 88
  and wastewater treatment, 798
Grout, around well casing, 88–90
Growth constants, bacterial, table of, 390
Growth requirements:
  algal, 324–325
  bacterial, 346–347
Gt_o values, 206
Gumbel paper, 77
G values, 203, 204–205

Hair cells, ear, 567, 569
Half-life ($T_{1/2}$):
  defined, 816
  naturally occurring radioisotopes, 819
  table of, 816
Hardness, 171
  bar chart of, 180–182
  carbonate (CH), 180
  defined, 179
  formation of, 179
  noncarbonate (NCH), 180
  polyvalent cations, 179
  total (TH), 180
Haul time, solid waste, 654–655
Hazard index for risk assessment, 724
Hazardous air pollutants (HAPs), 23–26, 464
  and health effects, 471
  origin and fate, 476
Hazardous and Solid Waste Amendments (HSWA), 30, 733–745
Hazardous Materials Transportation Act, 30
Hazardous Ranking System (HRS), 746–747, 793
Hazardous waste:
  abandoned sites, 29
  biological treatment, 755–758
  characteristics, 726–729
  chemical treatment, 758–762
  code, 726
  contained-in policy, 732
  corrosive, 728
  defined, 703, 726–732
  delisted, 732
  derived from rule, 732
  dilution rules, 732
  extraction procedure toxicity (EP), 728
  ignitable, 727
  immobilization of, 782–783
  incineration of, 773–782
  land disposal restrictions (LDR), 741–742
  landfill, 784–789
  legislation, 29–31
  listed, 727
  table of, Appendix C
  management, 750–755
  manifest, 733–735
  mixture rule, 732
  from MSW incinerator, 682
  physical/chemical treatment, 762–783
  reactive, 728
  small quantity generator (SQG), 734
  stabilization/solidification, 782–783

TCLP, 729, 783
toxicity characteristic (TC), 728
transporter regulations, 737–738
treatment, storage, and disposal (TSD), 733
treatment technologies, 755–783
Universal Treatment Standard, 742
waste audit, 751
waste-code carry through, 732
waste minimization, 750–755
Head in aquifer, 93–96
Head loss in filter, 235–240
Hearing:
frequency range, 567
impairment mechanism, 569
loss, 569–573
mechanism, 564–567
protection, 573–574, 622–623
threshold level (HTL), 569, 571–573
threshold shift, 571–573
Heat:
balance, DO sag, 306
island effect, meteorological, 498
treatment of sludge, 439
as water pollutant, 285, 287
Heating value of solid waste, 677, 690
Height of mass transfer unit (H_{og} or HTU), 516, 520
Helium-4, 813
Henry's law, 153, 513, 763
Henry's law constant, table of, Appendix A
Heterotrophic, 341, 342
Heuristic routing, 652
Hexachlorobenzene (HCB), 774
Hexachloropentadiene (HCCPD), 774
Higher heating value, 677
High frequency notch, 572
High level radioactive waste, 850
High lift pumps, 171
High pressure, meteorological, 491–493
High rate irrigation, wastewater treatment, 416
Highway vehicle noise, 584–585
Holland's formula for plume rise, 503
Home separation of solid waste, 636, 685
Horizontal flow
clarifier, 212, 216–219
grit chamber, 366–367
See also Settling tank
Hot soak, 541
Housing and Urban Development (HUD) noise assessment, 588
HRS, 746, 793
HSWA, 30, 733–745
Hydrated lime, 183
Hydraulic conductivity, 93, 96
of landfill liners, 673, 674
Hydraulic detention time, 389
Hydraulic gradient, 93–97, 786, 788
Hydraulic loading, 377, 416, 770
Hydrocarbon (HC) emissions and A/F ratio, 539, 540
Hydrogen chloride, hazardous waste emission of, 773, 778
Hydrofluorocarbons (HFCs), 489
Hydrograph:
abstraction, 63
base flow, 63
compound, 69–71
definition, 63
direct runoff, 63
excess rainfall, 63
lag time, 65
unit, definition of, 63
unit duration, 66
Hydrologic:
cycle, 47, 667
equation, 52
system:
closed, 52
open, 52
year, 77
Hydrologic Evaluation of Landfill Performance (HELP), 668
Hyetograph, 66
Hypochlorite, 241, 242
Hypochlorous acid, 146, 241–242
Hypolimnion, 321
dissolved oxygen, 321

ICR hazardous waste, 727
Ideal gas law, 461
Ignitable hazardous waste, 727
Immediate removal of hazardous waste, 794
Immobilization of hazardous waste, 782–783
Impact noise, 563

Impeller:
 constant, table of, 207
 types for mixing, 204
 power for mixing, 206
Impulse noise, 563
Incineration:
 of air pollutants, 522
 of hazardous waste, 773–782
 catalytic, 522
 modular, 680–682
 of sludge, 441
 of solid waste, 678–679
 water wall, 679
Indoor air pollution, 481–485
Indoor air quality model, 509–511
Industrial wastewater, 285
 characteristics, 285, 353–354
 pretreatment, 363–364
Inertial collection, 525
Inertial impaction parameter, 530
Infiltration, groundwater, 12, 48, 53–54
Infiltration and inflow (I & I), 12
Initial deficit (D_a), 307
Initial ultimate BOD, 305
In-line blender, 202, 203
Inspection/maintenance (I/M), 541
Integrated Risk Information System (IRIS), 718, 724
Intensity:
 rainfall, 58
 sound, 554
Intensity-duration-frequency curves (IDF), 58–62
Interception, 48
Interface:
 gas-liquid, 152–153
 liquid-solid, 412
Interference, well, 92
Interim primary drinking water regulations (IPDWRs), 159
Intermediate-level radioactive waste, 850
Intermittent noise, 563
 exposure, 574
Internal combustion engine noise, 588
International System of Units (SI), 5–7
 table of, *inside back cover*
Interparticle bridging, 178
Inverse square law:
 noise, 590, 592
 radiation, 841–842
Inversion:
 aloft, 508
 defined, 496
 form of dispersion equation, 508
 and noise, 594
Ion association, 140
Ion exchange:
 and hazardous waste, 765–770
 softening, 197–199
Ionic strength, 150
Ionization, chemical, 140–141
Iron, in drinking water, 157
Irrigation with wastewater, 416–418
Isobars, 491–493
Isolation for noise control, 614
Isotherm, adsorption, 519
Isothermal lapse rate, 496
Isotope, 813

Jar tests, 175–177, 225
Jet engine combustion, 538
Joint and several liability for hazardous waste, 748

KeV, 813
Kinematic viscosity,
 air, Appendix A
 Reynolds number, 221
 water, table of, Appendix A
Kinetic reactions, 40, 151–152
Kingdoms and microorganism classification, 340
Kjeldahl nitrogen, 352–353

Lagoons:
 sludge, 257–258
 wastewater treatment, 406–409
Lag phase, log growth curve, 347
Lag time:
 hydrograph, 65
 log-growth curve, 347
Lake:
 breeze, 498–499
 overturn, 321–322
 productivity, 323–324
 stratification, 321–322
Lakes, water quality management, 284, 320–330

Land ban, 741–742
Land disposal restrictions (LDR), 741–742, 783
Landfill:
  aerobic and anaerobic decomposition in, 668
  cell, 663, 664
  compaction ratios, 670
  density, 670
  depth of cover, 664
  designs, 662–664
  environmental considerations, 665
  equipment, 659–662
  final cover, 676
  gases, 665
  hazardous waste:
    cap, 786
    clay liner, 784, 787
    leachate collection, 787–788
    monitoring wells, 786, 789
    siting, 784–786
  leachate, 667–668
  leachate collection, 674–676
  liner, 673–674
  operation, 662–664
  site selection, 658–659
  volume, 668–673
Landfilling of sludge, 441, 442
Land/sea breeze, 498–499
Land spreading of sludge, 441
Land treatment:
  of hazardous waste, 783
  of wastewater, 415–418
Langmuir adsorption isotherm, 519
Lapse rate and atmospheric stability, 494–498
LD_{50}, 713, 714–715
Leachate:
  collection, 674–676
  hazardous waste, 787–788
  sanitary landfill, 667–668
Lead:
  in air, 465, 472, 541
  in drinking water, 157
  origin and fate of in air pollution, 476–477
  shielding, 846
Lean fuel mixtures (A/F ratio), 539
Level, sound:
  adding, 552–558
  averaging, 562
  defined, 555
  relative, 556
  statistic, 562, 581–582
L_{dn}, 590, 591, 610
L_{eq}, 582–584, 609
L_N, 581
L_{10} nomograph, 606
  blank for use, Appendix B
Liability for hazardous waste, 748
Liebig's law, 326
Lift station, 13
Light compensation level, 323
Limbus spiralis, 567
Lime:
  and calcium hydroxide, 178
  and flue gas desulfurization, 523
  and hazardous waste, 783
  and phosphorus removal, 413
  purity, 189
  slaking, 183
Lime-soda softening, 183–197
  calculations flow chart, 187
  pH, 183
  reactions, 184–185
Limestone, 466, 523
Limiting nutrient in lakes, 326
Limnology, 321
Linearized multistage model of toxicity, 718
Line-of-sight distance, noise barrier parameter, 556
Liner, landfill, 673–674
Liquid extraction, 772
Liquid injection hazardous waste incinerator, 775
Liquid ion exchange, 772
Liquid-liquid extraction, 772
Liquid scrubbing, 511, 529–532
Listed hazardous waste, 727
  table of, Appendix C
Liters per capita per day (Lpcd), 10
Lithotrophs, 341
Littoral zone in lakes, 323
Loading rate:
  anaerobic digester, 436, 477
  defined:
    filters, 228
    settling tanks, 216
    weirs, 215
  ion exchange column, 199

Loading rate (*cont.*)
oxidation ponds, 408
primary settling, 373
rapid sand filter, 229
secondary settling, 403
settling tank, surface, 216
slow sand filter, 229
solids, 256
trickling filter, 377
weir, 215, 373
Log growth curve, bacterial, 347
Log-Pearson Type III distribution, 78
Loudness, 568
Lower heating valve, 677
Lower respiratory tract (LRT), 470
Low fence, 573
Low flow, partial series, 81
Low-level radioactive waste (LLRW), 850
management, 853–863
Low-Level Waste Policy Act (LLWPA), 32, 853
amendments, 32
Low lift pumps, 171
Low pressure, meteorological, 491–492
Lung cancer, 471
LUST, 742

Macro air pollution, 481
Macronutrients, 346
Magnesium:
drinking water, 178–180
hardness, 178–180
scale, 186
softening, 184–186
Mainstream smoking, 484
Manifest, hazardous waste, 733, 734, 735
Market-based air pollution control, 486
Masking noise, 574
Mass balance:
absorption column, 513–516
activated sludge process, 383–390
approach, 35–41
conservative, 36, 38
definition, 35
diagram, 35, 36
and DO sag, 303–306
equalization basin, 87, 368–372
hazardous waste, 751–753
hydrologic, 52–53
and problem solving, 35–41
and rational method, 71–73
and reactor design, 200–201
and reservoir design, 84–87
and retention basin, 87
and return sludge, 396–397
and sludge, 253, 423–427
Mass defect, 812
Mass density, 134
Mass excess, 812
Mass transfer, 152, 249, 511
Materials balance, *see* Mass balance
Maturation ponds, 407
Matrix, 759
Maximum achievable control technology (MACT), 23, 464
Maximum contaminant level (MCL), 159, 725
tables of, 160, 166, 168
Maximum haul time, solid waste, 654
Maximum permissible concentrations of radionuclides (MPC), 834–836
Mean cell residence time (θ_c), 383
Media:
dual, 229
mono, 230
multi, 412
trickling filter, 375
Meltdown, nuclear reactor, 820
Membrane filter method, 167
Mercaptans, 346
Mesophiles, 342
Mesotrophic lakes, 324
Metabolism, 343
Metal recovery, 682, 755
Meterage, 9
Methane:
and anaerobic digestion, 346
as an end product, 346
fermentation, 346, 435
recovery from landfill, 689–690
Methemoglobinemia, 157
MeV, 813
mg/L as $CaCO_3$, 149
Micro air pollution, 480
Microcurie, 816
Micrograms per cubic meter ($\mu g/m^3$), 462–463
Micrometer, 462
Micron, 462

Micronutrients, 347
Migration velocity, 533
Millicurie, 816
Milliequivalents (meq), 139
Milligrams per liter (mg/L), 138
Million electron volts (MeV), 813
Minimal media test, 167
Mitigation, groundwater contamination, 795–798
Mixed liquor:
- activated sludge, 382
- suspended solids (MLSS), 394, 758
- volatile suspended solids (MLVSS), 388

Mixers, 202–205
Mixing:
- and DO sag, 305
- and water treatment, 201–211

Moderators, nuclear reactor, 819
Modular incinerator, 680–682
Molar concentration, 138
Molarity, 138
Molecular weight (MW), 138
Molecular sieve, 518
Monitoring wells, 740, 789, 860–861
Monochloramine, 243
Monod equation, 349–351, 387
Montreal Protocol on ozone, 488, 542
Most probable number (MPN), 444
Motor Vehicle Air Pollution Control Act, 20
Mottled soil, 357
Mottling of teeth, fluoride, 156
Muffler, noise control, 615
Multiclone, 528
Multimedia:
- filter, 412
- pollution, 7
- systems, 7, 15

Multiple emissions, radioactive, 815
Multiple-tube fermentation technique (MTF), 167
Multistage model of toxicity, 717
Municipal:
- collection of solid waste, 639
- sewage, 11, 285
- solid waste (MSW), 635, 679–680, 682–684
- wastewater treatment plant (WWTP), 11

Mutation, radiation, 832–833
National Acid Precipitation Assessment Program (NAPAP), 486
National ambient air quality standards (NAAQS), 22, 463, 465
National Contingency Plan (NCP), 31, 747–748
National Cooperative Highway Research Program (NCHRP), 599
National emission standards for hazardous pollutants (NESHAPs), 23, 464
National Institute for Occupational Safety and Health (NIOSH), 574
National Institute of Municipal Law Officers (NIMLO), 27
National Pollution Discharge Elimination System (NPDES), 17
National Priorities List (NPL), 745, 793
National Research Council (NRC), 378
National Safe Drinking Water Act, *see* Safe Drinking Water Act
Natural background radiation, 837–840
Navigable waters, 20
NCH_f, NCH_i, NCH_R, 188
NCP, 31, 747–748
Near field, noise, 592
Necrosis, 468
Nephlometric turbidity unit (NTU), 155
Net heating value of solid waste, 677
Network, weighting, for noise meter, 558–561
Neutral solution, 144
Neutron:
- activation, 819
- bombardment, 818

New source performance standards (NSPS), 22
Nitrate:
- and acid rain, 485–486
- in drinking water, 157
- as an electron acceptor, 344

Nitrification:
- ammonia oxidation, 301
- denitrification, 344, 360, 414–415

Nitrifying organisms, 301
Nitrogen:
- in activated sludge, 401–402
- and algal growth, 325
- and bacterial growth, 302
- and DO sag, 317
- effect on receiving body, 319–320

Nitrogen (*cont.*)
  as a nutrient, 286
  oxidation, 301–302
  treatment, 344, 360, 414–415
Nitrogen dioxide:
  and acid rain, 485–486
  control technologies, 524–525
  emission of, 477
  emissions and A/F ratio, 539–540
  health effects of, 472
  and indoor air pollution, 481
  origin and fate, 477
  standard, 465
Nitrogenous BOD (NBOD), 301–302
NOAEL, 713, 714
Noise:
  attenuation in atmosphere, 595
  attenuation estimate of X_0, 597
  construction, 585, 589
  control, 610–623
  defined, 551
  directivity, 593–594
  effects on people, 564–580
  and land use compatibility, 588–591
  paths, 610, 616, 617–619
  pollution control legislation, 26–28
  prediction worksheet, 603
    blank for use, Appendix B
  rating systems, 580–584
  receiver protection, 522–523
  reduction coefficient (NRC), 616–617
  shadow zone, 597
  spectra, 561
  traffic, prediction of, 599–610
  transportation, 584–585
  and zoning, 588
  See also Sound
Noise Abatement Act, 28
Noise Abatement Commission, 27
Nomographs for noise prediction, 600, 606, 607
  blank for use, Appendix B
Non-Aqueous Phase Liquids (NAPLs), 798
Nonattainment area, 464, 542
Noncarbonate hardness (NCH), 180
  precipitation of, 184–185
Nonflowing well, artesian, 49, 51
Nonhomogenous aquifer, 106–108
Non-point source of pollution, 285, 319
Normality, 139
NPL, 31, 745, 793
Nuclear reactor, 819
Nuclear Regulatory Commission (NRC), 32, 834
Nuclear Waste Policy Act, 32, 852
  amendments, 32
Nuclide, 811
Nuisance law municipal ordinances, 20
Number of gas transfer units (N_{og}), 515, 520
Nutrients in water pollution, 285–286, 319–320, 325, 415
  See also Nitrogen; Phosphorus

Occupational exposure, and air pollution, 469
Occupational noise exposure limits, 573–574
Occupational Safety and Health Act (OSHA), 28
Octave bands, table of, 562
Odor, 156, 351
Oil Pollution Act, 16
Oligotrophic lakes, 323
O & M costs for solid waste collection, 649
One-hit model of toxicity, 717
ONPG-MUG test, 167
On-site disposal systems, 354–361
Open system:
  hydrologic, 52
  microbial, 348
Order of reaction, 151
Organ of Corti, 566
Organic acids as electron acceptors, 346
Organotrophs, 341
Ossicles, in ear, 564
Outhouse, 354
Oval window of ear, 565–567
Overflow rate (v_o):
  defined, 216
  determination of, 220–228
  primary settling, 373
  secondary settling, 403
  water treatment, table of, 227
  weir, 215, 373
Overhead costs for solid waste collection, 649
Overland flow wastewater treatment, 416, 418

Overturn of lakes, 321
Oxidation ponds, 375, 406–409
Oxidation-reduction reactions, 141, 759
Oxyfuel, 541
Oxygen:
  deficit in rivers, 303, 307, 310
  demand, theoretical, 289
  demanding waste, 286–289, 302, 320
  as an electron acceptor, 344
Ozone:
  depletion, 486–489
  disinfection, 247–248
  effects on:
    people, 473
    plants, 468
  for odor control, 356
  legislation, 26
  origin and fate of, in air pollution, 477
  photoreactions, 478, 487
  standard for air pollution, 465

PAC, 249
Packed tower, 512–518
Paddle mixer, 205
  power for, 206
Palatable, 134
Parent isotope in radioactive decay, 814
Parshall flume, 203
Partial duration series, 78
  low flow, 81
Particle:
  counting, 136
  respirable, 470
  settling velocity, 220–222
  sizes, 137, 464
Particulate matter in air (PM_{10}, $PM_{2.5}$):
  control of, 525–535
  and episodes, 474, 475
  health effects of, 473
  from incinerators, 678, 774, 778
  origin and fate of, 480–481
  respirable, 470, 482, 483
  standard, 464, 465
Parts per million (ppm), 462
Pathogens:
  in drinking water, 158, 167, 171, 240
  in wastewater, 286
PCB, 31, 705–707, 726, 774, 777, 781–782
Peak flow, 9
Peak rate of demand, 9
Per capita, 9
Perched water table, 49, 50
Percolation test, 357
Periodic table, Inside front cover
Period of sound wave, 553
Permanent threshold shift (PTS), 571–572
Permits:
  air, 26
  NPDES, 17
Personal pollution and air pollution, 469
pH:
  and acid rain, 328–330, 485–486
  and ammonia stripping, 415
  and bulking sludge, 406
  and coagulation, 173, 175, 177
  defined, 144
  hazardous waste, 728, 759, 785
  and phosphorus removal, 413
  and softening, 183
  of wastewater, 352
Phons, 568
Phosphorus:
  and activated sludge, 401–402
  and algal growth, 325
  control of, 327–328, 413
  in detergents, 327
  in lakes, 325, 326–328
  as a nutrient, 286, 320
  in wastewater, 320, 327–328, 353, 360, 413, 434
Photochemical oxidants:
  health effects, 473
  origin and fate, 477–478
Photochemical reactions, 477, 478, 487
Photochemical smog, 478
Photon of light energy ($h\nu$), 477, 487
Photolysis, 476
Photosynthesis, 342
Photorophs, 341, 342
Physical/chemical treatment of hazardous waste, 762–783
PIC, 773, 774
Picocurie, 816
Piezometers, 51
Piezometric pressure, 51
  surfaces, 49
Pit privy, 354–355
pK_a, 145
pK_s, 142

pK_w, 144
Planck's constant, 475, 814
Planned removal of hazardous waste, 794
Plasmolysis, 468
Plate pressure filter for sludge, 262–264
Plate tower, 512
Plug flow reactor, 200, 390–391
Plume dispersion, 500–501
Pneumatic mixing, 207
POC, 773
POHC, 774, 777–779
Point precipitation analysis, 57–62
Point sources of pollution, 285
Polishing ponds, 407
Pollutant categories, 285
Polychlorinated biphenyls (PCBs), 31, 705–706, 726, 774, 777, 781–782
Polymers, 178
Polynuclear aromatic hydrocarbons (PAH), 774
Polyphosphate, 327
Polyvalent cations, 179
Population dynamics, microorganism, 348
Porosity, 93
  of rapid sand filter, 235–237
Positive crankcase ventilation (PCV), 541
Positron, 814
Post-Closure Liability Fund, 31
Potable, 134
Potentially responsible parties (PRP), 748
POTW, 11
Power, requirements, mixing, 206–207
Power, sound, 554
ppm, 462
Precipitation:
  chemical, 141–142, 760–762
  duration, 58
  equilibrium reactions, 141–142
  frequency, 59
  hazardous waste, 760–762
  intensity, 58
  interception, 48
  space, 57
Precoat filter for sludge, 261
Precursor:
  and acid rain, 485
  defined, 165
  trihalomethanes, 165
Preliminary assessment of hazardous waste contamination, 793
Presbycusis, 569
Presedimentation sludge, 251
Presence-absence coliform test, 167
Pressure:
  artesian piezometric, 49, 50–51
  atmospheric:
    gradient, 492
    high and low, 492–493
  head, 94
  sound, 554–557
  water, 8
Pretreatment of industrial waste, 363–364
Pretreatment unit operations, 361, 364–372
Prevention of Significant Deterioration (PSD), 22
Primary:
  settling, 372–374
  sludge, 420
  treatment, 361, 362, 372–374
Primary ambient air quality standards, 463, 465
Primary Drinking Water Regulations (IPDWRs):
  and hazardous waste, 789
  maximum contaminant levels (MCLs), 159, 160–164
Principal organic hazardous component (POHC), 777, 779
Priority pollutants, 18–19, 157
  table of, 19
Private collection of solid waste, 639
Products of complete combustion (POC), 773
Products of incomplete combustion (PIC), 773, 774
Profession, 2
Professional ethics, 3
Profundal zone, 323
Protista, 340
Proton (H^+), 143
Protoplasm, 467
Protozoa, 342
PRP, 748
Psychoacoustic, 568
Psychrophiles, 342
Public Health Service Act, 16
Publicly owned treatment works (POTW), 11
Public water supply systems, 88, 132
  defined, 159

Pulmonary emphysema, 471
Pumphouse, 89, 91
Pumping test curve, 105
Pumps:
  high lift, 171
  low lift, 171
  submersible, 88
Putrescible, 13

Quantitative flow diagram, 423–427
Quicklime, 183

Rad, 824
Radiation:
  attenuation, 843
  biological effects, 825–834
  dose, 823–825
  exposure, 825, 837–841
  natural background, 837–840
  protection, 841–850
  shielding, 842–847
  standards, 834–836
Radiation Control for Health and Safety Act, 33
Radiation fields of a sound source, 592–593
Radioactive:
  daughter, 814
  decay, 813–818
  defined, 815
  parent, 814
  waste, 850–863
  waste compacts, 853, 854
Radioactivity, 813
Radioisotope, 813, 818–819
Radon:
  control, 848–850
  indoor air pollution, 483
  standards, 834
  source, 483, 838–839
Radium, paint, 830
Radius of influence of well, 92
Raffinate, 772
Rainfall:
  analysis, 57–62
  spatial distribution, 57
Rainfall intensity:
  annual average probability, 59
  average recurrence interval, 59
  duration, 58
  frequency, 59
  probability, 59
  rank, 59
  return period, 59
Rain gage, 57
Rapid infiltration, wastewater treatment, 416, 418
Rapid mix, 170, 171, 202–204
  G values, 203
Rapid sand filter, 229
Rate:
  definition, 36
  of oxygen depletion, 291, 309
  rainfall, 58
  reaction, 40, 151–152
  transformation, 40
  See also Loading rate
Rate constant:
  air pollutant, 510
  base e, 291
  biological, table of, 390
  BOD, 291
  defined, 40, 151
  deoxygenation, 309, 310–311
  effect of temperature on, 293
  endogeneous decay, 350
  graphical determination of, 294–297
  growth rate, 350
  half saturation, 350
  nitrogenous deoxygenation, 317
  and reaction order, 151
  reaeration, 309, 311
  and temperature, 293–294, 300, 311
Rating systems, noise, 580–584
Rational method, 71–77
  frequency of occurrence, 76
  runoff coefficients, table of, 74
  steady state, 73
Raw:
  sewage, 351–353, 374
  sludge, 372, 420
  water, 171
RBE, 824
RCRA, 29, 658, 731, 732–733, 741
Reach of a river, 307, 317–319
Reaction:
  basin, 413
  heterogeneous, 151
  homogeneous, 151

Reaction (*cont.*)
kinetics, 40, 151–152
order, 151–152
rate, 40
rate constant, 40, 151–152, 309
Reactive hazardous waste, 728
Reactors, 199–211
batch, 199
completely mixed flow (CMF), 38, 200
completely stirred tank (CSTR), 200
design, 200
flow, 200
nuclear, 819
plug flow, 200
Reaeration:
and DO, 303
rate constant (k_r), 309, 311
Recarbonation, 171, 190, 194, 198
Receiver protection from noise, 622–623
Receiving body, 13
Recharge area, 49, 51
Recirculation ratio, 377, 379
Record of decision (ROD), 748, 793
Recurrence interval, 59
Recycling of:
hazardous waste, 754–755
solid waste, 682–691
Reductive dehalogenation, 756
Reference dose for toxicity (RfD), 718, 719
Refractory organics, 412
Refuse, 632, 634
Refuse-derived fuel (RDF), 679–680
Reg-neg, 165
Reid vapor pressure, 541
Relative biological effectiveness (RBE), 824
Rem, 825
Remedial:
design (RD), 794
investigation (RI), 747–748, 793
response, 794
Remediation of hazardous waste, 794
Removal of hazardous waste, 794
Reservoirs, 84–87
Residual:
BOD, 412
sludge, 441
suspended solids, 412
Resins, ion exchange, 197, 795
Resistivity, fly ash, 535
Resource recovery, 682–691
Resource Conservation and Recovery Act (RCRA), 29, 658, 731, 732–745, 777
permit application requirements, 733–742
Resource Recovery Act, 29
Respirable particles (RSP), 482, 483
Retention basin, 87
Retention time (t_o), 200
See also Detention time
Returnable beverage containers, 684–685
Return period, 59
mean annual flood, 78
Return rate, activated sludge, 396, 397
Return sludge, 382, 394–398
Reverberation of noise, 593, 616
Reverse flow cyclone, 526
Reverse osmosis treatment of hazardous waste, 771–772
Reynolds number (**R**), 221, 223, 235, 236
RI/FS, 747–748, 793
Right-to-know, 749
Rippl method, 84–87
Rising sludge, 344, 406
Risk:
assessment, 707, 712–725
characterization, 724–725
and chronic daily intake (CDI), 720–724
defined, 707
and exposure, 718–724
and hazard, 707
management, 707, 725–726
perception, 707–712
reasonable maximum exposure (RME), 720
and toxicity, 712–718
River and Harbor Act, 20
Roentgen (R), 823
Root mean square sound pressure, 554
Rotary kiln hazardous waste incinerator, 775–777
Rotating biological contactor, 375, 409–410
Rotifers, 342
Round window, ear, 565, 567
Routing, solid waste collection, 651–653
Rubbish, 633, 634
Run-of-bank sand, 235
Runoff:
analysis, 62–84
coefficients, 73, 74
defined, 47

direct, 48, 63
rational method, 71–77
stage, 62

Sabin, absorption coefficients, 616
Safe Drinking Water Act (SDWA), 16
amendments, 16
Salmonella in sludge, 444
Salt as a water pollutant, 287
Salvage, 685
Sample size, BOD, 298
Sand drying beds for sludge, 258–259
Sand filter,
See also Rapid sand filter
Sanitary:
landfill, 658–677
seal, well, 90
sewage, 11
See also Landfill
SARA, 733, 745–755
Saturation concentration:
and equilibrium, 152–153
oxygen in water, table of, Appendix A
Scala:
media, 566
tympani, 566
vestibuli, 566
Scale-up factors, settling tank, 224, 225
Schulze's equation, 380–381
Schultz-Hardy rule, 173, 174
Scouring in sedimentation tank, 213
Screen, well, 89
Scrubber, 511
Secondary air pollutants, 477
Secondary ambient air quality standards, 464, 465
Secondary clarifier, *see* Secondary settling tank
Secondary effluent, 331
Secondary maximum contaminant levels (SMCL), 168
Secondary settling tank, 375, 376, 382, 402–406
denitrification in, 406
side water depth, table of, 404
and SVI, 395
trickling filter, 375
Secondary sludge, 420
Secondary treatment, 361, 362, 374–411
Second stage digestion, 435
Sedimentation, 170, 171, 211–228
activated sludge, 220, 375, 376, 382, 402–406
basin, 170, 171, 212, 213, 214
grit, 219, 365–367
Type I, 219, 220, 366, 402
Type II, 220, 222, 372, 402
Type III, 220, 402
zone, 211–212
See also Settling tank
Sediment effect on DO sag, 319
Seepage spring, 50
Seepage velocity, 788
Selective catalytic reduction, 525
Senescent lakes, 324
Separation of solid waste, 636, 685, 688
Septic sewage, 351
Septic tank and tile field, 328, 356
Sequencing batch reactor (SBR), 757
Set-out, set-back pickup of solid waste, 640, 641
Settleability of activated sludge, 393
Settling column, 222–225
Settling tank:
baffles, 211, 212
cleaning, 212, 213
horizontal flow clarifier, 212, 216–219
particle trajectory, 217–218
percentage removal, 218–219, 223–225
inlet zone, 211, 212
outlet zone, 211, 212
overflow rate, 215–220, 223, 227
primary, wastewater, 372–374
removal efficiency, 216, 219, 223–225
scale-up factors, 224, 225
scouring, 213
secondary:
activated sludge, 375, 376, 382, 402–406
trickling filter, 375, 376
sludge:
collector, 212, 372, 373
hopper, 212, 373
storage, 211, 212
volume, 211–212, 382, 383, 392, 399, 422
zone, 212
surface loading rate, 216
upflow clarifier, 208, 216
weir arrangement, 215

Settling tank (*cont.*)
weir overflow rates, 215
zone, 212
Settling velocity, 216, 218, 219–222
Sewage:
domestic, 11, 351–353
flow, 12
industrial, 11, 353–354
lagoon, 406
raw, 351–353, 374
Sewers:
combined, 12, 285
sanitary, 12
storm, 12
Shockwave, noise, 578
Shredding of solid waste, 688
Side stream smoke, 484
Side water depth (SWD):
primary settling tank, 373
secondary settling tank, 403–404
Sieve, U.S. Standard, 232
Silica gel, 518
Sink, pollutant, 490, 509
Sinusoidal wave, 553
Site inspection, 793
SI units, 5–7, *inside back cover*
Slaking 183
Sleep interference, noise and, 578
Slope factor for risk assessment, 718, 724
Slow rate irrigation, wastewater treatment, 416
Slow sand filter, 229
Sludge:
age, 383
blanket, 207
bulking, 405–406
cake, 256
centrifuging, 260
characteristics of:
wastewater treatment, 419–420
water treatment, 250–253
conditioning, 255, 419, 439
continuous belt filter press (CBFP), 261, 441
disposal regulations, 442–444
dewatering, 254, 419, 439–441
drainage, 259
drying bed, 258–259, 439–440
filter backwash, 250, 253
filter press, 262–264
freeze treatment, 259–260
hazardous waste, 782
hopper, settling tank, 212
incineration, 441
lagoon, 257–258
management, 250–251
mass balance, 253, 423–427
minimization, 254
primary, 420
production, 344, 346, 399
raw, 372, 420
reduction, 419, 441
regulations, 442–444
retention time (SRT), 383
return, 382, 394–398
rising, 344
storage, settling tank, 211, 212
thickening, 254–256, 419, 427–433
treatment, 254–264, 418–441, 691
ultimate disposal, 264, 441, 691
vacuum filtration, 261, 440–441
volume, 251–253, 256
coagulation, 252
softening, 252
wasting, 382, 383, 392
water treatment plant, 171, 250–264
Sludge volume index (SVI), 394
Small quantity generator (SQG) of hazardous waste, 734
Smog, 475
photochemical, 478
Smoking and air pollution, 469, 482–483
Soda ash (soda), 178, 183–197, 523
purity, 189
Sodium carbonate, 178, 183, 523
Sodium in drinking water, 157
Softening:
and alkalinity, 183
and bicarbonate, 183
calculations flowchart, 187
and carbonate, 183
and carbonate hardness (CH), 184–197
cases, 196–197
floc, 201
ion exchange, 197–199
lime-soda, 183–197
mixing, 171
and noncarbonate hardness (NCH), 184–197
and pH, 183

plant, 8, 171
process, 171
reactions, 183–185
and split treatment, 194–196
sludge, 171, 252
Soil erosion as a water pollutant, 287, 328
Soil percolation test, 357
Solid bowl centrifuge, 260
Solidification of hazardous waste, 782–783
Solids:
computations, 256, 420–427
dissolved, 135
settleable, 353
suspended, 135
Solids retention time (SRT), 383
Solid waste:
characteristics, 632–635
collection, 639–658
collection cost, 641, 649–650
compaction density, 646, 670
compaction ratio, 646, 670
defined, 631
density, 635, 647
disposal alternatives, 638, 658–691
generation, 631–632
heating value, 677
legislation, 28–29
management, 14–15, 635–639
production rates, 631–635
Solid Waste Disposal Act, 28
Solubility constant, 142
table of, 142, Appendix A
Solvent:
extraction, 772
recovery, 755, 772
Sonic booms, 577–578
Sorption capacity, 785
Sound:
analyzer, 559
continuous, 563
directivity, 593–594
equivalent energy, 582–584
frequency, 553, 558–562, 567–574, 615
impact, 563
impulse, 563
intensity (I), 554
intermittent, 563
level:
adding, 557–558
averaging, 562
defined, 555
meter, 558
relative, 556
octave bands, 561–562
power (W), 554, 555
pressure, 554–557
speed of, 555
steady-state, 563
transmission, 590–610, 619
and stability, 594
waves, 552–554
Source separation, solid waste, 636, 685
Specific activity, 816
Specific gravity, 135
of microorganisms, 339
of sludge, 421
Specific weight of water, 134
table of, Appendix A
Specific yield of aquifer, 93
Speech interference level (SIL), 575
Speech interference, noise and, 574–576
Speed of sound (*c*), 555
Split treatment, 194–196
Spray chamber, 511, 512
Spray-dryer FGD, 523
Spring, 49–50
Stability of atmosphere, 494–498, 499, 501, 504
class, table of, 506
and noise, 594, 597
Stabilization of:
hazardous waste, 782–783
sludge, 419, 434
wastewater, 13, 339
Staff gage, 62
Stage, 62
Staged combustion, 524
Standard Methods for the Examination of Water and Wastewater, 297
State implementation plans (SIPs), 21
Static mixers, 207
Static water table, 92
Stationary phase, log growth curve, 347, 348
Steady state:
drawdown, 99–101
and mass balance, 38
noise, 563
time of concentration, 72
transmissibility, 104

Steam stripping of hazardous waste, 764
Stenothermophiles, 342
Sterilization, 240
Stilling well, 62
Stoichiometric, 136, 152
  defined, 539
Stokes law, 222, 366
Stoma, plant, 467
Storage coefficient:
  confined aquifer, 93
  unconfined aquifer, 93
Storage containers, solid waste, 636
Storage reservoir, 8, 84–87, 170, 171
Storm sewer, design frequency, 76
Stormwater, 12–13
Stratification of lakes, 321–322
Streamflow, 51–52, 62
Stream gages, 62–65
Streams Investigation Station, 16
Stripping, 147
Strong acids, 144
Subadiabatic lapse rate, 496
Substrate, waste as a, 343, 348
Subtitle C of RCRA, 732
Subtitle D of RCRA, 658, 664, 673, 674, 732
Sulfate:
  and acid rain, 485–486
  in drinking water, 157
  as an electron acceptor, 346
Sulfur oxides:
  and acid rain, 329, 485–486
  from combustion, 478–479
  effects on health, 473
  effects on materials, 466
  and electrostatic precipitators, 535
  and hazardous waste, 773
  origin and fate of, 478–480
  standard, 465
Superadiabatic lapse rate, 494
Superfund, 31, 733, 745
Superfund Amendments and Reauthorization Act (SARA), 31, 733, 745–755
Supersaturated solution, 142
Support media, filter, 230–231
Surface loading rate, 216, 229
Surface runoff, 47, 667
Surface water, 7, 47, 169
Surface water treatment rule (SWTR), 166
Surrogate POHC, 779
Susceptible population, 469
Suspended solids (SS), 135, 285–287, 353, 387
  digester supernatant, 434, 437
  filtration, 411–412
  removal efficiency, 361, 374, 412
  residual, 412
  as a water pollutant, 135
Synergism, 469
Synthetic organic chemicals (SOCs), 249
Systems approach, 7
Systems engineer, 7

Taste and odor, 156
TC, 727
TCDD, 705
TCLP, 729, 783
Tectoral membrane, ear, 567
Temperature, of drinking water, 156
Temperature coefficients:
  BOD rate constant, 294
  deoxygenation rate constant, 294, 311
  reaeration rate constant, 311
Temporary threshold shift (TTS), 571, 572
Terminal electron acceptor, 344, 346
Terminal head loss, 231
Terminal settling velocity (v_s):
  calculation of, 220–222
  and electrostatic precipitator, 533
  and particle deposition in lung, 470
Tertiary:
  ponds, 407
  sludge, 420
  treatment, 411
Theim equation, 99–100
Theoretical detention time (t_o), *see* Detention time; Retention time
Theoretical oxygen demand (ThOD), 289–290
Thermocline in lakes, 321
Thermophiles, 342
Thickener design, 430–432
Thickening, sludge, 256, 427–433
Thorium series, 818
Threshold dose, radiation, 827
Threshold planning quantity (TPQ), 749
Threshold shift, 571–573
Tile fields, table of application rates, 358

Time, to critical point in DO sag (t_c), 303, 312
Time of concentration (t_c), 65, 72, 75
Times Beach, Missouri, 703
Tipping bucket, rain gage, 57
Tissue weighting factor, 825
Title III of SARA, 749
Total dissolved solids (TDS), 135, 287, 353
Total hardness (TH), 180
Total Kjeldahl nitrogen (TKN):
- in digester supernatant, 434, 437–438
- in wastewater, 352–353

Total organic carbon (TOC), 166, 353, 354
Total suspended particulates (TSP), 464
Toxic hazardous waste, 726
Toxic inorganic substances:
- in drinking water, 156–157, 162–163
- as water pollutant, 287

Toxic organic substances in drinking water, 19, 157, 160–162
Toxic Release Inventory (TRI), 749
Toxic Substances Control Act (TSCA), 31, 732, 777, 781–782
Toxicity:
- and carcinogens, 715, 716, 724
- characteristic leaching procedure (TCLP), 729, 783
- and dose, 713–714
- and IRIS, 718
- and LD_{50}, 713, 714–715
- linearized multistage model, 718
- multistage model, 717
- and NOAEL, 713, 714
- and non-carcinogens, 715, 724
- one-hit model, 717
- reference dose (RfD), 718, 719
- slope factors, 718

Traffic noise model (TNM), 600
Traffic noise prediction, 599–610
Transfer station for solid waste, 654
Transient flow in aquifer, 98, 101–104
Transmissibility, 97
Transmission:
- airborne, of noise, 590–610
- loss, noise, 619, 621
- path and noise control, 610, 616, 617–619
- works, water, 8

Transpiration, 47
Transportation noise, 584–587, 599–610
Transporter regulations, hazardous waste, 737–738
Transuranic radioactive waste, 850
Trapping, precipitation, 48
Trash, 633, 634
Traveling media filter for sludge, 261
Treatment, storage or disposal (TSD), 733, 737, 738–742
- permits, 738, 739

Treatment ponds for wastewater, 406–409
Treatment works, 8
Trial burn, 775
Trichloramine, 243
Trickling filter, 375–381
Trihalomethanes (THMs), 165, 248, 249
Tritium, 812, 819
Truck, solid waste collection:
- capacity, 645–649
- cost, 649
- routing, 651–654

TSD, 733, 737
Turbidity, 136, 155, 167, 228, 319
Turbidity unit (TU), 136, 155, 228
Turbulence, atmospheric:
- mechanical, 493
- thermal, 493–494
- stability, 494–498

Turnover in lakes, 321
Tympanic membrane, in ear, 564
Tyndall effect, 136
Type I, II and III settling, *see* Sedimentation
Typhoid in water, 133–134

Ultimate BOD (L_0), 291, 295, 300, 305, 309
Ultimate disposal of sludge, 264, 441–442
Ultimate nitrogenous BOD (L_n), 317
Ultraviolet radiation:
- and disinfection, 248
- and ozone depletion, 487

Unconfined aquifer, 49
Underground storage tanks (UST), 30, 742–745
Uniform hazardous waste manifest, 733, 734, 735
Uniformity coefficient, 233
Unit duration of hydrograph, 66

United States Public Health Service (USPHS), 16, 18, 29
Unit hydrograph:
  definition, 63
  direct runoff, 63
  discharge, 63, 65
  duration, 66
  rainfall excess, 66
  separation, 66
Units of measure:
  air pollution, 462–463
  chemical, 136–139
  hydrologic, 51–52
  noise, 553–557
  radiation, 821–823
  SI, 6–7, *inside back cover*
  wastewater, *see* water
  water, 134–139
Universal gas constant, 460
Unsteady flow in aquifer, 101–106
Upflow clarifier, 212, 215–216
Upflow solids-contact, 207–208
Upper respiratory tract (URT), 470
Upwind attenuation of noise, 596
Uranium series, 818
Urban runoff, 285, 287
U.S. Environmental Protection Agency (USEPA),
  see Environmental Protection Agency
U.S. Geological Survey (USGS), 77, 667
U.S. Standard Sieve Series, 232
UST, 742–745
U.S. Weather Bureau standard rain gage, 57

Vacuum filter for sludge, 261, 440
Vadose zone, 49
Valley effect on meteorology, 499–500
Vault, radioactive waste storage, 860–862
Vault toilet, 355–356
Vectors, 444, 665
Velocity controlled grit chamber, 366
Velocity gradient (G), 201, 203, 206
Vent, well, 90
Venturi scrubber, 523, 529–532
Vibration isolation, noise, 614
Viscosity,
  of air, table of, Appendix A
  and mixing, 201–202
  and Reynolds number, 221–222
  and rapid sand filtration, 238
  and water, table of, Appendix A
Volatile acids, 346, 435
Volatile organic chemicals (VOCs), 249, 483, 485–486, 541, 665, 668, 790
Volatile suspended solids (VSS):
  defined, 387

Walsh-Healy Public Contracts Act, 27, 574
Wash troughs, filter, 231, 236
Waste activated sludge (WAS), 382, 383, 392
Waste audit, 751
Waste decomposition and products, table of, 345
Waste exchange, 754
Waste:
  hazardous, characteristics of, 726–729
  solid, characteristics of, 632–635
Waste minimization and:
  air pollution, 542
  hazardous waste, 30, 750–755
  multimedia systems, 16
  radioactive waste, 853–857
  sludge production, 254
  solid waste, 639, 682–691
  water conservation, 110
  water pollution, 285, 327, 328
Waste stabilization pond, 406
Waste to energy, 677–682
Wastewater:
  characteristics, 351–354
  disinfection, 411
  equalization basins, 87
  industrial, 353
  management, 10–13
  microbiology, 339–351
  sludge, 418–420
Wastewater treatment plant (WWTP), 11, 361
Wasting:
  activated sludge, 382, 383
  and food to microorganism ratio, 392
Water:
  degree of ionization, 144
  physical properties of, 134–136
    tables of, Appendix A
  raw, 171
Waterborne disease, 133

disinfection, 133
typhoid, 133
Water chemistry, 134–150
Water consumption, 8–10, 110
Water demand, 8–10, 110
Water Environment Federation (WEF), 379
Water filtration plant, 8, 170
Water mains, 8
Water Pollution Control Act, 17–20
Water pressure, 8
Water Quality Act, 17
Water quality, drinking, 154–168
Water quality goals, 168, 169
Water quality management:
in lakes, 284, 320–330
legislation, 16–20
in rivers, 284, 288–289, 302–320
Water quality standards:
contaminants, 154–168
maximum contaminant level, tables of, 160, 166, 168
Water resource management, 7–13
Watershed, 8, 48
Water softening plant, 8
Water systems, community, 88, 132
Water table:
aquifer, 49
perched, 49
well, 49
Water treatment systems, 168–172
Wavelength, 553
radiation, 814
Weak acid, 145–146
dissociation constants, 145, Appendix A
Weibull's formula, 59
Weighing bucket, 57
Weighting network, 558–561
Weight percent, 136–138
Weir:
grit chamber, 366, 367
loading rates (overflow rates), 215, 373, 403
plate, 62
settling tank, 215, 373, 403
stream gaging, 62
Well:
artesian, 50
casing, 88–90
cement grout, 88–90
cone-of-depression, 91–93
disinfection of, 91
drawdown, 92, 99–104
flowing artesian, 49, 51
hydraulics, 97–109
injection of hazardous waste, 783
interference, 92, 108–109
non-flowing artesian, 49, 51
piezometric surface, 49
pumphouse, 89, 91
radius of influence, 92
sanitary construction, 88–91
screen, 89
seal, 90
static water table, 92
system for contaminated groundwater, 796–797
vent, 90
water table, 49
Well function of u ($W(u)$), 101
table of, 102–103
Wet air oxidation, 760
Wind speed and pressure, 492
WIPP, 852
Works, defined, 8

X-rays:
defined, 815
exposure, 840
production, 820, 822
tube, 821

Yard waste, 635
Yield coefficient, 351, 388, 390
observed, 399
Yucca Flats, 852

Zeolites, 197, 518
Zinc in drinking water, 157
Zone of saturation, 49
Zone settling, 220
Zoning and noise, 588

Useful conversion factors

Multiply	By	To obtain
atmosphere (atm)	101.325	kilopascal (kPa)
Calorie (international)	4.1868	Joules (J)
centipoise	10^{-3}	Pa · s
centistoke	10^{-6}	m^2/s
cubic meter (m^3)	35.31	cubic feet (ft^3)
cubic meter	1.308	cubic yard (yd^3)
cubic meter	1,000.00	liter (L)
cubic meter/s	15,850.0	gallons/min (gpm)
cubic meter/s	22.8245	million gal/d (MGD)
cubic meter/m^2	24.545	gallons/sq ft (gal/ft^2)
cubic meter/d · m	80.52	gal/d · ft (gpd/ft)
cubic meter/d · m^2	24.545	gal/d · ft^2 (gpd/ft^2)
cubic meter/d · m^2	1.0	meters/d (m/d)
days (d)	24.00	hours (h)
days (d)	1,440.00	minutes (min)
days (d)	86,400.00	seconds (s)
dyne	10^{-5}	Newtons (N)
erg	10^{-7}	Joules (J)
grains (gr)	6.480×10^{-2}	grams (g)
grains/U.S. gallon	17.118	mg/L
grams (g)	2.205×10^{-3}	pounds mass (lb_m)
hectare (ha)	10^4	m^2
Hertz (Hz)	1	cycle/s
Joule (J)	1	N · m
J/m^3	2.684×10^{-5}	Btu/ft^3
kilogram/m^3 (kg/m^3)	8.346×10^{-3}	lb_m/gal
kilogram/m^3	1.6855	lb_m/yd^3
kilogram/ha (kg/ha)	8.922×10^{-1}	lb_m/acre
kilogram/m^2 (kg/m^2)	2.0482×10^{-1}	lb_m/ft^2
kilometers (km)	6.2150×10^{-1}	miles (mi)
kilowatt (kW)	1.3410	horsepower (hp)
kilowatt-hour	3.600	megajoules (MJ)
liters (L)	10^{-3}	cubic meters (m^3)
liters	1,000.00	milliliters (mL)
liters	2.642×10^{-1}	U.S. gallons
megagrams (Mg)	1.1023	U.S. short tons
meters (m)	3.281	feet (ft)
meters/d (m/d)	2.2785×10^{-3}	ft/min
meters/d	3.7975×10^{-5}	meters/s (m/s)
meters/s (m/s)	196.85	ft/min
meters/s	3.600	km/h
meters/s	2.237	miles/h (mph)
micron (μ)	10^{-6}	meters
milligrams (mg)	10^{-3}	grams (g)
milligrams/L	1	g/m^3
milligrams/L	10^{-3}	kg/m^3
Newton (N)	1	kg · m/s^2
Pascal (Pa)	1	N/m^2
Poise (P)	10^{-1}	Pa · s
square meter (m^2)	2.471×10^{-4}	acres
square meter (m^2)	10.7639	sq ft (ft^2)
square meter/s	6.9589×10^6	gpd/ft
Stoke (St)	10^{-4}	m^2/s
Watt (W)	1	J/s
Watt/cu meter (W/m^3)	3.7978×10^{-2}	hp/1,000 ft^3
Watt/sq meter · °C (W/m^2 · °C)	1.761×10^{-1}	Btu/h · ft^2 · °F

Unit Conversion Booklet

for use with

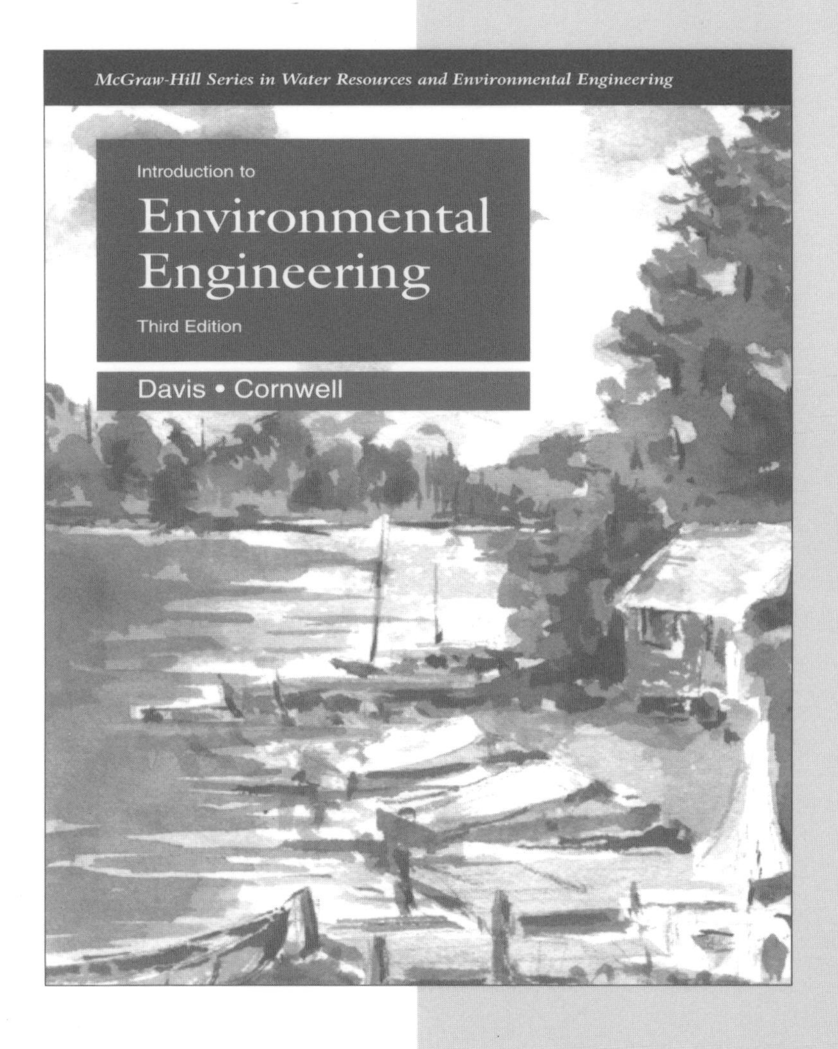

Engineering and Scientific Conversion Booklet

July, 1999 Edition

for use with

Introduction to Environmental Engineering

Third Edition

Dr. Mackensie L. Davis
Michigan State University

Dr. David A. Cornwell
Environmental Engineering and Technology, Inc.

Prepared by
Environmental Engineering Student Society
A Student Chapter of
Air & Waste Management Association and
Michigan Water Environment Association

Special Thanks to:
Dr. M. L. Davis, Stephen Jan Callister
and EESS members that have contributed to this project throughout the years

Boston Burr Ridge, IL Dubuque, IA Madison, WI New York San Francisco St. Louis
Bangkok Bogotá Caracas Lisbon London Madrid
Mexico City Milan New Delhi Seoul Singapore Sydney Taipei Toronto

McGraw-Hill Higher Education

A Division of The McGraw-Hill Companies

Unit Conversion Booklet for use with
INTRODUCTION TO ENVIRONMENTAL ENGINEERING

5 6 7 8 9 0 HAM/HAM 0 9 8 7 6 5 4 3 2 1 0

ISBN 0-07-237821-2

http://www.mhhe.com

Table of Contents

Page

Abbreviations 1

Atomic Weights and Atomic Masses of Elements 2

Conversion Table 3

- acres through bushel 4
- bushel through ft of water 5
- ft of water through ft^3/min 6
- ft^3/min through hp-h 7
- hp-h through kg-cal/min 8
- kg/m^3 through knots 9
- knots through microns 10
- microns through lb water 11
- lb water/min through rev/min/min 12
- rev/min/min through yd^2 13
- yd^2 through years 14

Greek Alphabet and Multiples 15

Gas Constant 16

Water Properties 17

Complimentary Error Function 18

Theis Solution 19

Periodic Table of the Elements 20

Functional Groups of Organic Compounds 21

Families of Organic Compounds 22

Calculus Integrals and Log Properties 23

Geometric Formulas-Areas 24

Geometric Formulas-Volumes 25

Abbreviations

Abs	....	absolute
Atm	....	Atmosphere
Bhp	....	brake horsepower unit
Br	....	British
Btu	....	British thermal unit
C	....	Coulomb
°C	....	degree Celsius
cm	....	centimeter
d	....	day
deg	....	degree
°F	....	degree Fahrenheit
ft	....	feet
g	....	gram
gal	....	gallons per minute
gf	....	gram force
gpd	....	gallons per day
gpm	....	gallons per minute
ha	....	hectare
hp	....	horsepower
Hg	....	mercury
h.	....	inch
J	....	Joule
K	....	degree Kelvin
Kg	....	kilogram
kg-cal	....	kilogram-calorie
km	....	kilometer
kpa	....	kilo-pascal
Kw	....	kilowatt
L	....	liter
lb	....	pound
lbf	....	pound force
m	....	meter
m	....	millibar
mi	....	mile
min	....	minute
mm	....	millimeter
mph	....	miles per hour
mg/l	....	milligrams per liter
MGD	....	million gallons per day
N	....	Newton
oz	....	ounce
Pa	....	Pascal
ppb	....	parts per billion
ppm	....	parts per million
psf	....	pounds per square foot
psi	....	pounds per square inch
°R	....	degree Rankine
rev	....	revolution
rpm	....	revolutions per minute
s	....	second
yd	....	yard
y	....	year
W	....	watt

Atomic Weights, and Atomic Masses of the Elements

Actinium	Ac	89	227.0278	Mercury	Hg	80	200.59
Aluminum	Al	13	26.98154	Molybdenum	Mo	42	95.94
Americium	Am	95	(243)	Neodymium	Nd	60	144.24
Antimony	Sb	51	121.75	Neptunium	Np	93	237.0482
Arsenic	As	33	74.9216	Nickel	Ni	28	58.70
Astatine	At	85	(210)	Niobium	Nb	41	92.9064
Barium	Ba	56	137.33	Nitrogen	N	71	14.00
Berkelium	Bk	97	(247)	Nobelium	No	102	(259)
Beryllium	Be	4	9.01218	Osmium	Os	71	90.2
Bismuth	Bi	83	208.9804	Oxygen	0	8	15.9994
Boron	B	5	10.81	Palladium	Pd	46	106.4
Bromine	Br	35	79.904	Phosphorous	P	15	30.97376
Cadmium	Cd	48	112.41	Platinum	Pt	78	195.09
Calcium	Ca	20	40.08	Plutonium	Pu	94	(244)
Californium	Ct	98	(251)	Polonium	Po	84	(209)
Carbon	C	6	12.011	Potassium	K	19	39.0983
Cerium	Ce	58	140.12	Praseodymium	Pr	59	140.9077
Cesium	Cs	55	132.9054	Promethium	Pm	61	(145)
Chlorine	Cl	17	35.453	Protactinium	Pa	91	231.0359
Chromium	Cr	24	51.996	Radium	Ra	88	226.0254
Cobalt	Co	27	58.9332	Radon	Rn	86	(222)
Copper	Cu	29	63.546	Rhenium	Rc	75	186.207
Curium	Cm	96	(247)	Rhodium	Rh	45	102.9055
Dysprosium	Dy	66	162.50	Rubidium	Rb	37	85.4678
Einsteinium	Es	99	(254)	Ruthenium	Ru	44	101.07
Erbium	Er	68	167.26	Samarium	Sm	62	150.4
Europium	Eu	63	151.96	Scandium	Sc	21	44.9559
Fermium	Fm	100	(257)	Selenium	Se	34	78.96
Fluorine	F	9	18.99840	Silicon	Si	14	28.0855
Francium	Fr	87	(223)	Silver	Ag	47	107.868
Gadolinium	Gd	64	157.25	Sodium	Na	11	22.98977
Gallium	Ga	31	69.72	Strontium	Sr	38	87.62
Germanium	Gc	32	72.59	Sulfur	S	16	32.06
Gold	Au	79	196.9665	Tantalum	Ta	73	180.9479
Hafnium	Hf	72	178.49	Technetium	Tc	43	(97)
Helium	He	2	4.00260	Tellurium	Te	52	127.60
Holmium	Ho	67	164.9304	Terbium	Tb	65	158.9254
Hydrogen	H	1	1.0079	Thallium	Tl	81	204.37
Indium	In	49	114.82	Thorium	Th	90	232.0.381
Iodine	I	53	126.9045	Thulium	Tm	69	168.9342
Iridium	Ir	77	192.22	Tin	Sn	50	118.69
Iron	Fe	26	55.847	Titanium	Ti	22	47.90
Krypton	Kr	36	83.80	Tungsten	W	74	183.85
Lanthanum	La	57	138.9055	Uranium	U	92	238.029
Lawrencium	Lr	103	(260)	Vanadium	V	23	50.44
Lead	Lu	71	174.97	Yttrium	Y	39	88.9059
Magnesium	Mg	12	24.305	Zinc	Zn	30	65.38
Manganese	Mn	25	54.9380	Zirconium	Zr	40	91.22
Mendelevium	Md	101	(258)				

Atomic number and atomic weights From Pure Appl Chem. 47,75 (1976).
A value in parenthesis is the mass number of the longest-lived isotope of the element.

Designation	***Symbol***	***ValueUnits***	
Atomic Mass Unit	u	1.660566×10^{-27}	kg
Avogadro's Constant	NA	6.022045×10^{26}	$kmol^{-1}$
Boltzmann's Constant	k	1.380662×10^{-23}	J/K
Bohr Radius		5.2918×10^{-11}	m
Centipoise		10^{-3}	Pa•s
Earth Avg. Density		5.517	g/cm^3
Earth Mass		5.9763×10^{24}	kg
Earth Avg. Radius		6.37×10^{6}	m
Earth Surface Area		5.10068×10^{8}	km
Earth-Moon Avg. Distance		3.844×10^{8}	m
Earth-Sun Avg. Distance		1.496×10^{11}	m
Electron Charge	qe, e-	1.60219×10^{-19}	C
Electron Rest Mass	me	9.1096×10^{-31}	kg
		5.4859×10^{-4}	u
Electron Volt	eV	1.602190×10^{-19}	J
Faraday Constant	F	9.648456×10^{7}	C/kmol
Gravitational Acceleration	g	9.80665	m/s^2
		32.17	ft/s^2
Joule	J	1	N-m
		1	$kg\text{-}s^2$
		9.48×10^{-4}	Btu
		0.2389	cal
		0.73756	ft•lb
Magnetic Field Constant	mo	1.256640×10^{-6}	N/A^2
Neutron rest mass	Mn	1.674954×10^{-27}	kg
Newton	N	1	$kg\text{-}m/s^2$
		0.22481	lbf
Pascal	Pa	1	N/m^2
Permeability Constant	mo	1.26×10^{-6}	$N\text{-}s^2/c^2$
Permittivity Const.	eo	8.854×10^{-12}	$C^2/N\text{-}m^2$
Planck's Constant	h	6.6262×10^{-34}	J-s
Poise	P	1	gm/cm-s
		1	$dyne\text{-}s/cm^2$
		0.1	Pa-s
Proton rest mass	mp	1.672649×10^{-27}	kg
Slug		1	$lbf\text{-}s^2/ft$
Speed of Light in Vacuum	c	2.997925×10^{8}	m/s
Speed of Sound in Air (0^0C)	cs	3.313621×10^{2}	m/s
Standard Atmosphere	atm	1.013250×10^{5}	Pa
Standard Kilomole Volume	Vo	22.41383	$m^3/kmole$
Stoke	St	1.000×10^{-4}	m^2/s
		1.075×10^{-3}	ft^2/s
		1	$poise\text{-}cm^3/g$
Sun Mass		1.991×10^{30}	kg
Sun Avg. Radius		6.9595×10^{8}	m
Sun's Radiation Intensity at Earth		0.032	$cal/cm^2\text{-}s$
		0.134	$J/cm^2\text{-}s$
Thermochemical Calorie	cal	4.184000	J
Watt	W	1	J/s

Multiply	***By***	***To Obtain***
acres	43,560	ft^2
acres	4047	m^2
acres	0.4047	ha
acres	1.563×10^{-3}	mi^2
acres	4840	yd^2
acre-ft	43,560	ft^3
acre-ft	325,851(U.S.)	gal
acre-ft	1233.49	m^3
angstrom	1.00×10^{-10}	m
atm	760	mm Hg (0 °C)
atm	29.92	in. Hg (0 °C)
atm	33.90	ft water (4°C)
atm	10,333	kg force/m^2
atm	101.325	kPa
atm	1.0584	short ton/ft^2
atm	14.70	lbf/in^2
atm	1.01325×10^5	N/m^2
bars	0.9869	atm
bars	1.0000×10^6	dynes/cm^2
bars	1.000×10^3	mb
bars	1.020×10^4	kg/m^2
bars	2,089	lb/ft^2
bars	14.50	lb/in^2
barrels-oil	42	gal-oil (U.S.)
barrels-cement	376	lb-cement
barrels-beer	2.0	kegs (1/2 barrels)
barrels-liquid	31.5	gal
bags-cement	94	lb-cement
Bhp	42.44	Btu/min
Bhp	3.30133×10^4	ft-1bf/min
Bhp	550	ft-1bf /s
Bhp	1.1014	hp (metric)
Bhp	10.69	kg-cal/min
Bhp	0.7457	kW
board-ft	144 in^2 x 1 in	in^3
Btu	0.2520	kg-cal
Btu	777.7	ft-lbf
Btu	3.929×10^{-4}	hp-h
Btu	1055	J
Btu	107.5	kg-m
Btu	2.928×10^{-4}	kW-h
Btu/ft^3	37.253	kJ/m^3
Btu/lb	2.3259×10^{-3}	MJ/kg
Btu/min	12.96	ft-1bf /s
Btu/min	0.02357	hp
Btu/min	17.5978	J/s (W)
Btu/min	17.5978	W
bushel	1.244	ft^3
bushel	2150	in^3
bushel	0.03524	m^3

Multiply	*By*	*To Obtain*
bushel	4	pecks
bushel	64	pints (dry)
bushel	32	quarts (dry)
calorie	4.184	J
centipoise	1×10^{-2}	g mass/cm-s
centipoise	6.72×10^{-4}	lb mass/ft-s
centipoise	2.088×10^{-5}	lbf-s/ft^2
centipoise	1.000×10^{-3}	Pa • s
cm^2	1.076×10^{-3}	ft^2
cm^2	0.1550	in^2
cm^2	1.00×10^{-4}	m^2
cm^2	100.0	mm^2
cm^3	3.531x10-5	ft^3
cm^3	6.102×10^{-2}	in^3
cm^3	3.381×10^{-2}	oz.
cm^3	1.000×10^{-6}	m^3
cm^3	1.000×10^{-3}	L
chain (surveyors)	66	ft
chain	0.1	Furlong
chain	100	links
chain	4	Rods
cord-feet	4 ft x 4 ft x 1 ft	ft^3
cords	8 ft x 4 ft x 4 ft	ft^3
days	24	hr.
days	1440	min
days	86400	sec.
degrees (angle)	60	min.
degrees (angle)	0.01745	radians
degrees (angle)	3600	sec.
degrees/s	0.01745	rad/s
degrees/s	0.1667	rpm
degrees/s	2.778×10^{-3}	rev/s
drams (av.)	1.772	grams
drams (av.)	0.0625	ounces
dynes	1.020×10^{-3}	gram-force
dynes	1.0000×10^{-5}	N
dynes	2.248×10^{-6}	lb
ergs	9.485×10^{-11}	Btu
ergs	1.0000	dyne-cm
ergs	7.376×10^{-8}	ft force-lb
ergs	1.020×10^{-3}	g-cm
ergs	1.0000×10^{-7}	J
ergs	2.389×10^{-11}	kg force-cal
ergs	1.020×10^{-8}	kg-m
fathoms (linear)	6	ft
ft	30.48	cm
ft	12	in.
ft	0.3048	m
ft	1.894×10^{-4}	miles (statute)
ft	0.3333	yd
ft of water	2.950×10^{-2}	atm

Multiply	*By*	*To Obtain*
ft of water	0.8826	in. Hg
ft of water	304.8	kg force/m^2
ft of water	2.98898	kPa
ft of water	62.42	psf
ft of water	0.4335	psi
ft/min	0.5080	cm/s
ft/min	0.01667	ft/s
ft/min	0.01829	km/h
ft/min	0.3048	m/min
ft/min	0.005080	m/s
ft/min	0.01136	mph
ft/s	0.3048	m/s
ft/s	1.097	km/h
ft/s	0.5921	knots
ft/s	18.29	m/min
ft/s	0.6818	mph
ft/s	0.01136	mi/min
ft-lb force	1.286×10^{-3}	Btu
ft-lb f	5.051×10^{-7}	Bhp-h
ft-lb f	3.241×10^{-4}	kg_f-cal
ft-lb f	0.1383	kg_f-m
ft-lb f	3.766×10^{-7}	kW-h
ft-lb f	1.356×10^{7}	ergs
ft-lb f	1.356	J
ft-lb f/min	1.286×10^{-3}	Btu/min
ft-lb f/min	0.01667	ft-lbf/s
ft-lb f/min	3.030×10^{-5}	Bhp
ft-lb f/min	0.0225970	J/s
ft-lb f/min	3.240×10^{-4}	kg_f-cal/min
ft-lb f/min	2.26×10^{-5}	kW
ft-lb f/s	1.818×10^{-3}	Bhp
ft-lbf/s	1.356	J/s
ft-lbf/s	1.944×10^{-2}	kg-cal/min
ft-lbf/s	1.356×10^{-3}	kW
ft^2	2.296×10^{-5}	acres
ft^2	929.0	cm^2
ft^2	144.0	in^2
ft^2	0.09290304	m^2
ft^2	3.587x10-8	mi^2
ft^2	0.1111	yd^2
ft^3	2.832×10^{4}	cm^3
ft^3	1728	in^3
ft^3	0.02832	m^3
ft^3	0.03704	yd^3
ft^3	7.48052	gal
ft^3	28.32	L
ft^3 of water (39.2·F)	62.43	lb of water
ft^3/min	4.72×10^{-4}	m^3/s
ft^3/min	0.1247	gal/sec
ft^3/min	0.4720	L/s
ft^3/min	62.43	lb water/min

Multiply	*By*	*To Obtain*
ft^3/min	7.48052	gpm
ft^3/s	0.646317	MGD
ft^3/s	448.831	gpm
ft^3/s	0.02832	m^3/s
furlongs	40	rods
gal (U.S. liquid)	3785.4	cm^3
gal	0.1337	ft^3
gal	231.0	in^3
gal	3.785×10^{-3}	m^3
gal	4.951×10^{-3}	yd^3
gal	3.7854	L
gal	8	pints (liq)
gal	4	quarts (liq)
gal (U.S.)	0.83267	British gal
gal water @ 4°C	8.33585	lb water gal
gpm	2.228×10^{-3}	ft^3/s
gpm	0.06309	L/s
gpm	6.309×10^{-5}	m^3/s
gpm	8.020	ft^3/h
gpm	5.451	m^3/d
gpm	1.440×10^{-3}	MGD
gpm/ft^2	58.68	m^3/d-m^2
gpd	1.54733×10^{-6}	ft^3/s
gpd/ft	1.242×10^{-2}	m^3/d-m
gpd/ft2	4.074×10^{-2}	m^3/d-m^2
grains/U.S. gal	17.119	ppm
grains/U.S. gal	142.86	lb/million gal
grains/Imp. gal	14.254	ppm
g	15.43	grains
g	2.205×10^{-3}	lb
g/cm^3	62.43	lbs/ft^3
g/cm^3	0.03613	lbs/in^3
g/L	58.417	grains/gal
g/L	8.345	lb/1000gal
g/L	0.062427	lb/ft^3
g/L	0.1335	oz/gal
g/L	1000	ppm
g/m3	1.000	mg/L
g/m3	1	ppm
h	3600	sec.
h	4.16667×10^{-2}	day
hectare	2.471	acres
hectare	1.076×10^5	ft^2
hectare	10,000	m^2
hertz	1	1/s
hp(boiler)	3.345×10^4	Btu/h
hp(boiler	9.810	kW
hp-h	2546	Btu
hp-h	1.98×10^6	ft-lbf
hp-h	641.3	kg-cal
hp-h	2.737×10^5	kg force-m

Multiply	*By*	*To Obtain*
hp-h	0.7457	kW-h
hp-h	2.684×10^{6}	J
inches	2.540	cm
in. Hg @ 0^{0}C	0.03342	atm
in Hg	1.133	ft water @ 4 ^{0}C
in Hg	345.316	kg/sq m
in. Hg	3.38639	kPa
in. Hg	33.8639	millibar
in. Hg	70.73	psf
in. Hg	0.4912	psi
in water @ 4^{0}C	0.002458	atm
in water	0.07355	in. Hg
in. water	25.4	kg/m^2
in. water	0.249	kPa
in. water	5.202	psf
in. water	0.03613	psi
in^2	6.452	cm^2
in^2	645.16	mm^2
in^3	16.39	cm^3
in^3	5.787×10^{-4}	ft^3
in^3	1.639×10^{-5}	m^3
in^3	2.143×10^{-5}	yd^3
in^3	4.329×10^{-3}	gal
in^3	1.639×10^{-2}	L
J	9.4845×10^{-4}	Btu
J	0.73756	ft-lbf
J	2.39×10^{-4}	kg-cal
J	2.7778×10^{-7}	kW-h
J	2.7778×10^{-4}	W-h
J	1.000×10^{7}	erg
J	0.1020	kg force-m
J/s	0.056907	Btu/min
J/s	0.73756	ft-lb/s
J/s	0.001341	hp Br
J/s	0.01434	kg-cal/min
J/s	0.001	kW
J/s	1.000	W
keg (beer)	64	quarts (liq)
kg	2.205	lb
kg	1.102×10^{-3}	tons (short)
kg	0.001	tons (metric)
kg-cal	3.968	Btu
kg-cal	3085.96	ft-lb
kg-cal	1.559×10^{-3}	hp-h
kg-cal	4184	J
kg-cal	1.162×10^{-3}	kW-h
kg-cal/min	51.46	ft-lb force/s
kg-cal/min	0.09351	hp
kg-cal/min	69.73	J/s
kg-cal/min	0.06973	kW

Multiply	***By***	***To Obtain***
kg/m^3	0.06243	lb/ft^3
kg/m^3	3.613×10^{-5}	lb/in^3
kg/m^3	1.6856	lb/yd^3
kg/m	0.6720	lb force/ft
kgf/ha	0.892	lb/acre
kgf/ha	4.46×10^{-4}	tons/acre
kgf/m^2	9.678×10^{-5}	atm
kgf/m^2	3.281×10^{-3}	ft. water @ 4°C (39.2°F)
kgf/m^2	2.896×10^{-3}	in. Hg @ 0°C (32°F)
kgf/m^2	0.2048	psf
kgf/m^2	1.422×10^{-3}	psi
kgf/m^2	7.3556×10^{-2}	mm Hg @ 0°C
km	3281	ft
km	0.6214	mi
km	1094	yd
km/h	0.2778	m/s
km/h	54.68	ft/min
km/h	0.9113	ft/s
km/h	0.5399	knots
km/h	16.67	m/min
km/h	0.6214	mph
km/L	2.3520	mi/gal
km^2	247.1	acres
km^2	100.0	hectares
km^2	1.076×10^7	ft^2
km^2	1.000×10^6	m^2
km^2	0.3861	mi^2
km^2	1.196×10^6	yd^2
kPa	9.869×10^{-3}	atm
kPa	0.33456	ft. water @ 4°C
kPa	0.29530	in. Hg @ 0°C
kPa	4.016	in. water
kPa	7.50064	mm Hg
kPa	101.966	mm water
kPa	1.000×10^3	N/m^2
kPa	20.886	psf
kPa	0.145033	psi
kW	56.825	Btu/min
kW	4.425×10^4	ft-lbf/min
kW	737.6	ft-lbf/s
kW	1.341	hp (British)
kW-h	3414	Btu
kW-h	2.655×10^6	ft-lbf
kW-h	1.341	hp-h (Brake)
kW-h	3.6×10^6	J
kW-h	859.1	kg-cal
knots	1.852	km/h

Multiply	*By*	*To Obtain*
knots	1.150779	mph
links (surveyor's)	7.92	in.
L	1.00×10^3	cm^3
L	0.035315	ft^3
L	61.024	in^3
L	1.00×10^{-3}	m^3
L	0.2642	gal (US liq)
L	1.057	quarts (liq)
L/min	5.885×10^{-4}	ft^3/s
L/min	1.667×10^{-5}	m^3/s
L/min	4.403×10^{-3}	gal/s
L/min	0.2642	gpm
meters	100	cm
meters	3.281	ft
meters	39.37	in
meters	1.00×10^{-3}	km
meters	1.00×10^3	mm
meters	1.094	yd
meters/min	0.01667	m/s
meters/min	3.281	ft/min
meters/min	0.05468	ft/s
meters/min	0.06	km/h
meters/min	0.03728	mph
meters/s	196.85	ft/min
meters/s	3.281	ft/s
meters/s	3.6	km/h
meters/s	0.06	km/min
meters/s	2.237	mph
meters/s	0.03728	mi/min
m^2	2.471×10^{-4}	acres
m^2	1.000×10^{-4}	hectares
m^2	10.763910	ft^2
m^2	3.861×10^{-7}	mi^2
m^2	1.196	yd^2
m^3	1.000×10^6	cm^3
m^3	35.31	ft^3
m^3	61,023	in^3
m^3	1.308	yd^3
m^3	264.2	gal (U.S.)
m^3	1.0×10^3	L
m^3/d-m	80.52	gpd/ft
m^3/d-m^2	24.545	gpd/ft^2
m^3/d-m^2	0.0170	gpm/ ft^2
m^3/s	2118.6	ft^3/min
m^3/s	35.31	ft^3/s
m^3/s	15852	gpm
m^3/s	264.2	gal/s
m^3/s	6.00×10^4	L/min
m^3/s	22.8245	MGD
microns	1.00×10^{-6}	meters

Multiply	***By***	***To Obtain***
microns	1.0	micrometers
miles	5280	ft
miles	1.609344	km
miles	1760	yd
miles/gal	0.42517	km/L
mph	0.44704	m/s
mph	88	ft/min
mph	1.467	ft/s
mph	1.609	km/h
mph	0.868976	knots
miles/min	2682	cm/s
miles/min	88	ft/s
miles/min	1.609	km/min
miles/min	60	mph
mi^2	640.0	acres
mi^2	259.00	hectares
mi^2	2.788×10^7	ft^2
mi^2	2.590	km^2
mi^2	3.098×10^6	yd^2
millibar	0.100	kPa
millibar	0.029530	in. Hg
millibar	0.75006	mm Hg
mm	0.03937	in.
mm Hg (@ 0^0C)	0.001316	atm
mm Hg	0.04461	ft water
mm Hg	13.595	kgf/m^2
mm Hg	0.133322	kPa
mm Hg	2.785	lbf/ft^2
mm Hg	0.01934	lbf/in^2
mm^2	0.01	cm^2
mm^2	1.55×10^{-3}	in^2
mg/L	1	ppm in water
MGD	1.54723	ft^3/s
MGD	4.3813×10^{-2}	m^3/s
min (angle)	2.909×10^{-4}	radians
N/m^2	1.000×10^{-3}	kPa
ounces	16	drams
ounces	437.5	grains
ounces	28.35	grams
ounces	0.0625	lb.
oz/gal	7.489	g/L
ounces (fluid)	1. 805	in^3
ounces (fluid)	0.02957	L
ppm	8.35	lb/million gal
pounds	16	ounces
pounds	7000	grains
pounds	5.0×10^{-4}	tons(short)
pounds	453.5924	grams
lb water	0.01602	ft^3
lb water	0.1198	gal

Multiply	*By*	*To Obtain*
lb water/min	2.67×10^{-4}	ft^3/s
lb/acre	1.1208	kg/ha
lb/gal	119.827	g/L
lb/ft^3	0.01602	$grams/cm^3$
lb/ft^3	16.019	kg/m^3
lb/ft^3	5.787×10^{-4}	lb/in^3
lb/ft^3	27.000	lb/yd^3
lb/in^3	27.68	$grams/cm^3$
lb/in^3	2.768×10^{4}	kg/m^3
lb/in^3	1728	lb/ft^3
lb/yd^3	0.5932	kg/m^3
lb/ft	1.488	kg/m
lb/in.	178.6	grams/cm
Pa	1.0000	N/m^2
ppb	1	μg/kg, μg/L
ppm	1	mg/kg, mg/L
psf	0.01602	ft water (@ 4°C)
psf	4.882428	kg/m^2
psf	6.945×10^{-3}	psi
lbf/ft^2	47.88	N/m^2
lbf/in^2	6895	N/m^2
psi	0.068046	atm
psi	2.307	ft water (@ 4°C)
psi	2.036	in. Hg (@ 0°C)
psi	51.7149	mm Hg
psi	703.1	kgf/m^2
psi	6.8948	kPa
quarts (US dry)	67.20	in^3
quarts (liq)	57.75	in^3
quarts (liq)	32	ounces (US)
quarts (liq)	2	pints (US)
quarts (US liq)	9.464×10^{-4}	m^3
radians	57.30	degrees
radians	3438	minutes
radians	0.6366	quadrants
rad/s	57.30	degrees/s
rad/s	0.1592	rev/s
rad/s	9.549	rpm
rad/s/s	572.958	rev/min/min
rad/s/s	0.1592	rev/s/s
reams	500	sheets of paper
revolutions	360	degrees
revolutions	6.283	radians
rpm	6.000	degrees/s
rpm	0.1047	rad/s
rpm	0.01667	rev/s
rev/min/min	1.745×10^{-3}	rad/s/s

Multiply	*By*	*To Obtain*
rev/min/min	2.778×10^{-4}	rev/s/s
rev/s	360	degrees/s
rev/s	6.283	rad/s
rev/s	60.000	rpm
rev/s/s	6.283	rad/s/s
rev/s/s	3600	rev/min/min
rod	0.25	Chain(survey)
rod	16.25	ft
rod	0.025	Furlongs
rod	25	links(survey)
seconds (time)	1.1574×10^{-5}	day
seconds (time)	2.7778×10^{-4}	hr
seconds (angle)	4.848×10^{-6}	rad
slugs	14.59	kg
slugs/ft^3	515.4	kg/m^3
surveyor's chain	66	ft
temp (°C)+273.15	1	abs temp (K)
temp (°C)+17.48	1.8	temp (°F)
temp (°F)+460	1	abs temp (°R)
temp (°F)-32	0.5555	temp (°C)
tons (long)	1016	kg
tons (long)	2240	lb
tons (long)	1.12000	tons (short)
tons (metric)	1.00×10^{3}	kg
tons (metric)	2205	lb
tons (short)	2000	lb
tons (short)	907.18486	kg
tons (short)	0.89287	tons (long)
tons (short)	0.90718	tons (metric)
tons/acre	2.2416×10^{3}	kg/ha
US $/Liter	3.785	US $/gal
US $/gal	0.2642	US $/Liter
W	0.0568	Btu/min
W	44.25	ft-lb/min
W	0.7376	ft-lb/s
W	1.341×10^{-3}	hp
W	1.000	J/s
W	14.34	cal/min
W	1.00×10^{-3}	kW
W-h	3.4144	Btu
W-h	2655.22	ft-lbf
W-h	1.341×10^{-3}	hp-h
W-h	0.859	kg-cal
W-h	367.1	kg-m
yd	91.44	cm
yd	3	ft
yd	36	in.
yd	0.9144	m
yd^2	2.066×10^{-4}	acres
yd^2	9.000	ft^2
yd^2	0.8361	m^2

Multiply	*By*	*To Obtain*
yd^2	3.228×10^{-7}	mi^2
yd^3	7.646×10^5	cm^3
yd^3	27	ft^3
yd^3	4.6656×10^4	in^3
years (common)	365	days
years (common)	8760	hr
years (leap)	366	days
years	5.2596×10^5	min
years	12	months
years	52.142857	weeks

Add your own conversion factors below:

Multiply	*By*	*To Obtain*

Greek Alphabet

Α	α	Alpha	Ν	ν	Nu
Β	β	Beta	Ξ	ξ	Xi
Γ	γ	Gamma	Ο	ο	Omicron
Δ	δ	Delta	Π	π	Pi
Ε	ε	Epsa	Σ	σ	Sigma
Η	η	Eilon	Ρ	ρ	Rho
Ζ	ζ	Zetta	Τ	τ	Tau
Θ	θ	Theta	Υ	υ	Upsilon
Ι	ι	Iota	Φ	ϕ	Phi
Κ	κ	Kappa	Χ	χ	Chi
Λ	λ	Lambda	Ψ	ψ	Psi
Μ	μ	Mu	Ω	ω	Omega

Multiples

Amount	*Submultiples*	*Prefixes*	*Symbols*
1 000 000 000 000 000 000	10^{18}	exa	E
1 000 000 000 000 000	10^{15}	peta	P
1 000 000 000 000	10^{12}	tera	T
1 000 000 000	10^{9}	giga	G
1 000 000	10^{6}	mega	M
1 000	10^{3}	kilo	k
1 00	10^{2}	hecto	h
1 0	10	deka	da
0.1	10^{-1}	deci	d
0.01	10^{-2}	centi	c
0.001	10^{-3}	milli	m
0.000 001	10^{-6}	micro	μ
0.000 000 001	10^{-9}	nano	n
0.000 000 000 001	10^{-12}	pico	p
0.000 000 000 000 001	10^{-15}	femto	f
0.000 000 000 000 000 001	10^{-18}	lto	a

Gas Constant (Universal), R

$$R = \frac{PV}{nT}$$

Volume	Temp.	moles	Atm	psia	mm Hg	in Hg	in H_2O
Liters	K	gm	0.08205	1.206	62.4	2.45	33.4
		lb	37.2	547	28,300	1113	15,140
cm^3	K	gm	82.05	1206	62,400	2450	33,400
		lb	37,200	547,000	$2.83X10^7$	$1.11X10^6$	$1.51X10^7$
ft^3	K	gm	0.00290	0.0426	2.20	0.00867	1.18
		lb	1.31	19.31	999	39.3	535

Some commonly used values:

8.3144	**kJ/kg mole-K**
8.314	**kPa-m^3/kg mole-K**
$8.20562x10^{-2}$	**m^3-atm/k mole-K**
$1.716x10^3$	**ft-lbf/slug-^{0}R**
21.9	**in Hg-ft^3/lb mole-^{0}R**
0.7302	**ft^3-atm/lb mole-^{0}R**
1.9872	**kcal/kg mole-K**

Other useful conversions for the gas constant:

1 Atm = 76 cm Hg
1 Atm = 33.90 ft H_2O
1 Atm = 1.01325 Pa
Std. Temp. = 273.15°K or 491.67°R
1 cm^3 = $1X10^{-6}$ m^3
1 ft^3 = 7.48052 gal

Physical Properties of Water at 1 atm

Temperature °C	Density kg/m³	Specific weight kN/m³	Dynamic Visc. mPa•s	KinematicVisc. μm²/s
0	999.842	9.805	1.787	1.787
3.98	1,000.000	9.807	1.567	1.567
5	999.967	9.807	1.519	1.519
10	999.703	9.804	1.307	1.307
12	999.500	9.802	1.235	1.236
15	999.103	9.798	1.139	1.140
17	998.778	9.795	1.081	1.082
18	998.599	9.793	1.053	1.054
19	998.408	9.791	1.027	1.029
20	998.207	9.789	1.002	1.004
21	997.996	9.787	0.998	1.000
22	997.774	9.785	0.955	0.957
23	997.542	9.783	0.932	0.934
24	997.300	9.781	0.911	0.913
25	997.048	9.778	0.890	0.893
26	996.787	9.775	0.870	0.873
27	996.516	9.773	0.851	0.854
28	996.236	9.770	0.833	0.836
29	995.948	9.767	0.815	0.818
30	995.650	9.764	0.798	0.801
35	994.035	9.749	0.719	0.723
40	992.219	9.731	0.653	0.658
45	990.216	9.711	0.596	0.602
50	988.039	9.690	0.547	0.554
60	983.202	9.642	0.466	0.474
70	977.773	9.589	0.404	0.413
80	971.801	9.530	0.355	0.365
90	965.323	9.467	0.315	0.326
100	958.366	9.399	0.282	0.294

Complimentary Error Function (erfc)

β	erf(β)	erfc(β)	β	erf(β)	erfc(β)
0	0	1.0	1.0	0.842701	0.157299
0.05	0.056372	0.943628	1.1	0.880205	0.119795
0.1	0.112463	0.887537	1.2	0.910314	0.089686
0.15	0.167996	0.832004	1.3	0.934008	0.065992
0.2	0.222703	0.777297	1.4	0.952285	0.047715
0.25	0.276326	0.723674	1.5	0.966105	0.033895
0.3	0.328627	0.671373	1.6	0.976348	0.023652
0.35	0.379382	0.620618	1.7	0.983790	0.016210
0.4	0.428392	0.571608	1.8	0.989091	0.010909
0.45	0.475482	0.524518	1.9	0.992790	0.007210
0.5	0.520500	0.479500	2.0	0.995322	0.004678
0.55	0.563323	0.436677	2.1	0.997021	0.002979
0.6	0.603856	0.396144	2.2	0.998137	0.001863
0.65	0.642029	0.357971	2.3	0.998857	0.001143
0.7	0.677801	0.322199	2.4	0.999311	0.000689
0.75	0.711156	0.288844	2.5	0.999593	0.000407
0.8	0.742101	0.257899	2.6	0.999764	0.000236
0.85	0.770668	0.229332	2.7	0.999866	0.000134
0.9	0.796908	0.203092	2.8	0.999925	0.000075
0.95	0.820891	0.179109	2.9	0.999959	0.000041

$$erf(\beta) = \frac{2}{\sqrt{\mu}} \int_0^{\beta} e^{-\text{э}^2} \, d\,\text{э}$$

$$erf(-\beta) = -erf\beta$$

$$erfc(\beta) = 1 - erf(\beta)$$

Theis Solution

TO PREDICT DRAWDOWN IN HYDRAULIC HEAD IN A CONFINED AQUIFER

Exponential Integral:

$$h_0 - h(r,t) = \frac{Q}{4 \cdot \pi \cdot T} \cdot \int_u^{\infty} \frac{e^{-u}}{u} \, du$$

Where:

$$u = \frac{r^2 \cdot S}{4 \cdot T \cdot t}$$

By substitution:

$$h_0 - h = \frac{Q}{4 \cdot \pi \cdot T} \cdot W(u)$$

Where:

W(u) = well function
Q= pumping rate
T = transmissivity
S = storativity
h_0-h = drawdown
r = radial distance
t = time after start of pumping

Values of W(u) for Various Values of u

u	1.0	2.0	3.0	4.0	5.0	6.0	7.0	8.0	9.0
x1	0.219	0.049	0.013	0.0038	0.0011	0.00036	0.00012	0.000038	0.000012
$x10^{-1}$	1.82	1.22	0.91	0.70	0.56	0.45	0.37	0.31	0.26
$x10^{-2}$	4.04	3.35	2.96	2.68	2.47	2.30	2.15	2.03	1.92
$x10^{-3}$	6.33	5.64	5.23	4.95	4.73	4.54	4.39	4.26	4.14
$x10^{-4}$	8.63	7.94	7.53	7.25	7.02	6.84	6.69	6.55	6.44
$x10^{-5}$	10.94	10.24	9.84	9.55	9.33	9.14	8.99	8.86	8.74
$x10^{-6}$	13.24	12.55	12.14	11.85	11.63	11.45	11.29	11.16	11.04
$x10^{-7}$	15.54	14.85	14.44	14.15	13.93	13.75	13.60	13.46	13.34
$x10^{-8}$	17.84	17.15	16.74	16.46	16.23	16.05	15.90	15.76	15.65
$x10^{-9}$	20.15	19.45	19.05	18.76	18.54	18.35	18.20	18.07	17.95
$x10^{-10}$	22.45	21.76	21.35	21.06	20.84	20.66	20.50	20.37	20.25
$x10^{-11}$	24.75	24.06	23.65	23.36	23.14	22.96	22.81	22.67	22.55
$x10^{-12}$	27.05	26.36	25.96	25.67	25.44	25.26	25.11	24.97	24.86
$x10^{-13}$	29.36	28.66	28.26	27.97	27.75	27.56	27.41	27.28	27.16
$x10^{-14}$	31.66	30.97	30.56	30.27	30.05	29.87	29.71	29.58	29.46
$x10^{-15}$	33.96	33.27	32.86	32.58	32.35	32.17	32.02	31.88	31.76

Periodic Table of the Elements

IA																VIIA	0
1 **H** 1.008	IIA											IIIA	IVA	VA	VIA	1 **H** 1.008	2 **He** 4.003
3 **Li** 6.941	4 **Be** 9.012											5 **B** 10.81	6 **C** 12.01	7 **N** 14.01	8 **O** 16.00	9 **F** 19.00	10 **Ne** 20.18
11 **Na** 22.99	12 **Mg** 24.31	IIIB	IVB	VB	VIB	VIIB	VIIIB			IB	IIB	13 **Al** 26.98	14 **Si** 28.09	15 **P** 30.97	16 **S** 32.07	17 **Cl** 35.45	18 **Ar** 39.95
19 **K** 39.09	20 **Ca** 40.08	21 **Sc** 44.96	22 **Ti** 47.87	23 **V** 50.94	24 **Cr** 52.00	25 **Mn** 54.94	26 **Fe** 55.85	27 **Co** 58.93	28 **Ni** 58.69	29 **Cu** 63.55	30 **Zn** 65.39	31 **Ga** 69.72	32 **Ge** 72.61	33 **As** 74.92	34 **Se** 78.96	35 **Br** 79.90	36 **Kr** 83.80
37 **Rb** 85.47	38 **Sr** 87.62	39 **Y** 88.91	40 **Zr** 91.22	41 **Nb** 92.91	42 **Mo** 95.94	43 **Tc** (98)	44 **Ru** 101.1	45 **Rh** 102.9	46 **Pd** 106.4	47 **Ag** 107.9	48 **Cd** 112.4	49 **In** 114.8	50 **Sn** 118.7	51 **Sb** 121.8	52 **Te** 127.6	53 **Br** 126.9	54 **Xe** 131.3
55 **Cs** 132.9	56 **Ba** 137.3	57 **La** 138.9	72 **Hf** 178.5	73 **Ta** 180.9	74 **W** 183.8	75 **Re** 186.2	76 **Os** 190.2	77 **Ir** 192.2	78 **Pt** 195.1	79 **Au** 197.0	80 **Hg** 200.6	81 **Tl** 204.4	82 **Pb** 207.2	83 **Bi** 209.0	84 **Po** (209)	85 **At** (210)	86 **Rn** (222)
87 **Fr** (223)	88 **Ra** (226)	89 **Ac** (227)	104 **Rf** (261)	105 **Db** (262)	106 **Sg** (263)	107 **Bh** (262)	108 **Hs** (265)	109 **Mt** (266)	110 (269)	111 (272)	112 (277)						

58 **Ce** 140.1	59 **Pr** 140.9	60 **Nd** 144.2	61 **Pm** (145)	62 **Sm** 150.4	63 **Eu** 152.0	64 **Gd** 157.3	65 **Tb** 158.9	66 **Dy** 162.5	67 **Ho** 164.9	68 **Er** 167.3	69 **Tm** 168.9	70 **Yb** 173.0	71 **Lu** 175.0
90 **Th** 232.0	91 **Pa** 231.0	92 **U** 238.0	93 **Np** (237)	94 **Pu** (244)	95 **Am** (243)	96 **Cm** (247)	97 **Bk** (247)	98 **Cf** (251)	99 **Es** (252)	100 **Fm** (257)	101 **Md** (258)	102 **No** (259)	103 **Lr** (262)

IA Alkali metals
IIA Alkaline-earth metals
IIIB-IIB Transition elements
VIA Chalcogens
VIIA Halogens
0 Noble gases

FunctionalGroups of Organic Compounds

Name	Symbol	Formula	Bonding Sites
aldehyde		CHO	1
alkyl	[R]	C_nH_{2n+1}	1
alkoxy	[RO]	$C_nH_{2n+1}O$	1
amine		NH_n	3-n [n=0,1,2]
aryl (benzene ring)	[Ar]	C_6H_5	1
carbinol		COH	3
carbonyl (keto)		CO	2
carboxyl		COOH	1
ester		COO	1
ether		O	2
halogen (halide)	[X]	Cl, Br, I, or F	1
hydroxyl		OH	1
nitrile		CN	1
nitro		NO_2	1

Families of Organic Compounds

Family	Structure	Example
acids		
carboxylic acids	[R]-COOH	acetic acid ($(CH_3)COOH$)
fatty acids	[Ar]-COOH	benzoic acid (C_6H_5COOH)
alcohols		
aliphatic	[R]-OH	methanol (CH_3OH)
aromatic	[Ar]-[R]-OH	benzyl alcohol ($C_6H_5CH_2OH$)
aldehydes	[R]-CHO	formaldehyde (HCHO)
alkyl halides	[R]-[X]	chloromethane (CH_3Cl)
amides	[R]-CO-NH_n	β-methylbutyramide ($C_4H_9CONH_2$)
amines	$[R]_{3-n}$-NH_n	methylamine (CH_3NH_2)
	$[Ar]_{3-n}$-NH_n	aniline ($C_6H_5NH_2$)
anhydrides	[R]-CO-O-CO-[R']	acetic anhydride $(CH_3CO)_2O$
aromatics	C_nH_n	benzene (C_6H_6)
aryl halides	[Ar]-[X]	fluorobenzene (C_6H_5F)
esters	[R]-COO-[R']	methyl acetate (CH_3COOCH_3)
ethers	[R]-O-[R']	diethyl ether ($C_2H_5OC_2H_5$)
	[Ar]-O-[Ar]	diphenyl ether ($C_6H_5OC_6H_5$)
hydrocarbons		
alkanes	C_nH_{2n+2}	octane (C_8H_{18})
alkenes	C_nH_{2n}	ethylene (C_2H_4)
alkynes	C_nH_{2n-2}	acetylene (C_2H_2)
ketones	[R]-[CO]-[R]	acetone ($(CH_3)_2CO$)

Useful Calculus Integrals

$$\int a^u\,du = \frac{1}{\ln(a)}\cdot a^u + C$$

$$\int e^u\,du = e^u + C$$

$$\int e^{au}\,du = \frac{e^{au}}{a} + C$$

$$\int \ln(u)du = u\cdot\ln(u) - u + C$$

$$\int u^{-1}\,du = \ln u^{-1} + C$$

$$\int u^n\,du = \frac{u^{n+1}}{n+1} + C$$

$$\int \sqrt{a^2 + u^2}\,du = \frac{u}{a}\cdot\sqrt{a^2 - u^2} + \frac{a^2}{2}\sin^{-1}\left(\frac{u}{a}\right) + C$$

$$\int u\,dv = uv - \int v\,du \text{ (integratio n by parts)}$$

$$\int \sin(u)du = -\cos(u) + C$$

$$\int \cos(u)du = \sin(u) + C$$

$$\int \tan(u)du = \ln|\sec(u)| + C$$

$$\int \cot(u)du = \ln|\sin(u)| + C$$

$$\int \sec(u)du = \ln|\sec(u) + \tan(u)| + C$$

$$\int \csc(u)du = \ln|\csc(u) - \cot(u)| + C$$

$$\int \sin^{-1}(u)du = u\cdot\sin^{-1}(u) + \sqrt{(1-u^2)} + C$$

$$\int \cos^{-1}(u)du = u\cdot\cos^{-1}(u) - \sqrt{1-u^2} + C$$

$$\int \sinh(u)du = \cosh(u) + C$$

$$\int \cosh(u)du = \sinh(u) + C$$

$$\int \tanh(u)du = \ln(\cosh(u)) + C$$

$$\int \coth(u)du = \ln|\sinh(u)| + C$$

$$\int \operatorname{csch}(u)du = \ln\left|\tanh\left(\frac{1}{2}\cdot u\right)\right| + C$$

$$\int \operatorname{sech}(u)du = \tan^{-1}|\sinh(u)| + C$$

$$\int \operatorname{sech}^2(u)du = \tanh(u) + C$$

$$\int \operatorname{csch}^2(u)du = -\coth(u) + C$$

$$\int u\cdot\sin(u)du = (\sin(u) - u\cdot\cos(u)) + C$$

$$\int u\cdot\cos(u)du = (\cos(u) + u\cdot\sin(u)) + C$$

Log Properties

$$Ln(1) = 0$$

$$Ln(e) = 1$$

$$Ln(a\cdot c) = Ln(a) + Ln(c)$$

$$Ln\left(\frac{a}{c}\right) = Ln(a) - Ln(c)$$

$$Ln(a^r) = r\,Ln(a)$$

$$Ln\left(\frac{1}{c}\right) = -Ln(c)$$

Geometric Formulas-Areas

Sketch		Area	Centroid	Second Moment
Rectangle	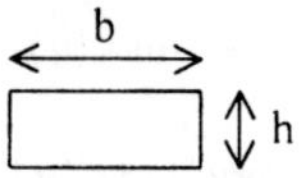	bh	$y_{bar} = h/2$	$I_{bar} = bh^3/12$
Triangle	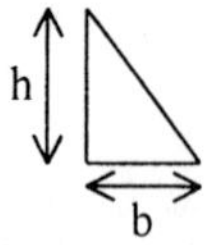	bh/2	$y_{bar} = h/3$	$I_{bar} = bh^3/36$
Circle	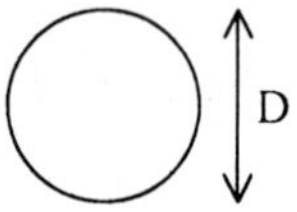	$\Pi D^2/4$	$y_{bar} = r$	$I_{bar} = \Pi D^4/64$
Semicircle	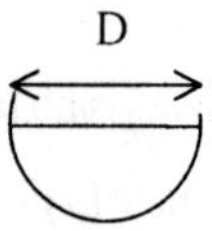	$\Pi D^2/8$	$y_{bar} = 4r/3\Pi$	$I_x = \Pi D^4/128$
Ellipse	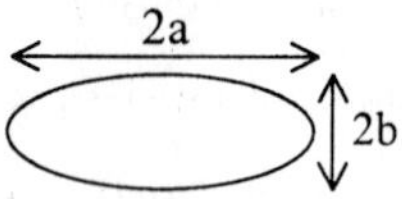	Πab	$y_{bar} = b$	$I_{bar} = \Pi ab^3/4$
Semiellipse	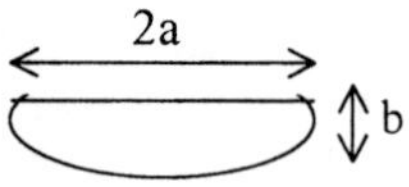	$\Pi ab/2$	$y_{bar} = 4b/3\Pi$	$I_x = \Pi ab^3/8$

Geometric Formulas-Volumes

	Sketch	Surface Area	Volume	Centroid
Cylinder	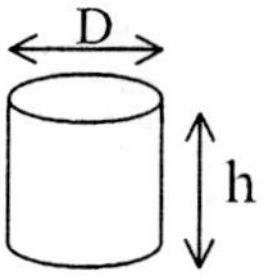 D, h	$\Pi Dh + \Pi D^2/2$	$\Pi D^2 h/4$	$y_{bar} = h/2$
Sphere	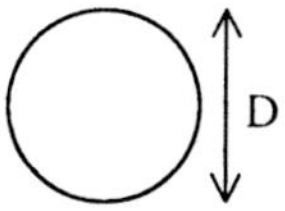 D	ΠD^2	$\Pi D^3/6$	$y_{bar} = r$
Cone	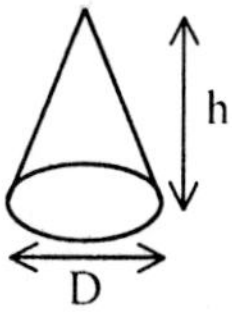 h, D	$\Pi[r^2 + r\,(r^2 + h^2)^{1/2}]$	$\Pi D^2 h/12$	$y_{bar} = h/4$
Hemisphere	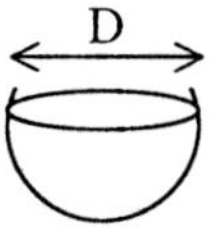 D	$3\Pi D^2/4$	$\Pi D^2/12$	$y_{bar} = 3r/8$

McGraw-Hill Higher Education

A Division of The McGraw-Hill Companies

www.mhhe.com